LOO-KENG HUA
SELECTED PAPERS

华罗庚

Loo-Keng Hua
Selected Papers

Edited by H. Halberstam

Springer-Verlag
New York Heidelberg Berlin

Loo-Keng Hua
Institute of Applied Mathematics
Academia Sinica
Beijing
China

Editor

H. Halberstam
University of Illinois at Urbana-Champaign
Department of Mathematics
Urbana, Illinois 61801
U.S.A.

AMS Subject Classification: 01A75, 10-XX, 30-XX

Library of Congress Cataloging in Publication Data
Hua, Loo-keng, 1911–
Loo-keng Hua selected papers.
Contents: Number theory/by Wang Yuan—Algebra and geometry/by Z. X. Wan—Function theory/by S. Kung and K. H. Look.
1. Mathematics—Collected works. I. Halberstam, H. (Heini) II. Title.
QA3.H7725 1982 510s [510] 82-19164

With 14 Illustrations

Printed and bound by Malloy Lithographing, Inc., Ann Arbor, MI.
Printed in the United States of America.

9 8 7 6 5 4 3 2 1

ISBN 0-387-**90744**-0 Springer-Verlag New York Heidelberg Berlin
ISBN 3-540-**90744**-0 Springer-Verlag Berlin Heidelberg New York

PREFACE

The unexpected arrival of Loo-Keng Hua in Europe in the fall of 1978 was for many of us a romantic event, a legend come to life. What had long been (and had seemed destined to remain) merely a revered name in the mathematical annals of our times assumed suddenly the handsome presence of the man himself, dignified yet jovial, youthful yet wise, serene yet ever questing for new achievements; and we realized how very much we had missed by his thirty years' absence from the international scene. While the publication of these "Selecta" from his writings needs no justification beyond what is in them, it will, I hope, serve also as a way of saying a most cordial "welcome back".

It has been an honor for me to play a small rôle in producing the Selecta. To select only parts from the imposing whole is automatically to be wrong, and it may well seem in the long run (to quote loosely from a poem of Hua himself) that I have repaid gifts in jade with artifacts of wood. Nevertheless, I bear sole responsibility for this selection, and my final purpose here is only to acknowledge all the kind help I have received: from Professor Hua himself and his assistant Mr. Na; from Chinese colleagues who provided chapter introductions (which I freely adapted); from Provost Timothy O'Meara of the University of Notre Dame, and Professor Raghavan Narasimhan of the University of Chicago; and from colleagues here in Illinois, Professors John D'Angelo, Everett Dade, Irving Reiner, and Joseph Rotman.

H. Halberstam

CONTENTS

*Numbers in brackets refer to bibliography pages 883–887.

PERMISSIONS

Springer-Verlag would like to thank the original publishers of Loo-Keng Hua's papers for granting permission to reprint specific papers in this volume (numbers in brackets refer to bibliography pp. 883–887):

Reprinted by permission from Akademie-Verlag:
[122] On the Riemannian curvature in the space of several complex variables. *Schriftenreihe des Instituts für Mathematik* **1** (1957), 245–263.

Reprinted by permission from the American Mathematical Society:
[62] On the least solution of Pell's equation. *Bull. AMS* **48** (1942), 731–735.
[68] A remark on a result due to Blichfeldt. *Bull. AMS* **51** (1945), 537–539.
[101] Supplement to the paper of Dieudonné on the automorphisms of classical groups. *Mem. AMS* **2** (1951), 96–122.
[60] On the number of partitions of a number into unequal parts. *Trans. AMS* **51** (1942), 194–201.
[64] On the distribution of quadratic non-residues and the Euclidean algorithm in real quadratic fields. I–II. Part II coauthored by Szu-Hoa Min. *Trans. AMS* **56** (1944), 527–546, 547–569.
[67] Geometries of matrices. I. Generalizations of von Staudt's Theorem; I_1. Arithmetical construction. *Trans. AMS* **57** (1945), 441–481, 482–490.
[76] Orthogonal classification of Hermitian matrices. *Trans. AMS* **59** (1946), 508–523.
[77] Geometries of matrices. II. Study of involutions in the geometry of symmetric matrices. *Trans. AMS* **61** (1947), 193–228.
[78] Geometries of matrices. III. Fundamental theorems in the geometries of symmetric matrices. *Trans. AMS* **61** (1947), 229–255.
[92] On the generators of the symplectic modular group. Coauthored by Irving Reiner. *Trans. AMS* **65** (1949), 415–426.
[102] Automorphisms of the unimodular group. Coauthored by Irving Reiner. *Trans. AMS* **71** (1951), 331–348.
[106] Automorphisms of the projective unimodular group. Coauthored by Irving Reiner. *Trans. AMS* **72** (1952), 467–473.

Reprinted by permission from *Annals of Mathematics*:
[74] On the theory of Fuchsian functions of several variables. *Ann. Math.* **47** (1946), 167–191.
[85] On the automorphisms of the symplectic group over any field. *Ann. Math.* **49** (1948), 739–759.
[93] Geometry of symmetric matrices over any field with characteristic other than two. *Ann. Math.* **50** (1949), 8–31.

Reprinted by permission from the *Canadian Journal of Mathematics*:
[100] On exponential sums over an algebraic number field. *Canadian J. Math.* **3** (1951), 44–51.

Reprinted by permission from The Johns Hopkins University Press:
[66] On the theory of automorphic functions of a matrix variable. I. Geometrical basis; II. The classification of hypercircles under the symplectic group. *Am. J. Math.* **66** (1944), 470–488, 531–563.

Reprinted by permission from *The Journal of Indian Mathematical Society*:
[47] On Waring's problem with cubic polynomial summands. *J. Indian Math. Soc.* **4** (1940), 127–135.

Reprinted by permission from C. F. Hodgson & Son Ltd.:
[28] A problem in the additive theory of numbers of several variables. *J. London Math. Soc.* **12** (1937), 257–261.
[42] A remark on the moment problem. *J. London Math. Soc.* **14** (1939), 84–86.
[94] Improvement of a result of Wright. *J. London Math. Soc.* **24** (1949), 157–159.

Reprinted by permission from Oxford University Press:
[30] On a generalized Waring problem. *Proc. London Math. Soc.* Series 2, **43** (1937), 161–182.
[36] On Waring's problem. *Q.J.M.* **9** (1938), 199–202.
[37] On Tarry's problem. *Q.J.M.* **9** (1938), 315–320.
[48] On a theorem due to Vinogradow. *Q.J.M.* **11** (1940), 161–176.
[63] The lattice-points in a circle. *Q.J.M.* **13** (1942), 18–29.
[75] On the extended space of several complex variables I. The space of complex spheres. *Q.J.M.* **17** (1946), 214–222.
[95] An improvement of Vinogradov's mean-value theorem and several applications. *Q.J.M.* **20** (1949), 48–61.

Reprinted by permission from *Studies in Applied Mathematics*:
[27] On Fourier transforms in L^p in the complex domain. Coauthored by Shien-siu Shü. *J. Math. Phys.* **15** (1936), 249–263.

A BIOGRAPHICAL NOTE

Hua Loo-Keng was born in November 1910, into a poor family in Jintar County, Jiangsu, China. He went to school for only nine years in all and did not get a university degree until the University of Nancy in France made him Docteur Honoris Causa in 1979. At the age of fifteen he left school in order to help his father to look after a small grocery store and he studied mathematics in his spare time. In 1930 he came to Tsing Hua University to work, and during six years he moved from being just a clerk, to assistant, then lecturer, and finally a research fellow of the China Foundation. Between 1936 and 1938 Hua lived in Cambridge, England at the invitation of G. H. Hardy to whom Hua had been recommended by Norbert Wiener. As soon as the Sino-Japanese war broke out, Hua returned to China and, from 1938 to 1946, taught in the South-West United University at Kunming as a professor. In the Spring of 1946 he was invited by the Institute of Mathematics of the Academy of Sciences of the USSR to visit the Soviet Union for three months. From 1946 to 1947 he worked as a member of the Institute for Advanced Study at Princeton, USA, and taught number theory in Princeton University. From 1948 to 1950 Hua was a professor in the University of Illinois at Urbana. In 1950, as soon as the People's Republic of China was founded, Hua returned once again, with his family, to China to take up permanent residence there.

After his return, he was appointed Professor at Tsing Hua University, director of the Institute of Mathematics of Academia Sinica, Vice President of the University of Science and Technology of China, and Vice President of Academia Sinica.

Hua is one of the earliest directors of research in China. Among those who participated in his seminars in Kunming in the forties and later became well known were Tuan Hsio-Fu, Min Szu-Hoa, Fan Ky and Shü Shien-Siu. After his return to China in 1950, he organized numerous seminars in different subjects and brought on many mathematicians of a younger generation. For example, in number theory there emerged a group of workers with Chen Jin-Run and Wang Yuan among them, in algebra there was Wan Zhe Xian, and in functions of several complex variables Look K. H. (Lu Qi Keng) and Kung Sheng. Hua's method for training students was quite remarkable. A fluent writer—even something of a poet—he wrote "Introduction to number theory", "Classical groups" and "Harmonic analysis of functions of several complex variables in classical domains" for the young people

Chen Jin-Run and Hua Loo-Keng at work.

who attended the seminars. Some participants in his seminars diversified later into other branches of mathematics. In the last fifteen years Hua has worked also in applied mathematics in response to his country's needs. Hua has long concerned himself with the teaching of mathematics in school, and has written for schools a series of supplementary texts. He was the first to organize mathematical Olympiads in China, and he has taken a particular interest in popularizing mathematics among the citizens of his country. In all these activities he has been conspicuously successful.

In recent times he has led Chinese scientific delegations to Europe and the United States, and has acted as ambassador of good will on behalf of the Chinese academic community during extended visits to the United States, Japan, Great Britain, France and Germany. In 1980 he was one of the principal speakers at the fourth International Congress on Mathematical Education in Berkeley, California. In 1982 he was elected a foreign associate of the United States National Academy of Sciences.

The present selection of his writings, compiled with his approval and assistance, is testimony to his highly influential contributions to many diverse areas of mathematics and mathematical education over more than half a century. As good a guide as any to his standing in contemporary mathematics as scientist and educator is the fact that his classical texts on 'Additive theory of prime numbers' and 'Harmonic Analysis of functions of several complex variables in the classical domains', first published in 1947 and 1958 respectively, continue to be maintained in print by the American Mathematical Society, while two other and quite different books, one with Wang Yuan on 'Application of Number theory to Numerical Analysis' and the other a massive 'Introduction to the Theory of Numbers', have been translated and published recently by Springer-Verlag.

H. Halberstam

NUMBER THEORY

WANG YUAN

The Estimation of Exponential Sums. Let q be a positive integer and $f(x)$ be a polynomial with integer coefficients

$$f(x) = a_k x^k + \cdots + a_1 x,$$

where $(a_k, \ldots, a_1, q) = 1$. Consider the complete exponential sum

$$S(q, f(x)) = \sum_{x=1}^{q} e^{2\pi i f(x)/q}.$$

If $f(x) = x^2$, $S(q, x^2)$ is the well-known Gaussian sum. It was proved by Gauss that

$$|S(q, x^2)| = q^{1/2}.$$

The problem of estimating $S(q, f(x))$ has a long history and until Hua [49A] solved it in 1940 it had been settled only for special polynomials. Hua proved in a characteristically elegant way that

$$|S(q, f(x))| = 0(q^{1-1/k+\epsilon}) \tag{1}$$

where ϵ denotes any pre-assigned positive number and the constant implied by the symbol "0" depends only on k and ϵ. In 1938, Hua [38] had obtained this result but the implied constant then depended also on the coefficients a_i $(1 \leqslant i \leqslant h)$ of $f(x)$. It is easily seen that (1) is best possible apart from possible improvement of the factor q^ϵ (for example, $|S(p^k, x^k)| = p^{k(1-1/k)}$) where p always denotes a prime and $p \nmid k$.

Hua later [100] generalized (1) to any algebraic number field $Q(\alpha)$ of degree n.

Concerning the corresponding complete exponential sum in two variables, Hua and Min [82] established that

$$\sum_{x=1}^{p} \sum_{y=1}^{p} e^{2\pi i f(x,y)/p} = 0(p^{2-2/k}),$$

where $f(x, y)$ cannot be reduced to a k-th power of a polynomial in a single variable. This was the first result of its kind. Recently the general problem has been settled by Deligne's proof of Weil's conjecture.

For the incomplete exponential sum, Hua [119] proved that

$$\sum_{x=1}^{P} e^{2\pi i h x^k/q} = \frac{P}{q} \sum_{x=1}^{q} e^{2\pi i h x^k/q} + 0(q^{1/2+\epsilon}). \tag{2}$$

This result has an important application to Waring's problem.

Hua improved and simplified Vinogradov's method for the estimation of Weyl's sums by pointing out that the essence of Vinogradov's method is the following mean value theorem [95]:

Let $f(x) = \alpha_k x^k + \cdots + \alpha_1 x$ and

$$C_k = C_k(P) = \sum_{x=a+1}^{a+P} e^{2\pi i f(x)}.$$

Let $t_1 = t_1(k) \geqslant \frac{1}{4}k(k+1) + lk$. Then

$$\int_0^1 \cdots \int_0^1 |C_k|^{2t_1} d\alpha_1 \ldots d\alpha_k \leqslant (7t_1)^{4t_1 l} P^{2t_1 - (1/2)k(k+1) + \delta} (\log P)^{2l},$$

where $\delta = \delta(k) = \frac{1}{2}k(k+1)(1 - 1/k)^l$.

From this one derives immediately the following theorem:

Suppose that $k \geqslant 12$, $2 \leqslant r \leqslant k$ and

$$\left|\alpha_r - \frac{h}{q}\right| \leqslant \frac{1}{q^2}, \qquad (h,q) = 1, \qquad 1 \leqslant q \leqslant P^r.$$

Then for $P \leqslant q \leqslant P^{r-1}$, we have

$$S = \sum_{x=1}^{P} e^{2\pi i f(x)} \ll P^{1 - 1/\sigma_k + \epsilon}$$

where $\sigma_k = 2k^2(2\log k + \log\log k + 3)$.

In all current monographs on analytic number theory, Vinogradov's method is stated according to Hua's formulation (See for example R. C. Vaughan, The Hardy–Littlewood Method, Cambridge Tracts in Mathematics, Vol. 80, 1981).

Concerning character sums, Hua [61] proved in 1942 that the estimation

$$\frac{1}{A+1}\left|\sum_{a=1}^{A} \sum_{n=-a}^{a} \chi(n)\right| \leqslant \sqrt{p} - \frac{A+1}{\sqrt{p}}, \qquad 1 \leqslant A < p,$$

holds for all non-principal characters modulo p. This led to improvements of the estimation of the least primitive root modulo p and of the least solution of Pell's equation [62].

Waring's Problem. In 1770, Waring conjectured that, for any integer $k \geqslant 2$, there always exists an integer $s = s(k)$ depending only on k such that every positive integer can be expressed as a sum of s k-th powers of non-negative integers. Waring's conjecture was proved by Hilbert in 1900. In the 20's of this century, Hardy and Littlewood created and developed a very powerful new analytic method in additive number theory, the so-called "circle method", which leads to much more precise results on Waring's problem. Let $G(k)$ denote the least integer s such that every sufficiently large integer N can be represented by

$$N = x_1^k + \cdots + x_s^k, \tag{3}$$

where x_i $(1 \leqslant i \leqslant s)$ are non-negative integers, Hardy and Littlewood established that $G(k) \leqslant (k-2)2^{k-1} + 5$; more than that, they obtained an asymptotic formula for the number of solutions of (3) when $s \geqslant (k-2)2^{k-1} + 5$. The result of Hardy and Littlewood was improved by Hua [36] in 1938 to

$$G(k) \leqslant 2^k + 1$$

and he proved also that the asymptotic formula for the number of solutions of (3) holds for $s \geqslant 2^k + 1$. For this he used what has come to be called Hua's inequality

$$\int_0^1 \left| \sum_{x=1}^{P} e^{2\pi i \alpha x^k} \right|^{2^k} d\alpha \ll P^{2^k - k + \epsilon}. \tag{4}$$

It may be of interest to observe that no improvement on the condition $s \geqslant 2^k + 1$ has yet been made for small values of k so far as the asymptotic formula itself is concerned. A new "elementary" proof of Hilbert's theorem was given by Linnik in 1943, and was selected by Khinchine as one of his 'three pearls'. Commenting on this proof, Davenport wrote: "The underlying ideas of this proof were undoubtedly suggested by certain features of the Hardy–Littlewood method, and in particular by Hua's inequality".

In the 30's of this century, many mathematicians studied the generalization of the x^k in Waring's problem to a polynomial of degree k. The main difficulty was removed by Hua, using his estimate (1). A very general formulation of the Hilbert theorem is due to Hua. Suppose that $f(x)$ denotes an integral polynomial of degree k, i.e.;

$$f(x) = a_k\binom{x}{k} + a_{k-1}\binom{x}{k-1} + \cdots + a_1\binom{x}{1},$$

where $\binom{x}{r} = x(x-1)\cdots(x-r+1)/r!$ and a_i $(1 \leqslant i \leqslant k)$ are integers which we assume always to be positive. Hua proved between 1937 and 1940 that the number of solutions of the equation

$$N = f(x_1) + \cdots + f(x_s) \tag{5}$$

has an asymptotic formula for $s \geqslant 2^k + 1$ $(1 \leqslant k \leqslant 10)$ [36] and [implicitly from Lemma 12.3 of 104] for $s \geqslant 2k^2(2\log k + \log\log k + 2.5)$ $(k > 10)$. The result is true also if (5) is replaced by

$$N = f_1(x_1) + \cdots + f_s(x_s),$$

where the $f_i(x)$ $(1 \leqslant i \leqslant s)$ are any given s integral polynomials of degree k. Let $G(f)$ denote the least integer s such that (5) is solvable for all sufficiently large N. Let $\partial^\circ f$ denote the degree of f. Hua [30], [45] established that

$$G(f \mid \partial^\circ f = k) \leqslant (k-1)2^{k+1},$$

and [47]

$$G(f \mid \partial^\circ f = 3) \leqslant 8.$$

He also proved [45] that

$$\max_{f(x)} G(f) \geqslant 2^k - 1,$$

where $f(x)$ runs over all integral polynomials of degree k, in which $k \geqslant 5$.

The essence of the circle method of Hardy and Littlewood may be sketched as follows. The number of solutions of the equation (3) may be represented by an integral

$$r_s(N) = \int_0^1 (T(\alpha))^s e^{-2\pi i N\alpha}\, d\alpha, \qquad T(\alpha) = \sum_{x=1}^{P} e^{2\pi i \alpha x^k}$$

where $P = [N^{1/k}]$. Divide the unit interval (0, 1) into two parts, the major arcs and the minor arcs having unions $\mathfrak{M}$ and $\mathfrak{m}$ respectively. Roughly speaking $\mathfrak{M}$ consists of disjoint intervals centering on the rational number k/q with small denominators and $\mathfrak{m}$ comprises the rest. Hardy and Littlewood proved that the asymptotic formula

$$\int_{\mathfrak{M}} (T(\alpha))^s e^{-2\pi i N\alpha}\, d\alpha = \mathfrak{S}(N) N^{s/k-1} \frac{\Gamma^s(1+1/k)}{\Gamma\left(\frac{s}{k}\right)} + (P^{s-k-1/k}),$$

holds for $s \geqslant 2k+1$, where $\mathfrak{S}(N)$ denotes the singular series. Hua [120] improved the result to $s \geqslant k+1$ in 1957. This result is best possible and the proof is based on his estimate (2).

Goldbach's Problem. The problem was raised in a letter from Goldbach to Euler in which two conjectures were proposed: (A) Every even integer $\geqslant 6$ is a sum of two odd primes and (B) every odd integer $\geqslant 9$ can be represented as a sum of three odd primes. Evidently (B) is a consequence of (A) and (B) was solved by Vinogradov in 1937 for all sufficiently large integers. By the use of Vinogradov's method, several mathematicians, Hua among them, proved that the conjecture (A) holds for almost all even integers. Specifically, Hua [32] established more generally that almost all even integers can be represented as sums of a prime and a k-th power of prime, where k is a given integer $\geqslant 1$.

The Waring–Goldbach Problem. Hua went on to study systematically the so called Waring–Goldbach problem, of the solvability of (3) and its generalizations in which the variables x_i $(1 \leqslant i \leqslant s)$ are restricted to prime numbers. The fruits of his researches are given in the well-known monograph "Additive prime number theory". For example, he proves there [cf. 39] that the number of solutions of the equation

$$N = p_1^k + \cdots + p_s^k \tag{6}$$

has an asymptotic formula for $s \geqslant 2^k + 1$ $(k \leqslant 10)$ and $s \geqslant 2k^2(2\log k + \log\log k + 2.5)$ $(k > 10)$; and he obtains analogues for prime arguments of many of the results sketched above.

Tarry's Problem. Let $N(k)$ denote the least integer t such that the system of equations

$$x_1^h + \cdots + x_t^h = y_1^t + \cdots + y_t^h, \qquad 1 \leqslant h \leqslant k \tag{8}$$

has a non-trivial solution, i.e., $x_1, \ldots, x_t, y_1, \ldots, y_t$ are positive integers but the set of $x_1, \ldots, x_t$ is not a permutation of $y_1, \ldots, y_t$. Let $M(k)$ be the least integer t such that (8) is solvable and

$$x_1^{k+1} + \cdots + x_t^{k+1} \neq y_1^{k+1} + \cdots + y_t^{k+1}.$$

Evidently, we have

$$k+1 \leqslant N(k) \leqslant M(k).$$

Hua [37] proved that, for $k \geqslant 12$,

$$M(k) \leqslant (k \pm 1)\left[\left[\frac{\log \frac{1}{2}(k+2)}{\log\left(1+\frac{1}{k}\right)}\right]+1\right] \sim k^2 \log k$$

which improves significantly on the early result $M(k) < 7k^2(k-11)(k+3)/216$ due to Wright. Hua's argument is elementary and straightforward.

It was pointed out by Hua that Vinogradov's method may be used to treat Tarry's problem. In this way Hua [104] obtained the following results: Let t_0 be given by the table

k	2	3	4	5	6	7	8	9	10	$k \geqslant 11$
t_0	3	8	23	55	120	207	336	540	885	$[k^2(3\log k + \log\log k + 9]$

Let $r_t(p)$ denote the number of solutions of (8) satisfying

$$1 \leqslant x_i, \quad y_i \leqslant P, \quad 1 \leqslant i \leqslant t.$$

Then

$$\lim_{p\to\infty} P^{1/2k(k+1)-2t} r_t(p) = c(t)$$

holds for $t > t_0$, where $c(t)$ is a positive constant depending on t only.

Partitions. Let $q(n)$ be the number of partitions of a positive integer n into unequal parts, or into odd parts. Hua [60] established an exact formula for $q(n)$ by modifying Hardy and Ramanujan's method:

$$q(n) = \frac{1}{2^{1/2}} \sum_{\substack{k=1 \\ k \text{ odd}}}^{\infty} \sum_{\substack{(h,k)=1 \\ 0<h\leqslant k}} \omega_{h,k} e^{-2\pi i h n/k} \frac{d}{dn} J_0\left(\frac{i\pi}{k}\left\{\frac{2}{3}\left(n+\frac{1}{24}\right)\right\}^{1/2}\right),$$

where $J_0(x)$ is the Bessel function of 0th order.

Circle Problem. Let $A(x)$ denote the number of lattice points (u,v) in the circle $u^2+v^2 \leqslant x$. To find the least θ such that the relation

$$A(x) = \pi x + O(x^{\theta+\epsilon})$$

holds for any positive ϵ, where the constant implied by the symbol "O" depends on ϵ, is called the Gauss circle problem. It was established by Gauss himself that $\theta = 1/2$ is admissible. In 1942, Hua [63] refined Titchmarsh's method for the circle problem and thereby improved his result $\theta = 15/46$ to $\theta = 13/40$. Moreover, Hua's paper pointed out a gap in Vinogradov's work on the same problem.

Reprinted from the
MATHEMATISCHE ZEITSCHRIFT
Vol. 41, No. 5, pp. 708–712, 1936

A Problem on the additive Theory of Number of several variables.

By

Loo-keng Hua *) in Peiping (China).

In this note the following question is answered:

Wheter there exists an integer s such that the system of diophantine equations

$$(1) \qquad \begin{cases} \sum_{\nu=1}^{s} \varepsilon_\nu P(x_\nu, y_\nu) = n, \\ \sum_{\nu=1}^{s} \varepsilon_\nu P'(x_\nu, y_\nu) = n', \end{cases} \qquad (\varepsilon_\nu = \pm 1)$$

is solvable simultaneously for all integers n and n', where P and P' are two integral-valued polynomials of the variables x and y?

The question is answered affirmatively provided that P and P' satisfy *a* certain restriction. The upper bound of s will be given here, it is equal to $2^k k - 1$, k being the greater of the degrees of P and P'.

We assume: $P(x, y)$ and $P'(x, y)$ are two integral-valued polynomials of k-th and k'-th degree respectively. For any prime p there do not exist integers q and q', $(q, q') = 1$, and l such that

$$(2) \qquad q\, P(x, y) + q'\, P'(x, y) \equiv l \pmod{p}$$

identically. Without loss of generality we may assume $k \geqq k'$. In the study of this problem, the condition stated here cannot be removed; for otherwise, n and n' cannot be two arbitrary integers.

Lemma 1. *Every integral valued polynomial $P(x, y)$ can be put in the form*

$$P(x, y) = \Sigma\, a_{\mu\nu}\, P_\mu(x)\, P_\nu(y),$$

where the a's are integers and

$$P_\mu(z) = \frac{z(z-1)\ldots(z-\mu+1)}{\mu!} \text{ for } \mu > 0; \; P_0(z) = 1.$$

Lemma 2. *Let*

$$P(x, y) = \Sigma\, a_{\mu\nu}\, P_\mu(x)\, P_\nu(y)$$

---

*) Research fellow of the China Foundation of the Promotion of Education and Culture.

and

$$P'(x, y) = \Sigma a'_{\mu\nu} P_\mu(x) P_\nu(y).$$

The necessary and sufficient condition that for a given prime p there exist integers q and q' and l such that

$$qP + q'P' \equiv l \pmod p$$

identically is that

$$q a_{\mu\nu} + q' a'_{\mu\nu} \equiv 0 \pmod p$$

for

$$q a_{00} + q' a'_{00} \equiv l \pmod p.$$

These Lemmas can be obtained easily in the same way as those of the author's former papers [1]).

Let $H(P, P', m, m')$ be the least integer of s for which the system of congruences

$$\sum_{\nu=1}^{s} \varepsilon_\nu P(x_\nu, y_\nu) \equiv n \pmod m$$

and

$$\sum_{\nu=1}^{s} \varepsilon_\nu P'(x_\nu, y_\nu) \equiv n' \pmod{m'}$$

is solvable simultaneously for any integers n and n'. Let $H(P, P', 0, 0)$ be the least integer of s for which (1) is solvable.

Let $\Delta_t^x(P(x, y))$ and $\Delta_t^y(P(x, y))$ be the t-th difference of $P(x, y)$ with respect to x and y respectively. $\Delta_t^x(P(x, y))$ can be considered as the sum of 2^t terms of $\pm P(x, y)$.

Lemma 3. *Let α, β, γ and δ be integers with the relation*

$$\begin{vmatrix} \alpha, & \beta \\ \gamma, & \delta \end{vmatrix} = \pm 1.$$

Then

$$H(P, P', 0, 0) = H(\alpha P + \beta P, \gamma P + \delta P', 0, 0).$$

The Lemma follows immediately.

Lemma 4. *Let $a_{uv} P_u(x) P_v(y)$ be a term of k-th degree in $P(x, y)$. If $a'_{uv} = 0$, and $P' \not\equiv 0 \pmod m$ then*

$$H(P, P', m, m') \leqq 2^{k-1} + H(P, P', (m, a_{uv}), m').$$

(Here m, m' may be zeros).

Proof. Here

$$\Delta_{u-1}^x \Delta_v^y P(x, y) = a_{u,v} x + a_{u-1,v} + a_{u-1,v+1} y,$$

$$\Delta_{u-1}^x \Delta_v^y P'(x, y) = a'_{u-1,v} + a'_{u-1,v+1} y.$$

---

[1]) Journ. of London Math. Soc. **11**, 1935, pp. 4—5. Science Report of Tsing Hua University, Ser. A. **3**, 1935, 247—260.

Let s be an integer $\geqq H(P, P', (m, a_{n\,v}), m')$, then

$$\sum_{\nu=1}^{s} \varepsilon_\nu P(x_\nu, y_\nu) \equiv n \quad (\operatorname{mod}(m, a_{u\,v})),$$

$$\sum_{\nu=1}^{s} \varepsilon_\nu P'(x_\nu, y_\nu) \equiv n' \quad (\operatorname{mod} m')$$

is solvable for all integers n and n', and thence

$$(3) \qquad \begin{cases} \sum_{\nu=1}^{s} \varepsilon_\nu P(x_\nu, y_\nu) + \Delta_{u-1}^{x} \Delta_{v}^{y} P(x, y) \equiv N \quad (\operatorname{mod} m), \\ \sum_{\nu=1}^{s} \varepsilon_\nu P'(x_\nu, y_\nu) + \Delta_{u-1}^{x} \Delta_{v}^{y} P(x, y) \equiv N' \quad (\operatorname{mod} m'), \end{cases}$$

(is solvable for all integers N and N'). In fact (3) can be written in the form or

$$\sum_{\nu=1}^{s} \varepsilon_\nu P(x_\nu, y_\nu) + a_{u,\,v}\, x + a_{u-1,\,v} + a_{u-1,\,v+1}\, y \equiv N \quad (\operatorname{mod} m),$$

$$\sum_{\nu=1}^{s} \varepsilon_\nu P'(x_\nu, y_\nu) + a'_{u-1,\,\nu} + a'_{u-1,\,v+1}\, y \equiv N' \quad (\operatorname{mod} m'),$$

From (3) the Lemma follows with the remark before the Lemma 3.

Lemma 5.

$$H(P, P', 0, 0) \geqq H(P, P', m, m').$$

This Lemma is evident.

Lemma 6. *Let*

$$R(x, y) = a\,x + b\,y + c,$$

$$R'(x, y) = a'\,x + b'\,y + c.$$

If for any prime p there do not exist integers q and q', $(q, q') = 1$, and $l(>1)$ such that

$$q\,R + q'\,R' \equiv l \quad (\operatorname{mod} p),$$

then

$$\begin{vmatrix} a & a' \\ b & b' \end{vmatrix} = \pm 1.$$

Thus

$$H(R, R', 0, 0) = H(x, y, 0, 0) = 1.$$

Proof. Let μ and μ' be two integers such that

$$a\,\mu + a'\,\mu' = 0, \qquad (\mu, \mu') = 1.$$

Then

$$b\,\mu + b\,\mu' = \pm 1.$$

for otherwise, let p be a prime divisor of $b\mu + b'\mu'$, then

$$\mu R + \mu' R' \equiv \mu c + \mu' c' \pmod{p}$$

identically. But this contradicts our hypothesis. Thus $(b, b') = 1$. Similarly $(a, a') = 1$. Therefore

$$\begin{vmatrix} a & a' \\ b & b' \end{vmatrix} = \pm 1.$$

Theorem 1. $H(P, P', 0, 0) \leqq 2^k k - 1$.

Proof. Let $a_{uv} P_u(x) P_v(y)$ be a term of $P(x, y)$ where $u + v = k$, $a_{uv} \neq 0$ and u is the greatest one of the u's satisfying the conditions $u + v = k$.

1. $a'_{uv} = 0$. By Lemmas 4 and 5

$$H(P, P', 0, 0) \leqq 2^{u+v-1} + H(P, P', a_{uv}, 0)$$
$$\leqq 2^{k-1} + H(Q, Q', 0, 0)$$

where Q and Q' are two Polynomials having the same properties as P and P' and

$$P \equiv Q, \quad P' \equiv Q' \pmod{a_{uv}}$$

and the coefficients of $P_u(x) P_v(y)$ in Q and Q' are zeros.

2. $a'_{uv} \neq 0$. Let r and r' be two integers having the properties

$$(r, r') = 1$$

and

$$r a_{uv} + r' a'_{uv} = 0.$$

We have integers s and s' such that

$$\begin{vmatrix} r & r' \\ s & s' \end{vmatrix} = \pm 1.$$

By Lemma 3, we have

$$H(P, P', 0, 0) = H(sP + s'P', rP + r'P', 0, 0)$$

and the coefficient of $P_u(x) P_v(y)$ in $rP + r'P'$ is zero. By 1. we have, thus, two polynomials Q and Q' having the same properties as P and P', and then coefficients of $P_u(x) P_v(y)$ being zero, such that

$$H(P, P', 0, 0) \leqq 2^{u+v-1} + H(Q, Q', 0, 0)$$
$$\leqq 2^{k-1} + H(Q, Q', 0, 0).$$

This is true in both cases.

Repeating this process, we have finally

$$H(P, P', 0, 0) \leqq \sum_{\nu=2}^{k} (\nu + 1) 2^{\nu-1} + H(R, R', 0, 0)$$

where R and R' are two linear polynomials in x and y and have the properties mentioned in Lemma 6. Thus we have

$$H(P, P', 0, 0) \leqq \sum_{\nu=1}^{k} (\nu + 1) 2^{\nu-1} - 1 = k 2^k - 1.$$

This result can be extended to t variables:

Theorem 2. *Let $P_1, \ldots, P_t$ be t integral-valued polynomials of $x_1, \ldots, x_t$, the highest of their degrees being equal to k, and suppose for any prime p there do not exist integers $q_1, \ldots, q_t$ and l (> 1) such that*

$$q_1 P_1 + \ldots + q_t P_t \equiv l \pmod{p}, \qquad (q_1, \ldots, q_t) = 1$$

Then

$$H(P_1, \ldots, P_s, 0, \ldots, 0) = O(2^k k^{t-1}).$$

(Eingegangen am 20. April 1936.)

Reprinted from the
JOURNAL OF THE LONDON MATHEMATICAL SOCIETY
Vol. 12, pp. 257–261, 1937

A PROBLEM IN THE ADDITIVE THEORY OF NUMBERS OF SEVERAL VARIABLES

LOO-KENG HUA‡§.

In an attempt to solve a problem suggested by the work of Prof. Mordell‖, I have obtained the following results.

Let¶

$$F_1(x, y),\ F_2(x, y),\ \ldots,\ F_{k+1}\,(x, y)$$

† *Cf.* S. Bochner, *op. cit.*, 191, Theorem 57.

‡ Research Fellow of the China Foundation for the Promotion of Education and Culture.

§ Received 1 November, 1936; read 12 November, 1936.

‖ *Journal London Math. Soc.*, 11 (1936), 204–208.

¶ When the number of the *F*'s is two, the author has obtained the more precise result given in *Math. Zeitschrift*, 41 (1936), 708–712.

be $k+1$ integral-valued polynomials of degree k. It is well-known that $F_i(x, y)$ can be expressed in the form

$$F_i(x, y) = \sum_{\substack{\mu+\nu \leq k \\ \mu \geq 0 \\ \nu \geq 0}} a^{(i)}_{\mu\nu} P_\mu(x) P_\nu(y),$$

where
$$\mu! P_\mu(x) = x(x-1) \dots (x-\mu+1) \quad [P_0(x) = 1]$$

and the a's are integers. Let

$$\begin{vmatrix} a^{(i)}_{0k} & \dots & a^{(i)}_{k0} \\ \dots & \dots & \dots \\ a^{(k+1)}_{0k} & \dots & a^{(k+1)}_{k0} \end{vmatrix} = \Delta.$$

We then have

THEOREM 1. *If $\Delta \neq 0$, then there exist integers $m_1, \dots, m_{k+1}$, and N such that, for any set of integers $n_i \equiv m_i \pmod{|\Delta|}$, we can find integers x_i, y_i and appropriate signs $\epsilon_i = \pm 1$ such that*

$$\sum_{i=1}^{N} \epsilon_i F_j(x_i, y_i) = n_j \quad (j = 1, \dots, k+1).$$

We can take N to be $[\frac{1}{2}k+1]\,2^{k-1}$.

Proof. There exist integers b_{ij} $(1 \leq i, j \leq k+1)$ such that the polynomials

$$Q_i(x, y) = \sum_{j=1}^{k+1} b_{ij} F_j(x, y) \quad (|b_{ij}| = \pm 1)$$

have the following special form:

$$Q_i(x, y) = \sum_{\substack{\mu+\nu \leq k \\ \mu \geq 0 \\ \nu \geq 0}} A^{(i)}_{\mu\nu} P_\mu(x) P_\nu(y),$$

$$A^{(i)}_{j, k-j} = 0, \quad \text{for} \quad j+1 < i,$$

$$A^{(1)}_{0k} \dots A^{(k+1)}_{k0} = \Delta.$$

It is easy to verify that, if the theorem is true for $Q_i(x, y)$, it holds for $F_i(x, y)$ also.

For any integral-valued polynomial $R(x, y)$, let $\Delta_t^x\big(R(x, y)\big)$ and $\Delta_t^y\big(R(x, y)\big)$ denote the t-th differences of $R(x, y)$ with respect to x and y. $\Delta_t^x\big(R(x, y)\big)$ is the sum of 2^t terms of the type $\pm R(x, y)$.

(1) k odd. Consider the $k+1$ expressions

$$\Delta_{k-1}^y Q_i(x_1, y_1)+\Delta_{k-3}^y \Delta_2^x Q_i(x_2, y_2)+\ldots+\Delta_{k-1}^x Q_i\big(x_{\frac12(k+1)}, y_{\frac12(k+1)}\big),$$

corresponding to $i=1, 2, \ldots, k+1$. These reduce to the system of linear forms

$$A_{0,k}^{(1)} y_1+A_{1,k-1}^{(1)} x_1+A_{2,k-2}^{(1)} y_2+A_{3,k-3}^{(1)} x_2+\ldots+m_1,$$

$$A_{1,k-1}^{(2)} x_1+A_{2,k-2}^{(2)} y_2+A_{3,k-3}^{(2)} x_2+\ldots+m_2,$$

$$\ldots \quad \ldots \quad \ldots \quad \ldots \quad \ldots \quad \ldots$$

where $m_1, m_2, \ldots$ are constant terms so arising. Since

$$A_{0,k}^{(1)} A_{1,k-1}^{(2)} \ldots A_{k,0}^{(k+1)}=\Delta,$$

for any system of integers n_i which are congruent to m_i (mod $|\Delta|$) respectively, the system of linear equations

$$\Delta_{k-1}^y Q_i(x_1, y_1)+\ldots+\Delta_{k-1}^x Q_i(x_{\frac12(k-1)}, y_{\frac12(k-1)})=n_i \quad (i=1, \ldots, k+1)$$

is solvable in integers. Hence we can solve the system of equations given in the theorem with $N=(k+1)\,2^{k-2}$.

(2) k even. By considering

$$\Delta_{k-1}^y Q_i(x_1, y_1)+\Delta_{k-3}^y \Delta_2^x Q_i(x_2, y_2)+\ldots+\Delta_1^y \Delta_{k-2}^x Q_i(x_{\frac12 k}, y_{\frac12 k})$$
$$+\Delta_{k-1}^x Q_i(x_{\frac12 k+1}, 0)$$

we can obtain the result in just the same way as in (1), the number of Q's now being $(k+2)\,2^{k-2}$.

COROLLARY. *If $\Delta=\pm 1$, then there exists an integer N such that*

$$\sum_{i=1}^{N} \epsilon_i F_j(x_i, y_i)=n_j \quad (\epsilon_i=\pm 1, \quad j=1, \ldots, k+1),$$

is solvable for every set of integers n_i.

s 2

For example, the set of polynomials

$$G_i(x, y) = P_i(x)\, P_{k-i}(y) \quad (i = 1, \ldots, k+1)$$

satisfies this condition.

The problem of finding N so that

$$n_1 X^k + n_2 X^{k-1} Y + \ldots + n_{k+1} Y^k = \sum_{i=1}^{N} \epsilon_i (x_i X + y_i Y)^k$$

is solvable, is equivalent to the problem discussed above with

$$F_1(x, y) = x^k, \quad F_2(x, y) = k x^{k-1} y, \ \ldots, \quad F_{k+1}(x, y) = y^k.$$

Notice that in this case

$$\Delta^x_{k-1} \Delta^y_{i-1} F_j(x, y) = \frac{k!}{(j-1)!\,(k-j+1)!} \Delta^x_{k-i} \Delta^y_{i-1} (x^{k-j+1} y^{j-1})$$

$$= \begin{cases} k!\,x, & \text{for } j = i, \\ k!\,y, & \text{for } j = i+1, \\ 0, & \text{otherwise,} \end{cases}$$

and, for k odd,

$$\sum_{i=1}^{\frac{1}{2}(k+1)} \Delta^x_{k-2i+1} \Delta^y_{2i-2} F_j(x_i, y_i) = \begin{cases} k!\,x_{\frac{1}{2}(j+1)}, & \text{for odd } j, \\ k!\,y_{\frac{1}{2}j}, & \text{for even } j; \end{cases}$$

and, for k even,

$$\sum_{i=1}^{\frac{1}{2}k} \Delta^x_{k-2i+1} \Delta^y_{2i-2} F_j(x_i, y_i) + \Delta^y_{k-1} F_j(x_{\frac{1}{2}k+1}, 0) = \begin{cases} k!\,x_{\frac{1}{2}(j+1)}, & \text{for odd } j, \\ k!\,y_{\frac{1}{2}j} & \text{for even } j. \end{cases}$$

Applying the method used before, we have

THEOREM 2. *Every homogeneous binary form of degree k, whose coefficients are multiples of $k!$, can be expressed as the sum or difference of at most $[\frac{1}{2}k+1]\,2^{k-2}$ k-th powers of linear forms with integral coefficients.*

Theorem 2 shows that it is sufficient to examine the forms whose coefficients lie between 0 and $k!$ in order to obtain a result for the number of positive or negative k-th powers of linear forms which suffice to represent a given form. Since the upper bound here given is not the least possible, the problem stated by Prof. Mordell can be partly solved by examining the forms whose coefficients are less than $k!$.

It does not seem easy to extend the method to more than two variables. The following method is simpler (but the upper bound obtained is larger) and can be extended to more than two variables.

Since

$$\Delta^y_{k-1} F_i(0, y_1)+\Delta^y_{k-2}\Delta_1 F_i(0, y_2)+\ldots+\Delta_1{}^x \Delta^y_{k-2} F_i(0, y_{k-1})+\Delta^x_{k-1} F_i(x, y_k)$$
$$= a^{(i)}_{1,k} y_1+a^{(i)}_{1,k-1} y_2+\ldots+a^{(i)}_{k-1,1} y_k+a^{(i)}_{k,0} x+m_i,$$

where the m's are integers, the system of Diophantine equations

$$\sum_{i=1}^{N} \epsilon_i F_j(x_i, y_i) = n_j \quad (j = 1, \ldots, k+1)$$

is solvable for $n_i = m_i \pmod{|\Delta|}$ with $N = k2^{k-1}$.

The corresponding upper bound for the problem of n variables is

$$\frac{1}{n!} k(k+1)\ldots(k+n) 2^{k-1}.$$

Applying this method we can get the following results :

THEOREM 3. *Let* $\mathfrak{R}$ *be a finite ring without nilpotent element and let its basis be* $\omega_1, \ldots, \omega_n$. *Then there exist integers* m, N, *and an element* ω *of the ring such that every element of the ring congruent to* ω (mod m) *can be expressed as the sum or difference of* N *values of* X^k, *where the* X'*s are elements of the ring.*

Let
$$X = x_1\omega_1+\ldots+x_n\omega_n,$$
and
$$X^k = y_1\omega_1+\ldots+y_n\omega_n,$$

where the y's are n homogeneous polynomials of degree k in $x_1, \ldots, x_n$. Let R be the resultant of the y's. Then a necessary and sufficient condition for

$$y_1 = 0, \ldots, y_n = 0$$

to have a non-trivial solution is that $R = 0$. Since the ring has no nilpotent element, we have $R \neq 0$. Therefore there do not exist integers $l_1, \ldots, l_n$ such that

$$l_1 y_1+\ldots+l_n y_n = 0$$

identically. By our previous argument the theorem follows.

I should like to express my thanks to Prof. Mordell for his encouragement.

36 Belvoir Road,
Cambridge.

ON A GENERALIZED WARING PROBLEM

By LOO-KENG HUA†.

[*Extracted from the Proceedings of the London Mathematical Society, Ser.* 2, *Vol.* 43, 1937.]

[Received 29 June, 1936.—Read 12 November, 1936.—Received in revised form 23 March, 1937.]

1. *Introduction.*

In the first part of this paper we wish to find a generalization for polynomials of the Hardy-Littlewood asymptotic formula in Waring's problem, that is to say a formula for the number of solutions of the Diophantine equation

$$(1) \qquad N = P_1(h_1) + \ldots + P_s(h_s) \quad (h_\nu \geqslant 0),$$

where the P's are integral-valued polynomials of the k-th degree with positive highest coefficient. The method used is essentially due to Vinogradow (**3**)‡. For major arcs, Heilbronn's method in (**1**) is adopted.

Let $G\{P(h)\}$ be the least integer s such that (1) has a solution for $P_1(h) = P_2(h) = \ldots = P_s(h) = P(h)$§ when N is sufficiently large. Then $P(h)$ must satisfy the condition that, if q is an integer greater than 1, there exists no integer d such that

$$P(h) \equiv d \pmod q$$

identically.

The principal results for $G\{P(h)\}$ are as follows.

THEOREM A. *If $P(h)$ is an odd integral-valued polynomial of the k-th degree, where $k > 20$, then*

$$G\{P(h)\} \leqslant \tfrac{1}{3}(2^{k+1} - 1);$$

† Research Fellow of the China Foundation for the Promotion of Education and Culture.

‡ See the list of references at the end of the paper.

§ A study of the problem with different P's will be given later.

if, further, the highest coefficient of $P(h)$ is $a/k!$, where $(a, k!) = 1$, then

$$G\{P(h)\} = O(2^{\epsilon k}),$$

for any $\epsilon > 0$.

This is equivalent to Theorems 5 and 6 together.

THEOREM B. *If $P(h)$ is an integral-valued polynomial of the k-th degree, where $k > 20$, then*

$$G\{P(h)\} \leqslant 3^{\frac{1}{2}k}\, 2^{\frac{1}{2}(k-1)+7}\, k,$$

and

$$G\{P(h)\} = O(k^3\, 2^{k-1});$$

if, further, the highest coefficient of $P(h)$ is $a/k!$, where $(a, k!) = 1$, then

$$G\{P(h)\} = O(3^{k(\frac{1}{2}+\epsilon)}).$$

This is equivalent to Theorems 7, 8, and 9.

The existence of the upper bound of $G\{P(h)\}$ was first proved by Kamke (**1**), but no one has given a definite upper bound for $G\{P(h)\}$.

The main difficulties of the work are arithmetical and not analytical, and in some cases I have quoted analytic theorems of Vinogradow or other writers in slightly generalized form. The alterations necessary in their analysis are not difficult.

I wish to take this opportunity of expressing my gratitude to Dr. Heilbronn for his encouragement and advice.

2. *Notation.*

$r_{P_1, \ldots, P_s}(N) = r(N)$ is the number of solutions of the Diophantine equation

$$N = P_1(h_1) + P_2(h_2) + \ldots + P_s(h_s) \quad (h_\nu \geqslant 0,\ \nu = 1, 2, \ldots, s),$$

where $P_\nu(h)$ $(\nu = 1, 2, \ldots, s)$ are integral-valued polynomials of the k-th degree with positive coefficients and constant term zero†.

N_ν is the greatest positive root of $P_\nu(h) = N (> 0)$. It exists and is positive and tends to infinity with N.

k is an integer.

† The only essential restriction is that the highest coefficient of each polynomial is positive. For we can always choose an integer l large enough to make all the coefficients of $Q_\nu(h) = P_\nu(h+l)$ positive.

$b = k/(k+1)$.

q_ν^* denotes the connected number† of q with respect to $P_\nu(h)$.

We write

$$e^{2\pi i x} = e(x), \quad \rho = e(l/q), \quad (l, q) = 1,$$

$$S_{\rho\nu} = \sum_{h=0}^{q_\nu^*} \rho^{P_\nu(h)} \quad (\nu = 1, 2, \ldots, s),$$

$$A(q) = A_s(q, j) = \sum_{\rho} \left(\prod_{\nu=1}^{s} \frac{S_{\rho\nu}}{q_\nu^*} \right) \rho^{-j},$$

where ρ runs through all primitive q-th roots of unity. In what follows Π and Σ will mean $\prod_{\nu=1}^{s}$ and $\sum_{\nu=1}^{s}$ respectively, unless the contrary is stated.

$$\mathfrak{S}(n) = \sum_{q=1}^{\infty} A_s(q, n),$$

$$T_\nu = T_\nu(\alpha) = \sum_{1 \leqslant \mu \leqslant P} e\{P_\nu(\mu)\,\alpha\},$$

$$I_\nu = I_\nu(\alpha) = \int_0^P e(\beta a_\nu v^k)\, dv,$$

where a_ν is the highest coefficient in $P_\nu(h)$.

N is the integer to be represented. We can define a constant c such that

$$\max_{1 \leqslant \nu \leqslant s} N_\nu \leqslant c N^{1/k}.$$

P is a number satisfying the condition

$$c N^{1/k} \leqslant P = O(N^{1/k}).$$

3. *Lemmas for exponential sums.*

LEMMA 3.1. *If $Q(h)$ is a polynomial with integral coefficients, and $(l, q) = 1$, then*

$$\sum_{h=1}^{q} e\left(\frac{l}{q}\, Q(h)\right) = O(q^{1-1/k+\epsilon}),$$

where the O depends only on ϵ and the coefficients of $Q(h)$ [Hua (**5**)].

† Hua (**3**), (**4**).

Lemma 3.2. *Let $k > 20$ and let N be a sufficiently large integer. Then we can choose P and H, depending only on k, so that*

$$cN^{1/k} \leqslant P = O(N^{1/k}), \quad 0 < H < 9k^3 \log k,$$

$$\int_0^1 |S|^H \, d\alpha = O(P^{H-k+\frac{1}{20}}),$$

where

$$S = \sum_{x=1}^{P} e\{Q(x)\}$$

and $Q(x)$ is an integral-valued polynomial of the k-th degree.

[This is a slight generalization of Lemma K of Vinogradow (**3**).]

Lemma 3.3. *If $\alpha_0, \alpha_1, \ldots, \alpha_k$ are real numbers, $k > 20$, and*

$$\left| \alpha_0 - \frac{l}{q} \right| < \frac{1}{q^2}, \quad f(x) = \alpha_0 x^k + \ldots + \alpha_k \quad [(l, q) = 1],$$

then

$$\left| \sum_{x=R+1}^{R+P} e\{f(x)\} \right| = O(P^{1-\frac{1}{18}k^{-3}(\log k)^{-1}}),$$

provided that

$$P^b < q \leqslant P^{k-b}.$$

[Vinogradow (**3**)].

4. *Farey dissection.*

We divide the interval $0 \leqslant \alpha \leqslant 1$ in the usual way into Farey arcs belonging to all rational points l/q with $1 \leqslant q \leqslant P^{k-b}$, $0 \leqslant l < q$, $(l, q) = 1$. We divide these arcs into major arcs $\mathfrak{M}(1 \leqslant q \leqslant P^b)$ and minor arcs $\mathfrak{m}(P^b < q \leqslant P^{k-b})$. In each case the arc has the form

$$\alpha = \frac{l}{q} + \beta, \quad -\vartheta_1 q^{-1} P^{b-k} \leqslant \beta \leqslant \vartheta_2 q^{-1} P^{b-k}, \quad \tfrac{1}{2} \leqslant \vartheta_1 \leqslant 1, \quad \tfrac{1}{2} \leqslant \vartheta_2 \leqslant 1.$$

5. *Lemmas for major arcs.*

Lemma 5.1. $$S_{\rho\nu} = O(q^{1-1/k}).$$

Proof. Let d be the least common denominator of the coefficients of $P_\nu(h)$. Then, by the definition of the connected number q_ν^*, $qd = q_\nu^* d_1$ for some integer d_1, and we have

$$S_{\rho\nu} = \sum_{h=1}^{q_\nu^*} e\left(\frac{dP_\nu(h)}{q_\nu^* d_1}\right) = \frac{1}{d_1} \sum_{h=1}^{q_\nu^* d_1} e\left(\frac{dP_\nu(h)}{q_\nu^* d_1}\right).$$

Since $dP_\nu(h)$ is a polynomial with integral coefficients and $d \leqslant k!$, we have

$$S_{\rho\nu} = O\{(q_\nu{}^* d_1)^{1-1/k} d_1^{-1}\} = O(q^{1-1/k}),$$

by Lemma 3.1.

LEMMA 5.2. *If* $s > 2k$, *then*

$$\mathfrak{S} = \mathfrak{S}(n)$$

converges absolutely.

Proof. By Lemma 5.1,

$$\left| \sum_\rho \left(\Pi \frac{S_{\rho\nu}}{q_\nu{}^*} \right) \rho^{-n} \right| \leqslant q, \; \Pi \left| \frac{S_{\rho\nu}}{q_\nu{}^*} \right| = O(q^{-1-1/k}),$$

which proves the lemma.

LEMMA 5.3.

$$I_\nu = O\{\min(P, |\beta|^{-1/k})\}$$

[Heilbronn (**1**), formula (13)].

LEMMA 5.4. *On* $\mathfrak{M}$ *we have*

$$T_\nu = \frac{S_{\rho\nu}}{q_\nu{}^*} I_\nu + O(P^b)$$

[Heilbronn (**1**), Lemma 4; Landau (**3**)].

LEMMA 5.5. *If* $|A_\nu - B_\nu| \leqslant C$ *and* $|B_\nu| \leqslant D$, *then*

$$\left| \prod_{\nu=1}^{t} A_\nu - \prod_{\nu=1}^{t} B_\nu \right| \leqslant K(t)(C^t + CD^{t-1}).$$

Proof. The lemma is evident for $t = 1$, and the general case follows by induction.

LEMMA 5.6. *If* $s > 2k+1$, *then*

$$\sum_{\mathfrak{M}} \int_{\mathfrak{M}} \left| \Pi T_\nu - \Pi \frac{S_{\rho\nu}}{q_\nu{}^*} I_\nu \right| d\alpha = O(P^{s-k+1+b})$$

[Lemma 5.5 and Heilbronn (**1**), Lemma 5].

LEMMA 5.7. *If* $0 < N \leqslant P^k$, *then*

$$\int_{-\infty}^{\infty} e^{-2\pi i N\beta} \Pi I_\nu \, d\beta = \frac{\Gamma^s(1+1/k)}{\Gamma(s/k)} \Pi a_\nu^{-1/k} N^{s/k-1}$$

[Landau (**3**)].

Lemma 5.8. *If* $0 < N \leqslant P^k$, *then*

$$\int_{\mathfrak{M}} e^{-2\pi i N\beta} \Pi I_\nu d\beta = \frac{\Gamma^s(1+1/k)}{\Gamma(s/k)} \Pi a_\nu^{-1/k} N^{(s/k)-1} + O(P^{s-k-b\{(s/k)-1\}} q^{(s/k)-1})$$

[Heilbronn (1), Lemma 7].

Lemma 5.9. *If* $0 < N \leqslant P^k$, *then*

$$\sum_{\mathfrak{M}} \int_{\mathfrak{M}} e^{-2\pi i N a} \Pi I_\nu da = \Pi a_\nu^{-1/k} \frac{\Gamma^s(1+1/k)}{\Gamma(s/k)} \mathfrak{S}(N) N^{(s/k)-1} + O(P^{s-k-(1-b)}).$$

Proof. By Lemma 5.6, the left-hand side of the formula is

$$O(P^{s-k-1+b}) + \sum_{\mathfrak{M}} \left(\Pi \frac{S_{\rho\nu}}{q_\nu{}^*}\right) \int_{\mathfrak{M}} e^{-2\pi i N a} \Pi I_\nu da$$

$$= O(P^{s-k-1+b}) + \frac{\Gamma^s(1+1/k)}{\Gamma(s/k)} \Pi a_\nu^{-1/k} \sum_{\mathfrak{M}} \Pi \frac{S_{\rho\nu}}{q_\nu{}^*} e^{-2\pi i N(a/q)}$$

$$+ O\left(\sum_{\mathfrak{M}} q^{-s/k} P^{-s-k-b\{(s/k)-1\}} q^{(s/k)-1}\right)$$

(Lemmas 5.8 and 5.9)

$$= O(P^{s-k-(1-b)}) + \frac{\Gamma^s(1+1/k)}{\Gamma(s/k)} \Pi a_\nu^{-1/k} N^{(s/k)-1} \{\mathfrak{S} + \sum_{q > P^b} O(q^{1-(s/k)})\}$$

$$+ O(P^{s-k-b(s/k-1)}) \sum_{\mathfrak{M}} q^{-1}$$

$$= O(P^{s-k-(1-b)}) + \frac{\Gamma^s(1+1/k)}{\Gamma(s/k)} \mathfrak{S} N^{(s/k)-1} + O(P^{s-k-b\{(s/k)-2\}}).$$

6. *Lemmas for minor arcs.*

Lemma 6.1. *If* $s > 10k^3 \log k$ *and* $k > 20$, *then*

$$\sum_{\mathfrak{m}} \int_{\mathfrak{m}} \prod_{\nu=1}^{s} T_\nu e^{-2\pi i N a} da = O(P^{s-k-(1-b)}).$$

Proof. Choose H and P to satisfy the requirements of Lemma 3.2. Since the geometric mean is less than the arithmetic mean, we have

$$\int_0^1 \prod_{\nu=1}^{H} |T_\nu| da \leqslant \frac{1}{H} \sum_{\nu=1}^{H} \left(\int_0^1 |T_\nu|^H da\right) = O(P^{H-k+\frac{1}{20}}).$$

Thus, by Lemma 3.3,

$$\sum_{\mathfrak{m}} \int_{\mathfrak{m}} \prod_{\nu=1}^{s} |T_\nu| \, d\alpha = O\left(\max_{\text{on } \mathfrak{m}} \prod_{\nu=H+1}^{s} |T_\nu| \int_0^1 \prod_{\nu=1}^{H} |T_\nu| \, d\alpha \right)$$

$$= O(P^{\{1-\frac{1}{18}k^{-3}(\log k)^{-1}\}(s-H)+H-k+\frac{1}{20}})$$

$$= O(P^{s-k-(1-b)}).$$

THEOREM 1. *Let $r(N)$ be the number of solutions of the Diophantine equation*

$$N = P_1(x_1) + \dots + P_s(x_s), \quad x_\nu \geqslant 0.$$

Then, if $k > 20$ and $s > 10k^3 \log k$, we have

$$r(N) = \left(\prod_{\nu=1}^{s} a_\nu^{-1/k} \right) \frac{\Gamma^s(1+1/k)}{\Gamma(s/k)} \mathfrak{S}(N) N^{(s/k)-1} + O(P^{s-k-(1-b)}).$$

The theorem follows from the Lemmas 5.9 and 6.1.

7. *Lemmas concerning congruences and singular series.*

The following notation will be used throughout the remaining part of this paper:

d_ν is the least common denominator of the coefficients of $P_\nu(h)$.

$d_\nu P_\nu(h) = \phi_\nu(h)$.

θ_ν is the highest power of p such that

$$\phi_\nu'(h) \equiv 0 \pmod{p^{\theta_\nu}}$$

for all values of h.

$P_\nu^*(h) = p^{-\theta_\nu} \phi_\nu'(h)$.

$M(n) = M(P_1, \dots, P_s, m, n)$ is the number of solutions of

$$(7.1) \quad n \equiv P_1(h_1) + \dots + P_s(h_s) \pmod{m}, \quad 0 \leqslant h_\nu \leqslant m_\nu^*, \quad \nu = 1, 2, \dots, s.$$

When $m = p^l$, $N(p^l, n) = N(P_1, \dots, P_s;\ p^l, n)$ denotes the number of solutions of the congruence (7.1) in which at least one of the numbers $P_1^*(h_1), \dots, P_s^*(h_s)$ is prime to p.

$$M(s.P, m, n) = M(P_1, \dots, P_s;\ m, n)$$

and

$$N(s.P, p^l, n) = N(P_1, \dots, P_s, p^l, n),$$

provided that $P_1, \dots, P_s$ are the same polynomial.

δ is the greatest integer satisfying $p^\delta \leqslant k-1$.

If we define t_ν by $p_\nu^* = p^{1+t_\nu}$, then $d_\nu = p^{t_\nu} d_\nu'$ and $(d_\nu', p) = 1$.

$$\gamma_\nu = \begin{cases} \theta_\nu + 2 - t_\nu + \delta & \text{for} \quad p = 2, \\ \theta_\nu + 1 - t_\nu + \delta & \text{for} \quad p \neq 2. \end{cases}$$

$\gamma = \max(\gamma_1, \ldots, \gamma_s)$.

LEMMA 7.1. *If* $(m^{(1)}, m^{(2)}) = 1$, *then*

$$M(m^{(1)}) M(m^{(2)}) = M(m^{(1)} m^{(2)}).$$

LEMMA 7.2.

$$\sum_{q/m} A(q) = \frac{m}{\Pi m_\nu^*} M(m).$$

LEMMA 7.3. *If* $(m^{(1)}, m^{(2)}) = 1$, *then*

$$\sum_{q/m^{(1)} m^{(2)}} A(q) = \sum_{q/m^{(1)}} A(q) \sum_{q/m^{(2)}} A(q).$$

These lemmas can be proved by the method used by Landau (**4**).

LEMMA 7.4. *If* $Q(h)$ *is a residual polynomial of the k-th degree* (mod p^α), *and* $Q'(h)$ *is a residual polynomial* (mod p^β), *then*

$$p^{\alpha-\beta} \leqslant k.$$

Proof. (1) First let us consider the special case

$$Q(h) = Q_r(h) = h(h-1) \ldots (h-r+1).$$

In this case, $\alpha = \sum_{l=1}^{\infty} \left[\frac{r}{p^l}\right]$. Since $|Q_r'(i)| = i!(r-i-1)!$, we have

$$\beta = \min_{0 \leqslant i \leqslant r} \left\{ \sum_{l=1}^{\infty} \left(\left[\frac{r-1-i}{p^l}\right] + \left[\frac{r}{p^l}\right] \right) \right\}.$$

If $p^\sigma | r$ and $p^{\sigma+1} \nmid r$, then

$$\alpha - \sum_{l=1}^{\infty} \left[\frac{r-1}{p^l}\right] = \sigma$$

and
$$\sum_{l=1}^{\infty} \left[\frac{r-1}{p^l}\right] - \sum_{l=1}^{\infty} \left(\left[\frac{r-1-i}{p^l}\right] + \left[\frac{i}{p^l}\right] \right) \leqslant \delta_1 - \sigma,$$

where δ_1 is the greatest integer satisfying $p^{\delta_1} \leqslant r$. Hence the lemma is true for $Q_r(h)$.

(2) Since $Q(h)$ can be put into the form

$$Q(h) = \sum_{r=1}^{k} a_r Q_r(h),$$

and $a_r Q_r(h)$ $(r=1, 2, \ldots, k)$ are residual polynomials $(\text{mod } p^{\alpha})$, the lemma follows from (1).

LEMMA 7.5. *If* $l \geqslant \gamma_\nu + 1$ *and* $h = y + p^{l+t_\nu-\theta_\nu-1} z$, *then*

$$P_\nu(h) \equiv P_\nu(y) + zp^{l-1} P_\nu^*(y) \pmod{p^l}$$

and

$$P_\nu^*(h) \equiv P_\nu^*(y) \pmod{p}.$$

Proof. By Taylor's expansion, we have

$$\phi_\nu(h) \equiv \phi_\nu(y) + zp^{l+t_\nu-\theta_\nu-1} \phi_\nu'(y) + \frac{1}{2!} z^2 p^{2(l+t_\nu-\theta_\nu-1)} \phi_\nu''(y) + \frac{1}{3!} z^3 p^{3(l+t_\nu-\theta_\nu-1)} \phi_\nu'''(y) + \ldots.$$

(1) $p \neq 2$. Since $\phi_\nu''(y)$ is a residual polynomial $(\text{mod } p^{\theta_\nu-\delta})$, it follows that

$$\phi_\nu(h) \equiv \phi_\nu(y) + zp^{l+t_\nu-\theta_\nu-1} \phi_\nu'(y) \pmod{p^{l+t_\nu}}.$$

(Notice that

$$p^{i(l+t_\nu-\theta_\nu-1)} \frac{\phi_\nu^{(i)}(y)}{i!} \equiv 0 \pmod{p^{l+t_\nu}} \quad \text{for} \quad i \geqslant z).$$

(2) $p=2$. In this case, $\frac{1}{2}\phi_\nu''(y)$ is a residual polynomial $(\text{mod } p^{\theta_\nu-\delta-1})$, and we have the lemma in the same way as in the case (1).

Using the same argument, we get the second conclusion.

LEMMA 7.6. *If* $l \geqslant \gamma_\nu + 1$, *then*

$$N(p^l) = p^{s-1} N(p^{l-1}).$$

Proof. In (7.1) let $h_\nu = y_\nu + p^{l+t_\nu-\theta_\nu-1} z_\nu$, where $0 \leqslant y_\nu < p^{l+t_\nu-\theta_\nu-1}$, $0 \leqslant z_\nu < p^{\theta_\nu+1}$, and p does not divide every $P_\nu^*(y_\nu)$. We obtain

$$\sum_{\nu=1}^{s} P_\nu(y_\nu) + p^{l-1} \sum_{\nu=1}^{s} z_\nu P_\nu^*(y_\nu) \equiv n \pmod{p^l}. \quad \begin{cases} 0 \leqslant y_\nu < p^{l+t_\nu-\theta_\nu-1}, \\ 0 \leqslant z_\nu < p^{\theta_\nu+1}). \end{cases}$$

Then to each solution of this congruence there corresponds a solution of the two congruences

(7.6.1) $\sum_{\nu=1}^{s} P_\nu(y_\nu) \equiv n \pmod{p^{l-1}}$, $0 \leqslant y_\nu < p^{l+t_\nu-\theta_\nu-1}$, $p \nmid$ every $P_\nu^*(y_\nu)$,

(7.6.2) $\sum_{\nu=1}^{s} z_\nu P_\nu^*(y_\nu) \equiv p^{-(l-1)}\left(n - \sum_{\nu=1}^{s} P_\nu(y_\nu)\right) \pmod{p}$, $0 \leqslant z_\nu < p^{\theta_\nu+1}$;

and conversely.

We shall prove next that the former has $\Pi p^{-\theta_\nu} N(p^{l-1})$ and the latter $p^{-1}\Pi p^{\theta_\nu+1}$ solutions. It follows from Lemma 7.5 that, if $x_\nu \equiv y_\nu \pmod{p^{l+t_\nu-\theta_\nu-1}}$, then $P_\nu(x_\nu) \equiv P_\nu(y_\nu) \pmod{p^{l-1}}$.

Hence, each solution of (7.6.1) gives Πp^{θ_ν} solutions of

$$\sum_{\nu=1}^{s} P_\nu(y_\nu) \equiv n \pmod{p^{l-1}}, \quad 0 \leqslant y_\nu < p^{l+t_\nu-1}, \quad p \nmid \text{every } P_\nu^*(y_\nu).$$

Since p does not divide every $P_\nu^*(y_\nu)$, we may assume that $p \nmid P_1^*(y_1)$. Then $z_\nu (\nu = 2, 3, \ldots, s)$ may be chosen arbitrarily $(\bmod\, p^{\theta_\nu+1})$ in (7.6.2) and z_1 is determined uniquely $(\bmod\, p)$. Thus, for any set of $z_\nu \pmod{p^{\theta_\nu+1}}$ $(\nu = 2, 3, \ldots, s)$, there are p^{θ_1} solutions z_1 for which $0 \leqslant z_1 < p^{\theta_1+1}$. Hence (7.6.2) has $p^{-1} \prod_{\nu=1}^{s} p^{\theta_\nu+1}$ solutions.

As a consequence of the above lemma we have

LEMMA 7.7. *If $l \geqslant \gamma$, then*

$$N(p^l) = p^{(l-\gamma)(s-1)} N(p^\gamma).$$

LEMMA 7.8. *If $M(P_1, \ldots, P_{s-1}, p^l, n) > 0$ for all integers n, then*

$$N(P_1, \ldots, P_{s-1}, P_s, p^l, n) > 0$$

for all integers n.

Proof. Since $P_s^*(h)$ is not congruent to zero $(\bmod\, p)$ identically, there is an integer h_s such that $p \nmid P_s^*(h_s)$, $0 \leqslant h < p^{l+t_s}$. By hypothesis, we see that the equation

$$\sum_{\nu=1}^{s-1} P_\nu(h_\nu) = n - P_s(h_s) \pmod{p^l}$$

has at least one solution and hence the lemma is true.

Let H denote the hypothesis that

$$N(P_1, \ldots, P_{s_0}, n, p^\gamma) > 0$$

for all primes p and all integers n.

LEMMA 7.9. *If H is true, and if $s \geqslant \max(s_0, 2k)$, then*

$$\mathfrak{S}(n) \geqslant D_1 > 0,$$

where D_1 (and later $D_2, \ldots, D_4$) is a positive constant depending on the coefficients of $P_\nu(h)$ $(\nu = 1, 2, \ldots, s)$.

Proof. By Lemma 7.2, we have

$$\sum_{q/p^l} A(q) = \frac{p^l}{\Pi p^{l+t_\nu}} M(p^l).$$

By Lemma 7.7,

$$\sum_{q/p} A(q) \geqslant p^{-l(s-1)-\Sigma t_\nu} N(p^l) \tag{7.9.1}$$
$$\geqslant p^{-\gamma(s-1)-\Sigma t_\nu},$$

which is independent of l. By Lemma 5.1,

$$\sum_{q/p^l} A(q) = 1 + \sum_{\lambda=1}^{l} A(q^\lambda) > 1 - D_2 \sum_{\lambda=1}^{\infty} p^{-\lambda/b}$$
$$= 1 - D_2/(p^{1/b}-1).$$

If $p > D_3 = (D_2+1)^b$, then

$$\prod_{D_3 < p \leqslant p^l} \sum_{q/p^l} A(q) > \prod_{D_3 < p} \{1 - D_2/(p^{1/b}-1)\},$$

where p_l is the l-th prime. The right-hand side converges. Further, by (7.9.1), we have

$$\prod_{p \leqslant D_3} \sum_{q/p^l} A(q) > D_4,$$

which is independent of l.

Let l tend to infinity. We have, by Lemma 7.3,

$$\mathfrak{S}(n) \geqslant D_1 > 0.$$

Thus we have

THEOREM 2. *If H is true, and $s > \max(s_0 - 1,\ 10k^3 \log k)$, then*

$$r_{P_1, \dots, P_s}(n) > 0.$$

We know [Hardy-Littlewood, (**1**)] that hypothesis H is satisfied when $s_0 = 4k$ and $P_1 = P_2 = \dots = P_{4k} = x^k$. *A fortiori*, it is satisfied when $P_1 = P_2 = \dots = P_{4k} = x^k$ and $P_{4k+1}, \dots, P_{s_0}$ are arbitrary. Consequently we have

THEOREM 3. *If $P_{4k+1}(h), \dots, P_s(h)$ are any integral-valued polynomials of the k-th degree, and $s > 10k^3 \log k$, then, if N is a sufficiently large positive integer, we can choose positive integers $h_1, h_2, \dots, h_s$ such that*

$$N = h_1^k + \dots + h_{4k}^k + P_{4k+1}(h_{4k+1}) + \dots + P_s(h_s) \quad (h_\nu \geqslant 0).$$

More generally, we have

THEOREM 4. *If $a_1, a_2, \dots, a_t$ is an admissible set as defined by Huston (**1**), and $P_{t+1}(h), \dots, P_s(h)$ are any integral-valued polynomials, then, if N is* *sufficiently large and* $s > 10k^3 \log k$, *the Diophantine equation*

$$N = a_1 h_1^k + \ldots + a_t h_t^k + P_{t+1}(h_{t+1}) + \ldots + P_s(h_s)$$

has a solution in positive integers.

In the remaining part of this paper the indices ν of θ_ν, t_ν, ϕ_ν, ... are suppressed.

8. *Lemmas for* θ.

LEMMA 8.1. *If* $p > k$, *then* $\theta = 0$ *and* $t = 0$.

LEMMA 8.2. *If* $p \leqslant k$, *then*

$$p^\theta \leqslant k 2^{k-1}.$$

Proof. Let

$$\phi(h) = c_0 h^k + \ldots + c_{k-1} h.$$

Then
$$p \nmid (c_0, \ldots, c_{k-1}).$$

Let δ_1 be the greatest power of p for which

$$p^{\delta_1} \mid \{k c_0, (k-1) c_1, \ldots, 2c_{k-2}, c_{k-1}\},$$

and let δ_2 be the greatest power of p for which there may exist a residual polynomial (mod p^{δ_2}) of the $(k-1)$-th degree of which at least one coefficient is not divisible by p. Then

$$\theta \leqslant \delta_1 + \delta_2.$$

Since $p \nmid (c_0, c_1, \ldots, c_{k-1})$, p^{δ_1} must divide one of the integers $1, 2, \ldots, k$. Therefore $p^{\delta_1} \leqslant k$.

Since
$$\delta_2 \leqslant \sum_{l=1}^{\infty} \left[\frac{k-1}{p^l}\right] \leqslant \frac{k-1}{p-1},$$

we have
$$p^{\delta_2} \leqslant p^{(k-1)/(p-1)} = e^{(k-1)\log p/(p-1)}.$$

When x takes all positive integral values greater than or equal to 2, the greatest value of $\log x/(x-1)$ is given by $x = 2$. Thus

$$p^\theta \leqslant p^{\delta_1+\delta_2} \leqslant k 2^{k-1}.$$

Consequently we have

LEMMA 8.3. *If* $p < k$ *and* $k \geqslant 4$, *then*

$$p^\gamma \leqslant k^3 2^{k-1}.$$

9. *The relation between Tarry's problem and Waring's problem for odd polynomials.*

Let $P(h)$ be an odd polynomial. Then

$$P(h) = b_0 Q_{2l+1}(h) + b_1 Q_{2l-1}(h) + \ldots + b_l Q_1(h),$$

where
$$Q_{2\mu+1}(h) = \frac{h(h^2-1)\ldots(h^2-\mu^2)}{(2\mu+1)!},$$

the b's are integers, and

$$(b_0, b_1, \ldots, b_l) = 1$$

[*cf.* Hua (**3**), Lemmas 2 and 4].

LEMMA 9.1. *If* $s \geqslant \frac{1}{3}(2^{k+1}-1)$, *then*

$$N(s.P, p^l, n) > 0$$

[Hua (**3**), Lemma 40].

In §§ 9-10 it will be shown that, if we have found a particular solution (non-trivial) of Tarry's problem for a definite $k = k_0$, then we can derive from it an upper bound of $G\{P(h)\}$ for all $k \geqslant k_0$.

Let us suppose that the particular solution is $a_1, \ldots, a_{j_0}; b_1, \ldots, b_{j_0}$, that is to say

$$(9.1) \quad \begin{cases} a_1{}^h + \ldots + a_{j_0}^h = b_1{}^h + \ldots + b_{j_0}^h \quad (1 \leqslant h \leqslant k_0 - 2), \\ a_1^{k_0-1} + \ldots + a_{j_0}^{k_0-1} \neq b_1^{k_0-1} + \ldots + b_{j_0}^{k_0-1}. \end{cases}$$

Let
$$a_1^{k_0-1} + \ldots + a_{j_0}^{k_0-1} - b_1^{k_0-1} - \ldots - b_{j_0}^{k_0-1} = X(a, b, k_0),$$

and let the greatest prime factor of X be $\mathfrak{p}$. For every solution of (9.1) we have a corresponding number $q = \max(j_0, \mathfrak{p})$. The least of the q's is denoted by $T(k_0)$.

LEMMA 9.2. *Let* $k \geqslant k_0$,

$$k-1 = l(k_0-1)+m, \quad l > 0, \quad 0 \leqslant m < k_0 - 1,$$

$j \geqslant 2^{m+l-1} j_0{}^l$. *Then there is a set of integers* $x_1, \ldots, x_j;\ y_1, \ldots, y_j$ *such that*

$$x_1{}^h + \ldots + x_j{}^h = y_1{}^h + \ldots + y_j{}^h \quad (1 \leqslant h \leqslant k-2),$$
$$x_1^{k-1} + \ldots + x_j^{k-1} \neq y_1^{k-1} + \ldots + y_j^{k-1}$$

and the greatest prime factor of

$$X(k) + x_1^{k-1} + \ldots + x_j^{k-1} - y_1^{k-1} - \ldots - y_j^{k-1}$$

does not exceed the greater of k *and* $\mathfrak{p}$.

Proof. By the same argument as that used by Maitland Wright (**3**), we can construct a polynomial

$$f_0(y) = (y-1)^{k_0-1} h_0(y) = \sum_{i=1}^{j_0} y^{a_i} - \sum_{i=1}^{j_0} y^{b_i};$$

and it is easy to verify that

$$\left\{ \left(y \frac{d}{dy} \right)^{k_0-1} f_0(y) \right\}_{y=1} = (k_0-1)!\, h_0(1) = X(k_0).$$

Construct

$$f(y) = (y-1)^m \{f_0(y)\}^l = (y-1)^{k-1} \{k_0(y)\}^l.$$

Then $f(y)$ contains at most $2^m(2j_0{}^l)$ terms, each of the type $\pm y^a$. Further,

$$X(k) = \left\{ \left(y \frac{d}{dy} \right)^{k-1} f(y) \right\}_{y=1} = (k-1)!\, \{h_0(1)\}^l,$$

and the lemma is proved.

LEMMA 9.3. *If* $p > k$ *and* $s \geqslant \max\{\mathfrak{p},\ 2^{m+l-1} j_0{}^l,\ \frac{1}{3}(2^{k_0+1}-1)\}$, *then*

$$N(s\,.\,P,\ p^\gamma,\ n) > 0$$

(*actually* $\gamma = 1$) *for all odd polynomials* $P(h)$.

It is worth while to notice that when k increases, only $2^{m+l-1} j_0{}^l$ increases, the other terms remain unchanged.

Proof. (1) If $p \leqslant \max(\mathfrak{p}, k)$, the conclusion is evident.

(2) If $p > \max(k, \mathfrak{p})$ (here $\gamma = 1$), then $p \nmid X(k)$. If $p \nmid a$ [a, which is not necessarily an integer, is the first coefficient in $P(h)$ and its denominator is relatively prime to p], then, if n is any integer, we can choose x_i, y_i, and x so that $j \leqslant 2^{m+l-1} j_0{}^l$ and

$$\sum_{i=1}^{j} \{P(x+x_i) - P(x+y_i)\} = aX(k)x + Y(k) \equiv n \pmod p.$$

If $p \mid a$, then $P(h) \equiv Q(h) \pmod p$, where $Q(h)$ is a polynomial whose degree is less than that of $P(h)$, and if its degree exceeds k_0, we can apply the same argument to $Q(h)$. If $Q(h)$ is a polynomial of the k_0-th degree or less, then, by Lemma 9.1, if n is any integer and $s \geqslant \frac{1}{3}(2^{k_0+1}-1)$, the congruence

$$\sum_{\nu=1}^{s} Q(h_\nu) \equiv n \pmod p$$

has a solution. The lemma is therefore proved.

LEMMA 9.4. *If* $\epsilon > 0$, *then*

$$2^{m+l-1} j_0^{\,l} = O(2^{\epsilon h}) \quad \text{as} \quad k \to \infty.$$

Proof. If k_0 is large, it follows from Maitland Wright's result (**4**) that

$$2^{m+l-1} j_0^{\,l} \leqslant 2^{m+l-1} \left(\frac{7k_0^{\,2}(k_0-11)^2}{72}+56\right)^l \leqslant 2^{m+l-1} k_0^{4l}$$

$$\leqslant 2^{k_0-1-(k-1)/(k_0-1)-1} k_0^{4(k-1)/(k_0-1)}$$

$$\leqslant 2^{k_0-2-(k-1)(1+4\log k_0/\log 2)/(k_0-1)}.$$

Since
$$\lim_{h_0\to\infty} \frac{1}{k_0-1}(1+4\log k_0/\log 2) = 0,$$

for any given $\epsilon > 0$, there is a k_0 such that

$$\frac{1}{k_0-1}(1+4\log k_0/\log 2) < \epsilon,$$

and this proves our lemma.

LEMMA 9.5. *If* $k \geqslant 10$, $p > k$, *and* $s \geqslant 2^{\frac{1}{2}(k-1)+7}$, *and* $P(h)$ *is an odd polynomial, then*

$$N(s\,.\,P,\, p^{\gamma},\, n) > 0.$$

Proof. Since

$$1^h+5^h+10^h+24^h+28^h+42^h+47^h+51^h$$
$$= 2^h+3^h+12^h+21^h+31^h+40^h+49^h+50^h \quad (1 \leqslant h \leqslant 7),$$

and

$$1^8+5^8+10^8+24^8+28^8+42^8+47^8+51^8$$
$$-2^8-3^8-12^8-21^8-31^8-40^8-49^8-50^8 = 2^{11}.3^3.5^2.7^2.11.13,$$

it follows that

$$k_0 = 9, \quad j_0 = 8, \quad \mathfrak{p} = 13.$$

Here
$$\max\{\mathfrak{p},\, \tfrac{1}{3}(2^{k_0+1})\} = 341$$

and
$$2^{m+j-1} j_0^{\,l} = 2^{m+4l-1} \leqslant 2^{k_0-1-4(h-1)/(k_0-1)-1} = 2^{\frac{1}{2}(k-1)+7}.$$

The conclusion follows from Lemma 9.3.

By Theorem 2 and Lemma 9.1 we have

THEOREM 5. *If $k > 20$, every sufficiently large integer can be expressed as the sum of $\frac{1}{3}(2^{k+1}-1)$ values, for $h \geqslant 0$, of an odd polynomial $P(h)$ of the k-th degree.*

THEOREM 6. *If the highest coefficient of $P(h)$ is $a/k!$, $(a, k!) = 1$, then*

$$G\{P(h)\} = O(2^{\epsilon k}).$$

This theorem follows from Lemma 9.4 and the fact that $\gamma \leqslant \delta + 2$.

10. *Lemmas for general polynomials.*

Let $K(2l)$ be the least s such that the system of congruences

$$x_1^{2\nu} + \dots + x_s^{2\nu} \equiv 0 \pmod{p}, \quad p + x_1, \quad \nu = 1, \dots, l$$

has a solution. We assume that $p > k$ and $2l \leqslant k$.

LEMMA 10.1. *If*

$$x_1{}^{\mu} + \dots + x_{s_1}^{\mu} \equiv 0 \pmod{p}, \quad p + x_1$$

and

$$y_1{}^{\nu} + \dots + y_{s_2}^{\nu} \equiv 0 \pmod{p}, \quad p + y_1,$$

then

$$\sum_{t_2=1}^{s_2} \sum_{t_1=1}^{s_1} (x_{t_1} y_{t_2})^{\nu} \equiv 0 \pmod{p}$$

and

$$\sum_{t_2=1}^{s_2} \sum_{t_1=1}^{s_1} (x_{t_1} y_{t_2})^{\mu} \equiv 0 \pmod{p}.$$

LEMMA 10.2. *There is an integer $m \leqslant 2l$ such that $m \mid p-1$, and there is no integer $\nu \leqslant l$ for which*

$$m \mid (p-1, 2\nu) \quad \textit{and} \quad \frac{(p-1, 2\nu)}{m} > 1.$$

These two lemmas are evident.

LEMMA 10.3. *If m is chosen as in Lemma 10.2 and $s \geqslant m(m+1)^{[2l/m]}$, then the system of congruences*

$$x_1^{2\nu} + \dots + x_s^{2\nu} \equiv 0 \pmod{p}, \quad p + x_1, \quad \nu = 1, \dots, l \tag{10.3}$$

has a solution.

Proof. Since $m \mid (p-1)$, there exists an integer g belonging to m (mod p). Thus

$$1^{2\nu}+g^{2\nu}+(g^2)^{2\nu}+\ldots+(g^{m-1})^{2\nu} \equiv 0 \pmod{p} \tag{10.3.1}$$

for all ν's satisfying $m \nmid 2\nu$. The other ν's which have not been discussed are those which satisfy

$$(2\nu,\ p-1) = m.$$

By Satz 301 of Landau (1), we can choose $x_1, \ldots, x_{m+1}$ so that

$$x_1^{2\nu}+\ldots+x_{m+1}^{2\nu} \equiv 0 \pmod{p}, \quad p \nmid x_1.$$

There are at most $[2\nu/m]$ ν's having this property. Therefore, by Lemma 10.1, the lemma is proved.

LEMMA 10.4.

$$K(2l) \leqslant 2l \,.\, 3^l.$$

If $l \geqslant 22$, *then*
$$K(2l) < 3^l.$$

Proof. (1) Here

$$(m+1)^{[2l/m]} \leqslant (m+1)^{2l/m} = e^{2l \log(m+1)/m}.$$

Since $\log(x+1)/x$ decreases as x increases for $x > e-1$, we have

$$(m+1)^{[2l/m]} \leqslant 3^l.$$

(2) If $m = 2$, since there is no congruence of the type (10.3.1), we can prove that

$$K(2l) < 3^l.$$

If $m \geqslant 3$,

$$K(2l) < m(m+1)^{[2l/m]} \leqslant 2l \,.\, 4^{\frac{2}{3}l};$$

and if $l \geqslant 22$,

$$2l \,.\, 4^{\frac{2}{3}l} \leqslant 3^l.$$

LEMMA 10.5. *If* $s > K(2l)$, *and* t *is any odd integer, then there are integers* $h_1, \ldots, h_s$ *such that*

$$h_1^{2r}+\ldots+h_s^{2r} \equiv 0 \pmod{p}, \quad l \geqslant r \geqslant 1, \tag{10.5.1}$$

and

$$h_1^{t}+\ldots+h_s^{t} \not\equiv 0 \pmod{p}. \tag{10.5.2}$$

Proof. Since, by Lemma 10.4, we can choose $h_1, \ldots, h_s$ satisfying (10.3) and

$$h_1^t+h_2^t+\ldots+h_s^t \not\equiv (-h_1)^t+h_2^t+\ldots+h_s^t \pmod{p},$$

the conclusion follows immediately.

Every integral-valued polynomial $R(h)$ with no constant term can be written in the form

$$R(h) = \tfrac{1}{2}\{h\,R_1(h)+R_2(h)\},$$

where $R_1(h)$ and $R_2(h)$ are two odd polynomials. It is evident that $hR_1(h)$, $R_2(h)$ are both integral-valued. Let the degrees of $hR_1(h)$ and $R_2(h)$ be $2t_1$ and $2t_2+1$ respectively. We can write

$$R_2(h) = \sum_{\nu=1}^{t_2} c_\nu\, Q_{2\nu+1}(h),$$

where the c's are integers.

Lemma 10.6. *If there is no integer a such that $P(h) \equiv a \pmod{p}$, then there is an integer q such that the coefficients c_ν of $R_2(h)$ deduced from $R(h) = P(h+q)$ satisfy the condition*

$$p \nmid (c_1, \ldots, c_{t_2}) \quad (p \neq 2).$$

Proof. Evidently

$$P(h)-R(-h) = R_2(h).$$

If the conclusion is false, we have, since $R_2(0) = 0$,

$$P(h+q)-P(-h+q) \equiv 0 \pmod{p}$$

for all values of h and q. In particular, we take $h = q$. Then

$$P(2h) \equiv 0 \pmod{p}$$

for all values of h. This contradicts our hypothesis.

Let $\Gamma\{P(h), p\}$ be the least integer s for which

$$\sum_{\nu=1}^{s} P(h_\nu) \equiv n \pmod{p^\gamma}$$

has a solution for all n.

LEMMA 10.7. *If $p > k$, there exists an odd polynomial $Q(h) \not\equiv 0 \pmod{p}$, of degree not greater than $2t_1+1$, such that, if*

$$s \geqslant K(2t_1)\,\Gamma\{Q(h),\, p\},$$

then the congruence
$$\sum_{\nu=1}^{s} P(h_\nu) \equiv n \pmod{p}$$

has a solution for every integer n.

Proof. By Lemma 10.6, there exists an integer q such that

$$R_2(h) \not\equiv 0 \pmod{p}.$$

Thus we can define an integer τ such that

$$p \mid (c_{t_2}, \ldots, c_{t_2-\tau+1}), \quad p \nmid c_{t_2-\tau}, \quad 0 \leqslant \tau < t_2.$$

Let $h_1, \ldots, h_{s_1}$ $\{s_1 = K(2t_1)\}$ be the integers satisfying

$$h_1^{2\nu}+\ldots+h_s^{2\nu} \equiv 0 \pmod{p}, \quad \nu = 1, \ldots, t_1,$$

$$h_1^{2(t_2-\tau)+1}+\ldots+h_s^{2(t_2-\tau)+1} \not\equiv 0 \pmod{p}. \tag{10.7.1}$$

Consider
$$\Psi(h) = \sum_{\nu=1}^{s_1} R(h_\nu h) \pmod{p}.$$

Here $\Psi(h)$ is an odd polynomial, and, by (10.7.1),

$$\Psi(h) \not\equiv 0 \pmod{p}.$$

The lemma follows immediately.

By Lemmas 8.3, 9.3, 9.4, 10.4, and 10.7, we obtain

LEMMA 10.8. *If $\epsilon > 0$, then*

$$\Gamma\{P(h),\, p\} = O(3^{k(\frac{1}{2}+\epsilon)}), \quad \textit{for all} \quad p > k.$$

THEOREM 7.

$$G\{P(h)\} = O(k^3\, 2^{k-1}).$$

By Lemmas 8.3, 9.5, 10.4, and 10.7, we obtain

LEMMA 10.9. *If $k \geqslant 15$, then*

$$\Gamma\{P(h),\, p\} \leqslant \max\,(k 3^{[\frac{1}{3}k]}\, 2^{\frac{1}{2}(k-1)+7},\; k^3\, 2^{k-1}) \leqslant k 3^{\frac{1}{3}k}\, 2^{\frac{1}{2}(k-1)+7}.$$

Consequently we have

THEOREM 8. *If* $k > 20$, *every sufficiently large integer is the sum of* $[k 3^{\frac{1}{2}k} 2^{\frac{1}{2}(k-1)+7}]$ *values, for non-negative values of* h, *of a polynomial* $P(h)$ *of the* k-*th degree.*

THEOREM 9. *If the coefficient of* $P(h)$ *is* $a/k!$, $(a, k!) = 1$, *then*

$$G\{P(h)\} = O(3^{k(\frac{1}{2}+\epsilon)}).$$

11. *Some further results.*

In this section we describe some methods which can be applied to special polynomials and indicate briefly the more interesting results obtained by them.

(1) We can prove by Heilbronn's method that, if $4k$ of the integers

$$a_1, a_2, \ldots, a_s$$

form an admissible set as defined by Huston, then

$$a_1 h_1^k + \ldots + a_s h_s^k = n$$

has a solution in non-negative integers for all large integers n, provided that

$$s \geqslant 6k \log k + \left\{4 + 3 \log \left(3 + \frac{2}{k}\right)\right\} k + 3.$$

This includes one of Dickson's results [Dickson (**1**)] on the generalized Waring problem.

(2) If we replace Lemmas 3.2 and 3.3 by Weyl's approximation and use Wright's method [Wright (**2**), (**3**)], we can show that a generalization of Wright's asymptotic formula holds for polynomials, since

$$s \geqslant 2^{k-1}(k-2) + 5 \quad (k \geqslant 3).$$

We can deduce from this the following results for polynomials of lower degree.

Let $G^*\{P(h)\}$ be the best possible integer for which the Waring-Kamke problem is solved with proportional conditions. Here certainly $P(h)$ has the property that there exist no integers c and d $(d > 1)$ such that

$$P(h) \equiv c \pmod{d}.$$

The results can be summarized as follows.

1. If $P(h)$ is a cubic integral-valued polynomial, then

$$G^*\{P(h)\} \leqslant 9.$$

This is better than James's result [James (**1**)], since the restriction upon the coefficients is removed. *Cf.* also James (**2**), Hua (**1**).

2. $$G^*\{\tfrac{1}{12}A(h^4-h^2)+Bh^2\} \leqslant 26 \quad [(A,\, 6B)=1].$$

3. $$G^*\{\tfrac{1}{12}A(h^4-h^2)+Bh^2\} \leqslant 31 \quad [(A,\, B)=1].$$

4. If $P(h)$ is a quartic integral-valued polynomial, then

$$G^*\{P(h)\} \leqslant 65.$$

5. If $P(h)$ is a quintic integral-valued polynomial, then

$$G^*\{P(h)\} \leqslant 81.$$

6. If $P(h)$ is an even integral-valued polynomial of the 6-th degree, then

$$G^*\{P(h)\} \leqslant 2304.$$

7. If $P(h)$ is an integral-valued polynomial of the 6-th degree, then

$$G^*\{P(h)\} \leqslant 4627.$$

8. If $P(h)$ is an integral-valued polynomial of the 7-th degree, then

$$G^*\{P(h)\} \leqslant 4691.$$

(3) In the case when $P(h)$ is an odd polynomial of the k-th degree, it can be shown by the method used by Hardy and Littlewood (**2**) that

$$G\{P(h)\} \leqslant (k-2)\,2^{k-2}+k+5+\left[\frac{(k-2)\log 2-\log k+\log(k-2)}{\log k-\log(k-1)}\right].$$

Although this upper bound is not very good for large k, it gives us some interesting results for $k \leqslant 20$ [Hua (**3**)].

(4) The upper bound in Lemma 8.3 can be improved by choosing special polynomials. There are a large number of them, for example, for which $\phi(0)$ (defined on p. 167) is equal to 1. In particular, it is easy to see that

$$G\{\tfrac{1}{2}(x^h+x)\} \leqslant 10k^3 \log k.$$

References.

L. E. Dickson (1), "On Waring's problem and its generalization", *Annals of Math.*, 37 (1936), 293–316.

G. H. Hardy and J. E. Littlewood (1), "Sôme problems of 'Partitio Numerorum' (IV): The singular series in Waring's problem, and the values of the number $G(k)$", *Math. Zeitschrift*, 12 (1922), 161–188.

——— (2), "Some problems of 'Partitio Numerorum' (VI): Further researches in Waring's problem", *Math. Zeitschrift*, 23 (1925), 1–37.

H. Heilbronn (1), "Über das Waringsche Problem", *Acta Arith.*, 1 (1935), 212–221.

L. K. Hua (1), "On Waring theorems with cubic polynomial summands", *Math. Annalen*, 111 (1935), 622–631.

——— (2), "An easier Waring-Kamke problem", *Journal London Math. Soc.*, 11 (1936), 2–3.

——— (3), "On Waring's problem with polynomial summands", *Amer. J. of Math.*, 58 (1936), 553–562.

——— (4), "On Waring's problem with polynomial summands", *Journal Chinese Math. Soc.*, 1 (1936), 23–61.

——— (5), "On an exponential sum". (In the Press.)

Ralph E. Huston (1), "Asymptotic generalization of Waring's theorem", *Proc. London Math. Soc.* (2), 39 (1935), 82–115.

R. D. James (1), "The representation of integers as sums of values of cubic polynomials", *Amer. J. of Math.*, 56 (1934), 303–315.

——— (2), "The representation of integers as sums of pyramidal numbers", *Math. Annalen*, 109 (1934), 196–189.

E. Kamke (1), "Verallgemeinerungen des Waring-Hilbertschen Satzes", *Math. Annalen*, 83 (1921), 85–112.

E. Landau (1), "Vorlesungen über Zahlentheorie", *Bd.* 1.

——— (2), "Zum Waringschen Problem", *Math. Zeitschrift*, 12 (1938), 219–247.

——— (3), "Über die Winogradoffsche Behandlung des Waringschen Problem", *Math. Zeitschrift*, 31 (1929), 318–338.

——— (4), "Zum Waringschen Problem, Dritte Abhandlung", *Math. Zeitschrift*, 32 (1930), 699–702.

I. Vinogradow (1), "On Waring's problem", *Annals of Math.*, 36 (1935), 395–405.

——— (2), "On asymptotic formula in Waring's problem", *Receuil Math.*, 1 (43), (1936), 169–174.

——— (3), "A new method of estimation of trigonometric sums", *Receuil Math.*, 1 (43), (1936), 175–188.

E. Maitland Wright (1), "An extension of Waring's problem", *Phil. Trans. Royal Soc.*, 232 (1933), 1–26.

——— (2), "Proportional conditions in Waring's problem", *Math. Zeitschrift*, 38 (1934), 728–746.

——— (3), "An easier Waring's problem", *Journal London Math. Soc.*, 11 (1934), 267–272.

——— (4), "On Tarry's problem I", *Quarterly Journal*, 6 (1935), 261–267.

36 Belvoir Road,
Cambridge.

Printed by C. F. Hodgson & Son, Ltd., Newton St., London, W.C.2.

Reprinted from the
QUARTERLY JOURNAL OF MATHEMATICS
Oxford Series 9, pp. 199–202, 1938

ON WARING'S PROBLEM

By LOO-KENG HUA (*Cambridge*)

[Received 1 February 1938]

THE object of the present paper is to give a proof that Hardy and Littlewood's asymptotic formula for the number of solutions of the Diophantine equation

$$N = x_1^k + \ldots + x_s^k \quad (x_\nu \geqslant 0)$$

is true for $s \geqslant 2^k+1$. This result is new only for $k < 14$. For $k \geqslant 14$, Vinogradow's contribution is much better than this. The most interesting particular case is $k = 4$, for which Estermann† and Davenport and Heilbronn‡ proved that every sufficiently large integer is a sum of 17 fourth powers, but they gave no asymptotic formula for the number of solutions.

More precisely, what I am going to prove is the following more general theorem:

Let $P_1(x),\ldots, P_s(x)$ be s integral-valued polynomials of the kth degree and let their first coefficients be positive numbers $a_1,\ldots, a_s$ respectively. Let $r(N)$ be the number of solutions of the Diophantine equation

$$N = P_1(x_1) + \ldots + P_s(x_s) \quad (x_\nu \geqslant 0).$$

Then, for $s \geqslant 2^k+1$, we have

$$r(N) = \prod_{\nu=1}^{s} a_\nu^{-1/k} \frac{\Gamma^s(1+1/k)}{\Gamma(s/k)} \mathfrak{S}(N) N^{s/k-1} + O(N^{s/k-1-\delta}),$$

where $\delta = 2^{1-k}s - z - \epsilon$ and ϵ is an arbitrary small positive number and $\mathfrak{S}(N)$ is defined in one of my previous papers.§

I shall give elsewhere an application of this theorem to prove that

$$G\{P(x)\} \leqslant 17,$$

provided that $P(x)$ is a quartic polynomial with coefficient of x^4 positive and that there does not exist an integer q (> 1) such that

$$P(x) \equiv P(0) \pmod q$$

for all x.

The proof of the theorem depends essentially on the following lemma which seems to have some interest in itself.

† *Proc. London Math. Soc.* 41 (1938), 127–42.
‡ Ibid. 41 (1938), 143–50.
§ Ibid. 43 (1937), 161–82.

MAIN LEMMA. *Let $P(x)$ be an integral-valued polynomial of the kth degree, and*

$$f(\alpha) = \sum_{x=1}^{p} \exp\{2\pi i\, P(x)\alpha\}.$$

Then

$$\int_0^1 |f(\alpha)|^\lambda \, d\alpha = O(p^{\mu(\lambda)}),$$

where $\{\lambda, \mu(\lambda)\}$ lies on a polygonal line with vertices $(2^\nu, 2^\nu - \nu + \epsilon)$ $(\nu = 1, \ldots, k)$, and the constants implied by the symbol O depend only on the coefficients of $P(x)$ and ϵ.

An improvement of this lemma for the particular case $P(x) = x^k$ and its application to the additive prime-number theory will appear elsewhere later.

Proof of the main lemma

Since

$$\log\left(\int_0^1 |f(\alpha)|^\nu \, d\alpha\right)$$

is a convex function of ν,† we only need to prove that

$$\int_0^1 |f(\alpha)|^{2^\nu} \, d\alpha = O(p^{2^\nu - \nu + \epsilon}) \quad \text{for} \quad \nu = 1, 2, \ldots, k. \tag{1}$$

Without loss of generality we assume that $P(x)$ is a polynomial with integer coefficients. In fact, let q be the least common denominator of the coefficients of $P(x)$, then, by Hölder's inequality,

$$\int_0^1 |f(\alpha)|^\lambda \, d\alpha = \int_0^1 \left| \sum_{a=1}^{q} \sum_{x=0}^{[(p-a)/q]} \exp\{2\pi i\, P(qx+a)\alpha\} \right|^\lambda d\alpha$$

$$\leqslant q^{\lambda-1} \sum_{a=1}^{q} \int_0^1 \left| \sum_{x=0}^{[(p-a)/q]} \exp[2\pi i \{P(qx+a) - P(a)\}\alpha] \right|^\lambda d\alpha,$$

where $P(qx+a) - P(a)$ is a polynomial with integer coefficients.

Then (1) is trivial for $\nu = 1$ and well known for $\nu = 2$.§ We are going to prove (1) by induction.

We use the abbreviation

$$\underset{y}{\Delta}\, Q(x) = \frac{1}{y}\{Q(x+y) - Q(x)\}.$$

† See, for example, Hardy, Littlewood, and Pólya, *Inequalities*, § 6.12.

§ See, for example, Landau, *Vorlesungen über Zahlentheorie*, Bd. 1, Satz 262, 37. There he deals only with the particular case $P(x) = x^k$. For the general case see Hua, *J. of Chinese Math. Soc.* 1 (1936), 23–61, Lemma 11.

Then $\underset{y}{\Delta} Q(x)$ is a polynomial of the $(h-1)$th degree in x, provided that $Q(x)$ is a polynomial of the hth degree. Let $\sum_{x}^{p}$ denote a summation with variable x whose number of terms is $O(p)$.

Consider

$$|f(\alpha)|^2 = \sum_{x_1=1}^{p} \sum_{x_2=1}^{p} \exp[2\pi i\{P(x_1)-P(x_2)\}\alpha]$$
$$= \sum_{x_2}^{p}\sum_{y_1}^{p} \exp[2\pi i\{P(x_2+y_1)-P(x_2)\}\alpha]$$
$$= \sum_{y_1}^{p}\sum_{x_2}^{p} \exp\{2\pi i y_1 \underset{y_1}{\Delta} P(x_2)\alpha\}.$$

By Schwarz's inequality, we have

$$|f(\alpha)|^4 \ll p \sum_{y_1}^{p} \Big| \sum_{x_2}^{p} \exp\{2\pi i y_1 \underset{y_1}{\Delta} P(x_2)\alpha\}\Big|^2$$
$$\ll p \sum_{y_1}^{p}\sum_{y_2}^{p}\sum_{x_3}^{p} \exp\{2\pi i y_1 y_2 \underset{y_2}{\Delta}\underset{y_1}{\Delta} P(x_3)\alpha\},$$

where $A \ll B$ means $A = O(B)$. Repeating this process, we obtain, in general,

$$|f(\alpha)|^{2^\mu} \ll p^{2^\mu-\mu-1} \sum_{y_1}^{p} \cdots \sum_{y_\mu}^{p} \sum_{x_{\mu+1}}^{p} \exp\{2\pi i y_1 \cdots y_\mu \underset{y_\mu}{\Delta}\underset{y_{\mu-1}}{\Delta} \cdots \underset{y_1}{\Delta} P(x_{\mu+1})\alpha\}$$
$$\ll p^{2^\mu-1}+p^{2^\mu-\mu-1} \sum_{y_1}^{p} \cdots \sum_{y_\mu}^{p} \sum_{x_{\mu+1}}^{p} {}^{*} \exp\{2\pi i y_1 \cdots y_\mu \underset{y_\mu}{\Delta} \cdots \underset{y_1}{\Delta} P(x_{\mu+1})\alpha\} \quad (2)$$

for $\mu = 1, 2,\ldots, k-1$, where * denotes the condition

$$y_1 \cdots y_\mu \underset{y_\mu}{\Delta} \cdots \underset{y_1}{\Delta} P(x_{\mu+1}) \neq 0.$$

We have, therefore,

$$\int_0^1 |f(\alpha)|^{2^\nu} d\alpha \ll p^{2^{\nu-1}-1} \int_0^1 |f(\alpha)|^{2^{\nu-1}} d\alpha + \quad (3)$$
$$+p^{2^{\nu-1}-\nu} \int_0^1 \sum_{y_1}^{p} \cdots \sum_{y_{\nu-1}}^{p} \sum_{x_\nu}^{p} {}^{*} \exp\{2\pi i y_1 \cdots y_{\nu-1} \underset{y_{\nu-1}}{\Delta} \cdots \underset{y_1}{\Delta} P(x_\nu)\alpha\} |f(\alpha)|^{2^{\nu-1}} d\alpha.$$

By the hypothesis of the induction, the first term on the right of (3) is

$$O(p^{2^{\nu-1}-1}.p^{2^{\nu-1}-\nu+1+\epsilon}) = O(p^{2^\nu-\nu+\epsilon}).$$

The second term on the right of (3) equals

$$p^{2^{\nu-1}-\nu} \int_0^1 \sum_{y_1}^{p} \cdots \sum_{y_\mu}^{p} \sum_{x_{\mu+1}}^{p} {}^{*} \sum_{z_1}^{p} \cdots \sum_{z_{2^{\nu-1}}}^{p} \exp[2\pi i\{y_1 \cdots y_{\nu-1} \underset{y_{\nu-1}}{\Delta} \cdots \underset{y_1}{\Delta} P(x_\nu)-$$
$$-P(z_1)+P(z_2)-\ldots+P(z_{2^{\nu-1}})\}\alpha]\, d\alpha = p^{2^{\nu-1}-\nu} R,$$

where R denotes the number of solutions of

$$y_1 \dots y_{\nu-1} \underset{y_{\nu-1}}{\Delta} \underset{y_1}{\Delta} \dots P(x_\nu) = P(z_1) - P(z_2) + \dots - P(z_{2^{\nu-1}}),$$

$$y_1 \dots y_{\nu-1} \underset{y_{\nu-1}}{\Delta} \dots \underset{y_1}{\Delta} P(x_\nu) \neq 0, \qquad z_\mu,\, y_\mu,\, x_{\nu+1} \ll P. \tag{4}$$

For given $z_1,\dots, z_{2^{\nu-1}}$, the number of solutions of (4) is

$$O\{d^{\nu-1}[P(z_1) - P(z_2) + \dots - P(z_{2^{\nu-1}})]\}.\dagger$$

Since $d(n) = O(n^\epsilon)$, we have

$$R \ll \sum_{z_1} \dots \sum_{z_{2^{\nu-1}}} {\ddagger}\, d^{\nu-1}\{P(z_1) - P(z_2) + \dots - P(z_{2^{\nu-1}})\}$$

$$\ll p^{2^{\nu-1}+\epsilon},$$

where ‡ denotes the condition $P(z_1) - P(z_2) + \dots - P(z_{2^{\nu-1}}) \neq 0$. The lemma is therefore proved.

Proof of the theorem

Let $p = N^{1/k}$ and

$$S_r(\alpha) = \sum_{x=1}^{p_r} \exp\{2\pi i\, P_r(x)\alpha\},$$

where p_r is the greatest root of $P_r(x) = N$. It is evident that when N is sufficiently large, p_r always exists and $p \ll p_r \ll p$. Then

$$r(N) = \int_0^1 \prod_{r=1}^{s} S_r(\alpha) \exp(-2\pi i \alpha N)\, d\alpha.$$

It is sufficient to estimate the part $\overline{W}$ of the integral corresponding to the minor arcs, since the remaining part can be treated, without any difficulty, by the method used in my previous paper.§

By Weyl's theorem‖ and the main lemma we have

$$\overline{W} \ll p^{(1-2^{1-k}+\epsilon)(s-2^k)} \int_0^1 \prod_{r=1}^{2^k} |S_r(\alpha)|\, d\alpha$$

$$\ll p^{(1-2^{1-k}+\epsilon)(s-2^k)} \Big(\prod_{r=1}^{2^k} \int_0^1 |S_r(\alpha)|^{2^k}\, d\alpha \Big)^{2^{-k}}$$

$$\ll p^{s-k-\delta}.$$

The theorem is proved.

In closing, I should like to express my warmest thanks to the referee for his valuable advice.

† $d(n)$ denotes number of divisors of n.

§ *Proc. London Math. Soc.* 43 (1937), 161–82.

‖ Landau, *Vorlesungen über Zahlentheorie*, Satz 267.

Reprinted from the
QUARTERLY JOURNAL OF MATHEMATICS
Oxford Series 9, pp. 315–320, 1938

ON TARRY'S PROBLEM

By LOO-KENG HUA (*Kunming, China*)

[Received 31 May 1938]

LET $M(k)$ be the least value of s such that the equations

$$a_1^h+...+a_s^h = b_1^h+...+b_s^h \quad (1 \leqslant h \leqslant k) \tag{1}$$

and

$$a_1^{k+1}+...+a_s^{k+1} \neq b_1^{k+1}+...+b_s^{k+1} \tag{2}$$

have a solution in integers. In this paper I shall prove that

$$M(k) \leqslant (k+1)\left(\left[\frac{\log\frac{1}{2}(k+2)}{\log(1+1/k)}\right]+1\right).$$

The method used is very elementary and no previous knowledge is assumed.*

Throughout the paper, $c_1, c_2,...$ denote positive numbers depending on k only.

LEMMA 1. *Given any positive H, there exists a set of positive integers $a_1,..., a_k$ (depending only on k, H) such that the product of the principal diagonal of the determinant*

$$D_k = \begin{vmatrix} 1 & . & . & . & . & 1 \\ a_1 & . & . & . & . & a_k \\ & & . & . & . & . \\ a_1^{k-1} & & . & . & . & a_k^{k-1} \end{vmatrix}$$

is greater than H times the sum of the absolute values of all the other terms in the expansion of the determinant.

Proof. We prove the lemma by induction. If $\phi_j(a_1,..., a_j)$ denotes the product of the principal diagonal minus H times the sum of the absolute values of the other terms in the determinant D_j $(j \leqslant k)$, then

$$\phi_j(a_1,..., a_j) = a_j^{j-1}\phi_{j-1}(a_1,..., a_{j-1})-H\psi(a_1,..., a_j),$$

where ψ is a polynomial of degree $j-2$ in a_j. Thus, if $a_1,..., a_{j-1}$ have been chosen to make ϕ_{j-1} positive, we can further choose a_j so large that ϕ_j is positive. But initially $\phi_1 = 1$. Thus the induction is established.

* By using a lemma of I. Vinogradoff's we can obtain clearer information about the number of solutions of (1) and (2) with $0 \leqslant a_i \leqslant P$, $0 \leqslant b_i \leqslant P$. More precisely, by that method we are able to prove that the number $r(P)$ of these solutions satisfies

$$cP^{2s-\frac{1}{2}k(k+1)} \leqslant r(P) \leqslant c'P^{2s-\frac{1}{2}k(k+1)},$$

where c and c' are two numbers depending only on k.

LEMMA 2. *Let $X_1,\ldots, X_k$ be integers lying in the intervals*

$$a_i Q \leqslant X_i \leqslant 2a_i Q,$$

where the a's are those defined in Lemma 1. Let N be the number of sets $(X_1,\ldots, X_k)$ for which

$$X_1^k+\ldots+X_k^k, \quad X_1^{k-1}+\ldots+X_k^{k-1}, \quad \ldots, \quad X_1+\ldots+X_k$$

lie in given intervals of lengths

$$O(Q^{k-1}), \quad O(Q^{k-2}), \quad \ldots, \quad O(Q), \quad O(1)$$

respectively. Then

$$N = O(1).$$

Proof. If $(X_1,\ldots, X_k)$ and $(X_1',\ldots, X_k')$ are two sets satisfying the requirements of the lemma, then

$$X_1^k-X_1'^k+\ldots+X_k^k-X_k'^k = O(Q^{k-1}),$$

$$\cdot \quad \cdot \quad \cdot \quad \cdot \quad \cdot \quad \cdot \quad \cdot \quad \cdot \quad \cdot \quad \cdot \quad \cdot$$

$$X_1-X_1'+\ldots+X_k-X_k' = O(1).$$

Let $Y_i = X_i-X_i'$. Then we have

$$A_{11} Y_1+\ldots+A_{1k} Y_k = O(Q^{k-1}),$$

$$\cdot \quad \cdot \quad \cdot \quad \cdot \quad \cdot \quad \cdot \quad \cdot \quad \cdot \quad \cdot$$

$$A_{k1} Y_1+\ldots+A_{hk} Y_k = O(1),$$

where
$$A_{ij} = X_j^{k-i}+X_j^{k-i-1}X_j'+\ldots+X_j'^{h-i}.$$

Then
$$(k-i+1)(a_j Q)^{k-i} \leqslant A_{ij} \leqslant (k-i+1)(2a_j Q)^{k-i}.$$

Consider the determinant $|A_{ij}|$. The product of the elements in the principal diagonal divided by the corresponding term of D_k is greater than

$$k!\, Q^{k-1+k-2+\ldots+2+1} = k!Q^{\frac{1}{2}k(k-1)}.$$

Further, the absolute values of the other terms in the expansion of $|A_{ij}|$ are less than

$$2^{\frac{1}{2}k(k-1)}k!\, Q^{\frac{1}{2}k(k-1)}$$

times the absolute values of the corresponding terms in D_k. By Lemma 1, with $H = 2^{\frac{1}{2}k(k-1)}$, we have

$$|A_{ij}| \geqslant c_1 Q^{\frac{1}{2}k(k-1)}.$$

Further,

$$\begin{vmatrix} O(Q^{k-1}) & a_{12} & \cdot & \cdot & \cdot & a_{1k} \\ \cdot & \cdot & \cdot & \cdot & \cdot & \cdot \\ O(1) & a_{k2} & \cdot & \cdot & \cdot & a_{kk} \end{vmatrix} = O(Q^{\frac{1}{2}k(k-1)}).$$

Thus
$$Y_1 = O(1).$$

Similarly, $Y_2 = O(1), \quad \ldots, \quad Y_k = O(1).$

Thus we have the lemma.

Let R_k be the number of solutions of

$$\sum_{j=1}^{n}\sum_{i=1}^{k} \chi_{ij}^{h} = \sum_{j=1}^{n}\sum_{i=1}^{k} \chi_{ij}'^{h} \quad (1 \leqslant h \leqslant k) \tag{3}$$

where the χ satisfy the conditions

$$a_i P^{(1-1/k)^{j-1}} \leqslant \chi_{ij} \leqslant 2a_i P^{(1-1/k)} \quad (i = 1, 2, \ldots, k), \tag{4}$$

and the χ' satisfy the same conditions. Let R_k' be the number of solutions of the equations

$$\sum_{j=1}^{n}\sum_{i=1}^{k} \chi_{ij}^{h} = \sum_{j=1}^{n}\sum_{i=1}^{k} \chi_{ij}'^{h} \quad (1 \leqslant h \leqslant k-1),$$

where the χ and χ' satisfy the same conditions.

LEMMA 3. $R_k' \geqslant c_2 P^{2k^2\{1-(1-1/k)^n\}-\frac{1}{2}k(k-1)}.$

Proof. Let $r(n_1, \ldots, n_{k-1})$ be the number of solutions of the equations

$$\sum_{j=1}^{n}\sum_{i=1}^{k} \chi_{ij}^{h} = n_h \quad (1 \leqslant h \leqslant k-1)$$

satisfying the conditions (4). Then, evidently,

$$\sum_{n_1} \cdots \sum_{n_{k-1}} r(n_1, \ldots, n_{k-1}) \geqslant c_3 P^{k\{1+(1-1/k)+\ldots+(1-1/k)^{n-1}\}}$$
$$= c_3 P^{k^2\{1-(1-1/k)^n\}}$$

where the summation runs over all possible sets $n_1, \ldots, n_k$. By Schwarz's inequality we have

$$\sum_{n_1} \cdots \sum_{n_{k-1}} r(n_1, \ldots, n_{k-1}) \leqslant \sqrt{\left\{\sum_{n_1} \cdots \sum_{n_{k-1}} 1 \sum_{n_1} \cdots \sum_{n_{k-1}} r^2(n_1, \ldots, n_{k-1})\right\}}$$
$$\leqslant \sqrt{\left\{c_4 P^{1+2+\ldots+k-1} \sum_{n_1} \cdots \sum_{n_{k-1}} r^2(n_1, \ldots, n_{k-1})\right\}},$$

since $c_5 P^h \leqslant n_h \leqslant c_6 P^h$. Thus

$$R_k' = \sum_{n_1} \cdots \sum_{n_{k-1}} r^2(n_1, \ldots, n_{k-1})$$
$$\geqslant c_2 P^{2k^2\{1-(1-1/k)^n\}-\frac{1}{2}k(k-1)}.$$

LEMMA 4. $R_k = O(P^{\{2k^2-\frac{1}{2}k(k+1)^n\}\{1-(1-1/k)^n\}}).$

Proof. From (3) and (4) we have

$$\sum_{i=1}^{k} \chi_{i1}^{h} - \sum_{i=1}^{k} \chi_{i1}'^{k} = O(P^{h(1-1/k)}) \quad (1 \leqslant h \leqslant k).$$

Then, for fixed χ_{i1}' $(i = 1, \ldots, k)$,

$$\sum_{i=1}^{k} \chi_{i1}^{k}, \quad \sum_{i=1}^{k} \chi_{i1}^{k-1}, \quad \ldots, \quad \sum_{i=1}^{k} x_{i1}$$

lie in intervals of the lengths

$$O(P^{k(1-1/k)}), \quad O(P^{(k-1)(1-1/k)}), \quad \ldots, \quad O(P^{(1-1/k)}) \tag{5}$$

respectively. Since the system of intervals (5) can be divided into

$$O\left(\frac{P^{k(1-1/k)}}{P^{k-1}} \frac{P^{(k-1)(1-1/k)}}{P^{k-2}} \cdots \frac{P^{2(1-1/k)}}{P} \frac{P^{1-1/k}}{1}\right) = O(P^{k-\frac{1}{2}(k+1)})$$

systems of intervals of lengths

$$O(P^{k-1}), \quad O(P^{k-2}), \quad \ldots, \quad O(P), \quad O(1),$$

by Lemma 2 (with $Q = P$), the number of systems of χ_{i1} $(i = 1,\ldots, k)$ is

$$O(P^{k-\frac{1}{2}(k+1)}).$$

Therefore the number of systems of χ_{i1} and χ'_{i1} $(i = 1,\ldots, k)$ is

$$O(P^{2k-\frac{1}{2}(k+1)}).$$

Further, for fixed χ_{ij}, χ'_{ij} $(1 \leqslant i \leqslant k;\ 1 \leqslant j \leqslant l-1)$ and $\chi'_{i,l}$ $(1 \leqslant i \leqslant k)$, we see by (3) and (4) that

$$\sum_{i=1}^{k} \chi_{il}^{k}, \quad \sum_{i=1}^{k} \chi_{il}^{k-1}, \quad \ldots, \quad \sum_{i=1}^{k} \chi_{il}$$

lie in intervals of the lengths

$$O(P^{k(1-1/k)^l}), \quad O(P^{(k-1)(1-1/k)^l}), \quad \ldots, \quad O(P^{(1-1/k)^l}) \tag{6}$$

respectively. Since

$$O\left(\frac{P^{k(1-1/k)^l}}{P^{(k-1)(1-1/k)^{l-1}}} \frac{P^{(k-1)(1-1/k)^l}}{P^{(k-2)(1-1/k)^{l-1}}} \cdots \frac{P^{(1-1/k)^l}}{1}\right) = O(P^{\{k-\frac{1}{2}(k+1)\}(1-1/k)^{l-1}}),$$

by Lemma 2 (with $Q = P^{(1-1/k)^{l-1}}$), the number of systems $\chi_{i,l}$ $(1 \leqslant i \leqslant k)$ is

$$O(P^{\{k-\frac{1}{2}(k+1)\}(1-1/k)^{l-1}}).$$

Therefore, for fixed χ_{ij}, χ'_{ij} $(1 \leqslant i \leqslant k;\ 1 \leqslant j \leqslant l-1)$, the number of systems of χ_{il} and χ'_{il}, is

$$O(P^{\{2k-\frac{1}{2}(k+1)\}(1-1/k)^{l-1}}).$$

Thus the total number of solutions of (3) with the restriction (4) is

$$O(P^{\{2k-\frac{1}{2}(k+1)\}\{1+(1-1/k)+\ldots+(1-1/k)^{n-1}\}})$$
$$= O(P^{\{2k^2-\frac{1}{2}k(k+1)\}\{1-(1-1/k)^n\}}).$$

THEOREM. *If* $n > \log \frac{1}{2}(k+1)/\{\log k - \log(k-1)\}$, *then there are infinitely many sets of integers satisfying*

$$\sum_{j=1}^{n} \sum_{i=1}^{k} \chi_{ij}^{h} = \sum_{j=1}^{n} \sum_{i=1}^{k} \chi_{ij}^{\prime h} \quad (1 \leqslant h \leqslant k-1)$$

and

$$\sum_{j=1}^{n} \sum_{i=1}^{k} \chi_{ij}^{k} \neq \sum_{j=1}^{n} \sum_{i=1}^{k} \chi_{ij}^{\prime k}.$$

Proof. We consider those χ_{ij} and χ'_{ij} satisfying (4). The number of solutions of the equations in the theorem is evidently equal to

$$R'_k - R_k.$$

By Lemmas 3 and 4,

$$R'_k - R_k \geqslant c_2\, P^{2k^2\{1-(1-1/k)^n\}-\frac{1}{2}k(k-1)} - O\left(P^{\{2k^2-\frac{1}{2}k(k+1)\}\{1-(1-1/k)^n\}}\right)$$
$$\geqslant c_7\, P^{2k^2\{1-(1-1/k)^n\}-\frac{1}{2}k(k-1)}$$

for sufficiently large P, since

$$n > \frac{\log \frac{1}{2}(k+1)}{\log k - \log(k-1)}.$$

Hence the theorem follows immediately.

Replacing k by $k+1$ in the theorem, we have consequently

$$M(k) \leqslant (k+1)\left(\left[\frac{\log \frac{1}{2}(k+2)}{\log(1+1/k)}\right]+1\right).$$

This asserts that $$\overline{\lim_{k\to\infty}}\, \frac{M(k)}{k^2 \log k} \leqslant 1.$$

In what follows I shall indicate a method of improving the constant on the right-hand side, but I am unable to obtain anything better than that
$$M(k) = O(k^2 \log k).$$

Let $k = 2l-1$ be an odd integer. Let $J(l)$ be the least integer s for which
$$\sum_{i=1}^{s} \chi_i^{2h} = \sum_{i=1}^{s} \chi_i'^{2h} \quad (h = 1,\ldots,l-1)$$
and
$$\sum_{i=1}^{s} \chi^{2l} \neq \sum_{i=1}^{s} \chi_i'^{2l}$$
is solvable. By the same method as before, we can show that

$$J(l) \leqslant l\left[\frac{\log \frac{1}{2}(l+1)}{\log l - \log(l-1)}+1\right].$$

Evidently, if
$$a_1^{2h}+\ldots+a_s^{2h} = b_1^{2h}+\ldots+b_s^{2h} \quad (1 \leqslant h \leqslant l-1),$$
$$a_1^{2l}+\ldots+a_s^{2l} \neq b_1^{2l}+\ldots+b_s^{2l},$$
then
$$\sum_{i=1}^{s} \{(x+a_i)^t+(x-a_i)^t\} = \sum_{i=1}^{s} \{(x+b_i)^t+(x-b_i)^t\} \quad (1 \leqslant t \leqslant 2l-1)$$
and
$$\sum_{i=1}^{s} \{(x+a_i)^{2l}+(x-a_i)^{2l}\} \neq \sum_{i=1}^{s} \{(x+b_i)^{2l}+(x-b_i)^{2l}\}.$$

Thus we have

$$M(k) \leqslant 2J(l) \leqslant 2l\left[\frac{\log \frac{1}{2}(l+1)}{\log l - \log(l-1)} + 1\right]$$

$$\leqslant (k+1)\left[\frac{\log \frac{1}{4}(k+3)}{\log(k+1) - \log(k-1)} + 1\right] \sim \tfrac{1}{2}k^2 \log k.$$

The other method is the reconsideration of the 'tails' of (3), i.e. the parts corresponding to χ_{in} and χ'_{in}. Such a method will sharpen the result by subtracting a number ($> \frac{1}{2}k$).

Remarks. 1. For $k \geqslant 15$, the result here given is better than E. M. Wright's result that, if $k \geqslant 12$, then

$$M(k) < \frac{7k^2(k-11)(k+3)}{216}.$$

2. The theorem holds good, if we make the restriction that the a and b are primes.

Reprinted from the
MATHEMATISCHE ZEITSCHRIFT
Vol. 44, No. 3, pp. 335–346, 1938

On the representation of numbers as the sums of the powers of primes[1].

By

Loo-keng Hua in Cambridge[2].

Let k be an integer $\geqq 4$, $a = 1/k$,

$$b = \begin{cases} k^3(\log k + 1{,}25 \log\log k^2) & \text{for } k \geqq 15, \\ 2^{k-1} & \text{for } k < 15, \end{cases}$$

and

$$m = \left[\frac{\log \frac{1}{2} b + \log(1 - 2a)}{\log k - \log(k-1)}\right].$$

Let[3] $p^\theta \| k$ and

$$\gamma = \begin{cases} \theta + 2 & \text{for } p = 2,\ 2 \mid k, \\ \theta + 1 & \text{otherwise} \end{cases}$$

$$K = \prod_{(p-1) \mid k} p^\gamma.$$

Theorem. *Every sufficiently large integer $N \equiv s \pmod K$ is a sum of s k-th powers of prims, provided $s \geqq s_0$, where $s_0 = s_0(k) = 2k + 2m + 7$.*

It is worthwhile to note that for large k, the order of s_0 is $6 k \log k$ which is as good as Vinogradow's result[4] for Waring's problem; for small k, $s_0(4) = 19$, $s_0(5) = 31$ which is very close to the results of Davenport, Estermann and Heilbronn[5]. The present method can also be used to prove that almost all integers $N \equiv s \pmod K$ are sums of s k-th powers of primes, provided $s \geqq k + m + 4$ ($\sim 3 k \log k$).

As to $k = 3$, the present method gives $s_0(3) = 11$[6].

---

[1] A preliminary account of the results has been published in Comptes Rendus de l'USSR, xvii, no. 5, 1937.

[2] Research fellow of the China Foundation for the promotion of Education and Culture.

[3] $p^t \| n$ means that $p^t \mid n$ and $p^{t+1} \nmid n$.

[4] On the upper bound of $G(n)$ in Waring's problem, Bull. de l'Acad. de l'URSS, VII Serie, Classe des sciences math. naturelles, 1934, p. 1455—1469; Une nouvelle variante de la demonstration du theorem de Waring, Compte Rendus **200** (1935), p. 182—184; On Waring's problem, Annals of Math. **36** (1935), p. 395—405.

[5] Proc. of London Math. Soc. (2) **41** (1936), p. 26—150.

[6] The author has obtained a new method by which he can prove that every sufficiently large odd integer is a sum of nine cubes of the primes. It will be published elsewhere seperately.

The present paper consists of three parts:

1. Study of singular series.
2. A lemma on Waring's problem.
3. Proof of the theorem.

1. Study of singular series.

Let

$$W_{h,q} = \sum_{\substack{l=1 \\ (l,q)=1}}^{q} e_q(h\,l^k), \quad e_q(x) = e^{\frac{2\pi i x}{q}},$$

$$B_s(N,q) = \sum_{\substack{h=1 \\ (h,q)=1}}^{q} \left(\frac{W_{h,q}}{\varphi(q)}\right)^s e_q(-aN),$$

$$\mathfrak{S}(n) = \sum_{q=1}^{\infty} B_s(N,q).$$

Lemma 1.1. *If* $(q_1, q_2) = 1$, *then*

$$W_{h, q_1 q_2} = W_{h q_1^{k-1}, q_2} W_{h q_2^{k-1}, q_1}$$

and

$$B_s(N, q_1 q_2) = B_s(N, q_1) B_s(N, q_2).$$

Proof. Let $l = l_1 q_2 + l_2 q_1$. Then

$$W_{h, q_1 q_2} = \sum_{\substack{l_1=1 \\ (l_1,q_1)=1}}^{q_1} \sum_{\substack{l_2=1 \\ (l_2,q_2)=1}}^{q_1} e_{q_1 q_2}(h\, q_2^h l_1^k + h\, q_1^k l_2^k)$$
$$= W_{h q_1^{k-1}, q_2} W_{h q_2^{k-1}, q_1}.$$

The second is an immediate consequence of the first.

Lemma 1.2. *If* $t > \gamma$, *then*

$$W_{h, p^t} = 0.$$

Proof. Let $l = l_1 + l_2 p^{t-\theta-1}$. By Landau[7]), Satz 290 (with an easy modification for $p = 2$) we have

$$W_{h, p^t} = \sum_{\substack{l_1=1 \\ (l_1,p)=1}}^{p^{t-\theta-1}} \sum_{l_2=1}^{p^{\theta+1}} e_{p^t}(h(l_1^k + p^{t-\theta-1} k\, l_1^{k-1} l_2)) = 0$$

since $p \nmid l_1 k p^{-\theta}$.

Lemma 1.3.

$$W_{h,q} = O(q^{\frac{1}{2}+\varepsilon})$$

[7]) Vorlesungen über Zahlentheorie, Bd. 1, This foot-note will not be repeated on similar occasions.

where the constants implied by the symbol O depend on k and ε.

Proof. By lemma 1.2

$$W_{h,p^t} = O(1) \quad \text{for } p \mid k,$$
$$W_{h,p^t} = 0 \quad \text{for } p \nmid k \text{ and } t > 1.$$

Further, by Landau, Satz 311, we have

$$|W_{h,p}| \leqq k\sqrt{p} \quad \text{for all } p,$$
$$\leqq p^{\frac{1}{2}+\varepsilon} \quad \text{for } p \geqq k^{1/\varepsilon}.$$

Let $q = p_1^{t_1} \ldots p_\nu^{t_\nu}$ and $p_1 < p_2 < \ldots < p_\nu$. Then by lemma 1.1

$$|W_{h,q}| = \prod_{p_i \leqq k^{1/\varepsilon}} \left|W_{h_i, p_i^{t_i}}\right| \prod_{p_i > k^{1/\varepsilon}} \left|W_{h_i, p_i^{t_i}}\right|$$
$$= O(q^{\frac{1}{2}+\varepsilon}).$$

Lemma 1.4. *If $s > 4$, then*

$$\mathfrak{S}(N) = \prod_p \chi_p(N),$$

where

$$\chi_p(N) = 1 + \sum_{t=1}^{\gamma} B_s(N, p^t).$$

Proof. Lemmas 1.1, 1.2 and 1.3.

Lemma 1.5. *Let $M_s(p^t, N)$ be the number of solutions of the congruence*

$$x_1^k + \ldots + x_s^k \equiv N \pmod{p^t}, \quad p \nmid x_1 \ldots x_s, \quad 0 < x_i < p^t.$$

Then

$$\varphi(p^t)^{-s} p^t M_s(p^t, N) = 1 + \sum_{d=1}^{t} B_s(N, p^d).$$

Proof. We have

$$M_s(p^t, N) = p^{-t} \sum_{\substack{l_1=1 \\ (l_1,p)=1}}^{p^t} \ldots \sum_{\substack{l_s=1 \\ (l_s,p)=1}}^{p^t} \sum_{h=1}^{p^t} e_{p^t}\left(h\left(l_1^k + \ldots + l_s^k - N\right)\right)$$

$$= p^{-t} \sum_{h=1}^{p^t} W_{h,p^t}^s \, e_{p^t}(-hN)$$

$$= p^{-t} \varphi^s(p^t)\left(1 + \sum_{d=1}^{t} B_s(N, p^d)\right).$$

Lemma 1.6. *Let $x_1, \ldots, x_m$ belong to m different residue classes,* mod p^l, *and $y_1, \ldots, y_n$ belong to n different residue classes,* mod p^l *and no two y's are congruent each other, mod p. Then the number of different residue classes represented by*

$$x_i + y_j \quad (1 \leqq i \leqq m, \; 1 \leqq j \leqq n)$$

is greater than or equal to

$$Min\,(m + n - 1,\ p^l).$$

(I. Chowla and Davenport)[8]).

Lemma 1.7. *For* $s \geqq 4k$ *and* $(p-1) \nmid k$, *then*

$$M_s(p^\gamma, N) > 0.$$

Proof. 1) $p \nmid k$, then $\gamma = 1$. Since $(p-1) \nmid k$, x^k gives

$$d = \frac{p-1}{(k, p-1)} > 0$$

different values, mod p, when x runs over $1, 2, \ldots, p-1 \pmod{p}$. By lemma 1.6, $x_1^k + \ldots + x_s^k$, $p \nmid x_1, \ldots, x_s$, gives

$$Min\,(d + (d-1)(s-1), p) = p$$

different values, mod p, since

$$s \geqq 4k \geqq \frac{p^\gamma - 1}{\frac{1}{2}d} \geqq \frac{p^\gamma - 1}{d-1}.$$

2) $k = p^\theta k_0$, $p \nmid k_0$. Then x^k gives at least $p-1$ different values no two of which are congruent each other, mod p, since

$$x^{p^\theta k_0} \equiv x^{k_0} \pmod{p} \quad \text{and} \quad (k_0, p-1) = 1.$$

Therefore $x_1^k + \ldots + x_s^k$ gives

$$Min\,(p - 1 + (p-2)(s-1),\ p^\gamma) = p^\gamma$$

different values, mod p^γ, since

$$s \geqq 4k \geqq 4p^\theta \geqq \frac{p^\gamma - p + 1}{p-2} + 1.$$

Lemma 1.8. *If* $s \equiv N \pmod{p^\gamma}$ *and* $(p-1) \mid k$, *then*

$$M_s(p^\gamma, N) > 0.$$

Evident.

Lemma 1.9. *If* $s \geqq 4k$ *and* $s \equiv N \pmod{p^\gamma}$ *for all* p *satisfying* $(p-1) \mid k$, *then*

$$\mathfrak{S}(N) \geqq A \text{ (independent of } N) > 0.$$

Proof. Using lemmas 1.5, 1.7 and 1.8 we obtain

$$\chi_p > 0 \quad \text{for all } p.$$

By Landau, Satz 311, we have

$$|B_s(N, p)| \leqq \frac{(k\sqrt{p})^s}{(p-1)^{s-1}} \leqq (2k)^s p^{-\frac{s}{2}+1}.$$

---

[8]) See, for example, Landau, Über einige neuere Fortschritte der additiven Zahlentheorie, Cambridge Tracts, no. 35, p. 8.

Therefore, for $p > (2\,k)^{4s}$,

$$\chi_p > 1 - p^{-\frac{s}{2}+1+\frac{1}{4}}.$$

Consequently

$$\mathfrak{S}(N) \geqq \prod_{p \leqq (2k)^{4s}} \chi_p \prod_{p > (2k)^{4s}} \left(1 - p^{-\frac{5}{4}}\right) \geqq A > 0.$$

Remark. We can prove in the same way that if $k = 3$ and $s = 11$ or 9, then

$$\mathfrak{S}(N) \geqq A > 0$$

for all odd N.

2. A lemma on Waring's problem.

Le N be a large integer and $P = \frac{1}{2} N^a$.

$$T(\alpha, P) = \sum_{P < n < 2P} e(n^k \alpha), \quad e(x) = e^{2\pi i x}.$$

$$T_i(\alpha) = T(\alpha, \alpha^{-i} P^{(1-a)^i}), \quad i = 0, \ldots, m+1.$$

$$Q(\alpha) = T_1(\alpha) \ldots T_m(\alpha)\, T_{m+1}^2(\alpha)$$
$$= \sum_n r_{m+2}(n)\, e(n\,\alpha).$$

$$R(\alpha) = T_0(\alpha) Q(\alpha)$$
$$= \sum_n r_{m+3}(n)\, e(n\,\alpha),$$

$$T_0^k(\alpha)\, R(\alpha) = \sum_n r_{k+m+3}(n)\, e(n\,\alpha).$$

Let $c_1, c_2, \ldots$ be numbers depending only on k. Then

$$c_1\, P^{k-1-(k-2)(1-a)^{m+1}} \leqq Q(0) \leqq c_2\, P^{k-1-(k-2)(1-a)^{m+1}}.$$

The constants implied by the symbol O depend only on k and ε, where ε is an arbitrary small positive number.

The purpose of this section is to prove that

$$\sum_n r_{k+m+3}^2(n) = O\left(P^{k+2} Q^2(0)\right).$$

We divide the interval $0 \leqq \alpha \leqq 1$ in the usual way into Farey arcs belonging to all rational points h/q with $1 \leqq q \leqq P^{h-1+\varepsilon}$, $0 < h < q$, $(h, q) = 1$. Each arc has the form

$$\alpha = h/q + \beta, \quad -\vartheta_1 q^{-1} P^{-k+1-\varepsilon} \leqq \beta \leqq \vartheta_2 q^{-1} P^{-k+1-\varepsilon},$$

where $\frac{1}{2} \leqq \vartheta_i \leqq 1$.

Let $\mathfrak{M}_{h,q}$ denote an arc (or a part of an arc) with

$$q \leqq P^{k/b}, \quad |\beta| \leqq q^{-1} P^{-k(1-1/b)}.$$

Let E be the set of the points which do not belong to any $\mathfrak{M}_{h,q}$.

22*

Let

$$S_{h,q} = \sum_{v=1}^{q} e_q(v^k h),$$

and

$$T^*(\alpha, h, q) = q^{-1} S_{h,q} \int_{P}^{2P} e^{2\pi i y^k \beta}\, dy.$$

Lemma 2.1[9]. *If* $(h, q) = 1$, *then*

$$S_{h,q} = O(q^{1-a}).$$

Lemma 2.2 (Davenport and Heilbronn)[10]

$$\sum_{x=1}^{m} e_q(h x^k) = \frac{m}{q} S_{h,q} + O(q^{3/4+\varepsilon}).$$

Consequently, if $m \leqq q$, *then, for* $k \geqq 4$,

$$\sum_{x=1}^{m} e_q(h x^k) = O(q^{1-a+\varepsilon}).$$

Lemma 2.3 (Weyl[11]) and Vinogradow[12]). *If* α *belongs to the arc corresponding* $P^{1-\varepsilon} < q \leqq P^{k-1+\varepsilon}$, *then*

$$T(\alpha) = O\left(P^{1-\frac{1}{b}+\varepsilon}\right).$$

Lemma 2.4.

$$\int_0^1 |R(\alpha)|^2 d\alpha = \sum_n r_{m+3}^2(n) = O(P^{1+\varepsilon} Q(0)).$$

Proof. $\sum\limits_n r_{m+3}^2(n)$ is the number of solutions of

$$x_0^k + \ldots + x_m^k + x_{m+1}^k + x_{m+1}'^k = y_0^k + \ldots + y_m^k + y_{m+1}^k + y_{m+1}'^k$$

where

$$2^{-} P^{(1-a)^i} \leqq x_i, y_i \leqq 2^{1-i} P^{(1-a)^i}.$$

(1) implies $x_i = y_i$ (for $i = 0, \ldots, m$) when P is large enough. For, if v is the first one of the suffixes for which $x_v \neq y_v$, then

$$|x_v^k - y_v^k| \geqq k(P^{(1-a)^v} 2^{-v})^{k-1}$$

is certainly greater than

$$y_{v+1}^2 + \ldots + y_{m+1}'^k$$

as P large enough. It is well-known that the number of solutions of

$$x_{m+1}^k + x_{m+1}'^k = y_{m+1}^k + y_{m+1}'^k$$

---

9) Landau, Satz 315.

10) Proc. of London Math. Soc. (2) 41 (1936), S. 449–453.

11) Landau, Satz 267.

I am indebted to Prof. Vinogradow for communicating this new result to me.

is $O(P^{2(1-a)^{m}+1+\varepsilon})$. The lemma is proved.

Lemma 2.5.

$$T^*(\alpha, h, q) = O(q^{-a}\, Min\,(P, |\beta|^{-a})).$$

Proof. Since

$$S_{h,q} = O(q^{1-a}),$$

and

$$\int_P^{2P} e^{2\pi i \beta y^k}\, dy = O(P)$$

it suffices to prove that

$$\int_P^{2P} e^{2\pi i \beta y^k}\, dy = O(|\beta|^{-a}).$$

This inequality follows at once by changing variables.

Lemma 2.6. *If* $q \leqq P^{1-\varepsilon}$, $|\beta| \leqq q^{-1} P^{-k+1-\varepsilon}$, *then*

$$T(\alpha) - T^*(\alpha, h, q) = O(q^{1-a+\varepsilon}).$$

Proof. The lemma can be proved by Estermann's argument (§§ 2.4—2.7)[13]) by replacing Weyl's estimation by lemma 2.2.

Lemma 2.7.

$$\int_E |T_0^k(\alpha)\, R(\alpha)|^2\, d\alpha = O(P^{k+2-c_3} Q^2(0)).$$

Proof. Since α is not on $\mathfrak{M}_{h,q}$, at least one of the following three conditions is satisfied:

(i) $$P^{1-\varepsilon} < q \leqq P^{k-1+\varepsilon},$$

(ii) $$P^{k/b} < q \leqq P^{1-\varepsilon},$$

(iii) $$q \leqq P^{k/b}, \quad |\beta| > q^{-1} P^{-k(1-1/b)}.$$

By lemmas 2.3, 2.5 and 2.6 we have on E

$$T_0(\alpha) = O\left(P^{1-\frac{1}{b}}\right).$$

By lemma 2.4, we have

$$\int_E |T_0^{2k}(\alpha)\, R^2(\alpha)|\, d\alpha = O\left(P^{2k\left(1-\frac{1}{b}\right)} \int_0^1 |R(\alpha)|^2\, d\alpha\right)$$

$$= O\left(P^{2k\left(1-\frac{1}{b}\right)+1+\varepsilon} Q(0)\right),$$

$$= O\left(P^{2k\left(1-\frac{1}{b}\right)+2-k+(k-2)(1-a)^{m}+1} Q^2(0)\right) = O(P^{k+2-c_3} Q^2(0)),$$

[13]) Proc. of London Math. Soc. (2) **41** (1936), S. 126—143.

since

$$-\frac{2k}{b}+(k-2)(1-a)^{m+1}<-\frac{2k}{b}+(k-2)(1-a)^{\log\frac{1}{2}b(1-2a)/\log(1-a)^{-1}}$$
$$=-\frac{2k}{b}+(k-2)\left(\frac{1}{2}b(1-2a)\right)^{-1}=0.$$

Lemma 2. 8.

$$\sum_{\mathfrak{M}}\int_{\mathfrak{M}}|T_0^k(\alpha)\,R(\alpha)|^2\,d\alpha=O\left(P^{k+2}Q^2(0)\right).$$

Proof. By lemmas 2. 5 and 2. 6, we have, on $\mathfrak{M}$,

$$T_0(\alpha)=O\left(q^{-a}\,Min\,(P,|\beta|^{-a})\right)+O\left(q^{1-a+\varepsilon}\right)$$
$$=O\left(q^{-a}\,Min\,(P,|\beta|^{-a})\right).$$

The summation inquestion does not exceed

$$O\left(\sum_{\mathfrak{M}}\int_{\mathfrak{M}}q^{-2(k+1)a}\,Min\,(P^{2(k+1)},|\beta|^{-2(1+a)})\,Q^2(0)\,d\beta\right)$$
$$=O\left(\sum_q q^{-1-2a}\,P^{k+2}Q^2(0)\right)=O\left(P^{k+2}Q^2(0)\right).$$

Lemma 2. 9.

$$\sum_n r_{k+m+3}^2(n)=O\left(P^{k+2}Q^2(0)\right).$$

Proof. The lemma follows from lemmas 2. 7 and 2. 8.

3. Proof of the theorem.

Let

$$\mathfrak{T}(\alpha,P)=\sum_{P<p<2P}e(p^k\alpha).$$

Correspondingly we define $\mathfrak{T}_i(\alpha)$ $(i=0,1,2,\ldots,m,m+1)$, $\mathfrak{Q}(a)$ and $\mathfrak{R}(\alpha)$. Let

$$\mathfrak{T}_0^{k+1}(\alpha)\,\mathfrak{Q}(\alpha)=\sum_n r'_{m+k+3}(n)\,e(n\alpha)$$

and

$$\mathfrak{T}_0^{2k+3}(\alpha)\,\mathfrak{Q}^2(\alpha)=\sum_n r'_{2m+2k+7}(n)\,e(n\alpha),$$

so that $r'_{2m+2k+7}(n)$ is the number of the solutions of

$$n=p_1^k+\ldots+p_{2m+2k+7}^k$$

where p's satisfy certain conditions.

Let

$$\mathfrak{T}_0^*(\alpha,h,q)=\frac{W_{h,q}}{\varphi(q)}\sum_{p^k\leqq n\leqq(2P)^k}\frac{e(n\beta)}{n^{1-a}\log n}.$$

We divide the interval $0\leqq\alpha\leqq 1$ into Farey arcs belonging to all rational points h/q with $1\leqq q\leqq NL^{-\sigma_0}$, $0\leqq h\leqq q$, $(h,q)=1$, where

$L = \log N$ and σ_0 is the number so chosen that lemma 3. 1 holds. We divide these arcs into major arcs $\mathfrak{M}\,(1 \leqq q \leqq L^{\sigma_0})$ and minor arcs $\mathfrak{m}\,(L^{\sigma_0} < q \leqq N L^{-\sigma_0})$. In either case the arc has the form

$$\alpha = \frac{h}{q} + \beta, \qquad -\vartheta_1 q^{-1} N^{-1} L^{\sigma_0} \leqq \beta \leqq \vartheta_2 q^{-1} N^{-1} L^{\sigma_0}$$

where $\frac{1}{2} \leqq \vartheta_i \leqq 1$.

Lemma 3. 1 (Vinogradow) [14]. *There exists an integer* σ_0 *such that on* $\mathfrak{m}$

$$\mathfrak{T}_0(\alpha) = O(P L^{-\sigma}).$$

In particular we can choose σ_0 *so that* $\sigma \geqq 2k + 2m + 8$.

Lemma 3. 2.

$$\sum_{\mathfrak{m}} \int_{\mathfrak{m}} |\mathfrak{T}_0^{2k+3}(\alpha)\, \mathfrak{Q}^2(\alpha)|\, d\alpha = O(P^{k+3} Q^2(0) L^{-\sigma}).$$

Proof. By lemma 2. 9 and 3. 1 we have

$$\sum_{\mathfrak{m}} \int_{\mathfrak{m}} |\mathfrak{T}_0^{2k+3}(\alpha)\, \mathfrak{Q}^2(\alpha)|\, d\alpha = O\Big(P L^{-\sigma} \int_0^1 |\mathfrak{T}_0^{k+1}(\alpha)\, \mathfrak{Q}(\alpha)|^2\, d\alpha\Big)$$
$$= O\big(P L^{-\sigma} \sum_n r'^{\,2}_{m+k+3}(n)\big) = Q\big(P L^{-\sigma} \sum_n r^2_{m+k+3}(n)\big)$$
$$= O(P^{k+2} Q^2(0)\, P L^{-\sigma}).$$

Lemma 3. 3 (Siegel-Walfisz) [15]. *If* $q \leqq L^{\sigma_0}$, $(l, q) = 1$, $n \leqq N$, *then*

$$\pi(n; l, q) = \frac{1}{\varphi(q)} \operatorname{li} n + O(N e^{-c_4 \sqrt{L}}).$$

Lemma 3. 4. On $\mathfrak{M}$

$$\mathfrak{T}_0(\alpha) - \mathfrak{T}_0^*(\alpha, a, q) = O(P e^{-c_5 \sqrt{L}}).$$

This is a consequence of lemma 3. 3 [16].

Lemma 3. 5.

$$\mathfrak{T}_0^*(\alpha, h, q) = O\big(q^{-\frac{1}{2}+\varepsilon} \mathit{Min}(P, |\beta|^{-a})\big).$$

This is a consequence of lemma 1. 2.

Lemma 3. 6.

$$\sum_{\mathfrak{M}} \int_{\mathfrak{M}} |\mathfrak{T}_0^{2k+3}(\alpha) - \mathfrak{T}_0^{*\,2k+3}(\alpha, h, q)|\, \mathfrak{Q}^2(\alpha)\, d\alpha = O(P^{k+3} Q^2(0)\, e^{-c_6 \sqrt{L}}).$$

Proof. By lemma 3. 4 and 3. 5 on $\mathfrak{M}$

$$\mathfrak{T}_0^{2k+3}(\alpha) - \mathfrak{T}_0^{*\,2k+3}(\alpha, h, q) = O\Big(\big(P L^{-\frac{1}{2}\sigma_0}\big)^{2k+2} P e^{-c_5 \sqrt{L}}\Big)$$
$$= O(P^{2k+3} e^{-c_7 \sqrt{L}}).$$

---

[14]) Comptes Rendus de l'URSS, XVI, No. 3, 1937.

[15]) Math. Zeitschr. **40** (1936), S. 592—601, Hilfssatz 3.

[16]) See, for example, Davenport and Heilbronn, Proc. of London Math. Soc. (2) **43** (1931), S. 142—151, Lemma 2.

Therefore the summation in question does not exceed

$$O\Big(\sum_{q \leqq L^{\sigma_0}} q \cdot P^{2k+3} e^{-c_7\sqrt{L}} \cdot q^{-1} N^{-1} L^{\sigma_0} Q^2(0)\Big) = O\big(P^{k+3} Q^2(0)\, e^{-c_6\sqrt{L}}\big).$$

Lemma 3. 7.

$$\sum_{\mathfrak{M}} \int_{\mathfrak{M}} |\mathfrak{T}_0^{*\,2k+3}(\alpha, h, q)| \left| \mathfrak{Q}^2(\alpha) - \Lambda^2 \Big(\frac{W_{a,q}}{\varphi(q)}\Big)^{2m+4} \right| d\alpha = O\big(P^{k+3} Q^2(0)\, e^{-c_8\sqrt{L}}\big)$$

where

$$c_9 \frac{Q(0)}{L^{m+2}} \leqq \Lambda \leqq c_{10} \frac{Q(0)}{L^{m+2}}.$$

Proof. We have

$$\left|\mathfrak{Q}(\alpha) - \mathfrak{Q}\Big(\frac{h}{q}\Big)\right| \leqq \sum_n r'_{m+2}(n)\, |e(n\alpha) - e(nh/q)|$$

$$\leqq |\beta| \sum_n n\, r'_{m+2}(n) = O(P^{k-1} |\beta| Q(0))$$

$$= O(P^{-1} L^{\sigma_0} Q(0)).$$

Further by lemma 3. 3,

$$\mathfrak{T}_i\Big(\frac{h}{q}\Big) = \sum_{2^{-i} P^{(1-a)^i} \leq p \leqq 2^{-i+1} P^{(1-a)^i}} e_q(h p^k)$$

$$= \sum_{\substack{l=1 \\ (l,q)=1}}^{q} e_q(h l^k)\big(\pi(2^{-i} P^{(1-a)^i}; l, q) - \pi(2^{-i+1} P^{(1-a)^i}; l, q)\big) + O(q^{\varepsilon})$$

$$= \frac{W_{h,q}}{\varphi(q)} \int_{2^{-i} P^{(1-a)^i}}^{2^{-i+1} P^{(1-a)^i}} \frac{dx}{\log x} + O\big(P^{(1-a)^i} e^{-c_{11}\sqrt{L}}\big).$$

Therefore

$$\mathfrak{Q}\Big(\frac{h}{q}\Big) = \Big(\frac{W_{h,q}}{\varphi(q)}\Big)^{m+2} \Lambda + O\big(Q(0)\, e^{-c_{12}\sqrt{L}}\big),$$

where

$$\Lambda = \left(\prod_{i=1}^{m+1} \int_{2^{-i} P^{(1-a)^i}}^{2^{-i+1} P^{(1-a)^i}} \frac{dx}{\log x}\right)\left(\int_{2^{-(m+1)} P^{(1-a)^{m+1}}}^{2^{-m} P^{(1-a)^{m+1}}} \frac{dx}{\log x}\right).$$

It is evident that Λ satisfies the inequality in the lemma.

Thus

$$\left|\mathfrak{Q}^2(\alpha) - \Lambda^2 \Big(\frac{W_{h,q}}{\varphi(q)}\Big)^{2(m+2)}\right| \leqq \left|Q(\alpha) - \Lambda \Big(\frac{W_{h,q}}{\varphi(q)}\Big)^{m+2}\right| Q(0)$$

$$= O\big(Q^2(0)\, e^{-c_{12}\sqrt{L}}\big).$$

The sum in question, by lemma 3. 5, does not exceed

$$O\Big(\sum_{q \leq L^h} q \cdot Q^2(0)\, e^{-c_{12}\sqrt{L}}\, q^{-\frac{2k+3}{2}+\varepsilon}\Big)\Big(\int_0^{P^{-k}} P^{2k+3}\, d\beta + \int_{P^{-k}} |\beta|^{-\frac{2k+3}{k}}\, d\beta\Big)$$

$$= O\big(P^{k+3} Q^2(0)\, e^{-c_8\sqrt{L}}\big).$$

Lemma 3. 8.

$$\sum_{\mathfrak{M}}\Big(\int_0^1 - \int_{\mathfrak{M}}\Big)\Big|\mathfrak{T}_0^{*\,2k+3}(\alpha, h, q)\, A^2\Big(\frac{W_{h,q}}{\varphi(q)}\Big)^{2m+4}\Big|\, d\alpha = O\big(P^{k+3} Q^2(0)\, L^{-\sigma_0(1-3a)}\big).$$

Proof. The left hand side does not exceed

$$O\Big(\sum_{q \leq L^{\sigma_0}} q \cdot Q^2(0)\, q^{-m-2} \cdot q^{-\frac{1}{2}(2k+3)} \int_{q^{-1}N^{-1}L^{\sigma_0}} |\beta|^{-\frac{2k+3}{k}}\, d\beta\Big)$$

$$= O\big(P^{k+3} Q^2(0)\, L^{-\sigma_0(1-3a)}\big).$$

Lemma 3. 9.

$$r_{2k+2m+7}(N) = A^2\,\mathfrak{T}(N)\,\psi(N) + O\big(P^{k+3} L^{-\sigma} Q^2(0)\big)$$

where

$$c_{13} P^{k+3} L^{-2k-2m-7} < \psi(N) < c_{14} P^{k+3} L^{-2k-2m-7}.$$

Proof. By lemmas 3. 2, 3. 6, 3. 7 and 3. 8, we have

$$r'_{2m+2k+7}(N) = \int_0^1 \mathfrak{T}_0^{2k+3}(\alpha)\, \mathfrak{Q}^2(\alpha)\, e^{-2\pi i N\alpha}\, d\alpha$$

$$= \sum_{q \leq L^{\sigma_0}} \sum_{\substack{h=1\\(h,q)=1}}^{q} A^2\Big(\frac{W_{h,q}}{\varphi(q)}\Big)^{2k+2m+7} e_q(-Nh)\, \Psi(N)$$

$$+ O\big(P^{k+3} L^{-\sigma} Q^2(0)\big)$$

where

$$\Psi(N) = \sum_{n_1 + \ldots + n_{2k+3} = N} \frac{1}{\prod_{i-1}^{2k+3} n_i^{1-a} \log n_i}.$$

It is obviously satisfy the inequality stated in the lemma.

The lemma follows, since

$$\sum_{q > L^{\sigma_0}} \sum_{\substack{h=1\\(h,q)=1}}^{q} \Big(\frac{W_{h,q}}{\varphi(q)}\Big)^{2k+2m+7} e_q(-Nh) = 0\Big(\sum_{q > L^{\sigma_0}} q^{-k-m-2}\Big)$$

$$= O\big(L^{-(k+m+1)\sigma_0}\big).$$

The theorem is proved by lemmas 1. 9 and 3. 9.

Appendix. Distribution of K.

The congruence condition in the theorem is essential and can not replaced by weaker one.

If k is an odd integer, then $K = 2$. Consequently we have

Theorem. *Let k be an odd integer. Every sufficiently large odd integer is a sum of $2k + 2m + 7$ k-th powers of primes. Every sufficiently large integer can be expressed as a sum of at most $2k + 2m + 8$ k-th powers of primes.*

When k is even, the distribution of K is very irregular and its order of magnitude is very large. To illustrate this, I give the following table.

k	K	k	K
2	$2^3 \cdot 3$	52	$2^4 \cdot 3 \cdot 5 \cdot 53$
4	$2^4 \cdot 3 \cdot 5$	54	$2^3 \cdot 3^4 \cdot 7 \cdot 19$
6	$2^3 \cdot 3^2 \cdot 7$	56	$2^5 \cdot 3 \cdot 5 \cdot 29$
8	$2^5 \cdot 3 \cdot 5$	58	$2^3 \cdot 3 \cdot 59$
10	$2^3 \cdot 3 \cdot 11$	60	$2^4 \cdot 3^2 \cdot 5^2 \cdot 7 \cdot 11 \cdot 13 \cdot 31 \cdot 61$
12	$2^4 \cdot 3^2 \cdot 5 \cdot 7 \cdot 13$	62	$2^3 \cdot 3$
14	$2^3 \cdot 3$	64	$2^8 \cdot 3 \cdot 5 \cdot 17$
16	$2^6 \cdot 3 \cdot 5 \cdot 17$	66	$2^3 \cdot 3^2 \cdot 7 \cdot 23 \cdot 67$
18	$2^3 \cdot 3^3 \cdot 7 \cdot 19$	68	$2^4 \cdot 3 \cdot 5$
20	$2^4 \cdot 3 \cdot 5^2 \cdot 11$	70	$2^3 \cdot 3 \cdot 11 \cdot 71$
22	$2^3 \cdot 3 \cdot 23$	72	$2^5 \cdot 3^3 \cdot 5 \cdot 7 \cdot 13 \cdot 19 \cdot 37 \cdot 73$
24	$2^5 \cdot 3^2 \cdot 5 \cdot 7 \cdot 13$	74	$2^3 \cdot 3$
26	$2^3 \cdot 3$	76	$2^4 \cdot 3 \cdot 5$
28	$2^4 \cdot 3 \cdot 5 \cdot 29$	78	$2^3 \cdot 3^2 \cdot 7 \cdot 79$
30	$2^3 \cdot 3^2 \cdot 7 \cdot 11 \cdot 31$	80	$2^6 \cdot 3 \cdot 5^2 \cdot 11 \cdot 17 \cdot 41$
32	$2^7 \cdot 3 \cdot 5 \cdot 17$	82	$2^3 \cdot 3 \cdot 83$
34	$2^3 \cdot 3$	84	$2^4 \cdot 3^2 \cdot 5 \cdot 7^2 \cdot 13 \cdot 29 \cdot 43$
36	$2^4 \cdot 3^3 \cdot 5 \cdot 7 \cdot 13 \cdot 19 \cdot 37$	86	$2^3 \cdot 3$
38	$2^3 \cdot 3$	88	$2^5 \cdot 3 \cdot 5 \cdot 23 \cdot 89$
40	$2^5 \cdot 3 \cdot 5^2 \cdot 11 \cdot 41$	90	$2^3 \cdot 3^3 \cdot 7 \cdot 11 \cdot 19 \cdot 31$
42	$2^3 \cdot 3^2 , 7^2 \cdot 43$	92	$2^4 \cdot 3 \cdot 5 \cdot 47$
44	$2^4 \cdot 3 \cdot 5 \cdot 23$	94	$2^3 \cdot 3$
46	$2^3 \cdot 3 \cdot 47$	96	$2^7 \cdot 3^2 \cdot 5 \cdot 7 \cdot 13 \cdot 17 \cdot 97$
48	$2^6 \cdot 3^2 \cdot 5 \cdot 7 \cdot 13 \cdot 17$	98	$2^3 \cdot 3$
50	$2^3 \cdot 3 \cdot 11$	100	$2^4 \cdot 3 \cdot 5^3 \cdot 11 \cdot 101$

(Eingegangen am 28. Februar 1938.)

Reprinted from the
JOURNAL OF THE CHINESE MATHEMATICAL SOCIETY
Vol. 2, pp. 175–191, 1940

ON A GENERALIZED WARING PROBLEM II

LOO-KENG HUA

Introduction

Let $P(h)$ be an integral-valued polynomial of the k-th degree and with positive first coefficient. Let $G(P(h))$ be the least value of integers such that the Diophantine equation

$$P(h_1) + \cdots + P(h_s) = N, \qquad h_\nu \geqslant 0,$$

is solvable for all sufficiently large integers N. In this paper we shall prove that

$$G(P(h)) \leqslant 2^{k+1}(k-1), \tag{1}$$

which is much better than the result of I([1]).

On the other hand, it will be shown in §4 that

$$G(P(h)) = \begin{cases} 2^k - 1 & \text{for } k \text{ odd} \\ 2^k & \text{for } k \text{ even,} \end{cases} \tag{2}$$

provided that $k \geqslant 5$ and

$$P(h) = 2^{k-1}F_k(h) - 2^{k-2}F_{k-1}(h) + \cdots + (-1)^{k-1}F_1(h),$$

where

$$i!\,F_i(h) = h(h-1)\cdots(h-i+1).$$

The only theorems in the whole theory of Waring's problem which give definite values of $G(P(h))$ are those involving quadratic polynomials, namely,

$$G(h^2) = 4, \qquad G\big(\tfrac{1}{2}(h^2 - h)\big) = 3.$$

There is some numerical evidence for the view that

$$G(P(h)) \leqslant \begin{cases} 2^k - 1 & \text{for odd } k, \\ 2^k & \text{for even } k, \end{cases}$$

holds generally.

Without loss of generality, we may assume that the constant term of $P(h)$ is zero and all its coefficients positive.

([1]) Proc. London Math. Soc. **43** (1937), 161–182. It will be referred to as I.

1. Let $r(N)$ be the number of solutions of the Diophantine equation

$$N = P(x_1) + \cdots + P(x_s), \qquad x_\nu \geqslant 0.$$

In I and elsewhere(²) the author proved the following

Theorem 1. *If*

$$s \geqslant \begin{cases} 10k^3 \log k & \text{for} \quad k > 15, \\ 2^k + 1 & \text{for} \quad 3 \leqslant k \leqslant 15, \end{cases}$$

then

$$r(N) + a^{-s/k} \frac{\Gamma^s(1 + 1/k)}{\Gamma(s/k)} \mathfrak{S}(N) N^{s/k - 1} + O(N^{s/k - 1 - \rho}),$$

where a is the first coefficient in $P(x)$, ρ *is a positive number independent of* N, *and* $\mathfrak{S}(N)$ *is the singular series defined in I.*

Consequently, in order to establish the result (1), we only need to prove that

$$\mathfrak{S}(N) \geqslant D > 0, \tag{3}$$

for $s \geqslant (k-1)2^{k+1}$, where D is a number independent of N.

Let d be the least common denominator of the coefficients of $P(h)$. Let p be a prime, $p^t \| d$ and $p^t P(h) = \Phi(h)$.(³) Then the denominators of the coefficients of $\Phi(h)$ are not divisible by p. Let $\theta^{(i)}$ be the greatest integer such that the i-th derivative of $\Phi(h)$ satisfies

$$\Phi^{(i)}(h) \equiv 0 \qquad (\text{mod } p^{\theta^{(i)}})(^4)$$

for all h, and let $P^*(h) = p^{-\theta'} \phi'(h)$. Let

$$\delta = \max_{1 \leqslant i \leqslant k-1} (\theta^{(i)} - \theta^{(i+1)}),$$

and

$$\gamma' = \begin{cases} \theta' + 2 - t + \delta & \text{for} \quad p = 2, \\ \theta' + 1 - t + \delta & \text{for} \quad p \neq 2. \end{cases}$$

Lemma 1.1. *If* $l \geqslant \gamma' + 1$ *and* $x = y + p^{l + t - \theta' + 1} z$, *then*

$$P(x) \equiv P(y) + z p^{l-1} P^*(y) \qquad (\text{mod } p^l)$$

(²) *Jour. of Chinese Math. Soc.* **1** (1936), 21–61, *and Quar. Jour. of Math.* **9** (1938), 199–202.

(³) The definition of $\Phi(h)$ differs from that of I by a factor relatively prime to p. We should correct $P_\nu^*(y)$ into $\alpha_\nu P^*(y)$ in lemma 7.5 of I, where $p \nmid \alpha_\nu$. However, this is not essential for the argument.

(⁴) $\Phi(h)$ is not necessarily a polynomial with integer coefficients. Hence $\Phi^{(i)}(h)$ may not be an integer. Yet it may be as an integer in the congruence $\Phi^{(i)}(h) \equiv 0$ (mod $p\theta^{(i)}$), since the denominators of all the coefficients are prime to p.

and

$$P^*(x) \equiv P^*(y) \qquad (\bmod p)$$

(This is sharper than I, lemma 7.5).

Proof. By Taylor's expansion, we have

$$\Phi(x) = \Phi(y) + zp^{l+t-\theta'-1}\Phi'(y) + \frac{1}{2!}z^2p^{2(l+t-\theta'-1)}\Phi''(y)$$
$$+ \frac{1}{3!}z^3p^{3(l+t-\theta'-1)}\Phi'''(y) + \cdots$$

1) $p \neq 2$. For $i \geqslant 2$, the power of p in

$$\frac{1}{i!}p^{i(l+t-\theta'-1)}\Phi^{(i)}(y)$$

is not less than $i(l+t-\theta'-1)+\theta^{(i)}-\Lambda_i$, where $p^{\Lambda_i}\|i!$ Since

$$\theta^{(i)} \geqslant \theta^{(i-1)} - \delta \geqslant \cdots \geqslant \theta' - (i-1)\delta$$

and $\Lambda_i \leqslant i-2$, we have

$$i(l+t-\theta'-1) - \theta^{(i)} - \Lambda_i \geqslant i(l+t-\theta'-1) + \theta' - (i-1)\delta - \Lambda_i$$
$$\geqslant l+t+2(i-1)-i-\Lambda_i \geqslant l+t.$$

It follows that

$$\Phi(x) \equiv \Phi(y) + zp^{l+t-\theta'-1}\Phi'(y) \qquad (\bmod p^{l+t}).$$

Therefore

$$P(x) \equiv P(y) + p^{l-1}zp^*(y) \qquad (\bmod p^l).$$

2) $p = 2$. For $i \geqslant 2$, the power of 2 in

$$\frac{1}{i!}\alpha^{i(l+t-\theta'-1)}\Phi^{(i)}(y)$$

is not less than

$$i(l+t-\theta'-1)+\theta^{(i)}-\Lambda_i \geqslant l+t+3(i-1)-i-\Lambda_i$$
$$\geqslant l+t,$$

where $2^{\Lambda_i}\|i!$. And the result for $p=2$ follows as before.

The remaining part of the lemma can also be proved by Taylor's expansion. From lemma 1.1 we can deduce, by the method used in I, lemma 7.6,

Lemma 1.2. *If* $l \geqslant 1+\gamma$, *then*

$$N(p^t) = p^{s-1}N(p^{t-1}).$$

Consequently (Cf. I §7) we have

Theorem 2. *If there exists an integer* s_0 *such that*

$$N(s_0 \cdot P, n, p\gamma') > 0$$

for all primes p and all integers n, then, for $s \geqslant \operatorname{Max}(s_0, 2k+1)$, *we have*

$$\mathfrak{S}(N) \geqslant D > 0.$$

Thus, in order to prove (1), we need only prove that

$$N(s_0 \cdot P, n, p\gamma') > 0 \tag{4}$$

for $s_0 \geqslant (k-1)2^{k+1}$. This will be proved in §§2 and 3.

2. Lemma 2.1 (DAVENPORT AND CHOWLA).[5] *Let* $\alpha_1, \ldots, \alpha_m$ *be m different residue classes,* mod h, *and* $\beta_1, \ldots, \beta_n$ *be n different residue classes,* mod h, *and* $(\beta_1, \ldots, \beta_n, h) = 1$. *Then the number of different residue classes represented by*

$$\alpha_i \quad \textit{or} \quad \alpha_i + \beta_j \qquad (1 \leqslant i \leqslant m, \quad 1 \leqslant j \leqslant n)$$

is greater than or equal to $\operatorname{Min}(m + n, h)$.

Lemma 2.2. *Let* $P(x)$ *be a polynomial of the k-th degree with integer coefficients,* mod p, *and* $\not\equiv P(0)$ (mod p). *If* $p > k$, *then the number of different values given by* $P(x)$, mod p, *is greater than or equal to*

$$[p/h] + 1.$$

Proof. Since, for any given integer α, the number of solutions of the congruence

$$P(x) \equiv a \qquad (\text{mod } p)$$

is at most k, the lemma follows immediately from the fact that $k \nmid p$.

Lemma 2.3. *For* $s_0 \geqslant 2k$,

$$N(s_0 \cdot P, n, p\gamma') > 0$$

for all integers n and $p > k$.

Proof. Now $\gamma' = 1$, since $p > k$. Firstly, we are going to prove that, for $s \geqslant 2k - 1$,

$$\sum_{\nu=1}^{s} P(x_\nu) \equiv n \qquad (\text{mod } p)$$

is solvable for all integers n. This statement is evident for $p \leqslant 2k - 1$. So we can assume $p > 2k - 1$. Since $P(x)$ gives at least $[p/k] + 1$ different values, mod p, and at least $[p/k]$ of them are not divisible by p, and

$$\begin{aligned}[p/k] + 1 + (s-1)[p/k] &\geqslant s(p/k - 1) + 1 \\ &\geqslant (2k-1)(p/k - 1) + 1 = p + (k-1)(p/k - 2) \geqslant p_1\end{aligned}$$

by lemma 2.1 the congruence is solvable for all integers n.

Next,[6] since $P^*(x) \not\equiv 0$ (mod p) for all x, we can choose an integer x' such that

[5] See E. Landau, *Über einige neuere Fortschritte der additiven Zahlentheorie*, Cambridge tract, p. 8.

[6] The argument of this paragraph will not be repeated in lemma 3.4.

$P^*(x') \not\equiv 0 \pmod p$. By the previous part of the proof, the congruence

$$\sum_{\nu=1}^{s} P(x_\nu) \equiv n - P(x') \qquad (\bmod\ p)$$

is solvable for $s \geqslant 2k-1$. Consequently we have the lemma.

3. In this section we shall use the following conventional notation: Let $Q(x)$ be a polynomial with rational coefficients and α be the greatest integer such that the least common denominator of $Q(x)$ is divisible by p^α. Then

$$Q(x) \equiv 0 \qquad (\bmod\ p^\alpha) \quad (\alpha \gtrless 0)$$

means that

$$p^\alpha Q(x) \equiv 0 \qquad (\bmod\ p^{a+\alpha})$$

for all integers x. For example

$$\frac{x^3 - x}{3^7} \equiv 0 \qquad (\bmod\ 3^{-6}).$$

We must note that "mod p^0" is different from our ordinary notation "mod 1." For example

$$\frac{x^3 - x}{3 \cdot 7} \equiv 0 \qquad (\bmod\ 3^0)$$

but

$$\frac{x^3 - x}{3 \cdot 7} \not\equiv 0 \qquad (\bmod\ 1),$$

since $1/F$ is to be interpreted as an integer, mod 3.

Lemma 3.1. *A necessary and sufficient condition for*

$$Q(x) \equiv 0 \qquad (\bmod\ p^\alpha)$$

is that

$$Q(x) = a_1 F_k(x) + \cdots + a_k F_1(x), p^\alpha / (a_1, \ldots, a_k)\text{[7]}$$

where $F_i(x)$ is defined in the introduction.

This can be proved easily by considering $p^\alpha Q(x)$.

Lemma 3.2. *If a is the greatest integer such that*

$$Q(x) \equiv 0 \qquad (\bmod\ p^\alpha),$$

and if

$$Q'(x) \equiv 0 \qquad (\bmod\ p^b),$$

(7) The previous conventional notation is used.

then

$$b - a \leqslant [k/(p-1)] - 1.$$

Proof. Since $b - a$ is unaltered if we replace $Q(x)$ by $p^t Q(x)$, we may assume without loss of generality that $a = 0$, i.e.

$$p^0 | (a_1, \ldots, a_k), \qquad p \nmid (a_1, \ldots, a_k)$$

by lemma 3.1. (The above formulae signify that all the denominators of the a's in their reduced forms are relatively prime to p). Write

$$Q'(x) = b_1 F_{k-1}(x) + b_2 F_{k-2}(x) + \cdots b_{k-1} F_1(x) + b_k .$$

Let $\Delta^i(P(x))$ be the i-th difference of $P(x)$. Then

$$b_1 = \Delta^{k-1} Q'(x) = (\Delta^{k-1} Q(x))' = a_1,$$

$$b_2 = (\Delta^{k-2} Q'(x))_{x=0} = (\Delta^{k-2} Q(x))'_{x=0} = -\frac{a_1}{2} + a_2,$$

$$\cdots\cdots\cdots\cdots\cdots\cdots\cdots\cdots\cdots\cdots\cdots\cdots\cdots\cdots$$

$$b_k = (Q'(x))_{x=0} = (-1)^k \left(\frac{a_1}{k} - \frac{a_2}{k-1} + \frac{a_s}{k-2} - \cdots + (-1)^k a_k \right).$$

From the hypothesis and lemma 3.1, we have

$$p^b | (b_1, \ldots, b_k).$$

If $b > [k/(p-1)] - 1$, then

$$p^b | (a_1, a_2, \ldots, a_{p-1})$$
$$p^{b-1} | (a_p, a_{p+1}, \ldots, a_{2p-2}),$$
$$p^{b-2} | (a_{2p-1}, \ldots),$$
$$\cdots \quad \cdots \quad \cdots$$
$$p | p^{b-([k/(p-1)]-1)} | (a_{([k/(p-1)]-1)(p-1)+1, \ldots,} a_k)$$

That is to say, $p | (a_1, \ldots, a_k)$. But this is a contradiction.

Lemma 3.3. *If* $p = 2$, *then*

$$p^{\gamma'} \leqslant (k-1) 2^{k+1};$$

if $p > 2$, *then*

$$p^{\gamma'} \leqslant (k-1) p^{[k/(p-1)]} < (k-1) 2^{k+1}.$$

Proof. By I, lemma 7.4, we have

$$p^{\delta} \leqslant k - 1.$$

In connection with lemma 3.2, we have, for $p = 2$,

$$p^{\gamma'} = p^{\theta' - t + 2 + \delta} \leqslant (k-1) 2^{k+1};$$

and for $p > 2$,

$$p^{\gamma'} \leqslant (k-1)p^{[k/(p-1)]}.$$

Consequently we have

Lemma 3.4. *If* $s \geqslant (k-1)2^{k+1} - 1$, *and* $P(x) \not\equiv 0 \pmod p$, *then*

$$\sum_{\nu=1}^{s} P(x_\nu) \equiv n \pmod{p\gamma'}$$

is solvable for all integers n. Therefore, for $s \geqslant (k-1)2^{k+1}$, *we have*

$$N(s_0 \cdot P, n, p\gamma') > 0$$

for all integers n and primes p.

Thus (4) is true and by theorem 1 and 2 we have the following

Theorem 3. *If $P(x)$ is an integral-valued polynomial of the k-th degree with positive first coefficient and is not congruent to $P(0)$ identically*, mod p, *where p is any prime, then*

$$G(P(x)) \leqslant (k-1)2^{k+1}.$$

4. In this section we shall consider the special polynomials.

$$H_k(x) = 2^{k-1}F_k(x) - 2^{k-2}F_{k-1}(x) + \cdots + (-1)^{k-1}F_1(x).$$

Lemma 4.1. *Let τ be the integer satisfying $2^\tau \| k!$ and σ the integer satisfying $2^\sigma \leqslant k < 2^{\sigma-1}$. Denote by M the aggregate of integers which consists of all the products of any $k-2$ integers taken among any k consecutive integers. Then every element of M is divisible by $2^{\tau-2\sigma+1}$*

Proof. First, let us consider those elements of M for which one of the k consecutive integers is zero. It is sufficient to prove that, (i) if e and $k-1-e$ are both less than 2^σ, then

$$2^{-\sigma+1}e!\,(k-1-e)!$$

is divisible by $2^{\tau-2\sigma+1}$; and (ii) if either e or $k-1-e$ is greater than or equal to 2^σ, then

$$2^{-\sigma}e!\,(k-1-e)!$$

is divisible by $2^{\tau-2\sigma+1}$. Evidently (i) and (ii) can be expressed alternatively as follows: (i)

$$\sum_{\lambda=1}^{\infty}\left(\left[\frac{k}{2^\lambda}\right] - \left[\frac{e}{2^\lambda}\right] - \left[\frac{k-1-e}{2^\lambda}\right]\right) \leqslant \sigma;$$

and (ii) if $2^{\sigma+1} > e \geqslant 2^{\sigma}$ (as we may suppose without loss of generality), then

$$\sum_{\lambda=1}^{\infty}\left(\left[\frac{k}{2^\lambda}\right]-\left[\frac{e}{2^\lambda}\right]-\left[\frac{k-1-e}{2^\lambda}\right]\right) \leqslant \sigma - 1.$$

We shall only give a proof of (ii), as (i) can be easily proved by the same argument.
It is evident that

$$\sum_{\lambda=\sigma}^{\infty}\left(\left[\frac{k}{2^\lambda}\right]-\left[\frac{e}{2^\lambda}\right]-\left[\frac{k-1-e}{2^\lambda}\right]\right) = 0.$$

Let $2^d \| k$. If $\lambda > d$, then

$$\left[\frac{k}{2^\lambda}\right]-\left[\frac{e}{2^\lambda}\right]-\left[\frac{k-1-e}{2^\lambda}\right] < \frac{k-1}{2^\lambda} - \left(\frac{e}{2^\lambda}-1\right)-\left(\frac{k-1-e}{2^\lambda}-1\right) = 2,$$

and consequently

$$\left[\frac{k}{2^\lambda}\right]-\left[\frac{e}{2^\lambda}\right]-\left[\frac{k-1-e}{2^\lambda}\right] \leqslant 1.$$

If $\lambda \leqslant d$, we write $k = 2^d k_0$, and

$$e = 2^d e_0 + 2^{d-1}e_1 + \cdots + e_d, \qquad 0 \leqslant e_i \leqslant 1, \quad i = 1, 2, \ldots, d.$$

Then

$$k - 1 - e = 2^d(k_0 - e_0 - 1) + 2^{d-1}(1-e_1) + \cdots + (1 - e_d).$$

It is easy to see that

$$\left[\frac{k}{2^\lambda}\right]-\left[\frac{e}{2^\lambda}\right]-\left[\frac{k-1-e}{2^\lambda}\right] \leqslant \sigma.$$

Hence

$$\sum_{\lambda=1}^{\infty}\left(\left[\frac{k}{2^\lambda}\right]-\left[\frac{e}{2^\lambda}\right]-\left[\frac{k-1-e}{2^\lambda}\right]\right) \leqslant \sigma - 1.$$

Next we consider the elements of M taken from among the consecutive numbers

$$a+1, a+2, \ldots, a+k. \tag{A}$$

Let Λ be the integer of (A) containing the greatest power of 2. (It is easy to verify that Λ is unique). Form the set

$$a+1-\Lambda, a+2-\Lambda, \ldots, a+k-\Lambda. \tag{B}$$

The power of 2 in any term of (A) is equal to that in the corresponding term of (B), the only exception being when Λ corresponds to zero. The most unfavourable element of M taken from (A) is

$$\frac{(a+1)(a+2)\cdots(a+k)}{\Lambda\cdot\Lambda_1},$$

where Λ_1 is a number of (A) with next highest power of 2. The power of 2 for this element is equal to that for

$$\frac{(\Lambda-(a+1)!(a+h-\Lambda)!}{\Lambda_1-\Lambda}.$$

Now applying the first part of the proof, we complete the proof of the lemma.

Lemma 4.2. *Let*

$$K_k(x)=x(x-1)\cdots(x-k+1).$$

Then $K_k(x)$ *is a residual polynomial,* mod $2^{\tau-\sigma}$, *and* $K_k''(x)$ *is a residual polynomial,* mod $2^{\tau-2\sigma+1}$.

Proof. The second part of the lemma follows immediately from lemma 4.1, and the first part is easily proved by the same argument.

Lemma 4.3. *There is an integer* x_1 *such that*

$$K_k(x_1)\equiv 0 \pmod{2^{\tau+1}}$$

and

$$2^{\tau-\sigma+1}\nmid K_k'(x_1).$$

There is an integer x_2 *such that*

$$K_k(x_2)\equiv 2^\tau \pmod{2^{\tau+1}}$$

and

$$2^{\tau-\sigma+1}\nmid K_k'(x_2).$$

Proof. Take $x_1=2^\sigma$. Then $K_k(x_2)=0$ and

$$K_k'(x_1)=2^{-\sigma}k!,$$

which is not divisible by $2^{\tau-\sigma+1}$.

Next take $x_2=k$, then

$$K_k(x_2)=k!\equiv 2^\tau \pmod{2^{\tau+1}}$$

and

$$K_k'(x_2)\equiv 2^{-\sigma}k! \pmod{2^{\tau-\sigma+1}}.$$

Theorem 4. *Let*

$$k!\,F_k(x)=K_k(x).$$

Then

$$F_k(x)\equiv a \pmod{2^l},\quad 2^{\tau-\sigma+1}\nmid K_k'(x)$$

is solvable for all values of a *and* $l(>0)$.

Proof. By lemma 4.3, the theorem is evident for $l = 1$. We shall prove it by induction. Let x_1 be an integer such that

$$K_k(x_1) \equiv k!\,a \qquad (\text{mod}\, 2^{l+\tau}), \qquad 2^{\tau-\sigma+1} \nmid K(x_1)$$

By the same method as in lemma 1.1, we have, for $l \geqslant 1$,

$$K_k(x + 2^{l+\sigma}y) \equiv K_k(x) + 2^{l+\sigma}yK_k'(x) \qquad (\text{mod}\, 2^{l+\tau+1}),$$

and

$$K_k'(x + 2^{l+\sigma}y) \equiv K_k'(x) \qquad (\text{mod}\, 2^{\tau-\sigma+1}).$$

Therefore we can find a y such that

$$K_k(x_1 + 2^{l+\sigma}y) \equiv k!\,a \qquad (\text{mod}\, 2^{l+\tau+1}),$$

and

$$2^{\tau-\sigma+1} \nmid K_k'(x_1 + 2^{l+\sigma}y).$$

Lemma 4.4.

$$H_k(y+1) + H_k(y) = 2^k F_k(y) + (-1)^{k+1}.$$

Proof. The lemma is evident for $k = 1$. It can be proved by induction, since

$$H_k(y) = 2^{k-1}F_k(y) - H_{k-1}(y).$$

Consequently we have

Lemma 4.5.

$$H_k(y+2) - H_k(y) = 2^k F_{k-1}(y).$$

Theorem 5.

$$G(H_k(x)) \geqslant \begin{cases} 2^k - 1 & \textit{for } k \textit{ odd}, \\ 2^k & \textit{for } k \textit{ even}. \end{cases}$$

Proof. By lemma 4.5, $H_k(x)$ gives only two different values, 0 and $(-1)^{k-1}$, $(\text{mod}\, 2^k)$. Therefore

$$G(H_k(x)) \geqslant 2^k - 1.$$

Furthermore if k is even, then, by lemma 4.5, $H_k(x)$ gives only three different values, 0, $(-1)^{k-1}$ and $(-1)^{k-1} + 2^k$, $(\text{mod}\, 2^{k+1})$. The congruence

$$\sum_{\nu=1}^{2^k-1} H_k(h_\nu) \equiv 2^k \qquad (\text{mod}\, 2^{k+1})$$

has no solution; therefore

$$G(H_k(x)) \geqslant 2^k.$$

Now we shall prove that

$$G(H_k(x)) \leqslant \begin{cases} 2^k - 1 & \text{for } k \text{ odd,} \\ 2^k & \text{for } k \text{ even.} \end{cases}$$

By the same argument as in §1, we only need to prove that

$$N(s_0 \cdot H_k, n, p\gamma') > 0$$

for

$$s_0 = \begin{cases} 2^k - 1 & \text{for } k \text{ odd,} \\ 2^k & \text{for } k \text{ even.} \end{cases}$$

CASE I. *k is odd.*
We define

$$E_k(y) = 2^{-k} H_k(2y) \quad \text{and} \quad O_k(y) = 2^{-k}(H_k(2y+1) - 1).$$

Lemma 4.6.

$$E_k(y) \not\equiv E_k(0) \quad \textit{and} \quad O_k(y) \not\equiv O_k(0) \qquad (\text{mod}\, 2).$$

Proof. 1) By lemma 4.5 we have

$$E_k(y+1) - E_k(y) = F_{k-1}(2y).$$

Since $F_{k-1}(k-1) = 1$, the first part of the lemma is proved.
2) Similarly, we have

$$O_k(y+1) - O_k(y) = F_{k-1}(2y+1).$$

Since $F_{k-1}(k) = k$, the second part of the lemma follows.

Lemma 4.7. *Let $p = 2$. The γ' (for the definition, see §1) of $E_k(x)$ is less than or equal to* 3.

Proof. It is sufficient to prove that $2^\tau E_k'(x)$ is not a residual polynomial, mod $2^{\tau-\sigma+2}$, since $\delta \leqslant \sigma$ and

$$\gamma' = \theta - t + \delta + 2 \leqslant \tau - \sigma + 1 - \tau + \sigma + 2 = 3.$$

If it were false, we should have

$$2^\tau(E_k'(y+2) - E_k'(y)) = 2^{\tau+1}F_{k-1}'(2y) \equiv 0 \qquad (\text{mod}\, 2^{\tau-\sigma+2}),$$

while

$$2^{\tau-\sigma+2} \nmid 2^{\tau+1}F_{k-1}'(k-1).$$

Lemma 4.8. *Let $p = 2$. The γ' of $O_k(x)$ is less than or equal to* 3.

The proof of the lemma is similar to that of lemma 4.7, except that the last statement is now to be replaced by

$$2^{\tau-\sigma+2} \nmid 2^{\tau+1}F_{k-1}'(k).$$

Lemma 4.9. *If $k \geqslant 5$ and $s \geqslant 2^k - 1$, then*

$$N(s \cdot H_k, 2^{\gamma'}, n) > 0.$$

for all values of n.

Proof. Let n' be the integer satisfying

$$n \equiv n' (\operatorname{mod} 2^k), \qquad 0 \leqslant n' < 2^k.$$

Then, by lemmas 4.7 and 4.8 ($2^{\gamma'} \leqslant 8$),

$$\sum_{\nu=1}^{n'} O_k(x_\nu) + \sum_{\nu=n'+1}^{2^k-1} E_k(x_\nu) \equiv m (\operatorname{mod} 2^l), \qquad 2 \nmid \text{one of } O_k^*(x_\nu) \text{ or } E_k^*(x_\nu),$$

is solvable for all integers m and $l > 0$. Consequently

$$\sum_{\nu=1}^{2^k-1} H_k(x_\nu) \equiv n \qquad (\operatorname{mod} 2^{\gamma'}), \qquad 2 \nmid \text{ one of } H_k^*(x_\nu),$$

is solvable for all integers n.

Lemma 4.10. *If $s \geqslant 2^k - 1$ and $k > 15$, then*

$$N(s \cdot H_k, p^{\gamma'}, n) > 0$$

for all integers n and primes p.

Proof. 1) If $p > k$, the lemma is proved in lemma 2.3.

2) If $k \geqslant p > 2$, the lemma follows from lemma 3.3 together with the inequalities

$$p^{\gamma'} \leqslant (k-1) p^{[k/(p-1)]} \leqslant (k-1) 3^{[1/2k]} < 2^k - 1$$

for $k > 15$.

3) If $p = 2$, the lemma is proved in lemma 4.9.

By theorems 1 and 2, we obtain, then,

$$G(H_k(x)) \leqslant 2^k - 1.$$

for odd $k > 15$. Consequently we have

Theorem 6. *If k is an odd integer > 15, then*

$$G(H_k(x)) = 2^k - 1.$$

CASE II. *k is even.*

Lemma 4.11. *If $s \geqslant 2^k$, $k \geqslant 6$, then*

$$N(s \cdot H_k, 2^{\gamma'}, n) > 0$$

for all values of p.

Proof. Let n' be the integer satisfying

$$n \equiv (-1)^{k-1} n' \qquad (\operatorname{mod} 2^k), \qquad 0 < n' \leqslant 2^k.$$

1) $n' > 8$. Define $O_k(x)$ as in case I. By lemma 4.8 (which holds also for even k),

$$\sum_{\nu=1}^{n'} O_k(x_\nu) \equiv m \pmod{2^l}, \qquad 2 \nmid \text{one of } O_k^*(x_\nu)$$

is solvable for all integers m and $l > 0$. Consequently

$$\sum_{\nu=1}^{2^k} H_k(x_\nu) \equiv n \pmod{2^{\gamma'}}, \qquad 2 \nmid \text{one of } H_k^*(x_\nu)$$

is solvable for all integers n.

2) $n' \leqslant 8$. By lemma 4.4 and theorem 4,

$$\sum_{\nu=2}^{n'} O_k(x_\nu) + 2^{-k}(H_k(x+1) + H_k(x)) \equiv m \pmod{2^k}$$

is solvable for all integers m and $l > 0$. And the lemma follows from the fact that $9 < 2^k$.

Lemma 4.12. *If* $k > 16$ *and* $s \geqslant 2^k$, *then*

$$N\left(s \cdot H_k, p^{\gamma'}, n\right) > 0$$

for all integers n and all primes p.

The proof of the lemma is the same as that of lemma 4.10.

Consequently we have

Theorem 7. *If k is an even integer* > 16, *then*

$$G(H_k(x)) = 2^k.$$

(Received 10 Feb., 1939)

Added on 5 March, 1939. The analytic part of the proof can be very much improved. Combining the arithmetical results of this paper, we can obtain, for example, that

$$G\left(\frac{1}{k!} x(x-1) \cdots (x-k+1)\right) \leqslant 2k + 2m + 5$$

where

$$m = \left[\frac{\log \frac{1}{2} b + \log(1 - 2/k)}{\log k - \log(k-1)} \right]$$

and

$$b = \begin{cases} k^3(\log k + 1.25 \log\log k^2) & \text{for } k \geqslant 15, \\ 2^{k-1} & \text{for } k < 15. \end{cases}$$

Further, theorems 6 and 7 are also true for $k \geqslant 4$.

Reprinted from the
JOURNAL OF INDIAN MATHEMATICAL SOCIETY
New Series, Vol. 4, No. 4, pp. 127–135, 1940

ON WARING'S PROBLEM WITH CUBIC POLYNOMIAL SUMMANDS*

By LOO-KENG HUA, *National Tsing Hua University, China.*

[Received 25 June 1940]

Let a cubic integral-valued polynomial be represented by

$$P(x)=\tfrac{1}{6}a(x^3-x)+\tfrac{1}{2}b(x^2-x)+cx+d,$$

where a, b, c and d are integers with $(a, b, c)=1$ and $a>0$. The object of this paper is to prove that the Diophantine equation

$$P(x_1)+\ldots+P(x_8)=N \qquad x_\nu \geqslant 0$$

is soluble for all sufficiently large integers N. This result is better than my previous one, where we require nine values of $P(x)$, $x\geqslant 0$.

1. Notation.

Let N be any sufficiently large integer, i.e. $N\geqslant c_1$, where c_1 and later $c_2, \ldots$ denote positive numbers depending only on the coefficients of $P(x)$. Let $2P$ be the greatest positive root of $P(x)=N$. It exists and is positive and tends to infinity with N.

Let d be the least common denominator of $P(x)$ and q be a positive integer. Let

$$q^*=(q, d)q$$

$$e(x)=e^{2\pi ix}$$

$$S_{a,q}=\sum_{h=1}^{q^*} e\Big(\frac{a}{q}P(h)\Big)$$

$$A(q)=\sum_{\substack{a=1\\(a,q)=1}}^{q}(S_{a,q}/q^*)^7 e\Big(-\frac{a}{q}n\Big)$$

and

$$\mathfrak{S}(n)=\sum_{q=1}^{\infty}A(q),\quad \mathfrak{S}'(n)=\sum_{q\leqslant P^{8/9}}A(q).$$

* This paper was sent to *Acta Arith.* before the war.
IV—18

Let

$$T(\alpha) = T(\alpha, P) = \sum_{P \leqslant \mu \leqslant 2P} e\Big(P(\mu)\alpha\Big),$$

$$T_1(\alpha) = T(\alpha, P^{4/5}),$$

$$T^7(\alpha)T_1(\alpha) = \sum_n r_8'(n)e(n\alpha)$$

and

$$I(\beta) = \int_P^{2P} e(\tfrac{1}{6}\beta a v^3)dv$$

We use ε to denote an arbitrarily small positive number.

We divide the interval $0 \leqslant \alpha \leqslant 1$ in the usual way into Farey arcs belonging to all rational points a/q with $1 \leqslant q \leqslant P^{2-\varepsilon}$, $0 \leqslant a \leqslant q$, $(a, q) = 1$. We classify these arcs into major arcs $M(1 \leqslant q \leqslant P^{1+\varepsilon})$ and minor arcs $m(P^{1+\varepsilon} < q \leqslant P^{2-\varepsilon})$. In either c se the arc has the form

$$\alpha = a/q + \beta, \ -\theta_1 q^{-1}P^{-2+\varepsilon} \leqslant \beta \leqslant \theta_2 q^{-1}P^{-2+\varepsilon},$$

where

$$\tfrac{1}{2} \leqslant \theta_1 \leqslant 1, \qquad \tfrac{1}{2} \leqslant \theta_2 \leqslant 1.$$

Further we divide major arcs into $M_1(1 \leqslant q \leqslant P^{8/9})$ and $M_2(P^{8/9} < q \leqslant P^{1+\varepsilon})$.

In order to avoid the repetition of many well-known arguments, we make frequent references to two of my previous papers*.

2. Minor arcs.

LEMMA 1†.

$$\int_0^1 |T(\alpha)|^6 d\alpha = O(P^{6-2-\frac{1}{2}+\varepsilon}).$$

LEMMA 2.

$$\int_0^1 |T^2(\alpha)T_1^4(\alpha)|\, d\alpha = O(P^{1+\frac{8}{5}+\varepsilon}).$$

The proof of the lemma is analogous to that due to Davenport‡ for $P(x) = x^3$.

* On a generalized Waring Problem, *Proc. London Math. Soc.* **43** (1937), 161-2; On Waring's Problem for fifth powers, *Proc. London. Math. Soc.* **45** (1939), 144-60. They will be referred to as I, II respectively.

† Hua, *Quarterly Jour.* **9** (1938), 199-202

‡ Davenport, *C. R.* **207** (1938), 1366.

LEMMA 3.

$$\int_0^1 |T^5(\alpha)T_1(\alpha)|d\alpha = O(P^{5+\frac45-2-\frac12-\frac1{40}+\varepsilon}).$$

Proof. By Hölder's inequality and Lemmas 1 and 2, we have

$$\int_0^1 |T^5(\alpha)T_1(\alpha)|d\alpha = \int_0^1 |T(\alpha)|^{9/2}|T(\alpha)|^{\frac12}|T_1(\alpha)|d\alpha$$

$$\leqslant \left(\int_0^1 |T(\alpha)|^{\frac92\cdot\frac43}d\alpha\right)^{\frac34}\left(\int_0^1 |T(\alpha)|^2|T_1(\alpha)|^4 d\alpha\right)^{\frac14}$$

$$= O(P^{\frac34(6-\frac52)+\frac14(1+\frac85)+\varepsilon}).$$

LEMMA 4.

$$\sum_m \int_m |T^7(\alpha)T_1(\alpha)|d\alpha = O(P^{5+\frac45-3-\frac1{40}+\varepsilon}).$$

Proof. By a well-known argument*, we have, on m,

$$T(\alpha) = O(P^{\frac34+\varepsilon}).$$

Hence

$$\sum_m \int_m |T^7(\alpha)T_1(\alpha)|d\alpha = O\left(P^{\frac32+\varepsilon}\int_0^1 |T^5(\alpha)T_1(\alpha)|d\alpha\right)$$

$$= O(P^{5+\frac45-3-\frac1{40}+\varepsilon}).$$

3. Major arcs.

LEMMA 5. *If* $|\beta| \leqslant \frac12$, *then*

$$q^{*-1}S_{a,q}I(\beta) = O(q^{-1/3+\varepsilon}\mathrm{Min}[P_1, |\beta|^{-\frac13}]).$$

Proof. Use Lemma 2.7 in II with the result

$$S_{a,q} = O(q^{\frac23+\varepsilon}).$$

LEMMA 6.‡ *If* $\alpha = a/q+\beta$, $q \leqslant P^{1+\varepsilon}$ *and* $|\beta| \leqslant q^{-1}P^{-2-\varepsilon}$, *then*

$$I(\alpha) - q^{*-1}S_{a,q}I(\beta) = O(q^{\frac34+\varepsilon}),$$

(II, Lemma 3.2).

LEMMA 7.

$$\sum_{M_2} \int_{M_2} |T^7(\alpha)T_1(\alpha)|d\alpha = O(P^{5+\frac45-3-\frac1{40}+\varepsilon}).$$

Proof. By Lemmas 5 and 6, we have

$$T(\alpha) = O(q^{\frac34+\varepsilon}) + O(q^{-\frac13+\varepsilon}P) = O(P^{3/4+\varepsilon}).$$

* See, e.g. Gebbcke, Satz 10, *Math. Annalen*, **105** (1931), 637-52.

‡ According to my present contributions on exponential sums, which will appear in *Chinese Jour. of Math.* Vol. 2, we can improve Lemma 6 so that the divisions of M_1 and M_2 are unnecessary. But for the convenience of the reader, I give that result in its present form.

The rest of the proof is similar to that of Lemma 4.

LEMMA 8.

$$\sum_{M_1}\int_{M_1}|T^7(\alpha)-q^{*-7}S^7_{a,q}I^7(\beta)|d\beta = O(P^{4-\frac{1}{3}+\varepsilon}).$$

Proof. By Lemmas 5 and 6, we have

$$|T^7(\alpha)-q^{*-7}S^7_{a,q}I^7(\beta)| = O(q^{\frac{3}{4}+\varepsilon}q^{-2}\operatorname{Min}[P^6, |\beta|^{-2}]),$$

since

$$q^{\frac{3}{4}+\varepsilon} \leqslant \begin{cases} q^{-\frac{1}{3}+\varepsilon}P^{\frac{13}{12}\cdot\frac{8}{9}} \leqslant q^{-\frac{1}{3}+\varepsilon}P, \\ P^{\frac{3}{4}\cdot\frac{8}{9}+\varepsilon} = P^{\frac{2}{3}+\varepsilon} \leqslant q^{-\frac{1}{3}+\varepsilon}|\beta|^{-\frac{1}{3}} \end{cases}.$$

Thus, the left side of the equation in the lemma does not exceed

$$O\Big[\sum_{q\leqslant P^{8/9}} q\cdot q^{\frac{3}{4}+\varepsilon}q^{-2}\Big(\int_0^{P^{-3}} P^6 d\beta+\int_{P^{-3}}|\beta|^{-2}d\beta\Big)\Big]$$

$$= O\Big(P^3\sum_{q\leqslant P^{8/9}} q^{-\frac{1}{4}}\Big) = O(P^{4-\frac{1}{3}+\varepsilon}).$$

LEMMA 9.

$$\sum_{M_1}\int_{\overline{M}_1}|q^{*-7}S^7_{a,q}I^7(\beta)|d\beta = O(P^{4-\frac{4}{9}+\varepsilon}),$$

where $\overline{M}_1$ is the complementary set of M_1 in $(-\infty, \infty)$.

Proof. The left side does not exceed

$$O\Big(\sum_{q\leqslant P^{8/9}} q\cdot q^{-\frac{7}{3}+\varepsilon}\int_{q^{-1}P^{-2+\varepsilon}}|\beta|^{-\frac{7}{3}}d\beta\Big)$$

$$= O\Big(P^{8/3}\sum_{q\leqslant P^{8/9}} q\cdot q^{-\frac{7}{3}}q^{\frac{4}{3}+\varepsilon}\Big) = O(P^{8/3+8/9+\varepsilon})$$

$$= O(P^{4-4/9+\varepsilon}).$$

LEMMA 10. *Let*

$$F(n) = \int_{-\infty}^{\infty} I^7(\beta)\, e^{-2\pi i n\beta}\, d\beta.$$

If $\frac{1}{10}N \leqslant n \leqslant N$, we have

$$c_2P^4 \leqslant F(n) \leqslant c_3P^4.$$

Proof. We have

$$F(n) = \int_{-\infty}^{\infty}\Big(\int_P^{2P} e(\tfrac{1}{6}ax^3\beta)\,dx\Big)^7 e(-n\beta)d\beta$$

$$= (2P)^4\int_{-\infty}^{\infty}\Big(\int_{\frac{1}{2}}^1 e(x^3\beta)\,dx\Big)^7 e(-c\beta)d\beta,$$

where $0 < c \leqslant 1+\varepsilon$. As in II, Lemma 1.8, we can prove that

$$\int_{-\infty}^{\infty}\left(\int_{\frac{1}{2}}^{1} e(x^3\beta)\,dx\right)^7 e(-c\beta)\,d\beta \geqslant c_4.$$

LEMMA 11.

$$\sum_{M_1}\int_{M_1} T^7(\alpha)e(-n\alpha)d\alpha = F(n)\ \mathfrak{S}'(n)+O(P^{4-\frac{1}{3}}).$$

Proof. By Lemmas 8, 9 and 10, we have

$$\sum_{M_1}\int_{M_1} T^7(\alpha)e(-n\alpha)$$

$$=\sum_{M_1} q^{*-7}S_{a,q}^7 e\left(-\frac{na}{q}\right)\int_{-\infty}^{\infty} I^7(\beta)e(-n\beta)d\beta + O(P^{4-\frac{1}{3}+\varepsilon})$$

$$=\mathfrak{S}'(n)F(n)+O(P^{4-\frac{1}{3}+\varepsilon}).$$

4. Singular series.

Let d be the least common denominator of the coefficients of $P(x)$, and $dP(x)=\phi(x)$. Let p denote a prime and $p^t \| d$ and let θ be the highest power of p such that $\phi'(x)\equiv 0 \pmod{p^\theta}$ for all values of x. Let $P^*(x)=p^{-\theta}\phi'(x)$.

Let

$$\gamma=\begin{cases}\theta+3-t & \text{for } p=2,\\ \theta+1-t & \text{for } p=2.\end{cases}$$

Let $N(p^\gamma, n)$ be the number of solutions of

$$P(x_1)+\ldots+P(x_7)\equiv n \pmod{p^\gamma},\ 0\leqslant x_i\leqslant p^\gamma,\ p \nmid \{P^*(x_1),\ldots, P^*(x_7)\}.$$

LEMMA 12. *If $N(n, p^\gamma)>0$ for all integers n and all primes p, then*

$$\mathfrak{S}(n)\geqslant c_5>0,$$

(I, Lemma 7.9).

LEMMA 13 (Davenport and Chowla).* *Let $\alpha_1,\ldots,\alpha_m$ be m different residue classes* mod h, *and $\beta_1,\ldots,\beta_n$ be n different residue classes* mod h, *and $(\beta_1\ldots\beta_n, h)=1$. Then the number of different residue classes of the form α_i and $\alpha_i+\beta_j$, $(1\leqslant i\leqslant m$, $1\leqslant j\leqslant n)$ is* $\geqslant \mathrm{Min}(m+n, h)$.

LEMMA 14. *If $p>3$, then $N(n, p^\gamma)>0$.*

* See, e.g. Landau, *Uber einige Neuere Fortschritte der additiven Zahlentheorie*, p. 79.

Proof. For $p>3$, we have $\theta=0$. Therefore $\gamma=1$. The number of solutions of $P(x)\equiv a(\bmod p)$ is at most 3. Thus $P(x)$ gives at least $[p/3]+1$ different values mod p. Thus, by Lemma 13, $P(x_1)+\ldots.+P(x_6)$ gives at least

$$\operatorname{Min}\left(p, 5\left[\frac{p}{3}\right]+\left[\frac{p}{3}\right]+1\right) \geqslant \operatorname{Min}\left[p, 6\left(\frac{p}{3}-1\right)+1\right] \geqslant p$$

different values mod p. That is

$$P(x_1)+\ldots.+P(x_6)\equiv n(\bmod p)$$

is solvable for all integers n and all primes p. We can choose x_7 such that $p \nmid P^*(x_7)$. And we have integers $y_1,\ldots, y_6$ such that

$$P(y_1)+\ldots.+P(y_6)\equiv n-P(x_7) \pmod p.$$

Then we have the lemma.

LEMMA 15. *If $p=3$, then $N(n, p^\gamma)>0$.*

Proof. For $p=3$, we have $\theta=0$ or 1.

(1) If $\theta=0$, then $\gamma=1$, the lemma is trivial.

(2) If $\theta=1$, then $\gamma=2$. If $3 \nmid a$, then $\theta=0$. Now we suppose that $3|a$. If $\theta=1$, then $3|(b,-\frac{1}{6}a+c)$. Then

$$P(x)\equiv\frac{a}{3}x^3+d\equiv\frac{a}{3}x+d \pmod 3.$$

Thus $P(x)$ gives 0, 1 and 2 (mod 3). By Lemma 13, we have the lemma, since $3+5.2>9$.

LEMMA 16. $N(n, 2^\gamma)>0$.

The proof of the lemma being very complicated, we divide it into several lemmas.

We may write

$$P(x)=\frac{a}{2}(x^3-x)+\frac{b}{2}(x^2-x)+cx,\ 2 \nmid (a, b, c)$$

$$P'(x)=3a\frac{x^2-x}{2}+\left(b+\frac{3a}{2}\right)x-\frac{a}{2}-\frac{b}{2}+c,$$

$$P''(x)=3ax+b,$$

since $\frac{1}{3}$ can be considered as an integer mod 2^γ and the constant term is insignificant here.

LEMMA 16.1. *If $\theta=0$, $P(x)$ gives either at least 2^{l-2} different odd numbers with $2 \nmid P^*(x)$ or at least 2^{l-2} different even numbers with $2 \nmid P^*(x)$.*

Proof. Since $P^*(x)\not\equiv 0 \pmod 2$ for all integers x, the lemma is evident for $l=2$. Let $l\geqslant 3$. Then, by the same method as in II, Lemma 7.6, we have

$$P(x+2^{l+t-1}y)\equiv P(x)+2^{l-1}yP^*(x) \bmod 2^l \qquad (1)$$

and

$$P^*(x) \equiv P^*(y) \pmod 2.$$

Let $x_0 \equiv x + 2^{t+2}$. Then $P(x_0)$ and $P(x_1)$ give two different values mod 2^3, and they are either both even or both odd, and they satisfy $2 \nmid P^*(x)$. Thus the lemma is true for $l = 3$.

Let $x_1, \ldots, x_{2^{l-3}}$ be integers, by which $P(x)$ gives 2^{l-3} different values mod 2^{l-1} and $2 \nmid P^*(x_i)$, $i = 1, \ldots, 2^{l-3}$ and they are either all odd or all even. Then, by (1), we have

$$x_1, \ldots, x_{2^{l-3}}, \; x_1 + 2^{l+t-1}, \ldots, x_{2^{l-3}} + 2^{l+t-1}$$

satisfying our requirement.

LEMMA 16.2. *If $\theta = 0$, then $N(n, 2^\gamma) > 0$.*

Proof. (1) If the 2^{l-2} different values given by Lemma 16.1 are all odd, then by Lemma 13, we have the result.

(2) If the 2^{l-2} different values given by Lemma 16.1 are even, we can obtain the lemma by considering $P(x)+1$ instead of $P(x)$.

LEMMA 16.3. *If $\theta > 0$, then $t = 0$.*

Proof. Suppose that $t \neq 0$. Then $t = 1$, and

$$\phi'(x) = 2P'(x) = 3a(x^2 - x) + (2b + 3a)x + 2c - b - a.$$

(One insignificant factor 3 of $\phi'(x)$ may be omitted here.) By the hypothesis, we have $2 \mid (2b+3a, 2c-b-a)$, then $2 \mid (a, b)$. This contradicts $t \neq 0$.

Hereafter we assume that $t = 0$, then $2 \mid a$, $2 \mid b$. Thus $2 \nmid c$. Without loss of generality we may assume that $c = 1$. For otherwise, let c' be an integer satisfying $cc' \equiv 1 \pmod{2^\gamma}$. If

$$c'P(x_1) + \ldots + c'P(x_7) \equiv n \pmod{2^\gamma}, \; 2 \nmid P^*(x_1)$$

is solvable for all integers n, then it certainly implies that $N(n, 2^\gamma) > 0$.

Next, for $\theta \geqslant 1$, we let

$$2 \mid a, 2 \mid b, 2 \mid \left(3a, b + \frac{3a}{2}, 1 - \frac{b}{2} - \frac{a}{2} \right)$$

then

$$2^2 \mid a \text{ and } 2 \| b.$$

Hereafter we write

$$a = 2^2 a', \quad b = 2(2b'+1).$$

LEMMA 16.4. *If* $\theta = 1$, *then* $N(n, 2^\gamma) > 0$.

Proof. Evidently $\gamma \leqslant 4$.

(1) If $2|a'$, $2|b'$, then $P(x)$ gives $1, 1+2^3$, 0 and $l_1 \pmod{2^4}$ where $2^2||l_1$. In fact, $P(x) \equiv x^2 \pmod{2^3}$, then $P(0) = 0$, $P(2) = l_1$, $2^2||l_1$; further $P(1) = 1$ and $P(1+2^2) \equiv 1+2^3 \pmod{2^4}$. In this case the lemma can be verified directly.

(2) If $2|a'$, $2 \nmid b'$, then $P(x)$ gives $0, 2^3, 1, 1+l_2 \pmod{2^4}$, where $2^2||l_2$. The proof of the statement is analogous to (1).

(3) If $2 \nmid a'$, then $P(x)$ gives $0, 2^3, 1, 1+2^3 \pmod{2^4}$ and each with $2 \nmid P^*(x)$. The proof of this statement is also easy. (But it is worthwhile to remark here that seven is best possible.)

LEMMA. 16.5. *If* $\theta = 2$, *then* $N(n, 2^\gamma) > 0$.

Proof. Then $\gamma = 5$. Since $t = 0$ and $\theta = 2$, we have

$$2|(a, b), 2^2|(3a, b+\tfrac{3}{2}a, 1-\tfrac{1}{2}b-\tfrac{1}{2}a).$$

Consequently

$$a = 2^2(2a''+1), \quad b = 2(4b''-1).$$

Then, using the relation $x^2 - x \equiv 0 \pmod 2$, we have

$$P(2x') \equiv -2^2x^2 + 2^3(2-a''+b'')x \pmod{2^5}$$
$$\equiv -2^2y^2 + 2^2(2-a''+b'')^2 \pmod{2^5},$$

where $$y = x-2+a''-b'';$$

and $$P(2x+1) \equiv 2^2 \,.\, 5x^2 + 3.2^3(1+b'')x + 1 \pmod{2^5}.$$

We can choose λ and μ such that

$$h \equiv \lambda + 3.2^2(2-a''+b'')^2 - 2^2\mu \pmod{2^5}, \; 0 \leqslant \lambda \leqslant 3, \; 0 \leqslant \mu < 8.$$

(1) If $\mu \neq 7$, then we have y_1, y_2 and y_3 such that

$$n \equiv \lambda + 3.2^2(2-a''+b'')^2 - 2^2(y_1^2+y_2^2+y_3^2) \pmod{2^5}$$
$$\equiv \lambda P(1) + P(x_1) + P(x_2) + P(x_3).$$

Since $\lambda \leqslant 3$, the lemma follows immediately.

(1) If $\lambda \neq 3$, then we have y_1, y_2, y_3 and y_4 such that

$$n \equiv \lambda + 4.2^2(2-a''+b'')^2 - 2^2(y_1^2+y_2^2+y_3^2+y_4^2) \pmod{2^5}$$

and we have the lemma.

(3) Suppose that $\lambda = 3$ and $\mu = 7$.

Since $P(3) \not\equiv P(1) \pmod{2^5}$, we have

$$n \equiv 2P(1) + P(3) + 3.2^2(2-a''+b'')^2 - 2^2\mu' \pmod{2^5}$$

and $\mu' \not\equiv 7 \pmod 8$. The result follows in the same way as in §1.

Lemma 16 now follows since $\theta \leqslant 2$.

LEMMA 17. *$\mathfrak{S}(n) \geqslant c_6 > 0$ for all n and $P > c_7$.*

Proof. We have

$$|\mathfrak{S}(n) - \mathfrak{S}'(n)| \leqslant \sum_{q \leqslant P^{8/9}} q.q^{-\frac{7}{3}+\varepsilon} = O(P^{-\frac{8}{27}+\varepsilon}).$$

By Lemmas 12, 14, 15 and 16, we have

$$\mathfrak{S}(n) \geqslant c_5 > 0.$$

Then

$$\mathfrak{S}'(n) \geqslant c_5 + O(P^{-\frac{8}{27}+\varepsilon}) \geqslant c_6.$$

5. Proof of the theorem.

THEOREM. *For $N \geqslant c_1$, we have $r'_8(N) > 0$.*

Proof. Write

$$T_1(\alpha) = \sum_n e(n\alpha).$$

By Lemmas 4 and 11, we have

$$\begin{aligned} r'_8(N) &= \int_0^1 T^7(\alpha) T_1(\alpha) e(-N\alpha) d\alpha \\ &= \sum_n \int_0^1 T^7(\alpha) e\{-(N-n)\alpha\} d\alpha \\ &= \sum_n \sum_{m_1} q^{*-1} S_{aq} e\{-(N-n)\frac{a}{q}\} \\ &\qquad \times \int_{-\infty}^{\infty} I^7(\beta) e\{-(N-n)\}\beta d\beta + O(P^{4+\frac{4}{5}-\frac{1}{40}-\varepsilon}) \\ &= \sum_n \mathfrak{S}'(N-n) F(N-n) + O(P^{4+\frac{4}{5}-\frac{1}{40}+\varepsilon}) \\ &\geqslant c_7 P^4 \sum_n 1 + O(P^{4+\frac{4}{5}-\frac{1}{40}+\varepsilon}) \\ &\geqslant c_8 P^{4+\frac{4}{5}} + O(P^{4+\frac{4}{5}-\frac{1}{40}+\varepsilon}). \end{aligned}$$

The theorem is therefore proved.

Reprinted from the
QUARTERLY JOURNAL OF MATHEMATICS
Oxford Series 11, pp. 161–176, 1940

ON A THEOREM DUE TO VINOGRADOW

By LOO-KENG HUA (*Kunming*)

[Received 26 September 1939]

1. Introduction

IT was proved by Vinogradow† that, *if P is a large positive integer, then*

$$\int_0^1 \dots \int_0^1 \left| \sum_{x=1}^{P} e^{2\pi i(\alpha_k x^k + \dots + \alpha_1 x)} \right|^r d\alpha_1 \dots d\alpha_k = O(P^{r-\frac{1}{2}k(k+1)}),$$

for the greatest even integer r satisfying

$$r < Lk(k+1)(k+2)\log k,$$

where the value of L, which depends on k, is given by the following table:

k	2	3	4	5	6	7	8	9	10	11	12	13	$\geqslant 14$
L	4·81	4·45	4·34	4·30	4·29	4·23	4·22	4·18	4·16	4·15	4·14	4·12	4·10

The result is very important, and has numerous applications, e.g. to Waring's problem, to the estimation of trigonometrical sums as given by Vinogradow himself,‡ to Tarry's problem and to the simultaneous Waring's problem as given by the author.§ Now I am able to improve the result for small k: namely, *for the s given by the table*||

k	2	3	4	5	6	7	8	9	10
s	6	16	46	124	312	760	1778	4068	9190
r	80	293	721	1453	2582	4148	6318	9092	12644

we have

$$\int_0^1 \dots \int_0^1 \left| \sum_{x=1}^{P} e^{2\pi i(\alpha_k x^k + \dots + \alpha_1 x)} \right|^s d\alpha_1 \dots d\alpha_k = O(P^{s-\frac{1}{2}k(k+1)+\epsilon}),$$

where the constant implied by the symbol O (and later by the equivalent symbol $\ll$) depends on k and ϵ only.

The method of proof is similar, in principle, to that used in my previous paper,†† but with a complicated modification.

We use $y\Delta f(x)$ to denote $f(x+y)-f(x)$; evidently $\Delta f(x)$ is a poly-

† *Recueil Math.* new series, 3 (1938), 435–71.

‡ Loc. cit., and *Travaux de l'Institut Math. de Tbilissi*, 5 (1938), 167–80.

§ Cf. the end of this paper.

|| The last line gives the values due to Vinogradow.

†† *Quart. J. of Math.* (Oxford), 9 (1938), 199–202.

3695.11 M

nomial in x and y. The Δ operation will only be applied to the variable x. It is easy to verify that

$$y_1 \dots y_\mu \Delta^\mu x^\nu = 0 \quad (\mu > \nu),$$

$$y_1 \dots y_\mu \Delta^\mu x^\mu = \mu!\, y_1 \dots y_\mu,$$

and

$$y_1 \dots y_\mu \Delta^\mu x^{\mu+1} = y_1 \dots y_\mu w,$$

where w is a linear form in $y_1, \dots, y_\mu$ and x with integer coefficients.

LEMMA 1. *Let $g_1(x), \dots, g_s(x)$ be polynomials in x, and let*

$$g(x) = \alpha_1 g_1(x) + \dots + \alpha_s g_s(x)$$

be of degree k. Let

$$F = \sum_{x=1}^{P} e^{2\pi i g(x)}$$

Then

$$F^{2^\mu} \leqslant P^{2^\mu - 1} + P^{2^\mu - \mu - 1} \sum_{y_1}^{P} \dots \sum_{y_\mu}^{P} \sum_{x_{\mu+1}}^{P}{}^{*} e^{2\pi i y_1 \dots y_\mu \Delta^\mu g(x_{\mu+1})}$$

for $\mu = 1, 2, \dots, k-1$, where $\sum_{y}^{P}$ denotes a sum† *with $O(P)$ terms, and where $*$ denotes the conditions $y_1 \dots y_\mu \Delta^\mu g_r(x_{\mu+1}) \neq 0$, for all those values of r for which $g_r(x)$ is of degree greater than μ.*

The proof is the same as that given in my previous paper.‡

THEOREM $A(k)$. *Let*

$$f(x) = a_0 x^k + a_1 x^{k-1} + \dots,$$

where a_0 is an integer $\leqslant 1$ and a_1 is an integer $\leqslant P$. Let

$$S_k = \sum_{x=1}^{P} e^{2\pi i (\alpha_k f(x) + \alpha_{k-2} x^{k-2} + \dots + \alpha_1 x)}.$$

Then we have

$$\int_0^1 \dots \int_0^1 |S_k|^\lambda \, d\alpha_1 \dots d\alpha_{k-2} \, d\alpha_k = O(P^{\lambda - \frac{1}{2}(k^2 - k + 2) + \epsilon}),$$

where the value of λ is given by the table

k	3	4	5	6	7	8	9	10
λ	10	32	86	220	536	1272	2930	6628.

THEOREM $B(k)$. *Let*

$$C_k = \sum_{x=1}^{P} e^{2\pi i (\alpha_k x^k + \dots + \alpha_1 x)}.$$

Then we have

$$\int_0^1 \dots \int_0^1 |C_k|^s \, d\alpha_1 \dots d\alpha_k = O(P^{s - \frac{1}{2}k(k+1) + \epsilon}),$$

where s is as given previously.

† The conditions of summation for y_2 may depend on the value of y_1, and so on.

‡ Loc. cit. 201.

The proofs of both theorems depend on one another; more precisely, I shall prove the truth of theorem $A(k)$ by using results in the proofs of $A(l_1)$ and $B(l_2)$ for $l_1 \leqslant k-1$ and $l_2 \leqslant k-1$, and the truth of theorem $B(k)$ by using results in the proofs of $A(l_1)$ and $B(l_2)$ for $l_1 \leqslant k$ and $l_2 \leqslant k-1$. The methods used differ for different values of k. Certainly there exists a uniform method for the induction, but such a method gives a worse result.

2. Proof of the theorems

2.01. We may assume that $f(x) = x^k$ in $A(k)$, provided that we make a certain slight modification in the enunciation. In fact,

$$\int_0^1 \cdots \int_0^1 |S_k|^{2\mu}\, d\alpha_1 \ldots d\alpha_{k-2}\, d\alpha_k$$

is equal to the number of solutions of

$$\begin{aligned} f(x_1)+\ldots+f(x_\mu) &= f(y_1)+\ldots+f(y_\mu),\\ x_1^{k-2}+\ldots+x_\mu^{k-2} &= y_1^{k-2}+\ldots+y_\mu^{k-2},\\ &\cdots\cdots\\ x_1+\ldots+x_\mu &= y_1+\ldots+y_\mu, \end{aligned}$$

subject to

$$1 \leqslant x_\nu \leqslant P, \quad 1 \leqslant y_\nu \leqslant P \qquad (\nu = 1, 2, \ldots, \mu).$$

Multiplying the equations by $k^k a_0^{k-1}$, $k^{k-2} a_0^{k-2}, \ldots, k a_0$ respectively, they become

$$\sum_{\nu=1}^{\mu} \{(ka_0 x_\nu)^k + ka_1(ka_0 x_\nu)^{k-1}+\ldots\} = \sum_{\nu=1}^{\mu} \{(ka_0 y_\nu)^k + ka_1(ka_0 y_\nu)^{k-1}+\ldots\},$$

$$\sum_{\nu=1}^{\mu} (ka_0 x_\nu)^{k-2} = \sum_{\nu=1}^{\mu} (ka_0 y_\nu)^{k-2},$$

$$\cdots\cdots$$

$$\sum_{\nu=1}^{\mu} (ka_0 x_\nu) = \sum_{\nu=1}^{\mu} (ka_0 y_\nu).$$

Put $x'_\nu = ka_0 x_\nu + a_1$, $y'_\nu = ka_0 y_\nu + a_1$; then we have the system of equations

$$\left.\begin{aligned} x_1'^k+\ldots+x_\mu'^k &= y_1'^k+\ldots+y_\mu'^k\\ x_1'^{k-2}+\ldots+x_\mu'^{k-2} &= y_1'^{k-2}+\ldots+y_\mu'^{k-2}\\ &\cdots\cdots\\ x'_1+\ldots+x'_\mu &= y'_1+\ldots+y'_\mu \end{aligned}\right\} \tag{1}$$

subject to the conditions

$$a_0 k \mid (x'_\nu - a_1), \tag{2}$$

$$a_1+1 \leqslant x'_\nu \leqslant a_1+P. \tag{3}$$

Certainly $$\int_0^1 \dots \int_0^1 |S_k|^{2\mu}\, d\alpha_1 \dots d\alpha_k$$

does not exceed the number of solutions of (1) with the condition (3) only. Thus we need only discuss the case with $f(x) = x^k$, provided that in Theorems $A(k)$, $B(k)$ we allow the range of summation to be any interval of length P with both end-points $\ll P$. From now onwards we suppose them enunciated in this slightly more general form.

2.02. We shall now prove that

$$\int_0^1 \dots \int_0^1 |S_k|^{2k}\, d\alpha_1 \dots d\alpha_k \ll P^{k+\epsilon}. \tag{4}$$

For this we require the following

LEMMA 2. *Let*

$$s_i = x_1^i + \dots + x_k^i \quad (i = 1, 2, \dots, k).$$

Then the symmetrical function

$$f = (s_1 - x_1) \dots (s_1 - x_k)$$

of $x_1, \dots, x_k$ *can be expressed as a function of* $s_1, \dots, s_{k-2}$ *and* s_k *only.*

Proof. We write

$$f = s_1^k - s_1^{k-1}\sigma_1 + \dots + (-1)^k \sigma_k,$$

where σ_i is the ith elementary symmetrical function of $x_1, \dots, x_k$. By a well-known result on symmetrical functions, we have

$$f = (-1)^k \sigma_k + (-1)^{k-1} \sigma_{k-1} s_1 + f_1(s_1, \dots, s_{k-2}). \tag{5}$$

By Newton's formulae, we have

$$(-1)^k k \sigma_k = -s_k + \sigma_1 s_{k-1} + (-1)^k \sigma_{k-1} s_1 + f_2(s_1, \dots, s_{k-2}),$$

and $$(-1)^{k-1}(k-1)\sigma_{k-1} = -s_{k-1} + f_3(s_1, \dots, s_{k-2}).$$

Consequently,

$$\begin{aligned} k\{(-1)^k \sigma_k + (-1)^{k-1} \sigma_{k-1} s_1\} &= -s_k + \sigma_1 s_{k-1} - s_1 s_{k-1} + f_4(s_1, \dots, s_{k-2}) \\ &= -s_k + f_4(s_1, \dots, s_{k-2}). \end{aligned}$$

Combining this with (5) we have the lemma.

2.03. *Proof of* (4). The integral on the left of (4) does not exceed the number of solutions of

$$\left.\begin{aligned} x_1^k + \dots + x_k^k &= y_1^k + \dots + y_k^k \\ x_1^{k-2} + \dots + x_k^{k-2} &= y_1^{k-2} + \dots + y_k^{k-2} \\ \cdots &\cdots \\ x_1 + \dots + x_k &= y_1 + \dots + y_k \end{aligned}\right\} \tag{6}$$

subject to $x_\nu = O(P)$, $y_\nu = O(P)$.

Let $$l = x_1+\ldots+x_k = y_1+\ldots+y_k.$$

By Lemma 2 the equations (6) imply that

$$(l-x_1)\ldots(l-x_k) = (l-y_1)\ldots(l-y_k).$$

For given $y_1,\ldots, y_k$ with the condition

$$(l-y_1)\ldots(l-y_k) \neq 0, \tag{7}$$

the number of sets of values for $x_1,\ldots, x_k$ does not exceed

$$d^k\{(l-y_1)\ldots(l-y_k)\} = O(P^\epsilon),$$

where $d(n)$ denotes the number of divisors of n.

If (7) does not hold, then one of the y's and also one of the x's is equal to l. Suppose $x_k = y_k = l$. There are $O(P)$ choices for x_k, and the equations (6) imply that

$$x_1^{k-2}+\ldots+x_{k-1}^{k-2} = y_1^{k-2}+\ldots+y_{k-1}^{k-2},$$
$$\cdot \quad \cdot \quad \cdot \quad \cdot \quad \cdot \quad \cdot \quad \cdot \quad \cdot \quad \cdot \quad \cdot$$
$$x_1+\ldots+x_{k-1} = y_1+\ldots+y_{k-1} = 0.$$

There are $O(P^{k-1})$ choices for x_{k-1}, $y_2,\ldots, y_{k-1}$, and these determine y_1 uniquely, and determine $x_1,\ldots, x_{k-2}$ with not more than $(k-2)!$ $(= O(1))$ possibilities. Hence in this case also there are $O(P^{k+\epsilon})$ solutions for (6). This proves (4).

2.04. We next prove that

$$\int_0^1 \ldots \int_0^1 |C_k|^{2(k+1)}\, d\alpha_1\ldots d\alpha_k \ll P^{k+1+\epsilon}. \tag{8}$$

For this we require the following

LEMMA 3. *The system of equations*

$$x_1^i+\ldots+x_{k+1}^i = y_1^i+\ldots+y_{k+1}^i \quad (i = 1, 2,\ldots, k) \tag{9}$$

implies a relation of the form

$$(x_{k+1}-y_{k+1})g(y_1,\ldots, y_k, x_k, y_{k+1}, x_{k+1}) = (x_k-y_k)h(y_1,\ldots, y_k, x_k),$$

where g and h are homogeneous polynomials of degree $k-1$ in the variables indicated. The homogeneous polynomial of degree $k-1$ in x_{k+1} and y_{k+1} only, contained in g, is not divisible by $x_{k+1}-y_{k+1}$, and the coefficients of x_k^{k-1} in h and x_{k+1}^{k-1} in g are constants different from zero.

Proof. Let

$$s_i = x_1^i+...+x_{k-1}^i, \qquad t_i = y_1^i+...+y_{k-1}^i.$$

We can write (9) in the form

$$s_\nu = t_\nu-(x_k^\nu-y_k^\nu)-(x_{k+1}^\nu-y_{k+1}^\nu) \quad (\nu = 1,..., k). \tag{10}$$

It is well known that

$$s_k-\sigma_1 s_{k-1}+\sigma_2 s_{k-2}+...+(-1)^{k-1}\sigma_{k-1} s_1 = 0, \tag{11}$$

where σ_i is the ith elementary symmetrical function of $x_1,..., x_{k-1}$. Since σ_i is expressible as a polynomial in $s_1,..., s_i$, we can write (11) in the form

$$s_k-s_1 s_{k-1}+\sigma_2(s_1, s_2)s_{k-2}+...+(-1)^{k-1}\sigma_{k-1}(s_1,..., s_{k-1})s_1 = 0. \tag{12}$$

Similarly,

$$t_k-t_1 t_{k-1}+\sigma_2(t_1, t_2)t_{k-2}+...+(-1)^{k-1}\sigma_{k-1}(t_1,..., t_{k-1})t_1 = 0. \tag{13}$$

If we substitute from (10) in (12), we obtain the left-hand side of (13), plus a number of terms, each of which is a product of powers of the t's and of factors of the type $x_k^\nu-y_k^\nu$, $x_{k+1}^\nu-y_{k+1}^\nu$. When we subtract from this the equation (13), and take over to the right-hand side all terms which do not contain any factor of the type $x_{k+1}^\nu-y_{k+1}^\nu$, we obtain a relation which is of the required form.

The part of g containing only x_{k+1} and y_{k+1} arises from those terms which contain only factors of the type $x_{k+1}^\nu-y_{k+1}^\nu$. All such terms contain at least two factors, except that term $-(x_{k+1}^k-y_{k+1}^k)$ which comes from s_k. Now this term is not divisible by $(x_{k+1}-y_{k+1})^2$, whereas all others are. This proves the assertion about g.

The coefficient of x_{k+1}^k in the expression (12) after substituting from (10) can be found by taking $s_\nu = -x_{k+1}^\nu$. This makes $\sigma_\nu = (-1)^\nu x_{k+1}^\nu$, and so the coefficient is

$$-1-(-1)(-1)+(1)(-1)-...+(-1)^{k-1}(-1)^{k-1}(-1) = -k. \tag{14}$$

Thus the coefficient of x_{k+1}^{k-1} in g is $-k$. Similarly for the coefficient of x_k^{k-1} in h.

2.05. *Proof of* (8). The integral on the left is the number of solutions of the equations (9) subject to $x_\nu = O(P)$, $y_\nu = O(P)$. By Lemma 3 the equations imply

$$(x_{k+1}-y_{k+1})g(y_1,..., y_k, x_k, y_{k+1}, x_{k+1}) = (x_k-y_k)h(y_1,..., y_k, x_k). \tag{15}$$

For given $y_1,\ldots, y_k, x_k$ satisfying

$$(x_k-y_k)h(y_1,\ldots,y_k,x_k) \neq 0, \tag{16}$$

the number of sets of x_{k+1}, y_{k+1} does not exceed the divisor function

$$d\{(x_k-y_k)h(y_1,\ldots,y_k,x_k)\} = O(P^\epsilon).$$

This follows from the fact that, by the second assertion of Lemma 3, the number of solutions of

$$x_{k+1}-y_{k+1} = c, \qquad g(y_1,\ldots,y_k,x_k,y_{k+1},x_{k+1}) = d$$

(c, d being non-zero integers) does not exceed the degree $k-1$ of g.

Further, if $\qquad (x_k-y_k)h(y_1,\ldots,y_k,x_k) = 0,$

then $\qquad (x_{k+1}-y_{k+1})g(y_1,\ldots,y_k,x_k,y_{k+1},x_{k+1}) = 0.$

In virtue of the last assertion of Lemma 3, these equations imply that, for given $y_1,\ldots, y_k$ and y_{k+1}, there are only $O(1)$ possible values for x_k and for x_{k+1}.

In either case, when $y_1,\ldots, y_k, x_k, x_{k+1}, y_{k+1}$ are known, there are only $O(1)$ possible sets of values for $x_1,\ldots, x_{k-1}$, by (9). Hence the number of solutions of (9) subject to $x_\nu \ll P$, $y_\nu \ll P$, is $O(P^{k+1+\epsilon})$. This proves (8).

2.06. $B(2)$ is the particular case $k = 2$ of (8).

2.07. *Proof of* $A(3)$. By Lemma 1 we have

$$|S_3|^4 \ll P^3+P\sum_{y_1}^{P}\sum_{y_2}^{P}\sum_{x_3}^{P}{}^{*}\, e^{2\pi i y_1 y_2 \Delta^2(\alpha_3 x_3^3+\alpha_1 x_3)},$$

where the * means that the variables are subject to $y_1 y_2 \Delta^2 x_3^3 \neq 0$. Multiplying this inequality throughout by $|S_3|^6$, and integrating with respect to α_1 and α_3, we have

$$\int_0^1\int_0^1 |S_3|^{10}\,d\alpha_1\,d\alpha_3 \ll P^3\int_0^1\int_0^1 |S_3|^6\,d\alpha_1\,d\alpha_3+PR,$$

where R denotes the number of solutions of

$$y_1 y_2 \Delta^2(x_3^3) = z_1^3+\ldots+z_3^3-z_4^3-\ldots-z_6^3,$$
$$y_1 y_2 \Delta^2(x_3^3) \neq 0,$$
$$0 = z_1+\ldots+z_3-z_4-\ldots-z_6,$$

with $z_\nu = O(P)$. Since $z_1,\ldots, z_5$ determine the other variables with only $O(P^\epsilon)$ possibilities, we have $R = O(P^{5+\epsilon})$. Thus, using (4) with $k = 3$, we have

$$\int_0^1\int_0^1 |S_3|^{10}\,d\alpha_1\,d\alpha_3 \ll P^{6+\epsilon}. \tag{17}$$

2.08. *Proof of* $B(3)$. By Lemma 1

$$|C_3|^4 \ll P^3 + P \sum_{y_1}^{P} \sum_{y_2}^{P} \sum_{x_3}^{P}{}^{*} e^{2\pi i y_1 y_2 \Delta^2(\alpha_3 x_3^3 + \alpha_2 x_3^2)}, \tag{18}$$

where the * means that the variables are subject to

$$y_1 y_2 \Delta^2(x_3^3) \neq 0, \qquad y_1 y_2 \Delta^2(x_3^2) \neq 0.$$

Multiplying this inequality throughout by $|C_3|^8$, and integrating with respect to α_1, α_2, α_3, using (8) with $k = 3$, we have

$$\int_0^1\int_0^1\int_0^1 |C_3|^{12}\, d\alpha_1\, d\alpha_2\, d\alpha_3 \ll P^{7+\epsilon} + PR,$$

where R is the number of solutions of

$$\begin{aligned} y_1 y_2 w &= z_1^3 + \ldots + z_4^3 - z_5^3 - \ldots - z_8^3 \quad (y_1 y_2 w \neq 0), \\ 2 y_1 y_2 &= z_1^2 + \ldots + z_4^2 - z_5^2 - \ldots - z_8^2, \\ 0 &= z_1 + \ldots + z_4 - z_5 - \ldots - z_8, \end{aligned}$$

where $w = \Delta^2(x_3^3) \ll P$ and $z_\nu \ll P$.

For any fixed w the number of solutions of

$$(2z_1^3 - wz_1^2) + \ldots + (2z_4^3 - wz_4^2) = (2z_5^3 - wz_5^2) + \ldots + (2z_8^3 - wz_8^2),$$
$$z_1 + \ldots + z_4 = z_5 + \ldots + z_8$$

is $O(P^{5+\epsilon})$, by (4) with $k = 3$ and $f(z) = 2z^3 - wz^2$. Therefore, $R = O(P^{6+\epsilon})$, and so

$$\int_0^1\int_0^1\int_0^1 |C_3|^{12}\, d\alpha_1\, d\alpha_2\, d\alpha_3 \ll P^{7+\epsilon}. \tag{19}$$

In the same way, multiplying (18) by $|C_3|^{12}$ and using (19) and

$$\int_0^1\int_0^1 |S_3|^{12}\, d\alpha_1\, d\alpha_3 \ll P^{8+\epsilon},$$

which is a trivial consequence of (17), we obtain

$$\int_0^1\int_0^1\int_0^1 |C_3|^{16}\, d\alpha_1\, d\alpha_2\, d\alpha_3 \ll P^{10+\epsilon}.$$

This is $B(3)$.

2.09. *Proof of* $A(4)$. By Lemma 1 we have

$$|S_4|^8 \ll P^7 + P^4 \sum_{y_1}^{P} \sum_{y_2}^{P} \sum_{y_3}^{P} \sum_{x_4}^{P}{}^{*} e^{2\pi i y_1 y_2 y_3 \Delta^3(\alpha_4 x_4^4 + \alpha_2 x_4^2 + \alpha_1 x_4)},$$

where the * denotes that $y_1 y_2 y_3 \Delta^3 x_4^4 \neq 0$. Multiplying by $|S_4|^8$ and integrating, and using (4) with $k = 4$, we have

$$\int_0^1\int_0^1\int_0^1 |S_4|^{16}\, d\alpha_1\, d\alpha_2\, d\alpha_4 \ll P^{11+\epsilon}+P^4 R,$$

where R is the number of solutions of

$$y_1 y_2 y_3 \Delta^3 x_4^4 = z_1^4+\ldots+z_4^4-z_5^4-\ldots-z_8^4 \quad (y_1 y_2 y_3 \Delta^3 x_4^4 \neq 0),$$

$$0 = z_1^2+\ldots+z_4^2-z_5^2-\ldots-z_8^2,$$

$$0 = z_1+\ldots+z_4-z_5-\ldots-z_8,$$

with $z_\nu \ll P$. Using

$$\int_0^1\int_0^1 |C_2|^8\, d\alpha_1\, d\alpha_2 \ll P^{5+\epsilon},$$

which is a trivial consequence of $B(2)$, we have $R \ll P^{5+\epsilon}$. Thus

$$\int_0^1\int_0^1\int_0^1 |S_4|^{16}\, d\alpha_1\, d\alpha_2\, d\alpha_4 \ll P^{11+\epsilon}. \tag{20}$$

Repeating the argument, but multiplying by $|S_4|^{16}$ and $|S_4|^{24}$, we obtain successively

$$\int_0^1\int_0^1\int_0^1 |S_4|^{24}\, d\alpha_1\, d\alpha_2\, d\alpha_4 \ll P^{18+\epsilon}, \tag{21}$$

$$\int_0^1\int_0^1\int_0^1 |S_4|^{32}\, d\alpha_1\, d\alpha_2\, d\alpha_4 \ll P^{25+\epsilon}. \tag{22}$$

This is $A(4)$.

2.10. *Proof of* $B(4)$. By Lemma 1 we have

$$|C_4|^4 \ll P^3+P \sum_{y_1}^{P} \sum_{y_2}^{P} \sum_{x_3}^{P}{}^{*}\, e^{2\pi i y_1 y_2 \Delta^2(\alpha_4 x_3^4+\alpha_3 x_3^3+\ldots)},$$

where the * denotes that the summation is subject to

$$y_1 y_2 \Delta^2 x_3^4 \neq 0, \qquad y_1 y_2 \Delta^2 x_3^3 \neq 0.$$

Multiplying the inequality by $|C_4|^{10}$ and integrating, we have, using (8),

$$\int_0^1\int_0^1\int_0^1\int_0^1 |C_4|^{14}\, d\alpha_1\, d\alpha_2\, d\alpha_3\, d\alpha_4 \ll P^{8+\epsilon}+PR,$$

where R denotes the number of solutions of

$$y_1 y_2 \Delta^2 x_3^4 = z_1^4 + \ldots - z_{10}^4 \quad (y_1 y_2 \Delta^2 x_3^4 \neq 0),$$
$$y_1 y_2 w = z_1^3 + \ldots - z_{10}^3 \quad (y_1 y_2 w \neq 0),$$
$$2y_1 y_2 = z_1^2 + \ldots - z_{10}^2,$$
$$0 = z_1 + \ldots - z_{10}.$$

(The usual conditions on the magnitude of w and the z's are to be understood.) For fixed w the number of solutions of

$$0 = (2z_1^3 - wz_1^2) + \ldots - (2z_{10}^3 - wz_{10}^2),$$
$$0 = z_1 + \ldots - z_{10}$$

is $\ll P^{6+\epsilon}$ by (17). Consequently, $R \ll P^{7+\epsilon}$, and

$$\int_0^1 \int_0^1 \int_0^1 \int_0^1 |C_4|^{14}\, d\alpha_1\, d\alpha_2\, d\alpha_3\, d\alpha_4 \ll P^{8+\epsilon}. \tag{23}$$

By Lemma 1 we have

$$|C_4|^8 \ll P^7 + P^4 \sum_{y_1}^{P} \sum_{y_2}^{P} \sum_{y_3}^{P} \sum_{x_4}^{P}{}^{*} e^{2\pi i y_1 y_2 y_3 \Delta^3 (\alpha_4 x_4^4 + \ldots)},$$

where the * denotes that $y_1 y_2 y_3 \Delta^3 x_4^4 \neq 0$. Multiplying by $|C_4|^{14}$ and integrating, we have, by (23),

$$\int_0^1 \int_0^1 \int_0^1 \int_0^1 |C_4|^{22}\, d\alpha_1 \ldots d\alpha_4 \ll P^{15+\epsilon} + P^4 R,$$

where R denotes the number of solutions of

$$y_1 y_2 y_3 w = z_1^4 + \ldots - z_{14}^4 \quad (y_1 y_2 y_3 w \neq 0),$$
$$6y_1 y_2 y_3 = z_1^3 + \ldots - z_{14}^3,$$
$$0 = z_1^2 + \ldots - z_{14}^2,$$
$$0 = z_1 + \ldots - z_{14}.$$

Clearly R does not exceed $P^{1+\epsilon}$ times the number of solutions of

$$(6z_1^4 - wz_1^3) + \ldots - (6z_{14}^4 - wz_{14}^3) = 0,$$
$$z_1^2 + \ldots - z_{14}^2 = 0,$$
$$z_1 + \ldots - z_{14} = 0$$

for a fixed w. By a trivial consequence of (4) we have

$$\int_0^1 \int_0^1 \int_0^1 |S_4|^{14}\, d\alpha_1\, d\alpha_2\, d\alpha_4 \ll P^{10+\epsilon},$$

and hence $R \ll P^{11+\epsilon}$. Thus,

$$\int_0^1\int_0^1\int_0^1\int_0^1 |C_4|^{22}\, d\alpha_1 \dots d\alpha_4 \ll P^{15+\epsilon}. \tag{24}$$

Using the same method, but appealing to (20), (21), (22) instead of to (4), we obtain successively

$$\int_0^1\int_0^1\int_0^1\int_0^1 |C_4|^{30}\, d\alpha_1 \dots d\alpha_4 \ll P^{22+\epsilon}, \tag{25}$$

$$\int_0^1\int_0^1\int_0^1\int_0^1 |C_4|^{38}\, d\alpha_1 \dots d\alpha_4 \ll P^{29+\epsilon}, \tag{26}$$

$$\int_0^1\int_0^1\int_0^1\int_0^1 |C_4|^{46}\, d\alpha_1 \dots d\alpha_4 \ll P^{36+\epsilon}. \tag{27}$$

This last is $B(4)$.

From now onwards we use the abbreviation

$$\int f\, d\alpha \quad \text{for} \quad \int_0^1 \dots \int_0^1 f(\alpha_1,\dots,\alpha_n)\, d\alpha_1 \dots d\alpha_n.$$

2.11. *Proof of* $A(5)$. By Lemma 1 we have

$$|S_5|^4 \ll P^3 + P \sum_{y_1}^{P} \sum_{y_2}^{P} \sum_{x_3}^{P}{}^{*}\, e^{2\pi i y_1 y_2 \Delta^2 g(x_3)},$$

where $g(x) = \alpha_5 x^5 + \alpha_3 x^3 + \alpha_2 x^2 + \alpha_1 x$. Multiplying by $|S_5|^{10}$ and integrating, we have, by (4),

$$\int |S_5|^{14}\, d\alpha \ll P^{8+\epsilon} + PR,$$

where R denotes the number of solutions of

$$\begin{aligned} y_1 y_2 \Delta^2 x_3^5 &= z_1^5 + \dots - z_{10}^5 \quad (y_1 y_2 \Delta^2 x_3^5 \neq 0),\\ y_1 y_2 w &= z_1^3 + \dots - z_{10}^3 \quad (y_1 y_2 w \neq 0),\\ 2 y_1 y_2 &= z_1^2 + \dots - z_{10}^2,\\ 0 &= z_1 + \dots - z_{10}. \end{aligned}$$

For a fixed w the number of solutions of

$$\begin{aligned} (2z_1^3 - w z_1^2) + \dots - (2z_{10}^3 - w z_{10}^2) &= 0,\\ z_1 + \dots - z_{10} &= 0 \end{aligned}$$

is $\ll P^{6+\epsilon}$, by $A(3)$. Hence $R \ll P^{7+\epsilon}$, and

$$\int |S_5|^{14}\, d\alpha \ll P^{8+\epsilon}. \tag{28}$$

By Lemma 1 we have

$$|S_5|^8 \ll P^7+P^4 \sum_{y_1}^{P}\sum_{y_2}^{P}\sum_{y_3}^{P}\sum_{x_4}^{P}{}^{*} e^{2\pi i y_1 y_2 y_3 \Delta^3 g(x_4)}.$$

Multiplying by $|S_5|^{14}$ and integrating, we have, by (28),

$$\int |S_5|^{22}\, d\alpha \ll P^{15+\epsilon}+P^4 R,$$

where R denotes the number of solutions of

$$\begin{aligned} y_1 y_2 y_3 \Delta^3 x_3^5 &= z_1^5+\ldots-z_{14}^5 \quad (y_1 y_2 y_3 \Delta^3 x_3^5 \neq 0),\\ 6y_1 y_2 y_3 &= z_1^3+\ldots-z_{14}^3,\\ 0 &= z_1^2+\ldots-z_{14}^2,\\ 0 &= z_1+\ldots-z_{14}. \end{aligned}$$

Clearly, $$R \ll P^{\epsilon} \int |C_2|^{14}\, d\alpha \ll P^{11+\epsilon}$$

by a trivial consequence of $B(2)$. Hence

$$\int |S_5|^{22}\, d\alpha \ll P^{15+\epsilon}. \tag{29}$$

By Lemma 1 we have

$$|S_5|^{16} \ll P^{15}+P^{11} \sum_{y_1}^{P}\sum_{y_2}^{P}\sum_{y_3}^{P}\sum_{y_4}^{P}\sum_{x_5}^{P}{}^{*} e^{2\pi i y_1 y_2 y_3 y_4 \Delta^4 g(x_5)}.$$

Multiplying by $|S_5|^{22}$ and integrating, we obtain

$$\int |S_5|^{38}\, d\alpha \ll P^{30+\epsilon}+P^{11} R,$$

where it is easily seen that

$$R \ll P^{\epsilon} \int |C_3|^{22}\, d\alpha \ll P^{16+\epsilon}$$

by a trivial consequence of $B(3)$. Thus

$$\int |S_5|^{38}\, d\alpha \ll P^{30+\epsilon}.$$

Repeating this process we obtain

$$\int |S_5|^{38+16\lambda}\, d\alpha \ll P^{30+15\lambda+\epsilon} \quad (\lambda = 0,\, 1,\, 2,\, 3). \tag{30}$$

The case $\lambda = 3$ is $A(5)$.

2.12. *Proof of* $B(5)$. By Lemma 1 we have

$$|C_5|^8 \ll P^7+P^4 \sum_{y_1}^{P}\sum_{y_2}^{P}\sum_{y_3}^{P}\sum_{x_4}^{P}{}^{*} e^{2\pi i y_1 y_2 y_3 \Delta^3(\alpha_5 x_4^5+\alpha_4 x_4^4+\ldots)}.$$

Multiplying this by $|C_5|^{12}$ and integrating, we have, by (8),

$$\int |C_5|^{20}\, d\alpha \ll P^{13+\epsilon}+P^4 R;$$

and it is easily seen that

$$R \ll P^{1+\epsilon} \int |S_4|^{12}\, d\alpha \ll P^{9+\epsilon}$$

by a trivial consequence of (4). Hence

$$\int |C_5|^{20}\, d\alpha \ll P^{13+\epsilon}.$$

Repeating the argument, but appealing to (20), (21), (22) instead of to (4), we obtain

$$\int |C_5|^{20+8\lambda}\, d\alpha \ll P^{13+7\lambda+\epsilon} \quad (\lambda = 1, 2, 3). \tag{31}$$

Using Lemma 1 with $\mu = 4$, and appealing to (30), we obtain

$$\int |C_5|^{44+16\lambda}\, d\alpha \ll P^{34+15\lambda+\epsilon} \quad (\lambda = 1, 2, 3, 4, 5). \tag{32}$$

The case $\lambda = 5$ is $B(5)$.

2.13. *Proof of* $A(6)$. Using Lemma 1 with $\mu = 3$, and the results in the proof of $A(4)$, we obtain

$$\int |S_6|^{12+8\lambda}\, d\alpha \ll P^{6+7\lambda+\epsilon} \quad (\lambda = 1, 2, 3, 4). \tag{33}$$

Using Lemma 1 with $\mu = 4$, and $B(3)$, we obtain

$$\int |S_6|^{60}\, d\alpha \ll P^{49+\epsilon}. \tag{34}$$

Using Lemma 1 with $\mu = 5$, and the results in the proof of $B(4)$, we obtain

$$\int |S_6|^{60+32\lambda}\, d\alpha \ll P^{49+31\lambda+\epsilon} \quad (\lambda = 1, 2, 3, 4, 5). \tag{35}$$

The case $\lambda = 5$ is $A(6)$.

2.14. *Proof of* $B(6)$. Here we depart from the procedure of starting with (8), but start instead with

$$\int |C_6|^{16}\, d\alpha \ll P^{9+\epsilon},$$

which is an evident consequence of (8). Using Lemma 1 with $\mu = 3$, and the results in the proof of $A(4)$, we obtain

$$\int |C_6|^{16+8\lambda}\, d\alpha \ll P^{9+7\lambda+\epsilon} \quad (\lambda = 1, 2, 3). \tag{36}$$

By Lemma 1 with $\mu = 4$, and the results in the proof of $A(5)$,

$$\int |C_6|^{40+16\lambda}\, d\alpha \ll P^{30+15\lambda+\epsilon} \quad (\lambda = 1, 2, 3, 4, 5). \tag{37}$$

By Lemma 1 with $\mu = 5$, and the results in the proof of $A(6)$,

$$\int |C_6|^{120+32\lambda}\, d\alpha \ll P^{105+31\lambda+\epsilon} \quad (\lambda = 1, 2, 3, 4, 5, 6). \tag{38}$$

2.15. *Proof of* $A(7)$. As an evident consequence of (4), we have

$$\int |S_7|^{16}\, d\alpha \ll P^{9+\epsilon}.$$

By Lemma 1 with $\mu = 3$, and the results in the proof of $A(4)$, we have

$$\int |S_7|^{16+8\lambda}\, d\alpha \ll P^{9+7\lambda+\epsilon} \quad (\lambda = 1, 2, 3). \tag{39}$$

By Lemma 1 with $\mu = 4$, and the results in the proof of $A(5)$,

$$\int |S_7|^{40+16\lambda}\, d\alpha \ll P^{30+15\lambda+\epsilon} \quad (\lambda = 1, 2, 3, 4, 5). \tag{40}$$

By Lemma 1 with $\mu = 5$, and (27),

$$\int |S_7|^{152}\, d\alpha \ll P^{136+\epsilon}. \tag{41}$$

By Lemma 1 with $\mu = 6$, and the results in the proof of $B(5)$,

$$\int |S_7|^{152+64\lambda}\, d\alpha \ll P^{136+63\lambda+\epsilon} \quad (\lambda = 1, 2, 3, 4, 5, 6). \tag{42}$$

2.16. *Proof of* $B(7)$. As an evident consequence of (8), we have

$$\int |C_7|^{24}\, d\alpha \ll P^{16+\epsilon}.$$

As in the proof of $A(7)$, we obtain

$$\int |C_7|^{24+8\lambda}\, d\alpha \ll P^{16+7\lambda+\epsilon} \quad (\lambda = 1, 2), \tag{43}$$

$$\int |C_7|^{40+16\lambda}\, d\alpha \ll P^{30+15\lambda+\epsilon} \quad (\lambda = 1, 2, 3, 4, 5), \tag{44}$$

$$\int |C_7|^{120+32\lambda}\, d\alpha \ll P^{105+31\lambda+\epsilon} \quad (\lambda = 1, 2, 3, 4, 5, 6), \tag{45}$$

$$\int |C_7|^{312+64\lambda}\, d\alpha \ll P^{291+63\lambda+\epsilon} \quad (\lambda = 1, 2, 3, 4, 5, 6, 7). \tag{46}$$

2.17. *Proof of* $A(8)$. As an evident consequence of (4), we have

$$\int |S_8|^{24}\, d\alpha \ll P^{16+\epsilon}.$$

We have then

$$\int |S_8|^{24+8\lambda}\, d\alpha \ll P^{16+7\lambda+\epsilon} \quad (\lambda = 1, 2), \tag{47}$$

$$\int |S_8|^{40+16\lambda}\, d\alpha \ll P^{30+15\lambda+\epsilon} \quad (\lambda = 1, 2, 3, 4, 5), \tag{48}$$

$$\int |S_8|^{120+32\lambda}\, d\alpha \ll P^{105+31\lambda+\epsilon} \quad (\lambda = 1, 2, 3, 4, 5, 6), \tag{49}$$

$$\int |S_8|^{376}\, d\alpha \ll P^{354+\epsilon}, \tag{50}$$

$$\int |S_8|^{376+128\lambda}\, d\alpha \ll P^{354+127\lambda+\epsilon} \quad (\lambda = 1, 2, \ldots, 7). \tag{51}$$

2.18. *Proof of* $B(8)$. We have

$$\int |C_8|^{18+16\lambda}\,d\alpha \ll P^{9+15\lambda+\epsilon} \qquad (\lambda = 1, 2, 3, 4, 5, 6), \tag{52}$$

$$\int |C_8|^{114+32\lambda}\,d\alpha \ll P^{99+31\lambda+\epsilon} \qquad (\lambda = 1, 2, 3, 4, 5, 6), \tag{53}$$

$$\int |C_8|^{306+64\lambda}\,d\alpha \ll P^{285+63\lambda+\epsilon} \qquad (\lambda = 1, 2, \ldots, 7), \tag{54}$$

$$\int |C_8|^{754+128\lambda}\,d\alpha \ll P^{726+127\lambda+\epsilon} \qquad (\lambda = 1, 2, \ldots, 8). \tag{55}$$

2.19. *Proof of* $A(9)$. We have

$$\int |S_9|^{18+16\lambda}\,d\alpha \ll P^{9+15\lambda+\epsilon} \qquad (\lambda = 1, 2, 3, 4, 5, 6), \tag{56}$$

$$\int |S_9|^{114+32\lambda}\,d\alpha \ll P^{99+31\lambda+\epsilon} \qquad (\lambda = 1, 2, 3, 4, 5, 6), \tag{57}$$

$$\int |S_9|^{306+64\lambda}\,d\alpha \ll P^{285+63\lambda+\epsilon} \qquad (\lambda = 1, 2, \ldots, 7), \tag{58}$$

$$\int |S_9|^{882}\,d\alpha \ll P^{853+\epsilon}, \tag{59}$$

$$\int |S_9|^{882+256\lambda}\,d\alpha \ll P^{853+255\lambda+\epsilon} \qquad (\lambda = 1, 2, \ldots, 8). \tag{60}$$

2.20. *Proof of* $B(9)$. We have

$$\int |C_9|^{20+16\lambda}\,d\alpha \ll P^{10+15\lambda+\epsilon} \qquad (\lambda = 1, \ldots, 5), \tag{61}$$

$$\int |C_9|^{100+32\lambda}\,d\alpha \ll P^{85+31\lambda+\epsilon} \qquad (\lambda = 1, \ldots, 6), \tag{62}$$

$$\int |C_9|^{292+64\lambda}\,d\alpha \ll P^{271+63\lambda+\epsilon} \qquad (\lambda = 1, \ldots, 7), \tag{63}$$

$$\int |C_9|^{740+128\lambda}\,d\alpha \ll P^{712+127\lambda+\epsilon} \qquad (\lambda = 1, \ldots, 8), \tag{64}$$

$$\int |C_9|^{1764+256\lambda}\,d\alpha \ll P^{1738+255\lambda+\epsilon} \qquad (\lambda = 1, \ldots, 9). \tag{65}$$

2.21. *Proof of* $A(10)$. We have

$$\int |S_{10}|^{20+16\lambda}\,d\alpha \ll P^{10+15\lambda+\epsilon} \qquad (\lambda = 1, \ldots, 5), \tag{66}$$

$$\int |S_{10}|^{100+32\lambda}\,d\alpha \ll P^{85+31\lambda+\epsilon} \qquad (\lambda = 1, \ldots, 6), \tag{67}$$

$$\int |S_{10}|^{292+64\lambda}\,d\alpha \ll P^{271+63\lambda+\epsilon} \qquad (\lambda = 1, \ldots, 7), \tag{68}$$

$$\int |S_{10}|^{740+128\lambda}\,d\alpha \ll P^{712+127\lambda+\epsilon} \qquad (\lambda = 1, \ldots, 8), \tag{69}$$

$$\int |S_{10}|^{2020}\,d\alpha \ll P^{1983+\epsilon}, \tag{70}$$

$$\int |S_{10}|^{2020+512\lambda}\,d\alpha \ll P^{1983+511\lambda+\epsilon} \qquad (\lambda = 1, \ldots, 9). \tag{71}$$

2.22. *Proof of* $B(10)$. As an evident consequence of (8) we have

$$\int |C_{10}|^{38}\,d\alpha \ll P^{27+\epsilon}.$$

Then

$$\int |C_{10}|^{38+16\lambda}\,d\alpha \ll P^{27+15\lambda+\epsilon} \quad (\lambda = 1,\ldots, 4), \tag{72}$$

$$\int |C_{10}|^{102+32\lambda}\,d\alpha \ll P^{87+31\lambda+\epsilon} \quad (\lambda = 1,\ldots, 6), \tag{73}$$

$$\int |C_{10}|^{294+64\lambda}\,d\alpha \ll P^{273+63\lambda+\epsilon} \quad (\lambda = 1,\ldots, 7), \tag{74}$$

$$\int |C_{10}|^{742+128\lambda}\,d\alpha \ll P^{714+127\lambda+\epsilon} \quad (\lambda = 1,\ldots, 8), \tag{75}$$

$$\int |C_{10}|^{1766+256\lambda}\,d\alpha \ll P^{1730+255\lambda+\epsilon} \quad (\lambda = 1,\ldots, 9), \tag{76}$$

$$\int |C_{10}|^{4070+512\lambda}\,d\alpha \ll P^{4025+511\lambda+\epsilon} \quad (\lambda = 1,\ldots,.10). \tag{77}$$

3. Applications

In this section I would like to mention some of the possible applications of the theorem.

1. *Tarry's problem*. I hope to discuss this in a later paper.

2. *Waring's problem*. Let $G(k)$ be the least integer s such that

$$N = x_1^k + \ldots + x_s^k$$

is solvable in non-negative integers $x_1,\ldots, x_s$ for all sufficiently large positive integers N. It was proved by Vinogradow that

$$G(k) < 4k\log k + 8k\operatorname{loglog} k + 12k.$$

If we introduce the result of this paper, and follow the same method as Vinogradow, we can obtain a result which is better than his for $k \leqslant 15{,}000$.

3. *Estimation of trigonometrical sums*. An improvement upon some of the results† due to Vinogradow has been obtained; the details will be given elsewhere later.

4. *Simultaneous Waring's problem*. Let T be the number of solutions of

$$x_1^h + \ldots + x_s^h = N_h \quad (h = 1, 2,\ldots, k),$$

in positive integers. The result of this paper implies that

$$T = O(P^{s-\frac{1}{2}k(k+1)+\epsilon}),$$

provided that s has the value given in the table in the Introduction. The exact relation between T and a multiple Fourier integral, and some theorems concerning congruences, will be given elsewhere later.

† The results in the papers cited in footnotes † and ‡ on p. 161, and *Bull. de l'Acad. des Sciences de l'U.R.S.S.*, 1938, 399–416.

Reprinted from the
JOURNAL OF THE CHINESE MATHEMATICAL SOCIETY
Vol. 2, pp. 301–312, Feb. 1940

ON AN EXPONENTIAL SUM*

LOO-KENG HUA

The main object of this paper is to prove the following theorem:

Let $f(x)$ be a polynomial of the k-th degree with integer coefficients,

$$f(x) = a_k x^k + \cdots + a_1 x$$

and let $(a_k, \ldots, a_1, q) = 1$. Then

$$S(q, f(x)) = \sum_{x=1}^{q} e_q(f(x)) = O(q^{1-1/k+\epsilon}), \qquad e_q(z) = e^{2\pi i z/q},$$

where the constant implied by the symbol O depends only on k and ϵ.

This result is better than my previous one[1] in which the constant implied by O depends also on the coefficients of the polynomial.

In §§3, 4 some easy applications of the theorem will be given. Another application of the theorem to a problem studied by Vinogradow will be given elsewhere.

§1. The theorem is a deduction of the following lemma:

Main Lemma. *Let $l > 1$ and p be a prime, and let*

$$f(x) = a_k x^k + \cdots + a_1 x$$

and $p \nmid (a_1, \ldots, a_k)$. Then

$$S(p^l, f(x)) = O(p^{l(1-1/k)}),$$

where the constant implied by the symbol O depends on k only.

The proof of the lemma will be given in the next section.

Lemma 1 (MORDELL).

$$S(p, f(x)) = O(p^{1-1/k}).$$

Lemma 2. *If $(q_1, q_2) = 1$ and $f(o) = 0$, then*

$$S(q_1 q_2, f(x)) = S(q_1, f(q_2 x)/q_2) S(q_2, f(q_1 x)/q_1).$$

* Received April 14, 1939.

(1) *Jour. of London Math. Soc.* **13** (1938), 54–61.

(2) *Quarterly Jour.* **3** (1932), 161–167.

Proof. Writing $x = q_1 y + q_2 z$, then as y and z run over the complete sets of residue systems mod q_1 and mod q_2 respectively, x runs over a complete set of residue system, mod $q_1 q_2$. Further we have evidently

$$e_{q_1q_2}(f(q_1 y + q_2 z)) = q_1(f(e_{q_1}z)/q_2)e_{q_2}(f(q_1 z)/q_1).$$

Thus

$$S(q_1q_2, f(x)) = \sum_{x=0}^{q_1q_2-1} e_{q_1q_2}(f(x)) = \sum_{y=0}^{q_2-1} e_{q_2}(f(q_1 y)/q_1) \sum_{z=0}^{q_1-1} e_{q_1}(f(q_2 z)/q_2)$$
$$= S(q_1, f(q_2 x)/q_2)S(q_2, f(q_1 x)/q_1).$$

Theorem 1. *If* $(a_1, \ldots, a_k, q) = 1$, *then we have*

$$S(q, f(x)) = O(q^{1-1/k+\epsilon}).$$

Proof. By the main lemma and the lemmas 1 and 2, we have

$$|S(q_1 f(x))| \leqq (c(k))^{\nu(q)} q^{1-1/k},$$

where $\nu(q)$ is the number of distinct prime factors of q. Since

$$c(k)^{\nu(q)} = O(q^{\epsilon}).$$

the theorem is proved.

§2. **Definition.** Let

$$f(x) = a_k x^k + \cdots + a_1 x$$

$p l_s \| s a_s, t = \operatorname{Min}(l_1, \ldots, l_k)$, $t \geqslant 0$. Let s be the greatest integer such that $p^t \| s a_s$. This integer is then defined to be the index of $f(x)$, and we write $s = \operatorname{ind} f(x)$. Immediately we have the following lemmas:

Lemma 1. $\operatorname{ind} f(x) = \operatorname{ind} f(x + \lambda)$.

Lemma 2. $\operatorname{ind} f(x) \geqq \operatorname{ind} f(px)$.

Lemma 3. *If* $\operatorname{ind} f(x) = \operatorname{ind} f(px)$, *then*

$$f'(x) \equiv 0 \pmod{p^{t+1}}$$

implies $p \mid x$.

Proof. By definition $l_s \leqslant l_{s'}$, for any s', and also $l_s + s \leqslant l_{s'} + s'$. Therefore

$$l_s < l_{s'} \qquad \text{for} \quad s \neq s';$$

in fact, if $s < s'$, the result is a trivial consequence of the definition, while if $s > s'$, then $l_s \leqslant l_{s'} + s' - s < l_{s'}$. Thus $f'(x) \equiv 0 \pmod{p^{t+1}}$ implies

$$s a_s x^{s-1} \equiv 0 \pmod{p^{t+1}},$$

i.e. $p \mid x$.

Proof of the main lemma. The lemma is immediate for $l \leqq t+1$, since

$$|S(p^{l}, f(x))| \leqslant p^{l} \leqslant p^{t+1} \leqslant p^{2t} \leqslant k^{2}, \qquad \text{for} \quad t > 0,$$

and, by a result due to Mordell

$$S(p, f(x)) = 0(p^{1-1/k}) \qquad \text{for} \quad t = 0.$$

Therefore we may assume that $l \geqq t+2$. Let

$$\lambda_1, \ldots, \lambda_e$$

be the distinct roots of the congruence

$$f'(x) \equiv 0 \pmod{p^{t+1}}.$$

Then evidently $e \leqslant p^t k \leqslant k^2$, and

$$\sum_{x=1}^{p^l} e_{p^l}(f(x)) = \sum_{i=1}^{p^{t+1}} \sum_{\substack{x=1 \\ x \equiv i\ (p^{t+1})}}^{p^l} e_{p^l}(f(x)).$$

If i is not equal to any one of the λ's then, letting $x = y + p^{l-t-1}z$, we have

$$\sum_{\substack{x=1 \\ x \equiv i,\ (p^{t+1})}}^{p^l} e_{p^l}(f(x)) = \sum_{\substack{y=1 \\ y \equiv i,\ (p^{t+1})}}^{p^{l-t-1}} e_{p^l}(f(y)) \sum_{\substack{z=1 \\ f'(y) \equiv 0,\ (p^{t+1})}}^{p^{t+1}} e_{p^{t+1}}(zf'(y)) = 0.$$

Therefore

$$\left| \sum_{x=1}^{p^l} e_{p^l}(f(x)) \right| = \left| \sum_{\substack{i=1 \\ x \equiv \lambda_i,\ (p^{t+1})}}^{e} \sum_{x=1}^{p^l} e_{p^l}(f(x)) \right|$$

$$\leqq e \operatorname*{Max}_{1 \leqslant i \leqslant e} \left| \sum_{y=1}^{p^{l-t-1}} e_{p^l}\big(f(\lambda_i + p^{t+1}y) - f(\lambda_i)\big) \right|$$

$$\leqq e \operatorname*{Max}_{2 \leqslant i \leqslant e} \left| \sum_{x=1}^{p^{l-t-1}} e_{p^{l-\mu_i}}(g_i(x)) \right|,$$

where p^{μ_i} is the highest power of p which divides all the coefficients of $f(\lambda_i + p^{t+1}y) - f(\lambda_i)$. Therefore

$$\left| \sum_{x=1}^{p^l} e_{p^l}(f(x)) \right| \leqslant e \operatorname*{Max}_{1 \leqslant i \leqslant e} p^{\mu_i - t - 1} \left| \sum_{x=1}^{p^{l-\mu_i}} e_{p^{l-\mu_i}}(g_i(x)) \right|$$

$$\leqslant k^2 \operatorname*{Max}_{1 \leqslant i \leqslant e} p^{\mu_i(1-1/k)} \left| \sum_{x=1}^{p^{l-\mu_i}} e_{p^{l-\mu_i}}(g_i(x)) \right|, \tag{1}$$

since $\mu_i \leqslant k(1+t)$.

If ind $f(x) = $ ind $f(px)$, then by lemma 3,

$$\sum_{x=1}^{p^l} e_{p^l}(f(x)) = p^{t+1} \sum_{y=1}^{p^{l-t-1}} e_{p^l}(f(y)) = p^{t+1} \sum_{y=1}^{p^{l-t-2}} e_{p^l}(f(py))$$

$$= p^{t+1} \sum_{y=1}^{p^{l-t-2}} e_{p^l-\mu}(g(x)) = p^{\mu-1} \sum_{y=1}^{p^{l-\mu}} e_{p^l-\mu}(g(y)),$$

where p^μ is the highest power of p divides all the coefficients of $f(py)$ and $f(py) = p^\mu g(y)$. We have then

$$\left| \sum_{x=1}^{p^l} e_{p^l}(f(x)) \right| \leqslant p^{\mu(1-1/k)} \left| \sum_{x=1}^{p^{l-\mu}} e_{p^l-\mu}(g(x)) \right|. \tag{2}$$

If we apply this method repeatedly, then there are at most k steps each giving a factor less than k^2 (using (1)), the other ones giving factor 1 only (using (2)). Thus

$$S(p^l, f(x)) = O(p^{l(1-1/k)}).$$

Remark. The ϵ in the theorem may be omitted in most cases. More precisely, by a little modification of the proof of the main lemma and a theorem due to Davenport([1]), we have

$$S(q^l, f(x)) = O(q^{1-1/k})$$

provided that k is not of the form 2^g or 3.2^g.

§3. The object of this section is to prove the following theorem:

Theorem 1. *Let*

$$f(x) = a_k x^k + \cdots + a_1 x, \qquad (a_k, \ldots, a_1, q) = 1,$$

then

$$\sum_{x=1}^{m} e_q(f(x)) = \frac{m}{q} S(q, f(x)) + O(q^{1-1/k+\epsilon}).$$

Evidently it is sufficient to prove that, if $0 < m < q$, we have

$$\sum_{x=1}^{m} e_q(f(x)) = O(q^{1-1/k+\epsilon}).$$

First, we shall find a function $g(x)$ with period q such that

$$g(x) = \begin{cases} 1 & \text{for} \quad 0 < x < m, \\ 0 & \text{for} \quad m < x < q. \end{cases}$$

If we assume $g(0) = g(m) = \frac{1}{2}$, then $g(x)$ can be represented by the Fourier series:

$$g(x) = \frac{m}{q} + \sum_{n=-\infty}^{\infty}{}' \frac{1}{2\pi i n} (e_q(nx) - e_q(n(x-m))),$$

([1]) *Jour. für Math.* **169** (1933), 158–176.

where in the summation the term $n = 0$ is excluded. Let

$$S_{q'} = \sum_{n=q+1}^{q'} e_q(nx).$$

It is well-known that if x is not a multiple of q, then

$$S_{q'} \leqslant \tfrac{1}{2}\{x/q\}^{-1}$$

where $\{t\}$ denotes the distance of t from the nearest integer. Consequently, by the method of partial summation, we have

$$\sum_{n=q+1}^{q'} \frac{1}{n} e_q(\pm nx) = O\left(\frac{1}{q\{x/q\}}\right).$$

Similarly, if $x \neq m$ and $0 < x < q$, then

$$\sum_{n=q+1}^{q'} \frac{1}{n} e_q(\pm(x-m)n) = O\left(\frac{1}{q\{(x-m)/q\}}\right).$$

Thus, for $x \neq m$ and $0 < x < q$, we have

$$g(x) = \frac{m}{q} + \sum_{n=-q}^{q}{}' \frac{1}{2\pi i n}(e_q(nx) - e_q(n(x-m)))$$
$$+ O\left(\frac{1}{q\{x/q\}}\right) + O\left(\frac{1}{q\{(x-m)/q\}}\right). \tag{3}$$

Next

$$\sum_{x=1}^{m} e_q(f(x)) = \sum_{x=1}^{q}{}^{*} e_q(f(x))g(x) + O(1)$$

where $\sum^*$ denotes a sum excluding $x = m$ and $x = q$. By (3), we have immediately

$$\sum_{x=1}^{m} e_q(f(x)) = \frac{m}{q}\sum_{x=1}^{q}{}^{*} e_q(f(x)) + \frac{1}{2\pi i}\sum_{n=-q}^{q}{}' \frac{1}{n}$$
$$\times\left(\sum_{x=1}^{q}{}^{*} e_q(f(x)+nx) - \sum_{x=1}^{q}{}^{*} e_q(f(x)+nx-mn)\right)$$
$$+ O\left(\sum_{x=1}^{q}{}^{*} \frac{1}{q\{x/q\}}\right) + O\left(\sum_{x=1}^{q}{}^{*} \frac{1}{q\{(x-m)/q\}}\right)$$
$$= I_1 + I_2 + I_3 + I_4 + I_5, \textit{ say}.$$

We have

$$I_4 = \sum_{x=1}^{q}{}^{*} \frac{1}{q\{x/q\}} \leqslant \frac{1}{q}\sum_{x=1}^{q/2} \frac{2q}{x} = O(\log q),$$

and the same result holds for I_5.

Finally we consider

$$\sum_{n=1}^{q} \frac{1}{n} \sum_{x=1}^{q} e_q(f(x) + nx).$$

Let $(a_k, \ldots, a_2, q) = q'$ and q'' be any factor of q'. We collect the terms of the sum for which n satisfies the condition

$$(a_k, \ldots, a_2, a_1 + n, q) = q''.$$

$$\begin{aligned}
&\sum_{n=1}^{q} \frac{1}{n}\left|\sum_{x=1}^{q} e_q(f(x)+nx)\right| \\
&\leqslant \sum_{q''/q} \sum_{\substack{n=1 \\ a_1+n\equiv 0,\,(q'')}}^{q} \frac{1}{n}\left|\sum_{x=1}^{q} e_{q/q''}\left(\frac{1}{q''}(f(x)+nx)\right)\right| \\
&= O\left[\sum_{q''/q} \sum_{\substack{n=1 \\ a_1+n\equiv 0,\,(q'')}}^{q} \frac{1}{n} q''(q/q'')^{1-1/k+\epsilon}\right] \\
&= O\left(\sum_{q''/q} \sum_{m=1}^{q/q''} \frac{1}{mq''} q''(q/q'')^{1-1/k+\epsilon}\right) \\
&= O\left(q^{1-1/k+\epsilon}\log q \sum_{q''/q} q''^{-1+1/k+\epsilon}\right) \\
&= O(q^{1-1/k+\epsilon})
\end{aligned}$$

This method gives

$$I_2 = O(q^{1-1/k+\epsilon}), \qquad I_3 = O(q^{1-1/k+\epsilon}).$$

Evidently

$$I_1 = O(q^{1-1/k+\epsilon}).$$

Combining all these results we obtain theorem 1.

Since the denominator of an integral-valued polynomial of the k-th degree is $\leqslant k!$, the theorem 1 is still true, if we assume only that $f(x)$ is an integral-valued polynomial of the k-th degree and $f(x) \not\equiv f(0)$, (mod p), where p is any factor of q.

§4. Finally I shall prove a theorem which has an interesting application to the problem of the "major arc" in Waring's problem.

Theorem 2. *Let $f(x)$ be an integral-valued polynomial. Let*

$$S(a) = \sum_{x=0}^{P} e^{2\pi i f(x)\alpha}, \qquad \alpha = \frac{a}{q} + \beta,$$

$$I(\beta) = \int_0^P e^{2\pi i f(x)\beta}\,dx.$$

Then, if $q = O(P^{1-\epsilon})$ and $|\beta| = O(q^{-1}P^{-k+1-\epsilon})$, we have

$$S(\alpha) = \bar{q}^{-1} S_{a,q} I(\beta) + O(q^{1-1/k+\epsilon}),$$

where $\bar{q} = q(q,d)$ and d is the least common denominator of the coefficients of $f(x)$,

and

$$S_{a,q} = \sum_{x=1}^{q} e_q(af(x))$$

and the constant implied by the symbol O depends on the coefficients of $f(x)$.

To prove this theorem we shall make use of the well-known Euler's summation formula:

We define

$$b_1(x) = x - [x] - \tfrac{1}{2},$$

where $[x]$ denotes the greatest integer which does not exceed x. We define $b_l(x)$ by induction

$$b_l(x+1) = b_l(x) \tag{1}$$

and

$$\int_0^x b_l(y)dy = b_{l+1}(x) - b_{l+1}(0). \tag{2}$$

Let $b > a$, and let $g(x)$ and its derivatives (as far as they occur below) be continuous for $a \leqslant x \leqslant b$. Then, for any t,

$$\begin{aligned}\sum_{\substack{m \\ a \leqslant m+t<b}} g(m+t) = &\int_a^b g(x)\,dx \\ &+ \sum_{r=0}^{l-1} \left\{ g^{(r)}(b) b_{r+1}(t-b) - g^{(r)}(a) b_{r+1}(t-a) \right\} \\ &- \int_a^b g^{(l)}(x) b_l(t-x)\,dx.\end{aligned}$$

Proof of the theorem.

First step.

$$\begin{aligned}S(\alpha) = \sum_{x=0}^{P} e^{2\pi i f(x)\alpha} &= \sum_{v=1}^{\bar{q}} \sum_{\substack{0 \leqslant r \leqslant P \\ r \equiv v,\, (\bar{q})}} e_q(af()v) e^{2\pi i \beta f(r)} \\ &= \sum_{v=1}^{\bar{q}} e_q(af(v))\, d_v,\end{aligned}$$

where

$$d_v = \sum_{\substack{j \\ 0 \leqslant \bar{q} j + v \leqslant P}} e^{2\pi i \beta f(\bar{q} j + v)} = \sum_{\substack{j \\ 0 \leqslant j + v/\bar{q} \leqslant P/\bar{q}}} \Phi(j + v/\bar{q}),$$

$$\Phi(x) = e^{2\pi i \beta f(\bar{q} x)}$$

By Euler's summation formula, we have

$$d_v = \int_0^{P/\bar{q}} \Phi(x)\,dx + \sum_{r=1}^{l-1} \left\{ \Phi^{(r)}\left(\frac{P}{\bar{q}}\right) b_{r+1}\left(\frac{v}{\bar{q}} - \frac{P}{\bar{q}}\right) - \Phi^{(r)}(0) b_{r+1}\left(\frac{v}{\bar{q}}\right) \right\}$$

$$- \int_0^{P/\bar{q}} \Phi^{(l)}(x) b_l\left(\frac{v}{\bar{q}} - x\right) dx. \tag{5}$$

Since

$$\int_0^{P/\bar{q}} \Phi(x)\,dx = \int_0^{P/\bar{q}} e^{2\pi i\beta f(\bar{q}x)}\,dx = \frac{1}{\bar{q}} \int_0^P e^{2\pi i\beta f(y)}\,dy,$$

we have, from (4) and (5)

$$S(a) = \frac{S_{aq}}{\bar{q}} I(p) + \sum_{r=1}^{l} \left\{ \Phi^{(r)}\left(\frac{P}{\bar{q}}\right) a_{r+1}\left(\frac{v}{\bar{q}} - \frac{P}{\bar{q}}\right) - \Phi^{(r)}(0) b_{r+1}\left(\frac{v}{\bar{q}}\right) \right\} - R$$

where

$$a_{r+1}\left(\frac{v}{\bar{q}} - t\right) = \sum_{v=1}^{\bar{q}} e_q(af(v)) b_{r+1}\left(\frac{v}{\bar{q}} - t\right),$$

$$R = \sum_{v=1}^{\bar{q}} e_q(af(v)) \int_0^{P/\bar{q}} \Phi^{(l)}(x) b_l\left(\frac{v}{\bar{q}} - x\right) dx.$$

Second step. If $q = O(P^{1-\epsilon})$, $\beta = O(q^{-1}P^{-k+1-\epsilon})$ and $0 < x \leqslant P/\bar{q}$, then

$$\Phi^{(r)}(x) = O(P^{-r\epsilon}). \tag{6}$$

Suppose $f(v)$ have only one term, namely $f(v) = Av^k$. Let

$$\psi(x) = e^{2\pi i\beta A(\bar{q}x)^k}.$$

First, we shall prove that

$$\psi^{(r)}(x) = O(P^{-r\epsilon}).$$

Let $\psi_1(z) = e^{z^k}$, then

$$\psi_1^{(r)}(z) = e^{z^k} F_r(z),$$

where $F_r(z)$ is a polynomial of the $r(k-1)$-th degree. Therefore

$$\psi^{(r)}(x) = e^{2\pi i\beta A(\bar{q}x)^k} F_r\big((2\pi i\beta A)^{1/k}\bar{q}x\big)\big((2\pi i\beta A)^{1/k}\bar{q}\big)^r.$$

Consequently

$$\psi^{(r)}(x) = O\big(1 + (|\beta|^{1/k}qx)^{(k-1)r}\big)(|\beta|^{1/k}q)^r = O(P^{-r\epsilon})$$

Next, we suppose $f(x)$ to be a polynomial with the first coefficient A. Let

$$\Phi(x) = \psi(x)\psi_1(x), \qquad \psi_1(x) = e^{2\pi i\beta(f(\bar{q}x) - A(\bar{q}x)^k)}.$$

Suppose (6) to be true for $k-1$, i.e. when $|\beta| \leqslant q^{-1}P^{-k+2-\epsilon}$, we have

$$\psi_1^{(r)}(x) = O(P^{-r\epsilon}).$$

Since $q^{-1}P^{-k+2-\epsilon} > q^{-1}P^{-k+1-\epsilon}$, we have

$$\psi_1^{(r)}(x) = O(P^{-r\epsilon}),$$

for $|\beta| = O(q^{-1}P^{-k+1-\epsilon})$. Further, since

$$\Phi^{(r)}(x) = \psi^{(r)}(x)\psi_1(x) + \binom{r}{1}\psi^{(r-1)}(x)\psi_1'(x) + \cdots + \psi(x)\psi_1^{(r)}(x),$$

we have

$$\Phi^{(r)}(x) = O\Big(\operatorname*{Max}_{0\leqslant i\leqslant r}\big(\psi^{(r-i)}(x)\psi_1^{(i)}(x)\big)\Big) = O(P^{-r\epsilon}).$$

Third step. Take

$$l = [1/\epsilon] + 1,$$

then

$$\Phi^{(l)}(x) = O(P^{-1}).$$

Therefore

$$|R| = O\Big(\frac{1}{q}\int_0^{P/\bar{q}} P^{-1}\,dx\Big) = O(1).$$

Fourth step. Let

$$S_v = \sum_{h=1}^{v} e_q(af(h)).$$

By the definition of $a_v(t)$, we have

$$\begin{aligned} a_r\Big(\frac{v}{\bar{q}} - t\Big) &= S_1 b_{r+1}\Big(\frac{1}{\bar{q}} - t\Big) + \sum_{v=2}^{\bar{q}} (S_v - S_{v-1}) b_{r+1}\Big(\frac{v}{\bar{q}} - t\Big) \\ &= \sum_{m=1}^{\bar{q}-1} S_m \Big\{ b_{r+1}\Big(\frac{m}{\bar{q}} - t\Big) - b_{r+1}\Big(\frac{m+1}{\bar{q}} - t\Big)\Big\} + S_q b_{r+1}(1-t). \end{aligned}$$

By theorem 1,

$$S_v = O(\bar{q}^{1-1/k+\epsilon}) \qquad \text{for} \quad 0 < v \leqslant q.$$

Thus

$$a_r\Big(\frac{v}{\bar{q}} - t\Big) = 0\Big(\bar{q}^{1-1/k+\epsilon}\Big\{\sum_{m=1}^{\bar{q}-1}\Big| b_{r+1}\Big(\frac{m}{\bar{q}} - t\Big) - b_{r+1}\Big(\frac{m+1}{\bar{q}} - t\Big)\Big| + 1\Big\}\Big).$$

Since b_{r+1} is a function of bounded variation, we have

$$a_r\Big(\frac{v}{\bar{q}} - t\Big) = O(\bar{q}^{1-1/k+\epsilon}).$$

Fifth step. Combining the results of the 2nd, 3rd and 4th steps, we have, in conclusion, that

$$S(\alpha) - \bar{q}^{-1}S_{aq}I(p) = O\Big(\bar{q}^{1-1/k+\epsilon}\sum_{r=1}^{l-1} P^{-r\epsilon} + 1\Big) = O(\bar{q}^{1-/k+\epsilon}).$$

Reprinted from the
TRANSACTIONS OF THE AMERICAN MATHEMATICAL SOCIETY
Vol. 51, pp. 194–201, Jan. 1942

ON THE NUMBER OF PARTITIONS OF A NUMBER INTO UNEQUAL PARTS[1]

BY
LOO-KENG HUA

1. **Introduction.** Let $q(n)$ be the number of partitions of an integer n into unequal parts, or into odd parts[2]. Then

$$(1.1)\qquad \begin{aligned} f(x) &= 1 + \sum_{n=1}^{\infty} q(n)x^n = (1+x)(1+x^2)(1+x^3)\cdots \\ &= \frac{1}{(1-x)(1-x^3)(1-x^5)\cdots}. \end{aligned}$$

Hardy and Ramanujan[3] indicated that by their fundamental analytic method one can obtain the following result:

$$\begin{aligned} q(n) &= \frac{1}{2^{1/2}}\frac{d}{dn}J_0\left[i\pi\left\{\tfrac{1}{3}(n+\tfrac{1}{24})\right\}^{1/2}\right] \\ &\quad + 2^{1/2}\cos\left(\tfrac{2}{3}\pi n - \tfrac{1}{9}\pi\right)\frac{d}{dn}J_2\left[\tfrac{1}{3}i\pi\left\{\tfrac{1}{3}(n+\tfrac{1}{24})\right\}^{1/2}\right] + \cdots \\ &\quad + \text{to } \left[\alpha n^{1/2}\right] \text{ terms} + O(1) \end{aligned}$$

where α is an arbitrary constant. This result is less satisfactory than that concerning the number $p(n)$ of partitions (unrestricted) of n, since in the latter case the error term approaches zero with increasing n. Recently Rademacher[4] obtained an equality for $p(n)$. The object of the present paper is to find an equality for $q(n)$. The work of this paper is a straightforward application of Hardy-Ramanujan's method with two modifications. These modifications are Kloosterman's sum and Rademacher's "Farey dissection of infinite order."

The present method may also be applied to find the explicit formula for

$$\sum_{x=1}^{[n^{1/2}]} p(n-x^2)$$

where $p(n)$ is the number of unrestricted partitions of n.

Presented to the Society, April 27, 1940; received by the editors January 9, 1941.

(1) This paper was accepted by Acta Arithmetica before the war.

(2) Cf. MacMahon, *Combinatory Analysis*, vol. 2, 1916, p. 11.

(3) Proceedings of the London Mathematical Society, (2), vol. 17 (1918), pp. 75–115.

(4) Proceedings of the London Mathematical Society, (2), vol. 43 (1937), pp. 241–254.

2. **Statement of the result.** Let

$$\epsilon_{h,k} = \begin{cases} \exp\left(-\pi i\left(\dfrac{(h'^2-1)}{8}\left(\dfrac{1-hh'}{k}-1\right)+\dfrac{h'(1-hh')}{8k} \right.\right. \\ \qquad \left.\left. +\dfrac{1}{24}\left(k+\dfrac{1-hh'}{k}\right)(hh'^2-h'-h)\right)\right), & \text{for } 2\mid k, \\ \exp\left(\dfrac{\pi i}{24}\left(k+\dfrac{1-hh'}{k}\right)(h+h'-h^2h')\right), & \text{for } 2\nmid k,\ 2\nmid h, \\ \exp\left(-\dfrac{\pi i}{8}\left(k^2-1-hk+\tfrac{1}{3}(h+h')\left(hh'k-\dfrac{hh'-1}{k}\right)\right)\right), & \\ & \text{for } 2\nmid k,\ 2\mid h, \end{cases}$$

and

$$\omega_{h,k} = \begin{cases} \epsilon_{h,k}\exp\left(-\dfrac{\pi i}{12k}(h+h')\right), & \text{for } 2\mid k, \\ \epsilon_{h,k}\exp\left(-\dfrac{\pi i}{24k}(2h-h')\right), & \text{for } 2\nmid k, \end{cases}$$

where $hh'\equiv 1 \pmod{k}$, $h\equiv h' \pmod{2}$.

THEOREM. *The number of partitions of an integer n into unequal parts is given by*

$$q(n) = \frac{1}{2^{1/2}}\sum_{k=1,\,k\text{ odd}}^{\infty}\ \sum_{(h,k)=1,\,0<h\leq k}\omega_{h,k}e^{-2\pi i hn/k}\frac{d}{dn}J_0\left(\frac{i\pi}{k}\left\{\tfrac{2}{3}\left(n+\tfrac{1}{24}\right)\right\}^{1/2}\right),$$

where $J_0(x)$ is the Bessel function of the 0th order.

3. **Farey dissection.** By means of Cauchy's integral formula we obtain for (1.1)

$$q(n) = \frac{1}{2\pi i}\int_C \frac{f(x)}{x^{n+1}}\,dx.$$

The path of integration may be the circle defined as $|x| = e^{-2\pi N^{-2}}$ where N is a certain positive integer at our disposal. In the usual way we divide the circle into Farey arcs $\xi_{h,k}$ of order N. The Farey arc $\xi_{h,k}$ is defined by

(3.1) $$x = \exp(2\pi ih/k - 2\pi N^{-2} + 2\pi i\vartheta), \qquad (h,k)=1,$$

and

(3.2) $$-\vartheta_1(h,k) = \frac{h+h_1}{k+k_1}-\frac{h}{k}\leq\vartheta\leq\frac{h+h_2}{k+k_2}-\frac{h}{k} = \vartheta_2(h,k)$$

where h_1/k_1, h/k, h_2/k_2 are three consecutive fractions in the Farey sequence of order N. It is well known that

$$(3.3)\qquad \frac{1}{k(N+k)} \leqq \vartheta_1(h,k) < \frac{1}{k(N+1)},\quad \frac{1}{k(N+k)} \leqq \vartheta_2(h,k) < \frac{1}{k(N+1)}.$$

We obtain then

$$(3.4)\qquad q(n) = \frac{1}{2\pi i} \sum_{(h,k)=1,0<h\leqq k\leqq N} \int_{\xi_{h,k}} \frac{f(x)}{x^{n+1}}\, dx.$$

Let I_1 and I_2 denote the sums of those terms satisfying $2|k$, and $2\nmid k$, respectively. Then, by (3.4), we have

$$(3.5)\qquad q(n) = I_1 + I_2.$$

4. **Lemmas on Kloosterman's sums.**

LEMMA 4.1 (5). *Let*

$$g(N,\vartheta,h,k) = \begin{cases} 1 & \text{for } -\vartheta_1(h,k) \leqq \vartheta \leqq \vartheta_2(h,k), \\ 0 & \text{otherwise.} \end{cases}$$

Then

$$g = \sum_{r=1}^{k} b_r e^{2\pi i r h'/k}$$

where h' iz an integer satisfying

$$hh' \equiv 1 \pmod{k},$$

and b_r is independent of h and

$$\sum_{r=1}^{k} |b_r| < \log 4k.$$

LEMMA 4.2. *Let a be an absolute constant. Then*

$$\sum_{0<h\leqq ak,(h,ak)=1,h\equiv l(a)} \exp\left(\frac{2\pi i}{ak}(nk+mh')\right) = O(k^{2/3+\epsilon}(n,k)^{1/3}).$$

LEMMA 4.3. *If k is even and $\omega_{h,k}$ as defined in §2, then*

$$S_k = \sum_{1\leqq h\leqq k,(h,k)=1,hh'\equiv 1\,(k)} \omega_{h,k} e^{2\pi i(nh+mh')/k} = O(k^{2/3+\epsilon}(n,k)^{1/3}).$$

Proof. For the sake of simplicity I give here only the proof of the case $24|k$.

(5) T. Estermann, Abhandlungen aus dem Mathematischen Seminar der Hamburgischen Universität, vol. 7 (1929), pp. 93, 94.

Then

$$S_k = \sum_{1\leq l\leq 24,(l,24)=1} \sum_{1\leq h\leq k,(h,k)=1,hh'\equiv 1,h\equiv l\ (24)} \omega_{h,k}e^{2\pi i(nh+mh')/k}.$$

The inner sum becomes a Kloosterman's sum as in Lemma 4.2. Therefore we have

$$S_k = O(k^{2/3+\epsilon}(n, k)^{1/3}).$$

As to the proof of the other cases, nothing is difficult but a little complicated, and the following fact is used: let

$$F(h, k) = \omega_{h,k}e^{2\pi i(nh+mh')/k};$$

then $F(h+k, k)=F(h, k)$.

LEMMA 4.4. *Let* $2\nmid k$ *and* $\omega_{h,k}$ *be as defined in* §2, *then*

$$S = \sum_{1\leq h\leq k,(h,k)=1,hh'\equiv 1\ (k),h'\ \mathrm{odd}} \omega_{h,k}e^{\pi i(2nh+mh')/k} = O(k^{2/3+\epsilon}(h, k)^{1/3}).$$

The proof is similar to that of Lemma 4.3, only notice that

$$S = \sum_{1\leq h<2k,(h,2k)=1,hh'\equiv 1\ (2k)}$$

5. Lemmas from the theory of the linear transformation of the elliptic modular functions.

LEMMA 5.1. *Suppose that* $2\nmid h$, $2|k$; *that* h' *is a positive integer satisfying* $hh'\equiv 1 \pmod k$; *that* $\omega_{h,k}$ *is defined in* §2; *and that*

$$x = \exp\left(-\frac{2\pi z}{k}+\frac{2h\pi i}{k}\right), \qquad x' = \exp\left(-\frac{2\pi}{kz}-\frac{2h'\pi i}{k}\right),$$

where the real part of z *is positive. Then*

$$f(x) = \omega_{h,k}\exp\left(-\frac{\pi}{12kz}+\frac{\pi z}{12k}\right)f(x').$$

Proof. If we take $a=h$, $b=-k$, $c=(1-hh')/k$, $d=h'$, so that $ad-bc=1$, and write

$$x = q^2 = e^{2\pi i\tau}, \qquad x' = Q^2 = e^{2\pi iT},$$
$$\tau = (h+iz)/k, \qquad T = (-h'+i/z)/k,$$

then we can easily verify that

$$T = \frac{c+d\tau}{a+b\tau}.$$

Also, in the notation of Tannery and Molk, we obtain

$$f(x) = \frac{1}{2^{1/3}} q^{-1/12} \frac{\phi(\tau)}{\chi(\tau)}, \qquad f(x') = \frac{1}{2^{1/3}} Q^{-1/12} \frac{\phi(T)}{\chi(T)}.$$

Then

$$\begin{aligned} f(x') &= \frac{1}{2^{1/3}} Q^{-1/12} \frac{\phi(T)}{\chi(T)} = \exp\left(\pi i\left(\tfrac{1}{8}(d^2-1)(c-1) + \frac{cd}{8}\right.\right. \\ &\qquad \left.\left. - \frac{(b-c)(bcd-a)}{24}\right)\right) \frac{1}{2^{1/3}} Q^{-1/12} \frac{\phi(\tau)}{\chi(\tau)} \\ &= \exp\left(\pi i\left(\tfrac{1}{8}(d^2-1)(c-1) + \frac{cd}{8} - \frac{(b-c)(bcd-a)}{24}\right)\right) q^{1/12} Q^{-1/12} f(x) \\ &= \exp\left(\pi i\left(\tfrac{1}{8}(d^2-1)(c-1) + \frac{cd}{8} - \frac{(b-c)(bcd-a)}{24}\right)\right) \\ &\quad \cdot \exp\left(\frac{\pi}{12k}\left(\frac{1}{z} - z\right)\right) \exp\left(\frac{\pi i}{12k}(h+h')\right) f(x). \end{aligned}$$

LEMMA 5.2. *Suppose that* $2 \nmid hk$ *and* $hh' \equiv 1 \pmod{2k}$, *that*

$$f_1(x) = \prod_1^\infty (1 + x^{n-1/2}) = 1 + \sum_{n=1}^\infty q_1(n) x^{n/2}.$$

Then

$$f(x) = \frac{\omega_{h,k}}{2^{1/2}} \exp\left(\frac{\pi}{12k}\left(z + \frac{1}{2z}\right)\right) f_1(x').$$

Proof. As in Lemma 5.1, we have

$$\begin{aligned} f_1(x) &= f_1(q^2) = \prod (1 + q^{2n-1}) = 2^{1/6} q^{1/24} \frac{1}{\chi(\tau)}, \\ f_1(x') &= 2^{1/6} Q^{1/24} \frac{1}{\chi(T)} = 2^{1/6} Q^{1/24} \exp\left(-\frac{(b-c)(abc-d)}{24}\pi i\right) \frac{\phi(\tau)}{\chi(\tau)} \\ &= 2^{1/6} Q^{1/24} \exp\left(-\frac{(b-c)(abc-d)}{24}\pi i\right) 2^{1/3} q^{1/12} f(x) \\ &= \exp\left(-\frac{(b-c)(abc-d)}{24}\pi i\right) 2^{1/2} \\ &\quad \cdot \exp\left(\frac{\pi i}{24}\left(-\frac{h'}{k} + \frac{i}{kz} + \frac{2h}{k} + \frac{2iz}{k}\right)\right) f(x). \end{aligned}$$

LEMMA 5.3. *Suppose that* $2|h$, $2\nmid k$, $hh' \equiv 1 \pmod{k}$, $2|h'$ *and suppose that*

$$f_2(x) = \prod_1^\infty (1 - x^{n-1/2}) = 1 + \sum q_2(n) x^{n/2}.$$

Then

$$f(x) = \frac{\omega_{h,k}}{2^{1/2}} \exp\left(\frac{\pi}{12k}\left(z + \frac{1}{2z}\right)\right) f_2(x').$$

Proof. We take

$$a = -h, \qquad b = k, \qquad c = (hh' - 1)/k, \qquad d = -h'.$$

Then

$$\begin{aligned} f_2(x') = f_2(Q^2) &= 2^{1/6} Q^{1/24} \frac{\psi(T)}{\chi(T)} \\ &= 2^{1/6} Q^{1/24} \exp\left(\frac{\pi i}{2}\left(\frac{b^2-1}{4} + \frac{ab}{4} - \frac{(a+d)(abd-c)}{12}\right)\right) \frac{\phi(\tau)}{\chi(\tau)} \\ &= 2^{1/2} \exp\left(\frac{\pi i}{2}\left(\frac{b^2-1}{4} + \frac{ab}{4} - \frac{(a+d)(abd-c)}{12}\right)\right) Q^{1/24} q^{1/12} f(x). \end{aligned}$$

6. **Approximation of the integrand.** Let

$$z = k(N^{-2} - i\vartheta).$$

Then

$$\begin{aligned} I_1 &= \sum_{1 \leq k \leq N, 2|k} \ \sum_{(h,k)=1, 0<h<k} \int_{-k^{-1}(N+1)^{-1}}^{k^{-1}(N+1)^{-1}} g(\vartheta) f(e^{(2\pi i h - 2\pi z)/k}) e^{-2\pi i h n/k + 2\pi z n/k} d\vartheta \\ &= \sum_{1 \leq k \leq N, 2|k} \ \sum_{(h,k)=1, 0<h<k} \int_{-k^{-1}(N+1)^{-1}}^{k^{-1}(N+1)^{-1}} g(\vartheta) \omega_{h,k} e^{(\pi/12k)(z - 1/z)} \\ &\qquad \cdot f(x') e^{-2\pi i h n/k + 2\pi z n/k} d\vartheta \\ &= \sum_{1 \leq k \leq N, 2|k} \ \sum_{(h,k)=1, 0<h<k} \int_{-k^{-1}(N+1)^{-1}}^{k^{-1}(N+1)^{-1}} g(\vartheta) \omega_{h,k} \\ &\qquad e^{(\pi/12k)(z-1/z) - 2\pi i h n/k + 2\pi z n/k} \sum_{\nu=0}^\infty q(\nu) e^{-(2\pi/kz + 2h'\pi i/k)\nu} d\vartheta \qquad (6.1) \\ &= \sum_{1 \leq k \leq N, 2|k} \ \sum_{(h,k)=1, 0<h<k} \int_{-k^{-1}(N+1)^{-1}}^{k^{-1}(N+1)^{-1}} \sum_{\nu=0}^\infty q(\nu) e^{-(2\pi/kz)(\nu + 1/24) + (2\pi z/k)(n + 1/24)} \\ &\qquad \cdot \sum_{r=1}^k b_r e^{2\pi i r h'/k} \omega_{h,k} e^{-2\pi i h n/k - 2\pi i h' \nu/k} d\vartheta. \end{aligned}$$

Since $(1/k)\mathcal{R}(1/z) \geqq \frac{1}{2}$, we have

$$|I_1| \leqq \sum_{1\leqq k\leqq N,2|k} \int_{-k^{-1}(N+1)^{-1}}^{k^{-1}(N+1)^{-1}} \sum_{\nu=0}^{\infty} q(\nu) \cdot \exp\left\{-\frac{2\pi}{k}\left(\nu+\frac{1}{24}\right)\mathcal{R}\frac{1}{z}+\frac{2\pi}{k}\left(n+\frac{1}{24}\right)\mathcal{R}z\right\}$$

$$\sum_{r=1}^{k} |b_r| \left| \sum_{(h,k)=1} \omega_{h,k} e^{-2\pi i h n/k+2h'(r-\nu)\pi i/k}\right| d\vartheta$$

$$= O\left(\sum_{k=1}^{N} \int_{-k^{-1}(N+1)^{-1}}^{k^{-1}(N+1)^{-1}} \sum_{\nu=0}^{\infty} q(\nu)e^{-\pi(\nu+1/24)} \sum_{r=1}^{k} |b_r| k^{2/3} d\vartheta\right)$$

$$= O\left(\sum_{k=1}^{N} \log k \cdot k^{2/3} \frac{1}{kN}\right) = O\left(\frac{1}{N}\sum_{k=1}^{N} k^{-1/3+\epsilon}\right)$$

$$= O(N^{-1/3+\epsilon}).$$

Let

$$J = \frac{1}{2^{1/2}} \sum_{k=1,k \text{ odd}}^{N} \sum_{(h,k)=1,0<h\leqq k} \int_{-k^{-1}(N+1)^{-1}}^{k^{-1}(N+1)^{-1}} g(\vartheta)\omega_{h,k} \cdot e^{(\pi/24k)(2z+1/z)-2\pi i h n/k+2\pi z n/k} d\vartheta.$$

The same method will give us that $|I_2-J| = O(N^{-1/3+\epsilon})$.

7. **A contour integration.** Let $w=N^{-2}-i\vartheta$. Then

$$J = \frac{-i}{2^{1/2}} \sum_{1\leqq k\leqq N,k \text{ odd}} \sum_{(h,k)=1,0<h\leqq k} \omega_{h,k}e^{-2\pi i h n/k} \int_{N^{-2}-i\vartheta_2}^{N^{-2}+i\vartheta_1} e^{2\pi w(n+1/24)+\pi/24k^2w} dw$$

$$= \frac{i}{2^{1/2}} \sum_{1\leqq k\leqq N,k \text{ odd}} \sum_{(h,k)=1,0<h\leqq k} \omega_{h,k}e^{-2\pi i h n/k}\left(\int_{N^{-2}+i\vartheta_1}^{N^{-2}+ik^{-1}(N+1)^{-1}} + \int_{N^{-2}+ik^{-1}(N+1)^{-1}}^{-N^{-2}+ik^{-1}(N+1)^{-1}} + \int_{-N^{-2}+ik^{-1}(N+1)^{-1}}^{-N^{-2}-ik^{-1}(N+1)^{-1}} + \int_{-N^{-2}-ik^{-1}(N+1)^{-1}}^{N^{-2}+ik^{-1}(N+1)^{-}} + \int_{N^{-2}-ik^{-1}(N+1)}^{N^{-2}-i\vartheta_2} - 2\pi i \text{ Residue at } 0\right)$$

$$= K_1 + K_2 + K_3 + K_4 + K_5 + L \text{ (say)}.$$

We have

$$K_1 = \frac{i}{2^{1/2}} \sum_{1\leqq k\leqq N,k \text{ odd}} \sum_{(h,k)=1,0<h\leqq k} \omega_{h,k}e^{-2\pi i h n/k} \cdot \int_{N^{-2}+ik^{-1}(N+k)^{-1}}^{N^{-2}+ik^{-1}(N+1)^{-1}} g(\vartheta)e^{2\pi w(n+1/24)+\pi/24k^2w} dw.$$

By Lemma 3.1, we have

$$K_1 = O\left(\sum_{1\leqq k\leqq N,k \text{ odd}} k^{2/3+\epsilon}\int_{k^{-1}(N+k)^{-1}}^{k^{-1}(N+1)^{-}} \exp\left\{2\pi\left(n+\frac{1}{24}\right)\mathcal{R}w+\frac{\pi}{24k^2}\mathcal{R}\frac{1}{w}\right\}dw\right.$$

$$= O\left(\sum_{k=1}^{N} k^{2/3+\epsilon}e^{-2\pi nN^{-2}}\int_{k^{-1}(N+k)^{-1}}^{k^{-1}(N+1)^{-1}} d\vartheta\right)$$

$$= O(N^{-1/3+\epsilon}).$$

Similar result holds for K_5.

We have

$$\mathcal{R}\frac{1}{k^2w} = \frac{N^{-2}}{k^2N^{-2}+N^2}, \qquad K_2 = O\left(\sum_{k=1}^{N} N^{-2}k^{2/3+\epsilon}\right) = O(N^{-1/3+\epsilon}).$$

Similar result holds for K_4.

Applying again Kloosterman's argument to K_3, we have also $K_3=O(N^{-1/3})$.

Finally we find the residue of exp $(2\pi w(n+1/24)+\pi/24k^2w)$ at $w=0$. We have the expansion

$$e^{2\pi w(n+1/24)} = \sum_{\nu=1}^{\infty}\frac{(2\pi w(n+1/24))^\nu}{\nu!},$$

$$e^{\pi/24k^2w} = \sum_{\mu=1}^{\infty}\frac{1}{\mu!}\left(\frac{\pi}{24k^2w}\right)^\mu.$$

The residue is, therefore,

$$\sum_{\mu=1}^{\infty}\frac{1}{\mu!}\left(\frac{\pi}{24k^2}\right)^\mu\frac{1}{(\mu-1)!}(2\pi(n+\tfrac{1}{24}))^{\mu-1}$$

$$= \frac{1}{2\pi}\frac{d}{dn}\sum_{\mu=1}^{\infty}\frac{1}{(\mu!)^2}\left(\frac{\pi}{24k^2}\right)^\mu(2\pi(n+\tfrac{1}{24}))^\mu$$

$$= \frac{1}{2\pi}\frac{d}{dn}\sum_{\mu=1}^{\infty}\frac{1}{2^{2\mu}(\mu!)^2}\left(\frac{\pi}{k}\left\{\tfrac{1}{3}(n+\tfrac{1}{24})\right\}^{1/2}\right)^{2\mu}$$

$$= \frac{1}{2\pi}\frac{d}{dn}J_0\left(\frac{i\pi}{k}\left\{\tfrac{1}{3}(n+\tfrac{1}{24})\right\}^{1/2}\right).$$

Therefore

$$q(n) = \frac{1}{2^{1/2}}\sum_{k=1,k \text{ odd}}^{N}\sum_{(h,k)=1,0<h\leqq k}\omega_{h,k}e^{-2\pi ihn/k}\frac{d}{dn}J_0\left(\frac{i\pi}{k}\left\{\tfrac{1}{3}(n+\tfrac{1}{24})\right\}^{1/2}\right)$$

$$+ O(N^{-1/3+\epsilon}).$$

Let $N\to\infty$; we obtain the theorem.

NATIONAL TSING HUA UNIVERSITY,
KUNMING, YÜNNAN, CHINA

Reprinted from the
BULLETIN OF THE AMERICAN MATHEMATICAL SOCIETY
Vol. 48, No. 10, pp. 731–735, 1942

ON THE LEAST SOLUTION OF PELL'S EQUATION

LOO-KENG HUA

Let x_0, y_0 be the least positive solution of Pell's equation

$$x^2 - dy^2 = 4,$$

where d is a positive integer, not a square, congruent to 0 or 1 (mod 4). Let $\epsilon = (x_0 + d^{1/2}y_0)/2$. It was proved by Schur[1] that

$$\epsilon < d^{d^{1/2}}, \tag{1}$$

or, more precisely,

$$\log \epsilon < d^{1/2}((1/2) \log d + (1/2) \log \log d + 1) . \tag{2}$$

He deduced (1) from (2) by the property that

$$d^{1/2}((1/2) \log d + (1/2) \log \log d + 1) < d^{1/2} \log d$$

for $d > 244.69 \cdots$, and, for $d \leqq 244$, (1) is established by direct computation. It is the object of the present note to establish a slightly better result that

$$\log \epsilon < d^{1/2}((1/2) \log d + 1). \tag{3}$$

Thus (1) follows immediately without any calculation. The method used is that described in the preceding paper.

Let $(d \mid r)$ be Kronecker's symbol. (We extend the definition to include negative values of r by the relation $(d \mid r_1) = (d \mid r_2)$ for $r_1 \equiv r_2 \pmod{d}$.)

Let f denote the fundamental discriminant related to d, that is,

$$d = m^2 f,$$

where f is not divisible by a square of odd prime and is either odd, or congruent to 8 or congruent to 12 (mod 16).

LEMMA 1. *For $d > 0$, we have*

$$\left(\frac{d}{r}\right) = \left(\frac{d}{-r}\right).$$

PROOF. Landau, *Vorlesungen über Zahlentheorie*, vol. 1, Theorem 101.

LEMMA 2. *We have*

Received by the editors December 3, 1941.

[1] Göttingen Nachrichter, 1918, pp. 30–36.

$$\sum_r \left(\frac{f}{r}\right) e^{2\pi i n r/f} = \left(\frac{f}{n}\right) f^{1/2},$$

where r runs over a complete residue system, mod f.

PROOF. Landau, loc. cit., Theorem 215.

LEMMA 3. *We have*

$$\frac{1}{A^*+1}\left|\sum_{a=1}^{A}\sum_{n=1}^{a}\left(\frac{f}{n}\right)\right| \leqq \frac{1}{2}\left(f^{1/2} - \frac{A^*+1}{f^{1/2}}\right),$$

where A^* is the least positive residue of A, mod f.

PROOF. (See Lemma 1 of the preceding paper.) We have, by Lemma 2,

$$\begin{aligned} f^{1/2}\sum_{a=1}^{A}\sum_{n=1}^{a}\left(\frac{f}{n}\right) &= \frac{1}{2} f^{1/2}\sum_{a=0}^{A}\sum_{n=-a}^{a}\left(\frac{f}{n}\right) \\ &= \frac{1}{2}\sum_{a=0}^{A}\sum_{n=-a}^{a}\sum_{r=1}^{f}\left(\frac{f}{r}\right)e^{2\pi i n r/f} \\ &= \frac{1}{2}\sum_{r=1}^{f}\left(\frac{f}{r}\right)\sum_{a=0}^{A}\sum_{n=-a}^{a} e^{2\pi i n r/f}. \end{aligned}$$

Then

$$\begin{aligned} f^{1/2}\left|\sum_{a=1}^{A}\sum_{n=1}^{a}\left(\frac{f}{n}\right)\right| &\leqq \frac{1}{2}\sum_{r=1}^{f-1}\left|\sum_{a=0}^{A}\sum_{n=-a}^{a} e^{2\pi i n r/f}\right| \\ &= \frac{1}{2}\sum_{r=1}^{f-1}\left(\frac{\sin (A+1)\pi r/f}{\sin \pi r/f}\right)^2 \\ &= \frac{1}{2}\sum_{r=1}^{f-1}\left(\frac{\sin (A^*+1)\pi r/f}{\sin \pi r/f}\right)^2 \\ &= \frac{1}{2}\sum_{r=1}^{f-1}\sum_{a=0}^{A^*}\sum_{n=-a}^{a} e^{2\pi i n r/f} \\ &= \frac{1}{2}((A^*+1)f - (A^*+1)^2), \end{aligned}$$

since

$$\sum_{r=1}^{f-1} e^{2\pi i n r/f} = \sum_{r=1}^{f} e^{2\pi i n r/f} - 1 = \begin{cases} -1 & \text{if } f \nmid n, \\ f-1 & \text{if } f \mid n. \end{cases}$$

LEMMA 4. *For any discriminant $d>0$ and $A>d^{1/2}$, we have*

$$\left|\sum_{a=1}^{A}\sum_{n=1}^{a}\left(\frac{d}{n}\right)\right| \leqq \frac{1}{2}Ad^{1/2}.$$

PROOF. It is well known that[2]

$$\left(\frac{d}{n}\right)=\left(\frac{f}{n}\right)\sum_{r\mid(m,n)}\mu(r).$$

Then

$$\begin{aligned}\sum_{a=1}^{A}\sum_{n=1}^{a}\left(\frac{d}{n}\right) &= \sum_{a=1}^{A}\sum_{n=1}^{a}\left(\frac{f}{n}\right)\sum_{r\mid(m,n)}\mu(r)\\ &= \sum_{r\mid m}\mu(r)\sum_{a=1}^{A}\sum_{n=1,\, r\mid n}^{a}\left(\frac{f}{n}\right) = \sum_{r\mid m}\mu(r)\sum_{a=1}^{A}\sum_{n=1}^{[a/r]}\left(\frac{f}{rn}\right)\\ &= \sum_{r\mid m}\mu(r)\left(\frac{f}{r}\right)\sum_{a=1}^{A}\sum_{n=1}^{[a/r]}\left(\frac{f}{n}\right).\end{aligned}$$

Then, by Lemma 2,

$$\begin{aligned}\left|\sum_{a=1}^{A}\sum_{n=1}^{a}\left(\frac{d}{n}\right)\right| &\leqq \frac{1}{2}\sum_{r\mid m}\left|\sum_{a=1}^{A}\sum_{n=1}^{[a/r]}\left(\frac{f}{n}\right)\right|\\ &\leqq \frac{1}{2}\sum_{r\mid m}r\left|\sum_{b=1}^{[A/r]}\sum_{n=1}^{b}\left(\frac{f}{n}\right)\right|\\ &\leqq \frac{1}{2}\sum_{r\mid m}r\left(\left(\left[\frac{A}{r}\right]+1\right)f^{1/2}-\frac{1}{f^{1/2}}\left(\left[\frac{A}{r}\right]+1\right)^2\right)\\ &\leqq \frac{1}{2}\sum_{r\mid m}r\cdot\frac{A}{r}f^{1/2} \leqq \frac{1}{2}Af^{1/2}m = \frac{1}{2}Ad^{1/2},\end{aligned}$$

since we have $f^{1/2}r<f^{1/2}m<A$,

$$f^{1/2}-\frac{1}{f^{1/2}}\left(\left[\frac{A}{r}\right]+1\right)^2 < f^{1/2}-\frac{1}{f^{1/2}}\cdot f = 0$$

and

$$\sum_{r\mid m}1 \leqq m.$$

LEMMA 5. *We have*

$$\sum_{n=1}^{\infty}\left(\frac{d}{n}\right)\frac{1}{n} < \frac{1}{2}\log d + 1.$$

[2] This follows from the fact that $\sum_{d\mid a}\mu(d)=0$ or 1 according as $a>1$ or $a=1$.

PROOF. For $n \geqq 1$ let

$$S(n) = \sum_{a=1}^{n} \sum_{m=1}^{a} \left(\frac{d}{m}\right),$$

and let $S(0) = S(-1) = 0$. Then we have

$$S(n) - 2S(n-1) + S(n-2) = \left(\frac{d}{n}\right), \qquad n \geqq 1,$$

and

$$\sum_{n=1}^{\infty} \left(\frac{d}{n}\right) \frac{1}{n} = \sum_{n=1}^{\infty} \{S(n) - 2S(n-1) + S(n-2)\} \frac{1}{n}$$
$$= \sum_{n=1}^{\infty} S(n) \left(\frac{1}{n} - \frac{2}{n+1} + \frac{1}{n+2}\right)$$
$$= \sum_{n=1}^{\infty} \frac{2S(n)}{n(n+1)(n+2)}.$$

We divide the series into two parts

$$S_1 = \sum_{n=1}^{A-1}, \qquad S_2 = \sum_{n=A}^{\infty}.$$

Since

$$|S(n)| \leqq \sum_{a=1}^{n} \sum_{m=1}^{a} 1 = \frac{n(n+1)}{2},$$

it follows that

$$|S_1| \leqq \sum_{n=1}^{A-1} \frac{1}{n+2}.$$

If $A > d^{1/2}$ we have by Lemma 4

$$|S_2| < \sum_{n=A}^{\infty} \frac{nd^{1/2}}{n(n+1)(n+2)} = \frac{d^{1/2}}{A+1}.$$

Hence

$$\left|\sum_{n=1}^{\infty} \left(\frac{d}{n}\right) \frac{1}{n}\right| \leqq \sum_{n=1}^{A-1} \frac{1}{n+2} + \frac{d^{1/2}}{A+1}$$
$$= \sum_{m=1}^{A-1} \frac{1}{m} - 1 - \frac{1}{2} + \frac{1}{A} + \frac{1}{A+1} + \frac{d^{1/2}}{A+1}$$
$$\leqq \log (A-1) - \frac{1}{2} + \frac{1}{A} + \frac{d^{1/2}+1}{A+1}.$$

Taking $A=[d^{1/2}]+1$ we have

$$\left|\sum_{n=1}^{\infty}\left(\frac{d}{n}\right)\frac{1}{n}\right| \leqq \log d^{1/2}-\frac{1}{2}+\frac{1}{d^{1/2}}+\frac{d^{1/2}+1}{d^{1/2}+1}$$

$$=\frac{1}{2}\log d+\frac{1}{2}+\frac{1}{d^{1/2}}<\frac{1}{2}\log d+1$$

since $d\geqq 5$.

THEOREM 1. *We have*

$$\log \epsilon < d^{1/2}((1/2)\log d+1).$$

PROOF. It is known that the number $h(d)$ of classes of non-equivalent quadratic forms with determinant $d>0$, is given by

$$h(d)=\frac{d^{1/2}}{\log \epsilon}\sum_{n=1}^{\infty}\left(\frac{d}{n}\right)\frac{1}{n}\cdot$$

Since $h(d)\geqq 1$, we have the theorem.

THEOREM 2 (Schur). *We have*

$$\log \epsilon \leqq d^{1/2}\log d.$$

PROOF. For $d>e^2$, the theorem follows from Theorem 1. If $d<e^2$, then $d=5$. Evidently $\epsilon=(3+5^{1/2})/2$ and

$$\log \epsilon < 5^{1/2}\log 5.$$

NATIONAL TSING HUA UNIVERSITY

Reprinted from the
QUARTERLY JOURNAL OF MATHEMATICS
Oxford Series 13, pp. 18–29, 1942

THE LATTICE-POINTS IN A CIRCLE

[Received 9 January 1942]

LET $R(x)$ denote the number of lattice-points inside and on the circle $u^2+v^2 = x$. It is easily proved that, as $x \to \infty$, $R(x) \sim \pi x$, and in fact that

$$R(x) = \pi x + O(x^{\alpha}) \tag{1}$$

for some values of α less than 1. It is a question of finding the lower bound, ϑ say, of the numbers α for which (1) is true. The best result hitherto obtained is that $\vartheta \leqslant \frac{15}{46}$. This was proved by Titchmarsh* in 1933. It is the purpose of the paper to prove that $\vartheta \leqslant \frac{13}{40}$. Titchmarsh's proof depends essentially on the fact that a quadratic form he uses is positive definite. In trying to sharpen the result one arrives at the difficulty that a certain quadratic form is not positive definite. But, on examination, it is found that the variables of the quadratic form are not perfectly general. For these variables so restricted we have fortunately that the values of the form are always positive.

1. Lemmas quoted from Titchmarsh's paper

LEMMA 1. *Let $a_{\mu\nu}$ be any numbers, real or complex, such that, if $s_{m,n} = \sum_{\mu=1}^{m} \sum_{\nu=1}^{n} a_{\mu\nu}$, then $|s_{m,n}| \leqslant G$ $(1 \leqslant m \leqslant M;\ 1 \leqslant n \leqslant N)$. Let $b_{m,n}$ denote real numbers, $0 \leqslant b_{m,n} \leqslant H$, and let each of the expressions*

$$b_{m,n}-b_{m,n+1}, \qquad b_{m,n}-b_{m+1,n}, \qquad b_{m,n}-b_{m+1,n}-b_{m,n+1}+b_{m+1,n+1}$$

be of constant sign for values of m and n in question. Then

$$\left|\sum_{m=1}^{M} \sum_{n=1}^{N} a_{m,n} b_{m,n}\right| \leqslant 5GH.$$

LEMMA 2. *Let $f(x,y)$ be a real function of x and y, and*

$$S = \sum\sum e^{2\pi i f(m,n)},$$

* *Proc. London Math. Soc.* (2), 38 (1935), 96–115. I must also refer to a paper of I. Vinogradow, *Bull. Acad. Sci. U.R.S.S.* 7 (1932), 313–36, in which the error-term $O(x^{17/53+\epsilon})$ is claimed. Unfortunately there seems to be an incurable mistake contained in the proof, namely, in § 3 F and G of his paper. For he states that, after a bulky calculation, he obtained the result $\sum_m \sum_n \sum \min\{P, (E)^{-1}\} \min\{P, (F)^{-1}) \ll p_1^4 p_2^4 p_3^4 M^{2+\epsilon} P^{-2}$. But this is apparently false as we see by considering the sum of those terms with $r_1 = s_1 = 0$ (actually the partial sum formed by these terms $\gg (M/P)^2 p_1^2 p_2^4 p_3^4 P^2$).

the sum being taken over the lattice points of a region D included in the rectangle $a \leqslant x \leqslant b$, $\alpha \leqslant y \leqslant \beta$. *Let*

$$S' = \sum\sum e^{2\pi i\{f(m+\mu,n+\nu)-f(m,n)\}}, \qquad S'' = \sum\sum e^{2\pi i\{f(m+\mu,n-\nu)-f(m,n)\}},$$

where μ *and* ν *are integers, and* S' *is taken over values of* m *and* n *such that both* (m,n) *and* $(m+\mu,n+\nu)$ *belong to* D; *and similarly for* S''. *Let* ρ *be a positive integer not exceeding* $b-a$, *and let* ρ' *be a positive integer not exceeding* $\beta-\alpha$. *Then*

$$S = O\left\{\frac{(b-a)(\beta-\alpha)}{(\rho\rho')^{\frac{1}{2}}}\right\}+O\left[\left\{\frac{(b-a)(\beta-\alpha)}{\rho\rho'}\sum_{\mu=1}^{\rho-1}\sum_{\nu=0}^{\rho'-1}|S'|\right\}^{\frac{1}{2}}\right]+$$

$$+O\left[\left\{\frac{(b-a)(\beta-\alpha)}{\rho\rho'}\sum_{\mu=0}^{\rho-1}\sum_{\nu=1}^{\rho'-1}|S''|\right\}^{\frac{1}{2}}\right].$$

LEMMA 2′. *If* $0 < \rho \leqslant b-a$, *then*

$$S = O\left\{\frac{(b-a)(\beta-\alpha)}{\rho^{\frac{1}{2}}}\right\}+O\left[\left\{\frac{(b-a)(\beta-\alpha)}{\rho}\sum_{\mu=1}^{\rho-1}|S'''|\right\}^{\frac{1}{2}}\right],$$

where
$$S''' = \sum\sum e^{2\pi i\{f(m+\mu,n)-f(m,n)\}}.$$

LEMMA 3. *Let* $f(x,y)$ *be a real differentiable function of* x *and* y. *Let* $f_x(x,y)$ *be a monotonic function of* x *for each value of* y *considered, and* $f_y(x,y)$ *be a monotonic function of* y *for each value of* x *considered. Let* $|f_x| \leqslant \frac{3}{4}$, $|f_y| \leqslant \frac{3}{4}$, *for* $a \leqslant x \leqslant b$, $\alpha \leqslant y \leqslant \beta$, *where* $b-a \leqslant l$, $\beta-\alpha \leqslant l$ $(l \geqslant 1)$. *Let* D *be the rectangle* $(a,b;\alpha,\beta)$, *or part of the rectangle cut off by a continuous monotonic curve. Then*

$$\sum\sum\nolimits_D e^{2\pi i f(m,n)} = \iint\limits_D e^{2\pi i f(x,y)}\,dxdy + O(l). \tag{2}$$

LEMMA 4. *Suppose that* $f(x,y)$ *is a real function of* x *and* y *with continuous partial derivatives of as many orders as may be required in the rectangle* $(a,b;\alpha,\beta)$, *and also that any curve defined by equating to zero a polynomial of given degree in these derivatives has* $O(1)$ *intersections with any other such curve, or with any straight line. Let* $b-a \leqslant l$, $\beta-\alpha \leqslant l$. *Let*

$$|f_{xx}| < AR, \qquad |f_{yy}| < AR, \qquad |f_{xy}| < AR \tag{3}$$

(A *denotes a positive absolute constant, not necessarily the same one at each occurrence) and*

$$|f_{xx}f_{yy}-f_{xy}^2| \geqslant r^2, \tag{4}$$

where $0 < r \leqslant R$, *throughout the rectangle. Let* $|f_x| \leqslant r_1$, $|f_y| \leqslant r_1$, $|f_{xxy}| \leqslant r_3$, $|f_{xyy}| \leqslant r_3$, $|f_{yyy}| \leqslant r_3$, *and let*

$$r_1 r_3 < K_1 r^2 \tag{5}$$

and

$$l r_3 < K_2 r, \tag{6}$$

where K_1 *and* K_2 *are sufficiently small constants. Then*

$$\int_a^b \int_\alpha^\beta e^{2\pi i f(x,y)}\, dx dy = O\left(\frac{1+|\log l|+|\log R|}{r}\right).$$

The lemmas 1, 2, 3, 4 correspond to the lemmas α, β, γ, ζ of Titchmarsh respectively.

2. Let

$$\Delta f(u,v) = f(u+m_1+m_2+m_3, v+n_1+n_2+n_3) - \\ - \sum f(u+m_1+m_2, v+n_1+n_2) + \sum f(u+m_1, v+n_1) - f(u,v),$$

and let

$$X = 6m_1 m_2 m_3, \qquad Y = 2\sum m_1 m_2 n_3,$$
$$Z = 2\sum m_1 n_2 n_3, \qquad W = 6n_1 n_2 n_3.$$

Then we have

$$\Delta u = \Delta v = 0, \qquad \Delta u^2 = \Delta uv = \Delta v^2 = 0,$$
$$\Delta u^3 = X, \qquad \Delta u^2 v = Y, \qquad \Delta uv^2 = Z, \qquad \Delta v^3 = W.$$

It is easy to prove that

$$\Delta(u^\lambda v^{k-\lambda})_{u=0,\, v=0} = O\{\eta^{k-3}(|X|+|Y|+|Z|+|W|)\}, \quad \text{for } 0 \leqslant \lambda \leqslant k,$$

if $m_i = O(\eta)$ and $n_i = O(\eta)$ for $i = 1, 2, 3$.

Let $|u_1| \leqslant \eta$, $|u_2| \leqslant \eta$, $\max(u,v) \geqslant L$. Formally, we have

$$\frac{\partial^2}{\partial u^2}\sqrt{\{(u+u_1)^2+(v+v_1)^2\}} = \frac{(v+v_1)^2}{\{(u+u_1)^2+(v+v_1)^2\}^{\frac{3}{2}}}$$
$$= \frac{v^2}{(u^2+v^2)^{\frac{3}{2}}}\left(1+\frac{2v_1}{v}+\frac{v_1^2}{v^2}\right)\left(1+2\frac{uu_1+vv_1}{u^2+v^2}+\frac{u_1^2+v_1^2}{u^2+v^2}\right)^{-\frac{3}{2}}$$
$$= \frac{v^2}{(u^2+v^2)^{\frac{3}{2}}}\left(1+\frac{2v_1}{v}+\frac{v_1^2}{v^2}\right)\left\{1-3\frac{uu_1+vv_1}{u^2+v^2}-\frac{3}{2}\frac{u_1^2+v_1^2}{u^2+v^2}+\right.$$
$$+\frac{15}{2}\left(\frac{uu_1+vv_1}{u^2+v^2}\right)^2+\frac{15}{2}\frac{(uu_1+vv_1)(u_1^2+v_1^2)}{(u^2+v^2)^2}-$$
$$-\frac{35}{2}\left(\frac{uu_1+vv_1}{u^2+v^2}\right)^3+\frac{15}{8}\left(\frac{u_1^2+v_1^2}{u^2+v^2}\right)^2-\frac{105}{4}\frac{(uu_1+vv_1)^2(u_1^2+v_1^2)}{(u^2+v^2)^3}+$$
$$\left.+\frac{315}{8}\left(\frac{uu_1+vv_1}{u^2+v^2}\right)^4+\dots\right\}. \tag{7}$$

Let $G(u,v) = \Delta\{\sqrt{(u^2+v^2)}\}$. Then, by (7), we have

$$G_{uu} = \frac{v^2}{(u^2+v^2)^{\frac{5}{2}}}\left\{\frac{15}{2}\frac{(X+Z)u+(Y+W)v}{u^2+v^2}-\right.$$

$$-\frac{35}{2}\frac{Xu^3+3Yu^2v+3Zuv^2+Wv^3}{(u^2+v^2)^2}-3\frac{Y+W}{v}+$$

$$\left.+15\frac{Yu^2+2Zuv+Wv^2}{v(u^2+v^2)}-3\frac{Zu+Wv}{v^2}\right\}+O\left(\frac{(|X|+|Y|+|Z|+|W|)\eta}{L^5}\right).$$

Similarly, we have

$$G_{vv} = \frac{u^2}{(u^2+v^2)^{\frac{5}{2}}}\left\{\frac{15}{2}\frac{(X+Z)u+(Y+W)v}{u^2+v^2}-\right.$$

$$-\frac{35}{2}\frac{Xu^3+3Yu^2v+3Zuv^2+Wv^3}{(u^2+v^2)^2}-3\frac{X+Z}{u}+$$

$$\left.+15\frac{Xu^2+2Yuv+Zv^2}{u(u^2+v^2)}-3\frac{Xu+Yv}{u^2}\right\}+O\left(\frac{(|X|+|Y|+|Z|+|W|)\eta}{L^5}\right)$$

and

$$G_{uv} = -\frac{uv}{(u^2+v^2)^{\frac{5}{2}}}\left\{\frac{15}{2}\frac{(X+Z)u+(Y+W)v}{u^2+v^2}-\right.$$

$$-\frac{35}{2}\frac{Xu^3+3Yu^2v+3Zuv^2+Wv^3}{(u^2+v^2)^2}-\frac{3}{2}\left(\frac{X+Z}{u}+\frac{Y+W}{v}\right)+$$

$$\left.+\frac{15}{2}\left(\frac{Xu^2+2Yuv+Zv^2}{u(u^2+v^2)}+\frac{Yu^2+2Zuv+Wv^2}{v(u^2+v^2)}\right)-3\frac{Yu+Zv}{uv}\right\}+$$

$$+O\left(\frac{(|X|+|Y|+|Z|+|W|)\eta}{L^5}\right).$$

Hence we have

$$G_{uu}G_{vv}-G_{uv}^2 = \frac{u^2v^2}{(u^2+v^2)^5}\left[\left\{\frac{15}{2}\frac{(X+Z)u+(Y+W)v}{u^2+v^2}-\right.\right.$$

$$-\frac{35}{2}\frac{Xu^3+3Yu^2v+3Zuv^2+Wv^3}{(u^2+v^2)^2}-3\frac{Y+W}{v}+$$

$$\left.+15\frac{Yu^2+2Zuv+Wv^2}{v(u^2+v^2)}-3\frac{Zu+Wv}{v^2}\right\}\times$$

$$\times\left\{\frac{15}{2}\frac{(X+Z)u+(Y+W)v}{u^2+v^2}-\frac{35}{2}\frac{Xu^3+3Yu^2v+3Zuv^2+Wv^3}{(u^2+v^2)^2}-\right.$$

$$\left.-3\frac{X+Z}{u}+15\frac{Xu^2+2Yuv+Zv^2}{u(u^2+v^2)}-3\frac{Xu+Yv}{u^2}\right\}-$$

$$-\left\{\frac{15}{2}\frac{(X+Z)u+(Y+W)v}{u^2+v^2}-\frac{35}{2}\frac{Xu^3+3Yu^2v+3Zuv^2+Wv^3}{(u^2+v^2)^2}-\right.$$

$$-\frac{3}{2}\left(\frac{X+Z}{u}+\frac{Y+W}{v}\right)+$$

$$\left.+\frac{15}{2}\left(\frac{Xu^2+2Yuv+Zv^2}{u(u^2+v^2)}+\frac{Yu^2+2Zuv+Wv^2}{v(u^2+v^2)}\right)-3\frac{Yu+Zv}{uv}\right\}^2\Bigg]+$$

$$+O\left(\frac{(X^2+Y^2+Z^2+W^2)\eta}{L^9}\right)$$

$$=\frac{u^2v^2}{(u^2+v^2)^5}\Bigg[\left(\frac{15}{2}\frac{(X+Z)u+(Y+W)v}{u^2+v^2}-\right.$$

$$\left.-\frac{35}{2}\frac{Xu^3+3Yu^2v+3Zuv^2+Wv^3}{(u^2+v^2)^2}\right)\times$$

$$\times\left(-3\frac{Xu+Yv}{u^2}-3\frac{Zu+Wv}{v^2}+6\frac{Yu+Zv}{uv}\right)+$$

$$+9\left(\frac{X+Z}{u}-5\frac{Xu^2+2Yuv+Zv^2}{u(u^2+v^2)}+\frac{Xu+Yv}{u^2}\right)\times$$

$$\times\left(\frac{Y+W}{v}-5\frac{Yu^2+2Zuv+Wv^2}{v(u^2+v^2)}+\frac{Zu+Wv}{v^2}\right)-$$

$$-\frac{9}{4}\left\{\frac{X+Z}{u}+\frac{Y+W}{v}-5\left(\frac{Xu^2+2Yuv+Zv^2}{u(u^2+v^2)}+\frac{Yu^2+2Zuv+Wv^2}{v(u^2+v^2)}\right)+\right.$$

$$\left.+2\frac{Yu+Zv}{uv}\right\}^2\Bigg]+O\left(\frac{(X^2+Y^2+Z^2+W^2)\eta}{L^9}\right)$$

$$=-\frac{3}{4(u^2+v^2)^7}Q(X,Y,Z,W)+O\left(\frac{(X^2+Y^2+Z^2+W^2)\eta}{L^9}\right),\quad\text{say.}$$

Then

$$-\tfrac{3}{4}Q(X,Y,Z,W)=-3[\tfrac{15}{2}(u^2+v^2)\{(X+Z)u+(Y+W)v\}-$$
$$-\tfrac{35}{2}(Xu^3+3Yu^2v+3Zuv^2+Wv^3)]\times$$
$$\times\{v^2(Xu+Yv)+u^2(Zu+Wv)-2uv(Yu+Zv)\}+$$
$$+9\{u(u^2+v^2)(X+Z)-5u(Xu^2+2Yuv+Zv^2)+(u^2+v^2)(Xu+Yv)\}\times$$
$$\times\{v(u^2+v^2)(Y+W)-5v(Yu^2+2Zuv+Wv^2)+(u^2+v^2)(Zu+Wv)\}-$$
$$-\tfrac{9}{4}\{v(u^2+v^2)(X+Z)+u(u^2+v^2)(Y+W)-$$
$$-5v(Xu^2+2Yuv+Zv^2)-5u(Yu^2+2Zuv+Wv^2)+$$
$$+2(u^2+v^2)(Yu+Zv)\}^2$$

$$=-3\{(-10u^3+\tfrac{15}{2}uv^2)X+(-45u^2v+\tfrac{15}{2}v^3)Y+$$
$$+(\tfrac{15}{2}u^3-45uv^2)Z+(\tfrac{15}{2}u^2v-10v^3)W\}\times$$

$$\times\{uv^2X-(2u^2v-v^3)Y-(-u^3+2uv^2)Z+u^2vW\}+$$

$$+9\{(-3u^3+2uv^2)X+(-9u^2v+v^3)Y+(u^3-4uv^2)Z\}\times$$

$$\times\{(-4u^2v+v^3)Y+(u^3-9uv^2)Z+(2u^2v-3v^3)W\}-$$

$$-\tfrac{9}{4}\{(-4u^2v+v^3)X+(-2u^3-7uv^2)Y+(-7u^2v-2v^3)Z+$$

$$+(u^3-4uv^2)W\}^2$$

$$=-\tfrac{3}{4}(8u^4+6u^2v^2+3v^4)v^2X^2-\tfrac{9}{4}(4u^6+4u^4v^2+21u^2v^4+6v^6)Y^2-$$

$$-\tfrac{9}{4}(6u^6+21u^4v^2+4u^2v^4+4v^6)Z^2-\tfrac{3}{4}(3u^4+6u^2v^2+8v^4)u^2W^2-$$

$$-\tfrac{3}{2}uv(-8u^4+4u^2v^2-3v^4)XY-\tfrac{3}{2}uv(-3u^4+4u^2v^2-8v^4)ZW+$$

$$+\tfrac{3}{2}(2u^6+20u^4v^2+9u^2v^4+6v^6)XZ+$$

$$+\tfrac{3}{2}(6u^6+9u^4v^2+20u^2v^4+2v^6)YW-$$

$$-\tfrac{3}{2}(4u^4+3u^2v^2+4v^4)uvXW+\tfrac{135}{2}u^3v^3YZ.$$

Therefore, we have

$$Q(X,Y,Z,W)$$

$$=(8u^4+6u^2v^2+3v^4)v^2X^2+3(4u^6+4u^4v^2+21u^2v^4+6v^6)Y^2+$$

$$+3(6u^6+21u^4v^2+4u^2v^4+4v^6)Z^2+(3u^4+6u^2v^2+8v^4)u^2W^2-$$

$$-2uv(8u^4-4u^2v^2+3v^4)XY-2uv(3u^4-4u^2v^2+8v^4)ZW-$$

$$-2(2u^6+20u^4v^2+9u^2v^4+6v^6)XZ-$$

$$-2(6u^6+9u^4v^2+20u^2v^4+2v^6)YW+$$

$$+2(4u^4+3u^2v^2+4v^4)uvXW-90u^3v^3YZ. \tag{8}$$

3. We put $n_1=0$. Then $W=0$. We have then

$$Q(X,Y,Z,0)$$

$$=(8u^4+6u^2v^2+3v^4)v^2X^2+3(4u^6+4u^4v^2+21u^2v^4+6v^6)Y^2+$$

$$+3(6u^6+21u^4v^2+4u^2v^4+4v^6)Z^2-2uv(8u^4-4u^2v^2+3v^4)XY-$$

$$-2(2u^6+20u^4v^2+9u^2v^4+6v^6)XZ-90u^3v^3YZ.$$

It is the object of the section to prove that

$$Q(X,Y,Z,0)\geqslant\tfrac{1}{10}\{(u^2+v^2)^2v^2X^2+(u^2+v^2)^3(Y^2+Z^2)\}.$$

Thus, for $v\geqslant u$,

$$Q(X,Y,Z,0)\geqslant\tfrac{1}{20}(u^2+v^2)^3(X^2+Y^2+Z^2).$$

Evidently we have

$$Y^2=4m_1^2(m_2n_3+m_3n_2)^2\geqslant 16m_1^2m_2m_3n_2n_3=\tfrac{4}{3}XZ. \tag{9}$$

Therefore

$$Q(X,Y,Z,0)$$

$$\geqslant (8u^4+6u^2v^2+3v^4)v^2X^2+9(u^6-2u^4v^2+\tfrac{11}{2}u^2v^4+v^6)Y^2+$$

$$+3(6u^6+21u^4v^2+4u^2v^4+4v^6)Z^2-$$

$$-2uv(8u^4-4u^2v^2+3v^4)XY-90u^3v^3YZ$$

$$\geqslant \tfrac{1}{10}\{(u^2+v^2)^2v^2X^2+(u^2+v^2)^3(Y^2+Z^2)\}+$$

$$+(\tfrac{79}{10}u^4+\tfrac{58}{10}u^2v^2+\tfrac{29}{10}v^4)v^2X^2-2uv(8u^4-4u^2v^2+3v^4)XY+$$

$$+(\tfrac{89}{10}u^4-\tfrac{183}{10}u^2v^2+\tfrac{312}{10}v^4)u^2Y^2+$$

$$+(18u^2+\tfrac{89}{10}v^2)v^4Y^2+(\tfrac{179}{10}u^2+\tfrac{627}{10}v^2)u^4Z^2-90u^3v^3YZ$$

$$\geqslant \tfrac{1}{10}\{(u^2+v^2)^2v^2X^2+(u^2+v^2)^3(Y^2+Z^2)\},$$

since

$$(79u^4+58u^2v^2+29v^4)(89u^4-183u^2v^2+312v^4)-10^2(8u^4-4u^2v^2+3v^4)^2$$

$$\geqslant 600u^8-3000u^6v^2+10000u^4v^4 \geqslant 0,$$

and

$$(180u^2+89v^2)(179u^2+627v^2)-450^2u^2v^2$$

$$\geqslant \{2\surd(180.179.89.627)+179.89+180.627-450^2)u^2v^2 \geqslant 0.$$

4. It is known that

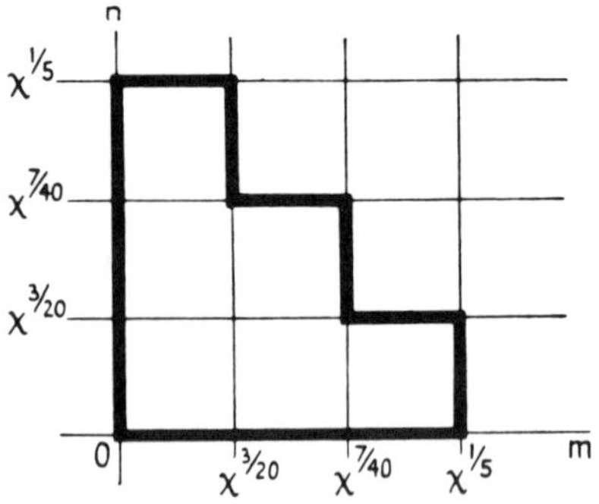

$$\int_0^x \{R(y)-\pi y\}\,dy = \frac{x}{\pi}\sum_{\nu=1}^{\infty}\frac{r(\nu)}{\nu}J_2\{2\pi\surd(\nu x)\},$$

where $r(\nu)$ is the number of solutions of the Diophantine equation

$$x^2+y^2=\nu.$$

Evidently, we have

$$\int_0^x \{R(y)-\pi y\}\,dy = \frac{4x}{\pi}\sum_{m=1}^{\infty}\sum_{n=0}^{\infty}\frac{J_2\{2\pi\surd[(m^2+n^2)x]\}}{m^2+n^2}.$$

Let C denote the region bounded by the heavy lines in the figure, and C' denote the remaining part in the first quadrant. It is easy to deduce that, if $0 < \alpha < 1$ (we shall put $\alpha = \frac{13}{40}$ later),

$$\int_x^{x\pm x^\alpha} \{R(y)-\pi y\}\,dy = 4\int_x^{x\pm x^\alpha} \sum_C\sum \frac{\sqrt{y}J_1\{2\pi\sqrt{[(m^2+n^2)y]}\}}{(m^2+n^2)^{\frac{1}{2}}}\,dy +$$

$$+\frac{4}{\pi}\left\{\sum_{C'}\sum \frac{yJ_2\{2\pi\sqrt{[(m^2+n^2)y]}\}}{m^2+n^2}\right\}_x^{x\pm x^\alpha}$$

$$= \sum\nolimits_1 + \{\sum\nolimits_2\}_x^{x\pm x^\alpha}, \quad \text{say.}$$

Now $$J_1\{2\pi\sqrt{(\nu y)}\} = \frac{\sin\{2\pi\sqrt{(\nu y)}-\frac{1}{4}\pi\}}{\pi(\nu y)^{\frac{1}{4}}} + O\left(\frac{1}{(\nu y)^{\frac{3}{4}}}\right).$$

Hence $$\sum\nolimits_1 = O\left(\int_x^{x\pm x^\alpha} |\phi(y)|y^{\frac{1}{4}}\,dy\right) + O(x^{\alpha-\frac{1}{4}}),$$

where

$$\phi(y) = \sum_C\sum \frac{e^{2\pi i\sqrt{\{(m^2+n^2)y\}}}}{(m^2+n^2)^{\frac{3}{4}}} \quad (x-x^\alpha \leqslant y \leqslant x+x^\alpha). \tag{10}$$

Similarly, we have

$$\sum\nolimits_2 = O\{y^{\frac{3}{4}}|\psi(y)|\} + O(x^{\frac{1}{4}}),$$

where

$$\psi(y) = \sum_{C'}\sum \frac{e^{2\pi i\sqrt{\{(m^2+n^2)y\}}}}{(m^2+n^2)^{\frac{5}{4}}} \quad (x-x^\alpha \leqslant y \leqslant x+x^\alpha). \tag{11}$$

If $m \leqslant x^{\frac{3}{20}}$, then

$$y^{\frac{1}{4}}\left|\sum_{\substack{C'\\ m\leqslant x^{\frac{3}{20}}}}\sum \frac{e^{2\pi i\sqrt{\{(m^2+n^2)y\}}}}{(m^2+n^2)^{\frac{3}{4}}}\right| \leqslant y^{\frac{1}{4}} \sum_{m\leqslant x^{\frac{3}{20}}} \sum_{n=1}^{\infty} \frac{1}{(m^2+n^2)^{\frac{3}{4}}}$$

$$\leqslant y^{\frac{1}{4}} \sum_{m\leqslant x^{\frac{3}{20}}} \left(\sum_{n=1}^{m} \frac{1}{n^{\frac{3}{2}}} + \sum_{n=m+1}^{\infty} \frac{1}{n^{\frac{3}{2}}}\right)$$

$$= O\left(y^{\frac{1}{4}} \sum_{m\leqslant x^{\frac{3}{20}}} m^{-\frac{1}{2}}\right) = O\left(x^{\frac{1}{4}+\frac{3}{40}}\right) = O(x^\alpha).$$

The same result holds for $n \leqslant x^{\frac{3}{20}}$.

If $m \geqslant x^{\frac{1}{5}}$, then

$$y^{\frac{3}{4}}\left|\sum_{\substack{C'\\ m>x^{\frac{1}{5}}}}\sum \frac{e^{2\pi i\sqrt{\{(m^2+n^2)y\}}}}{(m^2+n^2)^{\frac{5}{4}}}\right| \leqslant y^{\frac{3}{4}} \sum_{m>x^{\frac{1}{5}}} O\left(\frac{1}{m^{\frac{3}{2}}}\right) = O\left(x^{\frac{3}{4}-\frac{1}{10}}\right) = O(x^{2\alpha}).$$

The same result holds for $n \geqslant x^{\frac{1}{5}}$.

Let D be the region common to C and the square

$$x^{\frac{3}{20}} \leqslant m, n \leqslant x^{\frac{1}{5}}$$

and D' be the remaining part of the square. Thus

$$y^{\frac{1}{4}}|\phi(y)| = y^{\frac{1}{4}}\left|\sum_D\sum \frac{e^{2\pi i\sqrt{\{(m^2+n^2)y\}}}}{(m^2+n^2)^{\frac{3}{4}}}\right| + O(x^{\alpha}) \tag{12}$$

and

$$y^{\frac{3}{4}}|\psi(y)| = y^{\frac{3}{4}}\left|\sum_{D'}\sum \frac{e^{2\pi i\sqrt{\{(m^2+n^2)y\}}}}{(m^2+n^2)^{\frac{5}{4}}}\right| + O(x^{2\alpha}). \tag{13}$$

5. Now we consider a sum of the form

$$S = \sum_{m=M}^{M'}\sum_{n=N}^{N'} e^{2\pi i f(m,n)}, \qquad M \leqslant M' \leqslant 2M;\, N \leqslant N' \leqslant 2N$$

where $f(m,n) = \sqrt{\{(m^2+n^2)y\}}$. Let R denote the domain of summation and let $L = \max(m,n)$. Now the terms considered satisfy

$$x^{\frac{3}{20}} < L < x^{\frac{1}{5}}. \tag{14}$$

If $M'-M < x^{\frac{1}{8}}$, we have $S = O(Lx^{\frac{1}{8}})$, so that, by Lemma 1,

$$x^{\frac{1}{4}}\sum_R\sum \frac{e^{2\pi i f(m,n)}}{(m^2+n^2)^{\frac{3}{4}}} = O\left(x^{\frac{1}{4}+\frac{1}{8}}L^{-\frac{1}{2}}\right) = O\left(x^{\frac{1}{4}+\frac{1}{8}-\frac{3}{40}}\right) = O(x^{\alpha}) \tag{15}$$

and

$$x^{\frac{3}{4}}\sum_R\sum \frac{e^{2\pi i f(m,n)}}{(m^2+n^2)^{\frac{5}{4}}} = O\left(x^{\frac{3}{4}+\frac{1}{8}}L^{-\frac{3}{2}}\right) = O\left(x^{\frac{3}{4}+\frac{1}{8}-\frac{9}{40}}\right) = O(x^{2\alpha}). \tag{16}$$

A similar result holds if $N'-N < x^{\frac{1}{8}}$. We may therefore suppose that

$$M'-M > x^{\frac{1}{8}}, \qquad N'-N > x^{\frac{1}{8}}.$$

Applying Lemma 2′ once and Lemma 2 twice, we have

$$\begin{aligned} S &= O(L^2\rho^{-\frac{1}{2}}) + O\left[L\rho^{-\frac{1}{2}}\left(\sum_{m_1=1}^{\rho-1}|S_1|\right)^{\frac{1}{2}}\right] \\ &= O(L^2\rho^{-\frac{1}{2}}) + O\left[L^{\frac{3}{2}}\rho^{-1}\sum_{i=1}^{2}\left\{\sum_{m_1=1}^{\rho-1}\left(\sum_{m_2=1}^{\rho-1}\sum_{m_3=0}^{\rho-1}|S_2^{(i)}|\right)^{\frac{1}{2}}\right\}^{\frac{1}{2}}\right] \\ &= O(L^2\rho^{-\frac{1}{2}}) + O\left[L^{\frac{7}{4}}\rho^{-\frac{3}{2}}\sum_{i=1}^{4}\left\{\sum_{m_1=1}^{\rho-1}\left[\sum_{m_2=1}^{\rho-1}\sum_{n_2=0}^{\rho-1}\left(\sum_{m_3=1}^{\rho^2-1}\sum_{n_3=0}^{\rho^2-1}|S_3^{(i)}|\right)^{\frac{1}{2}}\right]^{\frac{1}{2}}\right\}^{\frac{1}{2}}\right], \end{aligned} \tag{17}$$

provided that

$$1 \leqslant \rho^2 \leqslant \tfrac{1}{2}x^{\frac{1}{8}}, \tag{18}$$

where

$$S_3^{(1)} = S_3 = \sum\sum e^{2\pi i g(m,n)}, \qquad g(m,n) = \sqrt{x}\, G(m,n)$$

with $W = 0$, and $S_3^{(2)}$, $S_3^{(3)}$, $S_3^{(4)}$ are corresponding sums with $(-m_3, n_3)$,

$(m_3, -n_3)$, $(-m_3, -n_3)$ for (m_3, n_3) respectively. We may suppose that $v > u$ (by symmetry). By § 3, we have

$$g_{uu}g_{vv}-g_{uv}^2 \geqslant \frac{x}{L^8}(X^2+Y^2+Z^2)+O\left(\frac{(X^2+Y^2+Z^2)x\eta}{L^9}\right)$$

$$\geqslant A\frac{x}{L^8}(X^2+Y^2+Z^2),$$

where A is a certain constant, provided that

$$\rho = o\left(L^{\frac{1}{2}}\right). \tag{19}$$

In fact, since $\eta = O(\rho^2)$,

$$\frac{(X^2+Y^2+Z^2)x\eta}{L^9} = O\left(\frac{x(X^2+Y^2+Z^2)\rho^2}{L^9}\right) = o\left(\frac{x(X^2+Y^2+Z^2)}{L^8}\right).$$

6. Since $X = O(\rho^4)$, $Y = O(\rho^4)$, $Z = O(\rho^4)$, we have

$$g_{uu} = O\left(\frac{x^{\frac{1}{2}}\rho^4}{L^4}\right)+O\left(\frac{x^{\frac{1}{2}}\rho^4\eta}{L^5}\right) = O\left(\frac{x^{\frac{1}{2}}\rho^4}{L^4}\right),$$

since $\eta = O(\rho^2) = O(L)$. A similar result holds for g_{vv} and g_{uv}. Hence, if

$$l = aL^4x^{-\frac{1}{2}}\rho^{-4}$$

with a sufficiently small a, the variation of g_u and g_v in a square of side l is less than $\frac{1}{2}$. Suppose the region of summation S_3 divided into such squares or parts of such squares. Then to each square correspond integers μ, ν such that, if

$$h(u,v) = g(u,v)-\mu u-\nu v,$$

then $|h_u| \leqslant \frac{3}{4}$, $|h_v| \leqslant \frac{3}{4}$. Hence, by Lemma 3, for each square

$$\sum\sum e^{2\pi i g(m,n)} = \iint e^{2\pi i h(u,v)}\,dudv + O(l).$$

By § 5, we can take (in Lemma 4),

$$r^2 = A\frac{x}{L^8}(X^2+Y^2+Z^2).$$

Also $$r_1 = O(1), \qquad r_3 = O\left(x^{\frac{1}{2}}\rho^4L^{-5}\right),$$

so that the condition (5) of Lemma 4 is satisfied if

$$L^3\rho^4 < K_1Ax^{\frac{1}{2}}(X^2+Y^2+Z^2). \tag{20}$$

Lemma 4 also requires that $lr_3 < K_2r$,

i.e. $$L^6 < K_2(X^2+Y^2+Z^2)x. \tag{21}$$

Since $X^2+Y^2+Z^2 = O(\rho^8)$, (21) is satisfied if (20) is satisfied. Hence, if (20) holds, Lemma 4 gives

$$\iint e^{2\pi i h(u,v)}\,du\,dv = O\left(\frac{L^4 \log x}{x^{\frac{1}{2}}(X^2+Y^2+Z^2)^{\frac{1}{2}}}\right),$$

assuming that $L = O(x^A)$.

The number of such terms is $O(L^2/l^2)$, provided that $l \leqslant L$. Hence

$$S_3 = O\left(\frac{L^6 \log x}{l^2 x^{\frac{1}{2}}(X^2+Y^2+Z^2)^{\frac{1}{2}}}\right)+O\left(\frac{L^2}{l}\right)$$

$$= O\left(\frac{x^{\frac{1}{2}}\rho^8 \log x}{L^2(X^2+Y^2+Z^2)^{\frac{1}{2}}}\right) = O\left(\frac{x^{\frac{1}{2}}\rho^8 \log x}{L^2 m_1 m_2 m_3}\right), \tag{22}$$

with the restriction (20). The result, of course, holds for $S_3^{(i)}$ $(i = 2, 3, 4)$.

7. Substituting (22) in (17), we have, for $v \geqslant u$,

$$O\left(\frac{L^{\frac{7}{4}}}{\rho^{\frac{3}{2}}}\left\{\sum_{m_1=1}^{\rho-1}\left[\sum_{m_2=1}^{\rho-1}\sum_{n_2=0}^{\rho-1}\left(\sum_{m_3=1}^{\rho^2-1}\sum_{n_3=0}^{\rho^2-1}|S_3|\right)^{\frac{1}{2}}\right]^{\frac{1}{2}}\right\}^{\frac{1}{2}}\right)$$

$$= O\left(\frac{L^{\frac{7}{4}}}{\rho^{\frac{3}{2}}}\left(\frac{x^{\frac{1}{2}}\rho^8 \log x}{L^2}\right)^{\frac{1}{8}}\left\{\sum_{m_1=1}^{\rho-1}\left[\sum_{m_2=1}^{\rho-1}\sum_{n_2=0}^{\rho-1}\left(\sum_{m_3=1}^{\rho^2-1}\sum_{m_3=0}^{\rho^2-1}\frac{1}{m_1 m_2 m_3}\right)^{\frac{1}{2}}\right]^{\frac{1}{2}}\right\}^{\frac{1}{2}}\right)$$

$$= O\left\{L^{\frac{3}{2}}\rho^{-\frac{1}{2}}x^{\frac{1}{16}}(\log x)^{\frac{1}{8}}.\rho(\log \rho)^{\frac{1}{8}}\right\}$$

$$= O\left\{L^{\frac{3}{2}}\rho^{\frac{1}{2}}x^{\frac{1}{16}}(\log x)^{\frac{1}{4}}\right\},$$

provided that (20) holds.

Next we consider the sum of those terms which do not satisfy (20); we have the inequality

$$X^2+Y^2+Z^2 = O\left(L^3\rho^4 x^{-\frac{1}{2}}\right).$$

Consequently we have $m_1 m_2 m_3 = O\left(L^3\rho^4 x^{-\frac{1}{2}}\right)$, $m_1 n_2 n_3 = O\left(L^3\rho^4 x^{-\frac{1}{2}}\right)$. Since $S_3 = O(L^2)$, the terms for which (20) is not satisfied contribute

$$O\left(L^{\frac{7}{4}}\rho^{-\frac{3}{2}}\left\{\sum_{m_1=1}^{\rho-1}\left[\sum_{m_2=1}^{\rho-1}\sum_{n_2=1}^{\rho-1}\left(\sum_{m_3=O(L^{\frac{3}{2}}\rho^2 x^{-\frac{1}{4}}m_1^{-1}m_2^{-1})}\;\sum_{n_3=O(L^{\frac{3}{2}}\rho^2 x^{-\frac{1}{4}}m_1^{-1}n_2^{-1})} 1\right)^{\frac{1}{2}}\right]^{\frac{1}{2}}\right\}^{\frac{1}{2}}\right)$$

$$= O\left(L^{\frac{7}{4}}\rho^{-\frac{3}{2}}\left\{\sum_{m_1=1}^{\rho-1}\left[\sum_{m_2=1}^{\rho-1}\sum_{n_2=1}^{\rho-1} L^{\frac{3}{2}}\rho^2 x^{-\frac{1}{4}}m_1^{-1}m_2^{-\frac{1}{2}}n_2^{-\frac{1}{2}}\right]^{\frac{1}{2}}\right\}^{\frac{1}{2}}\right)$$

$$= O\left(L^{\frac{17}{8}}\rho^{-1}x^{-\frac{1}{16}}\left[\sum_{m_1=1}^{\rho-1}\left(\sum_{m_2=1}^{\rho-1}\sum_{n_2=1}^{\rho-1}\frac{1}{(m_2 n_2)^{\frac{1}{2}}}\right)^{\frac{1}{2}}\frac{1}{m_1^{\frac{1}{2}}}\right]^{\frac{1}{2}}\right)$$

$$= O\left(L^{\frac{17}{8}}\rho^{-1}x^{-\frac{1}{16}}\rho^{\frac{1}{2}}\right) = O\left(L^{\frac{17}{8}}\rho^{-\frac{1}{2}}x^{-\frac{1}{16}}\right).$$

(The terms with $n_2 = 0$ contribute an insignificant order.) A similar result holds for $v < u$. Therefore

$$S = O\left(L^2\rho^{-\frac{1}{2}}\right)+o\left\{L^{\frac{3}{2}}\rho^{\frac{1}{2}}x^{\frac{1}{16}}(\log x)^{\frac{1}{4}}\right\}+O\left(L^{\frac{17}{8}}\rho^{-\frac{1}{2}}x^{-\frac{1}{16}}\right).$$

The first two terms are of the same form if

$$\rho = \left[L^{\frac{1}{2}}x^{-\frac{1}{16}}(\log x)^{-\frac{1}{4}}\right],$$

and, if this is a permissible value of ρ, we have

$$S = O\left\{L^{\frac{7}{4}}x^{\frac{1}{32}}(\log x)^{\frac{1}{8}}\right\}+O\left\{L^{\frac{15}{8}}x^{-\frac{1}{32}}(\log x)^{\frac{1}{8}}\right\} = O\left\{L^{\frac{7}{4}}x^{\frac{1}{32}}(\log x)\right\}^{\frac{1}{8}}. \quad (23)$$

Now we shall verify all the conditions. The condition (18)

$$1 \leqslant \rho^2 = Lx^{-\frac{1}{8}}(\log x)^{-\frac{1}{2}} \leqslant \tfrac{1}{2}x^{\frac{1}{8}}$$

can be written $\qquad x^{\frac{1}{8}}(\log x)^{\frac{1}{2}} \leqslant L \leqslant x^{\frac{1}{4}}(\log x)^{\frac{1}{2}}.$

This is always satisfied, since $\frac{1}{8} < \frac{3}{20} < \frac{1}{5} < \frac{1}{4}$. The condition (19) is

$$L^{\frac{1}{2}}x^{-\frac{1}{16}}(\log x)^{-\frac{1}{4}}-1 \leqslant \rho = o\left(L^{\frac{1}{2}}\right).$$

This is always satisfied.

8. By Lemma 1 and (23), we have

$$\sum_R\sum \frac{e^{2\pi i\sqrt{\{(m^2+n^2)y\}}}}{(m^2+n^2)^{\frac{3}{4}}} = O\left\{L^{\frac{7}{4}-\frac{3}{2}}x^{\frac{1}{32}}(\log x)^{\frac{1}{8}}\right\} \quad (24)$$

and

$$\sum_R\sum \frac{e^{2\pi i\sqrt{\{(m^2+n^2)y\}}}}{(m^2+n^2)^{\frac{5}{4}}} = O\left\{L^{\frac{7}{4}-\frac{5}{2}}x^{\frac{1}{32}}(\log x)^{\frac{1}{8}}\right\}. \quad (25)$$

Now we divide the sum

$$\sum_D\sum \frac{e^{2\pi i\sqrt{\{(m^2+n^2)y\}}}}{(m^2+n^2)^{\frac{3}{4}}} = \sum_{p=1}^{P}\sum_{q=1}^{Q}\left\{\sum_{2^{p-1}}^{2^p}\sum_{2^{q-1}}^{2^q} \frac{e^{2\pi i\sqrt{\{(m^2+n^2)y\}}}}{(m^2+n^2)^{\frac{3}{4}}}\right\}.$$

By (12) and (24) we have, taking $L_0 = x^{\frac{7}{40}}$,

$$\begin{aligned} y^{\frac{1}{4}}|\phi(y)| &= O\left(x^{\frac{1}{4}}\sum_{p=1}^{P}\sum_{q=1}^{Q}\{\max(2^p, 2^q)\}^{\frac{1}{4}}x^{\frac{1}{32}}(\log x)^{\frac{1}{8}}\right)+O(x^{\alpha}) \\ &= O\left\{x^{\frac{9}{32}}L_0^{\frac{1}{4}}(\log x)^{\frac{9}{8}}\right\}+O(x^{\alpha}) \\ &= O\{x^{\alpha}(\log x)^{\frac{9}{8}}\}. \end{aligned}$$

Similarly, by (13) and (25), we have

$$\begin{aligned} y^{\frac{3}{4}}|\psi(y)| &= O\left(x^{\frac{3}{4}}\sum_{p=1}^{P}\sum_{q=1}^{Q}\{\max(2^p, 2^q)\}^{-\frac{3}{4}}x^{\frac{1}{32}}(\log x)^{\frac{1}{8}}\right)+O\left(x^{\frac{1}{4}+\alpha}\right) \\ &= O\left\{x^{\frac{26}{40}}(\log x)^{\frac{9}{8}}\right\}+O\left(x^{\frac{1}{4}+\alpha}\right) \\ &= O\left\{x^{2\alpha}(\log x)^{\frac{9}{8}}\right\}. \end{aligned}$$

Thus we have

$$\int_x^{x\pm x^{\alpha}} \{R(y)-\pi y\}\,dy = O\left\{x^{2\alpha}(\log x)^{\frac{9}{8}}\right\}.$$

Hence, in the usual way, we deduce easily

$$R(x) = \pi x+O\left\{x^{\frac{13}{40}}(\log x)^{\frac{9}{8}}\right\}.$$

Reprinted from the
TRANSACTIONS OF THE AMERICAN MATHEMATICAL SOCIETY
Vol. 56, No. 3, pp. 537–546, 1944

ON THE DISTRIBUTION OF QUADRATIC NON-RESIDUES AND THE EUCLIDEAN ALGORITHM IN REAL QUADRATIC FIELDS. I

BY
LOO-KENG HUA

1. **Introduction.** One of the aims of this paper is to establish an explicit upper bound for the least quadratic non-residue, mod p. The bound is not the best(1) which the author can obtain. The author gives such a result owing to the following facts: in the present procedure we may adopt some known results due to Rosser(2) and it is sufficient to establish some typical results in the study of the E.A. (abbreviation of Euclidean algorithm) of real quadratic fields(3).

As to the results in the study of the E.A., we have the following theorem.

THEOREM. *For $d>e^{250}$, there is no E.A. in the quadratic field $R(d^{1/2})$, where d is a square-free integer.*

There are three ways to sharpen the result, (i) by means of Euler's summation formula to improve an estimate of a sum, (ii) reconsideration of the estimate of certain character sums, and (iii) by means of higher order "average" of Riemann-Mangoldt's formula to smooth some results concerning distribution of primes(4).

2. **Lemmas quoted from Rosser's paper.**

LEMMA 1. *Let*

$$\vartheta(x) = \sum_{p\leqq x} \log p$$

Presented to the Society, October 28, 1944; received by the editors November 29, 1943.

(1) A better result has been obtained, for example, we may have $d>e^{160}$ in the theorem. But the proof of it is at least ten times more difficult than the present one.

(2) Amer. J. Math. vol. 63 (1941) pp. 211–232.

(3) As to a detailed description of the history of this problem, see a paper by A. Brauer, Amer. J. Math. vol. 62 (1940) pp. 697–713.

(4) When the paper mentioned in footnote 3 appeared, it was unknown only in the following cases whether the E.A. exists or not:

I. $d=p$ where p is a prime of form $8n+1$ or $p=61$ and 109.

II. $d=p_1p_2\equiv 1$ (mod 24) where p_1 and p_2 are primes and $p_1\equiv p_2\equiv 3$ (mod 4).

In both cases it was known that the algorithm does not exist if d is sufficiently large. But in the meantime it was proved by Rédei (*Über den Euklidischen Algorithmus in reellquadratischen Zahlkörpern*, Mat Fiz. Lapok vol. 47 (1940) pp. 78–90) that the algorithm does not exist in the case II. The paper of Rédei was unknown to the author; therefore he considered the cases I and II in the original version of this paper. But the case II is now without any interest. In order to accelerate the publishing under the present conditions this paper was changed a little without the knowledge of the author such that only the case I is considered. A. Brauer.

where p runs over all primes not greater than x. Then we have, for $x \geqq 1$,

$$\vartheta(x) < (1 + 0.0376)x$$

and, for $x \geqq 51^2$, $\vartheta(x) > (1-0.0393)x$.

Proof. (1) By (10) of Rosser, we have, for $x \geqq e^{13.8}$, $\vartheta(x) < (1+0.0376)x$. As to $x < e^{13.8}$, we have $\vartheta(x) < x < (1+0.0376)x$ by Theorem 2 of Rosser.

(2) By (10) of Rosser, we have, for $x \geqq e^{13.8}$, $\vartheta(x) > (1-0.0393)x$. For $71^2 \leqq x < e^{13.8}$, we have, by Theorem 7 of Rosser,

$$\vartheta(x) > x - 2.78x^{1/2} > (1 - 0.0393)x.$$

For $51^2 \leqq x < 71^2$, we have, by Theorem 5 of Rosser,

$$\vartheta(x) > x - 2x^{1/2} > (1 - 0.0393)x.$$

LEMMA 2. *Let*

$$\pi(x) = \sum_{p \leqq x} 1, \qquad li\ x = \lim_{\epsilon \to 0} \left(\int_0^{1-\epsilon} + \int_{1+\epsilon}^{x} \right) \frac{dy}{\log y} \cdot$$

Then, for $x \geqq 2$, we have $\pi(x) < (1+0.0376)(li\ x + 1.85)$, and, for $x \geqq 51^2$, $\pi(x) > (1-0.0393) li\ x - 1.7$.

$$\pi(x) > (1 - 0.0393) li\ x - 1.7.$$

Proof. We have the identity

$$\pi(x) = \frac{\vartheta(x)}{\log x} + \int_2^x \frac{\vartheta(y)dy}{y \log^2 y} \cdot$$

(1) By Lemma 1, we have

$$\begin{aligned} \pi(x) &< 1.0376 \left(\frac{x}{\log x} + \int_2^x \frac{dy}{\log^2 y} \right) \\ &= 1.0376 \left(\int_0^x \frac{dy}{\log y} + \frac{2}{\log 2} - \int_0^2 \frac{dy}{\log y} \right) \\ &< 1.0376(li\ x + 1.85), \end{aligned}$$

using the value $li\ 2 = 1.04$.

(2) By the identity, we have, for $x \geqq K \geqq 51^2$,

$$\begin{aligned} \pi(x) - \pi(K) &= \frac{\vartheta(x)}{\log x} - \frac{\vartheta(K)}{\log K} + \int_K^x \frac{\vartheta(y)dy}{y(\log y)^2} \\ &> \frac{(1 - 0.0393)x}{\log x} - \frac{\vartheta(K)}{\log K} + (1 - 0.0393) \int_K^x \frac{dy}{(\log y)^2} \\ &= (1 - 0.0393) \left(li\ x + \frac{K}{\log K} - li\ K \right) - \frac{\vartheta(K)}{\log K} \cdot \end{aligned}$$

Thus

$$\pi(x) > (1 - 0.0393)(li\, x + K/\log K - li\, K) - (\vartheta(K)/\log K - \pi(K)).$$

Taking $K=51^2$, we have the lemma, since

$$\vartheta(K) = 2519.887, \qquad \pi(K) = 378, \qquad li\,(K) = 392.48.$$

LEMMA 3. *For* $x \geqq 2$,

$$\vartheta(x)/x \geqq 0.3465735.$$

Proof. For $x \geqq 16$, this follows from Lemma 1 and Rosser's Theorem 6. For $x \leqq 16$, the lemma is proved by the direct verifications

$$\vartheta(2)/2 = 0.3465735, \qquad \vartheta(3)/3 > \vartheta(4)/4 \geqq 0.44794,$$
$$\vartheta(5)/5 > \vartheta(6)/6 = 0.56686,$$
$$\vartheta(7)/7 > \vartheta(8)/8 > \vartheta(9)/9 > \vartheta(10)/10 = 0.53471,$$
$$\vartheta(11)/11 > \vartheta(12)/12 = 0.64542,$$
$$\vartheta(13)/13 > \vartheta(14)/14 > \vartheta(15)/15 > \vartheta(16)/16 = 0.64437.$$

3. A lemma concerning series.

LEMMA 4. *For* $q < A$,

$$\sum_{\nu=1}^{A/q} li \frac{A}{\nu} \leqq A \log \frac{\log A}{\log q} + \frac{A}{q} li\, q.$$

Proof. Since $d\, li(A/x)/dx \leqq 0$, we have

$$\sum_{\nu=1}^{A/q} li \frac{A}{\nu} \leqq \int_1^{A/q} li \frac{A}{x} dx + li\, A = \int_1^{A/q} dx \int_0^{A/x} \frac{dy}{\log y} + li\, A$$
$$= (A/q) li\, q + A \log\log A - A \log\log q.$$

REMARK. The inequality in the lemma may be sharpened by means of Euler's summation formula.

4. Lemmas concerning character sums.

LEMMA 5. *Let p be a prime and $p \equiv 1$ (mod 4). Then, for $A < p$, we have*

$$\sum_{a=1}^{A} \sum_{n=1}^{a} \left(\frac{n}{p}\right) \leqq \frac{1}{2} A p^{1/2}$$

where $(\frac{n}{p})$ is Legendre's symbol.

Proof. We may assume that $p < (A+1)^2$. For otherwise, we have

$$\left| \sum_{a=1}^{A} \sum_{n=1}^{a} \left(\frac{n}{p}\right) \right| \leqq \sum_{a=1}^{A} \sum_{n=1}^{a} 1 = \frac{1}{2} A(A+1) \leqq \frac{1}{2} A p^{1/2}.$$

It is well known that

$$\sum_{r=1}^{p}\left(\frac{r}{p}\right)e^{2\pi irn/p}=\left(\frac{n}{p}\right)p^{1/2}.$$

We have

$$p^{1/2}\sum_{a=1}^{A}\sum_{n=1}^{a}\left(\frac{n}{p}\right)=\frac{1}{2}p^{1/2}\sum_{a=0}^{A}\sum_{n=-a}^{a}\left(\frac{n}{p}\right)=\frac{1}{2}\sum_{a=0}^{A}\sum_{n=-a}^{a}\sum_{r=1}^{p}\left(\frac{r}{p}\right)e^{2\pi irn/p}$$

$$=\frac{1}{2}\sum_{r=1}^{p}\left(\frac{r}{p}\right)\sum_{a=0}^{A}\sum_{n=-a}^{a}e^{2\pi irn/p}$$

$$=\frac{1}{2}\sum_{r=1}^{p}\left(\frac{r}{p}\right)\frac{\sin^2\pi r(A+1)/p}{\sin^2\pi r/p}.$$

Therefore, we obtain

$$p^{1/2}\left|\sum_{a=1}^{A}\sum_{n=1}^{a}\left(\frac{n}{p}\right)\right|\leqq\frac{1}{2}\sum_{r=1}^{p-1}\frac{\sin^2\pi r(A+1)/p}{\sin^2\pi r/p}=\frac{1}{2}\sum_{r=1}^{p-1}\sum_{a=0}^{A}\sum_{n=-a}^{a}e^{2\pi irn/p}$$

$$=p(A+1)/2-(A+1)^2/2\leqq pA/2.$$

LEMMA 6. *Let $r_1, r_2, \cdots, r_s$ be s distinct primes different from p. Then*

$$\left|\sum_{a=1}^{A}\sum_{n=1,(n,r_1r_2\cdots r_s)=1}^{a}\left(\frac{n}{p}\right)\right|\leqq 2^{s-1}Ap^{1/2}.$$

Proof. The sum may be written as

$$\sum_{a=1}^{A}\sum_{n=1}^{a}\left(\frac{n}{p}\right)-\sum_{\nu=1}^{s}\sum_{a=1}^{A}\sum_{n=1,r_\nu|n}^{a}\left(\frac{n}{p}\right)+\sum_{1\leqq\nu<\mu\leqq s}\sum_{a=1}^{A}\sum_{n=1,r_\nu r_\mu|n}^{a}\left(\frac{n}{p}\right)-\cdots+\cdots.$$

There are 2^s sums each of the form

$$\sum_{a=1}^{A}\sum_{n=1,m|n}^{a}\left(\frac{n}{p}\right).$$

Now we have

$$\left|\sum_{a=1}^{A}\sum_{n=1,m|n}^{a}\left(\frac{n}{p}\right)\right|=\left|\sum_{a=1}^{A}\sum_{\lambda=1}^{[a/m]}\left(\frac{m\lambda}{p}\right)\right| \quad (5)$$

$$=\left|\sum_{a=1}^{A}\sum_{\lambda=1}^{[a/m]}\left(\frac{\lambda}{p}\right)\right|\leqq m\left|\sum_{b=1}^{A/m}\sum_{\lambda=1}^{b}\left(\frac{\lambda}{p}\right)\right|$$

$$\leqq m\frac{A}{m}p^{1/2}/2=Ap^{1/2}/2$$

by Lemma 5. Thus we have the lemma.

(5) $[x]$ denotes the integral part of x.

LEMMA 7. *Let r_1, r_2 and r_3 be the least three positive primes which are quadratic non-residues* mod p. *Then*

$$r_1 \leqq p^{1/2}, \qquad r_2 \leqq \frac{2}{1-\dfrac{1}{r_1}}\, p^{1/2}, \qquad r_3 \leqq \frac{4}{\left(1-\dfrac{1}{r_1}\right)\left(1-\dfrac{1}{r_2}\right)}\, p^{1/2}.$$

Proof. The first inequality(6) follows immediately from Lemma 5, for otherwise, taking $A = p^{1/2}$,

$$\frac{1}{2}\, p \geqq \sum_{a=1}^{A} \sum_{n=1}^{a} \left(\frac{n}{p}\right) = \sum_{a=1}^{A} \sum_{n=1}^{a} 1 = \sum_{a=1}^{A} a = \frac{1}{2} A(A+1);$$

this is impossible.

We have, by Lemma 6, with $A = r_2 - 1$,

$$\sum_{a=1}^{A} \sum_{n=1,\,(n,r_1)=1}^{a} 1 \leqq A p^{1/2}.$$

Consequently, we have

$$\left(1 - \frac{1}{r_1}\right) \sum_{a=1}^{A} a \leqq A p^{1/2},$$

that is,

$$(1 - 1/r_1) A(A+1)/2 \leqq A p^{1/2},$$

$$r_2 \leqq \frac{2}{1 - 1/r_1}\, p^{1/2}.$$

The third inequality follows similarly, since

$$\left[\frac{a}{r_1}\right] + \left[\frac{a}{r_2}\right] - \left[\frac{a}{r_1 r_2}\right] \leqq \frac{a}{r_1} + \frac{a}{r_2} - \frac{a}{r_1 r_2}.$$

5. The growth of the least quadratic non-residue.

LEMMA 8 (VINOGRADOV)(7). *Let $q_1, \cdots, q_s$ be all the primes not exceeding A which are quadratic non-residues* mod p. *Then, we have*

$$\frac{1}{2} \sum_{n=1}^{A} \left(1 - \left(\frac{n}{p}\right)\right) - \frac{1}{2} \sum_{n=1,\,(n,p)\neq 1}^{A} 1 \leqq \sum_{\nu=1}^{s} \left(\frac{A}{q_\nu}\right).$$

Proof. The left-hand side is the number of non-residues $n \leqq A$. Evidently each such n is divisible by one of the q's.

(6) See A. Brauer, *Über den kleinsten quadratischen Nichtrest*, Math. Zeit. vol. 33 (1931) pp. 161–176.

(7) Trans. Amer. Math. Soc. vol. 29 (1927) pp. 216–226.

LEMMA 9. *Under the same assumption as in Lemma 8, we have*

$$\frac{1}{2}\sum_{a=1}^{A}\sum_{n=1}^{a}\left(1-\left(\frac{n}{p}\right)\right)-\frac{1}{2}\sum_{a=1}^{A}\sum_{n=1}^{a}1 \leqq \sum_{a=1}^{A}\sum_{q_1\leqq q_\nu\leqq a}\left[\frac{a}{q_\nu}\right].$$

Proof. Summing up the formula in Lemma 7, we have the above lemma immediately.

LEMMA 10. *We have*

$$\sum_{q_1\leqq q\leqq A}\left[\frac{A}{q}\right]=\sum_{\nu=1}^{[A/q_1]}\pi\left(\frac{A}{\nu}\right)-\left[\frac{A}{q_1}\right](\pi(q_1)-1)$$

where q runs over all primes satisfying the inequality

$$q_1 \leqq q \leqq A.$$

Proof. We have

$$\begin{aligned}\sum_{q_1\leqq q\leqq A}\left[\frac{A}{q}\right] &= \sum_{A\geqq q>A/2}1+\sum_{A/2\geqq q>A/3}2+\cdots+\sum_{A/[A/q_1]\geqq q\geqq q_1}\left[\frac{A}{q_1}\right]\\ &=\pi(A)-\pi(A/2)+2(\pi(A/2)-\pi(A/3))+\cdots\\ &\quad+\left[\frac{A}{q_1}\right]\left(\pi\left(A\Big/\left[\frac{A}{q_1}\right]\right)-\pi(q_1)+1\right)\\ &=\sum_{\nu=1}^{[A/q_1]}\pi\left(\frac{A}{\nu_1}\right)-\left[\frac{A}{q_1}\right](\pi(q_1)-1).\end{aligned}$$

LEMMA 11. *We have, for $q<A$,*

$$\sum_{a=1}^{A}\sum_{\nu=1}^{[a/q]}\pi\left(\frac{a}{\nu}\right)<1.0376\times A(A+1)/2\left(\log\frac{\log A}{\log q}+\frac{li\, q}{q}+\frac{1.85}{q}\right).$$

Proof. By Lemmas 2 and 4, we have, for $q<a<A$,

$$\begin{aligned}\sum_{\nu=1}^{a/q}\pi\left(\frac{a}{\nu}\right) &< 1.0376\sum_{\nu=1}^{a/q}\left(li\frac{a}{\nu}+1.85\right)\\ &< 1.0376\left(a\log\frac{\log a}{\log q}+\frac{a}{q}li\, q+1.85\frac{a}{q}\right)\\ &< 1.0376\left(a\log\frac{\log A}{\log q}+a\frac{li\, q}{q}+1.85\frac{a}{q}\right).\end{aligned}$$

The inequality holds evidently for $q>a$. Thus

$$\sum_{a=1}^{A}\sum_{\nu=1}^{a/q}\pi\left(\frac{a}{\nu}\right)<\frac{1.0376}{2}A(A+1)\left(\log\frac{\log A}{\log q}+\frac{li\, q}{q}+\frac{1.85}{q}\right).$$

THEOREM 1. *Let q_1 be the least quadratic non-residue*, mod p. *Then, for $p \geqq e^{250}$, we have*

$$q_1 \leqq (60p^{1/2})^{0.625}.$$

Proof. (1) For $q_1 \leqq e^{80}$, we have

$$q_1 < (60e^{125})^{0.625} \leqq (60p^{1/2})^{0.625}.$$

(2) We suppose that $q_1 > e^{80}$. By Lemmas 9, 10, and 11, we have

$$\begin{aligned}\frac{1}{2}\sum_{a=1}^{A}\sum_{n=1}^{a}\left(1-\left(\frac{n}{p}\right)\right) &\leqq \sum_{a=1}^{A}\sum_{q_1\leqq q\leqq a}\left[\frac{a}{q}\right]\\ &= \sum_{a=1}^{A}\sum_{\nu=1}^{[a/q_1]}\pi\left(\frac{a}{\nu}\right)-\sum_{a=1}^{A}\left[\frac{a}{q_1}\right](\pi(q_1)-1)\\ &< 1.0376\,\frac{A(A+1)}{2}\left(\log\frac{\log A}{\log q_1}+\frac{li\,q_1}{q_1}+\frac{1.85}{q_1}\right)\\ &\quad -\frac{A(A+1)}{2}\left(\frac{1}{q_1}-\frac{2}{A+1}\right)(\pi(q_1)-1).\end{aligned}$$

By Lemmas 5 and 2, we have

$$\begin{aligned}\frac{1}{2}\left(\frac{A(A+1)}{2}-\frac{Ap^{1/2}}{2}\right) &< 1.0376\,\frac{A(A+1)}{2}\left(\log\frac{\log A}{\log q_1}+\frac{li\,q_1}{q_1}+\frac{1.85}{q_1}\right)\\ &\quad -\frac{A(A+1)}{2}\left(\frac{1}{q_1}-\frac{2}{A+1}\right)(0.9607\,li\,q_1-2.7).\end{aligned}$$

Therefore

$$\begin{aligned}\log\log q_1 < \log\log A &+ \frac{li\,q_1}{q_1}+\frac{1.85}{q_1}-\frac{1}{1.0376}\left(\frac{1}{2}-\frac{p^{1/2}}{2(A+1)}\right)\\ &-\frac{1}{1.0376}\left(\frac{1}{q_1}-\frac{2}{A+1}\right)(0.9607\,li\,q_1-2.7).\end{aligned}$$

Taking

$$A+1 = 60p^{1/2},$$

we deduce easily that

$$\begin{aligned}\log\log q_1 < \log\log A &+ 0.07412\,li\,q_1/q_1 + 4.453/q_1\\ &- 0.48188 + 0.00804\\ &+ 0.03115\,li\,q_1/p^{1/2} - 0.0546(1/p^{1/2}).\end{aligned}$$

Hence

$$\log\log q_1 < \log\log A - 0.472$$

for $e^{80} < q_1 < p^{1/2}$ and

$$0.07412\, li\, q_1/q_1 < 0.07412\, li\, e^{80}/e^{80} < 0.00095,$$
$$4.453/q_1 < 10^{-33},$$
$$0.03115\, li\, q_1/p^{1/2} < 0.03115\, li\, p^{1/2}/p^{1/2} < 0.03115/38 = 0.00082.$$

Therefore

$$q_1 < A^{e^{-0.472}} < A^{0.625}.$$

We have also:

THEOREM 2. *Let q_1, q_2 and q_3 be the least three prime quadratic non-residues,* mod p. *Then, for $p \geqq e^{250}$, we have*

$$q_2 \leqq (240 p^{1/2})^{0.625}$$

and

$$q_3 \leqq (720 p^{1/2})^{0.625}.$$

The proof of the theorem is similar to that of Theorem 1, but we start with the inequalities

$$\frac{1}{2}\sum_{a=1}^{A}\sum_{n=1,\, q_1 \nmid n}^{a}\left(1-\left(\frac{n}{p}\right)\right) \leqq \sum_{a=1}^{A}\left(\sum_{q_2 \leqq q \leqq a}\left[\frac{a}{q}\right] - \sum_{q_2 \leqq q \leqq a/q_1}\left[\frac{a}{q_1 q}\right]\right)$$

and

$$\frac{1}{2}\sum_{a=1}^{A}\sum_{n=1,\,(q_1q_2,n)=1}^{a}\left(1-\left(\frac{n}{p}\right)\right) \leqq \sum_{a=1}^{A}\left(\sum_{q_3 \leqq q \leqq a}\left[\frac{a}{q}\right] - \sum_{q_3 \leqq q \leqq a/q_1}\left[\frac{a}{q_1 q}\right] - \sum_{q_3 \leqq q \leqq a/q_2}\left[\frac{a}{q_2 q}\right] - \sum_{q_3 \leqq q \leqq a/q_1q_2}\left[\frac{a}{q_1 q_2 q}\right]\right),$$

respectively, instead of

$$\frac{1}{2}\sum_{a=1}^{A}\sum_{n=1}^{a}\left(1-\left(\frac{n}{p}\right)\right) \leqq \sum_{a=1}^{A}\sum_{q_1 \leqq q \leqq a}\left[\frac{a}{q}\right].$$

The corresponding estimates are given in Lemma 6.

6. **A necessary condition for the existence of E.A. in a quadratic field.**

LEMMA 12(8). *For a prime p of form $4n+1$, the E.A. cannot exist in $R(p^{1/2})$ if p can be written in the form*

$$p = q_1 n_1 + q_2 n_2,$$

where n_1, n_2, q_1, q_2 are all positive and quadratic non-residues (mod p), *and where the q_i are odd primes which divide $q_i n_i$ to an odd power for $i = 1, 2$.*

(8) P. Erdös and Ch. Ko, *Note on the Euclidean algorithm*, J. London Math. Soc. vol. 13 (1938) pp. 3–8.

LEMMA 13. *Suppose that* $s<q_1$. *Let* p_0 *be the least prime not dividing* s. *Then*

$$p_0 \leqq (1/0.346) \log q_1.$$

Proof. By Lemma 3,

$$\vartheta((1/0.346) \log q_1) \geqq \log q_1 > \log s.$$

Thus, there is a prime not greater than $(1/0.346) \log q_1$ not dividing s.

LEMMA 14. *Let* p *be a prime of form* $4n+1$. *Let* q_1, q_2, *and* q_3 *be the least three primes which are quadratic non-residues* mod p. *Suppose that* $q_1>3$. *If*

$$p > (1/0.346)q_1q_2q_3 \log q_1,$$

then we can find two positive numbers s *and* t *such that*

$$p = sq_2q_3 + tq_1$$

where $(\frac{s}{p})=1$ *and* $(s, q_2q_3)=(t, q_1)=1$.

Proof. We have

$$p = sq_2q_3 + tq_1, \qquad 0 < s < q_1$$

If $q_1 \nmid t$, the lemma follows from Lemma 12 since

$$sq_2q_3 < q_1q_2q_3 < p.$$

The other conditions are evident.

If $q_1 | t$, let p_0 be the least prime not dividing s, then there exists an integer μ such that

$$s + \mu q_1 \equiv 0 \pmod{p_0}, \qquad 0 < \mu < p_0 < q_1.$$

Hence

$$p = ((s + \mu q_1)/p_0)p_0q_2q_3 + (t - \mu q_2q_3)q_1.$$

Since

$$s + \mu q_1/p_0 < (1 + \mu)q_1/p_0 \leqq q_1$$

and, by Lemma 13,

$$((s + \mu q_1)/p_0)p_0q_2q_3 < p_0q_1q_2q_3 < p,$$

we have the lemma by Lemma 12.

LEMMA 15. *If* $q_1>3$ *and*

$$(1/0.346)q_1q_2q_3 \log q_1 < p,$$

then there is no E.A. in $R(p^{1/2})$.

Proof. The lemma is a consequence of Lemma 14.

LEMMA 16. *If* $q_1=3$, *there is no E.A. in* $R(p^{1/2})$ *providing that*

$$5q_2q_3 < p \quad \textit{for} \quad \left(\frac{5}{p}\right) = 1,$$

and

$$40q_3 < p \quad \textit{for} \quad \left(\frac{5}{p}\right) = -1.$$

Consequently Lemma 15 *holds also for* $q_1=3$.

Proof. (1) $(\frac{5}{p})=1$. We may write

$$p = sq_2q_3 + 3t, \quad \text{where} \quad s = 1 \text{ or } 2.$$

If $3\nmid t$, then this gives us a required decomposition; if $3\mid t$, then

$$p = (s+3)q_2q_3 + 3(t - q_2q_3)$$

will give us the same result.

(2) $(\frac{5}{p})=-1$. Then we may write

$$p = 5sq_3 + 3t, \quad \text{where} \quad s = 1 \text{ or } 2.$$

For $s=1$, the method in (1) gives us a required decomposition. If $s=2$ and $3\mid t$, we write

$$p = 40q_3 + 3(t - 10q_3).$$

7. Proof of the theorem for the E.A.

THEOREM 3. *For* $d>e^{250}$ *and square-free, there is no E.A. in the quadratic field* $R(d^{1/2})$.

Proof. According to the results which are already known, it is sufficient to consider the case $d=p\equiv 1 \pmod 4$. By Theorem 2 we have

$$(1/0.346)q_1q_2q_3 \log q_1 < (1/0.346)(60\cdot 240\cdot 720p^{3/2})^{0.625} \log (60p^{1/2})^{0.625} < p.$$

We have the theorem by Lemmas 15 and 16.

NATIONAL TSING HUA UNIVERSITY,
KUNMING, YUNNAN, CHINA.

Reprinted from the
TRANSACTIONS OF THE AMERICAN MATHEMATICAL SOCIETY
Vol. 56, No. 3, pp. 547–569, 1944

ON THE DISTRIBUTION OF QUADRATIC NON-RESIDUES AND THE EUCLIDEAN ALGORITHM IN REAL QUADRATIC FIELDS. II

BY
LOO-KENG HUA AND SZU-HOA MIN

1. **Introduction.** It is the purpose of the paper to establish the following result(1):

For a prime $p\equiv 17 \pmod{24}$, *there is no E.A. in* $R(p^{1/2})$ *except possibly for* $p=17, 41, 89, 113$, *and* 137.

It is known that for $p=17, 41$ the field is Euclidean. The only doubtful cases are $p=89$, 113, and 137.

The method used is a refinement of the previous paper, by introducing Euler's summation formula and a new estimation of character sum.

The numbering of theorems and lemmas continues from I, and the notations of I are preserved, for example p denotes a prime, $[\xi]$ denotes the integral part of ξ and $q_1<q_2<q_3$ denote the least three primes which are quadratic non-residues, mod p.

2. **Lemmas involving primes.**

LEMMA 17. *For* $11\leqq x\leqq 10^6$,

$$li\, x - li\, x^{1/2} < \pi(x) < li\, x$$

(Rosser, Theorems 1 and 3).

LEMMA 18. *For* $x\geqq 400$, $\vartheta(x)>(1-0.139)x$.

Proof. For $x<10^6$ use Theorem 7(2) of Rosser, and for $x>10^6$, use (10)(3) of Rosser.

LEMMA 19. *For* $401\leqq a\leqq b\leqq a^2$,

$$\sum_{a\leqq q\leqq b}\frac{1}{q^2}<\frac{1.2142}{a\log a}$$

where q runs over primes.

Presented to the Society, October 28, 1944; received by the editors November 29, 1943.

(1) Editor's footnote. The proof given here is entirely different from that given in L. Rédei, *Zur Frage des Euklidischen Algorithmus in quadratischen Zahlkörpern*, Math. Ann. vol. 118 (1942) pp. 588–608. In that paper Rédei proved that the Euclidean algorithm does not exist in quadratic fields $R(\mu^{1/2})$ if $\mu>41$ has the form $24n+17$. The paper is unknown to the author and arrived only recently in this country.

(2) For $0<x\leqq 10^6$, $x-2.78x^{1/2}<\vartheta(x)$.

(3) For $x>e^{13.8}$, $\vartheta(x)>(1-0.0393)x$.

Proof. By Lemma 1 (of I) and Lemma 18,

$$\sum_{a\leqq q\leqq b} \frac{1}{q^2} \leqq \sum_{a\leqq n\leqq b} \frac{\vartheta(n)-\vartheta(n-1)}{n^2\log n}$$
$$= \sum_{a\leqq n\leqq b} \vartheta(n)\left(\frac{1}{n^2\log n} - \frac{1}{(n+1)^2\log(n+1)}\right)$$
$$+ \frac{\vartheta(b)}{(b+1)^2\log(b+1)} - \frac{\vartheta(a-1)}{a^2\log a}$$
$$< 1.0376\left(\sum_{a\leqq n\leqq b} n\left(\frac{1}{n^2\log n} - \frac{1}{(n+1)^2\log(n+1)}\right)\right.$$
$$\left.+ \frac{b}{(b+1)^2\log(b+1)}\right) - \frac{(1-0.139)(a-1)}{a^2\log a}$$
$$= 1.0376\sum_{a\leqq n\leqq b}\frac{1}{n^2\log n} + \frac{0.1766(a-1)}{a^2\log a}$$
$$< \frac{1.0376}{\log a}\left(\frac{1}{a} - \frac{1}{b} + \frac{1}{a^2}\right) + \frac{0.1766}{a\log a}$$
$$< \frac{1.2142}{a\log a}.$$

3. Some results concerning trigonometrical sums.

LEMMA 20. *If A is an integer not less than* 1, $0<y\leqq\pi/2A$, *and* $|x|\leqq 1$, *we have*

$$\left|\sum_{n=1}^{A}(-1)^n\frac{\sin xyn}{\sin yn}\right| \leqq 1.$$

Proof. Without loss of generality we assume $x>0$. The function $\sin xz/\sin z$ increases monotonically for $0<z\leqq\pi/2$. In fact

$$\frac{d}{dz}\frac{\sin xz}{\sin z} = \frac{x\sin z\cos xz - \cos z\sin xz}{\sin^2 z} \geqq 0,$$

since the numerator of the expression vanishes at $z=0$ and its derivative is $(1-x^2)\sin z\sin xz\geqq 0$ for $0<z\leqq\pi/2$.

Thus the sum is an alternating series with increasing absolute values of terms. Since $0<x\leqq 1$ and $0<yn\leqq\pi/2$, we have

$$0 < \sin xyn/\sin yn \leqq 1;$$

the lemma follows immediately.

LEMMA 21. *Let A be an integer not less than* 1 *and m be a positive integer not greater than $p/2A$ and let*

$$S_m = S_m(A) = 2 \sum_{x=1}^{[(p-1)/2m]} \left(\frac{\sin \pi(A+1)mx/p}{\sin \pi mx/p}\right)^2.$$

Then

$$\left| S_m - p(A+1)/m + (A+1)^2 \right| < 3A + 1$$

and

$$S_1 = p(A+1) - (A+1)^2.$$

Proof. For the sake of brevity, we write

$$e(x) = e^{2\pi i x},$$

and we use $\mathcal{R}x$ to denote the real part of x.

We have

$$\begin{aligned}
S_m &= 2 \sum_{x=1}^{[(p-1)/2m]} \sum_{a=0}^{A} \sum_{n=-a}^{a} e(mnx/p) \\
&= 2\left[\frac{p-1}{2m}\right](A+1) + 4\sum_{a=1}^{A}\sum_{n=1}^{a} \\
&\qquad \cdot \mathcal{R}\left(\frac{2e((mn/p)([(p-1)/2m] + 1/2)) - e(mn/2p) + e(-mn/2p)}{2(e(mn/2p) - e(-mn/2p))}\right) \\
&= 2\left[\frac{p-1}{2m}\right](A+1) - A(A+1) \\
&\qquad + 4\sum_{a=1}^{A}\sum_{n=1}^{a} \mathcal{R}\,\frac{e((mn/p)([(p-1)/2m] + 1/2))}{e(mn/2p) - e(-mn/2p)} \\
&= 2\left[\frac{p-1}{2m}\right](A+1) - A(A+1) \\
&\qquad + 2\sum_{a=1}^{A}\sum_{n=1}^{a} \frac{\sin\left([(p-1)/2m] + 1/2\right)2\pi mn/p}{\sin \pi mn/p} \\
&= 2\left[\frac{p-1}{2m}\right](A+1) - A(A+1) \\
&\qquad + 2\sum_{a=1}^{A}\sum_{n=1}^{a} (-1)^n \frac{\sin\left([(p-1)/2m] - p/2m + 1/2\right)2\pi mn/p}{\sin \pi mn/p} \\
&= 2\left[\frac{p-1}{2m}\right](A+1) - A(A+1) + 2A\theta,
\end{aligned}$$

where $|\theta| \leqq 1$, by Lemma 20, since $m \leqq p/2A \leqq p/2a$ and

$$-1 + 1/2m \leqq [(p-1)/2m] - (p-1)/2m \leqq 0.$$

Since

$$-2(A+1) \leqq 2[(p-1)/2m](A+1) - p(A+1)/m \leqq -(A+1)/m < 0$$

we have

$$-4A - 2 < S_m - p(A+1)/m + A(A+1) < 2A.$$

Hence the inequality of the lemma follows.

In particular, when $m=1$, we have $\theta = 0$. Hence the equality follows.

LEMMA 22. *For* $\alpha \geqq 0$,

$$\sum_{n=1}^{\infty} \frac{\sin^2 \pi n \alpha}{n^2} = \frac{\pi^2}{2}(\{\alpha\} - \{\alpha\}^2),$$

where $\{\alpha\} = \alpha - [\alpha]$.

(See, for example, Whittaker-Watson, *Modern Analysis*, 4th ed., Example 3, p. 163.)

LEMMA 23. *Let* $A \geqq 0$, $m > 0$, *then*

$$-\frac{4p}{\pi^2 m} - \frac{8}{\pi^2} < S_m - \frac{p^2}{m^2}\left(\left\{\frac{(A+1)m}{p}\right\} - \left\{\frac{(A+1)m}{p}\right\}^2\right) < \frac{5p}{6m}.$$

Proof. (1) Using $\sin x \leqq x$ for $0 \leqq x \leqq \pi/2$, we have

$$\begin{aligned} S_m &> 2 \sum_{x=1}^{[(p-1)/2m]} \frac{\sin^2 \pi(A+1)mx/p}{(\pi m x/p)^2} \\ &> \frac{2p^2}{\pi^2 m^2}\left(\sum_{x=1}^{\infty} \frac{\sin^2 \pi(A+1)mx/p}{x^2} - \sum_{x=[(p-1)/2m]+1}^{\infty} \frac{1}{x^2}\right) \\ &> \frac{2p^2}{\pi^2 m^2}\left(\sum_{x=1}^{\infty} \frac{\sin^2 \pi(A+1)mx/p}{x^2} - \frac{2m}{p} - \frac{4m^2}{p^2}\right), \end{aligned}$$

since $[(p-1)/2m]+1 > (p-1)/2m - (2m-1)/2m + 1 = p/2m$ and

$$\sum_{x=[(p-1)/2m]+1}^{\infty} \frac{1}{x^2} < \int_{p/2m}^{\infty} \frac{dx}{x^2} + \frac{4m^2}{p^2} = \frac{2m}{p} + \frac{4m^2}{p^2}.$$

By Lemma 22,

$$S_m > \frac{p^2}{m^2}\left(\left\{\frac{(A+1)m}{p}\right\} - \left\{\frac{(A+1)m}{p}\right\}^2\right) - \frac{4p}{\pi^2 m} - \frac{8}{\pi^2}.$$

(2) For $0 \leqq x \leqq 0.44$, we have

$$(1 - x)^{-2} \leqq 1 + 5x$$

and for $0 < x \leqq \pi/2$, we have

$$\sin x > x - x^3/6.$$

Therefore, for $1 < x \leqq (p-1)/2m$,

$$\frac{1}{(\sin \pi m x/p)^2} < \frac{1}{(\pi^2 m^2 x^2/p^2)(1 - \pi^2 m^2 x^2/6p^2)^2} < \frac{p^2}{\pi^2 m^2 x^2}\left(1 + \frac{5\pi^2 m^2 x^2}{6p^2}\right)$$

since

$$\pi^2 m^2 x^2/6p^2 < \pi^2/24 < 0.44.$$

Hence, by Lemma 22,

$$\begin{aligned} S_m &< 2 \sum_{x=1}^{[(p-1)/2m]} \frac{p^2}{\pi^2 m^2} \frac{\sin^2 \pi(A+1)mx/p}{x^2}\left(1 + \frac{5\pi^2 m^2 x^2}{6p^2}\right) \\ &< \frac{2p^2}{\pi^2 m^2}\left(\sum_{x=1}^{\infty} \frac{\sin^2 \pi(A+1)mx/p}{x^2} + \sum_{x=1}^{[(p-1)/2m]} \frac{5\pi^2 m^2}{6p^2}\right) \\ &< \frac{p^2}{m^2}\left(\left\{\frac{(A+1)m}{p}\right\} - \left\{\frac{(A+1)m}{p}\right\}^2\right) + \frac{5p}{6m}. \end{aligned}$$

4. **Result concerning character sums.**

LEMMA 24. *Let $p > 10^6$ and let*

$$N = N(A) = 2 \sum_{x=1}^{(p-1)/2}{}' \left(\frac{\sin \pi(A+1)x/p}{\sin \pi x/p}\right)^2,$$

where the prime denotes that the sum runs over quadratic non-residues mod p. *If $q_1 = 3$, $q_2 > p^{1/2}/37.5$, $q_3 \geqq (p/5)^{1/2}$, then, for $1 \leqq A < 10p^{1/2}$, we have*

$$N > 20p(A+1)/81 - 0.05314pp^{1/2}.$$

Proof. We have

$$N > (S_3 - S_9) + (S_{27} - S_{81}) - S_{3q_2} - \sum_{(p/5)^{1/2} \leqq q \leqq (p-1)/6} S_{3q},$$

where q runs over the primes, since all positive integers less than $3(p/5)^{1/2}$ which are not divisible by q_2 but are exactly divisible by either 3 or $3^3 = 27$ are quadratic non-residues, mod p.

Since $2A \times 27 < 540p^{1/2} < p$, we have, by Lemma 21,

$$\begin{aligned} S_3 - S_9 + S_{27} &> (1/3 - 1/9 + 1/27)p(A+1) - (A+1)^2 - 9A - 3 \\ &= (7/27)p(A+1) - (A+1)^2 - 9A - 3. \end{aligned} \tag{1}$$

Since $81(A+1) < 810(p^{1/2}+1) < p$, we have, by Lemma 23,

$$S_{81} < p(A+1)/81 - (A+1)^2 + 5p/486. \tag{2}$$

By the same lemma, we have

$$S_{3q_2} < \frac{p^2}{9q_2^2} \times \frac{1}{4} + \frac{5p}{18q_2} = \frac{p^2}{36q_2^2} + \frac{5p}{18q_2} < 0.03909pp^{1/2}, \tag{3}$$

since $\{x\} - \{x\}^2 \leqq 1/4$ for all x. Similarly,

$$\sum_{(p/5)^{1/2} \leqq q \leqq (p-1)/6} S_{3q} < \sum_{(p/5)^{1/2} \leqq q \leqq (p-1)/6} \left(\frac{p^2}{36q^2} + \frac{5p}{18q}\right)$$
$$< \frac{p^2}{36} \frac{1.2142}{(p/5)^{1/2} \log (p/5)^{1/2}} + \frac{5p}{18}\left(\log \frac{(5p)^{1/2}}{6} + \left(\frac{5}{p}\right)^{1/2}\right),$$

by Lemma 19. Hence

$$\sum_{(p/5)^{1/2} \leqq q \leqq (p-1)/6} S_{3q} < \frac{1.2142(5)^{1/2}}{36 \times 6.103} pp^{1/2} + \frac{5pp^{1/2}}{18000}(5.9208 + 0.0023)$$
$$= \frac{2.71508}{219.708} pp^{1/2} + \frac{5 \times 5.9231}{18000} pp^{1/2} \tag{4}$$
$$= (0.01236 + 0.00165)pp^{1/2}$$
$$= 0.01401pp^{1/2}.$$

By (1), (2), (3) and (4), we have

$$N > 20p(A+1)/81 - 9A - 3 - 5p/486 - (0.03909 + 0.01401)pp^{1/2}$$
$$> 20p(A+1)/81 - 9A - 3 - 0.00002pp^{1/2} - 0.0531)pp^{1/2}$$
$$> 20p(A+1)/81 - 0.05314pp^{1/2}.$$

LEMMA 25. *Let* $p \equiv 1 \pmod 4$ *and*

$$S = \sum_{a=1}^{A} \sum_{n=1, 3 \nmid n}^{a} \left(\frac{n}{p}\right).$$

Then under the conditions of Lemma 24, we have, for $p^{1/2} < A < 10p^{1/2}$,

$$S < 41Ap^{1/2}/81 + 0.21256p.$$

Proof. We have

$$S = \sum_{a=1}^{A} \sum_{n=1}^{a} \left(\frac{n}{p}\right) + \sum_{a=1}^{A} \sum_{n=1}^{[a/3]} \left(\frac{n}{p}\right)$$
$$\leqq \sum_{a=1}^{A} \sum_{n=1}^{a} \left(\frac{n}{p}\right) + 2 \sum_{a=1}^{[A/3-1]} \sum_{n=1}^{a} \left(\frac{n}{p}\right) + 3 \left| \sum_{n=1}^{[A/3]} \left(\frac{n}{p}\right) \right|.$$

As in the proof of Lemma 5 of I, we have

$$p^{1/2}\sum_{a=1}^{A}\sum_{n=1}^{a}\left(\frac{n}{p}\right) = \sum_{x=1}^{(p-1)/2}\left(\frac{x}{p}\right)\left(\frac{\sin \pi(A+1)x/p}{\sin \pi x/p}\right)^2$$
$$= S_1(A)/2 - N(A).$$

Therefore, by Lemmas 21 and 24,

$$\begin{aligned} S &\leqq \frac{1}{2p^{1/2}}S_1(A) - \frac{1}{p^{1/2}}N(A) \\ &\quad + \frac{3}{2p^{1/2}}S_1\left(\left[\frac{A}{3}-1\right]\right) - \frac{3}{p^{1/2}}N\left(\left[\frac{A}{3}-1\right]\right) + A \\ &< \frac{1}{2p^{1/2}}\left(p(A+1) - (A+1)^2 - \frac{40}{81}p(A+1) + 0.10628pp^{1/2}\right) \\ &\quad + \frac{3}{2p^{1/2}}\left(p\left[\frac{A}{3}\right] - \left[\frac{A}{3}\right]^2 - \frac{40}{81}p\left[\frac{A}{3}\right] + 0.10628pp^{1/2}\right) + A \\ &< \frac{1}{2p^{1/2}}\left(\frac{41}{81}p(A+1) - (A+1)^2 + 0.10628pp^{1/2}\right) \\ &\quad + \frac{3}{2p^{1/2}}\left(\frac{41}{81}p\frac{A}{3} + 0.10628pp^{1/2}\right) \\ &< \frac{41}{81}Ap^{1/2} + 0.21256p. \end{aligned}$$

LEMMA 26. *Let* $p \equiv 1 \pmod 4$, *then under the conditions of Lemma* 24, *we have*

$$q_3 < 3p^{1/2}.$$

Proof. We have, for $A < q_3$,

$$\sum_{a=1}^{A}\sum_{n=1,3\nmid n}^{a}\left(\frac{n}{p}\right) > \sum_{a=1}^{A}\sum_{n=1,3\nmid n}^{a}1 - \sum_{a=1}^{A}\sum_{n=1,q_2|n}^{a}1 + \sum_{a=1}^{A}\sum_{n=1,3q_2|n}^{a}1$$
$$> A(A+1)/3 - (1/3q_2)A(A+1) - 2A.$$

By Lemma 7 of I, we have $q_3 < 7.5p^{1/2} < 10p^{1/2}$, therefore for $A = q_3 - 1$ we have, by Lemma 25 (if $A < p^{1/2}$, the result is trivial),

$$\begin{aligned} (A(A+1)/3)(1-1/q_2) - 2A &< 41Ap^{1/2}/81 + 0.21256p, \\ (A(A+1)/3)(1-0.0375) - 2A &< 0.5062Ap^{1/2} + 0.21256p, \\ 0.3208A^2 - 1.6792A &< 0.5062Ap^{1/2} + 0.21256p, \\ 0.3208A^2 - 0.5079Ap^{1/2} - 0.21256p &< 0. \end{aligned}$$

Hence

$$A < (1/0.6416)(0.5079 + (0.5079^2 + 4 \times 0.3208 \times 0.21256)^{1/2})p^{1/2},$$

which implies

$$q_3 < 3p^{1/2}.$$

5. **Lemmas concerning sums.**

LEMMA 27. *We have, for $q>0$,*

$$\frac{(A+2)(A-q)}{2q} < \sum_{a=1}^{A}\left[\frac{a}{q}\right] < \frac{1}{2q}\left(A - \frac{(q-2)}{2}\right)^2.$$

Proof. We have

$$\sum_{a=1}^{A}\left[\frac{a}{q}\right] = \sum_{q\leqq a<2q} + \sum_{2q\leqq a<3q} + \cdots + \sum_{([A/q]-1)q\leqq a<[A/q]q}\left[\frac{A}{q}-1\right] + \sum_{[A/q]q\leqq a\leqq A}\left[\frac{A}{q}\right]$$

$$= q\left(1+2+\cdots+\left[\frac{A}{q}-1\right]\right) + \left(A - \left[\frac{A}{q}\right]q+1\right)\left[\frac{A}{q}\right]$$

$$= \frac{1}{2}q\left[\frac{A}{q}\right]\left(\left[\frac{A}{q}\right]-1\right) + \left(A - \left[\frac{A}{q}\right]q+1\right)\left[\frac{A}{q}\right]$$

$$= \frac{1}{2}\left[\frac{A}{q}\right]\left(2A+2-\left[\frac{A}{q}\right]q - q\right)$$

$$= \frac{1}{2q}(A-\vartheta q)(A-q+\vartheta q+2)$$

$$= \frac{1}{2q}\left(\left(A-\frac{q}{2}+1\right)^2 - \left(\vartheta q+1-\frac{q}{2}\right)^2\right), \qquad \vartheta = \frac{A}{q} - \left[\frac{A}{q}\right],$$

which lies between $(A+2)(A-q)/2q$ and $(A-q/2+1)^2/2q$.

LEMMA 28 (EULER'S SUMMATION FORMULA). *Let $b_1(x)=x-[x]-1/2$ and*

$$b_2(x) = \frac{1}{12} + \int_0^x b_1(x)dx.$$

Let $b>a$ and let $g(x)$ with its first two derivatives be continuous for $a\leqq x<b$. Then

$$\sum_{a\leqq m<b} g(m) = \int_a^b g(x)dx + g(b)b_1(-b) - g(a)b_1(-a) + g'(b)b_2(-b) - g'(a)b_2(-a) - \int_a^b g''(x)b_2(-x)dx.$$

LEMMA 29. *For* $a-\lambda q>3q\geqq 9$,

$$\sum_{\nu=1}^{[(a-\lambda q)/3q]} li\,\frac{a}{3\nu+\lambda}<\frac{a}{3}\log\frac{\log(a/(3+\lambda))}{\log q}+\frac{(3\lambda+10.7322)a}{12(3+\lambda)^3\log(a/(3+\lambda))}$$
$$-\frac{3q^2b_2(-(a-\lambda q)/3q)}{a\log q}+\frac{a}{3q}\,li\,q$$
$$-\left(\frac{1}{2}+\frac{\lambda}{3}\right)li\,\frac{a}{3+\lambda}+li\,qb_1\left(-\frac{a-\lambda q}{3q}\right)+R,$$

where $R=li\,q$ *or* 0 *according as* $3q$ *does or does not divide* $a-\lambda q$.

Proof. Let $g(x)=li\,a/(3x+\lambda)$. Then

$$g'(x)=-\frac{3a}{(3x+\lambda)^2\log(a/(3x+\lambda))},$$
$$g''(x)=\frac{9a(2\log(a/(3x+\lambda))-1)}{(3x+\lambda)^3\log^2(a/(3x+\lambda))},$$

and for $1\leqq 3x+\lambda\leqq(1/e)a$, $g''(x)$ is monotonically decreasing, since the function $(2u-1)/u^2$ is monotonically decreasing for $u>1$.

By Lemma 28,

$$\sum_{\nu=1}^{[(a-\lambda q)/3q]} li\,\frac{a}{3\nu+\lambda}=\int_1^{(a-\lambda q)/3q} li\,\frac{a}{3t+\lambda}\,dt+\frac{1}{2}\,li\,\frac{a}{3+\lambda}$$
$$+li\,qb_1\left(-\frac{a-\lambda q}{3q}\right)+\frac{a}{4(3+\lambda)^2\log(a/(3+\lambda))}$$
$$-\frac{3ab_2(-(a-\lambda q)/3q)}{(a/q)^2\log q}$$
$$-\int_1^{(a-\lambda q)/3q}\frac{9a(2\log(a/(3x+\lambda))-1)b_2(-x)}{(3x+\lambda)^3\log^2(a/(3x+\lambda))}\,dx+R$$
$$=\frac{a}{3}\log\frac{\log(a/(3+\lambda))}{\log q}+\frac{a}{3q}\,li\,q-\frac{1}{3}(3+\lambda)\,li\,\frac{a}{3+\lambda}$$
$$+\frac{1}{2}\,li\,\frac{a}{3+\lambda}+li\,qb_1\left(-\frac{a-\lambda q}{3q}\right)$$
$$+\frac{a}{4(3+\lambda)^2\log(3/(3+\lambda))}-\frac{3q^2b_2(-(a-\lambda q)/3q)}{a\log q}$$
$$+\frac{3^{1/2}a\theta}{12(3+\lambda)^3\log(a/(3+\lambda))}+R,$$

where $\theta\leqq 1$, since

$$\int_1^{(a-\lambda q)/3q} li\,\frac{a}{3t+\lambda}\,dt$$

$$= \frac{1}{3}\left((3t+\lambda)\,li\,\frac{a}{3t+\lambda}\right)_1^{(a-\lambda q)/3q}$$

$$- \frac{a}{3}\int_1^{(a-\lambda q)/3q} \frac{d(a/(3t+\lambda))/dt}{a/(3t+\lambda)\log(a/(3t+\lambda))}\,dt$$

$$= \frac{1}{3}\left((3t+\lambda)\,li\,\frac{a}{3t+\lambda}\right)_1^{(a-\lambda q)/3q} - \frac{a}{3}\left(\log\log\frac{a}{3t+\lambda}\right)_1^{(a-\lambda q)/3q}$$

$$= \frac{a}{3q}\,li\,q - \frac{1}{3}(3+\lambda)\,li\,\frac{a}{3+\lambda} + \frac{a}{3}\log\frac{\log(a/(3+\lambda))}{\log q}$$

and

$$\left|\int_0^x b_2(-x)dx\right| \leqq \frac{3^{1/2}}{216}\;(^4).$$

The lemma follows immediately.

LEMMA 30. *Under the conditions of Lemma* 29, *and the condition* $\lambda > 0$, *we have*

$$\sum_{a=\lambda q}^{A}\sum_{\nu=0}^{[(a-\lambda q)/3q]} li\,\frac{a}{3\nu+\lambda} < \frac{1}{6}A(A+2)\log\log\frac{A}{3+\lambda}$$

$$-\frac{1}{6}(A^2+A+(3+\lambda)q)\log\log q + (A+1)\,li\,\frac{A}{\lambda} - \frac{(2\lambda+3)A}{6}\,li\,\frac{A}{3+\lambda}$$

$$+\frac{li\,q}{24q}(4A^2+8A+(2\lambda-3)^2q^2+4) + \frac{3\lambda+10.7322}{12(\lambda+3)^3}\,\frac{A}{\log A/(3+\lambda)}$$

$$+\frac{q^2}{8\log q}\left(\log\frac{A}{(3+\lambda)q} + \frac{1}{(3+\lambda)q}\right) - \lambda\int_q^{A/\lambda}\frac{x\,dx}{\log x}$$

$$+\frac{2\lambda(3+\lambda)^2+3\lambda+10.7322}{12(3+\lambda)}\int_q^{A/(3+\lambda)}\frac{x\,dx}{\log x}.$$

Proof. By Lemma 29, we have

$$\sum_{a=\lambda q}^{A}\sum_{\nu=0}^{[(a-\lambda q)/3q]} li\,\frac{a}{3\nu+\lambda} = \sum_{a=\lambda q}^{A} li\,\frac{a}{\lambda} + \sum_{a=(3+\lambda)q}^{A}\sum_{\nu=1}^{[(a-\lambda q)/3q]} li\,\frac{a}{3\nu+\lambda}$$

---

(4) For $0<x\leqq 1$, $\int_0^x b_2(-x)dx = \int_0^x((x^2-x)/2+1/12)dx = (x/12)(2x^2-3x+1)$, which takes its extrema at $x=(3\pm 3^{1/2})/6$. For $x=1$, the integral equals zero. Since $b_2(x-1)=b_2(x)$, we have the required formula.

$$< \int_{\lambda q}^{A} li\frac{x}{\lambda}dx + li\frac{A}{\lambda} + \sum_{a=(3+\lambda)q}^{A} \frac{a}{3}\log\frac{\log(a/(3+\lambda))}{\log q}$$

$$+ \frac{3\lambda + 10.7322}{12(3+\lambda)^3}\left(\int_{(3+\lambda)q}^{A} \frac{xdx}{\log(x/(3+\lambda))} + \frac{A}{\log(A/(3+\lambda))}\right)$$

$$- \sum_{a=(3+\lambda)q}^{A} \frac{3q^2}{a\log q} b_2\left(-\frac{a-\lambda q}{3q}\right)$$

$$+ \frac{1}{6q}(A+\lambda q+3q)(A-\lambda q-3q+1)\, li\, q$$

$$- \left(\frac{1}{2}+\frac{\lambda}{3}\right)\int_{(3+\lambda)q}^{A} li\frac{x}{3+\lambda}dx$$

$$+ li\, q \sum_{a=(3+\lambda)q}^{A}\left(b_1\left(-\frac{a-\lambda q}{3q}\right) + R'\right)$$

where $R'=1$ or 0 according as $3q$ does or does not divide $a-\lambda q$.

It is plain that

$$\int_{\lambda q}^{A} li\frac{x}{\lambda}dx = A\, li\frac{A}{\lambda} - \lambda q\, li\, q - \int_{\lambda/q}^{A}\frac{xdx}{\lambda\log x/\lambda},$$

$$\sum_{a=(3+\lambda)q}^{A}\frac{a}{3}\log\log\frac{a}{3+\lambda} < \frac{1}{3}\int_{(3+\lambda)q}^{A} x\log\log\frac{x}{3+\lambda}dx + \frac{A}{3}\log\log\frac{A}{3+\lambda}$$

$$= \frac{1}{6}\left(A^2\log\log\frac{A}{3+\lambda} - (3+\lambda)^2q^2\log\log q\right)$$

$$+ \frac{A}{3}\log\log\frac{A}{3+\lambda} - \frac{1}{6}\int_{(3+\lambda)q}^{A}\frac{xdx}{\log x/(3+\lambda)}$$

$$- \sum_{a=(3+\lambda)q}^{A}\frac{b_2(-(a-\lambda q)/3q)}{a} < \frac{1}{24}\sum_{a=(3+\lambda)q}^{A}\frac{1}{a}$$

$$< \frac{1}{24}\left(\log\frac{A}{(3+\lambda)q} + \frac{1}{(3+\lambda)q}\right),$$

since $b_2(x) \geqq -1/24$,

$$\int_{(3+\lambda)q}^{A} li\frac{x}{3+\lambda}dx = A\, li\frac{A}{3+\lambda} - (3+\lambda)q\, li\, q - \int_{(3+\lambda)q}^{A}\frac{xdx}{(3+\lambda)\log x/(3+\lambda)}$$

and

$$\sum_{a=(3+\lambda)q}^{A}\left(b_1\left(-\frac{a-\lambda q}{3q}\right)+R'\right) = \sum_{n=0}^{A-(3+\lambda)q}\left(b_1\left(-\frac{n}{3q}\right)+R''\right) = S,$$

say, where $R''=1$ or 0 according as $3q|n$ or $3q\nmid n$. Clearly,

$$S = \sum_{n=0}^{A-(3+\lambda)q}\left(-\frac{n}{3q}-\left[-\frac{n}{3q}\right]-\frac{1}{2}+R''\right)$$
$$= \sum_{n=0}^{A-(3+\lambda)q}\left(-\frac{n}{3q}+\left[\frac{n}{3q}\right]+\frac{1}{2}\right),$$

since $[-n/3q]=-[n/3q]+R''-1$. By Lemma 27,

$$S < -\frac{1}{6q}(A-(3+\lambda)q)(A-(3+\lambda)q+1)$$
$$+\frac{1}{6q}\left(A-(3+\lambda)q-\frac{3q-2}{2}\right)^2+\frac{1}{2}(A-(3+\lambda)q+1)$$
$$=\frac{1}{6q}(A-\lambda q)+\frac{(3q-2)^2}{24q}.$$

Therefore

$$\sum_{a=\lambda q}^{A}\sum_{\nu=0}^{[(a-\lambda q)/3q]} li\frac{a}{3\nu+\lambda} < A\,li\frac{A}{\lambda}-\lambda q\,li\,q-\int_{\lambda q}^{A}\frac{xdx}{\lambda\log x/\lambda}+li\frac{A}{\lambda}$$
$$-\frac{1}{6}(A+\lambda q+3q)(A-\lambda q-3q+1)\log\log q$$
$$+\frac{1}{6}\left(A^2\log\log\frac{A}{3+\lambda}-(3+\lambda)^2q^2\log\log q\right)+\frac{A}{3}\log\log\frac{A}{3+\lambda}$$
$$-\frac{1}{6}\int_{(3+\lambda)q}^{A}\frac{xdx}{\log x/(3+\lambda)}$$
$$+\frac{3\lambda+10.7322}{12(3+\lambda)^3}\left(\int_{(3+\lambda)q}^{A}\frac{xdx}{\log x/(3+\lambda)}+\frac{A}{\log A/(3+\lambda)}\right)$$
$$+\frac{q^2}{8\log q}\left(\log\frac{A}{(3+\lambda)q}+\frac{1}{(3+\lambda)q}\right)$$
$$+\frac{1}{6q}(A+\lambda q+3q)(A-\lambda q-3q+1)\,li\,q$$
$$-\left(\frac{1}{2}+\frac{\lambda}{3}\right)\left(A\,li\frac{A}{3+\lambda}\right)-(3+\lambda)q\,li\,q$$
$$-\int_{(3+\lambda)q}^{A}\frac{xdx}{(3+\lambda)\log x/(3+\lambda)}+li\,q\left(\frac{A-\lambda q}{6q}+\frac{(3q-2)^2}{24q}\right)$$

$$= \frac{1}{6} A(A+2) \log\log \frac{A}{3+\lambda} - \frac{1}{6}(A^2 + A + (3+\lambda)q) \log\log q + (A+1)\, li \frac{A}{\lambda}$$
$$- \frac{2\lambda+3}{6} A\, li \frac{A}{3+\lambda} + \frac{li\, q}{24q}(4A^2 + 8A + (2\lambda-3)^2 q^2 + 4)$$
$$+ \frac{3\lambda + 10.7322}{12(3+\lambda)^3} \frac{A}{\log (A/(3+\lambda))} + \frac{q^2}{8 \log q}\left(\log \frac{A}{(3+\lambda)q} + \frac{1}{(3+\lambda)q}\right)$$
$$- \int_{\lambda q}^{A} \frac{x dx}{\lambda \log x/\lambda} + \frac{2\lambda(3+\lambda)^2 + 3\lambda + 10.7322}{12(3+\lambda)} \int_q^{A/(3+\lambda)} \frac{x dx}{\log x}.$$

LEMMA 31. *Under the conditions of Lemma* 29,

$$\sum_{a=q}^{A}\left(\sum_{\nu=0}^{[(a-q)/3q]} li \frac{a}{3\nu+1} + \sum_{\nu=0}^{[(a-2q)/3q]} li \frac{a}{3\nu+2}\right)$$
$$< \frac{1}{6} A(A+2) \log\left(\log \frac{A}{4} \log \frac{A}{5}\right) - \frac{1}{6}(2A^2 + 2A + 9q) \log\log q$$
$$+ \frac{li\, q}{12q}(4A^2 + 8A + q^2 + 4) + \frac{0.03A}{\log A/5}$$
$$+ \frac{q^2}{8 \log q}\left(\log \frac{A^2}{20q^2} + \frac{q}{20q}\right) + A^2\left(\int_{1/2}^{1} \frac{(1-x)dx}{\log Ax} + \int_{1/4}^{1/2} \frac{(2-3x)dx}{\log Ax}\right.$$
$$\left.+ \int_{1/5}^{1/4} \frac{(1.1667 - 2.0472x)dx}{\log Ax} + 0.1016 \int_{q/A}^{1/5} \frac{x dx}{\log Ax}\right) + li\, A + li \frac{A}{2}.$$

Proof. By Lemma 30, the sum in question is less than

$$\frac{1}{6} A(A+2) \log\left(\log \frac{A}{4} \log \frac{A}{5}\right) - \frac{1}{6}(2A^2 + 2A + 9q) \log\log q$$
$$+ (A+1)\, li\, A + (A+1)\, li \frac{A}{2} - \frac{5}{6} A\, li \frac{A}{4} - \frac{7}{6} A\, li \frac{A}{5}$$
$$+ \frac{li\, q}{12q}(4A^2 + 8A + q^2 + 4) + \frac{0.03A}{\log (A/5)}$$
$$+ \frac{q^2}{8 \log q}\left(\log \frac{A^2}{20q^2} + \frac{9}{20q}\right) - \int_q^A \frac{x dx}{\log x} - \int_{2q}^{A} \frac{x dx}{2 \log x/2}$$
$$+ \frac{45.7322}{48} \int_q^{A/4} \frac{x dx}{\log x} + \frac{116.7322}{60} \int_q^{A/5} \frac{x dx}{\log x}.$$

Since

$$A\,li\,A + A\,li\frac{A}{2} - \frac{5}{6}A\,li\frac{A}{4} - \frac{7}{6}A\,li\frac{A}{5}$$
$$= A\int_{A/2}^{A}\frac{dx}{\log x} + 2A\int_{A/4}^{A/2}\frac{dx}{\log x} + \frac{7}{6}A\int_{A/5}^{A/4}\frac{dx}{\log x}$$

and

$$-\int_{q}^{A}\frac{xdx}{\log x} - \int_{2q}^{A}\frac{xdx}{2\log x/2} + \frac{45.7322}{48}\int_{q}^{A/4}\frac{xdx}{\log x} + \frac{116.7322}{60}\int_{q}^{A/5}\frac{xdx}{\log x}$$
$$= -\int_{A/2}^{A}\frac{xdx}{\log x} - 3\int_{A/4}^{A/2}\frac{xdx}{\log x} - 2.0472\int_{A/5}^{A/4}\frac{xdx}{\log x}$$
$$- 0.1016\int_{q}^{A/5}\frac{xdx}{\log x},$$

we have the lemma by replacing x in the integrals by Ax.

6. **The growth of smaller quadratic non-residues.**

LEMMA 32. *Let* $p\equiv 1 \pmod 4$, $p>10^6$, $q_1=3$. *Then* $5q_2q_3<p$.

Proof. Suppose on the contrary that $5q_2q_3\geqq p$. Then

(1) $$q_3 > (p/5)^{1/2}.$$

By Lemma 7 of I, we have $q_3\leqq 4p^{1/2}/(1-1/3)(1-1/5)=7.5p^{1/2}$.

If $q_2<p^{1/2}/37.5$, the lemma is immediate. If $q_2\geqq p^{1/2}/37.5$, by Lemma 26,

(2) $$q_3 < 3p^{1/2}.$$

The lemma follows for $q_2\leqq p^{1/2}/15$. Then we may assume that

(3) $$q_2 > p^{1/2}/15.$$

By Lemma 10 of I, we have

$$\frac{1}{2}\sum_{a=1}^{A}\sum_{n=1,3\nmid n}^{a}\left(1-\left(\frac{n}{p}\right)\right) \leqq \sum_{a=q_3}^{A}\left(\sum_{q_3\leqq q\leqq a}\left[\frac{a}{q}\right] - \sum_{q_3\leqq q\leqq a/3}\left[\frac{a}{3q}\right]\right)$$
$$+ \sum_{a=q_2}^{A}\left(\left[\frac{a}{q_2}\right]-\left[\frac{a}{3q_2}\right]\right)$$
$$= \sum_{a=q_3}^{A}\left(\sum_{\nu=1}^{[a/q_3]}\pi\left(\frac{a}{\nu}\right) - \left[\frac{a}{q_3}\right](\pi(q_3)-1)\right.$$
$$\left. - \sum_{\nu=1}^{[a/3q_3]}\pi\left(\frac{a}{3\nu}\right) + \left[\frac{a}{3q_3}\right](\pi(q_3)-1)\right)$$
$$+ \sum_{a=q_2}^{A}\left(\left[\frac{a}{q_2}\right]-\left[\frac{a}{3q_2}\right]\right).$$

Denoting the first side of the inequality by S_0, we have

$$(4)\quad \begin{aligned} S_0 &< \sum_{a=q_2}^{A} \sum_{\nu=1,3\nmid\nu}^{[a/q_3]} \pi\left(\frac{a}{\nu}\right) - \sum_{a=q_3}^{A}\left(\left[\frac{a}{q_3}\right]-\left[\frac{a}{3q_3}\right]\right)(\pi(q_3)-1) \\ &\quad + \sum_{a=q_2}^{A}\left(\left[\frac{a}{q_2}\right]-\left[\frac{a}{3q_2}\right]\right) = S' - S'' + S''', \text{ say.} \end{aligned}$$

(1) Suppose that $p>10^8$. Let us take $A=[10p^{1/2}]$. By Lemma 25,

$$(5)\quad \begin{aligned} S_0 &> (A^2/3 - 41Ap^{1/2}/81 - 0.21256p)/2 \\ &> (33.32 - 5.1 - 0.22)p/2 = 14p \end{aligned}$$

since

$$A^2/3 > (100p - 20p^{1/2})/3 = 33.32p + 4p/300 - 20p^{1/2}/3 > 33.32p.$$

By (3),

$$(6)\quad \begin{aligned} S''' &< \sum_{a=q_2}^{A}\left(\frac{a}{q_2} - \frac{a}{3q_2} + 1\right) < \frac{A(A+1)}{3q_2} + A \\ &< 50(10p^{1/2}+1) + 10p^{1/2} = 510p^{1/2} + 50 \\ &< 0.06p. \end{aligned}$$

By Lemma 2 of I, we have, since $q_3 > (p/5)^{1/2} \geqq 10^4/5^{1/2} > 51^2$,

$$S'' > (0.9607\, li\, q_3 - 2.7)\sum_{a=q_3}^{A}\left(\left[\frac{a}{q_3}\right]-\left[\frac{a}{3q_3}\right]\right).$$

By Lemma 27 and (2),

$$\begin{aligned} &\sum_{a=q_3}^{A}\left(\left[\frac{a}{q_3}\right]-\left[\frac{a}{3q_3}\right]\right) \\ &\quad > \frac{1}{2q_3}\left((A+2)(A-q_3) - \frac{1}{3}\left(A+1-\frac{3q_3}{2}\right)^2\right) \\ &\quad = \frac{1}{2q_3}\left(\frac{2}{3}A^2 + \frac{4}{3}A - \frac{3}{4}q_3^2 - q_3 - \frac{1}{3}\right) \\ &\quad > \frac{p}{q_3}\left(\frac{100}{3} - \frac{1}{3\times 10^8} - \frac{3^3}{8} - \frac{3}{2\times 10^4} - \frac{1}{6\times 10^8}\right) \\ &\quad > \frac{29.957}{q_3}p. \end{aligned}$$

Then

$$(7)\quad S'' > (29.957/q_3)(0.9607\, li\, q_3 - 2.7) > (28.779\, li\, q_3/q_3 - 0.03)p.$$

Finally, by Lemma 2 of I and Lemma 31,

$$S' < 1.0376 \sum_{a=q_3}^{A} \sum_{\nu=1,3\nmid\nu}^{a/q_3} (li\, a/\nu + 1.85)$$

$$< 1.0376\Big\{\frac{1}{6} A(A+2) \log\left(\log\frac{A}{4}\log\frac{A}{5}\right)$$

$$-\frac{1}{6}(2A^2+2A+9q_3)\log\log q_3 + \frac{li\, q_3}{12q_3}(4A^2+8A+q_3^2+4)$$

$$+\frac{0.03A}{\log A/5} + \frac{q_3^2}{8\log q_3}\left(\log\frac{A^2}{20q_3^2}+\frac{9}{20q_3}\right) + A^2\left(\int_{1/2}^{1}\frac{(1-x)dx}{\log Ax}\right.$$

$$+\int_{1/4}^{1/2}\frac{(2-3x)dx}{\log Ax} + \int_{1/5}^{1/4}\frac{(1.1667-2.0472)dx}{\log Ax}$$

$$\left.+0.1016\int_{q_3/A}^{1/5}\frac{xdx}{\log Ax}\right) + li\, A + li\frac{A}{2}\Big\}$$

$$+1.91956\sum_{a=q_3}^{A}\left(\left[\frac{a}{q_3}\right]-\left[\frac{a}{3q_3}\right]\right).$$

Since

$$\frac{1}{6}A(A+2)\log\left(\log\frac{A}{4}\log\frac{A}{5}\right) - \frac{1}{6}(2A^2+2A+9q_3)\log\log q_3$$

$$< \frac{1.0001}{6}A^2\left(\log\frac{\log(5/2)p^{1/2}\log 2p^{1/2}}{(\log p^{1/2}-(\log 5)/2)^2} + \frac{2\log\log(p/5)^{1/2}}{[10p^{1/2}]}\right)$$

$$< \frac{1.0001}{6}A^2\left(\log\frac{(4\log 10+\log 2.5)(4\log 10+\log 2)}{(4\log 10-(\log 5)/2)^2}+0.0002\right)$$

$$< \frac{1.0001}{6}A^2\left(\log\frac{10.267\times 9.9035}{8.40562^2}+0.0002\right)$$

$$= \frac{1.0001}{6}A^2(\log 1.42+0.0002)$$

$$< \frac{0.351}{6}A^2 < 5.85p.$$

$$\frac{li\, q_3}{12q_3}(4A^2+8A+q_3^2+4) < \frac{\log q_3}{q_3}p\left(\frac{100}{3}+\frac{2\times 10}{3\times 10^4}+\frac{3^2}{12}+\frac{1}{3\times 10^8}\right)$$

$$< 34.0841\, li\, q_3 p/q_3,$$

$$\frac{0.03A}{\log A/5} < \frac{0.3p^{1/2}}{\log(2p^{1/2}-1)} = \frac{0.3p}{10^4\log(2\times 10^4-1)} < 0.00001p,$$

$$\frac{q_3^2}{8\log q_3}\left(\log\frac{A^2}{20q_3^2}+\frac{9}{20q_3}\right)<\frac{9p}{8\log(3p^{1/2})}\left(\log\frac{10^2\times 5}{20}+\frac{9(5)^{1/2}}{20\times 10^4}\right)$$
$$<\frac{9}{8\times 10.30895}(3.21889+0.0002)p$$
$$<0.36p$$

(since $(p/5)^{1/2}<q_3<3p^{1/2}$).

$$\int_{1/2}^{1}\frac{(1-x)dx}{\log Ax}<\frac{1}{\log A/2}\int_{1/2}^{1}(1-x)dx=\frac{1}{8\times 10.81977}<0.0116,$$

$$\int_{1/4}^{1/2}\frac{(2-3x)dx}{\log Ax}<\frac{7}{32\log A/4}<\frac{7}{32\times 10.126635}<0.0217,$$

$$\int_{1/5}^{1/4}\frac{(1.1667-2.0472x)dx}{\log Ax}<\frac{1}{\log A/5}\left(\frac{1.1667}{20}-\frac{9\times 2.0472}{800}\right)<0.00357,$$

$$0.1016\int_{q_3/A}^{1/5}\frac{xdx}{\log Ax}<\frac{0.00204}{\log q_3}<0.0001,$$

$$li\,A+li\,A/2<2\,li\,A<2\times\frac{10p^{1/2}}{\log 10p^{1/2}-2}\,(^5)<\frac{2p}{10^3(\log 10^5-2)}$$
$$<0.0003p$$

and

$$\sum_{a=q_3}^{A}\left(\left[\frac{a}{q_3}\right]-\left[\frac{a}{3q_3}\right]\right)\leqq\frac{A(A+1)}{3q_3}+A<0.0085p.$$

We have

$$\begin{aligned}S'&<1.0376p\{5.85+34.0841\,li\,q_3/q_3+0.00001p+0.36\\&\quad+10^2(0.0116+0.0217+0.0036+0.0001)+0.0003\}\\&\quad+1.91956\times 0.0085p\\&<p(1.0376\times 10+0.02+1.0376\times 34.084)\,li\,q_3/q_3)\\&<p(10.40+35.366\,li\,q_3/q_3).\end{aligned}\tag{8}$$

By (4), (5), (6), (7), and (8), we have $14p<p(10.40+35.366\,li\,q_3/q_3)-(28.779\,li\,q_3/q_3-0.03)p+0.06p$, that is,

$$14<10.40+0.03+0.06+6.59\,li\,q_3/q_3$$
$$<10.49+6.59\times 1/(\log(10^4/5^{1/2})-2)<12,$$

which is absurd.

(5) For $e^4\leqq x$, $li\,x<x/(\log x-2)$. In fact, the inequality is true for $x=e^4$, and $d\,li\,x/dx\leqq d(x/(\log x-2))/dx$.

(2) Suppose $10^6 < p < 10^8$. We take $A = [6p^{1/2}]$. By Lemma 25,

$$(5') \quad \begin{aligned} S_0 &> (A^2/3 - 41Ap^{1/2}/81 - 0.21256p)/2 \\ &> (12 - 0.004 - 3.0371 - 0.21256)p/2 > 4.3731p. \end{aligned}$$

By Lemma 27

$$\begin{aligned} S''' &< \frac{1}{2q_2}\left(\left(A - \frac{q_2}{2} + 1\right)^2 - \frac{(A+2)(A-3q_2)}{3}\right) \\ &= \frac{1}{2q_2}\left(\frac{2}{3}A^2 + \frac{4}{3}A + \frac{(q_2+2)^2}{4}\right) \\ &= \frac{1}{q_2}\left(\frac{1}{3}A^2 + \frac{2}{3}A + \frac{(q_2+2)^2}{8}\right) < \frac{p}{q_2}\left(12 + \frac{4}{10^3} + \frac{(q_2+2)^2}{8p}\right) \\ &< 5q_3\left(12.004 + \frac{(p+10q_3)^2}{200pq_3^2}\right), \end{aligned}$$

for $q_2 > p/5q_3$, and the function is decreasing for $q_2 < q_3 < 3p^{1/2}$ (by (2)). Since

$$\frac{(p+10q_3)^2}{200pq_3^2} = \frac{1}{200p}\left(\frac{p}{q_3} + 10\right)^2 < \frac{1}{200}\left(5^{1/2} + \frac{10}{p^{1/2}}\right)^2 < \frac{1}{200}\left(5^{1/2} + \frac{1}{100}\right)^2 < 0.0253,$$

$$(6') \qquad S''' < 60.1465q_3.$$

Since $11 < (p/5)^{1/2} < q_3 < 3p^{1/2} < 10^6$, we have, by Lemma 18, $S'' > (li\, q_3 - li\, q_3^{1/2} - 1)\sum_{a=q_3}^{A}([a/q_3] - [a/3q_3])$.

By Lemma 27 and (2), we have as in (1)

$$\begin{aligned} \sum_{a=q_3}^{A}\left(\left[\frac{a}{q_3}\right] - \left[\frac{a}{3q_3}\right]\right) &> \frac{1}{2q_3}\left(\frac{2}{3}A^2 + \frac{4}{3}A - \frac{3}{4}q_3^2 - q_3 - \frac{1}{3}\right) \\ &> \frac{p}{2q_3}\left(\frac{2}{3}\cdot 6^2 - \frac{8}{p^{1/2}} + \frac{4}{3}\frac{6}{p^{1/2}} - \frac{3}{4}\frac{q_3^2}{p} - \frac{q_3}{p^{1/2}} - \frac{1}{p}\right) \\ &> \frac{p}{2q_3}\left(24 - \frac{3}{4}\frac{q_3^2}{p} - \frac{3}{p^{1/2}} - \frac{1}{p}\right) \\ &> \frac{p}{2q_3}\left(23.9969 - \frac{3}{4}\frac{q_3^2}{p}\right) > \frac{p}{q_3}\left(11.9984 - \frac{3}{8}\frac{q_3^2}{p}\right), \end{aligned}$$

$$\frac{1}{q_3} < \left(\frac{5}{p}\right)^{1/2} < \frac{p^{1/2}}{10^3} = 0.002237,$$

$$\frac{li\, q_3^{1/2}}{q_3} < \frac{li\,(10^3/5^{1/2})^{1/2}}{10^3/5^{1/2}} < \frac{li\, 21.176}{447.21} < \frac{10.75}{447.21} < 0.02405.$$

Therefore

$$(7') \qquad S'' > p(li\, q_3/q_3 - 0.0263)(11.9984 - 3q_3^2/8p).$$

Finally, by Lemmas 17 and 31, we have

$$S' = \sum_{a=q_3}^{A} \sum_{\nu=1,3\nmid\nu}^{a/q_3} \pi\left(\frac{a}{\nu}\right) < \sum_{a=q_3}^{A} \sum_{\nu=1,3\nmid\nu}^{a/q_3} li\frac{a}{\nu}$$

$$< \frac{1}{6}A(A+2)\log\left(\log\frac{A}{4}\log\frac{A}{5}\right) - \frac{1}{6}(2A^2+2A+9q_3)\log\log q_3$$

$$+ \frac{li\, q_3}{12q_3}(4A^2+8A+q_3^2+4) + \frac{0.03A}{\log A/5} + \frac{q_3^2}{8\log q_3}\left(\log\frac{A^2}{20q_3^2} + \frac{9}{20q_3}\right)$$

$$+ A^2\left(\int_{1/2}^{1}\frac{(1-x)dx}{\log Ax} + \int_{1/4}^{1/2}\frac{(2-3x)dx}{\log Ax}\right.$$

$$\left. + \int_{1/5}^{1/4}\frac{(1.1667-2.0472x)dx}{\log Ax} + 0.1016\int_{q_3/A}^{1/5}\frac{xdx}{\log Ax}\right) + li\, A + li\frac{A}{2}$$

$$< \frac{A^2}{3}\left\{\frac{1}{2}\left(1+\frac{2}{A}\right)\log\left(\log\frac{A}{4}\log\frac{A}{5}\right) - \log\log q_3 + \frac{6\, li\, A}{A^2}\right.$$

$$+ \frac{li\, q_3}{q_3}\left(1+\frac{2}{A}+\frac{q_3^2}{4A^2}+\frac{1}{A^2}\right) + \frac{0.09}{A\log A/5}$$

$$+ \frac{3q_3^2}{8A^2\log q_3}\left(\log\frac{A^2}{20q_3^2}+\frac{9}{20q_3}\right) + 3\left(\int_{1/2}^{1}\frac{(1-x)dx}{\log Ax}\right.$$

$$+ \int_{1/4}^{1/2}\frac{(2-3x)dx}{\log Ax} + \int_{1/5}^{1/4}\frac{(1.1667-2.0472x)dx}{\log Ax}$$

$$\left.\left. + 0.1016\int_{q_3/A}^{1/5}\frac{xdx}{\log Ax}\right)\right\}.$$

Since

$$\frac{1}{A}\log\left(\log\frac{A}{4}\log\frac{A}{5}\right) < \frac{1}{5999}\log(\log 1500\log 1200)$$

$$< \frac{1}{5999}\log(7.31323\times 7.09008) < 0.0007,$$

$$\frac{6\, li\, A}{A^2} < \frac{6}{5999(\log 5999 - 2)} < 0.00016,$$

$$\frac{0.09}{A\log A/5} < \frac{0.09}{5999} < 0.00002,$$

$$\frac{3p}{8A^2\log q_3}\left(\log\frac{A^2}{20q_3^2}+\frac{9}{20q_3}\right)<\frac{3}{286\ \log 10^3/5^{1/2}}\left(\log\frac{36}{4}+\frac{9(5)^{1/2}}{20\times 10^3}\right)<0.0038,$$

$$\int_{1/2}^{1}\frac{(1-x)dx}{\log Ax}<\frac{1}{\log A/2}\int_{1/2}^{1}(1-x)dx<\frac{1}{8.006}\times\frac{1}{8}<0.0157,$$

$$\int_{1/4}^{1/2}\frac{(2-3x)dx}{\log Ax}<\frac{1}{\log A/4}\int_{1/4}^{1/2}(2-3x)dx<0.03,$$

$$\int_{1/5}^{1/4}\frac{(1.1667-2.0472x)dx}{\log Ax}<\frac{1}{\log A/5}\int_{1/5}^{1/4}(1.1667-2.0472x)dx<0.005,$$

$$0.1016\int_{q_3/A}^{1/5}\frac{xdx}{\log Ax}<0.1016\times\frac{0.002}{\log q_3}<\frac{0.1016\times 0.002}{6.103}<0.0001,$$

and the total sum of the last four integrals is less than 0.0508, we have

$$\begin{aligned}S' < 12p\Big\{&\frac{1}{2}\log\left(\log\frac{A}{4}\log\frac{A}{5}\right)-\log\log q_3\\ (8')\qquad &+\frac{li\, q_3}{q_3}\left(1+\frac{2}{5.999p^{1/2}}+\frac{q_3^2}{143.952p}+0.00001\right)\\ &+0.0038\,\frac{q_3^2}{p}+0.15328\Big\}\end{aligned}$$

(for $0.0007+0.00016+0.00002+3\times 0.0508=0.15328$).

By (4), (5′), (6′), (7′) and (8′), we have

$$\begin{aligned}4.3731 < 12\Big\{&\frac{1}{2}\log\left(\log\frac{A}{4}\log\frac{A}{5}\right)-\log\log q_3\\ &+\frac{li\, q_3}{q_3}\left(1+\frac{2}{5.999p^{1/2}}+\frac{q_3^2}{143.952p}+0.00001\right)\\ &+0.0038\,\frac{q_3^2}{p}+0.15328\Big\}\\ &-(li\, q_3/q_3-0.0263)(11.9984-3q_3^2/8p)+60.1465q_3/p,\end{aligned}$$

$$\begin{aligned}\frac{4.3731}{12}<&\frac{1}{2}\log\left(\log\frac{A}{4}\log\frac{A}{5}\right)-\log\log q_3\\ &+\frac{li\, q_3}{q_3}(0.00014+0.00034+0.03825q_3^2/p+0.00001)\\ &+0.0263\left(1-\frac{q_3^2}{32p}\right)+0.0038q_3^2/p+\frac{60.1465}{12}\,\frac{q_3}{p}+0.15328.\end{aligned}$$

Since

$$\frac{li\, q_3}{q_3} < \frac{1}{\log q_3 - 2} < \frac{1}{4.103} < 0.24373,$$

we have

$$\begin{aligned} 0.3644 < \frac{1}{2} \log \left(\log \frac{A}{4} \log \frac{A}{5} \right) &- \log \log q_3 \\ &+ 0.24373(0.00049 + 0.03825 q_3^2/p) \\ &+ 0.0263(1 - q_3^2/32p) + 0.0038 q_3^2/p \\ &+ 5.013 q_3/p + 0.15328, \end{aligned}$$

that is,

$$\begin{aligned} 0.3644 < \frac{1}{2} \log \left(\log \frac{A}{4} \log \frac{A}{5} \right) &- \log \log q_3 + 0.0123 q_3^2/p \\ &+ 5.013 q_3/p + 0.1797. \end{aligned}$$

Then

$$\begin{aligned} \log \log q_3 < \frac{1}{2} \log \left(\log \frac{A}{4} \log \frac{A}{5} \right) &+ 0.0123 q_3^2/p \\ &+ 5.013 q_3/p - 0.1847. \end{aligned} \tag{9}$$

Since $(p/5)^{1/2} < q_3 < 3p^{1/2}$, we have

$$\begin{aligned} \log \log q_3 &< \frac{1}{2} \log \left(\log \frac{A}{4} \log \frac{A}{5} \right) + 0.0123 \times 9 \\ &\quad + \frac{5.013 \times 3}{10^3} - 0.1847 \\ &< \frac{1}{2} \log \left(\log \frac{A}{4} \right)^2, \end{aligned}$$

that is,

$$q_3 < (3/2) p^{1/2}. \tag{10}$$

For $p < 10^8$, we have by (10),

$$\begin{aligned} \frac{d}{dq_3} &\left(\log \log q_3 - \frac{0.0123}{p} q_3^2 - \frac{5.013 q_3}{p} \right) \\ &= \frac{1}{q_3 \log q_3} - \frac{0.0246 q_3}{p} - \frac{5.013}{p} \\ &> \frac{2}{3p^{1/2} \log ((3/2) \times 10^4)} - \frac{0.0246}{p^{1/2}} \frac{3}{2} - \frac{5.013}{p} > 0. \end{aligned}$$

Therefore (9) gives

$$\log\log\left(\frac{p}{5}\right)^{1/2} < \frac{1}{2}\log\left(\log\frac{3}{2}p^{1/2}\log\frac{6}{5}p^{1/2}\right)$$

$$+\frac{0.0123}{p}\left(\left(\frac{p}{5}\right)^{1/2}\right)^2+\frac{5.013}{p}\left(\frac{p}{5}\right)^{1/2} - 0.1847$$

$$< \frac{1}{2}\log\left(\log\frac{3}{2}p^{1/2}\log\frac{6}{5}p^{1/2}\right)$$

$$+ 0.00246 + \frac{5.013}{10^3(5)^{1/2}} - 0.1847$$

$$< \frac{1}{2}\log\left(\log\frac{3}{2}p^{1/2}\log\frac{6}{5}p^{1/2}\right) - 0.1799$$

or

$$0.35 < 0.1799 \times 2 < \log\frac{\log(3p^{1/2}/2)\log(6p^{1/2}/5)}{(\log(p/5)^{1/2})^2}$$

$$< \log\frac{\log(3/2\times 10^3)\log(6/5\times 10^3)}{(\log 10^3/5^{1/2})^2}$$

$$< \log 1.393 < 0.332,$$

which is absurd.

Thus the lemma is established.

7. Proof of the theorem for E. A.

THEOREM 4. *For a prime* $p\equiv 17 \pmod{24}$, *there is no E.A. in* $R(p^{1/2})$ *except possibly for* $p=17$, 41, 89, 113 *and* 137.

Proof. (1) Suppose $q_2=5$. Since

$$40q_3 < 7.5\times 40p^{1/2} = 300p^{1/2} < p$$

if $p>300^2=90000$, the theorem is true for $p>90000$. For $p<90000$, it is verified directly that the condition

$$40q_3 < p$$

holds except for $p=17$, 113, 233 and 257. But

$$233 = 2\cdot 5\cdot q_3 + 41.3 \qquad (q_3 = 11),$$

$$257 = 5q_3 + 74.3 \qquad (q_3 = 7),$$

by Lemma 15 (of I) the field $R(p^{1/2})$ has no E.A. except possibly for $p=17$, 113 and 137.

(2) Suppose $q_2\neq 5$. For $p>10^6$, the theorem follows from Lemma 32 and Lemma 16 (of I). For $p\leqq 10^6$, it is verified directly that the condition

$$5q_2q_3 < p$$

is satisfied except for $p=41$, 89 and 271. But

$$271 = 3q_3 + q_2 \cdot 34 \qquad (q_2 = 7,\ q_3 = 11),$$

by Lemma **14** (of I) we have the theorem.

NATIONAL TSING HUA UNIVERSITY,
KUNMING, YUNNAN, CHINA.

Reprinted from the
BULLETIN OF THE AMERICAN MATHEMATICAL SOCIETY
Vol. 51, No. 8, pp. 537–539, 1945

A REMARK ON A RESULT DUE TO BLICHFELDT

LOO-KENG HUA

Let $\sigma \geqq 1$ and $\xi_1, \cdots, \xi_n$ be $n \geqq 3$ linear forms of the real variables $x_1, \cdots, x_n$ of nonvanishing determinant Δ. For simplicity's sake we assume $|\Delta| = 1$. Let $2s$ of the forms be pairwise conjugate complex and the remaining $n - 2s$ be real. Then

$$|\xi_1|^\sigma + \cdots + |\xi_n|^\sigma \leqq 1$$

defines a symmetric convex body in the x-space, the volume $V(\sigma)$ of which equals

$$2^n \cdot \frac{\{\Gamma(1 + \alpha)\}^{n-2s}\{\pi\Gamma(1 + 2\alpha)/2^{1+2\alpha}\}^s}{\Gamma(1 + n\alpha)} \qquad (\alpha = 1/\sigma).$$

Minkowski's principle states that there is a lattice point $(x_1, \cdots, x_n) \neq (0, \cdots, 0)$ satisfying the inequality

$$|\xi_1|^\sigma + \cdots + |\xi_n|^\sigma \leqq r^\sigma$$

provided

$$r^n \geqq 2^n V^{-1}(\sigma). \tag{1}$$

By means of Blichfeldt's method, van der Corput and Schaake[1] obtained a sharpening of this result for $\sigma \geqq 2$. Decisive in this procedure is an inequality of the following form

$$\sum_{p,q=1}^{k} |z_p - z_q|^\sigma \leqq \epsilon(\sigma)k \cdot \sum_{p=1}^{k} |z_p|^\sigma, \tag{2}$$

where the factor $\epsilon(\sigma)$ depends neither on the arbitrary complex numbers z_p nor on k. Once such an inequality is known, (1) may be replaced by

$$r^n \geqq (\epsilon(\sigma))^{n/\sigma} \cdot \frac{n + \sigma}{\sigma} \cdot V^{-1}(\sigma). \tag{3}$$

The elementary relation

$$|u - v|^\sigma \leqq 2^{\sigma-1}(|u|^\sigma + |v|^\sigma)$$

(following from the fact that x^σ is a convex function of $x > 0$) implies (2) with $\epsilon(\sigma) = 2^\sigma$. Substituted in (3) this does not improve, but on

Received by the editors February 24, 1945.

[1] Acta Arithmetica vol. 2 (1936) pp. 152–160.

the contrary worsens, Minkowski's inequality. However, van der Corput and Schaake obtained the better value $2^{\sigma-1}$ for $\sigma \geqq 2$. I shall show here that $\epsilon(\sigma)=2$ is a legitimate choice for $1 \leqq \sigma \leqq 2$ and that both facts follow almost immediately from Marcel Riesz's convexity theorem.

Indeed, specialize this theorem (Theorem 296 on p. 219 of Hardy, Littlewood and Pólya's *Inequalities*) by taking $\gamma=\alpha$ and the X as the linear forms $X_{pq}=z_p-z_q$. It then turns out that the logarithm of the maximum $M_k(\alpha)$ of

$$\left\{\sum_{p,q=1}^{k} |z_p - z_q|^{1/\alpha} / k \sum_{p=1}^{k} |z_p|^{1/\alpha}\right\}^{\alpha}$$

for fixed k and variable $z_1, \cdots, z_k$ is a convex function of α in the interval $0 \leqq \alpha \leqq 1$. One readily verifies that

$$M_k(0) = 2, \qquad M_k(1/2) = 2^{1/2}, \qquad M_k(1) = 2(1 - 1/k) \leqq 2.$$

As

$$\left\{\sum_{p} |z_p|^{1/\alpha}\right\}^{\alpha} \to \max |z_p| \quad \text{for} \quad \alpha \to 0,$$

the first equation follows from $\max |z_p - z_q| \leqq 2 \cdot \max |z_p|$ together with the observation that the upper bound 2 is attained for $z_1=1$, $z_2=-1$, $z_3=\cdots=z_k=0$. Similarly the two other equations are immediate consequences of the elementary inequalities

$$\sum_{p,q} |z_p - z_q|^2 = 2k \sum_{p} |z_p|^2 - 2\left|\sum_{p} z_p\right|^2 \leqq 2k \sum_{p} |z_p|^2,$$

$$\sum_{p \neq q} |z_p - z_q| \leqq \sum_{p \neq q} (|z_p| + |z_q|) = 2(k-1) \sum_{p} |z_p|,$$

and the corresponding obvious observations about the z_p for which the upper bound is reached.

Let us use 2 as the basis of our logarithms. Then the values of $\log_2 M_k(\alpha)$ are 1, 1/2 and less than or equal to 1 for $\alpha=0, 1/2, 1$ respectively, and hence the broken line consisting of $1-\alpha$ for $0 \leqq \alpha \leqq 1/2$ and α for $1/2 \leqq \alpha \leqq 1$ gives an upper bound for the convex function $\log_2 M_k(\alpha)$. We thus obtain the promised result that (2) holds with

$$\epsilon(\sigma) = 2^{\sigma-1} \quad \text{for} \quad \sigma \geqq 2 \qquad \text{and} \qquad \epsilon(\sigma) = 2 \quad \text{for} \quad 1 \leqq \sigma \leqq 2. \tag{4}$$

Both choices are the best possible of their kinds, as, for $0 \leqq \alpha \leqq 1/2$, is shown by the example $k=2$, $z_1=-z_2=1$, and, for $1/2 \leqq \alpha \leqq 1$, by the example $z_1=-z_2=1$, $z_3=\cdots=z_k=0$, with large k.

Consider the case $1 \leqq \sigma \leqq 2$. If we substitute the value $\epsilon(\sigma) = 2$ in (3), we shall find that it does not always improve Blichfeldt's known inequality, in particular not for the most interesting case $\sigma = 1$. We observe that

$$\left(\frac{|\xi_1|^\sigma + \cdots + |\xi_n|^\sigma}{n}\right)^{1/\sigma}$$

is an increasing function of the exponent σ, while the upper bound for its lattice minimum as derived from (3), namely,

$$\left(\frac{2}{n}\right)^{1/\sigma}\left(\frac{n+\sigma}{\sigma}\right)^{1/n}(V(\sigma))^{-1/n}, \tag{5}$$

is not. For $s = 0$ the expression (5) tends to a limit with $n \to \infty$, namely

$$\frac{1}{2}\left(\frac{2}{\sigma e}\right)^{1/\sigma} \Big/ \Gamma\left(1 + \frac{1}{\sigma}\right) = 2^{\alpha-1}\left(\frac{\alpha}{e}\right)^{\alpha} \Big/ \Gamma(1+\alpha).$$

The logarithmic derivative of this function with respect to α is negative for $\alpha = 1/2$ and positive for $\alpha = 2/3$, and hence this function has a minimum between $\sigma = 2$ and $\sigma = 1.5$; numerical computation gives as its location $\sigma = \sigma_0 = 1.8653 \cdots$.[2] At this point the value of the function is

$$\leqq 1/(3.146e)^{1/2}$$

which is slightly better than the constant

$$1/(\pi e)^{1/2}$$

due to Blichfeldt.[3]

In conclusion, for $2 \geqq \sigma \geqq \sigma_0$, (1) may be replaced by

$$r^n \geqq 2^{n/\sigma}\left(\frac{n+\sigma}{\sigma}\right)V^{-1}(\sigma),$$

and, for $1 \leqq \sigma \leqq \sigma_0$, (1) may be replaced by

$$r^n \geqq 2^{n/\sigma_0}\left(\frac{n+\sigma_0}{\sigma_0}\right)V^{-1}(\sigma_0).$$

This would be true however σ_0 were chosen within the limits $1 \leqq \sigma_0 \leqq 2$; our special choice approaches the best possible for $n \to \infty$ (and $s = 0$) and is sharp enough to beat Blichfeldt's record by a slight margin, even for small n.

National Tsing Hua University

[2] The author is indebted to Mr. Sze for this numerical value.

[3] Trans. Amer. Math. Soc. vol. 15 (1914) pp. 227–235.

IMPROVEMENT OF A RESULT OF WRIGHT

LOO-KENG HUA*.

[*Extracted from the Journal of the London Mathematical Society, Vol. 24, 1949.*]

E. M. Wright, in his paper entitled "The Prouhet-Lehmer problem", [1], stated "I have not yet found an upper bound for $W(k, s)$ independent of s". Here I give one. $W(k, s)$ is the least integer j for which there are integers x_{iu} $(1 \leqslant i \leqslant j,\ 1 \leqslant u \leqslant s)$ such that

$$\sum_{i=1}^{j} x_{i1}^{h} = \sum_{i=1}^{j} x_{i2}^{h} = \ldots = \sum_{i=1}^{j} x_{is}^{h} \quad (1 \leqslant h \leqslant k) \tag{1}$$

and

$$\sum_{i=1} x_{ip}^{k+1} \neq \sum_{i=1}^{j} x_{iq}^{k+1} \cdot \tag{2}$$

for all p, q for which $p \neq q$. Another function $L(k, s)$, named after Lehmer [2], is defined in the same way as $W(k, s)$, except that the phrase "for all p, q for which $p \neq q$" is replaced by "for at least one pair of values p, q". Evidently $L(k, 2) = W(k, 2)$ and $L(k, s) \leqslant W(k, s)$. By an adaptation of my method [3], Wright [1] proved that my upper bound

$$j_0 = (k+1)\left(\left[\frac{\log \frac{1}{2}(k+2)}{\log(1+1/k)}\right]+1\right) \tag{3}$$

for $L(k, 2)$ holds good for $L(k, s)$ in general and that

$$W(k, s) \leqslant (k+1)\left(\left[\frac{\log \frac{1}{2}(k(s-1)+2)}{\log(1+1/k)}\right]+1\right).$$

In this note, I shall prove that my upper bound (3) holds for $W(k, s)$ for any s, which gives an upper bound independent of s. The result is evidently a consequence of the following

THEOREM. *Let $j \geqslant j_0$. For any given s, there exist integers*

$$N_1, \ldots, N_k;\quad M_1, \ldots, M_s \quad (M_{t_1} \neq M_{t_2} \text{ when } t_1 \neq t_2)$$

such that the s systems of Diophantine equations

$$R_t \ (1 \leqslant t \leqslant s): \quad \begin{cases} \sum_{i=1}^{j} x_{it}^{h} = N_h & (1 \leqslant h \leqslant k), \\ \sum_{i=1}^{j} x_{it}^{k+1} = M_t & (x_{it} \geqslant 0) \end{cases}$$

are soluble.

---

* Received 10 August, 1948; read 16 December, 1948.

Let $a_1, \ldots, a_{k+1}$ be the set of positive numbers in Lemma 1 of Hua [3] with $k+1$ for k. We suppose throughout that

$$a_u P^{\beta^{v-1}} \leqslant y_{uv} \leqslant 2a_u P^{\beta^{v-1}} \quad (1 \leqslant u \leqslant k+1,\ 1 \leqslant v \leqslant l),$$

where $\beta = k/(k+1)$. Let $r(n_1, \ldots, n_k)$ be the number of solutions of

$$\sum_{u=1}^{k+1} \sum_{v=1}^{l} y_{uv}^h = n_h \quad (1 \leqslant h \leqslant k).$$

There exists a set of integers $N_1, \ldots, N_h$ such that

$$r(N_1, \ldots, N_k) \geqslant c_1 P^{(k+1)^2(1-\beta^l)-\frac{1}{2}k(k+1)}, \tag{4}$$

where c_1 (and later $c_2, \ldots$) are positive and depend only on k. In fact, the number of different y-sets is not less than

$$\tfrac{1}{2} \prod_{u=1}^{k+1} \prod_{v=1}^{l} a_u P^{\beta^{v-1}} \geqslant c_2 P^{(k+1)(1+\beta+\ldots+\beta^{l-1})} = c_2 P^{(k+1)^2(1-\beta^l)}.$$

Since $|n_h| \leqslant c_3 P^h$, the number of different n-sets is

$$\leqslant c_4 P^{1+2+\ldots+k} = c_4 P^{\frac{1}{2}k(k+1)}.$$

Therefore there exists an n-set, say $N_1, \ldots, N_k$, such that

$$r(N_1, \ldots, N_k) \geqslant \frac{c_2}{c_4} P^{(k+1)^2(1-\beta^l)-\frac{1}{2}k(k+1)}.$$

The number of solutions of the system of equations

$$\sum_{u=1}^{k+1} \sum_{v=1}^{l} y_{uv}^h = N_h \quad (1 \leqslant h \leqslant k+1)$$

is $\leqslant c_5 P^{\frac{1}{2}k(k+1)(1-\beta^l)}$. This result is Lemma 2 of Wright [1].

Now we consider all the solutions of

$$\sum_{u=1}^{k+1} \sum_{v=1}^{l} y_{uv}^h = N_h \quad (1 \leqslant h \leqslant k).$$

To each solution, we have an integer M such that

$$\sum_{u=1}^{k+1} \sum_{v=1}^{l} y_{uv}^{k+1} = M.$$

If the set M contains only e ($\leqslant s-1$) different elements, say, $M_1, \ldots, M_e$, then the total number of solutions of e systems of equations

$$\Pi_i\ (1 \leqslant i \leqslant e): \quad \begin{cases} \sum_{u=1}^{k+1} \sum_{v=1}^{l} y_{uv}^h = N_h \quad (1 \leqslant h \leqslant k) \\ \sum_{u=1}^{k+1} \sum_{v=1}^{l} y_{uv}^{k+1} = M \end{cases}$$

is $\leqslant c_5 e P^{\frac{1}{2}k(k+1)(1-\beta^l)}$. For $l > \{\log \frac{1}{2}(k+2)\}/\log(1+1/k)$, we have

$$(5) \qquad c_5 e P^{\frac{1}{2}k(k+1)(1-\beta^l)} < c_1 P^{(k+1)^2(1-\beta^l)-\frac{1}{2}k(k+1)} \leqslant r(N_1, \ldots, N_k)$$

for P large. This is a contradiction.

The proof tells us a great deal more. For example, a relation between s and the size of the unknown x_{ij} can be obtained.

I take this opportunity to correct an error in the foot-note of my paper [**3**]. In the last line of the foot-note, the phrase "provided $s \geqslant ck^3 \log k$" is missing. This assertion was proved later in Hua [**5**] as Theorem 16, p. 133. Elsewhere, [**6**], I have announced a better result in this direction.

Literature.

1. E. M. Wright, *Journal London Math. Soc.*, 23 (1948), 279–285.
2. D. H. Lehmer, *Scripta Math.*, 13 (1947), 37–41.
3. L. K. Hua, *Quarterly J.*, 9 (1938), 315–320.
4. L. K. Hua, *Quarterly J.*, 11 (1940), 161–176.
5. L. K. Hua, *Travaux de l'Institute Math. Stekloff*, 22 (1947).
6. L. K. Hua, *Proc. Nat. Acad. Sci., U.S.A.*

University of Illinois,
Urbana, Illinois, U.S.A.

Reprinted from the
QUARTERLY JOURNAL OF MATHEMATICS
Vol. 20, pp. 48–61, March 1949

AN IMPROVEMENT OF VINOGRADOV'S MEAN-VALUE THEOREM AND SEVERAL APPLICATIONS

By LOO-KENG HUA

[Received 21 July 1948]

1. Introduction. IN 1940 I showed† that Vinogradov's estimation (1) of Weyl's sum depends essentially on a result which I called 'Vinogradov's mean-value theorem'. The purpose of the present paper is to improve this mean-value theorem. Since Vinogradov's method seems to have reached a final stage, any improvement of the constant in the exponent is worthy of consideration. More definitely, in this paper I shall establish a mean-value theorem by means of which we can establish sharper results about Waring's problem, distribution of primes, etc. The method used here seems to be much simpler than that originally used by Vinogradov.

The form of Vinogradov's mean-value theorem which will be proved here is the following.

THEOREM 1. *Let P and T be integers and $P \geqslant 2$,*

$$f(x) = \alpha_k x^k + \ldots + \alpha_1 x, \tag{1}$$

and let

$$C_k = C_k(P) = \sum_{T<x\leqslant T+P} e^{2\pi i f(x)}. \tag{2}$$

Then, when

$$s \geqslant \tfrac{1}{4}k(k+1)+lk, \tag{3}$$

we have

$$\int_0^1 \ldots \int_0^1 |C_k(P)|^{2s}\, d\alpha_1 \ldots d\alpha_2 \leqslant (7s)^{4sl}(\log P)^{2l} P^{2s-\frac{1}{2}k(k+1)+\delta}, \tag{4}$$

where

$$\delta = \tfrac{1}{2}k(k+1)(1-1/k)^l. \tag{5}$$

Arithmetically, the value of the integral in (4) is equal to the number of solutions of the system of equations

$$x_1^h + \ldots + x_s^h = y_1^h + \ldots + y_s^h \quad (1 \leqslant h \leqslant k), \tag{6}$$

where

$$T < x_i, \quad y_i \leqslant T+P.$$

† Hua, *Additive Prime Number Theory*. This booklet was accepted for publication by the Academy of U.S.S.R. in 1940, but its appearance was delayed by the war.

Setting $X_i = x_i - T$, $Y_i = y_i - T$, we obtain from (6)

$$\sum_{i=1}^{s} (X_i+T)^h = \sum_{i=1}^{s} (Y_i+T)^h \quad (1 \leqslant h \leqslant k). \tag{7}$$

Expanding the hth powers, we see that the system of equations (7) is equivalent to

$$\sum_{i=1}^{s} X_i^h = \sum_{i=1}^{s} Y_i^h \quad (1 \leqslant h \leqslant k), \tag{8}$$

where $0 < X_i, Y_i \leqslant P$. This establishes that the left-hand side of (4) is really independent of T.

Since Vinogradov's paper and my booklet are both published in Russian, the paper is set forth *ab initio*.

2. Lemmas

LEMMA 1. *Let $Q = RH$, $R > 1$, $H > 1$ and let $g_1,..., g_k$ be integers satisfying*

$$1 < g_1 < g_2 < ... < g_k \leqslant H, \qquad g_\nu - g_{\nu-1} > 1. \tag{9}$$

For each value of ν $(1 \leqslant \nu \leqslant k)$ let x_ν be a variable lying in the interval

$$-\omega+(g_\nu-1)R < x_\nu \leqslant -\omega+g_\nu R \quad (0 \leqslant \omega \leqslant Q). \tag{10}$$

The number of sets of such integers $x_1,..., x_k$ for which the values of

$$x_1^h+...+x_k^h \tag{11}$$

lie in intervals of lengths not exceeding Q^{h-1} $(1 \leqslant h \leqslant k)$ respectively, is less than or equal to

$$(2kH)^{\frac{1}{2}k(k-1)}. \tag{12}$$

Proof. Let $x_1,..., x_k$ and $y_1,..., y_k$ be two sets of integers satisfying the requirements of the lemma; let

$$s_h = \sum_{\nu=1}^{k} x_\nu^h, \qquad s'_h = \sum_{\nu=1}^{k} y_\nu^h,$$

and let σ_h be the hth elementary symmetric function of $x_1,..., x_k$ and σ'_h that of $y_1,..., y_k$. Then, by (10), we have

$$|s_h| \leqslant \sum_{\nu=1}^{k} |x_\nu|^h \leqslant kQ^h, \qquad |s'_h| \leqslant kQ^h \tag{13}$$

and

$$|\sigma_h| \leqslant \binom{k}{h} Q^h, \qquad |\sigma'_h| \leqslant \binom{k}{h} Q^h. \tag{14}$$

By the hypotheses, we have

$$|s_h - s'_h| \leqslant Q^{h-1} \quad (1 \leqslant h \leqslant k). \tag{15}$$

From (15), we shall deduce that

$$|\sigma_h - \sigma'_h| \leqslant \tfrac{3}{4}(2kQ)^{h-1} \quad \text{for} \quad 2 \leqslant h \leqslant k, \tag{16}$$

and therefore

$$|\sigma_h-\sigma'_h| \leqslant (2kQ)^{h-1} \quad \text{for} \quad 1 \leqslant h \leqslant k. \tag{17}$$

Since $\sigma_2 = \frac{1}{2}(s_1^2-s_2)$, we have by (13)

$$|\sigma_2-\sigma'_2| \leqslant \tfrac{1}{2}\{|s_1-s'_1|(|s_1|+|s'_1|)+2|s_2-s'_2|\} \leqslant \tfrac{1}{2}(2k+1)Q \leqslant \tfrac{3}{4}(2kQ), \tag{18}$$

so that (16) holds for $h = 2$. We use induction and suppose that (16) is true for $2 \leqslant h \leqslant t-1$. We then deduce from (13), (14), (15), and (16) that, for $1 \leqslant \nu \leqslant t-1$,

$$\begin{aligned}|\sigma_\nu s_{t-\nu}-\sigma'_\nu s'_{t-\nu}| &\leqslant |\sigma_\nu-\sigma'_\nu||s_{t-\nu}|+|\sigma'_\nu||s_{t-\nu}-s'_{t-\nu}| \\ &\leqslant \left\{(2k)^\nu k+\binom{k}{\nu}\right\}Q^{t-1} \leqslant \left(1+\frac{1}{\nu!}\right)(2k)^{\nu-1}kQ^{t-1}. \end{aligned}\tag{19}$$

By a well-known theorem on symmetric functions, however, we have

$$s_t-\sigma_1 s_{t-1}+\sigma_2 s_{t-2}-...+(-1)^t t\sigma_t = 0 \tag{20}$$

and

$$s'_t-\sigma'_1 s'_{t-1}+\sigma'_2 s'_{t-2}-...+(-1)^t t\sigma'_t = 0. \tag{21}$$

Combining (19), (20), and (21) we obtain

$$\begin{aligned}|\sigma_t-\sigma'_t| &\leqslant \frac{1}{t}\left(1+2k+\tfrac{3}{2}k\sum_{\nu=2}^{t-1}(2k)^{\nu-1}\right)Q^{t-1} \\ &\leqslant \tfrac{1}{2}[1+2k+\tfrac{3}{2}k\{(2k)^{t-1}-2k\}/(2k-1)]Q^{t-1} \\ &\leqslant \tfrac{1}{2}\{1+\tfrac{1}{2}k+\tfrac{3}{2}k(2k)^{t-1}/(2k-1)\}Q^{t-1} \\ &\leqslant \tfrac{3}{4}(2k)^{t-1}Q^{t-1}. \end{aligned}\tag{22}$$

Consequently, we have, for $|X| \leqslant Q$, that

$$\begin{aligned}|(X-x_1)...(X-x_k)-(X-y_1)...(X-y_k)| &\leqslant \sum_{h=1}^{k}|\sigma_h-\sigma'_h||X|^{k-h} \\ &\leqslant \left\{1+\tfrac{3}{4}\sum_{h=2}^{k}(2k)^{h-1}\right\}Q^{k-1} \\ &\leqslant \left\{1+\frac{6k}{4(2k-1)}\{(2k)^{k-1}-1\}\right\}Q^{k-1} \\ &\leqslant (2kQ)^{k-1}, \end{aligned}\tag{23}$$

since $2k/(2k-1) \leqslant \frac{4}{3}$.

But $|y_k-x_\nu| \geqslant R$ for $\nu = 1, 2,..., k-1$, so, if we set $X = y_k$ in (23), we obtain

$$R^{k-1}|y_k-x_k| \leqslant (2kQ)^{k-1}.$$

Therefore the number of x_k satisfying the requirements of our theorem does not exceed $(2kQ)^{k-1}$. Next, for fixed x_k, the numbers

$$x_1^h+...+x_{k-1}^h \quad (1 \leqslant h \leqslant k-1) \tag{24}$$

lie in intervals of lengths at most Q^{h-1} $(1 \leqslant h \leqslant k-1)$ respectively. This reduces to the exact formulation of our lemma with $k-1$ instead of k. The lemma is evident for $k = 1$. We suppose that it holds for smaller k; then the number of sets of integers $x_1,..., x_{k-1}$ satisfying the requirements imposed on (24) does not exceed

$$\{2(k-1)H\}^{\frac{1}{2}(k-1)(k-2)}.$$

Therefore the number of sets of integers satisfying the requirements imposed on (11) is less than or equal to

$$\{2(k-1)H\}^{\frac{1}{2}(k-1)(k-2)}(2kH)^{k-1} \leqslant (2kH)^{\frac{1}{2}k(k-1)}.$$

LEMMA 2. *Let* $c \geqslant 1$. *Under the same hypothesis as in Lemma* 1, *the number of sets of integers* $x_1,..., x_k$ *for which*

$$x_1^h+...+x_k^h \quad (1 \leqslant h \leqslant k)$$

lies in intervals of lengths not exceeding $cQ^{(1-1/k)h}$ *respectively* $(1 \leqslant h \leqslant k)$ *does not exceed*

$$(2c)^k(2kH)^{\frac{1}{2}k(k-1)}Q^{\frac{1}{2}(k-1)}. \tag{25}$$

Proof. We divide the hth interval into

$$\{cQ^{h(1-1/k)}/Q^{h-1}\}+1$$

parts and apply Lemma 1. Since

$$\prod_{h=1}^{k}\{(cQ^{h(1-1/k)}/Q^{h-1})+1\} \leqslant \prod_{h=1}^{k}(2cQ^{h(1-1/k)-(h-1)}) = (2c)^kQ^{\frac{1}{2}(k-1)},$$

we have at most $(2c)^kQ^{\frac{1}{2}(k-1)}$ sets of sub-intervals, each of them satisfying the hypothesis of Lemma 1. Therefore we have at most $(2kH)^{\frac{1}{2}k(k-1)}$ solutions for each set, and the theorem follows.

LEMMA 3. *The set of integers* $(g_1,..., g_b)$ *with* $1 \leqslant g_\nu \leqslant H$ *is said to be 'well-spaced' if there are at least* k *of them, say* $g_{j_1},..., g_{j_k}$, *satisfying*

$$g_{j_{\nu+1}}-g_{j_\nu} > 1 \quad (1 \leqslant \nu \leqslant k-1). \tag{26}$$

The number of not well-spaced sets is at most

$$b!\,3^bH^{k-1}. \tag{27}$$

Proof. We arrange $g_1,..., g_b$ in order of increasing magnitude

$$1 \leqslant g_1' \leqslant g_2' \leqslant ... \leqslant g_b', \tag{28}$$

and set $f_\nu = g_{\nu+1}'-g_\nu'$. If the set is not well-spaced, there are at most $k-2$ of the f's for which $f_\nu > 1$.

Consider now these sets with exactly σ $(0 \leqslant \sigma \leqslant k-2)$ f's with $f_\nu > 1$. The number of different positions of these σ f's is $\binom{b-1}{\sigma}$. Thus the number of different sets is at most

$$\binom{b-1}{\sigma} H^{\sigma+1} 2^{b-1-\sigma}$$

since $0 \leqslant f_\nu \leqslant H-1$ and $1 \leqslant g_1' \leqslant H$. The total number of not well-spaced sets is therefore

$$\leqslant \sum_{\sigma=0}^{k-2} \binom{b-1}{\sigma} H^{\sigma+1} 2^{b-1-\sigma} \leqslant (1+2)^{b-1} H^{k-1} \leqslant 3^b H^{k-1}.$$

The theorem now follows since the number of sets $(g_1, \ldots, g_b)$ corresponding to $(g_1', \ldots, g_b')$ is $b!$.

3. Recurrence formula

THEOREM 2. *Let b be an integer $\geqslant \frac{1}{4}k(k+1)+k$ and let η be the greatest integer not exceeding*

$$\frac{1}{k} \log Q / \log 2. \tag{29}$$

Then

$$\int_0^1 \cdots \int_0^1 |C_k(Q)|^{2b}\, d\alpha_1 \ldots d\alpha_k \leqslant (7b)^{4b} \max(1, \eta^2) Q^{2k - \frac{1}{2}(k+1) + 2(b-k)/k} \times$$

$$\times \int_0^1 \cdots \int_0^1 |C_k(Q^{1-1/k})|^{2(b-k)}\, d\alpha_1 \ldots d\alpha_k. \tag{30}$$

Proof. (i) We defined

$$C_k(Q) = \sum_{T < x \leqslant T+Q} e^{2\pi i f(x)}$$

in Theorem 1. From the remarks there, we see that without loss of generality we may assume that $T = 0$ hereafter. Suppose that $\eta \geqslant 2$, and let s be an integer satisfying $1 \leqslant s \leqslant \eta - 1$. We divide $C_k(Q)$ into 2^s parts, each of length $R_s = Q\, 2^{-s}$:

$$C_k(Q) = \sum_{g=1}^{2^s} \sum_{(g-1)R_s < x \leqslant g R_s} e^{2\pi i f(x)}$$

$$= \sum_{g=1}^{2^s} Z_{sg}, \quad \text{say.}$$

Let $Z = \{C_k(Q)\}^b$. Then

$$Z = \sum^{2^{sb}} Z_{sg_1} \ldots Z_{sg_b}, \tag{31}$$

where $\sum^{M}$ denotes a sum of at most M terms (I shall use this convention throughout the paper). I use the further abbreviation

$$Z_s = Z_{s;g_1,\ldots,g_b} = Z_{sg_1}\ldots Z_{sg_b}.$$

Those $Z_{s;g_1,\ldots,g_b}$ with well-spaced $g_1,\ldots,g_b$ are called *well-spaced sums* and are denoted by Z'_s. By Lemma 3, the number of not well-spaced sums does not exceed $b!\,3^b 2^{s(k-1)}$. Those Z_s which are not well-spaced sums are now decomposed further by dividing each factor into two parts, so that from each not well-spaced sum Z_s we obtain 2^b sums of the type Z_{s+1}. The number of well-spaced Z_{s+1} obtained from all of the not well-spaced Z_s clearly does not exceed

$$b!\,3^b 2^{s(k-1)}.2^b = b!\,6^b 2^{s(k-1)}.$$

Those Z_{s+1} thus obtained which are well-spaced are denoted by Z'_{s+1}, and we decompose the others as above. Since these Z_1 are always not well-spaced, we have no difficulty to begin with. We repeat this process for $s = 1, 2,\ldots, \eta-1$, and use Z'_η to denote *all* those Z_η obtained from those not well-spaced $Z_{\eta-1}$. We have therefore

$$Z = \sum_{s=1}^{\eta} \sum^{M_s} Z'_s, \tag{32}$$

where $M_s = b!\,6^b 2^{s(k-1)}$.

(ii) By Schwarz's inequality, we obtain

$$|C(Q)|^{2b} = |Z|^2 \leqslant \eta \sum_{s=1}^{\eta} \Big| \sum^{M_s} Z'_s \Big|^2 \leqslant \eta \sum_{s=1}^{\eta} M_s \sum^{M_s} |Z'_s|^2. \tag{33}$$

Suppose $g_1,\ldots, g_k$ of $Z'_{s;g_1\ldots g_b}$ $(1 \leqslant s \leqslant \eta-1)$ satisfy (9); otherwise, we can rearrange the subscripts. Since the geometrical mean does not exceed the arithmetical mean, we have

$$|Z_{sg_{k+1}}\ldots Z_{sg_b}|^2 \leqslant \frac{1}{b-k} \sum_{i=k+1}^{b} |Z_{sg_i}|^{2(b-k)}. \tag{34}$$

We divide Z_{sg_i} $(k+1 \leqslant i \leqslant b)$ into

$$[Q\,2^{-s}/(Q^{1-1/k}-1)]+1 \leqslant Q\,2^{-s}(Q^{1-1/k}-1)^{-1}+Q^{1/k}2^{-\eta}$$
$$\leqslant Q\,2^{-s}(\tfrac{3}{4}Q^{1-1/k})^{-1}+Q^{1/k}2^{-s-1} \leqslant Q^{1/k}2^{1-s}$$

(since $4 \leqslant 2^\eta \leqslant Q^{1/k} \leqslant Q^{1-1/k}$) parts, each of the form

$$C^* = \sum_x e^{2\pi i f(x)},$$

where x runs over an interval of length $\leqslant Q^{1-1/k}-1$; namely, we have an integer ω such that

$$\omega < x \leqslant \omega+Q', \qquad 0 < Q' \leqslant Q^{1-1/k}, \qquad 0 \leqslant \omega \leqslant g_i R_s \leqslant Q.$$

Then, by Hölder's inequality,

$$|Z_{sg_i}|^{2(b-k)} \leqslant \Big(\sum^{Q^{1/k}2^{1-s}} |C^*|\Big)^{2(b-k)} \leqslant (Q^{1/k}2^{1-s})^{2(b-k)-1} \sum^{Q^{1/k}2^{1-s}} |C^*|^{2(b-k)}. \tag{35}$$

From (33), (34), and (35) we then obtain

$$|Z|^2 \leqslant \frac{\eta}{b-k} \sum_{s=1}^{\eta} M_s (Q^{1/k}2^{1-s})^{2(b-k)-1} \sum^{N_s} |Z_{sg_1}|^2 \ldots |Z_{sg_k}|^2 |C^*|^{2(b-k)}, \tag{36}$$

where $N_s = M_s(b-k)Q^{1/k}2^{1-s} = b!\, 6^b . 2^{s(k-1)}(b-k)Q^{1/k}2^{1-s}$. Integrating over the unit hypercube $(0 \leqslant \alpha_1 \leqslant 1, \ldots, 0 \leqslant \alpha_k \leqslant 1)$, we have

$$\int_0^1 \ldots \int_0^1 |Z|^2\, d\alpha_1 \ldots d\alpha_k \leqslant \frac{\eta}{b-k} \sum_{s=1}^{\eta} M_s (Q^{1/k}2^{1-s})^{2(b-k)-1} \times$$

$$\times \sum^{N_s} \int_0^1 \ldots \int_0^1 |Z_{sg_1}|^2 \ldots |Z_{sg_k}|^2 |C^*|^{2(b-k)}\, d\alpha_1 \ldots d\alpha_k. \tag{37}$$

(iii) The expression

$$\int_0^1 \ldots \int_0^1 |Z_{sg_1}|^2 \ldots |Z_{sg_k}|^2 |C^*|^{2(b-k)}\, d\alpha_1 \ldots d\alpha_k \tag{38}$$

is equal to the number of solutions of the system of Diophantine equations

$$x_1^h + \ldots + x_k^h + y_1^h + \ldots + y_{b-k}^h = x_1'^h + \ldots + x_k'^h + y_1'^h + \ldots + y_{b-k}'^h \quad (1 \leqslant h \leqslant k),$$

where the y's lie in an interval of the form

$$\omega < y,\ y' \leqslant \omega + Q' \quad (0 < Q' \leqslant Q^{1-1/k};\ 0 \leqslant \omega \leqslant Q),$$

and the x_i and x_i' lie in intervals

$$(g_i - 1)R_s < x_i,\ x_i' \leqslant g_i R_s,$$

where, for $s \leqslant \eta - 1$, the integers $g_1, \ldots, g_k$ satisfy the condition (9).

We replace x by $X+\omega$ and y by $Y+\omega$. Then (38) is also the number of solutions of the system of equations

$$X_1^h + \ldots + X_k^h + Y_1^h + \ldots + Y_{b-k}^h = X_1'^h + \ldots + X_k'^h + Y_1'^h + \ldots + Y_{b-k}'^h \quad (1 \leqslant h \leqslant k), \tag{39}$$

where the Y's lie in the interval $(0, Q')$ and X_i and X_i' lie in

$$-\omega + (g_i - 1)R_s < X_i,\ X_i' \leqslant -\omega + g_i R_s \quad (0 \leqslant \omega \leqslant Q). \tag{40}$$

If now the X' are fixed arbitrarily, the conditions on the X satisfy the requirements of Lemma 1 with $R = R_s$ and Lemma 2 with $c = 2(b-k)$ and $H = 2^s$. Thus the number of sets of X and X' does not exceed

$$R_s^k\{4(b-k)\}^k(2k\,2^s)^{\frac{1}{2}k(k-1)}Q^{\frac{1}{2}(k-1)}$$
$$= \{4(b-k)\}^k(2k)^{\frac{1}{2}k(k-1)}2^{\frac{1}{2}sk(k-1)-sk}Q^{2k-\frac{1}{2}(k+1)}. \quad (41)$$

Further, for a fixed set of X and X', the number of sets of Y and Y' does not exceed

$$\int_0^1 \cdots \int_0^1 |C_k(Q^{1-1/k})|^{2(b-k)}\,d\alpha_1 \dots d\alpha_k,$$

since
$$\left|\int_0^1 f(x)e^{ixy}\,dx\right| \leqslant \int_0^1 |f(x)|\,dx.$$

Therefore, we have

$$\int_0^1 \cdots \int_0^1 |Z_{sg_1} \dots Z_{sg_k}|^2 |C^*|^{2(b-k)}\,d\alpha_1 \dots d\alpha_k$$
$$\leqslant \{4(b-k)\}^k(2k)^{\frac{1}{2}k(k-1)}2^{\frac{1}{2}sk(k+1)-2sk}Q^{2k-\frac{1}{2}(k+1)}\times$$
$$\times \int_0^1 \cdots \int_0^1 |C_k(Q^{1-1/k})|^{2(b-k)}\,d\alpha_1 \dots d\alpha_k, \quad (42)$$

for $1 \leqslant s \leqslant \eta-1$.

For $s = \eta$, we use the trivial inequality

$$\int_0^1 \cdots \int_0^1 |Z_{sg_1} \dots Z_{sg_k}|^2 |C^*|^{2(b-k)}\,d\alpha_1 \dots d\alpha_k$$
$$\leqslant R_\eta^{2k} \int_0^1 \cdots \int_0^1 |C_k(Q^{1-1/k})|^{2(b-k)}\,d\alpha_1 \dots d\alpha_k. \quad (43)$$

Since
$$R_\eta^{2k} = Q^{2k}2^{-2k\eta}$$
$$\leqslant 2^{-\eta[2k-\frac{1}{2}k(k+1)]}Q^{2k-\frac{1}{2}(k+1)}(Q\,2^{-\eta k})^{\frac{1}{2}(k+1)}$$
$$\leqslant 2^{-\eta[2k-\frac{1}{2}k(k+1)]}Q^{2k-\frac{1}{2}(k+1)}2^{\frac{1}{2}k(k+1)},\dagger$$

then (42) holds also for $s = \eta$.

† Since $\eta \geqslant \log Q/k \log 2 - 1$, $\log 2^\eta \geqslant \log Q^{1/k} - \log 2 = \log \frac{1}{2}Q^{1/k}$, then $Q\,2^{-\eta k} \leqslant 2^k$.

(iv) Combining (37) and (42) with $s = 1,\dots, \eta$, we have

$$\int_0^1 \dots \int_0^1 |C(Q)|^{2b}\, d\alpha_1 \dots d\alpha_k \leqslant \eta \sum_{s=1}^{\eta} M_s(Q^{1/k}2^{1-s})^{2(b-k)-1} N_s\{4(b-k)\}^k \times$$

$$\times (2k)^{\frac{1}{2}k(k-1)} 2^{\frac{1}{2}sk(k+1)-2sk} Q^{2k-\frac{1}{2}(k+1)} \int_0^1 \dots \int_0^1 |C_k(Q^{1-1/k})|^{2(b-k)}\, d\alpha_1 \dots d\alpha_k$$

$$\leqslant \eta c \sum_{s=1}^{\eta} 2^{-s\{2b-\frac{1}{2}k(k+1)-2k\}} Q^{2k-\frac{1}{2}(k+1)+2(b-k)/k} \times$$

$$\times \int_0^1 \dots \int_0^1 |C_k(Q^{1-1/k})|^{2(b-k)}\, d\alpha_1 \dots d\alpha_k$$

$$\leqslant \eta^2 c Q^{2k-\frac{1}{2}(k+1)+2(b-k)/k} \int_0^1 \dots \int_0^1 |C_k(Q^{1-1/k})|^{2(b-k)}\, d\alpha_1 \dots d\alpha_k, \tag{44}$$

since $2b \geqslant \frac{1}{2}k(k+1)+2k$, where

$$c = (b!\, 6^b)^2 2^{2(b-k)} (4b)^k (2k)^{\frac{1}{2}k(k-1)}.$$

Since $\quad c < (12b)^{2b}(4b)^b(2k)^b \leqslant \{(12b)^2 . 4b . 2b\}^b \leqslant (7b)^{4b},$

we have the theorem for $\eta \geqslant 2$.

(v) *The case* $\eta < 2$. Then

$$\frac{1}{k}\log Q/\log 2 < 2, \quad \text{i.e.} \quad Q < 4^k.$$

We divide $C_k(Q)$ into four parts, each of them of the form

$$C^* = \sum_{\omega < x \leqslant \omega + Q'} e^{2\pi i f(x)} \quad (0 < Q' \leqslant \tfrac{1}{4}Q \leqslant Q^{1-1/k}).$$

By Hölder's inequality, we have

$$|C_k(Q)|^{2b} \leqslant 4^{2b-1} \sum^{4} |C^*|^{2b} \leqslant 4^{2b-1} Q^{2k(1-1/k)} \sum^{4} |C^*|^{2(b-k)}.$$

Integrating over the unit hypercube, we have

$$\int_0^1 \dots \int_0^1 |C_k(Q)|^{2b}\, d\alpha_1 \dots d\alpha_k$$

$$\leqslant 4^{2b-1} Q^{2k(1-1/k)} \sum^{4} \int_0^1 \dots \int_0^1 |C^*|^{2(b-k)}\, d\alpha_1 \dots d\alpha_k$$

$$\leqslant 4^{2b} Q^{2k(1-1/k)} \int_0^1 \dots \int_0^1 |C_k(Q^{1-1/k})|^{2(b-k)}\, d\alpha_1 \dots d\alpha_k$$

$$\leqslant 4^{2b} Q^{2b-\frac{1}{2}(k+1)+2(b-k)/k} \int_0^1 \dots \int_0^1 |C_k(Q^{1-1/k})|^{2(b-k)}\, d\alpha_1 \dots d\alpha_k,$$

since $2b > \frac{1}{2}k(k+1)$, and we have the theorem.

4. Proof of Theorem 1. If $P^{1-1/k} \leqslant 3$, then $P \leqslant 9$, and the theorem is trivial. Accordingly we assume that $P^{1-1/k} > 3$, and consequently $P > e$.

The theorem is trivial for $l = 0$. We use induction on l, and we assume that it is true for $l-1$. By Theorem 2, we have

$$\int_0^1 \dots \int_0^1 |C_k(P)|^{2s}\, d\alpha_1 \dots d\alpha_k \leqslant (7s)^{4s} P^{2k-\frac{1}{2}(k+1)+2(s-k)/k} \times$$
$$\times (\log P)^2 \int_0^1 \dots \int_0^1 |C_k(P^{1-1/k})|^{2(s-k)}\, d\alpha_1 \dots d\alpha_k. \quad (45)$$

By the inductive hypothesis, with $l-1$, $s-k$, and $P^{1-1/k}$ instead of l, s, and P in the statement of Theorem 1, we have, for $P^{1-1/k} > 3 > 2$,

$$\int_0^1 \dots \int_0^1 |C_k(P^{1-1/k})|^{2(s-k)}\, d\alpha_1 \dots d\alpha_k \leqslant (7s)^{4s(l-1)} (\log P)^{2(l-1)} \times$$
$$\times P^{(1-1/k)\{2s-2k-\frac{1}{2}k(k+1)+\frac{1}{2}k(k+1)(1-1/k)^{l-1}\}}. \quad (46)$$

Combining (45) and (46), we have the theorem.

As a consequence of Theorem 1, we have

THEOREM 3. *Let $P \geqslant 2$ and $s \geqslant \frac{1}{4}k(k+1)+lk$, then we have*

$$\int_0^1 \Big|\sum_{x=1}^{P} e^{2\pi i \alpha x^k}\Big|^{2s} d\alpha_1 \leqslant s^k (7s)^{4sl} (\log P)^{2l} P^{2s-k+\delta},$$

where $\delta = \frac{1}{2}k(k+1)(1-1/k)^l$.

Proof. Let $r(N_1, \dots, N_k)$ be the number of solutions of

$$x_1^h + \dots + x_s^h - y_1^h - \dots - y_s^h = N_h \quad (1 \leqslant h \leqslant k;\ 1 \leqslant x, y \leqslant P).$$

Evidently, we have

$$\int_0^1 \Big|\sum_{x=1}^{P} e^{2\pi i \alpha x^k}\Big|^{2s} d\alpha \leqslant \sum_{|N_1| \leqslant sP} \dots \sum_{|N_{k-1}| \leqslant sP^{k-1}} r(N_1, \dots, N_{k-1}, 0),$$

since

$$r(N_1, \dots, N_{k-1}, N_k) = \int_0^1 \dots \int_0^1 |C_k(P)|^{2s} e^{2\pi i (N_1\alpha_1 + \dots + N_k \alpha_k)}\, d\alpha_1 \dots d\alpha_k$$
$$\leqslant \int_0^1 \dots \int_0^1 |C_k(P)|^{2s}\, d\alpha_1 \dots d\alpha_k.$$

The theorem is therefore an immediate consequence of Theorem 1.

5. Estimation of exponential sums

LEMMA 4. *Let* $$\left|\alpha-\frac{h}{q}\right| \leqslant \frac{1}{q^2}, \quad (h,q)=1.$$

Then $$\sum_{y=1}^{Y}\left|\sum_{n=f+1}^{f+N} e^{2\pi i\alpha yn}\right| \leqslant \left(\frac{Y}{q}+1\right)(N+q\log q).$$

The lemma is well known; for a proof see, for instance, Landau (**2**).

THEOREM 4. *Suppose that the number of solutions of the system of Diophantine equations*

$$x_1^h+\ldots+x_{2t}^h = y_1^h+\ldots+y_{2t}^h \quad (1 \leqslant h \leqslant k;\ 1 \leqslant x,y \leqslant Q) \tag{47}$$

does not exceed $$c_1(k,t)P^{4t-\frac{1}{2}k(k+1)+\delta'}. \tag{48}$$

Let $F(x) = \alpha_{k+1}x^{k+1}+\ldots+\alpha_1 x+\alpha_0$, *and*

$$S = \sum_{x=1}^{P} e^{2\pi iF(x)}, \tag{49}$$

and $$|\alpha_{k+1}-h/q| \leqslant q^{-2}, \quad (h,q)=1, \quad P \leqslant q \leqslant P^k.$$

Then we have

$$|S| \leqslant c_2(k,t)P^{1-\rho}, \qquad \rho = (1-\delta)/(4t+k-\delta). \tag{50}$$

Proof. Let p_1 be an integer, $1 \leqslant p_1 \leqslant P$, and let

$$S(y) = \sum_{x=1}^{p_1} e^{2\pi iF(x+y)}.$$

Then

$$S = \frac{1}{p_1}\sum_{x=1}^{p_1}\sum_{z=1}^{P} e^{2\pi iF(z)} = \frac{1}{p_1}\sum_{x=1}^{p_1}\sum_{y=1-x}^{P-x} e^{2\pi iF(x+y)}$$

$$= \frac{1}{p_1}\sum_{y=1}^{P} S(y)+Q_1 p_1, \quad \text{where } |Q_1| \leqslant 1. \tag{51}$$

Write $$F(x+y) = A_{k+1}x^{k+1}+A_k x^k+\ldots+A_0.$$

Then $$A_{k+1}=\alpha_{k+1}, \qquad A_k = \alpha_k+(k+1)\alpha_{k+1}y, \qquad \ldots. \tag{52}$$

By Hölder's inequality, we have

$$\left|\sum_{y=1}^{P} S(y)\right|^{2t} \leqslant P^{2t-1}\sum_{y=1}^{P}|S(y)|^{2t} = P^{2t-1}\sum_{y=1}^{P}\{S(y)\}^t\{\overline{S(y)}\}^t$$

$$= P^{2t-1}\sum_{y=1}^{P}\left(\sum_{x_1=1}^{p_1}\cdots\sum_{x_t=1}^{p_1}\sum_{x_1'=1}^{p_1}\cdots\sum_{x_t'=1}^{p_1} e^{2\pi i\phi}\right), \tag{53}$$

where

$$\phi = f(x_1+y)+\ldots+f(x_t+y)-f(x_1'+y)-\ldots-f(x_t'+y)$$

$$= A_{k+1}\left(\sum_{i=1}^{t} x_i^{k+1}-\sum_{i=1}^{t} x_i'^{k+1}\right)+A_k\left(\sum_{i=1}^{t} x_i^k-\sum_{i=1}^{t} x_i'^k\right)+\ldots.$$

Let $\psi(N_k,\ldots,N_1)$ be the number of solutions of

$$x_1^h+\ldots+x_t^h-x_1'^h-\ldots-x_t'^h = N_h \quad (1 \leqslant h \leqslant k;\ 1 \leqslant x,x' \leqslant p_1).$$

Then we have

$$\sum_{y=1}^{P} |S(y)|^{2t} \leqslant \sum_{x_1=1}^{p_1} \cdots \sum_{x_t=1}^{p_1} \sum_{x_1'=1}^{p_1} \cdots \sum_{x_t'=1}^{p_1} \Big| \sum_{y=1}^{P} e^{2\pi i \phi} \Big|$$

$$\leqslant \sum_{|N_1|\leqslant tp_1} \cdots \sum_{|N_k|\leqslant tp_1^k} \psi(N_k,\ldots,N_1) \Big| \sum_{y} \exp(A_k N_k+\ldots+A_1 N_1) \Big|$$

$$\leqslant \sqrt{\Big\{ \sum_{N_1} \cdots \sum_{N_k} \psi^2(N_k,\ldots,N_1) \sum_{N_1} \cdots \sum_{N_k} \Big| \sum_{y} \exp(A_k N_k+\ldots+A_1 N_1) \Big|^2 \Big\}} \tag{54}$$

by Schwarz's inequality.

First the expression

$$\sum_{N_1} \cdots \sum_{N_k} \psi^2(N_k,\ldots,N_1) = \sum_{N_1} \cdots \sum_{N_k} \Big| \int_0^1 \cdots \int_0^1 \Big| \sum_{x=1}^{p_1} \exp(\alpha_k x^k+\ldots+\alpha_1 x) \Big|^{2t} \times$$

$$\times \exp(-N_k \alpha_k - \ldots - N_1 \alpha_1)\, d\alpha_1 \ldots d\alpha_k \Big|$$

$$= \int_0^1 \cdots \int_0^1 \Big| \sum_{x=1}^{p_1} \exp(\alpha_k x^k+\ldots+\alpha_1 x) \Big|^{4t} d\alpha_1 \ldots d\alpha_k,$$

by the Parseval relation. By (48), we have

$$\sum_{N_1} \cdots \sum_{N_k} r^2(N_1,\ldots,N_k) \leqslant c_1(k,t) p_1^{4t-\frac{1}{2}k(k+1)+\delta'}. \tag{55}$$

Next, we have, by (52),

$$\sum_{|N_1|\leqslant tp_1} \cdots \sum_{|N_k|\leqslant tp_1^k} \Big| \sum_{y} e^{2\pi i (A_k N_k+\ldots+A_1 N_1)} \Big|^2$$

$$\leqslant t^k p_1^{1+\ldots+k-1} \sum_{y_1=1}^{P} \sum_{y_2=1}^{P} \Big| \sum_{N_k} e^{2\pi i (k+1)\alpha_{k+1}(y_1-y_2)N_k} \Big|$$

$$\leqslant t^k p_1^{\frac{1}{2}k(k-1)} P \sum_{Y=1}^{P} \Big| \sum_{N_k} e^{2\pi i \alpha_{k+1} Y N_k} \Big|,$$

(since the number of solutions of $(k+1)(y_1-y_2) = Y$ does not exceed P)

$$\leqslant c_3(k,t) p_1^{\frac{1}{2}k(k-1)} P \Big(\frac{P}{q}+1\Big)(p_1^k+q\log q) \tag{56}$$

by Lemma 4.

Combining (54), (55), (56) we have

$$\sum_{y} |S(y)|^{2t} \leqslant c_4(k,t) p_1^{2t+\frac{1}{2}\delta} P^{\frac{1}{2}} (1+p_1^{-k} q \log q)^{\frac{1}{2}}.$$

Consequently, from (53), we deduce

$$|S| \leqslant c_5(k,t) p_1^{\delta/4t} P^{1-1/4t} (1+p_1^{-k} q \log q)^{1/4t} + p_1.$$

For $P \leqslant q \leqslant P^k$, we have

$$|S| \leqslant c_5(k,t)(p_1^{(\delta-k)/4t}P^{1+(k-1)/4t}+p_1)\log P.$$

Taking
$$p_1 = P^{1-\rho}, \qquad \rho = \frac{1-\delta}{4t+k-\delta},$$
we have the theorem.

THEOREM 5. *Let* $k > 10$, *and*

$$\left|\alpha-\frac{h}{q}\right| \leqslant \frac{1}{q^2}, \quad (h,q)=1, \quad P \leqslant q \leqslant P^{k-1},$$

then

$$\left|\sum_{x=1}^{P} e^{2\pi i\alpha x^k}\right| \leqslant c_6(k)P^{1-1/\sigma}, \quad \sigma = 4k^2(\log k+\tfrac{1}{2}\log\log k+3).$$

Proof. Taking

$$t = \left[\tfrac{1}{8}k(k-1)+\frac{l(k-1)}{2}\right]+1, \qquad l = \left[\frac{\log\{\frac{1}{2}k(k-1)\}+\log\log k}{-\log\{1-1/(k-1)\}}\right]+1;$$

we have
$$\delta = \tfrac{1}{2}k(k-1)\left(1-\frac{1}{k-1}\right)^l < 1/\log k \quad (<\tfrac{1}{2}).$$

By Theorem 1, the hypotheses of Theorem 4 are true, with $\delta' = 1/\log k$.

Since
$$-1\bigg/\log\left(1-\frac{1}{k-1}\right) \leqslant k,$$

$$2l+\tfrac{1}{2}k \leqslant 2k(\log\tfrac{1}{2}h^2+\log\log k)+\tfrac{1}{2}k+1 < 2k(\log k^2+\log\log k),$$

then we have

$$\begin{aligned}\frac{4t-k-1-\delta}{1-\delta} &\leqslant \frac{(2l+\frac{1}{2}k)(k-1)-k+3-\delta}{1-\delta}\\ &< (2l+\tfrac{1}{2}k)k(1+2\delta)\\ &\leqslant 4k^2(\log k+\tfrac{1}{2}\log\log k+2+\log\log k/\log k)\\ &\leqslant 4k^2(\log k+\tfrac{1}{2}\log\log k+3). \qquad (57)\end{aligned}$$

The theorem follows from Theorem 4, since (57) is an open sign.

6. Applications

In this section I shall only indicate several applications which require merely straightforward alternation of the known methods.

(i) *The Waring–Goldbach problem.* Let $H(n)$ be the least integer s such that

$$p_1^k+\ldots+p_s^k = N \qquad (58)$$

is soluble for larger N, provided that N satisfies certain congruence conditions. By the method used in one of the author's papers (**3**), with the new exponent of Theorem 5 instead of the old one, we have

$$H(k) \leqslant s_0 \quad (\sim 4k \log k,\ k \text{ large}).$$

(ii) The asymptotic formula for the number of solutions of (58) is true when $s \geqslant s_0 = 4k^2(\log k + \frac{1}{2}\log\log k + 8)$. Certainly, we can also prove that the Hardy–Littlewood asymptotic formula for the number of decompositions of an integer into s positive kth powers holds also for this bound, which is sharper than Vinogradov's (**4**) bound $s \geqslant s_0 = 10k^2 \log k$.

REFERENCES

1. The latest version of Vinogradov's method, see *Bull. de l'Acad. des Sci. de l'URSS*, 6 (1942), 33–40.
2. Landau, *Vorlesungen über Zahlentheorie*, Bd. 1, p. 256.
3. Hua, *Math. Zeits.* 44 (1939), 335–46.
4. Vinogradov, *Comptes Rendus (Doklady)*.

[*Added* 20 *July* 1949.] I am indebted to Dr. J. L. B. Cooper for the information that my booklet was published in 1947 as No. 22 of the *Travaux de l'Institut math. Stekloff*. Also he showed me a copy of Vinogradov's booklet which contains the proof of the result of (ii) § 6 with $s \geqslant 10k^2 \log k$. It seems also worthy of mention that the exponent of Theorem 5 is slightly better than his result. He has

$$\sigma = 3k^2 \log\{12k(k+1)\} \quad (\sim 6k^2 \log k)$$

instead of $$\sigma = 4k^2(\log k + \tfrac{1}{2}\log\log k + 3)$$

in this paper.

Reprinted from the
CANADIAN JOURNAL OF MATHEMATICS
Vol. 3, No. 1, pp. 44–51, 1951

ON EXPONENTIAL SUMS OVER AN ALGEBRAIC NUMBER FIELD

LOO-KENG HUA

1. Introduction

LET K be an algebraic field of degree n over the rational field, and let $\mathfrak{d}$ be the ground ideal (differente) of the field. Let

$$f(x) = a_k x^k + \ldots + a_1 x + a_0$$

be a polynomial of the kth degree with coefficients in the field K, and let $\mathfrak{a}$ be the fractional ideal generated by $a_k, \ldots, a_1$, that is, $\mathfrak{a} = (a_k, \ldots, a_1)$. Suppose $\mathfrak{a}\mathfrak{d} = \mathfrak{r}/\mathfrak{q}$, where $\mathfrak{r}$ and $\mathfrak{q}$ are two relatively prime integral ideals, and

$$S(f(x), \mathfrak{q}) = S(f(x)) = S(\mathfrak{q}) = \sum_{x \bmod \mathfrak{q}} e^{2\pi i \operatorname{tr}(f(x))},$$

where x runs over a complete residue system, mod $\mathfrak{q}$. It is the aim of the paper to prove the following:

THEOREM 1. *For any given $\epsilon > 0$, we have*

$$S(f(x), \mathfrak{q}) = O(N(\mathfrak{q})^{1-1/k+\epsilon})$$

where the constant implied by the symbol O depends only on k, n and ϵ.

As usual, we use $\operatorname{tr}(\alpha)$ and $N(\mathfrak{q})$ to denote the trace of a number α and the norm of an ideal $\mathfrak{q}$ of K respectively.

This is a generalization of a theorem of the author's [1] over the rational field. The method used here is simpler and quite different from the original one.

2. A theorem on congruences

THEOREM 2. *Let $\mathfrak{p}$ be a prime ideal and let $s(x)$ be a polynomial with integral coefficients,* mod $\mathfrak{p}$. *Let α be a root of multiplicity m of the congruence*

$$s(x) \equiv 0 \pmod{\mathfrak{p}}.$$

Let λ be an integer, divisible by $\mathfrak{p}$ but not by $\mathfrak{p}^2$, and let u be the greatest integer such that $\mathfrak{p}^u$ divides all the coefficients of $s(\lambda x + \alpha) - s(\alpha)$. Let

$$t(x) \equiv \lambda^{-u}(s(\lambda x + \alpha) - s(\alpha)) \pmod{\mathfrak{p}}$$

be a polynomial with integral coefficients. Then $u \leqslant m$, and the congruence

$$t(x) \equiv 0 \pmod{\mathfrak{p}}$$

has at most m solutions.

Received September 14, 1949.

Proof. Without loss of generality, we may assume that $a = 0$. Then

$$s(x) = x^m s_1(x) + s_2(x),\ s_1(0) \not\equiv 0 \pmod{\mathfrak{p}}$$

where $s_2(x)$ is a polynomial of degree less than m and all its coefficients are divisible by $\mathfrak{p}$. Now we have

$$s(\lambda x) = \lambda^m x^m s_1(\lambda x) + s_2(\lambda x).$$

Since the coefficient of x^m is equal to $\lambda^m s_1(0)$ which is not divisible by $\mathfrak{p}^{m+1}$, we have $u \leqslant m$.

Since $\lambda^{-u}s(\lambda x)$ is congruent to a polynomial of degree not exceeding m, mod $\mathfrak{p}$, the theorem follows.

Remark. u is independent of the choice of λ. In fact, let λ' be another integer having the same property, then we have an integer τ such that

$$\lambda \equiv \lambda'\tau \pmod{\mathfrak{p}^{u+1}},\ \mathfrak{p} \nmid \tau.$$

Then

$$s(\lambda x + a) - s(a) \equiv s(\lambda'(\tau x) + a) - s(a) \qquad \pmod{\mathfrak{p}^{u+1}}$$

3. Several lemmas concerning algebraic numbers

Let $\mathfrak{g}$ be an ideal, fractional or integral, and $\mathfrak{a}$ be an integral ideal. It is clear that $\mathfrak{g}|\mathfrak{g}\mathfrak{a}$.

Now we divide the elements of $\mathfrak{g}$ into residue classes according to the modulus $\mathfrak{g}\mathfrak{a}$. The number of different classes is known to be $N(\mathfrak{a})$. We take an element from each class; the set so formed is called a complete residue system of $\mathfrak{g}$, mod $\mathfrak{g}\mathfrak{a}$.

The definition of the ground ideal $\mathfrak{d}$ can be stated in the following way:

$$\mathfrak{d}^{-1} \text{ is the aggregate of all numbers } \xi \text{ of } K$$

such that

$$e^{2\pi i \operatorname{tr}(\xi a)} = 1$$

for all integers a of K. Consequently, if β belongs to $(\mathfrak{q}\mathfrak{d})^{-1}$ and $a_1 \equiv a_2 \pmod{\mathfrak{q}}$, then

$$e^{2\pi i \operatorname{tr}(\beta a_1)} = e^{2\pi i \operatorname{tr}(\beta a_2)}.$$

This asserts that the sum $S(f(x), \mathfrak{q})$, which was defined at the beginning of the paper, is independent of the choice of the residue system, mod $\mathfrak{q}$.

THEOREM 3. *Let $\mathfrak{q}$ be an integral ideal. As ξ runs over a complete residue system of* $(\mathfrak{q}\mathfrak{d})^{-1}$, mod $\mathfrak{d}^{-1}$, *we have, for integral a,*

$$\sum_{\xi} e^{2\pi i \operatorname{tr}(\xi a)} = \begin{cases} N(\mathfrak{q}) & \text{if } \mathfrak{q} \mid a, \\ 0 & \text{if } \mathfrak{q} \nmid a. \end{cases}$$

Proof. If $\mathfrak{q} \mid a$, then ξa belongs to $\mathfrak{d}^{-1}$. Then $e^{2\pi i \operatorname{tr}(\xi a)} = 1$ for all ξ. Hence, we have the first conclusion.

If $\mathfrak{q} \nmid \alpha$, there is an element ξ_0, which belongs to $(\mathfrak{d}\mathfrak{q})^{-1}$, but $\xi_0\alpha$ does not belong to $\mathfrak{d}^{-1}$. In fact, if for all ξ_0 belonging to $(\mathfrak{d}\mathfrak{q})^{-1}$ we have $\xi_0\alpha$ belonging to $\mathfrak{d}^{-1}$, then we have

$$\mathfrak{d}^{-1} \mid \alpha(\mathfrak{d}\mathfrak{q})^{-1}.$$

Consequently $\mathfrak{q} \mid \alpha$. This is impossible. By the definition of $\mathfrak{d}^{-1}$ there is an integer γ such that

$$e^{2\pi i \operatorname{tr}(\gamma\xi_0\alpha)} \neq 1.$$

Since $\gamma\xi_0$ belongs to $(\mathfrak{d}\mathfrak{q})^{-1}$, we can write $\gamma\xi_0 = \xi_1$. Then

$$\sum_{\xi} e^{2\pi i \operatorname{tr}(\xi\alpha)} = \sum_{\xi} e^{2\pi i \operatorname{tr}((\xi+\xi_1)\alpha)} = e^{2\pi i \operatorname{tr}(\xi_1\alpha)} \cdot \sum_{\xi} e^{2\pi i \operatorname{tr}(\xi\alpha)}.$$

Thus we have the second conclusion of our theorem.

4. Proof of the theorem for $\mathfrak{q} = \mathfrak{p}$

In case $\mathfrak{q}$ is a prime ideal $\mathfrak{p}$, the exponential sum considered here reduces to a type of exponential sum over a finite field which has been discussed before [2]. But the author could not find an easy way to establish an explicit relationship between the exponential sums considered here and those over a finite field. Also, for the sake of completeness, the following proof is included here. The method is an adaptation of one due to Mordell [3].

THEOREM 4. *We have*

$$|S(f(x), \mathfrak{p})| \leqslant k^n N(\mathfrak{p})^{1-1/k}.$$

Proof. Without loss of generality, we may assume that α_k does not belong to $\mathfrak{d}^{-1}$, for otherwise

$$S(f(x), \mathfrak{p}) = S(f(x) - \alpha_k x^k, \mathfrak{p}),$$

since $e^{2\pi i \operatorname{tr}(\alpha_k x^k)} = 1$ for all integral x. Thus we now assume that α_k belongs to $(\mathfrak{p}\mathfrak{d})^{-1}$ but not to $\mathfrak{d}^{-1}$. The theorem is trivial for $N(\mathfrak{p}) \leqslant k^n$, since

$$|S(f(x), \mathfrak{p})| \leqslant N(\mathfrak{p}) \leqslant k^n N(\mathfrak{p})^{1-1/k}.$$

Now we assume $N(\mathfrak{p}) > k^n$ and consequently $\mathfrak{p} \nmid k!$ We have

$$|S(f(x))|^{2k} = \frac{1}{N(\mathfrak{p})(N(\mathfrak{p}) - 1)} \sum_{\lambda \bmod \mathfrak{p}}{}' \sum_{\mu \bmod \mathfrak{p}} |S(f(\lambda x + \mu))|^{2k},$$

where λ runs over a reduced residue system, mod $\mathfrak{p}$. Write

$$f(\lambda x + \mu) = \beta_k x^k + \ldots + \beta_0,$$

where

$$(1) \qquad \beta_k \equiv \alpha_k \lambda^k \pmod{\mathfrak{d}^{-1}},$$

$$(2) \qquad \beta_{k-1} \equiv k\alpha_k \lambda^{k-1} + \alpha_{k-1}\lambda^{k-1} \pmod{\mathfrak{d}^{-1}},$$

and so on.

For fixed $\beta_k, \beta_{k-1}, \ldots$ belonging to $(\mathfrak{p}\mathfrak{d})^{-1}$, the number of integers λ and μ does not exceed k. In fact, (1) asserts that $\beta_k - \alpha_k\lambda^k$ belongs to $\mathfrak{d}^{-1}$. (β_k and α_k belong to $(\mathfrak{p}\mathfrak{d})^{-1}$.) There is an integer τ belonging to $\mathfrak{p}\mathfrak{d}$ but not to $\mathfrak{p}$. Consequently $\tau\alpha_k$ and $\tau\beta_k$ are integers and $\mathfrak{p} \nmid \tau\alpha_k$; the congruence $\tau\beta_k \equiv \tau\alpha_k\lambda^k$ (mod $\mathfrak{p}$) has evidently at most k solutions. For a fixed λ, the same argument proves that μ is uniquely determined by (2), since $\mathfrak{p} \nmid k$.

Therefore, we have

$$\left| S(f(x), \mathfrak{p}) \right|^{2k} \leqslant \frac{k}{N(\mathfrak{p})\,(N(\mathfrak{p})-1)} \sum_{\beta_k} \cdots \sum_{\beta_1} \left| S(\beta_k x^k + \ldots + \beta_1 x) \right|^{2k},$$

where each β runs over a complete residue system of $(\mathfrak{p}\mathfrak{d})^{-1}$, mod $\mathfrak{d}^{-1}$.

We have

$$\sum_{\beta_k} \cdots \sum_{\beta_1} \left| S(\beta_k x^k + \ldots + \beta_1 x) \right|^{2k} = \sum_{\beta_k} \cdots \sum_{\beta_1} \sum_{x_1} \cdots \sum_{x_k} \sum_{y_1} \cdots \sum_{y_k} e^{2\pi i \operatorname{tr}(\psi)}$$
$$= N(\mathfrak{p})^k M,$$

where

$$\psi = \beta_k(x_1^k + \ldots + x_k^k - y_1^k - \ldots - y_k^k) + \beta_{k-1}(x_1^{k-1} + \ldots + x_k^{k-1} - y_1^{k-1} - \ldots - y_k^{k-1}) + \ldots + \beta_1(x_1 + \ldots + x_k - y_1 - \ldots - y_k),$$

and, by Theorem 3, M is equal to the number of solutions of the system of congruences

$$x_1^h + \ldots + x_k^h \equiv y_1^h + \ldots + y_k^h, \text{ mod } \mathfrak{p}, \qquad 1 \leqslant h \leqslant k.$$

By a theorem on symmetric functions, we deduce immediately

$$(X - x_1) \ldots (X - x_k) \equiv (X - y_1) \ldots (X - y_k), \text{ mod } \mathfrak{p},$$

since $\mathfrak{p} \nmid k!$ Then we have that $x_1, \ldots, x_k$ are a permutation of $y_1, \ldots, y_k$ and then

$$M \leqslant k! N(\mathfrak{p})^k.$$

Consequently, we have

$$\left| S(f(x), \mathfrak{p}) \right|^{2k} \leqslant \frac{k \cdot k!}{N(\mathfrak{p})(N(\mathfrak{p})-1)} N(\mathfrak{p})^{2k}$$
$$\leqslant 2k \cdot k! N(\mathfrak{p})^{2k-2}$$
$$\leqslant k^{2k} N(\mathfrak{p})^{2k-2}$$

and the theorem follows.

5. Proof of the theorem for $\mathfrak{q} = \mathfrak{p}^l$

THEOREM 5. *If* $\mathfrak{q} = \mathfrak{p}^l$, *and* $\mathfrak{p}$ *is a prime ideal, then*

(1) $$|S(f(x), \mathfrak{p}^l)| \leqslant k^{2n+1} N(\mathfrak{p}^l)^{1-1/k}.$$

Proof. Let

$$\mathfrak{b} = (ka_k, (k-1)a_{k-1}, \ldots, 2a_2, a_1).$$

Evidently $\mathfrak{a} \mid \mathfrak{b}$. Let t be the highest exponent of $\mathfrak{p}$ dividing $\mathfrak{b}\mathfrak{a}^{-1}$. Let m be the number of solutions, multiplicities being counted, of the congruence

$$f'(x) \equiv 0 \pmod{\mathfrak{p}^{t+1-l}} \tag{2}$$

as x runs over a complete residue system, mod $\mathfrak{p}$. (We have $m \leqslant k-1$.)

Evidently, (1) is a conseqeunce of the sharper result

$$|S(f(x), \mathfrak{p}^l)| \leqslant k^{2n} \max(1, m)\, N(\mathfrak{p}^l)^{1-1/k}. \tag{3}$$

If $t \geqslant 1$, then $\mathfrak{p}^t$ divides at least one of the integers $k, k-1, \ldots, 1$. Then

$$N(\mathfrak{p}^t) \leqslant k^n,$$

that is

$$N(\mathfrak{p}) \leqslant k^{n/t}. \tag{4}$$

Suppose that $l < 2(t+1)$. For $t = 0$, we have $l = 1$ and (3) follows from Theorem 4. If $t \geqslant 1$, then, by (4)

$$\begin{aligned} |S(f(x), \mathfrak{p}^l)| &\leqslant N(\mathfrak{p})^l \leqslant (N(\mathfrak{p}))^{l(1-1/k)} (N(\mathfrak{p}))^{(2t+1)/k} \\ &\leqslant N(\mathfrak{p})^{l(1-1/k)} k^{n(2+1/t)/k} \\ &\leqslant k^{2n} \cdot N(\mathfrak{p})^{l(1-1/k)}. \end{aligned}$$

Therefore (3) is true for $l \leqslant 2t+1$. Now we assume that $l \geqslant 2(t+1)$ and that (3) is true for smaller l.

Let $\mu_1, \ldots, \mu_r$ be the distinct roots of (2) with multiplicities $m_1, \ldots, m_r$ respectively. Then $m_1 + \ldots + m_r = m$. Evidently

$$S(f(x)) = \sum_x e^{2\pi i \operatorname{tr}(f(x))} = \sum_\nu \sum_{\substack{x \\ x \equiv \nu \pmod{\mathfrak{p}}}} e^{2\pi i \operatorname{tr}(f(x))} = \sum_\nu S_\nu$$

say, where ν runs over a complete residue system, mod $\mathfrak{p}$. If ν is not one of the μ's then, letting

$$x = y + \lambda^{l-t-1} z,$$

where λ is an integer belonging to $\mathfrak{p}$ but not to $\mathfrak{p}^2$, we have

$$\begin{aligned} S_\nu &= \sum_{\substack{y \bmod \mathfrak{p}^{l-t-1} \\ y \equiv \nu \pmod{\mathfrak{p}}}} \sum_{z \bmod \mathfrak{p}^{t+1}} e^{2\pi i \operatorname{tr}(f(y) + \lambda^{l-t-1} z f'(y))} \\ &= \sum e^{2\pi i \operatorname{tr}(f(y))} \sum_{z \bmod \mathfrak{p}^{t+1}} e^{2\pi i \operatorname{tr}(\lambda^{l-t-1} z f'(y))} \\ &= 0 \end{aligned}$$

by Theorem 3, since $\mathfrak{p}^{t+1-l} \nmid f'(y)$.

Therefore

$$| S(f(x)) | \leqslant \sum_{s=1}^{r} \left| \sum_{x \bmod \mathfrak{p}^{l-1}} e^{2\pi i \operatorname{tr}(f(\mu_s + \lambda y))} \right|$$

$$= \sum_{s=1}^{r} \left| \sum_{x \bmod \mathfrak{p}^{l-1}} e^{2\pi i \operatorname{tr}(f(\mu_s + \lambda y) - f(\mu_s))} \right|$$

$$= \sum_{s=1}^{r} N(\mathfrak{p})^{\sigma_s - 1} S(f(\mu_s + \lambda y) - f(\mu_s), \mathfrak{p}^{l-\sigma_s}), \tag{5}$$

where σ_s is defined in the following way: Let $\mathfrak{c}$ be the ideal generated by the coefficients of

$$f_s(y) = f(\mu_s + \lambda y) - f(\mu_s).$$

Evidently $\mathfrak{a}$ divides $\mathfrak{c}$, and σ_s is the highest power of $\mathfrak{p}$ dividing $\mathfrak{c}\mathfrak{a}^{-1}$. Also, if $l \leqslant \sigma_s$, we use the conventional meaning

$$S(f(\mu_s + \lambda y) - f(\mu_s), \mathfrak{p}^{l-\sigma_s}) = \mathfrak{p}^{l-\sigma_s}.$$

Now we are going to prove that

$$1 \leqslant \sigma_s \leqslant k. \tag{6}$$

If (6) is not true, then $\mathfrak{p}^{-l+k+1}$ divides all the coefficients of $f(\mu_s + \lambda y) - f(\mu_s)$; that is

$$\mathfrak{p}^{-l+k+1} \left| \frac{f^{(r)}(\mu_s)}{r!} \lambda^r, \right. \qquad 1 \leqslant r \leqslant k.$$

Consequently

$$\mathfrak{p}^{-l+1} \left| \frac{f^{(r)}(\mu_s)}{r!}, \right.$$

which is equal to a_r plus a linear combination of $a_k, \ldots, a_{r-1}$ with integral coefficients. Thus we deduce successively $\mathfrak{p}^{-l+1} | a_k$, $\mathfrak{p}^{-l+1} | a_{k-1}, \ldots, \mathfrak{p}^{-l+1} | a_1$. This contradicts $\mathfrak{q} = \mathfrak{p}^l$.

From (5) and (6), we have, for $l \geqslant \max(\sigma_1, \ldots, \sigma_r)$,

$$| S(f(x), \mathfrak{p}^l) | \leqslant \sum_{s=1}^{r} N(\mathfrak{p})^{\sigma_s(1-1/k)} | S(f_s(y), \mathfrak{p}^{l-\sigma_s}) |.$$

By the hypothesis of induction, we have

$$| S(f(x), \mathfrak{p}^l) | \leqslant k^{2n} \sum_{s=1}^{r} N(\mathfrak{p})^{\sigma_s(1-1/k)} m_s N(\mathfrak{p})^{(l-\sigma_s)(1-1/k)}$$

$$= k^{2n} m N(\mathfrak{p})^{l(1-1/k)}.$$

In case $l \leqslant \max(\sigma_1, \ldots, \sigma_r)$, we have $l \leqslant k$ and, by (5)

$$| S(f(x)) | \leqslant r \mathfrak{p}^{l-1} \leqslant m \mathfrak{p}^{l(1-1/k)}.$$

We have (3) and consequently (1). (Notice that if $\sum_{s=1}^{r} m_s = 0$, the method shows that $S(f(x)) = 0$, if $l \geqslant 2(t+1)$.)

THEOREM 6. *If* $(\mathfrak{q}_1, \mathfrak{q}_2) = 1$ *and* $f(0) = 0$, *then there are polynomials* $f_1(x)$ *and* $f_2(x)$ *each of degree k such that*

$$S(f(x), \mathfrak{q}_1\mathfrak{q}_2) = S(f_1(x), \mathfrak{q}_1)\ S(f_2(x), \mathfrak{q}_2).$$

Proof. We can find two integers λ_1 and λ_2 such that

$$(\lambda_1, \mathfrak{q}_1\mathfrak{q}_2) = \mathfrak{q}_2, \quad (\lambda_2, \mathfrak{q}_1\mathfrak{q}_2) = \mathfrak{q}_1.$$

Putting

$$x = \lambda_1 y_2 + \lambda_2 y_1,$$

then, as y_1 and y_2 run over complete residue systems mod $\mathfrak{q}_1$ and mod $\mathfrak{q}_2$ respectively, x runs over a complete residue system, mod $\mathfrak{q}_1\mathfrak{q}_2$. Then

$$\begin{aligned} S(f(x), \mathfrak{q}_1\mathfrak{q}_2) &= \sum_{y_1 \bmod \mathfrak{q}_1} \sum_{y_2 \bmod \mathfrak{q}_2} e^{2\pi i \operatorname{tr}(f(\lambda_1 y_2 + \lambda_2 y_1))} \\ &= \sum_{y_1 \bmod \mathfrak{q}_1} e^{2\pi i \operatorname{tr}(f(\lambda_2 y_1))} \sum_{y_2 \bmod \mathfrak{q}_2} e^{2\pi i \operatorname{tr}(f(\lambda_1 y_2))} \\ &= S(f_1(x), \mathfrak{q}_1))\ S(f_2(x), \mathfrak{q}_2)\ , \end{aligned}$$

where $f_1(x) = f(\lambda_2 x)$ and $f_2(x) = f(\lambda_1 x)$. Now we have to verify that the ideal generated by the coefficients of $f_1(x)$ can be expressed as $\mathfrak{r}(\mathfrak{d}\mathfrak{q}_1)^{-1}$, where $\mathfrak{r}$, $\mathfrak{q}$ are relatively prime integral ideals, but this is quite evident.

6. Proof of Theorem 1

Let
$$\mathfrak{q} = \mathfrak{p}_1^{l_1} \ldots \mathfrak{p}_s^{l_s}.$$

Then we have, by repeated application of Theorem 6,

$$S(f(x), \mathfrak{q}) = \prod_{i=1}^{s} S(f_i(x), \mathfrak{p}_i^{l_i})\ .$$

By Theorem 5, we have

$$\begin{aligned} |\ S(f(x), \mathfrak{q})\ | &\leqslant \sum_{i=1}^{s} k^{2n+1} N(\mathfrak{p}_i^{l_i})^{1-1/k} \\ &\leqslant \sum_{i=1}^{s} (1 + l_i)^{(2n+1)\log k/\log 2} N(\mathfrak{p}_i^{l_i})^{(1-1/k)} \\ &= d(\mathfrak{q})^{(2n+1)\log k/\log 2} N(\mathfrak{q})^{1-1/k} \\ &= O(N(\mathfrak{q})^{1-1/k+\epsilon}) \end{aligned}$$

where $d(\mathfrak{q})$ denotes the number of divisors of $\mathfrak{q}$.

Remarks. The previous method is practically an algorithm; more precisely, for a given polynomial, if we know the value of $S(f(x), \mathfrak{p}^l)$, $l \leqslant 2t + 1$, then we can find the value of $S(f(x), \mathfrak{p}^l)$.

References

[1] L. K. Hua, *On an exponential sum*, Jour. of Chinese Math. Soc., vol. 2 (1940), 301-312.
[2] L. K. Hua and S. H. Min, *On a double exponential sum*, Acad. Sinica Sci. Record, vol. 1 (1942), 23-25.
[3] L. J. Mordell, *On a sum analogous to a Gauss's sum*, Quart. J. Math. (Oxford), vol. 3 (1932), 161-167.

Tsing Hua University,
Peking, China

Reprinted from the
ACTA SCIENTIA SINICA
Vol. 1, pp. 1–76, 1952

ON THE NUMBER OF SOLUTIONS OF TARRY'S PROBLEM*

By LOO-KENG HUA
Tsing Hua University and Academia Sinica

I. INTRODUCTION

Let $k \geqslant 2$ and let t_0 be an integer, depending on k, defined by the following table:

k	2	3	4	5	6	7	8	9	10	>10
t_o	3	8	23	62	156	380	889	2034	4595	$[k^2(3 \log k + \log\log k + 4)]$

Let $r_t(P)$ be the number of solutions of the system of Diophantine equations

$$\begin{aligned} x_1^k + \cdots + x_t^k &= y_1^k + \cdots + y_t^k \\ &\cdots\cdots\cdots\cdots \\ x_1 + \cdots + x_t &= y_1 + \cdots + y_t, \end{aligned} \tag{1.1}$$

with the restriction that

$$1 \leqslant x_i, \quad y_i \leqslant P. \tag{1.2}$$

One of the purposes of the paper is to prove that, for $t > t_o$ we have

$$\lim_{P\to\infty} P^{\frac{1}{2}k(k+1)-2t} r_t(P) = \vartheta_o \mathfrak{S}, \tag{1.3}$$

where

*Received May 31, 1952.

$$(1.4)\qquad \vartheta_0 = \int_{-\infty}^{\infty} \cdots \int_{-\infty}^{\infty} \left| \int_0^1 e^{2\pi i(\beta_k x^k + \cdots + \beta_1 x)} dx \right|^{2t} d\beta_k \cdots d\beta_1 ,$$

$$(1.5)\qquad \mathfrak{S} = \sum_{q_1=1}^{\infty} \cdots \sum_{q_k=1}^{\infty} \sum_{\substack{h_1=1 \\ (h_1, q_1)=1}}^{q_1} \cdots \sum_{\substack{h_k=1 \\ (h_k, q_k)=1}}^{q_k} \left| q_1^{-1} \cdots q_k^{-1} \sum_{x=1}^{q_1 \cdots q_k} e\left(\frac{h_k}{q_k} x^k + \cdots + \frac{h_1}{q_1} x\right) \right|^{2t} .$$

We use the abbreviation $e(x)$ for $e^{2\pi i x}$ throughout the paper.

In 1939, the author[1] proved that (1.3) holds for

$$(1.6)\qquad t > \frac{1}{4} k(k+1)\left(\left[\frac{6 \log k + 2 \log \log 2k + \log 23.2}{-\log\left(1 - \frac{1}{k}\right)}\right] + k + 1\right)$$

which is asymptotically equal to $1.5\, k^3 \log k$ for large k (It was published in 1947). This new improvement is made possible by a new method of Vinogradow[5] and a result of the author[2].

Let $M(k)$ be the least possible integer t such that the system of equations

$$x_1^h + \cdots + x_t^h = y_1^h + \cdots + y_t^h , \qquad 1 \leqslant h \leqslant k ,$$

$$x_1^{k+1} + \cdots + x_t^{k+1} \neq y_1^{k+1} + \cdots + y_t^{k+1}$$

is solvable. Then, by (1.3), we evidently deduce

$$M(k) \leqslant t_0 + 1 .$$

A previous result of the author[3] was

$$M(k) \geqslant (k+1)\left(\left[\frac{\log \frac{1}{2}(k+2)}{\log(k+1) - \log k}\right] + 1\right) \qquad (\sim k^2 \log k, \text{ for } k \text{ large});$$

the difference in the two estimates is only a constant factor. This reveals the sharpness of the present method.

Particular attentions are also paid to the convergence problem of the real density ∂_0 and the p-adic density ∂_p which is a factor of our singular series. To handle the convergence problem of ∂_0 the author uses Young and Hausdorff's theorem on multiple Fourier integrals. The exponent of convergence so obtained is sharper than that obtained previously from the estimation of exponential integrals (Vinogradow [6]). The best possible exponent is still an open question.

For the p-adic density and the convergence of $\mathfrak{S}$, the author introduces a new method by means of which he obtains the best possible value of the convergence exponent.

The present paper is independent of the author's Russian book (Hua [1]). Aside from two previous papers (Hua [3] and [4]), the present paper is self-contained.

Throughout the paper, we use ε to denote an arbitrarily small positive number which may be different at each occurrence. We use $c_1, c_2, \ldots$ to denote constants depending only on k, ε and sometimes t. The constants implied by the symbol O are such c' s.

II. The Constant ϑ_0

Let D_0 be the domain

$$0 \leqslant x_1 < x_2 < x_3 < \cdots < x_k \leqslant 1 .$$

The domain D of the variables $y_1, \cdots, y_k$ is the image of D_0 under the mapping

$$y_h = x_1^h + \cdots + x_k^h , \qquad h = 1, \cdots, k . \tag{2.1}$$

The Jacobian of the mapping is equal to

$$\frac{\partial (y_1, \cdots, y_k)}{\partial (x_1, \cdots, x_k)} = k! \prod_{1 \leq i < j \leq k} (x_j - x_i) , \tag{2.2}$$

which is positive when $(x_1, \cdots, x_k)$ belongs to D_0.

We define $y=(y, \cdots, y_k)$ and

$$f(y_1, \cdots, y_k) = \frac{1}{k! \prod\limits_{1 \leq i < j \leq k} (x_j - x_i)} , \qquad \text{for } y \in D \tag{2.3}$$

$$= 0 , \qquad \text{otherwise} .$$

Evidently, $f(y_1, \ldots, y_k)$ is a function vanishing outside a finite domain (Namely it vanishes outside

$$0 \leqslant y_h \leqslant k, \qquad \text{for } h = 1, 2, \cdots, k).$$

The Fourier transform of the function $f(y_1, \ldots, y_k)$ is equal to[1]

$$\int_{-\infty}^{\infty} \cdots \int_{-\infty}^{\infty} f(y_1, \cdots, y_k)\, e(\gamma_1 y_1 + \cdots + \gamma_k y_k)\, dy_1 \cdots dy_k$$

$$= \int \cdots \int_D f(y_1, \cdots, y_k)\, e(\gamma_1 y_1 + \cdots + \gamma_k y_k)\, dy_1 \cdots dy_k$$

$$(2.4) \qquad = \int \cdots \int_{D_0} e(\gamma_k (x_1^k + \cdots + x_k^k) + \cdots + \gamma_1 (x_1 + \cdots + x_k))\, dx_1 \cdots dx_k$$

$$= \frac{1}{k!} \int_0^1 \cdots \int_0^1 e(\gamma_k (x_1^k + \cdots + x_k^k) + \cdots + \gamma_1 (x_1 + \cdots + x_k))\, dx_1 \cdots dx_k$$

$$= \frac{1}{k!} \left(\int_0^1 e(\gamma_k x^k + \cdots + \gamma_1 x)\, dx \right)^k .$$

That is,

$$(2.5) \qquad \frac{1}{k!} \left(\int_0^1 e(\gamma_k x^k + \cdots + \gamma_1 x)\, dx \right)^k$$

is the Fourier transform of a function vanishing outside of a finite domain. Unfortunately, $f(y_1, \ldots, y_k)$ does not belong to L^2. If it did, we could use Parseval's relation to obtain the value of D_0 for $t = k$. Nevertheless, we have the following result:

Lemma 2.1. *The function $f(y_1, \cdots, y_k)$ belongs to L^σ for $\sigma < 1 + \frac{2}{k}$, and it does not belong to L^σ for $\sigma > 1 + \frac{2}{k}$.*

Proof. We have

$$J = \int_{-\infty}^{\infty} \cdots \int_{-\infty}^{\infty} | f(y_1, \cdots, y_k) |^\sigma\, dy_1 \cdots dy_k$$

---

[1] A little different from the usual notation, we use $g(x) = \int_{-\infty}^{\infty} f(y)e(xy)dy$ to denote the Fourier transform of $f(y)$, the inversion formula now takes the form $f(y) = \int_{-\infty}^{\infty} g(x)e(-yx)dx$.

$$= \int_{D_0} \cdots \int (k! \prod_{1 \le i < j \le k} (x_j - x_i))^{1-\sigma} dx_1 \cdots dx_k . \tag{2.6}$$

The lemma is trivial for $\sigma \leqslant 1$ and the integral (2.6) is evidently divergent for $\sigma \geqslant 2$. Now we consider the range

$$1 < \sigma < 2 . \tag{2.7}$$

Let

$$x_k = x_{k-1} + \delta_{k-1}, \quad x_{k-1} = x_{k-2} + \delta_{k-2}, \quad \cdots, \quad x_2 = x_1 + \delta_1 .$$

The domain D_0 now becomes

$$0 \leqslant \delta_1, \ 0 \leqslant \delta_2, \ \cdots, \ 0 \leqslant \delta_{k-1}, \ \delta_1 + \cdots + \delta_{k-1} \leqslant 1 \tag{2.8}$$

and

$$0 \leqslant x_1 \leqslant 1 - \delta_1 - \cdots - \delta_{k-1},$$

since $x_k = \delta_1 + \cdots + \delta_{k-1} + x_1$. Let D' be the $(k-1)$-dimensional domain (2.8). Then we have

$$J = k!^{1-\sigma} \int_{D'} \cdots \int (\delta_1 \cdots \delta_{k-1} (\delta_1 + \delta_2)(\delta_2 + \delta_3) \cdots$$

$$\cdots (\delta_{k-2} + \delta_{k-1})(\delta_1 + \delta_2 + \delta_3) \cdots (\delta_1 + \cdots + \delta_{k-1}))^{1-\sigma} \times$$

$$\times (1 - \delta_1 - \cdots - \delta_{k-1}) \, d\delta_1 \cdots d\delta_{k-1}$$

$$\leqslant k!^{1-\sigma} \sum_{\lambda_1 + \cdots + \lambda_{k-1} = \frac{1}{2}k(k-1)} \int_{D'} \cdots \int (\delta_1^{\lambda_1} \cdots \delta_{k-1}^{\lambda_{k-1}})^{1-\sigma} \times \tag{2.9}$$

$$\times (1 - \delta_1 - \cdots - \delta_{k-1}) \, d\delta_1 \cdots d\delta_{k-1},$$

since $(A+B)^{1-\sigma} \leqslant A^{1-\sigma} + B^{1-\sigma}$ for $1 \leqslant \sigma < 2$ and $A \geqslant 0$, $B \geqslant 0$.

Each integral in (2.9) is a Dirichlet integral, by changing the variables, it becomes the simple integral

$$\int_0^1 \tau^{\frac{1}{2}k(k-1)(1-\sigma)+k-2} (1-\tau)\, d\tau$$

which converges for

$$\frac{1}{2} k(k-1)(1-\sigma) + k - 1 > 0,$$

or

$$\sigma < 1 + \frac{2}{k}.$$

Now we are going to prove that (2.6) diverges for $\sigma > 1 + \frac{2}{k}$. Let $0 < \delta < \frac{1}{k}$. Evidently the integral is greater than

$$k!^{1-\sigma} \int_{k\delta}^1 dx_k \int_{x_k-\delta}^{x_k} dx_{k-1} \cdots \int_{x_2-\delta}^{x_2} \Big(\prod_{1 \le i < j \le k} (x_j - x_i) \Big)^{1-\sigma} dx_1$$

$$\geqslant (k!(k-1)!(k-2)! \cdots 1)^{1-\sigma} \int_{k\delta}^1 dx_k \int_{x_k-\delta}^{x_k} dx_{k-1} \cdots \int_{x_2-\delta}^{x_2} \delta^{\frac{1}{2}(1-\sigma)k(k-1)} dx_1,$$

since $x_i - x_j \leqslant (i-j)\delta$ for $i > j$. Therefore

$$J > c_1 \delta^{k-1+\frac{1}{2}(1-\sigma)k(k-1)}$$

which tends to infinity as $\delta \to 0$, if $\sigma > 1 + \frac{2}{k}$.

Remark. It seems to be true that $f(y_1, \ldots, y_k)$ does not belong to $L^{1+2/k}$ The values of J can be evaluated for $k=2$ and $k=3$, and they are

$$\frac{2^{1-\sigma}}{(2-\sigma)(3-\sigma)} = 2^{1-\sigma} \frac{\Gamma(2-\sigma)}{\Gamma(4-\sigma)}, \qquad (3\cdot 2)^{1-\sigma} \frac{(\Gamma(2-\sigma))^2}{\Gamma(4-2\sigma)} \frac{\Gamma(5-3\sigma)}{\Gamma(7-3\sigma)}$$

respectively.

By means of Young and Hausdorff's theorem on Fourier integrals, we can deduce

Lemma 2.2. *The integral*

$$\int_0^1 e(\gamma_k x^k + \cdots + \gamma_1 x)\, dx \tag{2.10}$$

belongs to $L^{\frac{1}{2}k(k+2)+\epsilon}$.

Proof. Young and Hausdorff's theorem asserts that if f belongs to $L^p(1 < p \leqslant 2)$, its Fourier transform belongs to $L^{p'}$, where $p' = p/(p-1)$. Now we take $p = \sigma < 1 + \frac{2}{k}$, then $p' > \frac{k}{2} + 1$. That is, (2.5) belongs to $L^{\frac{k}{2}+1+\epsilon}$. Thus, (2.10) belongs to $L^{\frac{1}{2}k(k+2)+\epsilon}$.

Remarks. 1. It is an open question what the best possible exponent is.

2. The same method gives us that if (2.10) is considered as a function of a single variable $\gamma_u (1 \leqslant u \leqslant k)$, then (2.10) belongs to $L^{u+\epsilon}$. The proof is quite similar, since

$$f(y) = \begin{cases} e(\gamma_k y^{\frac{k}{u}} + \cdots + \gamma_{u-1} y^{\frac{u-1}{u}} + \gamma_{u+1} y^{\frac{u+1}{u}} + \cdots + \gamma_1 y^{\frac{1}{u}}) y^{\frac{1}{u}-1}, & \text{for } 0 \leqslant y \leqslant 1, \\ 0, & \text{otherwise} \end{cases} \tag{2.11}$$

is the Fourier transform of (2.10) with respect to γ_u. Evidently $f(y)$ of (2.11) belongs to L^σ for $\sigma < 1 + \frac{1}{u-1}$. By Young and Hausdorff's theorem, (2.10) belongs to $L^{u+\epsilon}$ with respect to γ_u. This suggests some result about the behaviour of (2.10) for γ large. This is the purpose of the next section.

III. Estimations of Exponential Integrals

We need a lemma concerning simple integral.

Lemma 3.1. *Let* $\varphi(x)$ *be a continuous real function having a finite*[2] *number of maximum and minimum in the interval* (a, b). *Then, we have*

$$\int_a^b \varphi(x)\, e(x)\, dx = O\left(\max_{0 \leq \xi \leq 1} \max_{a \leq v \leq b-\xi} \int_v^{v+\xi} |\varphi(x)|\, dx \right). \tag{3.1}$$

[2] By finite number, we mean that the number depends only on k.

Proof. The lemma is trivial for $b-a\leqslant 1$.

Now we suppose that $a<b-1$. It is sufficient to prove that

$$(3.2)\qquad \int_a^b \varphi(x)\, e(x)\, dx = O\Big(\max_{a\leq v\leq b-1}\int_v^{v+1}|\varphi(x)|\,dx\Big).$$

Without loss of generality, we may assume that $\varphi(x)$ is monotone, otherwise, we can divide the interval into finite number of pieces, in each of them the function is monotone. Since the treatments are entirely similar, we now assume that $\varphi(x)$ is decreasing. Since

$$\Big|\int_a^b \varphi(x)e(x)dx\Big|\leqslant\int_a^{[a]+1}|\varphi(x)|\,dx+\Big|\int_{[a]+1}^{[b]}\varphi(x)e(x)dx\Big|+\int_{[b]}^b|\varphi(x)|\,dx,$$

it is enough to prove (3.2) for the case in which a and b are integers.

Now we have

$$\int_a^b\varphi(x)\sin 2\pi x\,dx=\int_a^{a+\frac{1}{2}}\varphi(x)\sin 2\pi x\,dx+\int_{a+\frac{1}{2}}^{a+1}\varphi(x)\sin 2\pi x\,dx+\cdots$$

$$=\int_0^{\frac{1}{2}}(\varphi(x+a)-\varphi(x+a+\tfrac{1}{2})+\varphi(x+a+1)-\cdots-\varphi(x+b-\tfrac{1}{2}))\sin 2\pi x\,dx,$$

since $\varphi(x)$ is decreasing, we have

$$0\leqslant\varphi(x+a)-\varphi(x+a+\tfrac{1}{2})+\varphi(x+a+1)-\cdots-\varphi(x+b-\tfrac{1}{2})\leqslant\varphi(x+a).$$

Consequently, we obtain

$$0\leqslant\int_a^b\varphi(x)\sin 2\pi x\,dx\leqslant\int_0^{\frac{1}{2}}\varphi(x+a)\sin 2\pi x\,dx$$

$$\leqslant\int_0^{\frac{1}{2}}|\varphi(x+a)|\,dx=\int_a^{a+\frac{1}{2}}|\varphi(x)|\,dx.$$

Similar result can be established for

$$\int_a^b\varphi(x)\cos 2\pi x\,dx.$$

Combining these results, we have our lemma.

Lemma 3.2. *Let $\gamma_k, \cdots, \gamma_1$ be k real numbers and let*

$$I = \int_0^1 e(\gamma_k x^k + \cdots + \gamma_1 x)\, dx . \tag{3.3}$$

Then

$$I = O(\min(1, |\gamma_1|^{-\frac{1}{k}}, \cdots, |\gamma_k|^{-\frac{1}{k}})) . \tag{3.4}$$

Proof. It is evident that $|I| \leqslant 1$. Thus we may assume that $|\gamma_i| \geqslant 1$. We have

$$I^k = \int_0^1 \cdots \int_0^1 e(\psi)\, dx_1 \cdots dx_k$$

where

$$\psi = (x_1^k + \cdots + x_k^k)\gamma_k + \cdots + (x_1 + \cdots + x_k)\gamma_1 .$$

Evidently

$$I^k \leqslant k! \int \cdots \int_{0 \leq x_k \leq x_{k-1} \leq \cdots \leq x_1 \leq 1} e(\psi)\, dx_1 \cdots dx_k . \tag{3.5}$$

Now we consider the mapping

$$\begin{gathered} (x_1^k + \cdots + x_k^k)\,|\gamma_k| = y_k \\ \cdots\cdots\cdots\cdots\cdots\cdots \\ (x_1 + \cdots + x_k)\,|\gamma_1| = y_1 , \end{gathered} \tag{3.6}$$

and let R be the domain of y's. The Jacobian $\dfrac{\partial(x_1, \cdots, x_k)}{\partial(y_1, \cdots, y_k)} = g(y_1, \cdots, y_k)$ of (3.6) is always positive. Then

$$|k!^{-1} I^k| = \Big| \int \underset{R}{\cdots} \int e(y_k + \cdots + y_1)\, g(y_1, \cdots, y_k)\, dy_1 \cdots dy_k \Big|$$

$$\leqslant \int \cdots \int dy_1 \cdots dy_{h-1}\, dy_{h+1} \cdots dy_k \Big| \int e(y_h)\, g(y_1, \cdots, y_k)\, dy_h \Big|$$

$$= O\Big(\max_{\substack{v \\ 0 \le \xi \le 1}} \int \cdots \int dy_1 \cdots dy_{h-1}\, dy_{h+1} \cdots dy_k \int_{v \le y_h \le v+\xi} g(y_1, \cdots, y_k)\, dy_k \Big)$$

$$= O\Big(\max_{0 \le \xi \le 1} \underset{v \le y_h \le v+\xi}{\int_0^1 \cdots \int_0^1} dx_1 \cdots dx_k \Big), \tag{3.7}$$

by lemma 3.1.

Now we have

$$\underset{v \le y_h \le v+\xi}{\int_0^1 \cdots \int_0^1} dx_1 \cdots dx_k \leqslant \underset{v \le y_h \le v+\xi}{\int \cdots \int} dx_1 \cdots dx_k \leqslant \underset{v \le y_h \le v+1}{\int \cdots \int} dx_1 \cdots dx_k$$

$$= V(v+1) - V(v), \tag{3.8}$$

where $V(v)$ is the volume of the domain

$$y_h = |\gamma_h| (x_1^h + \cdots + x_k^h) \leqslant v, \qquad x_\nu \geqslant 0.$$

Since

$$V(v) = \eta \Big(\frac{v}{|\gamma_h|} \Big)^{k/h},$$

where η is a constant depending only on k and h, we have

$$V(v+1) - V(v) = O\Big(\Big(\frac{v+1}{|\gamma_h|} \Big)^{k/h} - \Big(\frac{v}{|\gamma_h|} \Big)^{k/h} \Big)$$

$$= O\Big(|\gamma_h|^{-k/h} \int_v^{v+1} t^{k/h-1}\, dt \Big)$$

$$= O\big(|\gamma_h|^{-k/h} (v+1)^{k/h-1} \big)$$

$$= O(|\gamma_h|^{-k/h}(|\gamma_h| k+1)^{k/h-1}), \tag{3.9}$$

since $v \leqslant k|\gamma_h|$. Combining (3.7), (3.8) and (3.9), we have

$$I^k = O(|\gamma_h|^{-k/h}(|\gamma_h| k+1)^{k/h-1})$$

$$= O(|\gamma_h|^{-k/h-1+k/h}) = O(|\gamma_h|^{-1}).$$

This proves our lemma.

Remark. The lemma is due to Vinogradow, but the proof here given is new, which suggests some possible improvements in the result. By this lemma, we can also establish a theorem concerning the convergence of the integral (2.10), but it is not so good as lemma 2.2. Since min $(A_k, \cdots, A_1) \leqslant (A_k \cdots A_1)^{1/k}$ for $A_v > 0$, we have

$$I = O(\min(1, |\gamma_k \cdots \gamma_1|^{-1/k^2})). \tag{3.10}$$

Consequently, for $2t \geq k^2 + \epsilon$, $\epsilon > 0$,

$$\int_{-\infty}^{\infty} \cdots \int_{-\infty}^{\infty} |I|^{2t} d\gamma_1 \cdots d\gamma_k$$

converges. That is, I belongs to $L^{k^2+\epsilon}$.

IV. Singular Series

$\mathfrak{S}$ is called the singular series of Tarry's problem. It is evidently positive, if it converges. The convergence of the series can be deduced from the following lemma.

Lemma 4.1 (Hua [4]). *Let a_k, ..., a_0 and q be integers and $q > 0$, and let*

$$g(x) = a_k x^k + \cdots + a_1 x + a_0, \tag{4.1}$$

$(a_k, \cdots, a_1, q) = 1$. *Then*

$$\sum_{x=1}^{q} e(g(x)/q) = O(q^{1-1/k+\epsilon}). \tag{4.2}$$

For the proof of this lemma see Hua [1] and [4]. Certainly, we can construct a proof of this lemma, by an adoption of the method of §§ 6-8. Actually, the proof given later is a refinement of the previous one.

Lemma 4.2. *The singular series* $\mathfrak{S}$ *converges for* $t > k^2$.

Proof. Let Q be the least common multiple of $q_1, \ldots, q_k$. Then, we have

$$\left(\frac{h_k Q}{q_k}, \ldots, \frac{h_1 Q}{q_1}, Q\right) = 1, \tag{4.3}$$

for otherwise, there exists a prime p such that

$$p \mid \left(\frac{h_k Q}{q_k}, \ldots, \frac{h_1 Q}{q_1}, Q\right). \tag{4.4}$$

Let b be the greatest integer such that p^b divides Q. By the definition of Q, we have one of the q's, say q_l, such that p^b divides q_l. But, from (4.4), p divides $\frac{h_l}{q_l} Q$ which implies p dividing h_l. This contradicts $(h_l, q_l) = 1$.

By lemma 4.1, we have

$$\sum_{x=1}^{q_1 \cdots q_k} e\left(\frac{h_k}{q_k} x^k + \cdots + \frac{h_1}{q_1} x\right) \tag{4.5}$$

$$= \frac{q_1 \cdots q_k}{Q} \sum_{x=1}^{Q} e(g(x)/Q)$$

$$= O(q_1 \cdots q_k Q^{-1/k+\epsilon}).$$

Consequently, we have

$$\mathfrak{S} = O\left(\sum_{q_1=1}^{\infty} \cdots \sum_{q_k=1}^{\infty} \sum_{h_1} \cdots \sum_{h_k} Q^{-2t/k+\epsilon}\right) \tag{4.6}$$

$$= O\left(\sum_{q_1=1}^{\infty} \cdots \sum_{q_k=1}^{\infty} q_1 \cdots q_k Q^{-2t/k+\epsilon}\right).$$

Since $Q \geq \max(q_1, \cdots, q_k) \geq (q_1 \cdots q_k)^{1/k}$, we have

$$\mathfrak{S} = O\left(\sum_{q_1=1}^{\infty} \cdots \sum_{q_k=1}^{\infty} (q_1 \cdots q_k)^{1-2t/k^2+\epsilon} \right),$$

which converges for $t > k^2$.

Remark. Actually the previous method can give us a slightly better result, since we did not fully utilize the properties of the series

(4.7) $$\tau = \sum_{q_1=1}^{\infty} \cdots \sum_{q_k=1}^{\infty} q_1 \cdots q_k \, Q^{-2t/k+\epsilon}.$$

For a fixed Q, we consider the sum

(4.8) $$\sigma(Q) = \sum \cdots \sum q_1 \cdots q_k$$

where the sum runs over all systems of integers $q_1, \ldots, q_k$ having the least common multiple Q.

Let $Q = p_1^{l_1} \cdots p_s^{l_s}$. Then

$$\sigma(Q) = \sum \cdots \sum (p_1^{l_{11}} \cdots p_s^{l_{1s}})(p_1^{l_{21}} \cdots p_s^{l_{2s}}) \cdots (p_1^{l_{k1}} \cdots p_s^{l_{ks}})$$

where the sum runs over all non-negative integers l_{ij} satisfying

$$\max(l_{1j}, \cdots, l_{kj}) = l_j.$$

Evidently, we have

(4.9) $$\sigma(Q) = k\,Q\,(S(Q))^{k-1},$$

where $S(Q)$ is the sum of divisors of Q. Since $S(Q) = O(Q^{1+\epsilon})$, we have

$$\tau = \sum_{Q=1}^{\infty} \sigma(Q)\, Q^{-2t/k+\epsilon} = k \sum_{Q=1}^{\infty} (S(Q))^{k-1} Q^{1-2t/k+\epsilon}$$

$$= O\left(\sum_{Q=1}^{\infty} Q^{k-2t/k+\epsilon} \right),$$

which converges for $t > \frac{1}{2} k(k+1)$.

But this is still not the best possible result, in fact, by a more elaborate method, we can prove that the series converges for $t>\frac{1}{4}k(k+1)+1$. On the contrary, we prove first that

Lemma 4.3. *The series* $\mathfrak{S}$ *diverges for* $t=\frac{1}{4}k(k+1)+1$.

Proof. The following series is a part of our singular series.

$$\mathfrak{S}_1=\sum_{p>k}\sum_{\substack{h_k=1\\ p\nmid h_k}}^{p^k}\cdots\sum_{\substack{h_1=1\\ p\nmid h_1}}^{p}\sum_{c=1}^{p}\left|p^{-1-\cdots-k}\sum_{x=1}^{p^{1+\cdots+k}}e\left(\frac{h_k}{p^k}(x+c)^k+\cdots+\frac{h_1}{p}(x+c)\right)\right|^{2t}$$

$$=\sum_p\sum_{h_k}\cdots\sum_{h_1}p\left|p^{-k}\sum_{x=1}^{p^k}e\left(\frac{h_k}{p^k}x^k+\cdots+\frac{h_1}{p}x\right)\right|^{2t}.$$

If we can prove that, for $p>k$,

$$\sum_{x=1}^{p^k}e\left(\frac{h_k}{p^k}x^k+\cdots+\frac{h_1}{p}x\right)=p^{k-1}, \tag{4.10}$$

then

$$\mathfrak{S}_1=\sum_{p>k}p^{\frac{1}{2}k(k-1)}(p-1)^k p^{-2t+1}$$

$$\geqslant\frac{1}{2^k}\sum_{p>k}p^{\frac{1}{2}k(k+1)-2t+1}$$

$$\geqslant\frac{1}{2^k}\sum_p p^{-1},$$

which diverges.

Now we are going to prove (4.10). Let $X=p^{k-1}y+z$, then

$$\sum_{x=1}^{p^k}e\left(\frac{h_k}{p^k}x^k+\cdots+\frac{h_1}{p}x\right)$$

$$=\sum_{z=1}^{p^{k-1}}e\left(\frac{h_k}{p^k}z^k+\cdots+\frac{h_1}{p}z\right)\sum_{y=1}^{p}e\left(\frac{k\,h_k}{p}y\,z^{k-1}\right).$$

The inner sum equals zero except when p divides z. Replacing $z = pw$, we have (4.10).

In order to establish the best upper bound of t, we require a more compact form of the singular series which tells us the arithmetical meaning of the singular series.

V. The Arithmetical Nature of the Singular Series

Lemma 5.1. *We have*

$$\mathfrak{S} = \sum_{Q=1}^{\infty} \sum_{(a_k, \cdots, a_1, Q)=1} \left| Q^{-1} \sum_{x=1}^{Q} e\,(a_k x^k + \cdots + a_1 x)/Q \right|^{2t}, \tag{5.1}$$

where each of $a_1, \ldots, a_k$ run over complete residue systems, mod Q, satisfying the condition $(a_k, \ldots, a_1, Q) = 1$.

Proof. The lemma is an immediate consequence of the following fact: The correspondence

$$\frac{h_l Q}{q_l} = a_l, \qquad l = 1, 2, \cdots, k,$$

where Q is the least common multiple of $q_1, \ldots, q_k$, is one to one between the range of summations of (1.5):

$$q_l = 1, 2, 3, \cdots, (h_l, q_l) = 1, \ 1 \leqslant h_l \leqslant q_l, \ l = 1, 2, \cdots, k$$

and the range of summations of (5.1)

$$Q = 1, 2, 3, \cdots, (a_k, \cdots, a_1, Q) = 1, \ 1 \leqslant a_l \leqslant Q, \ l = 1, \cdots, k.$$

This fact is almost self-evident. The lemma follows, since

$$q_1^{-1} \cdots q_k^{-1} \sum_{x=1}^{q_1 \cdots q_k} = Q^{-1} \sum_{x=1}^{Q}.$$

Definition. Let $N(p^\gamma)$ be the number of solutions of

$$x_1^h + \cdots + x_t^h \equiv y_1^h + \cdots + y_t^h \pmod{p^\gamma}$$

$$1 \leqslant h \leqslant k, \qquad 1 \leqslant x, y \leqslant p^{\gamma}.$$

Lemma 5.2. *We have*

$$(5.2) \qquad p^{-\gamma(2t-k)} N(p^{\gamma}) = \sum_{l=0}^{\gamma} \sum_{\substack{a_1=1 \\ p \nmid (a_1, \cdots, a_k)}}^{p^l} \cdots \sum_{a_k=1}^{p^l} \Big| p^l \sum_{x=1}^{p^l} e\,(a_k x^k + \cdots + a_1 x)/p^l) \Big|^{2t}.$$

Proof. Evidently, we have

$$N(p^{\gamma}) = \frac{1}{p^{\gamma k}} \sum_{a_1=1}^{p^{\gamma}} \cdots \sum_{a_k=1}^{p^{\gamma}} \Big| \sum_{x=1}^{p^{\gamma}} e\,((a_k x^k + \cdots + a_1 x)/p^{\gamma}) \Big|^{2t}$$

$$= \frac{1}{p^{\gamma k}} \sum_{m=0}^{\gamma} \sum_{\substack{a_1=1 \\ (a_1, \cdots, a_k, p^{\gamma}) = p^m}}^{p^{\gamma}} \cdots \sum_{a_k=1}^{p^{\gamma}} \Big| \sum_{x=1}^{p^{\gamma}} e\,((a_k x^k + \cdots + a_1 x)/p^{\gamma}) \Big|^{2t}$$

$$= \frac{1}{p^{\gamma k}} \sum_{m=0}^{\gamma} \sum_{\substack{b_1=1 \\ p \nmid (b_1, \cdots, b_k)}}^{p^{\gamma-m}} \cdots \sum_{b_k=1}^{p^{\gamma-m}} \Big| p^m \sum_{x=1}^{p^{\gamma-m}} ((b_k x^k + \cdots + b_1 x)/p^{\gamma-m}) \Big|^{2t}$$

$$= p^{2t\gamma-\gamma k} \sum_{l=0}^{\gamma} \sum_{\substack{b_1=1 \\ p \nmid (b_1, \cdots, b_k)}}^{p^l} \cdots \sum_{b_k=1}^{p^l} \Big| p^l \sum_{x=1}^{p^l} e\,((b_k x^k + \cdots + b_1 x)/p^l) \Big|^{2t}.$$

This proves lemma 5.2.

By lemma 4.2, since the right hand side of (5.2) is a part of the singular series, we have that the limit

$$\partial_p = \lim_{r \to \infty} p^{-\gamma(2t-k)} N(p^{\gamma})$$

exists for $t > k^2$. This is called the p-adic density of Tarry's problem. Owing to the absolute convergence of $\mathfrak{S}$, we can verify easily that

$$(5.3) \qquad \mathfrak{S} = \prod_p \partial p$$

if $t > k^2$. This is the arithmetic nature of the singular series. In §§ 6-9, we shall establish (5.3) with the best possible lower bound for t.

VI. p-adic Expansion of a Solution of a Congruence

Let p be a fixed prime and l a fixed integer > 0. Let $g_1(x)$ be a polynomial of the n-th degree with integer coefficients, mod p^l. Let r_0 ($\geqslant 0$) be the largest integer such that p^{r_0} divides all the coefficients of $g_1(x)$. Let x' be a solution of

(6.1) $$p^{-r_0} g_1(x) \equiv 0 \pmod{p}, \qquad 0 \leqslant x' < p.$$

Write

(6.2) $$g_2(x) = p^{-r_0} g_1(px + x').$$

We consider $g_2(x)$ instead of $g_1(x)$ and modulo p^{l-r_0} instead of modulo p^l. Then we have r_1, which is the highest power of p such that p^{r_1} divides all the coefficients of $g_2(x)$, evidently $r_1 \geqslant 1$. Let x'' be a solution of

(6.3) $$p^{-r_1} g_2(x) \equiv 0 \pmod{p}, \qquad 0 \leqslant x'' < p$$

and $g_3(x) = p^{-r_1} g_2(px + x'')$, mod $p^{l-r_0-r_1}$, etc.

Continuing this process, until, after e steps, we have $r_0 + \cdots + r_{e-1} = l$, and all the coefficients of $g_e(x)$ are divisible by $p^{l-r_1-\cdots-r_{e-2}}$.

Symbolically, we use

(6.4) $$x' + px'' + \cdots + p^{e-1} x^{(e)}$$

to denote the considered solution of

(6.5) $$g_1(x) \equiv 0 \pmod{p^l}.$$

Lemma 6.1. *The number of solutions of* (6.5) *in the sense previously described is at most equal to* n.

Proof. Let x' be a root of (6.1) of multiplicity m, it is sufficient to prove that (6.3) has at most m solutions, multiplicities being counted. In fact, let $x'_1, x'_2, \cdots, x'_v$ be v solutions of (6.1) with the multiplicities

$m_1, \ldots, m_v$ respectively. Then, we have v $g_2(x)$'s, the total number of roots of (6.3) is $\leqslant m_1 + \cdots\cdots + m_v \leqslant n$.

We may assume that $x' = 0$. By definition, we may write

$$p^{-r_0} g_1(x) = x^m h(x) + p j(x),$$

where $j(x)$ is a polynomial of the degree $< m$, and $p \nmid h(0)$. Thus, we have

$$g_2(x) = p^m x^m h(px) + p j(px).$$

Incidentally we proved $r_1 \leqslant m$ and

$$p^{-r_1} g_2(x) = p^{m-r_1} x^m h(px) + p^{1-r_1} j(px)$$

$$\equiv p^{m-r_1} x^m h(0) + p^{1-r_1} j(px) \pmod{p},$$

which is a polynomial of degree $\leqslant m$.

VII. A Chain of Characteristic Exponents of a Polynomial

Let

$$f(x) = a_k x^k + \cdots + a_1 x + a_0$$

and $p \nmid (a_k, \cdots, a_1)$. Let

(7.1) $$x = x' + p x'' + \cdots + p^{e-1} x^{(e)}$$

be a solution of

(7.2) $$f'(x) \equiv 0 \pmod{p^l}$$

in the sense of §6.

Let u_1 be the greatest integer such that p^{u_1} divides all the coefficients of

$$f(px + x') - f(x') \qquad (= p^{u_1} f_1(x), \text{ say}).$$

Let u_2 be the greatest integer such that p^{u_2} divides all the coefficients of

$$f_1(px+x'')-f_1(x'') \qquad (=p^{u_2}f_2(x), \text{ say}).$$

Since

$$f_1(px+x'')-f_1(x'')=p^{-u_1}(f(p^2x+px''+x')-f(px''+x')),$$

then $p^{u_1+u_2}$ is the highest power of p dividing all the coefficients of

$$f(p^2x+x'+px'')-f(x'+px'').$$

Similarly, we define $u_1, u_2, \ldots$ up to u_e which is called a chain of characteristic exponents with respect to the root (7.1) of the polynomial $f'(x)$ modulo p^l.

Lemma 7.1. *We have*

$$k \geqslant u_1 \geqslant u_2 \geqslant \cdots \geqslant u_e \geqslant 2. \tag{7.3}$$

Proof. By definition, we have evidently $u_i \geqslant 2$. It is only needed to prove that $k \geqslant u_1 \geqslant u_2$. Without loss of generality, we assume that $x'=x''=0$. Then

$$f(px)-f(0)=a_k p^k x^k+\cdots+a_1 px.$$

Evidently $u_1 \leqslant k$, since $p \nmid (a_k, \cdots, a_1)$. From the definition of u_1, we have

$$p^{u_1-1} \mid a_1,\ p^{u_1-2} \mid a_2,\ \cdots,\ p \mid a_{u_1-1}. \tag{7.4}$$

If $u_2 > u_1$, then by the definition of u_2, that is $p^{u_1+u_2}$ divides all the coefficients of $f(p^2x)-f(0)$, we have

$$p^{u_1+u_2-2} \mid a_1,\ p^{u_1+u_2-4} \mid a_2, \cdots,\ p^{u_1+u_2-2u_1} \mid a_{u_1}.$$

Consequently, we have

$$p^{u_1} \mid a_1,\ p^{u_1-1} \mid a_2,\ \cdots,\ p^2 \mid a_{u_1-1},\ p \mid a_{u_1},$$

that means, p^{u_1+1} divides all the coefficients of $f(px)-f(0)$, this contradicts our definition of u_1.

Lemma 7.2. *We have*

$$l + e - (u_1 + \cdots + u_e) \leqslant \left[\frac{\log k}{\log p}\right].$$

Proof. Without loss of generality, we assume

$$x' + x'' p + \cdots + x^{(e)} p^{e-1} = 0 .$$

Then, by definition, we deduce that all the coefficients of $f'(p^e x)$ are divisible by p^l. That is

$$p^l \mid v\, a_v\, p^{e(v-1)}, \qquad v = 1, 2, \cdots, k .$$

That is,

$$\frac{p^{l+e}}{(p^l, v)} \Bigm| a_v\, p^{ev},$$

all the coefficients of $f(p^e x) - f(0)$ are divisible by $p^{l+e-\nu}$, where $p^\nu = \max\limits_{1 \leq v \leq k} (p^l, v)$. Therefore

$$u_1 + \cdots + u_e \geqslant l + e - \left[\frac{\log k}{\log p}\right],$$

since $(p^l, v) \leqslant p^\nu \leqslant k$.

Lemma 7.3. *Let $e' \leqslant e$. The number of polynomials which have a root of the systems of the given first e' characteristic exponents*

$$k \geqslant u_1 \geqslant u_2 \geqslant \cdots \geqslant u_{e'} > 1$$

is equal to

$$p^{e'}\, p^{lk - \frac{1}{2}u_1(u_1-1) - \cdots - \frac{1}{2}u_{e'}(u_{e'}-1)} .$$

Proof. The possible choice of the root is $p^{e'}$. We assume that the root is 0, then apply (7.4) successively we have the lemma, since the condition of (7.4) restricts the possibility of $a_1, \ldots, a_k$ by a factor

$$p^{-1-2-\cdots-(u_1-1)} = p^{-\frac{1}{2}u_1(u_1-1)} .$$

VIII. Exponential Sum

Let $p \nmid (a_k, \cdots, a_1)$ and

$$f(x) = a_k x^k + \cdots + a_1 x + a_0 ,$$

and let

$$W = \left[\frac{\log k}{\log p}\right].$$

Let r be the greatest integer such that p^r divides all the coefficients of $f'(x)$. Evidently we have $r \leqslant W$.

Lemma 8.1. *If ξ is not a solution of*

$$p^{-r} f'(x) \equiv 0 \qquad (\mathrm{mod}\ p) , \tag{8.1}$$

then, for $l > 2W + 1$, we have

$$\sum_{x=1}^{p^{l-1}} e(f(\xi + px)/p^l) = 0 .$$

Proof. Let $x = y + p^{l-r-2} z$, then

$$\sum_{x=1}^{p^{l-1}} e(f(\xi + px)/p^l)$$

$$= \sum_{y=1}^{p^{l-r-2}} \sum_{z=0}^{p^{r+1}-1} e(f(\xi + py + p^{l-r-1} z)/p^l)$$

$$= \sum_{y=1}^{p^{l-r-2}} e(f(\xi + py)/p^l) \cdot \sum_{z=0}^{p^{r+1}-1} e(f'(\xi + py) z / p^{r+1})$$

$$= 0 ,$$

since $p^{r+1} \nmid f'(\xi)$.

Lemma 8.2. *Let ξ be a solution of* (8.1). *Let p^{u_1} be the highest power of p dividing all the coefficients of $f(px+\xi)-f(\xi)$ $(=p^{u_1}\, g(x)$, say). Then*

$$(8.2)\qquad \sum_{x=1}^{p^{l-1}} e(f(\xi+px)/p^l) = p^{u_1-1}\, e(f(\xi)/p^l) \sum_{x=1}^{p^{l-u_1}} e(g(x)/p^{l-u_1}).$$

Proof. We have

$$\sum_{x=1}^{p^{l-1}} e((f(\xi+px)-f(\xi))/p^l)$$

$$=\sum_{x=1}^{p^{l-1}} e(g(x)/p^{l-u_1}) = p^{u_1-1}\sum_{x=1}^{p^{l-u_1}} e(g(x)/p^{l-u_1}).$$

According to lemmas 8.1 and 8.2, we have that

$$\sum_{x=1}^{p^l} e(f(x)/p^l) = \sum_{\xi} p^{u_1-1}\, e(f(\xi)/p^l) \sum_{x=1}^{p^{l-u_1}} e(g(x)/p^{l-u_1}),$$

where ξ runs over the roots of (8.1) and u_1 and g depend on ξ.

If we use this method repeatedly, the exponential sum breaks into at most $k-l$ partial sums each of which corresponds to a root in the sense described in § 6. Certainly, we have to notice the fact $l>2W+1$.

Let e' be the greatest integer such that

$$l-u_1-\cdots-u_{e'-1} > 2W+1 \geqslant l-u_1-\cdots-u_{e'}.$$

The existence of e' is assured by lemma 7.2. Let $\xi=x'+px''+\cdots+p^{e'-1}x^{(e')}$ denote typically one of the roots. Then the partial sum corresponding to this root is equal to

$$(8.3)\qquad e^{2\pi i f(\xi)/p^l}\, p^{u_1+\cdots+u_{e'}-e'} \sum_{x=1}^{p^{l_0}} e^{2\pi i h(x)/p^{l_0}},$$

where

$$l_0 = l - u_1 - \cdots - u_{e'}, \qquad 0 \leqslant l_0 \leqslant 2W + 1.$$

Consequently, the absolute value of the partial sum corresponding to one of these roots with chain $(u_1, \ldots, u_{e'})$ is

$$(8.4) \qquad \leqslant p^{u_1 + \cdots + u_{e'} - e'} \cdot p^{l_0} = p^{l-e'}.$$

Let $$A(p^l) = \sum_{\substack{a_k=1 \\ p \nmid (a_k, \cdots, a_1)}}^{p^l} \cdots \sum_{a_1=1}^{p^l} \left| \frac{1}{p^l} \sum_{x=1}^{p^l} e((a_k x^k + \cdots + a_1 x)/p^l) \right|^{2t}.$$

We divide the inner sum $\sum_{x=1}^{p^l}$ according to the roots of $f'(x) \equiv 0 \pmod{p^l}$ in the sense of §§ 6-7, say

$$\sum\nolimits_1, \ \cdots, \ \sum\nolimits_m, \qquad m \leqslant k-1.$$

By Hölder's inequality, we have

$$\left| \sum_{x=1}^{p^l} e((a_k x^k + \cdots + a_1 x)/p^l) \right|^{2t} \leqslant k^{2t-1} \sum_{i=1}^{m} \left| \Sigma_i \right|^{2t}.$$

Thus $A(p^l)$ is dominated by the sum

$$(8.5) \qquad p^{-2tl} \sum_{\xi} |S(\xi)|^{2t},$$

where $S(\xi)$ corresponds to a partial sum corresponding to a polynomial with ξ as a root. Corresponding to a fixed chain

$$u_1, u_2, \cdots, u_{e'},$$

the number of polynomials is

$$\leqslant p^{e'} \cdot p^{lk - \frac{1}{2}u_1(u_1-1) - \cdots - \frac{1}{2}u_{e'}(u_{e'}-1)},$$

and each sum is

$$\leqslant p^{l-e'}$$

by (8.4). Thus the sum of all those partial sums each of them corresponding to a root with chain $u_1, \cdots, u_{e'}$ is

$$\leqslant p^{e'+lk-\frac{1}{2}u_1(u_1-1)-\cdots-\frac{1}{2}u_{e'}(u_{e'}-1)} \cdot p^{-2e't}.$$

Summing over all possible values of $u_1, \ldots, u_{e'}$ we have

$$A(p^l) = O\left(\sum p^{e'+lk-\frac{1}{2}u_1(u_1-1)-\cdots-\frac{1}{2}u_{e'}(u_{e'}-1)-2e't}\right), \tag{8.6}$$

where the sum runs over

$$k \geqslant u_1 \geqslant u_2 \geqslant \cdots \geqslant u_{e'} \geqslant 2,$$

and

$$2W + 1 \geqslant l - u_1 - \cdots - u_{e'} \geqslant 0.$$

The number of terms in the right of (8.6) is $O(l^k)$. In fact, we consider the number of solutions of

$$u_1 + \cdots + u_{e'} = l'. \tag{8.7}$$

Suppose there are e_k of the u's equal k, e_{k-1} equal $k-1, \ldots$, and e_2 of the u's equal 2. Then

$$0 \leqslant k e_k \leqslant l', \ 0 \leqslant (k-1) e_{k-1} \leqslant l', \cdots, \ 0 \leqslant 2 e_2 \leqslant l'.$$

Thus, there are

$$\leqslant \frac{(l'+1)^{k-1}}{k!} = O(l'^{k-1})$$

of different $u_1, \ldots, u_{e'}$ satisfying (8.7). Summing over from $l' = l - 2W - 1, \cdots, l$, we have the assertion.

Now we are going to find the term with the highest exponent in (8.6). For a fixed e', we claim that

$$(8.8)\quad u_1 = \cdots = u_{l'-\left[\frac{l'}{e'}\right]e'} = \left[\frac{l'}{e'}\right]+1,\ u_{l'-\left[\frac{l'}{e'}\right]e'+1} = \cdots = u_{e'} = \left[\frac{l'}{e'}\right]$$

gives us the term with the biggest exponent in (8.6). This is a consequence of the inequality: if $u > v+1$, then

$$\frac{u(u-1)}{2} + \frac{v(v-1)}{2} > \frac{(u-1)(u-2)}{2} + \frac{v(v+1)}{2},$$

that is, if two of the terms in the exponent corresponding to u and v having a difference $\geqslant 2$, we can replace them by $u-1$ and $v+1$, and the term so obtained gives a bigger exponent.

In case (8.8), the exponent is

$$e' + l'k - \frac{1}{2}\left(\left[\frac{l'}{e'}\right]+1\right)\left[\frac{l'}{e'}\right]\left(l' - \left[\frac{l'}{e'}\right]e'\right)$$
$$- \frac{1}{2}\left[\frac{l'}{e'}\right]\left(\left[\frac{l'}{e'}\right]-1\right)\left(e' - l' + \left[\frac{l'}{e'}\right]e'\right) - 2e't$$

$$(8.9)\quad = e' + l'k + \frac{e'}{2}\left[\frac{l'}{e'}\right]\left(\left[\frac{l'}{e'}\right]-1\right) - \left[\frac{l'}{e'}\right]\left(l' - \left[\frac{l'}{e'}\right]e'\right) - 2e't$$

$$(8.10)\quad \leqslant e' + l'k - \frac{e'}{2}\left[\frac{l'}{e'}\right]\left(\left[\frac{l'}{e'}\right]-1\right) - 2e't = \phi(e').$$

Let

$$(8.11)\quad t = \frac{1}{4}k(k+1) + 1 + \delta.$$

For $e' \mid l'$, let $l' = e'm$, $m \leqslant k$, then

$$\phi(e') = e' + e'mk - \frac{e'}{2}m(m-1) - 2e't$$

$$= e'\left(1 + \frac{1}{2} m (2k - m + 1) - 2t\right)$$

$$\leqslant e'\left(1 + \frac{1}{2} k (k + 1) - 2t\right) \leqslant -(1 + 2\delta)\, e'$$

$$= -(1 + 2\delta) \frac{l'}{m} \leqslant -(1 + 2\delta) \frac{l'}{k}. \tag{8.12}$$

In case $e' \nmid l'$, let $l' = e' m + e_1$, $1 \leqslant e_1 < e'$, $m + 1 \leqslant k$,

we have

$$\phi(e') = e' + (e' m + e_1) k - \frac{e'}{2} m (m - 1) - 2 e' t$$

$$\leqslant e'\left(1 + \frac{1}{2} k (k + 1) - 2t\right) = -(1 + 2\delta)\, e'$$

$$< -(1 + 2\delta) \frac{l'}{m + 1}$$

(8.13)

$$\leqslant -(1 + 2\delta) \frac{l'}{k}.$$

Thus we have

$$\phi(e') \leqslant -(1 + 2\delta) \frac{l - 2W - 1}{k}.$$

Combining these results, we have

Lemma 8.3. *For* $t = \frac{1}{4} k (k + 1) + 1 + \delta$;

$$A(p^l) = O\left(l^k p^{-(1+2\delta)\frac{l-1}{k}}\right).$$

In fact, if $W \geqslant 1$,

$$p^{\frac{2w+1}{k}} \leqslant p^{\frac{3w}{k}} = O(1).$$

Lemma 8.4. *The p-adic density ∂_p converges for* $t > \frac{1}{4} k(k+1)+1$.

Proof.

$$\partial_p = \sum_{l=0}^{\infty} A(p^l) = O\left(\sum_{l=0}^{\infty} l^k p^{-(1+2\delta)\frac{l}{k}}\right) = O(1).$$

Remark. It is easy to establish that the p-adic density ∂_p converges for $t > \frac{1}{4} k(k+1)$ and it diverges for $t = \frac{1}{4} k(k+1)$.

IX. Convergence of Singular Series

Lemma 9.1. *We have, for* $t \geq \frac{1}{4} k(k+1)+1+\delta$,

$$\sum_{l=k+1}^{\infty} A(p^l) = O(p^{-(1+2\delta)}).$$

Proof. By lemma 8.3, we have

$$\sum_{l=k+1}^{\infty} A(p^l) = O\left(\sum_{l=k+1}^{\infty} l^k p^{-(1+2\delta)\frac{l-1}{k}}\right)$$

$$= O\left(\sum_{m=0}^{\infty} (m+k+1) p^{-(1+2\delta)\frac{m}{k}} \cdot p^{-(1+2\delta)}\right).$$

Lemma 9.2. *We have, for* $p > k$, $t \geqslant (\frac{3}{2}+\delta) k$,

$$A(p) = O(p^{-(1+2\delta)}).$$

Proof. By Mordell's theorem[7], we have

$$\sum_{a_1=1}^{p} \cdots \sum_{a_k=1}^{p} \left| \frac{1}{p} \sum_{x=1}^{p} e(a_k x^k + \cdots + a_1 x)/p) \right|^{2t}$$

$$\leqslant O\left(p^{(2t-2k)(1-\frac{1}{k})-2t+k} \frac{1}{p^k} \sum_{a_1=1}^{p} \cdots \sum_{a_k=1}^{p} \left| \sum_{x=1}^{p} e((a_k x^k + \cdots + a_1 x)/p) \right|^{2k}\right)$$

$$= O(p^{2(t-k)(1-\frac{1}{k})-2t+2k})$$

$$= O(p^{-2(t-k)\frac{1}{k}})$$

$$= O(p^{-(1+2\delta)}).$$

Lemma 9.3. *We have, for* $t=\frac{1}{4}k(k+1)+1+\delta$, $p>k$,

$$A(p^2)+\cdots+A(p^k)=O(p^{-(1+2\delta)}).$$

Proof. Now we have $W=0$. The possible values of l' is l or $l-1$. By (8.12) and (8.13),

$$\phi(e')\leqslant-(1+2\delta)\frac{l'}{m}\leqslant-(1+2\delta), \qquad \text{if } e\,/\,l,$$

$$\leqslant-(1+2\delta)\frac{l'}{m+1}\leqslant-(1+2\delta), \qquad \text{if } e\nmid l,$$

since now we have $m\leqslant l'$ or $m+1\leqslant l'$ respectively. Thus, we have

$$A(p^2)+\cdots+A(p^k)=O(p^{-(1+2\delta)}).$$

Combining lemmas 9.1, 9.2 and 9.3 we have

Lemma 9.4. *For* $k\geqslant3$, *we have*

$$\partial_p-1=O(p^{-(1+2\delta)}).$$

Lemma 9.5. *Let* $k\geqslant3$. *For* $t>\frac{1}{4}k(k+1)+1$, *the infinite product*

$$\prod_p \partial_p$$

converges absolutely and

$$\mathfrak{S}=\prod_p \partial_p.$$

Remark. By a similar method we can prove that $\mathfrak{S}$ converges for $t>3$ as $k=2$. But for this case, we can find the explicit value of $\mathfrak{S}$, namely

$$\mathfrak{S} = 2\frac{\zeta(t-2)}{\zeta(t-1)}.$$

It is quite lucky to find the best possible exponent for $k \geqslant 3$, in fact, for $k=2$, $A(p)$ contributes the most essential part of the result. It is very difficult to find the best possible order of $A(p)$ for $k \geq 3$. Fortunately the dominating term is shifted to $A(p^t)$, for $t>1$, for which we can find the best possible estimation.

X. Several Lemmas

Let $<\xi> = \min(\xi - [\xi], [\xi]+1-\xi)$, $\{\xi\} = \xi - [\xi]$.

Lemma 10.1. *Let* $\tau \geqslant 1$,

$$\left|\alpha - \frac{h}{q}\right| \leqslant \frac{\tau}{q^2}, \quad (h, q) = 1, \quad 0 < q < p^s.$$

The number of solutions in integers y satisfying

$$<\alpha y> \leqslant \frac{V}{q}, \qquad f \leqslant y \leqslant f+N \tag{10.1}$$

is

$$\leqslant 2(V+2\tau)(Nq^{-1}+1). \tag{10.2}$$

Proof. It is sufficient to prove that the number of solutions of

$$<\alpha y> \leqslant \frac{V}{q}, \qquad f \leqslant y \leqslant f+q \tag{10.3}$$

is $\leqslant 2(V+2\tau)$. We write

$$y = f+z, \qquad \alpha = \frac{h}{q} + \frac{\tau\vartheta}{q^2}, \qquad |\vartheta| \leqslant 1.$$

Then

$$\alpha y = \frac{hz}{q} + \frac{\tau \vartheta z}{q^2} + \frac{hf}{q} + \frac{\tau \vartheta f}{q^2}$$

$$= \frac{hz + [c] + \{c\} + \tau \vartheta z/q}{q}, \qquad |\tau \vartheta z/q| \leqslant \tau,$$

where

$$c = hf + \tau \vartheta f/q.$$

The conclusion is evident for $q \leq 2(V + 2\tau)$. As z runs over a complete residue system modulo q, so does $\mathrm{w} = hz + [c]$. Thus

$$\alpha y = \frac{\mathrm{w} + \sigma(\mathrm{w})}{q},$$

where

$$-\tau \leqslant \{c\} - \tau \leqslant \sigma(\mathrm{w}) \leqslant \{c\} + \tau < 1 + \tau.$$

Evidently, those integers w satisfying

$$(10.4) \qquad V + \tau \leqslant \mathrm{w} < q - \tau - V - 1$$

gives

$$\frac{V}{q} \leqslant \frac{\mathrm{w} + \sigma(\mathrm{w})}{q} < 1 - \frac{V}{q},$$

which do not satisfy (10.3). The number of integers w satisfying (10.4) is $\geq q - 2V - 2\tau - 2$. Therefore the number of integers satisfying (10.3) is

$$\leqslant q - (q - 2V - 2\tau - 2) = 2V + 2\tau + 2 \leqslant 2(V + 2\tau).$$

Lemma 10.2. *Let* $Y \geq 1$ *and let* $A_0, A_1, \cdots, A_{k-1}$ *be integers satisfying*

$$A_0 = 1, \qquad |A_r| \leqslant (r+1) Y^r.$$

The system of equations

(10.5) $$v_r = \sum_{s=r}^{k} \binom{s+1}{r} A_{s-r} u_s$$

implies that $(k+1)\,k \cdots (r+1)\, u_r$ *is a linear combination of* $v_k, \cdots, v_r$ *with integral coefficients, namely*

(10.6) $$(k+1)\,k \cdots (r+1)\, u_r = \sum_{s=r}^{k} a_{rs} v_s ,$$

moreover

(10.7) $$a_{rs} = O\,(Y^{s-r}) .$$

Proof. The lemma is true for $r = k$. Suppose the theorem is true for k, $(k-1)$, . . . , $r+1$. Then, by (10.5)

$$(k+1) \cdots (r+1)\, u_r = (k+1) \cdots (r+2)\, (v_r - \sum_{s=r+1}^{k} \binom{s+1}{r} A_{s-r} u_s)$$

$$= (k+1) \cdots (r+2)\, v_r - \sum_{s=r+1}^{k} \binom{s+1}{r} A_{s-r} \frac{s\,!}{(r+1)!} (k+1) \cdots (s+1)\, u_s$$

$$= (k+1) \cdots (r+2)\, v_r - \sum_{s=r+1}^{k} \binom{s+1}{r} A_{s-r} \frac{s\,!}{(r+1)!} \sum_{u=s}^{k} a_{su} v_u$$

by the hypothesis of induction. Then, we have, for $k \geq u > r$,

$$a_{su} = \sum_{u \geq s \geq r+1}^{k} \binom{s+1}{r} A_{s-r} \frac{s\,!}{(r+1)!} a_{su}$$

which is evidently an integer and

$$a_{ru} = O\,(\sum_{u \geq s \geq r+1} | A_{s-r} |\, | a_{su} |)$$

$$= O\,(Y^{s-r} \cdot Y^{u-s}) = O\,(Y^{u-r}) .$$

In case $u = r$, it is evident that

$$a_{rr} = O(1).$$

Lemma 10.3. *Let $\xi_1, \cdots, \xi_n$ be real numbers. Then, for integers $l_1, \cdots, l_n$, we have*

$$< \sum_{i=1}^{n} l_i \xi_i > \leqslant \sum_{i=1}^{n} | l_i | < \xi_i > .$$

The lemma is a consequence of the following easy fact

$$< \xi_1 \pm \xi_2 > \leqslant < \xi_1 > + < \xi_2 > .$$

XI. A Bridge between an Individual Sum and an Average Value

Lemma 11.1. *Let $\alpha_k, \cdots, \alpha_1$ be real, and*

$$f(x) = \alpha_k x^k + \cdots + \alpha_1 x .$$

Suppose that $0 < \delta_1 < 1$ and T any integer,

$$(11.1) \qquad \int_0^1 \cdots \int_0^1 \Big| \sum_{x=T+1}^{T+P} e(f(x)) \Big|^{2t_1} d\alpha_1 \cdots d\alpha_k = O(P^{2t_1 - \frac{1}{2}k(k+1) + \delta_1}),$$

the constant implied by the symbol O may depend on t_1 and δ_1, but later we shall take t_1 and δ_1 as functions of k.

Let $\beta_{k+1}, \cdots, \beta_1$ be real numbers and

$$F(x) = \beta_{k+1} x^{k+1} + \cdots + \beta_1 x .$$

Let r be an integer, $k + 1 \geq r \geq 2$, suppose that

$$(11.2) \qquad \Big| \beta_r - \frac{h}{q} \Big| \leqslant \frac{1}{q^2}, \quad (h, q) = 1, \quad 1 \leqslant q \leqslant P^r .$$

Then, we have, for any T,

(11.3) $$S = \sum_{x=T+1}^{T+P} e(F(x)) = O\begin{cases} (P^{1-\rho} \quad), & \text{for } \tfrac{1}{2}P \leqslant q \leqslant P^{r-1}, \\ \left(P^{1-\rho}\left(\dfrac{q}{P^{r-1}}\right)^{\frac{1}{2t_1+k+1}}\right), & \text{for } P^{r-1} \leqslant q \leqslant P^r, \end{cases}$$

where

(11.4) $$\rho = \frac{1-\delta_1}{2t_1+k+1}.$$

Proof. (This important technique is due to Vinogradow.) For $0 < y \leqslant Y < P$, we write

$$S_0 = \sum_{x=T+1}^{T+P} e(F(x+y) - F(y)) = \sum_{x=T+1}^{T+P} e(\phi(x)),$$

where

$$\phi(x) = Y_1 x + \cdots + Y_{k+1} x^{k+1},$$

and

(11.5) $$Y_j = Y_j(y) = \frac{1}{j!}\frac{d^j}{dy^j}F(y)$$
$$= \binom{k+1}{j}\beta_{k+1} y^{k+1-j} + \cdots + \binom{j+1}{j}\beta_{j+1} y + \beta_j.$$

Evidently, we have $|S_0| = |S| + 2\vartheta y$, $|\vartheta| \leq 1$. Consequently,

$$|S| \leqslant Y^{-1}\sum_{y=1}^{Y} |S_0| + Y.$$

Applying Hölder's inequality twice, we have

$$|S|^{2t_1} \leqslant 2^{2t_1-1}\left((Y^{-1}\sum_{y=1}^{Y} |S_0|)^{2t_1} + Y^{2t_1}\right)$$

$$(11.6) \qquad = O\left(Y^{-1}\sum_{y=1}^{Y} |S_0|^{2t_1} + Y^{2t_1}\right).$$

Let

$$S_1 = \sum_{x=T+1}^{T+P} e(\alpha_1 x + \cdots + \alpha_k x^k + \beta_{k+1} x^{k+1}).$$

For a fixed y, then $Y_1, \cdots, Y_k$ are fixed; we consider those α's satisfying

$$\{\alpha_1 - Y_1\} \leqslant \frac{1}{2} P^{-2} Y, \cdots, \quad \{\alpha_k - Y_k\} \leqslant \frac{1}{2} P^{-k-1} Y, \quad 0 \leqslant \alpha_i < 1.$$

The domain so formed by $(\alpha_1, \cdots, \alpha_k)$ is denoted by $\Omega(y)$. If $\alpha_1, \cdots, \alpha_k$ belong to $\Omega(y)$, then we have

$$S_0 = S_1 + O(Y),$$

and consequently

$$(11.7) \qquad |S_0|^{2t_1} = O(|S_1|^{2t_1}) + O(Y^{2t_1}).$$

Combining (11.6) and (11.7), we have

$$|S|^{2t_1} = O(|S_1|^{2t_1}) + O(Y^{2t_1}).$$

Integrating over the domain $\Omega(y)$, we have

$$(11.8) \quad |S|^{2t_1} = O\left(P^{\frac{1}{2}k(k+1)+k} Y^{-k} \int \cdots \int_{\Omega(Y)} |S_1|^{2t_1} d\alpha_1 \cdots d\alpha_k\right) + O(Y^{2t_1}),$$

since

$$\int \cdots \int_{\Omega(y)} d\alpha_1 \cdots d\alpha_k \geqq \prod_{i=1}^{k} \left(\frac{1}{2} P^{-(i+1)} Y\right) = P^{-\frac{1}{2}k(k+1)-k} Y^k.$$

Now we are going to estimate the number of overlapping $\Omega(y)$'s. Suppose that $\Omega(y)$ and $\Omega(y_0)$ are overlapping. Then, we have

$$< Y_r(y) - Y_r(y_0) > \leqslant P^{-r-1} Y, \quad 1 \leqslant r \leqslant k.$$

Putting $v_r = Y_r(y) - Y_r(y_0)$, $u_s = \beta_{s+1}(y - y_0)$, and

$$A_{s-r} = \frac{y^{s-r+1} - y_0^{s-r+1}}{y - y_0},$$

then, from (11.5), we have

$$v_r = \sum_{s=r}^{k} \binom{s+1}{r} \beta_{s+1} (y^{s-r+1} - y_0^{s-r+1}),$$

$$= \sum_{s=r}^{k} \binom{s+1}{r} A_{s-r} u_s.$$

By lemmas 10.2 and 10.3, we have

$$< \frac{(k+1)!}{r!} u_r > \leqslant \sum_{s=r}^{k} | a_{rs} | < v_s >$$

$$= O(\sum_{s=r}^{k} Y^{s-r} P^{-s-1} Y)$$

$$= O(Y P^{-r-1}), \qquad 1 \leqslant r \leqslant k.$$

Replacing r by r–1, we have

(11.9) $$< \frac{(k+1)!}{(r-1)!} \beta_r (y - y_0) > = O(Y P^{-r}), \qquad 2 \leqslant r \leqslant k+1,$$

where

(11.10) $$1 \leqslant y \leqslant Y.$$

By lemma 10.1, the number of integers y satisfying (11.9) and (11.10) is

$$(11.11) \qquad O\left(\left(\frac{Yq}{p^r}+1\right)\left(1+\frac{Y}{q}\right)\right)=O\left(\frac{qY}{p^r}+1\right)=O\left(\frac{q}{p^{r-1}}+1\right),$$

since

$$q \geqslant \frac{1}{2}P \quad \text{and} \quad Y \leqslant P.$$

Therefore, any point $(\alpha_1, \ldots, \alpha_k)$ in the unit hypercube

$$0 \leqslant \alpha_1 \leqslant 1, \cdots, 0 \leqslant \alpha_k \leqslant 1$$

(it is better to say an n-dimensional torus, that is, we identify the side $\alpha_i=0$ with $\alpha_i=1$) is covered by at most $O\left(\frac{Yq}{p^r}+1\right)$ times by $\Omega(y)$, $y=1,\cdots,Y$. Hence, we have, by (11.8),

$$|S|^{2t_1} = O\left(P^{\frac{1}{2}k(k+1)+k} Y^{-k} \frac{1}{Y}\sum_{y=1}^{Y}\int\cdots\int_{\Omega}|S_1|^{2t_1}\,d\alpha_1\cdots d\alpha_k\right)+O(Y^{2t_1})$$

$$(11.12) \qquad = O\left(P^{\frac{1}{2}k(k+1)+k} Y^{-k-1}\left(\frac{q}{p^{r-1}}+1\right)\int_0^1\cdots\int_0^1|S_1|^{2t_1}\,d\alpha_1\cdots d\alpha_k\right)+O(Y^{2t_1}).$$

Since

$$\int_0^1\cdots\int_0^1|S_1|^{2t_1}\,d\alpha_1\cdots d\alpha_k \leqslant \int_0^1\cdots\int_0^1\left|\sum_{x=T+1}^{P+T}e(f(x))\right|^{2t_1}d\alpha_1\cdots d\alpha_k,$$

we have, by (11.1)

$$|S|^{2t_1} = O\left(P^{\frac{1}{2}k(k+1)+k} Y^{-k-1}\left(\frac{q}{p^{r-1}}+1\right)P^{2t_1-\frac{1}{2}k(k+1)+\delta_1}\right)+O(Y^{2t_1})$$

$$= O\left(P^{2t_1+k+\delta_1} Y^{-k-1}\left(\frac{q}{p^{r-1}}+1\right)\right)+O(Y^{2t_1}).$$

We take

$$Y = \begin{cases} \left[P^{1-\frac{1-\delta_1}{2t_1+k+1}}\right] + 1\ , & \text{for } \frac{P}{2} \leqslant q \leqslant P^{r-1}, \\ \left[P^{1-\frac{1-\delta_1}{2t_1+k+1}}\left(\frac{q}{P^{r-1}}\right)^{\frac{1}{2t_1+k+1}}\right] + 1, & \text{for } P^{r-1} \leqslant q \leqslant P^r, \end{cases}$$

the lemma follows.

Lemma 11.2. *Under the same assumption as lemma* 11.1, *we have*

$$S = \sum_{x=T+1}^{T+P} e(F(x)) = O(P\, q^{-\rho}), \quad \text{for } 1 \leqslant q \leqslant P.$$

Proof. We divide the sum into no more than $\frac{P}{q}$ parts, each of them is a sum of length Q which lies in

$$q \leqslant Q \leqslant 2q.$$

By lemma 11.1, we have

$$S = O\left(\frac{P}{q} \max_f \left|\sum_{x=f+1}^{f+Q} e(F(x))\right|\right)$$

$$= O\left(\frac{P}{q} Q^{1-1/\rho}\right) = O(P\, q^{-1/\rho}).$$

XII. Estimation of Exponential Sums

Lemma 12.1. (Hua [1]). *Let*

$$f(x) = \alpha_k x^k + \cdots + \alpha_1 x.$$

Let $t_1 = t_1(k)$ be the integer defined by the following table:

k	1	2	3	4	5	6	7	8	9	10
t_1	1	3	8	23	62	380	656	889	2034	4595

we have

$$\int_0^1 \cdots \int_0^1 \left| \sum_{x=0}^{P} e^{2\pi i f(x)} \right|^{2t_1} d\alpha_1 \cdots d\alpha_k = O(P^{2t_1 - \frac{1}{2}k(k+1) + \epsilon}).$$

Lemma 12.2. (Hua [2], Vinogradow). *Under the same assumption as in lemma* 12.1, *we have*

$$\int_0^1 \cdots \int_0^1 \left| \sum_{x=0}^{P} e^{2\pi i f(x)} \right|^{2t_1} d\alpha_1 \cdots d\alpha_k = O(P^{2t_1 - \frac{1}{2}k(k+1) + \delta + \epsilon}),$$

where

$$t_1(k) = t_1 = \begin{cases} \frac{1}{4} k(k+1) + lk, & \text{for } k \equiv 0,3 \pmod 4, \\ \frac{1}{4}(k^2 + k + 2) + lk, & \text{for } k \equiv 1,2 \pmod 4, \end{cases}$$

and

$$\delta(k) = \delta = \frac{1}{2} k(k+1)(1 - \frac{1}{k})^l, \ l \geqslant k.$$

Combining these two results with lemma 11.1, we have

Lemma 12.3. *Let* σ_k *be the number defined by the following table:*

k	2	3	4	5	6	7	8	9	10	11	> 11
σ_k	4	9	20	51	130	319	768	1781	4078	9201	$2k^2(2\log k + \log\log k + 3)$

Let $k \geq r \geq 2$ *and*

(12.1) $$|\alpha_r - \frac{h}{q}| \leqslant \frac{1}{q^2}, \quad (h, q) = 1, \quad 1 \leqslant q \leqslant P^r;$$

and let

$$f(x) = \alpha_k x^k + \cdots + \alpha_1 x.$$

Then we have

$$(12.2)\qquad \sum_{x=1}^{P} e(f(x)) = O(P^{1-\frac{1}{\sigma_k}+\varepsilon}), \qquad \text{for } P \leqslant q \leqslant P^{r-1},$$

$$(12.3)\qquad = O(P\, q^{-\frac{1}{\sigma_k}+\varepsilon}), \qquad \text{for } 1 \leqslant q \leqslant P.$$

Proof. For $k \leqslant 11$, the result is a direct consequence of lemmas 11.1, 11.2 and 12.1 with $\delta_1 = \varepsilon$ and

$$\sigma_k = 2t_1 + k + 1.$$

For $k > 11$, we use lemma 11.1 with

$$\delta_1 = \delta(k-1),$$

which is defined in lemma 12.2. Then we have, by lemma 12.2,

$$\frac{1}{\rho} = (2t_1(k-1)+k)\Big/(1 - \frac{1}{2}k(k-1)(1-\frac{1}{k-1})^l)$$

$$\leqslant (\frac{1}{2}k^2 + 2lk)\Big/(1 - \frac{1}{2}k(k-1)(1-\frac{1}{k-1})^l)$$

$$(12.4)\qquad \leqslant (\frac{1}{2}k^2 + 2lk)(1 + k^2(1-\frac{1}{k})^l),$$

if

$$(12.5)\qquad \frac{1}{2}k(k-1)(1-\frac{1}{k-1})^l \leqslant \frac{1}{2},$$

since $(1-x)^{-1} \leqslant 1 + 2x$ for $0 \leqslant x \leqslant \frac{1}{2}$. Taking

$$l = \left[\frac{2\log k + \log\log k}{-\log(1-\frac{1}{k})}\right] + 1,$$

we have then

$$l < k(2\log k + \log\log k) + 1$$

and

$$k^2\left(1-\frac{1}{k}\right)^l \leqslant \frac{1}{\log k}, \tag{12.6}$$

since

$$\frac{1}{-\log\left(1-\frac{1}{k}\right)} = \left(\frac{1}{k}+\frac{1}{2k^2}+\frac{1}{3k^3}+\cdots\right)^{-1} \leqslant k.$$

Therefore

$$\frac{1}{\rho} \leqslant \left(\frac{1}{2}k^2 + 2k^2(2\log k + \log\log k) + 2k\right)\left(1+\frac{1}{\log k}\right)$$

$$\leqslant 2k^2\left(2\log k + \log\log k + \frac{1}{4} + \frac{1}{k} + \frac{\log\log k}{\log k} + \frac{1}{4\log k} + \frac{1}{k\log k}\right)$$

$$\leqslant 2k^2(2\log k + \log\log k + 3)\ (=\sigma_k)$$

for $k > 11$. We have (12.2), since by (12.6), (12.5) is really satisfied, (12.3) is a consequence of lemma 11.2.

Another consequence of lemma 11.1 is

Lemma 12.4. *Under the same condition as lemma* 12.3 *and* $k \leqslant 11$, *we have*

$$\sum_{x=1}^{P} e(f(x)) = O\left(P^{1-\frac{1}{\sigma_k}+\varepsilon}\left(\frac{q}{P^{r-1}}\right)^{\frac{1}{\sigma_k}}\right), \quad P^{r-1} \leqslant q \leqslant P^r. \tag{12.7}$$

For $k > 11$, we could not get so precise a result as at this moment, we give the following partial result which is needed later.

Lemma 12.5. *Under the same conditions as lemma* 12.3 *and* $k > 11$, *we have*

$$\sum_{x=1}^{P} e(f(x)) = O\left(P^{1-\frac{1}{\rho'}}\right), \quad \text{for} \quad P^{r-1} \leqslant q \leqslant P^{r-\frac{1}{4k}}, \tag{12.8}$$

where

$$\rho' = 40k^3 \log k .$$

Proof. By lemma 11.1, we have

$$\sum_{x=1}^{P} e(f(x)) = O\left(P^{1-\rho}\left(P^{1-\frac{1}{4k}}\right)^{\frac{1}{2t_1+k+1}}\right)$$

$$= O\left(P^{1-\frac{\frac{1}{4k}-\delta_1}{2t_1+k+1}}\right).$$

Taking $\delta_1 = \delta(k-1)$, and

$$l = \left[\frac{4 \log k}{-\log\left(1-\frac{1}{k}\right)}\right] + 1 ,$$

we have then

$$l \leqslant 4k \log k + 1 ,$$

and

$$\delta_1 \leqslant k^2 \left(1-\frac{1}{k}\right)^l \leqslant \frac{1}{k^2} .$$

As inequality (12.4), we have

$$(2t_1(k-1)+k+1)\Big/\left(\frac{1}{4k}-\delta_1\right)$$

$$\leqslant \left(\frac{1}{2}k^2 + 2lk\right)\Big/\left(\frac{1}{4k} - \frac{1}{k^2}\right)$$

$$\leqslant \left(\frac{1}{2}k^2 + 8k^2 \log k + 2k\right) 4k \left(1-\frac{4}{k}\right)^{-1}$$

$$\leqslant 40k^3 \log k .$$

Remark. Since we shall have a better result later, so we did not fully utilize our method.

XIII. Remarks Concerning the Estimation of Exponential Sums

The previous estimations of exponential sums depend on the arithmetical nature of an individual coefficient α_r $(2 \leqslant r \leqslant k)$ of the polynomial. It seems to be quite likely that a satisfactory result should depend on the arithmetical nature of the totality of $\alpha_k, \cdots, \alpha_2$. Such a view point is supported by the following results (lemmas 13.2 and 13.3).

Lemma 13.1. *Let* $<\xi>$ *denote the distance from* ξ *to its nearest integer, and let* $V_1, \ldots, V_k$ *be integers. Let* $(h_i, q_i) = 1$. *Then the number of solutions* x *of the system of inequalities*

$$(13.1) \qquad < \frac{x h_i}{q_i} > \leqslant \frac{V_i}{q_i}, \qquad 1 \leqslant i \leqslant k, \quad f < x \leqslant f + N$$

is $\leq \left(\frac{N}{Q}+1\right)\prod_{i=1}^{k} \min(2V_i + 1, q_i)$, *where* Q *is the least common multiple of* $q_1, \ldots, q_k$.

Proof. The lemma is an evident consequence of the following statement: The number T of solutions x of the system of inequalities

$$< \frac{x h_i}{q_i} > \leqslant \frac{V_i}{q_i}, \qquad 1 \leqslant i \leqslant k, \quad f < x \leqslant f + Q$$

is $\leqslant \prod_{i=1}^{k} \min(2V_i + 1, q_i)$. Inasmuch as x runs over a complete residue system, mod Q, so does $x-f$, thus it is without loss of generality to assume $f=0$. Moreover, let y be the integer defined by

$$y \equiv x h_i \quad (\bmod q_i), \qquad 1 \leqslant i \leqslant k,$$

which is uniquely determined, mod Q. Also, it is evident that as x runs over a complete residue system, mod Q, so y does. Thus T is equal to the number of solutions of

$$< \frac{y}{q_i} > \leqslant \frac{V_i}{q_i}, \qquad 1 \leqslant i \leqslant k, \quad 0 < y \leqslant Q.$$

Evidently, this is equivalent to

$$y \equiv 1, \cdots, V_i;\ q_i - V_i, \cdots, q_i \ (\bmod q_i)$$

if $q_i \geqslant 2V_i+1$. For modulo Q, the number of y's is $\leqslant \prod_{i=1}^{k} \min(2V_i+1, q_i)$, since the number of solutions, mod $\frac{q_1 q_2}{(q_1, q_2)}$, of

$$x \equiv r_1 \pmod{q_1}, \quad x \equiv r_2 \pmod{q_2}$$

is $\geqslant 1$.

Lemma 13.2. *Let*

$$f(x) = \frac{h_{k+1}}{q_{k+1}} x^{k+1} + \cdots + \frac{h_1}{q_1} x .$$

Under the same hypothesis as lemma 11.1, *we have*

$$S = O(P^{l-\rho}), \tag{13.2}$$

provided

$$1 \leqslant q_r \leqslant P^{r-1}, \qquad (2 \leqslant r \leqslant k+1), \ P \leqslant Q_1,$$

where Q_1 *is the least common multiple of* $q_k, \ldots, q_2$.

Proof. We use the same method as lemma 11.1. Instead of (11.9), we now have

$$< \frac{(k+1)!}{r!} \frac{h_{r+1}}{q_{r+1}} (y-y_0) > \leqslant C_1 Y P^{-r}, \quad 1 \leqslant y \leqslant Y, \quad 1 \leqslant r \leqslant k. \tag{13.3}$$

By lemma 13.1, with $V_r = [C_1 q_r Y P^{-r}] + 1$, the number of solutions of (13.3) is

$$O\left(\left(\frac{Y}{Q_1} + 1\right) \prod_{r=1}^{k} \min(q_r Y P^{-r} + 1, \ q_r) \right)$$

$$= O\left(\left(\frac{Y}{Q_1} + 1\right) \prod_{r=1}^{k} \min(q_r P^{-r+1} + 1, \ q_r) \right)$$

$$= O(1).$$

In the same way as we did for lemma 11.1, we have (13.2).

We can use the method of lemma 11.2 to establish that

$$S = O\left(P\, Q_1^{-1/p}\right) \tag{13.4}$$

for $Q_1 \leqslant P$. But sometimes, we can use the following sharper results.

Lemma 13.3. *Let*

$$f(x) = \frac{h_k}{q_k} x^k + \cdots + \frac{h_1}{q_1} x, \qquad (h_i, q_i) = 1, \; q_i \geqslant 1,$$

and let Q_1 *be the least common multiple of* $q_k, \ldots, q_2$, *and* Q *be the least common multiple of* q_1 *and* Q_1. *Then, we have*

$$\sum_{x=1}^{P} e(f(x)) - \frac{P}{Q} \sum_{x=1}^{Q} e(f(x)) = O\left(Q \cdot Q_1^{-1/k+\varepsilon}\right). \tag{13.5}$$

Proof. By lemma 4.1, we have

$$\frac{P}{Q} \sum_{x=1}^{Q} e(f(x)) = O\left(P\, Q^{-1/k+\varepsilon}\right),$$

it is enough to prove that, for $1 \leqslant P \leqslant Q$,

$$\sum_{x=1}^{P} e(f(x)) = O\left(Q \cdot Q_1^{-1/k+\varepsilon}\right). \tag{13.6}$$

Let $Q = d\, Q_1$. We write

$$Q\, f(x) = g(x) = a_k x^k + \cdots + a_2 x^2 + a_1 x.$$

It is easy to verify that

$$(a_k, \cdots, a_2) = d. \tag{13.7}$$

Since

$$\frac{1}{Q}\sum_{n=1}^{Q}\sum_{m=1}^{P} e(n(m-x)/Q) = \begin{cases} 1, & \text{for } 1 \leqslant x \leqslant P, \\ 0, & \text{for } P+1 \leqslant x \leqslant Q, \end{cases}$$

we have

$$\sum_{x=1}^{P} e(g(x)/Q) = \sum_{x=1}^{Q} e(g(x)/Q)\frac{1}{Q}\sum_{n=1}^{Q}\sum_{m=1}^{P} e(n(m-x)/Q)$$

$$(13.8)\qquad = \frac{P}{Q}\sum_{x=1}^{Q} e(g(x)/Q) + \frac{1}{Q}\sum_{n=1}^{Q-1}\Big(\sum_{m=1}^{P} e(mn/Q)\Big) \times \sum_{x=1}^{Q} e(g(x)-nx)/Q).$$

The absolute value of the first term is $O(PQ^{-1/k+\varepsilon}) = O(QQ_1^{-1/k+\varepsilon})$ by lemma 4.1. Since

$$\Big|\sum_{m=1}^{P} e(nm/Q)\Big| = \left|\frac{1-e(nP/Q)}{1-e(n/Q)}\right| \leqslant \left|\frac{1}{\operatorname{Sin}\pi n/Q}\right| \leqslant \frac{1}{\langle n/Q\rangle},$$

and by lemma 4.1,

$$\sum_{x=1}^{Q} e((g(x)-nx)/Q) = O(QQ_1^{-1/k+\varepsilon}),$$

we have

$$\sum_{x=1}^{P} e(g(x)/Q) = O(PQ_1^{-1/k+\varepsilon}) + O\Big(Q_1^{-1/k+\varepsilon}\sum_{n=1}^{Q-1}\frac{1}{\langle n/Q\rangle}\Big)$$

$$= O(QQ_1^{-1/k+\varepsilon}) + O\Big(Q_1^{-1/k+\varepsilon}\sum_{n=1}^{1/2Q}\frac{Q}{n}\Big)$$

$$= O(QQ_1^{-1/k+\varepsilon}).$$

The lemma is now proved.

For $Q \leqslant P$, we have

$$\sum_{x=1}^{P} e(f(x)) = O(P\,Q^{-1/k}),$$

which is sharper than the result (13.4). Both of them are handicapped for small Q_1. The following easy lemma sometimes fills the gap.

Lemma 13.4. *If* $Q_1 < Q$, *we have*

$$\sum_{x=1}^{P} e(f(x)) = O(Q).$$

Proof. The lemma is trivial for $P \leqslant Q$. We suppose $Q < P$. Let $x = Q_1\, y + z$, where

$$1 \leqslant z \leqslant Q_1, \qquad 0 \leqslant y \leqslant \frac{P-z}{Q_1}.$$

Then, since $q_1 \nmid Q_1$, we have

$$\Big| \sum_{x=1}^{P} e(f(x)) \Big| = \Big| \sum_{z=1}^{Q_1} e(f(z)) \sum_{y=1}^{(P-z)/Q_1} e^{2\pi i\, h_1 Q_1 y/q_1} \Big|$$

$$\leqslant Q_1 \max_{z} \Big| \sum_{y} e^{2\pi i\, h_1 Q_1 y/q_1} \Big|$$

$$\leqslant Q_1 \frac{1}{\langle h_1 Q_1/q_1 \rangle} \leqslant Q_1 \frac{q_1}{(Q_1, q_1)} = Q.$$

XIV. An Analytic Formulation of the Theorem

Let

$$(14.1) \qquad T(\alpha_1, \cdots, \alpha_k) = \sum_{x=1}^{P} e(\alpha_k x^k + \cdots + \alpha_1 x).$$

By the orthogonality of the system $e(nx)$, we have

$$r_t(P) = \int_0^1 \cdots \int_0^1 | T(\alpha_1, \cdots, \alpha_k) |^{2t} d\alpha_1 \cdots d\alpha_k .$$

By the periodicity of the function $T(\alpha_1, \cdots, \alpha_k)$, we may write

$$(14.2) \qquad r_t(P) = \int_{-\frac{1}{\tau_1}}^{1-\frac{1}{\tau_1}} \cdots \int_{-\frac{1}{\tau_k}}^{1-\frac{1}{\tau_k}} | T(\alpha_1, \cdots, \alpha_k) |^{2t} d\alpha_1, \cdots d\alpha_k ,$$

where

$$\tau_1 = P^{\frac{1}{2}}, \quad \tau_v = P^{v-\frac{1}{2k}+\sigma}, \quad 2 \leqslant v \leqslant k,$$

and

$$\sigma = \frac{1}{k^3}.$$

It is known that to every point in the k-dimensional space $(\alpha_1, \cdots, \alpha_k)$, we have a rational point $\left(\frac{h_1}{q_1}, \cdots, \frac{h_k}{q_k}\right)$ such that

$$(14.3) \qquad \alpha_v = \frac{h_v}{q_v} + \beta_v, \quad (h_v, q_v) = 1, \quad | \beta_v | \leqslant \frac{1}{q_v \tau_v}, \quad 0 < q_v \leqslant \tau_v .$$

Let

$$\mathfrak{M} = \mathfrak{M}\left(\frac{h_k}{q_k}, \cdots, \frac{h_1}{q_1}\right)$$

be the domain defined by $|\beta_v| \leqslant \frac{1}{q_v \tau_v}$ $(1 \leqslant v \leqslant k)$ with

$$(14.4) \qquad 1 \leqslant q_v \leqslant P^{\frac{1}{2k}-2\sigma}, (2 \leqslant v \leqslant k), \quad 1 \leqslant q_1 \leqslant P^{\frac{1}{2}-\frac{1}{2k}+\sigma}.$$

It is easy to prove that no two of $\mathfrak{M}$'s are overlapping. In fact, suppose that

$$\mathfrak{M}\left(\frac{h_k}{q_k}, \cdots, \frac{h_1}{q_1}\right) \text{ and } \mathfrak{M}\left(\frac{h'_k}{q'_k}, \cdots, \frac{h'_1}{q'_1}\right)$$

are overlapping, then, we have, except the case $\frac{h_\nu}{q_\nu} = \frac{h'_\nu}{q'_\nu}$ for all ν, that there is a ν such that $\frac{h_\nu}{q_\nu} \neq \frac{h'_\nu}{q'_\nu}$, then

$$\frac{1}{q_\nu q'_\nu} \leqslant \frac{|h_\nu q'_\nu - h'_\nu q_\nu|}{q_\nu q'_\nu} = \left|\frac{h_\nu}{q_\nu} - \frac{h'_\nu}{q'_\nu}\right| \leqslant \frac{1}{q_\nu \tau_\nu} + \frac{1}{q'_\nu \tau_\nu}$$

$$\leqslant \frac{2}{\tau_\nu} \max\left(\frac{1}{q_\nu}, \frac{1}{q'_\nu}\right),$$

that is

$$\tau_\nu \leqslant 2 \max(q_\nu, q'_\nu) \begin{cases} \leqslant 2P^{\frac{1}{2k}-2\sigma}, & \text{for } \nu > 1, \\ \leqslant 2P^{\frac{1}{2}-\frac{1}{2k}-\sigma}, & \text{for } \nu = 1, \end{cases}$$

which are impossible.

Let E denote the remaining part of

$$-\frac{1}{\tau_\nu} \leqslant \alpha_\nu \leqslant 1 - \frac{1}{\tau_\nu}$$

after removing all these $\mathfrak{M}$'s. We let

$$(14.5) \qquad T_{(1)} = \int \cdots \int_E |T(\alpha_1, \cdots, \alpha_k)|^{2t} d\alpha_1 \cdots d\alpha_k$$

and

$$(14.6) \qquad T_{(2)} = \sum_{\mathfrak{M}} K\left(\frac{h_k}{q_k}, \cdots, \frac{h_1}{q_1}\right),$$

where

$$K\left(\frac{h_k}{q_k}, \cdots, \frac{h_1}{q_1}\right) = \int \cdots \int_{\mathfrak{M}} | T |^{2t} d\alpha_1 \cdots d\alpha_k . \tag{14.7}$$

Then

$$r_t(P) = T_{(1)} + T_{(2)} . \tag{14.8}$$

XV. Estimation of $T_{(1)}$

Lemma 15.1. *Suppose that* $(\alpha_1, \cdots, \alpha_k)$ *belongs to* E *and it satisfies* (14.3) *with*

$$1 \leqslant q_\nu \leqslant P^{\frac{1}{2k} - 2\sigma}, \qquad (2 \leqslant \nu \leqslant k), \tag{15.1}$$

then

$$T(\alpha_1, \cdots, \alpha_k) = O(P^{1-\sigma}) . \tag{15.2}$$

Proof. By lemma 13.4, we have, for any n

$$\begin{aligned} S_n &= \sum_{x=1}^{n} e\left(\frac{h_k}{q_k} x^k + \cdots + \frac{h_1}{q_1} x\right) = O(q_1 \cdots q_k) \\ &= O(P^{\frac{1}{2}} (P^{\frac{1}{2k} - 2\sigma})^{k-1}) = O(P^{1 - \frac{1}{2k} - 2\sigma(k-1)}) . \end{aligned} \tag{15.3}$$

Then

$$\begin{aligned} T(\alpha_1, \cdots, \alpha_k) &= \sum_{x=1}^{P} (S_x - S_{x-1}) e(\beta_k x^k + \cdots + \beta_1 x) \\ &= \sum_{x=1}^{P} S_x \Big(e(\beta_k x^k + \cdots + \beta_1 x) - e(\beta_k (x+1)^k + \cdots + \beta_1 (x+1))\Big) \\ &\quad + S_P e\Big(\beta_k (P+1)^k + \cdots + \beta_1 (P+1)\Big). \end{aligned} \tag{15.4}$$

Since

$$
\begin{aligned}
& e(\beta_k x^k + \cdots + \beta_1 x) - e(\beta_k (x+1)^k + \cdots + \beta_1 (x+1)) \\
& = O(|\beta_k| P^{k-1} + |\beta_{k-1}| P^{k-2} + \cdots + |\beta_1|) \\
& = O\left(\frac{P^{k-1}}{\tau_k} + \frac{P^{k-2}}{\tau_{k-1}} + \cdots + \frac{P}{\tau_2} + \frac{1}{q_1 \tau_1}\right) \\
& = O(P^{-1+\frac{1}{2k}-\sigma} + P^{-\frac{1}{2}-\frac{1}{2}+\frac{1}{2k}-\sigma}) \\
& = O(P^{-1+\frac{1}{2k}-\sigma}),
\end{aligned}
$$

since from (14.4) and (15.1) and $(\alpha_1, \cdots, \alpha_k)$ belonging to E, we deduce that $q_1 \geqslant P^{\frac{1}{2}-\frac{1}{2k}+\sigma}$. Therefore, by (15.3) and (15.4), we have

$$
\begin{aligned}
T(\alpha_1, \cdots, \alpha_k) &= O(\sum P^{1-\frac{1}{2k}-2\sigma(k-1)} \cdot P^{-1+\frac{1}{2k}-\sigma}) \\
&= O(P^{1-\sigma(2k-3)}) = O(P^{1-\sigma}).
\end{aligned}
$$

Lemma 15.2. *For $(\alpha_1, \cdots, \alpha_k)$ belonging to E, we have*

$$
T(\alpha_1, \cdots, \alpha_k) = O(P^{1-1/\lambda}), \tag{15.5}
$$

where

$$
\lambda = 40 k^3 \log k .
$$

Proof. Either one of the following k conditions holds

$$
P^{\frac{1}{2k}-2\sigma} < q_\nu \leqslant P^{\nu-\frac{1}{2k}+\sigma}, \qquad 1 < \nu \leqslant k, \tag{15.6}
$$

$$
P^{\frac{1}{2}-\frac{1}{2k}+\sigma} < q_1 \leqslant P^{\frac{1}{2}}. \tag{15.7}
$$

For the first case, we have, by lemmas 12.3, 12.4 and 12.5, for $k > 11$,

$$
T(\alpha_1, \cdots, \alpha_k) = O(P^{1-\frac{1}{40 k^3 \log k}}),
$$

since for $P^{\frac{1}{2k}-2\sigma} \leqslant q \leqslant P$,

$$T(\alpha_1, \cdots, \alpha_k) = O(P \cdot q^{-\frac{1}{\sigma_k}+\epsilon})$$

$$= O(P^{1-\frac{1}{40 k^3 \log k}}),$$

and for $k \leqslant 11$,

$$T(\alpha_1, \cdots, \alpha_k) = O(P^{1-\frac{1}{\sigma_k}+\epsilon}).$$

For the second case the result follows from lemma 15.1.

Lemma 15.3. *For*

$$t \geq k^2(3 \log k + \log \log k + 4), \quad k \geq 11,$$

$$t > t_0, \qquad\qquad k \leq 10,$$

we have

$$T_{(1)} = O(P^{2t-\frac{1}{2}k(k+1)-c_1}).$$

Proof. For $k \leq 10$ the proof is simpler, now we assume $k \geq 11$. Let $t_1 = t_1(k)$ be defined in lemma 12.2, let $t = t_1 + k^2$. Then, by lemma 12.2, we have

$$\int_E \cdots \int |T|^{2t} d\alpha_1 \cdots d\alpha_k \leq \max_{\alpha \in E} |T|^{2k^2} \int_0^1 \cdots \int_0^1 |T|^{2t_1} d\alpha_1 \cdots d\alpha_k$$

$$= O(P^{2k^2(1-1/40 k^3 \log k)} \cdot P^{2t_1-\frac{1}{2}k(k+1)+\frac{1}{2}k(k+1)(1-1/k)^l})$$

$$= O(P^{2t-\frac{1}{2}k(k+1)-\lambda}),$$

where

$$\lambda = \frac{1}{20\,k\log k} - \frac{1}{2}\,k\,(k+1)\,(1-1/k)^{l}.$$

We choose

$$l = \left[\log\,(10\,k^2\,(k+1)\log k)\,/\,(-\log\,(1-\frac{1}{k}))\right] + 1,$$

then $\lambda > 0$. Since, for $k \geq 11$,

$$l \leq k\log\,(22\,k^3\log k) + 1\,,$$

we have

$$t = t_1 + k^2 \leq l\,k + \frac{1}{4}\,(k^2 + k + 2) + k^2$$

$$\leq k^2\,(3\log k + \log\log k + \log 22) + \frac{5}{4}\,(k^2 + k + 2)$$

$$= k^2\left(3\log k + \log\log k + \log 22 + \frac{5}{4}\,(1 + \frac{1}{k} + \frac{2}{k^2})\right)$$

$$\leq k^2\,(3\log k + \log\log k + 4)\,.$$

XVI. Approximation Formula

Lemma 16.1. (van der Corput) *Let $f(x)$ be a real polynomial defined in interval (a, b), and its derivative satisfying*

$$|\,f'(x)\,| \leq \frac{1}{2}\,.$$

Then we have

$$\sum_{a\leq x\leq b} e\,(f\,(x)) = \int_a^b e\,(f\,(x))\,d\,x + O\,(1)\,.$$

Lemma 16.2. *If* $(\alpha_1, \cdots, \alpha_k)$ *is on* $\mathfrak{M}\left(\frac{h_1}{q_1}, \cdots, \frac{h_k}{q_k}\right)$, *we have then*

$$T(\alpha_1, \cdots, \alpha_k) = B\left(\frac{h_k}{q_k}, \cdots, \frac{h_1}{q_1}\right) R + O(H),$$

where, letting $H = q_1 \cdots q_k$,

$$B\left(\frac{h_k}{q_k}, \cdots, \frac{h_1}{q_1}\right) = \frac{1}{H} \sum_{y=1}^{H} e\left(\frac{h_k}{q_k} y^k + \cdots + \frac{h_1}{q_1} y\right)$$

and

$$R = \int_0^P e(\beta_k x^k + \cdots + \beta_1 x)\, dx.$$

Proof. Let $H = q_1 \cdots q_k$. Write

$$x = Hz + y, \qquad y = 1, \cdots, H; \qquad -\frac{y}{H} \leqslant z \leqslant \frac{P-y}{H}.$$

Then we have

$$T(\alpha_1, \cdots, \alpha_k) = \sum_{y=1}^{H} e\left(\frac{h_k}{q_k} y^k + \cdots + \frac{h_1}{q_1} y\right) W_y,$$

where

$$W_y = \sum_{-\frac{y}{H} \leq z \leq \frac{P-y}{H}} e(\varphi(z))$$

and

$$\varphi(z) = \beta_k (Hz + y)^k + \cdots + \beta_1 (Hz + y).$$

Since

$$\varphi'(z) = k\beta_k (Hz + y)^{k-1} H + \cdots + \beta_1 H$$

$$= O(|\beta_k| P^{k-1}H + |\beta_{k-1}| P^{k-2}H + \cdots + |\beta_1| H)$$

$$= O\left(\sum_{\nu=2}^{k} \frac{HP^{\nu-1}}{q_\nu \tau_\nu} + \frac{H}{q_1 \tau_1}\right)$$

$$= O\left(\sum_{\nu=2}^{k} P^{\left(\frac{1}{2k}-2\sigma\right)(k-2)} \cdot P^{\frac{1}{2}-\frac{1}{2k}+\sigma} \cdot \frac{P^{\nu-1}}{P^{\nu-\frac{1}{2k}+\sigma}} + P^{\left(\frac{1}{2k}-2\sigma\right)(k-1)} \cdot \frac{1}{P^{\frac{1}{2}}}\right)$$

$$= O\left(P^{-\frac{1}{k}-2\sigma(k-2)} + P^{-\frac{1}{2k}-2\sigma(k-1)}\right) = o(1)$$

as P tends to infinity, that is, $|\varphi'(z)| \leq \frac{1}{2}$ for P being large. By lamma 16.1, we have

$$W_y = \int_{-\frac{y}{H}}^{(P-y)/H} e(\varphi(z))\, dz + O(1)$$

$$= \frac{1}{H} \int_0^P e(\beta_k x^k + \cdots + \beta_1 x)\, dx + O(1).$$

Therefore we have

$$T(\alpha_1, \cdots, \alpha_k) = \frac{1}{H} \sum_{y=1}^{H} e\left(\frac{h_k}{q_k} x^k + \cdots + \frac{h_1}{q_1} x\right) \int_0^P e(\beta_k x^k + \cdots + \beta_1 x)\, dx + O(H).$$

Lemma 16.3. *Let* $\gamma_r = \beta_r P^r$ $(1 \leqslant r \leqslant k)$, *we have*

$$B\left(\frac{h_k}{q_k}, \cdots, \frac{h_1}{q_1}\right) R = O\left(PH^{-\frac{1}{k^2}+\epsilon} Z\right),$$

where

$$Z = \min\left(1, |\gamma_1|^{-\frac{1}{k}}, \cdots, |\gamma_k|^{-\frac{1}{k}}\right).$$

Proof. Let Q be the least common multiple of $q_1, \cdots, q_k$. By lemma 4.1, we have

$$B\left(\frac{h_k}{q_k},\cdots,\frac{h_1}{q_1}\right)=\frac{1}{Q}\sum_{y=1}^{Q} e\left(Q\left(\frac{h_k}{q_k}x^k+\cdots+\frac{h_1}{q_1}x\right)\Big/Q\right)$$

$$=O\left(Q^{-\frac{1}{k}+\epsilon}\right).$$

Since

$$Q \geqslant \max(q_1,\cdots,q_k) \geqslant (q_1\cdots q_k)^{\frac{1}{k}} = H^{\frac{1}{k}},$$

we have

$$B\left(\frac{h_k}{q_k},\cdots,\frac{h_1}{q_1}\right)=O\left(H^{-\frac{1}{k^2}+\epsilon}\right).$$

Further

$$R = P\int_0^1 e(\gamma_k x^k+\cdots+\gamma_1 x)\,dx.$$

The lemma follows from lemma 3.2.

Lemma 16.4. *On* $\mathfrak{M}$, *we have*

$$T(\alpha_1,\cdots,\alpha_k)=O\left(PH^{-\frac{1}{k^2}+\epsilon}Z\right).$$

Proof. Since α is on $\mathfrak{M}$, we have

$$H = q_1\cdots q_k \leqslant P^{\frac{1}{2}-\frac{1}{2k}+\sigma+(k-1)\left(\frac{1}{2k}-2\sigma\right)}$$

$$= P^{1-\frac{1}{k}-(2k-3)\sigma} \leqslant P^{1-\frac{1}{k}},$$

and

$$Z = \min\left(1, |\gamma_1|^{-\frac{1}{k}},\cdots,|\gamma_k|^{-\frac{1}{k}}\right)$$

$$= \min\left(1, (P\,|\,\beta_1\,|)^{-\frac{1}{k}}, \cdots, (P^k\,|\,\beta_k\,|)^{-\frac{1}{k}}\right)$$

$$\geqslant \min\left(1, \left(\frac{P}{\tau_1}\right)^{-\frac{1}{k}}, \cdots, \left(\frac{P^k}{\tau_k}\right)^{-\frac{1}{k}}\right)$$

$$\geqslant \min\left(1, P^{-\frac{1}{2k}}\right) = P^{-\frac{1}{2k}}.$$

Therefore

$$H = H^{-\frac{1}{k^2}+\epsilon} \cdot H^{1+\frac{1}{k^2}-\epsilon}$$

$$\leqslant H^{-\frac{1}{k^2}+\epsilon} P^{(1+\frac{1}{k^2}-\epsilon)(1-\frac{1}{k})}$$

$$\leqslant H^{-\frac{1}{k^2}+\epsilon} P \cdot P^{-\frac{1}{2k}} = O(P H^{-\frac{1}{k^2}+\epsilon} Z).$$

The lemma follows from lemma 16.3.

XVII. The Prinicipal Term

Lemma 17.1. *We have*

$$(17.1) \qquad K\left(\frac{h_k}{q_k}, \cdots, \frac{h_1}{q_1}\right)$$

$$= \left| B\left(\frac{h_k}{q_k}, \cdots, \frac{h_1}{q_1}\right)\right|^{2t} \int_{-q_k^{-1}\tau_k^{-1}}^{q_k^{-1}\tau_k^{-1}} \cdots \int_{-q_1^{-1}\tau_1^{-1}}^{q_1^{-1}\tau_1^{-1}} |R|^{2t}\, d\beta_1 \cdots d\beta_k$$

$$+ O\left(P^{2t-\frac{1}{2}k(k+1)-1} H^{1-(2t-1)/k^2+\varepsilon}\right).$$

Proof. Using the inequality

$$\big|\,|\xi|^{2t} - |\eta|^{2t}\big| \leqslant 2t\,|\xi-\eta|\,(|\xi|^{2t-1} + |\eta|^{2t-1}),$$

we have, on $\mathfrak{M}\left(\frac{h_k}{q_k}, \cdots, \frac{h_1}{q_1}\right)$,

$$|T(\alpha_1, \cdots, \alpha_k)|^{2t} - \left| B\left(\frac{h_k}{q_k}, \cdots, \frac{h_1}{q_1}\right)\right|^{2t} |R|^{2t}$$

$$= O\left(H\left(PH^{-\frac{1}{k^2}+\epsilon}Z\right)^{2t-1}\right).$$

Integrating over $\mathfrak{M}\left(\frac{h_k}{q_k}, \cdots, \frac{h_1}{q_1}\right)$, we have

(17.2) $K\left(\frac{h_k}{q_k}, \cdots, \frac{h_1}{q_1}\right)$

$$= \left| B\left(\frac{h_k}{q_k}, \cdots, \frac{h_1}{q_1}\right)\right|^{2t} \int_{-q_k^{-1}\tau_k^{-1}}^{q_k^{-1}\tau_k^{-1}} \cdots \int_{-q_1^{-1}\tau_1^{-1}}^{q_1^{-1}\tau_1^{-1}} |R|^{2t}\, d\beta_1 \cdots d\beta_k$$

$$+ O\left(HP^{2t-1}H^{-\frac{2t-1}{k^2}+\varepsilon} \int_{-\infty}^{\infty} \cdots \int_{-\infty}^{\infty} Z^{2t-1}\, d\beta_1 \cdots d\beta_k\right).$$

Let $\delta_\nu = \max(1, |\gamma_\nu|)$. Evidently we have

$$\prod_{\nu=1}^{k} \delta_\nu = \prod_{\nu=1}^{k} \max(1, |\gamma_\nu|) \leqslant \max(1, |\gamma_1|, \cdots, |\gamma_k|)^k,$$

and then

(17.3) $$Z \leqslant \prod_{\nu=1}^{k} \delta_\nu^{-\frac{1}{k^2}}.$$

Therefore we have

$$\int_{-\infty}^{\infty} \cdots \int_{-\infty}^{\infty} Z^{2t-1}\, d\beta_1 \cdots d\beta_k = P^{-\frac{1}{2}k(k+1)} \int_{-\infty}^{\infty} \cdots \int_{-\infty}^{\infty} Z^{2t-1}\, d\gamma_1 \cdots d\gamma_k$$

$$\leqslant P^{-\frac{1}{2}k(k+1)} \int_{-\infty}^{\infty} \cdots \int_{-\infty}^{\infty} \frac{d\gamma_1 \cdots d\gamma_k}{(\delta_1 \cdots \delta_k)^{(2t-1)/k^2}}$$

$$= O\left(P^{-\frac{1}{2}k(k+1)}\right),$$

since $\frac{2t-1}{k^2} > 1$ (*i.e.*, $2t > k^2+1$). The lemma is therefore proved.

Lemma 17.2. *We have*

$$(17.4)\quad P^{-2t+\frac{1}{2}k(k+1)}\int_{-q_k^{-1}\tau_k^{-1}}^{q_k^{-1}\tau_k^{-1}}\cdots\int_{-q_1^{-1}\tau_1^{-1}}^{q_1^{-1}\tau_1^{-1}}|R|^{2t}\,d\beta_k\cdots d\beta_1=\vartheta_0+O(P^{-(\frac{2t}{k^2}-1)\sigma}).$$

Proof. By changing of variables $\beta_\nu = P^{-\nu}\gamma_\nu$, we have that the left-hand side equals

$$\int_{-q_k^{-1}\tau_k^{-1}P^k}^{q_k^{-1}\tau_k^{-1}P^k}\cdots\int_{-q_1^{-1}\tau_1^{-1}P}^{q_1^{-1}\tau_1^{-1}P}\left|\int_0^1 e(\gamma_k x^k+\cdots+\gamma_1 x)\,dx\right|^{2t}d\gamma_k\cdots d\gamma_1$$

$$=\vartheta_0+O\Big(\sum_{i=1}^{k}\int_{-\infty}^{\infty}d\gamma_1\cdots\int_{-\infty}^{\infty}d\gamma_{j-1}\int_{\tau_j^{-1}q_j^{-1}P^j}^{\infty}d\gamma_j\int_{-\infty}^{\infty}d\gamma_{j+1}\cdots\int_{-\infty}^{\infty}|Z|^{2t}\,d\gamma_k\Big)$$

$$=\vartheta_0+O\Big(\sum_{j=1}^{k}\int_{\tau_j^{-1}q_j^{-1}P^j}^{\infty}\delta_j^{-2t/k^2}\,d\gamma_j\Big).$$

Since

$$\frac{P^j}{q_j\tau_j}\geqslant\frac{P^j}{P^{\frac{1}{2k}-2\sigma}P^{j-\frac{1}{2k}+\sigma}}=P^\sigma,\qquad \text{for } 2\le j\le k,$$

and

$$\frac{P}{q_1\tau_1}\geqslant\frac{P}{P^{\frac{1}{2}-\frac{1}{2k}+\sigma}P^{\frac{1}{2}}}=P^{\frac{1}{2k}-\sigma}\geqslant P^\sigma,$$

therefore the left-hand side is equal to

$$\vartheta_0+O\Big(\int_{P^\sigma}^{\infty}\delta^{-\frac{2t}{k^2}}\,d\gamma\Big)$$

$$=\vartheta_0+O\Big(\int_{P^\sigma}^{\infty}\gamma^{-\frac{2t}{k^2}}\,d\gamma\Big)$$

$$=\vartheta_0+O\Big(P^{-(\frac{2t}{k^2}-1)\sigma}\Big).$$

Lemma 17.3. *We have*

(17.5)

$$\mathfrak{S} - \sum_{q_1 \le P^{\frac{1}{2}-\frac{1}{2k}+\sigma}} \sum_{q_2 \le P^{2-\frac{1}{2k}+\sigma}} \cdots \sum_{q_k \le P^{k-\frac{1}{2k}+\sigma}} \sum_{\substack{h_1=1 \\ (h_1, q_1)=1}}^{q_1} \cdots \sum_{\substack{h_k=1 \\ (h_k, q_k)=1}}^{q_k} \left| B\left(\frac{h_k}{q_k}, \cdots, \frac{h_1}{q_1}\right) \right|^{2t}$$

$$= O\left(P^{-\frac{1}{k}(\frac{1}{2}-\frac{1}{2k}+\sigma)+\varepsilon}\right).$$

Proof. The right hand side is dominated by

$$F \sum_{q_1=1}^{\infty} \cdots \sum_{q_k=1}^{\infty} \sum_{h_1} \cdots \sum_{h_k} \left| B\left(\frac{h_k}{q_k}, \cdots, \frac{h_1}{q_1}\right) \right|^{2t-1} = O(F),$$

where

$$F = \max_{Q \ge P^{\frac{1}{2}-\frac{1}{2k}+\sigma}} \left| B\left(\frac{h_k}{q_k}, \cdots, \frac{h_1}{q_1}\right) \right|$$

and Q is the least common multiple of $q_1, \ldots, q_k$. By lemma 4.1, we have

$$F = O(Q^{-\frac{1}{k}+\varepsilon}) = O\left(P^{-\frac{1}{k}(\frac{1}{2}-\frac{1}{2k}+\sigma)+\varepsilon}\right).$$

Lemma 17.4. *We have*

$$T_{(2)} \sim \vartheta_0 \mathfrak{S} P^{2t-\frac{1}{2}k(k+1)}.$$

Proof. Summing up (17.1) with respect to h and q, we have

(17.6)

$$T_{(2)} = \sum_{h, q} \left| B\left(\frac{h_k}{q_k}, \cdots, \frac{h_1}{q_1}\right) \right|^{2t} \int_{-q_k^{-1}\tau_k^{-1}}^{q_k^{-1}\tau_k^{-1}} \cdots \int_{-q_1^{-1}\tau_1^{-1}}^{q_1^{-1}\tau_1^{-1}} |R|^{2t} \, d\beta_1 \cdots d\beta_k$$

$$+\,O\left(P^{2t-\frac{1}{2}k(k+1)-1}\sum_{q,h} H^{1-\frac{2t-1}{k^2}+\varepsilon}\right).$$

The series

$$\sum_{q,h} H^{1-(2t-1)/k^2+\varepsilon} = \sum_{q} H^{2-(2t-1)/k^2+\varepsilon}$$

converges for $2t > 3k^2+1$. That is, for $k \geqslant 5$,

$$\sum_{q,h} H^{1-(2t-1)/k^2+\varepsilon} = O(1).$$

In case $2 \leqslant k \leqslant 4$, we may verify directly that

$$\sum_{q,h} H^{1-(2t-1)/k^2+\varepsilon} = O(P^{1-c_k}), \qquad c_k > 0, \; c_2 = \frac{3}{8}.$$

For example, as $k=2$, we have

$$\sum_{q_1 \leq P^{\frac{1}{2}-\frac{1}{4}+\sigma}} \; \sum_{q_2 \leq P^{\frac{1}{4}-2\sigma}} (q_1 q_2)^{2-(8-1)/4+\varepsilon}$$

$$= \sum_{q_1 \leq P^{\frac{1}{4}+\sigma}} q_1^{\frac{1}{4}+\varepsilon} \sum_{q_2 \leq P^{\frac{1}{4}-2\sigma}} q_2^{\frac{1}{4}+\varepsilon}$$

$$= O\left(P^{\frac{5}{4}(\frac{1}{4}+\sigma)+\frac{5}{4}(\frac{1}{4}-2\sigma)+\varepsilon}\right) = O\left(P^{\frac{5}{8}}\right).$$

For $k=3$,

$$\sum_{q_1 \leq P^{\frac{1}{2}-\frac{1}{6}+\sigma}} \; \sum_{q_2 \leq P^{\frac{1}{6}-2\sigma}} \; \sum_{q_3 \leq P^{\frac{1}{6}-2\sigma}} (q_1 q_2 q_3)^{2-(18-1)/9+\varepsilon}$$

$$\leqslant \sum_{q_1 \leq P^{\frac{1}{3}+\sigma}} q_1^{\frac{1}{9}+\varepsilon} \sum_{q_2 \leq P^{\frac{1}{6}-2\sigma}} q_2^{\frac{1}{9}+\varepsilon} \sum_{q_3 \leq P^{\frac{1}{6}-2\sigma}} q_3^{\frac{1}{9}+\varepsilon}$$

$$= O\left(P^{\frac{10}{9}(\frac{1}{3}+\sigma+\frac{1}{6}-2\sigma+\frac{1}{6}-2\sigma)}\right) = O\left(P^{\frac{20}{27}}\right).$$

For $k=4$,

$$\sum_{q_1 \leq P^{\frac{1}{2}-\frac{1}{8}+\sigma}} \sum_{q_2 \leq P^{\frac{1}{8}-2\sigma}} \sum_{q_3 \leq P^{\frac{1}{8}-2\sigma}} \sum_{q_4 \leq P^{\frac{1}{8}-2\sigma}} (q_1 q_2 q_3 q_4)^{2-(48-1)/16+\epsilon}$$

$$= O\left(P^{\frac{1}{16}\left(\frac{3}{8}+\frac{3}{8}\right)}\right) = O\left(P^{\frac{3}{64}}\right).$$

Therefore the error term in (17.6) can be replaced by

$$O\left(P^{2t-\frac{1}{2}k(k+1)-\frac{3}{8}}\right).$$

By lemmas 17.2 and 17.3, we have

$$T_{(2)} = \sum_{h,q} \left| B\left(\frac{h_k}{q_k}, \cdots, \frac{h_1}{q_1}\right) \right|^{2t} \vartheta_0 P^{2t-\frac{k}{2}(k+1)}$$

$$+ O\left(P^{2t-\frac{1}{2}k(k+1)-(2t/k^2-1)\sigma}\right) + O\left(P^{2t-\frac{1}{2}k(k+1)-\frac{3}{8}}\right)$$

$$= \mathfrak{S}(P)\, \vartheta_0 P^{2t-\frac{1}{2}k(k+1)} + O\left(P^{2t-\frac{1}{2}k(k+1)-c_1}\right).$$

Proof of (1.3). Combining lemmas 15.3 and 17.4, we have the theorem which represented by (1.3).

XVIII. Estimation of an Exponential Sum

Now we are ready to answer the problem arrived at in § 12. Instead of lemma 12.5, we have the following comparatively satisfactory result.

Lemma 18.1. *Under the same assumption as lemma* 12.5, *we have, for* $k \geqslant 11$,

$$\sum_{x=1}^{P} e(f(x)) = O\left(P^{1-\frac{1}{\sigma'}}\left(\frac{q}{P^{r-1}}\right)^{\frac{1}{\sigma'}}\right), \qquad \text{for } P^{r-1} \leq q \leq P^r,$$

where

$$\sigma' = k^2 (3 \log k + \log \log k + 5) .$$

It is a direct consequence of lemma 11.1, with $\delta=0$.

XIX. A Remark Concerning Prouhet's Problem

Let $r_t^{(l)}(P)$ be the number of solutions of l sets of unknowns $x_1, \cdots, x_t; y_1, \cdots, y_t; \cdots; \omega_1, \cdots, \omega_t$ of the system of Diophantine equations:

$$(19.1) \quad \begin{cases} x_1^k + \cdots + x_t^k = y_1^k + \cdots + y_t^k = \cdots = \omega_1^k + \cdots + \omega_t^k, \\ x_1^{k-1} + \cdots + x_t^{k-1} = y_1^{k-1} + \cdots + y_t^{k-1} = \cdots = \omega_1^{k-1} + \cdots + \omega_t^{k-1}, \\ \cdots\cdots\cdots\cdots\cdots\cdots\cdots\cdots\cdots\cdots\cdots\cdots \\ x_1 + \cdots + x_t = y_1 + \cdots + y_t = \cdots = \omega_1 + \cdots + \omega_t \end{cases}$$

with the restrictions that

$$(19.2) \qquad 1 \leqslant x_i, y_i, \cdots, \omega_i \leqslant P, \qquad i = 1, 2, \cdots, t.$$

In particular, we have

$$r_t^{(2)}(P) = r_t(P).$$

The analytic formulation of $r_t^{(l)}(P)$ is the following:
Let

$$T(\alpha) = \sum_{x=1}^{P} e(\alpha_k x^k + \cdots + \alpha_1 x).$$

Then

$$(19.3) \qquad r_t^{(l)}(P) = \int_0^1 \cdots \int_0^1 \{T_P(\alpha^{(1)}) T_P(\alpha^{(2)}) \cdots T_P(\alpha^{(l)})\}^t \times$$
$$\times \tau(-\alpha^{(1)} - \cdots - \alpha^{(l)}) \, d\alpha^{(1)} \cdots d\alpha^{(l)}.$$

Certainly we can get an asymptotic formula by the method of Farey dissection. We are now going to indicate another way to handle this problem.

Let $r_s(N_k, \cdots, N_1)$ be the number of solutions of the system of Diophantine equations

$$(19.4)\qquad \begin{cases} x_1^k + \cdots + x_s^k = N_k, \\ \cdots\cdots\cdots\cdots\cdots\cdots \\ x_1 + \cdots + x_s = N_1, \end{cases} \qquad 1 \le x_i \le P.$$

Then, for $s > 2t_0$, which is defined in § 1, we have

$$(19.5)\qquad r_s(N_k, \cdots, N_1) = b_0 P^{s-\frac{1}{2}k(k+1)} \mathfrak{S}(N_k, \cdots, N_1) + O(P^{s-\frac{1}{2}k(k+1)-c}),$$

where

$$(19.6)\qquad b_0 = \int_{-\infty}^{\infty} \cdots \int_{-\infty}^{\infty} \left(\int_0^1 e(\gamma_k x^k + \cdots + \gamma_1 x)\, dx \right)^s \times$$

$$\times\, e\left(-\frac{N_k}{P^k} \gamma_k - \cdots - \frac{N_1}{P} \gamma_1 \right) d\gamma_k \cdots d\gamma_1$$

and

$$(19.7)\qquad \mathfrak{S}(N_k, \cdots, N_1) = \sum_{q_1, \cdots, q_k = 1}^{\infty} A(q_k, \cdots, q_1),$$

$$A(q_k, \cdots, q_1) = \sum_{\substack{h_1=1 \\ (h_1, q_1)=1}}^{q_1} \cdots \sum_{\substack{h_k=1 \\ (h_k, q_k)=1}}^{q_k} S^s e\left(-\frac{h_k}{q_k} N_k - \cdots - \frac{h_1}{q_1} N_1 \right)$$

and

$$S = \frac{1}{Q} \sum_{x=1}^{Q} e\left(\frac{h_k}{q_k} x^k + \cdots + \frac{h_1}{q_1} x \right)$$

and Q is the least common multiple of $q_k, \ldots, q_1$.

The proof is omitted, since it is entirely similar to that of (1.3). Nevertheless there are several interesting problems about convergence which arise here, since the convergence problems which confronted us in this paper, were problems of absolute convergence. The exponents of convergence of

(19.6) and (19.7) can be reduced considerably. For example, (19.6) converges for $s \geqslant k$. But the author will not go into the detailed discussions here.

After we obtain (19.5), the situation about Prouhet's and Tarry's problems can be easily visualized through the light of Fourier series. In fact, let

$$G(\alpha_k, \cdots, \alpha_1) = \left(\sum_{x=1}^{P} e(\alpha_k x^k + \cdots + \alpha_1 x) \right)^s .$$

(19.5) asserts the behaviour of the Fourier coefficients of the function $G(\alpha_k, \cdots, \alpha_1)$. Tarry's problem is equivalent to the Parseval relation that

$$\sum_{N_k} \cdots \sum_{N_1} r_s^2 (N_k, \cdots, N_1) = \int_0^1 \cdots \int_0^1 | G(\alpha_k, \cdots, \alpha_1) |^2 d\alpha_k \cdots d\alpha_1 .$$

Prouhet's problem is to investigate the behaviour of the sum

$$r_s^{(l)}(P) = \sum_{N_k} \cdots \sum_{N_1} r_s^l (N_k, \cdots, N_1) .$$

If we appeal directly to the theory of Fourier series, we can easily find out by Young and Hausdorff's theorem that

$$\gamma_s^{(l)}(P) \leqslant \left(\int_0^1 \cdots \int_0^1 | G(\alpha_k, \cdots, \alpha_1) |^{l/(l-1)} d\alpha_k \cdots d\alpha_1 \right)^{l-1}$$

$$= O(P^{ls - \frac{1}{2}k(k+1)(l-1)}) . \tag{19.8}$$

But by means of (19.5), we can find an asymptotic formula for $r_s^{(l)}(P)$ instead of the order-result (19.8).

XX. Remarks

1. Using the analogous between Fourier series and almost periodic functions, we can treat the following problem: Let

$$0 < e_1 < e_2 < \cdots < e_k$$

be k real numbers. Let $r(N, \Delta)$ be the number of solutions, in integers, of

$$|x_1^{e_1} + \cdots + x_s^{e_1} - N_1| \leqslant \Delta_1$$

$$\cdots\cdots\cdots\cdots\cdots\cdots$$

$$|x_1^{e_k} + \cdots + x_s^{e_k} - N_k| \leqslant \Delta_k, \qquad 1 \leqslant x_i \leqslant P,$$

where Δ_i approaches zero as P tends to infinity.

2. By the method in Hua [1], we can find the asymptotic formula for the number of (1.1) and (19.5) even if we restrict our unknowns to be primes, with the lower bound replacing t_0 by $t_0 + 1$.

XXI. In Case $k = 2$, We Can Push Our Theorem to the Best Form

Theorem. *Let $r'(N)$ be the number of solutions of the system of Diophantine equations*

$$(21.1) \qquad \begin{cases} x_1 + x_2 + x_3 = y_1 + y_2 + y_3 & x_1^2 + x_2^2 + x_3^2 \leqslant N, \\ x_1^2 + x_2^2 + x_3^2 = y_1^2 + y_2^2 + y_3^2 & y_1^2 + y_2^2 + y_3^2 \leqslant N. \end{cases}$$

Then, we have, as $N \to \infty$,

$$(21.2) \qquad r'(N) \sim \frac{35\sqrt{3}}{2} N^{\frac{3}{2}} \log N.$$

Proof. 1) Let $r(N_1, N_2)$ be the number of solutions of the system of Diophantine equations

$$(21.3) \qquad \begin{cases} x_1 + x_2 + x_3 = N_1 \\ x_1^2 + x_2^2 + x_3^2 = N_2, & N_2 \leqslant N. \end{cases}$$

Then, we have, by Schwartz's inequality,

(21.4) $$N_1^2 \leqslant 3N_2,$$

and also the congruence relation

(21.5) $$N_1 \equiv N_2 \pmod 2.$$

From (21.3), we have

$$x_1^2 + x_2^2 + (N_1 - x_1 - x_2)^2 = N_2,$$

that is

(21.6) $$(3x_1 - N_1)^2 + (3x_1 - N_1)(3x_2 - N_1)^2 + (3x_2 - N_1)^2 = \frac{3}{2}(3N_2 - N_1^2).$$

Now we consider the number $\psi(m)$ of solutions of

(21.7) $$X_1^2 + X_1 X_2 + X_2^2 = m,$$

which is known to be

(21.8) $$\psi(m) = 6\sum_{l/m}\left(\frac{-3}{l}\right).$$

If m is divisible by 9, from (21.7), we have consequently that $3 | (X_1, X_2)$, thus, for $3 | N_1$, (21.6) is always soluble and its number of solutions is equal to $\psi(\frac{3}{2}(3N_2 - N_1^2))$. Then, we have for $3 | N_1$, that

(21.9) $$r(N_1, N_2) = \psi\left(\frac{3}{2}(3N_2 - N_1^2)\right).$$

If $3 \nmid N_1$, we have

$$\frac{3}{2}(3N_2 - N_1^2) \equiv 3 \pmod 9.$$

From

$$X^2 + XY + Y^2 \equiv 3 \pmod 9,$$

we have $X \equiv Y \equiv \pm 1 \pmod 3$. If (21.7) has a solution (X_1, X_2), then $(-X_1, -X_2)$ is also a solution. Therefore

$$3x_1 - N_1 = \pm X_1, \qquad 3x_2 - N_1 = \pm X_2$$

is also always soluble by an appropriate choice of the sign, and the choice is unique. Therefore

$$r(N_1, N_2) = \frac{1}{2}\psi\left(\frac{3}{2}(3N_2 - N_1^2)\right) \tag{21.10}$$

if $3 \nmid N_1$.

Since

$$r'(N) = \sum_{N_1, N_2} r^2(N_1, N_2),$$

where N_1 and N_2 run over integers satisfying (21.4) and (21.5), we have, by (21.9) and (21.10),

$$\begin{aligned} r'(N) &= \sum_{3 \mid N_1} \sum_{N_2} \psi^2\left(\frac{3}{2}(3N_2 - N_1^2)\right) \\ &\qquad + \frac{1}{4} \sum_{3 \nmid N_1} \sum_{N_2} \psi^2\left(\frac{3}{2}(3N_2 - N_1^2)\right) \\ &= S_1 + \frac{1}{4} S_2, \end{aligned} \tag{21.11}$$

where

$$S_1 = \sum_{3 \mid N_1} \sum_{N_2} \psi^2\left(\frac{3}{2}(3N_2 - N_1^2)\right)$$

and

$$S_2 = \sum_{3 \nmid N_1} \sum_{N_2} \psi^2 \left(\frac{3}{2} (3 N_2 - N_1^2)\right).$$

2) Since $\psi(m) = \psi(3m)$, we have

$$S_2 = \sum_{3 \nmid N_1} \sum_{N_2} \psi^2 \left(\frac{1}{2} (3 N_2 - N_1^2)\right) \tag{21.12}$$

where $m = \frac{1}{2}(3N_2 - N_1^2)$. The number of terms which makes $\frac{1}{2}(3N_2 - N_1^2)$ to be a square is equal to the number of solutions of

$$3 N_2 = N_1^2 + 2 M^2,$$

which is $O(N_2^\epsilon)$. Since $\psi(N) = O(N^\epsilon)$, the sum of all those term making $\frac{1}{2}(3N_2 - N_1^2)$ to be a square is

$$O\left(\sum_{N_2} O(N^\epsilon)\right) = O(N^{1+\epsilon}). \tag{21.13}$$

We use S_2^* to denote the sum of the part of S_2, of which $\frac{1}{2}(3N_2 - N_1^2)$ is not a square. We have

$$\psi(m) = 6 \sum_{l \mid m} \left(\frac{-3}{l}\right)$$

$$= 6 \sum_{\substack{l \mid m \\ l < \sqrt{m}}} \left(-\frac{3}{l}\right) + 6 \sum_{\substack{l \mid m \\ l < \sqrt{m}}} \left(\frac{-3}{m/l}\right)$$

$$= 6\left(1 + \left(\frac{-3}{m}\right)\right) \sum_{\substack{l \mid m \\ l < \sqrt{m}}} \left(\frac{-3}{l}\right).$$

Since, for $m = \frac{1}{2}(3N_2 - N_1^2)$,

$$\left(\frac{-3}{m}\right) = \left(\frac{-3}{N_1^2}\right) = 1,$$

we have

$$S_2^* = 144 \sum_{3 \nmid N_1} \sum_{N_2} \Bigg(\sum_{\substack{l \mid m \\ l < \sqrt{m}}} \left(\frac{-3}{l}\right)\Bigg)^2$$

(21.14)
$$= 144 \sum_{l < \sqrt{\frac{3N}{2}}} \sum_{l' < \sqrt{\frac{3N}{2}}} \left(\frac{-3}{ll'}\right) A(l, l'),$$

where $A(l, l')$ is the number of solutions of

$$\frac{1}{2}(3N_2 - N_1^2) \equiv 0 \pmod{\frac{ll'}{(l, l')}},$$

$$N_1^2 \leqslant 3N_2 \leqslant 3N.$$

Thus

$$A(l, l') = \sum_{N_1} \left(\frac{N - N_1^2/3}{2ll'/(l, l')} + O(1)\right)$$

(21.15)
$$= \sum_{N_1^2 \leq 3N} \frac{(N - N_1^2/3)(l, l')}{2ll'} + O(N^{\frac{1}{2}}),$$

$$= \frac{(l, l')}{2ll'} \frac{5}{\sqrt{3}} N^{\frac{3}{2}} + O\left(\frac{(l, l')}{ll'} N\right) + O(N^{\frac{1}{2}}),$$

since

$$\sum_{N_1^2 \leq 3N} N = N(2[\sqrt{3N}] + 1) = 2\sqrt{3}\, N^{\frac{3}{2}} + O(N)$$

and

$$\sum_{N_1^2 \leq 3N} \frac{N_1^2}{3} = \frac{1}{9} (\sqrt{3N})^3 + O(N).$$

Therefore, by (21.14) and (21.15), we have

$$S_2^* = 120\sqrt{3}\, N^{\frac{3}{2}} \sum_{l<\sqrt{\frac{3N}{2}}} \sum_{l'<\sqrt{\frac{3N}{2}}} \frac{(l,l')}{ll'}\left(\frac{-3}{ll'}\right)$$

$$(21.16) \qquad + O\Big(N \sum_{l<\sqrt{\frac{3N}{2}}}{}' \sum_{l'<\sqrt{\frac{3N}{2}}} \frac{(l,l')}{ll'}\Big) + O(N^{\frac{3}{2}})$$

$$= 120\sqrt{3}\, N^{\frac{3}{2}} \sum_{l<\sqrt{\frac{3N}{2}}} \sum_{l'<\sqrt{\frac{3N}{2}}} \frac{(l,l')}{ll'}\left(\frac{-3}{ll'}\right) + O(N^{\frac{3}{2}}),$$

since $(l,l') \leqslant \sqrt{ll'}$,

$$\sum_{l<\sqrt{\frac{3N}{2}}} \sum_{l'<\sqrt{\frac{3N}{2}}} \frac{(l,l')}{ll'} \leqslant \Big(\sum_{l<\sqrt{\frac{3N}{2}}} \frac{1}{l^{1/2}}\Big)^2 = O(N^{\frac{1}{2}}).$$

3) Now we are going to prove that

$$(21.17) \qquad \sum_{l<\sqrt{\frac{3N}{2}}} \sum_{l'<\sqrt{\frac{3N}{2}}} \frac{(l,l')}{ll'}\left(\frac{-3}{ll'}\right) = \frac{1}{12}\log N + O((\log N)^{\frac{1}{2}}).$$

In the proof we shall use the following facts:

$$(21.18) \qquad \sum_{a\leq m\leq b} \frac{1}{m}\left(\frac{-3}{m}\right) = O\left(\frac{1}{a}\right)$$

which is a consequence of partial summations, and

$$(21.19) \qquad L_{-3}(1) = \sum_{m=1}^{\infty} \frac{1}{m}\left(\frac{-3}{m}\right) = \frac{\pi}{3\sqrt{3}}$$

which is well known.

Let $l = md$, $l' = m'd$ and $(m, m') = 1$. We consider first the partial sum of (21.17) consisting of those terms with $d > \sqrt{N}/\log N$. It does not exceed

$$\sum_{\frac{\sqrt{N}}{\log N}<d<\sqrt{\frac{3N}{2}}} \sum_{\substack{m<\sqrt{\frac{3N}{2}}\frac{1}{d} \\ (m,m')=1}} \sum_{m'<\sqrt{\frac{3N}{2}}\frac{1}{d}} \frac{1}{d\,m\,m'}$$

$$\leqslant \sum_{\frac{\sqrt{N}}{\log N}<d<\sqrt{\frac{3N}{2}}} \frac{1}{d} \sum_{m<\sqrt{\frac{3}{2}}\log N} \frac{1}{m} \sum_{m'<\sqrt{\frac{3}{2}}\log N} \frac{1}{m'}$$

(21.20) $$= O\left((\log\log N)^3\right).$$

The remaining part of the sum (21.17) is equal to

(21.21) $$U = \sum_{\substack{d<\sqrt{N}/\log N \\ 3\nmid d}} \sum_{\substack{m<\sqrt{\frac{3N}{2}}\frac{1}{d} \\ (m,m')=1}} \sum_{m'<\sqrt{\frac{3N}{2}}\frac{1}{d}} \frac{1}{d\,m\,m'}\left(\frac{-3}{mm'}\right)$$

$$= \sum_{\substack{d<\sqrt{N}/\log N \\ 3\nmid d}} \sum_{m<\sqrt{\frac{3N}{2}}\frac{1}{d}} \sum_{m'<\sqrt{\frac{3N}{2}}\frac{1}{d}} \frac{1}{dmm'}\left(\frac{-3}{mm'}\right) \sum_{n\mid(m,m')} \mu(n).$$

Let $m=n\,q$, $m'=n\,q'$, we have then

$$U = \sum_{\substack{d<\sqrt{N}/\log N \\ 3\nmid d}} \frac{1}{d} \sum_{\substack{n<\sqrt{\frac{3N}{2}}\frac{1}{d} \\ 3\nmid n}} \frac{\mu(n)}{n^2} \sum_{q<\sqrt{\frac{3N}{2}}\frac{1}{dn}} \sum_{q'<\sqrt{\frac{3N}{2}}\frac{1}{dn}} \frac{1}{q\,q'}\left(\frac{-3}{q\,q'}\right)$$

(21.22) $$= U_1 + U_2,$$

where U_1 is the sum over those $n>\left(\sqrt{\frac{3N}{2}}\frac{1}{d}\right)^{\frac{1}{2}}$, and U_2 is the remaining part of U. Then, by (21.18),

$$U_1 \leqslant \sum_{d<\sqrt{N}/\log N} \frac{1}{d} \sum_{\left(\sqrt{\frac{3N}{2}}\frac{1}{d}\right)^{\frac{1}{2}}<n} \frac{1}{n^2} \left| \sum_{q<\sqrt{\frac{3N}{2}}\frac{1}{dn}} \frac{1}{q}\left(\frac{-3}{q}\right) \right| \times$$

(21.23) $$\times \left| \sum_{q'<\sqrt{\frac{3N}{2}}\frac{1}{dn}} \frac{1}{q'}\left(\frac{-3}{q'}\right) \right|$$

$$= O\Big(\log N \sum_{(\log N)^{\frac{1}{2}} < n} \frac{1}{n^2}\Big) = O\big((\log N)^{\frac{1}{2}}\big).$$

Now we consider

$$U_2 = \sum_{\substack{d < \sqrt{N}/\log N \\ 3 \nmid d}} \frac{1}{d} \sum_{\substack{n < \left(\sqrt{\frac{3N}{2}}\frac{1}{d}\right)^{\frac{1}{2}} \\ 3 \nmid n}} \frac{\mu(n)}{n^2} \Bigg(\sum_{q < \sqrt{\frac{3N}{2}}\frac{1}{dn}} \frac{1}{q}\left(\frac{-3}{q}\right)\Bigg)^2.$$

Since, for $d < \sqrt{N}/\log N$,

$$\sum_{n > \left(\sqrt{\frac{3N}{2}}\frac{1}{d}\right)^{\frac{1}{2}}} \frac{\mu(n)}{n^2} = O\Big(\Big(\sqrt{\frac{3N}{2}}\frac{1}{d}\Big)^{-\frac{1}{2}}\Big) = O\Big((\log N)^{-\frac{1}{2}}\Big)$$

and, by (21.19)

$$\sum_{q > \sqrt{\frac{3N}{2}}\frac{1}{dn}} \left(\frac{-3}{q}\right)\frac{1}{q} = O\Big(\Big(\sqrt{\frac{3N}{2}}\frac{1}{dn}\Big)^{-1}\Big) = O\Big((\log N)^{-\frac{1}{2}}\Big),$$

we have

$$U_2 = \sum_{\substack{d < \sqrt{N}/\log N \\ 3 \nmid d}} \frac{1}{d} \sum_{\substack{n=1 \\ 3 \nmid n}}^{\infty} \frac{\mu(n)}{n^2} L^2_{-3}(1) + O\Big((\log N)^{\frac{1}{2}}\Big)$$

(21.24)

$$= \frac{9}{4\pi^2} \log N \, L^2_{-3}(1) + O\Big((\log N)^{\frac{1}{2}}\Big),$$

since

$$\sum_{\substack{d < \sqrt{N}/\log N \\ 3 \nmid d}} \frac{1}{d} = \sum_{d < \sqrt{N}/\log N} \frac{1}{d} - \frac{1}{3} \sum_{d < \frac{1}{3}\sqrt{N}/\log N} \frac{1}{d}$$

$$= \frac{1}{3} \log N + O(\log \log N)$$

and

$$\sum_{\substack{n=1\\3\nmid n}}^{\infty}\frac{\mu(n)}{n^2}=\sum_{n=1}^{\infty}\frac{\mu(n)}{n^2}-\frac{1}{9}\sum_{\substack{m=1\\3\nmid m}}^{\infty}\frac{\mu(3m)}{m^2}$$

$$=\frac{1}{\zeta(2)}+\frac{1}{9}\sum_{\substack{m=1\\3\nmid m}}^{\infty}\frac{\mu(m)}{m^2}$$

and

$$\sum_{\substack{n=1\\3\nmid n}}^{\infty}\frac{\mu(n)}{n^2}=\frac{9}{8}\,\frac{1}{\zeta(2)}=\frac{27}{4\pi^2}\,.$$

Combining (21.20), (21.21), (21.22), (21.23) and (21.24), we have (21.17). Therefore, by (21.16) and (21.17)

$$S_2=10\sqrt{3}\,N^{\frac{3}{2}}\log N+O\Big(N^{\frac{3}{2}}(\log N)^{\frac{1}{2}}\Big).$$

4) Now we are going to find the asymptotic formula of S_1. Let τ be the highest power of 3 dividing $\frac{3}{2}(3N_2-N_1^2)$ where $\tau\geqslant 2$. We sum up S_1 according to τ. Then we have

$$S_1=\sum_{\tau=2}^{W}\sum_{N_1}\sum_{\substack{N_2\\3^{\tau}\,\|\,\frac{3}{2}(3N_2-N_1^2)}}\psi^2\Big(\frac{3}{2}(3N_2-N_1^2)\Big)$$

where

$$W=[\log\tfrac{9}{2}N/\log 3]\,.$$

Since $\psi(3N)=\psi(N)$, we have

$$S_1=\sum_{\tau=1}^{W}\sum_{N_1}\sum_{N_2}\psi^2(m),$$

where $m = \frac{3}{2}(3N_2 - N_1^2)\, 3^{-\tau}$. Let us consider the sum

$$S_1' = \sum_{\tau \geq W/2} \sum_{N_1} \sum_{N_2} \psi^2(m),$$

which is

$$O(N^\varepsilon \sum_{\tau \geq W/2} \sum_{N_1} \sum_{N_2} 1),$$

and the number of solutions of

$$\frac{3}{2}(3N_2 - N_1^2) \equiv 0 \pmod{3^\tau}$$

$$N_1^2 \leqslant 3N_2 \leqslant 3N$$

is

$$O\Big(\sum_{N_1} \big(\frac{N - \frac{1}{3}N_1^2}{3^\tau} + 1\big)\Big) = O\Big(\big(\frac{N^{3/2}}{3^\tau} + 1\big)\Big).$$

Thus

$$S_1' = O\Big(N^\varepsilon \sum_{\tau \geq \frac{1}{2}W} \big(\frac{N^{3/2}}{3^\tau} + 1\big)\Big) = O(N^{\frac{3}{2}+\varepsilon} / 3^{\frac{1}{2}W})$$
$$= O(N^{1+\varepsilon}).$$

As in the treatment of S_2, we can prove that the partial sum S_1'' of those terms with m being a square ($\geqslant 0$) is also of lower order.

Now we are going to consider

$$S_1 - S_1' + S_1'' = \sum_{\tau \leq \frac{1}{2}W} \sum_{N_1} \sum_{N_2} \psi^2(m) = S_1''', \text{ say}.$$

Since m is not a square, we have

$$\psi(m) = 6 \sum_{\substack{l<\sqrt{m} \\ l/m}} \left(\frac{-3}{l}\right) + 6 \sum_{l<\sqrt{m}} \left(\frac{-3}{m/l}\right)$$

$$= 6\left(1 + \left(\frac{-3}{m}\right)\right) \sum_{l<\sqrt{m}} \left(\frac{-3}{l}\right)$$

$$= \begin{cases} 0, & \text{if} \quad m \equiv 2 \pmod 3 \\ 12 \sum_{l<\sqrt{m}} \left(\frac{-3}{l}\right), & \text{if} \quad m \equiv 1 \pmod 3 . \end{cases}$$

We have therefore

$$S_1''' = 144 \sum_{\tau=1}^{W/2} \sum_{\substack{N_1 \\ m\equiv 1\,(3)}} \sum_{N_2} \Big(\sum_{l\leq\sqrt{m}} \left(\frac{-3}{l}\right)\Big)^2$$

$$= 144 \sum_{\tau=2}^{W/2} \sum_{l<\sqrt{m}} \sum_{l'<\sqrt{m}} \left(\frac{-3}{l}\right)\left(\frac{-3}{l'}\right) A(l, l'; \tau),$$

where $A(l, l'; \tau)$ is the number of solutions of

$$\frac{1}{2}(3N_2 - N_1^2) \equiv 0 \quad \left(\bmod\ 3^{\tau-1}\frac{l\,l'}{(l, l')}\right)$$

$$N_1^2 \leqslant 3N_2 \leqslant 3N .$$

Thus

$$A(l, l'; \tau) = \sum_{N_1^2 \leq 3N} \left(\frac{(N - \frac{1}{3}N_1^2)\,(l, l')}{2\cdot 3^{\tau-2}\, l\, l'} + O(1)\right)$$

$$= \frac{(l, l')}{2ll'}\,\frac{5}{\sqrt{3}}\,\frac{1}{3^{\tau-2}}\,N^{\frac{3}{2}} + O\left(\frac{(l, l')}{3^{\tau}\, l\, l'}\,N\right) + O\left(N^{\frac{1}{2}}\right).$$

We have, similar to 3)

$$S_1 = 120\sqrt{3}\,N^{\frac{3}{2}} \sum_{\tau=2}^{W/2} \sum_{l\leq\sqrt{\frac{3N}{2\cdot 3^{\tau}}}} \sum_{l'\leq\sqrt{\frac{3N}{2\cdot 3^{\tau}}}} \frac{(l, l')}{l\,l'} \left(\frac{-3}{l\,l'}\right) \frac{1}{3^{\tau-2}} + O(N^{\frac{3}{2}})$$

$$= 120\sqrt{3}\,\frac{3}{2}\,N^{\frac{3}{2}}\,\frac{\log N}{12} + O\left(N^{\frac{3}{2}}\left(\log N^{\frac{1}{2}}\right)\right).$$

Combining both estimation for S_1 and S_2, we have

$$r'(N) = 10\sqrt{3}\left(\frac{3}{2}+\frac{1}{4}\right)N^{\frac{3}{2}}\log N + O\left(N^{\frac{3}{2}}(\log N)^{\frac{1}{2}}\right)$$

$$= \frac{35\sqrt{3}}{2}\,N^{\frac{3}{2}}\log N + O\left(N^{\frac{3}{2}}(\log N)^{\frac{1}{2}}\right).$$

Another possible way to find the number of solutions of (21.1) is by considering the indefinite quinary quadratic form

$$x_1^2 + x_2^2 + x_3^2 - y_1^2 - y_2^2 - (x_1 + x_2 + x_3 - y_1 - y_2)^2.$$

But it seems to be more complicated to bridge over from the condition of primary solutions and the condition of the form (21.1).

References

[1] Hua, Loo-Keng, 1947, Аддитивная теория простых чисел, Труды математического института Стеклова, XXII.

[2] ———, 1949, An improvement of Vinogradow's mean-value theorem and several applications. Quart. Journ. of Math. (Oxford) **20**, 48-61.

[3] ———, 1928, On Tarry's problem, *ibid.* (Oxford) **9**, 315-320.

[4] ———, 1939, On an exponential sum. Journ. Chinese Math. Soc., vol. 2.

[5] Виноградов, И. М., 1947, Метод тригонометрических сумм в теории чисел. Труды математического института Стеклова, XXIII.

[6] ———, 1938, Математ. сборник, **3**, 435—471.

[7] Mordell, 1939, Quart. Journ. of Math., **3**, 161-167.

Reprinted from the
SCIENCE RECORD
Vol. 1, No. 1, pp. 1–4, 1957

MATHEMATICS

ON EXPONENTIAL SUMS *

HUA LOO-KENG (華罗庚)
Institute of Mathematics, Academia Sinica
(Member of Academia Sinica)

The aim of the present note is to prove the following theorem, but its generalizations and applications will not be given here.

Theorem. Let k be an integer $\geqslant 2$, and $(h, q) = 1$. Then, for any $\varepsilon > 0$, we have

$$\sum_{x=1}^{P} e^{2\pi i h x^k / q} = \frac{P}{q} \sum_{x=1}^{q} e^{2\pi i h x^k / q} + O\left(q^{\frac{1}{2}+\varepsilon}\right),$$

where the constant implied by the symbol O depends only on k and ε.

This theorem improves a result of the author[1] in 1940. In wider range, it gives improvement of a celebrated result of Vinogradow[2].

The proof of the present theorem relies on the following lemmas.

Lemma 1. (Weil[3], Carlitz-Uchiyama[4]). Let p be a prime and

$$f(x) = a_k x^k + \dots + a_1 x + a_0, \qquad p \nmid (a_k, \dots, a_1)$$

we have

$$\left| \sum_{x=1}^{p} e^{2\pi i f(x)/p} \right| \leqslant k \sqrt{p} \quad . \tag{1}$$

Lemma 2. Under the hypothesis of lemma 1, we assume further that $p^t \parallel (k a_k, \dots, 2a_2, a_1)$ and the congruence

$$p^{-t} f'(x) \equiv 0 \pmod{p}, \quad 0 \leqslant x < p \tag{2}$$

has no double root. Then we have

$$\left| \sum_{x=1}^{p^l} e^{2\pi i f(x)/p^l} \right| \leqslant k^{\frac{5}{2}} p^{\frac{l}{2}} \quad . \tag{3}$$

Proof. Let $\mu_1, \dots, \mu_r$ $(r \leqslant k-1)$ be the roots of (2). The lemma is evidently a consequence of the following inequality

* Received Nov. 29, 1956.

$$\left|\sum_{x=1}^{p^l} e^{2\pi i f(x)/p^l}\right| \leqslant k \max(1, r)\, p^{\frac{1}{2}(l+t)}, \tag{4}$$

since $p^t \leqslant k$.

1) For $l < 2(t+1)$, if $t=0$ then $l=1$, (4) follows from lemma 1; if $t \geqslant 1$, we have

$$\left|\sum_{x=1}^{p^l} e^{2\pi i f(x)/p^l}\right| \leqslant p^l \leqslant p^{\frac{1}{2}(l+t)}\, p^{\frac{1}{2}(l-t)} \leqslant p^{\frac{1}{2}(t+1)}\, p^{\frac{1}{2}(l+t)} \leqslant k p^{\frac{1}{2}(l+t)}$$

2) For $l \geqslant 2(t+1)$, we have

$$\left|\sum_{x=1}^{p^l} e^{2\pi i f(x)/p^l}\right| \leqslant \sum_{j=1}^{r} \left|\sum_{y=1}^{p^{l-1}} e^{2\pi i (f(\mu_j+py) - f(\mu_j))/p^l}\right|. \tag{5}$$

Notice that in case (2) has no solution, i. e. $r=0$, both hands of (5) are zero.

Now let $g_j(x) = p^{-\sigma_j}(f(\mu_j+px) - f(\mu_j))$, $j=1, \ldots, r$, where p does not divide all the coefficients of $g_j(x)$. Since $f(\mu_j) \not\equiv 0 \pmod{p^{t+1}}$, $(j=1, \ldots, r)$, we have $2 \leqslant \sigma_j \leqslant t+2$, and by(5)

$$\left|\sum_{x=1}^{p^l} e^{2\pi i f(x)/p^l}\right| \leqslant \begin{cases} rp^{l-1} \leqslant rp^{t+1} \leqslant rkp, & \text{if } l \leqslant \max(\sigma_1, \ldots, \sigma r_1), \\ \sum\limits_{j=1}^{r} p^{\sigma_j - 1} \left|\sum\limits_{x=1}^{p^{l-\sigma_j}} e^{2\pi i g_j(x)/p^{l-\sigma_j}}\right|, & \text{if } l > \max(\sigma_1, \ldots, \sigma_r). \end{cases} \tag{6}$$

For the later case we use induction. Let p^{δ_j} divide all the coefficients of $g'_j(x)$, but p^{δ_j+1} do not. From the expression

$$p^{\sigma_j} g'_j(x) = f'(\mu_j)p + f''(\mu_j)p^2 x + \ldots + \frac{f^{(k)}(\mu_j)}{(k-1)!} p^k x^{k-1},$$

we see that $\sigma_j + \delta_j = t+2$. The congruence

$$p^{-\delta_j} g'_j(x) \equiv f'(\mu_j)\, p^{-t-1} + f''(\mu_j)\, p^{-t} x \equiv 0 \pmod p,\ 0 \leqslant x < p$$

has only one root. Therefore by induction, we obtain

$$\left|\sum_{x=1}^{p^l} e^{2\pi i f(x)/p^l}\right| \leqslant \sum_{j=1}^{r} p^{\sigma_j - 1} k p^{\frac{1}{2}(l-\delta_j+\sigma_j)} = krp^{\frac{1}{2}(l+t)}.$$

The lemma is proved.

Lemma 3. Let $\sigma \geqslant 0$ and $S(p^l, p^\sigma) = \max\limits_{p \nmid hm} \left|\sum\limits_{x=1}^{p^l} e^{2\pi i(hx^k + p^\sigma m x)/p^l}\right|$. Then

$$S(p^l, p^\sigma) \leqslant k^{\frac{5}{2}} p^{\frac{1}{2}(l+\sigma)}$$

Proof. For $l=1$, it follows from lemma 1.

We suppose now that $l\geqslant 2$, $k=p^{\theta}k_1$, $p\nmid k_1$.

Let $\theta\geqslant\sigma$. If the congruence

$$f'(x)=p^{\theta}k_1hx^{k-1}+p^{\sigma}m\equiv 0 \pmod{p^{\sigma+1}},\ 0\leqslant x<p$$

has no solution or $(k-1,\ p-1)$ simple roots, then by lemma 2, we have

$$\left|\sum_{x=1}^{p^l} e^{2\pi i\,(hx^k+p^{\sigma}mx)/p^l}\right|\leqslant k^{\frac{5}{2}}p^{\frac{l}{2}}.$$

Otherwise, $p\mid k-1$, $\theta=\sigma=0$, the similar but more elaborate method gives the same conclusion.

Suppose that $\theta<\sigma$. For $l<2(\theta+1)$, we have evidently $S(p^l,p^{\sigma})\leqslant \leqslant k^{\frac{5}{2}}p^{\frac{l}{2}}$. For $l\geqslant 2(\theta+1)$, the congruence $k_1hx^{k-1}+p^{\sigma-\theta}m\equiv 0$ (mod p), $0\leqslant x<p$ has only a single root $x=0$. Then we have

$$\left|\sum_{x=1}^{p^l} e^{2\pi i\,(hx^k+p^{\sigma}mx)/p^l}\right|\leqslant\left|\sum_{x=1}^{p^{l-1}} e^{2\pi i\,(hp^kx^k+p^{\sigma+1}mx)/p^l}\right|\leqslant$$

$$\leqslant\begin{cases} p^{k-1}\left|\sum_{x=1}^{p^{l-k}} e^{2\pi i(hx^k+p^{\sigma+1-k}mx)/p^{l-k}}\right|, & \text{for } l>k,\ \sigma+1\geqslant k,\\ 0, & \text{for } l>k,\ \sigma+1<k,\\ 0, & \text{for } l\leqslant k,\ \sigma+1<l,\\ p^{l-1}\leqslant p^{\frac{1}{2}(l+\sigma-1)}, & \text{for } l\leqslant k,\ \sigma+1\geqslant l.\end{cases}$$

For the later three cases, the lemma is evident, and for the first case, by induction, we have

$$\left|\sum_{x=1}^{p^l} e^{2\pi i\,(hx^k+p^{\sigma}mx)/p^l}\right|\leqslant p^{k-1}k^{\frac{5}{2}}p^{\frac{1}{2}(l-k+\sigma+1-k)}\leqslant k^{\frac{5}{2}}p^{\frac{1}{2}(l-\sigma)}.$$

The lemma is proved.

Proof of the theorem. Evidently we may assume that $1\leqslant P\leqslant q$, then we have

$$\sum_{x=1}^{P} e^{2\pi ihx^k/q}=\frac{1}{q}\sum_{n=1}^{q}\sum_{x=1}^{q} e^{2\pi i(hx^k+nx)/q}\sum_{t=1}^{P} e^{-2\pi int/q}=$$

$$=\frac{P}{q}\sum_{x=1}^{q} e^{2\pi ihx^k/q}+O\left(\frac{1}{q}\sum_{n=1}^{q-1}\left|\sum_{t=1}^{P} e^{-2\pi int/q}\right|\cdot\left|\sum_{x=1}^{q} e^{2\pi i(hx^k+nx)/q}\right|\right)=$$

$$= \frac{P}{q}\sum_{x=1}^{q} e^{2\pi i h x^k/q} + O\left(\sum_{n=1}^{q-1}\frac{1}{n}\left|\sum_{x=1}^{q} e^{2\pi i(hx^k+nx)/q}\right|\right).$$

Let $q = p_1^{l_1} \dots p_s^{l_s}$ be the canonical decomposition of q, now consider the sum of the terms of $\sum_{n=1}^{q-1}\frac{1}{n}\left|\sum_{x=1}^{q} e^{2\pi i(hx^k+nx)/q}\right|$ with $p_1^{a_1} \dots p_s^{a_s} \mid n$ and $p_1^{a_1+1} \nmid n, \dots, p_s^{a_s+1} \nmid n$. If we can prove that the sum of these terms is $O\left(\frac{q^{\frac{1}{2}+\varepsilon}}{(p_1^{a_1}\dots p_s^{a_s})^{\frac{1}{2}}}\right)$, then the theorem follows. Let $n = p_1^{a_1} \dots p_s^{a_s} m$, $(m, p_1 \dots p_s) = 1$. The partial sum

$$\leqslant \sum_{\substack{m \geq q-1\\(m,q)=1}} \frac{1}{p_1^{a_1}\dots p_s^{a_s} m}\prod_{i=1}^{s} S\left(p_i^{l_i}, p_i^{a_i}\right) \leqslant$$

$$\leqslant k^{\frac{5}{2}s}\frac{p_1^{\frac{1}{2}(l_1+a_1)}\dots p_s^{\frac{1}{2}(l_s+a_s)}}{p_1^{a_1}\dots p_s^{a_s}}\log q = O\left(\frac{q^{\frac{1}{2}+\varepsilon}}{(p_1^{a_1}\dots p_s^{a_s})^{\frac{1}{2}}}\right).$$

The theorem is thus established.

References

[1] Hua, L. K., 1940. On an Exponential Sum, *Journ. of Chinese Math. Soc.*, **2**, 301-312. 1953. Additive Prime Number Theory, *Acad. Sinica Press*, 11-12.

[2] Виноградов, И. М., 1952. Избранные Труды, Издательство Акад. Наук СССР, Москва, 291-295.

[3] Weil, A., 1941. On the Riemann Hypothesis in Functional Fields, *Proc. of Nat. Acad. of Sci. of U. S. A.*, **27**, 345-347.

[4] Carlitz, L. and Uchiyama, S., Bounds for Exponential Sums (in print).

ALGEBRA AND GEOMETRY

Z. X. WAN

Finite Groups. Early in 1938, while Hua taught at the Southwest Association Associated University in Kunming, he conducted a seminar on finite groups; among the topics studied were p-groups. In [46, 81] he introduced the concept of the rank of a p-group. A p-group $\mathfrak{g}$ of order p^n is said to be of rank α, if the maximum of the orders of its elements is $p^{n-\alpha}$. Using this concept he proved that a pseudo-basis exists in p-groups, i.e., if $p \geqslant 3$ and $n \geqslant 2\alpha + 1$, then every element of $\mathfrak{g}$ can be expressed uniquely as

$$G = A^{\delta} A_{\alpha}^{\delta_\alpha} A_{\alpha-1}^{\delta_{\alpha-1}} \ldots A_1^{\delta_1}, \qquad 0 \leqslant \delta \leqslant p^{n-\alpha} - 1, \qquad 0 \leqslant \delta_i \leqslant p - 1,$$

where A is of order $p^{n-\alpha}$ and $A_i^{p^i} = 1$. With the aid of pseudo-bases, and of a modified form of the enumeration principle of P. Hall, he proved several "Anzahl" theorems. For instance, if $\mathfrak{g}$ is a group of order p^n and rank α ($p \geqslant 3$, $n \geqslant 2\alpha + 1$), then (i), $\mathfrak{g}$ contains one and only one subgroup of order p^m and rank α ($2\alpha + 1 \leqslant m \leqslant n$); (ii) $\mathfrak{g}$ contains p^α cyclic subgroups of order p^m ($\alpha < m < n - \alpha - 1$); (iii) the number of elements of order $\leqslant p^m$ ($\alpha \leqslant m \leqslant n - \alpha$) in $\mathfrak{g}$ is equal to $p^{m+\alpha}$. The second and third results improved theorems of G. A. Miller and A. A. Kulakoff respectively.

Skew Fields. Since Hamilton's first example of a non-commutative division algebra—the quaternion algebra—division algebras have received a great deal of attention. By comparison, infinite dimensional division algebras—skew fields—were neglected; until, around 1950, with his perceptive direct algebraic method Hua proved several remarkable theorems in this area.

First, in 1949, Hua [88] proved that "every semi-automorphism of a skew field is either an automorphism or an anti-automorphism". (By a semi-automorphism of a skew field we mean a one-to-one mapping σ from the skew field into itself with the properties $(a + b)^\sigma = a^\sigma + b^\sigma$, $(aba)^\sigma = a^\sigma b^\sigma a^\sigma$ and $1^\sigma = 1$.) This theorem was referred to as the beautiful theorem of Hua by E. Artin in his book "Geometric Algebra". From it Hua [88, 97] deduced also the fundamental theorem of 1-dimensional projective geometry over a skew field. In 1950 [97] he extended his theorem to semi-homomorphisms of rings without zero divisors.

Secondly, in 1949 L. K. Hua [91] gave a straightforward proof that "every proper normal subfield of a skew field is contained in its center". This result appears in the literature as the Cartan-Brauer-Hua theorem. Before the work of Hua and Richard Brauer, Henri Cartan's proof had used the complicated device of Galois extensions over subfields. By contrast, Hua's proof requires only the elementary identity: If $ab \neq ba$, then

$$a = \left(b^{-1} - (a-1)^{-1}b^{-1}(a-1)\right)\left(a^{-1}b^{-1}a - (a-1)^{-1}b^{-1}(a-1)\right)^{-1}.$$

In 1950, Hua [96] proved also that "if a skew field is not a field, then its multiplicative group is not meta-abelian".

Classical Groups. Early in 1946, L. K. Hua [73] published his first paper on automorphisms of classical groups, in which he determined the automorphisms of a real symplectic group. Subsequently, in 1948, he [85] determined the automorphisms of asymplectic group over any field of characteristic not 2. The method of Hua for determining the automorphisms of symplectic groups can be applied also to classical groups of other types; but since Dieudonné published his results on the automorphisms of classical groups in 1951, Hua [101] restricted himself to publishing only solutions, by his own method, to a series of problems left open by Dieudonné. The first of these was the determination of automorphisms of $GL_2(K)$, K being an arbitrary skew field of characteristic $\neq 2$.

Besides $GL_2(K)$, Hua [101] determined also the automorphisms of $SL_4(K)$ and $PSL_4(K)$, where K is a skew field of characteristic not 2, and the automorphism of $O_4^+(K, F)$, where K is a field of characteristic not 2 and f is a quadratic form of index 2. Afterwards, Hua and Z. X. Wan [105] determined the automorphisms of $SL_2(K)$ and $PSL_2(K)$, where K is a skew field of characteristic $\neq 0$, the automorphisms of $SL_4(K)$ and $PSL_4(K)$, where K is a skew field of characteristic 2, and they proved also the nonisomorphism of certain linear groups.

Hua's work on the automorphism of classical groups, shows mastery of the techniques of matrix calculation. The procedure is to start with the low-dimensional cases and to proceed to the higher dimensional cases by induction, as in [85], for $SP_{2n}(K)$.

About the structure of classical groups, Hua extended the usual unitary group to the case when the basic field is not necessarily commutative but has an involutive anti-automorphism. He proved that the group $TU_n(K_1S)$ generated by unitary transvections modulo its center is a simple group, if S has index $\geqslant 1$ and that $TU_n(K_1S)$ is the commutator subgroup of $U_n(K_1S)$, if the index of S satisfies $n \geqslant 2V \geqslant 4$.

Hua and I. Reiner [102, 106] also determined the automorphisms of $GL_n(\mathbb{Z})$ and $DGL_n(\mathbb{Z})$, which was the start of the work on the automorphisms of classical groups over rings. They [92] also proved that $GL_n(\mathbb{Z})$ is generated by three elements, $SL_n(\mathbb{Z})$ by two elements, and $Sp\,2n(\mathbb{Z})$ by four elements for $n \geqslant 2$. Formerly

Poincaré* had stated without proof that $\mathrm{Sp}\,2n(\mathbb{Z})$ is generated by elementary matrices of two simple types, and later Brahana† had proved this by showing that every element of $\mathrm{Sp}\,2n(\mathbb{Z})$ is expressible as a product of matrices taken from some finite set of matrices.

Geometry of Matrices. [67, 76, 77, 78, 93, 99] Study of this topic was initiated by Hua and relates to Siegel's work on fractional linear transformations. In it, the points of the space are matrices of a certain kind, for instance, rectangular matrices, symmetric matrices or skew-symmetric matrices of the same size. There is then a group of motions in this space, and the problem is to characterize the group of motions by as few geometric invariants as possible. First, he studied the geometry of matrices of various types over the complex or real fields. Later, he extended his results to the case when the basic field is not necessarily commutative and discovered that the invariant"coherence" is alone sufficient to characterize the group of motions of the space. Take his paper [99] as an example. He proved the fundamental theorem of affine geometry of rectangular matrices: *Let* $1 < n \leqslant m$. *Then the one-to-one mappings from the set of* $n \times m$ *matrices over a skew field* K *onto itself preserving coherence* (*two matrices* M *and* N *are said to be coherent, if the rank of* $M - N$ *is* 1) *is necessarily of the form*

$$Z_1 = PZ^{\sigma}Q + R \tag{1}$$

where $P = P^{(n)}$ *and* $Q = Q^{(m)}$ *are invertible matrices,* R *is an* $n \times m$ *matrix, and* σ *is an automorphism of* K; *if* $n = m$, *then besides* (1) *we have also*

$$Z_1 = PZ'^{\tau}Q + R,$$

where τ *is an anti-automorphism of* K. From this theorem he deduced the fundamental theorem of the projective geometry of rectangular matrices (the Grassmann space), and he determined the Jordan isomorphism of total matrix rings over skew fields of characteristic $\neq 2$ and the Lie isomorphism of total matrix rings over skew fields of characteristic $\neq 2, 3$.

Arising from the geometry of matrices and the theory of functions of several complex variables, Hua went on to study the classification problem of matrices; for instance, the classification of complex symmetric and skew-symmetric matrices under the unitary group, of a pair of Hermitian matrices under congruence [66], and of Hermitian matrices under the orthogonal group [76]. [*Editorial note*: By 'elementary divisors of a characteristic matrix' is meant, in current usage, 'Jordan blocks' in a Jordan normal form, in the sense that $(X - \alpha)^d$ is an elementary divisor of multiplicity m if and only if the Jordan form has exactly m blocks of $d \times d$

* Poincaré, H., *Rend. Circ. Mat. Palermo* **18** (1904), 45–110.

† Brahana, R. R., *Ann. of Math. (2)* **24** (1923), 265–270.

matrices

$$J_d(\alpha) = \begin{pmatrix} \alpha & 1 & & \\ & \alpha & 1 & 0 \\ & 0 & \ddots & 1 \\ & & & \alpha \end{pmatrix}.$$

In [76], on p. 509 four lines from the bottom read $\bar{\Gamma}'Q(\bar{\Gamma}')^{-1} = T$ for $\bar{\Gamma}Q(\bar{\Gamma}')^{-1} = T$, and on p. 512, line four, read $H = KT_0$ for $K = HT_0$.]

Reprinted from the
TRANSACTIONS OF THE AMERICAN MATHEMATICAL SOCIETY
Vol. 57, pp. 441–481, May 1945

GEOMETRIES OF MATRICES. I. GENERALIZATIONS OF VON STAUDT'S THEOREM

BY
LOO-KENG HUA

It was first shown in the author's recent investigations on the theory of automorphic functions of a matrix-variable that there are three types of geometry playing important roles. Besides their applications, the author obtained a great many results which seem to be interesting in themselves.

The main object of the paper is to generalize a theorem due to von Staudt, which is known as the fundamental theorem of the geometry in the complex domain. The statement of the theorem is:

Every topological transformation of the complex plane into itself, which leaves the relation of harmonic separation invariant, is either a collineation or an anti-collineation.

Since the fields and groups may be varied, several generalizations of von Staudt's theorem will be given. The proofs of the theorems have interesting corollaries.

The paper contains also some fundamental results which will be useful in succeeding papers.

The interest of the paper seems to be not only geometric but also algebraic; for example we shall establish the following purely algebraic theorem:

Let $\mathfrak{M}$ be the module formed by n-rowed symmetric matrices over the complex field. Let Γ be a continuous (additive) automorphism of $\mathfrak{M}$ leaving the rank unaltered and $\Gamma(iX)=i\Gamma(X)$. Then Γ is an inner automorphism of $\mathfrak{M}$, that is, we have a nonsingular matrix T such that

$$\Gamma(X) = TXT'.$$

The author makes the paper self-contained in the sense that no knowledge of the author's contributions to the theory of automorphic functions is assumed.

I. Geometry of symmetric matrices

Let Φ be any field. In I, II, and III, capital Latin letters denote $n\times n$ matrices unless the contrary is stated. But on the contrary, we use $M^{(n,m)}$ to denote an $n\times m$ matrix, and $M^{(n)}=M^{(n,n)}$. I and 0 denote the identity and zero matrices respectively.

Throughout I, we use

Presented to the Society, April 28, 1945; received by the editors November 20, 1944.

$$\mathfrak{F} = \begin{pmatrix} 0 & I \\ -I & 0 \end{pmatrix}, \qquad \mathfrak{T} = \begin{pmatrix} I & 0 \\ 0 & I \end{pmatrix},$$

which are $2n$-rowed matrices.

1. Definitions. We make the following definitions.

A pair of matrices (Z_1, Z_2) is said to be *symmetric* if

$$(Z_1, Z_2)\mathfrak{F}(Z_1, Z_2)' = 0,$$

that is, if $Z_1Z_2' = Z_2Z_1'$. The pair is said to be *nonsingular* if (Z_1, Z_2) is of rank n.

A $2n\times 2n$ matrix $\mathfrak{T}$ is said to be *symplectic* if

$$\mathfrak{T}\mathfrak{F}\mathfrak{T}' = \mathfrak{F}.$$

Explicitly, let

$$\mathfrak{T} = \begin{pmatrix} A & B \\ C & D \end{pmatrix},$$

then we have

$$AB' = BA', \qquad CD' = DC', \qquad AD' - BC' = I.$$

Further, it may be easily verified that

$$\mathfrak{T}^{-1} = \begin{pmatrix} D' & -B' \\ -C' & A' \end{pmatrix}$$

is also symplectic.

We define

$$(W_1, W_2) = Q(Z_1, Z_2)\mathfrak{T}$$

to be a *symplectic transformation*, where Q is nonsingular and $\mathfrak{T}$ is symplectic.

Since

$$(W_1, W_2)\mathfrak{F}(W_1, W_2)' = Q(Z_1, Z_2)\mathfrak{T}\mathfrak{F}\mathfrak{T}'(Z_1, Z_2)'Q',$$

a symplectic transformation carries symmetric (nonsingular) pairs into symmetric (nonsingular) pairs.

We identify two nonsingular symmetric pairs of matrices (Z_1, Z_2) and (W_1, W_2) by means of the relation

$$(Z_1, Z_2) = Q(W_1, W_2).$$

It is called a point of the space. The space so defined is unaltered under symplectic transformations, which may be considered as the motions of the space.

If Z_1 and W_1 are both nonsingular and if $(W_1, W_2) = Q(Z_1, Z_2)\mathfrak{T}$ let

$$W = -W_1^{-1}W_2, \qquad Z = -Z_1^{-1}Z_2,$$

then W and Z are both symmetric and

$$Z = (AW + B)(CW + D)^{-1}.$$

Thus a symmetric pair of matrices may be considered as homogeneous coordinates of a symmetric matrix. The terminology "geometry of symmetric matrices" is thus justified.

2. **Equivalence of points.**

THEOREM 1. *Any two nonsingular symmetric pairs of matrices are equivalent. Or what is the same thing: every nonsingular symmetric pair is equivalent to* $(I, 0)$.

Proof. Let (Z_1, Z_2) be a nonsingular symmetric pair.

(1) If Z_1 is nonsingular, we have

$$(Z_1, Z_2) = Z_1(I, Z_1^{-1}Z_2) = Z_1(I, 0)\begin{pmatrix} I & S \\ 0 & I \end{pmatrix},$$

where $S = Z_1^{-1}Z_2$ is symmetric, and then

$$\begin{pmatrix} I & S \\ 0 & I \end{pmatrix}$$

is symplectic.

(2) Suppose Z_1 to be singular. We have nonsingular matrices P and Q such that

$$W_1 = PZ_1Q = \begin{pmatrix} I^{(r)} & 0^{(r,n-r)} \\ 0^{(n-r,r)} & 0^{(n-r)} \end{pmatrix},$$

and

$$(W_1, W_2) = P(Z_1, Z_2)\begin{pmatrix} Q & 0 \\ 0 & Q'^{-1} \end{pmatrix},$$

and

$$W_2 = PZ_2Q'^{-1} = \begin{pmatrix} s^{(r)} & m^{(r,n-r)} \\ q^{(n-r,r)} & t^{(n-r)} \end{pmatrix}.$$

Since

$$\begin{pmatrix} Q & 0 \\ 0 & Q'^{-1} \end{pmatrix}$$

is symplectic, (W_1, W_2) is nonsingular and symmetric. Consequently s is symmetric and q is a zero matrix.

Let

$$(U_1, U_2) = (W_1, W_2)\begin{pmatrix} I & -S \\ 0 & I \end{pmatrix},$$

where

$$S = \begin{pmatrix} s^{(r)} & 0 \\ 0 & I^{(n-r)} \end{pmatrix}.$$

Then

$$U_1 = W_1, \qquad U_2 = -W_1S + W_2 = \begin{pmatrix} 0 & m \\ 0 & t \end{pmatrix}.$$

Since (U_1, U_2) is nonsingular, $t^{(n-r)}$ is nonsingular. Let

$$(V_1, V_2) = (U_1, U_2)\begin{pmatrix} I & 0 \\ I & I \end{pmatrix};$$

then

$$V_1 = U_1 + U_2 = \begin{pmatrix} I^{(r)} & m \\ 0 & t \end{pmatrix},$$

which is nonsingular. By (1), we have the theorem.

3. Equivalence of point-pairs.

DEFINITION. Let (Z_1, Z_2) and (W_1, W_2) be two nonsingular symmetric pairs of matrices. We define the rank of

$$(Z_1, Z_2)\mathfrak{F}(W_1, W_2)' = Z_1W_2' - Z_2W_1'$$

to be the *arithmetic distance* between the two points represented. Evidently, the notion is independent of the choice of representation. Further, it is invariant under symplectic transformations. In fact, let

$$(Z_1^*, Z_2^*) = Q(Z_1, Z_2)\mathfrak{T}, \qquad (W_1^*, W_2^*) = R(W_1, W_2)\mathfrak{T},$$

then

$$(Z_1^*, Z_2^*)\mathfrak{F}(W_1^*, W_2^*)' = Q(Z_1, Z_2)\mathfrak{T}\mathfrak{F}\mathfrak{T}'(W_1, W_2)'R' = Q(Z_1, Z_2)\mathfrak{F}(W_1, W_2)'R'.$$

In nonhomogeneous coordinates, the arithmetic distance between two symmetric matrices W, Z is equal to the rank of $W-Z$.

THEOREM 2. *Two point-pairs are equivalent if and only if they have the same arithmetic distance. What is the same thing: every point-pair with arithmetic distance r is equivalent to*

$$(I, 0), \qquad (I, I_r)$$

where

$$I_r = \begin{pmatrix} I^{(r)} & 0 \\ 0 & 0 \end{pmatrix}.$$

Proof. By Theorem 1, we may assume that the point-pairs are of the form

$$(I, 0), \qquad (Z_1, Z_2).$$

The arithmetic distance being r, it follows that Z_2 is of rank r. We have two nonsingular matrices P and Q such that

$$QZ_2P = \begin{pmatrix} I^{(r)} & 0 \\ 0 & 0 \end{pmatrix} = I_r.$$

Then

$$Q(Z_1, Z_2)\begin{pmatrix} P'^{-1} & 0 \\ 0 & P \end{pmatrix} = (T, I_r)$$

and

$$Q(I, 0)\begin{pmatrix} P'^{-1} & 0 \\ 0 & P \end{pmatrix} = QP'^{-1}(I, 0).$$

Since (T, I_r) is a nonsingular symmetric pair, we have, consequently,

$$T \doteq \begin{pmatrix} s^{(r)} & t \\ 0 & p^{(n-r)} \end{pmatrix},$$

where s is symmetric and p is nonsingular. Then

$$\begin{pmatrix} I^{(r)} & -tp^{-1} \\ 0 & p^{-1} \end{pmatrix}(T, I_r) = \left(\begin{pmatrix} s^{(r)} & 0 \\ 0 & I^{(n-r)} \end{pmatrix}, I_r\right).$$

Further,

$$\left(\begin{pmatrix} s^{(r)} & 0 \\ 0 & I^{(n-r)} \end{pmatrix}, I_r\right)\begin{pmatrix} I & 0 \\ \begin{pmatrix} I-s & 0 \\ 0 & 0 \end{pmatrix} & I \end{pmatrix} = (I, I_r)$$

and

$$(I, 0)\begin{pmatrix} I & 0 \\ \begin{pmatrix} I-s & 0 \\ 0 & 0 \end{pmatrix} & I \end{pmatrix} = (I, 0).$$

Since

$$\begin{pmatrix} I & 0 \\ \begin{pmatrix} I-s & 0 \\ 0 & 0 \end{pmatrix} & I \end{pmatrix}$$

is symplectic, we have the result.

DEFINITION. The points (X_1, X_2) with singular X_1 are called *points at infinity* (or symmetric matrices at infinity). Finite points are those with nonsingular X_1.

LEMMA. *Any finite number of points may be carried simultaneously into finite points by a symplectic transformation, if Φ is the field of complex numbers.*

Proof. (1) Given any symmetric pair of matrices (T_1, T_2), we have a symplectic matrix

$$\begin{pmatrix} P_1 & P_2 \\ T_1 & T_2 \end{pmatrix}.$$

In fact, by Theorem 2, we have a symplectic $\mathfrak{T}$ such that

$$(T_1, T_2) = Q(-I, 0)\mathfrak{T}.$$

Let

$$(P_1, P_2) = Q'^{-1}(0, I)\mathfrak{T}.$$

Then

$$\begin{pmatrix} P_1 & P_2 \\ T_1 & T_2 \end{pmatrix} = \begin{pmatrix} Q'^{-1} & 0 \\ 0 & Q \end{pmatrix} \begin{pmatrix} 0 & I \\ -I & 0 \end{pmatrix} \mathfrak{T}$$

which is evidently symplectic.

(2) For a fixed point (X_1, X_2), the manifold

$$\det((X_1, X_2)\mathfrak{F}(Z_1, Z_2)') = 0$$

is of dimension $n(n+1)-2$. Let

$$(A_1, A_2), \cdots, (L_1, L_2)$$

be p given points. Then we have p manifolds

$$\det((A_1, A_2)\mathfrak{F}(Z_1, Z_2)') = 0, \cdots, \det((L_1, L_2)\mathfrak{F}(Z_1, Z_2)') = 0.$$

In the space, there is a point (T_1, T_2) which is not on any one of the manifolds. The transformation

$$(Y_1, Y_2) = Q(X_1, X_2)\begin{pmatrix} P_1 & P_2 \\ T_1 & T_2 \end{pmatrix}^{-1} = Q(X_1, X_2)\begin{pmatrix} T_2' & -P_2' \\ -T_1' & P_1' \end{pmatrix}$$

carries evidently the p points into finite points simultaneously.

4. Equivalence of triples of points.

Definition 1. A subspace is said to be *normal* if it is equivalent to the subspace formed by symmetric matrices (in nonhomogeneous coordinates) of the form

$$\begin{pmatrix} Z_0^{(r)} & 0 \\ 0 & 0^{(n-r)} \end{pmatrix}.$$

The least possible r is defined to be the *rank of the subspace*.

Definition 2. A triple of points is said to be of *degeneracy* $d = n - r$ if it belongs to a normal subspace of rank r.

Evidently degeneracy is invariant under symplectic transformations.

Theorem 3. *In the complex field, two triples of points are equivalent if and only if they have the same degeneracy and the arithmetic distances between any two corresponding pairs of points are equal.*

Proof. Evidently, if two triples are equivalent, they have the same degeneracy and the arithmetic distances between any two corresponding pairs of points are equal.

We prove the converse in six steps.

(1) Every triple with arithmetic distances n, n, r is equivalent to

$$0,\ I,\ \begin{pmatrix} -I^{(r)} & 0 \\ 0 & 0 \end{pmatrix} \qquad \text{(in nonhomogeneous coordinates).}$$

(Notice that now the degeneracy is 0.) We use $r(A, B)$ to denote the arithmetic distance between A and B. Let A, B, C be the three points of the triple. Then

$$r(A, B) = r(A, C) = n.$$

By Theorem 2, we may write in homogeneous coordinates

$$A = (I, 0), \qquad B = (0, I), \qquad C = (Z_1, Z_2).$$

Since $r(A, C) = n$ and Z_2 is nonsingular, we may write C as

$$(S, I),$$

where S is a symmetric matrix of rank r. We have a nonsingular matrix Γ such that

$$\Gamma S \Gamma' = I_r = \begin{pmatrix} I^{(r)} & 0 \\ 0 & 0 \end{pmatrix},$$

then

$$\Gamma \begin{pmatrix} (I, 0) \\ (0, I) \\ (S, I) \end{pmatrix} \begin{pmatrix} \Gamma' & 0 \\ 0 & \Gamma^{-1} \end{pmatrix} = \begin{pmatrix} \Gamma\Gamma'(I, 0) \\ (0, I) \\ (I_r, I) \end{pmatrix}.$$

Thus the triple is equivalent to

$$(I, 0), \qquad (0, I), \qquad (I_r, I).$$

Since (in the nonhomogeneous coordinate-system)

$$0,\ I,\ -I_r$$

is a triple with distances n, n, r, we have the theorem.

(2) Every triple of points with arithmetic distances n, s, t is equivalent to

$$0,\ I,\ \begin{pmatrix} -I^{(p)} & 0 & 0 \\ 0 & 0 & 0 \\ 0 & 0 & I^{(q)} \end{pmatrix},$$

where $p+q=s$, $n-q=t$. (Obviously, $s+t \geqq n$.)

In fact, we may assume that

$$A = (I, 0), \qquad B = (I, I), \qquad C = (Z_1, Z_2).$$

We may determine two nonsingular matrices U, V such that

$$UZ_2V = \begin{pmatrix} I^{(r)} & 0 \\ 0 & 0 \end{pmatrix},$$

where r is the rank of Z_2. If we set

$$G = \begin{pmatrix} V'^{-1} & 0 \\ V - V'^{-1} & V \end{pmatrix}$$

the relations

$$U(I, 0)G = UV'^{-1}(I, 0),$$
$$U(I, I)G = UV(I, I),$$
$$U(Z_1, Z_2)G = \left(P, \begin{pmatrix} I^{(r)} & 0 \\ 0 & 0 \end{pmatrix}\right)$$

imply that we may assume that

$$Z_1 = P, \qquad Z_2 = \begin{pmatrix} I^{(r)} & 0 \\ 0 & 0 \end{pmatrix}.$$

Owing to the symmetry, we have

$$P = \begin{pmatrix} S^{(r)} & W \\ 0 & T \end{pmatrix},$$

where S is symmetric and T is nonsingular. Further, since

$$\begin{pmatrix} I & -WT^{-1} \\ 0 & T^{-1} \end{pmatrix}\left(\begin{pmatrix} S^{(r)} & W \\ 0 & T \end{pmatrix}, \begin{pmatrix} I^{(r)} & 0 \\ 0 & 0 \end{pmatrix}\right) = \left(\begin{pmatrix} S^{(r)} & 0 \\ 0 & I \end{pmatrix}, \begin{pmatrix} I^{(r)} & 0 \\ 0 & 0 \end{pmatrix}\right),$$

we may assume that

$$Z_1 = \begin{pmatrix} S^{(r)} & 0 \\ 0 & I \end{pmatrix}, \qquad Z_2 = \begin{pmatrix} I^{(r)} & 0 \\ 0 & 0 \end{pmatrix}.$$

In the normal subspace of rank r, the points $(I^{(r)}, 0^{(r)})$, $(I^{(r)}, I^{(r)})$, $(S^{(r)}, I^{(r)})$ are, by (1), equivalent to

$$(I^{(r)}, 0^{(r)}), \qquad (I^{(r)}, I^{(r)}), \qquad \left(I^{(r)}, \begin{pmatrix} -I^{(p)} & 0 \\ 0 & 0^{(r-p)} \end{pmatrix}\right).$$

Thus, we have, in nonhomogeneous coordinates,

$$\begin{pmatrix} I^{(r)} & 0 \\ 0 & 0^{(n-r)} \end{pmatrix}, \qquad \begin{pmatrix} 0^{(r)} & 0 \\ 0 & I^{(n-r)} \end{pmatrix}, \qquad \begin{pmatrix} -I^{(p)} & 0 & 0 \\ 0 & 0^{(r-p)} & 0 \\ 0 & 0 & 0^{(n-r)} \end{pmatrix}.$$

The transformation

$$\begin{pmatrix} I^{(r)} & 0 \\ 0 & iI^{(n-r)} \end{pmatrix}\left(Z - \begin{pmatrix} 0^{(r)} & 0 \\ 0 & I^{(n-r)} \end{pmatrix}\right)\begin{pmatrix} I^{(r)} & 0 \\ 0 & iI^{(n-r)} \end{pmatrix} = W$$

carries the three points to the required form.

(3) Now we are going to prove that any three points are equivalent to

$$A = 0, \qquad B = b_1 \dotplus \cdots \dotplus b_\lambda, \qquad C = c_1 \dotplus \cdots \dotplus c_\lambda\text{[1]},$$

where b_ν and c_ν are unit matrices of degree r_ν multiplied with a factor 1, 0, or -1. (1) and (2) are special cases of this. We shall consider another special case with

$$A = 0, \qquad B = \begin{pmatrix} 0 & M \\ M' & 0 \end{pmatrix}, \qquad C = \begin{pmatrix} 0 & N \\ N' & 0 \end{pmatrix},$$

where

$$M = \begin{pmatrix} 0 \cdots 0 \\ I^{(m)} \end{pmatrix}, \qquad N = \begin{pmatrix} I^{(m)} \\ 0 \cdots 0 \end{pmatrix}, \qquad n = 2m + 1.$$

They form a triple with distances $2m$, $2m$, $2m$.

Now we are going to establish that there exists a symmetric matrix S such that the transformation

$$W = Z(SZ + I)^{-1}$$

will carry the three points to

$$A = 0, \qquad B = \begin{pmatrix} 0 & 0 \\ 0 & B_1^{(n-1)} \end{pmatrix}, \qquad C = \begin{pmatrix} 1 & 0 \\ 0 & C_1^{(n-1)} \end{pmatrix},$$

where B_1 is nonsingular. In fact S is given by

$$\begin{bmatrix} 1 & 0 & -1 \\ 0 & 0 & 0 \\ -1 & 0 & 1 \end{bmatrix}, \qquad \begin{bmatrix} 1 & 0 & 0 & -1 & 0 \\ 0 & 0 & 0 & 0 & 0 \\ 0 & 0 & 0 & 0 & 0 \\ -1 & 0 & 0 & 1 & 0 \\ 0 & 0 & 0 & 0 & 0 \end{bmatrix},$$

and so on. The general form may be obtained easily. Applying the results obtained in (2) to

$$0^{(n-1)}, \qquad B_1^{(n-1)}, \qquad C_1^{(n-1)},$$

we have the conclusion.

(4) Let B, C be a nonsingular pair of symmetric matrices (in the ordinary sense), that is, we have λ and μ such that

$$\det(\lambda B + \mu C) \neq 0.$$

Suppose C is nonsingular; the conclusion announced in (3) is true by (2). Otherwise ($\lambda \neq 0$) we have Γ such that

(1) $\dotplus$ and $\sum'$ denote direct sums.

$$\Gamma'(\lambda B + \mu C)\Gamma = I,$$

$$\Gamma' C \Gamma = \begin{pmatrix} C_1^{(r)} & 0 \\ 0 & 0 \end{pmatrix}, \qquad C_1^{(r)} \text{ nonsingular.}$$

Then

$$\lambda \Gamma' B \Gamma = \begin{pmatrix} I^{(r)} - \mu C_1^{(r)} & 0 \\ 0 & I^{(n-r)} \end{pmatrix}.$$

Applying the results of (2) to

$$0, \qquad \frac{1}{\lambda}(I^{(r)} - \mu C_1^{(r)}), \qquad C_1^{(r)},$$

and

$$0, \qquad \frac{1}{\lambda} I^{(n-r)}, \qquad 0,$$

we have the result announced in (3).

(5) Finally, for any pair of symmetric matrices (cf. the lemma of §3)

$$B, \quad C$$

we have a nonsingular matrix Γ such that

$$\Gamma B \Gamma' = b_1 \dotplus \cdots \dotplus b_\lambda$$

and

$$\Gamma C \Gamma' = c_1 \dotplus \cdots \dotplus c_\lambda,$$

where

$$(b_\nu, c_\nu)$$

is either the pair discussed in (4) or the pair discussed in (3), hence the results in (3).

(6) By a rearrangement and some evident modifications, for a triple of points with degeneracy t, we have

$$\begin{aligned} A &= \quad 0^{(p)} \dotplus 0^{(q)} \dotplus 0^{(r)} \dotplus 0^{(s)} \dotplus 0^{(t)}, \\ B &= \quad I^{(p)} \dotplus 0^{(q)} \dotplus I^{(r)} \dotplus I^{(s)} \dotplus 0^{(t)}, \\ C &= -I^{(p)} \dotplus I^{(q)} \dotplus 0^{(r)} \dotplus I^{(s)} \dotplus 0^{(t)}, \end{aligned}$$

which is the only possible form. The arithmetic distances between two points are given by

$$\begin{aligned} a &= r(B, C) = p + q + r, \\ b &= r(C, A) = p + q + s, \\ c &= r(A, B) = p + r + s. \end{aligned}$$

Thus, for given t, a, b, c, if the equations are soluble, the solution is unique. We have therefore the theorem.

The conditions for solubility are

$$\begin{aligned} n - t &\geqq a, b, c, \\ a + b + c &\geqq 2(n - t). \end{aligned} \tag{1}$$

In terms of a "triangle" we have the following theorem.

THEOREM 4. *A triangle of degeneracy t with sides a, b, c exists if and only if* (1) *holds. If it exists, it is unique apart from equivalence.*

Incidentally, we have

$$a + b \geqq 2(n - t) - c \geqq c;$$

equality holds if and only if $c = a + b = n - t$.

The "triangle-relation"

$$a + b \geqq c, \qquad b + c \geqq a, \qquad c + a \geqq b$$

does not guarantee the existence of triangles with a given degeneracy, for example, $n = 2$, $t = 0$, $a = b = c = 1$. But we have the following theorem.

THEOREM 5. *Given the lengths of three sides a, b, c $(\leqq n)$, where the sum of every two is greater than the third one, there are λ non-equivalent triangles, where*

$$\lambda = \begin{cases} [(a + b + c)/2] - \max(a, b, c) + 1, & \text{for } n \geqq [(a + b + c)/2]\text{(2)}, \\ n - \max(a, b, c) + 1, & \text{for } n < [(a + b + c)/2]. \end{cases}$$

Proof. From $a+b \geqq c$, $b+c \geqq a$, $c+a \geqq b$, we have

$$a + b + c \geqq 2 \max(a, b, c).$$

There always exists a t such that

$$a + b + c \geqq 2(n - t) \geqq 2 \max(a, b, c).$$

Then

$$\max(0, n - [(a + b + c)/2]) \leqq t \leqq n - \max(a, b, c).$$

Thus, the number of t's is equal to

$$\begin{aligned} n - \max(a, b, c) - \max(0, n - [(a + b + c)/2]) + 1 \\ = \min(n, [(a + b + c)/2]) - \max(a, b, c) + 1. \end{aligned}$$

COROLLARY 1. *If one of the sides is of length n, the triangle is unique.*

COROLLARY 2. *If the sum of two sides is equal to the third, then the triangle is unique.*

(2) $[x]$ denotes the integral part of x.

5. Equivalence of quadruples of points.

DEFINITION. Let Z_1, Z_2, Z_3, Z_4 be four points in the nonhomogeneous coordinate-system. The matrix

$$(Z_1 - Z_3)(Z_1 - Z_4)^{-1}(Z_2 - Z_4)(Z_2 - Z_3)^{-1}$$

is defined to be the cross-ratio-matrix of the four points, and it is denoted by

$$(Z_1, Z_2; Z_3, Z_4).$$

It is defined only when $Z_1 - Z_4$ and $Z_2 - Z_3$ are nonsingular.

In the homogeneous coordinate-system, we let P_1, P_2, P_3, P_4 be four points with coordinates

$$(X_1, Y_1), \qquad (X_2, Y_2), \qquad (X_3, Y_3), \qquad (X_4, Y_4).$$

In terms of

$$\langle P_i, P_j\rangle = (X_j, Y_j)\mathfrak{F}(X_i, Y_i)',$$

the cross-ratio-matrix is defined by

$$(P_1, P_2; P_3, P_4) = \langle P_1, P_3\rangle\langle P_1, P_4\rangle^{-1}\langle P_2, P_4\rangle\langle P_2, P_3\rangle^{-1},$$

provided that it is not meaningless.

Let P_i^* be the point with coordinates

$$(X_i^*, Y_i^*) = Q_i(X_i, Y_i)\mathfrak{T},$$

where $\mathfrak{T}$ is symplectic; then

$$\begin{aligned}\langle P_i^*, P_j^*\rangle &= (X_j^*, Y_j^*)\mathfrak{F}(X_i^*, Y_i^*)' \\ &= Q_j(X_j, Y_j)\mathfrak{F}(X_i, Y_i)'Q_i' = Q_j\langle P_i, P_j\rangle Q_i'.\end{aligned}$$

Therefore

$$\begin{aligned}(P_1^*, P_2^*; P_3^*, P_4^*) &= \langle P_1^*, P_3^*\rangle\langle P_1^*, P_4^*\rangle^{-1}\langle P_2^*, P_4^*\rangle\langle P_2^*, P_3^*\rangle^{-1} \\ &= Q_3\langle P_1, P_3\rangle Q_1'Q_1'^{-1}\langle P_1, P_4\rangle Q_4 Q_4^{-1}\langle P_2, P_4\rangle Q_2'Q_2'^{-1}\langle P_2, P_3\rangle^{-1}Q_3^{-1} \\ &= Q_3(P_1, P_2; P_3, P_4)Q_3^{-1},\end{aligned}$$

and we now state the following theorem.

THEOREM 6. *In an algebraically closed field, two quadruples of points, no two of the points having arithmetic distance less than n, are equivalent if and only if their cross-ratio-matrices are equivalent.*

In order to prove Theorem 6, we need to establish the following theorem.

THEOREM 7. *In the algebraically closed field, any quadruple of points, no two of which have arithmetic distance less than n, is equivalent to*

$$0, \qquad \infty, \qquad \sum_{1\le i\le \nu}{}' a_i, \qquad \sum_{1\le i\le \nu}{}' b_i,$$

where

$$a_i = \begin{pmatrix} 0 \cdots 0 & 1 \\ 0 \cdots 1 & 0 \\ \cdots\cdots & . \\ 1 \cdots 0 & 0 \end{pmatrix}, \qquad b_i = \begin{pmatrix} 0 & 0 \cdots 0 & \lambda_i \\ 0 & 0 \cdots \lambda_i & 1 \\ . & \cdots\cdots & . \\ \lambda_i & 1 \cdots 0 & 0 \end{pmatrix}, \qquad \lambda_i \neq 0 \quad or \quad 1.$$

Proof. In homogeneous coordinates, we may write the four points as

$$(0, I), \qquad (I, 0), \qquad (Z_1, Z_2), \qquad (W_1, W_2).$$

Since no two of the arithmetic distances are less than n, Z_1, Z_2, W_1, W_2 are all nonsingular. We may write them in the nonhomogeneous coordinates as

$$0, \qquad \infty, \qquad S_1, \qquad S_2.$$

We have a nonsingular matrix T such that

$$TS_1T' = \sum{}' a_i, \qquad TS_2T' = \sum{}' b_i.$$

The theorem follows.

The proof of Theorem 6 is now evident.

REMARK. The equivalence of quadruples in any field seems to be more difficult. The condition in Theorem 6 is insufficient for the real case. (A signature system is required.)

DEFINITION. We define a quadruple of points satisfying

$$(P_1, P_2; P_3, P_4) = -I$$

to be a *harmonic range*.

Evidently a harmonic range is invariant under a symplectic transformation.

6. **Von Staudt's theorem in the complex number field.** Now we let Φ be the field formed by complex numbers.

We use $\overline{Z}$ to denote the conjugate complex matrix of Z. The transformation

$$(W_1, W_2) = Q(\overline{Z}_1, \overline{Z}_2)\mathfrak{T}$$

carrying a symmetric pair (W_1, W_2) into a symmetric pair (Z_1, Z_2) is called *anti-symplectic* if Q is nonsingular and $\mathfrak{T}$ symplectic.

THEOREM 8. *A transformation satisfying the following conditions:*

(1) *one-to-one and continuous,*

(2) *carrying symmetric matrices into symmetric matrices,*

(3) *keeping arithmetic distance invariant,*

(4) *keeping the harmonic relation invariant,*

is either a symplectic or an anti-symplectic transformation.

Proof. Let Γ be the transformation considered. Taking three points A, B,

C (symmetric matrices), no two of which have arithmetic distance less than n, let A_1, B_1, C_1 be their images. By (3), the arithmetic distance between any two of A_1, B_1, C_1 is n. Let $\mathfrak{T}_1$ and $\mathfrak{T}_2$ be two symplectic transformations carrying respectively A, B, C and A_1, B_1, C_1 into 0, I, ∞, in accordance with Theorem 3. Then, without loss of generality, we may assume that

$$0 = \Gamma(0), \qquad I = \Gamma(I), \qquad \infty = \Gamma(\infty).$$

Since

$$Z, \qquad Z_1, \qquad (Z + Z_1)/2, \qquad \infty$$

form a harmonic range, we have

$$\Gamma(Z) + \Gamma(Z_1) = \Gamma(Z + Z_1).$$

Consequently,

$$\Gamma(rZ) = r\Gamma(Z)$$

for all rational r. By continuity, this holds for all real r.

Now we introduce the following notations:

$$E_{ii} = (p_{st}), \qquad p_{st} = \begin{cases} 1 \text{ if } s = t = i, \\ 0 \text{ otherwise} \end{cases}$$

and

$$E_{ij} = (q_{st}), \qquad q_{st} = \begin{cases} 1 \text{ if } s = i,\ t = j \text{ or } s = j,\ t = i, \\ 0 \text{ otherwise.} \end{cases}$$

Let

$$\Gamma(E_{ii}) = M_i.$$

Since M_i is of rank 1 and symmetric, we have

$$M_i = (\lambda_{i1}, \cdots, \lambda_{in})'(\lambda_{i1}, \cdots, \lambda_{in}).$$

Let

$$\Lambda = (\lambda_{ij}).$$

Then

$$\begin{aligned} I = \Gamma(I) &= \sum_{i=1}^{n} \Gamma(E_{ii}) = \sum_{i=1}^{n} M_i \\ &= \sum_{i=1}^{n} (\lambda_{i1}, \cdots, \lambda_{in})'(\lambda_{i1}, \cdots, \lambda_{in}) \\ &= \sum_{i=1}^{n} (\lambda_{ij}\lambda_{ik}) = \left(\sum_{i=1}^{n} \lambda_{ij}\lambda_{ik}\right) \\ &= \Lambda'\Lambda. \end{aligned}$$

That is, Λ is an orthogonal matrix

$$(\lambda_{i1}, \cdots, \lambda_{in})\Lambda' = (\delta_{i1}, \cdots, \delta_{in}),$$

where δ_{ij} is Kronecker's delta. Thus

$$\Lambda\Gamma(E_{ii})\Lambda' = E_{ii}.$$

Let

$$\Delta(Z) = \Lambda\Gamma(Z)\Lambda',$$

then Δ has the same property as Γ, that is,

$$\Delta(Z + Z_1) = \Delta(Z) + \Delta(Z_1),$$
$$0 = \Delta(0), \qquad I = \Delta(I), \qquad \infty = \Delta(\infty)$$

and

$$E_{ii} = \Delta(E_{ii}).$$

Let

$$\Delta(E_{ij}) = M = (m_{st}), \qquad i \neq j,$$

M is of rank 2. Since

$$E_{ij} + \lambda E_{ii} + E_{jj}/\lambda$$

is of rank 1, owing to the invariance of arithmetic distance, the matrix

$$M + \lambda E_{ii} + E_{jj}/\lambda \tag{1}$$

is also of rank 1 for all λ. We are going to prove that $M = \pm E_{ij}$. In fact, we may assume that $i=1, j=2$. The two-rowed minor of (1)

$$\begin{vmatrix} m_{11} + \lambda & m_{12} \\ m_{12} & m_{22} + 1/\lambda \end{vmatrix} = 0$$

for any λ, that is

$$m_{11}m_{22} - m_{12}^2 + m_{11}/\lambda + m_{22}\lambda + 1 = 0,$$

that is

$$m_{11} = m_{22} = 0, \qquad m_{12} = \pm 1.$$

Further

$$\begin{vmatrix} m_{11} + \lambda & m_{1t} \\ m_{1t} & m_{tt} \end{vmatrix} = 0 \qquad t \geqq 3,$$

for all λ, then $m_{tt}=0$, $m_{1t}=0$ for all $t \geqq 3$. Finally

$$\begin{vmatrix} 0 & m_{st} \\ m_{st} & 0 \end{vmatrix} = 0 \quad \text{if} \quad (s, t) \neq (1, 2),$$

then $m_{st}=0$ for $(s, t) \neq (1, 2)$. Hence, we have

$$M = \pm E_{12}.$$

Thus

$$\Delta(E_{ij}) = \pm E_{ij}.$$

Let

$$D = [\epsilon_1, \cdots, \epsilon_n], \qquad \epsilon_\nu = \pm 1.$$

Then

$$D\Delta(E_{ij})D' = \pm \epsilon_i\epsilon_j E_{ij}.$$

Thus we may choose ϵ properly so that

$$D\Delta(E_{1i})D' = E_{1i}.$$

Let $D\Delta D' = \Pi$. Then Π has all properties of Δ and further

$$\Pi(E_{1i}) = E_{1i}. .$$

Now we consider E_{ij}. Without loss of generality we take $(i, j) = (2, 3)$. Then, if $\Pi(E_{23}) = -E_{23}$, we have

$$\Pi(E_{11} + E_{33} + E_{12} + E_{13} + E_{23}) = E_{11} + E_{33} + E_{12} + E_{13} - E_{23},$$

since

$$\begin{vmatrix} 1 & 1 & 1 \\ 1 & 0 & \epsilon \\ 1 & \epsilon & 1 \end{vmatrix} = -(\epsilon - 1)^2,$$

which is equal to zero for $\epsilon = 1$ and not zero for $\epsilon = -1$. Consequently the ranks of $E_{11}+E_{33}+E_{12}+E_{13}+E_{23}$ and $E_{11}+E_{33}+E_{12}+E_{13}-E_{23}$ are not equal. This is impossible.

Thus, we have

$$\Pi(X) = X$$

for all real X. (If we do not use continuity, it holds for all rational X.) We may assume Γ to be Π.

Further for real Y, the four points Yi, $-Yi$, Y, $-Y$, form a harmonic range, while

$$\Gamma(Y) = Y, \qquad \Gamma(-Y) = -Y,$$

thus we have

$$\Gamma(iY)Y^{-1} = -Y\Gamma(iY)^{-1}.$$

In particular,

$$(\Gamma(iI))^2 = -I.$$

Then

$$\Gamma(iI) = iJ$$

where J is an involutory symmetric matrix, that is $J^2 = I$ and $J = J'$. We have a matrix T (not necessarily orthogonal) such that

$$J = T'T.$$

Let

$$T'^{-1}\Gamma(Z)T^{-1} = \Phi(Z),$$

we have then

$$\Phi(iI) = iI.$$

Let

$$\nabla(Z) = - i\Phi(iZ).$$

Then

$$\nabla(0) = 0, \qquad \nabla(I) = I, \qquad \nabla(\infty) = \infty,$$

so the ranks of Z and $\nabla(Z)$ are equal. By the method used before, we have

$$\nabla_1 = B'\nabla B$$

such that

$$\nabla_1(X) = X,$$

for all real X. Thus, we have finally that

$$\Gamma(X + iY) = \Gamma(X) + \Gamma(iY) = X + iA'YA,$$

where A is independent of X and Y.

Now we have

$$A'YA\cdot Y^{-1} = Y(A'YA)^{-1},$$

that is,

$$(A'YAY^{-1})^2 = I,$$

for all real Y. Here we introduce a lemma.

Lemma. *Let A be a nonsingular matrix. If*

$$(A'YAY^{-1})^2 = I$$

for all symmetric Y then $A = \rho I$, where $\rho = \pm 1$ or $\pm i$.

If the lemma is true, then

$$\Gamma(X + iY) = X + iY \quad \text{or} \quad X - iY.$$

The theorem is proved.

Proof of the lemma. (1) We have a nonsingular matrix Γ such that

$$\Gamma^{-1}A\Gamma = B$$

and

$$B = J_1 \dotplus \cdots \dotplus J_\nu,$$

where J_i is a Jordan matrix of degree n_i. Evidently

$$(B'(\Gamma'Y\Gamma)B(\Gamma'Y\Gamma)^{-1})^2 = \Gamma'(A'YAY^{-1})^2\Gamma'^{-1} = I.$$

Thus it is sufficient to prove the theorem for B instead of A.

(2) We shall prove $n_i=1$. In fact, if

$$J_i = \begin{pmatrix} \lambda & 1 \cdots 0 \\ 0 & \lambda \cdots 0 \\ \cdot & \cdots\cdots \\ 0 & 0 \cdots \lambda \end{pmatrix},$$

then

$$(J_i'IJ_iI)_\cdot^2 = \begin{pmatrix} \lambda^2 & \lambda & 0 \cdots & 0 \\ \lambda & 1+\lambda^2 & \lambda \cdots & 0 \\ \cdot & \cdot & \cdots\cdots & \cdot \\ 0 & 0 & 0 \cdots & 1+\lambda^2 \end{pmatrix}^2 = \begin{pmatrix} \lambda^2(\lambda^2+1) & 2\lambda^3+\lambda & \lambda^2 \cdots \\ \cdot & \cdot & \cdots \\ \cdot & \cdot & \cdots \\ \cdot & \cdot & \cdots \end{pmatrix} = I,$$

which is impossible. Thus $n_i=1$, that is,

$$B = [\epsilon_1, \cdots, \epsilon_n], \qquad \epsilon_\nu \neq 0,$$

which is a diagonal matrix.

(3) Putting

$$Y = \begin{pmatrix} 1 & 1 & 0 & 0 \cdots 0 \\ 1 & 0 & 0 & 0 \cdots 0 \\ 0 & 0 & 1 & 0 \cdots 0 \\ 0 & 0 & 0 & 1 \cdots 0 \\ \cdot & \cdot & \cdot & \cdots\cdots \\ 0 & 0 & 0 & 0 \cdots 1 \end{pmatrix}$$

we have

$$\left(\begin{pmatrix} \epsilon_1 & 0 \\ 0 & \epsilon_2 \end{pmatrix}\begin{pmatrix} 1 & 1 \\ 1 & 0 \end{pmatrix}\begin{pmatrix} \epsilon_1 & 0 \\ 0 & \epsilon_2 \end{pmatrix}\begin{pmatrix} 0 & 1 \\ 1 & -1 \end{pmatrix}\right)^2 = \begin{pmatrix} \epsilon_1^2\epsilon_2^2 & 2\epsilon_1^2\epsilon_2(\epsilon_1-\epsilon_2) \\ 0 & \epsilon_1^2\epsilon_2^2 \end{pmatrix} = \begin{pmatrix} 1 & 0 \\ 0 & 1 \end{pmatrix},$$

which implies

$$\epsilon_1 = \epsilon_2.$$

Similarly

$$B = \epsilon I, \qquad \epsilon \neq 0.$$

Then $A=\epsilon I$. For $Y=I$, we have $\epsilon^4=1$, that is $\epsilon=\pm 1, \pm i$. The lemma is thus completely proved.

7. **Remarks.** The following results are contained in the proof of Theorem 8:

THEOREM 9. *Let Φ be the complex field, and let $\mathfrak{M}$ be a module formed by sym-*

metric matrices over Φ. Let Γ be an additive continuous automorphism of $\mathfrak{M}$ leaving the rank invariant, so that Γ satisfies

(i) $\Gamma(X)\in\mathfrak{M}$, *if* $X\in\mathfrak{M}$;

(ii) $\Gamma(X+Y)=\Gamma(X)+\Gamma(Y)$, *if* $X\in\mathfrak{M}$ *and* $Y\in\mathfrak{M}$;

(iii) $\Gamma(iX)=i\Gamma(X)$; *and*

(iv) $\Gamma(X)$ *has the same rank as* X.

Then $\Gamma(X)$ *is an inner automorphism, that is*

$$\Gamma(X) = TXT'$$

for certain T.

In the case of the real field the situation is more complicated. In Theorem 9, we require an additional condition that the signature of $\Gamma(X)$ is the same as that of X.

The analogue of Theorem 8 in the real field is more complicated. Since Theorem 7 is not true in case of the real field, degeneracy and lengths of sides do not characterize the equivalence of triples of points, for example, there does not exist a real symplectic transformation Γ satisfying

$$\Gamma(0) = 0, \qquad \Gamma(\infty) = \infty, \qquad \Gamma\begin{pmatrix}2 & 0\\ 0 & -2\end{pmatrix} = \begin{pmatrix}-2 & 0\\ 0 & -2\end{pmatrix}.$$

In fact the transformation satisfying the first two of these relations is of the form

$$\Gamma(Z) = CZC'$$

where C is nonsingular and real. It keeps the signature invariant. By means of the signature of a triple, we may obtain an analogue of Theorem 9 in the real field.

II. Geometry of skew-symmetric matrices

Throughout II, we use

$$\mathfrak{F}_1 = \begin{pmatrix}0 & I\\ I & 0\end{pmatrix}, \qquad \mathfrak{I} = \begin{pmatrix}I & 0\\ 0 & I\end{pmatrix}.$$

We let $n=2m$.

8. **Notions.** A pair of matrices (Z_1, Z_2) is said to be *skew-symmetric* if

$$(Z_1, Z_2)\mathfrak{F}_1(Z_1, Z_2)' = 0,$$

that is,

$$Z_1Z_2' = -Z_2Z_1'.$$

A $2n\times 2n$ matrix $\mathfrak{T}$ is said to be $\mathfrak{F}_1$-orthogonal, if

$$\mathfrak{T}\mathfrak{F}_1\mathfrak{T}' = \mathfrak{F}_1.$$

We define

$$(W_1, W_2) = Q(Z_1, Z_2)\mathfrak{T}$$

to be an $\mathfrak{F}_1$-orthogonal transformation if Q is nonsingular.

The transformation carries nonsingular skew-symmetric pairs into nonsingular skew-symmetric pairs.

The nonsingular skew-symmetric pair of matrices may be considered as the homogeneous coordinates of a skew-symmetric matrix.

It is easy to verify that the geometry so obtained (analogous to I) is *transitive*, that is, any two points of the space are equivalent.

We define the rank of

$$(Z_1, Z_2)\mathfrak{F}_1(W_1, W_2)'$$

to be the *arithmetic distance* between the two points represented by (Z_1, Z_2) and (W_1, W_2).

We have also:

Two point-pairs are equivalent if and only if they have the same arithmetic distance.

We may also define the cross-ratio-matrix of four points P_1, P_2, P_3, P_4, $P_i = (X_i, Y_i)$ $(i = 1, 2, 3, 4)$ by

$$\langle P_1, P_3\rangle\langle P_1, P_4\rangle^{-1}\langle P_2, P_4\rangle\langle P_2, P_3\rangle^{-1},$$

where

$$\langle P_i, P_j\rangle = (X_j, Y_j)\mathfrak{F}_1(X_i, Y_i)'.$$

The analogue of Theorem 6 is also true.

If the cross-ratio-matrix is equal to $-I$, we define P_1, P_2, P_3, P_4 to be a harmonic range.

9. **An algebraic theorem.** On the ground of similarity, the following statement seems to be true.

Let Γ be a continuous (additive) automorphism of the module formed by all skew-symmetric matrices, such that $\Gamma(iX) = i\Gamma(X)$, and that the rank is left invariant. Then $\Gamma(X)$ is an inner automorphism.

Unfortunately, this statement is false and so the situation becomes more complicated. For $n = 2$,

$$\Gamma\begin{pmatrix} 0 & a & b & c \\ -a & 0 & d & e \\ -b & -d & 0 & f \\ -c & -e & -f & 0 \end{pmatrix} = \begin{pmatrix} 0 & a & b & d \\ -a & 0 & c & e \\ -b & -c & 0 & f \\ -d & -e & -f & 0 \end{pmatrix}$$

is an automorphism but not an inner automorphism.

It is an automorphism of the required kind, since the principal minors form equal sets, say

$$(af - be + cd)^2;\ a^2, b^2, c^2, d^2, e^2, f^2.$$

It is not an inner automorphism. In fact, we write

$$\Gamma\begin{pmatrix} P & Q \\ -Q' & R \end{pmatrix} = \begin{pmatrix} P & Q' \\ -Q & R \end{pmatrix},$$

where P and R are two-rowed skew-symmetric matrices. Suppose it is an inner automorphism, that is, that there exists a nonsingular matrix

$$\begin{pmatrix} A & B \\ C & D \end{pmatrix}$$

such that

$$\begin{pmatrix} A & B \\ C & D \end{pmatrix}\begin{pmatrix} P & Q \\ -Q' & R \end{pmatrix}\begin{pmatrix} A & B \\ C & D \end{pmatrix}' = \begin{pmatrix} P & Q' \\ -Q & R \end{pmatrix}, \tag{1}$$

for all P, Q, R.

In particular, if $P=R=0$, $Q=I$, we have

$$\begin{pmatrix} A & B \\ C & D \end{pmatrix}\begin{pmatrix} 0 & I \\ -I & 0 \end{pmatrix}\begin{pmatrix} A & B \\ C & D \end{pmatrix}' = \begin{pmatrix} 0 & I \\ -I & 0 \end{pmatrix}. \tag{2}$$

Combining (1) and (2), we have

$$\begin{pmatrix} A & B \\ C & D \end{pmatrix}\begin{pmatrix} Q & -P \\ R & Q' \end{pmatrix} = \begin{pmatrix} Q' & -P \\ R & Q \end{pmatrix}\begin{pmatrix} A & B \\ C & D \end{pmatrix}.$$

Putting $P=R=0$, we have

$$AQ = Q'A, \qquad BQ' = Q'B, \qquad CQ = QC, \qquad DQ' = QD$$

for any Q. Consequently, we obtain

$$B = \beta I, \qquad C = \gamma I, \qquad A = D = 0.$$

But

$$\begin{pmatrix} 0 & \beta I \\ \gamma I & 0 \end{pmatrix}\begin{pmatrix} P & Q \\ -Q' & R \end{pmatrix}\begin{pmatrix} 0 & \gamma I \\ \beta I & 0 \end{pmatrix} = \begin{pmatrix} \beta^2 R & -\beta\gamma Q' \\ \beta\gamma Q & \gamma^2 P \end{pmatrix},$$

which, in general, is not equal to

$$\begin{pmatrix} P & Q' \\ -Q & R \end{pmatrix}.$$

Thus the automorphism is *not* an inner automorphism.

The above argument suggests that in general we might have $m-1$ basic automorphisms:

(i) $a_{14} \rightleftarrows a_{23}$, other elements invariant;

(ii) $a_{14} \rightleftarrows a_{23}$, $a_{16} \rightleftarrows a_{25}$, other elements invariant;

(iii) $a_{14}\rightleftarrows a_{23}$, $a_{16}\rightleftarrows a_{25}$, $a_{18}\rightleftarrows a_{27}$, other elements invariant;

. . . .

$(m-1)$ $a_{14}\rightleftarrows a_{23}$, $\cdots$, $a_{1,2m}\rightleftarrows a_{2,2m-1}$, other elements invariant.

Such a reasonable suggestion is a false one, since for $m\geqq 3$, "$a_{14}\rightleftarrows a_{23}$" does not keep the rank invariant, for example,

$$\Gamma\begin{pmatrix} 0 & 0 & 0 & 0 & 1 & 0 \\ 0 & 0 & 1 & 0 & 0 & 1 \\ 0 & -1 & 0 & 0 & -1 & 0 \\ 0 & 0 & 0 & 0 & 0 & 0 \\ -1 & 0 & 1 & 0 & 0 & 1 \\ 0 & -1 & 0 & 0 & -1 & 0 \end{pmatrix} = \begin{pmatrix} 0 & 0 & 0 & 1 & 1 & 0 \\ 0 & 0 & 0 & 0 & 0 & 1 \\ 0 & 0 & 0 & 0 & -1 & 0 \\ -1 & 0 & 0 & 0 & 0 & 0 \\ -1 & 0 & 1 & 0 & 0 & 1 \\ 0 & -1 & 0 & 0 & -1 & 0 \end{pmatrix}.$$

Here one matrix is singular while the other is nonsingular.

THEOREM 10. *Let Φ be the field of complex numbers. Let $\mathfrak{M}$ be the module formed by all skew-symmetric matrices over Φ. Let Γ be a continuous (additive) automorphism of Φ leaving the rank invariant and $\Gamma(iX)=i\Gamma(X)$. Then, for $m\neq 2$, Γ is an inner automorphism. For $m=2$ there exists a nonsingular matrix T such that*

$$\Gamma(X) = TX_\nu T'$$

where X_ν is either X or

$$\begin{pmatrix} 0 & a_{12} & a_{13} & a_{23} \\ -a_{12} & 0 & a_{14} & a_{24} \\ -a_{13} & -a_{14} & 0 & a_{34} \\ -a_{23} & -a_{24} & -a_{34} & 0 \end{pmatrix}.$$

Proof. (i) Evidently, the automorphisms

$$Y = TX_\nu T'$$

satisfy the requirement, where T is nonsingular.

(ii) The additive property may be stated as

$$\Gamma(X+Y) = \Gamma(X) + \Gamma(Y), \tag{1}$$

for any two X and Y belonging to $\mathfrak{M}$. Putting $X=Y=0$, we have

$$\Gamma(0) = 0. \tag{2}$$

It is also very easy to deduce that

$$\Gamma(rX) = r\Gamma(X) \tag{3}$$

for any rational r. By continuity, it holds for any real r. Since $\Gamma(iX)=i\Gamma(X)$, the relation holds for all complex r.

Let

$$A = \Gamma\left(\begin{pmatrix} 0 & 1 \\ -1 & 0 \end{pmatrix} \dotplus \cdots \dotplus \begin{pmatrix} 0 & 1 \\ -1 & 0 \end{pmatrix}\right),$$

where A is a nonsingular skew-symmetric matrix. There exists a matrix Q such that

$$QAQ' = \begin{pmatrix} 0 & 1 \\ -1 & 0 \end{pmatrix} \dotplus \cdots \dotplus \begin{pmatrix} 0 & 1 \\ -1 & 0 \end{pmatrix}.$$

Let $\Gamma_1(X) = Q\Gamma(X)Q'$, then Γ_1 is an automorphism satisfying the properties given in the theorem and

$$\Gamma_1(J) = J, \tag{4}$$

where

$$J = \begin{pmatrix} 0 & 1 \\ -1 & 0 \end{pmatrix} \dotplus \cdots \dotplus \begin{pmatrix} 0 & 1 \\ -1 & 0 \end{pmatrix}.$$

Write Γ instead of Γ_1. (In the following we shall repeat this procedure by the simple statement "we may let Γ satisfy (4).")

(iii) Let $\lambda_1, \cdots, \lambda_m$ be m distinct numbers, and let

$$A = \begin{pmatrix} 0 & \lambda_1 \\ -\lambda_1 & 0 \end{pmatrix} \dotplus \cdots \dotplus \begin{pmatrix} 0 & \lambda_m \\ -\lambda_m & 0 \end{pmatrix}.$$

Consider the two pairs of matrices A, J and $\Gamma(A)$, J. Since $\Gamma(A-\lambda J) = \Gamma(A) - \lambda J$, the characteristic roots of $\Gamma(A-\lambda J)$ are also $\lambda_1, \cdots, \lambda_m$. (Each is a double root.) We have a nonsingular matrix M such that

$$M\Gamma(A)M' = A, \qquad MJM' = J. \tag{5}$$

Now we are going to prove that M can be chosen independent of the λ's. Write $M = M_{\lambda_1, \cdots, \lambda_m}$. We have

$$\Gamma\left(\begin{pmatrix} 0 & 0 \\ 0 & 0 \end{pmatrix} \dotplus \cdots \dotplus \begin{pmatrix} 0 & 1 \\ -1 & 0 \end{pmatrix} \dotplus \cdots \dotplus \begin{pmatrix} 0 & 0 \\ 0 & 0 \end{pmatrix}\right)$$

$$= M_i\left(\begin{pmatrix} 0 & 0 \\ 0 & 0 \end{pmatrix} \dotplus \cdots \dotplus \begin{pmatrix} 0 & 1 \\ -1 & 0 \end{pmatrix} \dotplus \cdots \dotplus \begin{pmatrix} 0 & 0 \\ 0 & 0 \end{pmatrix}\right) M_i' \text{ [3]},$$

where $M_i = M_{\lambda_1, \cdots, \lambda_m}$ with $\lambda_i = 1$ and $\lambda_j = 0$ for $j \neq i$. In this expression, only the $(2i-1)$th and $2i$th columns are significant. Let P be a matrix having $(2i-1)$th and $2i$th columns in common with M_i for $i = 1, 2, \cdots, n$. Then

$$\Gamma(A) = PAP'.$$

Putting $\lambda_1 = \cdots = \lambda_m = 1$, we find that P is nonsingular.

---

(3) The term different from the zero-matrix is the ith term of the sum.

Now we may let

$$\Gamma(A) = A, \tag{6}$$

where

$$A = \begin{pmatrix} 0 & \lambda_1 \\ -\lambda_1 & 0 \end{pmatrix} \dotplus \cdots \dotplus \begin{pmatrix} 0 & \lambda_m \\ -\lambda_m & 0 \end{pmatrix}.$$

(iv) The theorem is evident for $m=1$.

Now we take $m=2$. Let

$$\Gamma \begin{pmatrix} 0 & a & b & c \\ -a & 0 & d & e \\ -b & -d & 0 & f \\ -c & -e & -f & 0 \end{pmatrix} = \begin{pmatrix} 0 & a' & b' & c' \\ -a' & 0 & d' & e' \\ -b' & -d' & 0 & f' \\ -c' & -e' & -f' & 0 \end{pmatrix}.$$

Since

$$\begin{vmatrix} 0 & a-\lambda & b & c \\ -a+\lambda & 0 & d & e \\ -b & -d & 0 & f-\mu \\ -c & -e & -f+\mu & 0 \end{vmatrix} = 0,$$

that is, $(a-\lambda)(f-\mu)-be+dc=0$, if and only if

$$\begin{vmatrix} 0 & a'-\lambda & b' & c' \\ -a'+\lambda & 0 & d' & e' \\ -b' & -d' & 0 & f'-\mu \\ -c' & -e' & -f'+\mu & 0 \end{vmatrix} = 0,$$

we have

$$a = a', \quad f = f', \tag{7}$$

$$be - cd = b'e' - c'd'. \tag{8}$$

Now we consider

$$\Gamma \begin{pmatrix} \cdot & \begin{matrix} 1 & 0 \\ 0 & 1 \end{matrix} \\ * & \cdot \end{pmatrix} = \begin{pmatrix} \cdot & M \\ * & \cdot \end{pmatrix},$$

$$\Gamma \begin{pmatrix} \cdot & \begin{matrix} 1 & 0 \\ 0 & 0 \end{matrix} \\ * & \cdot \end{pmatrix} = \begin{pmatrix} \cdot & M_1 \\ * & \cdot \end{pmatrix},$$

where "·" stands for zero-matrix and "*" stands for a matrix which either is evident or has no essential significance in the consideration. (This convention will be retained for the rest of the paper.) Then M is of rank 2, M_1 and $M-M_1$ are of ranks less than or equal to 1, that is

$$|M_1 - \lambda M| = 0$$

has two characteristic roots 0 and 1. We can find two matrices P and Q of determinant 1, such that

$$PMQ = \begin{pmatrix} 1 & 0 \\ 0 & 1 \end{pmatrix}, \qquad PM_1Q = \begin{pmatrix} 1 & 0 \\ 0 & 0 \end{pmatrix}.$$

Therefore

$$\begin{pmatrix} P & 0 \\ 0 & Q' \end{pmatrix}\begin{pmatrix} \cdot & M \\ * & \cdot \end{pmatrix}\begin{pmatrix} P' & 0 \\ 0 & Q \end{pmatrix} = \begin{bmatrix} \cdot & \begin{matrix} 1 & 0 \\ 0 & 1 \end{matrix} \\ * & \cdot \end{bmatrix},$$

and

$$\begin{pmatrix} P & 0 \\ 0 & Q' \end{pmatrix}\begin{pmatrix} \cdot & M_1 \\ * & \cdot \end{pmatrix}\begin{pmatrix} P' & 0 \\ 0 & Q \end{pmatrix} = \begin{bmatrix} \cdot & \begin{matrix} 1 & 0 \\ 0 & 0 \end{matrix} \\ * & \cdot \end{bmatrix}.$$

Since

$$\begin{pmatrix} P & 0 \\ 0 & Q' \end{pmatrix}\begin{bmatrix} \begin{matrix} 0 & \lambda \\ -\lambda & 0 \end{matrix} & \cdot \\ \cdot & \begin{matrix} 0 & \mu \\ -\mu & 0 \end{matrix} \end{bmatrix}\begin{pmatrix} P' & 0 \\ 0 & Q \end{pmatrix} = \begin{bmatrix} \begin{matrix} 0 & \lambda \\ -\lambda & 0 \end{matrix} & \cdot \\ \cdot & \begin{matrix} 0 & \mu \\ -\mu & 0 \end{matrix} \end{bmatrix},$$

we may let

$$\Gamma \begin{bmatrix} \cdot & \begin{matrix} b & 0 \\ 0 & e \end{matrix} \\ * & \cdot \end{bmatrix} = \begin{bmatrix} \cdot & \begin{matrix} b' & 0 \\ 0 & e' \end{matrix} \\ * & \cdot \end{bmatrix}.$$

We deduce easily that

(9) $$b = b', \qquad e = e'.$$

From (8) we have

$$cd = c'd'.$$

In particular, we have

$$\Gamma \begin{bmatrix} \cdot & \begin{matrix} 0 & 1 \\ 1 & 0 \end{matrix} \\ * & \cdot \end{bmatrix} = \begin{bmatrix} \cdot & \begin{matrix} 0 & c' \\ d' & 0 \end{matrix} \\ * & \cdot \end{bmatrix}, \qquad c'd' = 1.$$

Since

$$\begin{pmatrix} (d')^{1/2} & 0 & 0 & 0 \\ 0 & (d')^{-1/2} & 0 & 0 \\ 0 & 0 & (d')^{-1/2} & 0 \\ 0 & 0 & 0 & (d')^{1/2} \end{pmatrix} \begin{pmatrix} 0 & a & b & c' \\ -a & 0 & d' & e \\ -b & -d' & 0 & f \\ -c' & -e & -f & 0 \end{pmatrix} \begin{pmatrix} (d')^{1/2} & 0 & 0 & 0 \\ 0 & (d')^{-1/2} & 0 & 0 \\ 0 & 0 & (d')^{-1/2} & 0 \\ 0 & 0 & 0 & (d')^{1/2} \end{pmatrix}$$

$$= \begin{pmatrix} 0 & a & b & 1 \\ -a & 0 & 1 & e \\ -b & -1 & 0 & f \\ -1 & -e & -f & 0 \end{pmatrix}$$

we may let

$$\Gamma \begin{pmatrix} 0 & 0 & 0 & 1 \\ 0 & 0 & 1 & 0 \\ 0 & -1 & 0 & 0 \\ -1 & 0 & 0 & 0 \end{pmatrix} = \begin{pmatrix} 0 & 0 & 0 & 1 \\ 0 & 0 & 1 & 0 \\ 0 & -1 & 0 & 0 \\ -1 & 0 & 0 & 0 \end{pmatrix}.$$

Finally, we have

$$\Gamma \begin{pmatrix} 0 & 0 & 0 & 1 \\ 0 & 0 & 0 & 0 \\ 0 & 0 & 0 & 0 \\ -1 & 0 & 0 & 0 \end{pmatrix} = \begin{pmatrix} 0 & 0 & 0 & \lambda \\ 0 & 0 & \mu & 0 \\ 0 & -\mu & 0 & 0 \\ -\lambda & 0 & 0 & 0 \end{pmatrix}.$$

Then

$$\left| \begin{pmatrix} 0 & \lambda \\ \mu & 0 \end{pmatrix} - k \begin{pmatrix} 0 & 1 \\ 1 & 0 \end{pmatrix} \right| = 0$$

for $k=1$ and 0, so we have either

$$\Gamma(X) = X$$

or

$$\Gamma(X) = X_1.$$

(v) For the sake of simplicity, we give the proof for $m=3$. The method is valid for any m.

Let $M=(a_{ij})$, $M_1=(a'_{ij})$ and

$$\Gamma(M) = M_1.$$

Since, by (iii),

$$\Gamma \begin{pmatrix} \cdot & & \cdot & & \cdot & \\ & & 0 & \lambda & & \\ \cdot & & -\lambda & 0 & \cdot & \\ & & & & 0 & \mu \\ \cdot & & \cdot & & -\mu & 0 \end{pmatrix} = \begin{pmatrix} \cdot & & \cdot & & \cdot & \\ & & 0 & \lambda & & \\ \cdot & & -\lambda & 0 & \cdot & \\ & & & & 0 & \mu \\ \cdot & & \cdot & & -\mu & 0 \end{pmatrix},$$

the determinants of

$$M-\begin{pmatrix} \cdot & \cdot & \cdot \\ & \begin{matrix} 0 & \lambda \\ -\lambda & 0 \end{matrix} & \\ \cdot & & \cdot \\ & & \begin{matrix} 0 & \mu \\ -\mu & 0 \end{matrix} \\ \cdot & \cdot & \end{pmatrix} \quad \text{and} \quad M_1-\begin{pmatrix} \cdot & \cdot & \cdot \\ & \begin{matrix} 0 & \lambda \\ -\lambda & 0 \end{matrix} & \\ \cdot & & \cdot \\ & & \begin{matrix} 0 & \mu \\ -\mu & 0 \end{matrix} \\ \cdot & \cdot & \end{pmatrix}$$

are identically equal. Comparing the coefficients of $\lambda^2\mu^2$, we find $a_{12}=a'_{12}$. Similarly, we deduce

$$a_{34}=a'_{34}, \qquad a_{56}=a'_{56}.$$

Now we let

$$\Gamma\begin{pmatrix} & \begin{matrix} 1 & 0 \\ 0 & 1 \end{matrix} & \\ \cdot & & \cdot \\ * & \cdot & \cdot \\ \cdot & \cdot & \cdot \end{pmatrix} = M_1.$$

Since for $\lambda\mu=1$

$$\left|\begin{pmatrix} \cdot & \begin{pmatrix} 1 & 0 \\ 0 & 1 \end{pmatrix} \\ * & \cdot \end{pmatrix} \dot{-} \begin{pmatrix} \begin{matrix} 0 & \lambda \\ -\lambda & 0 \end{matrix} & \cdot \\ \cdot & \begin{matrix} 0 & \mu \\ -\mu & 0 \end{matrix} \end{pmatrix}\right| = 0,$$

it follows that

$$M_1-\begin{pmatrix} \begin{matrix} 0 & \lambda \\ -\lambda & 0 \end{matrix} & \cdot & \cdot \\ \cdot & \begin{matrix} 0 & \mu \\ -\mu & 0 \end{matrix} & \cdot \\ \cdot & \cdot & \cdot \end{pmatrix}$$

is of rank not greater than 2 for all λ, μ satisfying $\lambda\mu=1$. Thus we have

$$\begin{vmatrix} 0 & \lambda & a'_{13} & a'_{14} \\ -\lambda & 0 & a'_{23} & a'_{24} \\ & & 0 & \mu \\ & * & -\mu & 0 \end{vmatrix} = \begin{vmatrix} 0 & \lambda & a'_{13} & a'_{15} \\ -\lambda & 0 & a'_{23} & a'_{25} \\ & & 0 & a'_{35} \\ & * & -a'_{35} & 0 \end{vmatrix} = \begin{vmatrix} 0 & \lambda & a'_{13} & a'_{16} \\ -\lambda & 0 & a'_{23} & a'_{26} \\ & & 0 & a'_{36} \\ & * & -a'_{36} & 0 \end{vmatrix},$$

$$= \begin{vmatrix} 0 & \lambda & a'_{14} & a'_{15} \\ -\lambda & 0 & a'_{24} & a'_{25} \\ & & 0 & a'_{45} \\ & * & -a'_{45} & 0 \end{vmatrix} = \begin{vmatrix} 0 & \lambda & a'_{14} & a'_{16} \\ -\lambda & 0 & a'_{24} & a'_{26} \\ & & 0 & a'_{46} \\ & * & -a'_{46} & 0 \end{vmatrix} = \begin{vmatrix} 0 & \lambda & a'_{15} & a'_{16} \\ -\lambda & 0 & a'_{25} & a'_{26} \\ & * & & \cdot \end{vmatrix} = 0.$$

The first equation gives

$$\begin{vmatrix} a'_{13} & a'_{14} \\ a'_{23} & a'_{24} \end{vmatrix} = 1,$$

so we may take

$$\begin{pmatrix} a'_{13} & a'_{14} \\ a'_{23} & a'_{24} \end{pmatrix} = \begin{pmatrix} 1 & 0 \\ 0 & 1 \end{pmatrix}.$$

(In fact, we can choose a suitable Q such that

$$\begin{bmatrix} I & 0 & 0 \\ 0 & Q & 0 \\ 0 & 0 & I \end{bmatrix} M \begin{bmatrix} I & 0 & 0 \\ 0 & Q' & 0 \\ 0 & 0 & I \end{bmatrix}$$

has the required form.)

Then, from the system of equations, we have

$$a'_{35} = a'_{36} = a'_{45} = a'_{46} = a'_{15} = a'_{16} = a'_{25} = a'_{26} = 0.$$

Thus, we may let

$$\Gamma \begin{bmatrix} \cdot & \begin{matrix} 1 & 0 \\ 0 & 1 \end{matrix} & \cdot \\ * & \cdot & \cdot \\ \cdot & \cdot & \cdot \end{bmatrix} = \begin{bmatrix} \cdot & \begin{matrix} 1 & 0 \\ 0 & 1 \end{matrix} & \cdot \\ * & \cdot & \cdot \\ \cdot & \cdot & \cdot \end{bmatrix}.$$

Let

$$\Gamma \begin{bmatrix} \cdot & \begin{matrix} 1 & 0 \\ 0 & 0 \end{matrix} & \cdot \\ * & \cdot & \cdot \\ \cdot & \cdot & \cdot \end{bmatrix} = \begin{bmatrix} \cdot & P & * \\ * & \cdot & * \\ * & * & \cdot \end{bmatrix} = M_1.$$

Since M_1 is of rank 2, we have $|P| = 0$. Since

$$\begin{bmatrix} \cdot & \begin{matrix} 1 & 0 \\ 0 & 1 \end{matrix} & \cdot \\ * & \cdot & \cdot \\ \cdot & \cdot & \cdot \end{bmatrix} - M_1 \tag{10}$$

is of rank 2, we have $|P-I| = 0$. There is also a matrix Q such that

$$QPQ^{-1} = \begin{pmatrix} 1 & 0 \\ 0 & 0 \end{pmatrix}.$$

Thus we may assume that

$$\Gamma\begin{pmatrix} \cdot & \begin{matrix}1 & 0\\ 0 & 0\end{matrix} & \cdot \\ * & \cdot & \cdot \\ \cdot & \cdot & \cdot \end{pmatrix} = \begin{pmatrix} \cdot & \begin{matrix}1 & 0\\ 0 & 0\end{matrix} & \begin{matrix} a'_{15} & a'_{16} \\ a'_{25} & a'_{26} \end{matrix} \\ * & \cdot & \begin{matrix} a'_{35} & a'_{36} \\ a'_{45} & a'_{46} \end{matrix} \\ * & * & \cdot \end{pmatrix}.$$

By consideration of the ranks of the previous matrix and the matrix given in (10) we find

$$a'_{15} = a'_{16} = a'_{25} = a'_{26} = a'_{35} = a'_{36} = a'_{45} = a'_{46} = 0.$$

Consider again a general skew-symmetric matrix M and its image $M_1 = \Gamma(M)$. By the same method used for $m=2$, we have either

$$M_1 = \begin{pmatrix} 0 & a_{12} & a_{13} & a_{14} & a'_{15} & a'_{16} \\ & 0 & a_{23} & a_{24} & a'_{25} & a'_{26} \\ & & 0 & a_{34} & a'_{35} & a'_{36} \\ & & & 0 & a'_{45} & a'_{46} \\ & * & & & 0 & a'_{56} \\ & & & & & 0 \end{pmatrix}$$

or

$$M_1 = \begin{pmatrix} 0 & a_{12} & a_{13} & a_{23} & a'_{15} & a'_{16} \\ & 0 & a_{14} & a_{24} & a'_{25} & a'_{26} \\ & & 0 & a_{34} & a'_{35} & a'_{36} \\ & & & 0 & a'_{45} & a'_{46} \\ & * & & & 0 & a'_{56} \\ & & & & & 0 \end{pmatrix}.$$

Repeating the process for

$$\begin{pmatrix} a_{15} & a_{16} \\ a_{25} & a_{26} \end{pmatrix},$$

we obtain either

$$a_{25} \rightleftarrows a_{25}, \qquad a_{16} \rightleftarrows a_{16},$$

or

$$a_{25} \rightleftarrows a_{16}, \qquad a_{16} \rightleftarrows a_{25}.$$

For the equivalence of "$a_{14} \rightleftarrows a_{23}$" and "$a_{16} \rightleftarrows a_{25}$," we have three cases: (α) $\Gamma(M) = M_1$, M_1 is obtained by replacing a_{35}, a_{36}, a_{45}, a_{46} by a'_{35}, a'_{36}, a'_{45}, a'_{46} in M;

$$(\beta) \qquad \Gamma(M) = M_2 = \begin{pmatrix} 0 & a_{12} & a_{13} & a_{23} & a_{15} & a_{16} \\ & 0 & a_{14} & a_{24} & a_{25} & a_{26} \\ & & 0 & a_{34} & a'_{35} & a'_{36} \\ & & & 0 & a'_{45} & a'_{46} \\ & * & & & 0 & a_{56} \\ & & & & & 0 \end{pmatrix},$$

$$(\gamma) \qquad \Gamma(M) = M_3 = \begin{pmatrix} 0 & a_{12} & a_{13} & a_{23} & a_{15} & a_{25} \\ & 0 & a_{14} & a_{24} & a_{16} & a_{26} \\ & & 0 & a_{34} & a'_{35} & a'_{36} \\ & & & 0 & a'_{45} & a'_{46} \\ & * & & & 0 & a_{56} \\ & & & & & 0 \end{pmatrix}.$$

(α) We leave a_{35}, a_{36}, a_{45}, a_{46} arbitrary. Putting $a_{16}=a_{24}=a_{34}=1$, $a_{25}=x$, and the others equal to 0, we have $\det(M)=(x-a_{35})^2$, $\det(M_1)=(x-a'_{35})^2$. We have $\det(M)=0$ if and only if $\det(M_1)=0$, that is $a_{35}=a'_{35}$. Next putting $a_{16}=a_{23}=a_{34}=1$, $a_{25}=x$, and the others equal to 0, we obtain $a_{45}=a'_{45}$. Putting $a_{15}=a_{24}=a_{14}=1$, $a_{26}=x$, and the others equal to 0 and putting $a_{15}=a_{23}=a_{34}=1$, $a_{26}=x$ and the others equal to 0, we have respectively

$$a_{36} = a'_{36}, \qquad a_{46} = a'_{46}.$$

Thus we have

$$\Gamma(M) = M.$$

(β) and (γ) Putting $a_{24}=a_{34}=1$, $a_{16}=a_{25}=x$ and others equal to 0, we have $a_{35}=a'_{35}$. Further, if we put

$$a_{12} = a_{13} = 0, \quad a_{14} = x, \quad a_{15} = 1, \quad a_{16} = 0, \quad a_{23} = y, \quad a_{24} = a_{25} = 0, \quad a_{26} = 1,$$
$$a_{34} = 0, \quad a_{35} = -1, \quad a_{36} = a_{45} = a_{46} = 0, \quad a_{56} = 1,$$

then

$$d(M) = (x(y-1))^2, \qquad d(M_1) = (y(x-1))^2.$$

By putting $x=-1$, $y=+1$, we see that this is impossible.

The general proof may be arranged in the following steps:

(a) Dividing the matrix into m^2 2-rowed matrices.

(b) Choosing the first row of the small matrices as in the case $m=3$ and applying the analogous method as above to the image.

(c) Determining the other small matrices by the method given for $m=3$ (from (10) et seq.).

(d) Considering the 6-rowed minors we find that the exceptional case appearing for $m=2$ cannot exist for $m\geqq 3$.

10. Another generalization of von Staudt's theorem. The transformation

$$(W_1,W_2)=Q(\bar{Z}_1,\bar{Z}_2)\mathfrak{T}$$

is called *anti-orthogonal* if Q is nonsingular and $\mathfrak{T}$ is $\mathfrak{F}_1$-orthogonal.

LEMMA. *Let A be a nonsingular matrix. Suppose that*

$$(A'YAY^{-1})^2 = I$$

holds for all skew-symmetric Y. For $m=1$, A is a matrix of determinant ± 1. For $m>1$, then

$$A = \rho I, \quad \text{or} \quad A = \rho T[1,\cdots,1,-1]T^{-1},$$

where $\rho=\pm 1$, $\pm i$, and conversely.

Proof. (i) The result is evident for $m=1$.

For $m>1$, $A=\rho I$ evidently satisfies the equation. Now we prove that $A=\rho T[1,\cdots,1,-1]T^{-1}$ satisfies the equation.

We write

$$T'YT = \begin{pmatrix} Y_1^{(n-1)} & v \\ -v' & 0 \end{pmatrix}, \qquad (T'YT)^{-1} = \begin{pmatrix} Y_1^* & v^* \\ -v^{*\prime} & 0 \end{pmatrix}.$$

Then

$$\begin{pmatrix} I & 0 \\ 0 & I \end{pmatrix} = \begin{pmatrix} Y_1 & v \\ -v' & 0 \end{pmatrix}\begin{pmatrix} Y_1^* & v^* \\ -v^{*\prime} & 0 \end{pmatrix} = \begin{pmatrix} Y_1Y_1^* - vv^{*\prime} & Y_1v^* \\ -v'Y_1^* & -v'v^* \end{pmatrix}.$$

Further

$$\begin{aligned} T'A'YAY^{-1}T'^{-1} &= \rho^2[1,\cdots,1,-1]\begin{pmatrix} Y_1 & v \\ -v' & 0 \end{pmatrix}[1,\cdots,1,-1]\begin{pmatrix} Y_1^* & v^* \\ -v^{*\prime} & 0 \end{pmatrix} \\ &= \rho^2\begin{pmatrix} Y_1Y_1^* + vv^{*\prime} & Y_1v^* \\ v'Y_1^* & v'v^* \end{pmatrix} = \rho^2\begin{pmatrix} I + 2vv^{*\prime} & 0 \\ 0 & -1 \end{pmatrix}. \end{aligned}$$

Since $v^{*\prime}v=-1$, we have

$$(I^{(n-1)} + 2vv^{*\prime})^2 = I + 4vv^{*\prime} + 4vv^{*\prime}vv^{*\prime} = I^{(n-1)};$$

hence the result.

(ii) As in the proof of the lemma of §6, we may assume that

$$A = J_1 \dot{+} J_2 \dot{+} \cdots,$$

where J_ν is again of degree n_ν. The number of odd n_ν's is always even.

(iii) We consider the case

$$A = J^{(n)},$$

where n is even. Write

$$J = \begin{pmatrix} p & q \\ 0 & p \end{pmatrix}, \qquad p = p^{(n/2)}, \qquad q = q^{(n/2)},$$

where

$$p = \begin{bmatrix} \lambda & 1 & 0 \cdots 0 \\ 0 & \lambda & 1 \cdots 0 \\ \cdot & \cdot & \cdots\cdots \\ 0 & 0 & 0 \cdots \lambda \end{bmatrix}, \qquad q = \begin{bmatrix} 0 & 0 \cdots 0 \\ 0 & 0 \cdots 0 \\ \cdot & \cdots\cdots \\ 1 & 0 \cdots 0 \end{bmatrix}.$$

Take

$$Y = \begin{pmatrix} 0 & I^{(n/2)} \\ -I^{(n/2)} & 0 \end{pmatrix};$$

then

$$I = (A'YAY^{-1})^2 = \left(\begin{pmatrix} p' & 0 \\ q' & p' \end{pmatrix}\begin{pmatrix} 0 & I \\ -I & 0 \end{pmatrix}\begin{pmatrix} p & q \\ 0 & p \end{pmatrix}\begin{pmatrix} 0 & -I \\ I & 0 \end{pmatrix}\right)^2$$
$$= \begin{pmatrix} (p'p)^2 & * \\ * & * \end{pmatrix}$$

implies $(p'p)^2 = I^{(n/2)}$, which is impossible for $n > 2$ (that is $m > 1$).

(iv) Let

$$A = J_1 \dotplus J_2, \qquad n_1 + n_2 = n$$

where n_1 and n_2 are both odd. Let $n_1 = n_2$. Write

$$A = \begin{pmatrix} p & 0 \\ 0 & q \end{pmatrix}, \qquad Y = \begin{pmatrix} 0 & I^{(n/2)} \\ -I & 0 \end{pmatrix}.$$

Then, we have

$$(A'YAY^{-1})^2 = \begin{pmatrix} (p'q)^2 & * \\ * & * \end{pmatrix} = I,$$

which is possible only for $n_1 = n_2 = 1$. Further let $n_1 > n_2$. We write

$$A = \begin{pmatrix} p & q \\ 0 & r \end{pmatrix}, \qquad Y = \begin{pmatrix} 0 & I^{(n/2)} \\ -I & 0 \end{pmatrix}.$$

Then

$$(A'YAY^{-1})^2 = \begin{pmatrix} (p'r)^2 & * \\ * & * \end{pmatrix} = I,$$

which is impossible for $n_1>3$. For $n_1=3$ and $n_2=1$, we have consequently

$$A = \begin{pmatrix} \lambda & 1 & 0 & 0 \\ 0 & \lambda & 1 & 0 \\ 0 & 0 & \lambda & 0 \\ 0 & 0 & 0 & \mu \end{pmatrix}, \qquad \mu = -\lambda, \qquad \lambda^4 = 1.$$

Taking

$$Y = \begin{pmatrix} \cdot & & 2 & 1 \\ & & 1 & 1 \\ * & & \cdot & \end{pmatrix},$$

we find this also to be impossible.

Thus each of the numbers

$$n_1, n_2, \cdots$$

must be either 1 or 2.

(v) It is easily seen that no two of the n_ν's can be 2. In fact

$$A = \begin{pmatrix} \lambda_1 & 1 & 0 & 0 \\ 0 & \lambda_1 & 0 & 0 \\ 0 & 0 & \lambda_2 & 1 \\ 0 & 0 & 0 & \lambda_2 \end{pmatrix} = \begin{pmatrix} p & \cdot \\ \cdot & q \end{pmatrix},$$

say. Taking

$$Y = \begin{pmatrix} 0 & I \\ -I & 0 \end{pmatrix},$$

we find this to be impossible.

(vi) Further, if one of the n_ν's is 2, then n is equal to 2. In fact, suppose that

$$A = \begin{pmatrix} \lambda_1 & 1 & & \\ 0 & \lambda_1 & & \\ & & \lambda_2 & 0 \\ & & 0 & \lambda_3 \end{pmatrix}.$$

Taking

$$Y = \begin{pmatrix} 0 & I \\ -I & 0 \end{pmatrix},$$

we have

$$\lambda_2 = -\lambda_3, \qquad \lambda_1^2\lambda_2^2 = 1.$$

Taking further

$$Y=\begin{pmatrix} 0 & 0 & 1 & 1 \\ 0 & 0 & 1 & 0 \\ -1 & -1 & 0 & 0 \\ -1 & 0 & 0 & 0 \end{pmatrix},$$

we have $\lambda_2=\lambda_3$. Both results cannot hold simultaneously.

(vii) Suppose $n\geqq 4$. Let,

$$A = [\lambda_1, \lambda_2, \lambda_3, \lambda_4].$$

Taking

$$Y=\begin{pmatrix} 0 & 0 & 1 & 1 \\ 0 & 0 & 1 & 0 \\ -1 & -1 & 0 & 0 \\ -1 & 0 & 0 & 0 \end{pmatrix},$$

we have

$$(1)\ (A'YAY^{-1})^2=\begin{pmatrix} \lambda_1^2\lambda_4^2 & \lambda_1(\lambda_1\lambda_4+\lambda_2\lambda_3)(\lambda_3-\lambda_4) & \cdot & \cdot \\ \cdot & \lambda_2^2\lambda_3^2 & \cdot & \cdot \\ \cdot & \cdot & \lambda_2^2\lambda_3^2 & \lambda_3(\lambda_1\lambda_4+\lambda_2\lambda_3)(\lambda_1-\lambda_2) \\ \cdot & \cdot & \cdot & \lambda_1^2\lambda_4^2 \end{pmatrix}=I$$

which implies

$$\lambda_1^2\lambda_4^2 = \lambda_2^2\lambda_3^2 = 1.$$

Since $\lambda_1, \lambda_2, \lambda_3, \lambda_4$ can be permuted, we have

$$\lambda_i^2\lambda_j^2 = 1 \qquad \text{for all } i, j\ (i \neq j).$$

Thus

$$\lambda_1^4 = 1, \qquad \lambda_1^2 = \lambda_2^2 = \lambda_3^2 = \lambda_4^2 = \pm 1.$$

In general $A=[\lambda_1, \cdots, \lambda_n]$ where $\lambda_1^2= \cdots =\lambda_n^2=\pm 1$. By choosing a suitable ρ in the lemma, we may consider the case with

$$\lambda_1^2 = \lambda_2^2 = \cdots = \lambda_n^2 = 1.$$

If among the λ's there occur two positive and two negative numbers, we take $\lambda_1=\lambda_4=1$, $\lambda_2=\lambda_3=-1$. Then (1) is impossible. Thus we have the lemma.

THEOREM 11. *A transformation satisfying the following conditions:*

(1) *one-to-one and continuous,*

(2) *carrying skew-symmetric matrices into skew-symmetric matrices,*

(3) *keeping arithmetic distance invariant,*
and
(4) *keeping the harmonic relation invariant,*
is for $n\neq 4$ *either* $\mathfrak{F}_1$*-orthogonal or anti-orthogonal. In the case* $n=4$, *the transformation is either* $\mathfrak{F}_1$*-orthogonal or anti-orthogonal, or is equivalent to*

$$\Gamma(Z) = Z_1, \quad \text{where} \quad Z = \begin{pmatrix} p & q \\ -q' & r \end{pmatrix}, \qquad Z_1 = \begin{pmatrix} p & q' \\ -q & r \end{pmatrix},$$

or equivalent to

$$\Gamma(Z) = \bar{Z}_1.$$

Proof. (i) A triple of points, no two of which have arithmetic distance less less than n, is equivalent to

$$0, \quad \begin{pmatrix} 0 & I^{(m)} \\ -I^{(m)} & 0 \end{pmatrix}, \quad \infty.$$

We may let the transformation satisfy

$$\Gamma(0) = 0, \qquad \Gamma(\mathfrak{F}) = \mathfrak{F}, \qquad \Gamma(\infty) = \infty.$$

(ii) As in the symmetric case, we have

$$\Gamma(Z) + \Gamma(Z_1) = \Gamma(Z + Z_1).$$

Again, for $n\neq 4$, we have, analogous to the symmetric case,

$$\Gamma(X + iY) = X + iA'YA.$$

Since Y, $-Y$ are separated harmonically by iY and $-iY$ for real Y, we have

$$\Gamma(Yi)Y^{-1} = -Y(\Gamma(Yi))^{-1},$$

that is,

$$(A'YAY^{-1})^2 = I$$

for all real Y. Now we suppose $m\geqq 3$(4); then we have

$$\Gamma(X + iY) = X \pm iY$$

or

$$\Gamma(X + iY) = X \pm iT'^{-1}[1, \cdots 1, -1]T'YT[1, \cdots, 1, -1]T^{-1}.$$

The first case is what we require. Changing variables in the second case we may let

$$\Gamma(X + iY) = X \pm i[1, \cdots, 1, -1]Y[1, \cdots, 1, -1].$$

Since

(4) For $m=1$, the result is almost evident.

$$\begin{vmatrix} 0 & 0 & 1 & 1 \\ 0 & 0 & i & i \\ \cdot & & & \cdot \end{vmatrix} = 0 \quad \text{and} \quad \begin{vmatrix} 0 & 0 & 1 & 1 \\ 0 & 0 & i & -i \\ \cdot & & & \cdot \end{vmatrix} \neq 0$$

the rank is not invariant. The last case does not satisfy our requirement.

In case $m=2$, a great deal of special consideration is needed. Apart from the lemma, we require the solutions of

$$(A'Y_1AY^{-1})^2 = I,$$

where

$$Y = \begin{pmatrix} P & Q \\ -Q' & R \end{pmatrix}, \quad Y_1 = \begin{pmatrix} P & Q' \\ -Q & R \end{pmatrix}.$$

The proof of the lemma establishes that either

$$A = \rho I$$

or

$$A = \rho T[1, 1, 1, -1]T^{-1}.$$

(The Q used here is always symmetric.)

As in the preceding proof we have four cases,

$$\Gamma(X+iY) = X \pm iA'YA, \quad \text{or} \quad X_1 \pm iA'Y_1A',$$
$$\text{or} \quad X_1 \pm iA'YA, \quad \text{or} \quad X \pm iA'Y_1A.$$

By the previous argument, for $m \geqq 3$, we have, for the first two cases,

$$\Gamma(Z) = Z, \bar{Z}, Z_1, \bar{Z}_1,$$

where $Z=X+iY$, and Z_1 is obtained from Z by the process yielding Y_1 from Y.

Next, if $\Gamma(X+iY)=X_1 \pm iA'YA$, we have either

$$\Gamma(X+iY) = X_1 \pm iY \quad \text{or} \quad X_1 \pm i[1, 1, 1, -1]Y[1, 1, 1, -1].$$

Putting

$$Z = \begin{pmatrix} 0 & 0 & i & i \\ 0 & 0 & 1 & 1 \\ -i & -1 & 0 & 0 \\ -i & -1 & 0 & 0 \end{pmatrix},$$

$$\Gamma(Z) = \begin{pmatrix} 0 & 0 & i & i+1 \\ 0 & 0 & 0 & 1 \\ -i & 0 & 0 & 0 \\ -i-1 & -1 & 0 & 0 \end{pmatrix} \quad \text{or} \quad \begin{pmatrix} 0 & 0 & i & -i+1 \\ 0 & 0 & 0 & 1 \\ -i & 0 & 0 & 0 \\ i-1 & -1 & 0 & 0 \end{pmatrix},$$

we see that both these automorphisms could render a singular matrix nonsingular. Thus both these cases are ruled out.

Finally, the possibility of

$$\Gamma(X + iY) = X \pm iAY_1A'$$

may be treated in a similar way.

III. Geometry of Hermitian matrices

The geometry of symmetric matrices in the real domain is closely analogous to the present geometry.

11. **Notions.** We define an *Hermitian pair* (Z_1, Z_2) of matrices by

$$(\bar{Z}_1, \bar{Z}_2)\mathfrak{F}(Z_1, Z_2)' = 0.$$

A conjunctively symplectic matrix $\mathfrak{T}$ is defined by[5]

$$\mathfrak{T}^*\mathfrak{F}\mathfrak{T}' = \mathfrak{F}.$$

We define a conjunctively symplectic transformation by

$$(W_1, W_2) = Q(Z_1, Z_2)\mathfrak{T}.$$

We identify two nonsingular Hermitian pairs of matrices (Z_1, Z_2) and (W_1, W_2) by means of the relation

$$(Z_1, Z_2) = Q(W_1, W_2).$$

It is called a point of the space. Evidently, the space so defined is transitive. The rank of

$$(\bar{W}_1, \bar{W}_2)\mathfrak{F}(Z_1, Z_2)'$$

is defined to be the arithmetic distance of the points (W_1, W_2) and (Z_1, Z_2). Two pairs of points are equivalent if and only if they have the same arithmetic distance.

Let P_1, P_2, P_3 be three points no two of which have arithmetic distance less than n. Then they are equivalent to the three points

$$0, \infty, K(= [1, \cdots, 1, -1, \cdots, -1]).$$

The signature of K is defined to be the signature of the range P_1, P_2, P_3. (The order of points is significant.)

Evidently the signature is invariant under the group. We also may say that two triples of points are in the same sense if they have the same signature. We may prove that if two ranges are in the same sense, there is a conjunctively symplectic transformation carrying one into the other.

As to the equivalence of quadruples of points, a great deal of difficulty arises from the fact that the existing treatments of the theory of Hermitian

(5) $\mathfrak{T}^*$ denotes the conjugate complex matrix of $\mathfrak{T}$.

forms are incomplete. We shall give elsewhere a complete classification and then its application to the present problem will be immediate.

12. A further generalization of von Staudt's theorem.

THEOREM 12. *Let* Γ *be an additive continuous automorphism of the module formed by all Hermitian matrices keeping rank and signature invariant. Then* Γ *is either an inner automorphism or an anti-automorphism* $(Z \to \overline{PZP'})$.

Proof. (Cf. the results of §6.) (i) We have $\Gamma(0)=0$. Let

$$\Gamma(I) = H_0.$$

Since H_0 is positive definite, we may let

$$\Gamma(I) = I.$$

As in the proof of Theorem 8 (in the real field), we may let

$$\Gamma(X) = X$$

for all real symmetric X.

Let Y be any real skew-symmetric matrix, and let

$$\Gamma(iY) = H.$$

Since

$$\det (X + iY) = 0$$

if and only if

$$\det (X + H) = 0,$$

by Hilbert's theorem on polynomial ideals, we have an integer ρ such that

$$(\det (X + iY))^\rho \equiv 0 \ (\text{mod} \det (X + H))$$

and

$$(\det (X + H))^\rho \equiv 0 \ (\text{mod} \det (X + iY)),$$

in the polynomial ring formed from the real field by adjunction of the elements of X. Let $X = X' = (x_{ij})$. We write

$$\det (X + iY) = f_1 x_{11} + g_1,$$
$$\det (X + H) = f_2 x_{11} + g_2,$$

where f_1, f_2, g_1, g_2 are elements in the ring $\mathfrak{R}$ (generated by the elements of X omitting x_{11}). Since

$$(f_2(f_1 x_{11} + g_1) - f_1(f_2 x_{11} + g_2))^\rho \equiv 0 \ (\text{mod} \det (X + H)),$$

and since $f_2 g_1 - f_1 g_2$ is independent of x_{11}, we have

$$f_2 g_1 - f_1 g_2 = 0.$$

Since the determinant of a Hermitian matrix is an irreducible polynomial in its elements, f_1 and f_2 are irreducible and f_1 and g_1 have no common divisor. Consequently we have

$$f_1 = f_2, \qquad g_1 = g_2.$$

In this procedure, we have to compare one of the coefficients.

Thus

$$\det(X + iY) = \det(X + H).$$

Consequently each principal minor of iY equals the corresponding principal minor of H. We complete the proof with the aid of the following lemma.

LEMMA. *If two Hermitian matrices H and K have the same principal minors of orders* 1, 2, 3, *then* (*for* $\exp(x) = e^x$)

$$H = [\exp(i\theta_1), \cdots, \exp(i\theta_n)]K^*[\exp(-i\theta_1), \cdots, \exp(-i\theta_n)],$$

where K^* is obtained from K by replacing k_{rs} by either k_{rs} or $\bar{k}_{rs}$.

From the lemma, we may let

$$h_{rs} = \pm iy_{rs}.$$

Since one of

$$\begin{pmatrix} 2 & i & -i \\ -i & 0 & 1 \\ i & 1 & 0 \end{pmatrix} \quad \text{and} \quad \begin{pmatrix} 2 & i & -\bar{i} \\ -i & 0 & 1 \\ \bar{i} & 1 & 0 \end{pmatrix}$$

is singular and the other is not, we have

$$H = \pm iY.$$

The proof of the lemma is straightforward. Considering the 1-rowed principal minors of $H = (h_{rs})$, $K = (k_{rs})$, we have

$$h_{rr} = k_{rr}.$$

Since

$$\begin{vmatrix} h_{rr} & h_{rs} \\ \bar{h}_{rs} & h_{ss} \end{vmatrix} = \begin{vmatrix} h_{rr} & k_{rs} \\ \bar{k}_{rs} & h_{ss} \end{vmatrix},$$

we have

$$|h_{rs}|^2 = |k_{rs}|^2.$$

We may choose $\theta_1, \cdots, \theta_n$ such that the matrix

$$[\exp(i\theta_1), \cdots, \exp(i\theta_n)]H[\exp(-i\theta_1), \cdots, \exp(-i\theta_n)]$$

has real $h_{12}, h_{13}, \cdots, h_{1n}$. We may let

$$h_{1i} = k_{1i}$$

be real and positive. Consider

$$\begin{vmatrix} h_{11} & h_{1i} & h_{ij} \\ \bar{h}_{1i} & h_{ii} & h_{ij} \\ \bar{h}_{1j} & \bar{h}_{ij} & h_{jj} \end{vmatrix} = \begin{vmatrix} h_{11} & h_{1i} & h_{1j} \\ \bar{h}_{1i} & k_{ii} & k_{ij} \\ \bar{h}_{1j} & \bar{k}_{ij} & h_{jj} \end{vmatrix}.$$

We have

$$h_{ij} + \bar{h}_{ij} = k_{ij} + \bar{k}_{ij};$$

then letting $h_{ij}=\alpha+\beta i$, $k_{ij}=\gamma+\delta i$, we have

$$\alpha^2 + \beta^2 = \gamma^2 + \delta^2, \qquad \alpha = \gamma.$$

Thus $\beta = \pm\delta$ and we have

$$h_{ij} = k_{ij} \quad \text{or} \quad \bar{k}_{ij}. \qquad Q.E.D.$$

THEOREM 13. *A transformation satisfying the following conditions:*

(1) *one-to-one and continuous,*

(2) *carrying Hermitian matrices into Hermitian matrices,*

(3) *keeping arithmetic distance invariant,*

(4) *keeping sense of triples of points invariant,*

(5) *keeping the harmonic relation invariant,*

is either a conjunctively symplectic or a conjunctively anti-symplectic transformation.

The proof is omitted because of the similarity to the real analogue of Theorem 8 (cf. §7).

IV. GEOMETRY OF RECTANGULAR MATRICES

13. Subgeometries of the geometry of unitary matrics. The geometry studied in III may also be interpreted as the geometry of *unitary matrices*. Since the matrix

$$\mathfrak{F} = i\begin{pmatrix} 0 & I \\ -I & 0 \end{pmatrix}$$

is of signature 0, we may use

$$\mathfrak{F} = \begin{pmatrix} I & 0 \\ 0 & -I \end{pmatrix}$$

instead of it. Then the pair of matrices (Z_1, Z_2), satisfying

$$(\bar{Z}_1, \bar{Z}_2)\begin{pmatrix} I & 0 \\ 0 & -I \end{pmatrix}(Z_1, Z_2)' = 0,$$

that is

$$\bar{Z}_1 Z_1' - \bar{Z}_2 Z_2' = 0, \qquad I - (\bar{Z}_1^{-1}\bar{Z}_2)(Z_1^{-1}Z_2)' = 0,$$

is the homogeneous representation of a unitary matrix. We can generalize the idea a little further. Let

$$\mathfrak{F}_2 = \begin{pmatrix} I^{(n)} & 0 \\ 0 & -I^{(m)} \end{pmatrix}, \qquad n \geqq m.$$

The matrix $\Gamma^{(m+n)}$ satisfying

$$\bar{\Gamma}\mathfrak{F}_2\Gamma' = \mathfrak{F}_2$$

is called conjunctive with signature (n, m). The pair $(Z_1^{(n)}, Z_2^{(n,m)})$ of matrices satisfying

$$(\bar{Z}_1, \bar{Z}_2)\mathfrak{F}_2(Z_1, Z_2)' = 0$$

is called an (n, m)-unitary pair.

Instead of going into the details of this geometry we shall be content to make the following remark.

Let

$$W_1^{(n,m)} = Z_1^{-1} Z_2.$$

Then we have

$$\bar{W}_1 W_1' = I.$$

W_1 is formed by m columns of a unitary matrix. Thus the geometry may be considered as a subgeometry of the geometry of unitary matrices by identifying the elements with the same m columns as an element of the subgeometry. This may be described in short as "the process of projection."

14. **Remarks.**

(i) The condition "one-to-one" is redundant, since the invariance of arithmetic distance implies it.

(ii) The continuity for the real case is also very probably redundant. (Cf. Sierpinski's contribution to the solution of the functional equation $f(x+y) = f(x)+f(y)$.)

(iii) The geometry of pairs of matrices $(Z_1^{(n)}, Z_2^{(n,m)})$ with the group given in IV has interesting applications to the study of automorphic functions. It is *not* an analogue of projective geometry but of non-Euclidean geometry.

(iv) Analogous to IV, we may establish a geometry of real rectangular matrices.

Finally the author should like to express his warmest thanks to the referee for his help with the manuscript.

National Tsing Hua University,
Kunming, Yunnan, China.

Reprinted from the
TRANSACTIONS OF THE AMERICAN MATHEMATICAL SOCIETY
Vol. 57, pp. 482–490, May 1945

GEOMETRIES OF MATRICES. I_1
ARITHMETICAL CONSTRUCTION

BY
LOO-KENG HUA

A discussion of the redundancy of the conditions involved in the generalizations of von Staudt's Theorem should be preceded by a study of the involutions. As an illustration and a supplement to part I, we give here a discussion of the geometry of 2-rowed symmetric matrices; we intend to treat the general case at a later occasion. More definitely, the condition concerning the harmonic separation is a consequence of the invariance of arithmetic distance. For the real case, even continuity (as well as the condition concerning the signature) is redundant (a proof for even order symmetric matrices has been obtained). As to the general discussion, some knowledge concerning involutions seems to be indispensable; the author will come back to it later.

Throughout the paper, the notations in I are taken over and we assume that $n=2$.

1. Normal subspaces.

THEOREM 1. *Given two matrices Z_1 and Z_2 with arithmetic distance $r(Z_1, Z_2) = 1$, the points Z satisfying*

$$r(Z, Z_1) \leqq 1, \qquad r(Z, Z_2) \leqq 1$$

form a normal subspace.

Proof. Without loss of generality, we may take

$$Z_1 = \begin{pmatrix} 1 & 0 \\ 0 & 0 \end{pmatrix}, \qquad Z_2 = \begin{pmatrix} -1 & 0 \\ 0 & 0 \end{pmatrix}.$$

Let

$$Z = \begin{pmatrix} x & y \\ y & z \end{pmatrix}.$$

Then we have

$$(x \pm 1)z - y^2 = 0,$$

that is, $z=y=0$. The theorem is now evident.

DEFINITION. The normal subspace obtained in Theorem 1 is said to be spanned by Z_1 and Z_2.

DEFINITION. Two normal subspaces are said to be *complementary* if there is one and only one pair of matrices, one from each subspace, with arithmetic distance less than 2.

Presented to the Society, April 28, 1945; received by the editors January 2, 1945.

THEOREM 2. *Two complementary subspaces may be carried simultaneously to*

$$\begin{pmatrix} x & 0 \\ 0 & 0 \end{pmatrix}, \quad \begin{pmatrix} 0 & 0 \\ 0 & y \end{pmatrix}.$$

Proof. We may let

$$\begin{pmatrix} 1 & 0 \\ 0 & 0 \end{pmatrix}, \quad \begin{pmatrix} -1 & 0 \\ 0 & 0 \end{pmatrix}$$

be the matrices to span the first subspace, and

$$\begin{pmatrix} 0 & 0 \\ 0 & 1 \end{pmatrix}, \quad X$$

be the matrices to span the second subspace. In fact, by a theorem of I, we may carry any three points A, B, C with

$$r(A, B) = 1, \qquad r(A, C) = r(B, C) = 2,$$

into

$$A_1 = \begin{pmatrix} 1 & 0 \\ 0 & 0 \end{pmatrix}, \quad B_1 = \begin{pmatrix} -1 & 0 \\ 0 & 0 \end{pmatrix}, \quad C_1 = \begin{pmatrix} 0 & 0 \\ 0 & 1 \end{pmatrix}.$$

Since

$$r\left(\begin{pmatrix} 0 & 0 \\ 0 & 1 \end{pmatrix}, X\right) = 1,$$

we have

$$X = \begin{pmatrix} a^2 & ab \\ ab & b^2 + 1 \end{pmatrix}.$$

If $a \neq 0$, the matrix

$$\begin{pmatrix} x & 0 \\ 0 & 0 \end{pmatrix}, \qquad x = a^2/(b^2 + 1),$$

of the first subspace is at distance 1 from X. This contradicts our hypothesis. Thus the second is spanned by

$$\begin{pmatrix} 0 & 0 \\ 0 & 1 \end{pmatrix}, \quad \begin{pmatrix} 0 & 0 \\ 0 & b^2 + 1 \end{pmatrix}.$$

The theorem follows.

Consequently two complementary subspaces have a unique matrix in common.

THEOREM 3. *Let* $\mathfrak{S}$ *denote the set of matrices* S *such that*

$$r(S, P) = 2,$$

where P is the common matrix of two complementary normal subspaces. Then $\mathfrak{S}$ is transitive under the subgroup leaving the two subspaces invariant.

Proof. Let the two complementary subspaces be

$$\begin{pmatrix} x & 0 \\ 0 & 0 \end{pmatrix}, \quad \begin{pmatrix} 0 & 0 \\ 0 & y \end{pmatrix}$$

and S be the points in $\mathfrak{S}$. By the hypothesis with $P=0$, we know that S is nonsingular. The transformation

$$Z_1 = Z(-S^{-1}Z + I)^{-1}$$

carries S into ∞, and it leaves the subspaces

$$\begin{pmatrix} x & 0 \\ 0 & 0 \end{pmatrix}, \quad \begin{pmatrix} 0 & 0 \\ 0 & y \end{pmatrix}$$

invariant.

2. **Direct sum of complementary subspaces.**

Definition. The subspace formed by points of the form

$$\begin{pmatrix} x & 0 \\ 0 & y \end{pmatrix}$$

is called a (completely) *reducible subspace* which is the direct sum of the subspaces

$$\begin{pmatrix} x & 0 \\ 0 & 0 \end{pmatrix}, \quad \begin{pmatrix} 0 & 0 \\ 0 & y \end{pmatrix}.$$

Now we shall give its arithmetic construction. We take, by Theorem 2, I as any point of the set $\mathfrak{S}$. Find P satisfying

$$r\left(P, \begin{pmatrix} x & 0 \\ 0 & 0 \end{pmatrix}\right) = r\left(P, \begin{pmatrix} 0 & 0 \\ 0 & y \end{pmatrix}\right) = r(P, I) = 1.$$

Let

$$P = \begin{pmatrix} a^2+1 & ab \\ ab & b^2+1 \end{pmatrix};$$

we have

$$(a^2+1-x)(b^2+1) = a^2b^2, \qquad (a^2+1)(b^2+1-y) = a^2b^2.$$

Then, $a^2 = y(1-x)/(xy-x-y)$, $b^2 = x(1-y)/(xy-x-y)$.

Thus we have two solutions:

$$P_\pm = \begin{pmatrix} 1 & 0 \\ 0 & 1 \end{pmatrix} + \frac{1}{xy-x-y}\begin{pmatrix} y(1-x) & \pm(xy(1-x)(1-y))^{1/2} \\ \pm(xy(1-x)(1-y))^{1/2} & x(1-y) \end{pmatrix}.$$

Finally, we find all K satisfying

$$r(K, P_+) = r(K, P_-) = 1.$$

Putting

$$K = I + \frac{1}{(xy - x - y)}\begin{pmatrix} k_1 & k_2 \\ k_2 & k_3 \end{pmatrix},$$

we have

$$(k_1 - y(1 - x))(k_3 - x(1 - y)) - (k_2 \pm (xy(1 - x)(1 - y))^{1/2})^2 = 0.$$

Then $k_2 = 0$, $k_1 = (1+\rho)y(1-x)$, $k_3 = (1+1/\rho)x(1-y)$.

The matrices of the form

$$K = \begin{pmatrix} 1 + (1 + \rho)y(1 - x) & 0 \\ 0 & 1 + (1 + 1/\rho)x(1 - y) \end{pmatrix}$$

run over all matrices of the reducible space.

Since we use the arithmetical notion only, we have the following general definition.

DEFINITION. Given two complementary subspaces $\mathfrak{X}$ and $\mathfrak{Y}$, let Q be a point of the set $\mathfrak{S}$. The reducible space (or direct sum of both subspaces) is defined by the aggregate of points K such that

$$r(P_+, K) = r(P_-, K) = 1,$$

where P_+ and P_- are both solutions of

$$r(Q, P) = r(P, X) = r(P, Y) = 1,$$

and X and Y belong to $\mathfrak{X}$ and $\mathfrak{Y}$ respectively.

As a consequence of Theorems 2 and 3 we have the following theorem.

THEOREM 4. *The aggregate of reducible spaces is transitive.*

3. **Involutions.** Given

$$Z = \begin{pmatrix} z_1 & z_2 \\ z_2 & z_3 \end{pmatrix},$$

we wish to find all matrices

$$\begin{pmatrix} x & 0 \\ 0 & y \end{pmatrix}$$

of a reducible space such that

$$r\left(\begin{pmatrix} x & 0 \\ 0 & y \end{pmatrix}, \begin{pmatrix} z_1 & z_2 \\ z_2 & z_3 \end{pmatrix}\right) = 1,$$

consequently

$$(x - z_1)(y - z_3) - z_2^2 = 0.$$

This set is denoted by Σ. To each matrix Z we have a set Σ. Conversely, to each Σ, we have two matrices

$$Z \quad \text{and} \quad \begin{pmatrix} z_1 & -z_2 \\ -z_2 & z_3 \end{pmatrix} = \begin{pmatrix} 1 & 0 \\ 0 & -1 \end{pmatrix} Z \begin{pmatrix} 1 & 0 \\ 0 & -1 \end{pmatrix}.$$

Thus we obtain a transformation

$$(1) \qquad Z_1 = \begin{pmatrix} 1 & 0 \\ 0 & -1 \end{pmatrix} Z \begin{pmatrix} 1 & 0 \\ 0 & -1 \end{pmatrix},$$

which is called an *involution of the first kind*. Further, each point of the reducible space is a fixed point of (1); there are no other fixed points.

Since

$$\frac{1}{2}\begin{pmatrix} 1 & -1 \\ 1 & 1 \end{pmatrix}\begin{pmatrix} 1 & 0 \\ 0 & -1 \end{pmatrix}\begin{pmatrix} 1 & 1 \\ -1 & 1 \end{pmatrix} = \begin{pmatrix} 0 & 1 \\ 1 & 0 \end{pmatrix},$$

we have an equivalent involution

$$(2) \qquad Z_1 = \begin{pmatrix} 0 & 1 \\ 1 & 0 \end{pmatrix} Z \begin{pmatrix} 0 & 1 \\ 1 & 0 \end{pmatrix};$$

and, since

$$\begin{pmatrix} 1 & i \\ i & 1 \end{pmatrix}^{-1}\begin{pmatrix} 1 & 0 \\ 0 & -1 \end{pmatrix}\begin{pmatrix} 1 & i \\ i & 1 \end{pmatrix} = \begin{pmatrix} 0 & i \\ -i & 0 \end{pmatrix},$$

we have another equivalent involution

$$(3) \qquad Z_1 = -\begin{pmatrix} 0 & 1 \\ -1 & 0 \end{pmatrix} Z \begin{pmatrix} 0 & -1 \\ 1 & 0 \end{pmatrix}.$$

It may be shown that the most general form of involutions of the first kind(1) is

$$(4) \qquad Z_1 = (PZ - K_1)(K_2 Z + P')^{-1},$$

where K_1 and K_2 are skew symmetric and

$$\mathfrak{F} = \begin{pmatrix} P & -K_1 \\ K_2 & P' \end{pmatrix}$$

(1) The general definition of an involution of the first or second kind is that $\mathfrak{F}^2 = \mathfrak{I}$ or $-\mathfrak{I}$. It may be verified easily that they are equivalent symplectically to (1) or (5) respectively. A detailed study of involutions forms the subject of II, which will appear soon. For this reason, the author omits some details of the discussion. Certainly, for $n = 2$, the properties used can all be verified directly and easily.

is symplectic. We use $\mathfrak{F}_1$, $\mathfrak{F}_2$, $\mathfrak{F}_3$ to denote the matrices of (1), (2) and (3) respectively. We may easily verify that

$$\mathfrak{F}_i\mathfrak{F}_j = -\mathfrak{F}_j\mathfrak{F}_i,$$

and

$$\mathfrak{F}_1\mathfrak{F}_2\mathfrak{F}_3 = \begin{pmatrix} iI & 0 \\ 0 & -iI \end{pmatrix},$$

that is,

$$Z_1 = -Z. \tag{5}$$

This is called an involution of the second kind.

4. **Commutative involutions.** Let $\mathfrak{S}$ and $\mathfrak{S}_0$ be two reducible subspaces associated with two involutions of the first kind, $\mathfrak{F}$ and $\mathfrak{F}_0$ respectively. (It may be shown easily that, if $\mathfrak{F}\mathfrak{F}_0=\mathfrak{F}_0\mathfrak{F}$, then $\mathfrak{F}_0$ carries $\mathfrak{S}$ into itself pointwise.)

If $\mathfrak{F}$ carries $\mathfrak{S}_0$ into itself, but not pointwise, then

$$\mathfrak{F}_0\mathfrak{F} = -\mathfrak{F}\mathfrak{F}_0.$$

In fact, $\mathfrak{F}^{-1}\mathfrak{F}_0\mathfrak{F}$ leaves $\mathfrak{S}_0$ fixed pointwise. Thus

$$\mathfrak{F}^{-1}\mathfrak{F}_0\mathfrak{F}_1 = \pm\,\mathfrak{F}_0;$$

the upper sign is ruled out, since $\mathfrak{F}$ does not leave $\mathfrak{S}_0$ pointwise fixed.

THEOREM 5. *Any pair of commutative involutions of the first kind may be carried into $\mathfrak{F}_1$ and $\mathfrak{F}_2$ simultaneously.*

Proof. The first one may be assumed to be $\mathfrak{F}_1$. Let the second be given by (4). Then

$$\begin{pmatrix} P & -K_1 \\ K_2 & P' \end{pmatrix}\begin{bmatrix} 1 & 0 & 0 & 0 \\ 0 & -1 & 0 & 0 \\ 0 & 0 & 1 & 0 \\ 0 & 0 & 0 & -1 \end{bmatrix} = -\begin{bmatrix} 1 & 0 & 0 & 0 \\ 0 & -1 & 0 & 0 \\ 0 & 0 & 1 & 0 \\ 0 & 0 & 0 & -1 \end{bmatrix}\begin{pmatrix} P & -K_1 \\ K_2 & P' \end{pmatrix},$$

and we have

$$P = \begin{pmatrix} 0 & p_1 \\ p_2 & 0 \end{pmatrix}, \quad K_1 = \begin{pmatrix} 0 & k_1 \\ -k_1 & 0 \end{pmatrix}, \quad K_2 = \begin{pmatrix} 0 & k_2 \\ -k_2 & 0 \end{pmatrix},$$

where

$$p_1p_2 + k_1k_2 = 1.$$

If

$$\begin{pmatrix} p_1-\lambda & -k_1 \\ k_2 & p_2-\lambda \end{pmatrix}$$

has only simple elementary divisors, we have a, b, c, d such that

$$\begin{pmatrix} a & b \\ c & d \end{pmatrix}\begin{pmatrix} p_1 & -k_1 \\ k_2 & p_2 \end{pmatrix}\begin{pmatrix} a & b \\ c & d \end{pmatrix}^{-1} = \begin{pmatrix} \lambda & 0 \\ 0 & 1/\lambda \end{pmatrix}, \qquad ad - bc = 1.$$

Then

$$\begin{bmatrix} 0 & a & 0 & b \\ a & 0 & b & 0 \\ 0 & c & 0 & d \\ c & 0 & d & 0 \end{bmatrix}\begin{pmatrix} P & -K_1 \\ K_2 & P' \end{pmatrix}\begin{bmatrix} 0 & d & 0 & -b \\ d & 0 & -b & 0 \\ 0 & -c & 0 & a \\ -c & 0 & a & 0 \end{bmatrix}$$

$$= \begin{bmatrix} 0 & \lambda & 0 & 0 \\ 1/\lambda & 0 & 0 & 0 \\ 0 & 0 & 0 & 1/\lambda \\ 0 & 0 & \lambda & 0 \end{bmatrix}.$$

(Notice that the transformation carries $\mathfrak{F}_1$ into $-\mathfrak{F}_1$, but they denote the same transformation.) Further

$$\begin{bmatrix} 1 & 0 & 0 & 0 \\ 0 & \lambda & 0 & 0 \\ 0 & 0 & 1 & 0 \\ 0 & 0 & 0 & 1/\lambda \end{bmatrix}$$

carries $\mathfrak{F}$ into $\mathfrak{F}_2$. If

$$\begin{pmatrix} p_1 - \lambda & -k_1 \\ k_2 & p_2 - \lambda \end{pmatrix}$$

has a double elementary divisor, we may take

$$\begin{pmatrix} p_1 & -k_1 \\ k_2 & p_2 \end{pmatrix} = \pm \begin{pmatrix} 1 & 0 \\ 1 & 1 \end{pmatrix}.$$

Now we have to consider the case

$$\begin{bmatrix} 0 & 1 & 0 & 0 \\ 1 & 0 & 0 & 0 \\ 0 & 1 & 0 & 1 \\ -1 & 0 & 1 & 0 \end{bmatrix}.$$

The transformation

$$\begin{pmatrix} 1 & 0 & 0 & 0 \\ 0 & 1 & 0 & 0 \\ -1 & 0 & 1 & 0 \\ 0 & 0 & 0 & 1 \end{pmatrix}$$

carries $\mathfrak{F}$ into $\mathfrak{F}_2$. Thus, we have the theorem.

THEOREM 6. *Any triple of commutative involutions of the first kind may be carried into $\mathfrak{F}_1$, $\mathfrak{F}_2$ and $\mathfrak{F}_3$ simultaneously.*

Proof. We may assume that the first two are $\mathfrak{F}_1$ and $\mathfrak{F}_2$. Let the third one be $\mathfrak{F}$. Since $\mathfrak{F}_1\mathfrak{F} = -\mathfrak{F}\mathfrak{F}_1$, we have

$$\mathfrak{F} = \begin{pmatrix} 0 & p_1 & 0 & -k_1 \\ p_2 & 0 & k_1 & 0 \\ 0 & k_2 & 0 & p_2 \\ -k_2 & 0 & p_1 & 0 \end{pmatrix}, \qquad p_1p_2 + k_1k_2 = 1.$$

Since $\mathfrak{F}_2\mathfrak{F} = -\mathfrak{F}\mathfrak{F}_2$, we have $p_1 = -p_2$. We may change it into $\mathfrak{F}_3$, since

$$\begin{vmatrix} p_1 - \lambda & -k_1 \\ k_2 & -p_1 - \lambda. \end{vmatrix} = \lambda^2 + 1.$$

5. **Involution of the second kind.** An involution of the second kind has two isolated fixed points with arithmetic distance 2. Conversely, any two given points, with arithmetic distance 2, will serve as the isolated fixed points of an involution of the second kind, which is uniquely determined by them. In fact, let 0 and ∞ be fixed points, then the transformation takes the form

$$Z_1 = -AZA', \qquad A^2 = I.$$

We have P such that

$$PAP' = \begin{cases} \pm I, \\ \pm \begin{pmatrix} 1 & 0 \\ 0 & -1 \end{pmatrix}. \end{cases}$$

The latter case cannot happen, since then 0 would not be an isolated fixed point. Thus we have

$$Z_1 = -Z.$$

Now we may define harmonic ranges. Four points Z_1, Z_2, Z_3, Z_4, no two of them with arithmetic distance less than 2, are said to form a harmonic range, if the involution determined by Z_1, Z_2 permutes Z_3 and Z_4.

Analytically, we let

$$(Z_1, Z_2, Z_3, Z_4) = ((Z_1 - Z_3)(Z_1 - Z_4)^{-1})((Z_2 - Z_3)(Z_2 - Z_4)^{-1})^{-1}.$$

The involution

$$(Z - Z_1)(Z - Z_2)^{-1} = -(Z^* - Z_1)(Z^* - Z_2)^{-1}$$

carries Z_3 into Z_4. Evidently

$$(Z_1, Z_2, Z_3, Z_4) = -I.$$

This condition is sufficient as well as necessary.

Thus the "invariance of arithmetic distances" implies the invariance of "harmonic range."

NATIONAL TSING HUA UNIVERSITY,
KUNMING, YUNNAN, CHINA.

Reprinted from the
TRANSACTIONS OF THE AMERICAN MATHEMATICAL SOCIETY
Vol. 59, No. 3, pp. 508–523, 1946

ORTHOGONAL CLASSIFICATION OF HERMITIAN MATRICES

BY
LOO-KENG HUA

1. **Introduction.** Elsewhere(1) the author established the symplectic classification of Hermitian matrices, which has applications to the geometry of symmetric matrices. It is the purpose of the present paper to treat the analogous problem: the orthogonal classification of Hermitian matrices. In other words, it may also be described as the quasi-unitary classification of symmetric matrices. Besides their own interest, the results of the present paper have applications in the geometries of skew-symmetric and symmetric matrices, as well as in the theory of automorphic functions of a matrix variable.

It should be noted that the method previously used is only applicable to the present problem for matrices of even order. In order to establish the general solution, we introduce here a different method.

Unless the contrary is stated, throughout the paper, capital latin letters denote n-rowed square matrices. Further let M' and $\overline{M}$ respectively denote the transposed matrix and conjugate imaginary matrix of M. I denotes the identity matrix.

2. **Statement of the problems.** Two Hermitian matrices H and K are said to be *conjunctive orthogonally*, if there is an orthogonal matrix P such that

$$PHP' = K. \tag{1}$$

From (1) and the orthogonality of P we deduce immediately that

$$P\overline{H}HP' = \overline{K}K. \tag{2}$$

Therefore, if two Hermitian matrices H and K are conjunctive orthogonally, the elementary divisors of the characteristic matrices of $\overline{H}H$ and $\overline{K}K$ are the same.

In the converse, two problems arise:

1. Does $\overline{H}H$ take any prescribed elementary divisors? The answer is in the negative. Then we ask further:

2. For a given admissible set of elementary divisors, is there a unique H, apart from orthogonal conjunctiveness, such that $\overline{H}H$ takes the given set as its elementary divisors?

The answer is also in the negative. More definitely, the elementary divisors of $\overline{H}H$ characterize the orthogonal conjunctiveness of H with a reservation about the uncertainty of signs related to "signature."

Presented to the Society, February 23, 1946; received by the editors October 18, 1945.

(1) Amer. J. Math. vol. 66 (1944) pp. 531–563.

More generally, apparently, we have the following problem. Two pairs (H, S) and (H_1, S_1) of matrices, where H and H_1 are Hermitian and S and S_1 are symmetric, are said to be equivalent if there exists a nonsingular matrix P such that

$$\bar{P}HP' = H_1, \tag{3}$$

and

$$PSP' = S_1. \tag{4}$$

Consequently, we have

$$\bar{P}HS^{-1}\bar{H}\bar{S}^{-1}\bar{P}^{-1} = H_1S_1^{-1}\bar{H}_1\bar{S}_1^{-1}, \tag{5}$$

if S is nonsingular. Hereafter we discuss only the case with nonsingular S. Analogously, we have also two problems.

3. Solution of problem 1.

THEOREM 1. *Let T be a symmetric matrix. The equation*

$$\bar{H}H = T \tag{6}$$

is soluble in an Hermitian H, if and only if the elementary divisors of T have the following two properties:

(i) *Each elementary divisor corresponding to a negative characteristic root must occur an even number of times.*

(ii) *Complex elementary divisors must occur only in complex conjugate pairs.*

Analogously, we have

THEOREM 1'. *The equation*

$$HS^{-1}\bar{H}\bar{S}^{-1} = Q \tag{7}$$

is soluble in symmetric S and Hermitian H if and only if the elementary divisors of Q satisfy (i) *and* (ii) *of Theorem* 1.

Equivalence of both theorems. If $S^{-1} = \Gamma\Gamma'$ in (7), we have

$$(\bar{\Gamma}'H\Gamma)(\Gamma'\bar{H}\bar{\Gamma}) = \bar{\Gamma}'Q(\bar{\Gamma}')^{-1}. \tag{8}$$

Thus, Theorem 1' implies Theorem 1. Conversely, since we have Γ such that $\bar{\Gamma}Q(\bar{\Gamma}')^{-1} = T$, Theorem 1 also implies Theorem 1'.

The sole difficulty lies in proving the property (i).

Proof of the theorems. (1) Suppose that

$$T = \begin{pmatrix} q_1 & 0 \\ 0 & q_2 \end{pmatrix}, \tag{9}$$

where q_1 and $\bar{q}_2$ have no characteristic root in common. Let

$$H = \begin{pmatrix} h_1 & k \\ \bar{k}' & h_2 \end{pmatrix}. \tag{10}$$

From (6), (9), and (10), we have

$$\bar{h}_1 h_1 + \bar{k}\bar{k}' = q_1, \tag{11}$$

$$\bar{h}_1 k + \bar{k} h_2 = 0, \tag{12}$$

and

$$k' k + \bar{h}_2 h_2 = q_2. \tag{13}$$

From (11), (12) and (13), we deduce immediately

$$\begin{aligned} q_1 \bar{k} &= \bar{h}_1 h_1 \bar{k} + \bar{k}\bar{k}'\bar{k} \\ &= -\bar{h}_1 k \bar{h}_2 + \bar{k}\bar{k}'\bar{k} \\ &= \bar{k}(h_2 \bar{h}_2 + \bar{k}'\bar{k}) = \bar{k}\bar{q}_2. \end{aligned}$$

Since q_1 and $\bar{q}_2$ have no characteristic root in common, we have $k=0$. Thus (6) is soluble if and only if both equations

$$h_1 \bar{h}_1 = q_1, \qquad h_2 \bar{h}_2 = q_2$$

are soluble in Hermitian h_1 and h_2. Thus we have to consider only either that T has a real root or that T has a pair of conjugate imaginary roots.

(2) Now we introduce the notations:

$$j^{(p)} = \begin{pmatrix} 0 & 0 \cdots 0 & 1 \\ 0 & 0 \cdots 1 & 0 \\ \cdot & \cdot \quad \cdot \quad \cdot & \cdot \\ 1 & 0 \cdots 0 & 0 \end{pmatrix} = (a_{ij})$$

where $a_{ij}=1$ if $i+j=p+1$ and 0 otherwise. Let $j_\alpha^{(p)}$ be the ordinary Jordan's canonical form[2] of a matrix with elementary divisor $(x-\alpha)^p$. Then jj_α is symmetric.

It is well known that there exists a real polynomial $f(x)$ such that

$$f(j_1)^2 = j_1.$$

Let $s=\alpha^{1/2} jf(j_1)$ which is symmetric, since

$$jj_1^{\sigma} = j_1' j j_1^{\sigma-1} = \cdots = j_1'^{\sigma} j.$$

It can be verified easily that s satisfies

---

[2] For example, $j_\alpha^{(2)} = \begin{pmatrix} \alpha & 1 \\ 0 & \alpha \end{pmatrix}$.

$$sjs = \alpha jj_1. \tag{14}$$

(3) Now we consider the case that T has only a pair of conjugate imaginary characteristic roots. Evidently the elementary divisors of $\overline{H}H-\lambda I$ and $H\overline{H}-\lambda I$ are the same; the solvability, therefore, implies the property (ii) of the theorem.

Conversely, the pair of symmetric matrices (I, T) is congruent to a pair of matrices (T_1, T_2) which is a direct sum of the following pairs:

$$\begin{pmatrix} j & 0 \\ 0 & j \end{pmatrix}, \quad \begin{pmatrix} \alpha jj_1 & 0 \\ 0 & \bar{\alpha} jj_1 \end{pmatrix}.$$

Let

$$\Gamma\Gamma' = T_1, \qquad \Gamma T\Gamma' = T_2.$$

Then, (6) implies

$$(\Gamma\overline{H}\overline{\Gamma}')\overline{T}_1^{-1}(\overline{\Gamma}H\Gamma') = T_2, \tag{15}$$

and conversely.

Since T_1 is real and $T_1^2 = I$, the solvability of (6) is equivalent to that of

$$\overline{H}T_1H = T_2.$$

Since the equation

$$\begin{pmatrix} 0 & s \\ \bar{s} & 0 \end{pmatrix}\begin{pmatrix} j & 0 \\ 0 & j \end{pmatrix}\begin{pmatrix} 0 & \bar{s} \\ s & 0 \end{pmatrix} = \begin{pmatrix} \alpha jj_1 & 0 \\ 0 & \bar{\alpha} jj_1 \end{pmatrix}$$

is equivalent to $sjs = \alpha jj_1$, which is soluble in s by (14), the theorem is proved for the case with complex characteristic roots.

(4) The easiest case is that in which T has a positive root. Since we may assume that (T_1, T_2) of (3) is a direct sum of

$$(j, \alpha jj_1), \qquad \alpha > 0,$$

by (14), there exists a real s such that

$$sjs = \alpha jj_1,$$

and the theorem is now proved.

(5) Finally, there comes the difficult case that α is negative. Without loss of generality, we may assume that $\alpha = -1$.

Since, by (14), we have a symmetric matrix s such that

$$\begin{pmatrix} 0 & s \\ \bar{s} & 0 \end{pmatrix}\begin{pmatrix} j & 0 \\ 0 & j \end{pmatrix}\begin{pmatrix} 0 & \bar{s} \\ s & 0 \end{pmatrix} = \begin{pmatrix} -jj_1 & 0 \\ 0 & -jj_1 \end{pmatrix},$$

we see that if (i) is satisfied, (7) is soluble. The sole difficulty lies in the converse.

(6) Now we write

$$T_0 = \dot{j}^{(s_1)} \dotplus \cdots \dotplus \dot{j}^{(s_e)}, \tag{16}$$

$$T_1 = \dot{j}_{-1}^{(s_1)} \dotplus \cdots \dotplus \dot{j}_{-1}^{(s_e)}. \tag{17}$$

Let $K = HT_0$, then the equation

$$\overline{H}T_1H = T_0 \tag{18}$$

takes the form

$$\overline{K}T_0T_1K = I. \tag{19}$$

From (19), we deduce immediately that

$$KT_0T_1 = KT_0T_1\overline{K}T_0T_1K = T_0T_1K. \tag{20}$$

Let

$$K = (k_{ij}), \qquad k_{ij} = k_{ij}^{(s_i,s_j)}.$$

From (20), we deduce

$$k_{ij}\dot{j}_{-1}^{(s_j)} = \dot{j}_{-1}^{(s_i)} k_{ij}. \tag{21}$$

It is known that

$$k_{ij} = \begin{pmatrix} a_{ij} & b_{ij} & c_{ij} & \cdots \\ 0 & a_{ij} & b_{ij} & \cdots \\ \cdot & \cdot & \cdot & \cdot \\ \cdot & \cdot & \cdot & a_{ij} \end{pmatrix}, \qquad \text{if } s_i = s_j. \tag{22}$$

and in case $s_j > s_k$, we add $s_j - s_k$ rows of zeros below the matrix of the form (21) and in case $s_j < s_k$, we add $s_k - s_j$ columns of zeros before the matrix of the form (21).

From (19), we find

$$\sum_j \bar{k}_{ij}\dot{j}_{-1}^{(s_j)} k_{jk} = \delta_{ik} I^{(s_k)}.$$

Now we consider the element at the (1, 1) position of the case $s_i = s_k$. By (21),

$$\sum_j \bar{k}_{ij} k_{jk}\dot{j}_{-1}^{(s_j)} = \delta_{ik} I^{(s_k)}. \tag{23}$$

By the constitution of the first column of j_{-1}, we need only consider the element at the (1, 1) position of $k_{ij}k_{jk}$. It equals zero, if $s_j < s_k$ and if $s_j > s_k$. Thus the sum (23) runs only over all those j's with $s_j = s_k$.

Let s be any integer occurring in the set $s_1, \cdots, s_e$. Without loss of generality, we may assume that

$$s_i = s \quad \text{for} \quad 1 \leqq i \leqq m, \qquad s_i \neq s \quad \text{for} \quad m < i \leqq e.$$

Then, from (23), we deduce

$$-\sum_{j=1}^{m} \bar{a}_{ij}a_{jk} = \delta_{ik}, \qquad 1 \leqq i,\ k \leqq m,$$

that is $\bar{A}A = -I^{(m)}$ if we let $A = (a_{ij})_{1 \leqq i,\, j \leqq m}$. Taking determinants of both sides, we have $|d(A)|^2 = (-1)^m$, which can not hold for odd m.

The theorem is now proved.

4. Square root of an orthogonal matrix. Now we require a result very likely due to Hilton, but his paper is not available here. Accordingly we state the rediscovered result in the following without proof.

We say a matrix P is *orthogonal with respect to a nonsingular symmetric matrix* S, if

$$PSP' = S.$$

In particular, for $S = I$, we omit the phrase "with respect to I."

Two matrices A and B, orthogonal with respect to S, are said to be similar orthogonally, if there exists a matrix P orthogonal with respect to S such that $A = PBP^{-1}$.

THEOREM 2 (Hilton). *Every orthogonal matrix is similar orthogonally to a direct sum of orthogonal matrices with elementary divisors either of the form*

$$\text{(i)} \qquad (x - \alpha)^r, \qquad (x - 1/\alpha)^r,$$

or of the form

$$\text{(ii)} \qquad (x \pm 1)^r, \qquad \textit{for odd } r.$$

For the first case, the matrix orthogonal with respect to

$$S = \begin{pmatrix} 0 & I^{(r)} \\ I^{(r)} & 0 \end{pmatrix}$$

is similar to the matrix

$$M = M^{(2r)} = \begin{pmatrix} (\dot{j}_\alpha^{(r)})' & 0 \\ 0 & (\dot{j}_\alpha^{(r)})^{-1} \end{pmatrix} \tag{24}$$

orthogonally, and for the second case, the matrix orthogonal with respect to

$$S_1 = \begin{pmatrix} 0 & 0 & I^{(p)} \\ 0 & 1 & 0 \\ I^{(p)} & 0 & 0 \end{pmatrix}$$

is similar to

$$N = N^{(r)} = \begin{bmatrix} (\dot{j}_{\pm 1}^{(p)})' & 0 & 0 \\ v & 1 & 0 \\ -(\dot{j}_{\pm 1}^{(p)})^{-1} v'v/2 & -(\dot{j}_{\pm 1}^{(p)})^{-1} v' & (\dot{j}_{\pm 1}^{(p)})^{-1} \end{bmatrix} \tag{25}$$

orthogonally, where $v=(0, 0, \cdots, 0, 1)$ *and* $2p+1=r$.

Evidently, the squares of the matrices M and N are matrices M_1 and N_1, orthogonal with respect to S and S_1, with elementary divisors

$$\text{(i}'\text{)} \qquad (x-\alpha^2)^r, \qquad (x-\alpha^{-2})^r$$

and

$$\text{(ii}'\text{)} \qquad (x+1)^r$$

respectively.

With a slight modification we have the following assertion: For given orthogonal M_1 and N_1 with elementary divisors (i′) and (ii′) respectively, we can find orthogonal matrices M and N with the elementary divisors (i) and (ii) respectively and with α in the right half-plane in the plane of complex numbers. Thus, except when α is negative, an orthogonal matrix has a square root matrix.

THEOREM 3. *Let* R *be an orthogonal matrix without a negative root and suppose that we have an Hermitian matrix* H *such that*

$$RH = H\bar{R}'. \tag{26}$$

Then there exists an orthogonal matrix Q *with characteristic roots on the right half-plane such that* $Q^2=R$ *and*

$$QH = H\bar{Q}'. \tag{27}$$

Proof. We may assume that R is a direct sum of the matrices

$$R = r_1 \dotplus r_2 \dotplus \cdots \dotplus r_e$$

where r_i has elementary divisors either of the form (i′) or of the form (ii′). They are denoted typically by M_1 and N_1 respectively. We construct M and N accordingly; they are denoted by $q_1, \cdots, q_e$. Let

$$Q = q_1 \dotplus q_2 \dotplus \cdots \dotplus q_e.$$

Then, evidently, we have $Q^2=R$.

Further, from

$$RH = H\bar{R}',$$

we have

$$r_i h_{ij} = h_{ij}\bar{r}_j', \tag{28}$$

and if this implies

$$q_i h_{ij} = h_{ij}\bar{q}_j', \tag{29}$$

we have $QH=H\bar{Q}'$ and consequently we have the theorem.

In order to verify that (28) implies (29), we need only verify that

$$M^2X = XM^2 \quad \text{and} \quad N^2X = XN^2$$

implies

$$MX = XM \quad \text{and} \quad NX = XN,$$

respectively, where M and N are defined exactly as in (24) and (25). Straightforward calculation establishes both implications.

5. Solution of the second problem.

THEOREM 4. *If $H\overline{H}=K\overline{K}$, where H and K are both Hermitian, then there exist two matrices P and Q such that*

$$\overline{P}HP' = \begin{pmatrix} h_1 & 0 \\ 0 & h_2 \end{pmatrix}$$

and

$$\overline{Q}KQ' = \begin{pmatrix} -h_1 & 0 \\ 0 & h_2 \end{pmatrix}.$$

Proof. Since $K^{-1}H=R$ is orthogonal, we have an orthogonal matrix T such that

$$T^{-1}RT = \begin{pmatrix} r_1 & 0 \\ 0 & r_2 \end{pmatrix} = R_1,$$

where r_1 contains all negative characteristic roots of R_1. Then, we have

$$H_1 = K_1R_1 = \overline{R}_1' K_1,$$

where $H_1=\overline{T}'HT$, $K_1=\overline{T}'KT$ are both Hermitian. Since r_1 and $\bar{r}_2$ have no characteristic root in common, we have

$$K_1 = \begin{pmatrix} k_1 & 0 \\ 0 & k_2 \end{pmatrix}$$

and

$$H_1 = \begin{pmatrix} h_1 & 0 \\ 0 & h_2 \end{pmatrix},$$

where

$$k_1r_1 = \bar{r}_1' k_1$$

and

$$k_2r_2 = \bar{r}_2' k_2.$$

By Theorem 3, we have an orthogonal q_2 such that $q_2^2=r_2$ and

$$h_2 = k_2q_2^2 = \bar{q}_2' k_2q_2.$$

Further we have an orthogonal matrix q_1 such that $q_1^2=-r_1$ and

$$h_1 = -k_1q_1^2 = -\bar{q}_1' k_1q_1.$$

The theorem follows.

6. **Explicit result.** The result previously obtained may be concluded in the following theorem.

THEOREM 5. *Let H be a nonsingular Hermitian matrix and S be a nonsingular symmetrix matrix. The elementary divisors of the characteristic matrix of*

$$HS^{-1}\overline{HS}^{-1} \tag{30}$$

are of the following three types:

(i) $(x-\alpha)^\lambda$, *for real* $\alpha > 0$

(ii) $(x-\beta)^\lambda,\quad (x-\beta)^\lambda$, *for real* $\beta < 0$

and

(iii) $(x-\gamma)^\lambda,\quad (x-\bar{\gamma})^\lambda$, *for complex* γ.

Let $\alpha_1=\alpha^{1/2}$, $\beta_1=i(-\beta)^{1/2}$ *and* $\gamma_1=\gamma^{1/2}$; *the determination is taken on the right half-plane. Then the pair of matrices* (H, S) *is equivalent to*(3)

$$\begin{cases} H_1 = \sum_\alpha{}'(\pm\, jj_{\alpha 1}) + \sum_\beta{}'\begin{pmatrix} 0 & jj_{\beta 1}\\ j\bar{j}_{\beta 1} & 0\end{pmatrix} \dotplus \sum_\gamma{}'\pm\begin{pmatrix} 0 & jj_{\gamma 1}\\ j\bar{j}_{\gamma 1} & 0\end{pmatrix} \\ S_1 = \sum{}'(\pm\, j) \quad \dotplus \sum{}'\begin{pmatrix} j & 0\\ 0 & j\end{pmatrix} \quad \dotplus \sum{}'\pm\begin{pmatrix} j & 0\\ 0 & j\end{pmatrix}. \end{cases} \tag{31}$$

In the theorem we have to justify only a point that the pair of matrices

$$\left(\begin{pmatrix} 0 & jj_{\beta 1}\\ j\bar{j}_{\beta 1} & 0\end{pmatrix}, \begin{pmatrix} j & 0\\ 0 & j\end{pmatrix}\right)$$

is equivalent to

$$\left(-\begin{pmatrix} 0 & jj_{\beta 1}\\ j\bar{j}_{\beta 1} & 0\end{pmatrix}, -\begin{pmatrix} j & 0\\ 0 & j\end{pmatrix}\right).$$

In fact, we have a nonsingular matrix γ such that $\gamma j\bar{j}_{\beta 1}\gamma' = -jj_{\beta 1}$ and $\gamma j\gamma' = j$. Let

$$P = \begin{pmatrix} 0 & i\bar{\gamma}\\ i\gamma & 0\end{pmatrix}.$$

Then

$$\overline{P}\begin{pmatrix} 0 & jj_{\beta 1}\\ j\bar{j}_{\beta 1} & 0\end{pmatrix}P' = -\begin{pmatrix} 0 & jj_{\beta 1}\\ j\bar{j}_{\beta 1} & 0\end{pmatrix}$$

and

$$P\begin{pmatrix} j & 0\\ 0 & j\end{pmatrix}P' = -\begin{pmatrix} j & 0\\ 0 & j\end{pmatrix}.$$

The situation of the indeterminate signs corresponding to a positive root

(3) $\sum'$ denotes direct sum.

or to a pair of conjugate complex roots will be clarified by introducing the concept of signature system.

Suppose that $HS^{-1}\overline{H}\overline{S}^{-1}$ has an elementary divisor $(x-\alpha)^\lambda$ $(\alpha>0)$ repeated m times. Let p and q be the number of positive signs and negative signs appearing in the expression (31) corresponding to the elementary divisor $(x-\alpha)^\lambda$. Then (p, q), $p+q=m$, is called the signature corresponding to the elementary divisor $(x-\alpha)^\lambda$. Similarly we define the signature corresponding to

$$(x-\gamma)^\lambda(x-\bar{\gamma})^\lambda$$

for complex γ. The totality of elementary divisors and their corresponding signatures is called the elementary divisors with signature system of the pair of matrices (H, S). Evidently, (H, S) and (H_1, S_1) are equivalent, if they have the same elementary divisors with the same signature system. The converse of this statement is also true. The proof of this fact can be constructed by adapting the results to be obtained in §§8–10 and the method used in (6) of §3. Owing to the similarity, we give here no details of the proof.

7. **Anti-involutions.** An orthogonal matrix T satisfying

$$T\overline{T} = I \tag{32}$$

is called an *anti-involution of the first kind*, and that satisfying

$$T\overline{T} = -I \tag{33}$$

is called an *anti-involution of the second kind.*

Since $TT'=I$, it is evident that the involution T of the first kind is an Hermitian matrix. By Theorem 5, we have an orthogonal matrix P such that

$$PT\overline{P}' = [1, \cdots, 1, -1, \cdots, -1]$$

which is a diagonal matrix.

In case of the second kind, iT is Hermitian, the roots of

$$|(iT)(\overline{iT}) - \lambda I| = 0$$

are all equal to -1. The case can only happen for n even. There exists an orthogonal matrix P such that

$$P(iT)\overline{P}' = \begin{pmatrix} 0 & iI^{(p)} \\ -iI^{(p)} & 0 \end{pmatrix},$$

where $n=2p$.

8. **Automorphs.** Now we are going to study the group formed by all the matrices P satisfying

$$PSP' = S \tag{34}$$

and

$$\overline{P}HP' = H. \tag{35}$$

By Theorem 5, we may write

$$S = S_1^{(p)} \dotplus S_2^{(q)}, \qquad H = H_1^{(p)} \dotplus H_2^{(q)}, \qquad p + q = n,$$

where the two matrices

$$H_1 S_1^{-1} \overline{H}_1 \overline{S}_1^{-1} \quad \text{and} \quad H_2 S_2^{-1} \overline{H}_2 \overline{S}_2^{-1}$$

have no characteristic root in common. From (34) and (35), we have

$$\overline{P} H S^{-1} = H S^{-1} P. \tag{36}$$

It follows immediately that P is decomposable as

$$P = \begin{pmatrix} P_1^{(p)} & 0 \\ 0 & P_2^{(q)} \end{pmatrix}.$$

Therefore, without loss of generality, we treat the problem with (i) $HS^{-1}\overline{HS}^{-1}$ having only a pair of conjugate roots, (ii) $HS^{-1}\overline{HS}^{-1}$ having only a positive root, and (iii) $HS^{-1}\overline{HS}^{-1}$ having only a negative root.

We shall discuss the cases separately.

9. **The case with a pair of complex roots.** Now, without loss of generality, we take

$$S = \sum_{1 \leqq i \leqq e}{}' \pm \begin{pmatrix} j^{(s_i)} & 0 \\ 0 & j^{(s_i)} \end{pmatrix}$$

and

$$H = \sum_{1 \leqq i \leqq e}{}' \pm \begin{pmatrix} 0 & jj_\gamma^{(s_i)} \\ j\bar{j}_\gamma^{(s_i)} & 0 \end{pmatrix},$$

where $s_1 \geqq s_2 \geqq \cdots \geqq s_e$. Putting

$$J^{(p)} = \sum_{1 \leqq i \leqq e}{}' (\pm j^{(s_i)})$$

and

$$J_\gamma^{(p)} = \sum_{1 \leqq i \leqq e}{}' (\pm jj_\gamma^{(s_i)}),$$

we treat, without loss of generality, the case that

$$S = \begin{pmatrix} J & 0 \\ 0 & J \end{pmatrix}, \qquad H = \begin{pmatrix} 0 & J_\gamma \\ \overline{J}_\gamma & 0 \end{pmatrix}.$$

Let

$$P = \begin{pmatrix} A & B \\ C & D \end{pmatrix}, \qquad A = A^{(p)},$$

and so on. Then, from (36), we have

$$\begin{pmatrix} \bar{A} & \bar{B} \\ \bar{C} & \bar{D} \end{pmatrix} \begin{pmatrix} 0 & J_\gamma J \\ \overline{J_\gamma J} & 0 \end{pmatrix} = \begin{pmatrix} 0 & J_\gamma J \\ \overline{J_\gamma J} & 0 \end{pmatrix} \begin{pmatrix} A & B \\ C & D \end{pmatrix},$$

that is

$$\begin{aligned} \bar{B}\bar{J}_\gamma J &= J_\gamma JC, & \bar{A}J_\gamma J &= J_\gamma JD, \\ \bar{D}\bar{J}_\gamma J &= \bar{J}_\gamma JA, & \bar{C}J_\gamma J &= \bar{J}_\gamma JB. \end{aligned} \tag{37}$$

Then, we have

$$B(J_\gamma J)^2 = (\bar{J}_\gamma J)\bar{C}(J_\gamma J) = (\bar{J}_\gamma J)^2 B.$$

It follows that $B=0$ since the characteristic root of $(J_\gamma J)^2$ is equal to γ^2 and $\gamma^2 \neq \bar{\gamma}^2$. Consequently $C=0$.

Further, from (37), we have

$$A(\bar{J}_\gamma J)^2 = (\bar{J}_\gamma J)^2 A, \tag{38}$$

and

$$D = (J_\gamma J)^{-1}\bar{A}(J_\gamma J). \tag{39}$$

From (34) we deduce

$$AJA' = J. \tag{40}$$

Conversely, for any A satisfying (38) and (40), the matrix

$$P = \begin{pmatrix} A & 0 \\ 0 & (J_\gamma J)^{-1}\bar{A}(J_\gamma J) \end{pmatrix} \tag{41}$$

satisfies our requirement.

Then we need to find those A satisfying both (38) and (40).

Let

$$A = (I+Q)(I-Q)^{-1}$$

(the exceptional case $|I+A|=0$ is negligible in counting of parameters).

From (38) and (40), we have

$$Q(\bar{J}_\gamma J)^2 = (\bar{J}_\gamma J)^2 Q$$

and

$$QJ + JQ' = 0.$$

Let $QJ=K$, which is a skew symmetric matrix and which satisfies

$$K(J\bar{J}_\gamma)^2 = (J\bar{J}_\gamma)^2 K. \tag{42}$$

(Notice that $JJ_\gamma J = J'_{\gamma'}$.)

Putting

$$K = (k_{ij})_{1 \leq i,j \leq s}, \qquad k_{ij} = k_{ij}^{(s_i, s_j)},$$

and

$$k_{ij} = -k'_{ji},$$

we have

$$k_{ij}(\bar{j}_\gamma^{(s_j)})^2 = (\bar{j}_\gamma^{(s_j)})'^2 k_{ij}.$$

It follows immediately that

$$k_{ij} = \begin{pmatrix} x_1 & x_2 \cdots & x_{s_j-1} & x_{s_j} \\ x_2 & x_3 \cdots & x_{s_j} & 0 \\ \cdot & \cdot\ \cdot\ \cdot & \cdot & \cdot \\ x_{s_j} & 0 \cdots & 0 & 0 \\ 0 & 0 \cdots & 0 & 0 \\ \cdot & \cdot\ \cdot\ \cdot & \cdot & \cdot \end{pmatrix},$$

which contains s_i rows, if $s_j < s_i$ with addition of a sufficient number of zero rows, and which contains $2s_j$ parameters (a complex number is counted as two parameters). Further in case $i=j$ evidently k_{ii} equals zero.

Therefore K depends on

$$\begin{aligned} &2s_2 + 2s_3 + \cdots + 2s_e \\ &\qquad + 2s_3 + \cdots + 2s_e \\ &\qquad\qquad \cdots\cdots \\ &\qquad\qquad\qquad + 2s_e \\ &= 2s_2 + 4s_3 + \cdots + 2(e-1)s_e \end{aligned}$$

parameters.

10. **The case with a positive root.** Now we take

$$S = \sum_{1\leq i\leq e}{}' (\pm\, \dot{j}^{(s_i)}),$$

$$H = \sum_{1\leq i\leq e}{}' (\pm\, j\dot{j}_\alpha^{(s_i)}), \qquad \alpha > 0,$$

where $s_1 \geqq s_2 \geqq \cdots \geqq s_e$. Let

$$P = (I+Q)(I-Q)^{-1}.$$

We find, from (34) and (35), that

$$QS + SQ' = 0,$$

and

$$\bar{Q}H + HQ' = 0.$$

Putting $QS=K$, which is skew symmetric, we have

$$\overline{K}S^{-1}H - HS^{-1}K = 0. \tag{43}$$

Since $\alpha \neq -\alpha$, we find that K is real. Let

$$K = (k_{ij}), \qquad k_{ij} = k_{ij}^{(s_i,s_j)}.$$

Then

$$\bar{k}_{ij}\dot{j}_\alpha^{(s_j)} = (\dot{j}^{(s_j)}\dot{j}_\alpha^{(s_i)}\dot{j}^{(s_i)})k_{ij} = \dot{j}^{(s_i)'}k_{ij}.$$

In a manner similar to that of §9, it can be shown that the number of parameters is equal to

$$2s_2 + 4s_3 + \cdots + 2(e-1)s_e.$$

11. **The case with a negative root.** Now we may let

$$S = \sum_{1 \leq i \leq e}{}' \begin{pmatrix} \dot{j}^{(s_i)} & 0 \\ 0 & \dot{j}^{(s_i)} \end{pmatrix},$$

$$H = \sum_{1 \leq i \leq e}{}' i \begin{pmatrix} 0 & \dot{j}\dot{j}_\beta^{(s_i)} \\ -\dot{j}\dot{j}_\beta^{(s_i)} & 0 \end{pmatrix}$$

where $\beta > 0$. Let

$$J = \sum_{1 \leq i \leq e}{}' \dot{j}^{(s_i)}, \qquad J_\beta = \sum_{1 \leq i \leq e}{}' \dot{j}^{(s_i)}\dot{j}_\beta^{(s_i)}.$$

We may consider, without loss of generality, the case

$$S = \begin{pmatrix} J & 0 \\ 0 & J \end{pmatrix}, \qquad H = i\begin{pmatrix} 0 & J_\beta \\ -J_\beta & 0 \end{pmatrix}.$$

Letting

$$P = (I + Q)(I - Q)^{-1}$$

and

$$Q\begin{pmatrix} J & 0 \\ 0 & J \end{pmatrix} = K,$$

which is skew symmetric, we find, from (36), that

$$\overline{K}\begin{pmatrix} 0 & JJ_\beta \\ -JJ_\beta & 0 \end{pmatrix} = \begin{pmatrix} 0 & J_\beta J \\ -J_\beta J & 0 \end{pmatrix} K. \tag{44}$$

Let

$$K = \begin{pmatrix} K_1 & L \\ -L' & K_2 \end{pmatrix}. \tag{45}$$

From (44), we have

(46) $$\bar{L}JJ_\beta = J_\beta JL', \qquad \bar{L}'JJ_\beta = J_\beta JL,$$

(47) $$\bar{K}_1JJ_\beta = J_\beta JK_2, \qquad \bar{K}_2JJ_\beta = J_\beta JK_1.$$

Putting $\bar{L}JJ_\beta = T$, which is Hermitian by (46), we have

(48) $$T(JJ_\beta)^2 = (J_\beta J)^2T,$$

(49) $$K_1(JJ_\beta)^2 = (J_\beta J)^2K_1$$

and

(50) $$K_2 = (J_\beta J)^{-1}\bar{K}_1(JJ_\beta).$$

Consequently, for T and K_1 satisfying (48) and (49) and K_2 defined by (50), we have K defined by (45) to meet our requirement.

The number of parameters of K_1 is equal to

$$2s_2 + 4s_3 + \cdots + 2(e-1)s_e,$$

and the number of T is equal to

$$2s_2 + 4s_3 + \cdots + 2(e-1)s_e + s_1 + s_2 + \cdots + s_e.$$

The total number of parameters of K is, therefore, equal to

$$s_1 + 5s_2 + 9s_3 + \cdots + (4e-3)s_e.$$

12. **Automorphs (continuation).** As a consequence of the previous results we have the following statement: Let the roots of $HS^{-1}\bar{H}\bar{S}^{-1}$ be

$$\alpha_1, \cdots, \alpha_\lambda \qquad (\alpha > 0)$$

with the multiplicities

$$p_1, \cdots, p_\lambda;$$

$$\beta_1, \cdots, \beta_\mu; \qquad \bar{\beta}_1, \cdots, \bar{\beta}_\mu \qquad (\beta \text{ complex})$$

with the multiplicities

$$q_1, \cdots, q_\mu; \qquad q_1, \cdots, q_\mu$$

and

$$\gamma_1, \cdots, \gamma_\nu \qquad (\gamma < 0)$$

with the multiplicities

$$2r_1, \cdots, 2r_\nu$$

respectively. Then *the group of automorphs of* (H, S) *depends on at least*

(51) $$r_1 + r_2 + \cdots + r_\nu$$

parameters. (Evidently, this is a best possible constant.)

THEOREM 6. *Given* λ *and an Hermitian matrix* H, *the symmetric matrices* S *satisfying*

$$d(HS^{-1}\overline{HS}^{-1} - \lambda I) = 0 \tag{52}$$

depends on $n(n+1)-1$, $n(n+1)-2$, *and* $n(n+1)-3$ *parameters according as* λ *is positive, complex, and negative respectively.*

Proof. The group of conjunctive automorphs of H, that is, the group formed by Γ satisfying

$$\overline{\Gamma} H \Gamma' = H,$$

depends on n^2 parameters. Now we ask what is the number of parameters of distinct

$$\Gamma S \Gamma'.$$

Since the matrix Γ satisfying

$$\overline{\Gamma} H \Gamma' = H, \qquad \Gamma S \Gamma' = S$$

depends on not more than $r_1 + \cdots + r_\nu$ parameters, if the roots of $HS^{-1}\overline{HS}^{-1}$ are given at the beginning of the section, the totality of different symmetric matrices $\Gamma S \Gamma'$ depends on $n^2 - r_1 - \cdots - r_\nu$ parameters.

For a given positive λ, the manifold formed by S (varying all the other roots) depends on $n^2+(n-1)$ parameters. Similarly we have the result for a complex root.

For a given negative root, the number of parameters of other roots is equal to $n-2$. Further the different symmetric $\Gamma S \Gamma'$, for all roots being given, depends on at most n^2-1 parameters. (In case all the others are non-negative n^2-1 is the exact number.) Thus the total number of parameters is equal to n^2-n-3.

Finally, the author wishes to express his warmest thanks to the referee for his help with the manuscript.

NATIONAL TSING HUA UNIVERSITY,
KUNMING, YUNNAN, CHINA.

Reprinted from the
TRANSACTIONS OF THE AMERICAN MATHEMATICAL SOCIETY
Vol. 61, pp. 193–228, March 1947

GEOMETRIES OF MATRICES. II. STUDY OF INVOLUTIONS IN THE GEOMETRY OF SYMMETRIC MATRICES

BY
LOO-KENG HUA

1. **Introduction.** The paper contains a detailed study of the involutions in the geometry of symmetric matrices over the complex field. It is one of the aims of the paper to establish the following theorem:

A topological automorphism of the group formed by the symplectic transformations is either an inner automorphism or an anti-symplectic transformation.

More precisely, we identify two symplectic matrices $\mathfrak{T}$ and $-\mathfrak{T}$ as a symplectic transformation $\mathfrak{T}_0$. A continuous automorphism of the group formed by $\mathfrak{T}_0$ is either of the form $\mathfrak{P}_0\mathfrak{T}_0\mathfrak{P}_0^{-1}$ or $\mathfrak{P}_0\mathfrak{T}_0^*\mathfrak{P}_0^{-1}$, where $\mathfrak{P}_0$ denotes a symplectic transformation and $\mathfrak{T}_0^*$ is the conjugate complex of $\mathfrak{T}_0$.

The following result, which can also be derived from Mohr's results(¹) on the representations of the symplectic group, can be obtained as an immediate consequence of our present theorem: Every topological automorphism of the group formed by all symplectic matrices (that is, we do not identify $\mathfrak{T}$ and $-\mathfrak{T}$) is either an inner automorphism or the conjugate complex of an inner automorphism. Actually, by means of the method used in the paper, an independent proof of this result can be obtained which is much simpler than that of the first theorem, since the distinction between involutions of the first and the second kind now is apparent.

In the course of our discussion, we find the explicit normal forms of involutions and anti-involutions. The manifold of the fixed points of all sorts of involutions has also been determined completely.

As an introduction, several types of geometries keeping an involution or an anti-involution as absolute have been enumerated. Those obtained from anti-involutions are generalizations of non-Euclidean geometries and those obtained from involutions give us several new types of geometries, whose real analogy (which will be given elsewhere later) is a generalization of Möbius geometry of circles.

Furthermore, the author shows that every symplectic transformation is a product of *two* involutions and *four* anti-involutions, and that for $n=2^\sigma\tau$, τ odd, in the space of symmetric matrices of order n, we have at most $\sigma+3$ pairs of points of which any two pairs separate each other harmonically.

Presented to the Society, December 29, 1946; received by the editors February 27, 1946,

(¹) Göttingen Dissertation, 1933. The author is indebted to the referee for this reference, but unfortunately it is not available in China.

Algebraically speaking, the last result is equivalent to the following one: let $\mathfrak{T}_1, \cdots, \mathfrak{T}_s$ be symplectic matrices of order $2n$ satisfying

$$\mathfrak{T}_i^2 = -\mathfrak{J}, \qquad \mathfrak{T}_i\mathfrak{T}_j = -\mathfrak{T}_j\mathfrak{T}_i$$

where $\mathfrak{J}$ denotes the $2n$-rowed identity. Then $s \leqq \sigma+3$ and this maximum is attained. As a by-product the author establishes also that if $S_1, \cdots, S_s$ are n-rowed symmetric matrices satisfying $S_i^2 = -I$, $S_iS_j = -S_jS_i$, then $s \leqq \sigma+1$. In case of skew symmetric matrices, we have $s \leqq \sigma$. These maximums are all attained.

As in I([2]), capital latin letters denote $n \times n$ matrices unless the contrary is stated. On the other hand, we use $M^{(l,m)}$ to denote an $l \times m$ matrix and $M^{(m)} = M^{(m,m)}$. I and 0 denote the identity and zero matrices respectively. We use also

$$\mathfrak{F} = \begin{pmatrix} 0 & I \\ -I & 0 \end{pmatrix}, \qquad \mathfrak{J} = \begin{pmatrix} I & 0 \\ 0 & I \end{pmatrix}$$

which are $2n$-rowed matrices. p and q denote two integers satisfying $p+q=n$.

2. **Classification of involutions.** First of all we identify the transformations which have the same effect in the space of symmetric matrices.

THEOREM 1. *In the space of symmetric matrices (in homogeneous coordinates), two substitutions*

$$Q(W_1, W_2) = (Z_1, Z_2)\mathfrak{T}, \tag{1}$$

and

$$Q_0(W_1, W_2) = (Z_1, Z_2)\mathfrak{F}_0 \tag{2}$$

induce the same mapping of the space, if and only if

$$\mathfrak{T} = \pm \mathfrak{F}_0. \tag{3}$$

In particular, (1) *carries every point of the space into itself, if and only if*

$$\mathfrak{T} = \pm \mathfrak{J}.$$

Proof. It is sufficient to establish the second part of the theorem. Putting $Z_1 = 0$, we have, by the supposition, $W_1 = 0$. Hence $C = 0$, if we put

$$\mathfrak{T} = \begin{pmatrix} A & B \\ C & D \end{pmatrix}. \tag{4}$$

Similarly, putting $Z_2 = 0$, we find $B = 0$. Now the transformation becomes

$$W = AZA'$$

([2]) The first paper of the series will be referred to as I (Trans. Amer. Math. Soc. vol. 57 (1945) pp. 441–481).

in the nonhomogeneous coordinate system. By the assumption,

$$Z = AZA' \tag{5}$$

holds for all Z, consequently $A = \pm I$.

From Theorem 1, we deduce at once the following theorem.

THEOREM 2. *A symplectic transformation*

$$Q(W_1, W_2) = (Z_1, Z_2)\mathfrak{T}, \qquad \mathfrak{T} = \begin{pmatrix} A & B \\ C & D \end{pmatrix}$$

is an involution if and only if

$$\begin{pmatrix} A & B \\ C & D \end{pmatrix} = \pm \begin{pmatrix} D' & -B' \\ -C' & A' \end{pmatrix}, \tag{6}$$

that is, $\mathfrak{T}\mathfrak{F}$ *is either skew symmetric or symmetric according as* $\mathfrak{T}^2 = \mathfrak{J}$ *or* $-\mathfrak{J}$.

By an involution, we understand a symplectic transformation whose square induces the identity mapping. Similarly, we define an anti-involution as an anti-symplectic mapping whose square induces the identity mapping.

THEOREM 3. *An anti-symplectic transformation*

$$Q(W_1, W_2) = (\bar{Z}_1, \bar{Z}_2)\mathfrak{T}, \qquad \mathfrak{T} = \begin{pmatrix} A & B \\ C & D \end{pmatrix} \tag{7}$$

is an anti-involution, if and only if

$$\begin{pmatrix} A & B \\ C & D \end{pmatrix}^* = \pm \begin{pmatrix} D' & -B' \\ -C' & A' \end{pmatrix}, \tag{8}$$

that is, $\mathfrak{T}\mathfrak{F}$ *is either a skew Hermitian or an Hermitian matrix according as* $\mathfrak{T}\mathfrak{T}^* = \mathfrak{J}$ *or* $-\mathfrak{J}$.

Proof. From (7) and

$$Q^*(Z_1, Z_2) = (\bar{P}_1, \bar{P}_2)\mathfrak{T}, \tag{9}$$

we have

$$\bar{Q}^*Q(W_1, W_2) = (P_1, P_2)\mathfrak{T}^*\mathfrak{T}.$$

This represents the identical mapping, if and only if $\mathfrak{T}^*\mathfrak{T} = \pm\mathfrak{J}$. Further, since $\mathfrak{T}$ is symplectic, that is $\mathfrak{T}\mathfrak{F}\mathfrak{T}' = \mathfrak{F}$, we have

$$\mathfrak{T}^*\mathfrak{F} = \mathfrak{T}^*(\mathfrak{T}\mathfrak{F}\mathfrak{T}') = \pm\, \mathfrak{F}\mathfrak{T}' = \mp\, (\mathfrak{T}^*\mathfrak{F})^{*\prime}.$$

The theorem is now evident.

DEFINITION 1. The involutions satisfying $\mathfrak{T}^2 = \mathfrak{J}$ are called *involutions of the first kind* and those satisfying $\mathfrak{T}^2 = -\mathfrak{J}$ are called *involutions of the second kind.*

DEFINITION 2. The anti-involutions satisfying $\mathfrak{T}\mathfrak{T}^* = \mathfrak{J}$ are called *anti-involutions of the first kind* and those satisfying $\mathfrak{T}\mathfrak{T}^* = -\mathfrak{J}$ are called *anti-involutions of the second kind.*

3. **Normal form of involutions.** Suppose that $\mathfrak{T}$ is an involution of the first kind. By Theorem 2, $\mathfrak{T}\mathfrak{F}$ is skew symmetric. Now we consider the pair of skew symmetric matrices $(\mathfrak{T}\mathfrak{F}, \mathfrak{F})$. The elementary divisors of the matrix $\mathfrak{T}\mathfrak{F} - \lambda\mathfrak{F} = (\mathfrak{T} - \lambda\mathfrak{J})\mathfrak{F}$ are those of $\mathfrak{T} - \lambda\mathfrak{J}$. Since $\mathfrak{T}^2 = \mathfrak{J}$, the elementary divisors are all simple and the characteristic roots are ± 1. Since the determinant of $\mathfrak{T}$ is equal to 1, the multiplicity of the root -1 is even. The pair of skew symmetric matrices

$$\left(\begin{pmatrix} 0 & H \\ -H & 0 \end{pmatrix}, \mathfrak{F}\right), \qquad H = [1, \cdots, 1, -1, \cdots, -1]$$

has the same elementary divisors as the pair of matrices $(\mathfrak{T}\mathfrak{F}, \mathfrak{F})$. Here H is a diagonal matrix with p terms 1 and q terms -1. Hence we have a matrix $\mathfrak{P}$ such that

$$\mathfrak{P}(\mathfrak{T}\mathfrak{F}, \mathfrak{F})\mathfrak{P}' = \left(\begin{pmatrix} 0 & H \\ -H & 0 \end{pmatrix}, \mathfrak{F}\right).$$

Thus $\mathfrak{P}$ is symplectic, and

$$\mathfrak{P}\mathfrak{T}\mathfrak{P}^{-1} = -\mathfrak{P}\mathfrak{T}\mathfrak{F}\mathfrak{P}'\mathfrak{F} = -\begin{pmatrix} 0 & H \\ -H & 0 \end{pmatrix}\begin{pmatrix} 0 & I \\ -I & 0 \end{pmatrix} = \begin{pmatrix} H & 0 \\ 0 & H \end{pmatrix},$$

which gives the transformation

$$W = HZH. \tag{10}$$

Therefore we have the following theorem.

THEOREM 4. *Every involution of the first kind is equivalent to* (10) *symplectically, where H is a diagonal matrix with p positive* 1's *and q negative* 1's *and $p \leqq q$. Further, no two of these involutions are equivalent.*

The last sentence can be justified by considering the multiplicity of the characteristic root 1 of the symplectic matrix.

DEFINITION 3. This involution is said to be *of signature* (p, q).

Now we consider an involution $\mathfrak{T}$ of the second kind. Then we have a pair of matrices $(\mathfrak{T}\mathfrak{F}, \mathfrak{F})$; the first matrix is symmetric and the second skew symmetric. The characteristic roots of $\mathfrak{T}$ are $\pm i$ and the elementary divisors are all simple. Let(3) $d(\mathfrak{T} - \lambda\mathfrak{J}) = f(\lambda)$. Since $\mathfrak{T}\mathfrak{F} = \mathfrak{F}\mathfrak{T}'^{-1}$, we have

(3) $d(X)$ denotes the determinant of the matrix X.

$$f(\lambda) = d(\mathfrak{T} - \lambda\mathfrak{J})d(\mathfrak{F}) = d(\mathfrak{F}\mathfrak{T}'^{-1} - \lambda\mathfrak{F})$$
$$= \lambda^{2n}d\left(\mathfrak{T}' - \frac{1}{\lambda}\,\mathfrak{J}\right) = \lambda^{2n}f\left(\frac{1}{\lambda}\right).$$

Hence the multiplicities of the characteristic roots i and $-i$ are equal.

Since

$$\begin{pmatrix} I & 0 \\ 0 & I \end{pmatrix} - \lambda\begin{pmatrix} 0 & I \\ -I & 0 \end{pmatrix}$$

has the same elementary divisors as $\mathfrak{T}\mathfrak{F}-\lambda\mathfrak{F}$, we have a symplectic matrix $\mathfrak{P}$ carrying $\mathfrak{T}$ into

$$\begin{pmatrix} 0 & I \\ -I & 0 \end{pmatrix}$$

which corresponds to the transformation

$$W = -Z^{-1}. \tag{11}$$

Hence, we have the following theorem.

THEOREM 5. *Every involution of the second kind is equivalent to* (11) *symplectically.*

Remarks. 1. For $n=1$, no involution of the first kind exists.

2. The following normal form of an involution of the second kind is sometimes useful:

$$W = -Z. \tag{12}$$

3. In the case $n=2p$, we sometimes use

$$W = \begin{pmatrix} 0 & I^{(p)} \\ I^{(p)} & 0 \end{pmatrix} Z \begin{pmatrix} 0 & I^{(p)} \\ I^{(p)} & 0 \end{pmatrix} \tag{13}$$

as the normal form of an involution of the signature (p, p).

Evidently, two simplectic transformations with matrices $\mathfrak{T}_1$ and $\mathfrak{T}_2$ are commutative if and only if

$$\mathfrak{T}_1\mathfrak{T}_2 = \pm\, \mathfrak{T}_2\mathfrak{T}_1. \tag{14}$$

Consequently, we see that the product of two commutative involutions $\mathfrak{T}_1$ and $\mathfrak{T}_2$ of the first kind is an involution either of the first kind or of the second kind according as $\mathfrak{T}_1\mathfrak{T}_2=\mathfrak{T}_2\mathfrak{T}_1$ or $\mathfrak{T}_1\mathfrak{T}_2=-\mathfrak{T}_2\mathfrak{T}_1$. In particular, for n even, $n=2p$, say, (11) may be regarded as a product of two involutions of the first kind:

$$W = \begin{pmatrix} 0 & I^{(p)} \\ -I^{(p)} & 0 \end{pmatrix} Z \begin{pmatrix} 0 & -I^{(p)} \\ I & 0 \end{pmatrix},$$

and

$$W = -\begin{pmatrix} 0 & I^{(p)} \\ -I & 0 \end{pmatrix} Z^{-1} \begin{pmatrix} 0 & I^{(p)} \\ I & 0 \end{pmatrix}.$$

For n odd, such a decomposition does not exist.

Further, any involution of the first kind is a product of involutions of the first kind of the signature $(1, n-1)$, which are called *fundamental involutions.*

4. **Equivalence of anti-involutions.** A little attention should be paid to the equivalence of anti-symplectic transformations. Let $(W_1, W_2)=Q(\overline{Z}_1, \overline{Z}_2)\mathfrak{T}$. From $(U_1, U_2)=Q_1(Z_1, Z_2)\mathfrak{P}$, $(V_1, V_2)=Q_2(W_1, W_2)\mathfrak{P}$, we have

$$(V_1, V_2) = Q_2(W_1, W_2)\mathfrak{P} = Q_2Q(\overline{Z}_1, \overline{Z}_2)\mathfrak{T}\mathfrak{P} = Q_2Q\overline{Q}_1^{-1}(U_1, U_2)\mathfrak{P}^{*-1}\mathfrak{T}\mathfrak{P}.$$

Thus we have the following definition.

DEFINITION. *Two anti-symplectic transformations with matrices $\mathfrak{T}_1$ and $\mathfrak{T}_2$ are said to be equivalent, if there exists a symplectic matrix $\mathfrak{P}$ such that*

$$\mathfrak{P}^{*-1}\mathfrak{T}_1\mathfrak{P} = \mathfrak{T}_2. \tag{15}$$

Notice that if (U_1, U_2) and (Z_1, Z_2) (and (V_1, V_2) and (W_1, W_2)) are related anti-symplectically, then we have $\mathfrak{P}^{*-1}\mathfrak{T}_1^*\mathfrak{P}=\mathfrak{T}_2$ instead of (15).

5. **Relation between anti-involutions and hypercircles and the normal form of the anti-involution.** Let $\mathfrak{H}$ be an Hermitian matrix. The points (Z_1, Z_2) of the space for which

$$(Z_1, Z_2)\mathfrak{H}(\overline{Z}_1, \overline{Z}_2)' \tag{16}$$

is positive definite form a hypercircle. $\mathfrak{H}$ is called the matrix of the hypercircle. The skew matrix

$$\mathfrak{H}'\mathfrak{F}\mathfrak{H} \tag{17}$$

is called the discriminantal matrix of $\mathfrak{H}$.

For an anti-involution of the first kind $\mathfrak{T}$, we have a hypercircle with the matrix

$$i\mathfrak{T}\mathfrak{F}, \tag{18}$$

in fact, since $\mathfrak{T}$ is symplectic, $i\mathfrak{T}\mathfrak{F}=i\mathfrak{F}\mathfrak{T}'^{-1}=i\mathfrak{F}\mathfrak{T}^{*\prime}=(i\mathfrak{T}\mathfrak{F})^{*\prime}$; and for an anti-involution of the second kind $\mathfrak{T}$, we have a hypercircle

$$\mathfrak{T}\mathfrak{F}. \tag{19}$$

Their discriminantal matrices are, respectively,

$$(i\mathfrak{T}\mathfrak{F})'\mathfrak{F}(i\mathfrak{T}\mathfrak{F}) = \mathfrak{F}^3 = -\mathfrak{F}$$

and

$$(\mathfrak{T}\mathfrak{F})'\mathfrak{F}(\mathfrak{T}\mathfrak{F}) = -\mathfrak{F}^3 = \mathfrak{F}.$$

THEOREM 6. *To each hypercircle with discriminantal matrix* $-\mathfrak{F}$ *or* $\mathfrak{F}$, *there corresponds an anti-involution of the first and the second kind respectively.*

Proof. (1) Let $\mathfrak{H}$ be an Hermitian matrix satisfying $\mathfrak{H}'\mathfrak{F}\mathfrak{H} = -\mathfrak{F}$. Let

$$\mathfrak{T} = i\mathfrak{H}\mathfrak{F}.$$

Then

$$\mathfrak{T}'\mathfrak{F}\mathfrak{T} = \mathfrak{F}\mathfrak{H}'\mathfrak{F}\mathfrak{H}\mathfrak{F} = -\mathfrak{F}^3 = \mathfrak{F},$$

and

$$\mathfrak{T}^*\mathfrak{T} = -i\mathfrak{H}^*\mathfrak{F}i\mathfrak{H}\mathfrak{F} = -\mathfrak{F}^2 = \mathfrak{I}.$$

Hence $\mathfrak{T}$ is a symplectic matrix and we have an anti-involution of the first kind.

(2) If $\mathfrak{H}'\mathfrak{F}\mathfrak{H} = \mathfrak{F}$, set $\mathfrak{T} = -\mathfrak{H}\mathfrak{F}$. We obtain consequently

$$\mathfrak{T}'\mathfrak{F}\mathfrak{T} = -\mathfrak{F}\mathfrak{H}'\mathfrak{F}\mathfrak{H}\mathfrak{F} = -\mathfrak{F}^3 = \mathfrak{F},$$

$$\mathfrak{T}^*\mathfrak{T} = \mathfrak{H}\mathfrak{F}\mathfrak{H}\mathfrak{F} = -\mathfrak{I},$$

that is, $\mathfrak{T}$ defines an anti-involution of the second kind.

Further, from $\mathfrak{P}^*\mathfrak{T}\mathfrak{P}^{-1} = \mathfrak{T}_1$, we have

$$\mathfrak{P}^*\mathfrak{T}\mathfrak{F}\mathfrak{P}' = \mathfrak{P}^*\mathfrak{T}\mathfrak{P}^{-1}\mathfrak{F} = \mathfrak{T}_1\mathfrak{F}.$$

Therefore, in order to classify anti-involutions, we have to classify their corresponding hypercircles.

Since

$$\mathfrak{H}'\mathfrak{F}\mathfrak{H} - \lambda\mathfrak{F} = \pm\mathfrak{F} - \lambda\mathfrak{F}$$

has characteristic roots either all $+1$ or all -1 and has simple elementary divisors, we have[4] a symplectic matrix $\mathfrak{P}$ such that

$$\mathfrak{P}^*\mathfrak{H}\mathfrak{P}' = \begin{pmatrix} H_1 & 0 \\ 0 & H_2 \end{pmatrix}$$

where H_1 is a diagonal matrix with p terms 1 and q terms -1

Since the discriminantal matrix is $\pm\mathfrak{F}$, we have then

$$H_1'H_2 = -I, \qquad H_2'H_1 = -I$$

for anti-involutions of the first kind, and

$$H_1'H_2 = I, \qquad H_2'H_1 = I$$

for anti-involutions of the second kind.

Thus, for anti-involutions of the first kind, the hypercircle is symplectically conjunctive to[5]

[4] See the author's paper, *On the theory of automorphic functions of a matrix variable*, II, Amer. J. Math. vol. 66 (1944) pp. 531–563.

[5] Two Hermitian matrices H and K are said to be conjunctive, if there exists a matrix A such that $\bar{A}HA' = K$.

(20)
$$\begin{pmatrix} I & 0 \\ 0 & -I \end{pmatrix}$$

and for the second kind, it is symplectically conjunctive to

(21)
$$\begin{pmatrix} H_1 & 0 \\ 0 & H_1 \end{pmatrix}.$$

Therefore, we have the following theorem.

THEOREM 7. *An anti-involution of the first kind is equivalent to*

(22)
$$W = \bar{Z}^{-1}$$

and an anti-involution of the second kind is equivalent to

(23)
$$W = -H\bar{Z}^{-1}H,$$

where H denotes a diagonal matrix with p terms 1 and q terms -1, $p \leqq q$.

We may also prove that they are all non-equivalent, since their hypercircles are non-equivalent.

An anti-involution, equivalent to (23), is called an *anti-involution of signature (p, q).*

Remark. (22) is equivalent to

(24)
$$W = -\bar{Z}.$$

In the case $p=q$, (23) is equivalent to

(25)
$$W = \begin{pmatrix} 0 & I^{(p)} \\ -I & 0 \end{pmatrix} \bar{Z} \begin{pmatrix} 0 & -I \\ I & 0 \end{pmatrix}.$$

6. **Decomposition of involutions into anti-involutions.** Now we are going to express involution as product of anti-involutions.

THEOREM 8. *Every involution is a product of two commutative anti-involutions of the first kind.*

Before proving Theorem 8, we give the following rules concerning the multiplication of anti-symplectic transformations.

(1) The product of two anti-symplectic transformations with matrices $\mathfrak{T}_1$ and $\mathfrak{T}_2$ is a symplectic transformation with matrix $\mathfrak{T}_1^*\mathfrak{T}_2$.

(2) They are commutative if and only if $\mathfrak{T}_1^*\mathfrak{T}_2 = \pm\mathfrak{T}_2^*\mathfrak{T}_1$.

Proof of Theorem 8. (1) Let

$$\mathfrak{T}_1 = \begin{pmatrix} 0 & iI \\ iI & 0 \end{pmatrix}, \quad \mathfrak{T}_2 = \begin{pmatrix} 0 & H_i \\ H_i & 0 \end{pmatrix}.$$

Then

$$\mathfrak{T}_2^*\mathfrak{T}_1 = \begin{pmatrix} H & 0 \\ 0 & H \end{pmatrix} = \mathfrak{T}_1^*\mathfrak{T}_2,$$

which corresponds to the involution $W = HZH$ of the first kind. Further $\mathfrak{T}_1^*\mathfrak{T}_1 = \mathfrak{T}_2^*\mathfrak{T}_2 = \mathfrak{I}$, hence $\mathfrak{T}_1$ and $\mathfrak{T}_2$ represent two anti-involutions of the first kind.

(2) Let

$$\mathfrak{T}_1 = \begin{pmatrix} iI & 0 \\ 0 & -iI \end{pmatrix}, \qquad \mathfrak{T}_2 = \begin{pmatrix} 0 & Ii \\ Ii & 0 \end{pmatrix}.$$

Then $\mathfrak{T}_1^*\mathfrak{T}_1 = \mathfrak{T}_2^*\mathfrak{T}_2 = \mathfrak{I}$. Further

$$\mathfrak{T}_2^*\mathfrak{T}_1 = \begin{pmatrix} 0 & -I \\ I & 0 \end{pmatrix} = -\mathfrak{T}_1^*\mathfrak{T}_2$$

which corresponds to the involution $Z_1 = -Z^{-1}$ of the second kind.

THEOREM 9. *Every anti-involution of the second kind is a product of three mutually commutative anti-involutions of the first kind.*

Proof. The anti-involution of the second kind

$$Z_1 = -H\overline{Z}^{-1}H$$

is a product of $Z_1 = -Z_2^{-1}$ and $Z_2 = H\overline{Z}H$, and the former one is a product of two anti-involutions of the first kind:

$$Z_1 = -\overline{Z}_3, \qquad Z_3 = \overline{Z}_2^{-1}.$$

The three anti-involutions so obtained are evidently mutually commutative.

Remark. Theorems 8 and 9 (and later Theorem 10) suggest that anti-involutions of the first kind can be used as generators in the group of symplectic and anti-symplectic transformations. Does the anti-involution of the second kind play the same role? The answer seems to be negative. In fact, an involution of the second kind cannot be decomposed into a product of two anti-involutions of the second kind. Let $\mathfrak{T}$ be an involution of the second kind. Suppose the contrary, that is, suppose that we have two anti-involutions $\mathfrak{T}_1$ and $\mathfrak{T}_2$ of the second kind such that $\mathfrak{T} = \mathfrak{T}_1^*\mathfrak{T}_2$. Then, we have

$$-\mathfrak{I} = \mathfrak{T}^2 = (\mathfrak{T}_1^*\mathfrak{T}_2)(\mathfrak{T}_1^*\mathfrak{T}_2),$$

that is,

$$\mathfrak{T}_2\mathfrak{T}_1^* = -\mathfrak{T}_1\mathfrak{T}_2^*. \tag{26}$$

Now, we are going to show that, in particular, for $\mathfrak{T}_1 = \mathfrak{F}$ we have no $\mathfrak{T}_2$ satisfying this condition. (Notice that no generality is lost.) From

$$\mathfrak{T}_2\mathfrak{F} = -\mathfrak{F}\mathfrak{T}_2^*,$$

and (by Theorem 3) $\mathfrak{T}_2\mathfrak{F} = (\mathfrak{T}_2\mathfrak{F})^{*\prime} = -\mathfrak{F}\mathfrak{T}_2^{*\prime}$, we deduce that $\mathfrak{T}_2$ is symmetric. Further, the equation

$$\mathfrak{T}_2\mathfrak{T}_2^{*\prime} = \mathfrak{T}_2\mathfrak{T}_2^{*} = -\mathfrak{J}$$

is impossible, since the matrix on the left is a positive definite Hermitian matrix, while the matrix on the right is negative definite.

7. **Decomposition of symplectic transformations into involutions.** First of all we give the normal form of a symplectic matrix $\mathfrak{T}$. We have a symplectic matrix $\mathfrak{P}$ such that $\mathfrak{P}^{-1}\mathfrak{T}\mathfrak{P}$ is a direct sum of symplectic transformations of the forms

$$W^{(\sigma)} = J_{\pm 1}^{(\sigma)} Z^{(\sigma)} J_{\pm 1}^{(\sigma)\prime} + S^{(\sigma)} \tag{27}$$

and

$$W^{(\tau)} = J_{\alpha}^{(\tau)} Z^{(\tau)} J_{\alpha}^{(\tau)}, \qquad \alpha \neq \pm 1,\ \tau \text{ odd}, \tag{28}$$

where

$$J_{\alpha}^{(\tau)} = \begin{pmatrix} \alpha & 0 & \cdots & 0 \\ 1 & \alpha & \cdots & 0 \\ \cdot & \cdot & \cdot & \cdot \\ 0 & 0 & \cdots & \alpha \end{pmatrix}, \qquad S = \begin{pmatrix} 1 & 0 & \cdots & 0 \\ 0 & 0 & \cdots & 0 \\ \cdot & \cdot & \cdot & \cdot \\ 0 & 0 & \cdots & 0 \end{pmatrix}.$$

By a direct sum of two symplectic transformations with matrices

$$\begin{pmatrix} A^{(\sigma)} & B_1 \\ C_1 & D_1 \end{pmatrix}, \qquad \begin{pmatrix} A_2^{(\tau)} & B_2 \\ C_2 & D_2 \end{pmatrix}$$

we mean a symplectic transformation with the $2(\sigma+\tau)$-rowed matrix

$$\begin{pmatrix} \begin{pmatrix} A_1 & 0 \\ 0 & A_2 \end{pmatrix} & \begin{pmatrix} B_1 & 0 \\ 0 & B_2 \end{pmatrix} \\ \begin{pmatrix} C_1 & 0 \\ 0 & C_2 \end{pmatrix} & \begin{pmatrix} D_1 & 0 \\ 0 & D_2 \end{pmatrix} \end{pmatrix}.$$

This result, according to valuable information from Professor H. Weyl, is due to Williamson; however, it was also proved by the author independently. Since Williamson's paper is not available in China, the author is obliged to give the preceding result without a necessary quotation.

THEOREM 10. *Every symplectic transformation is a product of two involutions of the second kind; consequently, every symplectic transformation is a product of four anti-involutions of the first kind.*

Proof. (1) Let

$$J = \begin{pmatrix} 0 & 0 \cdots 0 & 1 \\ 0 & 0 \cdots 1 & 0 \\ \cdot & \cdot \cdot \cdot \cdot & \cdot \\ 1 & 0 \cdots 0 & 0 \end{pmatrix}, \qquad \mathfrak{P} = \begin{pmatrix} 0 & J \\ -J & 0 \end{pmatrix}.$$

Evidently, $\mathfrak{P}$ is an involution of the second kind. Let

$$\mathfrak{T} = \begin{pmatrix} J_\alpha & 0 \\ 0 & J_\alpha'^{-1} \end{pmatrix}.$$

Then

$$(\mathfrak{T}\mathfrak{P})^2 = \begin{pmatrix} 0 & J_\alpha J \\ -J_\alpha'^{-1} J & 0 \end{pmatrix}^2 = -\mathfrak{I},$$

since $J_\alpha J = J J_\alpha'$. Thus $\mathfrak{T}$ is a product of two involutions of the second kind.

(2) Next we consider the case

$$\mathfrak{T} = \begin{pmatrix} J_1 & S J_1'^{-1} \\ 0 & J_1'^{-1} \end{pmatrix}. \tag{29}$$

Let

$$M^{(n)} = M = (m_{ij}), \qquad m_{ij} = \begin{cases} (-1)^{i-1} \binom{i-1}{j-1} & \text{for } i \geqq j, \\ 0 & \text{for } i < j, \end{cases} \tag{30}$$

and $M^2 = (t_{ij})$. Then, by (30),

$$t_{ij} = \sum_{h=1}^{m} m_{ih} m_{hj} = (-1)^{i-1} \sum_{j \leqq k \leqq i} (-1)^{k-1} \binom{i-1}{k-1} \binom{k-1}{j-1} = \begin{cases} 0 & \text{for} \quad i \neq j, \\ 1 & \text{for} \quad i = j. \end{cases}$$

Therefore, we have

$$M^2 = I. \tag{31}$$

Consequently, the matrix

$$\mathfrak{P} = \begin{pmatrix} iM & 0 \\ 0 & -iM' \end{pmatrix}$$

denotes an involution of the second kind. We may verify directly that

$$J_1 M^{(n)} = \begin{pmatrix} 1 & 0 \\ 0 & -M^{(n-1)} \end{pmatrix}.$$

Thus

$$\mathfrak{T}\mathfrak{P} = i\begin{pmatrix} J_1 M & -SJ_1'^{-1}M' \\ 0 & -J_1'^{-1}M' \end{pmatrix}$$

is also an involution of the second kind, since $S = J_1 MSM'J_1'$.

(3) Similarly we treat the case

$$\begin{pmatrix} J_{-1} & SJ_{-1}'^{-1} \\ 0 & J_{-1}'^{-1} \end{pmatrix}.$$

On account of the result quoted at the beginning of the section, we have the theorem.

8. **Geometries induced by anti-involutions.** The results of sections 8, 9, 10, 11 and 12 can be used to introduce several types of geometry; the detailed study will be given elsewhere.

We take an anti-involution $\mathfrak{J}$ of the first kind as an absolute. The group G formed by all symplectic transformations $\mathfrak{T}$ commutative with $\mathfrak{J}$ is called the group of motion. Correspondingly, we have a hypercircle with the matrix $i\mathfrak{J}\mathfrak{F}$. The group of transformations $\mathfrak{T}$ with $\mathfrak{T}\mathfrak{J} = \mathfrak{J}\mathfrak{T}^*$ form a subgroup G_1 whose index in G is equal to 2.

In fact, from

$$\mathfrak{T}_1\mathfrak{J} = \mathfrak{J}\mathfrak{T}_1^*$$

and

$$\mathfrak{T}_2\mathfrak{J} = \mathfrak{J}\mathfrak{T}_2^*$$

we deduce $\mathfrak{T}_1\mathfrak{T}_2\mathfrak{J} = \mathfrak{T}_1\mathfrak{J}\mathfrak{T}_2^* = \mathfrak{J}\mathfrak{T}_1\mathfrak{T}_2^*$.

As the transformation

$$(W_1, W_2) = Q(Z_1, Z_2)\mathfrak{T} \tag{32}$$

of G_1 carries

$$(Z_1, Z_2)i\mathfrak{J}\mathfrak{F}(Z_1, Z_2)^{*\prime} \tag{33}$$

into

$$Q(W_1, W_2)i\mathfrak{T}\mathfrak{J}\mathfrak{F}\mathfrak{T}^{*\prime}(W_1, W_2)^{*\prime}\overline{Q}' = Q(W_1, W_2)i\mathfrak{J}\mathfrak{F}(W_1, W_2)^{*\prime}\overline{Q}'$$

for $\mathfrak{T}$ belonging to G_1, the rank and signature of the Hermitian matrix (33) classify the points of the space into transitive sets. This statement will be proved in the remark of the next section.

Now we take the set of points for which (33) is positive definite. This set is called a *hyperbolic space.* The corresponding geometry is called a *hyperbolic geometry of symmetric matrices.* By Theorem 7, we find that, apart from equivalence, there is one and only one type of hyperbolic geometry. The group G_1 is called the group of motion of the space.

Let P_1 and P_2 be two points of the hyperbolic space. Every motion evidently carries the cross-ratio matrix

$$\begin{aligned}\{P_1, P_2\} &= (P_1, \mathfrak{J}(P_1), P_2, \mathfrak{J}(P_2)) \\ &= (P_1 - P_2)(P_1 - \mathfrak{J}(P_2))^{-1}((\mathfrak{J}(P_1) - P_2)(\mathfrak{J}(P_1) - \mathfrak{J}(P_2))^{-1})^{-1}\end{aligned} \tag{34}$$

into a similar matrix. (34) is called the *distance-matrix* between two points. We have the following theorems.

THEOREM 11. *The hyperbolic space is transitive.*

THEOREM 12. *The distance matrix has simple elementary divisors and positive characteristic roots. Two point-pairs are equivalent if and only if their distance matrices have the same characteristic roots (multiplicities are counted).*

THEOREM 13. *The space is symmetric.*

By a symmetric space, we mean that to each point of the space there exists an involution with the point as its isolated fixed point and that every transformation of the space having the point as fixed point is commutative with the involution.

The proof of these three theorems will be given in the next section.

9. **Visualization of the hyperbolic geometry.** For the sake of concreteness, we take (cf. (24))

$$\mathfrak{J} = i\begin{pmatrix} I & 0 \\ 0 & -I \end{pmatrix}.$$

The hypercircle is now formed by those points $Z = X + iY$ where X and Y are real and Y is positive definite. The group G_1 is formed by the symplectic transformations

$$W = (AZ + B)(CZ + D)^{-1}, \tag{35}$$

where

$$\mathfrak{T} = \begin{pmatrix} A & B \\ C & D \end{pmatrix}$$

is real. This is Siegel's generalization of the Poincaré half-plane.

Sometimes we take, by Theorem 7,

$$\mathfrak{J} = i\begin{pmatrix} 0 & I \\ I & 0 \end{pmatrix}.$$

The space is formed by those points for which $I - Z\bar{Z}$ is positive definite. The group of motions consists of the transformations

$$W = (AZ + B)(\bar{B}Z + \bar{A})^{-1}. \tag{36}$$

This is a generalization of the unit circle in the complex plane.

Theorem 11 is evident, since any point $P = Q + iR$ with positive definite R may be carried into iR by the transformation $W = Z - Q$ of (35) and since there exists a real matrix C such that $CRC' = I$, then $W = C^{-1}ZC'^{-1}$ carries iR into iI.

To prove Theorem 12, we use the second representation. We may take one of the points to be zero. Then $\{Z, 0\} = (Z, \bar{Z}^{-1}, 0, \infty) = Z\bar{Z}$, which is positive definite. We have a unitary matrix U such that

$$UZ\bar{Z}\bar{U}' = [\lambda_1^2, \cdots, \lambda_n^2], \qquad \lambda_r \geqq 0,$$

and

$$UZU' = [\lambda_1, \cdots, \lambda_n].$$

This gives Theorem 12.

The space is transitive. In order to prove Theorem 13, we have to show that there exists one and only one involution which has 0 as its isolated fixed point and that the subgroup leaving 0 invariant is commutative with it.

The transformation leaving 0 invariant carries also ∞ into itself. Then it takes the form

$$W = AZA'. \tag{37}$$

(37) denotes an involution if and only if $A^2 = \pm I$. The equation $Z = AZA'$ has no other solution but zero if and only if the characteristic roots of the second power matrix of A are different from 1. Further the roots of A are ± 1 and $\pm i$; the only possibility is that $A = \pm iI$ by Theorem 44.4 of MacDuffee[6]. Now we have the unique involution

$$W = -Z. \tag{38}$$

It is evidently commutative with all transformations (37).

Remark. In order to prove the promised statement of the preceding section, it is sufficient to prove that the points

$$X + Yi, \qquad X, Y \text{ real},$$

are classified into transitive sets according to the ranks and signatures of Y. This is evident, since $Z_1 = Z - X$ carries the point into Yi and $Z_1 = AZA'$ carries Yi into a diagonal matrix of the form $i[1, \cdots, 1, -1, \cdots, -1, 0, \cdots, 0]$.

10. **Elliptic geometries.** Now we take $\mathfrak{J}$ to be an anti-involution of the second kind as an absolute. More definitely, we take

$$\mathfrak{J} = \begin{pmatrix} 0 & -H \\ H & 0 \end{pmatrix}$$

where

$$H = [1, \cdots, 1, -1, \cdots, 1],$$

where there are p 1's and q -1's, $q \geqq q$. The hypercircle has the matrix

$$\begin{pmatrix} H & 0 \\ 0 & H \end{pmatrix}.$$

(6) *Theory of matrices*, 1933.

The space is formed by the points for which

$$Z_1H\bar{Z}_1' + Z_2H\bar{Z}_2' = (Z_1, Z_2)\mathfrak{J}\mathfrak{F}(Z_1, Z_2)^{*\prime} \tag{39}$$

is of the same signature as H. This geometry will be called the *elliptic geometry of signature* (p, q). The group of motions is defined by the symplectic transformations

$$(Z_1, Z_2) = Q(W_1, W_2)\mathfrak{T}, \tag{40}$$

with

$$\mathfrak{T}\mathfrak{J} = \mathfrak{J}\mathfrak{T}^*, \tag{41}$$

since (40) carries (39) into

$$Q(W_1, W_2)\mathfrak{T}\mathfrak{J}\mathfrak{F}\mathfrak{T}^{*\prime}(W_1, W_2)'\bar{Q}' = Q(W_1, W_2)\mathfrak{J}\mathfrak{F}(W_1, W_2)^{*\prime}\bar{Q}'.$$

In the case $p=q$, the transformations (40) with $\mathfrak{T}\mathfrak{J}=-\mathfrak{J}\mathfrak{T}^*$ instead of (41) also leave the space invariant. Then the group of motions is now constituted by both kinds of transformations.

From (41), we have

$$\mathfrak{T} = \begin{pmatrix} A & B \\ -H\bar{B}H & H\bar{A}H \end{pmatrix}.$$

THEOREM 14. *The space is transitive.*

Proof. For any point (A, B) of the space making $\bar{A}HA'+\bar{B}HB'$ of signature (p, q) we can find a matrix P such that

$$\bar{P}(\bar{A}HA' + \bar{B}HB')P' = H.$$

Then

$$\mathfrak{P} = \begin{pmatrix} PA & PB \\ -H\bar{P}\bar{B}H & H\bar{P}\bar{A}H \end{pmatrix} \tag{42}$$

is symplectic, in fact

$$PA(PB)' = PAB'P' = (PB)(PA)',$$
$$(-H\bar{P}\bar{B}H)(H\bar{P}\bar{A}H)' = (H\bar{P}\bar{A}H)(-H\bar{P}\bar{B}H)',$$

and

$$PA(H\bar{P}\bar{A}H)' - PB(H\bar{P}\bar{B}H)' = P(AH\bar{A}' + BH\bar{B}')\bar{P}'\ H = I.$$

The transformation $(Z_1, Z_2)=Q(W_1, W_2)\mathfrak{P}$ is a motion of the space, since

$$\mathfrak{P}^*\begin{pmatrix} H & 0 \\ 0 & H \end{pmatrix}\mathfrak{P}' = \begin{pmatrix} H & 0 \\ 0 & H \end{pmatrix},$$

and it carries $(I, 0)$ into the point (A, B).

The group of stability at the origin, that is, the subgroup of transformations leaving 0 invariant, consists of the transformations of the form

$$W = AZA',$$

where A is a conjunctive automorphism of H, that is, $\bar{A}HA' = H$. We may prove also the following theorem.

THEOREM 15. *The elliptic space of signature (p, q) is symmetric.*

As in §9, we may introduce the distance matrix between two points P_1 and P_2 by $\{P_1, P_2\} = (P_1, \mathfrak{J}(P_1), P_2, \mathfrak{J}(P_2))$. Putting $P_1 = Z$ and $P_2 = 0$, we have

$$\{Z, 0\} = ZH\bar{Z}H. \tag{43}$$

Here the problem arises to classify symmetric matrices Z under the group of conjunctive automorphisms of an Hermitian matrix. It would require too much space to solve this problem: hence at present we content ourselves with the special case $H = I$, which is the direct generalization of elliptic geometry.

THEOREM 16. *The distance matrix has simple elementary divisors and positive characteristic roots in the elliptic space of signature $(n, 0)$. Two point-pairs are equivalent if and only if their distance matrices have the same characteristic roots (multiplicities are counted).*

Notice that in the elliptic space of signature $(n, 0)$ the group of motions is formed by all unitary symplectic matrices.

11. **Geometries induced by involutions.** Now we taken an involution $\mathfrak{J}$ of the second kind as the absolute; without loss of generality we may take

$$W = -Z. \tag{44}$$

From

$$\begin{pmatrix} A & B \\ C & D \end{pmatrix}\begin{pmatrix} iI & 0 \\ 0 & -iI \end{pmatrix} = \begin{pmatrix} iI & 0 \\ 0 & -iI \end{pmatrix}\begin{pmatrix} A & B \\ C & D \end{pmatrix},$$

we deduce that $B = C = 0$. Thus the transformations of the space are given by

$$Z_1 = AZA', \tag{45}$$

which form a group G_1. If the involution $Z_1 = -Z^{-1}$ is adjoined to G_1, a group G is obtained which contains G_1 as a subgroup of index 2. The geometry of the symmetric matrices under the group G_1 is equivalent to the algebraic problem "congruent classification of matrices." Evidently the space is not transitive. The rank of P is the characteristic invariant for a finite point.

For a pair of finite points P_1 and P_2, we have $(P_1, P_2, 0, \infty) = P_1P_2^{-1}$. Thus the elementary divisors of $P_1P_2^{-1}$ characterize the equivalence of a pair of finite points.

Now we take an involution $\mathfrak{J}$ of the first kind with signature (p, q) as an absolute, for example

$$Z_1 = HZH. \tag{46}$$

Now the matrix $\mathfrak{T}$ of the transformations satisfies

$$\begin{pmatrix} A & B \\ C & D \end{pmatrix}\begin{pmatrix} H & 0 \\ 0 & H \end{pmatrix} = \begin{pmatrix} H & 0 \\ 0 & H \end{pmatrix}\begin{pmatrix} A & B \\ C & D \end{pmatrix}.$$

Consequently we have

$$A = \begin{pmatrix} a_1^{(p)} & 0 \\ 0 & a_2^{(q)} \end{pmatrix}, \quad B = \begin{pmatrix} b_1^{(p)} & 0 \\ 0 & b_2^{(q)} \end{pmatrix}, \quad C = \begin{pmatrix} c_1^{(p)} & 0 \\ 0 & c_2^{(q)} \end{pmatrix}, \quad D = \begin{pmatrix} d_1^{(p)} & 0 \\ 0 & d_2^{(q)} \end{pmatrix}.$$

Then $\mathfrak{T}$ may be considered as a direct product of two symplectic matrices

$$\begin{pmatrix} a_1 & b_1 \\ c_1 & d_1 \end{pmatrix}, \qquad \begin{pmatrix} a_2 & b_2 \\ c_2 & d_2 \end{pmatrix}.$$

The group so formed is denoted by G_1.

The equation

$$\begin{pmatrix} A & B \\ C & D \end{pmatrix}\begin{pmatrix} H & 0 \\ 0 & H \end{pmatrix} = -\begin{pmatrix} H & 0 \\ 0 & H \end{pmatrix}\begin{pmatrix} A & B \\ C & D \end{pmatrix}$$

cannot hold, except when $p=q$, $n=2p$. Then we have a group G which is generated by G_1 and the additional transformation

$$W = \begin{pmatrix} 0 & I^{(p)} \\ I & 0 \end{pmatrix} Z \begin{pmatrix} 0 & I \\ I & 0 \end{pmatrix}.$$

For simplicity's sake, we take $p=n-1$, $q=1$, that is, the absolute is a fundamental involution.

THEOREM 17. *The arithmetic distance of P and $\mathfrak{J}(P)$ characterizes the equivalence of points. More definitely, every point is either equivalent to* 0 *or equivalent to*

$$\begin{bmatrix} 0 & 0 & \cdots & 0 & 1 \\ 0 & 0 & \cdots & 0 & 0 \\ \cdot & \cdot & \cdot & \cdot & \cdot \\ 0 & 0 & \cdots & 0 & 0 \\ 1 & 0 & \cdots & 0 & 0 \end{bmatrix}.$$

Proof. Let

$$P = \begin{pmatrix} P_1 & V' \\ V & P \end{pmatrix}, \qquad P_1 = P_1^{(n-1)}.$$

Then

$$W = Z - \begin{pmatrix} P_1 & 0 \\ 0 & P \end{pmatrix}$$

carries P into

$$\begin{pmatrix} 0 & v' \\ v & 0 \end{pmatrix}. \tag{47}$$

If $v=0$, there is nothing to be proved. Otherwise, we have a matrix M such that $vM=(1, 0, \cdots, 0)$, $M=M^{(n-1)}$ Then the transformation

$$W = \begin{pmatrix} M & 0 \\ 0 & 1 \end{pmatrix} Z \begin{pmatrix} M & 0 \\ 0 & 1 \end{pmatrix}$$

carries (47) into the required form.

The non-equivalence of the two points given in Theorem 17 is evident.

It is easy to extend the theorem for arbitrary p and q.

Remarks. 1. The geometries so obtained are generalizations of the complex analogy of the Möbius geometry of circles. There is a particularly great variety of geometries over the real field.

2. We may take several commutative involutions as absolutes; for example, we take $W=-Z$, $W=Z^{-1}$ as absolutes, then the transformations take the form $W=\Gamma Z\Gamma'$, where Γ is an orthogonal matrix. Thus "the orthogonal classification of symmetric matrices" may be considered as the geometry of symmetric matrices with two absolutes.

3. To each involution $\mathfrak{I}$ of the second kind, we have a symplectic symmetric matrix $\mathfrak{I}\mathfrak{F}$, and conversely. Thus, the geometry of symplectic symmetric matrices may be considered as a geometry of involutions of the second kind. In particular, for $n=1$, it gives the ordinary treatment of involutions in the complex plane. It may also be extended to the study of anti-involutions, and so on. But the author will not go into the detailed discussion of this problem.

12. **Laguerre's geometry of matrices.** The transformations

$$W = AZA' + S \tag{48}$$

form a subgroup, which leaves ∞ invariant. The geometry under this group is called *Laguerre's geometry of symmetric matrices*, or *affine geometry of symmetric matrices*.

THEOREM 18. *Finite points are transitive under the group. Two point-pairs are equivalent if and only if their arithmetic distances are the same.*

We may obtain invariants from the "projective" geometry of symmetric matrices by selecting a point to be infinity. For the discussion of the equivalence of the triples of points, we introduce a "simple ratio-matrix," that is,

$$(X_1, X_2, X_3, \infty) = (X_1 - X_2)(X_3 - X_2)^{-1}.$$

Further, the transformations (48) with unitary A form a subgroup. The geometry under this group may be called the Euclidean geometry of symmetric matrices. The transformations (48) with A of determinant I form also a group. The corresponding geometry may be called the special Laguerre geometry.

13. **Fixed points of an involution of the second kind.** In §§13, 14 we determine the fixed points of involutions.

THEOREM 19. *An involution of the second kind has two isolated fixed points; they form a nonspecial pair, that is, a pair of points with arithmetic distance n. Conversely, a nonspecial pair of points determines uniquely an involution of the second kind ha ing the points as its isolated fixed points.*

Proof. (1) It is sufficient to prove that $\pm iI$ are two isolated solutions of

$$Z^2 = -I. \tag{49}$$

If Z_0 is a non-scalar solution of (49), then $\Gamma Z_0 \Gamma'$ is also a solution for all orthogonal Γ. Thus Z_0 is not an isolated fixed point of $W = -Z^{-1}$.

If Z_0 is scalar, then $Z_0 = \pm iI$. From $(iI+\Delta)^2 = -I$ we have $\Delta(\Delta+2iI) = 0$. For sufficiently small Δ, $\Delta+2iI$ is nonsingular, consequently $\Delta = 0$. This establishes that $\pm iI$ are the two isolated fixed points.

(2) Conversely, let 0 and ∞ be two fixed points. Then the involution takes the form

$$W = AZA', \qquad A^2 = -I.$$

Using this in connection with (37), we have $W = -Z$.

THEOREM 20. *Let P_1 and P_2 be two isolated fixed points of an involution of the second kind; all its fixed points X are given by*

$$r(P_1, X) + r(X, P_2) = n \tag{50}$$

and conversely(7).

Proof. (1) All the solutions of (49) are given by

$$Z = \Gamma \begin{pmatrix} -I^{(p)}i & 0 \\ 0 & I^{(q)}i \end{pmatrix} \Gamma',$$

where Γ is orthogonal, since the elementary divisors of $Z - \lambda I$ are all simple. Further

$$r(Z, Ii) = r\left(\begin{pmatrix} -I^{(p)}i & 0 \\ 0 & I^{(q)}i \end{pmatrix}, Ii\right) = p$$

and $r(-Ii, Z) = q$.

(7) $r(A, B)$ denotes the rank of $A - B$; it is called the arithmetic distance between A and B.

(2) Now we let

$$r(Z, iI) = p \quad \text{and} \quad r(Z, -iI) = q.$$

There exists a matrix Q such that

$$QZQ' = \sum' J_k^{(C_k)}, \qquad QIQ' = \sum' J^{(C_k)}$$

where

$$J = \begin{pmatrix} 0 & 0 \cdots 0 & 1 \\ 0 & 0 \cdots 1 & 0 \\ \cdot & \cdot \cdots \cdot & \cdot \\ 1 & 0 \cdots 0 & 0 \end{pmatrix}, \qquad J_k = \begin{pmatrix} 0 & \cdots 0 & \lambda_k \\ 0 & \cdots \lambda_k & 1 \\ \cdot & \cdots \cdot & \cdot \\ \lambda_k & \cdots 0 & 0 \end{pmatrix}.$$

Direct verification shows that $C_k=1$ and that $d(Z-\lambda I)=0$ has i as root of multiplicity p and has $-i$ as root of multiplicity q. Thus we have an orthogonal Q such that

$$Z = Q\begin{pmatrix} I^{(p)}i & 0 \\ 0 & -I^{(q)}i \end{pmatrix}Q'.$$

THEOREM 21. *Let P_1 and P_2 be two isolated fixed points of an involution of the second kind. Let X be a point satisfying $r(P_1, X)=r(P_2, X)=n$, and let X_0 be the image of X under the involution. Then P_1, P_2, X, X_0 form a harmonic range.*

Proof. We may take $P_1=0$, $P_2=\infty$. Then $X_0=-X$. We have consequently $(P_1, P_2, X, X_0)=(0, \infty, X, -X)=-I$.

14. **Fixed elements of an involution of the first kind.** In order to determine the fixed elements of an involution of the first kind, we have to introduce the concept of decomposable subspace.

DEFINITION. A manifold in the space is said to form a *decomposable subspace of the type* (p, q), if we have a symplectic transformation carrying the manifold to a manifold formed by the points of the form

$$\begin{pmatrix} Z_1^{(p)} & 0 \\ 0 & Z_2^{(q)} \end{pmatrix}, \qquad p+q=n, \tag{51}$$

where Z_1 and Z_2 are called *components*.

THEOREM 22. *Let $p+q=n$. The fixed points of an involution of signature (p, q) form a decomposable subspace of type (p, q).*

Proof. We write

$$H = \begin{pmatrix} I^{(p)} & 0 \\ 0 & -I^{(q)} \end{pmatrix},$$

and

$$Z = \begin{pmatrix} Z_{11}^{(p)} & Z_{12} \\ Z'_{12} & Z_{22}^{(q)} \end{pmatrix}.$$

Then

$$HZH - Z = \begin{pmatrix} 0 & -2Z_{12} \\ -2Z'_{12} & 0 \end{pmatrix} = 0$$

which implies $Z_{12}=0$. Thus we have the theorem.

Theorems 22 and 2 give us a general definition of decomposable subspace:

Let $\mathfrak{K}$ be a skew symmetric symplectic matrix. The points (Z_1, Z_2) satisfying

$$(Z_1, Z_2)\mathfrak{K}(Z_1, Z_2)' = 0 \tag{52}$$

define a decomposable space. It is of the type (p, q), if $d(\mathfrak{K}-\lambda\mathfrak{F})=0$ has 1 and -1 as roots of multiplicity $2p$ and $2q$ respectively.

The justification is almost evident, since on multiplying $(W_1, W_2) = Q(Z_1, Z_2)\mathfrak{T}$ on the right by $\mathfrak{F}(Z_1, Z_2)'$, we obtain

$$(W_1, W_2)\mathfrak{F}(Z_1, Z_2)' = Q(Z_1, Z_2)\mathfrak{K}(Z_1, Z_2)'.$$

(Notice that (Z_1, Z_2) and (W_1, W_2) denote the same point, if and only if $(Z_1, Z_2)\mathfrak{F}(W_1, W_2)'=0$.)

Further we may prove that the involution is uniquely determined by the decomposable subspace.

In the same way we may define the manifold of the fixed points of an involution of the second kind.

15. **Fixed points of an anti-involution.** Now we are going to find the fixed elements of anti-involutions.

THEOREM 23. *The fixed points of an anti-involution of the first kind form a connected piece of dimension $n(n+1)/2$ (real parameters).*

In fact, the fixed points of

$$W = \bar{Z} \tag{24}$$

are the real Z. The theorem is now evident.

THEOREM 24. *An anti-involution with signature (p, q) $(p\neq q)$ has no fixed point and that with signature (p, p) has fixed points depending on $n(n+1)/2$ real parameters.*

Proof. We have (23). Its fixed points are given by $ZH\bar{Z}=-H$. This is impossible except for $p=q$.

In the case $p=q$, we take the normal form

$$W = \begin{pmatrix} 0 & I^{(p)} \\ I & 0 \end{pmatrix}\bar{Z}\begin{pmatrix} 0 & I \\ I & 0 \end{pmatrix}. \tag{25}$$

Set

$$Z = \begin{pmatrix} Z_1 & Z_2 \\ Z_2' & Z_3 \end{pmatrix}.$$

The fixed points of (25) are given by

$$Z_1 = \overline{Z}_3, \qquad Z_2 = \overline{Z}_2'.$$

The number of parameters of Z_1 and Z_2 are $p(p+1)$, p^2 respectively. Thus the total number of parameters is equal to

$$p(p+1) + p^2 = p(2p+1) = n(n+1)/2.$$

16. **Number of parameters of involutions.** Now we are going to determine the dimensions of the manifolds formed by involutions and anti-involutions.

THEOREM 25. *The number of parameters (complex) of involutions of the second kind is equal to $n(n+1)$ and that of involutions of the first kind of signature (p, q) is equal to $4pq$.*

Notice that $4pq \leqq n^2 < n(n+1)$.

Proof. (1) By Theorem 19, we have the first part of the theorem.

(2) Now we consider involutions of the signature (p, q). It is known that the group of $2n$-rowed symplectic matrices depends on $n(2n+1)$ parameters. We are going to find the number of parameters of the subgroup leaving the decomposable subspace

$$\begin{pmatrix} X_1^{(p)} & 0 \\ 0 & X_2^{(q)} \end{pmatrix}$$

invariant. By the result of §11 and Theorem 22, the number of parameters of the subgroup is equal to $p(2p+1)+q(2q+1)$. Therefore the number of parameters of involutions of the signature (p, q) is equal to

$$n(2n+1) - p(2p+1) - q(2q+1) = 2(n^2 - p^2 - q^2) = 4pq,$$

since $n = p+q$.

Now we give also the number of parameters of anti-involutions.

THEOREM 26. *Anti-involutions of the first or the second kind each form an $n(2n+1)$ parametric family (real parameters).*

Proof. We consider the anti-involution $\mathfrak{T}$ of the first kind. Put

$$\mathfrak{T} = (\mathfrak{I} - \mathfrak{S}\mathfrak{F})(\mathfrak{I} + \mathfrak{S}\mathfrak{F})^{-1}.$$

From $\mathfrak{T}^*\mathfrak{T} = \mathfrak{I}$ and $\mathfrak{T}\mathfrak{F}\mathfrak{T}' = \mathfrak{F}$ we deduce easily that

$$\mathfrak{S} + \mathfrak{S}^* = 0, \qquad \mathfrak{S} = \mathfrak{S}'.$$

That is, $\mathfrak{S}$ is pure imaginary and symmetric. The theorem follows.

In the case of anti-involutions of the second kind, we have correspondingly

$$\mathfrak{S}\mathfrak{F}\mathfrak{S}^* = \mathfrak{F}, \qquad \mathfrak{S} = \mathfrak{S}'.$$

Putting $\mathfrak{S}=(\mathfrak{J}-\mathfrak{P})(\mathfrak{J}+\mathfrak{P})^{-1}$ again, we have $\mathfrak{P}\mathfrak{F}+\mathfrak{F}\mathfrak{P}^*=0$, $\mathfrak{P}=\mathfrak{P}'$, that is,

$$\mathfrak{P} = \begin{pmatrix} S & H \\ H' & -S \end{pmatrix},$$

where S is symmetric and H is Hermitian. The number of parameters is $n(n+1)+n^2=n(2n+1)$.

17. **Dieder manifolds.** In this section we study the dimension of dieder manifold.

DEFINITION. Let A and B form a nonspecial pair. The locus of the points X satisfying

$$r(A, X) + r(X, B) = n \tag{50}$$

is called a *dieder manifold* of the point-pair (A, B).

As a consequence of Theorem 20, the fixed points of an involution of the second kind form a dieder manifold of its two isolated fixed points. It is called the *dieder manifold of the involution.*

THEOREM 27. *Let $\mathfrak{S}$ be an involution commutative with $\mathfrak{T}$, an involution of the second kind. Then $\mathfrak{S}$ carries the dieder manifold of $\mathfrak{T}$ into itself.*

Proof. $\mathfrak{S}$ carries the isolated fixed points (A, B) of $\mathfrak{T}$ to (C, D). Then (C, D) are isolated fixed points of $\mathfrak{S}\mathfrak{T}\mathfrak{S}^{-1}$. Since $\mathfrak{S}\mathfrak{T}\mathfrak{S}^{-1}=\pm\mathfrak{T}$, $\mathfrak{S}$ either keeps A and B fixed or permutes A and B. In both cases the dieder manifold is invariant.

DEFINITION. The part of the dieder manifold satisfying

$$r(A, X) = p \tag{53}$$

is called the *component of index p* with respect to A.

THEOREM 28. *The component of index p is of dimension $2pq$.*

Proof. It is now more convenient to use the homogeneous coordinate-system. We may assume that A and B are $(0, I)$ and $(I, 0)$ respectively. The component of index p is constituted by the points

$$(Z_1, Z_2) \tag{54}$$

where Z_1 is of rank p and Z_2 of rank q. Evidently

$$Z_1 = \begin{pmatrix} I^{(p)} & a \\ 0 & 0 \end{pmatrix}, \quad Z_2 = \begin{pmatrix} 0 & 0 \\ b & I^{(q)} \end{pmatrix} \tag{55}$$

are points on the component. The number of parameters is $2pq$. Thus we have to establish the following two facts:

(1) We have no Q, different from the identity, such that

$$(56)\qquad Q\left(\begin{pmatrix} I^{(p)} & a \\ 0 & 0 \end{pmatrix}, \begin{pmatrix} 0 & 0 \\ b & I^{(q)} \end{pmatrix}\right) = \left(\begin{pmatrix} I^{(p)} & a_1 \\ 0 & 0 \end{pmatrix}, \begin{pmatrix} 0 & 0 \\ b_1 & I^{(q)} \end{pmatrix}\right).$$

(2) In general, each point on the component is of the form (55).

To prove (1), we let

$$Q = \begin{pmatrix} q_{11} & q_{12} \\ q_{21} & q_{22} \end{pmatrix}.$$

Comparing the elements on both sides of (56), we have $q_{11}=I^{(p)}$, $q_{12}=0$, $q_{21}=0$, $q_{22}=I^{(q)}$. Thus (1) is true.

(2) The transformation leaving $(0, I)$ and $(I, 0)$ invariant of is the form

$$W_1 = QZ_1A, \qquad W_2 = QZ_2D, \qquad AD' = I.$$

Thus the points on the components are given by

$$W_1 = Q\begin{pmatrix} I^{(p)} & 0 \\ 0 & 0 \end{pmatrix} A = Q\begin{pmatrix} a_{11}^{(p)} & a_{12} \\ 0 & 0 \end{pmatrix},$$

$$W_2 = Q\begin{pmatrix} 0 & 0 \\ 0 & I^{(q)} \end{pmatrix} A'^{-1} = Q\begin{pmatrix} 0 & 0 \\ \alpha_{21} & \alpha_{22} \end{pmatrix},$$

where

$$A = \begin{pmatrix} a_{11} & a_{12} \\ a_{21} & a_{22} \end{pmatrix}, \qquad A'^{-1} = \begin{pmatrix} \alpha_{11} & \alpha_{12} \\ \alpha_{21} & \alpha_{22} \end{pmatrix}.$$

We take the general case $d(a_{11})\neq 0$, $d(d_{22})\neq 0$. Thus

$$\begin{pmatrix} a_{11}^{-1} & 0 \\ 0 & a_{22}^{-1} \end{pmatrix}\left(\begin{pmatrix} a_{11} & a_{12} \\ 0 & 0 \end{pmatrix}, \begin{pmatrix} 0 & 0 \\ \alpha_{21} & \alpha_{22} \end{pmatrix}\right)$$

gives the required form. The conditions of inequality are irrelevant in counting the parameters.

Consequently, two components of indices r and s of a dieder manifold are not equivalent topologically, if $r\neq s$ and $r+s\neq n$. As a consequence of Theorem 27, we have the following theorem.

THEOREM 29. *A dieder manifold is of dimension* $2[n/2](n-[n/2])$ *where* $[x]$ *denotes the integral part of* x.

18. **Commutative involutions of the second kind.** Without loss of generality the problem to be solved can be stated as follows:

Given an involution of the second kind, say (12), find all involutions of the second kind commutative with it.

Let $\mathfrak{T}$ be the matrix of the required involution, then $\mathfrak{T}^2 = -\mathfrak{T}$ and

$$\mathfrak{T}\begin{pmatrix} iI & 0 \\ 0 & -iI \end{pmatrix} = \pm \begin{pmatrix} iI & 0 \\ 0 & -iI \end{pmatrix}\mathfrak{T}. \tag{57}$$

(1) Taking the upper sign, we have

$$\mathfrak{T} = \begin{pmatrix} A & B \\ C & D \end{pmatrix}, \qquad B = C = 0.$$

That is, the involution takes the form

$$W = AZA', \qquad A^2 = -I. \tag{58}$$

Now we are going to find its isolated fixed points. There exists a nonsingular matrix Γ such that $\Gamma A \Gamma^{-1} = iH$, where $H = [1, \cdots, 1, -1, \cdots, -1]$, where there are p 1's and q -1's. Let

$$W_0 = \Gamma W \Gamma', \qquad Z_0 = \Gamma Z \Gamma';$$

we have

$$W_0 = -HZ_0H \tag{59}$$

and notice that (12) takes the form $W_0 = -Z_0$. We drop the subscript 0.

Using homogeneous coordinates, we have

$$(Z_1, Z_2) = Q(W_1, W_2)\begin{pmatrix} iH & 0 \\ 0 & -iH \end{pmatrix}. \tag{60}$$

The transformation

$$\mathfrak{T} = \begin{pmatrix} I^p & iI_q \\ iI_q & I^p \end{pmatrix}, \qquad I^p = \begin{pmatrix} I^{(p)} & 0 \\ 0 & 0 \end{pmatrix}, \qquad I_q = \begin{pmatrix} 0 & 0 \\ 0 & I^{(q)} \end{pmatrix} \tag{61}$$

carries (60) into

$$\mathfrak{T}\begin{pmatrix} iH & 0 \\ 0 & -iH \end{pmatrix}\mathfrak{T}^{-1} = \begin{pmatrix} I^p & iI_q \\ iI_q & I^p \end{pmatrix}\begin{pmatrix} iH & 0 \\ 0 & -iH \end{pmatrix}\begin{pmatrix} I^p & -iI_q \\ -iI_q & I^p \end{pmatrix}$$
$$= \begin{pmatrix} iI & 0 \\ 0 & -iI \end{pmatrix}.$$

The last matrix represents (12), and it has 0 and ∞ as its isolated fixed points. Thus the isolated fixed points of (60) are (I^p, iI_q), (iI_q, I^p) which lie on the dieder manifold of (12).

(2) Taking the lower sign of (57), we have $A = D = 0$. Then

$$\mathfrak{T} = \begin{pmatrix} 0 & B \\ -B'^{-1} & 0 \end{pmatrix}.$$

Since

$$\begin{pmatrix} 0 & B \\ -B'^{-1} & 0 \end{pmatrix}^2 = \begin{pmatrix} -BB'^{-1} & 0 \\ 0 & * \end{pmatrix} = -\mathfrak{I},$$

B is symmetric ($=S$, say). Then we have

$$W = SZ^{-1}S. \tag{62}$$

It has $\pm S$ as its two isolated fixed points. Since 0, ∞, S, $-S$ form a harmonic range, we have the following theorem.

THEOREM 30. *If an involution of the second kind is commutative with a fixed involution of the same kind, its isolated fixed points either lie on the dieder manifold of the fixed involution or separate the isolated fixed points of the fixed involution harmonically.*

The proof of Theorem 29 suggests also the following theorem.

THEOREM 31. *The involution of the second kind commutative with a fixed involution of the second kind depends on $n(n+1)/2$ parameters (complex).*

Proof. The case (2) gives us $n(n+1)/2$ parameters, as S is symmetric. The case (1) gives $2[n/2](n-[n/2])(<n(n+1)/2)$ parameters as shown by the following:

LEMMA. *The solution of the matrix equation*

$$A^2 = I \tag{63}$$

depends on $2[n/2](n-[n/2])$ parameters.

In fact, let A have 1 as characteristic root of multiplicity p, and -1 of multiplicity q. The most general solution is given by

$$\Gamma^{-1}[1, \cdots, 1, -1, \cdots, -1]\Gamma$$

where Γ is nonsingular.

Further, if $\Gamma^{-1}[1, \cdots, 1, -1, \cdots, -1]\Gamma = [1, \cdots, 1, -1, \cdots, -1]$ then

$$\Gamma = \begin{pmatrix} \overset{(p)}{\gamma_1} & 0 \\ 0 & \overset{(q)}{\gamma_2} \end{pmatrix},$$

which depends on p^2+q^2 variables. Thus for a fixed P, A depends on

$$n^2 - p^2 - q^2 = 2pq$$

parameters. This expression has its maximum for $p=[n/2]$.

Since

$$2[n/2](n - [n/2]) \leq 2(n/2)^2 < n(n+1)/2,$$

we have Theorem 30.

19. **Number of parameters of involutions commutative with a given involution.** Now we push Theorem 31 a little further.

THEOREM 32. *The involutions commutative with a fixed involution of the second kind depend on $n(n+1)/2$ parameters.*

Proof. We need only to consider the involutions of the first kind commutative $W=-Z$. Then, we have either

$$Z_1 = AZA', \qquad A^2 = I \tag{64}$$

or, for even n,

$$Z_1 = KZ^{-1}K \tag{65}$$

where K is skew symmetric. (64) depends on $[n/2](n-[n/2])$ parameters by the lemma of the last section. (65) depends on $n(n-1)/2$ parameters. Both do not exceed $n(n+1)/2$.

THEOREM 33. *The involutions commutative with a fixed involution of the first kind of signature (p, q) depend on $p(p+1)+q(q+1)$ parameters.*

Proof. We take $Z_1=HZH$, $H=[1, \cdots, 1, -1, \cdots, -1]$ as the fixed involution. Let

$$\mathfrak{T} = \begin{pmatrix} A & B \\ C & D \end{pmatrix}$$

be an involution commutative with it, then we have

$$\begin{pmatrix} A & B \\ C & D \end{pmatrix}\begin{pmatrix} H & 0 \\ 0 & H \end{pmatrix} = \pm \begin{pmatrix} H & 0 \\ 0 & H \end{pmatrix}\begin{pmatrix} A & B \\ C & D \end{pmatrix}. \tag{66}$$

(1) Taking the upper sign, we have

$$A = \begin{pmatrix} \overset{(p)}{a_1} & 0 \\ 0 & a_2 \end{pmatrix}, \quad B = \begin{pmatrix} \overset{(p)}{b_1} & 0 \\ 0 & b_2 \end{pmatrix}, \quad C = \begin{pmatrix} \overset{(p)}{c_1} & 0 \\ 0 & c_2 \end{pmatrix}, \quad D = \begin{pmatrix} \overset{(p)}{d_1} & 0 \\ 0 & d_2 \end{pmatrix},$$

where

$$\begin{pmatrix} a_1 & b_1 \\ c_1 & d_1 \end{pmatrix}, \quad \begin{pmatrix} a_2 & b_2 \\ c_2 & d_2 \end{pmatrix}$$

are involutions of the same kind. By Theorem 25, the number of parameters is

$$p(p+1) + q(q+1).$$

(2) We take the lower sign in (66). This is possible only when $p=q$. Then

$$A = \begin{pmatrix} 0^{(p)} & a_1 \\ a_2 & 0 \end{pmatrix}, \quad B = \begin{pmatrix} 0 & b_1 \\ b_2 & 0 \end{pmatrix}, \quad C = \begin{pmatrix} 0 & c_1 \\ c_2 & 0 \end{pmatrix}, \quad D = \begin{pmatrix} 0 & d_1 \\ d_2 & 0 \end{pmatrix}.$$

Let

$$T_1 = \begin{pmatrix} a_1 & b_1 \\ c_1 & d_1 \end{pmatrix}, \qquad T_2 = \begin{pmatrix} a_2 & b_2 \\ c_2 & d_2 \end{pmatrix}.$$

We have $T_1T_2 = I^{(p)}$ and T_1 is symplectic. Thus the number of parameters of T_1 is equal to $p(2p+1)$ and T_2 is determined uniquely. Since

$$p(2p + 1) < 2p(p + 1)$$

we have the theorem.

Notice that

$$(67) \qquad \begin{aligned} p(p + 1) + q(q + 1) &= p^2 + q^2 + p + q \\ &\geqq (p + q)^2/2 + p + q > n(n + 1)/2. \end{aligned}$$

20. **Further study of commutative involutions.** First of all, let us classify commutative involutions.

DEFINITION. Two commutative involutions of the second kind, $\mathfrak{T}_1$ and $\mathfrak{T}_2$, which have isolated fixed points forming a harmonic range, are said to be commutative nondegenerately. Otherwise, we say they are commutative degenerately. For the degenerate case, the arithmetic distance between an isolated fixed point of $\mathfrak{T}_1$ and the other of $\mathfrak{T}_2$ is either p or q. We assume that $0 < p \leqq n/2$. Then p is called the *arithmetic distance* of $\mathfrak{T}_1$ and $\mathfrak{T}_2$.

Analytically, $\mathfrak{T}_1$ and $\mathfrak{T}_2$ are commutative nondegenerately if and only if

$$(68) \qquad \mathfrak{T}_1\mathfrak{T}_2 = -\mathfrak{T}_2\mathfrak{T}_1$$

(see (2) of §18). In the present section we are going to establish the following theorem.

THEOREM 34. *Let $n = 2^\sigma\tau$, where τ is odd. There are $\sigma+3$ involutions of the second kind any two of which are commutative nondegenerately, and $\sigma+3$ is the maximal number.*

In other words, *there are $\sigma+3$ nonspecial pairs of points such that any two pairs form a harmonic range.*

We prove the present theorem together with the following:

THEOREM 35. *Let $n = 2^\sigma\tau$, where τ is odd. Let $\mathrm{P}(n)$, $\Sigma(n)$ and $\mathrm{K}(n)$ denote the greatest integer s for which there exist s n-rowed symplectic, symmetric and skew-symmetric matrices $T_1, \cdots, T_s$ such that*

$$(69) \qquad T_i^2 = -I, \qquad T_iT_j = -T_jT_i, \qquad i \neq j,$$

respectively. Then

$$(70) \qquad \mathrm{P}(n) = \sigma + 2 \qquad (\sigma \geqq 1),$$

$$(71) \qquad \Sigma(n) = \sigma + 1 \qquad (\sigma \geqq 0),$$

and

$$\mathrm{K}(n) = \sigma \qquad (\sigma \geqq 1). \tag{72}$$

Notice that (70) implies Theorem 35 and that $\mathrm{P}(n)$ and $\mathrm{K}(n)$ are defined only for even n.

We establish (70), (71) and (72) as consequences of the equalities

$$\mathrm{P}(2n) = 2 + \Sigma(n), \tag{73}$$

$$\Sigma(2n) = 2 + \mathrm{K}(n) \tag{74}$$

and

$$\mathrm{K}(2n) = 1 + \mathrm{K}(n) \tag{75}$$

and the three initial equalities

$$\mathrm{K}(2\tau) = 1, \tag{76}$$

$$\Sigma(\tau) = 1 \tag{77}$$

and

$$\Sigma(2\tau) = 2 \tag{78}$$

for odd τ.

We are going to prove (73) first. We start with the involution (12). By (2) of §18, the involutions commutative with (12) are of the form (62), where S is a symmetric matrix. There is a matrix Γ such that $\Gamma'\Gamma = S^{-1}$. We put $\Gamma' W\Gamma$ and $\Gamma' Z\Gamma$ instead of W and Z, then (12) takes its original form and (62) becomes

$$W = Z^{-1}. \tag{79}$$

Hence any pair of nondegenerately commutative involutions may be carried simultaneously to (12) and (79).

Now we consider an involution of the second kind commutative with both (12) and (79). Since it commutes with (12), it is of the form (62); since it commutes with (79), we have $S^2 = -I$. Let

$$W = S_i Z^{-1} S_i, \qquad 1 \leqq i \leqq \mathrm{P}(2n) - 2,$$

be involutions commutative nondegenerately. Then we have $S_i^2 = -I$, $S_i S_j^{-1} = -S_j S_i^{-1}$, that is,

$$S_i S_j = -S_j S_i, \qquad S_i^2 = -I. \tag{80}$$

Therefore, we have

$$\mathrm{P}(2n) - 2 \leqq \Sigma(n).$$

On the other hand, if we have $\Sigma(n)$ symmetric matrices satisfying (80), we may find $2+\Sigma(n)$ symplectic matrices with the required property. (In the following we shall not repeat this statement at similar occasions.) Therefore we have (73).

From (80), it follow that S_1 has simple elementary divisors and $\pm i$ are its characteristic roots. We have an orthogonal matrix Γ such that

$$\Gamma S_1 \Gamma' = \begin{pmatrix} I^{(a)}i & 0 \\ 0 & -I^{(b)}i \end{pmatrix}.$$

We may put

$$S_1 = \begin{pmatrix} I^{(a)}i & 0 \\ 0 & -I^{(b)}i \end{pmatrix}. \tag{81}$$

If there exists an S_2 such that

$$S_1S_2 = -S_2S_1, \qquad S_1 = -S_2S_1S_2^{-1},$$

then S_1 and $-S_1$ have the same characteristic equation. This is only possible for $a=b$. Therefore, we have (77). Now we take $n=2a$.

From $S_1S_l+S_lS_1=0$, $S_l^2=-I$ we deduce that

$$S_l = \begin{pmatrix} 0 & iA_l^{(a)} \\ iA_l^{-1} & 0 \end{pmatrix}, \qquad \geqq 2, \tag{82}$$

where

$$A_lA_l' = I. \tag{83}$$

Putting

$$\begin{pmatrix} A_2' & 0 \\ 0 & I^{(a)} \end{pmatrix} S_k \begin{pmatrix} A_2 & 0 \\ 0 & I^{(a)} \end{pmatrix}$$

instead of S_k, we find that (81) remains unaltered and S_2 takes the form

$$S_2 = \begin{pmatrix} 0 & iI \\ iI & 0 \end{pmatrix}. \tag{84}$$

If (82) is commutative with (84) nondegenerately, for $l \geqq 3$, then the A_l ($l \geqq 3$) are skew symmetric and satisfy $A_l^2=-I^{(a)}$, $A_lA_k=-A_kA_l$. In case a is odd, there is no such A, we have therefore (78). Otherwise, we have (74).

Now we start again with $(2n)$-rowed skew symmetric matrices

$$\mathfrak{K}_1, \cdots, \mathfrak{K}_s$$

satisfying

$$\mathfrak{K}_i\mathfrak{K}_j = -\mathfrak{K}_j\mathfrak{K}_i, \qquad \mathfrak{K}_i^2 = -\mathfrak{J}. \tag{85}$$

We may take

$$\mathfrak{K}_1 = \begin{pmatrix} 0 & I \\ -I & 0 \end{pmatrix} = \mathfrak{F}.$$

From $\mathfrak{K}_i\mathfrak{K}_j = -\mathfrak{K}_j\mathfrak{K}_i$, we find

$$\mathfrak{K}_k = \begin{pmatrix} K_{1k} & K_{2k} \\ K_{2k} & -K_{1k} \end{pmatrix},$$

where K_{1k} and K_{2k} are skew symmetric matrices.

We introduce the transformation

$$\mathfrak{K}_k\dagger = \frac{i}{2}\begin{pmatrix} I & iI \\ I & -iI \end{pmatrix}\mathfrak{K}_k\begin{pmatrix} I & I \\ iI & -iI \end{pmatrix}. \tag{86}$$

Then $\mathfrak{K}_1\dagger = \mathfrak{F}$ and

$$\mathfrak{K}_k\dagger = \begin{pmatrix} L_k & 0 \\ 0 & M_k \end{pmatrix},$$

where

$$L_k = -K_{2k} + iK_{1k}, \tag{87}$$

$$M_k = K_{2k} + iK_{1k}. \tag{88}$$

$\mathfrak{K}_k\dagger$ $(k \geqq 2)$ can exist only when n is even. Thus we have (76). From (85), we deduce that

$$L_iM_i = I \tag{89}$$

and

$$L_iM_j + L_jM_i = 0, \qquad i \geqq 2, j \geqq 2. \tag{90}$$

Consequently, we have $L_2, \cdots, L_s$ such that

$$L_iL_j^{-1} + L_jL_i^{-1} = 0. \tag{91}$$

We have Γ such that $\Gamma L_2\Gamma' = F$, where

$$F = \begin{pmatrix} 0 & I^{(a)} \\ -I & 0 \end{pmatrix}, \qquad a = 2n.$$

We may take $L_2 = F$. Let

$$N_i = F^{-1}L_i, \qquad 3 \leqq i \leqq s.$$

Then, from (91),

$$N_i^2 = L_2^{-1}L_iL_2^{-1}L_i = -I, \tag{92}$$

$$FN_i + N_iF = L_i + F^{-1}L_iF = 0 \tag{93}$$

and

(94) $N_iN_j + N_jN_i = L_2^{-1}L_jL_2^{-1}L_i + L_2^{-1}L_iL_2^{-1}L_j = -(L_j^{-1}L_i + L_i^{-1}L_j) = 0.$

Further

$$N_i' = L_i'F'^{-1} = L_iF^{-1} = -F^{-1}L_i = -N_i.$$

Thus we have $s-1$ n-rowed skew symmetric matrices $F, N_3, \cdots, N_s$ satisfying (92), (93) and (94). Thus, we have (75).

Remark. If we restrict the T's to be orthogonal in the theorem, the maximum s is equal to $K(n)=\sigma$. In fact, from

$$T^2 = -I, \qquad TT' = I,$$

we deduce immediately $T=-T'$.

21. **Parameters of involutions commutative with a pair of involutions.** Now we establish the characteristic distinction of all sorts of commutative involutions of the second kind.

THEOREM 36. *Given two nondegenerately commutative involutions of the second kind, the involutions of the second kind commutative with both depend on* $[n/2](n-[n/2])$ *parameters (complex).*

Proof. (1) Now we take, without loss of generality, the two fixed involutions as (12) and (11).

The involutions, nondegenerately commutative with both given involutions, are given by

(62) $$W = SZ^{-1}S, \qquad S^2 = -I.$$

We have an orthogonal matrix Γ such that

$$\Gamma S\Gamma' = Hi, \qquad H = [1, \cdots, 1, -1, \cdots, -1].$$

The totality of orthogonal matrices Γ depends on $n(n-1)/2$ parameters. Further from $\Gamma H\Gamma'=H$, we deduce

$$\Gamma = \begin{pmatrix} t^{(p)} & 0 \\ 0 & t^{(q)} \end{pmatrix},$$

which depends on $p(p-1)/2+q(q-1)/2$ parameters. Thus S depends on

$$(n(n-1) - p(p-1) - q(q-1))/2 = pq \leqq [n/2](n-[n/2])$$

parameters.

(2) Consider the involutions commutative degenerately with one of the fixed involutions, say $W=-Z$. Then they are of the form (58). If they commute with $W=-Z^{-1}$, either degenerately or not, then

$$W = AZ^{-1}A' = (AZA')^{-1} = A'^{-1}ZA^{-1},$$

and we deduce $A'A=\pm I$. Then

$$A' = -A'A^2 = \mp A,$$

that is, A is either symmetric or skew-symmetric. If it is symmetric, the number of symmetric involutions A has been counted in (1).

Now we assume that A is skew-symmetric and orthogonal. Consequently n is even, equal to $2p$, say. Since $A^2=-I$, A has $\pm i$ as its characteristic roots and its elementary divisors are simple. Since

$$d(A-\lambda I)=d(A-\lambda I)'=d(-A-\lambda I),$$

the multiplicities of i and $-i$ are equal.

The pair of matrices

$$\begin{pmatrix} 0 & I^{(p)} \\ -I^{(p)} & 0 \end{pmatrix}, \qquad \begin{pmatrix} I^{(p)} & 0 \\ 0 & I^{(p)} \end{pmatrix}$$

has the prescribed elementary divisors. Thus we have an orthogonal matrix Γ such that

$$\Gamma A \Gamma' = \begin{pmatrix} 0 & I \\ -I & 0 \end{pmatrix}.$$

The totality of orthogonal matrices depends on $n(n-1)/2$ parameters and symplectic orthogonal matrices of order $2p$ depend on p^2 parameters. Thus the number of parameters is

$$n(n-1)/2 - n^2/4 < ((n-1)/2)(n-((n-1)/2)) \leqq [n/2](n-[n/2]),$$

and the theorem follows.

THEOREM 37. *The involutions of the second kind commutative with two commutative involutions of the second kind of arithmetic distance p depend on $p(p+1)/2+q(q+1)/2$ parameters.*

Proof. We may write both fixed involutions as (12) and

$$W=-HZH \tag{59}$$

where $H=[1,\cdots,1,-1,\cdots,-1]$.

(1) The involutions commutative degenerately with (12) are given by (58). If they commute with (59), we have $AH=\pm HA$. Consequently, we have

$$A=\begin{pmatrix} a_1^{(p)} & 0 \\ 0 & a_2^{(q)} \end{pmatrix}, \quad \text{or} \quad A=\begin{pmatrix} 0 & b_1^{(p)} \\ b_2^{(p)} & 0 \end{pmatrix},$$

the second case can appear only when $p=n/2$.

(11) The number of parameters of

$$\begin{pmatrix} a_1 & 0 \\ 0 & a_2 \end{pmatrix}, \qquad a_1^2=-I^{(p)}, \qquad a_2^2=-I^{(q)}$$

is equal to $2[p/2](p-[p/2])+2[q/2](q-[q/2])$, by the lemma of §18.

(12) From $A^2=-1$, we deduce

$$\begin{pmatrix} 0 & b_1 \\ b_2 & 0 \end{pmatrix}, \qquad b_1 b_2 = -I^{(p)}.$$

Thus the number of parameters is equal to p^2, which is less than $p(p+1)$.

(2) The involutions commuting with (12) nondegenerately are of the form

$$W = SZ^{-1}S.$$

If they commute with (59), we have $HS = \pm SH$. Consequently, we obtain

$$S = \begin{pmatrix} s_1 & 0 \\ 0 & s_2 \end{pmatrix}, \quad \text{or} \quad S = \begin{pmatrix} 0 & t \\ t' & 0 \end{pmatrix}.$$

The number of parameters in the first case is

$$p(p+1)/2 + q(q+1)/2,$$

and in the second case, $n=2p$, is $p^2<p(p+1)$.

Finally, we have the theorem, since

$$2[p/2](p-[p/2]) + 2[q/2](q-[q/2]) < p(p+1)/2 + q(q+1)/2.$$

22. **Automorphisms of the group of symplectic transformations.** Finally, we are going to prove the following theorem.

THEOREM 38. *A topological automorphism of the group of symplectic transformations is either an inner automorphism or an anti-symplectic transformation.*

More explicitly: Let $\mathfrak{G}$ be the group of symplectic transformations. It may be visualized as the totality of symplectic matrices $\mathfrak{T}$, but we have to identify $\mathfrak{T}$ and $-\mathfrak{T}$. $\mathcal{A}$ is called an automorphism over $\mathfrak{G}$, written as

$$\mathcal{A}(\mathfrak{T}) = \mathfrak{T}^\dagger, \tag{95}$$

if it builds up a one-to-one and bi-continuous relation between $\mathfrak{T}$ and $\mathfrak{T}^\dagger$, and if

$$\mathcal{A}(\mathfrak{T}_1\mathfrak{T}_2) = \mathfrak{T}_1^\dagger\mathfrak{T}_2^\dagger. \tag{96}$$

The conclusion of the theorem is that we have either a symplectic matrix $\mathfrak{P}$ such that

$$\mathcal{A}(\mathfrak{T}) = \mathfrak{P}\mathfrak{T}\mathfrak{P}^{-1} \tag{97}$$

or a symplectic matrix $\mathfrak{P}$ such that

$$\mathcal{A}(\mathfrak{T}) = (\mathfrak{P}^\dagger\mathfrak{T})^\dagger\mathfrak{P}^{-1} = \mathfrak{P}\mathfrak{T}^\dagger\mathfrak{P}^{-1}, \tag{98}$$

for all $\mathfrak{T}$.

Proof. Evidently the automorphism $\mathcal{A}$ carries an involution into an involution and a pair of commutative involutions into a pair of commutative involutions. The involutions commutative with an involution of the first kind of signature (p, q) form a manifold of dimension $p(p+1)+q(q+1)$, by Theorem 33, and those commutative with an involution of the second kind form a manifold of dimension $n(n+1)/2$ by Theorem 32. Since the numbers

$$n(n+1)/2, \qquad p(p+1)+q(q+1), \qquad 1 \leqq q \leqq n/2,$$

are never equal, and since a topological transformation leaves the dimension invariant, $\mathcal{A}$ leaves the kind and signature of involutions invariant.

Further, the manifold formed by the involutions of the second kind permuting with a pair of commutative involutions of arithmetic distance p is of dimension

$$p(p+1)/2+q(q+1)/2$$

by Theorem 37. They are all different for $0<p\leqq n/2$. Moreover the manifold formed by the involutions of the second kind permuting with a pair of nondegenerate commutative involutions is of dimension $[n/2](n-[n/2])$, by Theorem 36. Therefore the "degeneracy" of a pair of commutative involutions is also invariant under $\mathcal{A}$.

Each involution of the second kind is determined uniquely by its two isolated fixed points. Thus, we use (A, B) to denote an involution of the second kind possessing A and B as its two isolated fixed points.

We now prove that two involutions with a common isolated fixed point, and the other two fixed points forming a nonspecial pair, are carried by $\mathcal{A}$ into involutions with the same property. In fact, let

$$(A, B), \qquad (A, C) \tag{99}$$

be the two involutions under consideration, and suppose that $\mathcal{A}$ carries them into

$$(P, Q), \qquad (R, S). \tag{100}$$

The transformations commutative degenerately with both involutions (99) form a group isomorphic to the orthogonal group. In fact we may take $A=0$, $B=\infty$ and $C=I$; the conclusion follows immediately.

Suppose that at least one of the arithmetic distances $r(P, R)$, $r(P, S)$, $r(Q, R)$ and $r(Q, S)$ is different from 0 and n. Then the subgroup formed by the transformations commutative degenerately with (P, Q) and (R, S) is reducible. In fact, we may let, in homogeneous coordinates,

$$P=(I, 0), \qquad Q=(0, I), \qquad R=\left\{\begin{pmatrix} I & 0 & 0 \\ 0 & I & 0 \\ 0 & 0 & 0 \end{pmatrix}, \begin{pmatrix} 0 & 0 & 0 \\ 0 & I & 0 \\ 0 & 0 & I \end{pmatrix}\right\}.$$

Direct verification asserts that the group is formed by the elements of the form:

$$\begin{pmatrix} A & 0 \\ 0 & A'^{-1} \end{pmatrix}, \qquad A = \begin{bmatrix} A_{11} & 0 & 0 \\ A_{21} & A_{22} & 0 \\ A_{31} & A_{32} & A_{33} \end{bmatrix}, \qquad A_{22}A'_{22} = I.$$

Since the orthogonal group cannot be isomorphic to a reducible group, we have the conclusion that either our assertion is true or that

$$r(P, R) = r(P, S) = r(Q, R) = r(Q, S) = n.$$

The latter case cannot happen. In fact, there is no pair of matrices (X, Y) to separate both (A, B) and (A, C) harmonically, and this may be justified easily by putting $A=0$, $B=\infty$ and $C=I$. On the other hand, there exists a nonspecial pair (X, Y) to separate (P, Q) and (R, S) both harmonically. In fact, we may take $P=0$, $Q=\infty$, $R=I$; then the pair $(S_1, -S_1)$ separates both pairs harmonically when $S_1^2=S$. Thus we have established our assertion.

Let $\Sigma(A)$ be the set of involutions having a common fixed point A. Then, by **A**, it is mapped into another set with the same property. In fact, if (A, B), (A, C), $(r(B, C)=n)$ are mapped into $(A\dagger, B\dagger)$, $(A\dagger, C\dagger)$; a third pair (A, D) must be mapped into either $(A\dagger, D\dagger)$ or $(B\dagger, C\dagger)$. The latter case cannot happen, since **A** cannot map a fourth one. We may remove the assumption $r(B, C)=n$ by means of a consideration of continuity.

Suppose $\Sigma(A)$ is carried into $\Sigma(A\dagger)$ by **A**. Therefore **A** induces a mapping of the space of symmetric matrices: $(A\to A\dagger)$.

Let $r(A, B)=p$. The points C satisfying

$$r(A, C) = n, \qquad r(B, C) = q$$

form a manifold of dimension $nq-q(q+1)/2$. Let D be the image of B under the involution (A, C). Then (A, C), (B, D) are two commutative involutions with the arithmetic distance q. Suppose **A** carries $\Sigma(A)$ and $\Sigma(B)$ into $\Sigma(A\dagger)$ and $\Sigma(B\dagger)$. Then we find that $r(A\dagger, B\dagger)=p$ or q. Since $np-p(p+1)/2 \neq nq-q(q+1)/2$ for $p\neq q$, we have $r(A\dagger, B\dagger)=p$.

Therefore the automorphism $\mathcal{A}$ induces a mapping on the space of symmetric matrices possessing the following properties:

(1) It is topological.

(2) It keeps harmonic separation invariant.

(3) It keeps arithmetic distance invariant.

The theorem follows from the generalization of von Staudt's theorem for symmetric matrices.

TSING HUA UNIVERSITY,
KUNMING, CHINA.

Reprinted from the
TRANSACTIONS OF THE AMERICAN MATHEMATICAL SOCIETY
Vol. 61, pp. 229–255, March 1947

GEOMETRIES OF MATRICES. III. FUNDAMENTAL THEOREMS IN THE GEOMETRIES OF SYMMETRIC MATRICES

BY
LOO-KENG HUA

1. **Introduction.** The author gave, in the paper I_1(1), a discussion of the redundancy of the conditions which appeared in the generalization of von Staudt's theorem for symmetric matrices of order 2. The author then could also give a proof for matrices of all even orders but was unable to prove the result in its full generality, and further the proof depended essentially on some results about involutions. The author has now found a simpler proof which holds for any order greater than 2 and which is independent of the results about involutions. Nevertheless, the proof is carried out by means of mathematical induction upon the result of I_1, which depends essentially on the theory of involutions.

Several theorems for geometries keeping an involution as an absolute have also been obtained.

The second part of the paper is concerned with analytic mappings. The projective space of symmetric matrices may also be considered as the extended space of several complex variables as defined by Osgood(2). So far as the author is aware, the completeness of the group of automorphic mappings of an extended space has been established only for two special cases, namely the space of function theory and the complex projective space. If we assume that the group is topological, we may deduce the result from a theorem of E. Cartan(3) for semi-simple groups. In this paper, the problem is solved without any restriction besides analyticity.

As an application of the previous result in combination with the continuity theorem due to Levi(4) and with a result due to the author(5), we solve also the corresponding problem for the group of automorphs of the elliptic space. It should be remarked that the corresponding problem for the hyperbolic space was solved by C. L. Siegel(6) in a recent important paper.

Presented to the Society, April 27, 1946; received by the editors May 30, 1945.

(1) Papers I, I_1, and II under the same title were published in Trans. Amer. Math. Soc. vol. 57 (1945) pp. 441–481, 482–490, vol. 61 (1947) pp. 193–228. They will be referred to in the present paper simply by their numbers.

(2) *Lehrbuch der Funktionentheorie*, vol. 2, 1929.

(3) *La théorie des groupes finis et continus et l'analysis situs*, Mémorial des sciences mathématiques, Paris, 1930, p. 51.

(4) E. E. Levi, Annales de Mathématiques Pures et Appliquées (3) vol. 17 (1910).

(5) L. K. Hua, Trans. Amer. Math. Soc. vol. 59 (1946) pp. 508–523.

(6) C. L. Siegel, Amer. J. Math. vol. 65 (1943) pp. 1–86.

2. **Arithmetic distance.** Let $\mathfrak{S}$ and $\mathfrak{T}$ be two sets of points in the "projective" space of symmetric matrices. The arithmetic distance between $\mathfrak{S}$ and $\mathfrak{T}$ is defined to be the least upper bound of the arithmetic distance between any two points S and T, where S and T belong to $\mathfrak{S}$ and $\mathfrak{T}$ respectively. It is denoted by $r(\mathfrak{S}, \mathfrak{T})$.

In particular for $\mathfrak{S}=\mathfrak{T}$, the distance $r(\mathfrak{S}, \mathfrak{S})$ is called the arithmetic diameter of the set $\mathfrak{S}$.

Given a positive integer ρ, a set $\mathfrak{S}$ is called a *maximal set of rank* ρ, if $\mathfrak{S}$ is of arithmetic diameter ρ, and if any set properly containing $\mathfrak{S}$ is of arithmetic diameter greater than ρ.

THEOREM 1. *A normal subspace of rank* ρ *is a maximal set of rank* ρ.

Proof. Since arithmetic distance is invariant, we may take the normal subspace to consist of the points

$$\begin{pmatrix} X^{(\rho)} & 0 \\ 0 & 0 \end{pmatrix}.$$

For any vector v and any number a, not both zero, we have $X^{(\rho)}$ such that

$$\begin{pmatrix} X^{(\rho)} & v' \\ v & a \end{pmatrix}$$

is a nonsingular $(\rho+1)$-rowed matrix. The theorem is now evident.

Is the converse of Theorem 1 true? It is true for $\rho=1$, but in general we have the following "gegenbeispiel":

The set of all matrices of rank 1 forms a maximal set $\mathfrak{S}$ of rank 2, but does not form a normal subspace. In fact, the equation

$$d\left(\begin{pmatrix} x & y & z \\ y & u & v \\ z & v & w \end{pmatrix} - \begin{pmatrix} a^2 & ab & ac \\ ab & b^2 & bc \\ ac & bc & c^2 \end{pmatrix}\right) = 0$$

for all a, b, c implies that

$$\begin{pmatrix} x & y & z \\ y & u & v \\ z & v & w \end{pmatrix}$$

is of rank 1. This fact shows that $\mathfrak{S}$ is a maximal set. On the other hand, we have $r(\mathfrak{S}, 0)=1$, which cannot hold in the normal subspace.

3. **Arithmetic property of the normal subspace.** Now we extend the concept of dieder manifold a little further:

DEFINITION. Let P_1 and P_2 be two points. The points X satisfying

$$r(P_1, X) + r(X, P_2) = r(P_1, P_2)$$

are said to form a *dieder manifold spanned by* P_1 and P_2.

THEOREM 2. *A maximal set of rank* ρ *is a normal space if and only if it contains all dieder manifolds spanned by any two points of the set.*

Proof. (1) Let

$$P = 0, \qquad Q = [1, \cdots, 1, 0, \cdots, 0]$$

(where the 1's are ρ in number); we shall prove that, if

$$r(P, X) + r(X, Q) = r(P, Q),$$

then X takes the form

$$\begin{pmatrix} X_1^{(\rho)} & 0 \\ 0 & 0 \end{pmatrix}.$$

We consider the pair of matrices Q and X. There exists a nonsingular matrix Γ(7) such that

$$\Gamma^{-1}X\Gamma'^{-1} = [1, \cdots, 1, 0, \cdots, 0], \qquad 0 \leqq p \leqq \rho$$

(where the 1's are p in number), and

$$\Gamma^{-1}Q\Gamma'^{-1} = Q.$$

From the second equation, we find that

$$\Gamma = \begin{pmatrix} A & B \\ 0 & C \end{pmatrix}, \qquad AA' = I.$$

Then

$$X = \Gamma \begin{pmatrix} I^{(p)} & 0 \\ 0 & 0 \end{pmatrix} \Gamma'.$$

Consequently, we have the assertion.

Therefore a normal subspace contains the dieder manifold spanned by any pair of points of the subspace.

(2) Without loss of generality, we may assume that the maximal set contains two points

$$P = 0, \qquad Q = [1, \cdots, 1, 0, \cdots, 0]$$

(where the 1's are ρ in number). So it contains

$$P_i = [1, \cdots, 1, 0, 1, \cdots, 1, 0, \cdots, 0], \qquad 1 \leqq i \leqq \rho$$

(where the first zero in the brackets appears in the ith place, the second in the $(\rho-1)$th), and

$$P_{ij} = P_iP_j, \qquad P_{ijk} = P_iP_jP_k,$$

and so on.

(7) See Turnbull and Aitken, *Theory of canonical matrices*, 1931, p. 135.

Let

$$X = (a_{rs})_{1 \leqq r,s \leqq n}.$$

be a point of the set. Then

$$r(X, P_i) \leqq \rho, \qquad r(X, P_{ij}) \leqq \rho, \qquad r(X, P_{ijk}) \leqq \rho,$$

and so on. Consequently, the matrix

$$\begin{pmatrix} a_{11}-\epsilon_1 & a_{12} & \cdots & a_{1\rho} & a_{1\,\rho+1} & \cdots & a_{1n} \\ a_{12} & a_{22}-\epsilon_2 & \cdots & a_{2\rho} & a_{2\,\rho+1} & \cdots & a_{2n} \\ \cdot & \cdot & \cdot & \cdot & \cdot & \cdot & \cdot \\ a_{1\rho} & a_{2\rho} & \cdots & a_{\rho\rho}-\epsilon_\rho & a_{\rho\,\rho+1} & \cdots & a_{\rho n} \\ a_{1\,\rho+1} & a_{2\,\rho+1} & \cdots & a_{\rho\,\rho+1} & a_{\rho+1\,\rho+1} & \cdots & a_{\rho+1\,n} \\ \cdot & \cdot & \cdot & \cdot & \cdot & \cdot & \cdot \\ a_{1n} & a_{2n} & \cdots & a_{\rho n} & a_{\rho+1\,n} & \cdots & a_{nn} \end{pmatrix}$$

is always of rank ρ for $\epsilon_1, \cdots, \epsilon_\rho$ arbitrarily taken from 0 and 1. In particular,

$$\begin{vmatrix} a_{11}-\epsilon_1 & a_{12} & \cdots & a_{1\rho} & a_{1i} \\ a_{12} & a_{22}-\epsilon_2 & \cdots & a_{2\rho} & a_{2i} \\ \cdot & \cdot & \cdot & \cdot & \cdot \\ a_{1\rho} & a_{2\rho} & \cdots & a_{\rho\rho}-\epsilon_\rho & a_{\rho i} \\ a_{1i} & a_{2i} & \cdots & a_{\rho i} & a_{ii} \end{vmatrix} = 0$$

for $\rho < i \leqq n$ and any choice of ϵ. Putting $\epsilon_1 = 0$ and $\epsilon_1 = 1$ and subtracting the results, we have

$$\begin{vmatrix} a_{22}-\epsilon_2 & \cdots & a_{2\rho} & a_{2i} \\ \cdot & \cdot & \cdot & \cdot \\ a_{2\rho} & \cdots & a_{\rho\rho}-\epsilon_\rho & a_{\rho i} \\ a_{2i} & \cdots & a_{\rho i} & a_{ii} \end{vmatrix} = 0.$$

Repeating the same process, we have finally

$$\begin{vmatrix} a_{\rho\rho}-\epsilon_\rho & a_{\rho i} \\ a_{\rho i} & a_{ii} \end{vmatrix} = 0$$

for $\epsilon_\rho = 0$ and 1. Consequently $a_{ii} = 0$ and $a_{\rho i} = 0$. Varying ρ, we find that $a_{ji} = 0$ for all $1 \leqq j \leqq \rho$. Further varying i from $\rho+1$ to n, we have the assertion.

Therefore the elements of the set are of the form

$$\begin{pmatrix} X^{(\rho)} & 0 \\ 0 & 0 \end{pmatrix}.$$

By Theorem 1 and (1) we have the second part of our theorem.

Since pairs of points with arithmetic distance ρ $(1 \leqq \rho \leqq n)$ form a transitive set, the proof of the previous theorem establishes also the following:

THEOREM 3. *Given two points with arithmetic distance ρ, there is one and only one normal subspace of rank ρ which contains both.*

4. Proof of the fundamental theorem.

THEOREM 4. *A continuous mapping carrying symmetric matrices into symmetric matrices and leaving arithmetic distance invariant is either a symplectic transformation or an anti-symplectic transformation.*

Proof. (1) The theorem was proved for matrices of order 2 in I_1. Now we shall establish the general theorem by induction on the order of matrices. We thus suppose that the order of matrices is not less than 3.

(2) Let

$$\Gamma(Z) = Z_1 \tag{1}$$

be a mapping satisfying our conditions. The points of the form

$$\begin{pmatrix} \overset{(n-1)}{*} & 0 \\ 0 & 0 \end{pmatrix}$$

form a normal subspace of rank $n-1$. Since the arithmetic distance is invariant, the set of points

$$\Gamma\begin{pmatrix} \overset{(n-1)}{*} & 0 \\ 0 & 0 \end{pmatrix}$$

forms also a normal subspace of rank $n-1$. Since the totality of all normal subspaces of rank $n-1$ form a transitive set under the group of symplectic transformations, we may therefore assume that Γ satisfies

$$\Gamma\begin{pmatrix} \overset{(n-1)}{W} & 0 \\ 0 & 0 \end{pmatrix} = \begin{pmatrix} \overset{(n-1)}{W_1} & 0 \\ 0 & 0 \end{pmatrix}. \tag{2}$$

It induces a continuous mapping on the $(n-1)$-rowed matrices W and it keeps arithmetic distance invariant. Therefore by the hypothesis of induction we find either a symplectic mapping

$$W = (\alpha W_1 + \beta)(\gamma W_1 + \delta)^{-1}, \qquad \alpha = \alpha^{(n-1)}, \tag{3}$$

and so on, or an anti-symplectic mapping

$$\overline{W} = (\alpha W_1 + \beta)(\gamma W_1 + \delta)^{-1}. \tag{4}$$

The symplectic transformation $Z = (AZ_1 + B)(CZ_1 + D)^{-1}$ with

$$A = \begin{pmatrix} \alpha & 0 \\ 0 & 1 \end{pmatrix}, \quad B = \begin{pmatrix} \beta & 0 \\ 0 & 0 \end{pmatrix}, \quad C = \begin{pmatrix} \gamma & 0 \\ 0 & 0 \end{pmatrix}, \quad D = \begin{pmatrix} \delta & 0 \\ 0 & 1 \end{pmatrix}$$

carries the transformation Γ inducing (3) into a new one satisfying

$$\Gamma\begin{pmatrix} W^{(n-1)} & 0 \\ 0 & 0 \end{pmatrix} = \begin{pmatrix} W^{(n-1)} & 0 \\ 0 & 0 \end{pmatrix}. \tag{5}$$

Applying the same method to those Γ inducing (4), we conclude, in both cases, that we may assume, without loss of generality, that Γ satisfies (5).

(3) Since

$$\Gamma\begin{pmatrix} 0^{(n-1)} & 0 \\ 0 & 1 \end{pmatrix}$$

is a matrix of rank 1, we may let

$$\Gamma\begin{pmatrix} 0^{(n-1)} & 0 \\ 0 & 1 \end{pmatrix} = (a_1, \cdots, a_n)'(a_1, \cdots, a_n).$$

Since the arithmetic distance between

$$\begin{pmatrix} I^{(n-1)} & 0 \\ 0 & 0 \end{pmatrix}, \quad \begin{pmatrix} 0^{(n-1)} & 0 \\ 0 & 1 \end{pmatrix}$$

is equal to n, we have $a_n \neq 0$. The transformation $Z_1 = A'ZA$ with

$$A = \begin{pmatrix} I^{(n-1)} & 0 \\ -va_n^{-1} & a_n^{-1} \end{pmatrix}, \qquad v = (a_1, \cdots, a_{n-1}),$$

carries (5) into itself and $(a_1, \cdots, a_n)'(a_1, \cdots, a_n)$ into

$$\begin{pmatrix} 0^{(n-1)} & 0 \\ 0 & 1 \end{pmatrix}.$$

Therefore, we may assume further that

$$\Gamma\begin{pmatrix} 0^{(n-1)} & 0 \\ 0 & 1 \end{pmatrix} = \begin{pmatrix} 0^{(n-1)} & 0 \\ 0 & 1 \end{pmatrix}. \tag{6}$$

(4) The transformation Γ leaves

$$\begin{bmatrix} 0^{(1)} & 0 & 0 \\ 0 & I^{(n-2)} & 0 \\ 0 & 0 & 0 \end{bmatrix}, \quad \begin{bmatrix} 0 & 0 & 0 \\ 0 & 0^{(n-2)} & 0 \\ 0 & 0 & 1 \end{bmatrix}$$

invariant, therefore, by Theorem 2, we find

$$\Gamma\begin{pmatrix} 0 & 0 \\ 0 & W^{(n-1)} \end{pmatrix} = \begin{pmatrix} 0 & 0 \\ 0 & W_1^{(n-1)} \end{pmatrix}.$$

By the supposition of induction, we have either

(7) $$W^{(n-1)} = (\alpha W_1 + \beta)(\gamma W_1 + \delta)^{-1}$$

or

(8) $$\overline{W}^{(n-1)} = (\alpha W_1 + \beta)(\gamma W_1 + \delta)^{-1},$$

where

$$\begin{pmatrix} \alpha & \beta \\ \gamma & \delta \end{pmatrix}$$

is a $2(n-1)$-rowed symplectic matrix.

According to (5), the transformation (7) and (8) leaves every point of the form

(9) $$W = \begin{pmatrix} X^{(n-2)} & 0 \\ 0 & 0 \end{pmatrix}$$

invariant; (8) cannot hold.

Since (7) leaves every point of the form (9) invariant, we have

$$\alpha = \begin{pmatrix} \pm I^{(n-2)} & a_2' \\ 0 & a_4 \end{pmatrix}, \qquad \beta = 0$$

$$\gamma = \begin{pmatrix} 0 & c_2' \\ c_3 & c_4 \end{pmatrix}, \qquad \delta = \begin{pmatrix} \pm I^{(n-2)} & 0 \\ \mp a_2 a_4^{-1} & a_4^{-1} \end{pmatrix}.$$

Further, (7) carries

$$\begin{pmatrix} 0 & 0 \\ 0 & 1^{(1)} \end{pmatrix}$$

into itself; we have $a_2 = 0$.

Let

$$A = \begin{pmatrix} \pm 1 & 0 \\ 0 & \alpha^{(n-1)} \end{pmatrix}, \qquad B = \begin{pmatrix} 0 & 0 \\ 0 & \beta^{(n-1)} \end{pmatrix},$$

$$C = \begin{pmatrix} 0 & 0 \\ 0 & \gamma^{(n-1)} \end{pmatrix}, \qquad D = \begin{pmatrix} \pm 1 & 0 \\ 0 & \delta^{(n-1)} \end{pmatrix}.$$

The transformation

$$Z = (AZ_1 + B)(CZ_1 + D)^{-1}$$

carries Γ into a new one satisfying (5) and

(10) $$\Gamma\begin{pmatrix} 0 & 0 \\ 0 & W^{(n-1)} \end{pmatrix} = \begin{pmatrix} 0 & 0 \\ 0 & W^{(n-1)} \end{pmatrix}$$

since the transformation leaves every point

$$\begin{pmatrix} X_1^{(n-1)} & 0 \\ 0 & 0 \end{pmatrix}$$

invariant.

(5) Since

$$\Gamma\begin{pmatrix}1 & 0\\ 0 & 0^{(n-1)}\end{pmatrix}=\begin{pmatrix}1 & 0\\ 0 & 0^{(n-1)}\end{pmatrix},\qquad \Gamma\begin{pmatrix}0^{(n-1)} & 0\\ 0 & 1\end{pmatrix}=\begin{pmatrix}0^{(n-1)} & 0\\ 0 & 1\end{pmatrix},$$

we have

$$\Gamma\begin{bmatrix}x_{11} & 0 & \cdots & 0 & x_{1n}\\ 0 & 0 & \cdots & 0 & 0\\ \cdot & \cdot & \cdots & \cdot & \cdot\\ 0 & 0 & \cdots & 0 & 0\\ x_{1n} & 0 & \cdots & 0 & x_{nn}\end{bmatrix}=\begin{bmatrix}x'_{11} & 0 & \cdots & 0 & x'_{1n}\\ 0 & 0 & \cdots & 0 & 0\\ \cdot & \cdot & \cdots & \cdot & \cdot\\ 0 & 0 & \cdots & 0 & 0\\ x'_{1n} & 0 & \cdots & 0 & x'_{nn}\end{bmatrix}$$

and by hypothesis of induction, we have

$$\begin{pmatrix}x_{11} & x_{1n}\\ x_{1n} & x_{nn}\end{pmatrix}=\left(\alpha\begin{pmatrix}x'_{11} & x'_{1n}\\ x'_{1n} & x'_{nn}\end{pmatrix}+\beta\right)\left(\gamma\begin{pmatrix}x'_{11} & x'_{1n}\\ x'_{1n} & x'_{nn}\end{pmatrix}+\delta\right)^{-1},$$

which is a symplectic transformation. (Notice that we omit the "anti-symplectic" case by the same reason given in (3).)

Since it leaves every point

$$\begin{pmatrix}x & 0\\ 0 & 0\end{pmatrix},\quad \begin{pmatrix}0 & 0\\ 0 & y\end{pmatrix}$$

invariant, we have either

$$\alpha=I^{(2)},\qquad \beta=0,\qquad \gamma=\begin{pmatrix}0 & c\\ c & 0\end{pmatrix},\qquad \delta=I^{(2)}$$

or

$$\alpha=\begin{pmatrix}1 & 0\\ 0 & -1\end{pmatrix},\qquad \beta=0,\qquad \gamma=\begin{pmatrix}0 & c\\ -c & 0\end{pmatrix},\qquad \delta=\begin{pmatrix}1 & 0\\ 0 & -1\end{pmatrix}.$$

For the first case the transformation with

$$A=I,\qquad B=0,\qquad C=\begin{bmatrix}0 & 0 & \cdots & 0 & c\\ 0 & 0 & \cdots & 0 & 0\\ \cdot & \cdot & \cdots & \cdot & \cdot\\ 0 & 0 & \cdots & 0 & 0\\ c & 0 & \cdots & 0 & 0\end{bmatrix},\qquad D=I$$

and for the second case the transformation with

$$A = [1, 1, \cdots, 1, -1], \quad B = 0, \quad C = \begin{pmatrix} 0 & 0 \cdots 0 & c \\ 0 & 0 \cdots 0 & 0 \\ \cdot & \cdot \cdots \cdot & \cdot \\ 0 & 0 \cdots 0 & 0 \\ -c & 0 \cdots 0 & 0 \end{pmatrix}, \quad D = A,$$

carry Γ into a new one which satisfies (5), (10) and

$$\text{(11)} \qquad \Gamma \begin{pmatrix} x_{11} & 0 \cdots 0 & x_{1n} \\ 0 & 0 \cdots 0 & 0 \\ \cdot & \cdot \cdots \cdot & \cdot \\ 0 & 0 \cdots 0 & 0 \\ x_{1n} & 0 \cdots 0 & x_{nn} \end{pmatrix} = \begin{pmatrix} x_{11} & 0 \cdots 0 & x_{1n} \\ 0 & 0 \cdots 0 & 0 \\ \cdot & \cdot \cdots \cdot & \cdot \\ 0 & 0 \cdots 0 & 0 \\ x_{1n} & 0 \cdots 0 & x_{nn} \end{pmatrix}.$$

(6) In particular, we conclude that we may assume that

$$\Gamma([\lambda_1, \cdots, \lambda_{n-1}, 0]) = [\lambda_1, \cdots, \lambda_{n-1}, 0],$$
$$\Gamma([0, \lambda_2, \cdots, \lambda_n]) = [0, \lambda_2, \cdots, \lambda_n],$$

and

$$\Gamma([\lambda_1, 0, \cdots, 0, \lambda_n]) = [\lambda_1, 0, \cdots, 0, \lambda_n].$$

Now we are going to prove that

$$\text{(12)} \qquad \Gamma([\lambda_1, \cdots, \lambda_n]) = [\lambda_1, \cdots, \lambda_n].$$

Since

$$r([\lambda_1, \cdots, \lambda_n], [\lambda_1, \cdots, \lambda_{n-1}, 0]) = 1$$

we have

$$\Gamma([\lambda_1, \cdots, \lambda_n]) = [\lambda_1, \cdots, \lambda_{n-1}, 0] + (a_1, \cdots, a_n)'(a_1, \cdots, a_n).$$

Since

$$r([\lambda_1, \cdots, \lambda_n], [0, \lambda_2, \cdots, \lambda_n]) = 1,$$

we have $a_2 = a_3 = \cdots = a_{n-1} = 0$. Further, since

$$r([\lambda_1, \lambda_2, \cdots, \lambda_n], [\lambda_1, 0, \cdots, 0, \lambda_n])$$
$$= r([\lambda_1, \cdots, \lambda_{n-1}, 0] + (a_1, 0, \cdots, 0, a_n)'(a_1, 0, \cdots, 0, a_n), [\lambda_1, 0, \cdots, 0, \lambda_n]),$$

we obtain that the matrix

$$\begin{pmatrix} a_1^2 & a_1 a_n \\ a_1 a_n & a_n^2 - \lambda_n \end{pmatrix}$$

is of rank zero, that is $a_1 = 0$ and $a_n^2 = \lambda_n$. Thus we have the assertion.

(7) Let

$$\Gamma(Z) = Z_1, \qquad Z = (z_{ij}), \qquad Z_1 = (z'_{ij}).$$

Since (12) holds, the equation

$$d(Z - [\lambda_1, \cdots, \lambda_n]) = 0$$

(consider Z as fixed and vary $\lambda_1, \cdots, \lambda_n$) implies

$$d(Z_1 - [\lambda_1, \cdots, \lambda_n]) = 0$$

and vice versa. This implies

$$d(Z - [\lambda_1, \cdots, \lambda_n]) = d(Z_1 - [\lambda_1, \cdots, \lambda_n])$$

identically in $\lambda_1, \cdots, \lambda_n$. Consequently, we have

$$z_{ii} = z'_{ii}, \qquad z_{ii}z_{jj} - z_{ij}^2 = z'_{ii}z'_{jj} - z'^2_{ij},$$

and

$$z_{ij} = \pm z'_{ij}.$$

The transformation

$$[1, -1, 1, \cdots, 1]Z[1, -1, 1, \cdots, 1]$$

carries z_{ii} into itself and z_{12} into $-z_{12}$. Thus we may assume that

$$z_{1i} = z'_{1i}.$$

By the same consideration given in the proof of Theorem 8 of I, we have

$$\Gamma(Z) = Z.$$

The theorem therefore follows.

5. **Affine geometry of symmetric matrices.**

THEOREM 5. *A continuous mapping carrying finite points into finite points, infinite points into infinite points, and keeping arithmetic distance invariant is either an affine mapping*

$$W = AZA'rS \tag{13}$$

or an anti-affine mapping

$$\overline{W} = AZA' + S. \tag{13'}$$

This is an immediate consequence of Theorem 4.

6. **Möbius geometry of symmetric matrices.** As in II, we let $\mathfrak{J}$ be a fundamental involution as an absolute, for example:

$$W = HZH, \tag{14}$$

where $H=[1, 1, \cdots, 1, -1]$. The arithmetic distance between P and $\mathfrak{J}(P)$ is either 0 or 2. In case $P=\mathfrak{J}(P)$ we say that P is a *point-matrix.*

THEOREM 6. *A continuous mapping carrying point-matrices into point-matrices, non-point-matrices into non-point-matrices and keeping arithmetic distance invariant is a Möbius mapping or an anti-Möbius mapping.*

Proof. We take $\mathfrak{J}$ in the form of (14), which carries matrices of the form

$$\begin{pmatrix} Z_1^{(n-1)} & 0 \\ 0 & z \end{pmatrix} \tag{15}$$

into themselves. Thus the mappings under consideration carry the set defined by (15) onto itself.

Without loss of generality we may assume that the mappings keep 0 and ∞ invariant; by Theorem 4, it takes either the form

$$W = AZA'$$

or

$$\overline{W} = AZA'.$$

For the first case, we put

$$A = \begin{pmatrix} A_1 & \alpha_1' \\ \alpha_2 & a \end{pmatrix},$$

and we have

$$\begin{pmatrix} W_1 & 0 \\ 0 & w \end{pmatrix} = \begin{pmatrix} A_1 & \alpha_1' \\ \alpha_2 & a \end{pmatrix}\begin{pmatrix} Z_1 & 0 \\ 0 & z \end{pmatrix}\begin{pmatrix} A_1' & \alpha_2' \\ \alpha_1 & a \end{pmatrix}.$$

Consequently, we have

$$A_1Z_1\alpha_2' + \alpha_1' za = 0$$

for all Z_1 and z. It follows that

(i) $$\alpha_1 = 0 \quad \text{or} \quad a = 0$$

and

(ii) $$\alpha_2 = 0 \quad \text{or} \quad A_1 = 0.$$

The case with $\alpha_1=0$, $\alpha_2=0$ is what we are looking for. The other cases cannot happen by the nonsingularity of A, except when $n=2$, $a=A_1=0$. For this case we have also a Möbius mapping.

A similar method may be used for the second case.

7. **Manifold at infinity.** Let M be an idempotent symmetric matrix, that is, $M^2=M$. Then $I-M$ is also an idempotent symmetric matrix, since $(I-M)^2=I-2M+M^2=I-M$. Further M and $I-M$ annihilate each other, namely

$$M(I-M) = (I-M)M = 0.$$

From an idempotent symmetric matrix M, we can construct a symplectic transformation

$$W = (MZ + (I - M))(-(I - M)Z + M)^{-1}.$$

Now we consider those transformations with diagonal M. They are 2^n in number, including the identity. The $2^n - 1$ non-identity transformations are called the *fundamental semi-involutions* of the space of symmetric matrices. They are of fundamental importance in the study of the extended space.

For the sake of later use, we give the explicit expression of the semi-involution with

$$M = \begin{pmatrix} I^{(r)} & 0 \\ 0 & 0 \end{pmatrix}.$$

Let

$$Z = \begin{pmatrix} Z_{11}^{(r)} & Z_{12} \\ Z_{12}' & Z_{22} \end{pmatrix};$$

we have

$$W = \begin{pmatrix} Z_{11} - Z_{12}Z_{22}^{-1}Z_{12}' & -Z_{12}Z_{22}^{-1} \\ -Z_{22}^{-1}Z_{12}' & -Z_{22}^{-1} \end{pmatrix}. \tag{16}$$

THEOREM 7. *Every point at infinity can be carried into a finite point by one of the* $2^n - 1$ *semi-involutions.*

Proof. Suppose that

$$(Z_1, Z_2)$$

be the homogeneous coordinate of a point at infinity, with Z_2 of rank r. There exist two permutation matrices(8) P and Q such that

$$PZ_2Q = \begin{pmatrix} Z_{11}^{(r)} & * \\ * & * \end{pmatrix}$$

where $Z_{11}^{(r)}$ is nonsingular. We have a nonsingular R such that

$$RPZ_2Q = \begin{pmatrix} Z_{11} & Z_{12} \\ 0 & 0 \end{pmatrix}.$$

Let

$$RP(Z_1, Z_2)\begin{pmatrix} Q'^{-1} & 0 \\ 0 & Q \end{pmatrix} = (W_1, W_2).$$

Then

(8) Corresponding to a permutation

$$\begin{pmatrix} 1, 2, \cdots, n \\ i_1, i_2, \cdots, i_n \end{pmatrix}$$

we have a permutation matrix which is the matrix of the linear transformation $x_p' = x_{i_p}$ $(1 \leq p \leq n)$. Evidently, we have $P' = P^{-1}$.

$$W_2 = \begin{pmatrix} Z_{11} & Z_{12} \\ 0 & 0 \end{pmatrix}.$$

Let

$$W_1 = \begin{pmatrix} W_{11} & W_{12} \\ W_{21} & W_{22} \end{pmatrix}$$

and

$$T = \begin{pmatrix} I & -\overset{-1}{Z}_{11} Z_{12} \\ 0 & I \end{pmatrix}.$$

We have immediately

$$W_2 T = \begin{pmatrix} Z_{11} & 0 \\ 0 & 0 \end{pmatrix}$$

and

$$W_1 T'^{-1} = \begin{pmatrix} * & * \\ 0 & W_{22} \end{pmatrix}.$$

The zero at the left lower corner is obtained because (W_1, W_2) is a symmetric pair. Then W_{22} is nonsingular, since (W_1, W_2) is a nonsingular pair.

Let

$$M = \begin{pmatrix} I^{(r)} & 0 \\ 0 & 0 \end{pmatrix}.$$

Then

$$(W_1, W_2)\begin{pmatrix} M & I - M \\ -(I - M) & M \end{pmatrix} = (P_1, P_2),$$

where

$$P_1 = W_1(I - M) + W_2 M = \begin{pmatrix} 0 & W_{12} \\ 0 & W_{22} \end{pmatrix} + \begin{pmatrix} Z_{11} & Z_{12} \\ 0 & 0 \end{pmatrix},$$

which is evidently nonsingular.

Now we consider

$$\begin{pmatrix} Q & 0 \\ 0 & Q \end{pmatrix}\begin{pmatrix} M & I - M \\ -(I - M) & M \end{pmatrix}\begin{pmatrix} Q' & 0 \\ 0 & Q' \end{pmatrix} = \begin{pmatrix} M_1 & I - M_1 \\ -(I - M_1) & M_1 \end{pmatrix},$$

$$M_1 = QMQ^{-1},$$

which is a semi-involution, since $Q' = Q^{-1}$. The theorem is therefore established.

Since

$$d(-(I - M)Z + M) \tag{17}$$

is a principal minor of Z, we have the following theorem.

THEOREM 8. *The manifold at infinity is carried by the* $2^n - 1$ *semi-involutions into manifolds defined by equating the principal minrrs of* Z *to zero.*

8. **Group of analytic automorphs of the space.** Now we come to the second part of the paper. The projective space of symmetric matrices may be regarded as an extended space of several complex variables as defined by Osgood. The aim of the present section is to establish the following theorem.

THEOREM 9. *An analytic mapping carrying the extended space onto itself is a symplectic mapping.*

Proof. (1) Let

$$W = \Gamma(Z) \tag{18}$$

be an analytic mapping carrying the projective space of symmetric matrices onto itself. By a theorem due to Osgood[9], the mapping is birational; consequently, (18) may be written as

$$w_{ij} = p_{ij}(Z)/q(Z) \tag{19}$$

where $p_{ij}(Z)$ and $q(Z)$ are $2^{-1}n(n+1)+1$ polynomials without common divisor.

(2) There is a point S at which

$$p_{ij}(S) \neq 0, \qquad q(S) \neq 0.$$

The transformation $Z=Z_1+S$ carries (18) into a new one with

$$p_{ij}(0) \neq 0, \qquad q(0) \neq 0. \tag{20}$$

The transformation

$$W = \Gamma(-Z_1^{-1}) \tag{21}$$

maps also the space onto itself, and

$$w_{ij} = p_{ij}(-Z_1^{-1})(d(Z_1))^{\lambda}/q(-Z_1^{-1})(d(Z_1))^{\lambda},$$

where λ is the least integer making all numerators and denominators integral. By (20), $p_{ij}(-Z_1^{-1})(d(Z_1))^{\lambda}$ and $q(-Z_1^{-1})(d(Z_1))^{\lambda}$ are of degree $n\lambda$. Consider the Jacobian of (21). Let Δ and Δ_1 be the inverses of the Jacobians of (18) and (21) respectively. Notice that Δ and Δ_1 are polynomials, for otherwise there would exist some point for which the Jacobians vanish. Then we have

$$\Delta_1(Z_1) = \Delta(-Z_1^{-1})(d(Z_1))^{n+1},$$

since the Jacobian of $Z=-Z_1^{-1}$ is $(d(Z_1))^{-(n+1)}$. Since $q(0)\neq 0$, we have $\Delta(0)\neq 0$. Consequently $\Delta_1(Z_1)$ is a polynomial of degree $n(n+1)$.

Without loss of generality, we may now assume that p_{ij} and q are polynomials of degree $n\lambda$ and that the inverse of the Jacobian of (18) is a polynomial of degree $n(n+1)$ and its term of highest degree is equal to a constant

(9) Ibid. p. 295.

multiple of $(d(Z))^{n+1}$. The highest terms of p_{ij} and q are also constant multiples of $(d(Z))^{\lambda}$.

(3) We decompose q into irreducible factors

$$q = q_1^{\lambda_1} \cdots q_l^{\lambda_l}. \tag{22}$$

Now we are going to establish that $(q_1 \cdots q_l)^{n+1}$ divide Δ. By some easy transformation we may suppose that q_1 does not divide p_{nn}. We consider the product of the transformation (16) with $r=n-1$ and (18). The inverse of the Jacobian of the new mapping is equal to

$$(p_{nn}/q)^{n+1}\Delta,$$

which is a polynomial. Thus q_1^{n+1} divides Δ. Therefore we have the assertion.

(4) Since $d(Z)$ is irreducible and

$$\Delta(Z) = c(d(Z))^{n+1} + \cdots,$$

we have immediately that $l=1$. Therefore

$$q(Z) = (ad(Z) + \cdots)^{\lambda}.$$

Further (16) with $(n-1)$ carries p_{nn} into the denominator, then we have also

$$p_{nn}(Z) = (a_{nn}d(Z) + \cdots)^{\lambda}.$$

The mapping cannot be one-to-one except when $\lambda=1$.

Now we may assume that

$$q(Z) = ad(Z) + \cdots, \tag{23}$$

$$p_{ij}(Z) = a_{ij}d(Z) + \cdots, \tag{24}$$

and

$$\Delta(Z) = c(q(Z))^{n+1}. \tag{25}$$

(5) Now we shall prove that $p_{ii}p_{jj}-p_{ij}^2$ is divisible by q. In fact, by (16) with $r=n-2$, we have

$$W = \begin{pmatrix} * & * \\ * & \dfrac{q}{p_{n-1\,n-1}p_{nn} - p_{n\,n-1}^2}\begin{pmatrix} p_{nn} & -p_{n\,n-1} \\ -p_{n\,n-1} & p_{n-1\,n-1}\end{pmatrix}\end{pmatrix}.$$

If q does not divide $p_{n-1\,n-1}p_{nn}-p_{n\,n-1}^2$, the manifold $q=0$ is mapped into a manifold of dimension not greater than $n(n+1)-3$, which is impossible. Similarly, every three-rowed principal minor of (p_{ij}) is divisible by q.

(6) Since

$$\Delta(-Z^{-1})(d(Z))^{n+1} = \Delta_1(Z),$$

we have

$$q(-Z^{-1})(d(Z)) = q_1(Z),$$

where $q_1(Z)$ is a polynomial. By means of (16) with $r=n-1$, we may find that

$$p_{ii}(-Z^{-1})d(Z)$$

is also a polynomial. Further, by means of (16) with $r=n-2$, we find that

$$\frac{p_{ii}(-Z^{-1})p_{jj}(-Z^{-1}) - p_{ij}^2(-Z^{-1})}{q(-Z^{-1})} d(z)$$

is also a polynomial. Thus

$$p_{ij}^2(-Z^{-1})(d(Z))^2$$

is a polynomial, and so is $p_{ij}(-Z^{-1})d(Z)$. Therefore we have to find the polynomial $p(Z)$ of degree n such that

$$p(-Z^{-1})d(Z)$$

is a polynomial.

(7) The answer to the question raised in (6) is that if we put

$$p(Z) = \sum_{k=0}^{n} p^{(k)}(Z),$$

where $p^{(k)}$ is a homogeneous polynomial of degree k, then $p^{(k)}(Z)$ is a linear combination of the k-rowed minors of Z. This will be proved in the next section, owing to its independent interest.

(8) Now we have, instead of (23) and (24), the following expressions

$$q(Z) = ad(Z) + \sum_{k=0}^{n-1} q^{(k)}, \qquad a \neq 0,$$

$$p_{ij}(Z) = a_{ij}d(Z) + \sum_{k=0}^{n-1} p_{ij}^{(k)},$$

where $q^{(k)}$ and $p_{ij}^{(k)}$ are linear combinations of the k-rowed minors of Z. We may let $a=1$. Let z_{ij}^* be the cofactor of z_{ij}, then

$$q^{(n-1)} = \sum_{1\leq i\leq j\leq n} c_{ij}z_{ij}^*.$$

There exists S such that

$$q(Z - S)$$

contains no term of order $n-1$. Thus, we may assume that $q^{(n-1)}=0$.

Further the transformation

$$X = W - (a_{ij})$$

carries (18) into a new one with $a_{ij}=0$. Up to the present, (18) takes the form

$$q(Z) = d(Z) + \sum_{k=0}^{n-2} q^{(k)} \tag{26}$$

and

$$p_{ij}(Z) = \sum_{k=0}^{n-1} p_{ij}^{(k)}, \tag{27}$$

where $q^{(k)}$ and $p_{ij}^{(k)}$ are linear combinations of the k-rowed minors of Z. Δ is given by (25).

Since the singular matrices form a manifold of dimension $n(n+1)-2$, there is a nonsingular matrix Z carried into a nonsingular matrix W. Without loss of generality we may assume that

$$I = \Gamma(I). \tag{28}$$

(9) We write

$$p_{ij}^{(n-1)} = \sum_{1\leqq s\leqq t\leqq n} a_{ij,st} \overset{*}{z}_{st}, \qquad 1 \leqq i; j \leqq n. \tag{29}$$

We shall prove that (29) forms a system of independent equations. In fact, the degree of the Jacobian of (18) with (26) and (27) is, by direct verification, not greater than

$$(-2)\frac{n(n+1)}{2} - 1 = -n(n+1) - 1.$$

the last "-1" appears, as the highest terms are dependent. Then Δ is of degree not less than $n(n+1)+1$, which is impossible. We have therefore the assertion.

(10). Now we consider the matrix

$$(p_{ij}^{(n-1)}) = \sum_{1\leqq s,t\leqq n} A_{st} \overset{*}{z}_{st}, \qquad A'_{st} = A_{st}. \tag{30}$$

By (5), we have the consequence that $d(Z)$ divides

$$\begin{vmatrix} p_{ii}^{(n-1)} & p_{ij}^{(n-1)} \\ p_{ij}^{(n-1)} & p_{jj}^{(n-1)} \end{vmatrix} \quad \text{and} \quad \begin{vmatrix} p_{ii}^{(n-1)} & p_{ij}^{(n-1)} & p_{ik}^{(n-1)} \\ p_{ij}^{(n-1)} & p_{jj}^{(n-1)} & p_{jk}^{(n-1)} \\ p_{ik}^{(n-1)} & p_{jk}^{(n-1)} & p_{kk}^{(n-1)} \end{vmatrix}.$$

Therefore if $d(Z)=0$, we find that (30) is of rank not greater than 1. In particular A_{ii} are of rank 1, as that A_{ii} cannot be of rank 0 has been shown in (9).

We write

$$A_{ii} = (a_{i1}, \cdots, a_{in})'(a_{i1}, \cdots, a_{in}).$$

By (28), we have

$$\sum_{i=1}^{n} A_{ii} = I.$$

Let

$$A = (a_{ij}),$$

then

$$A'A = I.$$

Thus $A\Gamma(Z)A'$ carries (30) into a new one with

$$A_{ii} = [0, \cdots, 0, 1, 0, \cdots, 0]$$

(where there are $i-1$ zeros preceding the one).

We write

$$p_{11}^{(n-1)} = \overset{*}{z}_{11} + \sum_{1 \leqq i < j \leqq n} a_{ij}\overset{*}{z}_{ij}, \qquad p_{22}^{(n-1)} = \overset{*}{z}_{22} + \sum_{1 \leqq i < j \leqq n} c_{ij}\overset{*}{z}_{ij}.$$

Since $d(Z)$ divides

$$p_{11}^{(n-1)} p_{22}^{(n-1)} - (p_{12}^{(n-1)})^2$$

we write

$$p_{11}^{(n-1)} p_{22}^{(n-1)} - (p_{12}^{(n-1)})^2 = d(Z)g(Z).$$

By the lemma given in the next section, $p_{11}^{(n-1)}p_{22}^{(n-1)} - (p_{12}^{(n-1)})^2$ is a linear combination of the two-rowed minors of

$$(\overset{*}{z}_{ij}).$$

We write

$$p_{12}^{(n-1)} = \sum_{1 \leqq i < j \leqq n} b_{ij}\overset{*}{z}_{ij}.$$

If p_{12} contains $z_{ij}^*(ij \neq 1, 2)$ then $p_{11}p_{22}$ must contain $x_{ii}^*x_{jj}^*$, which is impossible. Therefore

$$p_{12}^{(n-1)} = \pm \overset{*}{z}_{12}.$$

Consequently,

$$p_{11}^{(n-1)} p_{22}^{(n-1)} - (p_{12}^{(n-1)})^2 = \overset{*}{z}_{11} \sum c_{ij}\overset{*}{z}_{ij} + \overset{*}{z}_{ij} \sum a_{ij}\overset{*}{z}_{ij} + (\sum a_{ij}\overset{*}{z}_{ij})(\sum c_{ij}\overset{*}{z}_{ij})$$

is a multiple of $d(Z)$. This is possible only when $c = a = 0$. Thus we have

$$p_{ii}^{(n-1)} = \overset{*}{z}_{ii}, \qquad p_{ij}^{(n-1)} = \pm \overset{*}{z}_{ij}.$$

The transformation

$$[1, -1, 1, \cdots, 1]Z^*[1, -1, \cdots, 1]$$

carries z_{12}^* into $-z_{12}^*$, therefore we may assume that

$$\overset{(n-1)}{p_{1i}} = \overset{*}{z}_{1i}.$$

Since

$$\begin{vmatrix} \overset{*}{z}_{11} & \overset{*}{z}_{12} & \overset{*}{z}_{13} \\ \overset{*}{z}_{12} & \overset{*}{z}_{22} & -\overset{*}{z}_{23} \\ \overset{*}{z}_{13} & -\overset{*}{z}_{23} & \overset{*}{z}_{33} \end{vmatrix}$$

is not divisible by $d(Z)$, we have

$$\overset{(n-1)}{p_{ij}} = \overset{*}{z}_{ij}.$$

Up to the present, we arrive at the conclusion that we may let

$$q(Z) = d(Z) + \sum_{k=0}^{n-2} q^{(k)} \tag{31}$$

and

$$p_{ij}(z) = \overset{*}{z}_{ij} + \sum_{k=0}^{n-2} \overset{(k)}{p_{ij}}. \tag{32}$$

(11) We write

$$p_{11}p_{22} - p_{12}^2 \equiv 0 \pmod q$$

in a more precise form

$$p_{11}p_{22} - p_{12}^2 = q\psi, \tag{33}$$

where

$$\psi = \psi^{(n-2)} + \psi^{(n-3)} + \cdots,$$

and $\psi^{(k)}$ is a homogeneous polynomial of degree k.

Comparing the terms of degree $2n-3$ in (33), we have

$$\overset{*}{z}_{11}\overset{(n-2)}{p_{22}} + \overset{*}{z}_{22}\overset{(n-2)}{p_{11}} - 2\overset{*}{z}_{12}\overset{(n-2)}{p_{12}} = d(Z)\overset{(n-3)}{\psi}.$$

By the result which will be proved in (12) we have

$$\overset{(n-2)}{p_{11}} = \overset{(n-2)}{p_{12}} = \overset{(n-2)}{p_{22}} = \overset{(n-3)}{\psi} = 0.$$

Now we are going to prove that $p_{11}^{(\sigma)}$, $p_{12}^{(\sigma)}$, $p_{22}^{(\sigma)}$ and $\psi^{(\sigma-1)}$ are all zero. Suppose that the assertion is true for $\sigma > \rho$. Then

$$\overset{*}{z}_{11}\overset{(\rho)}{p_{22}} + \overset{*}{z}_{22}\overset{(\rho)}{p_{11}} - 2\overset{*}{z}_{12}\overset{(\rho)}{p_{12}} = d(Z)\overset{(\rho-1)}{\psi}.$$

By (12) we have $p_{11}^{(\rho)} = p_{12}^{(\rho)} = p_{22}^{(\rho)} = \psi^{(\sigma-1)} = 0$. Thus we have finally

$$q(Z) = d(Z), \qquad p_{ij}(Z) = \overset{*}{z}_{ij}.$$

The theorem is proved completely, except the verification of the following assertion.

(12) For $n-2\geqq\sigma>0$, from

$$\overset{*}{z}_{11}\overset{(\sigma)}{p}_{22}+\overset{*}{z}_{22}\overset{(\sigma)}{p}_{11}-2\overset{*}{z}_{12}\overset{(\sigma)}{p}_{12}=d(Z)\overset{(\sigma-1)}{\psi},$$

we can deduce that $p_{11}^{(\sigma)}=p_{12}^{(\sigma)}=p_{22}^{(\sigma)}=\psi^{(\sigma-1)}=0$.

To prove this assertion, we make a transformation $Z=W^{-1}$. Then

$$\overset{*}{z}_{ij}=w_{ij}/d(W),\qquad \overset{(\sigma)}{p}_{ij}(Z)=\overset{(n-\sigma)}{r}_{ij}(W)/d(W),$$
$$\overset{(\sigma-1)}{\psi}(Z)=\overset{(n-\sigma+1)}{\phi}(W)/d(W);$$

we have

$$w_{11}\overset{(n-\sigma)}{r}_{22}+w_{22}\overset{(n-\sigma)}{r}_{11}-2w_{12}\overset{(n-\sigma)}{r}_{12}=\overset{(\tau)}{\phi}(W),$$

where $\phi^{(\tau)}$ is a sum of $\tau=(n-\sigma+1)$-rowed minors. Let ϕ_1 be a τ-rowed minor contained in ϕ. Putting all elements of W other than those contained in ϕ equal to zero, we find that ϕ_1, a determinant of order τ, may be expressed as

$$w_{12}r_{22}+w_{22}r_{11}-2w_{12}r_{12}$$

where r_{11}, r_{12}, r_{22} are sums of $(\tau-1)$-rowed minors of ϕ_1. This is impossible for $n>3$ as shown by the fact that the determinant vanishes identically for $w_{11}=w_{12}=w_{22}=0$. Thus we suppose that $\tau=3$. Since w_{22} contains no term with factors w_{11} and w_{13}, we have

$$r_{22}=c_1(w_{22}w_{33}-w_{23}^2),$$

and similarly

$$r_{11}=c_2(w_{11}w_{33}-w_{13}^2).$$

The equality

$$c_1(w_{22}w_{33}-w_{23}^2)w_{11}+c_2(w_{11}w_{33}-w_{13}^2)w_{22}-2w_{12}r_{12}$$
$$=w_{11}w_{22}w_{33}-w_{23}^2w_{11}-w_{13}^2w_{22}+\cdots$$

implies

$$c_1(w_{22}w_{33}-w_{23}^2)w_{11}+c_2(w_{11}w_{33}-w_{13})^2w_{23}$$
$$\equiv w_{11}w_{22}w_{33}-w_{23}^2w_{11}-w_{13}^2w_{22}\pmod{w_{12}},$$

which is evidently impossible.

9. **A result concerning adjugate.** Now we are going to verify the assertion stated in (7) of the previous section.

Let

$$X=(x_{ij})$$

denote an n-rowed symmetric matrix and let

$$Y = (y_{ij})$$

be its inverse. Now we consider the following problem: find the polynomial $f(Y)$ in y_{ij} satisfying

$$f(Y)d(X) = g(X), \tag{34}$$

where $g(X)$ is a polynomial in x_{ij}. Since $d(X)$ is a homogeneous polynomial, we need only consider the homogeneous part of $f(Y)$ as well as $g(X)$. Thus our problem is reduced to finding homogeneous $f(Y)$ satisfying (34). Let l be the degree of $f(Y)$ and

$$f(Y) = \sum_{p_{11}+\cdots+p_{nn}=l} p(p_{11}, \cdots, p_{nn}) y_{11}^{p_{11}} \cdots y_{nn}^{p_{nn}}. \tag{35}$$

Then $g(X)$ is of degree $n-l$ and can be written as

$$g(X) = \sum_{q_{11}+\cdots+q_{nn}=n-l} q(q_{11}, \cdots, q_{nn}) x_{11}^{q_{11}} \cdots x_{nn}^{q_{nn}}. \tag{36}$$

Let

$$X^* = (x^*_{ij}) = d(X)Y.$$

Then (34) may be written as

$$\begin{aligned} &\sum_{p_{11}+\cdots+p_{nn}=k} p(p_{11}, \cdots, p_{nn}) x_{11}^{*p_{11}} \cdots x_{nn}^{*p_{nn}} \\ &\qquad = (d(X))^{l-1} \sum_{q_{11}+\cdots+q_{nn}=n-l} q(q_{11}, \cdots, q_{nn}) x_{11}^{q_{11}} \cdots x_{nn}^{q_{nn}}. \end{aligned} \tag{37}$$

Notice that the relation (37) is a reciprocal one, in fact, from (37), we deduce also

$$\begin{aligned} &\sum_{q_{11}+\cdots+q_{nn}=n-l} q(q_{11}, \cdots, q_{nn}) x_{11}^{*q_{11}} \cdots x_{nn}^{*q_{nn}} \\ &\qquad = d(X)^{n-l-1} \sum_{p_{11}+\cdots+p_{nn}=n} p(p_{11}, \cdots, p_{nn}) x_{11}^{p_{11}} \cdots x_{nn}^{p_{nn}}. \end{aligned} \tag{38}$$

Does (37) have a solution? It is known([10]) that a minor of order l satisfies our requirement. The purpose of the present section is to establish the converse, namely:

THEOREM 10. (37) *holds if and only if* (36) *is a linear combination of the* $(n-l)$*-rowed minors. Consequently,* (34) *holds if and only if* $g(X)$ *is a linear combination of the minors of* X.

([10]) Wedderburn, *Lectures on matrices*, Amer. Math. Soc. Colloquium Publications, vol. 17, p. 67, formula (20).

Proof. The theorem is evidently true for $l=0, 1, n$ and $n-1$. Consequently the theorem is true for $n=1, 2$ and 3.

Let $n\geqq 4$ and $n/2\leqq l\leqq n-2$. We write

$$g(X) = g_1(X) + g_0(X),$$

where $g_1(X)$ vanishes for $x_{11}=x_{12}=\cdots=x_{1n}=0$. Putting

$$X = \begin{pmatrix} 1 & 0 \\ 0 & X_1 \end{pmatrix},$$

we find, by comparing the homogeneous parts, that $g_0(X)=g_0(X_1)$ satisfies (37) with $l-1, n-1$ instead of l, n. By hypothesis of induction, $g_0(X)$ is a linear combination of $(n-l)$-rowed minors of X_1.

Let

$$g_1(X) = g_2(X) + g_{1,0}(X)$$

where $g_2(X)$ vanishes for

$$x_{11} = x_{12} = \cdots = x_{1n} = 0$$

and for

$$x_{12} = x_{22} = \cdots = x_{2n} = 0.$$

Then $g_{10}(X)$ is also a linear combination of the $(n-l)$-rowed minors of X.

Proceeding successively, we have

$$g(X) = g_0(X) + g_{10}(X) + \cdots + g_{n0}(X) + \psi(X),$$

where $\psi(X)$ vanishes for

$$x_{11} = x_{12} = \cdots = x_{1n} = 0,$$

for

$$x_{12} = x_{22} = \cdots = x_{2n} = 0,$$

$$\cdots\cdots\cdots\cdots\cdots\cdots,$$

and for

$$x_{1n} = x_{2n} = \cdots = x_{nn} = 0.$$

Since the degree of $\psi(X)$ is $n-l\leqq n/2$, it is only possible for $l=n/2$, and $\psi(X)$ contains only terms of the form

$$cx_{12}x_{34}\cdots x_{n-1n}.$$

The term $(x_{11}\cdots x_{nn})^{l-1}(x_{12}x_{34}\cdots x_{n-1n})$ cannot appear on the left side of (37). Therefore we have the theorem.

10. **Elliptic geometry.** Now we let

$$\xi = \begin{pmatrix} H & 0 \\ 0 & H \end{pmatrix},$$

where $H=[1, \cdots, 1, -1, \cdots, -1]$ with p positive 1's and q negative 1's, $p+q=n$.

The points (W_1, W_2) making

$$\overline{(W_1, W_2)}\xi(W_1, W_2)' \tag{43}$$

of the same signature as H form the elliptic space with signature (p, q). The symplectic transformation

$$(Z_1, Z_2) = Q(W_1, W_2)F, \qquad \overline{F}\xi F' = \xi$$

is called a motion of the space.

THEOREM 11. *Any nonsingular symmetric pair of matrices (W_1, W_2) making (43) nonsingular belongs to the elliptic space of signature (p, q).*

Proof. We have a nonsingular matrix Γ such that

$$\overline{\Gamma}\overline{(W_1, W_2)}\xi(W_1, W_2)'\Gamma' = H_0 \tag{44}$$

where H_0 is a diagonal matrix with p' positive 1's and q' negative 1's.

Let

$$(W_1^*, W_2^*) = \Gamma(W_1, W_2).$$

Construct the symplectic matrix

$$F = \begin{pmatrix} W_1^* & W_2^* \\ -H_0\overline{W}_2^*H & H_0\overline{W}_1^*H \end{pmatrix}.$$

We may verify directly that

$$\overline{F}\begin{pmatrix} H & 0 \\ 0 & H \end{pmatrix}F' = \begin{pmatrix} H_0 & 0 \\ 0 & H_0 \end{pmatrix}.$$

Owing to the invariance of signature, we have $H=H_0$. The theorem follows.

From Theorem 11, we have that the elliptic space of signature (p, q) is formed by all nonsingular symmetric pairs of matrices (W_1, W_2) except those lying on

$$d(\overline{(W_1, W_2)}\xi(W_1, W_2)') = 0.$$

From Theorem 11, we may easily find that

$$d(\overline{(W_1, W_2)}\xi(W_1, W_2)'H) \geqq 0 \tag{45}$$

for any nonsingular pair of matrices (W_1, W_2) and that the equality holds on a manifold of dimension not greater than $n(n+1)-2$. Now we shall go further to establish the following theorem.

THEOREM 12. *If*

$$d((\overline{W_1, W_2})\xi(W_1, W_2)') = 0$$

then (W_1, W_2) *lies on a manifold of dimension not greater than* $n(n+1)-3$.

Proof. By the consideration of semi-involutions, we have only to consider the nonhomogeneous expression, that is, we are going to prove that the symmetric matrices Z satisfying

$$d(H + ZH\overline{Z}) = 0 \tag{46}$$

form a manifold of dimension not greater than $n(n+1)-3$.

The equation (46) implies that $HZ^{-1}H\overline{Z}^{-1}$ has at least a negative root, if Z is nonsingular. By a theorem due to the author(11), those Z form a manifold of dimension not greater than $n(n+1)-3$.

The equation $d(Z)=0$ does not imply $d(H + ZH\overline{Z}) = 0$ identically. The manifold defined by

$$d(Z) = d(H + ZH\overline{Z}) = 0$$

is therefore of dimension not greater than $n(n+1)-3$.

THEOREM 13 (FUNDAMENTAL THEOREM OF THE ELLIPTIC SPACE). *An analytic automorph of the elliptic space of signature* (p, q) *is a motion of the space.*

Proof. Let

$$W = \Gamma(Z), \qquad \Gamma = (f_{ij})$$

by an analytic mapping of the elliptic space onto itself. Then $f_{ij}(Z)$ is analytic in the whole extended space of symmetric matrices except possibly on a manifold of dimension not greater than $n(n+1)-3$. By the continuity theorem(12) of functions of several complex variables, $f_{ij}(Z)$ is analytic in the whole extended space. Its inverse mapping is also analytic everywhere. The theorem follows from Theorem 12.

11. **Hyperbolic space.** The hyperbolic space is formed by the symmetric matrices Z making $I-Z\overline{Z}$ positive definite. A symplectic mapping carrying the space onto itself is called a motion of the space.

THEOREM 14 (SIEGEL). *An analytic mapping carrying the hyperbolic space onto itself is a motion of the space.*

For completeness, we give here a proof which is different from that due to Siegel.

Proof. (1) The space is transitive, hence we need only to consider the trans-

(11) Ibid. Theorem 6.

(12) Levi, ibid., or Satz 17, Folgerung 1 of Behnke and Thullen, *Theorie der Funktionen mehrerer komplexer Veränderlichen*, Julius Springer, 1934.

formations keeping 0 invariant. By a theorem for circular regions proved by H. Cartan([13]), the mapping is a linear one. Let

$$W = \sum_{1 \leq r \leq s \leq h} A_{rs} z_{rs}, \qquad A_{rs} = A'_{rs}, \tag{47}$$

be the mapping. Since $I - Z\bar{Z} = 0$ is the intersection of all the algebraic surfaces bounding the space, the mapping (47), therefore, carries unitary symmetric matrices into unitary symmetric matrices.

Since any unitary symmetric matrix can be expressed as UU' where U is unitary, we may assume that (47) carries I into itself. Consequently, we have

$$\sum_{r=1}^{n} A_{rr} = I. \tag{48}$$

Putting $Z = [e^{i\theta_1}, \cdots, e^{i\theta_n}]$, we have

$$I = W\bar{W} = \left(\sum_{r=1}^{n} A_{rr} e^{i\theta_r}\right)\left(\sum_{r=1}^{n} \bar{A}_{rr} e^{-i\theta_r}\right)$$

for any real θ. Consequently we obtain

$$A_{rr}\bar{A}_{ss} = 0 \qquad \text{for } r \neq s. \tag{49}$$

(3) From (48) and (49), we have

$$A_{rr} = A_{rr}\left(\sum_{s=1}^{n} \bar{A}_{ss}\right) = A_{rr}\bar{A}_{rr} = \left(\sum_{s=1}^{n} A_{ss}\right)\bar{A}_{rr} = \bar{A}_{rr}.$$

Then A_{rr} are real and $A_{rr}A_{ss} = 0$, $A_{rr}^2 = A_{rr}$. Therefore, we have a real orthogonal T such that

$$TA_{rr}T' = [0, \cdots, 0, 1, 0, \cdots, 0]$$

(where the 1 in the brackets is in the rth place). Without loss of generality we assume that

$$A_{rr} = [0, \cdots, 0, 1, 0, \cdots, 0] \tag{50}$$

(where the 1 in brackets is in the rth place).

(4) Now we put

$$Z = \begin{pmatrix} 0 & e^{i\theta} \\ e^{i\theta} & 0 \end{pmatrix} + [1, \cdots, 1].$$

From (48) and $I = W\bar{W}$, we have

$$A_{12}\bar{A}_{12} + e^{i\theta}A_{12}(I - A_{11} - A_{22}) + (I - A_{11} - A_{22})e^{-i\theta}\bar{A}_{12} + (I - A_{11} - A_{22})^2 = I,$$

for all θ. Thus we have

([13]) Journal de Mathématique (9) vol. 10 (1931) Theorem 6.

$$A_{12} = \alpha^{(2)} + 0,$$

where $\alpha\bar{\alpha} = I^{(2)}$.

(5) Further we let

$$Z = \begin{pmatrix} e^{i\tau}/2^{1/2} & 1/2^{1/2} \\ 1/2^{1/2} & -e^{-i\tau}/2^{1/2} \end{pmatrix} + 0^{(n-2)}.$$

From (48) and $I - W\overline{W} = 0$, we deduce immediately that

$$\begin{aligned} I^{(2)} &= 2^{-1}\left(\begin{pmatrix} e^{i\tau} & 0 \\ 0 & -e^{-i\tau} \end{pmatrix} + \alpha\right)\left(\begin{pmatrix} e^{-i\tau} & 0 \\ 0 & -e^{i\tau} \end{pmatrix} + \bar{\alpha}\right) \\ &= 2^{-1}I^{(2)} + 2^{-1}\left(\alpha\begin{pmatrix} e^{-i\tau} & 0 \\ 0 & -e^{i\tau} \end{pmatrix} + \begin{pmatrix} e^{i\tau} & 0 \\ 0 & -e^{-i\tau} \end{pmatrix}\bar{\alpha}\right) + 2^{-1}\alpha\bar{\alpha}. \end{aligned}$$

Consequently

$$\alpha\begin{pmatrix} e^{-i\tau} & 0 \\ 0 & -e^{i\tau} \end{pmatrix} + \begin{pmatrix} e^{i\tau} & 0 \\ 0 & -e^{-i\tau} \end{pmatrix}\bar{\alpha} = 0$$

for all τ. Then

$$\alpha = \pm\begin{pmatrix} 0 & 1 \\ 1 & 0 \end{pmatrix}.$$

Similarly A_{rs} is determined completely apart from a sign. We may assume that A_{1s} takes the positive sign. Since

$$\begin{pmatrix} z_{11} & z_{12} & z_{13} \\ z_{12} & z_{22} & z_{23} \\ z_{13} & z_{23} & z_{33} \end{pmatrix}$$

being unitary does not imply that

$$\begin{pmatrix} z_{11} & z_{12} & z_{13} \\ z_{12} & z_{22} & -z_{23} \\ z_{13} & -z_{23} & z_{33} \end{pmatrix}$$

is unitary, the theorem is now established.

Finally, the author gives the following theorem which shows the importance of the notion of the characteristic roots of the distance-matrix.

THEOREM 15. *A mapping carrying the hyperbolic space "$I - Z\overline{Z}$ being positive definite" into itself, and keeping the characteristic roots of the distance matrix between two points invariant, is either a hyperbolic motion or a hyperbolic motion combined with a reflexion $Z = \overline{W}$.*

Proof. The proof is comparatively simple, and it contains some repetition of our old argument. The author gives only the main procedure of the proof.

(1) The distance matrix $\mathfrak{D}(A, B)$ of two points A and B of the space is defined by

$$\mathfrak{D}(A, B) = (A - B)(A - \bar{B}^{-1})^{-1}((\bar{A}^{-1} - B)(\bar{A}^{-1} - \bar{B}^{-1})^{-1})^{-1}.$$

In particular for $B=0$, we have

$$\mathfrak{D}(A, 0) = A\bar{A}'.$$

(2) Owing to the transitivity, we need only consider the mapping $W=\Gamma(Z)$ keeping 0 invariant. Thus both Hermitian matrices $W\bar{W}$ and $Z\bar{Z}$ are conjunctive. In particular the mapping keeps the rank of Z invariant.

(3) Let

$$\Gamma([1/2, 0, \cdots, 0]) = (a_1, \cdots, a_n)'(a_1, \cdots, a_n).$$

By a slight modification, we may let

$$\Gamma([1/2, 0, \cdots, 0]) = [1/2, 0, \cdots, 0].$$

From the invariance of the characteristic roots of distance-matrix we find

$$\Gamma([\lambda_1, 0, \cdots, 0]) = [\lambda_1, 0, \cdots, 0],$$

for all real λ_1. Without loss of generality, we may modify Γ such that

$$\Gamma([0, \lambda_2, 0, \cdots, 0]) = [0, \lambda_2, 0, \cdots, 0],$$

$$\cdots\cdots\cdots\cdots\cdots\cdots\cdots,$$

$$\Gamma([0, 0, \cdots, 0, \lambda_n]) = [0, 0, \cdots, 0, \lambda_n])$$

for all real $\lambda_2, \cdots, \lambda_\Lambda$.

Next, we may show that for any real diagonal Λ we have

$$\Gamma(\Lambda) = \Lambda,$$

and that for all real X, we have

$$\Gamma(X) = X.$$

(4) Since $W=\Gamma(Z)$ and

$$r(X, W) = r(X, Z),$$

we find that

$$| d(X - W) |^2 = | d(X - Z) |^2$$

for all real X. We deduce that

$$W = Z \quad \text{or} \quad \bar{Z}.$$

National Tsing Hua University,
Kunming, Yunnan, China.

Reprinted from the
THE SCIENCE REPORTS OF NATIONAL TSING HUA UNIVERSITY
Series A, Vol. 4, Nos. 4–6, pp. 313–327, 1947

SOME "ANZAHL" THEOREMS FOR GROUPS OF PRIME POWER ORDERS.

By Loo-keng Hua (華羅庚)

Department of Mathematics

(*Received October 5, 1940*)

ABSTRACT.

We define a group $\mathcal{G}$ *of order* p^n (*p being a prime*) *to be of rank* α *if the maximum of the orders of all elements of* $\mathcal{G}$ *is equal to* $p^{n-\alpha}$. *First the author introduces a "pseudo base", which is the main weapon of the paper. Then the author proves the "Anzahl" theorem that if* $p\geq 3$, $n\geq 2\alpha+1$, *then* (*i*) $\mathcal{G}$ *contains one and only one subgroup of order* p^m ($2\alpha+1\leq m\leq n$) *of rank* α; (*ii*) $\mathcal{G}$ *contains* p^α *cyclic subgroups of order* p^m ($\alpha<m<n-\alpha+1$); *and* (*iii*) $\mathcal{G}$ *contains* $p^{m+\alpha}$ *elements satisfying* $G^{p^m}=1$. *Next the author determines completely the ranks of all major subgroups and the exact number of the major subgroups of each rank. More precisely, the rank of a major subgroup of index* p^β *is either* $\alpha-\beta+1$ *or* $\alpha-\beta$, *and for* $p\geq 3$ *and* $n\geq 2\alpha+1$, *the number of those of rank* $\alpha-\beta+1$ *and the number of those of rank* $\alpha-\beta$ *are equal to* $\phi_{d-1,\beta-1}$ *and* $p^\beta\phi_{d-1,\beta}$ *respectively, where d is the number of mutually independent generators of the group and for* $0\leq r\leq\nu$

$$\phi_{\nu,r}=\frac{(p^\nu-1)....(p^{\nu-r+1}-1)}{(p^r-1)....(p-1)}$$

Then the author deduces a more precise formula than the enumeration principle due to Hall. This is the second weapon of the paper. By these two weapons, the following Anzahl theorem is obtained: If $p\geq 3$ *and* $n\geq 2\alpha+1$, *then the number of its subgroups of order* p^m ($2\alpha-\beta<m<n-\beta+1$) *and of rank* $\alpha-\beta$ *with* $0\leq\beta\leq d$ *is congruent to* $p^\beta\phi_{d-1,\beta}$ (*mod* $p^{d-\beta+1}$). *For* $\beta=d$, $p^\beta\phi_{d-1,\beta}$ *is to be replaced by zero. In the later part of the paper, the author selects the following typical one: For* $p\geq 3$ *if* $\mathcal{G}$ *contains* p^β *cyclic subgroups of order* p^m *for a fixed m satisfying* $m>\beta+1$, *it is of rank* β.

INTRODUCTION.

For convenience, we shall define a group $\mathcal{G}$ of order p^n (p being a prime) to be of rank δ if the maximum of the orders of all elements G of $\mathcal{G}$ is equal to $p^{n-\delta}$.

Throughout this paper, we shall always denote by p an odd prime.

In §1, by introducing a "pseudo-base", we shall prove the following "Anzahl" theorem:

If $\mathcal{G}$ is of order p^n ($p\geq 3$, $n\geq 2\alpha+1$) of rank α, then (i) $\mathcal{G}$ contains one and only one subgroup of order p^m ($2\alpha+1\leq m\leq n$) of rank α; (ii) $\mathcal{G}$ contains p^α cyclic subgroups of order p^m ($\alpha<m<n-\alpha+1$); and (iii) $\mathcal{G}$ contains $p^{m+\alpha}$ elements satisfying

$$G^{p^m}=1.$$

The second and the third statement of the above theorem yield better results than the well-known theorems due to Miller and Kulakoff respectively.

In §2, the ranks of all the major subgroups and the exact number of the major subgroups of each rank will be determined completely; more precisely, *in a group $\mathcal{G}$ of order p^n of rank α, the rank of a major subgroup of index p^β is either $\alpha-\beta+1$ or $\alpha-\beta$, and for $p\geq 3$ and $n\geq 2\alpha+1$, the number of those of rank $\alpha-\beta+1$ and the number of those of rank $\alpha-\beta$ are equal to $\phi_{d-1,\beta-1}$ and $p^\beta\phi_{d-1,\beta}$ respectively, where d is the number of generators of the group and for $o\leq\tau\leq\nu$*

$$\phi_{\nu,\tau}=\frac{(p^\nu-1)\dots(p^{\nu-\tau+1}-1)}{(p^\tau-1)\dots(p-1)}=\frac{(p^\nu-1)\dots(p^\nu-p^{\tau-1})}{(p^\tau-1)\dots(p^\tau-p^{\tau-1})}.$$

By means of these results we shall establish a more precise formula than the enumeration principle due to Hall.

Some precise results concerning the number of subgroups of order p^m in $\mathcal{G}$ will be given in §3, the chief theorem is the following:

Let $\mathcal{G}$ be a group of order p^n and of rank α with $p\geq 3$ and $n\geq 2\alpha+1$. then the number of its subgroups of order p^m ($2\alpha-\beta<m<n-\beta+1$) and of rank $\alpha-\beta$ with $o\leq\beta\leq d$ is congruent to $p^\beta\phi_{d-1,\beta}$ (mod $p^{d-\beta+1}$). For $\beta=d$, *$p^\beta\phi_{d-1,\beta}$ has to be replaced by* 0.

Among the theorems of §4, we shall state the following typical one which is a generalization of a result due to Miller and is in a certain sense a converse of a result of §1:

A group of order p^n ($p\geq 3$) which contains p^β cyclic subgroups of order p^m for a fixed m satisfying $m>\beta+1$ is of rank β.

For $\beta=0$, the theorem is trivial, and for $\beta=1$, we have the case discussed by Miller.

The author is indebted to Mr. H, F, Tuan for his revision of the manuscripts. Some new results of Kulakoff's type have been obtained by him.

§1. **Theorem** 1. *Let $\mathcal{G}$ be a group of order p^n and of rank α with $p\geq 3$ and $n\geq 2\alpha+1$, then we have the following conclusions:*

(i) *there exist $\alpha+1$ elements $A, A_1, \ldots\ldots, A_\alpha$ of $\mathcal{G}$ such that every element G of $\mathcal{G}$ can be expressed uniquely as*

$$G=A_\alpha^{\lambda_\alpha} \ldots A_1^{\lambda_1} A^{\lambda} \quad 1\leq\lambda_\alpha\leq p, \ldots, 1\leq\lambda_1\leq p. \ 1\leq\lambda\leq p^{n-\alpha},$$

where the order of A_δ does not exceed p^δ and that of A is exactly $p^{n-\alpha}$ (Pseudo-base);

(ii) *the equality*

$$(B_1B_2)^{p^\alpha}=B_1^{p^\alpha} B_2^{p^\alpha}$$

holds for any pair of elements B_1 and B_2 of $\mathcal{G}$;

(iii) *the commutators of $\mathcal{G}$ are of order $\leq p^\alpha$, and the simple 3-fold commutators of $\mathcal{G}$ are of order $\leq p^{\alpha-1}$; and*

(iv) *A^{p^α} belongs to the central of $\mathcal{G}$.*

Proof For $\alpha=0$ the theorem is trivial. For $\alpha=1$, the theorem is true. [1] Assume that $\alpha>1$ and that the theorem is true for $\alpha-1$. $\mathcal{G}$ has a maximal subgroup $\mathfrak{M}$ of rank $\alpha-1$. By hypothesis of induction, since $(n-1)\geq 2(\alpha-1)+1$, the elements M of $\mathfrak{M}$ are uniquely expressible as

(i)′ $$M=A_{\alpha-1}^{\lambda_{\alpha-1}}\ldots A_1^{\lambda_1} A^{\lambda}, \quad 1\leq\lambda_{\alpha-1}\leq p, \ldots, 1\leq\lambda_1\leq p, \cdots, 1\leq\lambda\leq p^{n-\alpha},$$

where A_δ is of order $\leq p^\delta$ and A is of order $p^{n-\alpha}$, and the following relations

(ii)′ $$(M_1M_2)^{p^{\alpha-1}}=M_1^{p^{\alpha-1}} M_2^{p^{\alpha-1}},$$

(iii′) $$(M_1, M_2)^{p^{\alpha-1}}=1, \text{ and } (M_1, M_2, M_3)^{p^{\alpha-2}}=1$$

(1) See, *e.g.*, H. Zassenhaus, Lehrbuch der Gruppentheorie I, 114-5 (1937).

hold for arbitrary elemcnts M_1, M_2 and M_3 of $\mathfrak{M}$, and further

(iv′) $A^{p^{\alpha-1}}$ belongs to the central of $\mathfrak{M}$.

Let B be any element of $\mathcal{G}$ but not in $\mathfrak{M}$. Then $\mathcal{G}=\{\mathfrak{M}, B\}$. Since B^p belongs to $\mathfrak{M}$, by (iv′) we have

$$A^{p^{\alpha-1}} B^p = B^p A^{p^{\alpha-1}}.$$

Let

$$B^{-1}A B = AC,$$

where C belengs to $\mathfrak{M}$, since $\mathfrak{M}$ is normal in $\mathcal{G}$, By (ii′), we have

$$B^{-1} A^{p^{\alpha-1}} B = A^{p^{\alpha-1}} C^{p^{\alpha-1}}$$

Further, $C^{p^{\alpha-1}}$ belongs to the cyclic subgroup $\{A\}$, for otherwise the group $\{C,A\}$ would be greater than $\mathfrak{M}$. We can let

$$C^{p^{\alpha-1}} = A^{p^{\alpha-1} b},$$

since the order of C cannot exceed that of A, Then

$$B^{-1} A^{p^{\alpha-1}} B = A^{p^{\alpha-1}(1+b)}.$$

Consequently by (iv′)

$$A^{p^{\alpha-1}} = B^{-p} A^{p^{\alpha-1}} B^p = A^{p^{\alpha-1}(1+b)^p}.$$

Therefore

$$(1+b)^p \equiv 1 \pmod{p^{n-2\alpha+1}}. \tag{1}$$

For $n \geq 2\alpha+1$, we have then

$$b \equiv 0 \pmod{p^{n-2\alpha}}. \tag{2}$$

Thus

$$C^{p^{\alpha-1}} = A^{p^{n-\alpha-1}c},$$

i.e. $(A,B)=C$ is of order $\leq p^\alpha$, and it follows that (B,A^p) is of order $\leq p^{\alpha-1}$. We then obtain (iv) by $(B, A^{p^\alpha})=1$ and (iv′).

Further, the orders of the commutators (B, A_i), $i=1,\dots,\alpha-1$, are $\leq p^{\alpha-1}$, since let

$$A_i^{-1} B^{-1} A_i B = C_i, \quad (C_i \text{ in } \mathfrak{M}),$$

we have

$$1=B^{-1}A_i^{p^{\alpha-1}}B=(A_iC_i)^{p^{\alpha-1}}=C_i^{p^{\alpha-1}}$$

by (ii') and (i'). Let M be any element of $\mathfrak{M}$, and let $M=M_1A^{\lambda}$ with $M_1=A_{\alpha-1}^{\lambda_{\alpha-1}}\dots A_1^{\lambda_1}$. Then since

$$(B,M)=(B,M_1A^{\lambda})=(B,A^{\lambda})A^{-\lambda}(B,M_1)A^{\lambda}$$

we have

$$(B,M)^{p^\alpha}=(B,A^{\lambda})^{p^\alpha}A^{-p^\alpha\lambda}(B,M_1)^{p^\alpha}A^{p^\alpha\lambda}=1.$$

Further, if both B and B' are in $\mathfrak{G}$ but not in $\mathfrak{M}$, then we have an element M of $\mathfrak{M}$ such that $B'=B^{\lambda}M\cdot$ Therefore

$$(B,B')=B^{-1}M^{-1}B^{-\lambda}BB^{\lambda}M=(B,M)$$

is of order $\leq p^\alpha$.

Now let K be any commutator, then K belongs to $\mathfrak{M}$ and is of order $\leq p^\alpha$. Thus (K,C) is of order $\leq p^{\alpha-1}$ for any C of $\mathfrak{M}$ by (iii'). Finally we can write

$$K=M_1A^{p^{n-2\alpha}\lambda'}\quad\text{with}\quad M_1=A_{\alpha-1}^{\lambda_{\alpha-1}}\dots A_1^{\lambda_1}.$$

Then, for any B of $\mathfrak{G}$ not of $\mathfrak{M}$, we have

$$(K,B)=A^{-p^{n-2\alpha}\lambda'}(M_1,B)A^{p^{n-2\alpha}\lambda'}(A^{p^{n-2\alpha}\lambda'},B).$$

By (ii'). and since $(M,B)^{p^{\alpha-1}}=1$. we have

$$(K,B)^{p^{\alpha-1}}=1,\quad\text{for } n\geq 2\alpha+1.$$

Hence the proof of (iii) is complete.

It is easy to see by mathematical induction that

$$(B_1B_2)^p=B_1^pB_2^pK_1K_2\dots K_t,$$

where among these K's $\frac{1}{2}t(p-1)$ are (B_1, B_2) and the others are higher fold commutators. By (ii′) we then have for $p \geq 3$

$$(B_1 B_2)^{p^\alpha} = B_1^{p^\alpha} B_2^{p^\alpha} K_1^{t^{\alpha-1}} \ldots K_t^{t^{\alpha-1}} = B_1^{t^\alpha} B_2^{t^\alpha} (B_1, B_2)^{\frac{1}{2}t^\alpha (p-1)}$$

$$= B_1^{t^\alpha} B_2^{t^\alpha}.$$

Therefore (ii) is true.

If B is of order $\leq p^\alpha$, then $B = A_\alpha$ satisfies our requirement of (i). If B is of order p^β $(\beta > \alpha)$, then

$$B^p = A_{\alpha-1}^{\mu_{\alpha-1}} \ldots A_1^{\mu_1} A^{p^{n-\alpha-\beta+1}\mu} \quad \text{with } p+\mu.$$

Therefore

$$B^{p^\alpha} = A^{p^{n-\beta}\mu}$$

by (ii′) and (i′). Let $A_\alpha = B^{-1} A^{p^{n-\alpha-\beta}\mu}$, then

$$A_\alpha^{p^\alpha} = B^{-p^\alpha} A^{p^{n-\beta}\mu} = 1$$

by (ii). (i) is thus proved, since the uniqueness is a trivial consequence of the relative orders of A_α's.

Remark. For $n = 2\alpha$, (1) implies $b \equiv 0 \pmod p$. The theorem is still ture.

Theorem 2. (*Generalization of a result due to Kulakoff*) *The number of elements of order* $\leq p^m$ $(\alpha \leq m \leq n-\alpha)$ *of a group of order* p^n $(p \geq 3,\ n \geq 2\alpha+1)$ *of rank* α *iz equal to* $p^{m+\alpha}$.

Proof. By theorem 1 (i) and (ii), the elements of order $\leq p^m$ are of the form

$$A_\alpha^{\lambda_\alpha} \ldots A_1^{\lambda_1} A^{p^{n-\alpha-m}\mu},\ 1 \leq \lambda_\alpha \leq p, \ldots,\ 1 \leq \lambda_1 \leq p,\ 1 \leq \mu \leq p^m.$$

There are $p^{m+\alpha}$ such elements

Theorem 3. (*Generalization of a result due to Miller*) *The number of cyclic subgroups of order* p^m $(\alpha < m < n-\alpha+1)$ *of a group of order* p^n $(p \geq 3,\ n \geq 2\alpha+1)$ *of rank* α *is equal to* p^α.

Proof. The number of elements of order p^m of the group $\mathfrak{G}$ is $p^{\alpha}\phi(p^m)$. Therefore we have the theorem.

Theorem 4. *The group of order p^n ($p\geq 3$, $n\geq 2\alpha+1$) of rank α contains one and only one subgroup of order p^m ($n\geq m\geq 2\alpha$) of rank α,*

Proof. The set of elements

$$M=A_{\alpha}^{\lambda_\alpha}\ldots A_1^{\lambda_1}A^{p\lambda'}\quad 0\leq\lambda_\alpha\leq p-1,\ldots,0\leq\lambda_1\leq p-1,0\leq\lambda'\leq p^{n-\alpha-1}-1. \qquad (3)$$

forms a group, since (M, A^p) is of order $\leq p^{\alpha-1}$, *i.e.* $(M,A^p)=M_1A^{\lambda p}$ with $M_1=A_{\alpha-1}^{\lambda_\alpha}\ldots A_1^{\lambda_1}$. Further (3) contains all elements of order$\leq p^{n-\alpha-1}$. The theorem is then proved by induction.

Remark. For $n=2\alpha$, theorem 4 is then trivial.

§2. Let $\mathfrak{G}$ be any p-group of order p^n and of rank α and let $\mathfrak{M}_\beta$ denote a typical major subgroup of index p^β in $\mathfrak{G}$ with $0\leq\beta\leq d=d(\mathfrak{G})$, where d is the number of mutually independent generating elements of $\mathfrak{G}$. In particular $\mathfrak{G}=\mathfrak{M}_0$, $D=\mathfrak{M}_d=$the principal subgroup of $\mathfrak{G}$.

Theorem 5, *If $\mathfrak{G}$ is of rank α, then the rank of $\mathfrak{H}$, which is a subgroup of $\mathfrak{G}$ of index p^β, is $\geq\alpha-\beta$* and $\leq\alpha$.

Proof. 1) Evidently

$$H^{p^{n-\alpha}}=1$$

for all elements H of $\mathfrak{H}$. Thus the rank of $\mathfrak{H}$ is $\geq(n-\beta)-(n-\alpha)=\alpha-\beta$.

2) Evidently G^{p^β} belongs to $\mathfrak{H}$ for all elements G of $\mathfrak{G}$. If $\mathfrak{H}$ is of rank γ, then

$$\left(G^{p^\beta}\right)^{p^{n-\beta-\gamma}}=G^{p^{n-\gamma}}=1$$

Therefore $n-\alpha\leq n-\gamma$ and $\gamma\leq\alpha$.

Theorem 6, *If $\mathfrak{G}$ is of rank α, then the rank of any $\mathfrak{M}_\beta$ is $\leq\alpha-\beta+1$.*

Proof. Suppose $\mathfrak{M}_\beta$ be of rank γ. Then

$$\left(G^{p}\right)^{p^{n-\beta-\gamma}}=1$$

for all elements G of $\mathcal{G}$, since G^p belongs to $\mathfrak{M}_\beta$. Therefore $n-\beta-\gamma+1\geq n-\alpha$ and $\gamma\leq\alpha-\beta+1$.

According to theorems 5 and 6, we can classify the major subgroups $\mathfrak{M}_\beta$ into two classes.

I. those $\mathfrak{M}_\beta$ of rank $\alpha-\beta+1$; the typical one will be denoted by $\mathfrak{M}_{\beta 1}$;

II. those $\mathfrak{M}_\beta$ of rank $\alpha-\beta$; the typical one will be denoted by $\mathfrak{M}_{\beta 2}$.

Theorem 4 asserts that there is one and only one $\mathfrak{M}_{1,1}$ provided $p\geq 3$, $n\geq 2\alpha+1$.

Theorem 7. *Let $p\geq 3$ and $n\geq 2\alpha+1$. The subgroups of G which lie between $\mathfrak{M}_{1,1}$ and $\mathcal{G}$ are those of I and conversely.*

Proof. If $\mathfrak{M}_\beta$ lies in $\mathfrak{M}_{1,1}$, which is of rank α, the rank of $\mathfrak{M}_\beta$ is $\geq\alpha-(\beta-1)=\alpha-\beta+1$, by theorem 5. By theorem 6, the rank of $\mathfrak{M}_\beta$ is equal to $\alpha-\beta+1$. Hence $\mathfrak{M}_\beta$ lies always in I.

Next we shall prove that every $\mathfrak{M}_{\beta 1}$ lies between $\mathfrak{M}_{11}$ aud $\mathcal{G}$. This is trivial for $\beta=1$. Hence we suppose $\beta\geq 2$. There exists an $\mathfrak{M}_{\beta-1}$ which contains $\mathfrak{M}_{\beta,1}$. If $\mathfrak{M}_{\beta-1}$ belongs to I, *i,e.* $\mathfrak{M}_{\beta-1}=\mathfrak{M}_{\beta-1,1}$ the theorem follows by induction. Hence suppose that $\mathfrak{M}_{\beta-1}$ belongs to II and we write $\mathfrak{M}_{\beta-1}=\mathfrak{M}_{\beta-1,2}$. The intersection of $\mathfrak{M}_{1,1}$ and $\mathfrak{M}_{\beta-1,2}$ is of order $p^{n-\beta}$ and of rank $\alpha-\beta+1$, since $\mathfrak{M}_{\beta-1,2}$ is not a subgroup of $\mathfrak{M}_{1,1}$ by the direct part of the present theorem and

$$\frac{(\text{Order of } \mathfrak{M}_{1,1})\ (\text{Order of } \mathfrak{M}_{\beta-1,2})}{\text{Order of } \mathfrak{M}_{11}\wedge\mathfrak{M}_{\beta-1,2}}=\text{Order of } \mathcal{G}$$

or

$$\frac{p^{n-1}\cdot p^{n-\beta+1}}{\text{Order of } \mathfrak{M}_{11}\wedge\mathfrak{M}_{\beta-1,2}}=p^n$$

Now $\mathfrak{M}_{\beta-1,2}$, of order $p^{n-\beta+1}$ and rank $\alpha-(\beta-1)$, has one and only one subgroup of rank $\alpha-\beta+1$ and of order $p^{n-\beta}$ by theorem 4 for $n-\beta+1\geq 2\alpha-\beta+2\geq 2(\alpha-\beta+1)+\beta-1\geq 2(\alpha-\beta+1)+1$, the intersection should be $\mathfrak{M}_{\beta,1}$. Thus

$$\mathfrak{M}_{11}\wedge\mathfrak{M}_{\beta-1,2}=\mathfrak{M}_{\beta,1}.$$

Consequently $\mathfrak{M}_{\beta,1}$ is a subgroup of $\mathfrak{M}_{1,1}$. The theorem is thus completely proved.

Corollary 7. 1. *If $p\geq 3$ and $n\geq 2\alpha+1$, then the principal subgroup of a group of order p^n and of rank α is of rank $\alpha-d+1$.*

Corollary 7. 2. *If $p \geq 3$ and $n \geq \alpha+1$, then a group of order p^n with a cyclic principal subgroup of index p^d is of rank $\alpha = d-1$. In other words, the number of members of a minimal base is equal to its rank plus one.*

Corollary 7. 3. *If $p \geq 3$ and $n \geq 2\alpha+1$, the number of elements of a minimal base of a group of order p^n and of rank α does not exceed $\alpha+1$.*

Theorem 8. *Let $p \geq 3$ and $n \geq 2\alpha+1$. The number of $\mathfrak{M}_{\beta,1}$ is equal to $\phi_{d-1,\ \beta-1}$ and that of $\mathfrak{M}_{\beta,2}$ is equal to $p^{\beta}\ \phi_{d-1,\ \beta}$.*

Proof. The number of $\mathfrak{M}_{\beta 1}$ is equal to the number of subgroups of order $p^{n-\beta}$ which lie between $\mathfrak{M}_{1,1}$ and $\mathfrak{D}$. Now $\dfrac{\mathfrak{M}_{1,1}}{\mathfrak{D}}$ is Abelian of type $\underbrace{(1,1,\ldots\ldots,1)}_{d-1}$, we have the first statement of the theorem. The second follows immediately since

$$\phi_{d,\ \beta} - \phi_{d-1,\ \beta-1} = p^{\beta}\phi_{d-1,\ \beta}.$$

Theorem 9. *Let $p \geq 3$ and $n > 2\alpha+1$. Then*

$$d\,(\mathfrak{M}_{\beta,1}) \geq d\,(\mathfrak{G}) - \beta + 1$$

and

$$d\,(\mathfrak{M}_{\beta,2}) \geq d(\mathfrak{G}) - \beta.$$

Proof. It is evident that $\mathfrak{D}\,(\mathfrak{G}) \geq \mathfrak{D}\,(\mathfrak{M}_{\beta})$, ; hence we have the inequality

$$d\,(\mathfrak{M}_{\beta}) \geq d\,(\mathfrak{G}) - \beta.$$

In particular, the last inequality is trivial and is true for arbitrary n. To prove the first inequality, we have merely to show that for $n > 2\alpha+1$

$$\mathfrak{D}\,(\mathfrak{G}) \neq \mathfrak{D}\,(\mathfrak{M}_{1,1}).$$

For suppose $\mathfrak{D}\,(\mathfrak{G}) = \mathfrak{D}(\mathfrak{M}_{1,1})$, then corollary 7.1 shows that $\mathfrak{D}\,(\mathfrak{M}_{1,1})$ would be of rank $\alpha-(d-1)+1 = \alpha-d+2$, since $n-1 \geq 2\alpha+1$ for $n > 2\alpha+1$. But the rank of $\mathfrak{D}\,(\mathfrak{G})$ is $\alpha-d+1$. Thus $\mathfrak{D}\,(\mathfrak{G}) \neq \mathfrak{D}\,(\mathfrak{M}_{1,1})$, and the theorem is proved.

Corollary 9. 1. *Let $\mathfrak{G}$ be a group of order p^n and of rank α with $p \geq 3$ and $n \geq 2\alpha+1$. If $\mathfrak{G}$ has a cyclic principal subgroup of index p^d, then each major subgroup $\mathfrak{M}_{\gamma i}$ has also a cyclic principal subgroup, and for $n > 2\alpha+1$, there is one and only one maximal subgroup which has also a cyclic principal subgroup of index p^d.*

This is an immediate consequence of corollaries 7.2 and 7.3 together with theorem 9. The converse of the corollary is not true as is shown by the following "Gegenbeispiel":

$$A_1^{p^{n-2}}=1,\ A_2^p=1,\ (A_1, A_2)^p=1,\ ((A_1, A_2), A_1)=1,\ ((A_1, A_2), A_2)=1.$$

Theorem 10. *(Modified form of Hall's enumeration principle.) Let $\mathcal{G}$, $\mathfrak{M}_{\beta,1}$ and $\mathfrak{M}_{\beta,2}$ have the same meaning as defined above. Let $\mathfrak{K}$ be any class whose members are elements or subgroups or (still more generally) sets of elements of $\mathcal{G}$; and let each member of $\mathfrak{K}$ belong to at least one maximal subgroup of $\mathcal{G}$. Let $n(\mathfrak{M}_{\beta i})$ denote the number of members of $\mathfrak{K}$ which belong to $\mathfrak{M}_{\beta i}$. Then*

$$n(\mathfrak{M}_0)-\sum_{(\mathfrak{M}_{1,1})} n(\mathfrak{M}_{1,1})-\sum_{(\mathfrak{M}_{1,2})} n(\mathfrak{M}_{1,2})+p\left(\sum_{(\mathfrak{M}_{2,1})} n(\mathfrak{M}_{2,1})+\sum_{(\mathfrak{M}_{2,2})} n(\mathfrak{M}_{2,2})\right)$$

$$-p^3\left(\sum_{(\mathfrak{M}_{31})} n(\mathfrak{M}_{31})-\sum_{(\mathfrak{M}_{32})} n(\mathfrak{M}_{32})\right)+\dots+(-1)^d p^{\frac{1}{2}(d-1)} n(\mathfrak{M}_d)=0,$$

the sum $\sum\limits_{(\mathfrak{M}_{\beta i})}$ being taken over all the subgroups $\mathfrak{M}_{\beta i}$ of $\mathcal{G}$.

The derivation of this theorem from that due to Hall is immediate.

§3. **Theorem** 11. *Let $\mathcal{G}$ be a group of order p^n and of rank $\alpha \geq 1$, (consequently $d \geq 2$), with $p \geq 3$ and $n \geq 2\alpha+1$. Then the number of its subgroups of order p^m $(n > m \geq 2\alpha)$ and of rank $\alpha-1$ is congruent to*

$$p\,\phi_{d-1,1}=p+p^2+\dots+p^{d-1} \pmod{p^d}.$$

Proof. For $m=n-1$, the subgroups considered are those of $\mathfrak{M}_{1,2}$, which are $p\,\phi_{d-1,1}$ in number, hence the theorem is true for $m=n-1$. In particular, the theorem is true for $n=2\alpha+1$. We shall therefore assume that $n > 2\alpha+1$ and $n-1 > m \geq 2\alpha$ and apply mathematical induction on n.

Taking $\mathfrak{K}$ to be the set of subgroups of order p^m and of rank $\alpha-1$, we have, by theorems 10, 4 and 5

$$n(\mathfrak{M}_0)=n(\mathfrak{M}_{11})+\sum_{(\mathfrak{M}_{12})} n(\mathfrak{M}_{12})-p\sum_{(\mathfrak{M}_{21})} n(\mathfrak{M}_{21}).$$

By theorems 8 and 9 and hypothesis of induction, we have

$$n(\mathfrak{M}_0)\equiv 1\cdot p\,\phi_{d(\mathfrak{M}_{1,1})-1,1}+p\,\phi_{d-1,1}\cdot 1-p\,\phi_{d-1,1}\cdot 1 \left(\bmod\ p^{d(\mathfrak{M}_{11})}\right)$$

$$\equiv p\,\phi_{d-1,1} \pmod{p^d}$$

Theorem 12. *In a group $\mathcal{G}$ of order p^n and of rank α ($p \geq 3$ and $n \geq 2\alpha+1$), the number of subgroups of order $p^{n-\beta}$ $(d \geq \beta \geq 0)$ and of rank $\alpha-\beta$ is congruent* *to* $p^{\beta}\phi_{d-1,\,\beta}$ (mod $p^{d-\beta+1}$).—*Thus the number of nonmajor subgroups of order* $p^{n-\beta}$ *and of rank* $\alpha-\beta$ *in* $\mathfrak{G}$ *is divisible by* $p^{\alpha-\beta+1}$.

(For $\beta=d$, $p^{\beta}\phi_{d-1,\,\beta}$ is to be replaced by 0.)

Proof. There is no subgroup of rank $\alpha-\beta$ of order greater than $p^{n-\beta}$ by theorem 5.

The theorem is trivial for $\beta=0$ and it is true for $\beta=1$ by theorem 8. We shall therefore assume $\beta>1$ and apply induction on β. Taking $\mathfrak{K}$ to be the set of all subgroups of order $p^{n-\beta}$ and of rank $\alpha-\beta$, we have, by theorems 10 and 5 (hence $\mathfrak{M}_{\gamma 1}$ contains no subgroup of rank $(\alpha-\gamma+1)-(\beta-\gamma+1)$ of order $> p^{(n-\gamma)-(\beta-\gamma+1)}=p^{n-\beta-1}$),

$$n(\mathfrak{M}_0)=\sum_{(\mathfrak{M}_1)} n(\mathfrak{M}_1)-p\sum_{(\mathfrak{M}_2)} n(\mathfrak{M}_2)+p^3\sum_{(\mathfrak{M}_3)} n(\mathfrak{M}_3)-\dots$$

$$-(-1)^{\beta+1}p^{\frac12\beta(\beta+1)}\sum_{(\mathfrak{M}_{\beta+1})} n(\mathfrak{M}_{\beta+1})$$

$$=\sum_{(\mathfrak{M}_{12})} n(\mathfrak{M}_{12})-p\sum_{(\mathfrak{M}_{22})} n(\mathfrak{M}_{22})+p^3\sum_{(\mathfrak{M}_{32})} n(\mathfrak{M}_{32})-\dots$$

$$-(-1)^{\beta}p^{\frac12\beta(\beta-1)}\sum_{(\mathfrak{M}_{\beta,2})} n(\mathfrak{M}_{\beta 2}).$$

Now $\mathfrak{M}_{\gamma 2}$ is a group of order $p^{n-\gamma}$ and of rank $\alpha-\gamma$, and $n(\mathfrak{M}_{\gamma 2})$ is equal to the number of subgroups of order $p^{(n-\gamma)-(\beta-\gamma)}$ and of rank $(\alpha-\gamma)-(\beta-\gamma)$ with $\beta-\gamma<\beta$, further $n-\gamma\geq 2(\alpha-\gamma)+1$ and $0\leq\beta-\gamma\leq d-\gamma\leq d(\mathfrak{M}_{\gamma 2})$, we have, by hpyothesis of induction and by theorems 8 and 9.

$$-(-1)^{\gamma}p^{\frac12\gamma(\gamma-1)}\sum_{(\mathfrak{M}_{\gamma 2})} n(\mathfrak{M}_{\gamma 2})$$

$$\equiv-(-1)^{\gamma}p^{\frac12\gamma(\gamma-1)}\sum_{(\mathfrak{M}_{\gamma 2})} p^{\beta-\gamma}\phi_{d(\mathfrak{M}_{\gamma 2})-1,\,\beta-\gamma}\pmod{p^{\delta}}$$

$$\left(\delta=\min_{(\mathfrak{M}_{\gamma 2})} d(\mathfrak{M}_{\gamma 2})-\beta+\gamma+1+\tfrac12\gamma(\gamma-1)\right)$$

$$\equiv-(-1)^{\gamma}p^{\frac12\gamma(\gamma-1)}\sum_{(\mathfrak{M}_{\gamma 2})} p^{\beta-\gamma}\phi_{d-\gamma-1,\,\beta-\gamma}\left(\bmod\ p^{d-\beta+1+\frac12\gamma(\gamma-1)}\right)$$

$$\equiv -(-1)^\gamma p^{\frac{1}{2}\gamma(\gamma-1)} \sum_{(\mathfrak{M}_{\gamma_2})} p^{\beta-\gamma} \phi_{d-\gamma-1,\,\beta-\gamma} \pmod{p^{d-\beta+1}}$$

$$\equiv -(-1)^\gamma p^{\frac{1}{2}\gamma(\gamma-1)} p^\gamma \phi_{d-1,\,\gamma}\, p^{\beta-\gamma} \phi_{d-\gamma-1,\,\beta-\gamma} \pmod{p^{d-\beta+1}}$$

$$\equiv -(-1)^\gamma p^{\frac{1}{2}\gamma(\gamma-1)} p^\beta \phi_{d-1,\,\gamma}\, \phi_{d-\gamma-1,\,\beta-\gamma} \pmod{p^{d-\beta+1}}.$$

Consequently

$$n(\mathfrak{M}_0) \equiv p^\beta \left(\phi_{d-1,1}\,\phi_{d-2,\,\beta-1} - p\,\phi_{d-1,\,2}\,\phi_{d-3,\,\beta-2} + \dots - (-1)^\beta p^{\frac{1}{2}\beta(\beta-1)} \phi_{d-1,\,\beta}\right)$$

$$\equiv p^\beta \phi_{d-1,\,\beta} \pmod{p^{d-\beta+1}},$$

and the theorem is thus proved.

Corollary 12. 1. *Let $\mathfrak{G}$ have a cyclic principal subgroup of index p^a (consequently it is of rank $\alpha = d-1$). Then the number of subgroups of order $p^{n-\beta}$ $(d \geq \beta \geq 0)$ and of rank $\alpha-\beta$ is equal to $p^\beta \phi_{d-1,\,\beta}$.*

Proof. Since for this case, $d(\mathfrak{M}_{\gamma_2}) = d-\gamma$, we have everywhere equalities in place of former congruences.

Theorem 13. *Let $\mathfrak{G}$ be a group of order p^n and of rank α with $p \geq 3$ and $n \geq 2\alpha+1$. The number of its subgroups of order p^m $(2\alpha-\beta < m < n-\beta+1)$ and of rank $\alpha-\beta$ $(0 \leq \beta \leq d)$ is congruent to $p^\beta \phi_{d-1,\,\beta}$ (mod $p^{d-\beta+1}$). (For $\beta = d$, $p^\beta \phi_{d-1,\,\beta}$ has to be replaced by 0.)*

Proof. Theorem 12 asserts that the theorem is true for $m = n-\beta$. Thus, in particular, the theorem is true for $n = 2\alpha+1$. By theorems 4 and 11, the theorem is true for $\beta = 0$ and $\beta = 1$ respectively. Hence suppose $n > 2\ \ +1$, $m < n-\beta$ and $\beta > 1$, and prove by induction on n and β. Let $\mathfrak{K}$ be the set of all subgroups of order p^m of rank $\alpha-\beta$. By theorems 8, 9 and 10 and hypothesis of induction, we then have as before

$$n(\mathfrak{M}_0) \equiv n(\mathfrak{M}_{1,\,1}) + \sum_{(\mathfrak{M}_{1,\,2})} n(\mathfrak{M}_{1,\,2}) - p \sum_{(\mathfrak{M}_{2,\,1})} n(\mathfrak{M}_{2,\,1}) - p \sum_{(\mathfrak{M}_{2,\,2})} n(\mathfrak{M}_{2,\,2}) + \dots$$

$$\equiv p^\beta (\phi_{d-1,\,\beta} + \phi_{d-1,1}\,\phi_{d-1,\,\beta-1} - \phi_{d-1,\,1}\,\phi_{d-1,\,\beta-1} + \dots$$

$$-(-1)^\gamma p^{\frac{1}{2}\gamma(\gamma-1)} \phi_{d-\gamma-1,\,\beta-\gamma}\,\phi_{d-1,\,\gamma} - (-1)^{\gamma+1} p^{\frac{1}{2}(\gamma-1)\gamma} \times$$

$$\phi_{d-\gamma-1,\,\beta-\gamma}\,\phi_{d-1,\,\gamma} + \dots)$$

$$\equiv p^\beta \phi_{d-1,\beta} \pmod{p^{d-\beta+1}},$$

hence the theorem follows.

Corollary 13. 1. *Let $\mathcal{G}$ have a cyclic principal subgroup of index p^d (consequently it is of rank $\alpha = d-1$). Then the number of its subgroups of order p^m with $2\alpha - \beta < m < n - \beta + 1$ and of rank $\alpha - \beta$ with $0 \leq \beta \leq d$ is equal to $p^\beta \phi_{d-1,\beta}$.*

Proof. Since $d(\mathfrak{M}_{\gamma 1}) = \alpha - \gamma + 2 = d - \gamma + 1$ and $d(\mathfrak{M}_{\gamma 2}) = d - \gamma$, we have everywhere equalities in place of the former congruences.

Corollary 13. 2. *In a group of order p^n and of rank α with $p \geq 3$ and $n \geq 2\alpha + 1$, the number of subgroups of order p^m and of rank $\alpha - \beta$ with $0 \leq \beta \leq \alpha$ and $2\alpha - \beta < m < n - \beta + 1$ is congruent to $p^\beta \pmod{p^{\beta+1}}$.*

Proof. For $0 \leq \beta \leq \left[\frac{d}{2}\right]$, the corollary is true by theorem 13. Hence we can assume that $\beta > \left[\frac{d}{2}\right]$ and apply induction on .

For $m = n - \beta$, we have then

$$n(\mathfrak{M}_0) = n(\mathfrak{M}_{1,1}) + \sum n(\mathfrak{M}_{12}) - p \sum n(\mathfrak{M}_{21}) - p \sum n(\mathfrak{M}_{22}) + \ldots$$

$$= 0 + p\,\phi_{d-1,1}\,(p^{\beta-1} + \ldots) - 0 - p \cdot p^2\,\phi_{d-2,2}\,(p^{\beta-2} + \ldots) + \ldots$$

$$\equiv p^\beta \pmod{p^{\beta+1}}$$

Hence the corollary is true for $m = n - \beta$ and is thus true for $n = 2\alpha + 1$. We can then assume $n > 2\alpha + 1$, also $m < n - \beta$, and apply induction on n.

Now as before

$$n(\mathfrak{M}_0) = n(\mathfrak{M}_{1,1}) + \sum n(\mathfrak{M}_{12}) - p \sum n(\mathfrak{M}_{21}) - p \sum n(\mathfrak{M}_{22}) + \ldots$$

$$\equiv p^\beta + p\,\phi_{d-1,1}\,p^{\beta-1} - p\,\phi_{d-1,1}\,p^{\beta-1} - p \cdot p^2\,\phi_{d-1,2}\,p^{\beta-2} + \ldots \pmod{p^{\beta+1}}$$

$$\equiv p^\beta \pmod{p^{\beta+1}},$$

hence the corollary.

By the above corollary and theorem 5, we have.

Corollary 13. 3. *In a group of order p^n and of rank α with $p \geq 3$, the number of subgroups of order p^m and of rank $\alpha - \beta$ with $2\alpha - \beta < n$ is always congruent to $0 \pmod{p^\beta}$.*

§4. **Theorem** 14. *The number of subgroups of order p^m of rank β ($p \geq 3$, $m > 2\beta+1$) in a group of order p^n ($n \geq m$) of rank α ($\alpha > \beta$) is congruent to* 0 (mod p).

Proof. For $m=n$, the theorem is trivial, hence we assume $m < n$. For $n \leq \alpha+\beta+1$, the theorem is true, since the group would then contain no subgroup of order p^m of rank β ($m > 2\beta+1$) by theorem 5. For $\alpha=\beta+1$, we have $n > m > 2\beta+1=2\alpha-1$, and therefore the theorem is true by theorem 11. Hence assume $n > \alpha+\beta+1$ and $\alpha > \beta+1$ and apply induction on n and β. By the enumeration principle, we have therefore

$$n(\mathfrak{M}_0) \equiv \sum n(\mathfrak{M}_{1,1}) + \sum n(\mathfrak{M}_{1,2}) \pmod p;$$

then by hypothesis of induction, $n(\mathfrak{M}_0) \equiv 0 \pmod p$.

Theorem 15. *The number of cyclic subgroups of order p^m ($p \geq 3$, $m > \beta$) in a group of order p^n ($n \geq m$) of rank $\alpha \geq \beta$ is congruent to* 0 (mod p^β).

Proof. For $m=n$, the theorem is trival, hence we assume $m < n$. For $n \leq \alpha + \beta$, the theorem is true since the group would then contain no cyclic subgroups of such orders by theorem 5. For $\alpha = \beta$, the theorem is true for $\beta < m \leq n-\alpha$ (then $n \geq m+\alpha > \alpha+\beta \geq 2\alpha$) by theorem 3 and for $m > n-\alpha$ by theorem 5. Hence assume $n > \alpha+\beta$ and $\alpha > \beta$ and apply induction on n and β. By the enumeration principle, we have

$$n(\mathfrak{M}_0) \equiv \sum n(\mathfrak{M}_{1,1}) + \sum n(\mathfrak{M}_{12}) - p\sum n(\mathfrak{M}_{21}) - p\sum n(\mathfrak{M}_{22}) + \dots\dots$$

By hypothesis of induction, we have then

$$n(\mathfrak{M}_0) \equiv -(-1)^{\alpha-\beta+1} p^{\frac{1}{2}(\alpha-\beta+1)(\alpha-\beta)} \sum n(\mathfrak{M}_{\alpha-\beta+1,2}) - (-1)^{\alpha-\beta+2} \times$$
$$p^{\frac{1}{2}(\alpha-\beta+2)(\alpha-\beta+1)} \sum n(\mathfrak{M}_{\alpha-\beta+2,1}) + \dots \pmod{p^\beta}$$

Now for $\gamma \geq 2\alpha-\beta+1$, we have $m >$ (rank of $\mathfrak{M}_{\gamma i}$); hence by theorems 3 and 5,

$$n(\mathfrak{M}_{\gamma 1}) \equiv 0 \pmod{p^{\alpha-\gamma+1}},\quad n(\mathfrak{M}_{\gamma 2}) \equiv 0 \pmod{p^{\alpha-\gamma}}.$$

And for $\gamma \geq 2\alpha-\beta+1$, we have also

$$\tfrac{1}{2}\gamma(\gamma-1)+\alpha-\gamma=\tfrac{1}{2}\gamma(\gamma-3)+\alpha \geq \beta,$$

so the theorem is proved.

Theorem 16. *If a group of order p^n ($p \geq 3$) contains one and only one subgroup of rank β of order p^m for a fixed m, $n \geq m > 2\beta+1$, then it is of rank β.*

Proof. If the rank of the group is less than β, then by theorem 5, it contains no subgroup of rank β. And if the rank of the group is greater than β, then by theorem 14, the number of subgroups of rank β is divisible by p. Hence the theorem.

Theorem 17. *If a group of order p^n $(p \geq 3)$ contains exactly p^β cyclic subgroups of order p^m for a fixed m, $\beta+1 < m \leq n$, then it is of rank β.*

Proof. Let γ be the rank of the group considered.

1) If $\gamma < \beta$, then $\gamma < \gamma+1 < \beta+1 < m$. By theorems 3 and 5, the group would contain either p^γ or no cyclic subgroups of order p^m, according as $m \leq n-\gamma$ (then $n \geq m+\gamma > 2\gamma$) or $m > n-\gamma$. Hence we get a contradiction.

2) If $\gamma > \beta$, then $\gamma \geq \beta+1$. By theorem 15, the number of cyclic subgroups of order p^m $(\beta+1 < m \leq n)$ would be divisible by $p^{\beta+1}$. Hence we get a contradiction.

Thus the theorem is proved.

REFERENCES:

1. G. A. Miller, Proc. London Math. Soc. (2) **2,** 142—143 (1904).
2. A. Kulakoff, Math. Annalen, **104,** 778—793 (1931).
3. P. Hall, Proc. London Math. Soc. (2) **36,** 29—95 (1933-1934).
4. H. Zassenhaus, *Lehrbuch der Gruppentheorie* I (1937).
5. L. K. Hua—H. F. Tuan, Sci. Rep. Nat. Tsing-Hua Univ. (A) **4** 145—154 (1940).
6. L. K. Hua—H. F. Tuan, J. Chinese Math. Soc. **2** 313—319 (1940).

Annals of Mathematics
Vol. 49, No. 4, October, 1948

ON THE AUTOMORPHISMS OF THE SYMPLECTIC GROUP OVER ANY FIELD

By Loo-Keng Hua

(Received November 7, 1946, Revised August 15, 1947)

1. Introduction

Let Φ be any field with characteristic $\neq 2$. Unless the contrary is stated, we use small Latin letters to denote elements of the field and capital Latin letters to denote n-rowed matrices over Φ. For matrices which are not $n \times n$, we use $M^{(m)}$ to denote an $m \times m$ matrix. I and 0 denote the n-rowed identity and zero matrices respectively, and $[a_1, \cdots, a_n]$ will denote a diagonal matrix with $a_1, \cdots, a_n$ on its diagonal. We use small Greek letters for automorphisms of the field, i.e.

$$(a + b)^\sigma = a^\sigma + b^\sigma, \qquad (ab)^\sigma = a^\sigma b^\sigma$$

for a and b belonging to Φ.

Let

$$\mathfrak{F} = \begin{pmatrix} 0 & I \\ -I & 0 \end{pmatrix}.$$

The group formed by the $2n$-rowed matrices $\mathfrak{T}$ satisfying

$$\mathfrak{T}\mathfrak{F}\mathfrak{T}' = \mathfrak{F}, \tag{1}$$

where $\mathfrak{T}'$ denotes the transposed matrix of $\mathfrak{T}$, is called symplectic and is denoted by Sp $(\Phi, 2n)$.

Apparently, there are three different types of automorphisms $\mathcal{A}(\mathfrak{T})$ of Sp $(\Phi, 2n)$:

1) Inner automorphisms. For a symplectic $\mathfrak{P}$, we have

$$\mathcal{A}(\mathfrak{T}) = \mathfrak{P}\mathfrak{T}\mathfrak{P}^{-1}. \tag{2}$$

2) Automorphisms induced by those of the field:

$$\mathcal{A}(\mathfrak{T}) = \mathfrak{T}^\sigma. \tag{3}$$

3) Semi-inner automorphisms: We define the group formed by the matrices $\mathfrak{P}$ satisfying

$$\mathfrak{P}\mathfrak{F}\mathfrak{P}' = a\mathfrak{F} \tag{4}$$

to be the extended symplectic group. Then, for $\mathfrak{P}$ satisfying (4), we have also an automorphism

$$\mathcal{A}(\mathfrak{T}) = \mathfrak{P}^{-1}\mathfrak{T}\mathfrak{P}.$$

Since

$$(b\mathfrak{P})^{-1}\mathfrak{T}(b\mathfrak{P}) = \mathfrak{P}^{-1}\mathfrak{T}\mathfrak{P}$$

and

$$(b\mathfrak{P})\mathfrak{F}(b\mathfrak{P})' = b^2a\mathfrak{F},$$

we have only to consider those a which represent different cosets of the factor group of the multiplicative group of Φ by the group Φ_2 of square elements of Φ. More precisely, let

$$a, b, \cdots$$

be a representative system of Φ/Φ_2. Then the elements of the extended symplectic group can be expressed as

$$\begin{pmatrix} I & 0 \\ 0 & aI \end{pmatrix} \mathfrak{P},$$

where $\mathfrak{P}$ is symplectic.

It is the purpose of this paper to establish that the group generated by these three types of automorphisms is the largest group of automorphisms of Sp $(\Phi, 2n)$, provided that the characteristic of Φ is not equal to 2.

In other words, analytically, we have the following

THEOREM. *All the automorphisms* $\mathcal{A}(\mathfrak{P})$ *of* Sp $(\Phi, 2n)$ *can be expressed as*

$$\mathcal{A}(\mathfrak{T}) = \mathfrak{P}\mathfrak{R}\mathfrak{T}^{\sigma}\mathfrak{R}^{-1}\mathfrak{P}^{-1}, \tag{5}$$

where σ *runs over all automorphisms of the field,* $\mathfrak{P}$ *runs over all symplectic matrices and*

$$\mathfrak{R} = \begin{pmatrix} I & 0 \\ 0 & aI \end{pmatrix} \tag{6}$$

where a *runs over a complete residue system of* Φ/Φ_2.

The corresponding problem for the special linear group was solved by O. Schreier and B. L. van der Waerden.[1] (See Appendix.)

2. Involutions

We now consider involutions, that is, matrices $\mathfrak{T}$ of Sp $(\Phi, 2n)$ such that

$$\mathfrak{T}^2 = \mathfrak{J}, \tag{7}$$

where $\mathfrak{J}$ is the $2n$-rowed identity.

THEOREM 1. *Every involution in* Sp$(\Phi, 2n)$ *is equivalent sympletically to the normal form*

[1] *Die Automorphism der projektiven Gruppen*, Hamburg. Univ. Math. Seminar, Abh. Bd. 6, 1928, 303–322. See appendix.

$$\begin{pmatrix} J & 0 \\ 0 & J \end{pmatrix} \tag{8}$$

where $J = [1, \cdots, 1, -1, \cdots, -1] = \{p, q\}$, *say, consists of* p *ones and* q *minus ones. Precisely, there exists a symplectic matrix* $\mathfrak{P}$ *such that*

$$\mathfrak{P}\mathfrak{T}\mathfrak{P}^{-1} = \begin{pmatrix} J & 0 \\ 0 & J \end{pmatrix}.$$

The theorem is evident for $n = 1$; in fact, let

$$\mathfrak{T} = \begin{pmatrix} a & b \\ c & d \end{pmatrix} \qquad ad - bc = 1,$$

(7) implies

$$a^2 + bc = 1, \qquad (a + d)\, b = (a + d)\, c = 0.$$

Consequently, we have

$$\mathfrak{T} = \pm\mathfrak{I}.$$

Now we suppose that the theorem holds for symplectic matrices of order less than $2n$.

From (1) and (7), we have

$$\mathfrak{T}\mathfrak{F} = \mathfrak{F}\mathfrak{T}'^{-1} = \mathfrak{F}\mathfrak{T}' = -(\mathfrak{T}\mathfrak{F})',$$

that is, $\mathfrak{T}\mathfrak{F}$ is skew symmetric, more precisely, $\mathfrak{T}$ has the form

$$\mathfrak{T} = \begin{pmatrix} A & K_1 \\ K_2 & A' \end{pmatrix}, \tag{9}$$

where K_1, and K_2 are skew symmetric. Since $\mathfrak{T}$ is symplectic we have also that

$$AK_1, \qquad K_2A \tag{10}$$

are symmetric and that

$$A^2 + K_1K_2 = I. \tag{11}$$

There is a non-singular matrix Q such that

$$QK_1Q' = \begin{pmatrix} k_1^{(r)} & 0 \\ 0 & 0 \end{pmatrix} = K_1^0$$

where k_1 is a non-singular skew symmetric r-rowed matrix. (In case $r = n$, there are no zeros in the expression.) The transformation

$$\begin{pmatrix} Q & 0 \\ 0 & Q'^{-1} \end{pmatrix} \begin{pmatrix} A & K_1 \\ K_2 & A' \end{pmatrix} \begin{pmatrix} Q & 0 \\ 0 & Q'^{-1} \end{pmatrix}^{-1} \tag{12}$$

carries K_1 into K_1^0, where

$$\begin{pmatrix} Q & 0 \\ 0 & Q'^{-1} \end{pmatrix}$$

is clearly symplectic. Thus we may assume at the beginning that

$$K_1 = \begin{pmatrix} k_1^{(r)} & 0 \\ 0 & 0 \end{pmatrix}.$$

From (10), we have

$$A = \begin{pmatrix} a_1^{(r)} & a_2^{(r,n-r)} \\ a_3^{(n-r,r)} & a_4^{(n-r)} \end{pmatrix} \qquad a_3 = 0, \qquad a_1 k_1 \text{ symmetric}$$

The symplectic matrix

$$\mathfrak{P} = \begin{pmatrix} I & 0 \\ S & I \end{pmatrix}$$

carries (9) into

$$(13) \qquad \mathfrak{P}\mathfrak{T}\mathfrak{P}^{-1} = \begin{pmatrix} I & 0 \\ S & I \end{pmatrix}\begin{pmatrix} A & K_1 \\ K_2 & A' \end{pmatrix}\begin{pmatrix} I & 0 \\ -S & I \end{pmatrix} = \begin{pmatrix} A - K_1 S & K_1 \\ * & * \end{pmatrix}.$$

We choose

$$S = \begin{pmatrix} k_1^{-1} a_1, & k_1^{-1} a_2 \\ (k_1^{-1} a_2)', & 0 \end{pmatrix}$$

then the A of $\mathfrak{P}\mathfrak{T}\mathfrak{P}^{-1}$ takes the form

$$A = \begin{pmatrix} 0 & 0 \\ 0 & a_4^{(n-r)} \end{pmatrix}$$

where a_4 is non-singular. Now we assume at the beginning that A has this form. From (10) and (11), we deduce that

$$K_2 = \begin{pmatrix} k_1^{-1} & 0 \\ 0 & k_2 \end{pmatrix}.$$

Then, every involution is equivalent to

$$\mathfrak{T} = \begin{pmatrix} 0 & 0 & k_1 & 0 \\ 0 & a & 0 & 0 \\ k_1^{-1} & 0 & 0 & 0 \\ 0 & k_2 & 0 & a' \end{pmatrix}.$$

This induces two symplectic involutions

$$\mathfrak{T}^{(2r)} = \begin{pmatrix} 0 & k_1 \\ k_1^{-1} & 0 \end{pmatrix}, \qquad \mathfrak{T}^{(2n-2r)} = \begin{pmatrix} a & 0 \\ k_2 & a' \end{pmatrix}.$$

By hypothesis of induction, the theorem is true for $r \neq 0$ and $\neq n$. A similar result holds if the rank of k_2 is neither 0 nor n.

If the ranks of k_1 and k_2 are both 0, then $A^2 = I$, it is known that we have Q such that $QAQ^{-1} = J$. The theorem is now evident, since

$$\begin{pmatrix} Q & 0 \\ 0 & Q^{l-1} \end{pmatrix}\begin{pmatrix} A & 0 \\ 0 & A' \end{pmatrix}\begin{pmatrix} Q & 0 \\ 0 & Q^{l-1} \end{pmatrix}^{-1} = \begin{pmatrix} J & 0 \\ 0 & J \end{pmatrix}.$$

If $r = n$, then we have a suitable S such that (13) carries $\mathfrak{T}$ into the form (9) with $A = 0$. That is

$$\mathfrak{T} = \begin{pmatrix} 0 & K \\ K^{-1} & 0 \end{pmatrix}. \tag{14}$$

Note that this case can only happen for even n, $= 2m$ say. We can choose Q such that (12) carries (14) into the same form with

$$K = \begin{pmatrix} 0 & I^{(m)} \\ -I^{(m)} & 0 \end{pmatrix}. \tag{15}$$

If (15) is satisfied, the symplectic matrix

$$\begin{pmatrix} 0 & 0 & I^{(m)} & 0 \\ 0 & I^{(m)} & 0 & 0 \\ -I^{(m)} & 0 & 0 & 0 \\ 0 & 0 & 0 & I^{(m)} \end{pmatrix}$$

carries (14) into

$$\begin{pmatrix} 0 & 0 & I & 0 \\ 0 & I & 0 & 0 \\ -I & 0 & 0 & 0 \\ 0 & 0 & 0 & I \end{pmatrix}\begin{pmatrix} 0 & 0 & 0 & I \\ 0 & 0 & -I & 0 \\ 0 & -I & 0 & 0 \\ I & 0 & 0 & 0 \end{pmatrix}\begin{pmatrix} 0 & 0 & -I & 0 \\ 0 & I & 0 & 0 \\ I & 0 & 0 & 0 \\ 0 & 0 & 0 & I \end{pmatrix} = \begin{pmatrix} 0 & -I & 0 & 0 \\ -I & 0 & 0 & 0 \\ 0 & 0 & 0 & -I \\ 0 & 0 & -I & 0 \end{pmatrix}.$$

This gives $\mathfrak{T}$ with zero K_1 and the theorem follows from the previous case.

DEFINITION. An involution equivalent to (8) is called an *involution of signature* $\{p, q\}$.

3. Non equivalence of symplectic groups

Now we consider a set of symplectic matrices $\mathfrak{T}_i (1 \leqq i \leqq l)$ satisfying

$$\mathfrak{T}_i\mathfrak{T}_j = \mathfrak{T}_j\mathfrak{T}_i, \qquad \mathfrak{T}_i^2 = \mathfrak{J}. \tag{16}$$

THEOREM 2. *We have* $l \leqq 2^n$ *and the maximum is attained. More precisely, any set of commutative symplectic involutions is equivalent symplectically to a subset of the set of matrices of the form*

$$\begin{pmatrix} J & 0 \\ 0 & J \end{pmatrix},$$

where J is a diagonal matrix with ± 1 on its diagonal.

The proof of the theorem is evident for $n = 1$ (Cf. the first paragraph of the proof of Theorem 1). Now we apply induction on n.

Let $\mathfrak{T}_1$ be an involution different from $\pm\mathfrak{J}$. By Theorem 1, we may assume that it has the form

$$\mathfrak{T}_1 = \begin{pmatrix} J & 0 \\ 0 & J \end{pmatrix}, \qquad J = \{p, q\}, \qquad p > 0, q > 0.$$

These matrices commutative with $\mathfrak{T}_1$ are of the form

$$\mathfrak{T}_i = \begin{pmatrix} a_1^{(p)} & 0 & b_1^{(p)} & 0 \\ 0 & a_2^{(q)} & 0 & b_2^{(q)} \\ c_1^{(p)} & 0 & d_1^{(p)} & 0 \\ 0 & c_2^{(q)} & 0 & d_2^{(q)} \end{pmatrix}_i.$$

Thus we have

$$\begin{pmatrix} a_1 & b_1 \\ c_1 & d_1 \end{pmatrix}_i \begin{pmatrix} a_2 & b_2 \\ c_2 & d_2 \end{pmatrix}_i$$

two sets of commutative involutions of Sp $(\Phi, 2p)$ and Sp $(\Phi, 2q)$, respectively. By the hypothesis of induction we have Theorem 2.

Since an isomorphic mapping carries involution into involution and commutative involutions into commutative involutions, we have

THEOREM 3. *For a fixed field Φ, no two* Sp $(\Phi, 2n)$ *with different n are isomorphic.*

4. Generators of Sp $(\Phi, 2n)$

THEOREM 4. *The group* Sp $(\Phi, 2n)$ *is generated by the elements*

$$\begin{pmatrix} A & 0 \\ 0 & A^{-1} \end{pmatrix}, \quad \begin{pmatrix} I & S \\ 0 & I \end{pmatrix}, \quad S = S'$$

and

$$\begin{pmatrix} M & I - M \\ -(I - M) & M \end{pmatrix},$$

where M is a diagonal matrix with zero and one on its diagonal.

PROOF. For a given symplectic matrix $\mathfrak{T}$, we can find a suitable[2] M such that

$$\mathfrak{T} = \begin{pmatrix} M & I - M \\ -(I - M) & M \end{pmatrix} = \begin{pmatrix} A & B \\ C & D \end{pmatrix}$$

with non-singular A. Then

$$\begin{pmatrix} A & B \\ C & D \end{pmatrix} \begin{pmatrix} A^{-1} & 0 \\ 0 & A' \end{pmatrix} = \begin{pmatrix} I & S \\ P & Q \end{pmatrix}$$

where $S = S'$. Further, since

$$\begin{pmatrix} I & S \\ P & Q \end{pmatrix} \begin{pmatrix} I & -S \\ 0 & I \end{pmatrix} = \begin{pmatrix} I & 0 \\ R & I \end{pmatrix},$$

where R is symmetric and

$$\begin{pmatrix} M & I - M \\ -(I - M) & M \end{pmatrix}^{-1} \begin{pmatrix} I & 0 \\ R & I \end{pmatrix} \begin{pmatrix} M & I - M \\ -(I - M) & M \end{pmatrix} = \begin{pmatrix} I & -R \\ 0 & I \end{pmatrix}$$

with $M = 0$, we have the theorem.

THEOREM 5. *The group* Γ *generated by all square elements of* Sp $(\Phi, 2n)$ *is identical with* Sp $(\Phi, 2n)$.

PROOF. Since the field is of characteristic $\neq 2$, we deduce that as S runs over all symmetric matrices, so does $2S$. From

$$\begin{pmatrix} I & S \\ 0 & I \end{pmatrix}^2 = \begin{pmatrix} I & 2S \\ 0 & I \end{pmatrix},$$

we see that Γ contains all elements of the form

$$(*) \qquad \mathfrak{P} \begin{pmatrix} I & S \\ 0 & I \end{pmatrix} \mathfrak{P}^{-1}$$

where $\mathfrak{P}$ is any symplectic matrix and S is any symmetric matrix. In particular Γ contains

$$\begin{pmatrix} I & 0 \\ -J & I \end{pmatrix} = \begin{pmatrix} 0 & I \\ -I & 0 \end{pmatrix} \begin{pmatrix} I & S \\ 0 & I \end{pmatrix} \begin{pmatrix} 0 & I \\ -I & 0 \end{pmatrix}^{-1}.$$

For non-singular S, we have

$$\begin{pmatrix} S & 0 \\ 0 & S^{-1} \end{pmatrix} = \begin{pmatrix} I & -S \\ 0 & I \end{pmatrix} \begin{pmatrix} I & 0 \\ S^{-1} - I & I \end{pmatrix} \begin{pmatrix} I & I \\ 0 & I \end{pmatrix} \begin{pmatrix} I & 0 \\ -(I - S) & I \end{pmatrix},$$

thus, Γ contains elements of the form

[2] L. K. HUA, *Geometrics of Matrices* III, Trans., Amer. Math. Soc. 61(1947), 229-255, Theorem 7. Notice that the proof holds for any field.

$$\begin{pmatrix} S & 0 \\ 0 & S^{-1} \end{pmatrix}$$

where $S = S'$.

We shall justify later that any non-singular matrix A can be expressed as product of symmetric matrices, the group contains

$$\begin{pmatrix} A & 0 \\ 0 & A^{-1} \end{pmatrix}.$$

Since

$$\begin{pmatrix} 0 & 1 \\ -1 & 0 \end{pmatrix} = \begin{pmatrix} 1 & 2 \\ 0 & 1 \end{pmatrix} \left(\begin{pmatrix} 1 & 0 \\ 1 & 1 \end{pmatrix} \begin{pmatrix} 1 & 1 \\ 0 & 1 \end{pmatrix} \begin{pmatrix} 1 & 0 \\ 1 & 1 \end{pmatrix}^{-1} \right),$$

the group contains also

$$\begin{pmatrix} M, & I - M \\ -(I - M) & M \end{pmatrix}.$$

By Theorem 4, we have the theorem.

Now we are going to justify that every non-singular matrix A is a product of symmetric matrices.

Let $J_{ij}(i \neq j)$ denote a matrix with 1 at (ij) and (ji) position and with 1 at (k, k) position for $k \neq i$ and j and with zero elsewhere. It is known that the group of non-singular matrices A is generated by J_{ij} and

$$T(d) = \begin{pmatrix} 1 & d & 0 & 0 & \cdots \\ 0 & 1 & 0 & 0 & \cdots \\ 0 & 0 & 1 & 0 & \cdots \\ \cdots & \cdots & \cdots & \cdots & \cdots \end{pmatrix} \qquad H(a) = \begin{pmatrix} a & 0 & 0 & \cdots \\ 0 & 1 & 0 & \cdots \\ 0 & 0 & 1 & \cdots \\ \cdots & \cdots & \cdots & \cdots \end{pmatrix}.$$

Evidently J_{ij} and $H(a)$ are symmetric and since

$$\begin{pmatrix} 1 & d \\ 0 & 1 \end{pmatrix} = \begin{pmatrix} 0 & 1 \\ 1 & 0 \end{pmatrix} \begin{pmatrix} 0 & 1 \\ 1 & d \end{pmatrix},$$

we have our assertion.

5. Proof of the theorem

1) The theorem is true for $n = 1$.[3] We suppose that the theorem holds for symplectic group of order less than $2n$.

2) Since the normalisator of an involution of signature $\{p, q\}$ is a direct product of two symplectic groups of matrices of orders $2p$ and $2q$, we see that

[3] SCHREIER AND V. D. WAERDEN, *ibid.* Note that, in their theorem, they identified two matrices—$\mathfrak{T}$ and $\mathfrak{T}$ as a single element. But we can find the present conclusion easily. See appendix.

the automorphism $\mathcal{A}(\mathfrak{T})$ carries an involution of signature $\{p, q\}$ into an involution of signature $\{p, q\}$ or $\{q, p\}$. By Theorem we may assume, without loss of generality, that

$$\mathfrak{A}\begin{pmatrix}\{1, n-1\} & 0 \\ 0 & \{1, n-1\}\end{pmatrix} = \begin{pmatrix}\{1, n-1\} & 0 \\ 0 & \{1, n-1\}\end{pmatrix}$$

$$\text{or } = \begin{pmatrix}-\{1, n-1\} & 0 \\ 0 & -\{1, n-1\}\end{pmatrix}.$$

The aggregate of symplectic matrices commutative with

$$\begin{pmatrix}\pm\{1, n-1\}, & 0 \\ 0 & \pm\{1, n-1\}\end{pmatrix}$$

consists of all elements of the form

$$(17)\qquad \mathfrak{T} = \begin{pmatrix} a & 0 & b & 0 \\ 0 & A_1 & 0 & B_1 \\ c & 0 & d & 0 \\ 0 & C_1 & 0 & D_1 \end{pmatrix}, \qquad A_1 = A_1^{(n-1)}, \qquad B_1 = B_1^{(n-1)}, \quad \text{etc.}$$

We shall call $\mathfrak{T}$ the direct product of two matrices

$$\mathfrak{t} = \mathfrak{t}^{(2)} = \begin{pmatrix} a & b \\ c & d \end{pmatrix}, \qquad \mathfrak{T}_1 = \mathfrak{T}_1^{(2n-2)} = \begin{pmatrix} A_1 & B_1 \\ C_1 & D_1 \end{pmatrix}$$

which are symplectic. We use the notation $\mathfrak{T} = \mathfrak{t} \times \mathfrak{T}_1$.

Consider the group Λ formed by the elements of the form

$$\begin{pmatrix} a & 0 & b & 0 \\ 0 & \pm I & 0 & 0 \\ c & 0 & d & 0 \\ 0 & 0 & 0 & \pm I \end{pmatrix}.$$

$\mathcal{A}$ carries Λ into

$$\begin{pmatrix} a^* & 0 & b^* & 0 \\ 0 & A_1^* & 0 & B_1^* \\ c^* & 0 & d^* & 0 \\ 0 & C_1^* & 0 & D_1^* \end{pmatrix}$$

where

$$\begin{pmatrix} A_1^* & B_1^* \\ C_1^* & D_1^* \end{pmatrix}$$

for a group $\mathfrak{T}_1^*$. It is clear that $\mathfrak{T}_1^*$ is a normal subgroup of $\mathfrak{T}_1$. But the factor group Sp $(\Phi, 2n-2)$ over its centrum $(\mathfrak{J}^{(2n-2)}, -\mathfrak{J}^{(2n)-2})$ is simple,[4] and

[4] VAN DER WAERDEN, *Gruppen von Linearen Transforman Theorem*, 1935, Ergebnisse der Math. Bd. 4, heft 2, p. 10–11.

then $\mathfrak{T}_1^*$ is either $\mathfrak{T}_1$ or identity or the centrum. In case $n \neq 2$, the first case can never happen (Theorem 3). The second case also cannot happen, for otherwise, the inverse of $\mathcal{A}$ will cease to carry elements of the form

$$\begin{pmatrix} a & 0 & b & 0 \\ 0 & -I & 0 & 0 \\ c & 0 & d & 0 \\ 0 & 0 & 0 & -I \end{pmatrix}$$

into elements of the form (17). Squaring all the elements of Λ, we conclude, by Theorem 5, that $\mathcal{A}$ induces an automorphism on $\mathfrak{t}$. Similarly, $\mathcal{A}$ induces an automorphism on $\mathfrak{T}_1$. In case $n = 2$, there is the additional possibility of interchanging both factors. But the inner automorphism

$$\begin{pmatrix} 0 & 1 & 0 & 0 \\ 1 & 0 & 0 & 0 \\ 0 & 0 & 0 & 1 \\ 0 & 0 & 1 & 0 \end{pmatrix} \mathfrak{T} \begin{pmatrix} 0 & 1 & 0 & 0 \\ 1 & 0 & 0 & 0 \\ 0 & 0 & 0 & 1 \\ 0 & 0 & 1 & 0 \end{pmatrix}$$

carries the factors into the right order. By the hypothesis of induction, we have a two-rowed symplectic matrix $\mathfrak{p}$, a matrix

$$\mathfrak{r} = \begin{pmatrix} 1 & 0 \\ 0 & a_0 \end{pmatrix}$$

and an automorphism α of Φ such that

$$\mathcal{A}(\mathfrak{t}) = \mathfrak{p}\mathfrak{r}\mathfrak{t}^{\alpha}\mathfrak{r}^{-1}\mathfrak{p}^{-1};$$

and we have a $(2n - 2)$-rowed symplectic matrix $\mathfrak{P}_1$, a matrix

$$\mathfrak{R}_1 = \begin{pmatrix} I^{(n-1)} & 0 \\ 0 & b_0 I^{(n-1)} \end{pmatrix}$$

and an automorphism β of Φ such that

$$\mathcal{A}(\mathfrak{T}_1) = \mathfrak{P}_1\mathfrak{R}_1\mathfrak{T}^{\beta}\mathfrak{R}_1^{-1}\mathfrak{P}_1^{-1}.$$

We define

$$\mathfrak{P} = \mathfrak{p} \times \mathfrak{P}_1$$

and

$$\mathfrak{R} = \begin{pmatrix} 1 & 0 \\ 0 & b_0 I \end{pmatrix}.$$

Then the automorphism

$$(\mathfrak{R}^{-1}\mathfrak{P}^{-1}\mathcal{A}(\mathfrak{T})\mathfrak{P}\mathfrak{R})^{\beta^{-1}}$$

carries $\mathfrak{T}$ given by (17) into

$$\begin{pmatrix} a^{\tau} & 0 & pb^{\tau} & 0 \\ 0 & A_1 & 0 & B_1 \\ (1/p)c^{\tau} & 0 & d^{\tau} & 0 \\ 0 & C_1 & 0 & D_1 \end{pmatrix},$$

where $\tau = \alpha\beta^{-1}$ and $p = a_0 b_0^{-1}$.

Therefore, without loss of generality, we may assume that

$$\mathcal{A}\begin{pmatrix} a & 0 & b & 0 \\ 0 & A_1 & 0 & B_1 \\ c & 0 & d & 0 \\ 0 & C_1 & 0 & D_1 \end{pmatrix} = \begin{pmatrix} a^{\tau} & 0 & pb^{\tau} & 0 \\ 0 & A_1 & 0 & B_1 \\ (1/p)c^{\tau} & 0 & d^{\tau} & 0 \\ 0 & C_1 & 0 & D_1 \end{pmatrix} \tag{18}$$

where p is either 1 or not a square element of Φ.

3) Let Δ be the sub-group formed by matrices of the form

$$\begin{pmatrix} I & [a_1, \cdots, a_n] \\ 0 & I \end{pmatrix}.$$

Then the equation (18) asserts that $\mathcal{A}$ induces a mapping carrying Δ onto itself. Now we consider all the symplectic matrices commutative with every element of Δ. These matrices form a group Σ_2. The elements of Σ_2 are of the form

$$\begin{pmatrix} E & S \\ 0 & E \end{pmatrix}, \qquad S = S'. \qquad E = [\pm 1, \cdots, \pm 1].$$

In fact, from

$$\begin{pmatrix} A & B \\ C & D \end{pmatrix}\begin{pmatrix} I & [a_1, \cdots, a_n] \\ 0 & I \end{pmatrix} = \begin{pmatrix} I & [a_1, \cdots, a_n] \\ 0 & I \end{pmatrix}\begin{pmatrix} A & B \\ C & D \end{pmatrix} \tag{19}$$

we find immediately

$$[a_1, \cdots, a_n]\, C = 0,$$

that is $C = 0$. Then $D = A'^{-1}$ and $B = AS$. (19) implies also

$$A[a_1, \cdots, a_n] + B = B + [a_1, \cdots, a_n]D,$$

that is,

$$A[a_1, \cdots, a_n]A' = [a_1, \cdots, a_n]$$

for all $[a_1, \cdots, a_n]$. It follows immediately that $A = [\pm 1, \pm 1, \cdots, \pm 1]$.

Therefore $\mathcal{A}$ induces a mapping carrying Σ_2 onto itself. All the square elements form also a group Σ. Then $\mathcal{A}$ induces also an automorphism on Σ. The elements of Σ takes the form

$$\begin{pmatrix} I & S \\ 0 & I \end{pmatrix}$$

and conversely every element of this form belongs to Σ since the characteristic of Φ is not equal to 2. Thus we have

$$\mathcal{A}\begin{pmatrix} I & S \\ 0 & I \end{pmatrix} = \begin{pmatrix} I & S^* \\ 0 & I \end{pmatrix}. \tag{20}$$

This induces a mapping carrying symmetric matrices onto symmetric matrices.

By (18), we have

$$\mathcal{A}\begin{pmatrix} 0 & I \\ -I & 0 \end{pmatrix} = \begin{pmatrix} 0 & D \\ -D^{-1} & 0 \end{pmatrix}, \qquad D = [p, 1, \cdots, 1]. \tag{21}$$

Combining (20) and (21), we have

$$\begin{aligned} \mathcal{A}\begin{pmatrix} I & 0 \\ S & I \end{pmatrix} &= \mathcal{A}\left(\begin{pmatrix} 0 & I \\ -I & 0 \end{pmatrix}\begin{pmatrix} I & -S \\ 0 & I \end{pmatrix}\begin{pmatrix} 0 & -I \\ I & 0 \end{pmatrix}\right) \\ &= \begin{pmatrix} 0 & D \\ -D^{-1} & 0 \end{pmatrix}\begin{pmatrix} I & -S^* \\ 0 & I \end{pmatrix}\begin{pmatrix} 0 & -D \\ D^{-1} & 0 \end{pmatrix} \\ &= \begin{pmatrix} I & 0 \\ D^{-1}S^*D^{-1} & I \end{pmatrix}. \end{aligned}$$

Then we have

$$\mathcal{A}\begin{pmatrix} I & 0 \\ T & I \end{pmatrix} = \begin{pmatrix} I & 0 \\ T^{**} & I \end{pmatrix}, \tag{22}$$

where T and T^{**} are symmetric. The group form by elements of the form

$$\begin{pmatrix} I & 0 \\ T & I \end{pmatrix}$$

is denoted by Σ_0.

Consider those elements carrying the groups Σ and Σ_0 onto themselves. They form a group with elements of the form

$$\begin{pmatrix} A & 0 \\ 0 & A'^{-1} \end{pmatrix}.$$

Therefore we have

$$\mathcal{A}\begin{pmatrix} A & 0 \\ 0 & A'^{-1} \end{pmatrix} = \begin{pmatrix} A^* & 0 \\ 0 & A^{*\prime -1} \end{pmatrix}, \tag{23}$$

that is $\mathcal{A}$ induces an automorphism of the general linear group ($A \leftrightarrow A^*$).

4) In (18), we take

$$A_1 = aI, \qquad b = c = 0, \qquad B_1 = 0, \qquad C_1 = 0,$$

then we have

$$\mathcal{A}\begin{pmatrix} aI^{(n)} & 0 \\ 0 & a^{-1}I^{(n)} \end{pmatrix} = \begin{pmatrix} [a^\tau, a, \cdots, a] & 0 \\ 0 & [a^\tau, a, \cdots, a]^{-1} \end{pmatrix}.$$

If the field contains more than three elements, we can find an element $a \neq \pm 1$. If $a^\tau \neq a$ and a^{-1}, the normalizer of

$$\begin{pmatrix} aI & 0 \\ 0 & a^{-1}I^{(n)} \end{pmatrix} \quad \text{and} \quad \begin{pmatrix} [a^\tau, a, \cdots, a] & 0 \\ 0 & [a^\tau, a, \cdots, a]^{-1} \end{pmatrix}$$

cannot be isomorphic. This is absurd. Therefore $a^\tau = a$ or a^{-1} for all a belonging to Φ. If there is an element a in Φ such that

$$a^\tau = a^{-1}$$

then, if Φ is of characteristic $\neq 3$, we have

$$2a^{-1} = a^{-1} + a^{-1} = a^\tau + a^\tau = (2a)^\tau = \begin{cases} (2a)^{-1} \\ \text{or } 2a. \end{cases}$$

Both are impossible for $a \neq \pm 1$. If Φ is of characteristic 3, we have

$$1 + a^{-1} = 1 + a^\tau = (1 + a)^\tau = \begin{cases} 1 + a \\ \text{or } (1 + a)^{-1}. \end{cases}$$

Both are also impossible for $a \neq \pm 1$. Therefore for $a \neq \pm 1$, we have

$$a^\tau = a.$$

This holds also for $a = \pm 1$. Therefore we have $\tau = 1$, the identity automorphism.

The field having only three elements is the field:

$$-1, 0, 1 \qquad (\text{mod } 3).$$

The only automorphism of the field is the identity automorphism. We have also $\tau = 1$.

Therefore we have $\tau = 1$ in (18).

5) Another particular case of (18) is

$$\mathcal{A}\begin{pmatrix} I & [b_1, \cdots, b_n] \\ 0 & I \end{pmatrix} = \begin{pmatrix} I & [pb_1, \cdots, b_n] \\ 0 & I \end{pmatrix}. \tag{24}$$

Combining (23) and (24), we have

$$\mathcal{A}\begin{pmatrix} I & A[b_1, \cdots, b_n]A' \\ 0 & I \end{pmatrix} = \mathcal{A}\left(\begin{pmatrix} A & 0 \\ 0 & A'^{-1} \end{pmatrix}\begin{pmatrix} I & [b_1, \cdots, b_n] \\ 0 & I \end{pmatrix}\begin{pmatrix} A & 0 \\ 0 & A'^{-1} \end{pmatrix}^{-1}\right)$$

$$= \begin{pmatrix} I & A^*[pb_1, b_2, \cdots, b_n]A^{*\prime} \\ 0 & I \end{pmatrix}.$$

This asserts that (20) induces a mapping carrying the set of symmetric matrices onto itself and the ranks of S and S^* are equal.

Combining (20) and (24), we have

$$\mathcal{A}\begin{pmatrix} I & S - [b_1, \cdots, b_n] \\ 0 & I \end{pmatrix} = \mathcal{A}\left(\begin{pmatrix} I & S \\ 0 & I \end{pmatrix}\begin{pmatrix} I & [b_1, \cdots, b_n] \\ 0 & I \end{pmatrix}^{-1}\right)$$
$$= \begin{pmatrix} I & S^* - [pb_1, b_2, \cdots, b_n] \\ 0 & I \end{pmatrix}.$$

Consequently, if we have a set of values $b_1, \cdots, b_n$ such that

$$\det (S - [b_1, \cdots, b_n]) = 0$$

then it satisfies also

$$\det (S^* - [pb_1, b_2, \cdots, b_n]) = 0$$

and vice versa.

Since these equations are linear in each of the variables $b_1, \cdots, b_n$, and the characteristic of Φ is not equal to 2,[5] the coefficients are proportional, that is, if we put $S = (s_{ij})$ and $S^* = (\overset{*}{s}_{ij})$

$$\frac{1}{p} = \frac{s_{11}}{\overset{*}{s}_{11}} = \frac{s_{ii}}{p\overset{*}{s}_{ii}} \qquad \text{for } 1 < i$$

$$= \frac{s_{11}s_{jj} - s_{1j}^2}{\overset{*}{s}_{11}\overset{*}{s}_{jj} - \overset{*}{s}{}_{1j}^2} \qquad \text{for } 1 < j$$

$$= \frac{s_{ii}s_{jj} - s_{ij}^2}{p(\overset{*}{s}_{ii}\overset{*}{s}_{jj} - \overset{*}{s}{}_{ij}^2)} \qquad \text{for } 1 < i < j$$

$$= \begin{vmatrix} s_{11} & s_{12} & s_{13} \\ s_{12} & s_{22} & s_{23} \\ s_{13} & s_{23} & s_{33} \end{vmatrix} \Bigg/ \begin{vmatrix} \overset{*}{s}_{11} & \overset{*}{s}_{12} & \overset{*}{s}_{13} \\ \overset{*}{s}_{12} & \overset{*}{s}_{22} & \overset{*}{s}_{23} \\ \overset{*}{s}_{13} & \overset{*}{s}_{23} & \overset{*}{s}_{33} \end{vmatrix}, \tag{25}$$

that is

$$\overset{*}{s}_{11} = s_{11}p, \qquad \overset{*}{s}_{ii} = s_{ii}, \qquad 1 < i,$$
$$s_{ij}^2 = \overset{*}{s}{}_{ij}^2, \qquad 1 < i < j,$$
$$p(s_{11}s_{jj} - s_{1j}^2) = \overset{*}{s}_{11}\overset{*}{s}_{jj} - \overset{*}{s}{}_{ij}^2$$
$$= ps_{11}s_{jj} - \overset{*}{s}{}_{ij}^2.$$

[5] Notice that by eliminating b_1, we have an equation which is quadratic in each of $b_2, \ldots, b_n$ and equal to zero for all $b_2, \ldots, b_n$. Thus the coefficients of the equation are identical zero. Consequently we have the assertion.

Then we have

$$p s_{ij}^2 = s_{ij}^{*2} .$$

This is only possible when p is a square element. Then, we have

$$p = 1.$$

Consequently, we have

$$s_{ii} = s_{ii}^* , \qquad s_{ij} = \pm s_{ij}^* .$$

Another special case of (18) is

$$\mathcal{A}\begin{pmatrix} I & \begin{pmatrix} 1 & 0 \\ 0 & B \end{pmatrix} \\ 0 & I \end{pmatrix} = \begin{pmatrix} I & \begin{pmatrix} 1 & 0 \\ 0 & B \end{pmatrix} \\ 0 & I \end{pmatrix}. \tag{26}$$

Using (26) instead of (24), we have, by the same argument, that

$$s_{ij} = s_{ij}^* \qquad \text{for } 1 < i < j.$$

Let $G_{ij}(i \neq j)$ denote a matrix with 1 at (ij) and (ji)-position with 0 elsewhere. We have already proved that

$$G_{ij}^* = G_{ij} \qquad \text{for } 1 < i < j,$$

$$G_{ii}^* = G_{ii} .$$

We have a matrix $A = [-1, 1, \cdots, 1]$ such that

$$A G_{12} A' = -G_{12} .$$

Thus without loss of generality, we may assume that $\mathcal{A}$ satisfies (18) with $p = 1$ and $\tau = 1$ and $G_{12}^* = G_{12}$.

If $G_{13}^* = -G_{13}$, by (25), we have

$$0 = \begin{vmatrix} 1 & 1 & 1 \\ 1 & 0 & 1 \\ 1 & 1 & 1 \end{vmatrix} = \begin{vmatrix} 1 & 1 & -1 \\ 1 & 0 & 1 \\ -1 & 1 & 1 \end{vmatrix} = 4$$

this is impossible. Therefore

$$G_{1i}^* = G_{1i} .$$

Since

$$[a, 1, \cdots, 1] G_{12} [a, 1, \cdots, 1]' = a G_{12} ,$$

the additive property of the module formed by symmetric matrices implies

$$S^* = S$$

for all symmetric S. Therefore

$$\mathcal{A}\begin{pmatrix} I & S \\ 0 & I \end{pmatrix} = \begin{pmatrix} I & S \\ 0 & I \end{pmatrix}. \tag{27}$$

6) Combining (23) and (27), we have

$$ASA' = A^*SA^{*\prime}$$

for all symmetric matrices S. Then we have

$$A^* = \pm A.$$

Those A satisfying $A^* = A$ form a subgroup Γ. The index of Γ in the full linear group is either 1 or 2. Now we are going to prove that it cannot be two, therefore Γ is the full linear group. By (18), all these A of the form

$$\begin{pmatrix} a_1 & 0 \\ 0 & A_1(n-1) \end{pmatrix} \tag{28}$$

belong to Γ. For $n \geqq 3$ the full linear group is generated by the elements (28) and J_{12}, where J_{ij} denotes a matrix with 1 at (i, j) and (j, i) position and with 1 at (k, k) position for $k \neq i, k = j$ and with 0 elsewhere. If the index is two then

$$J_{12}^* = -J_{12}.$$

By (28), we have

$$J_{ij}^* = J_{ij} \qquad \text{for } i > 1, j > 1. \tag{29}$$

Then, from

$$J_{13} = J_{23}J_{12}J_{23}, \qquad J_{23} = J_{13}J_{12}J_{13},$$

we deduce that

$$J_{13}^* = -J_{13}, \qquad J_{23}^* = -J_{23}.$$

The last equation contradicts (29). Therefore

$$A = A^*$$

for all non-singular A, that is

$$\mathcal{A}\begin{pmatrix} A & 0 \\ 0 & A'^{-1} \end{pmatrix} = \begin{pmatrix} A & 0 \\ 0 & A'^{-1} \end{pmatrix}. \tag{30}$$

In case $n = 2$, by (18), we have that all those A of the form

$$\begin{pmatrix} a_1 & 0 \\ 0 & a_2 \end{pmatrix}$$

belong to Γ. Further we have

$$\begin{pmatrix} 1 & s \\ 0 & 1 \end{pmatrix}^* = \pm\begin{pmatrix} 1 & s \\ 0 & 1 \end{pmatrix}.$$

Squaring both sides, we have

$$\begin{pmatrix}1 & 2s\\ 0 & 1\end{pmatrix}^* = \begin{pmatrix}1 & 2s\\ 0 & 1\end{pmatrix}$$

for all s, then we have

$$\begin{pmatrix}1 & s\\ 0 & 1\end{pmatrix}^* = \begin{pmatrix}1 & s\\ 0 & 1\end{pmatrix}.$$

Similarly, we have

$$\begin{pmatrix}1 & 0\\ t & 1\end{pmatrix}^* = \begin{pmatrix}1 & 0\\ t & 1\end{pmatrix}.$$

Since

$$\begin{pmatrix}0 & 1\\ -1 & 0\end{pmatrix} = \begin{pmatrix}1 & 1\\ 0 & 1\end{pmatrix}\begin{pmatrix}1 & 0\\ -1 & 1\end{pmatrix}\begin{pmatrix}1 & 1\\ 0 & 1\end{pmatrix},$$

we have

$$\begin{pmatrix}0 & 1\\ 1 & 0\end{pmatrix}^* = \begin{pmatrix}0 & 1\\ 1 & 0\end{pmatrix}.$$

Evidently the full linear group is generated by elements of the form

$$\begin{pmatrix}a_1 & 0\\ 0 & a_2\end{pmatrix}, \quad \begin{pmatrix}1 & s\\ 0 & 1\end{pmatrix}, \quad \begin{pmatrix}0 & 1\\ 1 & 0\end{pmatrix}.$$

The assertion is also true for $n = 2$.

7) From (18), we have

$$\mathscr{A}\begin{pmatrix}M & I - M\\ -(I - M) & M\end{pmatrix} = \begin{pmatrix}M & I - M\\ -(I - M) & M\end{pmatrix}. \tag{31}$$

Combining (27), (30) and (31) and using Theorem 4 we have the theorem.

Appendix

Automorphisms of unimodular group

Since the proof of O. Schreier and B. v. d. Waerden contains an unjustifiable point,[6] for completeness we include the following appendix.

[6] Loc. cit., p. 313. Let S be a linear fractional transformation on Φ. Let $\mathfrak{N}_s$ be the normalizer of S and $\overline{\mathfrak{N}}_s$ be the normalizer of $\mathfrak{N}_s$. They assert that

$$\overline{\mathfrak{N}}_s : \mathfrak{N}_s = \begin{cases}1 & \text{if } S \text{ is parabolic,}\\ 2 & \text{if } S \text{ is elliptic and hyperbolic.}\end{cases}$$

However this is not true for S being elliptic. Since the fixed elements of an elliptic transformation may not be in the field, sometimes there does not exist a transformation in Φ to permit them.

In the appendix, we do not assume that the characteristic p of the field is not equal to 2 and we only consider two rowed matrices. The theorem to be proved is the following.

THEOREM. *Every automorphism of the unimodular group of order two is of the form*

$$PT^{\sigma}P^{-1}$$

where P is non-singular and σ is an automorphism of the field.

DEFINITION. For $p \neq 0$, a unimodular matrix A is called *parabolic*, if $A^{2p} = I$. For $p = 0$, a unimodular matrix A is called *parabolic*, if there are infinitely many unimodular matrices similar to A and commutative with A.

LEMMA 1. *A necessary and sufficient condition for a matrix A to be parabolic is that the characteristic equation has a double root ± 1, and consequently, every parabolic element is similar to*

$$\pm\begin{pmatrix}1 & s\\ 0 & 1\end{pmatrix}$$

under the unimodular group.

PROOF. 1) $p \neq 0$. In an extended field, the characteristic roots of A are α and $1/\alpha$. From $A^{2p} = I$, we deduce

$$\alpha^{2p} = 1.$$

Consequently $\alpha^2 = 1$ and $\alpha = \alpha^{-1} = \pm 1$.

2) $p = 0$. Suppose $\alpha \neq \pm 1$, we have in an extended field a matrix P such that

$$PAP^{-1} = \begin{pmatrix}\alpha & 0\\ 0 & \alpha^{-1}\end{pmatrix}.$$

If B is commutative with A, we have

$$PBP^{-1} = \begin{pmatrix}\beta & 0\\ 0 & \beta^{-1}\end{pmatrix}.$$

Since A and B are similar, we have either $\alpha = \beta$ or $\alpha = \beta^{-1}$. That is, we have only finite number of B commutative with and similar to A.

3) If the matrices A has a double characteristic root, we have a matrix P in the field such that

$$PAP^{-1} = \pm\begin{pmatrix}1 & s\\ 0 & 1\end{pmatrix}.$$

Evidently, all the matrices of the form

$$\pm\begin{pmatrix}1 & sa^2\\ 0 & 1\end{pmatrix} = \pm\begin{pmatrix}a & 0\\ 0 & a^{-1}\end{pmatrix}\begin{pmatrix}1 & s\\ 0 & 1\end{pmatrix}\begin{pmatrix}a & 0\\ 0 & a^{-1}\end{pmatrix}^{-1}$$

are similar to and commutative with $\pm\begin{pmatrix}1 & s\\0 & 1\end{pmatrix}$.

LEMMA 2. *Any two non-commutative parabolic elements can be carried simultaneously into*

$$\pm\begin{pmatrix}1 & s\\0 & 1\end{pmatrix}\quad\pm\begin{pmatrix}1 & 0\\t & 1\end{pmatrix}$$

by a unimodular transformation.

PROOF. By Lemma 1, we may assume that they take the form

$$\pm\begin{pmatrix}1 & s\\0 & 1\end{pmatrix},\quad\begin{pmatrix}a & b\\c & d\end{pmatrix},\qquad a+d=\pm 2,\qquad ad-bc=1,\qquad c\neq 0.$$

We can always find a unique t in the field such that

$$-ct^2+(d-a)t+b=0.$$

Then

$$\begin{pmatrix}1 & t\\0 & 1\end{pmatrix}\begin{pmatrix}a & b\\c & d\end{pmatrix}\begin{pmatrix}1 & t\\0 & 1\end{pmatrix}^{-1}$$

carries the second matrix into our required form.

PROOF OF THE THEOREM. By Lemma 2, we may assume that after subjecting the given automorphism to an inner automorphism we have s_0, s_0^*, t_0, t_0^*, none of them being zero and a choice of $\pm$ such that

$$\mathcal{A}\left(\pm\begin{pmatrix}1 & s_0\\0 & 1\end{pmatrix}\right)=\pm\begin{pmatrix}1 & s_0^*\\0 & 1\end{pmatrix}$$

and

$$\mathcal{A}\left(\pm\begin{pmatrix}1 & 0\\t_0 & 1\end{pmatrix}\right)=\pm\begin{pmatrix}1 & 0\\t_0^* & 1\end{pmatrix}.$$

The normalisator of $\pm\begin{pmatrix}1 & s_0\\0 & 1\end{pmatrix}$ is the group formed by elements of the form

$$\pm\begin{pmatrix}1 & s\\0 & 1\end{pmatrix}.$$

Squaring, we have

$$\begin{pmatrix}1 & 2s\\0 & 1\end{pmatrix}.$$

Thus, for $p \neq 2$, $\mathcal{A}$ carries the group Γ_1

$$\begin{pmatrix} 1 & s \\ 0 & 1 \end{pmatrix}$$

onto itself. For $p = 2$, $\pm\begin{pmatrix} 1 & s \\ 0 & 1 \end{pmatrix} = \begin{pmatrix} 1 & s \\ 0 & 1 \end{pmatrix}$, the statement is also true. Similarly $\mathcal{A}$ carries the group Γ_2 formed by matrices

$$\begin{pmatrix} 1 & 0 \\ t & 1 \end{pmatrix}$$

onto itself.

Consider those elements which carry Γ_1 and Γ_2 onto themselves, they take the form

$$\begin{pmatrix} a & 0 \\ 0 & a^{-1} \end{pmatrix},$$

and constitute a group Σ.

Further consider those elements which permute the groups Γ_1 and Γ_2, they take the form

$$\begin{pmatrix} 0 & d \\ -d^{-1} & 0 \end{pmatrix}.$$

Since

$$\begin{pmatrix} d^{-1} & 0 \\ 0 & d \end{pmatrix} P \begin{pmatrix} d^{-1} & 0 \\ 0 & d \end{pmatrix}^{-1}$$

keeps Γ_1, Γ_2 and Σ unaltered and

$$\begin{pmatrix} d^{-1} & 0 \\ 0 & d \end{pmatrix}\begin{pmatrix} 0 & d \\ -d^{-1} & 0 \end{pmatrix}\begin{pmatrix} d^{-1} & 0 \\ 0 & d \end{pmatrix}^{-1} = \begin{pmatrix} 0 & 1 \\ -1 & 0 \end{pmatrix},$$

therefore, we may assume that after another inner automorphism,

$$\mathcal{A}\begin{pmatrix} 0 & 1 \\ -1 & 0 \end{pmatrix} = \begin{pmatrix} 0 & 1 \\ -1 & 0 \end{pmatrix}$$

and

$$\mathcal{A}\begin{pmatrix} 1 & s \\ 0 & 1 \end{pmatrix} = \begin{pmatrix} 1 & \sigma(s) \\ 0 & 1 \end{pmatrix}, \qquad \sigma(s_1 + s_2) = \sigma(s_1) + \sigma(s_2)$$

and

$$\mathcal{A}\begin{pmatrix} a & 0 \\ 0 & a^{-1} \end{pmatrix} = \begin{pmatrix} \rho(a) & 0 \\ 0 & \rho(a)^{-1} \end{pmatrix}, \qquad \rho(a_1a_2) = \rho(a_1)\rho(a_2).$$

$$\mathcal{A}\begin{pmatrix} a & 0 \\ 0 & a^{-1} \end{pmatrix} = \begin{pmatrix} \rho(a) & 0 \\ 0 & \rho(a)^{-1} \end{pmatrix}, \qquad \rho(a_1a_2) = \rho(a_1)\rho(a_2).$$

Consequently

$$\mathcal{A}\begin{pmatrix} 1 & 0 \\ t & 1 \end{pmatrix} = \mathcal{A}\left(\begin{pmatrix} 0 & 1 \\ -1 & 0 \end{pmatrix}\begin{pmatrix} 1 & -t \\ 0 & 1 \end{pmatrix}\begin{pmatrix} 0 & 1 \\ -1 & 0 \end{pmatrix}^{-1}\right) = \begin{pmatrix} 1 & 0 \\ \sigma(t) & 1 \end{pmatrix}.$$

Since

$$\left(\begin{pmatrix} 1 & a \\ 0 & 1 \end{pmatrix}\begin{pmatrix} 0 & 1 \\ -1 & 0 \end{pmatrix}\right)^3 = I$$

if and only if $a = 1$, we have

$$\sigma(1) = 1.$$

Since

$$\begin{pmatrix} x & 0 \\ 0 & x^{-1} \end{pmatrix} = \begin{pmatrix} 1 & -x \\ 0 & 1 \end{pmatrix}\begin{pmatrix} 1 & 0 \\ x^{-1}-1 & 1 \end{pmatrix}\begin{pmatrix} 1 & 1 \\ 0 & 1 \end{pmatrix}\begin{pmatrix} 1 & 0 \\ -(1-x) & 1 \end{pmatrix},$$

we have

$$\begin{pmatrix} \rho(x) & 0 \\ 0 & \rho(x)^{-1} \end{pmatrix} = \begin{pmatrix} 1 & -\sigma(x) \\ 0 & 1 \end{pmatrix}\begin{pmatrix} 1 & 0 \\ \sigma(x^{-1}) - 1 & 1 \end{pmatrix}\begin{pmatrix} 1 & 1 \\ 0 & 1 \end{pmatrix}\begin{pmatrix} 1 & 0 \\ -1 + \sigma(x) & 1 \end{pmatrix}$$

$$= \begin{pmatrix} * & 1 - \sigma(x)\sigma(x^{-1}) \\ * & \sigma(x^{-1}) \end{pmatrix}.$$

We deduce

$$1 - \sigma(x)\sigma(x^{-1}) = 0, \qquad \rho(x)^{-1} = \sigma(x^{-1}),$$

that is, for $x \neq 0$,

$$\sigma(x) = \rho(x).$$

This asserts that $\sigma(x)$ is an autmorphism of the field.

INSTITUTE FOR ADVANCED STUDY

Reprinted from the Proceedings of the NATIONAL ACADEMY OF SCIENCES,
Vol. 34, No. 6, pp. 258–263. June, 1948

ON THE EXISTENCE OF SOLUTIONS OF CERTAIN EQUATIONS IN A FINITE FIELD

BY L. K. HUA AND H. S. VANDIVER

INSTITUTE FOR ADVANCED STUDY, PRINCETON, NEW JERSEY, AND DEPARTMENT OF APPLIED MATHEMATICS, UNIVERSITY OF TEXAS

Communicated April 7, 1948

In another paper[1] one of the authors stated that he had arrived at limits, both inferior and superior, for the number of solutions of the equation

$$c_1x_1^{a_1} + c_2x_2^{a_2} + \ldots + c_sx_s^{a_s} + c_{s+1} = 0 \tag{1}$$

in the x's where a's are integers such that $0 < a < p^n - 1$; $s \geqq 2$ for $c_{s+1} \neq 0$ and $s > 2$ for $c_{s+1} = 0$, the c's being given elements of a finite field of order p^n, p prime, which will be designated by $F(p^n)$; and

$$c_1 \ldots c_sx_1 \ldots x_s \neq 0. \tag{2}$$

As a consequence of this result, one can obtain

THEOREM I. *The equation (1) with the restriction (2) always has at least k solutions in the x's for k any given positive integer provided p^n exceeds a certain limit.*

In this paper we shall give two quite different approaches to establish this theorem. The first is closely related to the one previously mentioned and the other argument, although subject to the limitation $s > 2$ when $c_{s+1} \neq 0$, is far simpler and is based on a method[1a] which was introduced by one of the authors in the study of generalized Gaussian sums over a finite field.

The limit given here can be sharpened, and the proof of this will be published later.

Elsewhere[2] it was shown that the exact number of solutions of (1) may be determined directly if we know the exact number of solutions of $\theta^{m\alpha}$, $\theta^{m\beta}$. of the equation

$$\theta^{i+m\alpha} + \theta^{j+m\beta} + 1 = 0 \tag{3}$$

for each i and j; θ being a primitive root of $F(p^n)$; $(p^n - 1, a_i) = d_i$, $i = 1, 2, \ldots, s$; m is the L.C.M. of $d_1, d_2, \ldots, d_s$. Further in this argument some or all of the a's may equal $p^n - 1$. To prove this we employed formulae (5), (12) and (15) of the paper just mentioned and all of these contain positive terms only.

We may adapt this method, however, to finding limits from the number of solutions of (1) by first finding limits for the number of solutions of (3) and then employing (5), (12) and (15) of the previous paper.[2]

In the latter the formulae

$$\begin{aligned} A_{oo} &= (c-1)^2 + c(m-1); \qquad A_{ho} = A_{ok} = c^2 - c, \\ A_{hk} &= c^2,\ h \not\equiv k; \qquad A_{hh} = c^2 - c \\ h &\not\equiv 0 \pmod{m},\ k \not\equiv 0 \pmod{m},\ p^n = 1 + mc \end{aligned} \tag{4}$$

are derived; where

$$A_{hk} = \sum_{i,j}^{0 \text{ to } m-1} [i, j][i + h, j + k]$$

and $[i, j]$ is the number of different pairs r and t in the set $0, 1, \ldots, c - 1$ which satisfy the equation

$$1 + \theta^{i+rm} = \theta^{j+tm}; \qquad p^n - 1 = cm. \tag{5}$$

If $\{i, j\}$ denotes the number of solutions of (3) then we note that

$$[i, j] = \{i, j + \epsilon\}, \tag{6}$$

where $\epsilon = \text{ind}\,(-1)$, with $x = \theta^{\text{ind}\, x}$. We have

$$\sum_{i,j}^{0 \text{ to } m-1} (m^2[i, j] - p^n)^2 = \sum_{i,j}^{0 \text{ to } m-1} (m^4[i, j]^2 - 2m^2p^n[i, j]) + p^{2n}m^2$$

which gives, after employing (4) on the right as well as the known relations

$$\begin{aligned} [i, j] = [j + \epsilon, i + \epsilon] = [-j + \epsilon, i - j] = [i - j + \epsilon, -j] = [-i, \\ j - i] = [j - i + \epsilon, -i + \epsilon, -i], \end{aligned} \tag{7}$$

$$\left(\frac{p^n}{m^2} + \frac{D}{\sqrt{d}}\right) \geqq [i, j] \geqq \left(\frac{p^n}{m^2} - \frac{D}{\sqrt{d}}\right), \tag{8}$$

where

$$D = \frac{1}{m}\sqrt{p^n(m^2 - 3m + 2) + 3m + 1}.$$

Here d is the number of different pairs in (7) in the sense that $[h, k]$ is different from $[h_1, k_1]$ if $h \not\equiv h_1$, or $k \not\equiv k_1 \pmod{m}$.

The use of (8) with (6) of this paper and (5), (12) and (15) of the former paper yield superior and inferior limits for the number of solutions of (1), and we obtain a proof of Theorem I.

For the second proof of Theorem I (except for $s = 2$ with $c_{s+1} \neq 0$) we proceed as follows:

Any element α of $F(p^n)$ may be written uniquely in the form, if θ is a primitive root in $F(p^n)$,

$$\alpha = d_0 + d_1\theta + \ldots + d_{n-1}\theta^{n-1} = h(\theta) \tag{9}$$

where the d's are in the included field $F(p)$. Now θ satisfies an irreducible equation $f(x) = 0$ of degree n with coefficients in $F(p)$ and whose roots are $\theta_1, \theta_2 \ldots, \theta_n$, with $\theta_1 = \theta$. Set (here tr is short for trace)

$$\sum_{i=1}^{n} h(\theta_i) = tr\ h(\theta).$$

We shall now prove

LEMMA 1. *If* $\zeta = e^{2i\pi/p}$, *then, for* $n < p$,

$$\sum_{\alpha \epsilon F(p^n)} \zeta^{tr(\alpha t)} = \begin{cases} 0 \text{ for } t \neq 0 \\ p^n \text{ for } t = 0. \end{cases} \tag{10}$$

This statement is obvious for $t = 0$. For $t \neq 0$, αt ranges over all the elements of $F(p^n)$. Hence it is sufficient to prove that

$$\sum_{\alpha \epsilon K} \zeta^{tr(\alpha t)} = \sum_{d_0=0}^{p-1} \ldots \sum_{d_{n-1}=0}^{p-1} \zeta^{T} = \prod_{i=0}^{n-1} \left(\sum_{d_i=0}^{p-1} \zeta^{d_i tr(\theta^i)} \right) = 0, \tag{11}$$

where

$$T = tr(d_o + d_1\theta + \ldots + d_{n-1}\theta^{n-1}).$$

There is one of the i's such that p does not divide $tr(\theta^i)$. For, if the contrary is true then $tr(\theta^i) = 0$ for all $0 \leqq i \leqq n - 1$. Since $\theta_1 = \theta$, satisfies $(x) = 0$ of degree n as do also $\theta_2, \ldots, \theta_n$, then Newton's relations between the elementary symmetric functions of these roots and the sum of the $f$lth powers of said roots give easily $\theta^n = c$, with c in $F(p)$. This yields $\theta^{n(p-1)} = 1$, contradicting the definition of θ as a generator of $F(p^n)$ for $n > 1$, since $p^n - 1 = (p - 1)(p^{n-1} + \ldots + 1) > n(p - 1)$. This proves (10).

We now prove

LEMMA 2. *Write* K *for* $F(p^n)$. *If*

$$\psi(t, a, \zeta) = \sum_{\substack{x \epsilon K \\ x \neq 0}} \zeta^{tr(tx^a)}$$

then for $t \neq 0$,

$$|\psi(t, a, \zeta)| \leqq (p^n - 1, a)p^{n/2}. \tag{12}$$

Now

$$\frac{1}{p^n}\sum_{t \epsilon K} (\psi(t, a, \zeta)\psi(t, a, \zeta^{-1})) = \sum_{\epsilon K} \sum_{\substack{x,\, y \,\epsilon\, K \\ x,\, y \neq 0}} \zeta^{tr(tx^a - ty^a)}. \tag{13}$$

Using Lemma 1, the right-hand member reduces to $(p^n - 1)(p^n - 1, a)$, since this is the number of different solutions in x and y, $\neq 0$, of

$$x^a = y^a,$$

in $F(p^n)$. Since $|\psi(t, a, \zeta)| = |\psi(t, a, \zeta^{-1})|$, then (12) gives, if we set $\psi(t, a, \zeta) = \psi(t, a)$,

$$\frac{1}{p^n}\sum_{t \epsilon K} |\psi(t, a)|^2 = (p^n - 1)(p^n - 1, a). \tag{14}$$

Since, if $t \neq 0$, the number of solutions of $ty^a = q$ is $\leqq (p^n - 1, a)$, then

$$(p^n - 1)|\psi(t, a)|^2 = \sum_{\substack{y \,\epsilon\, K \\ y \neq 0}} |\psi(ty^a, a)|^2,$$

$$\leqq (p^n - 1, a) \sum_{q \,\epsilon\, K} |\psi(q, a)|^2,$$

$$\leqq (p^n - 1, a)^2 p^n (p^n - 1),$$

and by (14), we then have Lemma 2. Then for the proof of Theorem I, we have

$$N = \frac{1}{p^n}\sum_{t \epsilon K} \psi(c_1 t, a_1)\ldots\psi(c_s t, a_s)\zeta^{tr(c_{s+1} t)}$$

$$= \frac{(p^n - 1)^s}{p^n} + \frac{1}{p^n}\sum_{\substack{t \,\epsilon\, K \\ t \neq 0}} \psi(c_1 t, a_1)\ldots\psi(c_s t, a_s)\, \zeta^{tr(c_{s+1} t)}, \tag{15}$$

where N is the number of solutions of (1), and now we count all the solutions under the restriction (2). (In connection with (5) we did not do this exactly, that is, all the solutions of (5) would be $m^2[i, j]$.) Then

$$\left|N - \frac{(p^n - 1^s}{p^n}\right| \leqq \frac{1}{p^n} \sum_{\substack{t \,\epsilon\, K \\ t \neq 0}} |\psi(c_1 t, a_1)|\ldots|\psi(c_s t, a_s)|,$$

and by Lemma 2, we have

$$\left|N - \frac{(p^n - 1)^s}{p^n}\right| \leqq \prod_{i=1}^{s} (p^n - 1.\ a_i)\cdot p^{ns/} .$$

whence

$$\frac{(p^n-1)^s}{p^n} - Ap^{ns/2} \leqq N \leqq \frac{(p^n-1)^s}{p^n} + Ap^{ns/2}; \qquad (16)$$

$$A = \prod_{i=1}^{s} (p^n - 1, a_i).$$

The quantity $A > 0$ never exceeds the fixed value $\prod_{i=1}^{s} a_i$. Set $p^n - 1 = h$. Then the left-hand member of (16) becomes, if $r = (s+2)/2$,

$$\frac{h^r\left(h^{(s-r)} - A\left(1+\frac{1}{h}\right)^r\right)}{p^n}, \qquad (17)$$

which for $s > 2$ and p^n sufficiently large is obviously $>k$ for k any integer. This proves Theorem I, for $s > 2$. From (16) we may also write

$$N = \frac{(p^n-1)^s}{p^n} + 0\,(p^{ns/2}). \qquad (18)$$

We may now apply Theorem I to congruences with respect to an ideal prime modulus in any commutative ring R with a unity element. We may consider the congruence

$$\alpha_1 x_1^{a_1} + \alpha_2 x_2^{a_2} + \ldots + \alpha_s x_s^{a_s} + \alpha_{s+1} \equiv 0 \pmod{\mathfrak{p}}, \qquad (19)$$

where now the α's are fixed elements in R, that is, we may fix the a's and α's and consider the above congruence for various values of $\mathfrak{p}$. This is a bit different from the situation in Theorem I where the domain of the coefficients changes with each $\mathfrak{p}$. But if the number of incongruent residues (norm) in R of the ideal $\mathfrak{p}$ is finite then the residue classes form a field, since a finite integral domain is a field. Hence we may apply Theorem I and obtain

THEOREM II. *The congruence (19) always has at least k solutions in the x's for k any positive integer provided*

$$\alpha_1\alpha_2\ldots\alpha_s x_1 x_2 \ldots x_s \not\equiv 0 \pmod{\mathfrak{p}} \qquad (20)$$

$s \geqq 2$ for $c_{s+1} \not\equiv 0 \pmod{\mathfrak{p}}$ and $s > 2$ for $c_{s+1} \equiv 0 \pmod{\mathfrak{p}}$; $\mathfrak{p}$ is a prime ideal of finite norm in a commutative ring R with a unity element, the a's are fixed positive integers, the α's are fixed elements in R, and the norm of $\mathfrak{p}$ is sufficiently large.

If R is the ring of integers in an algebraic field then every ideal has a finite norm so that the Theorem II holds for any algebraic ring of this type with $\mathfrak{p}$ any prime ideal in the ring. If R is the ring of rational integers, then $n = 1$ and R is the system of residue classes modulo p, and we have in particular the

COROLLARY. *If in (19) the x's are rational integers and* $\mathfrak{p} = (p)$ *with* p *a prime rational integer then this congruence always has* k *solutions for* p *sufficiently large with the other conditions in Theorem II holding,* k *any integer.*

This corollary was given by Mordell[1] when $\alpha_{s+1} \not\equiv 0$, and we include *all*[3] solutions of (19). We considered only solutions prime to $\mathfrak{p}$ (primitive solutions) in our work, following the conditions (2) and (20).

[1] These PROCEEDINGS, **33**, 236–242 (1947). In this paper reference to previous results was made, but an important paper by Mordell, *Mathematische Zeitschrift*, **37**, 207 (1933) which bore more directly on the contents, was, unfortunately, not mentioned. Other relevant references are Pellet, *Bull. Math. Soc. France*, **15**, 80–93 (1886); Dickson, *Crelle*, **135**, 181–188 (1909); Hurwitz, *Crelle*, **136**, 272–292 (1909); Mitchell, *Ann. Math., II*, **18**, 120 (1917); Davenport, *Jour. London Math. Soc.*, **6**, 49–54 (1931); Schur, I., *Jahresber. Deutsch. Math. Verein.*, **25**, 114 (1916).

[1a] Cf. a paper by Hua, Loo-Keng, "On a Double Exponential Sum," soon to appear in the *Science Reports of Tsing Hua University*. An abstract is given in the *Science Record of the Acad. Sinica*, **1**, Nos. 1–2.

[2] These PROCEEDINGS, **32**, 47–52 (1946).

[3] Mordell's method of proof is different from either of those used in the present paper. In some ways there is quite a distinction between finding the primitive solutions of an equation in a finite field and finding all solutions. The congruence

$$x_1^2 + x_2^2 + \ldots + x_s^2 \equiv 0 \pmod 2$$

has no solutions in integers prime to 2, if s is odd, but evidently has solutions for some of the x's even. Again the congruence $x^7 + y^7 + 1 \equiv 0 \pmod{491}$ has no solutions in integers x and y prime to 491. But the congruence $x^7 + 1 \equiv 0 \pmod{491}$ obviously has solutions.

Reprinted from the Proceedings of the NATIONAL ACADEMY OF SCIENCES,
Vol. 35, No. 2, pp. 94–99. February, 1949

CHARACTERS OVER CERTAIN TYPES OF RINGS WITH APPLICATIONS TO THE THEORY OF EQUATIONS IN A FINITE FIELD

BY L. K. HUA AND H. S. VANDIVER

DEPARTMENT OF MATHEMATICS, UNIVERSITY OF ILLINOIS, AND DEPARTMENT OF APPLIED MATHEMATICS, UNIVERSITY OF TEXAS

Communicated December 17, 1948

Characters, modulo m, have been defined (for example by Landau[1]) and have been employed extensively in Group Theory and Number Theory, particularly in some of the most important parts of Analytic Number Theory. However, we may define them by means of any ring with a unity element and which contains only a finite number of elements which are not zero divisors, instead of the special ring consisting of the residue classes modulo m, and a number of extensions of known results concerning said special case are immediate. We will set up this concept and then specialize it for the finite field using results which are then applied to the theory of certain equations in a finite field which were considered in another paper[2] by the authors. This type of equation is

$$c_1x_1^{a_1} + c_2x_2^{a_2} + \ldots + c_sx_s^{a_s} + c_{s+1} = 0 \qquad (1)$$

where the a's are integers such that $0 < a < p^n - 1$; $s \geqq 2$ for $c_{s+1} \neq 0$ and $s > 2$ for $c_{s+1} = 0$, the c's being given elements of a finite field of order p^n, p prime, which will be designated by $F(p^n)$ and

$$c_1 \ldots c_sx_1 \ldots x_s \neq 0. \qquad (2)$$

In the present paper we obtain a new explicit expression for the number of solutions of (1) in terms of generalized Lagrange resolvents and (see the τ number defined in Theorem I) and using this we find better limits for the number of solutions of (1) than those derived in E.

Let R be a ring with a unity element which has the property of containing only a finite number of elements which are not divisors of zero. A set of such elements forms a group. For, since it does not contain zero or any divisors of zero, the cancellation holds and a finite semi-group having this property is known to be a group. Or, otherwise expressed, these elements are units in R.

Let an arithmetical function $\chi(a)$ be defined for any element a of R so that

I. $\chi(a)$ is a complex number.
II. $\chi(1) \neq 0$.
III. $\chi(a) = 0$ when a is zero or any divisor of zero.
IV. If $a_1 = a_2$ in R, then $\chi(a_1) = \chi(a_2)$.
V. $\chi(a_1 a_2) = \chi(a_1)\chi(a_2)$.

The arguments employed by Landau[1] for the special ring formed by the residue classes modulo m by which he obtains his Satz 127 to Satz 133 inclusive may be easily extended to yield the following:

For each character χ we have

$$\chi(1) = 1. \tag{3}$$

There is a character designated by χ_1 such that

$$\chi_0(a) = 1, \tag{3a}$$

for each a in R. Let $\phi(R) = h$ be the number of distinct units in R We shall call h the *indicator*[3] of R. Then since the units u form a group, $u^h = 1$ in R, and by I and V,

$$\chi(a) = e^{2i\pi b/h}; \tag{4}$$

for $0 \leqq b < h$.

We also obtain

LEMMA I.

$$\sum_{a \in R} \chi(a) = \begin{cases} h \text{ for } \chi = \chi_0, \\ 0 \text{ for } \chi \neq \chi_0. \end{cases} \tag{5}$$

Suppose that we have, if $h \equiv 0 \pmod{k}$, a character such that $(\chi(a))^k = 1$ for each unit a in R, this will be called a special character and designated by χ_k. In particular χ_0 is a special character for any k.

How let R be a finite field of order p^n, p prime, and designate the same by $F(p^n)$. Let θ be a primitive root of this field, and let β be a primitive $(p^n - 1)$th root of unity. Then by (4)

$$\theta = \beta^b; \; 0 \leqq b < p^n - 1,$$

and since all the elements, $\neq 0$, of $F(p^n)$ are given by powers of θ then the whole character, by V, is given by powers of β^b. Now consider χ_k. Then

$$\chi_k(\theta) = \beta^d \tag{6}$$

and by definition of χ_k, $(\chi_k(\theta))^k = \beta^{dk} = 1$ which gives $d \equiv 0 \pmod{(p^n-1)/k}$. It follows from this that there are exactly k special characters, and in particular there are h distinct characters defined by R. We now have

LEMMA II.

$$\sum_{\chi_k} \chi_k(a) = \begin{cases} k \text{ for } a \text{ a } k\text{th. power,} \\ 0 \text{ otherwise.} \end{cases} \tag{7}$$

This follows easily from (6) and the remarks following it.

We also have[5]

LEMMA III. *If* $\zeta = e^{2i\pi/p}$, *then in* $F(p^n)$,

$$\sum_{a \in k} \zeta^{tr(at)} = \begin{cases} 0 \text{ for } t \neq 0, \\ p^n \text{ for } t = 0. \end{cases} \tag{8}$$

We now proceed to[6]

THEOREM I. *The number of solutions N of (1) with the restriction (2) is*

$$N = \frac{(p^n - 1)^s}{p^n} + \frac{T}{p^n}$$

where

$$T = \sum_{\substack{a \in k \\ a \neq 0}} \prod_{i=1}^{s} \sum_{\chi k i} \chi_{k_i}(a^{-1}c_i^{-1})\tau(\chi_{k_i})\zeta^{tr(ac_s + 1)};$$

$$\tau(\chi_{k_i}) = \sum_{b \in k} \chi_{k_i}(b)\zeta^{tr(b)}$$

and $k_i = (a_i, p^n - 1)$. *Also K stands for* $F(p^n)$.

To prove this we note that

$$N = \frac{1}{p^n} \sum_{a \in k} \sum_{i=1}^{s} \sum_{\substack{x_i \in K \\ x_i \neq 0}} \zeta^{tr(aG(x))}; \tag{9}$$

$$G(x) = c_1 x_2^{k_1} + c_2 x_3^{k_2} + \ldots + c_s x_s^{k_s} + c_{s+1}.$$

First it is known that the number of solutions of (1) and the number of solutions of $G(x) = 0$ are the same if $(p^n - 1, a_i) = k_i$. Also when we carry out the summation in (9a) for the x_i's all the terms are unity when we have a set of x's which satisfy $G(x) = 0$ with the restriction (2) and when we sum with respect to a we obtain p^n according to Lemma III, but when we have a set of x_i's such that (1) is not satisfied then by the same lemma the sum with respect to a gives zero. (The relation (15) in E, p. 261, is a transformation of (9).) In the right-hand member, if we separate the zero and non-zero values of a, we obtain from (9),

$$N = \frac{(p^n - 1)^s}{p^n} + \frac{T}{p^n}, \tag{10}$$

where

$$T = \frac{1}{p^n} \sum_{\substack{a \in K \\ a \neq 0}} \sum_{i=1}^{s} \sum_{\substack{x_i \in K \\ x_i \neq 0}} \zeta^{tr(aG(x))}$$

Now employing Lemma II,

$$\sum_{\substack{x \in K \\ x \neq 0}} \zeta^{tr(cx^k)} = \sum_{y \in K} \sum_{\chi_k} \chi_k (c^{-1}y) \zeta^{tr(y)}$$

$$= \sum_{\chi_k} \chi_k(c^{-1})\tau(\chi_k)$$

and then (10) gives

$$T = \sum_{\substack{a \in K \\ a \neq 0}} \prod_{i=1}^{s} \sum_{\chi_{k_i}} \chi_{k_i} (c_i^{-1} a^{-1})\tau(\chi_{k_i})\zeta^{tr(ac_{s+1})}. \tag{11}$$

We now employ (10) and (11) to find limits for N. We see from (4) that for a fixed a

$$\prod_{i=1}^{s} \chi_{u_i}(a) = \chi_v(a), \tag{11a}$$

for some v. Also it is known[6] (p. 152) that

$$|\tau(\chi)| = p^{n/2}, \qquad \chi \neq \chi_0; \tag{11b}$$

$$|\tau(\chi_0)| = 1.$$

If we write by (11a),

$$\prod_{i=1}^{s} \chi_{k_i}(a^{-1}) = \chi(a);$$

$$\sum_{a \in K} \chi(ac_{s+1}^{-1})\zeta^{tr(a)} = \chi(c_{s+1}^{-1}) \sum_{a \in K} \chi(a)\zeta^{tr(a)} = \chi(c_{s+1}^{-1})\tau(\chi); \; c_{s+1} \neq 0,$$

and employ (11) and (11b) noting that there are k characters χ_k, and that χ_0 is a χ_k,

$$|T| \leqq \prod_{i=1}^{s} \sum_{\chi_{k_i}} |\chi_{k_i}(c_i^{-1}) \, ||\tau(\chi)|| \, \tau(\chi_{k_i})|$$

$$|T| \leqq \prod_{i=1}^{s} (1 + (k_i - 1)p^{n/2})p^{n/2}. \tag{12}$$

Let us now suppose that $c_{s+1} = 0$. Then we have $\zeta^{tr(ac_s+1)} = 1$ in (11) and

$$\sum_{\substack{a \epsilon K \\ a \neq 0}} \chi(c_i^{-1}a^{-1}) \leqq p^n - 1$$

and then

$$|T| \leqq (p^n - 1) \sum_{i=1}^{s} (1 + (k_i - 1)p^{n/2}) \tag{13}$$

Hence (10) gives for $c_{s+1} \neq 0$,

$$\left| N - \frac{(p^n - 1)^s}{p^n} \right| \leqq \prod_{i=1}^{s} (1 + (k_i - 1)\, p^{n/2})p^{-n/2}, \tag{14}$$

and for $c_{s+1} = 0$,

$$\left| N - \frac{(p^n - 1)^s}{p^n} \right| \leqq \frac{(p^n - 1)}{p^n} \prod_{i=1}^{s} (1 + (k_i - 1)\, p^{n/2}). \tag{15}$$

Whence we have

THEOREM II. *If N is the number of solutions of (1) under the condition (2), then for $c_{s+1} \neq 0$, $(p^n - 1, a_i) = k_i$,*

$$\frac{(p^n - 1)^s}{p^n} - D \leqq N \leqq \frac{(p^n - 1)^s}{p^n} + D, \tag{16}$$

and for $c_{s+1} = 0$,

$$\frac{(p^n - 1)^s}{p^n} - D_1 \leqq N \leqq \frac{(p^n - 1)^s}{p^n} + D_1, \tag{17}$$

where

$$D = p^{-n/2} \prod_{i=1}^{s} (1 + (k_i - 1)\, p^{n/2}),$$

and

$$D_1 = \frac{p^n - 1}{p^n} \prod_{i=1}^{s} (1 + (k_i - 1)\, p^{n/2}).$$

From the above theorem we infer immediately that

$$N = \frac{(p^n - 1)^s}{p^n} + 0(p^{\frac{n(s-v)}{2}}), \tag{18}$$

where $v = 1$ or 0 according as $c_{s+1} \neq 0$ or $c_{s+1} = 0$.

If we consider (1) without the restriction $x_1, x_2 \ldots x_s \neq 0$, we may find limits for the number of solutions by considering one or more of the c's in (1) as being zero in turn and employ for each of the resulting equations the limits (16) and (17) and adding the results we obtain inferior and superior limits of somewhat the same type as (16) and (17) and in particular we infer that if N_1 is said number then

$$N_1 = p^{n(s-1)} + 0(p^{n(s-v)/2}); \qquad (19)$$

v having the same meaning as in (18).

[1] *Vorlesungen über Zahlentheorie*, Leipzig, Hirzel, 1927, Bd. I, 83–87. Cf. also Dick son, *Modern Elementary Theory of Numbers*, University of Chicago Press, 1939, pp. 272–276.

[2] These PROCEEDINGS, **34**, 258–263 (1948). This article will be referred to as E. Characters defined by a general finite field have been considered by a number of writers. Cf. the paper referred to in our footnote 5 below and the references given therein as well as Davenport, *Acta Mathematica*, **71**, 99–121 (1939).

[3] This term is employed for the reason that the special R consisting of the residue classes modulo m, our units correspond to the integers less than m and prime to it.

[4] In this paper we are using the symbols 1 and 0 both for the unity and zero elements in the complex field and the unity and zero elements in R. This should not cause confusion if we keep in mind that $\chi(a)$ is always the complex field while a is always in R.

[5] This was proved for $n < p$ in E, and the argument employed there may be extended to the case $n \geqq p$. However, the result was previously known in general, in fact it follows from the results in Stickelberger, *Math. Annalen*, **37**, 321 (1890), § 1, paragraph 5, since any finite field may be represented by a complete set of residue classes with respect to a prime ideal modulus in an algebraic field.

[6] Davenport and Hasse, *J. f. Math.* (*Crelle*), **172**, 151–174 (1935) found, for the case $s = 2$ a formula related to our (10). See their relation (6.2), p. 173, and the value of N given at the top of p. 174. Their result will be discussed in a future paper by the authors.

Reprinted from the Proceedings of the NATIONAL ACADEMY OF SCIENCES,
Vol. 35, No. 7, pp. 386–389. July, 1949

ON THE AUTOMORPHISMS OF A SFIELD

BY LOO-KENG HUA

DEPARTMENT OF MATHEMATICS, UNIVERSITY OF ILLINOIS

Communicated by H. S. Vandiver, May 8, 1949

Let K be a sfield. A mapping $(a \to a^\sigma)$ of K onto itself is called a semi-automorphism if it satisfies

$$(a + b)^\sigma = a^\sigma + b^\sigma \tag{1}$$

$$(aba)^\sigma = a^\sigma b^\sigma a^\sigma \tag{2}$$

and

$$1^\sigma = 1. \tag{3}$$

The well-known examples of semi-automorphisms are automorphisms, which satisfy $(ab)^\sigma = a^\sigma b^\sigma$, and anti-automorphisms, which satisfy $(ab)^\sigma = b^\sigma a^\sigma$.

It is a known problem about the existence of semi-automorphism other than automorphisms and anti-automorphisms. It is the aim of the note to settle this problem, namely:

THEOREM 1. *Every semi-automorphism is either an automorphism or an anti-automorphism.*

To prove the theorem we need several simple consequences of (1), (2) and (3).

Putting $b = a^{-1}$ in (2), we have

$$(a^{-1})^\sigma = (a^\sigma)^{-1}, \tag{4}$$

and replacing a by $a + b$ and b by 1 in (2), we have, by (3),

$$(ab)^\sigma + (ba)^\sigma = a^\sigma b^\sigma + b^\sigma a^\sigma. \tag{5}$$

Applying (2) twice, we obtain, by (4),

$$\begin{aligned}(ba)^\sigma &= (ab(ab)^{-1}ba)^\sigma \\ &= a^\sigma b^\sigma (ab)^{\sigma-1} b^\sigma a^\sigma.\end{aligned} \tag{6}$$

Substituting (6) into (5), we have

$$(ab)^\sigma + a^\sigma b^\sigma (ab)^{\sigma-1} b^\sigma a^\sigma = a^\sigma b^\sigma + b^\sigma a^\sigma$$

which is equivalent to

$$((ab)^\sigma - a^\sigma b^\sigma)(1 - (ab)^{\sigma-1} b^\sigma a^\sigma) = 0. \tag{7}$$

We deduce immediately

$$(ab)^\sigma = \begin{cases} a^\sigma b^\sigma, & \text{or} \\ b^\sigma a^\sigma. \end{cases} \tag{8}$$

Suppose that we have a pair of elements a and b such that

$$(ab)^\sigma = b^\sigma a^\sigma \neq a^\sigma b^\sigma. \tag{9}$$

Then, for any c, we have

$$(ac)^\sigma = c^\sigma a^\sigma. \tag{10}$$

In fact, otherwise, we would have, by (8),

$$(ac)^\sigma = a^\sigma c^\sigma \neq c^\sigma a^\sigma,$$

and, by (9) and (1)

$$a^\sigma c^\sigma + b^\sigma a^\sigma = (ac)^\sigma + (ab)^\sigma = (a(b+c))^\sigma = \begin{cases} a^\sigma(b^\sigma + c^\sigma) & \text{or} \\ (b^\sigma + c^\sigma)a^\sigma \end{cases}$$

Both conclusions are impossible. Similarly, we prove that, for any d,

$$(db)^\sigma = b^\sigma d^\sigma. \tag{11}$$

Now we are going to prove that

$$(dc)^\sigma = c^\sigma d^\sigma.$$

Suppose the contrary, that is, by (8),

$$(dc)^\sigma = d^\sigma c^\sigma (\neq c^\sigma d^\sigma) \tag{12}$$

Similar to the previous argument, we can establish that

$$(ac)^\sigma = a^\sigma c^\sigma \tag{13}$$

and

$$(db)^\sigma = d^\sigma b^\sigma. \tag{14}$$

Now we consider the elements:

$$b^\sigma a^\sigma + (ac)^\sigma + (db)^\sigma + d^\sigma c^\sigma = ((a+d)(b+c))^\sigma =$$
$$\begin{cases} (a^\sigma + d^\sigma)(b^\sigma + c^\sigma) & \text{or} \\ (b^\sigma + c^\sigma)(a^\sigma + d^\sigma). \end{cases}$$

The first conclusion contradicts (9), by (10) and (11), and the second conclusion contradicts (12), by (13) and (14).

Therefore, if there is a pair of elements a,b such that $(ab)^\sigma = b^\sigma a^\sigma (\neq a^\sigma b^\sigma)$, then, for any pair of elements c and d, we have $(cd)^\sigma = d^\sigma c^\sigma$. This proves our theorem.

By means of the previous theorem, we also settle a problem in the study of projective geometry on a line over a sfield. Namely:

THEOREM 2. *Any one to one mapping carrying the projective line over a sfield of characteristic $\neq 2$ onto itself and keeping harmonic relation invariant is a semi-linear transformation induced by an automorphism or an anti-automorphism.*

This theorem was established for the quaternion algebra (Ancochea[1]) and later for division algebra (Ancochea[2]), characteristic $\neq 2$ and he left the general problem open.

As an easy consequence of the following theorem on geometry of matrices (the proof will be given elsewhere), we can extend theorem 1 to any semi-simple ring with a descending chain condition.

THEOREM 3. *Two $n \times m$ matrices Z and W are said to coherent, if the rank of their difference $Z - W$ is one. Suppose $1 < n \leq m$. Any one to* *one mapping carrying $n \times m$ matrices into $n \times m$ matrices and leaving the coherence relation invariant is of the form*

$$Z_1 = PZ^{\sigma}Q + R \tag{15}$$

where $P\ (=P^{(n)})$ and $Q(=Q^{(m)})$ are non-singular and $R = R^{(n,\,m)}$, and σ is an automorphism of the sfield. In case $n = m$, in addition to (15), we have also

$$Z_1 = PZ'^{\sigma}Q + R \tag{16}$$

where σ is an anti-automorphism of the sfield and Z' is the transposed matrix of Z.

By entirely different methods, Ancochea[2] and Kaplansky[3] treated the problem of semi-automorphisms under some restrictions. The former established the corresponding theorem for a simple algebra over a field of characteristic $\neq 2$, and the later extended it to semi-simple algebra over any field. Both of their methods rooted in the structure theory of linear algebras, therefore neither of them can be extended to the general case.

[1] Ancochea, G., *J. Math.*, **184**, 192–198 (1942).
[2] Ancochea, G., *Annals Math.*, **48**, 147–153 (1947).
[3] Kaplansky, I., *Duke Math. J.*, **14**, 521–525 (1947).

Reprinted from the Proceedings of the NATIONAL ACADEMY OF SCIENCES,
Vol. 35, No. 8, pp. 477–481. August, 1949

ON THE NUMBER OF SOLUTIONS OF SOME TRINOMIAL EQUATIONS IN A FINITE FIELD

BY LOO-KENG HUA AND H. S. VANDIVER

DEPARTMENT OF MATHEMATICS, UNIVERSITY OF ILLINOIS, AND DEPARTMENT OF APPLIED MATHEMATICS, UNIVERSITY OF TEXAS

Communicated July 1, 1949

In the present paper we shall use the results in two previous papers[1] by the authors to obtain (Theorem I) the exact number, explicitly in terms of p, n, k_1 and k_2, of solutions of the equation, if $0 < a_i < p^n - 1$, $(p^n - 1, a_i) = k_i$, $i = 1, 2$;

$$c_1x_1^{a_1} + c_2x_2^{a_2} + c_3 = 0, \tag{1}$$

in x_1 and x_2, non-zero elements of a finite field $F(p^n)$ of order p^n, p an odd prime with c_1, c_2, c_3 given elements of $F(p^n)$, $c_1c_2c_3 \neq 0$, in certain special cases. Secondly, we shall find limits (Theorem II) for the number of solutions of (1) which are better than those given in C for the solutions of this particular equation, and which have the unusual property that they agree with the exact values found for the number of solutions in the special cases treated in Theorem I. We shall employ the notation used in H (defined in (9b) just below relation (14) of that paper).

Let α be a primitive (ds_1s_2)-th root of unity, $m = ds_1s_2$ where $ds_1 = k_1$, $ds_2 = k_2$ where $(k_1, k_2) = d$, whence $(s_1, s_2) = 1$. Let χ_{k_1} be a special k_1-character and χ_{k_2} a special k_2-character in $F(p^n)$ and write, if K stands for $F(p^n)$,

$$\tau(\chi_{k_1}^{\mu_1}) = \sum_{a \in K} \alpha^{\mu_1 s_2 \text{ ind } (a)} \zeta^{tr(a)}$$

and

$$\tau(\chi_{k_2}^{\mu_2}) = \sum_{a \in K} \alpha^{\mu_2 s_1 \text{ ind } (a)} \zeta^{tr(a)},$$

where $tr(a)$ is the trace of a in $F(p^n)$ with respect to $F(p)$, and $g^{\text{ind } (a)}$ is defined by

$$g^{\text{ind } (a)} = a$$

in $F(p^n)$, g being a primitive root of $F(p^n)$ and we define ind (0) as zero. Hence, as is known, provided $\chi_{k_1}^{\mu_1}\chi_{k_2}^{\mu_2} \neq 1$,

$$\frac{\tau(\chi_{k_1}^{\mu_1})\tau(\chi_{k_2}^{\mu_2})}{\tau(\chi_{k_1}^{\mu_1}\ \chi_{k_2}^{\mu_2})} = \sum_a \alpha^{\mu_1 s_2 \text{ ind } (a) + \mu_2 s_1 \text{ ind } (1 - a)}.$$

Also, setting $(-a_1)$ for (a) we obtain

$$\sum_{a \in K} \alpha^{\mu_1 s_2 \text{ ind } (a) + \mu_2 s_1 \text{ ind } (1-a)}$$

$$= \sum_{a_1 \in K} \alpha^{\mu_1 s_2 \text{ ind } (-a_1) + \mu_2 s_1 \text{ ind } (1+a_1)}$$

$$= \sum_{a_1 \in K} \alpha^{\mu_1 s_2 \text{ ind } (-1)} \cdot \alpha^{\mu_1 s_2 \text{ ind } (a_1) + \mu_2 s_1 \text{ ind } (1+a_1)}.$$

Let $n = 2n_1$, then $p^{2n_1} - 1 \equiv 0 \pmod 4$ and if $(\mu_1 s_2(p^{2n_1} - 1))/2 = e$ so that $\alpha^{\mu_1 s_2 \text{ ind } (-1)} = (-1)^e = 1$,

$$\frac{\tau(\chi_{k_1}^{\mu_1})\tau(\chi_{k_2}^{\mu_2})}{\tau(\chi_{k_1}^{\mu_1}\chi_{k_2}^{\mu_2})} = \sum_{a \in K} \alpha^{\mu_1 s_2 \text{ ind } (a) + \mu_2 s_1 \text{ ind } (a+1)} = \psi(\alpha) \qquad (2)$$

Now Mitchell[2] showed that if $\mu_1 s_2 \not\equiv 0 \pmod m$, $\mu_2 s_1 \not\equiv 0 \pmod m$ and $\mu s_2 + \mu' s_1 \not\equiv 0 \pmod m$ then

$$\psi(\alpha) = (-1)^{r-1} p^{n_1}, \qquad (3)$$

where p belongs to the exponent $2t$, modulo m and $n_1 = rt$ with $p^t \equiv -1 \pmod m$. We now employ (18) of H, which may be written as follows:

$$N = p^n - 2 - \sum_{\mu_1 = 0}^{k_1 - 1} \chi_{k_1}^{\mu_1}\left(-\frac{c_3}{c_1}\right) - \sum_{\mu_2 = 0}^{k_2 - 1} \chi_{k_2}^{\mu_2}\left(-\frac{c_3}{c_2}\right) - \sum_{v = 1}^{d - 1} \chi_{k_1}^{s_1 v}\left(-\frac{c_2}{c_1}\right)$$

$$+ \sum_{\mu_1, \mu_2} \chi_{k_1}^{\mu_1}\left(-\frac{c_3}{c_1}\right) \chi_{k_2}^{\mu_2}\left(-\frac{c_3}{c_2}\right) \frac{\tau(\chi_{k_1}^{\mu_1})\tau(\chi_{k_2}^{\mu_2})}{\tau(\chi_{k_1}^{\mu_1}\chi_{k_2}^{\mu_2})}, \qquad (4)$$

where in the last term $\chi_{k_1}^{\mu_1} \neq 1$, $\chi_{k_2}^{\mu_2} \neq 1$, $\chi_1^{\mu_1}\chi_2^{\mu_2} \neq 1$, and divide the discussion into seven cases in order to express s in terms of integers only. Because of (3) and $\chi(\alpha) = -1$ for $\mu_1 = 0$, $\mu_2 \neq 0$, also for $\mu_1 \neq 0$, $\mu_2 = 0$, this is always possible.

The cases are:

I. $c_3 = c_2 = c_1$. II. $c_3 = c_1$, $c_2 \neq c_1$ with $\text{ind } c_2 \equiv \text{ind } c_1 \pmod d$. III. $c_3 = c_1$, $\text{ind } c_2 \not\equiv \text{ind } c_1 \pmod d$. IV. $c_3 = c_2$, $c_1 \neq c_2$, $\text{ind } c_2 \equiv \text{ind } c_1 \pmod d$. V. $c_3 = c_2$, $c_2 \not\equiv c_1 \pmod d$. VI. $c_3 \neq c_2$, $c_3 \neq c_1$, $\text{ind } c_2 \equiv \text{ind } c_1 \pmod d$. VII. $c_3 \neq c_1$, $c_3 \neq c_2$, $c_1 \not\equiv c_2 \pmod d$. Write

$$A = \sum_{\mu_1, \mu_2} \chi_{k_1}^{\mu_1}\left(-\frac{c_3}{c_1}\right) \chi_{k_2}^{\mu_2}\left(-\frac{c_3}{c_2}\right) \frac{\tau(\chi_{k_1}^{\mu_1})\tau(\chi_{k_2}^{\mu_2})}{\tau(\chi_{k_1}^{\mu_1}\chi_{k_2}^{\mu_2})} \qquad (5)$$

where $\chi_{k_1}^{\mu_1} \neq 1$, $\chi_{k_2}^{\mu_2} \neq 1$, $\chi_{k_1}^{\mu_1}\chi_{k_2}^{\mu_2} \neq 1$.

In case I, we find

$$\sum_{\mu_1 = 0}^{k_1 - 1} \chi_{k_1}^{\mu_1}\left(-\frac{c_3}{c_1}\right) = k_1,$$

$$\sum_{\mu_2 = 0}^{k_2 - 1} \chi_{k_2}^{\mu_2}\left(-\frac{c_3}{c_2}\right) = k_2,$$

$$\sum_{v=1}^{d-1} \chi_{k_1}^{s_1 v}\left(-\frac{c_2}{c_1}\right) = d - 1.$$

To reduce A in this case, we take all the terms in which $\chi_{k_1}^{\mu_1} \neq 1$, $\chi_{k_2}^{\mu_2} \neq 1$ and then subtract all the terms from the result in which $\chi_{k_1}^{\mu_1}\chi_{k_2}^{\mu_2} = 1$ with $\chi_{k_1}^{\mu_1} \neq 1$, $\chi_{k_2}^{\mu_2} \neq 1$ and employ (3). The terms of the first type give

$$(k_1 - 1)(k_2 - 1)p^{n_1}(-1)^{r-1}$$

As in the proof of (18) of s there are $(d - 1)$ terms of the second type just mentioned. Hence (4) reduces to

$$p^n + 1 - k_1 - k_2 - d + ((k_1 - 1)(k_2 - 1) - (d - 1))(-1)^{r-1}p^{n_1}. \tag{6}$$

For case II, we find the same value for the first summation in (4) but for the second, we have the value 0 since $c_3 \neq c_2$. In reducing A for this case and separating the summation in the same way it was separated in case I, we have for the first part $(k_1 - 1)p^{n_1}(-1)^r$ and for the second part $(d - 1)p^{n_1}(-1)^r$. Hence, we have for case II the value, if N_i denotes the number of solutions in case i, i = I, II, ..., VII.

$$N_2 = p^n + 1 - d - k_1 + (k_1 + d - 2)p^{n_1}(-1)^r. \tag{7}$$

Similarly for cases III, IV, etc., we have

$$N_3 = p^n + 1 - k_1 + (k_1 - 2)p^{n_1}(-1)^r. \tag{8}$$

$$N_4 = p^n + 1 - d - k_2 + (k_2 + d - 2)p^{n_1}(-1)^r. \tag{9}$$

$$N_5 = p^n + 1 - k_2 + (k_2 - 2)p^{n_1}(-1)^r. \tag{10}$$

$$N_6 = p^n + 1 - d + p^{n_1}(-1)^r(d - 2). \tag{11}$$

$$N_7 = p^n + 1 + (-1)^{r-1}2p^{n_1}. \tag{12}$$

THEOREM I. *If N_i is the number of solutions of the equation*

$$c_1x_1^{a_1} + c_2x_2^{a_2} + c_3 = 0$$

in x_1 and x_2, non-zero elements of a finite field $F(p^n)$ of order p^n, p an odd prime, with c_1, c_2, c_3 given non-zero elements of $F(p^n)$ with N_i, $i = 1, 2, \ldots, 7$, the number of solutions in the corresponding seven cases numbered I, II, ..., VII, given just above the relation (5); $0 < a_i < p^n - 1$, $(a_i, p^n - 1) = k_1$, $i = 1, 2$; $m = ds_1 s_2$, $ds_1 = k_1$, $ds_2 = k_2$, $(k_1, k_2) = d$; $n = 2n_1$, p belongs to the exponent $2t$ modulo m, $n_1 = rt$ with $p^t \equiv -1 \pmod m$, then the values of N_i are given by the relations (6)–(12), inclusive.

Mitchell[2] obtained these results for the special case when $k_1 = k_2$ by a more complicated method.

We now use (4) to find closer limits for N in the special case when $s = 2$, than those found in C for equation (1) of that paper. Set

$$M = p^n \sum_{\mu_1=0}^{k_1-1} \chi_{k_1}^{\mu_1}\left(-\frac{c_3}{c_1}\right) - \sum_{\mu_2=0}^{k_2-1} \chi_{k_2}^{\mu_2}\left(-\frac{c_3}{c_2}\right) - \sum_{v=1}^{d-1} \chi_{k_1}^{vs_1}\left(-\frac{c_2}{c_1}\right).$$

Now every term on the right of (4) is real so that we may write $N \leqq M + |A|$ and $N \geqq M - |A|$ and using the known result

$$|\tau(\chi)| = p^{n/2}, \chi \neq \chi_0, \tag{13}$$

we have, if

$$D = |((k_1 - 1)(k_2 - 1) - (d - 1))p^{n/2}|, \tag{13a}$$
$$N \leqq M + D, \tag{14}$$
$$N \geqq M - D. \tag{15}$$

To obtain limits in terms of rational integers only, divide the discussion into the seven cases I–VII defined in the proof of Theorem I. Then we can evaluate M exactly as in the determination of N in Theorem I in each of the seven cases.[4]

Write N_i; $i = 1, 2, \ldots, 7$ for the value of N corresponding to each of the seven cases referred to in Theorem I, respectively. Then from (14) and (15) we have the

THEOREM II. *If*

$$c_1x_1^{a_1} + c_2x_2^{a_2} + c_3 = 0,$$

where c_1, c_2, c_3 are given elements in a finite field $F(p^n)$ of order p^n, p prime, $c_1c_2c_3 \neq 0$, then we have the following limits for N_i, the number of solutions in the x's, neither zero, of the above equation,

$$M_i - D \leqq N_i \leqq M_i + D, \tag{16}$$

where $M_1 = p^n + 1 - k_1 - k_2 - d$; $M_2 = p^n + 1 - d - k_1$; $M_3 = p^n + 1 - k_1$; $M_4 = p^n + 1 - d - k_2$; $M_5 = p^n + 1 - k_2$; $M_6 = p^n + 1 - d$; $M_7 = p^n + 1$; $k_1 = (p^n - 1, a_1)$, $k_2 = (p^n - 1, a_2)$, $d = (k_1, k_2)$ *and D is defined in (13a), M_i; $i = 1, 2, \ldots, 7$, is the value of M corresponding to each of the seven cases, I, II, ..., VII, of Theorem I, respectively.*

For the case when $n = 1$ and $k_1 = k_2$ is an odd prime, Hurwitz[3] by a different method proved the above theorem using rational integers only. His argument is much longer than ours. For the particular cases treated in Theorem I the *exact value for N agrees with the limits found in Theorem II, when $c_1 = c_2 = c_3$, with the superior limit when r is odd and the inferior limit when r is even.* For the case $n = 1$, $k_1 = k_2$ an odd prime, this fact was noted by Mitchell,[2] who also stated that Hurwitz's method could be extended so as to obtain limits for the case when n is general, which had a similar property.

[1] These PROCEEDINGS, **35,** 94–99 (1949), this will be referred to as C; **35,** 451–457 (1949), this will be referred to as H.

[2] *Ann. Math.*, **II,** 18, 120 (1917).

[3] *Crelle,* **136,** 272–292 (1909).

[4] As a check on the accuracy of our work, the present problem was discussed in a mathematical seminar, using a different method of proof. During the resulting discussion, Mr. Olin B. Faircloth made a suggestion, which, when used in connection with the method employed in this paper yielded a better superior limit for N, than that which we originally had.

Reprinted from the Proceedings of the NATIONAL ACADEMY OF SCIENCES,
Vol. 35, No. 8, pp. 481–487. August, 1949

ON THE NATURE OF THE SOLUTIONS OF CERTAIN EQUATIONS IN A FINITE FIELD

BY L. K. HUA AND H. S. VANDIVER

DEPARTMENT OF MATHEMATICS, UNIVERSITY OF ILLINOIS, AND DEPARTMENT OF APPLIED MATHEMATICS, UNIVERSITY OF TEXAS

Communicated May 23, 1949

In two recent papers[1] the authors considered the equation

$$c_1x_1^{a_1} + c_2x_2^{a_2} + \ldots + c_sx_s^{a_s} + c_{s+1} = 0 \tag{1}$$

in the x's, where the a's are integers such that $0 < a < p^n - 1$; $s \geqq 2$ for $c_{s+1} \neq 0$, and $s > 2$ for $c_{s+1} = 0$ the c's being given elements of a finite field of order p^n, p an odd prime, which will be designated by $F(p^n)$; and

$$c_1c_2 \ldots c_sx_1x_2 \ldots x_s \neq 0 \tag{2}$$

in $F(p^n)$. Limits were found for the number of solutions of (1) and it was proved that for p^n sufficiently large (1) always had solutions under the restriction (2). In the statement above concerning (1) we made the provision that $s \geqq 2$ for $c_{s+1} \neq 0$. If we consider the case where $s = 1$, (1) reduces to the equation

$$c_1x_1^{a_1} + c_2 = 0, \tag{3}$$

with $c_1c_2 \neq 0$. Since any finite field can be represented by means of the residue classes with respect to some prime ideal in an algebraic field, then examination of the solutions of (3) is included in the theory of the congruence

$$x^n \equiv \alpha \pmod{\mathfrak{p}}, \tag{4}$$

where α is an integer in an algebraic field K, $\mathfrak{p}$ is a prime ideal in that field and $\alpha \not\equiv 0 \pmod{\mathfrak{p}}$. The study of the last relation led to the theory of the class field and the laws of reciprocity for nth powers.[2]

We wish to point out the sharp distinction between this problem and the problem of finding the solutions or the number of solutions of the equation (1) and when $s \geqq 2$ for $c_{s+1} \neq 0$.

In addition to this, we shall also obtain an expression (Theorem I) for the number of solutions of (1) when $c_{s+1} \neq 0$ and $s = 2$, which leads to the proof of Theorem I and which we shall also have occasion to make use of in another paper. Also we obtain Theorem II, which gives the exact number of solutions of (1) when the k's are prime each to each, $c_{s+1} = 0$ and $(a_i, p^n - 1) = k_i$; $i = 1, 2, \ldots, s$.

We consider equation (4) and confine ourselves to the case where K is an algebraic field which contains a primitive lth root of unity with l prime. Then

$$x^l - \alpha \equiv 0 \pmod{\mathfrak{p}} \qquad (5)$$

if and only if the lth power character of α, modulo $\mathfrak{p}$ is unity, or

$$\left(\frac{\alpha}{\mathfrak{p}}\right)_l = 1.$$

However, by a known theorem[3]

$$\left(\frac{\alpha}{\mathfrak{p}}\right)_l \neq 1$$

for an infinity of distinct prime ideals $\mathfrak{p}$ provided α does not equal an lth power of an integer in K. Now the last statement is contrary to what we have for $s > 1$ in (1) with $c_{s+1} \neq 0$. For, as we have shown in E (p. 262) the congruence

$$\alpha_1 x_1^{a_1} + \alpha_2 x_2^{a_2} + \ldots + \alpha_s x_s^{a_s} + \alpha_{s+1} \equiv 0 \pmod{\mathfrak{p}}, \qquad (6)$$

with

$$\alpha_1\alpha_2 \ldots \alpha_s x_1 x_2 \ldots x_s \not\equiv 0 \pmod{\mathfrak{p}}$$

for $s \geqq 2$ with $c_{s+1} \not\equiv 0 \pmod{\mathfrak{p}}$ always has solutions if the norm of $\mathfrak{p}$ exceeds a certain limit, in fact (6) always has at least k solutions in the x's for k any positive integer if the norm of p exceeds a certain limit. In particular when we consider the case where there are no solutions it is known[4] that $x^7 + y^7 + 1 \equiv 0 \pmod{\rho}$ has no solutions in integers prime to p for just four primes p, namely 29, 71, 113, 491, but has solutions for all the other primes p. As for $x^5 + y^5 + 1 \equiv 0 \pmod{p}$ it has no solutions for $p = 11, 41, 71, 101$. Also $x^3 + y^3 + 1 \equiv 0 \pmod{p}$ has no solutions only for $p = 7$ and 13. It is also known[5] that for $m < 10l$, and $p = 1 + ml$ and $h(l)$ is the number of primes p for which

$$x^l + y^l + 1 \equiv 0 \pmod{p} \qquad (7)$$

has no solutions for $xy \not\equiv 0 \pmod{p}$ then $h(11) = 3$, $h(13) = 4$, $h(17) = 4$,

$h(19) = 6$. In view of these facts it may happen that $h(l)$ increases as l increases. Note also from the cases 3, 5 and 7 that there is a range of values for p in which the corresponding congruences may or may not have solutions.

From the above considerations we may say that associated with any form of the type given by the left-hand member of (4) there is an integer $e(k)$ such that there are exactly $e(k)$ values of $\mathfrak{p}$ such that (4) has k solutions. Even in the case where k is zero, the class of forms satisfying the condition may be rather general. For example,[6] it was shown that if c is a prime and m is an integer such that $p = 1 + mc$ with p prime then

$$a_1x_1^m + a_2x_2^m + \ldots + a_sx_s^m \equiv 0 \pmod p$$

has only the solutions $x_1 \equiv x_2 \equiv \ldots \equiv x_s \equiv 0 \pmod p$ provided $s \leqq c - 2$ the sum of no n of the a's is $\equiv 0 \pmod p$ $0 < n \leqq s$, and

$$(|a_1| + |a_2| + \ldots + |a_s|)^{\phi(c)} < p,$$

where $\phi(c)$ is the indicator of c. Select $a_1 = a_2 = \ldots = a_{s-1} = 1$ and $a_s = -s$, and also an m and p with $p = 1 + mc$ with c fixed and satisfying the other conditions just mentioned then it is evident that

$$x_1^m + x_2^m + \ldots + x_{s-1}^m - sx_s^m \equiv 0 \pmod p$$

has no primitive solutions. (Incidentally, in view of Theorem III of the article just referred to, the equation

$$x_1^m + x_2^m + \ldots + x_{s-1}^m - sx_s^m = 0$$

has no solutions in rational integers unless each x is zero, if there is a prime p with $p = 1 + mc$ and $(2s - 1)^{\varphi(c)} < p$.)

We now recur to the subject of the number of solutions of equation (1), treated in E and C. The case $s = 2$ with $c_3 \neq 0$, has received considerable attention in the literature.[7] Let us examine

$$c_1x_1^{a_1} + c_2x_2^{a_2} + c_3 = 0, \tag{8}$$

with

$$x_1x_2c_1c_2c_3 \neq 0. \tag{8a}$$

Theorem I of C, we see will give, if N is the number of solutions of (8) in x_1 and x_2 under the conditions (8a),

$$N = \frac{(p^n - 1)^2}{p^n} + \frac{T}{p^n}, \tag{9}$$

where we may write T in the form, if $k_i = (a_i, p - 1)$, μ_1 ranges over 0, 1, ..., $k_1 - 1$; μ_2 over 0, 1, ..., $k_2 - 1$

$$\sum_{\substack{a \in K \\ a \neq 0}} \sum_{\mu_1, \mu_2} \tau(\chi_{k_1}^{\mu_1})\tau(\chi_{k_2}^{\mu_2})\chi_{k_1}^{\mu_1}(c_1^{-1})\chi_{k_2}^{\mu_2}(c_2^{-1})\chi_{k_1}^{\mu_1}(a^{-1})\chi_{k_2}^{\mu_2}(a^{-1})\zeta^{tr(ac_3)}, \tag{9a}$$

where χ_{k1} is a special character defined by k_1 or as we shall say, a k_1-character, and similarly χ_{k_2} is a special character defined by k_2 and where, if we write K for $F(p^n)$,

$$\tau(\chi, \zeta) = \tau(\chi) = \sum_{b\epsilon K} \chi(b)\, \zeta^{tr(b)};\ \zeta = e^{2i\pi/p}, \tag{9b}$$

with $\chi(0) = 0$. This is obtained from the fact that if α is a primitive $(p^n - 1)$th root of unity then $\alpha^{\mu \text{ ind } a}$, where a ranges over all elements of $F(p^n)$, $\neq 0$, gives a character as defined in C where ind a is given by

$$g^{\text{ind } a} = a,$$

and where g is a primitive root in $F(p^n)$, so we obtain all characters by taking $\mu = 0, 1, \ldots, p^n - 2$. We have from C the relations, if χ_0 is the principal character,

$$\left.\begin{aligned} |\tau(\chi)| &= p^{n/2};\ \chi \neq \chi_0 \\ \tau(\chi_0) &= -1 \end{aligned}\right\}. \tag{10}$$

To reduce (9a) we divide the discussion into four cases.

First Case. $\chi_{k_1}^{\mu_1} = 1, \chi_{k_2}^{\mu_2} = 1$. In this event $\chi_{k_1}^{\mu_1}(a^{-1}) = \chi_{k_2}^{\mu_2}(a^{-1}) = 1$, $\chi_{k_1}^{\mu_1}(c_1^{-1}) = \chi_{k_2}^{\mu_2}(c_2^{-1}) = 1$, which reduces (9a), using (10), to

$$\sum_a \zeta^{tr(ac_3)} = -1. \tag{11}$$

Second Case. $\chi_{k_1}^{\mu_1} = 1;\ \chi_{k_2}^{\mu_2} \neq 1$. This gives from (9a) and (10),

$$-\sum_a \tau(\chi_{k_2}^{\mu_2})\chi_{k_2}^{\mu_2}(c_2^{-1}a^{-1})\zeta^{tr(ac_3)} = \sum_a -\tau(\chi_{k_2})\chi_{k_2}(c_2^{-1})(\chi_{k_2}(a))^{-1}\zeta^{tr(ac_3)},$$

and setting $ac_3 = -a_1$, using $c_3 \neq 0$, this reduces to

$$-\sum_{a_1} \tau(\chi_{k_2}^{\mu_2})\chi_{k_2}^{\mu_2}(-c_2^{-1}c_3)(\chi_{k_2}^{\mu_3}(a_1))^{-1}\zeta^{-tr\, a_1} = -\sum \tau(\chi_{k_2})\overline{\tau(\chi_{k_2})}\chi_{k_2}\left(-\frac{c_3}{c_2}\right),$$

and using the known relation

$$\tau(\chi_{k_2}^{\mu_2})\overline{\tau(\chi_{k_2}^{\mu_2})} = p^n, \tag{11a}$$

this is seen to equal

$$-p^n \sum_{\mu_2} \chi_{k_2}^{\mu_2}\left(-\frac{c_3}{c_2}\right). \tag{12}$$

Also, if $\chi_{k_1}^{\mu_1} \neq 1$ and $\chi_{k_2}^{\mu_2} = 1$ we find in a similar way that the result is

$$-p^n \sum_{\mu_1} \chi_{k_1}^{\mu_1}\left(-\frac{c_3}{c_1}\right). \tag{13}$$

Third Case. $\chi_{k_1}^{\mu_1}\chi_{k_2}^{\mu_2} = 1$ with $\chi_{k_1}^{\mu_1} \neq 1$, $\chi_{k_2}^{\mu_2} \neq 1$, (9a) reduces to

$$\sum_{a}\sum_{\mu_1}\chi_{k_1}{}^{\mu_1}(c_1{}^{-1})\chi_{k_1}{}^{-\mu_1}(c_2{}^{-1})\tau(\chi_{k_1}{}^{\mu_1})\tau(\chi_{k_1}{}^{-\mu_1}, \zeta)\zeta^{tr(ac_3)}.$$

Now

$$\begin{aligned}\tau(\chi_{k_1}{}^{-\mu_1}, \zeta) &= \overline{\tau(\chi_{k_1}{}^{-\mu_1}, \zeta^{-1})}\chi_{k_1}(-1)\\ &= \tau(\chi_{k_1}{}^{\mu_1})\chi_{k_1}{}^{\mu_1}(-1).\end{aligned}$$

Hence for the admissible values of μ_1, (9a) reduces to

$$-\sum_{\mu_1}\chi_{k_1}{}^{\mu_1}\left(-\frac{c_2}{c_1}\right)p^n. \tag{14}$$

Now we must determine the possible values of μ_1 from $\chi_{k_1}{}^{\mu_1}\chi_{k_2}{}^{\mu_2} = 1$. Let α be a primitive (ds_1s_2)th root of unity where $ds_1 = k_1$, $ds_2 = k_2$ and where $(k_1, k_2) = d$, whence $(s_1, s_2) = 1$. Then $\chi_{k_1}{}^{\mu_1}\chi_{k_2}{}^{\mu_2} = 1$ gives integers μ_1 and μ_2 such that $\alpha^{s_1\mu_1 + s_2\mu_2} = 1$ or $s_2\mu_1 + s_1\mu_2 \equiv 0 \pmod{(ds_1s_2)}$ so that $\mu_1 = v_1s_1$, $\mu_2 = v_2s_2$. Hence the possible values of μ_1 are s_1v_1; $v_1 = 1, 2, \ldots, d - 1$, as the case $v_1 = 0$ was excepted. Hence (9a) reduces to, for this case,

$$-\sum_{v_1 = 1}^{d-1}\chi_{k_1}{}^{s_1v_1}\left(-\frac{c_2}{c_1}\right)p^n. \tag{15}$$

Fourth Case. Here $\chi_{k_1}{}^{\mu_1} \neq 1$, $\chi_{k_2}{}^{\mu_2} \neq 1$, $\chi_{k_1}{}^{\mu_1}\chi_{k_2}{}^{\mu_2} \neq 1$. Consider such terms in (9a). They may be written in the form

$$\sum_{\mu_1, \mu_2}\sum_{a \in K}\tau(\chi_{k_1}{}^{\mu_1})\tau(\chi_{k_2}{}^{\mu_2})\chi_{k_1}{}^{\mu_1}\chi_{k_2}{}^{\mu_2}(a^{-1})\zeta^{tr(ac_3)}\cdot\chi_{k_1}{}^{\mu_1}(c_1{}^{-1})\chi_{k_2}{}^{\mu_2}(c_2{}^{-1}). \tag{16}$$

Write $a_1 = ac_3$, then $a^{-1} = a_1{}^{-1}c_3$ and using $c_3 \neq 0$,

$$\begin{aligned}\sum_{\substack{a_1 \in K\\ a_1 \neq 0}}\chi_{k_1}{}^{\mu_1}\chi_{k_2}{}^{\mu_2}(a^{-1}) &= \sum_{a_1 \in K}\chi_{k_1}{}^{\mu_1}\chi_{k_2}{}^{\mu_2}(c_3)(\chi_{k_1}{}^{\mu_1}\chi_{k_2}{}^{\mu_2}(a))^{-1}\zeta^{tr(a_1)}\\ &= \sum_{a_1 \in K}\chi_{k_1}{}^{\mu_1}\chi_{k_2}{}^{\mu_2}(c_3)(\chi_{k_1}{}^{\mu_1}\chi_{k_2}{}^{\mu_2}(a_1))^{-1}\chi_{k_1}{}^{\mu_2}\chi_{k_2}{}^{\mu_2}(-1)\zeta^{-tr(a_1)}\end{aligned}$$

Hence using (11a), we find from (9a) that the terms in case 4 in T may be written as

$$p^n\sum_{\mu_1, \mu_2}\chi_{k_1}{}^{\mu_1}\left(-\frac{c_3}{c_1}\right)\chi_{k_2}{}^{\mu_2}\left(-\frac{c_3}{c_2}\right)\frac{\tau(\chi_{k_1}{}^{\mu_1})(\chi_{k_2}{}^{\mu_2})}{\tau(\chi_{k_1}{}^{\mu_1}\chi_{k_2}{}^{\mu_2})}, \tag{17}$$

$\chi_{k_1}{}^{\mu_1} \neq 1$, $\chi_{k_2}{}^{\mu_2} \neq 1$, $\chi_{k_1}{}^{\mu_1}\chi_{k_2}{}^{\mu_2} \neq 1$. Using (9), (9a), (11), (12), (13), (14), (15) and (17), we find

$$\begin{aligned}N = p^n &- \sum_{\mu_1=0}^{k_1-1}\chi_{k_1}^{\mu_1}\left(-\frac{c_3}{c_1}\right) - \sum_{\mu_2=0}^{k_2-1}\chi_{k_2}^{\mu_2}\left(-\frac{c_3}{c_2}\right) - \sum_{v=1}^{d-1}\chi_{k_1}{}^{s_1v}\left(-\frac{c_2}{c_1}\right)\\ &+ \sum_{\mu_1, \mu_2}{}'\chi_{k_1}{}^{\mu_1}\left(-\frac{c_3}{c_1}\right)\chi_{k_2}{}^{\mu_2}\left(-\frac{c_3}{c_2}\right)\frac{\tau(\chi_{k_1}{}^{\mu_1})\tau(\chi_{k_2}{}^{\mu_2})}{\tau(\chi_{k_1}{}^{\mu_1}\chi_{k_2}{}^{\mu_2})}\end{aligned} \tag{18}$$

where, in the last term, $\chi_{k_1}{}^{\mu_1} \neq 1$, $\chi_{k_2}{}^{\mu_2} \neq 1$, $\chi_{k_1}{}^{\mu_1}\chi_{k_2}{}^{\mu_2} \neq 1$.

Whence we have the (cf. Davenport-Hasse[8])

THEOREM I. *If N is the number of solutions in x_1 and x_2 of*

$$c_1x_1^{a_1} + c_2x_2^{a_2} + c_3 = 0,$$

with c_1, c_2, c_3, x_1, x_2 in a finite field $F(p^n)$ of order p^n with p an odd prime, $c_1c_2c_3x_1x_2 \neq 0$ in $F(p^n)$; $(a_1, p^n - 1) = k_1, 0 < k_1 < p^n - 1$, $(a_2, p^n - 1) = k_2, 0 < k_2 < p^n - 1$, χ_{k_1} is a special k_1-character, χ_{k_2} is a special k_2-character, $d = (k_1, k_2)$, $ds_1 = k_1$, $\tau(\chi)$ is defined as in (9b), then the relation (18) holds.

We shall now obtain N explicitly in terms of p, n and s, for special values of k_1 when $c_{s+1} = 0$. By the remark immediately following (9b), we may write

$$\chi_{k_i}^{\mu_i}(a^{-1}) = \alpha_i^{\mu_i \operatorname{ind}(a^{-1})} \tag{19}$$

for any i in the set $0, 1, \ldots, k_i - 1$, and where α_i is a primitive k_ith root of unity. Set in Theorem I of C, $c_{s+1} = 0$ then for this case we have

$$T = \sum_{\mu_i = 0}^{k_i - 1} \sum_{\substack{a \in K \\ a \neq 0}} (\alpha_1^{\mu_1}\alpha_2^{\mu_2} \ldots \alpha_s^{\mu_s})^{\operatorname{ind}(a^{-1})} \cdot \prod_{i=1}^{s} \alpha_i^{\mu_i \operatorname{ind}(c_i^{-1})}\tau(\alpha_i^{\mu_i}). \tag{20}$$

Assume that the k's are prime each to each and carry out the summation with respect to a we obtain

THEOREM II. *If N_s is the number of solutions of (1) in x's under the conditions (2) and the other conditions stated in connection with (1), and also $k_1, k_2, \ldots, k_s$ are prime each to each with $c_{s+1} = 0$, then*

$$N_s = \frac{p^n - 1}{p^n}((p^n - 1)^{s-1} + (-1)^s). \tag{21}$$

This result may be proved in a simpler way as we shall now show.

Consider the equation

$$c_1x_1^{k_1} + \ldots + c_sx_s^{k_s} = 0, \quad (k_i, k_j) = 1, \quad i \neq j, \tag{22}$$

and $A = k_1k_2 \ldots k_i$ dividing $p^n - 1$. Since $(k_1, k_2, \ldots k_s) = 1$, we have integers λ and μ such that

$$\lambda k_1 + \mu k_2 \ldots k_s = 1, \qquad (\lambda, p^n - 1) = 1.$$

Putting

$$x_1 = y_1^{\lambda}, \quad x_2 = y_1^{-\mu A/k_1k_2}y_2, \quad x_3 = y_1^{-\mu A/k_1k_3}y_3, \ldots.$$

Then we have

$$c_1y_1^{\lambda k_1} + y_1^{-\mu A/k_1}(c_2y_2^{k_2} + \ldots + c_sy_s^{k_s}) = 0$$

i.e.,

$$c_1y_1 + c_2y_2^{k_2} + \ldots + c_sy_s^{k_s} = 0$$

Since, for given elements of $F(p^n)$, $y_2, \ldots, y_s$ such that

$$c_2 y_2^{k_2} + \ldots + c_s y_s^{k_s} \neq 0$$

we have a unique y_1, therefore, we have

$$N_s = (p^n - 1)^{s-1} - N_{s-1}.$$

Consequently

$$N_s = (p^n - 1)^{s-1} - (p^n - 1)^{s-2} + \ldots + (-1)^s(p^n - 1)$$

$$= \frac{p^n - 1}{p^n}((p^n - 1)^{s-1} + (-1)^s).$$

We also have, since the residue classes with respect to a rational prime modulus p form a finite field of order p, the

COROLLARY I. *If N is the number of incongruent solutions in the y's of the congruence, with p an odd prime,*

$$d_1 y_1^{a_1} + d_2 y^{a_2} + \ldots + d_s y_s^{a_s} \equiv 0 \pmod p$$

where $d_1 d_2 \ldots d_s \not\equiv 0 \pmod p$, $y_1 y_2 \ldots y_s \not\equiv 0 \pmod p$; $k_1, k_2, \ldots, k_s$, prime each to each with $(a_i, p - 1) = k_i$; $0 < k_i < p - 1, i = 1, 2, \ldots, s$ then

$$N = \frac{p - 1}{p}((p - 1)^{s-1} + (-1)^s).$$

It may also be noted that the procedure used in the second proof of theorem II gives us a method for determining the solutions of (22).

[1] These PROCEEDINGS, **34,** 258–263 (1948), this will be referred to as E; **35,** 94–99 (1949), this will be referred to as C.

[2] Hasse, H., *Ber. Deut. Math. Verein.*, **35,** 1–55 (1926); **36,** 233–311 (1927); *Erganzungsband*, **VI** (1930).

[3] Takagi, T., *J. Coll. Sci., Imp. Univ. Tokyo*, **42,** 16 (1920).

[4] Dickson, L. E., *Messenger Math.*, **38,** 14–33 (1908).

[5] These results are due to N. G. W. H. Beeger, unpublished.

[6] Vandiver, H. S., these PROCEEDINGS, **32,** 101–106 (1946).

[7] Cf. footnote 1 of E.

[8] The proof of this theorem we gave as a direct application of our general theorem I of C to the special case $s = 2$. Another proof may be obtained by using the result of Davenport and Hasse, *Jour. für Math.* (Crelle) 172, p. 174 relation (6.5) which gives the number of solutions of $ax^m + by^n = c$, $abc \neq 0$ in $F(p^n)$, where x and y are not restricted, as they were in our work, to non-zero values in $F(p^n)$. To do this we consider the possible solutions involving zero and subtract the number of them from the right-hand member of the relation (6.5) just referred to. We note that the number of solutions of $ax^m = c$ and $by^n = c$ are, respectively,

$$\sum_{\mu=0}^{m-1} \overline{x}^n\left(\frac{c}{a}\right); \sum_{\nu=0}^{n-1} \psi\left(\frac{c}{b}\right)$$

which when taken in connection with (6.5) yields a formula equivalent to our relation (18).

Reprinted from the Proceedings of the NATIONAL ACADEMY OF SCIENCES,
Vol. 35, No. 9, pp. 533–537. September, 1949

SOME PROPERTIES OF A SFIELD

BY LOO-KENG HUA

DEPARTMENT OF MATHEMATICS, UNIVERSITY OF ILLINOIS

Communicated by H. S. Vandiver, July 12, 1949

All the results of this paper are initiated from the almost trivial identity that if $ab \neq ba$, then

$$a = (b^{-1} - (a - 1)^{-1}b^{-1}(a - 1))(a^{-1}b^{-1}a - (a - 1)^{-1}b^{-1}(a - 1))^{-1} \tag{1}$$

Among these results, there are two interesting ones which are the perfect form of two theorems due to H. Cartan[1] and J. Dieudonné,[2] respectively.

THEOREM 1. *Every sfield is generated by a non-central element and its conjugates.*

Let K be the sfield and L be the conjugate set. The identity (1) asserts that if there is an element b of L such that $ab \neq ba$, then a belongs to the sfield K, generated by L. That is, $K - K_1$ and K_1 are commutative elementwise.

Suppose our theorem is false; we have an element a of $K - K_1$ and two elements b and b' of K_1 such that $bb' \neq b'b$. Since a and ab belong to $K - K_1$, we have $(ab)b' = b'(ab) = ab'b$, which contradicts $bb' = b'b$. The theorem follows.

An immediate consequence of Theorem 1 is the following interesting result:

THEOREM 2. *Every proper normal subsfield of a sfield is contained in the center.*

H. Cartan proved the theorem under the assumption that the rank of the sfield over its center is finite. His proof is far more complicated than the present one.

More precise results can be obtained by specializing the subgroup.

THEOREM 3. *Every sfield, which is not a field, has no proper subsfield containing all its commutators.*

It is enough to prove the existence of a commutator which does not belong to the center. Suppose $ab \neq ba$. From a modification of the identity (1):

$$a = (1 - (a - 1)^{-1}b^{-1}(a - 1)b)(a^{-1}b^{-1}ab - (a - 1)^{-1}b^{-1}(a - 1)b)^{-1}, \tag{2}$$

we deduce that at least one of $a^{-1}b^{-1}ab$ and $(a - 1)^{-1}b^{-1}(a - 1)b$ does belong to the center.

As corollaries of Theorem 3, we have

THEOREM 4. *If the center of a sfield contains all the commutators of the sfield, then the sfield is commutative. An element of a sfield commutative with all the commutators of the sfield belongs to the center.*

Moreover, as a simple consequence of

$$a^{-1}c^{-1}ac = a^{-2}(ac^{-1})^2c^2 \tag{3}$$

and Theorem 3, we have

THEOREM 5. *Every sfield, which is not a field, has no proper subsfield containing all its square elements.*

THEOREM 6. *If the center of a sfield contains all the square elements of the sfield, then the sfield is commutative. An element of a sfield commutative with all the square elements of the sfield belongs to the center.*

By means of an identity in the theory of finite differences:

$$\Delta^{k-1}x^k = k!(x + {}^1/_2(k - 1)),$$

where $\Delta f(x) = f(x + 1) - f(x)$ and $\Delta^i f(x) = \Delta(\Delta^{i-1}f(x))$, we have

THEOREM 7. *Every sfield of characteristic $\geq k$, which is not commutative, has no proper subsfield containing all its k-th power elements.*

Let us denote $a^{-1}b^{-1}ab$ by (a, b). An r-commutator $(r \geq 2)$ is defined inductively by the relation

$$(a_1, \ldots, a_r) = (a_1, (a_2, \ldots, a_r)). \tag{4}$$

Let C_r be the set of all r-commutators of the sfield.

THEOREM 8. *Let $r \geq 2$. Every sfield, which is not commutative, has no proper subsfield containing all its r-commutators. An element of a sfield which permutes with all the r-commutators of the sfield belongs to the center.*

Let A_r and B_r denote the two propositions of our theorem. B_r follows from A_r immediately. Theorems 3 and 5 assert that both A_2 and B_2 are true. We shall prove the theorem by induction, we assume that A_{r-1} and B_{r-1} are both true. Let a be an element of the sfield, which does not belong to the center. By B_{r-1}, we have an element belonging to C_{r-1} such that a $b \neq ba$. The identity (2) asserts that a belongs to the sfield generated by C_r. Therefore A_r is true. The theorem is proved.

The theorem asserts that the lower central series of the multiplicative group of a sfield can never be ended at the identity. It is an almost trivial fact that the upper central series can never go up. It is a comparatively difficult question about the derived series. Elsewhere I shall prove

THEOREM 9. *Let $r \geq 1$. Every sfield is generated by all the elements of the r-th derived group.*

An application of Theorem 1 is to establish the simplicity of the projective special linear group $PSL_n(K)$[3] of dimension n over the sfield K.

THEOREM 10. *The projective special linear group $PSL_n(K)$ is simple except when $n = 2$ and K has 2 or 3 elements.*

In case K is a field of characteristic $\neq 2$ or K is a complete field, the theorem was proved by Burnside-Jordan-Dickson (see Dickson,[4] p. 85 and Van der Waerden,[5] p. 7). For the incomplete field, it was proved by Iwasawa.[6] For a sfield, Dieudonné[2] proved the theorem for $n > 2$, and for $n = 2$[7] except in the case when the center of K has 2, 3 or 5 elements. Since both exceptional cases in Theorem 10 are not simple, it is the final answer to the problem about the simplicity of $PSL_n(K)$.

Owing to the previous known results, we assume, from now on, that $n = 2$ and the sfield is not commutative. It is the same thing to prove that

THEOREM 11. *Every normal subgroup of $SL_2(K)$ is contained in the center, except when K has 2 or 3 elements.*

Let A be an element of $SL_2(K)$ which does not belong to the center. We are going to prove that A and its conjugates generate $SL_2(K)$. We shall prove that there is a matrix Q belonging to $SL_2(K)$ such that QAQ^{-1} has a non-zero element at (1, 2) position. In fact, let $A = \begin{pmatrix} \alpha & \beta \\ \gamma & \delta \end{pmatrix}$. If $\beta \neq 0$, $Q = I$ satisfies our requirement, and if $\gamma \neq 0$, then $Q = \begin{pmatrix} 0 & 1 \\ -1 & 0 \end{pmatrix}$ does. If $\beta = \gamma = 0$, we have an element x such that $\alpha x \neq x\delta$, since A does not belong to the center. Then

$$\begin{pmatrix} 1 & x \\ 0 & 1 \end{pmatrix}\begin{pmatrix} \alpha & 0 \\ 0 & \delta \end{pmatrix}\begin{pmatrix} 1 & x \\ 0 & 1 \end{pmatrix}^{-1} = \begin{pmatrix} * & -\alpha x + x\delta \\ * & * \end{pmatrix}.$$

We may assume that the subgroup generated by A and its conjugates contains an element with $\beta \neq 0$. Since

$$\begin{pmatrix} 1 & 0 \\ \beta^{-1}\alpha & 1 \end{pmatrix}\begin{pmatrix} \alpha & \beta \\ \gamma & \delta \end{pmatrix}\begin{pmatrix} 1 & 0 \\ \beta^{-1}\alpha & 1 \end{pmatrix}^{-1} = \begin{pmatrix} 0 & * \\ * & * \end{pmatrix},$$

we have assumed that the group contains an element of the form $A = \begin{pmatrix} 0 & \beta \\ \gamma & \delta \end{pmatrix}$. We take

$$B = \begin{pmatrix} -\gamma^{-1}\delta\beta^{-1} & \gamma^{-1} \\ -\gamma\kappa & 0 \end{pmatrix},$$

where κ belongs to the commutator subgroup of K, so that B belongs to $SL_2(K)$. We have

$$A(BA^{-1}B^{-1}) = AB(BA)^{-1} = \begin{pmatrix} -\beta\gamma\kappa & 0 \\ * & -(\gamma\kappa\beta)^{-1} \end{pmatrix}$$

Now we shall prove that we can choose κ such that $-\beta\gamma\kappa$ does not belong to the center. For otherwise $(-\beta\gamma\kappa_1)^{-1}(-\beta\gamma\kappa_2) = \kappa_1^{-1}\kappa_2$ belongs to the center for any two commutators κ_1 and κ_2, that is, the commutators of κ belong to the center of K. By Theorem 4, K is a field which contradicts our supposition. Therefore the subgroup contains an element of the form $C = ABA^{-1}B^{-1} = \begin{pmatrix} a & 0 \\ c & d \end{pmatrix}$, where a does not belong to the center. We have an element λ of K such that $d\lambda \neq \lambda a$, then the subgroup contains

$$\begin{pmatrix} a & 0 \\ c & d \end{pmatrix}\begin{pmatrix} 1 & 0 \\ \lambda & 1 \end{pmatrix}\begin{pmatrix} a & 0 \\ c & d \end{pmatrix}^{-1}\begin{pmatrix} 1 & 0 \\ -\lambda & 1 \end{pmatrix} = \begin{pmatrix} 1 & 0 \\ d\lambda a^{-1} - \lambda & 1 \end{pmatrix}.$$

We may suppose that our subgroup contains $\begin{pmatrix} 1 & 0 \\ c & 1 \end{pmatrix}$, $c \neq 0$.

We are going to prove that the subgroup contains $\begin{pmatrix} 1 & 0 \\ 1 & 1 \end{pmatrix}$. In fact, if $c = \pm 1$, our assertion is trivial. Let us suppose $c \neq 0, \neq \pm 1$. The field K_1, obtained from the center of K by adjunction of c, contains more than 3 elements. $SL_2(K_1)$ is generated by any one element, not in the center, and its conjugates; in particular, $\begin{pmatrix} 1 & 0 \\ c & 1 \end{pmatrix}$ and its conjugates generate $SL_2(K_1)$. Thus $\begin{pmatrix} 1 & 0 \\ 1 & 1 \end{pmatrix}$ is in the subgroup under consideration. Consequently the subgroup contains

$$\begin{pmatrix} \lambda^{-1} & 0 \\ 0 & \lambda \end{pmatrix}\begin{pmatrix} 1 & 0 \\ 1 & 1 \end{pmatrix}\begin{pmatrix} \lambda^{-1} & 0 \\ 0 & \lambda \end{pmatrix}^{-1} = \begin{pmatrix} 1 & 0 \\ \lambda^2 & 1 \end{pmatrix}$$

for all λ.

If the characteristic of the sfield is not equal to 2, then the subgroup contains all the elements of the form

$$\begin{pmatrix} 1 & 0 \\ (\lambda + 1)^2 & 1 \end{pmatrix}\begin{pmatrix} 1 & 0 \\ \lambda^2 & 1 \end{pmatrix}^{-1}\begin{pmatrix} 1 & 0 \\ 1 & 1 \end{pmatrix}^{-1} = \begin{pmatrix} 1 & 0 \\ 2\lambda & 1 \end{pmatrix}.$$

Therefore we have the theorem.

If the characteristic of the sfield equals 2, since

$$\begin{pmatrix} 1 & (\lambda + 1)^{-2} \\ 0 & 1 \end{pmatrix}\begin{pmatrix} 1 & 0 \\ \lambda^2 & 1 \end{pmatrix}\begin{pmatrix} 1 & 1 \\ 0 & 1 \end{pmatrix}\begin{pmatrix} 1 & 0 \\ (1 + \lambda^{-1})^{-2} & 1 \end{pmatrix} = \begin{pmatrix} (\lambda + 1)^{-2} & 0 \\ 0 & (\lambda + 1)^2 \end{pmatrix},$$

the subgroup contains elements of the form

$$\begin{pmatrix} a^2 & 0 \\ 0 & a^{-2} \end{pmatrix}.$$

for all a. Let a be any element of the sfield. From

$$\begin{pmatrix} 1 & a \\ 0 & 1 \end{pmatrix}\begin{pmatrix} a^2 0 & \\ 0 & a^{-2} \end{pmatrix}\begin{pmatrix} 1 & \lambda^2 \\ 0 & 1 \end{pmatrix}\begin{pmatrix} 1 & a \\ 0 & 1 \end{pmatrix}^{-1}\begin{pmatrix} a^{-2} & 0 \\ 0 & a^2 \end{pmatrix} = \begin{pmatrix} 1 & a^2\lambda^2 a^2 + a^5 + a \\ 0 & 1 \end{pmatrix}$$

$$= \begin{pmatrix} a^2 0 & \\ 0 & a^{-2} \end{pmatrix}\begin{pmatrix} 1 & \lambda^2 \\ 0 & 1 \end{pmatrix}\begin{pmatrix} a^2 0 & \\ 0 & a^{-2} \end{pmatrix}^{-1}\begin{pmatrix} (a + 1)^2 & 0 \\ 0 & (a + 1)^{-2} \end{pmatrix} \times$$

$$\begin{pmatrix} 1 & a \\ 0 & 1 \end{pmatrix}\begin{pmatrix} (a + 1)^2 & 0 \\ 0 & (a + 1)^{-2} \end{pmatrix}^{-1},$$

we deduce that $\begin{pmatrix} 1 & a \\ 0 & 1 \end{pmatrix}$ belongs to the subgroup. Since $\begin{pmatrix} 1 & a \\ 0 & 1 \end{pmatrix}$ and its conjugates generate $SL_2(K)$ as a runs over all elements of K, the theorem follows.

The previous method can be used to establish the easier

THEOREM 12. *Every normal subgroup of the general linear group $GL_n(K)$, which is not contained in the center, contains $SL_n(K)$, except the cases when $n = 2$ and K has 2 or 3 elements.*

[1] Cartan, H., *Ann. école normale superieure*, **64**, 59–77 (1947), Theorem 4.

[2] Dieudonnè, J., *Bull. Soc. math. France*, **71**, 27–45 (1943).

[3] For the definition and properties of $PSL_n(K)$, see Dieudonnè.[6]

[4] Dickson, L. E., *Linear Groups*, Leipzig, 1901.

[5] Van der Waerden, B. L., *Gruppen von linearen Transformationen*, Berlin, 1935, p. 7.

[6] Iwasawa, M., *Proc. Imp. Acad. Japan*, **17**, 57 (1941).

[7] The author had some difficulty in understanding Dieudonné's proof. In fact, all the parabolic elements of $PSL_2(K)$ do not form a single conjugate set in $PSL_2(K)$.

Reprinted from the
TRANSACTIONS OF THE AMERICAN MATHEMATICAL SOCIETY
Vol. 65, No. 3, pp. 415–426, 1949

ON THE GENERATORS OF THE SYMPLECTIC MODULAR GROUP

BY
L. K. HUA AND I. REINER

Introduction. Let n be a positive integer. Throughout this paper, unless the contrary is stated, we shall use capital Latin letters to denote n-rowed matrices and capital German letters to denote $2n$-rowed matrices. Furthermore, an r-rowed matrix R will be denoted by $R^{(r)}$. Let

$$\mathfrak{F} = \begin{pmatrix} 0 & I \\ -I & 0 \end{pmatrix},$$

where I and 0 denote the identity and zero matrices respectively. Let Γ be the group of all matrices $\mathfrak{M}$ with rational integral elements which satisfy

$$\mathfrak{M}\mathfrak{F}\mathfrak{M}' = \mathfrak{F}, \tag{1}$$

where $\mathfrak{M}'$ denotes the transpose of $\mathfrak{M}$. Let Γ_0 be the factor group of Γ over its centrum; Γ_0 is called the symplectic modular group. It can be thought of as being obtained from Γ by identifying the elements $\mathfrak{M}$ and $-\mathfrak{M}$. In applications to modular functions of the nth degree([1]) and to the projective geometry of matrices([2]) it is customary to identify $\mathfrak{M}$ and $-\mathfrak{M}$ as a single transformation. For this reason we have considered Γ_0 rather than Γ; it might be pointed out, however, that the generators of Γ_0 obtained in this paper happen to be a set of generators of Γ.

It is the aim of this paper to find the generators of the symplectic modular group. It will be proved here that this group is generated by two or four independent elements, according as $n=1$ or $n>1$. The method used here can be extended so as to give a set of generators for matrices with elements in any Euclidean ring. In particular, we give the details for the generalized Picard group at the end of this paper.

Problems of this type have been considered previously. Poincaré([3]) stated without proof that every matrix $\mathfrak{M}$ with integral elements for which $\mathfrak{M}\mathfrak{S}\mathfrak{M}' = \mathfrak{S}$, where $\mathfrak{S}$ is the direct sum of n two-rowed skew-symmetric matrices, is expressible as a product of elementary matrices of two simple types. Brahana([4]) proved this and extended the result to the case where $\mathfrak{S}$ is any skew-symmetric matrix by showing in this case that every such matrix $\mathfrak{M}$ is ex-

Presented to the Society, February 28, 1948; received by the editors March 6, 1948.

([1]) C. L. Siegel, Math. Ann. vol. 116 (1939) pp. 617–657.

([2]) L. K. Hua, Trans. Amer. Math. Soc. vol. 57 (1945) pp. 441–490.

([3]) H. Poincaré, Rend. Circ. Mat. Palermo vol. 18 (1904) pp. 45–110.

([4]) H. R. Brahana, Ann. of Math. (2) vol. 24 (1923) pp. 265–270.

pressible as a product of matrices taken from some finite set of matrices. From the results given in the present paper, a much stronger form of Brahana's result can be easily deduced.

1. If we set

$$\mathfrak{M} = \begin{pmatrix} A & B \\ C & D \end{pmatrix}, \tag{2}$$

(1) is equivalent to

$$AB' = BA', \qquad CD' = DC', \qquad AD' - BC' = I. \tag{3}$$

By taking inverses of both sides of (1) and using $\mathfrak{F}^{-1} = -\mathfrak{F}$, we can deduce that $\mathfrak{M}'\mathfrak{F}\mathfrak{M} = \mathfrak{F}$, so that

$$A'C = C'A, \qquad B'D = D'B, \qquad A'D - C'B = I. \tag{4}$$

We shall begin by showing in §3 that Γ_0 is generated by the following types of elements:

(I) *Translations*:

$$\mathfrak{T} = \begin{pmatrix} I & S \\ 0 & I \end{pmatrix},$$

where S is symmetric.

(II) *Rotations*:

$$\mathfrak{R} = \begin{pmatrix} U & 0 \\ 0 & U'^{-1} \end{pmatrix},$$

where U is unimodular, that is, abs $U=1$ (where abs U denotes the absolute value of the determinant of U).

(III) *Semi-involutions*:

$$\mathfrak{S} = \begin{pmatrix} J & I - J \\ J - I & J \end{pmatrix}$$

where J is a diagonal matrix whose diagonal elements are 0's and 1's, so that $J^2 = J$ and $(I-J)^2 = I-J$.

It is easily verified that matrices of types I, II and III satisfy (1).

2. In this section we prove two lemmas on matrices.

LEMMA 1. *Let m be a nonzero integer, and let T be an n-rowed symmetric matrix at least one of whose elements is not divisible by m. There exists a symmetric matrix S with integral elements such that*

$$0 < \text{abs}\,(T - mS) < |m|^n. \tag{5}$$

Proof. The lemma is evident for $n=1$. Consider next $n=2$; let $T=(t_{ij})$, $S=(s_{ij})$. Then

$$\text{abs } (T - mS) = \left| (t_{11} - ms_{11})(t_{22} - ms_{22}) - (t_{12} - ms_{12})^2 \right|. \tag{6}$$

If m divides both t_{11} and t_{22}, it cannot divide t_{12}; we can then choose S so that $t_{11}-ms_{11}=t_{22}-ms_{22}=0$ and $0<|t_{12}-ms_{12}|<|m|$. Suppose on the other hand that m does not divide one of the diagonal elements, say t_{11}. Fix s_{12} arbitrarily, and choose s_{11} so that $0<|t_{11}-ms_{11}|<|m|$. Since (6) can be written as

$$\text{abs } (T - mS) = \left| - m(t_{11} - ms_{11})s_{22} + \cdots \right|,$$

where $\cdots$ represents terms not involving s_{22}, we can choose an integer s_{22} by the Euclidean algorithm so that

$$0 < \text{abs } (T - mS) \leqq \left| m(t_{11} - ms_{11}) \right| < |m|^2.$$

Suppose now that the result has been established for $n=r-1$ with $r\geqq 3$; we shall deduce it for $n=r$. Let $T=T^{(r)}$ and let some element t_{ij} of T be not divisible by m. Since $r\geqq 3$, there exists a diagonal element t_{kk} of T which is not in the same row or column as t_{ij}. Let T_1 be the symmetric $(r-1)$-rowed matrix obtained from T by omitting the kth row and kth column; let S_1 be similarly related to S. By the induction hypothesis, we may choose S_1 symmetric so that

$$0 < \text{abs } (T_1 - mS_1) < |m|^{r-1}. \tag{7}$$

However, we have

$$\text{abs } (T - mS) = \left| (t_{kk} - ms_{kk}) \det (T_1 - mS_1) + \cdots \right|, \tag{8}$$

where $\cdots$ represents terms not involving s_{kk}. Choose s_{lk} arbitrarily for $l=1, 2, \cdots, k-1, k+1, \cdots, r$. Then by the Euclidean algorithm we can choose s_{kk} so that

$$0 < \text{abs } (T - mS) \leqq |m| \text{ abs } (T_1 - mS_1) < |m|^r.$$

This completes the proof of the lemma.

LEMMA 2. *Let A and B satisfy $AB'=BA'$ and let* det $A\neq 0$. *There exists a symmetric matrix S such that either*

$$B - AS = 0 \tag{9}$$

or

$$0 < \text{abs } (B - AS) < \text{abs } A. \tag{10}$$

Proof. From $AB'=BA'$ and det $A\neq 0$, we may deduce that A^*B is symmetric, where A^* denotes the adjoint of A. We apply Lemma 1 with $T=A^*B$ and $m=\det A$. Either every element of A^*B is divisible by m, in which case there exists a symmetric matrix S with $A^*B=mS$, or else there exist symmetric matrices R and S such that $A^*B=mS+R$ with $0<\text{abs } R<|m|^n$. In virtue of the relation $A^*A=mI$, these alternatives become: either $B=AS$ (in which case (9) holds), or $B-AS=AR/m$; however,

$$\operatorname{abs}\frac{AR}{m} = \frac{(\operatorname{abs} A)(\operatorname{abs} R)}{|m|^n} = \frac{\operatorname{abs} R}{|m|^{n-1}},$$

so that

$$0 < \operatorname{abs}(B - AS) < |m| = \operatorname{abs} A.$$

3. We are now ready to show that Γ_0 is generated by matrices of types I, II and III. Let $\mathfrak{M}$ given by (2) be an arbitrary element of Γ_0. It suffices to prove that by multiplying $\mathfrak{M}$ by matrices of types I, II and III, one obtains a product of matrices of those types.

(3) implies that not both A and B are 0. Since

$$\begin{pmatrix} A & B \\ C & D \end{pmatrix}\begin{pmatrix} 0 & I \\ -I & 0 \end{pmatrix} = \begin{pmatrix} -B & A \\ * & * \end{pmatrix},$$

we may assume that A has rank $r>0$. Furthermore,

$$\begin{pmatrix} U & 0 \\ 0 & U'^{-1} \end{pmatrix}\begin{pmatrix} A & B \\ C & D \end{pmatrix}\begin{pmatrix} V & 0 \\ 0 & V'^{-1} \end{pmatrix} = \begin{pmatrix} UAV & * \\ * & * \end{pmatrix},$$

so that A may be taken to be of the form

$$A = \begin{pmatrix} A_1 & 0 \\ A_2 & 0 \end{pmatrix}, \tag{11}$$

where A_1 is an r-rowed nonsingular matrix. We similarly decompose B as

$$B = \begin{pmatrix} B_1^{(r)} & * \\ * & * \end{pmatrix}.$$

From (3) it is easily seen that $A_1B_1' = B_1A_1'$. By Lemma 2, there exists a symmetric matrix S_1 with either $A_1S_1+B_1=0$ or $0<\operatorname{abs} R_1<\operatorname{abs} A_1$, where $R_1=A_1S_1+B_1$. Define

$$S = \begin{pmatrix} S_1^{(r)} & 0 \\ 0 & 0 \end{pmatrix}.$$

Then

$$\begin{pmatrix} A & B \\ C & D \end{pmatrix}\begin{pmatrix} I & S \\ 0 & I \end{pmatrix} = \begin{pmatrix} A & AS + B \\ * & * \end{pmatrix}, \tag{12}$$

so that A remains unaltered while B_1 of B is replaced by 0 or R_1. If the second alternative occurs, we proceed as follows: let

(13) $$J = \begin{pmatrix} 0 & 0 \\ 0 & I^{(n-r)} \end{pmatrix}.$$

Then

(14) $$\begin{pmatrix} A & B \\ C & D \end{pmatrix}\begin{pmatrix} J & I - J \\ J - I & J \end{pmatrix} = \begin{pmatrix} \bar{A} & * \\ * & * \end{pmatrix},$$

where

$$\bar{A} = AJ - B(I - J) = \begin{pmatrix} -R_1 & 0 \\ * & 0 \end{pmatrix}.$$

We now repeat the process as before, and so on. Since there are only finitely many positive integers less than abs A_1, this process eventually terminates. Thus, by multiplying $\mathfrak{M}$ by matrices of types I, II and III one arrives at a matrix

$$\begin{pmatrix} A_0 & B_0 \\ * & * \end{pmatrix}$$

with

$$A_0 = \begin{pmatrix} R & 0 \\ * & 0 \end{pmatrix}, \qquad B_0 = \begin{pmatrix} 0 & * \\ * & * \end{pmatrix}$$

and det $R \neq 0$. One readily deduces from $A_0 B_0' = B_0 A_0'$ that B_0 must be of the form

$$B_0 = \begin{pmatrix} 0 & * \\ 0 & * \end{pmatrix}.$$

But then

$$\begin{pmatrix} 0 & I \\ -I & 0 \end{pmatrix}\begin{pmatrix} A_0 & B_0 \\ C_0 & D_0 \end{pmatrix}\begin{pmatrix} J & I - J \\ J - I & J \end{pmatrix} = \begin{pmatrix} A^+ & B^+ \\ 0 & D^+ \end{pmatrix}$$

where J is given by (13). Finally we notice that for a matrix

$$\begin{pmatrix} A & B \\ 0 & D \end{pmatrix}$$

of Γ_0, we must have $A = U$ unimodular, $D = U'^{-1}$, and thence from (3), $B = SU'^{-1}$ with symmetric S. Therefore

$$\begin{pmatrix} A & B \\ 0 & D \end{pmatrix} = \begin{pmatrix} I & S \\ 0 & I \end{pmatrix}\begin{pmatrix} U & 0 \\ 0 & U'^{-1} \end{pmatrix}.$$

This completes the proof that Γ_0 is generated by the matrices of types I, II and III.

4. The set of matrices of types I, II and III which generate Γ_0 are certainly not independent generators. Let us reduce the number of generators as much as possible. Since

$$\begin{pmatrix} I & S_1 \\ 0 & I \end{pmatrix}\begin{pmatrix} I & S_2 \\ 0 & I \end{pmatrix} = \begin{pmatrix} I & S_1 + S_2 \\ 0 & I \end{pmatrix},$$

the subgroup formed by matrices of type I is generated by those type I matrices whose S's are given by

$$(15) \qquad S_0 = \begin{bmatrix} 1 & 0 \cdots 0 \\ 0 & 0 \cdots 0 \\ \cdot & \cdot \;\cdot\;\cdot \;\cdot \\ 0 & 0 \cdots 0 \end{bmatrix}, \qquad S_1 = \begin{bmatrix} 0 & 1 & 0 \cdots 0 \\ 1 & 0 & 0 \cdots 0 \\ 0 & 0 & 0 \cdots 0 \\ \cdot & \cdot & \cdot \;\cdot\;\cdot\;\cdot \\ 0 & 0 & 0 \cdots 0 \end{bmatrix}$$

and all matrices obtained from these by interchanging any two rows and the corresponding columns. Next we note that

$$\begin{pmatrix} U & 0 \\ 0 & U'^{-1} \end{pmatrix}\begin{pmatrix} I & S \\ 0 & I \end{pmatrix}\begin{pmatrix} U^{-1} & 0 \\ 0 & U' \end{pmatrix} = \begin{pmatrix} I & USU' \\ 0 & I \end{pmatrix},$$

so that the group generated by matrices of types I and II is the same as that generated by all type II matrices and the two translations whose S's are given by (15). However, we have

$$\begin{pmatrix} 1 & -1 \\ 0 & 1 \end{pmatrix}\begin{pmatrix} 1 & 0 \\ 0 & -1 \end{pmatrix}\begin{pmatrix} 1 & 0 \\ -1 & 1 \end{pmatrix} = \begin{pmatrix} 0 & 1 \\ 1 & -1 \end{pmatrix} = \begin{pmatrix} 0 & 1 \\ 1 & 0 \end{pmatrix} - \begin{pmatrix} 0 & 0 \\ 0 & 1 \end{pmatrix}.$$

Hence the translation with S_1 is obtainable from that with S_0 and the matrices of type II. Therefore Γ_0 is generated by the matrix

$$(16) \qquad \mathfrak{T}_0 = \begin{pmatrix} I & S_0 \\ 0 & I \end{pmatrix}$$

with S_0 given by (15), and all matrices of types II and III.

Since

$$\begin{pmatrix} U & 0 \\ 0 & U'^{-1} \end{pmatrix}\begin{pmatrix} V & 0 \\ 0 & V'^{-1} \end{pmatrix} = \begin{pmatrix} UV & 0 \\ 0 & (UV)'^{-1} \end{pmatrix},$$

in order to find the generators of the subgroup of rotations we have merely to find the generators of the group of unimodular matrices. These are given by the following theorem.

THEOREM 1. *Let $n \geqq 2$. Every unimodular matrix with rational integral elements is a product of the matrices U_1, U_2, U_3 and their inverses, where*

$$(17) \qquad U_1 = \begin{pmatrix} 0 & 0 & \cdots & 0 & 1 \\ 1 & 0 & \cdots & 0 & 0 \\ \cdot & \cdot & \cdot & \cdot & \cdot \\ 0 & 0 & \cdots & 0 & 0 \\ 0 & 0 & \cdots & 1 & 0 \end{pmatrix}, \qquad U_2 = \begin{pmatrix} 1 & 1 & \cdots & 0 & 0 \\ 0 & 1 & \cdots & 0 & 0 \\ \cdot & \cdot & \cdot & \cdot & \cdot \\ 0 & 0 & \cdots & 1 & 0 \\ 0 & 0 & \cdots & 0 & 1 \end{pmatrix},$$

$$U_3 = \begin{pmatrix} -1 & 0 & \cdots & 0 & 0 \\ 0 & 1 & \cdots & 0 & 0 \\ \cdot & \cdot & \cdot & \cdot & \cdot \\ 0 & 0 & \cdots & 1 & 0 \\ 0 & 0 & \cdots & 0 & 1 \end{pmatrix}.$$

Proof. It is known(5) that every unimodular matrix is a product of U_1, U_2, U_3, and

$$U_4 = = \begin{pmatrix} 0 & 1 & \cdots & 0 & 0 \\ 1 & 0 & \cdots & 0 & 0 \\ \cdot & \cdot & \cdot & \cdot & \cdot \\ 0 & 0 & \cdots & 1 & 0 \\ 0 & 0 & \cdots & 0 & 1 \end{pmatrix},$$

and their inverses. It is sufficient to show that U_4 is expressible as a product of U_1, U_2, U_3 and their inverses. We define $T = U_2U_1$ for the remainder of this proof, and let $\mathfrak{r} = (r_1, \cdots, r_n)'$ be a column vector. Then

$$T\mathfrak{r} = \begin{pmatrix} r_n + r_1 \\ r_1 \\ \cdot \\ \cdot \\ \cdot \\ r_{n-1} \end{pmatrix},$$

$$T^2\mathfrak{r} = \begin{pmatrix} r_{n-1} + r_n + r_1 \\ r_n + r_1 \\ r_1 \\ \cdot \\ \cdot \\ \cdot \\ r_{n-2} \end{pmatrix}, \cdots, T^{n-1}\mathfrak{r} = \begin{pmatrix} r_2 + r_3 + \cdots + r_n + r_1 \\ r_3 + \cdots + r_n + r_1 \\ \cdot \\ \cdot \\ \cdot \\ r_n + r_1 \\ r_1 \end{pmatrix}.$$

---

(5) See for example, C. C. MacDuffee, *The theory of matrices*, Berlin, 1933, p. 34, Theorem 22.5.

Therefore

$$U_1^{-1}T^{n-1}\mathfrak{r} = \begin{pmatrix} r_3 + \cdots + r_n + r_1 \\ \vdots \\ r_n + r_1 \\ r_1 \\ r_2 + r_3 + \cdots + r_n + r_1 \end{pmatrix},$$

so that

$$(T^{-1})^{n-2}U_1^{-1}T^{n-1}\mathfrak{r} = \begin{pmatrix} r_1 \\ r_2 + \cdots + r_n + r_1 \\ r_3 \\ \vdots \\ r_n \end{pmatrix},$$

$$U_1(T^{-1})^{n-2}U_1^{-1}T^{n-1}\mathfrak{r} = \begin{pmatrix} r_n \\ r_1 \\ r_2 + \cdots + r_n + r_1 \\ r_3 \\ \vdots \\ r_{n-1} \end{pmatrix}.$$

From this we see that

$$T^{n-3}U_1(T^{-1})^{n-2}U_1^{-1}T^{n-1}\mathfrak{r} = \begin{pmatrix} r_3 + r_4 + \cdots + r_n \\ r_4 + \cdots + r_n \\ \vdots \\ r_n \\ r_1 \\ r_2 + \cdots + r_n + r_1 \end{pmatrix}$$

and

$$(T^{-1})^{n-2}U_1T^{n-3}U_1(T^{-1})^{n-2}U_1^{-1}T^{n-1}\mathfrak{r} = \begin{pmatrix} r_n \\ r_1 \\ r_2 + r_1 \\ \vdots \\ r_{n-1} \end{pmatrix}.$$

Define

$$U\dagger = U_1^{-1}(T^{-1})^{n-2}U_1T^{n-3}U_1(T^{-1})^{n-2}U_1^{-1}T^{n-1}.$$

Then

$$U\dagger = \begin{pmatrix} 1 & 0 \\ 1 & 1 \end{pmatrix} \dotplus I^{(n-2)},$$

where $\dotplus$ denotes the direct sum of matrices. But from

$$U_3 = \begin{pmatrix} -1 & 0 \\ 0 & 1 \end{pmatrix} \dotplus I^{(n-2)} \quad \text{and} \quad U_2^{-1} = \begin{pmatrix} 1 & -1 \\ 0 & 1 \end{pmatrix} \dotplus I^{(n-2)}$$

we deduce

$$\begin{aligned} U_3U\dagger U_2^{-1}U\dagger &= \begin{pmatrix} -1 & 0 \\ 0 & 1 \end{pmatrix}\begin{pmatrix} 1 & 0 \\ 1 & 1 \end{pmatrix}\begin{pmatrix} 1 & -1 \\ 0 & 1 \end{pmatrix}\begin{pmatrix} 1 & 0 \\ 1 & 1 \end{pmatrix} \dotplus I^{(n-2)} \\ &= \begin{pmatrix} 0 & 1 \\ 1 & 0 \end{pmatrix} \dotplus I^{(n-2)} = U_4. \end{aligned}$$

This completes the proof of the theorem.

COROLLARY. *Let $n \geqq 2$. Every unimodular matrix with rational integral elements of determinant $+1$ is a product of powers of U_2 and*

$$U_5 = \begin{Bmatrix} 0 \cdots 0 & (-1)^{n-1} \\ 1 \cdots 0 & 0 \\ \cdots\cdots & \cdot \\ 0 \cdots 0 & 0 \\ 0 \cdots 1 & 0 \end{Bmatrix}.$$

By Theorem 1 we see now that Γ_0 is generated by $\mathfrak{T}_0$ and the set of all semi-involutions and the three rotations defined by

$$\mathfrak{R}_i = \begin{pmatrix} U_i & 0 \\ 0 & U_i'^{-1} \end{pmatrix}, \qquad i = 1, 2, 3. \tag{18}$$

We finally consider type III matrices. Let J_r be the diagonal matrix obtained from the identity matrix by replacing the rth 1 by 0. In that case, if $r \neq s$, we have

$$\begin{pmatrix} J_r & I - J_r \\ J_r - I & J_r \end{pmatrix}\begin{pmatrix} J_s & I - J_s \\ J_s - I & J_s \end{pmatrix} = \begin{pmatrix} J_{rs} & I - J_{rs} \\ J_{rs} - I & J_{rs} \end{pmatrix},$$

where J_{rs} is obtained from the identity matrix by replacing the rth and sth ones by 0's. Therefore, in order to obtain all type III matrices, we need only those semi-involutions

$$\begin{pmatrix} J_r & I - J_r \\ J_r - I & J_r \end{pmatrix}, \qquad r = 1, 2, \cdots, n, \tag{19}$$

with J_r defined above. Now, let U be that unimodular matrix obtained from I by interchanging the 1st and rth rows; then we have

$$\begin{pmatrix} U & 0 \\ 0 & U'^{-1} \end{pmatrix}\begin{pmatrix} J_r & I - J_r \\ J_r - I & J_r \end{pmatrix}\begin{pmatrix} U^{-1} & 0 \\ 0 & U' \end{pmatrix} = \begin{pmatrix} J_1 & I - J_1 \\ J_1 - I & J_1 \end{pmatrix}.$$

Therefore Γ_0 is generated by the matrices $\mathfrak{T}_0$, $\mathfrak{R}_i$ $(i=1, 2, 3)$ and the matrix

$$\mathfrak{S}_0 = \begin{pmatrix} J_1 & I - J_1 \\ J_1 - I & J_1 \end{pmatrix}, \tag{20}$$

with J, previously defined. But

$$\mathfrak{S}_0^2 = \mathfrak{R}_3,$$

so that $\mathfrak{R}_3$ may be dropped from the list of generators. Therefore we have the following theorem.

THEOREM 2. *Γ_0 is generated by the four matrices $\mathfrak{T}_0$, $\mathfrak{R}_1$, $\mathfrak{R}_2$ and $\mathfrak{S}_0$ given by* (15), (18) *and* (20), *for $n>1$. For $n=1$, Γ_0 is generated by $\mathfrak{T}_0$ and $\mathfrak{S}_0$.*

5. In this section we shall prove the independence of the generators given in Theorem 2. For $n=1$, this is trivial because $\mathfrak{S}_0$ is of finite order while $\mathfrak{T}_0$ is not. Hereafter we suppose that $n>1$.

(1) *Independence of* $\mathfrak{T}_0$. We consider the transformation

$$(X_1, Y_1) = (X, Y)\mathfrak{M}; \tag{21}$$

if XY' is symmetric, it is easily verified that X_1Y_1' is also symmetric. We shall show that if the diagonal elements of XY' are even, those of X_1Y_1' will also be even if $\mathfrak{M}$ is $\mathfrak{R}_1$, $\mathfrak{R}_2$ or $\mathfrak{S}_0$, while if $\mathfrak{M}=\mathfrak{T}_0$, it is possible to choose X and Y so that some diagonal element of X_1Y_1' is odd. This will show that $\mathfrak{T}_0$ is not expressible as a product of $\mathfrak{R}_1$, $\mathfrak{R}_2$ and $\mathfrak{S}_0$ and their inverses.

Assume now that the diagonal elements of XY' are even. From (21) one readily deduces that if $\mathfrak{M}$ is a rotation, $X_1Y_1' = XY'$, so that the diagonal elements of X_1Y_1' are also even. If secondly $\mathfrak{M}$ is a semi-involution, we have

$$X_1 = XJ + Y(I - J), \qquad Y_1 = - X(I - J) + YJ,$$

so that

$$X_1Y_1' = XJY' - Y(I - J)X' = XJY' + YJX' - YX'.$$

Since XJY' is the transpose of YJX', it is again clear that the diagonal elements of X_1Y_1' are even. Finally, suppose $\mathfrak{M}=\mathfrak{T}_0$. Then we obtain

$$X_1Y_1' = X(XS_0+Y)' = XY' + XS_0X'$$

and for $X=I$ the first diagonal element of X_1Y_1' is odd. This completes the proof of the independence of $\mathfrak{T}_0$. We may remark in passing, however, that $\mathfrak{T}_0^2$ is expressible as a product of the powers of $\mathfrak{R}_1$, $\mathfrak{R}_2$ and $\mathfrak{S}_0$.

(2) *Independence of* $\mathfrak{R}_1$. Let $\mathfrak{r}=(r_1, \cdots, r_n, s_1, \cdots, s_n)'$ be a column vector with $2n$ components. It is clear that the second component r_2 is unaffected when $\mathfrak{r}$ is multiplied on the left by any of the matrices $\mathfrak{T}_0$, $\mathfrak{R}_2$, and $\mathfrak{S}_0$ and their inverses. Under multiplication on the left by $\mathfrak{R}_1$, however, r_2 is replaced by r_1. Hence $\mathfrak{R}_1$ cannot be expressed as a product of $\mathfrak{T}_0$, $\mathfrak{R}_2$ and $\mathfrak{S}_0$ and their inverses.

(3) *Independence of* $\mathfrak{R}_2$. Multiplying $\mathfrak{r}$ on the left by $\mathfrak{R}_1$ or $\mathfrak{S}_0$ or their inverses permutes components of $\mathfrak{r}$; under any such permutation, however, any r_i and its corresponding s_i remain n components apart. Since the effect of multiplying on the left by $\mathfrak{T}_0$ is to replace r_1 by r_1+s_1, it is clear that by multiplying $\mathfrak{r}$ on the left by a product of $\mathfrak{R}_1$, $\mathfrak{S}_0$ and $\mathfrak{T}_0$ and their inverses, r_1 may be replaced by a linear combination of r_1 and s_1 and its position may be changed. It is however impossible to replace r_1 by r_1+r_2 in this way, and this is exactly the effect of multiplication of $\mathfrak{r}$ on the left by $\mathfrak{R}_2$. This proves the independence of $\mathfrak{R}_2$.

(4) *Independence of* $\mathfrak{S}_0$. We note that

$$\begin{pmatrix} * & * \\ 0 & * \end{pmatrix}\begin{pmatrix} * & * \\ 0 & * \end{pmatrix} = \begin{pmatrix} * & * \\ 0 & * \end{pmatrix}.$$

Since $\mathfrak{T}_0$, $\mathfrak{R}_1$ and $\mathfrak{R}_2$ and their inverses are all of the form

$$\begin{pmatrix} * & * \\ 0 & * \end{pmatrix}$$

and $\mathfrak{S}_0$ is not of this form, it is clear that $\mathfrak{S}_0$ is not expressible as a product of $\mathfrak{T}_0$, $\mathfrak{R}_1$ and $\mathfrak{R}_2$ and their inverses.

6. Our previous method can be extended to any Euclidean ring; in particular, for the ring formed by the Gaussian integers, we have the following result:

THEOREM 3. *Let* Γ' *be the group of matrices* $\mathfrak{M}$ *with Gaussian integers as elements which satisfy* (1). *Let* Γ_0' *be obtained from* Γ' *by identifying the four elements* $\pm\mathfrak{M}$ *and* $\pm i\mathfrak{M}$. *Then for* $n>1$, Γ_0' *is generated by the matrices* $\mathfrak{T}_0$, $\mathfrak{R}_1$, $\mathfrak{R}_2$ *and* $\mathfrak{S}_0$ *defined previously, and the matrix*

$$\mathfrak{T}_1 = \begin{pmatrix} I & S_1 \\ 0 & I \end{pmatrix} \qquad \textit{where } S_1 = iS_0. \tag{22}$$

For $n=1$, Γ_0' *is generated by* $\mathfrak{T}_0$, $\mathfrak{T}_1$ *and* $\mathfrak{S}_0$.

The independence of the generators is shown as follows (with suitable modifications when $n=1$):

(1) *Independence of* $\mathfrak{T}_0$. We use the method of §5, (1). Let XY' be a symmetric matrix with Gaussian integers as elements, such that the real part of each diagonal element is even. This property is preserved when (X, Y) is subjected to the transformations $\mathfrak{T}_1$, $\mathfrak{R}_1$, $\mathfrak{R}_2$ and $\mathfrak{S}_0$ according to (21), but not for the transformation $\mathfrak{T}_0$.

(2) *Independence of* $\mathfrak{T}_1$. This is clear since $\mathfrak{T}_1$ is the only generator which is not real.

(3) The independence of $\mathfrak{R}_1$, $\mathfrak{R}_2$ and $\mathfrak{S}_0$ follow exactly as before.

TSING HUA UNIVERSITY,
PEIPING, CHINA.
INSTITUTE FOR ADVANCED STUDY,
PRINCETON, N. J.

ANNALS OF MATHEMATICS
Vol. 50, No. 1, January, 1949

GEOMETRY OF SYMMETRIC MATRICES OVER ANY FIELD WITH CHARACTERISTIC OTHER THAN TWO

BY LOO-KENG HUA

(Received February 6, 1947, Revised October 7, 1947.)

1. Introduction

It is the aim of this paper to establish a generalization of von Staudt's theorem for the geometry of symmetric matrices over any field with characteristic $\neq 2$. Before stating the theorem explicitly, we explain some notations which will be used throughout the paper.

Let Φ be a field with characteristic $\neq 2$. We use capital Latin letters to denote n-rowed matrices over Φ, and I and 0 to denote the n-rowed identity and zero matrices respectively. For matrices which are not $n \times n$, we use $M^{(l,m)}$ to denote an $l \times m$ matrix and $M^{(m)} = M^{(m,m)}$. M' denotes the transposed matrix of M. We use, also, small Greek letters σ and τ as superscripts to denote automorphisms of the field Φ, that is, for any two elements a and b of Φ, we have

$$(a + b)^\sigma = a^\sigma + b^\sigma \qquad (ab)^\sigma = a^\sigma b^\sigma$$

Let

$$\mathfrak{F} = \begin{pmatrix} 0 & I \\ -I & 0 \end{pmatrix} \qquad \mathfrak{J} = \begin{pmatrix} I & 0 \\ 0 & I \end{pmatrix}.$$

A pair of matrices (X, Y) is called nonsingular and symmetric, if (X, Y) is of rank n and

$$(1) \qquad (X, Y)\mathfrak{F}(X, Y)' = 0.$$

We shall say that two nonsingular symmetric pairs (X, Y) and (X_1, Y_1) belong to the same class, if and only if there exists a nonsingular Q such that

$$(2) \qquad (X_1, Y_1) = Q(X, Y)$$

holds. A class of nonsingular symmetric pairs of matrices will be called a point $P = \{(X, Y)\}$. The totality of these points forms *a projective space of symmetric matrices*.

The rank of

$$< P, P_1 > = (X, Y)\mathfrak{F}(X_1, Y_1)'$$

is called the *arithmetic distance* between the two points P and P_1 represented by (X, Y) and (X_1, Y_1) respectively. It is denoted by $r(P, P_1)$. Evidently this notion is independent of the particular choice of the elements from the same class. It is also obvious that two points coincide if and only if their arithmetic distance is zero.

Our problem is to find all the transformations carrying the projective space of symmetric matrices onto itself and keeping arithmetic distance invariant.

Evidently, the symplectic transformations

$$(3) \qquad (X_1, Y_1) = Q(X, Y)\mathfrak{T}, \qquad \mathfrak{T}\mathfrak{F}\mathfrak{T}' = \mathfrak{F},$$

satisfy our requirement. But this does not tell the whole truth; in fact, the transformations

$$(4) \qquad (X_1, Y_1) = Q(X^\sigma, Y^\sigma)$$

also satisfy our requirement.

In order to find the full truth, we have to extend the notion of the symplectic group. We consider the transformation

$$(5) \qquad (X_1, Y_1) = Q(X, Y)\mathfrak{T}, \qquad \mathfrak{T}\mathfrak{F}\mathfrak{T}' = a\mathfrak{F},$$

where a is an element $\neq 0$. Further, since

$$(b\mathfrak{T})\mathfrak{F}(b\mathfrak{T})' = b^2 a\mathfrak{F}$$

and

$$(X_1, Y_1) = Qb^{-1}(X, Y)b\mathfrak{T},$$

only those a which represent different classes of the factor group of the multiplicative group of Φ modulo the multiplicative group Φ of all square elements of Φ are essentially taken into consideration. The transformations (5) are called extended symplectic transformations, and they form a group to be called the extended symplectic group. It is apparent that the factor group of the extended symplectic group over the symplectic group is isomorphic to the factor group Φ/Φ_2.

It will be proved in this paper that, for $n > 1$,

Every transformation carrying the projective space of symmetric matrices of order $n > 1$ onto itself and keeping arithmetic distance invariant is of the form

$$(6) \qquad (X_1, Y_1) = Q(X_1^\sigma, Y_1^\sigma)\mathfrak{T}$$

where Q is a nonsingular $n \times n$ matrix, σ is an automorphism of Φ, $\mathfrak{T}$ is a $2n \times 2n$ matrix satisfying

$$\mathfrak{T}\mathfrak{F}\mathfrak{T}' = a\mathfrak{F},$$

and a runs over a complete residue system of Φ/Φ_2.

In nonhomogeneous coordinates $(Z = Y^{-1}X)$, (6) takes the form

$$(7) \qquad Z_1 = a(AZ^\sigma + B)(CZ^\sigma + D)^{-1},$$

where $AB' = BA'$, $CD' = DC'$, $AD' - BC' = I$.

If, in particular, Φ is the real field, then it has no automorphism other than identity, and $+1$ and -1 are the representatives of the factor group Φ/Φ_2. Therefore we deduce that any transformation carrying the projective space of real symmetric matrices of order $n > 1$ onto itself and keeping arithmetic distance invariant is of the form

$$Z_1 = (AZ + B)(CZ + D)^{-1},$$

where $AB' = BA'$, $CD' = DC'$, $AD' - BC' = \pm I$.

A comparison with the theorem stated in I^1 reveals that we have eliminated the conditions on "continuity," "sense" and "harmonic separation." This is the final goal of our algebraic treatment.

Notice that, for $n = 1$, the theorem is no longer true. An additional condition concerning harmonic separation is needed. For a field of characteristic 2, the theorem is not true for $n = 1$ even with the condition on harmonic separation.

Before proceeding to the text, it may be worthwhile to mention several connections of the theorem with various fields of mathematics.

Algebraists may have some interest in the following theorem which is suggested by the "affine geometry" of symmetric matrices. Consider only the "finite" points (X, Y); that is, pairs for which Y is nonsingular. They can be represented in nonhomogeneous coordinates $Z(=Y^{-1}X)$. The arithmetic distance between two points Z and Z_1 is simply the rank of the difference $Z - Z_1$. Then we have

Let Σ be the set of all symmetric matrices of order $n \geqq 2$ over the field Φ with characteristic $\neq 2$. Any mapping of Σ onto itself leaving the rank of the difference of two matrices invariant is of the form

$$Z_1 = aAZ^{\sigma}A' + S, \qquad S = S', \tag{8}$$

where $a \in \Phi$ and σ is an automorphism of Φ.

This theorem assures the algebraists that their study of symmetric matrices under the congruence group is really a very general one. We shall proceed to prove this theorem by purely algebraic methods, and then we shall deduce our fundamental theorem of the projective geometry of symmetric matrices from it.

From the geometrical point of view, the case $n = 2$ is very interesting. Let (x_1, x_2, x_3, x_4) and (y_1, y_2, y_3, y_4) be two distinct points of ordinary projective 3-space. We use

$$(X, Y) = \begin{pmatrix} x_1 & x_2 & x_3 & x_4 \\ y_1 & y_2 & y_3 & y_4 \end{pmatrix}$$

to denote the line passing through both points. Two pairs of matrices (X, Y) and (X_1, Y_1) represent the same line if and only if we have a nonsingular matrix Q such that

$$(X, Y) = Q(X_1, Y_1).$$

[1] *Geometrics of Matrices*, Trans., Amer. Math. Soc. 57 (1945) 441–481. In particular, p. 459. The papers under the same title published in Trans. of Amer. Math. Soc. 57 (1945), 441–481; 61 (1947), 193–228, 229–255 will be referred as I, II and III respectively. Compare also C. R. (Doklady) Acad. Sci. URSS (N.S.) 53, 95–97 (1946).

The lines (X, Y) which satisfy

$$(X, Y)\mathfrak{F}(X, Y)' = 0$$

form a line complex. Thus a point of the projective space of 2-rowed symmetric matrices denotes a line on the complex. For two lines (X, Y) and (X_1, Y_r) the rank of $(X, Y)\mathfrak{F}(X_1, Y_1)'$ is either 0 or 1 or 2; its significance is that the lines coincide, intersect, or do not intersect, respectively. Therefore the present statement in its most specific form gives us a rational treatment of the line geometry on a line complex. It seems also to be new. Notice that the line geometry on a line complex is identical with Lie's geometry of circles and that Laguerre's geometry of circles is the "affine geometry" of 2-rowed symmetric matrices. This indicates still other applications of both the homogeneous and nonhomogeneous forms of our theorem.

2. Subspaces

In §§2–6, we consider only the finite points of the projective space of symmetric matrices. Thus there will be no confusion if we use a symmetric matrix to denote a point. Throughout §§2–4, we assume that $n = 2$.

Let S and T be two symmetric matrices with arithmetic distance 1, that is, $S - T$ is of rank 1. The set of symmetric matrices P satisfying

$$r(P, S) \leqq 1, \qquad r(P, T) \leqq 1$$

forms a *subspace spanned by S and T* which will be denoted by $\Pi(S, T)$.

We have a transformation of the form (8) carrying S and T into

$$0, \qquad \begin{pmatrix} x_0 & 0 \\ 0 & 0 \end{pmatrix}, \qquad (x_0 \neq 0),$$

respectively. Thus $\Pi(S, T)$ is equivalent under (8) to $\Pi\left(0, \begin{pmatrix} x_0 & 0 \\ 0 & 0 \end{pmatrix}\right)$. Let

$$P = \begin{pmatrix} x & y \\ y & z \end{pmatrix}$$

be a point of $\Pi\left(0, \begin{pmatrix} x_0 & 0 \\ 0 & 0 \end{pmatrix}\right)$, then

$$r(0, P) \leqq 1, \qquad r\left(\begin{pmatrix} x_0 & 0 \\ 0 & 0 \end{pmatrix}, P\right) \leqq 1$$

implies

$$xz - y^2 = (x - x_0)z - y^2 = 0.$$

Consequently we have $z = 0$ and $y = 0$, that is, $\Pi\left(0, \begin{pmatrix} x_0 & 0 \\ 0 & 0 \end{pmatrix}\right)$ consists of all the matrices of the form

$$\begin{pmatrix} x & 0 \\ 0 & 0 \end{pmatrix}. \tag{9}$$

Therefore every subspace can be carried by a transformation of the form (8) *into* $\Pi\left(0, \begin{pmatrix} 1 & 0 \\ 0 & 0 \end{pmatrix}\right)$ *which has the explicit form given by* (9).

Two subspaces are said to be *complemented* if they have one and only one point in common.

THEOREM 1. *Every pair of complemented subspaces can be carried simultaneously into*

$$\Pi\left(0, \begin{pmatrix} 1 & 0 \\ 0 & 0 \end{pmatrix}\right), \qquad \Pi\left(0, \begin{pmatrix} 0 & 0 \\ 0 & 1 \end{pmatrix}\right)$$

by a transformation of the form (8).

PROOF. We may suppose that one of the subspaces is $\Pi\left(0, \begin{pmatrix} 1 & 0 \\ 0 & 0 \end{pmatrix}\right)$ and that the common point is 0. Let U be a point in the other subspace different from 0. Then, since $r(0, U) = 1$, we have

$$U = \epsilon \begin{pmatrix} a_2 & ab \\ ab & b^2 \end{pmatrix}, \qquad b \neq 0.$$

The transformation

$$Z_1 = \begin{pmatrix} 1 & 1 \\ 0 & -ab^{-1} \end{pmatrix} Z \begin{pmatrix} 1 & -ab^{-1} \\ 0 & 1 \end{pmatrix}^{1}$$

carries $\Pi\left(0, \begin{pmatrix} 1 & 0 \\ 0 & 0 \end{pmatrix}\right)$ onto itself and carries U into

$$\begin{pmatrix} 0 & 0 \\ 0 & u \end{pmatrix}, \qquad u \neq 0.$$

The space spanned by 0 and $\begin{pmatrix} 0 & 0 \\ 0 & u \end{pmatrix}$ is $\Pi\left(0, \begin{pmatrix} 0 & 0 \\ 0 & 1 \end{pmatrix}\right)$, which proves the theorem.

Consider a point Z, outside a pair of complemented subspaces Π_1 and Π_2, whose arithmetic distance from the common point of Π_1 and Π_2 is 2. Suppose that Z is so chosen that there is a point P_1 of Π_1 satisfying

$$r(Z, P_1) = 1$$

and that for each point Q of $\Pi(P_1, Z)$, there exists a point P_2 of Π_2 such that

$r(P_2, Q) = 1$. As Q runs over all elements of $\Pi(P_1, Z)$, the set of all points in the subspaces $\Pi(P_2, Q)$ form what we shall call a *reducible subspace* Σ. Π_1 and Π_2 will be called the *components* of Σ.

THEOREM 2. *Every reducible subspace can be carried by a transformation of* (8) *into the set of matrices*

$$\begin{pmatrix} x & 0 \\ 0 & z \end{pmatrix} \tag{10}$$

where x and z run over all elements of the field.

To prove the theorem, we may assume that

$$\Pi_1 = \Pi\left(0, \begin{pmatrix} 1 & 0 \\ 0 & 0 \end{pmatrix}\right), \qquad \Pi_2 = \Pi\left(0, \begin{pmatrix} 0 & 0 \\ 0 & 1 \end{pmatrix}\right).$$

Then,

$$Z = \begin{pmatrix} a & b \\ b & c \end{pmatrix}, \qquad ac - b^2 \neq 0.$$

To ensure the existence of P_1, we have to assume that $c \neq 0$ in which case

$$P_1 = \begin{pmatrix} a - b^2/c & 0 \\ 0 & 0 \end{pmatrix}.$$

Then $\Pi(P_1, Z)$ is the set of points:

$$Q = (1 - t)\begin{pmatrix} a - b^2/c & 0 \\ 0 & 0 \end{pmatrix} + t\begin{pmatrix} a & b \\ b & c \end{pmatrix}$$

where t runs over all elements of Φ.

The existence of P_2 in Π_2 asserts that

$$(1 - t)\left(a - \frac{b^2}{c}\right) + t + a \neq 0$$

i.e.

$$a - \frac{b^2}{c} + t\,\frac{b^2}{c} \neq 0$$

for all t. Consequently $b = 0$, $ac \neq 0$ and

$$P_2 = \begin{pmatrix} 0 & 0 \\ 0 & tc \end{pmatrix}.$$

Now Q takes the form

$$\begin{pmatrix} a & 0 \\ 0 & tc \end{pmatrix},$$

and the space spanned by Q and P_2 consists of the points

$$\begin{pmatrix} sa & 0 \\ 0 & tc \end{pmatrix}.$$

This proves the theorem.

3. Construction of involutions

We begin by defining an involution. Let Σ be a reducible subspace and Z a point not on Σ. Let Δ be the set of elements X on Σ for which

$$r(X, Z) \leqq 1.$$

Then there is one and only one symmetric matrix $\bar{Z}$, different from Z, such that

$$r(X, \bar{Z}) \leqq 1$$

for all X belonging to Δ. This defines $\bar{Z}$ for every Z not on Σ. If Z is on Σ, let $Z = \bar{Z}$. The mapping $Z \to \bar{Z}$ is called *an involution of the first kind.*

If Σ has the form (10) and

$$Z = \begin{pmatrix} a & b \\ b & c \end{pmatrix},$$

then

$$r\left(Z, \begin{pmatrix} x & 0 \\ 0 & z \end{pmatrix}\right) \leqq 1$$

implies

(11) $$(a - x)(c - z) - b^2 = 0.$$

That is, Δ is the set of elements satisfying (11). Let

$$\bar{Z} = \begin{pmatrix} \bar{a} & \bar{b} \\ \bar{b} & \bar{c} \end{pmatrix}.$$

then (11) implies

$$(\bar{a} - x)(\bar{c} - z) - \bar{b}^2 = 0.$$

Consequently $a = \bar{a}$, $c = \bar{c}$ and $b^2 = \bar{b}^2$. Then $\bar{b} = -b$.

Analytically, the mapping can therefore be expressed in the form

(12) $$\bar{Z} = \begin{pmatrix} 1 & 0 \\ 0 & -1 \end{pmatrix} Z \begin{pmatrix} 1 & 0 \\ 0 & -1 \end{pmatrix}.$$

THEOREM 3. *The mapping $Z \to \bar{Z}$ is an involution of the first kind if and only if*

(13) $$\bar{Z} = AZA' + S, \qquad A^2 = I, \qquad ASA' + S = 0.$$

Using (12), let

$$Z = BWB' + T$$

and

$$\bar{Z} = B\bar{W}B' + T.$$

Then

$$\begin{aligned}\bar{W} &= B^{-1}\begin{pmatrix}1 & 0\\ 0 & -1\end{pmatrix} BWB' \begin{pmatrix}1 & 0\\ 0 & -1\end{pmatrix} B'^{-1} \\ &\quad + B^{-1}\left(\begin{pmatrix}1 & 0\\ 0 & -1\end{pmatrix} T \begin{pmatrix}1 & 0\\ 0 & -1\end{pmatrix} - T\right) B'^{-1},\end{aligned} \tag{14}$$

which is evidently of the form (13).

Conversely, since $A^2 = I$, we can find B such that

$$BAB^{-1} = \begin{pmatrix}1 & 0\\ 0 & -1\end{pmatrix}$$

(or $\pm I$, but these cases give only the identity mapping). Further

$$S = \frac{1}{2}(S - ASA') = B^{-1}\left(\begin{pmatrix}1 & 0\\ 0 & -1\end{pmatrix} T \begin{pmatrix}1 & 0\\ 0 & -1\end{pmatrix} - T\right) B'^{-1}$$

where

$$T = -\tfrac{1}{2}BSB'.$$

That is, (13) can be expressed as (14), which proves the theorem.

Theorem 3 shows that an involution of the first kind is uniquely determined by a reducible subspace and that every point of this reducible space is invariant under the involution.

Two different involutions of the first kind are said to be commutative if one carries the reducible subspace of the other onto itself.

Theorem 4. *Any pair of commutative involutions of the first kind can be carried simultaneously into* (12) *and*

$$\bar{Z} = \begin{pmatrix}0 & 1\\ 1 & 0\end{pmatrix} Z \begin{pmatrix}0 & 1\\ 1 & 0\end{pmatrix}. \tag{15}$$

Proof. Let one of the involutions be (12). The other carries the reducible subspace

$$\begin{pmatrix}x & 0\\ 0 & z\end{pmatrix}$$

onto itself. Let

$$A = \begin{pmatrix} a_1 & a_2 \\ a_3 & a_4 \end{pmatrix}, \qquad S = \begin{pmatrix} s_1 & s_2 \\ s_2 & s_3 \end{pmatrix}$$

in (13) and $Z = \begin{pmatrix} x & y \\ y & z \end{pmatrix}$. Then we have

$$\bar{y} = a_1a_3x + (a_2a_3 + a_1a_4)y + a_2a_4z + s_2 = 0$$

for $y = 0$ and arbitrary x and z. Thus

$$a_1a_3 = a_2a_4 = s_2 = 0.$$

Since A is nonsingular, we have either

$$a_1 = a_4 = 0$$

or

$$a_2 = a_3 = 0.$$

The second case cannot happen, as otherwise the second involution coincides with the first. Thus we have $a_1 = a_4 = 0$, and

$$\bar{Z} = \begin{pmatrix} 0 & b \\ 1/b & 0 \end{pmatrix} Z \begin{pmatrix} 0 & b \\ 1/b & 0 \end{pmatrix}' + \begin{pmatrix} -sb^2 & 0 \\ 0 & s \end{pmatrix}, \tag{16}$$

Using

$$\bar{Z} = \begin{pmatrix} 0 & b \\ 1 & 0 \end{pmatrix} \overline{W} \begin{pmatrix} 0 & b \\ 1 & 0 \end{pmatrix}' + \begin{pmatrix} -b^2s & 0 \\ 0 & 0 \end{pmatrix}$$

$$Z = \begin{pmatrix} 0 & b \\ 1 & 0 \end{pmatrix} W \begin{pmatrix} 0 & b \\ 1 & 0 \end{pmatrix}' + \begin{pmatrix} -b^2s & 0 \\ 0 & 0 \end{pmatrix},$$

we have (12) and (15) with W instead of Z.

The product of two commutative involutions of the first kind is called *an involution of the second kind*. Therefore an involution of the second kind can be carried into

$$\bar{Z} = \begin{pmatrix} 0 & 1 \\ -1 & 0 \end{pmatrix} Z \begin{pmatrix} 0 & -1 \\ 1 & 0 \end{pmatrix}. \tag{17}$$

The fixed elements of this transformation are of the form

$$t\begin{pmatrix} 1 & 0 \\ 0 & 1 \end{pmatrix}. \tag{18}$$

The manifold of the fixed elements of an involution of the second kind is called a *chain*.

We are now going to examine all the chains in a reducible subspace. Notice that (16) represents all the involutions of the first kind which are commutative with (12). The product (12) and (16) is

$$\bar{Z} = \begin{pmatrix} 0 & b \\ -1/b & 0 \end{pmatrix} Z \begin{pmatrix} 0 & b \\ -1/b & 0 \end{pmatrix}' + \begin{pmatrix} -sb^2 & 0 \\ 0 & s \end{pmatrix}.$$

The chain in the reducible space $\begin{pmatrix} x & 0 \\ 0 & z \end{pmatrix}$ is therefore of the form

$$\begin{pmatrix} b^2(t-s) & 0 \\ 0 & t \end{pmatrix},$$

where t runs over all elements of Φ. We put in more parameters to make it similar to a straight line. The chain is also equivalent to

$$\begin{pmatrix} at+b & 0 \\ 0 & ct+d \end{pmatrix} \tag{19}$$

if $a = cx^2$ is soluble.

In the next section, we shall obtain harmonic separation by means of the construction of complete quadrateral. But we should notice that we have neither the fact that passing through any two points there is a "line" nor the fact that any two "lines" have a point of intersection.

The chains passing through a point $\begin{pmatrix} b & 0 \\ 0 & d \end{pmatrix}$ are of the form (19). The chain passing through two points

$$\begin{pmatrix} a & 0 \\ 0 & b \end{pmatrix} \quad \text{and} \quad \begin{pmatrix} c & 0 \\ 0 & d \end{pmatrix}$$

is of the form

$$\lambda \begin{pmatrix} a & 0 \\ 0 & b \end{pmatrix} + (1-\lambda) \begin{pmatrix} c & 0 \\ 0 & d \end{pmatrix}, \tag{20}$$

if it exists. The condition for existence is that

$$(a-c)/(b-d) \tag{21}$$

is a square element in Φ.

4. Proof of the theorem for $n = 2$.

THEOREM 5. *Every mapping carrying symmetric matrices into symmetric matrices and keeping the rank of differences invariant is a combination of a homothetic transformation*

$$Z_1 = aZ,$$

a translation

$$Z_1 = Z + S, \qquad S' = S,$$

an automorphism of the field

$$Z_1 = Z^\sigma$$

and a congruence relation

$$Z_1 = AZA', \qquad \det(A) \neq 0.$$

This is practically the theorem stated in the introduction. In this section, we shall give a proof for $n = 2$.

Notice that the notion of reduced subspaces and chains depends only on the notion of arithmetic distance.

Let $\Gamma(Z)$ be a mapping of the type under consideration. It is evidently one to one. Without loss of generality, we may assume that it carries the reducible subspace

$$\begin{pmatrix} x & 0 \\ 0 & z \end{pmatrix}$$

onto itself and the chain $t\begin{pmatrix} 1 & 0 \\ 0 & 1 \end{pmatrix}$ onto itself, and moreover, that

$$\Gamma(0) = 0, \qquad \Gamma(I) = I.$$

Let

$$\Gamma(aI) = a^\sigma I.$$

We shall prove that σ is an automorphism of the field Φ.

1) $(a + b)^\sigma = a^\sigma + b^\sigma$.

Consider only the chains in the reducible subspace

$$\begin{pmatrix} x & 0 \\ 0 & z \end{pmatrix}.$$

Any chain, different from tI, passing through 0 can be expressed as

$$\begin{pmatrix} 1 & 0 \\ 0 & p^2 \end{pmatrix} t, \qquad t \in \Phi. \tag{22}$$

Any chain which does not intersect tI can be expressed as

$$tI + \begin{pmatrix} q & 0 \\ 0 & 0 \end{pmatrix}, \qquad t \in \Phi. \tag{23}$$

The intersection of (22) and (23) is given by

$$\frac{q}{1 - p^2}\begin{pmatrix} 1 & 0 \\ 0 & p^2 \end{pmatrix}. \tag{24}$$

The chain passing through aI and (24) is of the form

$$atI + \frac{q}{1-p^2}\begin{pmatrix}1 & 0\\ 0 & p^2\end{pmatrix}(1-t), \qquad t \in \Phi, \tag{25}$$

provided that

$$\left(a - \frac{q}{1-p^2}\right) \Big/ \left(a - \frac{qp^2}{1-p^2}\right) \tag{26}$$

is a square element in Φ.

From bI, we draw a chain which does not intersect (22); it has the form

$$\begin{pmatrix}1 & 0\\ 0 & p^2\end{pmatrix}t + bI, \qquad t \in \Phi. \tag{27}$$

The intersection of (27) and (23) is given by

$$\frac{q}{1-p^2}\begin{pmatrix}1 & 0\\ 0 & p^2\end{pmatrix} + bI. \tag{28}$$

From (28), we draw a chain which does not intersect (25); it has the form

$$\left(aI - \begin{pmatrix}1 & 0\\ 0 & p^2\end{pmatrix}\frac{q}{1-p^2}\right)t + \frac{q}{1-p^2}\begin{pmatrix}1 & 0\\ 0 & p^2\end{pmatrix} + bI, \qquad t \in \Phi. \tag{29}$$

The existence of (29) is assured by (26). This chain intersects tI at the point $(a+b)I$.

That is, for any p and q satisfying (26), we can construct $(a+b)I$ by means of the processes of "meet" and "join." This property is certainly carried over by the operation Γ. The existence of p and q satisfying (26) is assured since (26) is linear in q. Therefore we have

$$\Gamma(aI + bI) = \Gamma(aI) + \Gamma(bI).$$

2) $(ab)^\sigma = a^\sigma b^\sigma$.

From 0 we draw a chain

$$\begin{pmatrix}1 & 0\\ 0 & p^2\end{pmatrix}t, \qquad t \in \Phi. \tag{30}$$

From I, we draw another chain

$$\begin{pmatrix}1 & 0\\ 0 & q^2\end{pmatrix}t + I \qquad t \in \Phi. \tag{31}$$

Their intersection is given by

$$\frac{q^2-1}{q^2-p^2}\begin{pmatrix}1 & 0\\ 0 & p^2\end{pmatrix}. \tag{32}$$

Connecting (32) and bI, we have

$$t\frac{q^2-1}{q^2-p^2}\begin{pmatrix}1 & 0\\ 0 & p^2\end{pmatrix}+(1-t)bI, \qquad t\in\Phi \tag{33}$$

provided that

$$\left(\frac{q^2-1}{q^2-p^2}-b\right)\Big/\left(\frac{q^2-1}{q^2-p^2}p^2-b\right) \tag{34}$$

is a square element r^2.

From aI we draw a chain not intersecting (31),

$$\begin{pmatrix}1 & 0\\ 0 & q^2\end{pmatrix}t+aI, \qquad t\in\Phi \tag{35}$$

which intersects (30) at

$$\frac{q^2-1}{q^2-p^2}a\begin{pmatrix}1 & 0\\ 0 & p^2\end{pmatrix}. \tag{36}$$

From (36), we draw a chain not intersecting (33),

$$t\left(\frac{q^2-1}{q^2-p^2}\begin{pmatrix}1 & 0\\ 0 & p^2\end{pmatrix}-bI\right)+\frac{q^2-1}{q^2-p^2}a\begin{pmatrix}1 & 0\\ 0 & p^2\end{pmatrix}. \tag{37}$$

This intersects the chain tI at abI. The existence of (37) is assured by (34).

Therefore we have

$$(ab)^\sigma = a^\sigma b^\sigma,$$

provided that we can justify that (34) is always soluble in p, q, r. That is, for any b, we need p, q, r in the field such that

$$b=\frac{q^2-1}{q^2-p^2}\,\frac{1-p^2r^2}{1-r^2}.$$

For any b, there exist two elements u and v of the field such that

$$b=uv, \qquad -2uv+u+v\neq 0$$

Thus, we have a solution

$$r=0, \qquad p=\frac{v-u}{u+v-2uv}, \qquad q=\frac{2-u-v}{u+v-2uv}.$$

We may assume without loss of generality that

$$\Gamma(aI)=aI \tag{38}$$

for all a belonging to Φ.

Let $\begin{pmatrix}x & 0\\ 0 & z\end{pmatrix}$ be any point of the reducible space which is not on the chain tI,

and

$$\Gamma\begin{pmatrix} x & 0 \\ 0 & z \end{pmatrix} = \begin{pmatrix} x^* & 0 \\ 0 & z^* \end{pmatrix}. \tag{39}$$

Since there are two and only two points on the chain tI, namely

$$\begin{pmatrix} x & 0 \\ 0 & x \end{pmatrix} \quad \text{and} \quad \begin{pmatrix} z & 0 \\ 0 & z \end{pmatrix}$$

such that

$$r\left(\begin{pmatrix} x & 0 \\ 0 & x \end{pmatrix}, \begin{pmatrix} x & 0 \\ 0 & z \end{pmatrix}\right) = r\left(\begin{pmatrix} z & 0 \\ 0 & z \end{pmatrix}, \begin{pmatrix} x & 0 \\ 0 & z \end{pmatrix}\right) = 1,$$

we have then

$$\Gamma\begin{pmatrix} x & 0 \\ 0 & z \end{pmatrix} = \begin{pmatrix} x & 0 \\ 0 & z \end{pmatrix} \quad \text{or} \quad \begin{pmatrix} z & 0 \\ 0 & x \end{pmatrix}. \tag{40}$$

We may assume that

$$\Gamma\begin{pmatrix} 1 & 0 \\ 0 & 0 \end{pmatrix} = \begin{pmatrix} 1 & 0 \\ 0 & 0 \end{pmatrix}$$

for otherwise, we can use

$$\Gamma_1(z) = \begin{pmatrix} 0 & 1 \\ 1 & 0 \end{pmatrix} \Gamma(z) \begin{pmatrix} 0 & 1 \\ 1 & 0 \end{pmatrix}$$

instead of Γ, and Γ_1 satisfies our requirement. Consequently

$$\Gamma\begin{pmatrix} 0 & 0 \\ 0 & 1 \end{pmatrix} = \begin{pmatrix} 0 & 0 \\ 0 & 1 \end{pmatrix}.$$

Since $\begin{pmatrix} x & 0 \\ 0 & 0 \end{pmatrix}$ lies on the subspace $\Pi\left(0, \begin{pmatrix} 1 & 0 \\ 0 & 0 \end{pmatrix}\right)$, we have

$$\Gamma\begin{pmatrix} x & 0 \\ 0 & 0 \end{pmatrix} = \begin{pmatrix} x & 0 \\ 0 & 0 \end{pmatrix}$$

for all $x \in \Phi$. Similarly, we have

$$\Gamma\begin{pmatrix} 0 & 0 \\ 0 & z \end{pmatrix} = \begin{pmatrix} 0 & 0 \\ 0 & z \end{pmatrix}.$$

Since

$$r\left(\begin{pmatrix} x & 0 \\ 0 & z \end{pmatrix}, \begin{pmatrix} x & 0 \\ 0 & 0 \end{pmatrix}\right) = 1,$$

but

$$r\left(\begin{pmatrix} z & 0 \\ 0 & x \end{pmatrix}, \begin{pmatrix} x & 0 \\ 0 & 0 \end{pmatrix}\right) \neq 1,$$

we have inmediately

$$\Gamma\begin{pmatrix} x & 0 \\ 0 & z \end{pmatrix} = \begin{pmatrix} x & 0 \\ 0 & z \end{pmatrix}. \tag{41}$$

Now we let

$$\Gamma\begin{pmatrix} a & b \\ b & c \end{pmatrix} = \begin{pmatrix} a^* & b^* \\ b^* & c^* \end{pmatrix}, \tag{42}$$

we consider these $\begin{pmatrix} x & 0 \\ 0 & z \end{pmatrix}$ satisfying

$$r\left(\begin{pmatrix} a & b \\ b & c \end{pmatrix}, \begin{pmatrix} x & 0 \\ 0 & z \end{pmatrix}\right) \leqq 1,$$

that is

$$(a - x)(c - z) - b^2 = 0.$$

By (41) and (42), this equation implies

$$(a^* - x)(c^* - z) - b^{*2} = 0$$

and vice versa. Thus we have

$$a = a^*, \qquad c = c^*, \qquad b = \pm b^*.$$

Since

$$\begin{pmatrix} 1 & 0 \\ 0 & -1 \end{pmatrix}\begin{pmatrix} a & b \\ b & c \end{pmatrix}\begin{pmatrix} 1 & 0 \\ 0 & -1 \end{pmatrix} = \begin{pmatrix} a & -b \\ -b & c \end{pmatrix}$$

we may assume that

$$\Gamma\begin{pmatrix} x & 1 \\ 1 & z \end{pmatrix} = \begin{pmatrix} x & 1 \\ 1 & z \end{pmatrix}. \tag{43}$$

Suppose that there exists $b \neq 0$ such that

$$\Gamma\begin{pmatrix} a & b \\ b & c \end{pmatrix} = \begin{pmatrix} a & -b \\ -b & c \end{pmatrix}.$$

By (43), we should then have

$$(x - a)(z - c) - (1 - b)^2 = 0$$

which implies

$$(x - a)(z - c) - (1 + b)^2 = 0.$$

This is impossible, since the characteristic of the field is different from 2. Therefore

$$\Gamma\begin{pmatrix} a & b \\ b & c \end{pmatrix} = \begin{pmatrix} a & b \\ b & c \end{pmatrix},$$

and the theorem is proved for $n = z$.

REMARK. The theorem seems still to be true for the field with characteristic 2. For the prime field with characteristic 2, the proof is quite simple. The space contains only eight points

$$\begin{pmatrix} 0 & 0 \\ 0 & 0 \end{pmatrix}, \quad \begin{pmatrix} 1 & 0 \\ 0 & 0 \end{pmatrix}, \quad \begin{pmatrix} 0 & 0 \\ 0 & 1 \end{pmatrix}, \quad \begin{pmatrix} 1 & 1 \\ 1 & 1 \end{pmatrix},$$
$$\begin{pmatrix} 1 & 0 \\ 0 & 1 \end{pmatrix}, \quad \begin{pmatrix} 1 & 1 \\ 1 & 0 \end{pmatrix}, \quad \begin{pmatrix} 0 & 1 \\ 1 & 1 \end{pmatrix}, \quad \begin{pmatrix} 0 & 1 \\ 1 & 0 \end{pmatrix}.$$

We may assume that

$$\Gamma\begin{pmatrix} 0 & 0 \\ 0 & 0 \end{pmatrix} = \begin{pmatrix} 0 & 0 \\ 0 & 0 \end{pmatrix}, \quad \Gamma\begin{pmatrix} 1 & 0 \\ 0 & 0 \end{pmatrix} = \begin{pmatrix} 1 & 0 \\ 0 & 0 \end{pmatrix}, \quad \Gamma\begin{pmatrix} 0 & 0 \\ 0 & 1 \end{pmatrix} = \begin{pmatrix} 0 & 0 \\ 0 & 1 \end{pmatrix}.$$

The only singular matrix which has not been taken into consideration is $\begin{pmatrix} 1 & 1 \\ 1 & 1 \end{pmatrix}$.

Therefore

$$\Gamma\begin{pmatrix} 1 & 1 \\ 1 & 1 \end{pmatrix} = \begin{pmatrix} 1 & 1 \\ 1 & 1 \end{pmatrix}.$$

Since $\begin{pmatrix} 1 & 0 \\ 0 & 1 \end{pmatrix}$ is the only element satisfying

$$r\left(\begin{pmatrix} 1 & 0 \\ 0 & 1 \end{pmatrix}, \begin{pmatrix} 1 & 0 \\ 0 & 0 \end{pmatrix}\right) = r\left(\begin{pmatrix} 1 & 0 \\ 0 & 1 \end{pmatrix}, \begin{pmatrix} 0 & 0 \\ 0 & 1 \end{pmatrix}\right) = 1$$

we have

$$\Gamma\begin{pmatrix} 1 & 0 \\ 0 & 1 \end{pmatrix} = \begin{pmatrix} 1 & 0 \\ 0 & 1 \end{pmatrix}.$$

Similarly,

$$\Gamma\begin{pmatrix} 1 & 1 \\ 1 & 0 \end{pmatrix} = \begin{pmatrix} 1 & 1 \\ 1 & 0 \end{pmatrix}, \quad \Gamma\begin{pmatrix} 0 & 1 \\ 1 & 1 \end{pmatrix} = \begin{pmatrix} 0 & 1 \\ 1 & 1 \end{pmatrix}.$$

Finally, we have

$$\Gamma\begin{pmatrix}0 & 1\\ 1 & 0\end{pmatrix} = \begin{pmatrix}0 & 0\\ 1 & 0\end{pmatrix}.$$

5. Subspaces.

In order to extend our notion of subspaces to higher dimensions, we introduce the concept of a dieder manifold.

DEFINITION. Let P and Q be two symmetric matrices of order n. The symmetric matrices X satisfying

$$r(P, X) + r(X, Q) = r(P, Q) \tag{44}$$

form a manifold which is called a *dieder manifold* spanned by the points P and Q. The arithmetic distance between P and Q is called the *extent* of the dieder manifold.

Since the pairs of points having a fixed arithmetic distance form a transitive set[2], the dieder manifolds of the same extent form a transitive set.

THEOREM 6. *If $p < n$, the points on the dieder manifold spanned by*

$$0, \quad \begin{pmatrix}I^{(p)} & 0\\ 0 & 0\end{pmatrix}$$

are of the form

$$\begin{pmatrix}X^{(p)} & 0\\ 0 & 0\end{pmatrix}. \tag{45}$$

PROOF. Let X be a point on the dieder manifold and

$$X = \begin{pmatrix}X_{11}^{(p)} & X_{12}\\ X_{12}^{1} & X_{22}\end{pmatrix}, \qquad X_{12} = X_{12}^{(p,n-p)}, \qquad X_{22} = X_{22}^{(n-p)}.$$

Since

$$p = r(P, X) + r(X, Q) \geqq r(0, X_{11}) + r(X_{11}, I^{(p)}) \geqq r(0, I^{(p)}) = p,$$

we have

$$r(0, X_{11}) + r(X_{11}, I^{(p)}) = p. \tag{46}$$

There is a nonsingular p-rowed matrix Γ such that $\Gamma X_{11}\Gamma^{-1}$ is of the normal form. In virtue of (46), we find that the normal form becomes

$$\Gamma X_{11}\Gamma^{-1} = \begin{pmatrix}I^{(p)} & 0\\ 0 & 0\end{pmatrix}, \qquad 0 \leqq q \leqq p. \tag{47}$$

We shall now prove that $X_{12} = 0$, $X_{22} = 0$. Let

$$X_0 = \begin{pmatrix}X_{11}^{(p)} & v'\\ v & a\end{pmatrix}$$

[2] I theorem 2.

be a $(p + 1)$-rowed principal minor of X; it is sufficient to prove that $v = 0$, $a = 0$. From (44), we deduce in the same way used to establish (46), that

$$p = r(p, X) + r(X, Q) \geqq r(0^{(p+1)}, X_0) + r\left(X_0, \begin{pmatrix} I^{(p)} & 0 \\ 0 & 0^{(1)} \end{pmatrix}\right) \geqq p.$$

Consequently, we have

$$r(0^{(p+1)}, X_0) = r(0^{(p)}, X_{11}) = q \tag{48}$$

and

$$r\left(X_0, \begin{pmatrix} I^{(p)} & 0 \\ 0 & 0 \end{pmatrix}\right) = r(X_{11}, I^{(p)}) = p - q. \tag{49}$$

From (48), it follows that

$$X_0 = \begin{pmatrix} \Gamma^{-1} & 0 \\ 0 & 0 \end{pmatrix} \begin{pmatrix} I^{(q)} & 0 & w_1' \\ 0 & 0^{(p-q)} & w_2' \\ u_1 & u_2 & 0 \end{pmatrix} \begin{pmatrix} \Gamma & 0 \\ 0 & 1 \end{pmatrix}$$

is of rank q, where

$$(u_1, u_2)\Gamma = v, \qquad (w_1, w_2)\Gamma'^{-1} = v.$$

Since

$$\begin{pmatrix} I^{(q)} & 0 & 0 \\ 0 & I^{(p-q)} & 0 \\ -u_1 & 0 & 1 \end{pmatrix} \begin{pmatrix} I^{(q)} & 0 & w_1' \\ 0 & 0 & w_2' \\ u_1 & u_2 & a \end{pmatrix} \begin{pmatrix} I^{(q)} & 0 & -w_1' \\ 0 & I & 0 \\ 0 & 0 & 1 \end{pmatrix} = \begin{pmatrix} I^{(q)} & 0 & 0 \\ 0 & 0 & w_2' \\ 0 & u_2 & a - u_1 w_1' \end{pmatrix}$$

is of rank q, we deduce immediately that $u_2 = 0$, $w_2 = 0$, and $a - u_1 w_1' = 0$.

Using (49), we see also that $u_1 = 0$, $w_1 = 0$ and $a - u_2 w_2' = 0$. Therefore $v = 0$ and $a = 0$. This establishes our theorem.

DEFINITION. Let $p < n$. A set of points is said to form a normal subspace of rank p, if every pair of points of the set has arithmetic distance $\leqq p$, if it contains two points of arithmetic distance p, and if it contains all dieder manifolds spanned by any two points of the set.

THEOREM 7. *Normal subspaces of the same rank form a transitive set; more precisely, every normal subspace of rank p can be transformed by* (8) *into the normal subspace*

$$\begin{pmatrix} X^{(p)} & 0 \\ 0 & 0 \end{pmatrix} \tag{50}$$

where $X^{(p)}$ runs over all p-rowed symmetric matrices.

For the proof of this theorem see III, theorem 2, p. 231 (2).

6. Proof of Theorem 5 in general.

We now suppose that $n \geqq 3$. Let

$$\Gamma(Z) = Z_1$$

be the mapping under consideration. Suppose that

$$\Gamma(0) = 0.$$

The points of the form

$$\begin{pmatrix} X_1^{(n-1)} & 0 \\ 0 & 0 \end{pmatrix}$$

form a normal subspace of rank $n - 1$. Since the arithmetic distance is invariant, the set of points

$$\Gamma\begin{pmatrix} X^{(n-1)} & 0 \\ 0 & 0 \end{pmatrix}$$

constitutes also a normal subspace of rank $n - 1$. Since the set of all normal subspaces forms a transitive set (Theorem 7), we may suppose that

$$\Gamma\begin{pmatrix} W^{(n-1)} & 0 \\ 0 & 0 \end{pmatrix} = \begin{pmatrix} W_1^{(n-1)} & 0 \\ 0 & 0 \end{pmatrix}.$$

Thus Γ induces a mapping on $(n - 1)$-rowed symmetric matrices, and it keeps arithmetic distance invariant. By the hypothesis of induction, we have

$$W = a\alpha W_1^{\sigma}\alpha^{1} + \mu, \qquad \alpha = \alpha^{(n-1)}, \qquad \mu^{1} = \mu = \mu^{(n-1)}.$$

Then the mapping

$$Z = aAZ_1^{\sigma}A' + S,$$

where

$$A = \begin{pmatrix} \alpha & 0 \\ 0 & 1 \end{pmatrix}, \qquad S = \begin{pmatrix} \mu & 0 \\ 0 & 0 \end{pmatrix}$$

carries the mapping $\Gamma(Z)$ into a new one with

$$\Gamma\begin{pmatrix} W^{(n-1)} & 0 \\ 0 & 0 \end{pmatrix} = \begin{pmatrix} W^{(n-1)} & 0 \\ 0 & 1 \end{pmatrix}. \tag{51}$$

Since

$$\begin{pmatrix} 0^{(n-1)} & 0 \\ 0 & 1 \end{pmatrix}$$

is of rank 1, we may let

$$\Gamma\begin{pmatrix} 0^{(n-1)} & 0 \\ 0 & 1 \end{pmatrix} = \epsilon(a_1, \cdots, a_n)'(a_1, \cdots, a_n), \tag{52}$$

where ϵ is either 1 or $x^2 = \epsilon$ is not soluble. Since the arithmetic distance between

$$\begin{pmatrix} I^{(n-1)} & 0 \\ 0 & 0 \end{pmatrix} \quad \text{and} \quad \begin{pmatrix} 0^{(n-1)} & 0 \\ 0 & 0 \end{pmatrix}$$

is n, by (51), we have $a_n \neq 0$. Let

$$A = \begin{pmatrix} I^{(n-1)} & 0 \\ -a_n^{-1} v & a_n^{-1} \end{pmatrix}, \qquad v = (a_1, \cdots, a_{n-1}).$$

Then

$$Z_1 = A'ZA$$

carries (51) into itself and

$$A'(a_1, \cdots, a_n)'(a_1, \cdots, a_n)A = (0, 0, \cdots, 1)'(0, 0, \cdots, 1).$$

Thus we may assume further that

$$\Gamma\begin{pmatrix} 0^{(n-1)} & 0 \\ 0 & 1 \end{pmatrix} = \epsilon\begin{pmatrix} 0^{(n-1)} & 0 \\ 0 & 1 \end{pmatrix}. \tag{53}$$

The points

$$\begin{pmatrix} 1 & 0 & \cdots & 0 \\ 0 & 0 & \cdots & 0 \\ \cdots & \cdots & \cdots & \cdots \\ 0 & 0 & \cdots & 0 \end{pmatrix} \quad \text{and} \quad \begin{pmatrix} 0 & 0 & \cdots & 0 \\ 0 & 0 & \cdots & 0 \\ \cdots & \cdots & \cdots & \cdots \\ 0 & 0 & \cdots & \epsilon \end{pmatrix}, \quad \epsilon \neq 0$$

span a normal subspace

$$\begin{pmatrix} a_{11} & 0 & \cdots & a_{1n} \\ 0 & 0 & \cdots & 0 \\ \cdots & \cdots & \cdots & \cdots \\ 0 & 0 & \cdots & 0 \\ a_{1n} & 0 & \cdots & a_{nn} \end{pmatrix}$$

of rank 2. The mapping Γ induces an automorphic mapping on two rowed symmetric matrices

$$\begin{pmatrix} a_{11} & a_{1n} \\ a_{1n} & a_{nn} \end{pmatrix}.$$

Let it be

$$\gamma\begin{pmatrix} a_{11} & a_{1n} \\ a_{1n} & a_{nn} \end{pmatrix} = \begin{pmatrix} \overset{*}{a}_{11} & \overset{*}{a}_{1n} \\ \overset{*}{a}_{1n} & \overset{*}{a}_{nn} \end{pmatrix},$$

and it satisfies

$$\gamma(0) = 0, \quad \gamma\begin{pmatrix} x & 0 \\ 0 & 0 \end{pmatrix} = \begin{pmatrix} x & 0 \\ 0 & 0 \end{pmatrix}, \quad \gamma\begin{pmatrix} 0 & 0 \\ 0 & 1 \end{pmatrix} = \begin{pmatrix} 0 & 0 \\ 0 & \epsilon \end{pmatrix}. \tag{54}$$

Since the theorem is true for $n = 2$, we have, by the first equation of (54),

$$\gamma\begin{pmatrix} a_{11} & a_{1n} \\ a_{1n} & a_{nn} \end{pmatrix} = aA\begin{pmatrix} a_{11} & a_{1n} \\ a_{1n} & a_{nn} \end{pmatrix}^{\sigma} A' \qquad A = A^{(2)},$$

where a is either 1 or a nonsquare element of Φ. From the second and third equations of (54), we deduce

$$A = \begin{pmatrix} b & 0 \\ 0 & c \end{pmatrix}$$

and

$$ab^2x^\sigma = x, \qquad ac^2 = \epsilon^\sigma$$

for all x. If in particular $x = 1$, we see that a and ϵ must be square elements, whence $a = \epsilon = 1$, and $b^2 = c^2 = 1$. Consequently $x^\sigma = x$, that is, σ is the identity automorphism. Substantially, we have only two cases: (1) $b = c = 1$ and (2) $b = -c = 1$.

Thus

$$(55) \qquad \Gamma\begin{pmatrix} a_{11} & 0 & \cdots, & 0, & a_{1n} \\ 0 & 0 & \cdots & 0 & 0 \\ \cdots & \cdots & \cdots & \cdots & \cdots \\ a_{1n} & 0 & \cdots & 0 & a_{nn} \end{pmatrix} = \begin{pmatrix} a_{11} & 0 & \cdots & 0 & \pm a_{1n} \\ 0 & 0 & \cdots & 0 & 0 \\ \cdots & \cdots & \cdots & \cdots & \cdots \\ \pm a_{1n} & 0 & \cdots & 0 & a_{nn} \end{pmatrix}.$$

The mapping Γ leaves

$$(56) \qquad \begin{pmatrix} 0^{(1)} & 0 & 0 \\ 0 & I^{(n-2)} & 0 \\ 0 & 0 & 1^{(1)} \end{pmatrix}, \quad \begin{pmatrix} 0^{(1)} & 0 & 0 \\ 0 & 0^{(n-2)} & 0 \\ 0 & 0 & 1^{(1)} \end{pmatrix}$$

invariant and

$$\Gamma\begin{pmatrix} 0 & 0 \\ 0 & W^{(n-1)} \end{pmatrix} = \begin{pmatrix} 0 & 0 \\ 0 & W^{(n-1)} \end{pmatrix},$$

since two points of (56) span the normal subspace in the bracket. We have by the hypothesis of induction that

$$W = aAW^\tau A' + S, \qquad A = A^{(n-1)}, \qquad S = S^{(n-1)} = S',$$

where a is either 1 or a non-square element. By (51) and (55), it leaves

$$W = \begin{pmatrix} X^{(n-2)} & 0 \\ 0 & 0 \end{pmatrix}, \quad \begin{pmatrix} 0^{(n-2)} & 0 \\ 0 & x \end{pmatrix}$$

pointwise invariant, whence $S = 0$, $a = 1$, $\tau = 1$ and

$$A^{(n-1)} = I \quad \text{or} \quad A^{(n-1)} = \begin{pmatrix} I^{(n-2)} & 0 \\ 0 & -1 \end{pmatrix}.$$

For the former case we have

$$(57) \qquad \Gamma\begin{pmatrix} 0 & 0 \\ 0 & W^{(n-1)} \end{pmatrix} = \begin{pmatrix} 0 & 0 \\ 0 & W^{(n-1)} \end{pmatrix}.$$

For the latter case

$$Z_1 = \begin{pmatrix} I^{(n-1)} & 0 \\ 0 & -1 \end{pmatrix} Z \begin{pmatrix} I^{(n-1)} & 0 \\ 0 & -1 \end{pmatrix}$$

carries Γ into a new transformation which satisfies (51), (55) and (57).

As in III, p. 238, we deduce

$$\Gamma((z_{ij})) = (z_{ij}^1), \qquad z_{ii} = z_{ii}^1, \qquad z_{1n} = \pm z_{1n}^1$$

and

$$z_{ij} = z_{ij}^1 \qquad \text{for } (i, j) \neq (1, n).$$

Suppose that there exists a $Z = (z_{ij})$ such that

$$\Gamma \begin{pmatrix} z_{11}, & \cdots, & z_{1n} \\ \cdots & \cdots & \cdots \\ z_{1n}, & \cdots, & z_{nn} \end{pmatrix} = \begin{pmatrix} z_{11}, & \cdots, & z_{1n-1}, & -z_{1n} \\ \cdots & \cdots & \cdots & \cdots \\ -z_{1n}, & \cdots, & z_{n-1,n} & -z_{nn} \end{pmatrix}, \qquad z_{1n} \neq 0,$$

Since

$$\Gamma \begin{pmatrix} z_{11} - z_{1n}, & \cdots, & z_{1n-1} - z_{1n}, & 0 \\ \cdots & \cdots & \cdots & \cdots \\ 0 & , \cdots, & z_{n-1,n} - z_{1n}, & z_{nn} - z_{1n} \end{pmatrix}$$
$$= \begin{pmatrix} z_{11} - z_{1n}, & \cdots, & z_{1n-1} - z_{1n}, & 0 \\ \cdots & \cdots & \cdots & \cdots \\ 0 & , \cdots, & z_{n-1,n} - z_{1n}, & z_{nn} - z_{1n} \end{pmatrix}$$

and since arithmetic distance is invariant, we see that the matrices

$$\begin{pmatrix} z_{1n} & \cdots & z_{1n} & z_{1n} \\ \cdots & \cdots & \cdots & \cdots \\ z_{1n} & \cdots & z_{1n} & z_{1n} \end{pmatrix} \quad \text{and} \quad \begin{pmatrix} z_{1n}, & \cdots, & z_{1n}, & -z_{1n} \\ z_{1n}, & \cdots, & z_{1n}, & z_{1n} \\ \cdots & \cdots & \cdots & \cdots \\ -z_{1n}, & \cdots, & z_{1n}, & z_{1n} \end{pmatrix}$$

have the same rank. This is a contradiction, since the former is of rank 1 and the latter is of rank 3. Thus $z_{1n}^1 = z_{1n}$. That is,

$$\Gamma(Z) = Z.$$

The theorem follows.

7. Proof of the fundamental theorem of the projective geometry of symmetric matrices.

A point (X, Y), in homogeneous coordinate, is called finite if Y is nonsingular, otherwise, it is called infinite. The patricular infinite point $(I, 0)$ is denoted by ∞. Evidently a necessary and sufficient condition for a point P to be finite is that

$$r(P, \infty) = n.$$

Since the projective space of symmetric matrices is transitive, we may assume that the automorphic mapping of the space carries ∞ into itself. If it keeps arithmetic distance invariant, the mapping carries finite points into finite points and keeps the arithmetic distance of finite points invariant. By Theorem 5, we have

THEOREM 8. *Let $n > 1$. An automorphic mapping of the projective space keeping arithmetic distance invariant is of the form*

$$(X_1, Y_1) = Q(X^\sigma, Y^\sigma)\mathfrak{T},$$

where

$$\mathfrak{T}\mathfrak{F}\mathfrak{T}' = a\mathfrak{F}.$$

8. Remarks.

1°. The condition stated in the theorem can be somewhat weakened since the invariance of the arithmetic distance 1 implies the invariance of arithmetic distances > 1. In fact we have

THEOREM 9. *Two points P and Q are of arithmetic distance p if and only if there exists $p - 1$ points $X_1, \cdots, X_{p-1}$ such that*

$$(58)\qquad r(P, X_1) = r(X, X_2) = \cdots = r(X_{p-2}, X_{p-1}) = r(X_{p-1}, Q) = 1,$$

p being the least integer with this property.

PROOF. 1) If P and Q have the arithmetic distance p, we may assume without loss of generality that they are

$$0, \quad \begin{pmatrix} I^{(p)} & 0 \\ 0 & 0 \end{pmatrix}$$

respectively. Then,

$$X_q : \begin{pmatrix} I^{(q)} & 0 \\ 0 & 0 \end{pmatrix}, \qquad q = 1, 2, \cdots, p - 1$$

are the points required.

2) By the triangle inequality for arithmetic distances, we have

$$\begin{aligned} r(P, Q) &\leqq r(P, X_1) + r(X_1, X_2) + \cdots + r(X_{p-1}, Q) \\ &\leqq P. \end{aligned}$$

If $r(P, Q) < p$, then we should have fewer X's such that (58) holds. Therefore $r(P, Q) = p$.

From this we deduce immediately the following sharper result:

THEOREM 10. *Let Σ be the set of all symmetric matrices of order $n \geqq 2$ over a field Φ with characteristic $\neq 2$. Any one to one mapping of Σ onto itself carrying a* *pair of symmetric matrices with arithmetic distance one into a pair with the same property is of the form*

$$Z_1 = aAZ^{\sigma}A' + S, \qquad S = S'$$

where $a \in \Phi$ *and* σ *is an automorphism of* Φ.

2°. The previous theorem seems to remain true for a field with characteristic 2.

3°. The method for the construction of involutions has not been used to its full capacity. In fact, it is very likely that we can construct the whole symplectic group by means of this procedure.

THE INSTITUTE FOR ADVANCED STUDY

Reprinted from the
CHINESE MATHEMATICAL SOCIETY
Vol. 1, No. 2, pp. 109–163, 1951

A THEOREM ON MATRICES OVER A SFIELD AND ITS APPLICATIONS

By

LOO-KENG HUA

Academia Sinica, Peking and Tsing-hua University, Peking

1. *Introduction.* Let Φ be a sfield (or a division ring). Let m and n be two positive integers such that $n \leqq m$. We use $M = M^{(n,m)}$ to denote an $n \times m$ matrix over Φ and $M^{(n)} = M^{(n,n)}$ and M' to denote the transposed matrix of M

Two $n \times m$ matrices Z and W are said to be *coherent*, if the rank of their difference $Z - W$ is one.

The aim of part I of the paper is to establish the following theorem:

THEOREM 1. *Suppose* $1 < n \leqq m$. *Any one-to-one mapping which carries* $n \times m$ *matrices into* $n \times m$ *matrices and leaves the coherence invariant is of the form*

$$Z_1 = P Z^{\sigma} Q + R \tag{1}$$

where $P(= P^{(n)})$ *and* $Q(= Q^{(m)})$ *are non-singular,* R *is an* $n \times m$ *matrix and* σ *is an automorphism of the sfield* Φ. *When* $n = m$, *in addition to* (1), *we have also*

$$Z_1 = P Z'^{\sigma} Q + R \tag{2}$$

where σ *is an anti-automorphism of the sfield.*

This asserts that the equivalence relation used in the theory of matrices has a unique position.

In a previous paper (Hua [1]) the author established a corressponding theorem for symmetric matrices by the construction of involutions. The method to be used in the present investigation is entirely different. The main idea is based upon the method of "meet and join". It suggests a possibility for the formulation of this geometry in terms of lattice theory.

In case $m=n=1$, the problem is closely related to the fundamental theorem of projective geometry over a sfield, which has been discussed, though very incompletely, by Ancochea [5]. In a short note the author [6] proved the fundamental theorem without restriction. There seems to have no essential difficulty to build up a theory of projective geometry over a simple ring with chain condition, but now instead of the harmonic relation, the weaker concept of coherence plays an important role. Later we shall treat the geometry of rectangular matrices which is practically more general than this.

In part II and part III, we shall give several comparatively more direct applications.

It is well-known that every automorphism of a simple algebra is an inner one (see, e. g. Albert [2], p.51). It was extended by Ancochea [3] and Kaplansky [4] to the so-called semi-automorphisms. As an easy consequence of our present theorem with $m=n(>1)$, we solve the problem about semi-automorphisms of a simple ring with descending chain condition for the left ideals.

It can be readily seen that a semi-automorphism is just a Jordan automorphism, provided that the charactaristic of the sfield is different from 2. Incidentally we solve the problem of Jordan isomorphism for simple ring with chain condition. How is the analogous problem for Lie isomorphism? The answer is not so perfect as the corresponding result for Jordan isomorphism. The author can only establish the result for sfield with characteristic different from 2 and 3. Nevertheless, in case the sfield is commutative our result is better than the known ones. (Jacobson [7])

Another application is to the Grassmann geometry, namely:

Let us consider the space formed by $(n-1)$-dimensional linear manifolds of the $(m+n-1)$-dimensional projective space. Two different $(n-1)$-dimensional manifold are said to be *coherent*, if both of them are contained in an n-dimensional linear manifold. By means of theorem 1, we can obtain all the one-to-one transformations keeping coherence invariant.

More precisely, let

$$(Z_{i\,1}, \cdots Z_{i\,m+n}) \qquad 1 \leqq i \leqq n$$

be n points in the $(m+n-1)$-dimensional space. Let the matrix

$$Z = (Z_{ij})_{1 \leqq i \leqq n,\, 1 \leqq j \leqq m+n}$$

denote the $(n-1)$-dimensional manifold spanned by these n points, where Z is assumed of rank n.

Two $n \times (m+n)$ matrices Z and Z_1, both of rank n, represent the same manifold if and only if there is a matrix $Q = Q^{(n)}$, naturally nonsingular, such that

$$Z_1 = QZ. \tag{3}$$

We shall study the space formed by all the $(n-1)$-dimensional linear manifolds. The space admits evidently the mapping

$$Z_1 = Q Z^{\sigma} T \tag{4}$$

where $Q = Q^{(n)}$ and $T = T^{(m+n)}$ are both nonsingular.

Analytically, two $(n-1)$-dimensional linear manifolds W and Z are coherent, if and only if the rank of the $2n \times (m+n)$-matrix

$$\begin{pmatrix} Z \\ W \end{pmatrix} \tag{5}$$

is equal to $n+1$.

It will be established as a consequence of theorem 1 that any one-to-one transformation which carries the space formed by all $(n-1)$-dimensional manifolds onto itself and leaves the coherence invariant is of the form (4). However, in case $m=n$, besides (4), it admits another type of transformations which will be described in the text.

Part I. Algebraic Part

2. Affine geometry. In this section we shall assume that $n=1$.

Theorem 1 fails for $n=1$, because the coherence is now merely a condition which asserts that the correspondence is one-to-one. For the sake of completeness, we give here the following corresponding theorem with an additional condition. Now we suppose $m>1$.

A *line* (or left line) is defined by those points given by

$$t(a_1, \cdots, a_m) + (b_1, \cdots, b_m), \tag{6}$$

where the a's and the b's are fixed and t runs over all the elements of the sfield Φ. Two lines are called intersecting if they have a point in common. This is called the *coherence relation between lines.*

Evidently, passing througth two points $(x_1, \cdots, x_m)$ and $(y_1, \cdots, y_m)$, we have a unique line

$$(1-t)(x_1, \cdots, x_m) + t(y_1, \cdots, y_m). \tag{7}$$

THEOREM 2. *Let $m>1$. Every one to-one mapping of the m-dimensional space (or left space) onto itself carrying lines into lines is of the form*

$$(z_1, \cdots, z_m) = (w_1, \cdots, w_m)^{\sigma} Q + (r_1, \cdots, r_m), \tag{8}$$

where $Q=Q^{(m)}$ is a non-singular matrix and σ is an automorphism of the sfield.

The converse of the theorem is also true, since it is evident that (8) carries lines into lines. The theorem should have been known to geometers, since its proof depends on the classical method of meet and join. However I could not find an exact reference for it. Nevertheless, the theorem is known at least for the real field (Veblen and Whitehead [8]). The proof is comparatively simpler there, since the identity is the only automorphism. For completeness a proof of theorem 2 is given in the next section

3. *Proof* of theorem 2.

1) $m=2$. Suppose that a pair of point (a_1, a_2) and (b_1, b_2) is carried into (c_1, c_2) and (d_1, d_2) respectively. Then the line joining the previous pair

$$(1-x)(a_1, a_2) + x(b_1, b_2)$$

is carried into the line joining the second pair

$$(1-x^\sigma)(c_1, c_2) + x^\sigma(d_1, d_2). \tag{16}$$

we shall now prove that $x \to x^\sigma$ is an automorphism of the sfield· Since there is a mapping of the form (8) carrying any pair of points, into $(0,0)$ and $(1,0)$ respectively, it is enough to prove our assertion for $(a_1, a_2) = (c_1, c_2) = (0,0)$ and $(b_1, b_2) = (d_1, d_2) = (1,0)$. That is, if the line $(t, 0)$ is carried into $(t^\sigma, 0)$ and $1^\sigma = 1$, $0^\sigma = 0$, then

$$(s+t)^\sigma = s^\sigma + t^\sigma, (st)^\sigma = s^\sigma t^\sigma.$$

These follow from the classical method of "meet and join" as illustrated respectively in the following figures:

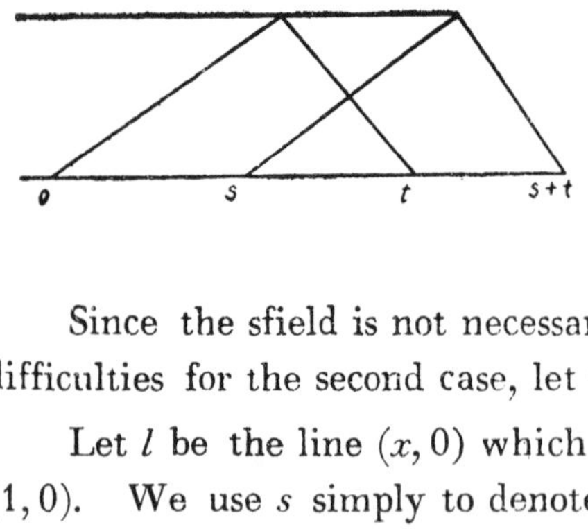

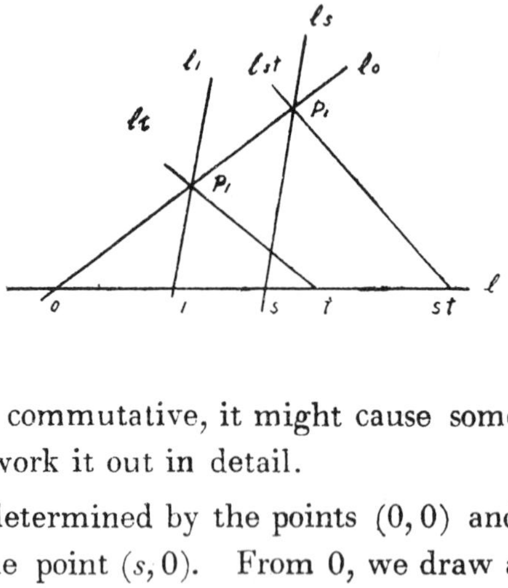

Since the sfield is not necessarily commutative, it might cause some difficulties for the second case, let us work it out in detail.

Let l be the line $(x, 0)$ which is determined by the points $(0,0)$ and $(1,0)$. We use s simply to denote the point $(s, 0)$. From 0, we draw a line l_0: $((xl_0, x), x \varepsilon \Phi)$ which is different from 1, and from 1, we draw a line l_1 $((xl_1 + 1, x), x \varepsilon \Phi)$ which is different from l and is not parallel* to l_0 $(i.e. l_0 \neq l_1)$ The intersection of the lines l_0 and l_1 is denoted by $p_1((l_0 - l_1)^{-1} l_0, (l_0 - l_1)^{-1})$. Parallel to l_1 and passing through s, we draw a line l_s $(xl_1 + s, x)$. The intersection of l_s and l_0 is denoted by $p_2(s(l_0 - l_1)^{-1} l_0, s(l_0 - l_1)^{-1})$. The line l_t joining p_1 and t is denoted by

*By parallel, we mean that the two lines do not intersect.

$$((1-y)((l_0-l_1)^{-1}l_0,(l_0-l_1)^{-1})+y(l,0)).$$

parallel to l_t and passing through p_2 we draw a line l_{st}:

$$((s(l_0-l_1)^{-1}l_0, s(l_0-l_1)^{-1})+x(t-(l_0-l_1)^{-1}l_0,-(l_0-l_1)^{-1})).$$

The intersection of l_{st} and the basic line l is given by $x=st$, that is, it is the point $(st,0)$. This completes the proof of the statement.

Let $A(x_1,x_2)$ be the mapping under consideration. Since any three points not on a line can be carried into $(0,0),(1,0)$ and $(0,1)$, we may assume that

$$A(0,0)=(0,0),\quad A(1,0)=(1,0),\quad A(0,\ 1)=(0,1).$$

After an automorphism, we may assume that, by the result proved before,

$$A(x,0)=A(x,0)$$

holds for all x belonging to Φ and,

$$A(0,y)=(0,y^\sigma)$$

for all y belonging to Φ. Then the point (x,y), being the intersection of a line passing through $(x,0)$ and parallel to y-axis with a line passing through $(0,y)$ and parallel to x-axis, is mapped into (x,y^σ), i. e.

$$A(x,y)=(x,y^\sigma). \tag{17}$$

By (16) with $(a_1,a_2)=(c_1\,c_2)=0$, $(b_1\,b_2)=(x,y)$ and $(d_1,d_2)=(x,y^\sigma)$ we have

$$A(t(x,y))=(t^\tau(x,y^\sigma)). \tag{18}$$

On the other hand, substituting tx and ty for x and y in (17) we obtain

$$A(tx,ty)=(tx,(ty)^\sigma). \tag{17_1}$$

Comparing (18) and (17_1), we have

$$\tau=\sigma=1.$$

Therefore

$$A(x,y)=(x,y)$$

for all x and y in Φ. The theorem is thus proved.

2) Let $m > 2$. The theorem can be proved by induction, and we shall leave to the reader to generalize it.

Remark. Without any essential difficulty, we can extend our theorem to the following form: Let Φ_m $(m > 1)$ be the m-dimensional affine (left or right) space over the sfield Φ. If there is a one-to-one mapping carrying Φ_m onto $\Phi'_{m'}$, and carrying lines into lines, then we have $m = m'$ and that either Φ and Φ' are isomorphic or they are anti-isomorphic. In the first case, Φ_m and Φ'_m are both left or right spaces; if we identify Φ' with Φ and suppose that Φ_m and Φ'_m are both left spaces, we obtain the transformation (2). For the second case, one of Φ_m and Φ'_m is the left and the other the right space; suppose Φ_m is the left space and Φ'_m is the right space the mapping can be written as

$$(w_1, \cdots w_m) \longrightarrow (Q(w_1 \cdots w_m)^{\tau})' + (r_1 \cdots r_m)'$$

where τ denotes the anti-isomorphism.

§ 4. Maximal set of rank 1.

Definition. A set of $n \times m$ matrices over Φ is called *maximal* (of rank 1), if any pair among them is coherent and if there is no other $n \times m$ matrix, outside of the set, which is coherent to each element of the set.

THEOREM 4. *Every maximal set is equivalent under the group* (1) *to either*

$$\begin{pmatrix} x_1 & x_2 & \cdots & x_m \\ 0 & 0 & \cdots & 0 \\ \cdot & \cdot & \cdot & \cdot \\ 0 & 0 & \cdots & 0 \end{pmatrix} \tag{19}$$

or

$$\begin{pmatrix} y_1 & 0 & \cdots & 0 \\ y_2 & 0 & \cdots & 0 \\ \cdot & \cdot & \cdot & \cdot \\ y_n & 0 & \cdots & 0 \end{pmatrix} \tag{20}$$

Proof. We may suppose that the set contains two points 0 and

$$Z_0=\begin{pmatrix} 1\,0\cdots 0 \\ 0\,0\cdots 0 \\ \cdots\cdots \\ 0\,0\cdots 0 \end{pmatrix}.$$

Let Z be another point of the set. Since Z and 0 are coherent, the rank of Z is 1, we have therefore

$$Z=\begin{pmatrix} a_1 b_1 & \cdots & a_1 b_m \\ \cdots & \cdots & \cdots \\ a_n b_1 & \cdots & a_n b_m \end{pmatrix}.$$

Since $Z-Z_0$ is of rank 1, we have

$$a_i b_j=0, \quad \text{for} \quad 2\leqq i\leqq n,\ 2\leqq j\leqq m.$$

If $a_2,\cdots,a_n$ are not all zero, we have $b_2=\cdots=b_m=0$, that is

(21)
$$Z=\begin{pmatrix} a_1 b_1\, 0 \cdots 0 \\ a_2 b_1\, 0 \cdots 0 \\ \cdots\cdots\cdots \\ a_n b_1\, 0 \cdots 0 \end{pmatrix}.$$

Otherwise, we have

(22)
$$Z=\begin{pmatrix} a_1 b_1 & \cdots & a_1 b_m \\ 0 & \cdots & 0 \\ \cdots & \cdots & \cdots \\ 0 & \cdots & 0 \end{pmatrix}.$$

The set of elements of the form

(23)
$$\begin{pmatrix} z\,0\cdots 0 \\ 0\,0\cdots 0 \\ \cdots\cdots \\ 0\,0\cdots 0 \end{pmatrix}$$

is evidently not a maximal set. If the set contains a point besides that of (23), say a point (21) with $a_2,\cdots,a_m$ not all zero, then none of the other elements of the set can be of the form (22). This asserts that the maximal set containing 0 and Z_0 is either (19) or (20), since evidently every pair of elements of (19) or (20) is coherent.

The above proof enables us also to establish the following theorem:

THEOREM 5. *Given any two coherent matrices, there are two and only two maximal sets containing both of them.*

§ 5. Intersections of maximal sets.

Now we consider the intersection of two maximal sets. Supposing that their intersection is not empty, we may take 0 to be one of their common point. A maximal set containing 0 is either of the form

$$P\begin{pmatrix} y_1 y_2 & \cdots & y_m \\ 0, 0 & \cdots, & 0 \\ \cdot & \cdots & \cdot \\ 0\ 0 & \cdots & 0 \end{pmatrix} Q,\quad P = P^{(n)},\ Q = Q^{(m)},$$

or

$$P\begin{pmatrix} x_1, 0, \cdots, 0 \\ x_2, 0, \cdots, 0 \\ \cdots\cdots \\ x_n, 0, \cdots, 0 \end{pmatrix} Q.$$

In fact, let Z be a point of rank 1, then we have non-singular P and Q such that

$$P^{-1} Z Q^{-1} = \begin{pmatrix} 1, 0, \cdots, 0 \\ 0, 0, \cdots, 0 \\ \cdots\cdots \\ 0, 0, \cdots, 0 \end{pmatrix}.$$

This establishes the assertion by the result of § 4. Therefore we may assume that the two maximal sets are either

$$\text{(i)}\qquad \begin{pmatrix} x_1, \cdots, x_m \\ 0, \cdots, 0 \\ \cdots\cdots \\ 0, \cdots, 0 \end{pmatrix},\quad P\begin{pmatrix} y_1, \cdots, y_m \\ 0, \cdots, 0 \\ \cdots\cdots \\ 0, \cdots, 0 \end{pmatrix} Q,$$

or

$$\text{(ii)}\qquad \begin{pmatrix} x_1, 0, \cdots, 0 \\ x_2, 0, \cdots, 0 \\ \cdots\cdots \\ x_n, 0, \cdots, 0 \end{pmatrix}, \quad P\begin{pmatrix} y_1, \cdots, y_m \\ 0, \cdots 0 \\ \cdots\cdots \\ 0, \cdots 0 \end{pmatrix}Q$$

or

$$\text{(iii)}\qquad \begin{pmatrix} x_1, 0, \cdots, 0 \\ x_2, 0, \cdots, 0 \\ \cdots\cdots \\ x_n, 0, \cdots, 0 \end{pmatrix}, \quad P\begin{pmatrix} y_1, 0, \cdots, 0 \\ y_2, 0, \cdots, 0 \\ \cdots\cdots \\ y_n, 0, \cdots, 0 \end{pmatrix}Q.$$

The first and the third cases can be treated similarly.

Now we consider the case (*i*). Without loss of generality we may assume that $Q=I$. Let

$$P=(p_{rs})_{1\leqq r,s\leqq n}.$$

The intersection is given by those y satisfying

$$(p_{21}, \cdots p_{n1})'(y_1, \cdots, y_m)=0.$$

If $(p_{21},\cdots,p_{n1})$ is a zero vector, both maximal sets are identical. If $(p_{21},\cdots,p_{n1})$ is not a zero vector, then $(y_1,\cdots,y_m)$ is a zero vector. That is, the maximal sets have only one point in common.

For the second case, the transformation $Z_1=P^{-1}Z$ carries both maximal sets of (ii) into those of (ii) with $P=I$ and $Q=I$. Therefore the intersection of two maximal sets is given by

$$\text{(24)}\qquad \begin{pmatrix} x, 0, 0, \cdots, 0 \\ 0, 0, 0, \cdots, 0 \\ \cdots\cdots \\ 0, 0, 0, \cdots, 0 \end{pmatrix}.$$

Therefore the intersection of two distinct maximal sets is either empty, a single point or a set equivalent to (24) under the group (1).

Definition. The intersection of two distinct maximal sets, which contains more than one point in common, is called a *line*.

THEOREM 6. *Let X and Y be a pair of coherent matrices. There is one and only one line passing through them.*

Proof. Without loss of generality, we may assume that

$$X = 0, \quad Y = \begin{pmatrix} 1, 0, \cdots, 0 \\ 0, 0, \cdots, 0 \\ \cdots\cdots \\ 0, 0, \cdots, 0 \end{pmatrix}.$$

By Theorem 5, there aro two (and only two) maximal sets containing both; they are (19) and (20). The intersection is given by

$$\begin{pmatrix} t, 0, \cdots, 0 \\ 0, 0, \cdots, 0 \\ \cdots\cdots \end{pmatrix}.$$

Therfore we have the theorem.

The general equation of a line can be put into the form

$$p' t q + W$$

where p and q are two vectors of dimension n and m respectively Furthermore, the general equation of a line in the maximal set (19) is of the form

$$t\begin{pmatrix} a_1\, a_2 \cdots a_m \\ 0 \end{pmatrix} + \begin{pmatrix} b_1\, b_2 \cdots b_m \\ 0 \end{pmatrix} \tag{25}$$

and in the maximal set (20) is of the form

$$\begin{pmatrix} a_1 & \\ a_2 & 0 \\ \vdots & \\ a_n & \end{pmatrix} t + \begin{pmatrix} b_1 & \\ b_2 & 0 \\ \vdots & \\ b_n & \end{pmatrix}. \tag{26}$$

§6. Linear subspaces.

Let us now consider the geometry in a maximal set. We call, sometimes, a line a linear subspace of dimension 1. Let l be a line of the maximal set. Let p be a point of the maximal set outside of l. Draw all lines in the maximal set passing through p and intersecting l. The set of points on all these lines is defined to be a linear subspace of dimension 2. Let q be a linear subspace of dimension 2, and let p be a point of the maximal space but not in q. We construct the lines passing through p and intersecting q. The set of all the points on all these lines is called a linear subspace of dimension 3, etc. Evidently we have

THEOREM 7. *Every linear subspace of dimension d can be constructed from any line of it by the previous process and the number of steps is exactly $d-1$. d is invariant under any mapping leaving the coherence invariant.*

Therefore, we have

THEOREM 8. *The maximal set* (19) *is an affine space of dimension m and* (20) *is of dimension n.*

Evidently, for $m=n$, (19) is equivalent to (20) under a transformation of (2) which carries right lines into left lines and it can occur only when the field has an anti-automorphism. Moreover, if $m \neq n$, any mapping leaving the coherence invariant cannot carry (19) into (20).

§ 7. Arithmetic distance.

Definition. Two matrices X_0 and X_r are said to be of arithmetic distance r, if there exist $r-1$ matrices

$$X_1, \cdots, X_{r-1}$$

such that X_i and X_{i-1} $(1 \leqq i \leqq r)$ are coherent, and r is the least integer having this property.

THEOREM 9. *Two matrices X and Y are of arithmetic distance r, if and only if $X-Y$ is of rank r.*

Proof. If $X-Y$ is of rank r, we may assume that $X=0$ and

$$Y = \begin{pmatrix} I^{(r)} & 0 \\ 0 & 0 \end{pmatrix}$$

where $I^{(r)}$ denotes the r-rowed identity. Then

$$X_\varrho = \begin{pmatrix} I^{(\varrho)} & 0 \\ 0 & 0 \end{pmatrix}, \quad 1 \leqq \varrho \leqq r-1,$$

are intermediate points, therefore the arithmetic distance between X and Y is $\leqq r$.

Further, from the well-known property that

$$\text{rank } (P+Q) \leqq \text{rank } P + \text{rank } Q$$

we deduce that, if X and Y are of arithmetic distance r, the rank of $X-Y$ is $\leqq r$. The theorem is thus proved.

THEOREM 10. *Any r maximal sets of rank 1, which have a unique point in common and such that no two of them have any other common point can be carried by* (1) *simultaneously into*

$$(27) \qquad \begin{pmatrix} x_{11}, \cdots, x_{1m} \\ 0, \cdots, 0 \\ \cdots\cdots \\ 0, \cdots, 0 \end{pmatrix}, \begin{pmatrix} 0, \cdots, 0 \\ x_{21}, \cdots, x_{2m} \\ 0, \cdots, 0 \\ \cdots\cdots \end{pmatrix}, \cdots, \begin{pmatrix} 0, \cdots, 0 \\ \cdots\cdots \\ x_{r1}, \cdots, x_{rm} \\ 0, \cdots, 0 \end{pmatrix}$$

or into

$$(28) \qquad \begin{pmatrix} y_{11}, 0, \cdots, 0 \\ y_{21}, 0, \cdots, 0 \\ \cdots\cdots \\ y_{n1}, 0, \cdots, 0 \end{pmatrix}, \begin{pmatrix} 0, y_{12}, 0, \cdots, 0 \\ 0, y_{22}, 0, \cdots, 0 \\ \cdots\cdots \\ 0, y_{n2}, 0, \cdots, 0 \end{pmatrix}, \cdots, \begin{pmatrix} 0, \cdots, 0, y_{1r}, 0, \cdots, 0 \\ 0, \cdots, 0, y_{2r}, 0, \cdots, 0 \\ \cdots\cdots \\ 0, \cdots, 0, y_{nr}, 0, \cdots, 0 \end{pmatrix}$$

Proof. In fact, let the common point be 0 and one of the maximal sets of rank 1 be

$$\begin{pmatrix} x_{11}, \cdots, x_{1m} \\ 0, \cdots, 0 \\ \cdots\cdots \\ 0, \cdots, 0 \end{pmatrix}$$

(Owing to similarity, we omit the other case).

Let

$$Z_1 = \begin{pmatrix} a_1 b_1, \cdots, a_1 b_m \\ \cdots\cdots\cdots \\ a_n b_1, \cdots, a_n b_m \end{pmatrix}$$

be a point of the second maximal set of rank 1. Then, we have $P^{(n-1)}$ and $Q^{(m)}$ such that

$$\begin{pmatrix} 1 & 0 \\ 0 & P^{(n-1)} \end{pmatrix} Z_1 Q^{(m)} = \begin{pmatrix} c_1 & c_2 & \cdots & c_m \\ 1 & 0 & \cdots & 0 \\ 0 & 0 & \cdots & 0 \\ \cdots & & & \\ 0 & 0 & \cdots & 0 \end{pmatrix}$$

Since it is of rank 1, we actually have $c_2 = \cdots = c_m = 0$. (Notice that such at transformation leaves the first maximal set invariant.) The transformation

$$Z_1 = \begin{pmatrix} 1 & -c_1 & 0 & \cdots & 0 \\ 0 & 1 & 0 & \cdots & 0 \\ 0 & 0 & 0 & \cdots & 0 \\ \cdots & & & & \\ 0 & 0 & 0 & \cdots & 0 \end{pmatrix} Z$$

leaves the first maximal set fixed and the previous point into

$$\begin{pmatrix} 0 & 0 & 0 & \cdots & 0 \\ 1 & 0 & 0 & \cdots & 0 \\ 0 & 0 & 0 & \cdots & 0 \\ \cdots & & & & \\ 0 & 0 & 0 & \cdots & 0 \end{pmatrix},$$

which spans with 0 a maximal set of rank 1:

$$\begin{pmatrix} 0, & 0 & \cdots & 0 \\ x_{21} & x_{22} & \cdots & x_{2m} \\ 0 & 0 & \cdots & 0 \\ \cdots & & & \end{pmatrix} \tag{29}$$

or

$$(30)\qquad \begin{pmatrix} y_{11} & 0 & \cdots & 0 \\ y_{21} & 0 & \cdots & 0 \\ \cdot & \cdot & \cdots & \cdot \\ y_{n1} & 0 & \cdots & 0 \end{pmatrix}$$

(30) cannot happen, since the intersection of the first and the second maximal sets contains more than one point.

Proceeding in this way, we have Theorem 10.

§8. *Proof* of theorem 1.

We write

$$X = \begin{pmatrix} x_1 \\ \cdot \\ \cdot \\ \cdot \\ x_n \end{pmatrix} = (y_1, \cdots, y_m)$$

where x_i is a vector $(x_{i1}, \cdots, x_{im})$ and y_j is a column

$$y_j = \begin{pmatrix} y_{1j} \\ \cdot \\ \cdot \\ \cdot \\ y_{nj} \end{pmatrix}$$

we use also X_i and Y_j to denote respectively the matrices

$$\begin{pmatrix} 0 \\ \cdot \\ \cdot \\ \cdot \\ x_i \\ 0 \\ \cdot \\ \cdot \\ \cdot \end{pmatrix}, (0, \cdots, 0, y_j, 0, \cdots, 0).$$

Let $A(X)$ be a mapping considered in Theorem 1. By Theorem 10, we may assume that

$$A(X_i) = X_i^*, \quad A(0) = 0.$$

By the result of § 6, A induces a mapping of x_i onto x_i^* carrying lines into lines. Therefore, by Theorem 2, we have

$$x_i^* = x_i^{\sigma} Q_i \tag{31}$$

where $Q_i = Q_i^{(m)}$ is non-singular and σ_i is an automorphism of Φ. Without loss of generality we can assume

$$\sigma_1 = 1, \quad Q_1 = I^{(m)}. \tag{32}$$

The maximal set of rank 1, other than X_1 containing both 0 and $(0, 0, \cdots 0, x_{1j}, 0, \cdots 0)$ is Y_j. Thus

$$A(Y_j) = Y_j^*.$$

By Theroem 2, we have

$$y_j^* = P_j y^{\tau_j} \tag{33}$$

where $P_j = P_j^{(n)}$ is non-singular and τ_j is an automorphism of Φ.

Let $x_{ij} E_{ij}$ be the intersection of X_i and Y_j, then we have

$$A(x_{ij} E_{ij}) = x_{ij}^* E_{ij}.$$

Consequently, Q_i and P_j are diagonal matrix. Write

$$Q_i = [q_{i1}, \cdots, q_{im}], \quad P_j = [p_{1j}, \cdots, p_{nj}].$$

Comparing (31) and (33), we have

$$x_{ij}^{\sigma_i} q_{ij} = p_{ij} x_{ij}^{\tau_j}, \tag{34}$$

which holds for all x_{ij} of Φ. In particular, for $x_{ij} = 1$, we have $q_{ij} = p_{ij}$. Therefore, for $i = 1$, $p_{ij} = 1$ and $\tau_j = 1$ for all $j = 1, \cdots, n$. And then σ_i's are inner automorphisms. If we use $P_1^{-1} A$ instead of A we may assume without loss of generality, that A satisfies (31), (32), (33) and

$$P_1 = I^{(n)}.$$

Then from (34) for $j = 1$ we deduce that $q_{i1} = 1$ and $\sigma_i = 1$ for $i = 1, \cdots, m$,

Further, the linear dependence of x_1 and x_i implies that of $x_1 = x_1 Q_1$ and $x_i Q_i$. Thus $Q_i = p_i I$. Since $Q_i = (1, q_{i2}, \cdots, q_{in})$, $Q_i = I$ for all i. Similarly, $P_i = 1$ for all j. Thus we have

$$A(X_i) = X_i, \quad A(Y_j) = Y_j. \tag{35}$$

Now we consider

$$A\begin{pmatrix} x_1 \\ \cdot \\ \cdot \\ \cdot \\ x_n \end{pmatrix} = \begin{pmatrix} x_1^* \\ \cdot \\ \cdot \\ \cdot \\ x_n^* \end{pmatrix}$$

Suppose that the rank of $\begin{pmatrix} x_1 \\ \cdot \\ \cdot \\ \cdot \\ x_n \end{pmatrix}$ is n, so is $\begin{pmatrix} x_1{}^* \\ \cdot \\ \cdot \\ \cdot \\ x_n{}^* \end{pmatrix}$, by Theorem 8. Since

$$A\begin{pmatrix} x_1 - \lambda_2 x_2 - \cdots - \lambda_n x_n \\ 0 \\ \cdot \\ \cdot \\ \cdot \\ 0 \end{pmatrix} = \begin{pmatrix} x_1 - \lambda_2 x_2 - \cdots - \lambda_n x_n \\ 0 \\ \cdot \\ \cdot \\ \cdot \\ 0 \end{pmatrix}.$$

the rank of

$$\begin{pmatrix} x_1^* - x_1 - \lambda_2 x_2 - \cdots - \lambda_n x_n \\ x_2^* \\ \cdot \\ \cdot \\ \cdot \\ x_n^* \end{pmatrix}$$

is less than n for any λ. Therefore

$$x_1 = x_1^* + \sum_{j=2}^{n} \mu_j x_j^*$$

$$x_k = \sum_{j=2}^{n} \mu_{kj} x_j^*, \quad 2 \leq k \leq n.$$

Applying the same method to the i-th row instead of the first row, we have

$$x_i = x_i^* + \sum_{\substack{j=1 \\ j \neq i}}^{n} v_{ij}^{(i)} x_j^*$$

$$x_k = \sum_{\substack{j=1 \\ j \neq i}}^{n} v_{kj}^{(i)} x_j^*, \quad k \neq i.$$

Eliminating x_1, we have

$$x_1^* = \sum_{\substack{j=1 \\ j \neq i}}^{n} v_{ij}^{(i)} x_j^* - \sum_{j=2}^{n} \mu_j x_j^*$$

Because of the independence of $x_1^*, \cdots, x_n^*$, we have

$$\mu_j = 0 \quad 2 \leqq i \leqq n.$$

Thus

$$x_1 = x_1^*.$$

Similarly, we have $x_i = x_i^*$, that is, if X is of rank n, we have

(36) $$AX = X.$$

As to the singular case, the theorem can be proved as follows: Let

$$A \begin{pmatrix} x_1 \\ \cdot \\ \cdot \\ \cdot \\ x_n \end{pmatrix} = \begin{pmatrix} x_1^* \\ \cdot \\ \cdot \\ \cdot \\ x_n^* \end{pmatrix}$$

If $x_{11} \neq 0$, we have, by (34), that

$$A \begin{pmatrix} x_{11} & x_{12} & x_{13} & \cdots & x_{1n} & x_{1n+1} & \cdots & x_{1m} \\ 0 & \lambda_2 & 0 & \cdots & 0 & 0 & \cdots & 0 \\ 0 & 0 & \lambda_3 & \cdots & 0 & 0 & \cdots & 0 \\ \cdot & \cdot & \cdot & \cdot & \cdot & \cdot & \cdot & \cdot \\ 0 & 0 & 0 & & \lambda_n & 0 & \cdots & 0 \end{pmatrix} = \begin{pmatrix} x_{11} & x_{12} & x_{13} & \cdots \\ 0 & \lambda_2 & 0 & \cdots \\ 0 & 0 & \lambda_3 & \cdots \\ \cdot & \cdot & \cdot & \cdot \\ 0 & 0 & & \cdots \end{pmatrix}$$

Then

$$\begin{pmatrix} x_{11}^* - x_{11} & x_{12}^* - x_{12} & x_{13}^* - x_{13} & \cdots \\ x_{21}^* & x_{22}^* - \lambda_2 & x_{23}^* & \cdots \\ x_{31}^* & x_{32}^* & x_{32}^* - \lambda_3 & \cdots \\ \cdots\cdots & \cdots\cdots & \cdots\cdots & \end{pmatrix}$$

is of rank $< n$ for any $\lambda_2 \lambda_3 \cdots \neq 0$. Therefore the only possible case is that

$$x_{11}^* - x_{11} = 0.$$

If $x_{11}^* \neq 0$, we proceed in the reverse way. Thus we have $x_{11}^* = x_{11}$ Similarly, we have

$$x_{ij}^* = x_{ij}.$$

The theorem is thus proved completely.

Remark. Without any essential difficulty, we can extend our theorem to the following form: Let $\Phi_{n,m}$ $(l < n \leqq m)$ be the totality of all the $n \times m$ matrices over the sfield Φ. If there is a one to one mapping carrying $\Phi_{n,m}$ onto $\Phi'_{n',m'}$ and carrying the coherence relation of the one into the other, then we have either $n = n'$ $m = m'$ or $n = m', m = n'$: For the first case, the sfields Φ and Φ' are isomorphic; if we identify them, we obtain the transformation (1). For the second case, the sfield Φ and Φ' are anti-isomorphisc, the mapping can be written as

$$(P Z^{\tau} Q + R)'$$

where τ denotes the anti-isomorphism.

Part II. Geometry of Rectangular Matrices

§8. Projective geometry of rectangular matrices.

We introduce the homogeneous coordinates of a rectangular matrix $Z(= Z^{(n,m)})$. We write

$$Z = (X, Y)$$

where $Y=Y^{(n)}$, $X=X^{(n,m)}$. The pair of matrices

$$(X, Y)$$

is called a homogeneous coordinate of the matrix Z. A necessary and sufficient condition that two pairs of matrices

$$(X, Y),\quad (X_1, Y_1)$$

represent the same matrix Z is that we have a non-singular $Q\,(=Q^{(n)})$ such that

$$Q(X, Y)=(X_1, Y_1). \tag{35}$$

Now we extend this notion to the pairs with singular Y.

We start with an $n\times(n+m)$-matrix W of rank n. Two matrices W and W_1 are said to be equivalent, if there is a matrix $Q\,(=Q^{(n)})$, (automatically non-singular) such that

$$W^*=QW. \tag{36}$$

This equivalence relation classifies $n\times(n+m)$ matrices into classes. We identify the matrices belonging to a class as a point. The totality of these points form the *projective space of rectangular matrices*. In case $n=1$, this reduces to the ordinary projective geometry.

The space admits the transformation

$$W^*=QWP \tag{37}$$

where $P=P^{(m+n)}$, $Q=Q^{(n)}$ are non-singular. The space is evidently transitive. Two different points W_1 and W_2 are said to be coherent, if the rank of

$$\begin{pmatrix} W_1 \\ W_2 \end{pmatrix} \tag{38}$$

is $n+1$. This is the smallest possible value for distinct W_1 and W_2. In fact, if (38) is of rank n, since W_1 is of rank n, each row of W_2 depends on those of W_1, and consequently

$$W_2 = QW_1$$

That is, W_1 and W_2 represent the same point.

Evidently, (37) leaves the coherence invariant, since

$$\begin{pmatrix} W_1^* \\ W_2^* \end{pmatrix} = \begin{pmatrix} Q_1 & O \\ O & Q_2 \end{pmatrix} \begin{pmatrix} W_1 \\ W_2 \end{pmatrix} P.$$

Let us break W up into two parts

$$W = (X, Y), \quad X = X^{(n,m)}, \quad Y = Y^{(n)}.$$

Those with non-singular Y are called finite points and those with singular Y are called points at infinity. If $W^* = (X^*, Y^*)$ and $W = (X, Y)$ are both finite points, we have the birational transformation

(39) $$Z^* = Y^{*-1} X^* = (ZB + D)^{-1}(ZA + C),$$
where

$$P = \begin{pmatrix} A & B \\ C & D \end{pmatrix}, A = A^{(m)}, B = B^{(m\ n)}, C = C^{(n,m)}, D = D^{(n)}.$$

Two finite points Z and Z_1 are coherent, if and only if the rank of $Z - Z_1$ is one. In fact, we have now

$$\begin{pmatrix} X & Y \\ X_1 & Y_1 \end{pmatrix} = \begin{pmatrix} Y & O \\ O & Y_1 \end{pmatrix} \begin{pmatrix} Z & I \\ Z_1 & I \end{pmatrix} = \begin{pmatrix} Y & O \\ O & Y_1 \end{pmatrix} \begin{pmatrix} I & I \\ O & I \end{pmatrix} \begin{pmatrix} Z - Z_1 & O \\ Z_1 & I \end{pmatrix}.$$

THEOREM 10. *Any coherent pair of points can be carried simultaneously, into*

$$(O, I^{(n)})$$

and

$$(N, I^{(n)})$$

where N is a matrix with 1 at $(1, 1)$-position and zero elsewhere.

Proof. Since the space is transitive, we may assume that one of the points is

$$(O, I^{(n)})$$

Break the coordinates of the other point into

$$(X^{(n,m)}, Y^{(n)}). \tag{40}$$

Then X is of rank 1. We have matrices $Q(=Q^{(n)})$ and $P(=P^{(m)})$ such that

$$QXP = N.$$

Then, we have

$$Q(X\,Y)\begin{pmatrix} P & O \\ O & I \end{pmatrix} = (N, QY)$$

and

$$(O\,I)\begin{pmatrix} P & O \\ O & I \end{pmatrix} = (O, I)$$

Therefore we may assume that $X=N$ in (40).

If Y is non-singular, then

$$(N, Y)\begin{pmatrix} I & O \\ O & Y^{-1} \end{pmatrix} = (N, I)$$

and

$$Y(O, I)\begin{pmatrix} I & O \\ O & Y^{-1} \end{pmatrix} = (O, I).$$

The theorem is proved.

If Y is singular, sinee (N, Y) is of rank n, we can write Y as

$$\begin{pmatrix} \lambda_2 y_2 + \cdots + \lambda_n y_n \\ y_2 \\ \cdot \\ \cdot \\ \cdot \\ y_n \end{pmatrix}$$

where $y_2, \cdots, y_n$ are independent vectors. There exists a vector z such that

$$\begin{pmatrix} z+\lambda_2 y_2+\cdots+\lambda_n y_n \\ y_2 \\ \cdot \\ \cdot \\ \cdot \\ y_n \end{pmatrix}$$

is of rank n. Let

$$C=\begin{pmatrix} z \\ 0 \\ \cdot \\ \cdot \\ \cdot \\ 0 \end{pmatrix}$$

Then

$$(O, I)\begin{pmatrix} I & C \\ O & I \end{pmatrix}=(O, I)$$

and

$$(N, Y)\begin{pmatrix} I & C \\ O & I \end{pmatrix}=(N, NC+Y)$$

where $NC+Y$ is non-singular. The theorem reduces to the cases considered previously.

§ 9. Projective geometry of square matrices.

In case $m=n$, there is an extra type of transformations which corresponds to the non-homogeneous transformation of the form

$$Z^*=(Z^{\tau\prime}B+D)^{-1}(Z^{\tau\prime}A+C) \tag{41}$$

where τ is an anti-automorphism of the sfield. Among such transformations the most essential one is

$$Z^*=Z^{\tau\prime}. \tag{42}$$

If homogeneous coordinates are again introduced, we have

$$Y^{*-1}X^*=X^{\tau\prime}Y^{\tau\prime-1},$$

that is

$$X^*Y^{\tau\prime}=Y^*X^{\tau\prime}.$$

we write this relation as

(43) $$(X^* Y^*)\mathfrak{F}(X, Y)^{\tau'} = 0.$$

where

(44) $$\mathfrak{F} = \begin{pmatrix} O & I \\ -I & O \end{pmatrix}.$$

Now we are going to prove that (43) defines a one-to-one mapping carrying the projective space of matrices onto itself. Let (X_1, Y_1) be another point which satisfies

$$(X_1, Y_1)\mathfrak{F}(X, T)^{\tau'} = 0.$$

Then, we have

$$\begin{pmatrix} X^* & Y^* \\ X_1 & Y_1 \end{pmatrix}\mathfrak{F}(X, Y)^{\tau'} = 0.$$

By Sylvester's law of nullity, we deduce that

$$\begin{pmatrix} X^* & Y^* \\ X_1 & Y_1 \end{pmatrix}$$

is of rank $\leqq n$. Since (X^*, Y^*) is of rank n, we have a non-singular Q such that

$$(X^*, Y^*) = Q(X_1, Y_1).$$

More generally, let $\mathfrak{R}$ be any $2n$-rowed non-singular matrix. Then

(45) $$(X^*, Y^*)\mathfrak{R}(X, Y)^{\tau'} = 0$$

defines a mapping from (X, Y) into (X^*, Y^*).

The product of two such relations is a transformation of the form (1). In fact, let $\sigma = \tau\tau_1$ which is an automorphism of the field, then from

$$(X_1, Y_1)\mathfrak{R}_1(X, Y)^{\tau'} = 0$$

and

$$(X_2, Y_2)\mathfrak{R}_2(X_1, Y_1)^{\tau_1'} = 0$$

we deduce

$$\begin{pmatrix} (X, Y)^\sigma \mathfrak{R}_1^{\tau_1}{}' \\ (X_2, Y_2)\mathfrak{R}_2 \end{pmatrix} (X_1, Y_1)^{\tau_1}{}' = 0$$

By Sylvester's law of nullity, we have

$$(X_2, Y_2)\mathfrak{R}_2 = Q(X, Y)^\sigma \mathfrak{R}_1^{\tau_1}{}'$$

i. e.

(46) $$(X_2, Y_2) = Q(X, Y)^\sigma \mathfrak{R}_1^{\tau_1}{}' \mathfrak{R}_2$$

which is a transformation of the form (1).

This establishes also the following

THEOREM 11. *The enlarged group generated by all the transformations of* (1) *and all the relations* (45) *is simply generated by* (1) *and one of relations of* (45), *say for example* (43).

Now we are going to state the fact that the coherence also remains unaltered under the enlarged group.

THEOREM 12. *The mapping* (45) *keeps the arithmetic distance invariant. More precisely, let*

$$(X^*, Y^*)\mathfrak{R}(X, Y)^{\tau\prime} = 0$$

and

$$(X_1^*, Y_1^*)\mathfrak{R}(X_1, Y_1)^{\tau\prime} = 0.$$

The two matrices

$$\begin{pmatrix} X & Y \\ X_1 & Y_1 \end{pmatrix} \quad \text{and} \quad \begin{pmatrix} X^* & Y^* \\ X_1^* & Y_1^* \end{pmatrix}$$

have the same rank

Proof. We have a transformation (1) carrying (X, Y) and (X_1, Y_1) into (O, I) and (L_r, I), where

$$L_r = \begin{pmatrix} I^{(r)} & O \\ O & O \end{pmatrix};$$

let the transformation be

$$(X, Y) = Q(U, V)\, T.$$

Let

$$(X^*, Y^*)\,\mathfrak{R}\, T^{\tau\prime} = (U^*, V^*)$$

then, we have

$$(U^*, V^*)(O, I)^{\tau\prime} = 0$$

and

$$(U_1^*, V_1^*)(L_r, I)^{\tau\prime} = 0.$$

Consequently

$$V^* = 0, \quad U_1^* L_r + V_1^* = 0.$$

The rank of

$$\begin{pmatrix} U^* & V^* \\ U_1^* & V_1^* \end{pmatrix} = \begin{pmatrix} U^* & O \\ U_1^* & -U_1^* L_r \end{pmatrix}$$

is equal to the rank of $U_1^* L_r$, i.e. $\leqq n+r$. Therefore we have

$$\text{rank of}\begin{pmatrix} X & Y \\ X_1 & Y_1 \end{pmatrix} = \text{rank of}\begin{pmatrix} U & V \\ U_1 & V_1 \end{pmatrix} \geqq \text{rank of}\begin{pmatrix} U^* & V^* \\ U_1^* & V_1^* \end{pmatrix} = \text{rank of}\begin{pmatrix} U^* & V^* \\ U_1^* & V_1^* \end{pmatrix}$$

By symmetry, we have the theorem.

From (46), we deduce easily that (45) is involutory, if and only if $\sigma = 1$ and $\mathfrak{R} = \varrho\, \mathfrak{R}^{\tau\prime}$ where ϱ belongs to the center of Φ. Consequently $\mathfrak{R} = \varrho\,(\varrho\,\mathfrak{R}^{\tau\prime})' = \varrho\,\varrho^{\tau}\,\mathfrak{R}$, that is $\varrho\,\varrho^{\tau} = 1$. Such an involutorial correspondence (45) is called a duality. The fixed elements (if exist) of a duality, i. e. those points (X, Y) satisfying

$$(X\ Y)\,\mathfrak{R}\,(X\ Y)^{\tau\prime} = 0$$

form a set which is called a *complex*. The coherence of a pair of points (X, Y) and (X_1, Y_1) both on the complex is equivalent to the statement that the rank of the matrix

$$(X\ Y)\,\mathfrak{R}\,(X_1\ Y_1)^{\tau\prime}$$

is one. In fact, the rank of

$$\begin{pmatrix} X & Y \\ X_1 & Y_1 \end{pmatrix}\mathfrak{R}\,(X_1\ Y_1)^{\tau\prime} = \begin{pmatrix} (X\ Y)\,\mathfrak{R}\,(X_1\ Y_1)^{\tau\prime} \\ 0 \end{pmatrix}$$

is equal to 1. By Sylvester's law of nullity, we have

$$\begin{pmatrix} X & Y \\ X_1 & Y_1 \end{pmatrix}$$

is of rank $\leqq n+1$. The assertion follows.

In case the complex is not an empty set and Φ is of characteristic $\neq 2$ $\mathfrak{R}$ can be transformed into a particularly convenient normal form. We have two n-rowed matrices S and T such that

$$\mathfrak{F} = \begin{pmatrix} X & Y \\ S & T \end{pmatrix}$$

is non-singular. Then

$$\mathfrak{F}\mathfrak{R}\mathfrak{F}^{\tau\prime} = \begin{pmatrix} O & B \\ \varrho^{\tau} B^{\tau\prime} & H \end{pmatrix}$$

where B is non-singular and $H^{\tau\prime} = \varrho H$. Since

$$\begin{pmatrix} B^{-1} & O \\ O & I \end{pmatrix} \begin{pmatrix} O & B \\ \varrho^{\tau} B^{\tau\prime} & H \end{pmatrix} \begin{pmatrix} (B^{\tau\prime})^{-1} & O \\ O & I \end{pmatrix} = \begin{pmatrix} O & I \\ \varrho^{\tau} I & H \end{pmatrix}.$$

and

$$\begin{pmatrix} I & O \\ -\frac{1}{2}H & I \end{pmatrix} \begin{pmatrix} O & I \\ \varrho^{\tau} I & H \end{pmatrix} \begin{pmatrix} I & -\frac{1}{2}H^{\tau\prime} \\ O & I \end{pmatrix} = \begin{pmatrix} O & I \\ \varrho^{\tau} I & O \end{pmatrix},$$

we have a matrix $\mathfrak{Y}$ such that

$$\mathfrak{Y}\mathfrak{R}\mathfrak{Y}^{\tau\prime} = \begin{pmatrix} O & I \\ \varrho^{\tau} I & O \end{pmatrix}.$$

Corresponding to this normal form, we have the transformation, in non-homogeneous coordinates

$$Z^{*} = -\varrho Z^{\tau\prime}.$$

It is an interesting problem to characterize the geometry of matrices on a complex by means of the property of coherence. In a previous occation, the author established a particular case with Φ commutative and $\tau = 1$ and $\varrho = -1$.

Remark. In the non-homogeneous coordinates the transformation takes the form

$$Z^* = (ZB + D)^{-1}(ZA + C)$$

and

$$Z^* = (Z^{\tau\prime} B + D)^{-1}(Z^{\tau\prime} A + C)$$

Apparently, it seems to be difficult to construct the product of two transformations. This difficulty can be avoided by means of the following identity: For any $Z (= Z^{(n)})$,

$$(ZB + D)^{-1}(ZA + C) = (SZ - R)(-QZ + P)^{-1}$$

where

$$\begin{pmatrix} A & B \\ C & D \end{pmatrix}\begin{pmatrix} P & Q \\ R & S \end{pmatrix} = \begin{pmatrix} I & O \\ O & I \end{pmatrix}$$

§ 9. Maximal set.

We define similarly that

A *maximal set* (of rank 1) is a set of points in the projective space of rectangular matrices, any pair of which are coherent, if there is no other point of the space can be added to the set such that the set still haye the same property.

Similar to Theorem 3, we have

THEOREM 13. *Under the group* (37), *every maximal set* (*of rank* 1) *is equivalent either to the set constituted by the finite points*

$$\begin{pmatrix} x_1 & \cdots & x_m & 1 & 0 & \cdots & 0 \\ 0 & \cdots & 0 & 0 & 1 & \cdots & 0 \\ \cdot & \cdot & \cdot & \cdot & \cdot & \cdot & \cdot \\ 0 & \cdots & 0, & 0 & 0 & \cdots & 1 \end{pmatrix} \tag{47}$$

and the infinite point,

$$\begin{pmatrix} x_1 & \cdots & x_m & 0 & 0 & \cdots & 0 \\ 0 & \cdots & 0 & 0 & 1 & \cdots & 0 \\ \cdot & \cdot & \cdot & \cdot & \cdot & \cdot & \cdot \\ 0 & \cdots & 0 & 0 & 0 & \cdots & 1 \end{pmatrix} \tag{48}$$

or to the set constituted by the finite points

$$\begin{pmatrix} y_1 0 \cdots 0 1 0 \cdots 0 \\ y_2 0 \cdots 0 0 1 \cdots 0 \\ \cdot \\ \cdot \\ \cdot \\ y_n 0 \cdots 0 0 0 \cdots 1 \end{pmatrix} \tag{49}$$

and the infinite points

$$\begin{pmatrix} y_1 0 \cdots 0 0 0 \cdots 0 \\ y_2 0 \cdots 0 0 1 \cdots 0 \\ \cdot \\ \cdot \\ \cdot \\ y_n 0 \cdots 0 0 0 \cdots 1 \end{pmatrix} \tag{50}$$

The proof is similar to that of Theorem 3 with some slight modifications.

Analogously, we define "lines" as the intersections of two distinct maximal sets which contain more than one point in common, and the arithmetic distance between two points as we did in § 6. By these concepts, we construct linear subspaces (cf. § 5). The linear subspaces of dimension $m-n$ are of primary importance. It can be proved without difficulty that every subspace of dimension $m-n$ can be carried into the following normal form: the set S is formed by the points

$$\begin{pmatrix} x_{n+1} & \cdots & x_m 1 0 \cdots 0 0 \cdots 0 \\ 0 & \cdots & 0 \quad 0 1 \cdots 0 0 \cdots 0 \\ \cdot & \cdots & \cdots \\ 0 & \cdots & 0 \quad 0 0 \cdots 1 0 \cdots 0 \end{pmatrix}$$

and

$$\begin{pmatrix} y_{n+1} & \cdots & y_m 0 0 \cdots 0 0 \cdots 0 \\ 0 & \cdots & 0 \quad 0 1 \cdots 0 0 \cdots 0 \\ \cdot & \cdots & \cdots \\ 0 & \cdots & 0 \quad 0 0 \cdots 1 0 \cdots 0 \end{pmatrix}$$

where at least one of the y's is not zero.

§ 10. Fundamental theorem in the projective geometry of rectangular matrices.

THEOREM 14. *Any one-to-one mapping carrying the projective space of rectangular matrices onto itself is of the form* (37). *In case* $m = n$, *we have an extra type as described in* § 8.

Proof. Since the linear subspaces of dimension $m-n$ are transitive, we may assume, without loss of generality that the mapping A under consideration leaves the manifold S invariant.

It is easy to verify that a point P is a point at infinity, if and only if, there is a point on S, of which the arithmetic distance from P is $< n$.

Thus the mapping A carries finite points into finite points and infinite points into infinite points. By Theorem 1, we may assume that all the finite points are left fixed elementwisely by A. Since any $m-n$ dimensional linear subspace may be regarded as a manifold S, applying the same argument as above to new S's and comparing the common finitc points arising from different manifolds S we know that all points, both finite and infinite are left fixed elementwisely by A. The theorem is thus proved.

Part III. Applications to Algebra.

§ 11. Module.

Now we consider the set of $n \times m$ matrices as a module, for it is closed with respect to addition. A one-to-one mapping $(A \to A^{\sigma})$ of a module onto itself is called a module automorphism, if it keeps the additive relation invariant, that is,

$$(A + B)^{\sigma} = A^{\sigma} + B^{\sigma} \tag{51}$$

An easy consequence of our theorem 1 is the following

THEOREM 15. *Every module automorphism of the set of* $n \times m$ *matrices carrying matrices of rank one into matrices of rank one is of the form*

$$Z_1 = P Z^{\sigma} Q \tag{52}$$

where $P\ (=P^{(n)})$ *and* $Q\ (=Q^{(m)})$ *are non-singular and* σ *is an automorphism of the sfield* Φ. *When* $m = n$, *in addition to* (52), *we have also*

(53) $$Z_1 = P Z'^{\tau} Q$$

where τ is an anti-automorphism of the sfield.

In fact, let $\Gamma(Z) = Z_1$ be the mapping under consideration. From (51). we deduce immediately $\Gamma(0)=0$, $\Gamma(-A) = -\Gamma(A)$. If $P-Q$ is of rank 1, then $\Gamma(P-Q) = \Gamma(P) - \Gamma(Q)$ is also of rank 1. Thus the theorem follows immediately from theorem 1.

Now we restrict ourselves to the case $m=n$. The set of all n-rowed matrices form a ring which is known as a total matrix ring. A semi-automorphism $(A \to A^\sigma)$ of a ring is a one to one mapping of the ring onto itself satisfying

(54) $$(A+B)^\sigma = A^\sigma + B^\sigma,$$

(55) $$(ABA)^\sigma = A^\sigma B^\sigma A^\sigma$$

and

(56) $$I^\sigma = I.$$

Evidently, automorphisms and anti-automorphisms are semi-automorphisms. From (55) and (54), we deduce

(57) $$(XY+YX)^\sigma = [(X+Y)^2 - X^2 - Y^2]^\sigma$$

$$= [(X+Y)^\sigma]^2 - (X^\sigma)^2 - (Y^\sigma)^2 = X^\sigma Y^\sigma + Y^\sigma X^\sigma$$

It is easy to deduce (55) and (56) from (57) and (54), provided that the characteristic of the sfield is different from two. (In fact, it follows from the identity

$$2(XYX) = (XY+YX)X + X(XY+YX) - (X^2Y + YX^2).$$

THEOREM 16. *Every semi-automorphism of the total matrix ring over a sfield is either of the form*

(58) $$Z_1 = P Z^{\tau} P^{-1}$$

where τ is an automorphism of the sfield, or of the form

(59) $$Z_1 = P Z'^{\tau} P^{-1}$$

where τ is an anti-automorphism of the sfield.

Proof. 1) By theorem 15, it is enough to prove that a semi-automorphism carries matrices of rank 1 into matrices of rank 1. Let $A \to A^\sigma$ be the mapping under consideration.

2) An element A satisfying $A^2 = A$ is called an idempotent matrix. From (55) with $B = 1$, we have $(A^\sigma)^2 = A^\sigma$. That is, a semi-automorphism carries idempotent elements into idempotent elements.

3) Let B an element orthogonal to an idempotent element A, that is, $AB = BA = O$. Then, by (57), we have

$$O = (AB + BA)^\sigma = A^\sigma B^\sigma + B^\sigma A^\sigma,$$

and by (55),

$$O = (ABA)^\sigma = A^\sigma B^\sigma A^\sigma = -(A^\sigma)^2 B^\sigma = -A^\sigma B^\sigma.$$

Therefore

$$A^\sigma B^\sigma = B^\sigma A^\sigma = 0.$$

4) An idempotent element A is called irreducible, if $A = B + C$, $B^2 = B$, $C^2 = C$, $BC = CB = O$ implies either $B = O$ or $C = O$. This property is evidently invariant under a semi-automorphism.

5) The theorem follows from theorem 1 and the following

Lemma. A matrix $M(\neq 0)$ is of rank 1 if and only if there exists an irreducible idempotent matrix E such that

$$(I - E)M(I - E) = 0$$

and

$$(M - EME)^2 = 0.$$

Notice that both relations are invariant under a semi-automorphism.

6) *Proof of the lemma.* If M is of rank 1, there is a matrix P such that

$$PMP^{-1} = \begin{pmatrix} u \\ 0 \\ \cdot \\ \cdot \\ \cdot \\ 0 \end{pmatrix}$$

where u is a row vector. Taking $E = E_1 = (a_{ij})$, $a_{ij} = 0$ except $a_{11} = 1$, we can verify both equations. Conversely, without loss of generality, we take $E=E_1$. Then from the first equation, we have

$$M=\begin{pmatrix} a & b \\ c & o \end{pmatrix}, \quad b=b^{(1,n-1)} \quad c=c^{(n-1,1)}$$

From the second equation

$$(M-EME)^2=\begin{pmatrix} o & b \\ c & o \end{pmatrix}^2=0$$

it follows that either $b=o$ or $c=o$. Therefore M is of rank 1.

Remark. It is known (Artin [9]) that a simple ring with descending chain condition for left ideals is simply a total matrix ring over a sfield. Therefore theorem 16 can be "dignified" as follows:

THEOREM 17. *A semi-automorphism of a simple ring with descending chain condition for left ideals is either an automorphism or an anti-automorphism.*

THEOREM 18. *An automorphism* (or *anti-automorphism*) *of a simple ring with descending chain condition for left ideals is a product of an inner automorphism and an automorphism (or an anti-automorphism) induced by an automorphism (or an anti-automorphism) of the sfield.*

The method of the proof can be easily extended to establish a corresponding theorem for semi-simple rings.

Let R be a semi-simple ring with descending chain condition for left ideals. It is known that it is isomorphic to the ring formed by all the matrices of the form

$$X_1^{(m_1)}+X_2^{(m_2)}+\cdots+X_r^{(m_r)}$$

where $X_i^{(mi)}$ runs through all (m_i, m_i) matrices with elements in a sfield Φ_i.

The steps 1)-8) are the same except that the E of 5) may be one of that form in a certain block. Let E belong to the i-th component, and let its image belong to the j-th component. The same method 7) asserts that the mapping carries all the elements of rank 1 in the i-th component into all the elements of rank 1 in the j-th components. Therefore our mapping carries the i-th components into the j-th components. The i-th components are matrices in Φ_i and contain m_i mutually orthogonal irreducible idempotent elements, and we deduce then that $m_i = m_j$ and Φ_i and Φ_j are isomorphic or anti-isomorphic. If we arrange the decomposition into a proper order, the semi-automorphism induces a semi-automorphism on each component. This is a generalization of a result of Kaplansky [4].

Certainly, we can apply the previous method to the problem of isomorphism of different Jordan ring.

§ 12. Jordan and Lie automorphisms.

A one-to-one mapping of a ring onto itself satisfying (54) and (57) is called a Jordan automorphism of the ring. Since (54) and (57) imply (55) and (56), in case that the basis sfield is of characteristic different from 2, we can restate theorem 15 as

THEOREM 19. *A Jordan automorphism of a simple ring with descending chain condition for left ideals is either an automorphism or an anti-automorphism, provided that the characteristic of the sfield is different from 2.*

An analogous, but much more difficult, problem is about the Lie automorphism.

A Lie automorphism $(A \to A^\sigma)$ is a one-to-one mapping of a ring onto itself satisfying the properties that

$$(A + B)^\sigma = A^\sigma + B^\sigma$$

and

$$(AB - BA)^\sigma = A^\sigma B^\sigma - B^\sigma A^\sigma. \tag{60}$$

For brevity, we introduce the notation

$$[A, B] = AB - BA \tag{61}$$

Therefore (60) can be written as

$$[A, B]^\sigma = [A^\sigma, B^\sigma].$$

The module generated by the elements of the form $[A, B]$ is called the derived Lie module of the ring, or simply the derived module. We use L to denote the derived Lie module of the sfield Φ.

Now we consider again the total matrix ring R over a sfield Φ. The elements of the form λI is isomorphic to the ground sfield, we identify these elements with those of the ground sfield. The results of Lie automorphism is not so perfact as that of Jordan automorphism; we have to impose several conditions in order to obtain a similar conclusion. Before we are going to prove the theorem, we need quite a number of lemmas.

§ 13.   Properties of the derived Lie module.

THEOREM 20.   *A necessary and sufficient condition for a matrix belonging to the derived Lie module of R is that its trace belongs to the derived Lie module L of Φ.*

Proof. 1)   Since

$$[A_1 + A_2, B] = [A_1, B] + [A_2, B]$$

it is enough to verify that, for $A = a E_{ij}$, the trace of $[A, B]$ belongs to the derived Lie module L of Φ, where E_{ij} is the matrix with zero element everywhere except 1 at the (i,j)-position.   Now we have

$$tr\, [a E_{ij}, B] = tr\, (a E_{ij} B - B E_{ij} a) = a b_{ji} - b_{ji} a$$

which evidently belongs to L.

2) we can readily verify that

$$[a E_{ij}, E_{ij}] = a E_{ij}. \quad \text{for} \quad i \neq j$$

and

$$[a E_{1i}, E_{i1}] = a E_{11} - a E_{ii}, \quad \text{for} \; i \neq 1.$$

Any matrix with trace belonging to L is a sum of the previous elements and

$$b E_{11}$$

where b belongs to L.   Since

$$b = \sum_i \left(a_i b_i - b_i a_i\right)$$

we deduce that

$$b E_{11} = \sum_i (a_i E_{11} b_i E_{11} - b_i E_{11} a_i E_{11});$$

the theorem is now proved.

Consequently, the property that the trace belongs to the derived module of K is invariant under similarity transformation, since

$$P[S\ T]P^{-1} = [PSP^{-1}, PTP^{-1}].$$

Incidentally, we obtain the following

THEOREM 21. *If S and T are similar to each other, then the trace of S is congruent to that of T, mod L, where L is the derived Lie module of Φ.*

Proof. Let

$$PSP^{-1} = T$$

It is sufficient to prove successively that the theorem is true for $n=2$ and for

$$P = \begin{pmatrix} 0 & 1 \\ 1 & 0 \end{pmatrix}, \begin{pmatrix} \lambda & 0 \\ 0 & 1 \end{pmatrix} \text{ and } \begin{pmatrix} 1 & \mu \\ 0 & 1 \end{pmatrix}$$

since the general linear group is generated by these elements. Since

$$\lambda a \lambda^{-1} - a = (\lambda a)\lambda^{-1} - \lambda^{-1}(\lambda a)$$

belongs to L, our theorem follows immediately.

§14. Lemmas. Now we assume that the characteristic of the sfield differs from 2 and 3.

A matrix A is calld an *I-matrix*, if it can be expressed as

$$A = \alpha I + E, \tag{62}$$

where α belongs to the center of Φ and E is an idempotent matrix, *i.e.* $E^2 = E$ and $E \neq 0$.

THEOREM 22. *A matrix A is an I-matrix, if and only if*

$$[A,[A,[A,B]]]=[A,B] \tag{63}$$

for all B.

Proof. We have

$$[A,[A,[A,B]]]=A^3B-3A^2BA+3ABA^2-BA^3$$

Since $[A,B]=[A-aI,B]$, evidently (62) implies (63).

Now we suppose that

$$A^3B-3A^2BA+3ABA^2-BA^3=AB-BA \tag{64}$$

for all B. We write

$$A=(a_{ij}),\ A^l=(a_{ij}^{(l)}),\ B=(b_{ij}).$$

Then, for all B, we have

$$\sum_{j=1}^{n}(a_{ij}^{(3)}b_{jk}-b_{ij}a_{jk}^{(3)})-3\sum_{j,l=1}^{n}(a_{ij}^{(2)}b_{jl}a_{lk}-a_{ij}b_{jl}a_{lk}^{(2)})=\sum_{j=1}^{n}(a_{ij}b_{jk}-b_{ij}a_{jk}).$$

Putting $B=xE_{jl}$, we have

$$a_{ij}^{(2)}xa_{lk}=a_{ij}xa_{lk}^{(2)}\quad \text{for}\quad j\neq i,k\neq l. \tag{65}$$

$$a_{ij}^{(3)}x-a_{ij}x-3a_{ij}^{(2)}xa_{kk}+3a_{ij}xa_{kk}^{(2)}=0\quad \text{for}\quad j\neq i,k=l, \tag{66}$$

$$xa_{lk}^{(3)}-xa_{lk}+3a_{ii}^{(2)}xa_{lk}-3a_{ii}xa_{lk}^{(2)}=0\quad \text{for}\quad j=i,k\neq l, \tag{67}$$

and

$$a_{ii}^{(3)}x-xa_{kk}^{(3)}-3a_{ii}^{(2)}xa_{kk}+3a_{ii}xa_{kk}^{(2)}=a_{ii}x-xa_{kk} \tag{68}$$

for $j=i,k=l$. (65)-(68) hold for all x.

If A is a central element, *i.e.* $A=cI$, where c is a central element of K, than $A=(c-1)I+I$ and A is an I-matrix. Thus there remains to consider the case that A is non-central. If A does not belong to the center, we have an invertible matrix P such that the element at $(1,2)$-

position of PAP^{-1} is different from zero. In fact, if there is a non-diagonal element of A different from zero, we have a permutation matrix P such that PAP^{-1} satisfies our requirement Next, if A is a diagonal matrix, the statement follows from the following formula

$$\begin{pmatrix} 1 & t \\ 0 & 1 \end{pmatrix} \begin{pmatrix} a & o \\ o & b \end{pmatrix} \begin{pmatrix} 1 & t \\ 0 & 1 \end{pmatrix}^{-1} = \begin{pmatrix} a & tb-at \\ o & b \end{pmatrix}$$

Since the relation (63) and the property of a matrix being an I-matrix are both invariant under similarity transformations Without loss of generality we can assume that $a_{12} \neq 0$.

From (65), we have

$$a_{12}^{-1} a_{12}^{(2)} x a_{lk} = x a_{lk}^{(2)},$$

for $k \neq l$ and all x. If $a_{12}^{(2)} \neq 0$ and $a_{lk} \neq 0$, then

$$a_{12}^{-1} a_{12}^{(2)} x = x a_{lk}^{(2)} a_{lk}^{-1},$$

for all x. Consequently, we have

$$a_{12}^{-1} a_{12}^{(2)} = a_{lk}^{(2)} a_{lk}^{-1} = \varrho$$

where ϱ belongs to the center. Thus

$$a_{lk}^{(2)} = \varrho\, a_{lk} \tag{69}$$

for $k \neq l$ when $a_{12}^{(2)} \neq 0$ and $a_{lk} \neq 0$. The element ϱ is independent of l and k. If $a_{12}^{(2)} \neq 0$ but $a_{lk} = 0$, then $a_{lk}^{(2)} = 0$, (69) is also true. Finally, if $a_{12}^{(2)} = 0$, then $a_{lk}^{(2)} = 0$; in this case (69) also holds for $\varrho = 0$. Therefore (69) holds for all $k \neq l$. From (66) with $x = 1$, we have

$$a_{12}^{(3)} - a_{12} = 3\, a_{12}^{(2)} a_{11} - 3\, a_{12} a_{11}^{(2)}$$

$$= 3\, a_{12}^{(2)} a_{kk} - 3\, a_{12} a_{kk}^{(2)}$$

then, since the characteristic of the sfield is $\neq 3$,

$$a_{12}^{(2)}(a_{11}-a_{kk})=a_{12}(a_{11}^{(2)}-a_{kk}^{(2)}).$$

From (69), we deduce

$$a_{kk}^{(2)}-a_{11}^{(2)}=\varrho(a_{kk}-a_{11}). \tag{70}$$

From (69) and (70), we have

$$A^2-a_{11}^{(2)}I=\varrho(A-a_{11}I).$$

That is

$$A^2=\varrho A+\lambda I, \tag{71}$$

where ϱ belongs to the center. We deduce immediately that λI is commutative with A. Repeating (71), we have

$$A^3=\varrho A^2+\lambda A=(\varrho^2+\lambda)A+\varrho\lambda I. \tag{72}$$

Substituting (71) and (72) into (64), we have

$$[(\varrho^2+\lambda)A+\varrho\lambda I]B-3(\varrho A+\lambda I)BA+3AB(\varrho A+\lambda I)$$
$$-3B[(\varrho^2+\lambda)A+\varrho\lambda I]=AB-BA$$

for all B, and then

$$(\varrho^2+4\lambda-1)(AB-BA)=0$$

for all matrix B which are commutative with λ. Since A does not belong to the center, we can choose a B which is commutative with λ but not with A. In fact, we can choose B as a matrix with elements in the prime field and $AB\neq BA$. Therefore we have

$$\varrho^2+4\lambda-1=0.$$

Then, from (71),

$$\left(A-\frac{\varrho-1}{2}I\right)^2=A-\frac{\varrho-1}{2}I,$$

that is, $A-\frac{\varrho-1}{2}I$ is an idempotent element.

Remark The theorem is not true for fields of characteristic 3. In fact $A^3 = A + \tau I$ where τ belongs to the center, is a solution, but A is not an I-matrix.

$$A = \begin{pmatrix} 0 & 1 & 0 \\ 0 & 0 & 1 \\ \tau & 1 & 0 \end{pmatrix}$$

is such a matrix. The theorem does not hold for fields of characteristic 2, since

$$A = \begin{pmatrix} 0 & 1 \\ 1 & 0 \end{pmatrix}$$

is a "Gegenbeispiel".

A matrix A is called an *N-matrix,* if it can be expressed as

$$A = \alpha I + N \tag{77}$$

where α belongs to the center of Φ and N is a nilpotent element that is, $N^2 = 0$.

THEOREM 23. *A matrix A is an N-matrix, if and only if*

$$[A,[A,[A,B]]] = 0 \tag{78}$$

for all B.

Proof. Instead of (64), we now have

$$A^3 B - 3A^2 BA + 3ABA^2 - BA^3 = 0 \tag{79}$$

Proceeding as in the proof of theorem 22, we prove that

$$A^2 = \varrho A + \lambda I,$$

where $\varrho^2 + 4\lambda = 0$. Consequently

$$\left(A - \frac{\varrho}{2} I\right)^2 = 0.$$

§15. Continuation.

THEOREM 24. *Suppose that* $A^2 = A$, $AC = CA$, $AD = DA$ *and* $A \neq 0$ *and* $\neq I$ *Then*

$$[D,[A,B]]+[C,[A,[C,[A,B]]]]=0 \tag{80}$$

for all B if and only if either

$$C=\lambda I+F,\quad D=\mu I+F^2 \tag{81}$$

where λ and μ belongs to the center and $AF=FA=0$, or

$$C=\lambda I-G,\quad D=\mu I-G^2 \tag{82}$$

where λ and μ belongs to the center, $AG=GA=G$

Proof. We expand (80) as

$$D(AB-BA)-(AB-BA)D+AC^2B+C^2BA+ABC^2+BAC^2$$
$$-2(AC^2BA+ABAC^2+ACBC+CBAC)+4ACBAC=0$$

which can be written as

$$(C^2+D)AB+BA(C^2+D)+(C^2-D-2C^2A)BA \tag{83}$$
$$+AB(C^2-D-2C^2A)+2C(2ABA-AB-BA)C=0.$$

In order to verify that (81) implies (80), we need only to prove that $C=F$ and $D=F^2$ satisfy (83). This is evident. In order to verify that (82) implies (80), we need only to prove that $C=-G$, $D=-G^2$ satisfy (83). This is also evident.

Conversely, without loss of generality, we may assume that

$$A=\begin{pmatrix} I^{(r)} & O \\ O & O \end{pmatrix}$$

since for every idempotent A, we have an invertible matrix P such that PAP^{-1} is of the prescribed form. We take

$$B=\begin{pmatrix} O^{(r)} & O \\ B_{21} & O \end{pmatrix},\quad B_{21}=B_{21}^{(n-r,r)}.$$

Then $AB=0$ and

$$BA=\begin{pmatrix} O & O \\ B_{21} & O \end{pmatrix}=B$$

Substituting into (85), we have

$$\begin{pmatrix} O & O \\ B_{21} & O \end{pmatrix}\left(\begin{pmatrix} C^*_{11} & C^*_{12} \\ C^*_{21} & C^*_{22} \end{pmatrix} + \begin{pmatrix} D_{11} & D_{12} \\ D_{21} & D_{22} \end{pmatrix}\right) + \left(\begin{pmatrix} C^*_{11} & C^*_{12} \\ C^*_{21} & C^*_{22} \end{pmatrix} - \begin{pmatrix} D_{11} & D_{12} \\ D_{21} & D_{22} \end{pmatrix}\right)\begin{pmatrix} O & O \\ B_{21} & O \end{pmatrix}$$

$$-2\begin{pmatrix} C_{11} & C_{12} \\ C_{21} & C_{22} \end{pmatrix}\begin{pmatrix} O & O \\ B_{21} & O \end{pmatrix}\begin{pmatrix} C_{11} & C_{12} \\ C_{21} & C_{22} \end{pmatrix} = 0,$$

where

$$C^2 = \begin{pmatrix} C^*_{11} & C^*_{12} \\ C^*_{21} & C^*_{22} \end{pmatrix} \quad D = \begin{pmatrix} D_{11} & D_{12} \\ D_{21} & D_{22} \end{pmatrix}, \text{ etc.}$$

Then, for all B_{21}, we have

(i) $$(C^*_{12} - D_{12})B_{21} = 2C_{12}B_{21}C_{11},$$

(ii) $$C_{12}B_{21}C_{12} = 0,$$

(iii) $$B_{21}(C^*_{11} + D_{11}) + (C^*_{22} - D_{22})B_{21} = 2C_{22}B_{21}C_{11}$$

and

(iv) $$B_{21}(C^*_{12} + D_{12}) = 2C_{22}B_{21}C_{12}.$$

From (ii), we deduce that $C_{12}=0$* and from (i) and (iv) we deduce $D_{12}=C^*_{12}=0$.

Putting

$$B = \begin{pmatrix} O & B_{12} \\ O & O \end{pmatrix}, \quad B_{12} = B_{12}^{(r,n-r)}$$

we deduce similarly $D_{21}=C^*_{12}=C_{21}=0$ and

(v) $$(C^*_{11} + D_{11})B_{12} + B_{12}(C^*_{22} - D_{22}) = 2C_{11}B_{12}C_{22}.$$

Since $C^*_{22}=C^2_{22}$, $C^*_{11}=C^2_{11}$, we write

$$C^2_{22} - D_{22} = (\alpha_{ij}), \quad C^2_{11} + D_{11} = (\beta_{ij})$$

---

*In fact, it is evident that if $AXB=0$ for all possible X, then we have either $A=0$ or $B=0$.

$$C_{22} = (\gamma_{ij}), \quad C_{11} = (\delta_{ij})$$

and $B_{21} = (b_{ij})$, then, from (iii), we have

$$\sum_{j=1}^{r} b_{ij}\beta_{jk} + \sum_{j=1}^{n-r} \alpha_{ij} b_{jk} = 2 \sum_{s=1}^{n-r} \sum_{t=1}^{r} \gamma_{is} b_{st} \delta_{tk}$$

for all b_{ij}. Consequently, we have

(vi) $$\gamma_{is}\delta_{tk} = 0 \quad \text{for} \quad t \neq k, i \neq s$$

(vii) $$2\gamma_{ii} b \delta_{tk} = b\beta_{tk} \quad \text{for} \quad t \neq k$$

(viii) $$2\gamma_{is} b \delta_{kk} = \alpha_{is} b \quad \text{for} \quad i \neq s,$$

(ix) $$2\gamma_{ii} b \delta_{kk} = \alpha_{ii} b + b\beta_{kk}$$

for all b.

If there is a non-diagonal element of C_{22} different from zero, say $\gamma_{is} \neq 0$ and $i \neq s$, then, from (vi) we have $\delta_{tk} = 0$ for all $t \neq k$ Then $\beta_{tk} = 0$ for all $t \neq k$, by (vii). Further, from (viii), we deduce that

$$\frac{1}{2}\gamma_{is}^{-1}\alpha_{is} = \delta$$

belongs to the center and $\delta_{kk} = \delta$. That is, $C_{11} = \delta I$.

From (ix), we have $\beta_{kk} = \beta$ for all k and

$$2\gamma_{ii}\delta b = \alpha_{ii} b + b\beta$$

for all b. In particular $b = 1$, then $\alpha_{ii} = 2\gamma_{ii}\delta - \beta$. Substituting into the previous equation, we have $b\beta - \beta\, b = 0$ for all b. Therefore β belongs to the center. Then

(x) $$C_{11}^2 + D_{11} = \beta I.$$

From (iii) and (ix), we have

$$C_{22}^2 - D_{22} = (\alpha_{ij}) = 2(\gamma_{ij})\delta - \beta I$$

$$= 2C_{22}\delta - \beta I.$$

That is

$$D_{22} = C_{22}^2 - 2C_{22}\delta + \beta I$$

$$= (C_{22} - \delta I)^2 - (\delta^2 - \beta) I.$$

Therefore

$$C = \begin{pmatrix} \delta I^{(r)} & O \\ O & C_{22} \end{pmatrix} = \delta I^{(n)} + \begin{pmatrix} O & O \\ O & C_{22} - \delta I \end{pmatrix},$$

and

$$D = \begin{pmatrix} D_{11} & O \\ O & D_{22} \end{pmatrix} = \begin{pmatrix} \beta I - C_{11}^2 & O \\ O & (C_{22} - \delta I)^2 - (\delta^2 - \beta) I \end{pmatrix}$$

$$= (\beta - \delta^2) I + \begin{pmatrix} O & O \\ O & (C_{22} - \delta I)^2 \end{pmatrix}$$

This proves the theorem.

By a slightly different method, we establish our theorem for the case C_{11} is not diagonal but notice that now we obtain the second conclusion instead of (8)

Suppose that C_{11} and C_{22} are both diagonal, then, from (vii) and (viii), both D_{11} and D_{22} are diagonal, and

(xi) $$2\gamma_i b \delta_k = \alpha_i b + b \beta_k$$

for all b, we write γ_i for γ_{ii} etc. If there is one of the differences $\gamma_i - \gamma$ different from zero, then we have

$$2(\gamma_i - \gamma_j) b \delta_k = (\alpha_i - \alpha_j) b$$

for all b. Then δ_k belongs to the center and

$$\delta_k = \delta = \frac{1}{2}(\gamma_i - \gamma_j)^{-1}(\alpha_i - \alpha_j)$$

which is independent of k. From (xi), we have

$$(2\gamma_i \delta - \alpha_i) b = b \beta_k$$

for all b. It follows that

$$\beta_k = \beta = 2\gamma_i \delta - \alpha_i$$

is a central element for all k. Therefore we have

$$C_{11}^2 + D_{11} = \beta I, \quad C_{11} = \delta I$$

Then

$$\begin{pmatrix} C_{11} & O \\ O & C_{22} \end{pmatrix} = \begin{pmatrix} \delta I & O \\ O & C_{22} \end{pmatrix} = \delta I + \begin{pmatrix} O & O \\ O & C_{22} - \delta I \end{pmatrix},$$

and

$$\begin{pmatrix} D_{11} & O \\ O & D_{22} \end{pmatrix} = \begin{pmatrix} (\beta - \delta^2) I & O \\ O & C_{22}^2 - 2\delta C_{22} + \beta I \end{pmatrix}$$

$$= (\beta - \delta)^2 I + \begin{pmatrix} O & O \\ O & (C_{22} - \delta I)^2 \end{pmatrix}$$

The theorem follows.

In case there is one of the difference $\delta_i - \delta_j$ different from zero, then we have

$$\gamma_i = \gamma, \quad \alpha_i = \alpha$$

which belong to the center, and

$$2\gamma \delta_k = \alpha + \beta_k.$$

Then

$$C_{22}^2 - D_{22} = \alpha I, \quad C_{22} = \gamma I$$

and

$$C_{11}^2 + D_{11} = -\alpha I + 2\gamma C_{11}.$$

Then

$$C = \begin{pmatrix} C_{11} & O \\ O & \gamma I \end{pmatrix} = \gamma I + \begin{pmatrix} C_{11} - \gamma I & O \\ O & O \end{pmatrix}$$

and

$$D = \begin{pmatrix} D_{11} & O \\ O & D_{22} \end{pmatrix} = (\gamma^2 - \alpha) I + \begin{pmatrix} -(C_{11} - \gamma I)^2 & O \\ O & O \end{pmatrix}.$$

That is,

$$C = \lambda I - G \quad D = \mu I - G^2$$

where

$$G = \begin{pmatrix} -(C_{11} - \gamma I) & O \\ O & O \end{pmatrix}$$

satisfies evidently $GA=AG=G$.

THEOREM 25. *Suppose that* $A^2=A(\neq 0, 1)$, $C^2=C$, $AC=CA$. Then

$$[C,[A,B]]+[C[A[C,[A,B]]]]=0 \tag{84}$$

for all B, *if and only if*

$$AC=0 \quad \text{or} \quad (I-A)(I-C)=0.$$

Proof. The theorem is an easy consequence of theorem 24. by putting $D=C^2=C$. From (81), we have

$$C^2=(\lambda I+F)^2=\mu I+F^2$$

then $\lambda^2=\mu$, $2\lambda=0$. Consequently $C=F$ and $AC=CA=0$

From (82) and $C^2=D=C$, we have

$$C^2=(\lambda I-G)^2=uI-G^2=\lambda I-G$$

then $\lambda=1$ and

$$(I-A)(I-C)=(I-A)G=G-G=0.$$

Remark. Theorem 25 can be proved very simply by the direct expansion of (84). In fact (84) is equivalent to

$$\begin{aligned} &CA(CAB-CBA-ABC+BAC)\\ &-C(CAB-CBA-ABC+BAC)A\\ &-A(CAB-CBA-ABC+BAC)C\\ &+(CAB-CBA-ABC+BAC)AC\\ &=-(CAB-CBA-ABC+BAC). \end{aligned} \tag{85}$$

Evidently $CA=O$ or $(I-C)(I-A)=O$ implies (85). Conversely multiplying (85) by AC on the left, we have

$$ACB(I-A)(I-C)=0$$

for all B. This implies either AC or $(I-A)(I-C)=O$. Therefore we have theorem 25.

§16. A theorem on Lie automorphism.

The situation about Lie automorphism of th matrix ring R differs greatly from that of Jordan automorphism.

Definition. A mapping $A \to A^\sigma$ of R into itself is called a Lie representation, if it satisfies

$$(A+B)^\sigma = A^\sigma + B^\sigma, \quad [A, B]^\sigma = [A^\sigma, B^\sigma].$$

The representation is called central, if it maps R into its center, that is

$$A \to \lambda(A) I. \tag{87}$$

where $\lambda(A)$ belongs to the center of the basic sfield Φ.

THEOREM 26. *Every central representation can be expressed as*

$$\lambda(A) = \mu(tr\, A),$$

where $tr\, A$ denotes the trace of A and $\mu(a)$ is a central Lie representation of the sfield Φ.

THEOREM 27. *A central Lie representation of a sfield is a representation of the factor group of the additive group of Φ over its derived Lie module into the center of Φ.*

Owing to their simplicity, we omit the proofs of both theorems.

If $A \to A^\sigma$ is a Lie representation, so is

$$A \to \lambda(A) I + A^\sigma \tag{88}$$

This very fact makes some trouble. In fact, sometimes, (88) ceases to be one-to-one.

First of all, let us investigate the condition under which (88) is one-to-one. That is, to find the condition under which the equality

$$\lambda(A) I + A^\sigma = 0 \tag{89}$$

implies $A = O$.

Evidently, if A belongs to the center, A^σ belongs to the center. Conversely, suppose that A^σ belongs to the center, but A does not. There exist a matrix B such that $C = [A, B] \neq 0$. Then C is mapped into

$$[\lambda(A) I + A^\sigma, \lambda(B) I + B^\sigma] = [A^\sigma, B^\sigma] = 0$$

which contradicts the assumption "one to one". Therefore, if A^σ belongs to the center, so does A.

Foom (89), we therefore deduce that A belongs to the center, say $A=\alpha I$, where α belongs to the center of Φ. From (89), we deduce the condition

$$n\mu(\alpha)+\alpha^{\sigma}=0$$

Therefore (88) is one to one, if and only if

(90) $$n\mu(\alpha)+\alpha^{\sigma}\neq 0$$

for all $\alpha\,(\neq 0)$ belonging to the center of Φ.

THEOREM 28. *Let Φ be a sfield with characteristic different from 2 and 3. Every Lie automorphism of R $(n>2)$ is either of the form*

(91) $$A\to\mu(tr\,A)I+PA^{\tau}P^{-1}$$

where $\mu(d)$ is a central Lie representation of Φ and τ is an automorphism of Φ and $n\mu(\alpha)-\alpha^{\tau}\neq 0$ for all $\alpha\,(\neq 0)$ belonging to the center of Φ; or of the form

(92) $$A\to\mu(tr\,A)I-PA^{\tau\prime}P^{-1}$$

where τ is an anti-automorphism of Φ and $n\mu(\alpha)-\alpha^{\tau}\neq 0$ for all $\alpha\,(\neq 0)$ belonging to the center of Φ.

Proof. 1) Idempotent elements. Let $E_1,\cdots,E_n$ be n elements different from zero such that

$$E_i^2=E_i \quad\text{for}\quad 1\leqq i\leqq n$$

and

$$E_iE_j=E_jE_i=0 \quad\text{for}\quad i\neq j.$$

By Theorem 22, E_i's are carried into

$$\lambda_i I+E_i^*,\quad E_i^{*2}=E_i^*$$

where λ_i belongs to the center. From Theorem 25, we have either

$$E_i^*E_j^*=E_j^*E_i^*=0$$

or

$$(I-E_i^*)(I-E_j^*)=(I-E_j^*)(I-E_i^*)=0.$$

Suppose that we have

$$E_1^* E_2^* = E_1^* E_3^* = 0,$$

but

$$(I - E_2^*)(I - E_3^*) = 0.$$

Then, we have

$$E_1^* = E_1^*(I - E_2^*) = E_1^*(I - E_2^*)(I - E_3^*) = 0,$$

which is impossible. Therefore we can only have either

(i) $$E_i^* E_j^* = E_j^* E_i^* = 0 \quad \text{for all} \quad i \neq j,$$

or

$$(I - E_i^*)(I - E_j^*) = (I - E_j^*)(I - E_i^*) = 0 \quad \text{for all} \quad i \neq j.$$

For the second case, we replace $I - E_i^*$ by E_i^* we have then

(ii) $$E_i \to \lambda_i I - E_i^*$$

where

$$E_i^{*2} = E_i^*, \quad E_i^* E_j^* = E_j^* E_i^* = 0 \quad \text{for} \quad i \neq j.$$

Without loss of generality, after subjecting the given Lie automorphism to an inner automorphism, we may assume that

$$E_i = E_i^* = (a_{st})$$

where $a_{st} = 0$ except $a_{ii} = 1$.

2) We have

(93) $$[E_i [E_j, P]] = p E_{ij} + q E_{ji} \quad \text{for} \quad i \neq j$$

where pE_{ij} denotes a matrix (a_{st}) with $a_{st} = 0$ except $a_{ij} = p$. Consider elements of the form (93) whose squares are zero. Then $pq = 0$. Therefore the matrix pE_{ij} is mapped into

$$\lambda I + N, \quad N^2 = 0, \quad \lambda \,\varepsilon\, \text{center}$$

by thecrem 23. Moreover, by 1), (93) is mapped into

$$p^* E_{ij} + q^* E_{ij} = \lambda I + N.$$

We deduce

$$O = N^2 = (p^* E_{ij} + q^* E_{ji} - \lambda I)^2,$$

that is

$$p^* q^* E_i + q^* p^* E_j - 2(p^* E_{ij} + q^* E_{ji}) \lambda + \lambda^2 I = 0$$

If $\lambda \neq 0$, then $p^* = q^* = 0$, which is impossible. For $\lambda = 0$, we have $p^* q^* = q^* p^* = 0$. Consequently, a matrix of the form pE_{ij} is mapped either into $p^* E_{ij}$ or into $q^* E_{ji}$

Next, from

$$\alpha E_{12} \to \alpha^* E_{12} \quad \beta E_{13} \to \beta^* E_{31}$$

we deduce

$$O = [\alpha E_{12}, \beta E_{13}] \to [\alpha^* E_{12}, \beta^* E_{31}] = -\beta^* \alpha^* E_{32}$$

which is absurd. Therefore we have either

(94) $\alpha E_{ij} \to \alpha^* E_{ij}$ for all $i \neq j$ and $\alpha \varepsilon \Phi$

or

(95) $\alpha E_{ij} \to \alpha^* E_{ji}$ for all $i \neq j$ and $\alpha \varepsilon \Phi$.

Suppose that we have (94), then

$$E_{1i} \to \gamma_i E_{1i}.$$

The mapping

$$[1, \gamma_2, \cdots, \gamma_n] X [1, \gamma_2, \cdots, \gamma_n]^{-1}$$

leaves E_i's invariant and carries $\gamma_i E_{1i}$ into E_{1i}, therefore after subjecting our Lie automorphism to an inner automorphism we can assume

$$E_{1i} \to E_{1i}, \quad i = 1, \cdots, n.$$

Further, let $E_{ij} \to \gamma E_{ij}$ then

$$[E_{1j}, E_{ij}] = E_{1j} \quad (j \neq 1)$$

is mapped into

$$[E_{1i}, \gamma E_{ij}] = \gamma E_{1j}$$

thus $\gamma = 1$. Further, for $j = 1$, the element

$$[E_{1i}, E_{i1}] = E_1 - E_i$$

is mapped into

$$[E_{1i}, \gamma E_{i1}] = (\lambda_1 - \lambda_i) I + E_1 - E_i$$

in case (i). Consequently

$$(\gamma - 1)(E_1 - E_i) = (\lambda_1 - \lambda_i) I$$

Then $\lambda_1 = \lambda_i$ and $\gamma = 1$. Consequently, we have

(i) $$E_i \to \lambda I + E_i, \quad E_{ij} \to E_{ij}.$$

In case (ii), it is mapped into

$$[E_{1i}, \gamma E_{i1}] = (\lambda_1 - \lambda_i) I - (E_1 - E_i),$$

then

$$(\gamma + 1)(E_1 - E_i) = (\lambda_1 - \lambda_i) I.$$

Then $\lambda_1 = \lambda_i$ and $\gamma = -1$. Consequently, $E_{i1} \to -E_{i1}$ for $i \neq 1$. For $i \neq 1$. we have

$$[E_{1i}, E_1] = E_{i1}$$

which is mapped into

$$[-E_{i1}, \lambda_1 I - E_1] = E_{i1} = -E_{i1}$$

which is impossible.

Similarly, with (95), we have

$$E_i \to \lambda I - E_i, \quad E_{1i} \to E_{i1}, \quad E_{j1} \to E_{1j}$$

$$E_{ij} \to -E_{ji} \quad \text{for} \quad i, j \neq 1.$$

Notice that the combination of (95) and (i) is impossible. The inner automorphism $[-1, 1, \cdots, 1] A [-1, 1, \cdots, 1]$ carries the automorphism into one satisfying

(ii) $$E_i \to \lambda I - E_i, \quad E_{ij} \to -E_{ji}.$$

3) Principal matrices of rank 1.

Let $P (\neq 0)$ be a matrix satisfying

$$PE_1 = E_1 P = P, \quad PE_i = E_i P = 0 \quad \text{for} \quad 2 \leqq i \leqq n.$$

For $PE_i = E_i P = O$ or P, we deduce from theorem 24 with $D = C^2$ and $C = P$ that P is mapped into $\mu I + P^*$ such that

$$P^* E_i = E_i P^* = 0 \quad \text{or} \quad P^*$$

If $P^* E_i = E_i P^* = O$ for all i, then $P^* = O$ which is impossible. From $P^* E_i = E_i P^* = P^*$, we deduce $P^* = \alpha E_i$. Therefore we have

$$\alpha E_1 \to \lambda I + \alpha^* E_i, \quad \lambda \in \text{center}.$$

If $i \neq 1$, we have

$$[\alpha E_1, E_{1k}] = \alpha E_{1k}, \quad k \neq 1, \quad k \neq i,$$

but

$$[\alpha^* E_1, E_{1k}] = 0 \quad \text{and} \quad [\alpha^* E_i, -E_{k1}] = 0,$$

both are impossible. Therefore, since $n \geqq 3$, we have

$$\alpha E_1 \to \lambda I + \alpha^* E_1.$$

Therefore, we may assume that,

(96) $$\alpha E_i \to \lambda_i I + \beta_i E_i.$$

4) In conclusion, we established that either

(i)
$$\alpha E_i \to \mu_i(\alpha) J + \alpha^*_{ii} E_i$$
$$\alpha E_{ij} \to \alpha^*_{ij} E_{ij}$$
$$1^*_{ij} = 1$$

or

(ii)
$$\alpha E_i \to \mu_i(\alpha) I = \alpha^*_{ii} E_i$$
$$\alpha E_{ij} \to -\alpha^*_{ij} E_{ij}$$
$$1^*_{ij} = 1$$

For (i), the element

$$\alpha E_i - \alpha E_j = [\alpha E_{ij}, E_{ji}]$$

is mapped into

$$(\mu_i(\alpha) - \mu_j(\alpha)) I + \alpha^*_{ii} E_i - \alpha^*_{jj} E_j = [\alpha^*_{ij} E_{ij}, E_{ji}]$$

$$= \alpha^*_{ij} E_i - \alpha^*_{ij} E_j.$$

We have then, since $n \geqq 3$,

$$\mu_i(\alpha) = \mu_j(\alpha), \quad \alpha^*_{ii} = \alpha^*_{ij} = \alpha^*_{jj}.$$

that is, we have

(i) $$\alpha E_i \to \mu(\alpha) I + \alpha^* E_i, \quad \alpha E_{ij} \to \alpha^* E_{ij}, \quad 1^* = 1$$

Similarly for case (ii), we have

(ii) $$\alpha E_i \to \mu(\alpha) I - \alpha^* E_i, \quad \alpha E_{ij} \to -\alpha^* E_{ij}, \quad 1^* = -1.$$

Notice that the mapping $\alpha \to \alpha^*$ is one-to-one, for otherwise from $\alpha \neq \alpha'$ and

$$\alpha E_i \to \mu(\alpha) \pm \alpha^* E_i, \quad \alpha' E_i \to \mu(\alpha') I \pm \alpha^* E_i,$$

we deduce

$$(\alpha - \alpha') E_i \to (\mu(\alpha) - \mu(\alpha')) I$$

which is impossible, since the right hand side belongs to the center but the left does not. Now the mapping (i) (or (ii)) is decomposed into two parts:

(97) $$A \to \mu(tr\, A)$$

and

(98) $$A = (\alpha_{ij}) \to A^* = (\alpha_{ij}^*).$$

The second mapping is a one-to-one mapping carrying R onto itself. It carries matrices of rank one into matrices of rank one, since the first mapping is independent of the choice of $E_1, \cdots, E_n$

By theorem 1, we have consequenently either

(99) $$A^* = P A^{\tau} Q$$

where τ is an automorphism of Φ or

(100) $$A^* = P A^{\tau\prime} Q$$

where τ is an anti-automorphism. Since the mapping brings central elements into central elements, we have $Q = \beta P^{-1}$ where β belongs to the center. From (99), we deduce $A^* = \beta P A^{\tau} P^{-1}$, then

$$[A, B]^* = \beta P [A, B] P^{-1} = [A^*, B^*] = \beta^2 P [A, B] P^{-1}.$$

Consequently $\beta^2 = \beta$, i. e. $\beta = 0$ or 1. From (100), we deduce $\beta = 0$ or -1. Therefore we have either

$$A^* = P A^{\tau} P^{-1}$$

or

$$A^* = P A^{\tau\prime} P^{-1}$$

the theorem is now proved.

Remark. For $n = 2$, the theorem is not true. For example, for the field of all complex numbers, we have a mapping

$$\begin{pmatrix} a_1 + i a_2 & b_1 + i b_2 \\ c_1 + i c_2 & d_1 + i d_2 \end{pmatrix} \rightarrow \begin{pmatrix} a - i d_2 & b_2 + \bar{i} b_2 \\ c_1 + i c_2 & d_1 - i a_2 \end{pmatrix}$$

which is not of the type described in the theorem.

REFERENCES

[1] Hua, Annals of Math., 50 (1949), 8-31.
[2] Albert, Structure of Algebras.
[3] Ancochea, Annals of Math., 48 (1947), 147-153.
[4] Kaplansky, Duke Math. Jour., 14 (1947), 521-525.
[5] Ancochea, Jour. für Math., 184 (1942), 192-198.
[6] Hua, Proc. of Nat. Aca. of Sc., U.S.A. 35 (1949), 396-389.
[7] Jacobson, Duke Math. Jour., 3 (1938) 534-551 and Amer. Jour. of Math., 63 (1941), 481-515.
[8] Veblen and Whitehead, Foundations of Differential geometry, Cambridge Tracts, 29 (1932), p.12.
[9] Artin and Whaples, Amer. Jour. of Math., 65 (1943) 87-107.
[10] Hua, Science Reports of Nat. Tsinghua Univ., Series A, 5 (1938), 150-181.

(Recieved Dec. 8, 1950)

Reprinted from MEMOIRS OF THE AMERICAN MATHEMATICAL SOCIETY, Number 2.

SUPPLEMENT TO THE PAPER OF DIEUDONNÉ ON THE AUTOMORPHISMS OF CLASSICAL GROUPS

By Loo-Keng Hua

1. Introduction. In a previous paper, the author [1] determined the group of automorphisms of the symplectic group $Sp_n(K)$ over a field K of characteristic different from two. The method used there applies equally well to the general linear group $GL_n(K)$ and after some complicated modifications to the orthogonal group $O_n(K,f)$, with a quadratic form f of index $\nu \geq 1$. Professor Dieudonné [2] used an entirely different approach and comprehensively worked out the group of automorphisms of classical groups with several exceptions. (See the last section of his paper). It is the aim of this paper to give solutions of some of the problems which he left open.

The first difficult problem cited in his paper is the determination of the group of automorphisms of $GL_2(K)$, where K is a sfield. For K of characteristic different from two he fails to characterize the transvections. Even if he were to succeed in characterizing the transvections, as he does when K is of characteristic two, the notion of semi-automorphisms of a sfield is still an obscure point of the final result. As a by-product of the present investigation, the author [3] proves that a semi-automorphism (in the sense of Ancochea [4] and Kaplansky [5]) of any sfield is either an automorphism or an anti-automorphism. Combining this with the result of Dieudonné, we can immediately solve the problem of automorphisms of $GL_2(K)$ (and $PGL_2(K)$), for K of characteristic 2. In this paper we shall solve the more difficult problem concerning the automorphisms of $GL_2(K)$, for K a sfield of characteristic different from two.

The second problem which he left unsolved concerns the automorphisms of the special linear groups $SL_4(K)$ and $PSL_4(K)$. Since the dimension is higher, the problem is much easier than the previous one. The group of automorphisms is determined in § 6.

The only problem left unsettled in the family of orthogonal groups with indefinite fundamental form is the determination of the group of automorphisms of $O_4^+(K,f)$ (and $PO_4^+(K,f)$), for the case where K is of characteristic different from two and the quadratic form f has index 2. This is closely related to the author's study of geometry of skew-symmetric matrices. Some new types of automorphisms appear; these surprising phenomena throw a new light on the study of the commutator subgroup $\Omega_n(K, f)$ of the orthogonal group, which will be a subject of our later discussion.

It seems worthwhile to mention that the nature of such problems show that the lower the dimension the harder the problem. Dieudonné adopted a method which worked smoothly for large n, and treated individually the cases with small n. As the author mentioned before, the difficulty increases as n diminishes; Dieudonné's method becomes very clumsy for smaller n, and sometimes he is unable to solve the case for smallest n. On the other hand the author's method starts with the least possible n, which is usually the most difficult case. Therefore the reader will have little difficulty in extending the special results of this paper to the general case by means of the inductive method used in [1]. Moreover, in contrast with that of Dieudonné, the author's method uses only the calculus of matrices.

I. LINEAR GROUPS

2. Automorphisms of the general linear group $GL_2(K)$. Let K be a sfield of characteristic $\neq 2$. An element M of $GL_2(K)$ is said to be an involution if $M^2=I$. Since $-I$ is the only involution commutative with every element of $GL_2(K)$, any automorphism of $GL_2(K)$ must leave $-I$ invariant.

Let A and B be two anti-commutative involutions, that is

$$A^2 = B^2 = I, \qquad AB = -BA. \tag{1}$$

These are invariant relations under any automorphism. Since A and B can be simultaneously carried into

$$\begin{pmatrix} 1 & 0 \\ 0 & -1 \end{pmatrix} \quad \text{and} \quad \begin{pmatrix} 0 & 1 \\ 1 & 0 \end{pmatrix} \tag{2}$$

by a similarity relation,* we may assume that the automorphism under consideration leaves both matrices of (2) unaltered.

It is easy to verify that a matrix commutative with both matrices of (2) takes the form

$$\begin{pmatrix} \lambda & 0 \\ 0 & \lambda \end{pmatrix} \tag{3}$$

where λ runs over all non-zero elements of the sfield K. Therefore the automorphism under consideration carries the group constituted by all the matrices of the form (3) onto itself.

Let K_0 be the center of the sfield K. Every matrix commutative with

* If there is a non singular P such that $PXP^{-1}=Y$, then we say that X and Y are similar.

all the elements of (3) belongs to $GL_2(K_0)$ and conversely. Therefore our automorphism induces an automorphism on $GL_2(K_0)$ which leaves both matrices of (2) unaltered.

The parabolic elements* are defined in the same way as before ([1], p. 756); namely, if the characteristic p of the sfield differs from 0, a matrix A is called parabolic if $A^{2p}=I$; and in case p=0, if there are infinitely many matrices similar to A and commutative with A. Since $SL_2(K_0)$ is generated by all the parabolic elements of $GL_2(K_0)$, the automorphism under consideration induces an automorphism on $SL_2(K_0)$. It was proved in [1] that every automorphism of $SL_2(K_0)$ is of the form

$$(4) \qquad A \to P\, A^{\sigma}\, P^{-1}$$

where P is a non-singular matrix and σ is an automorphism of the field K_0. This is defined for A belonging to $SL_2(K_0)$.

Since

$$\begin{pmatrix}1&0\\0&-1\end{pmatrix} A \begin{pmatrix}1&0\\0&-1\end{pmatrix}, \begin{pmatrix}0&1\\1&0\end{pmatrix} A \begin{pmatrix}0&1\\1&0\end{pmatrix}, \begin{pmatrix}0&1\\1&0\end{pmatrix} A \begin{pmatrix}1&0\\0&-1\end{pmatrix}$$

all belong to $SL_2(K_0)$ for A belonging to $SL_2(K_0)$, and our automorphism leaves (2) unaltered, we deduce immediately that P is either λI or $\mu\begin{pmatrix}0&1\\-1&0\end{pmatrix}$. That is, we have only two possible cases of (4), namely,

$$(5) \qquad A \to A^{\sigma}$$

and

$$(6) \qquad A \to \begin{pmatrix}0&1\\-1&0\end{pmatrix} A^{\sigma} \begin{pmatrix}0&1\\-1&0\end{pmatrix}^{-1}$$

We are going to discuss both cases separately. For the first case we have, in particular,

$$(7) \qquad \begin{pmatrix}1&0\\x&1\end{pmatrix} \to \begin{pmatrix}1&0\\x^{\sigma}&1\end{pmatrix}, \qquad \text{for } x \in K_0,$$

$$(8) \qquad \begin{pmatrix}a&0\\0&a^{-1}\end{pmatrix} \to \begin{pmatrix}a^{\sigma}&0\\0&(a^{\sigma})^{-1}\end{pmatrix}, \qquad \text{for } a(\neq 0) \in K_0$$

and

$$(9) \qquad \begin{pmatrix}0&1\\-1&0\end{pmatrix} \to \begin{pmatrix}0&1\\-1&0\end{pmatrix}$$

Notice that

$$\begin{pmatrix}0&1\\-1&0\end{pmatrix}\begin{pmatrix}1&0\\x&1\end{pmatrix}\begin{pmatrix}0&-1\\1&0\end{pmatrix} = \begin{pmatrix}1&-x\\0&1\end{pmatrix} \to \begin{pmatrix}1&-x^{\sigma}\\0&1\end{pmatrix}$$

---

* Notice that the definition does not coincide with the usual definition of parabolic element.

Next we are going to consider those parabolic elements of $GL_2(K)$ which are commutative with (7). By commutativity they are of the form

$$\begin{pmatrix} a & 0 \\ y & a \end{pmatrix};$$

since they are parabolic, we have $a = \pm 1$. Squaring all those elements, we deduce that our automorphism induces the mapping

$$(10) \qquad \begin{pmatrix} 1 & 0 \\ x & 1 \end{pmatrix} \to \begin{pmatrix} 1 & 0 \\ x^{\sigma} & 1 \end{pmatrix},$$

for all x belonging to K, noticing that the definition of x^{σ} is now extended to any element x of K. Evidently, we have

$$(11) \qquad (x + y)^{\sigma} = x^{\sigma} + y^{\sigma}$$

for all x and y of K.

Since

$$\begin{pmatrix} x & 0 \\ 0 & x^{-1} \end{pmatrix}, \qquad x\ (\neq 0) \in K$$

is commutative with (8), and

$$(12) \quad \begin{pmatrix} x & 0 \\ 0 & x^{-1} \end{pmatrix} = \begin{pmatrix} 1 & -x \\ 0 & 1 \end{pmatrix} \begin{pmatrix} 1 & 0 \\ x^{-1} & 1 \end{pmatrix} \begin{pmatrix} 1 & 0 \\ -1 & 1 \end{pmatrix} \begin{pmatrix} 1 & 1 \\ 0 & 1 \end{pmatrix} \begin{pmatrix} 1 & 0 \\ -1 & 1 \end{pmatrix} \begin{pmatrix} 1 & 0 \\ x & 1 \end{pmatrix},$$

it is mapped by our automorphism into the following form

$$\begin{pmatrix} \lambda & 0 \\ 0 & \mu \end{pmatrix} = \begin{pmatrix} 1 & -x^{\sigma} \\ 0 & 1 \end{pmatrix} \begin{pmatrix} 1 & 0 \\ x^{*} & 1 \end{pmatrix} \begin{pmatrix} 1 & 0 \\ -1 & 1 \end{pmatrix} \begin{pmatrix} 1 & 1 \\ 0 & 1 \end{pmatrix} \begin{pmatrix} 1 & 0 \\ -1 & 1 \end{pmatrix} \begin{pmatrix} 1 & 0 \\ x^{\sigma} & 1 \end{pmatrix}.$$

Comparing the elements in the (1, 2) position, we have

$$x^{\sigma}\, x^{*} = 1,$$

that is, $(x^{-1})^{\sigma} = (x^{\sigma})^{-1}$, and

$$\begin{pmatrix} x & 0 \\ 0 & x^{-1} \end{pmatrix} \to \begin{pmatrix} x^{\sigma} & 0 \\ 0 & x^{-\sigma} \end{pmatrix}$$

for all $x \in K$. Considering the element

$$\begin{pmatrix} x\,y\,x & 0 \\ 0 & (x\,y\,x)^{-1} \end{pmatrix} = \begin{pmatrix} x & 0 \\ 0 & x^{-1} \end{pmatrix} \begin{pmatrix} y & 0 \\ 0 & y^{-1} \end{pmatrix} \begin{pmatrix} x & 0 \\ 0 & x^{-1} \end{pmatrix}$$

we deduce that

$$(x\,y\,x)^{\sigma} = x^{\sigma}\, y^{\sigma}\, x^{\sigma}.$$

Evidently $1^{\sigma} = 1$; therefore, by a theorem of the author [3], we deduce that the mapping $x \to x^\sigma$ is either an automorphism or an anti-automorphism.

Since $SL_2(K)$ is generated by the elements of (9) and (10) (Dieudonné [6]), we have the following two possibilities:

1) If $x \to x^\sigma$ is an automorphism, the automorphism of $SL_2(K)$ under consideration is

$$A \to A^\sigma . \tag{13}$$

2) If $x \to x^\sigma$ is an anti-automorphism, since

$$\begin{pmatrix} 1 & x \\ 0 & 1 \end{pmatrix} \to \begin{pmatrix} 1 & x^\sigma \\ 0 & 1 \end{pmatrix} = \begin{pmatrix} 0 & 1 \\ -1 & 0 \end{pmatrix} \begin{pmatrix} 1 & 0 \\ x^\sigma & 1 \end{pmatrix}^{-1} \begin{pmatrix} 0 & 1 \\ -1 & 0 \end{pmatrix}^{-1}$$

$$\begin{pmatrix} x & 0 \\ 0 & x^{-1} \end{pmatrix} \to \begin{pmatrix} x & 0 \\ 0 & x^{-\sigma} \end{pmatrix} = \begin{pmatrix} 0 & 1 \\ -1 & 0 \end{pmatrix} \begin{pmatrix} x & 0 \\ 0 & x^{-\sigma} \end{pmatrix}^{-1} \begin{pmatrix} 0 & 1 \\ -1 & 0 \end{pmatrix}^{-1}$$

$$\begin{pmatrix} 0 & 1 \\ -1 & 0 \end{pmatrix} \to \begin{pmatrix} 0 & 1 \\ -1 & 0 \end{pmatrix} = \begin{pmatrix} 0 & 1 \\ -1 & 0 \end{pmatrix} \begin{pmatrix} 0 & -1 \\ 1 & 0 \end{pmatrix}^{-1} \begin{pmatrix} 0 & -1 \\ 1 & 0 \end{pmatrix} ,$$

we have the automorphism, after an inner automorphism,

$$A \to (A'^\sigma)^{-1} \tag{14}$$

where A' denotes the transpose of A.

By similar methods applied to (6), we obtain the same results (13) and (14). From now on we assume that the automorphism of $GL_2(K)$ under consideration satisfies (13) or (14) for A belonging to $SL_2(K)$. We now consider $GL_2(K)$; certainly the most essential part is to find the mapping of those elements of the form

$$\begin{pmatrix} 1 & 0 \\ 0 & \lambda \end{pmatrix} , \quad \lambda \in K. \tag{15}$$

Since such an element permutes with any element (12), with $x \in K_0$, it is mapped into an element of the form

$$\begin{pmatrix} a_\lambda & 0 \\ 0 & d_\lambda \end{pmatrix}$$

Since

$$\begin{pmatrix} 1 & 0 \\ 0 & \lambda \end{pmatrix} \begin{pmatrix} 1 & 0 \\ x & 1 \end{pmatrix} \begin{pmatrix} 1 & 0 \\ 0 & \lambda \end{pmatrix}^{-1} = \begin{pmatrix} 1 & 0 \\ \lambda x & 1 \end{pmatrix}$$

is mapped into

$$\begin{pmatrix} 1 & 0 \\ (\lambda x)^\sigma & 1 \end{pmatrix},$$

we deduce

$$(16) \qquad d_\lambda \; x^\sigma \; a_\lambda^{-1} = (\lambda x)^\sigma$$

for all x. In particular setting x = 1, we have

$$(17) \qquad d_\lambda = \lambda^\sigma \; a_\lambda \;;$$

that is, (15) is mapped into

$$(18) \qquad \begin{pmatrix} 1 & 0 \\ 0 & \lambda^\sigma \end{pmatrix} a_\lambda = d_\lambda \begin{pmatrix} \lambda^{-\sigma} & 0 \\ 0 & 1 \end{pmatrix} .$$

Combining (16) and (17), we have

$$\lambda^\sigma \; a_\lambda \; x^\sigma = (\lambda x)^\sigma \; a_\lambda .$$

In case σ is an automorphism, we have

$$a_\lambda \; x^\sigma = x^\sigma \; a_\lambda \;;$$

that is, a_λ belongs to the center, and (15) is mapped into

$$a_\lambda \begin{pmatrix} 1 & 0 \\ 0 & \lambda^\sigma \end{pmatrix} ,$$

where a_λ is a representation of the multiplicative group into the center.

In case σ is an anti-automorphism, we have

$$(\lambda^\sigma a_\lambda) \; x^\sigma = x^\sigma \; (\lambda^\sigma a_\lambda),$$

that is, $d_\lambda = \lambda^\sigma a_\lambda$ belongs to the center, and (15) is mapped into

$$d_\lambda \begin{pmatrix} \lambda^{-\sigma} & 0 \\ 0 & 1 \end{pmatrix}$$

Therefore we have

THEOREM 1. Every automorphism of $GL_2(K)$ is a product of an inner automorphism and one of the following two types: (i)

$$(19) \qquad A \to \chi(A) \; A^\sigma$$

where σ is an automorphism of K and $\chi(A)$ is a representation of $GL_n(K)$ into the multiplicative group of K_0; and (ii)

$$(20) \qquad A \to \chi(A) \; (A'^{-1})^\sigma$$

where σ is an anti-automorphism of K.

REMARK. The theorem also holds for sfield of characteristic 2.

3. Automorphisms of $SL_2^{\pm}(K)$. We now consider a group $SL_2^{\pm}(K)$ which differs slightly from $SL_2(K)$. The projective geometry on a line over a sfield suggests that the group $SL_2^{\pm}(K)$, obtained by successive application of harmonic separation, is of primary interest (Ancochea [7]). As we know, in the group $SL_2(K)$, involutions do not always exist. More precisely, if -1 does not belong to the commutator subgroup of the group of multiplication, there does not exist an involution (a relation obtained by harmonic conjugacy).

By an easy investigation, we find that if -1 belongs to the commutator subgroup of the multiplicative group of K, then $SL_2^{\pm}(K) = SL_2(K)$; otherwise, $SL_2^{\pm}(K)$ is obtained from $SL_2(K)$ by adjoining an involution, such as

$$\begin{pmatrix} 1 & 0 \\ 0 & -1 \end{pmatrix}.$$

The automorphisms of $SL_2(K)$ are still to be determined when -1 is not in the commutator subgroup.

The theorem which we shall prove in this section is the following:

THEOREM 2. Every automrophism of $SL_2^{\pm}(K)$ is either of the form

$$\chi(A) \quad P \quad A^{\sigma} \quad P^{-1}, \tag{21}$$

where P is a non singular matrix, σ is an automorphism of K, and $\chi(A)$ is a representation of $SL_2^{\pm}(K)$ into the multiplicative group of the center K_0, or of the form

$$\chi(A) \quad P \quad A'^{\sigma-1} P^{-1}, \tag{22}$$

where σ is an anti-automorphism.

REMARK concerning $\chi(A)$: In case -1 belongs to the commutator subgroup of the multiplicative group of K, then we have always $\chi(A) = 1$. Otherwise we have another possibility, namely

$$\chi(A) = \begin{cases} 1 & \text{if } A \text{ belongs to } SL_2(K) \\ -1 & \text{otherwise.} \end{cases}$$

PROOF of theorem 2. Let us start with a pair of anti-commutative involutions A and B, that is

$$A^2 = B^2 = I, \quad A B = -B A .$$

There exists a matrix P, which may not belong to $SL_2^{\pm}(K)$ such that

$$P A P^{-1} = \begin{pmatrix} 1 & 0 \\ 0 & -1 \end{pmatrix}, \quad P B P^{-1} = \begin{pmatrix} 0 & 1 \\ 1 & 0 \end{pmatrix}$$

An element of $SL_2^{\pm}(K)$ commuting with both matrices of (2) is of the form

$$\begin{pmatrix} \lambda & 0 \\ 0 & \lambda \end{pmatrix}, \tag{24}$$

where either λ^2 or $-\lambda^2$ belongs to the commutator subgroup of the group of multiplication of K. By theorem 4 of Hua [8], every element commutative with all λ belongs to the center K_0. Therefore a matrix commutative with all elements of (24) is of the form

$$\begin{pmatrix} a & b \\ c & d \end{pmatrix}, \tag{25}$$

where a, b, c, d belong to K_0, and the determinant of (25) (or its negative) belongs to the intersection of the center and the commutator subgroup of the multiplicative group. The remaining part of the proof is similar to that of theorem 1.

4. Automorphisms of $PGL_2(K)$. We identify all the elements of the form γM of $GL_2(K)$ as a single element of $PGL_2(K)$, where $\gamma (\neq 0)$ runs over all the elements of the center. A matrix M represents an involution if it satisfies

$$M^2 = \gamma I, \quad (\gamma \in K_0) \tag{26}$$

and if M does not represent the identity of $PGL_2(K)$. γ is called the scalar of the matrix M.

Since we have a matrix Q such that

$$Q \ M \ Q^{-1} = \begin{pmatrix} 0 & 1 \\ c & d \end{pmatrix}, \tag{27}$$

therefore an involution satisfying (26) is similar to the normal form

$$\begin{pmatrix} 0 & 1 \\ \gamma & 0 \end{pmatrix} \tag{28}$$

If $\begin{pmatrix} 0 & 1 \\ \gamma & 0 \end{pmatrix}$ and $\begin{pmatrix} 0 & 1 \\ \delta & 0 \end{pmatrix}$ are similar in $PGL_2(K)$, then from

$$\begin{pmatrix} 0 & 1 \\ \gamma & 0 \end{pmatrix} \begin{pmatrix} a & b \\ c & d \end{pmatrix} = \epsilon \begin{pmatrix} a & b \\ c & d \end{pmatrix} \begin{pmatrix} 0 & 1 \\ \delta & 0 \end{pmatrix},$$

where ϵ belongs to the center, we deduce

$$c = \epsilon \delta b, \quad d = \epsilon a, \quad \gamma a = \epsilon \delta b, \quad \gamma b = \epsilon c;$$

consequently

$$(\gamma - \epsilon^2 \delta) \ b = (\gamma - \epsilon^2 \delta) \ a = 0.$$

Since $\underline{a}$ and $\underline{b}$ cannot be both zero, we have $\gamma = \epsilon^2 \delta$. Conversely, we have

$$\begin{pmatrix} 0 & 1 \\ \gamma & 0 \end{pmatrix} = \epsilon \begin{pmatrix} 1 & 0 \\ 0 & \epsilon \end{pmatrix} \begin{pmatrix} 0 & 1 \\ \delta & 0 \end{pmatrix} \begin{pmatrix} 1 & 0 \\ 0 & \epsilon^{-1} \end{pmatrix}.$$

Therefore two involutions, represented by matrices with scalar γ and δ, are similar in $PGL_2(K)$ if and only if $\gamma\delta^{-1}$ is the square of a central element.

Therefore, all the involutions can be classified into conjugate sets Γ_α, where α runs over all the elements of the factor group of the multiplicative group of the center by its subgroup of all square elements.

We are now going to prove that the automorphism under consideration leaves Γ_1 invariant. Γ_1 is a conjugate set containing two distinct commuting elements, namely

$$\begin{pmatrix} 1 & 0 \\ 0 & -1 \end{pmatrix} \text{ and } \begin{pmatrix} 0 & 1 \\ 1 & 0 \end{pmatrix}.$$

It is enough to distinguish Γ_1 from those Γ_α containing two commutative involutions. Let Γ_α be a set containing two commuting involutions A and B with scalars α and $\alpha\delta^2$. From $A B = \epsilon B A$, ϵ belonging to the center, we deduce

$$B A^2 = B\alpha = \alpha B = A^2 B = \epsilon^2 B A^2, \tag{29}$$

that is, $\epsilon = \pm 1$. Therefore $(A B)^2 = \pm\alpha^2 \delta^2 I$, so that A B belongs either to Γ_1 and Γ_{-1}. Therefore our automorphism either leaves both Γ_1 and Γ_{-1} invariant, or permutes them.

1^0. If -1 is the square of a central element, then Γ_{-1} coincides with Γ_1.

2^0. Suppose that -1 is not a square of a central element but is a square of an element $\underline{a}$ in K. The product of any two distinct commuting elements of Γ_1 belongs to Γ_{-1}. In fact, let us fix one of them as $A = \begin{pmatrix} 1 & 0 \\ 0 & -1 \end{pmatrix}$; then, from $A B = B A$, we deduce that B is either the identity or γA, γ belonging to center. Thus, we always have $A B = -B A$ (the existence of B is shown by (2)), and $(A B)^2 = -I$. In Γ_{-1} we have two commuting elements

$$\begin{pmatrix} a & 0 \\ 0 & -a \end{pmatrix}, \quad \begin{pmatrix} 0 & a \\ a & 0 \end{pmatrix},$$

whose product is

$$a^2 \begin{pmatrix} 0 & 1 \\ -1 & 0 \end{pmatrix} = -\begin{pmatrix} 0 & 1 \\ -1 & 0 \end{pmatrix}$$

which belongs to Γ_{-1}. Therefore our automorphism cannot interchange Γ_1 and Γ_{-1}.

3^0. Next we suppose that -1 is not the square of an element in K. Let A and B be involutions of Γ_1 and Γ_{-1} respectively, and let them commute. We may fix A to be $\begin{pmatrix} 1 & 0 \\ 0 & -1 \end{pmatrix}$. Then B takes one of the forms: $\begin{pmatrix} \lambda & 0 \\ 0 & \mu \end{pmatrix}$ or $\begin{pmatrix} 0 & \lambda \\ \mu & 0 \end{pmatrix}$. However, the first is impossible, since $\lambda^2 = -1$ is insoluble. Therefore we have

$$B = \begin{pmatrix} 0 & \lambda \\ -\lambda^{-1} & 0 \end{pmatrix}.$$

The product A B belongs to Γ_1. Therefore the automorphism under consideration cannot interchange Γ_1 and Γ_{-1}.

Take a pair of commuting involutions of Γ_1. Without loss of generality, we assume that the automorphism leaves

$$(2) \qquad \begin{pmatrix} 1 & 0 \\ 0 & -1 \end{pmatrix} \quad \text{and} \quad \begin{pmatrix} 0 & 1 \\ 1 & 0 \end{pmatrix}$$

invariant. The elements commuting with both matrices of (2) are of the form

$$(30) \qquad \begin{pmatrix} a & 0 \\ 0 & a \end{pmatrix}, \begin{pmatrix} a & 0 \\ 0 & -a \end{pmatrix}, \begin{pmatrix} 0 & b \\ b & 0 \end{pmatrix}, \begin{pmatrix} 0 & b \\ -b & 0 \end{pmatrix},$$

by (29). Squaring all these elements, we obtain

$$(31) \qquad \begin{pmatrix} a^2 & 0 \\ 0 & a^2 \end{pmatrix}.$$

Consider the set of elements commuting with all these elements, that is

$$(32) \qquad \begin{pmatrix} p & q \\ r & s \end{pmatrix} \begin{pmatrix} a^2 & 0 \\ 0 & a^2 \end{pmatrix} = \rho \begin{pmatrix} a^2 & 0 \\ 0 & a^2 \end{pmatrix} \begin{pmatrix} p & q \\ r & s \end{pmatrix}$$

where ρ belongs to the center and may depend on a. Suppose $p \neq 0$; then from

$$p\, a^2 = \rho\, a^2\, p,$$

we have

$$\begin{pmatrix} 1 & qp^{-1} \\ rp^{-1} & sp^{-1} \end{pmatrix} \begin{pmatrix} a^2 & 0 \\ 0 & a^2 \end{pmatrix} p = \frac{1}{\rho} \begin{pmatrix} p & q \\ r & s \end{pmatrix} \begin{pmatrix} a^2 & 0 \\ 0 & a^2 \end{pmatrix} = \begin{pmatrix} a^2 & 0 \\ 0 & a^2 \end{pmatrix} \begin{pmatrix} p & q \\ r & s \end{pmatrix},$$

that is

$$\begin{pmatrix} 1 & qp^{-1} \\ rp^{-1} & sp^{-1} \end{pmatrix} \begin{pmatrix} a^2 & 0 \\ 0 & a^2 \end{pmatrix} = \begin{pmatrix} a^2 & 0 \\ 0 & a^2 \end{pmatrix} \begin{pmatrix} 1 & qp^{-1} \\ rp^{-1} & sp^{-1} \end{pmatrix}.$$

By theorem 6 of Hua [8], qp^{-1}, rp^{-1} and sp^{-1} all belong to the center. By similarly treating the case $p = 0$, $q \neq 0$, we reach the conclusion that

$$\begin{pmatrix} p & q \\ r & s \end{pmatrix} = \lambda M,$$

where $\underline{M}$ belongs to $PGL_2(K_0)$, and $\lambda a^2 \lambda^{-1} a^{-2}$ belongs to the center for all $\underline{a}$, by (32).

Consider those elements of the form

$$(\lambda M)\ (\mu N)^2\ (\lambda M)^{-1}\ (\mu N)^{-2} = \zeta\, MN^2 M^{-1} N^{-2}, \qquad \zeta \in K_0,$$

where $\underline{M}$ and $\underline{N}$ belong to $PGL_2(K_0)$. The elements $MN^2N^{-1}N^{-2}$ generate the group $PSL_2(K_0)$; in fact, they generate a normal subgroup of $PSL_2(K_0)$, but since $PSL_2(K_0)$ is simple,* we have the assertion. Therefore the automorphism of $PSL_2(K)$ induces an automorphism on $PSL_2(K_0)$.

Now we use non-homogeneous representations: an element of $PGL_2(K)$, represented by a matrix $\begin{pmatrix} a & b \\ c & d \end{pmatrix}$, can be realized as a transformation

$$z_1 = (a z + b)(c z + d)^{-1} \tag{33}$$

on a projective line over the sfield K. It is evident that the realization is one-to-one.

We define a parabolic element $\underline{t}$ by the following properties: if the characteristic $\underline{p}$ of our sfield is not zero, a transformation $\underline{t}$ is called parabolic, if $\underline{t}^p$ equals the identity of $PGL_2(K)$; and, for $\underline{p} = 0$, a transformation $\underline{t}$ is called parabolic, if there are infinitely many elements similar to $\underline{t}$ and commutative with $\underline{t}$. We shall now prove that every parabolic element of $PSL_2(K_0)$ can be carried into the form

$$z_1 = z + a \qquad (\text{where } \underline{a} \text{ belongs to } K_0).$$

In fact, suppose $\underline{p} = 0$, and that $\underline{t}$ is represented by a matrix $\underline{T}$ belonging to $SL_2(K_0)$, where

$$T^p = \alpha I .$$

Since $\underline{T}^p$ belongs to $SL_2(K_0)$, we have $\alpha = \pm 1$. As in [1], we have our assertion. In case $\underline{p} = 0$, the argument of [1] can still be applied.

After we have taken care of the parabolic elements, we can follow either the method of Schreier- van der Waerden [9] or the method of the author [1] to obtain the automorphism of $PSL_2(K_0)$. More precisely, we may assume that our automorphism of $PGL_2(K)$, induces an automorphism of $PSL_2(K)$ which carries the transformations

$$z_1 = z + t \quad \text{to} \quad z_1 = z + t^{\sigma}, \qquad t \in K_0, \tag{34}$$

and

$$z_1 = -1/z \quad \text{to} \quad z_1 = -1/z, \tag{35}$$

where $\underline{t}^{\sigma}$ is an automorphism of the field K_0.

---

* Notice that if K_0 has only 3 elements, the characteristic of K is equal to 2, the case which was excluded at the beginning.

Consider all the parabolic elements commuting with the translation group (34). It is easy to see that they are of the form

$$z_1 = z + t, \qquad t \in K. \tag{36}$$

Therefore, our automorphism carries (36) into

$$z_1 = z + t^{\sigma}$$

where t^{σ} is defined for all t belonging to K. From the identity (12), we deduce by the same argument following (12), that t^{σ} is either an automorphism or an anti-automorphism of the sfield; and

THEOREM 3. Every automorphism of $PGL_2(K)$ is induced by an automorphism of $GL_2(K)$.

REMARK. The theorem is still true for characteristic two.

5. Automorphisms of $PSL_2^{\pm}(K)$. As in §4, we classify the involutions into conjugate sets; the situation now becomes very much more complicated, however, since neither can the scalar be arbitrary, nor does the scalar determine a unique conjugate set. It is not always possible to find the matrix Q, belonging to $SL_2^{\pm}(K)$, such that (27) is true. A matrix of scalar γ is similar to a matrix of the form

$$\begin{pmatrix} 0 & q^{-1} \\ \gamma q & 0 \end{pmatrix} \tag{37}$$

under $SL_2(K)$. A conjugate set Γ_{α} of §4 may now break up into several conjugate sets. Nevertheless, the same argument gives us the result that the pair of sets Γ_1 and Γ_{-1} is invariant under automorphisms.

Fortunately Γ_1 does not split into several conjugate sets; in fact,

$$\begin{pmatrix} 0 & q^{-1} \\ q & 0 \end{pmatrix}\begin{pmatrix} 1 & -\frac{1}{2}q^{-1} \\ q & \frac{1}{2} \end{pmatrix} = \begin{pmatrix} 1 & -\frac{1}{2}q^{-1} \\ q & \frac{1}{2} \end{pmatrix}\begin{pmatrix} 1 & 0 \\ 0 & -1 \end{pmatrix}$$

and

$$\begin{pmatrix} 1 & -\frac{1}{2}q^{-1} \\ q & \frac{1}{2} \end{pmatrix} = \begin{pmatrix} 1 & 0 \\ q & 1 \end{pmatrix}\begin{pmatrix} 1 & -\frac{1}{2}q^{-1} \\ 0 & 1 \end{pmatrix}$$

which belongs to $SL_2(K)$.

As in §4, we divide our treatment into three cases: 1° and 3° can be treated analogously. The only difficult case is where (2°) -1 is a square of a non-central element a. Now Γ_{-1} splits into several conjugate sets. We shall now prove that every involution (namely (37) with $\gamma = -1$) of Γ_{-1} is similar to

$$(38)\qquad \begin{pmatrix} a & 0 \\ 0 & -qaq^{-1} \end{pmatrix}$$

under the group $SL_2(K)$. In fact this is a consequence of the following simple calculation:

$$(39)\qquad \begin{pmatrix} 1 & -\frac{1}{2}a \\ 0 & 1 \end{pmatrix} \begin{pmatrix} 1 & 0 \\ -a & 1 \end{pmatrix} \begin{pmatrix} 0 & 1 \\ -1 & 0 \end{pmatrix} \begin{pmatrix} 1 & 0 \\ a & 1 \end{pmatrix} \begin{pmatrix} 1 & \frac{1}{2}a \\ 0 & 1 \end{pmatrix} = \begin{pmatrix} a & 0 \\ 0 & -a \end{pmatrix}.$$

Then (37) with $\gamma = -1$ is similar to (38) under $SL_2(K)$, since

$$\begin{pmatrix} 1 & 0 \\ 0 & q \end{pmatrix} \begin{pmatrix} 1 & -\frac{1}{2}a \\ 0 & 1 \end{pmatrix} \begin{pmatrix} 1 & 0 \\ -a & 1 \end{pmatrix} \begin{pmatrix} 1 & 0 \\ 0 & q \end{pmatrix}^{-1}$$

belongs to $SL_2(K)$. Putting $\underline{q}\ \underline{a}$ instead of $\underline{q}$ in (37), we find that

$$(40)\qquad \begin{pmatrix} a & 0 \\ 0 & -qaq^{-1} \end{pmatrix} \quad \text{and} \quad \begin{pmatrix} 0 & (qa)^{-1} \\ -qa & 0 \end{pmatrix}$$

are similar under the group $SL_2(K)$. Their product is

$$\begin{pmatrix} 0 & q^{-1} \\ -q & 0 \end{pmatrix}$$

which is similar to (40) too. Therefore any set γ contains three mutually commuting elements.

As we proved beofre (§ 4), the product of two commuting elements of Γ_1 is not an element of Γ_1.

Therefore the automorphism carries Γ_1 onto itself. By an argument similar to that of § 4, we may assume that our automorphism leaves

$$\begin{pmatrix} 1 & 0 \\ 0 & -1 \end{pmatrix}, \begin{pmatrix} 0 & 1 \\ +1 & 0 \end{pmatrix}$$

invariant.

The elements of $PSL_2(K)$ commuting with both these elements are of the form

$$\begin{pmatrix} a & 0 \\ 0 & a \end{pmatrix}, \begin{pmatrix} a & 0 \\ 0 & -a \end{pmatrix}, \begin{pmatrix} 0 & b \\ b & 0 \end{pmatrix}, \begin{pmatrix} 0 & b \\ -b & 0 \end{pmatrix}.$$

Squaring all of these elements, we obtain

$$\begin{pmatrix} a^2 & 0 \\ 0 & a^2 \end{pmatrix}.$$

Since they belong to $PSL_2(K)$, then $\underline{a}^2$ or $-\underline{a}^2$ belongs to the commutator subgroup.

Using the lemma below, we proceed as in § 4, and obtain

THEOREM 4. Every automorphism of $PSL_2^{\pm}(K)$ is induced by an automorphism of $SL_2^{\pm}(K)$.

LEMMA. Let C be the commutator subgroup of the multiplicative group of K and let C^2 be the group generated by all the square elements of C. Then either K is commutative, or the sfield K is generated by all the elements of C^2. Consequently, if an element commutes with all the elements of C^2, then it belongs to center.

PROOF. If C^2 is not contained in the center, the lemma follows from theorem 1 of Hua [8].

Now we suppose C^2 is contained in the center K_0 and K is not commutative. Let

$$1, \quad a_1, \quad a_2, \ldots, (a_i \in C)$$

be the representative system of the factor group of C over C^2. Since C generates the sfield, by theorems 4 and 1 of Hua [8], we have therefore

$$K = K_0 (a_1, a_2, \ldots),$$

where

$$a_i^2 = \gamma_i, \qquad (a_i a_j)^2 = \gamma_i \gamma_j \delta_{ij}$$

and γ's and δ's belong to the field K_0.

Since K is not a field, there is a $\delta_{ij} \neq 1$. We may suppose that $\delta_{12} = \delta \neq 1$.

From

$$a_1(a_2 + 1) = \delta(a_2 + 1) a_1 + a_1 (1 - \delta),$$

we deduce

$$a_1(a_2 + 1) a_1^{-1} (a_2 + 1)^{-1} = \delta + (a_2 + 1)^{-1} (1 - \delta).$$

Suppose that the squares of commutators belong to the center. Then

$$(\delta + (a_2 + 1)^{-1} (1 - \delta))^2 = \beta$$

where β belongs to K_0. Consequently

$$(\delta a_2 + 1)^2 = \beta (a_2 + 1)^2;$$

we deduce immediately

$$\beta = \delta, \quad \delta \gamma_2 = 1.$$

Consider

$$(a_2 + 1)\, a_1 = \frac{1}{\delta}\, a_1\, (a_2 + 1) + a_1 \left(1 - \frac{1}{\delta} \right);$$

we deduce similarly

$$\frac{1}{\delta}\, \gamma_2 = 1.$$

Consequently $\delta = -1$ and $\gamma_2 = -1$. Similarly $\gamma_1 = -1$.

Suppose that K_0 has more than three elements (it is easy to prove that if K_0 contains less than 4 elements, there does not exist such a K). Take x of K_0 different from 0 and ± 1. Then

$$a_1\, (1 + x\, a_2)\, a_1^{-1}\, (1 + x\, a_2)^{-1} = (1 - x\, a_2)\, (1 + xa_2)^{-1} =$$

$$= \frac{1 - x^2}{1 + x^2} - \frac{2x}{1 + x^2}\, a_2,$$

its square is not in the center. This is a contradiction. We have therefore the lemma.

6. Automorphism of $SL_4(K)$. The case left open by Dieudonné concerning the automorphism of $SL_4(K)$ is that where K is not commutative, K has characteristic $\neq 2$ and -1 is not contained in the commutator subgroup of the multiplicative group.

Any two commuting involutions of $SL_4(K)$ can be brought into the form

$$(41) \qquad J_1 = \begin{pmatrix} 1 & 0 & 0 & 0 \\ 0 & 1 & 0 & 0 \\ 0 & 0 & -1 & 0 \\ 0 & 0 & 0 & -1 \end{pmatrix}, \qquad J_2 = \begin{pmatrix} 1 & 0 & 0 & 0 \\ 0 & -1 & 0 & 0 \\ 0 & 0 & 1 & 0 \\ 0 & 0 & 0 & -1 \end{pmatrix}.$$

Let

$$(42) \qquad J_3 = J_1\, J_2 = \begin{pmatrix} 1 & 0 & 0 & 0 \\ 0 & -1 & 0 & 0 \\ 0 & 0 & -1 & 0 \\ 0 & 0 & 0 & 1 \end{pmatrix}.$$

We may assume, that after a transformation, our automorphism leaves J_1, J_2 and J_3 invariant.

We define K_1 by means of the equation

$$(43) \qquad K_1\, J_1 = J_1\, K_1, \quad K_1\, J_2 = -J_3\, K_1, \quad K_1{}^2 = -J_1.$$

Solving (43), we obtain

$$K_1 = \begin{pmatrix} 0 & a & 0 & 0 \\ -a^{-1} & 0 & 0 & 0 \\ 0 & 0 & \pm 1 & 0 \\ 0 & 0 & 0 & \pm 1 \end{pmatrix},$$

where the signs $\pm$ are either both + or both -. By a transformation, we may assume without loss of generality that our automorphism leaves J_1, J_2, J_3 invariant, and either leaves

$$(44)\qquad K_1^+ = \begin{pmatrix} 0 & 1 & 0 & 0 \\ -1 & 0 & 0 & 0 \\ 0 & 0 & 1 & 0 \\ 0 & 0 & 0 & 1 \end{pmatrix}, \qquad K_1^- = \begin{pmatrix} 0 & 1 & 0 & 0 \\ -1 & 0 & 0 & 0 \\ 0 & 0 & -1 & 0 \\ 0 & 0 & 0 & -1 \end{pmatrix}$$

invariant, or permutes them. By similar considerations, we may assume that our automorphism leaves the following two pairs pair-wise invariant:

$$(45)\qquad K_2^+ = \begin{pmatrix} 0 & 0 & 1 & 0 \\ 0 & 1 & 0 & 0 \\ -1 & 0 & 0 & 0 \\ 0 & 0 & 0 & 1 \end{pmatrix}, \qquad K_2^- = \begin{pmatrix} 0 & 0 & 1 & 0 \\ 0 & -1 & 0 & 0 \\ -1 & 0 & 0 & 0 \\ 0 & 0 & 0 & -1 \end{pmatrix}$$

and

$$(46)\qquad K_3^+ = \begin{pmatrix} 0 & 0 & 0 & 1 \\ 0 & 1 & 0 & 0 \\ 0 & 0 & 1 & 0 \\ -1 & 0 & 0 & 0 \end{pmatrix}, \qquad K_3^- = \begin{pmatrix} 0 & 0 & 0 & 1 \\ 0 & -1 & 0 & 0 \\ 0 & 0 & -1 & 0 \\ -1 & 0 & 0 & 0 \end{pmatrix}.$$

($K_2 = K_2^{\pm}$ satisfies

$$K_2J_2 = J_2K_2, \quad K_2J_3 = -J_1K_2, \quad K_2^2 = -J_2,$$

and $K_3 = K^{\pm}{}_3$ satisfies

$$K_3J_3 = J_3K_3, \quad K_3J_1 = -J_2K_3, \quad K_3^2 = -J_3).$$

The subgroup commuting with J_1 consists of all elements of the form

$$(47)\qquad T = \begin{pmatrix} A & 0 \\ 0 & B \end{pmatrix},$$

where A and B are two-rowed matrices. Consider the group G formed by

$$(48)\qquad T\,K_1^{\pm}\,T^{-1} = \begin{pmatrix} A\begin{pmatrix} 0 & 1 \\ -1 & 0 \end{pmatrix}A^{-1} & 0 \\ 0 & \pm I^{(2)} \end{pmatrix}.$$

Since all of the elements of the form $A\begin{pmatrix} 0 & 1 \\ -1 & 0 \end{pmatrix}A^{-1}$ generate (Dieudonné [6]) the group $SL_2(K)$, all the elements of the form (48) generate a group formed by all elements of the form

$$(49)\qquad \begin{pmatrix} A & 0 \\ 0 & \pm I \end{pmatrix}$$

where <u>A</u> runs over all elements of $SL_2(K)$. Squaring all of the elements of (49), we obtain

$$\begin{pmatrix} A^2 & 0 \\ 0 & I \end{pmatrix}.$$

Since $SL_2(K)$ is also generated by all of its square elements (Dieudonné [6¶]) it follows that our automorphism induces an automorphism on the subgroup of elements

$$\begin{pmatrix} A & 0 \\ 0 & I \end{pmatrix}$$

where $\underline{A}$ runs over all the elements of $SL_2(K)$.

If we also use J_3, our automorphism induces an automorphism on

$$\begin{pmatrix} A & 0 \\ 0 & \begin{pmatrix} 1 & 0 \\ 0 & -1 \end{pmatrix}^{\ell} \end{pmatrix}$$

where $\underline{A}$ runs over all the elements of $SL_2^{\pm}(K)$. Notice that the integer ℓ is uniquely determined by $\underline{A}$. More precisely, in case -1 belongs to the commutator subgroup of the multiplicative group, then ℓ is always even, otherwise ℓ is even for $A \in SL_2(K)$ and ℓ is odd for $A \notin SL_2(K)$.

By theorem 2, our automorphism gives us the following mappings:

$$(50) \qquad \begin{pmatrix} A & 0 \\ 0 & I \end{pmatrix} \to \begin{pmatrix} P A^{\sigma} P^{-1} & 0 \\ 0 & I \end{pmatrix}$$

where σ is an automorphism of K or

$$(51) \qquad \begin{pmatrix} A & 0 \\ 0 & I \end{pmatrix} \to \begin{pmatrix} P A'^{\sigma^{-1}} P^{-1} & 0 \\ 0 & I \end{pmatrix}$$

where σ is an anti-automorphism of K, and where $\underline{A}$ runs over all the elements of $SL_2(K)$. Consequently our automorphism cannot permute $\underline{K}_1^+$ and $\underline{K}_1^{-1}$, and must keep $\underline{K}_1^+$ invariant. By an inner automorphism, keeping $\underline{K}_1^+$ invariant, we may assume that our automorphism induces either a mapping

$$(52) \qquad \begin{pmatrix} A & 0 \\ 0 & I \end{pmatrix} \to \begin{pmatrix} A^{\sigma} & 0 \\ 0 & I \end{pmatrix}$$

or

$$(53) \qquad \begin{pmatrix} A & 0 \\ 0 & I \end{pmatrix} \to \begin{pmatrix} A'^{\sigma^{-1}} & 0 \\ 0 & I \end{pmatrix} .$$

Using $\underline{J}_2$ and $\underline{K}_2$ instead of $\underline{J}_1$ and $\underline{K}_1$, we can prove that our automorphism keeps $\underline{K}_2^+$ invariant, and similarly $\underline{K}_3^+$. Since $SL_4(K)$ is generated by K_1^+, K_2^+, K_3^+ and

$$\begin{pmatrix} 1 & x & 0 & 0 \\ 0 & 1 & 0 & 0 \\ 0 & 0 & 1 & 0 \\ 0 & 0 & 0 & 1 \end{pmatrix} ,$$

we have the following theorem:

THEOREM 5. Every automorphism of $SL_4(K)$ is the restriction of an automorphism of $GL_4(K)$.

Since there is no essential difficulty in the case of $PSL_4(K)$, the author will not discuss it here.

II. ORTHOGONAL GROUPS

The case of the orthogonal group left open by Dieudonné is $O_4^+(K,f)$, where K is a field of characteristic $\neq 2$ and f is a quadratic form of index 2.

7. Preliminaries.

THEOREM 6 (Witt [10]). Every quaternary quadratic form of index 2 is equivalent to

$$f = x_1 x_3 + x_2 x_4,$$

whose matrix is

$$(54) \qquad \mathfrak{J} = \begin{pmatrix} 0 & I \\ I & 0 \end{pmatrix}$$

where I and 0 are two-rowed identity and zero matrices.

From now on, we assume that our fundamental quadratic form has the matrix $\mathfrak{J}$.

A four-rowed matrix $\mathfrak{T}$ satisfying

$$(55) \qquad \mathfrak{T}\mathfrak{J}\mathfrak{T}' = \mathfrak{J}$$

with determinant of $\mathfrak{T}$ equal to 1, is an element of $O_4^+(K)$. If we put

$$(56) \qquad \mathfrak{T} = \begin{pmatrix} A & B \\ C & D \end{pmatrix}, \qquad A = A^{(2)}, \text{ etc.},$$

then we have

$$(57) \qquad AB' + BA' = 0, \quad CD' + DC' = 0, \quad AD' + BC' = I.$$

Since $\mathfrak{T}'$ is also orthogonal, we deduce

$$(58) \qquad A'C + C'A = 0, \quad B'D + D'B = 0, \quad A'D + C'B = I.$$

Later we shall need another group containing O_4^+ as its subgroup and which is defined by the matrices $\mathfrak{P}$ satisfying

$$(59) \qquad \mathfrak{P}\mathfrak{J}\mathfrak{P}' = a\mathfrak{J},$$

where $\underline{a}$ is an element ($\neq 0$) of the field. This group is denoted by $GO_4 = GO_4(K,f)$.

THEOREM 7. If $\mathfrak{T}$ belongs to O_4^+, then in the expression (56), A is either zero or non-singular.

PROOF. Suppose that $\underline{A}$ is of rank 1, we have two matrices $P(=P^{(2)})$ and $Q(=Q^{(2)})$ such that

$$P\,A\,Q = \begin{pmatrix} 1 & 0 \\ 0 & 0 \end{pmatrix}$$

Since both

$$\begin{pmatrix} P & 0 \\ 0 & P'^{-1} \end{pmatrix} \quad \text{and} \quad \begin{pmatrix} Q & 0 \\ 0 & Q'^{-1} \end{pmatrix}$$

belongs to O_4^+, without loss of generality we may consider those T with $A = \begin{pmatrix} 1 & 0 \\ 0 & 0 \end{pmatrix}$. From (57), A B' is skew symmetric so we may deduce that

$$B = \begin{pmatrix} 0 & b_2 \\ 0 & b_4 \end{pmatrix}, \qquad b_4 \neq 0,$$

and from (58)

$$C = \begin{pmatrix} 0 & 0 \\ c_3 & c_4 \end{pmatrix}, \qquad c_4 \neq 0.$$

Again from (57)

$$D = \begin{pmatrix} 1 & d_2 \\ -b_2c_4 & d_4 \end{pmatrix}, \qquad b_4c_4 = 1;$$

in conclusion, we have

$$\mathfrak{T} = \begin{pmatrix} 1 & 0 & 0 & b_2 \\ 0 & 0 & 0 & b_4 \\ 0 & 0 & 1 & d_2 \\ 0_2 & c_4 & -b_2c_4 d_4 \end{pmatrix}$$

Its determinant is $-b_4c_4 = -1$. This proves theorem 2.

REMARK. The converse of the theorem is also true, i.e., any T satisfying (55) with non-singular $\underline{A}$ must belong to O_4^+. This is an easy consequence of theorem 8.

THEOREM 8. Write $K = \begin{pmatrix} 0 & 1 \\ -1 & 0 \end{pmatrix}$. Every element of O_4^+ can be expressed uniquely as one of the following forms:

$$(60) \quad \begin{pmatrix} I & 0 \\ xK & I \end{pmatrix} \begin{pmatrix} A & 0 \\ 0 & A'^{-1} \end{pmatrix} \begin{pmatrix} I & yK \\ 0 & I \end{pmatrix} = \begin{pmatrix} A & yAK \\ xKA & (1-xyd(A))A'^{-1} \end{pmatrix}^{*}$$

and

$$(61) \quad \begin{pmatrix} 0 & I \\ I & 0 \end{pmatrix} \begin{pmatrix} A & 0 \\ 0 & A'^{-1} \end{pmatrix} \begin{pmatrix} I & zK \\ 0 & I \end{pmatrix} = \begin{pmatrix} 0 & A'^{-1} \\ A & zAK \end{pmatrix}$$

where x, y and z are elemets of K and A is a non-singular two-rowed matrix.

---

* We use d(A) to denote the determinant of A.

PROOF. In case $\underline{A}$ is non-singular, from (57) and (58) we deduce

$$B = y\,A\,K, \quad C = x\,K\,A$$

and

$$D = x\,y\,K\,A\,K + A'^{-1} = (1 - x\,y\,d(A))\,A'^{-1}$$

since $A\,K\,A' = d(A)\,K$. The case $A = 0$ can easily be expressed as (61). The uniqueness can also be proved without any difficulty.

Definition. We use

$$(x, A, y), \quad (\infty, A, z)$$

to denote the matrices of (60) and (61) respectively, which may be called the coordinates of an element of O_4^+.

8. An extra type of automorphisms. From our previous experience, we might expect that the only automorphism is

$$\mathfrak{P}\,\mathfrak{U}\,\mathfrak{P}^{-1},$$

where $\mathfrak{P}$ belongs to $GO_4(K)$. But this is entirely false; there are many automorphisms of $O_4^+(K)$ which are not those of $O_4(K)$.

Let $a \to a^{\tau}$ be an automorphism of the field. Let $\varphi(a)$ be a representation of the multiplicative group of the field into itself, and let it satisfy

$$\varphi^2(a) = \frac{a^{\tau}}{a} \quad . \tag{62}$$

This is called a representation induced by the automorphism.

Example 1. Let K be a rational field. The identity automorphism is the only one. We write a rational number as

$$r = \pm\, 2^{\ell_1}\ 3^{\ell_2}\ \ldots\ p_n^{\ell_n}\ \ldots,$$

where p_n denotes the n-th prime, and the ℓ_n are integers. We fix a prime p_n. Then we define

$$\varphi(r) = (-1)^{\ell_n} \quad .$$

This evidently satisfies (62). There is exactly one other induced representation

$$\varphi(r) = \begin{cases} 1 & \text{if } r > 0 \\ -1 & \text{if } r < 0 \end{cases} \quad .$$

Example 2. Let K be a real field. The induced representation

$$\varphi(a) = \begin{cases} 1 & \text{if } a > 0 \\ -1 & \text{if } a < 0 \end{cases}$$

is the only one.

Example 3. Let K be a complex field, and $\underline{a}^{\tau}$ be the conjugate of $\underline{a}$. Then

$$\varphi(a) = |a|/a$$

is an induced mapping. Since

$$\frac{a^{\tau}}{a} = \frac{|a|^2}{a^2} = (\varphi(a))^2,$$

this does not exhaust all of the induced representations of the complex number field, which contains infinitely many automorphisms.

Now we express the previous notion in matrix form. We define

$$(63) \qquad \chi(A) = \varphi(\det A).$$

THEOREM 9. The mapping

$$(64) \qquad (x, A, y) \longrightarrow (x^{\tau}, \chi(A) A, y^{\tau})$$

defines an automorphism of O_4^+, where $\chi(A)$ is related to τ by (63) and (62).

PROOF. The theorem follows from theorem 8, if we can prove that (64) carries the following identities into their corresponding ones:

$$(65) \qquad (0, I, y)\ (x_1, I, 0) = (x_1(1-y\,x_1)^{-1},\ (1-y\,x_1)\,I,\ y(1-y\,x_1)^{-1}), \quad \text{if } x_1\, y = 1,$$

$$(66) \qquad (0, I, x_1^{-1})\ (x_1, I, 0) = (\infty,\ x_1 K,\ -x_1^{-1}),$$

$$(67) \qquad (0, I, x)\ (\infty, I, 0) = (-x^{-1},\ xK,\ -x^{-1})$$

and

$$(68) \qquad (0, A, 0)\ (\infty, I, 0) = (\infty,\ A'^{-1},\ 0).$$

To verify (65), we need to prove that

$$(y_1^{\tau}\ (1-y^{\tau}\, x_1^{\tau})^{-1},\ (1-y^{\tau}\, x_1^{\tau})I,\ y^{\tau}\,(1-y^{\tau}\, x_1^{\tau})^{-1} =$$
$$= ((x_1(1-y_1 x_1)^{-1})^{\tau},\ \chi(1-y\,x_1)I)\ (1-y\,x_1)I,\ (y(1-y\,x_1)^{-1})^{\tau}).$$

The first and the last terms are evidently equal to the corresponding ones, by the definition of an automorphism. The middle terms are equal, since

$$\chi((1-yx_1)\,I) = \varphi((1-yx_1)^2) = (\varphi(1-yx_1))^2 = (1-y^{\tau}x_1^{\tau})/(1-yx_1).$$

To verify (66), we need only establish that

$$x_1^{\tau} K = \chi(x_1 K) x_1 K.$$

This is also true, since

$$\chi(x_1 K) = \varphi(x_1{}^2) = x_1^{\tau}/x_1$$

The other two can be proved similarly.

9. Structure of the group $O_4^+(K,f)$.

THEOREM 10. The commutator subgroup $\Omega_4^+(K,f)$ of $O_4^+(K,f)$ consists of those elements

(69)
$$\begin{pmatrix} A & B \\ C & D \end{pmatrix},$$

where the determinants of A, B, C, D are square elements of the field.

PROOF. Notice that from the expressions (60) and (61) we deduce easily that if one of the determinants of A, B, C, D is a square element different from zero, then all the others are squares. By means of (60) and (61), we can easily verify that all the elements of (69) form a normal subgroup. The factor group of $O_4^+(K,f)$ by (69) is generated by

$$\begin{pmatrix} 1 & 0 & 0 & 0 \\ 0 & \lambda & 0 & 0 \\ 0 & 0 & 1 & 0 \\ 0 & 0 & 0 & \lambda^{-1} \end{pmatrix}, \quad \text{and} \quad \begin{pmatrix} 0 & I \\ I & 0 \end{pmatrix},$$

and it is abelian, since

$$\begin{pmatrix} 1 & 0 & 0 & 0 \\ 0 & \lambda & 0 & 0 \\ 0 & 0 & 1 & 0 \\ 0 & 0 & 0 & \lambda^{-1} \end{pmatrix} \begin{pmatrix} 0 & I \\ I & 0 \end{pmatrix} \begin{pmatrix} 1 & 0 & 0 & 0 \\ 0 & \lambda & 0 & 0 \\ 0 & 0 & 1 & 0 \\ 0 & 0 & 0 & \lambda^{-1} \end{pmatrix} \begin{pmatrix} 0 & I \\ I & 0 \end{pmatrix}$$

belongs to (69). Therefore $\Omega_4^+(K,f)$ is contained in (69), by the property of commuator subgroup.

Next we are going to prove that the subgroup (69) is generated by

(70)
$$\begin{pmatrix} I & xK \\ 0 & I \end{pmatrix}, \quad \begin{pmatrix} 0 & I \\ I & 0 \end{pmatrix}$$

and

(71)
$$\begin{pmatrix} A & 0 \\ 0 & A'^{-1} \end{pmatrix},$$

where $\underline{A}$ belongs to $SL_2(K)$. In fact, the group generated by (70) and (71) evidently contains

$$(72)\qquad \begin{pmatrix} I & 0 \\ yK & I \end{pmatrix} = \begin{pmatrix} 0 & I \\ I & 0 \end{pmatrix} \begin{pmatrix} I & yK \\ 0 & I \end{pmatrix} \begin{pmatrix} 0 & I \\ I & 0 \end{pmatrix}$$

and consequently it contains

$$\begin{pmatrix} I & (1-a^{-1})K \\ 0 & I \end{pmatrix} \begin{pmatrix} I & 0 \\ K & I \end{pmatrix} \begin{pmatrix} I & (1-a)K \\ 0 & I \end{pmatrix} \begin{pmatrix} K & 0 \\ -a^{-1}K & I \end{pmatrix} = \begin{pmatrix} a^{-1}I & 0 \\ 0 & aI \end{pmatrix} .$$

Since a matrix Q of determinant q^2 can be expressed as $\frac{1}{q}\, Q\, qI$, the group generated by (70) and (71) also contains

$$(73)\qquad \begin{pmatrix} Q & 0 \\ 0 & Q'^{-1} \end{pmatrix} = \begin{pmatrix} qI & 0 \\ 0 & q'^{-1}I \end{pmatrix} \begin{pmatrix} 1/q\, Q & 0 \\ 0 & qQ'^{-1} \end{pmatrix} .$$

Evidently (70), (72) and (73) generate the group formed by (69). Therefore (70) and (71) generate the group (69).

The group $\Omega_4^+(K,f)$ contains

$$\begin{pmatrix} A & 0 \\ 0 & A'^{-1} \end{pmatrix} \begin{pmatrix} B & 0 \\ 0 & B'^{-1} \end{pmatrix} \begin{pmatrix} A & 0 \\ 0 & A'^{-1} \end{pmatrix}^{-1} \begin{pmatrix} B & 0 \\ 0 & B'^{-1} \end{pmatrix}^{-1}$$

so it contains (71). Since

$$\begin{pmatrix} rI & 0 \\ 0 & r^{-1}I \end{pmatrix} \begin{pmatrix} I & sK \\ 0 & I \end{pmatrix} \begin{pmatrix} rI & 0 \\ 0 & r^{-1}I \end{pmatrix}^{-1} \begin{pmatrix} I & sK \\ 0 & I \end{pmatrix}^{-1} = \begin{pmatrix} I & s(r^2-1)K \\ 0 & I \end{pmatrix} ,$$

$\Omega_4^+(K,f)$ contains the first term of (70); similarly it contains $\begin{pmatrix} I & 0 \\ yK & I \end{pmatrix}$. Since

$$\begin{pmatrix} I & K \\ 0 & I \end{pmatrix} \begin{pmatrix} I & 0 \\ K & I \end{pmatrix} \begin{pmatrix} -K & 0 \\ 0 & -K \end{pmatrix} \begin{pmatrix} I & K \\ 0 & I \end{pmatrix} = \begin{pmatrix} 0 & I \\ I & 0 \end{pmatrix} ,$$

therefore $\Omega_4^+(K,f)$ contains (70) and (71), and consequently it contains the group (69). The theorem follows.

THEOREM 11. The group $P\Omega_4^+(K,f)$ is a direct product of two groups each of which is isomorphic to $PSL_2(K)$. More precisely, $\Omega_4^+(K,f)$ can be expressed as a direct product of three irreducible components: one formed by the identity and its negation, and two being isomorphic to $PSL_2(K)$.

PROOF. The two subgroups are the one formed by (71) and the one generated by (70). More precisely, we may write the second as

$$(74)\qquad \begin{pmatrix} qI & rK \\ -sK & tI \end{pmatrix}, \quad qt - rs = 1.$$

The correspondence of (74) and $\begin{pmatrix} q & r \\ -s & t \end{pmatrix}$ established the isomorphism of the group (74) and $SL_2(K)$.

The intersection of the groups (70) and (74) is given by

$$\pm \begin{pmatrix} I & 0 \\ 0 & I \end{pmatrix}$$

which is the identity of $P\Omega_4^+(K,f)$. Moreover, for determinant of $\underline{A}$ equal to $\underline{q}^2$, we have

$$\begin{pmatrix} A & x\,A\,K \\ y\,K\,A & (1-xyd(A))A'^{-1} \end{pmatrix} = \begin{pmatrix} 1/q\,A & 0 \\ 0 & qA'^{-1} \end{pmatrix} \begin{pmatrix} q\,I & q\,x\,K \\ q\,y\,K & (1-xyq^2)q^{-1}I \end{pmatrix}$$

and

$$\begin{pmatrix} 0 & A \\ A'^{-1} & zA'^{-1}K \end{pmatrix} = \begin{pmatrix} 1/q\,AK & 0 \\ 0 & q(AK)'^{-1} \end{pmatrix} \begin{pmatrix} 0 & -q\,K \\ -1/q\,K & z/q\,I \end{pmatrix};$$

theorem 11 follows from the definition of the direct product.

THEOREM 12. Every automorphism of $O_4^+(K,f)$ is a combination of

$$\mathfrak{X} \to \mathfrak{Q}\,\mathfrak{X}^{\sigma}\,\mathfrak{Q}^{-1}, \tag{75}$$

where $\mathfrak{Q}$ belongs to GO_4 and σ is an automorphism and that given by (64). Those of $PO_4^+(K,f)$ are induced from the automorphism of $O_4^+(K,f)$.

PROOF. Since $\Omega_4^+(K,f)$ (or $P\Omega_4^+(K,f)$ is a characteristic invariant subgroup, an automorphism of $O^+(K,f)$ (or PO_4^+) induces an automorphism on $\Omega_4^+(K,f)$ (or $P\Omega_4^+$). Since $\Omega_4^+(K,f)$ is a direct product of three groups G_0, G_1 and G_2, where G_1 and G_2 denote the subgroups of $PO_4^+(K,f)$ represented by (71) and (74), and since these are isomorphic to $PSL_2(K)$, it follows that the automorphism of $\Omega_4^+(K,f)$ can be obtained by those of G_1 and G_2 and by adjoining one which permutes G_1 and G_2. (Since G_0 has only identity automorphism).

Since we have, by putting $A = \begin{pmatrix} a & b \\ c & d \end{pmatrix}$,

$$\begin{pmatrix} 1 & 0 & 0 & 0 \\ 0 & 0 & 0 & 1 \\ 0 & 0 & 1 & 0 \\ 0 & 1 & 0 & 0 \end{pmatrix} \begin{pmatrix} A & B \\ C & D \end{pmatrix} \begin{pmatrix} 1 & 0 & 0 & 0 \\ 0 & 0 & 0 & 1 \\ 0 & 0 & 1 & 0 \\ 0 & 1 & 0 & 0 \end{pmatrix} = \begin{pmatrix} a\,I & b\,K \\ -c\,K & d\,I \end{pmatrix},$$

therefore, the permutation of the two components of G_1 and G_2 is merely an automorphism induced by an element of $O_4(K,f)$. Hence after an automorphism of the type (75) if necessary, we may assume that the automorphism of $\Omega_4^+(K,f)$ induces automorphisms on G_1 and G_2, say

$$\begin{pmatrix} A & 0 \\ 0 & A'^{-1} \end{pmatrix} \to \pm \begin{pmatrix} Q & 0 \\ 0 & Q'^{-1} \end{pmatrix} \begin{pmatrix} A & 0 \\ 0 & A'^{-1} \end{pmatrix}^{\sigma} \begin{pmatrix} Q & 0 \\ 0 & Q'^{-1} \end{pmatrix}^{-1} \tag{76}$$

and

$$(77)\quad \begin{pmatrix} qI & rK \\ -sK & tI \end{pmatrix} \longrightarrow \pm \begin{pmatrix} q_1 I & r_1 K \\ -s_1 K & t_1 I \end{pmatrix} \begin{pmatrix} qI & rK \\ -sK & tI \end{pmatrix}^{\tau} \begin{pmatrix} q_1 I & r_1 K \\ -s_1 K & t_1 K \end{pmatrix}^{-1}$$

where σ and τ are two automorphsims of the field. Squaring the elements on both sides of (75) and (77), and since $SL_2(K)$ is generated by all its square elements, we may omit the $\pm$ signs in (76) and (77).

We shall now construct a matrix $\mathfrak{X}$ belonging to $GO_4(K,f)$ which effects both (75) and (76) at the same time. In fact, let the determinant of Q be λ, and write $R = \begin{pmatrix} 1 & 0 \\ 0 & \lambda \end{pmatrix}^{-1} Q$. Then

$$\mathfrak{X} = \begin{pmatrix} I & 0 \\ 0 & \lambda I \end{pmatrix} \begin{pmatrix} 1 & 0 & 0 & 0 \\ 0 & \lambda & 0 & 0 \\ 0 & 0 & 1 & 0 \\ 0 & 0 & 0 & \lambda^{-1} \end{pmatrix} \begin{pmatrix} R & 0 \\ 0 & R'^{-1} \end{pmatrix} \begin{pmatrix} q_1 I & r_1 K \\ -s_1 K & t_1 I \end{pmatrix}$$

which evidently belongs to $GO_4(K,f)$, and

$$\mathfrak{X} \begin{pmatrix} A & 0 \\ 0 & A'^{-1} \end{pmatrix} \mathfrak{X}^{-1} = \begin{pmatrix} I & 0 \\ 0 & \lambda I \end{pmatrix} \begin{pmatrix} Q & 0 \\ 0 & Q'^{-1} \end{pmatrix} \begin{pmatrix} A & 0 \\ 0 & A'^{-1} \end{pmatrix} \begin{pmatrix} Q & 0 \\ 0 & Q'^{-1} \end{pmatrix} \begin{pmatrix} I & 0 \\ 0 & \lambda I \end{pmatrix}^{-1}$$

$$= \begin{pmatrix} Q & 0 \\ 0 & Q'^{-1} \end{pmatrix} \begin{pmatrix} A & 0 \\ 0 & A'^{-1} \end{pmatrix} \begin{pmatrix} Q & 0 \\ 0 & Q'^{-1} \end{pmatrix}^{-1}$$

since

$$\begin{pmatrix} A & 0 \\ 0 & A'^{-1} \end{pmatrix} \begin{pmatrix} q_1 I & r_1 K \\ -s_1 K & t_1 I \end{pmatrix} = \begin{pmatrix} q_1 I & r_1 K \\ -s_1 K & t_1 I \end{pmatrix} \begin{pmatrix} A & 0 \\ 0 & A'^{-1} \end{pmatrix};$$

and

$$\mathfrak{X} \begin{pmatrix} qI & rK \\ -sK & tI \end{pmatrix} \mathfrak{X}^{-1} = \begin{pmatrix} 1 & 0 & 0 & 0 \\ 0 & \lambda & 0 & 0 \\ 0 & 0 & \lambda & 0 \\ 0 & 0 & 0 & 1 \end{pmatrix} \begin{pmatrix} q_1 I & r_1 K \\ -s_1 K & t_1 I \end{pmatrix} \begin{pmatrix} R & 0 \\ 0 & R'^{-1} \end{pmatrix} \begin{pmatrix} qI & rK \\ -sK & tI \end{pmatrix} \times$$

$$\times \begin{pmatrix} R & 0 \\ 0 & R'^{-1} \end{pmatrix}^{-1} \begin{pmatrix} q_1 I & r_1 K \\ -s_1 K & t_1 I \end{pmatrix}^{-1} \begin{pmatrix} 1 & 0 & 0 & 0 \\ 0 & \lambda & 0 & 0 \\ 0 & 0 & \lambda & 0 \\ 0 & 0 & 0 & 1 \end{pmatrix}^{-1}$$

$$= \begin{pmatrix} q_1 I & r_1 K \\ -s_1 K & t_1 I \end{pmatrix} \begin{pmatrix} qI & rK \\ -sK & tI \end{pmatrix} \begin{pmatrix} q_1 I & r_1 K \\ -s_1 K & t_1 I \end{pmatrix}^{-1},$$

since

$$\begin{pmatrix} 1 & 0 & 0 & 0 \\ 0 & \lambda & 0 & 0 \\ 0 & 0 & \lambda & 0 \\ 0 & 0 & 0 & 1 \end{pmatrix} \quad \text{and} \quad \begin{pmatrix} q_1 & 0 & 0 & r_1 \\ 0 & q_1 & -r_1 & 0 \\ 0 & -s_1 & t_1 & 0 \\ s_1 & 0 & 0 & t_1 \end{pmatrix}$$

evidently commute. Therefore, after an automorphism of the type (75), we may assume that our automorphism under consideration induces the mapping

$$(78)\qquad \begin{pmatrix} A & 0 \\ 0 & A'^{-1} \end{pmatrix} \rightarrow \begin{pmatrix} A & 0 \\ 0 & A'^{-1} \end{pmatrix}$$

and

$$(79)\qquad \begin{pmatrix} q\,I & r\,K \\ -s\,K & t\,I \end{pmatrix} \rightarrow \begin{pmatrix} q^{\tau}\,I & r^{\tau}\,K \\ -s^{\tau}\,K & t^{\tau}\,I \end{pmatrix}$$

If K contains more than three elements, we have the following particular case of (79);

$$(80)\qquad \begin{pmatrix} a\,I & 0 \\ 0 & a^{-1}I \end{pmatrix} \rightarrow \begin{pmatrix} a^{\tau}\,I & 0 \\ 0 & a^{-\tau}I \end{pmatrix}, \qquad a^{\tau} \neq \pm 1.$$

Thus elements which commute with (80) are of the form

$$(81)\qquad \begin{pmatrix} P & 0 \\ 0 & P'^{-1} \end{pmatrix} \quad ;$$

by theorem 1, we have

$$\begin{pmatrix} P & 0 \\ 0 & p'^{-1} \end{pmatrix} \rightarrow \begin{pmatrix} \chi(P)\, P^{\mu} & 0 \\ 0 & \chi(P)^{-1} P^{\mu\,\prime -1} \end{pmatrix} ,$$

from (78), we have $\mu = 1$, and from (79),

$$\chi(q \quad I) = q^{\tau}/q,$$

Consequently

$$\left(\chi\begin{pmatrix} 1 & 0 \\ 0 & q \end{pmatrix}\right)^2 = \chi\begin{pmatrix} 1 & 0 \\ 0 & q \end{pmatrix} \chi\begin{pmatrix} q & 0 \\ 0 & 1 \end{pmatrix} = \chi(\,Q\,I\,) = q^{\tau}/q,$$

and

$$(\chi(A))^2 = \det A^{\tau} / \det A.$$

Our theorem is proved for K containing more than three elements (certainly a little detail is needed to complete the proofs of O_4^+ and PO_4^+). In case K contains only three elements, the factor group O_4^+/Ω_4^+ is of order two, and it is represented by

$$\begin{pmatrix} I & 0 \\ 0 & I \end{pmatrix}, \quad \mathfrak{U} = \begin{pmatrix} U & 0 \\ 0 & U^{-1} \end{pmatrix} \qquad U = \begin{pmatrix} 1 & 0 \\ 0 & -1 \end{pmatrix}.$$

Therefore we need only find the mapping of $\mathfrak{U}$. $\mathfrak{U}$ is mapped to $\mathfrak{U}\mathfrak{P}$, where $\mathfrak{P}$ belongs to Ω_4. There are two ways to map the element

$$\mathfrak{U}\begin{pmatrix} A & 0 \\ 0 & A'^{-1} \end{pmatrix}\begin{pmatrix} qI & rK \\ -sK & tI \end{pmatrix}\mathfrak{U}^{-1} = \begin{pmatrix} UAU^{-1} & 0 \\ 0 & (UAU^{-1})'^{-1} \end{pmatrix} X$$

$$X\begin{pmatrix} qI & -rK \\ sK & tI \end{pmatrix},$$

they give us the identity

$$\mathfrak{U}\mathfrak{P}\begin{pmatrix} A & 0 \\ 0 & A'^{-1} \end{pmatrix}\begin{pmatrix} q^\tau I & r^\tau K \\ s^\tau K & t^\tau I \end{pmatrix}\mathfrak{P}^{-1}\mathfrak{U}^{-1} = \begin{pmatrix} UAU^{-1} & 0 \\ 0 & (UAU^{-1})'^{-1} \end{pmatrix} X$$

$$X\begin{pmatrix} q^\tau I & -r^\tau K \\ s^\tau K & t^\tau I \end{pmatrix}$$

for all possible A, q, r, s and t. It follows that $\mathfrak{P}$ is either an identity or its negation, and we obtain the same conclusion.

REFERENCES

1. HUA, L. K., On the automorphisms of the symplectic group over any field, Annals of Math. 49(1948), pp.739-759.

2. DIEUDONNÉ, J., On the automorphisms of classical groups. This volume.

3. HUA, L. K., On the automorphisms of a sfield, Proc. Nat. Acad. Sci. 35 (1949), pp.386-389.

4. ANCOCHEA, G., On semi-automorphisms of division algebras, Annals of Math. 48(1947), pp.147-153.

5. KAPLANSKY, I., Semi-automorphisms of Rings, Duke Math. Jour. 14(1947), pp.521-525.

6. DIEUDONNÉ, J., Les déterminants sur un corps non commutatif, Bull. Soc. Math. Fr., 79(1943), pp.27-45.

6.* Ibid., p.41, theorem 2. Dieudonné, Compléments à trois articles antérieurs, Bull. Soc. Math. Fr., 74(1946), pp.59-68.

7. ANCOCHEA, G., Le théorème de van Staudt en géométrie projective quaternionne, Jour. für die reine und angewandte Math., 184(1942), pp.192-198.

8. HUA, L. K., Some properties of a sfield, Proc. of Nat. Aca. Sci., 35 (1949), pp.533-537.

9. SCHREIER O. and van der WAERDEN, B. L., Die Automorphismen der projektiven gruppen, Hamburg Univ. Math. Seminar, Abh. 6(1928), pp.303-322, cf. also appendix of (1).

10. WITT, E., Theorie der quadratische Formen in beliebigen Körpern, Jour. für Math., 176(1937), pp.21-44.

Reprinted from the
TRANSACTIONS OF THE AMERICAN MATHEMATICAL SOCIETY
Vol. 71, No. 3, pp. 331–348, 1951

AUTOMORPHISMS OF THE UNIMODULAR GROUP

BY
L. K. HUA AND I. REINER

Notation. Let $\mathfrak{M}_n$ denote the group of $n\times n$ integral matrices of determinant ± 1 (the unimodular group). By $\mathfrak{M}_n^+$ we denote that subset of $\mathfrak{M}_n$ where the determinant is $+1$; $\mathfrak{M}_n^-$ is correspondingly defined. Let $I^{(n)}$ (or briefly I) be the identity matrix in $\mathfrak{M}_n$, and let X' represent the transpose of X. The direct sum of the matrices A and B will be represented by $A\dotplus B$;

$$A \overset{s}{=} B$$

will mean that A is similar to B. In this paper, we shall find explicitly the generators of the group $\mathfrak{A}_n$ of all automorphisms of $\mathfrak{M}_n$.

1. **The commutator subgroup of $\mathfrak{M}_n$.** The following result is useful, and is of independent interest.

THEOREM 1. *Let $\mathfrak{N}_n$ be the commutator subgroup of $\mathfrak{M}_n$. Then trivially $\mathfrak{N}_n\subset\mathfrak{M}_n^+$. For $n=2$, $\mathfrak{N}_n$ is of index 2 in $\mathfrak{M}_n^+$, while for $n>2$, $\mathfrak{N}_n=\mathfrak{M}_n^+$.*

Proof. Consider first the case where $n=2$. Define

$$S = \begin{pmatrix} 0 & 1 \\ -1 & 0 \end{pmatrix}, \qquad T = \begin{pmatrix} 1 & 1 \\ 0 & 1 \end{pmatrix}. \tag{1}$$

It is well known that S and T generate $\mathfrak{M}_2^+$. An element X of $\mathfrak{M}_2^+$ is called *even* if, when X is expressed as a product of powers of S and T, the sum of the exponents is even; otherwise, X is called *odd*. Since all relations satisfied by S and T are consequences of

$$S^2 = -I, \qquad (ST)^3 = I,$$

it follows that the parity of $X\in\mathfrak{M}_2^+$ depends only on X, and not on the manner in which X is expressed as a product of powers of S and T. Let $\mathfrak{E}$ be the subgroup of $\mathfrak{M}_2^+$ consisting of all even elements; then clearly $\mathfrak{E}$ is of index 2 in $\mathfrak{M}_2^+$. It suffices to prove that $\mathfrak{E}=\mathfrak{N}_2$.

We prove first that $\mathfrak{N}_2\subset\mathfrak{E}$. Since the commutator subgroup of a group is always generated by squares, it suffices to show that $A\in\mathfrak{M}_2$ implies $A^2\in\mathfrak{E}$. For $A\in\mathfrak{M}_2^+$, this is clear. If $A\in\mathfrak{M}_2^-$, set $A=XJ=JY$, where

$$J = \begin{pmatrix} 1 & 0 \\ 0 & -1 \end{pmatrix}, \tag{2}$$

Presented to the Society, December 29, 1950; received by the editors January 8, 1951.

and X and $Y \in \mathfrak{M}_2^+$. Then $A^2 = XY = XJ^{-1}XJ$. Hence we need only prove that if $X \in \mathfrak{M}_2^+$, X and $J^{-1}XJ$ are of the same parity. This is easily verified for $X = S$ or T; since S and T generate $\mathfrak{M}_2^+$, and $J^{-1}X_1X_2J = J^{-1}X_1J \cdot J^{-1}X_2J$, the result follows.

On the other hand we can show that $\mathfrak{C} \subset \mathfrak{N}_2$. For, $\mathfrak{C}$ is generated by T^2 and ST, since $TS = (ST \cdot T^{-2})^2$. However, $T^2 = TJT^{-1}J^{-1} \in \mathfrak{N}_2$, and therefore also $(T')^{-2} \in \mathfrak{N}_2$. Furthermore, $ST = TST^{-1}S^{-1}(T')^{-2}T^2 \in \mathfrak{N}_2$. This completes the proof for $n = 2$.

Suppose now that $n > 2$, and define

$$R = \begin{pmatrix} 0 \cdots 0 & (-1)^{n-1} \\ 1 \cdots 0 & 0 \\ \cdots\cdots & \cdot \\ 0 \cdots 1 & 0 \end{pmatrix} \in \mathfrak{M}_n^+, \qquad S = \begin{pmatrix} 0 & 1 \\ -1 & 0 \end{pmatrix} \dotplus I^{(n-2)} \tag{3}$$

$$T = \begin{pmatrix} 1 & 1 \\ 0 & 1 \end{pmatrix} \dotplus I^{(n-2)}.$$

(The symbols S and T defined here are the analogues in $\mathfrak{M}_n^+$ of those defined by (1). It will be clear from the context which are meant.) For $n > 2$ we have(1)

$$T' = [R^{-1}(TR)^{-(n-2)}R(TR)^{n-2}](TR)^{-1}[R(TR)^{-(n-2)}R^{-1}(TR)^{n-2}](TR) \in \mathfrak{N}_n.$$

Further $S = TST^{-1}S^{-1}(T')^{-2}T \in \mathfrak{N}_n$. Finally, for odd n there exists a permutation matrix P such that $R^2 = P^{-1}RP$, whence $R = R^{-1}P^{-1}RP \in \mathfrak{N}_n$. For even n, R represents the monomial transformation

$$\begin{pmatrix} x_1 & x_2 \cdots x_{n-1} & x_n \\ x_2 & x_3 \cdots x_n & -x_1 \end{pmatrix},$$

which is a product of

$$\begin{pmatrix} x_1 & x_2 & x_3 \cdots x_{n-1} & x_n \\ x_2 & -x_1 & x_3 \cdots x_{n-1} & x_n \end{pmatrix}, \quad \begin{pmatrix} x_1 & x_2 & x_3 & x_4 \cdots x_n \\ -x_3 & x_2 & x_1 & x_4 \cdots x_n \end{pmatrix},$$

$$\begin{pmatrix} x_1 & x_2 & x_3 & x_4 \cdots x_n \\ x_4 & x_2 & x_3 & -x_1 \cdots x_n \end{pmatrix}, \cdots, \begin{pmatrix} x_1 & x_2 \cdots x_{n-1} & x_n \\ x_n & x_2 \cdots x_{n-1} & -x_1 \end{pmatrix},$$

each factor of which is similar to S (and hence is in $\mathfrak{N}_n$). Since T and R generate $\mathfrak{M}_n^+$, the theorem is proved.

COROLLARY 1. *In any automorphism of $\mathfrak{M}_n$, always $\mathfrak{M}_n^+ \to \mathfrak{M}_n^+$.*

Proof. For $n > 2$ this is an immediate corollary, since the commutator subgroup goes into itself in any automorphism. For $n = 2$, let $S \to S_1$ and

(1) L. K. Hua and I. Reiner, Trans. Amer. Math. Soc. vol. 65 (1949) p. 423.

$T \to T_1$. Then $ST \in \mathfrak{N}_2$ implies $S_1T_1 \in \mathfrak{N}_2$, so $\det(S_1T_1) = 1$. Further, $S^2 = -I$ implies $S_1^2 = -I$, so $\det S_1 = 1$, since the minimum function of S_1 is x^2+1, and the characteristic function must therefore be a power of x^2+1. This completes the proof when $n=2$.

2. **Automorphisms of $\mathfrak{M}_2^+$.** We wish to determine the automorphisms of $\mathfrak{M}_2$. Since every automorphism of $\mathfrak{M}_2$ takes $\mathfrak{M}_2^+$ into itself, we shall first determine all automorphisms of $\mathfrak{M}_2^+$. For $X \in \mathfrak{M}_2^+$, define $\epsilon(X) = +1$ or -1, according as X is even or odd.

THEOREM 2. *Every automorphism of $\mathfrak{M}_2^+$ is of one of the forms*

$$\text{(I)} \qquad X \in \mathfrak{M}_2^+ \to AXA^{-1} \qquad A \in \mathfrak{M}_2$$

or

$$\text{(II)} \qquad X \in \mathfrak{M}_2^+ \to \epsilon(X) \cdot AXA^{-1}, \qquad A \in \mathfrak{M}_2.$$

That is, the automorphism group of $\mathfrak{M}_2^+$ is generated by the set of "inner" automorphisms $X \to AXA^{-1}$ ($A \in \mathfrak{M}_2$) and the automorphism $X \to \epsilon(X) \cdot X$.

Proof. Let τ be an automorphism of $\mathfrak{M}_2^+$; it certainly leaves $I^{(2)}$ and $-I^{(2)}$ individually unaltered. Let S and T (as given by (1)) be mapped into S^τ and T^τ. Then $(S^\tau)^2 = -I$. Since all second order fixed points are equivalent, there exists a matrix $B \in \mathfrak{M}_2$ such that $BS^\tau B^{-1} = S$. Instead of τ, consider the automorphism τ': $X \to BX^\tau B^{-1}$, which leaves S unaltered. Assume hereafter that τ leaves S invariant. (It is this sort of replacement of τ by τ' which we shall mean when we refer to some property holding "after a suitable inner automorphism.") Set

$$T^\tau = \begin{pmatrix} a & b \\ c & d \end{pmatrix}.$$

From $(ST)^3 = I$ we obtain $(ST^\tau)^3 = I$, whence $b - c = 1$. Since $\det T^\tau = 1$, we get

$$ad = 1 + bc = c^2 + c + 1 > 0.$$

Set $N = |a+d|$. If $N \geqq 3$, consider the elements generated by S and T^τ (mod N). Since $a+d \equiv 0$ (mod N), we find that $(T^\tau)^2 \equiv I$ (mod N). Furthermore $(ST^\tau)^3 \equiv I$ (mod N); therefore S and T^τ generate (mod N) at most the 12 elements

$$\pm I, \ \pm S, \ \pm T^\tau, \ \pm ST^\tau, \ \pm T^\tau S, \ \pm ST^\tau S.$$

But if τ is an automorphism, S and T^τ generate $\mathfrak{M}_2^+$, which has more than 12 elements (mod N) for $N \geqq 3$.

Therefore $N \leqq 2$. Since $ad > 0$, either $a = d = 1$ or $a = d = -1$, and thence $b = 1$, $c = 0$ or $b = 0$, $c = -1$. There are 4 possibilities for T^τ:

$$T^{\tau} = \begin{cases} T_0 = \begin{pmatrix} 1 & 1 \\ 0 & 1 \end{pmatrix}, & T_2 = \begin{pmatrix} -1 & 1 \\ 0 & -1 \end{pmatrix}, \\ T_1 = \begin{pmatrix} 1 & 0 \\ -1 & 1 \end{pmatrix}, & T_3 = \begin{pmatrix} -1 & 0 \\ -1 & -1 \end{pmatrix}. \end{cases}$$

Since S and T generate $\mathfrak{M}_2^+$, to determine τ it is sufficient to specify S^τ and T^τ. Thus every automorphism of $\mathfrak{M}_2^+$ is of the form $S \to BSB^{-1}$, $T \to BT_iB^{-1}$ (for some i, $i=0, 1, 2, 3$), where $B \in \mathfrak{M}_2$. If J is given by (2), we have:

$$T_0 = T, \qquad T_1 = STS^{-1}, \qquad T_2 = -JTJ^{-1}, \qquad T_3 = -SJTJ^{-1}S^{-1},$$

and also $S = -JSJ^{-1}$. The possible automorphisms are:

$$\begin{array}{lll} i = 0: & S \to BSB^{-1}, & T \to BTB^{-1}. \\ i = 1: & S \to BS\cdot S\cdot S^{-1}B^{-1}, & T \to BS\cdot T\cdot S^{-1}B^{-1}. \\ i = 2: & S \to -BJ\cdot S\cdot J^{-1}B^{-1}, & T \to -BJ\cdot T\cdot J^{-1}B^{-1}. \\ i = 3: & S \to -BSJ\cdot S\cdot J^{-1}S^{-1}B^{-1}, & T \to -BSJ\cdot T\cdot J^{-1}S^{-1}B^{-1}. \end{array}$$

These automorphisms are of two types: for $i=0$ and 1, $S \to ASA^{-1}$, $T \to ATA^{-1}$, which imply that $X \in \mathfrak{M}_2^+ \to AXA^{-1}$; for $i=2$ and 3, $S \to -ASA^{-1}$, $T \to -ATA^{-1}$, which imply that $X \in \mathfrak{M}_2^+ \to \epsilon(X)\cdot AXA^{-1}$. This completes the proof.

3. **Automorphisms of $\mathfrak{M}_n^+$ and $\mathfrak{M}_n$.** We are now faced with the problem of determining the automorphisms of $\mathfrak{M}_2$ from those of $\mathfrak{M}_2^+$. We shall have the same problem for $\mathfrak{M}_n$ and $\mathfrak{M}_n^+$. As we shall see, the passage from $\mathfrak{M}_n^+$ to $\mathfrak{M}_n$ is trivial, and most of the difficulty lies in determining the automorphisms of $\mathfrak{M}_n^+$. In this paper we shall prove the following results:

THEOREM 3. *For $n>2$, the group of those automorphisms of $\mathfrak{M}_n^+$ which are induced by automorphisms of $\mathfrak{M}_n$ is generated by*

(i) *the set of all "inner" automorphisms*

$$X \in \mathfrak{M}_n^+ \to AXA^{-1} \qquad (A \in \mathfrak{M}_n),$$

and

(ii) *the automorphism*

$$X \in \mathfrak{M}_n^+ \to X'^{-1}.$$

REMARK. When $n=2$, the automorphism (ii) is the same as $X \to SXS^{-1}$, hence is included in (i). The automorphism $X \to \epsilon(X)\cdot X$ occurs only for $n=2$. Furthermore, for odd n all automorphisms of $\mathfrak{M}_n^+$ are induced by automorphisms of $\mathfrak{M}_n$.

THEOREM 4. *The generators of $\mathfrak{A}_n$ are*

(i) *the set of all inner automorphisms*

$$X \in \mathfrak{M}_n \to AXA^{-1} \qquad (A \in \mathfrak{M}_n),$$

(ii) *the automorphism* $X \in \mathfrak{M}_n \rightarrow X'^{-1}$,
(iii) *for even n only, the automorphism*

$$X \in \mathfrak{M}_n \rightarrow (\det X) \cdot X,$$

and
(iv) *for* $n=2$ *only, the automorphism*

$$X \in \mathfrak{M}_2^+ \rightarrow \epsilon(X) \cdot X, \qquad X \in \mathfrak{M}_2^- \rightarrow \epsilon(JX) \cdot X,$$

where J is given by (2).
Further, when $n=2$, *the automorphism* (ii) *may be omitted from this list.*

Let us show that Theorem 4 is a simple consequence of Theorem 3. Let τ be any automorphism of $\mathfrak{M}_n$. By Corollary 1, τ induces an automorphism on $\mathfrak{M}_n^+$ which, by Theorems 2 and 3, can be written as:

$$X \in \mathfrak{M}_n^+ \rightarrow \alpha(X) \cdot AX^*A^{-1},$$

where $A \in \mathfrak{M}_n$, $\alpha(X)=1$ for all X or $\alpha(X)=\epsilon(X)$ for all X (this can occur only when $n=2$), and where either $X^*=X$ for all X or $X^*=X'^{-1}$ for all X.
Let Y and $Z \in \mathfrak{M}_n^-$; then

$$Y^\tau Z^\tau = (YZ)^\tau = \alpha(YZ) \cdot A(YZ)^*A^{-1},$$

whence

$$Y^\tau = \alpha(YZ) \cdot AY^*Z^*A^{-1}(Z^\tau)^{-1}.$$

Let $Z \in \mathfrak{M}_n^-$ be fixed; then

$$Y^\tau = \alpha(YZ) \cdot AY^*B \qquad \text{for all } Y \in \mathfrak{M}_n^-,$$

where A and B are independent of Y. But then

$$AY^*B \cdot AY^*B = (Y^\tau)^2 = (Y^2)^\tau = \alpha(Y^2)A(Y^2)^*A^{-1},$$

so that

$$(BA)Y^*(BA) = \alpha(Y^2)Y^*.$$

Since this is valid for all $Y \in \mathfrak{M}_n^-$, we see that of necessity $\alpha(Y^2)=1$ for all Y, and $BA = \pm I$. This shows that either $Y^\tau = \alpha(YZ) \cdot AY^*A^{-1}$ for all $Y \in \mathfrak{M}_n^-$, or $Y^\tau = -\alpha(YZ) \cdot AY^*A^{-1}$ for all $Y \in \mathfrak{M}_n^-$. If $n=2$ and $\alpha(YZ)=\epsilon(YZ)$, it is trivial to verify that either $\epsilon(YZ)=\epsilon(JY)$ for all $Y \in \mathfrak{M}_2^-$ or $\epsilon(YZ)=-\epsilon(JY)$ for all $Y \in \mathfrak{M}_2^-$.

The remainder of the paper will be concerned with proving Theorem 3.

4. Canonical forms for involutions. In the proof of Theorem 3 we shall use certain canonical forms of involutions under similarity transformations.

LEMMA 1. *Under a similarity transformation, every involution* $X \in \mathfrak{M}_n$ *such that* $X^2 = I^{(n)}$ *can be brought into the form*

$$W(x, y, z) = \underset{(x \text{ terms})}{L \dotplus \cdots \dotplus L} \dotplus (-I)^{(y)} \dotplus I^{(z)}, \tag{4}$$

where $2x+y+z=n$ *and*

$$L = \begin{pmatrix} 1 & 0 \\ 1 & -1 \end{pmatrix}.$$

Proof. We prove first, by induction on n, that every $X \in \mathfrak{M}_n$ satisfying $X^2 = I$ is similar to a matrix of the form

$$\begin{pmatrix} I^{(l)} & 0 \\ M & -I^{(n-l)} \end{pmatrix}. \tag{5}$$

For $n=1$ and 2, this is trivial. Let the theorem be proved for n, and assume that $X^2 = I^{(n+1)}$, where $n \geqq 2$. Then $X^2 - I = 0$, or $(X-I)(X+I)=0$. If $X-I$ is nonsingular, then $X = -I$ and the result is obvious. Hence, supposing that $X - I$ is singular (so that $\lambda = 1$ is a characteristic root of X), there exists a primitive column vector $\mathfrak{t} = (t_1, \cdots, t_{n+1})'$ with integral elements such that $\mathfrak{t}'X = \mathfrak{t}'$. Choose $P \in \mathfrak{M}_{n+1}$ with first row $\mathfrak{t}'$. Then

$$PXP^{-1} = \begin{pmatrix} 1 & \mathfrak{n}' \\ \mathfrak{x} & X_1 \end{pmatrix},$$

where $\mathfrak{n}$ denotes a vector whose components are 0; thus

$$X \overset{s}{=} \begin{pmatrix} 1 & \mathfrak{n}' \\ \mathfrak{x} & X_1 \end{pmatrix}.$$

But

$$I^{(n+1)} = X^2 \overset{s}{=} \begin{pmatrix} 1 & \mathfrak{n}' \\ (I+X_1)\mathfrak{x} & X_1^2 \end{pmatrix}$$

shows that $X_1^2 = I^{(n)}$ and $(I+X_1)\mathfrak{x} = \mathfrak{n}$. By the induction hypothesis,

$$X_1 \overset{s}{=} \begin{pmatrix} I^{(m)} & 0 \\ M & -I^{(n-m)} \end{pmatrix},$$

and, after making the similarity transformation, we have (as a consequence of $(I+X_1)\mathfrak{x} = \mathfrak{n}$)

$$\begin{pmatrix} 2I^{(m)} & 0 \\ M & 0 \end{pmatrix} \mathfrak{x} = \mathfrak{n}.$$

Therefore

$$\mathfrak{x} = (\underbrace{0, \cdots, 0}_{(m \text{ terms})}, \underbrace{*, \cdots, *}_{(n-m \text{ terms})})',$$

where * denotes an arbitrary element. Thus

$$X \overset{s}{=} \begin{pmatrix} 1 & \mathfrak{n}' & \\ 0 & & \\ \cdot & I^{(m)} & 0 \\ \cdot & & \\ 0 & & \\ * & & \\ \cdot & M & -I^{(n-m)} \\ \cdot & & \\ * & & \end{pmatrix} = \begin{pmatrix} I^{(m+1)} & 0 \\ \overline{M} & -I^{(n-m)} \end{pmatrix}.$$

This completes the first part of the proof.

Suppose we now subject (5) to a further similarity transformation by

$$\begin{pmatrix} A^{(l)} & 0 \\ C & D^{(n-l)} \end{pmatrix} \in \mathfrak{M}_n.$$

A simple calculation shows that we obtain a matrix given by (5) with M replaced by $\overline{M}$, where $\overline{M} = 2CA^{-1} + DMA^{-1}$. Choosing firstly $C=0$, A and D unimodular, we find that $\overline{M} = DMA^{-1}$, and by proper choice of A and D we can make $\overline{M}$ diagonal. Supposing this done, secondly put $A=I$, $D=I$; we find that $\overline{M} = M + 2C$. Since C is arbitrary, we can bring $\overline{M}$ into the form

$$\begin{pmatrix} I^{(k)} & 0 \\ 0 & 0 \end{pmatrix},$$

where k is the rank of M. Since we can interchange two rows and simultaneously interchange the corresponding columns by means of a similarity transformation, the lemma follows.

It is easily seen that

$$W(x, y, z) \overset{s}{=} W(\bar{x}, \bar{y}, \bar{z})$$

only when $x = \bar{x}$, $y = \bar{y}$, and $z = \bar{z}$. Furthermore, changing the order of terms in the direct summation does not alter the similarity class. The number A_n of nonsimilar involutions in $\mathfrak{M}_n$ is therefore equal to the number of solutions of $2x + y + z = n$, $x \geqq 0$, $y \geqq 0$, $z \geqq 0$. This gives

$$A_n = \begin{cases} \left(\dfrac{n+2}{2}\right)^2, & n \text{ even}, \\ \dfrac{(n+1)(n+3)}{4}, & n \text{ odd}. \end{cases} \tag{6}$$

Let B_n be the number of nonsimilar involutions in $\mathfrak{M}_n^+$, where the similarity factors are in $\mathfrak{M}_n$. One easily obtains

$$(7) \qquad B_n = \begin{cases} (A_n - 1)/2, & \text{if } n \equiv 0 \pmod 4, \\ A_n/2, & \text{otherwise.} \end{cases}$$

5. **Automorphisms of $\mathfrak{M}_3^+$.** We shall now prove Theorem 3 for $n=3$. Let

$$I_1 = \begin{pmatrix} -1 & 0 & 0 \\ 0 & -1 & 0 \\ 0 & 0 & 1 \end{pmatrix}, \qquad I_2 = \begin{pmatrix} 1 & 0 & 0 \\ 1 & -1 & 0 \\ 0 & 0 & -1 \end{pmatrix} \in \mathfrak{M}_3^+.$$

Then $I_1^2 = I^{(3)}$. Let τ be any automorphism of $\mathfrak{M}_3^+$ and let $X = I_1^\tau$; then $X^2 = I^{(3)}$. By Lemma 1, the matrices I_1, I_2, and $I^{(3)}$ form a complete system of nonsimilar involutions in $\mathfrak{M}_3^+$. Therefore

$$X \overset{\varepsilon}{=} I_1 \text{ or } I_2.$$

After a suitable inner automorphism, we may assume that either $I_1 \to I_1$ or $I_1 \to I_2$. We shall show that this latter case is impossible by considering the normalizer groups of I_1 and I_2. The normalizer group of I_1, that is, the group of matrices $\in \mathfrak{M}_3^+$ which commute with I_1, consists of all elements of $\mathfrak{M}_3^+$ of the form

$$\begin{pmatrix} a & b & 0 \\ c & d & 0 \\ 0 & 0 & e \end{pmatrix},$$

and is isomorphic to $\mathfrak{M}_2$. That of I_2 consists of all elements of $\mathfrak{M}_3^+$ of the form

$$\begin{pmatrix} a & 0 & 0 \\ (a-e)/2 & e & f \\ -h/2 & h & i \end{pmatrix},$$

and is isomorphic to that subgroup $\mathfrak{G}$ of $\mathfrak{M}_2$ consisting of the elements

$$\begin{pmatrix} e & f \\ h & i \end{pmatrix} \in \mathfrak{M}_2, \qquad \text{where } \left.\begin{aligned} e &\equiv 1 \\ h &\equiv 0 \\ i &\equiv 1 \end{aligned}\right\} \pmod 2.$$

Since e and i are both odd, $\mathfrak{G}$ contains no element of order 3, and hence is not isomorphic to $\mathfrak{M}_2$. But then $I_1 \to I_2$ is impossible.

We may assume thus that after a suitable inner automorphism, I_1 is invariant. Thence elements of $\mathfrak{M}_3^+$ which commute with I_1 map into elements of the same kind, so that

$$\begin{pmatrix} X & \mathfrak{n}' \\ \mathfrak{n} & \pm 1 \end{pmatrix} \in \mathfrak{M}_3^+ \to \begin{pmatrix} X^\tau & \mathfrak{n}' \\ \mathfrak{n} & \pm 1 \end{pmatrix}.$$

Since this induces an automorphism $X \to X^\tau$ on $\mathfrak{M}_2$, we see that $\det X^\tau = \det X$, and hence the plus signs go together, and so do the minus signs. By Theorem 2 and that part of Theorem 4 which follows from Theorem 2, there exists a matrix $A \in \mathfrak{M}_2$ such that $X^\tau = \pm AXA^{-1}$; here, the plus sign certainly occurs when X is an even element of $\mathfrak{M}_2^+$, and if the minus sign occurs for one odd element of $\mathfrak{M}_2^+$, then it occurs for *every* odd element of $\mathfrak{M}_2^+$. By use of a further inner automorphism using the factor $A^{-1} \dotplus I^{(1)}$, we may assume that

$$\begin{pmatrix} X & \mathfrak{n}' \\ \mathfrak{n} & \pm 1 \end{pmatrix} \in \mathfrak{M}_3^+ \to \begin{pmatrix} \pm X & \mathfrak{n}' \\ \mathfrak{n} & \pm 1 \end{pmatrix}, \tag{8}$$

so that

$$M = \begin{pmatrix} 1 & 0 & 0 \\ 0 & -1 & 0 \\ 0 & 0 & -1 \end{pmatrix} \to M \quad \text{or} \quad M \to N = \begin{pmatrix} -1 & 0 & 0 \\ 0 & 1 & 0 \\ 0 & 0 & -1 \end{pmatrix}.$$

Since

$$N = \begin{pmatrix} 0 & -1 & 0 \\ 1 & 0 & 0 \\ 0 & 0 & 1 \end{pmatrix} \cdot M \cdot \begin{pmatrix} 0 & 1 & 0 \\ -1 & 0 & 0 \\ 0 & 0 & 1 \end{pmatrix},$$

we may assume (after a further inner automorphism, if necessary) that I_1, M, and N are all invariant under the automorphism (but (8) need not hold).

Thus, after a suitably chosen inner automorphism, we have I_1, M, and N invariant. Therefore there exist A, B, and $C \in \mathfrak{M}_2$ such that

$$\begin{aligned} &\begin{pmatrix} X & \mathfrak{n} \\ \mathfrak{n}' & \pm 1 \end{pmatrix} \in \mathfrak{M}_3^+ \to \begin{pmatrix} \pm AXA^{-1} & \mathfrak{n} \\ \mathfrak{n}' & \pm 1 \end{pmatrix}, \\ &\begin{pmatrix} \pm 1 & \mathfrak{n}' \\ \mathfrak{n} & X \end{pmatrix} \in \mathfrak{M}_3^+ \to \begin{pmatrix} \pm 1 & \mathfrak{n}' \\ \mathfrak{n} & \pm BXB^{-1} \end{pmatrix}, \\ &\begin{pmatrix} a & 0 & b \\ 0 & \pm 1 & 0 \\ c & 0 & d \end{pmatrix} \in \mathfrak{M}_3^+ \to \begin{pmatrix} \alpha & 0 & \beta \\ 0 & \pm 1 & 0 \\ \gamma & 0 & \delta \end{pmatrix}, \end{aligned} \tag{9}$$

where

$$\begin{pmatrix} \alpha & \beta \\ \gamma & \delta \end{pmatrix} = \pm C \begin{pmatrix} a & b \\ c & d \end{pmatrix} C^{-1},$$

and $\mathfrak{n} = (0, 0)'$. Here, the $+1$ on the left goes with the $+1$ on the right always (and the -1's go together); further, when X is an even element of $\mathfrak{M}_2^+$, the plus sign occurs before AXA^{-1}, BXB^{-1}, and CXC^{-1}, while if the minus sign occurs before one of these for any odd $X \in \mathfrak{M}_2^+$, it occurs there for every odd $X \in \mathfrak{M}_2^+$.

Now we may assume that at most one of A, B, and C has determinant -1; for if both A and B (say) have determinant -1, apply a further inner automorphism (with factor N) which leaves I_1, M, and N invariant and changes the signs of det A and det B. Suppose hereafter, without loss of generality, that det $A = \det B = 1$.

Next, N is invariant, but by (9) goes into

$$\begin{pmatrix} \pm A\begin{pmatrix} -1 & 0 \\ 0 & 1 \end{pmatrix}A^{-1} & \mathfrak{n}' \\ \mathfrak{n} & -1 \end{pmatrix},$$

so that

$$\pm A\begin{pmatrix} -1 & 0 \\ 0 & 1 \end{pmatrix}A^{-1} = \begin{pmatrix} -1 & 0 \\ 0 & 1 \end{pmatrix}.$$

This gives two possibilities:

$$A = I^{(2)} \quad \text{or} \quad \begin{pmatrix} 0 & 1 \\ -1 & 0 \end{pmatrix}.$$

The same holds true for B (but not necessarily for C, since det $C = \pm 1$).

Suppose firstly that either A or B is $I^{(2)}$, say $A = I^{(2)}$. Then

$$T = \begin{pmatrix} 1 & 1 & 0 \\ 0 & 1 & 0 \\ 0 & 0 & 1 \end{pmatrix} \to \begin{pmatrix} \pm\begin{pmatrix} 1 & 1 \\ 0 & 1 \end{pmatrix} & \begin{matrix} 0 \\ 0 \end{matrix} \\ \begin{matrix} 0 & 0 \end{matrix} & 1 \end{pmatrix}.$$

Case 1. T invariant. Then

$$\begin{pmatrix} 0 & 1 & 0 \\ -1 & 0 & 0 \\ 0 & 0 & 1 \end{pmatrix} \quad \text{and} \quad \begin{pmatrix} 0 & 1 & 0 \\ 1 & 0 & 0 \\ 0 & 0 & -1 \end{pmatrix}$$

are both invariant. (The first matrix is invariant in virtue of the remarks after (9); the second is invariant because it is M times the first.) For either possible choice of B we find that

$$\begin{pmatrix} -1 & 0 & 0 \\ 0 & 0 & 1 \\ 0 & 1 & 0 \end{pmatrix} \to \begin{pmatrix} -1 & \begin{matrix} 0 & 0 \end{matrix} \\ \begin{matrix} 0 \\ 0 \end{matrix} & \pm\begin{pmatrix} 0 & 1 \\ 1 & 0 \end{pmatrix} \end{pmatrix}.$$

Therefore

$$U = \begin{pmatrix} 0 & 1 & 0 \\ 0 & 0 & 1 \\ 1 & 0 & 0 \end{pmatrix} = \begin{pmatrix} -1 & 0 & 0 \\ 0 & -1 & 0 \\ 0 & 0 & 1 \end{pmatrix} \begin{pmatrix} -1 & 0 & 0 \\ 0 & 0 & 1 \\ 0 & 1 & 0 \end{pmatrix} \begin{pmatrix} 0 & 1 & 0 \\ 1 & 0 & 0 \\ 0 & 0 & -1 \end{pmatrix}$$

is mapped into

$$\begin{pmatrix} -1 & 0 & 0 \\ 0 & -1 & 0 \\ 0 & 0 & 1 \end{pmatrix} \begin{pmatrix} -1 & 0 \quad 0 \\ 0 & \\ 0 & \pm \begin{pmatrix} 0 & 1 \\ 1 & 0 \end{pmatrix} \end{pmatrix} \begin{pmatrix} 0 & 1 & 0 \\ 1 & 0 & 0 \\ 0 & 0 & -1 \end{pmatrix} = \begin{cases} U, & \text{if } + \text{ is used,} \\ V, & \text{if } - \text{ is used,} \end{cases}$$

where $V = I_1 U I_1^{-1}$. Thus, in this case, $T \to T = I_1 T I_1^{-1}$, and either $U \to U$ or $U \to I_1 U I_1^{-1}$. Since T and U generate(2) $\mathfrak{M}_3^+$, the automorphism is inner.

Case 2.

$$T \to \begin{pmatrix} -1 & -1 & 0 \\ 0 & -1 & 0 \\ 0 & 0 & 1 \end{pmatrix}.$$

Then

$$\begin{pmatrix} 0 & 1 & 0 \\ 1 & 0 & 0 \\ 0 & 0 & -1 \end{pmatrix} \to \begin{pmatrix} 0 & -1 & 0 \\ -1 & 0 & 0 \\ 0 & 0 & -1 \end{pmatrix},$$

and one finds in this case that

$$U \to \begin{pmatrix} 0 & -1 & 0 \\ 0 & 0 & 1 \\ -1 & 0 & 0 \end{pmatrix} \quad \text{or} \quad \begin{pmatrix} 0 & -1 & 0 \\ 0 & 0 & -1 \\ 1 & 0 & 0 \end{pmatrix}.$$

If we set $Z = TU^2$, then

$$\begin{pmatrix} 1 & 0 & 0 \\ 1 & 1 & 0 \\ 0 & 0 & 1 \end{pmatrix} = (UZ^{-1})^2 UZ^2. \tag{10}$$

Now certainly the left side of (10) maps into

$$\begin{pmatrix} -1 & 0 & 0 \\ -1 & -1 & 0 \\ 0 & 0 & 1 \end{pmatrix},$$

(2) L. K. Hua and I. Reiner, loc. cit.

whereas, knowing T^τ and U^τ, we can compute Z^τ and thence can find the image of the right side of (10). We readily find (for either value of U^τ) that the right side of (10) maps into

$$\begin{pmatrix} 1 & \cdot & \cdot \\ 3 & \cdot & \cdot \\ \cdot & \cdot & \cdot \end{pmatrix},$$

and hence we have a contradiction.

Therefore case 2 cannot occur, and so if either A or B equals $I^{(2)}$, the automorphism is inner. Suppose hereafter that

$$A = B = \begin{pmatrix} 0 & 1 \\ -1 & 0 \end{pmatrix}.$$

In this case we have

$$T \to \begin{pmatrix} \pm\begin{pmatrix} 1 & 0 \\ -1 & 1 \end{pmatrix} & \begin{matrix} 0 \\ 0 \end{matrix} \\ \begin{matrix} 0 & 0 \end{matrix} & 1 \end{pmatrix}.$$

Case 1*.

$$T \to \begin{pmatrix} 1 & 0 & 0 \\ -1 & 1 & 0 \\ 0 & 0 & 1 \end{pmatrix}.$$

Then as before

$$\begin{pmatrix} 0 & 1 & 0 \\ -1 & 0 & 0 \\ 0 & 0 & 1 \end{pmatrix} \quad \text{and} \quad \begin{pmatrix} 0 & 1 & 0 \\ 1 & 0 & 0 \\ 0 & 0 & -1 \end{pmatrix}$$

are invariant, and again $U^\tau = U$ or V. After a further inner automorphism by a factor of I_1 (in the latter case) we also have $U \to U$. But then

$$T \to T'^{-1}, \qquad U \to U'^{-1}.$$

(This automorphism is easily shown to be a non-inner automorphism.)

Case 2*.

$$T \to \begin{pmatrix} -1 & 0 & 0 \\ 1 & -1 & 0 \\ 0 & 0 & 1 \end{pmatrix}.$$

Then

$$\begin{pmatrix} 0 & 1 & 0 \\ 1 & 0 & 0 \\ 0 & 0 & -1 \end{pmatrix} \to \begin{pmatrix} 0 & -1 & 0 \\ -1 & 0 & 0 \\ 0 & 0 & -1 \end{pmatrix},$$

and again we find that there are two possibilities for U^τ, each of which leads to a contradiction, just as in case 2. Therefore Theorem 3 holds when $n=3$.

6. **A fundamental lemma.** Theorem 3 will be proved by induction on n; the result has already been established for $n=2$ and 3. In going from $n-1$ to n, the following lemma is basic:

LEMMA 2. *Let $n \geqq 4$, and define $J_1=(-1)\dotplus I^{(n-1)}$. In any automorphism τ of $\mathfrak{M}_n$, $J_1^\tau = \pm A J_1 A^{-1}$ for some $A \in \mathfrak{M}_n$.*

Proof. By Corollary 1, $J_1^\tau \in \mathfrak{M}_n^-$, and J_1^τ is an involution. After a suitable inner automorphism, we may assume that $J_1^\tau = W(x, y, z)$ (as defined by (4)), where $2x+y+z=n$ and $x+y$ is odd. Every element of $\mathfrak{M}_n$ which commutes with J_1 maps into an element of $\mathfrak{M}_n$ which commutes with W. Every matrix in $\mathfrak{M}_n^+$ maps into a matrix in $\mathfrak{M}_n^+$. Combining these facts, we see that the group $\mathfrak{G}_1$ consisting of those elements of $\mathfrak{M}_n^+$ which commute with J_1 is isomorphic to $\mathfrak{G}_2$, the corresponding group for W. If we prove that this can happen only for $x=0$, $y=1$, $z=n-1$ or $x=0$, $y=n-1$, $z=1$, the result will follow.

The group $\mathfrak{G}_1$ consists of the matrices in $\mathfrak{M}_n^+$ of the form $(\pm 1)\dotplus X_1$, $X_1 \in \mathfrak{M}_{n-1}$, and so clearly $\mathfrak{G}_1 \cong \mathfrak{M}_{n-1}$.

The group $\mathfrak{G}_2$ is easily found to consist of all matrices $C \in \mathfrak{M}_1^+$ of the form (we illustrate the case where $x=2$):

$$C = \left[\begin{array}{cccc|ccc|ccc} a_1 & 0 & a_2 & 0 & 0 & \cdots & 0 & 2\beta_1 & \cdots & 2\beta_z \\ \dfrac{a_1-d_1}{2} & d_1 & \dfrac{a_2-d_2}{2} & d_2 & \alpha_1 & \cdots & \alpha_y & \beta_1 & \cdots & \beta_z \\ a_3 & 0 & a_4 & 0 & 0 & \cdots & 0 & 2\delta_1 & \cdots & 2\delta_z \\ \dfrac{a_3-d_3}{2} & d_3 & \dfrac{a_4-d_4}{2} & d_4 & \gamma_1 & \cdots & \gamma_y & \delta_1 & \cdots & \delta_z \\ \hline \epsilon_1 & -2\epsilon_1 & \zeta_1 & -2\zeta_1 & & & & & & \\ \vdots & \vdots & \vdots & \vdots & & U & & & 0 & \\ \epsilon_y & -2\epsilon_y & \zeta_y & -2\zeta_y & & & & & & \\ \hline \eta_1 & 0 & \theta_1 & 0 & & & & & & \\ \vdots & \vdots & \vdots & \vdots & & 0 & & & V & \\ \eta_z & 0 & \theta_z & 0 & & & & & & \end{array}\right] \begin{array}{l} \\ 2x \text{ rows} \\ \\ \\ y \text{ rows} \\ \\ z \text{ rows} \\ \end{array}$$

$$\underbrace{\qquad\qquad}_{2x \text{ columns}} \quad \underbrace{\qquad}_{y \text{ columns}} \quad \underbrace{\qquad}_{z \text{ columns}}$$

For the moment put

$$K = \begin{pmatrix} 1 & 0 \\ -1/2 & 1 \end{pmatrix} \dotplus \underset{(x \text{ terms})}{\cdots} \dotplus \begin{pmatrix} 1 & 0 \\ -1/2 & 1 \end{pmatrix} \dotplus I^{(n-2x)}.$$

Then a simple calculation gives:

$$KCK^{-1} = \begin{bmatrix} a_1 & 0 & a_2 & 0 & 0 \cdots 0 & 2\beta_1 \cdots 2\beta_z \\ 0 & d_1 & 0 & d_2 & \alpha_1 \cdots \alpha_y & 0 \cdots 0 \\ a_3 & 0 & a_4 & 0 & 0 \cdots 0 & 2\delta_1 \cdots 2\delta_z \\ 0 & d_3 & 0 & d_4 & \gamma_1 \cdots \gamma_y & 0 \cdots 0 \\ 0 & -2\epsilon_1 & 0 & -2\zeta_1 & & \\ \vdots & \vdots & \vdots & \vdots & U & 0 \\ 0 & -2\epsilon_y & 0 & -2\zeta_y & & \\ \eta_1 & 0 & \theta_1 & 0 & & \\ \vdots & \vdots & \vdots & \vdots & 0 & V \\ \eta_z & 0 & \theta_z & 0 & & \end{bmatrix}$$

and so C is similar to

$$\begin{bmatrix} a_1 & a_2 & 2\beta_1 \cdots 2\beta_z \\ a_3 & a_4 & 2\delta_1 \cdots 2\delta_z \\ \eta_1 & \theta_1 & \\ \vdots & \vdots & V \\ \eta_z & \theta_z & \end{bmatrix} \dotplus \begin{bmatrix} d_1 & d_2 & \alpha_1 \cdots \alpha_y \\ d_3 & d_4 & \gamma_1 \cdots \gamma_y \\ -2\epsilon_1 & -2\zeta_1 & \\ \vdots & \vdots & U \\ -2\epsilon_y & -2\zeta_y & \end{bmatrix}$$

$$= \underset{x \qquad z}{\begin{bmatrix} S_1 & 2R_1 \\ Q_1 & T_1 \end{bmatrix}}\begin{matrix} x \\ z \end{matrix} \dotplus \underset{x \qquad y}{\begin{bmatrix} S_2 & Q_2 \\ 2R_2 & T_2 \end{bmatrix}}\begin{matrix} x \\ y \end{matrix},$$

with a fixed similarity factor depending only on W. Therefore $\mathfrak{G}_2 \cong \mathfrak{G}$, where $\mathfrak{G} = \mathfrak{G}(x, y, z)$ is the group of matrices in $\mathfrak{M}_n^+$ of the form

$$\underset{x \qquad z}{\begin{bmatrix} S_1 & 2R_1 \\ Q_1 & T_1 \end{bmatrix}}\begin{matrix} x \\ z \end{matrix} \dotplus \underset{x \qquad y}{\begin{bmatrix} S_2 & Q_2 \\ 2R_2 & T_2 \end{bmatrix}}\begin{matrix} x \\ y \end{matrix},$$

where $S_1 \equiv S_2 \pmod 2$. Here $2x+y+z=n$ and $x+y$ is odd.

We wish to prove that $\mathfrak{M}_{n-1} \cong \mathfrak{G}(x, y, z)$ only when $x=0$, $y=1$, $z=n-1$ or $x=0$, $y=n-1$, $z=1$. In order to establish this, we shall prove that in all other cases the number of involutions in $\mathfrak{G}$ which are nonsimilar in $\mathfrak{G}$ is greater than the number of involutions in $\mathfrak{M}_{n-1}$ which are nonsimilar in $\mathfrak{M}_{n-1}$;

this latter number is, of course, A_{n-1} (given by (6)).

We shall briefly denote the elements of $\mathfrak{G}$ by $A\dot{+}B$, where

$$A=\begin{pmatrix} S_1 & 2R_1 \\ Q_1 & T_1 \end{pmatrix} \quad \text{and} \quad B=\begin{pmatrix} S_2 & Q_2 \\ 2R_2 & T_2 \end{pmatrix}.$$

If $A_1\dot{+}B_1$ and $A_2\dot{+}B_2$ are two involutions in $\mathfrak{G}$, where either

$$A_1 \overset{s}{\neq} A_2$$

in $\mathfrak{M}_{x+z}$ or

$$B_1 \overset{s}{\neq} B_2$$

in $\mathfrak{M}_{x+y}$, then certainly

$$A_1\dot{+}B_1 \overset{s}{\neq} A_2\dot{+}B_2$$

in $\mathfrak{G}$ (these may be similar in $\mathfrak{M}_n$, however). Therefore, the matrices $A\dot{+}B$, where

$$A = I^{(a_1)} \dot{+} (-I)^{(b_1)} \dot{+} \underbrace{L \dot{+} \cdots \dot{+} L}_{(c_1 \text{ terms})},$$

$$B = I^{(a_2)} \dot{+} (-I)^{(b_2)} \dot{+} \underbrace{L \dot{+} \cdots \dot{+} L}_{(c_2 \text{ terms})},$$

obtained by taking different sets of values of $(a_1, b_1, c_1, a_2, b_2, c_2)$, if they lie in $\mathfrak{G}$, are certainly nonsimilar in $\mathfrak{G}$. Here we have

$$a_1+b_1+2c_1=x+z, \quad a_2+b_2+2c_2=x+y, \quad b_1+b_2+c_1+c_2 \text{ even}.$$

If $x\neq 0$, we impose the further restriction that $c_1\leqq(z+1)/2$, $c_2\leqq(y+1)/2$, and that in B instead of L we use L'. These conditions will insure that $A\dot{+}B\in\mathfrak{G}$. We certainly do not (in general) get all of the nonsimilar involutions of $\mathfrak{G}$ in this way, but instead we obtain only a subset thereof. Call the number of such matrices N.

For $x=0$, we have $N = B_yB_z + (A_y - B_y)(A_z - B_z)$. Since y is odd, $A_y=2B_y$, and therefore

$$N = B_yA_z = B_yA_{n-y}.$$

Case 1. n even. Then $N=(y+1)(y+3)(n-y+1)(n-y+3)/32$. If neither y nor $n-y$ is 1 (certainly neither can be zero), then

$$(y+1)(n-y+1) \geqq 4(n-2) \quad \text{and} \quad (y+3)(n-y+3) \geqq 6n,$$

so that

$$N \geqq (24/32)\, n(n-2).$$

For $n=4$, $x=0$, either $y=1$ or $z=1$. For $n\geqq 6$, we have $N>A_{n-1}$. Hence in this case $\mathfrak{G}$ is not isomorphic to $\mathfrak{M}_{n-1}$. (If either y or $n-y=1$, then $W(x, y, z) = \pm J_1$.)

Case 2. n odd. Then $N=(y+1)(y+3)(n-y+2)^2/32$. We find again that $N>A_{n-1}$ for $n\geqq 5$.

This settles the cases where $x=0$. Suppose that $x\neq 0$ hereafter. Then N is the number of solutions of

$$a_1+b_1+2c_1=x+z,\quad a_2+b_2+2c_2=x+y,\quad b_1+b_2+c_1+c_2 \text{ even},$$
$$0\leqq c_1\leqq \frac{z+1}{2},\qquad 0\leqq c_2\leqq \frac{y+1}{2}.$$

Using $[r]$ to denote the greatest integer less than or equal to r, we readily find that N is given by

$$\frac{1}{2}\left[\frac{z+3}{2}\right]\left[\frac{y+3}{2}\right]\left(x+z+1-\left[\frac{z+1}{2}\right]\right)\left(x+y+1-\left[\frac{y+1}{2}\right]\right).$$

By considering separately the cases where y and z are both even, one even and one odd, and so on, it is easy to prove that $N\geqq A_{n-1}$ in all cases except when both y and z are zero. Leaving aside this case for the moment, consider the matrix $A_0\dotplus I^{(x+y)}\in\mathfrak{G}$, where $A_0\in\mathfrak{M}_{x+z}$ is given by

$$A_0=\begin{pmatrix}1 & 2 & 2\cdots & 2\\ 0 & -1 & 0\cdots & 0\\ 0 & 0 & -1\cdots & 0\\ \cdot & \cdot & \cdots\cdot & \cdot\\ 0 & 0 & 0\cdots & -1\end{pmatrix}.$$

The matrix $A_0\dotplus I^{(x+y)}$ is certainly an involution in $\mathfrak{G}$. Since, in $\mathfrak{M}_{x+z}$,

$$A_0\overset{s}{=}\begin{pmatrix}1 & 0\cdots & 0\\ 0 & -1\cdots & 0\\ \cdot & \cdots\cdot & \cdot\\ 0 & 0\cdots & -1\end{pmatrix}=A_1,$$

$A_0\dotplus I^{(x+y)}$ can be similar (in $\mathfrak{G}$) only to that matrix (counted in the N matrices) of the form $A_1\dotplus I^{(x+y)}$. But from

$$A_1\cdot\begin{pmatrix}a_1 & a_2\cdots a_x & 2b_1\cdots 2b_z\\ \cdot\;\cdot\;\cdot\;\cdot\;\cdot\;\cdot\;\cdot\;\cdot\;\cdot\;\cdot\;\cdot\;\cdot\\ \cdot\;\cdot\;\cdot\;\cdot\;\cdot\;\cdot\;\cdot\;\cdot\;\cdot\;\cdot\;\cdot\;\cdot\\ \cdot\;\cdot\;\cdot\;\cdot\;\cdot\;\cdot\;\cdot\;\cdot\;\cdot\;\cdot\;\cdot\;\cdot\end{pmatrix}=\begin{pmatrix}a_1 & a_2\cdots a_x & 2b_1\cdots 2b_z\\ \cdot\;\cdot\;\cdot\;\cdot\;\cdot\;\cdot\;\cdot\;\cdot\;\cdot\;\cdot\;\cdot\;\cdot\\ \cdot\;\cdot\;\cdot\;\cdot\;\cdot\;\cdot\;\cdot\;\cdot\;\cdot\;\cdot\;\cdot\;\cdot\\ \cdot\;\cdot\;\cdot\;\cdot\;\cdot\;\cdot\;\cdot\;\cdot\;\cdot\;\cdot\;\cdot\;\cdot\end{pmatrix}\cdot A_0$$

we obtain

$$a_1 = a_2 = \cdots = a_x = 2b_1,$$

which is impossible. Hence $\mathfrak{G}$ contains at least $N+1$ nonsimilar involutions, and therefore $\mathfrak{G}$ is not isomorphic to $\mathfrak{M}_{n-1}$ in these cases.

We have left only the case $y=z=0$, $x=n/2$; then n is singly even. Here we may choose $A=W(c_1, b_1, a_1)$, $B=W(c_1, b_2, a_2)$, where

$$a_1 + b_1 + 2c_1 = x, \qquad a_2 + b_2 + 2c_1 = x, \qquad b_1 + b_2 \text{ even.}$$

Then $A\dotplus B\in\mathfrak{G}$, and the various matrices are nonsimilar. The number of such matrices is $(x+1)(x+2)(x+3)/12$, which is greater than A_{n-1} for $n\geqq 14$. For $n=6$, $\mathfrak{M}_{n-1}$ contains an element of order 5, while $\mathfrak{G}$ does not. For $n=10$, $\mathfrak{M}_{n-1}$ contains an element of order 7, while $\mathfrak{G}$ does not. This completes the proof of the lemma.

7. **Proof of Theorem 3.** We are now ready to give a proof of Theorem 3 by induction on n. Hereafter, let $n\geqq 4$ and suppose that Theorem 3 holds for $n-1$. If τ is any automorphism of $\mathfrak{M}_n$, by Corollary 1 and Lemma 2 we know that τ takes $\mathfrak{M}_n^+$ into itself, and $J_1^\tau=\pm AJ_1A^{-1}$. If we change τ by a suitable inner automorphism, then we may assume that $J_1\to\pm J_1$. When n is odd, certainly $J_1\to J_1$; when n is even, by multiplying τ by the automorphism $X\in\mathfrak{M}_n\to(\det X)\cdot X$ if necessary, we may again assume $J_1\to J_1$.

Therefore, every $M\in\mathfrak{M}_n^+$ which commutes with J_1 goes into another such element, that is,

$$\begin{pmatrix} \pm 1 & \mathfrak{n}' \\ \mathfrak{n} & X \end{pmatrix}^\tau = \begin{pmatrix} \pm 1 & \mathfrak{n}' \\ \mathfrak{n} & X^\tau \end{pmatrix}.$$

Since this induces an automorphism on $\mathfrak{M}_{n-1}$, we have $\det X^\tau=\det X$, so that the plus signs go together, as do the minus signs. Furthermore, by our induction hypothesis,

$$X^\tau = \pm AX^*A^{-1},$$

where $A\in\mathfrak{M}_{n-1}$ and either $X^*=X$ for all $X\in\mathfrak{M}_{n-1}$ or $X^*=X'^{-1}$ for all $X\in\mathfrak{M}_{n-1}$; here the minus sign can occur only for $X\in\mathfrak{M}_{n-1}^-$, and if it occurs for one such X, it occurs for all $X\in\mathfrak{M}_{n-1}^-$. After changing our original automorphism by a factor of $I^{(1)}\dotplus A^{-1}$, we may assume that $X^\tau=\pm X^*$.

Let J_ν be obtained from $I^{(n)}$ by replacing the νth diagonal element by -1. Then

$$J_1J_n = \begin{pmatrix} -1 & 0\cdots 0 & 0 \\ 0 & 1\cdots 0 & 0 \\ \cdot & \cdots\cdots & \cdot \\ 0 & 0\cdots 1 & 0 \\ 0 & 0\cdots 0 & -1 \end{pmatrix} \to \begin{pmatrix} -1 & \mathfrak{n}' \\ \mathfrak{n} & \pm\begin{pmatrix} 1\cdots 0 & 0 \\ \cdots\cdot 0 & 0 \\ 0\cdots 1 & 0 \\ 0\cdots 0 & -1 \end{pmatrix}^* \end{pmatrix}.$$

The minus sign here is impossible by Lemma 2, since $n \geqq 4$. Hence $J_1 J_n$ is invariant, and therefore so is J_n. By the same reasoning all of the J_ν $(\nu=1, \cdots, n)$ are invariant.

From the above remarks we see that for $X \in \mathfrak{M}_{n-1}^{+}$,

$$\begin{pmatrix} 1 & \mathfrak{n}' \\ \mathfrak{n} & X \end{pmatrix}^{\tau} = \begin{pmatrix} 1 & \mathfrak{n}' \\ \mathfrak{n} & A_1 X^* A_1^{-1} \end{pmatrix}, \cdots, \begin{pmatrix} X & \mathfrak{n} \\ \mathfrak{n}' & 1 \end{pmatrix}^{\tau} = \begin{pmatrix} A_n X^* A_n^{-1} & \mathfrak{n} \\ \mathfrak{n}' & 1 \end{pmatrix},$$

where $A_\nu \in \mathfrak{M}_{n-1}$, and in fact $A_1 = I$. Now suppose that $Z \in \mathfrak{M}_{n-2}^{+}$, and form $I^{(2)} \dotplus Z$. Since it commutes with both J_1 and J_2, its image must do likewise. But then

$$A_1 \begin{pmatrix} 1 & \mathfrak{n}' \\ \mathfrak{n} & Z \end{pmatrix} A_1^{-1} = \begin{pmatrix} 1 & \mathfrak{n}' \\ \mathfrak{n} & \overline{Z} \end{pmatrix}$$

for every $Z \in \mathfrak{M}_{n-2}^{+}$. Setting

$$A_1 = \begin{pmatrix} a & \mathfrak{x}' \\ \mathfrak{y} & A \end{pmatrix}$$

we obtain $\mathfrak{x}' Z = \mathfrak{x}'$, $\mathfrak{y} = \overline{Z} \mathfrak{y}$. Since this holds for all $Z \in \mathfrak{M}_{n-2}^{+}$, we must have $\mathfrak{x} = \mathfrak{y} = \mathfrak{n}$, so that A_1 is itself decomposable. A similar argument (considering the matrices commuting with both J_1 and J_ν, for $\nu = 3, \cdots, n$) shows that A_1 is diagonal. Correspondingly, all of the A_ν are diagonal. It is further clear that all of the A_ν $(\nu = 1, \cdots, n)$ are sections of a single diagonal matrix $D^{(n)}$. Using the further inner automorphism factor D^{-1}, we may henceforth assume that $X^\tau = X^*$ for every decomposable $X \in \mathfrak{M}_n^{+}$, where either $X^* = X$ always or $X^* = X'^{-1}$ always. Since $\mathfrak{M}_n^{+}$ is generated by the set of decomposable elements of $\mathfrak{M}_n^{+}$, the theorem is proved.

TSING HUA UNIVERSITY,
PEKING, CHINA.
UNIVERSITY OF ILLINOIS,
URBANA, ILL.

Reprinted from the
TRANSACTIONS OF THE AMERICAN MATHEMATICAL SOCIETY
Vol. 72, No. 3, pp. 467–473, 1952

AUTOMORPHISMS OF THE PROJECTIVE UNIMODULAR GROUP

BY
L. K. HUA AND I. REINER

Notation. Let $\mathfrak{M}_n$ denote the group of $n\times n$ integral matrices of determinant ± 1 (the unimodular group). By $\mathfrak{M}_n^+$ we denote that subset of $\mathfrak{M}_n$ where the determinant is $+1$; $\mathfrak{M}_n^-$ is correspondingly defined. Let $\mathfrak{P}_{2n}$ be obtained from $\mathfrak{M}_{2n}$ by identifying $+X$ and $-X$, $X\in\mathfrak{M}_{2n}$. (This is the same as considering the factor group of $\mathfrak{M}_{2n}$ by its centrum.) We correspondingly obtain $\mathfrak{P}_{2n}^+$ and $\mathfrak{P}_{2n}^-$ from $\mathfrak{M}_{2n}^+$ and $\mathfrak{M}_{2n}^-$. Let $I^{(n)}$ (or briefly I) be the identity matrix in $\mathfrak{M}_n$, and let X' denote the transpose of X. The direct sum of A and B is represented by $A\dotplus B$, while

$$A \overset{s}{=} B$$

means that A is similar to B.

In this paper we shall find explicitly the generators of the group $\mathfrak{B}_{2n}$ of all automorphisms of $\mathfrak{P}_{2n}$, thereby obtaining a complete description of these automorphisms. This generalizes the result due to Schreier([1]) for the case $n=1$.

We shall frequently refer to results of an earlier paper: *Automorphisms of the unimodular group*, L. K. Hua and I. Reiner, Trans. Amer. Math. Soc. vol. 71 (1951) pp. 331-348. We designate this paper by AUT.

1. **The commutator subgroup of $\mathfrak{P}_{2n}$.** The following useful result is an immediate consequence of the corresponding theorem for $\mathfrak{M}_{2n}$ (AUT, Theorem 1).

THEOREM 1. *Let $\mathfrak{S}_{2n}$ be the commutator subgroup of $\mathfrak{P}_{2n}$. Then clearly $\mathfrak{S}_{2n}\subset\mathfrak{P}_{2n}^+$. For $n=1$, $\mathfrak{S}_{2n}$ is of index 2 in $\mathfrak{P}_{2n}^+$, while for $n>1$, $\mathfrak{S}_{2n}=\mathfrak{P}_{2n}^+$.*

THEOREM 2. *In any automorphism of $\mathfrak{P}_{2n}$, always $\mathfrak{P}_{2n}^+$ goes into itself.*

Proof. This is a corollary to Theorem 1 when $n>1$, since the commutator subgroup goes into itself under any automorphism. For $n=1$, suppose that $\pm S\rightarrow\pm S_1$ and $\pm T\rightarrow\pm T_1$, where

$$S=\begin{pmatrix}0&1\\-1&0\end{pmatrix},\qquad T=\begin{pmatrix}1&1\\0&1\end{pmatrix}. \tag{1}$$

Since S and T generate $\mathfrak{M}_2^+$, it follows that $\pm S$ and $\pm T$ generate $\mathfrak{P}_2^+$,

Received by the editors May 18, 1951.

([1]) Abh. Math. Sem. Hamburgischen Univ. vol. 3 (1924) p. 167.

and hence so must $\pm S_1$ and $\pm T_1$. It is therefore sufficient to prove that $\det S_1 = \det T_1 = +1$. From $(ST)^3 = I$ we deduce $S_1T_1 = \pm T_1^{-1}S_1^{-1}T_1^{-1}S_1^{-1}$, so that $\det S_1T_1 = 1$. Hence either S_1 and T_1 are both in $\mathfrak{P}_2^+$ or both in $\mathfrak{P}_2^-$; we shall show that the latter alternative is impossible.

Suppose that $\det S_1 = \det T_1 = -1$. From $S^2 = I$ we deduce $S_1^2 = \pm I$; if $S_1^2 = -I$, then $S_1^2 + I = 0$ and the characteristic equation of S_1 is $\lambda^2 + 1 = 0$, from which it follows that $\det S_1 = 1$; this contradicts our assumption that $\det S_1 = -1$, so of necessity $S_1^2 = I$. But if this is the case, then it is easy to show that there exists a matrix $A \in \mathfrak{M}_2$ such that AS_1A^{-1} takes one of the two canonical forms

$$\begin{pmatrix} 1 & 0 \\ 0 & -1 \end{pmatrix} \quad \text{and} \quad \begin{pmatrix} 1 & 0 \\ 1 & -1 \end{pmatrix}.$$

By considering instead of the original automorphism τ, a new automorphism τ' defined by: $X^{\tau'} = AX^{\tau}A^{-1}$, we may hereafter assume that

$$S_1 = \pm \begin{pmatrix} 1 & 0 \\ 0 & -1 \end{pmatrix} \quad \text{or} \quad \pm \begin{pmatrix} 1 & 0 \\ 1 & -1 \end{pmatrix}.$$

Let

$$T_1 = \pm \begin{pmatrix} a & b \\ c & d \end{pmatrix};$$

then $ad - bc = -1$.

Now we observe that $J = (1) \dotplus (-1)$ is distinct from $\pm I$ and $\pm S$, that it commutes with S, and that JT is an involution. Hence there exists a matrix $M \in \mathfrak{P}_2$ distinct from $\pm I$ and $\pm S_1$, such that M commutes with S_1, and MT_1 is an involution.

Case 1.

$$S_1 = \pm \begin{pmatrix} 1 & 0 \\ 0 & -1 \end{pmatrix}.$$

Since $(S_1T_1)^3 = \pm I$, we find that $a - d = \pm 1$. The only matrices commuting with S_1 which are distinct from $\pm I$ and $\pm S_1$ are

$$\pm \begin{pmatrix} 0 & 1 \\ 1 & 0 \end{pmatrix} \quad \text{and} \quad \pm \begin{pmatrix} 0 & 1 \\ -1 & 0 \end{pmatrix}.$$

If M is either of the first two matrices, then the condition that MT_1 be an involution yields $b + c = 0$. Thus $a = d \pm 1$, $b = -c$, and $ad - bc = -1$. Combining these, we obtain $d(d \pm 1) + c^2 = -1$, which is impossible. The other two choices for M imply $b = c$, and therefore $d(d \pm 1) - c^2 = -1$. Hence $1 - 4(1 - c^2)$ is a perfect square; but $4c^2 - 3 = f^2$ implies $(2c + f)(2c - f) = 1$, whence $c = \pm 1$.

But then $ad=0$; from $a-d=\pm 1$ we deduce that $a^2-d^2=\pm 1$, whence $(S_1T_1^2)^3 = \pm I$, which is impossible.

Case 2.

$$S_1 = \pm \begin{pmatrix} 1 & 0 \\ 1 & -1 \end{pmatrix}.$$

From $(S_1T_1)^3=\pm I$ we obtain $a-d+b=\pm 1$. For M there are the four possibilities

$$\pm \begin{pmatrix} 1 & -2 \\ 0 & -1 \end{pmatrix} \quad \text{and} \quad \pm \begin{pmatrix} 1 & -2 \\ 1 & -1 \end{pmatrix}.$$

Since MT_1 is an involution, in the first two cases we have $a-2c-d=0$, whence

$$ad - bc = \{(a+d)^2 + (a-d\pm 1)^2 - 1\}/4 \neq -1.$$

In the second two cases we find that $a-2c+b-d=0$, so that $2c=a+b-d = \pm 1$, which is again a contradiction. This completes the proof of Theorem 2.

2. **Automorphisms of $\mathfrak{P}_2^+$.** Let us now determine all automorphisms of $\mathfrak{P}_2$. Since every such automorphism takes $\mathfrak{P}_2^+$ into itself, we begin by considering all automorphisms of $\mathfrak{P}_2^+$.

THEOREM 3. *Every automorphism of $\mathfrak{P}_2^+$ is of the form $X\in\mathfrak{P}_2^+\to AXA^{-1}$ for some $A\in\mathfrak{M}_2$; that is, all automorphisms of $\mathfrak{P}_2^+$ are "inner" (with $A\in\mathfrak{M}_2$ rather than $A\in\mathfrak{P}_2^+$.)*

Proof. Let τ be any automorphism of $\mathfrak{P}_2^+$, and define S and T as before; let $S_0\in\mathfrak{M}_2$ be a fixed representative of $\pm S^\tau$. By Theorem 2, $S_0\in\mathfrak{M}_2^+$, and therefore $S_0^2=-I$. Let T_0 be that representative of $\pm T^\tau$ for which $(S_0T_0)^3=I$ is valid. Then $S\to S_0$, $T\to T_0$ induces a mapping from $\mathfrak{M}_2^+$ onto itself. The mapping is one-to-one, for although an element of $\mathfrak{M}_2^+$ can be expressed in many different ways as a product of powers of S and T, these expressions can be gotten from one another by use of $S^2=-I$, $(ST)^3=I$; since S_0 and T_0 satisfy these same relations, the mapping is one-to-one. It is an automorphism because τ is one. Therefore (AUT, Theorem 2) there exists an $A\in\mathfrak{M}_2$ such that $S_0=\pm ASA^{-1}$, $T_0=\pm ATA^{-1}$. This proves the result.

COROLLARY. *Every automorphism of $\mathfrak{P}_2$ is of the form $X\in\mathfrak{P}_2\to AXA^{-1}$ for some $A\in\mathfrak{M}_2$.*

(This corollary is a simple consequence of Theorem 3, as is shown in AUT by the remarks following the statement of Theorem 4.)

3. **The generators of $\mathfrak{B}_{2n}$.** Our main result may be stated as follows:

THEOREM 4. *The generators of $\mathfrak{B}_{2n}$ are*

(i) *The set of all inner automorphisms*:

$$\pm X \in \mathfrak{P}_{2n} \rightarrow \pm AXA^{-1} \qquad (A \in \mathfrak{M}_{2n}),$$

and

(ii) *The automorphism* $\pm X \in \mathfrak{P}_{2n} \rightarrow \pm X'^{-1}$.

REMARK. For $n=1$, the automorphism (ii) is a special case of (i).

In the proof of Theorem 4 by induction on n, the following lemma (which has already been established for $n=1$) will be basic:

LEMMA 1. *Let* $J_1=(-1)\dotplus I^{(2n-1)}$. *In any automorphism* τ *of* $\mathfrak{P}_{2n}$, $J_1^\tau = \pm AJ_1A^{-1}$ *for some* $A \in \mathfrak{M}_{2n}$.

Proof. The result is already known for $n=1$. Hereafter let $n \geqq 2$. Certainly $(J_1^\tau)^2 = \pm I$ and det $J_1^\tau = -1$. If $(J_1^\tau)^2 = -I$, then the minimum function of J_1^τ is λ^2+1, and its characteristic function must be some power of λ^2+1, whence det $J_1^\tau=1$. Therefore $(J_1^\tau)^2=I$ is valid in $\mathfrak{M}_{2n}$. After a suitable inner automorphism, we may assume that

$$J_1^\tau = W(x, y, z) = L \dotplus \cdots \dotplus L \dotplus (-I)^{(y)} \dotplus I^{(z)},$$

where

$$L = \begin{pmatrix} 1 & 0 \\ 1 & -1 \end{pmatrix}$$

occurs x times, $2x+y+z=2n$, and $x+y$ is odd. (This follows from AUT, Lemma 1.)

Let $\mathfrak{G}_1$ be the group consisting of all elements of $\mathfrak{P}_{2n}$ which commute with J_1, and $\mathfrak{G}_2$ the corresponding group for J_1^τ. The lemma will be proved if we can show that $\mathfrak{G}_1$ is not isomorphic to $\mathfrak{G}_2$ unless $J_1^\tau = \pm J_1$. The group $\mathfrak{G}_1$ consists of the matrices $\pm(1 \dotplus X_1) \in \mathfrak{P}_{2n}$, so that $\mathfrak{G}_1 \cong \mathfrak{M}_{2n-1}$. The number of nonsimilar involutions in $\mathfrak{G}_1$ is therefore $n(n+1)$ (see AUT, §4). We shall prove that $\mathfrak{G}_2$ contains more than $n(n+1)$ involutions which are nonsimilar in $\mathfrak{G}_2$, except when $x=0$, $y=1$, $z=2n-1$ or $x=0$, $y=2n-1$, $z=1$.

Those elements $\pm C \in \mathfrak{P}_{2n}$ which commute with W must satisfy one of the two equations: $CW=WC$ or $CW=-WC$. The solutions of the first of these equations form a subgroup of $\mathfrak{G}_2$, and this subgroup is known (see AUT, proof of Lemma 2) to be isomorphic to $\mathfrak{G}_0=\mathfrak{G}_0(x, y, z)$ consisting of all matrices in $\mathfrak{P}_{2n}$ of the form

$$\begin{pmatrix} S_1 & 2R_1 \\ Q_1 & T_1 \end{pmatrix} \dotplus \begin{pmatrix} S_2 & Q_2 \\ 2R_2 & T_2 \end{pmatrix},$$

where S_1, S_2, T_1, and T_2 are square matrices of dimensions x, x, z, and y respectively, and where $S_1 \equiv S_2 \pmod 2$, $2x+y+z=2n$, and $x+y$ and $x+z$ are both odd.

Next we prove that $\overline{C}W=-W\overline{C}$ is solvable only when $y=z$. The space

$\mathfrak{U}$ of vectors $\mathfrak{u}$ such that $W\mathfrak{u}=\mathfrak{u}$ is of dimension $x+z$, while the space $\mathfrak{B}$ of vectors $\mathfrak{v}$ for which $W\mathfrak{v}=-\mathfrak{v}$ has dimension $x+y$. But if $\overline{C}W=-W\overline{C}$, then $W\overline{C}\mathfrak{u}=-\overline{C}\mathfrak{u}$ and $W\overline{C}^{-1}\mathfrak{v}=\overline{C}^{-1}\mathfrak{v}$, so the dimensions of $\mathfrak{U}$ and $\mathfrak{B}$ must be the same, whence $y=z$. Hence if $y\neq z$, there are no solutions of $\overline{C}W=-W\overline{C}$, $\overline{C}\in\mathfrak{M}_{2n}$.

We may now proceed to find a lower bound for the number of nonsimilar matrices in $\mathfrak{G}_0(x, y, z)$. We briefly denote the elements of $\mathfrak{G}_0$ by $A\dotplus B$, where

$$A = \begin{pmatrix} S_1 & 2R_1 \\ Q_1 & T_1 \end{pmatrix} \quad \text{and} \quad B = \begin{pmatrix} S_2 & Q_2 \\ 2R_2 & T_2 \end{pmatrix}.$$

If $A_1\dotplus B_1$ and $A_2\dotplus B_2$ are two distinct involutions in $\mathfrak{G}_0$, where either

$$A_1 \overset{S}{\neq} A_2 \quad \text{in} \quad M_{x+z} \quad \text{or} \quad B_1 \overset{S}{\neq} B_2 \quad \text{in} \quad M_{x+y},$$

then certainly

$$A_1 \dotplus B_1 \overset{S}{\neq} A_2 \dotplus B_2 \quad \text{in} \quad \mathfrak{G}_0.$$

Now let

$$A = I^{(a_1)} \dotplus (-I)^{(b_1)} \dotplus L \dotplus \cdots \dotplus L,$$
$$B = I^{(a_2)} \dotplus (-I)^{(b_2)} \dotplus L \dotplus \cdots \dotplus L,$$

where L occurs c_1 times in A and c_2 times in B; the various elements $A\dotplus B$ gotten by taking different sets of values of $(a_1, b_1, c_1, a_2, b_2, c_2)$, if they lie in $\mathfrak{G}_0$, are certainly nonsimilar in $\mathfrak{G}_0$, except that $A\dotplus B$ and $(-A)\dotplus(-B)$ are the same element of $\mathfrak{G}_0$. Hence the number N of nonsimilar involutions of $\mathfrak{G}_0$ is at least half of the number N_1 of solutions of

$$a_1 + b_1 + 2c_1 = x + z,$$
$$a_2 + b_2 + 2c_2 = x + y,$$

where if $x\neq 0$ we impose the restrictions that $c_1\leqq(z+1)/2$, $c_2\leqq(y+1)/2$, and that in B instead of L we use L'. (These conditions insure that $A\dotplus B\in\mathfrak{G}_0$.) As in the previous paper, one readily shows that $N>n(n+1)$ unless $J_1^\tau=\pm J_1$. We omit the details.

This leaves only the case where $y=z$. If $\overline{C}W=-W\overline{C}$, then $\overline{C}^kW=(-1)^kW\overline{C}^k$; therefore no odd power of $\overline{C}$ can be $\pm I$. Let p be a prime such that $n<p<2n$. Since $x+y=n$, certainly n is odd, and $p\geqq n+2$. Now $\mathfrak{G}_1$ (being isomorphic to $\mathfrak{M}_{2n-1}$) contains infinitely many elements of order p. However, $\mathfrak{G}_2$ contains only two such elements, since $\overline{C}^p\neq\pm I$ by the above argument, while if $C\in\mathfrak{G}_0$ and $C^p=\pm I$, then setting $C=A^{(n)}\dotplus B^{(n)}$ shows that $A^p=\pm I$ and $B^p=\pm I$. However, $A\in\mathfrak{M}_n$, and if $A^p=\pm I$, then the minimum function of A must divide $\lambda^p\mp 1$. But the degree of the minimum function is at most n, and therefore is less than $p-1$, whereas $\lambda^p\mp 1$ is the product of a linear factor $\lambda \mp 1$ and an irreducible factor of degree $p-1$; thence the minimum function of A is $\lambda \mp 1$, so $A = \pm I$. In the same way $B = \pm I$. Hence the only solutions are $C = I^{(n)} \dotplus I^{(n)}$ and $C = -I^{(n)} \dotplus I^{(n)}$. This completes the proof of the lemma. We remark that the use of the existence of the prime p could have been avoided, but the proof is much quicker this way.

4. **Proof of the main theorem.** We are now ready to prove Theorem 4 by induction on n. Hereafter, let $n \geqq 2$ and assume that Theorem 4 holds for $n-1$. Let τ be any automorphism of $\mathfrak{P}_{2n}$; then by Lemma 1, $J_1^\tau = \pm A J_1 A^{-1}$ for some $A \in \mathfrak{M}_{2n}$. If we change τ by a suitable inner automorphism, we may assume that $J_1^\tau = \pm J_1$.

Therefore, every $M \in \mathfrak{P}_{2n}$ which commutes with J_1 goes into another such element, that is,

$$\pm \begin{bmatrix} 1 & \mathfrak{n}' \\ \mathfrak{n} & X \end{bmatrix}^\tau = \pm \begin{bmatrix} 1 & \mathfrak{n}' \\ \mathfrak{n} & Y \end{bmatrix},$$

where $\mathfrak{n}$ denotes a column vector all of whose components are zero, and $X \in \mathfrak{M}_{2n-1}$. Thus, τ induces an automorphism on $\mathfrak{M}_{2n-1}$. Consequently (AUT, Theorem 4) there exists a matrix $A \in \mathfrak{M}_{2n-1}$ such that $Y = AX^*A^{-1}$ for all $X \in \mathfrak{M}_{2n-1}$, where either $X^* = X$ for all $X \in \mathfrak{M}_{2n-1}$ or $X^* = X'^{-1}$ for all $X \in \mathfrak{M}_{2n-1}$. After a further inner automorphism by a factor of $(1) \dotplus A^{-1}$, we may assume that $J_1^\tau = \pm J_1$ and also that $X^\tau = Y = X^*$ for all $X \in \mathfrak{M}_{2n-1}$.

Let J_ν be obtained from $I^{(2n)}$ by replacing the νth diagonal element by -1. Then

$$(J_1 J_{2n})^\tau = \pm \begin{bmatrix} 1 & 0 & \cdots & 0 & 0 \\ 0 & -1 & \cdots & 0 & 0 \\ \cdot & \cdot & \cdots & \cdot & \cdot \\ 0 & 0 & \cdots & -1 & 0 \\ 0 & 0 & \cdots & 0 & 1 \end{bmatrix}^\tau = \pm \begin{bmatrix} 1 & \mathfrak{n}' \\ \mathfrak{n} & \begin{bmatrix} -1 & \cdots & 0 & 0 \\ \cdot & \cdots & \cdot & \cdot \\ 0 & \cdots & -1 & 0 \\ 0 & \cdots & 0 & 1 \end{bmatrix}^* \end{bmatrix}$$

$$= \pm J_1 J_{2n},$$

so that $\pm J_{2n}$ is invariant. Similarly, all of the matrices $\pm J_\nu$ $(\nu = 1, \cdots, 2n)$ are invariant. Therefore for any $X \in \mathfrak{M}_{2n-1}$ we have

$$\pm \begin{pmatrix} 1 & \mathfrak{n}' \\ \mathfrak{n} & X \end{pmatrix}^\tau = \pm \begin{pmatrix} 1 & \mathfrak{n}' \\ \mathfrak{n} & A_1 X^* A_1^{-1} \end{pmatrix}, \cdots, \pm \begin{pmatrix} X & \mathfrak{n} \\ \mathfrak{n}' & 1 \end{pmatrix}^\tau = \pm \begin{pmatrix} A_{2n} X^* A_{2n}^{-1} & \mathfrak{n} \\ \mathfrak{n}' & 1 \end{pmatrix},$$

with $A_\nu \in \mathfrak{M}_{2n-1}$, and in fact $A_1 = I$.

Now suppose that $Z \in \mathfrak{M}_{2n-2}$, and consider $\pm(Z \dotplus I^{(2)})$; since it commutes with J_{2n-1} and J_{2n}, so does its image. But therefore

$$A_{2n}\begin{pmatrix} Z & \mathfrak{u} \\ \mathfrak{u}' & 1 \end{pmatrix} A_{2n}^{-1} = \begin{pmatrix} \bar{Z} & \mathfrak{u} \\ \mathfrak{u}' & 1 \end{pmatrix},$$

where $\bar{Z}$ denotes some matrix in $\mathfrak{M}_{2n-2}$. From this one easily deduces that A_{2n} must be of the form $B\dot{+}(1)$, with $B \in \mathfrak{M}_{2n-2}$. By considering the matrices commuting with J_ν and J_{2n} for $\nu=1, \cdots, 2n-2$ we see that A_{2n} must be diagonal. Furthermore, it is clear that all of the A_ν $(\nu=1, \cdots, 2n)$ must be diagonal, and all are sections of one diagonal matrix $D^{(2n)}$. Using the further inner automorphism factor D^{-1}, we find that $\pm X^\tau = \pm X^*$ for every decomposable matrix $\pm X \in \mathfrak{P}_{2n}$. Since $\mathfrak{P}_{2n}$ is generated by the set of its decomposable matrices, the theorem is proved.

TSING HUA UNIVERSITY,
PEKING, CHINA.
UNIVERSITY OF ILLINOIS,
URBANA, ILL.

FUNCTION THEORY

S. Kung and K. H. Look

In 1935 E. Cartan proved that there are precisely six types of irreducible homogeneous bounded symmetric domains. Two of these are exceptional in that they occur only in dimensions 16 and 27. The other four are the so-called "classical domains". These are defined by matrix inequalities such as

$$\{£ : £ \text{ is an } n \times n \text{ symmetric complex matrix such that } I - £\bar{£} \text{ is positive definite}\}.$$

This domain, and an unbounded equivalent form of it, were investigated extensively by Siegel in 1943 (see Vol. III of Siegel's "Topics in complex function theory," p. 120) in an attempt to study Abelian functions in several complex variables. At about the same time, in [66], Hua established many of the basic geometric results that underlie the theory of automorphic functions on the classical domains. Inevitably there is some overlap with Siegel's work. Here, as elsewhere in his mathematical writings, Hua's methods are distinguished by their direct style and concrete nature, and by a sure mastery of difficult computation. There is an excellent exposition of these and other results in his important monograph [3], which contains also material from the joint work [128, 129] with K. H. Look. Their papers on the Poisson and Bergman kernels still serve as models for the investigation of many questions in the theory of boundary behaviour of holomorphic functions (see Stein's monograph bearing this title). Monograph [3] remains in print as Volume 6 of the Translations series of the American Mathematical Society; in his foreword, the Editor stressed the importance of the subject matter for the representation theory of Lie groups, the theory of homogeneous spaces and the theory of automorphic functions of several complex variables; and he drew attention to the fact that most of the results in the book are due to Hua himself. A noteworthy feature of the book is the technical machinery developed by Hua (see Chapter I of [3]), e.g., a class of algebraic identities, or the computation of integrals of functions of a matrix argument, which were then, and still are, of independent interest.

[129] is now of historical interest, as a general theory does exist, J. J. Kohn's solution of the $\bar{\partial}$-Neumann problem. It is noteworthy that (p. 1035) Hua realized even then the importance of Hodge theory on open Hermitian manifolds.

In [122] Hua proved that the Riemannian curvature in the Bergman metric of a bounded domain is at most 2. (Incidentally, inequality (1.9) on the first page should read $R \leqslant 2$.) He showed here also that the holomorphy domains of constant curvature can be mapped onto the ball; general theorems of this kind continue to be of interest.

Hua was led by his work on functions of several complex variables to study some hard problems on partial differential equations. His researches, alone and jointly, are described in book [10] written in collaboration with his former students Z. Q. Hu and W. Lin.

Reprinted from the
AMERICAN JOURNAL OF MATHEMATICS
Vol. 66, pp. 470–488, 1944

ON THE THEORY OF AUTOMORPHIC FUNCTIONS OF A MATRIX VARIABLE I—GEOMETRICAL BASIS.*

(Dedicated to Professor K. L. Hiong, Chancellor of National Yunnan University, on his fiftieth birthday)

By Loo-Keng Hua.

The present paper is a revised form of another manuscript which the author had previously submitted for publication. The revision was necessary because the original manuscript contained some results (found independently by the author in some research begun in 1941) that have been recently published in Prof. C. L. Siegel's paper on Symplectic Geometry.[1] It is the aim of this paper to give a brief account of those results which are interfluent with Siegel's contributions. The remaining part of the author's research will be given later separately.

The paper is divided into two parts: the first part (**1-7**) is algebraic in nature and gives a very brief description of the main theory with which the author deals. In the second part (**8-10**) the author proves that the spaces which play the important rôles in the theory of analytic mappings have non-positive Riemannian curvature. Thus the geometries under consideration are sufficiently regular, and the development (in broad-line) of the theory of automorphic functions presents no serious difficulties.

The situation of the problem is well described by a statement due to Poincaré:

Note.—Because of the poor mail service between the U. S. and China, a number of minor changes in this paper have been made here, with the consent of the editors, by Prof. Hua's friend Dr. Hsio-Fu Tuan and Prof. C. L. Siegel.

* Received September 10, 1943.

[1] C. L. Siegel, "Symplectic Geometry," *American Journal of Mathematics*, vol. 65 (1943), pp. 1-86. Another important reference is:

C. L. Siegel, "Einführung in die Theorie der Modulfunktionen *n*-ten Grades, *Math. Annalen*, vol. 116 (1939), pp. 617-657.

The author is greatly indebted to Prof. H. Weyl for sending him a copy of Siegel's paper on Symplectic Geometry. The author would like also to express his thanks to Prof. P. C. Tang and Prof. S. S. Chern, for each sent to him one of the following two important references:

G. Giraud, *Leçons sur les fonctions automorphes*, Gauthier-Villars, Paris, 1920;

E. Cartan, "Sur les domaines bornés homogènes de l'espace de *n* variables complexes," *Hamb. Abh.*, vol. 11 (1935), pp. 116-162.

"La géomètrie non-euclidienne est la clef véritable du probleme qui nous occupe," *Acta. Math.*, vol. 39 (1923), p. 100.

1. Groups. Throughout the paper, capital Latin letters denote $n \times n$ matrices with complex elements unless the contrary is stated. A' denotes the transposed matrix of A and $\bar{A}$ denotes the conjugate complex matrix of A. I denotes the unit matrix and O denotes the zero matrix.

We use the notations

$$\mathfrak{F} = \begin{pmatrix} O & I \\ -I & O \end{pmatrix}, \qquad \mathfrak{F}_1 = \begin{pmatrix} O & I \\ I & O \end{pmatrix}$$

and

$$\mathfrak{J} = \begin{pmatrix} I & O \\ O & I \end{pmatrix}$$

Let

$$\mathfrak{T} = \begin{pmatrix} A & B \\ C & D \end{pmatrix}.$$

We shall consider three types of matrices $\mathfrak{T}$:

(i) Those $\mathfrak{T}$ satisfying

$$\mathfrak{T}\mathfrak{F}\mathfrak{T}' = \mathfrak{F}$$

are called *symplectic*. The condition may be written as

$$AB' = BA', \qquad CD' = DC', \qquad AD' - BC' = I;$$

(ii) Those $\mathfrak{T}$ satisfying

$$\mathfrak{T}\mathfrak{F}_1\mathfrak{T}' = \mathfrak{F}_1$$

are called *orthogonal*, or more definitely, $\mathfrak{F}_1$-*orthogonal*. The condition may be written as

$$AB' = -BA', \qquad CD' = -DC', \qquad AD' + BC' = I;$$

(iii) Those $\mathfrak{T}$ satisfying

$$\bar{\mathfrak{T}}\mathfrak{F}\mathfrak{T}' = \mathfrak{F}$$

are called *conjunctive-symplectic*. The condition may be written as

$$\bar{A}B' = \bar{B}A', \qquad \bar{C}D' = \bar{D}C', \qquad \bar{A}D' - \bar{B}C' = I.$$

Remark. Apparently we have a fourth type of matrices $\mathfrak{T}$ satisfying

$$\bar{\mathfrak{T}}\mathfrak{F}_1\mathfrak{T}' = \mathfrak{F}_1.$$

Since $\mathfrak{F}$ and $\mathfrak{F}_1$ are conjunctive the fourth type coincides with (iii).

THEOREM 1. *Each type of matrices form a group with respect to multiplication.*

Definition. The group formed by symplectic matrices is called the *symplectic group.* Similarly we define the $\mathfrak{F}_1$-orthogonal group and the *conjunctive-symplectic group.*

2. Spaces analogous to the projective space. (A detailed study will be given elsewhere later.)

DEFINITION 1. *A pair of matrices* (Z_1, Z_2) *is said to be symmetric or skew-symmetric if we have*

$$(Z_1, Z_2)\mathfrak{F}\,(Z_1, Z_2)' = 0$$

or

$$(Z_1, Z_2)\mathfrak{F}_1(Z_1, Z_2)' = 0$$

respectively.

(Certainly we might define Hermitian pairs, but this would be of no interest in the study of automorphic functions.)

DEFINITION 2. *A pair of matrices* (Z_1, Z_2) *is said to be non-singular, if the rank of the* $n \times 2n$ *matrix* (Z_1, Z_2) *is equal to* n.

DEFINITION 3. *A symplectic transformation is defined by*

$$(W_1, W_2) = Q(Z_1, Z_2)\mathfrak{T}$$

where Q *is non-singular and* $\mathfrak{T}$ *is symplectic. Similarly, we define* $\mathfrak{F}_1$*-orthogonal and conjunctive-symplectic transformations.*

THEOREM 2. *A symplectic transformation carries a non-singular symmetric pair of matrices into a non-singular symmetric pair. An* $\mathfrak{F}_1$*-orthogonal transformation carries a non-singular skew-symmetric pair of matrices into a non-singular skew-symmetric pair.*

Proof (for the symplectic case).

$$\begin{aligned}(W_1, W_2)\mathfrak{F}(W_1, W_2)' &= Q(Z_1, Z_2)\mathfrak{T}\mathfrak{F}\mathfrak{T}'(Z_1, Z_2)'Q' \\ &= Q(Z_1, Z_2)\mathfrak{F}(Z_1, Z_2)'Q' = 0.\end{aligned}$$

Thus we may take a non-singular symmetric pair (Z_1, Z_2)[2] as a point of the space and the symplectic group as the group of motions of the space. Then

[2] Identify (Z_1, Z_2) with (QZ_1, QZ_2) for any non-singular Q.

we obtain a geometry analogous to the projective geometry. A similar consideration holds for skew-symmetric pairs. A detailed treatment of independent interest will be given elsewhere.

We can also take a non-singular pair (Z_1, Z_2) as a point of our space and the conjunctive-symplectic group as the group of motions of the space. Then we also obtain a type of geometry.

We now indicate a general treatment which will be described for the symplectic case only.

Let $\mathfrak{H}$ be a Hermitian matrix. We define the symmetric pairs making

$$(\overline{Z_1, Z_2})\mathfrak{H}(Z_1, Z_2)'$$

positive definite to be a space.[3] The group of motions of the space is the subgroup of the symplectic group leaving $\mathfrak{H}$ invariant. Thus we establish a geometry analogous to non-euclidean geometry.

Thus the symplectic classification of Hermitian matrices is of the first importance. After the classification and the study of the structure of the group of automorphisms we arrive at the conclusion that there are three types of geometries of fundamental importance, namely those with

$$\mathfrak{H} = \begin{pmatrix} H & 0 \\ 0 & H \end{pmatrix}, \tag{1}$$

where H is a diagonal matrix $[1, \cdots, 1, -1, \cdots, -1]$; those with

$$\mathfrak{H} = \begin{pmatrix} H & 0 \\ 0 & 0 \end{pmatrix}; \tag{2}$$

and those with

$$\mathfrak{H} = \begin{pmatrix} I & 0 \\ 0 & -I \end{pmatrix}. \tag{3}$$

Correspondingly, we define the related geometries to be elliptic with signature H, parabolic with signature H and hyperbolic, respectively.

In this paper, we consider only the hyperbolic geometry. More definitely, all non-singular pairs (Z_1, Z_2) of matrices, making

$$(\overline{Z_1, Z_2}) \begin{pmatrix} I & 0 \\ 0 & -I \end{pmatrix} (Z_1, Z_2)'$$

positive definite, form a hyperbolic space. The symplectic transformations with matrix $\mathfrak{T}$ satisfying

[3] It is defined to be a hypercircle in the "projective" space.

$$\bar{\mathfrak{T}} \begin{pmatrix} I & O \\ O & -I \end{pmatrix} \mathfrak{T}' = \rho \begin{pmatrix} I & O \\ O & -I \end{pmatrix}, \qquad \rho = \pm 1$$

form the group of motions of the space.

Analogously, the non-singular skew-symmetric pairs (Z_1, Z_2) of matrices making

$$(\bar{Z}_1, \bar{Z}_2) \begin{pmatrix} I & O \\ O & -I \end{pmatrix} (Z_1, Z_2)'$$

positive definite form a space. The $\mathfrak{F}_1$-orthogonal transformations $\mathfrak{T}$ satisfying

$$\bar{\mathfrak{T}} \begin{pmatrix} I & O \\ O & -I \end{pmatrix} \mathfrak{T}' = \rho \begin{pmatrix} I & O \\ O & -I \end{pmatrix}$$

form the group of motions of the space.

There is a distinction between the symplectic and the conjunctive-symplectic cases, since the introduction of a "hypercircle" is not necessary in the latter case. The non-singular pairs of matrices (Z_1, Z_2) making

$$(\overline{Z_1, Z_2}) \begin{pmatrix} O & I \\ -I & O \end{pmatrix} (Z_1, Z_2)'$$

positive definite form the space, and the conjunctive-symplectic group is the group of motions of the space.

Notice that the transformation

$$(W_1, W_2) = Q(Z_1, Z_2) \begin{pmatrix} iI/\sqrt{2} & iI/\sqrt{2} \\ -iI/\sqrt{2} & iI/\sqrt{2} \end{pmatrix}$$

carries the space of points (Z_1, Z_2) such that

$$\text{"}(\overline{Z_1, Z_2}) \begin{pmatrix} I & O \\ O & -I \end{pmatrix} (Z_1, Z_2)' \text{ is positive definite"}$$

into the space of points (W_1, W_2) such that

$$\text{"}(\overline{W_1, W_2}) \begin{pmatrix} O & I \\ I & O \end{pmatrix} (W_1, W_2)' \text{ is positive definite."}$$

3. An extension of the conjunctive-symplectic group. We now consider the conjunctive-symplectic case with

$$\mathfrak{H} = \begin{pmatrix} I & O \\ O & -I \end{pmatrix}.$$

It is clear that there is no essential reason to restrict $\mathfrak{H}$ to be a Hermitian matrix with signature (n, n), except to have an analogy with the symplectic case. Thus we may extend the conjunctive-symplectic case much further. The procedure is as follows:

Let

$$\mathfrak{H}=\begin{pmatrix} I^{(n)} & 0 \\ 0 & -I^{(m)} \end{pmatrix}.$$

The points of the space are then given by matrices

$$(Z_1{}^{(n)}, \qquad Z_2{}^{(n,m)})\ [4]$$

such that

$$\overline{(Z_1{}^{(n)}, Z_2{}^{(n,m)})}\,\mathfrak{H}\,(Z_1{}^{(n)}, Z_2{}^{(n,m)})'$$

is positive definite. The group of motions consists of the transformations

$$(W_1{}^{(n)}, W_2{}^{(n,m)}) = Q(Z_1{}^{(n)}, Z_2{}^{(n,m)})\mathfrak{T}^{(n+m)}$$

where

$$\bar{\mathfrak{T}}^{(n+m)}\begin{pmatrix} I^{(n)} & 0 \\ 0 & -I^{(m)} \end{pmatrix}\mathfrak{T}' = \begin{pmatrix} I^{(n)} & 0 \\ 0 & -I^{(m)} \end{pmatrix}.$$

Writing

$$\mathfrak{T}=\begin{pmatrix} A^{(n)} & B^{(n,m)} \\ C^{(m,n)} & D^{(m)} \end{pmatrix},$$

we have the conditions:

$$\bar{A}A' - \bar{B}B' = I, \qquad \bar{A}C' = \bar{B}D', \qquad \bar{C}C' - \bar{D}D' = I.$$

The group so obtained is called the conjunctive group of signature (n, m).

4. Non-homogeneous coördinates. Let (W_1, W_2) and (Z_1, Z_2) be two symmetric (or skew) pairs connected by

$$(W_1, W_2) = Q(Z_1, Z_2)\mathfrak{T}.$$

If W_1 and Z_1 are both non-singular, let

$$W = -W_1^{-1}W_2, \qquad Z = -Z_1^{-1}Z_2;$$

then W and Z are symmetric (or skew) matrices connected by

$$W = (-A + ZC)^{-1}(B - ZD),$$

[4] $Z^{(n)}$ denotes an $n \times n$ matrix; $Z^{(n,m)}$ denotes an $n \times m$ matrix.

i. e.,

$$Z = (AW + B)(CW + D)^{-1}.$$

Thus a non-singular symmetric pair of matrices (W_1, W_2) may be considered as the homogeneous coördinates of a symmetric (or skew) matrix W.

Now in non-homogeneous coördinates, the geometries take the following forms:

(i) The space is formed by the symmetric matrices Z satisfying

$$I - Z\bar{Z} > 0.$$[5]

The group of motions is given by

$$W = (AZ + B)(CZ + D)^{-1}$$

and

$$\mathfrak{T} = \begin{pmatrix} A & B \\ C & D \end{pmatrix}$$

is symplectic satisfying

$$\bar{\mathfrak{T}} \begin{pmatrix} I & 0 \\ 0 & -I \end{pmatrix} \mathfrak{T}' = \begin{pmatrix} I & 0 \\ 0 & -I \end{pmatrix}.$$

(ii) The space is formed by the skew-symmetric matrices Z satisfying

$$I + Z\bar{Z} > 0.$$

The group of motions is given by

$$W = (AZ + B)(CZ + D)^{-1}$$

where $\mathfrak{T}$ is $\mathfrak{F}$-orthogonal leaving $I + Z\bar{Z} > 0$ invariant.

(iii) The space is formed by the $n \times m$ matrices Z satisfying

$$I - Z\bar{Z}' > 0.$$

The group of motions is the conjunctive group of signature (n, m).

As we transform $\begin{pmatrix} I & 0 \\ 0 & -I \end{pmatrix}$ into $\begin{pmatrix} 0 & I \\ I & 0 \end{pmatrix}$, we find that the symplectic and the conjunctive-symplectic cases have also the following equivalent expressions.

For the symplectic case, the space is formed by the symmetric matrices

$$Z = X + iY$$

[5] "> 0" means "being positive definite."

with Y positive definite, and the group of motions can be easily verified to be the real symplectic group.

For the conjunctive-symplectic case, we define

$$\frac{Z'+\bar{Z}}{2}, \qquad \frac{Z'-\bar{Z}}{2i}$$

to be the virtual real and imaginary parts of Z. The space is formed by the matrices Z with positive definite virtual imaginary parts. The group of motions is the conjunctive-symplectic group. Both correspond to the Poincaré half-plane.

Remark. After laying the foundation of the present theory, the author found in Cartan's paper that there are four general types (and two special types) of bounded symmetric spaces for analytic mappings. They are the previous types (i), (ii) and (iii), and a type studied by G. Giraud (in 1920) with the hyperabelian group. Thus the present treatment may be considered as complete in a certain sense.

5. Metrization of the space. Let $\mathfrak{H}$ be a Hermitian matrix

$$\mathfrak{H} = \begin{pmatrix} H_1 & \bar{L}' \\ L & H_2 \end{pmatrix}.$$

Let $\mathfrak{T} = \begin{pmatrix} A & B \\ C & D \end{pmatrix}$ be a symplectic matrix satisfying

$$\bar{\mathfrak{T}}\mathfrak{H}\mathfrak{T}' = \mathfrak{H}.$$

and set

$$H(Z) = \bar{Z}H_2Z + \bar{Z}L + \bar{L}'Z + H_1; \quad Z = (AW - B)(-CW + D)^{-1}.$$

Then

$$W = (D'Z + B')(C'Z + A')^{-1} = (ZC + A)^{-1}(ZD + B).$$

Therefore

$$\begin{aligned} H(W) &= \overline{(ZC+A)}^{-1}\{(\bar{Z}\bar{D} + \bar{B})H_2(D'Z + B') \\ &\quad + (\bar{Z}\bar{D} + \bar{B})L(C'Z + A') + (\bar{Z}\bar{C} + \bar{A})\bar{L}'(D'Z + B') \\ &\quad + (\bar{Z}\bar{C} + \bar{A})H_1(C'Z + A')\}(C'Z + A')^{-1} \\ &= (\overline{ZC + A})^{-1}H(Z)(C'Z + A')^{-1}. \end{aligned}$$

Further

$$\begin{aligned} dW &= (ZC + A)^{-1}dZD - (ZC + A)^{-1}dZC\,(ZC + A)^{-1}(ZD + B) \\ &= (ZC + A)^{-1}\,dZ(ZC + A)'^{-1}. \end{aligned}$$

Therefore we have

THEOREM 3. *The characteristic equation of the matrix*

$$dZ(H(Z))^{-1}d\bar{Z}(\overline{H(Z)})^{-1}$$

is invariant under the group of automorphisms of the hypercircle $H(Z)$. *In particular, we have*

$$\sigma((H(Z))^{-1}d\bar{Z}(\overline{H(Z)})^{-1}dZ)$$

as an invariant quadratic differential form under the group, where $\sigma(X)$ *denotes the trace of the matrix* X.

For the case corresponding to the Poincaré half-plane, we have that the quadratic differential form

$$\sigma(Y^{-1}dZY^{-1}d\bar{Z})$$

is invariant under all real symplectic transformations.

A similar result holds for the $\mathfrak{F}$-orthogonal case.

THEOREM 4. *For the conjunctive group with signature* (n, m), *we have the invariant quadratic differential form*

$$\sigma((I - Z\bar{Z}')^{-1}dZ(I - \bar{Z}'Z)^{-1}d\bar{Z}'),$$

where

$$Z = Z^{(n,m)}.$$

Proof. Let

$$\mathfrak{T}^{(m+n)} = \begin{pmatrix} A^{(n,n)}, & B^{(n,m)} \\ C^{(m,n)}, & D^{(m,n)} \end{pmatrix}$$

be a matrix satisfying the condition

$$\bar{\mathfrak{T}}\begin{pmatrix} I^{(n)} & 0 \\ 0 & -I^{(m)} \end{pmatrix}\mathfrak{T}' = \begin{pmatrix} I^{(n)} & 0 \\ 0 & -I^{(m)} \end{pmatrix}.$$

This condition may also be written as

$$\begin{pmatrix} \bar{A} & \bar{B} \\ \bar{C} & \bar{D} \end{pmatrix}\begin{pmatrix} A' & -C' \\ -B' & D' \end{pmatrix} = I^{(m+n)}.$$

Consequently

$$\begin{pmatrix} A' & -C' \\ -B' & D' \end{pmatrix}\begin{pmatrix} \bar{A} & \bar{B} \\ \bar{C} & \bar{D} \end{pmatrix} = I^{(m+n)};$$

i. e.,

$$A'\bar{A} - C'\bar{C} = I, \qquad A'\bar{B} = C'\bar{D}, \qquad -B'\bar{B} + D'\bar{D} = I.$$

On setting

$$W^{(n,m)} = (AZ^{(n,m)} + B)(CZ + D)^{-1},$$

we have

$$W = (Z\bar{B}' + \bar{A}')^{-1}(Z\bar{D}' + \bar{C}').$$

Then

$$\begin{aligned} I - W\bar{W}' &= I - (Z\bar{B}' + \bar{A}')^{-1}(Z\bar{D}' + \bar{C}')(D\bar{Z}' + C)(B\bar{Z}' + A)^{-1} \\ &= (Z\bar{B}' + \bar{A}')^{-1}((Z\bar{B}' + \bar{A}')(B\bar{Z}' + A) - (Z\bar{D}' + \bar{C}')(D\bar{Z}' + C))(B\bar{Z}' + A)^{-1} \\ &= (Z\bar{B}' + \bar{A}')^{-1}(I - Z\bar{Z}')(B\bar{Z}' + A)^{-1}. \end{aligned}$$

Furthermore

$$\begin{aligned} I - \bar{W}'W &= I - (\overline{CZ + D})'^{-1}(\overline{AZ + B})'(AZ + B)(CZ + D)^{-1} \\ &= (\overline{CZ + D})'^{-1}(I - \bar{Z}'Z)(CZ + D)^{-1}. \end{aligned}$$

Finally, we have

$$\begin{aligned} dW &= AdZ(CZ + D)^{-1} - (AZ + B)(CZ + D)^{-1}CdZ(CZ + D)^{-1} \\ &= (A - (Z\bar{B}' + \bar{A}')^{-1}(Z\bar{D}' + C')C)dZ(CZ + D)^{-1} \\ &= (Z\bar{B}' + \bar{A}')^{-1}dZ(CZ + D)^{-1}. \end{aligned}$$

Combining all these results, we have

$$\begin{aligned} &(I - W\bar{W}')^{-1}dW(I - \bar{W}'W)^{-1}d\bar{W}' \\ &\quad = (B\bar{Z}' + A)(I - Z\bar{Z}')^{-1}dZ(I - \bar{Z}'Z)^{-1}d\bar{Z}'(B\bar{Z}' + A)^{-1}. \end{aligned}$$

This completes the proof.

6. Unitary equivalence.

LEMMA. *Given a unitary matrix U, there exists a matrix V such that*

(a) $$V^2 = U, \ V\text{: unitary},$$

and

(b) $$\text{if } \ U'A = AU, \ \text{ then } \ V'A = AV.$$

Proof. There exists a unitary matrix Γ such that

$$\Gamma^{-1}U\Gamma = D,$$

where D is a diagonal matrix $[e^{i\theta_1}, \cdots, e^{i\theta_n}]$ and $0 \leq \theta_\nu < 2\pi$. Let

$$V = \Gamma D_{\frac{1}{2}}\Gamma^{-1},$$

where $D_{\frac{1}{2}} = [e^{\frac{1}{2}i\theta_1}, \cdots, e^{\frac{1}{2}i\theta_n}]$. Now we shall verify that V possesses the required properties. (a) is evident. For (b): If $U'A = AU$, then

$$\Gamma'^{-1}D\Gamma'A = A\Gamma D\Gamma^{-1},$$

i. e.,

$$D\Gamma'A\Gamma = \Gamma'A\Gamma D.$$

Thus D is commutative with $\Gamma'A\Gamma$; then so is $D_{\frac{1}{2}}$, i. e.,

$$D_{\frac{1}{2}}\Gamma'A\Gamma = \Gamma'A\Gamma D_{\frac{1}{2}}.$$

Consequently

$$V'A = AV.$$

THEOREM 5. *Let Z be a non-singular symmetric matrix with complex elements; then there exists a unitary matrix U such that*

$$UZU' = [\mu_1, \cdots, \mu_n]$$

where $\mu_1, \cdots \mu_n$ are the positive square roots of the characteristic roots of $Z\bar{Z}$.

Proof. Since $Z\bar{Z}$ is a positive definite Hermitian matrix, we have a unitary matrix V such that

$$VZ\bar{Z}\bar{V}' = [\mu_1{}^2, \cdots, \mu_n{}^2], \qquad \mu_\nu > 0;$$

i. e.,

$$(1) \qquad W\bar{W} = [\mu_1{}^2, \cdots, \mu_n{}^2],$$

where $W = VZV'$. Evidently, $W_0 = [\mu_1, \cdots, \mu_n]$ is a solution of (1).

Now

$$W\bar{W} = Wq_0\bar{W}_0, \quad \text{i. e.,} \quad (W_0{}^{-1}W)(\overline{W_0{}^{-1}W})' = I,$$

i. e., $W_0{}^{-1}W$ is unitary, say U_0; then

$$W = W_0U_0.$$

Since W and W_0 are both symmetric, we have

$$W_0U_0 = U'_0W_0.$$

By the lemma, we have a unitary matrix U such that $U^2 = U_0$ and $W_0U = U'W_0$. Then

$$W = W_0 U_0 = W_0 U^2 = U' W_0 U.$$

Thus

$$VZV' = U' W_0 U,$$

and we have proved the theorem.

THEOREM 6. *Let Z be a non-singular skew-symmetric matrix, then $Z\bar{Z}$ is a negative definite Hermitian matrix and its characteristic polynomial is a perfect square.*

Proof. Evidently, we have

$$Z\bar{Z} = -Z\bar{Z}',$$

hence $Z\bar{Z}$ is negative definite. Further

$$| Z\bar{Z} - \lambda I | = | \bar{Z} | \; | Z - \lambda \bar{Z}^{-1} |.$$

Since $Z - \lambda \bar{Z}^{-1}$ is a skew-symmetric matrix, its determinant is a perfect square.

THEOREM 7. *Let Z be a non-singular skew-symmetric matrix; then we have a unitary matrix U such that*

$$UZU' = \begin{pmatrix} 0 & d_1 \\ -d_1 & 0 \end{pmatrix} \dot{+} \cdots \dot{+} \begin{pmatrix} 0 & d_{n/2} \\ -d_{n/2} & 0 \end{pmatrix},$$

where $d_1^2, d_1^2, \cdots, d_{n/2}^2, d_{n/2}^2$ are the characteristic roots of $-Z\bar{Z}$.

Proof. By Theorem 6, we have a unitary matrix V such that

$$VZ\bar{Z}'\bar{V}' = [d_1^2, d_1^2, \cdots, d_{n/2}^2, d_{n/2}^2].$$

Let $VZV' = W$. Clearly

$$W_0 = \begin{pmatrix} 0 & d_1 \\ -d_1 & 0 \end{pmatrix} \dot{+} \cdots \dot{+} \begin{pmatrix} 0 & d_{n/2} \\ -d_{n/2} & 0 \end{pmatrix}$$

satisfies

$$W\bar{W}' = W_0 \bar{W}'_0.$$

The remaining part of the proof is the same as that of Theorem 5.

Remark. Both Theorems 5 and 7 may be extended to the singular case without any essential difficulty.

7. Existence and uniqueness of the geodesic passing through two given points.

THEOREM 8. *Passing through any two points of the hyperbolic space with the symplectic group, there is one and only one geodesic.*

Before proceeding to prove this fundamental theorem, we shall require a theorem concerning the equivalence of point-pairs.

THEOREM 9. *In the hyperbolic space with the symplectic group (in Poincaré's representation) any two points are equivalent to the two points* iI *and* iD, *where* D *is a diagonal matrix*

$$[d_1, \cdots, d_n], \qquad d_\nu > 0.$$

Proof. 1) The group is evidently transitive. Thus we can let one of the two points be iI.

2) The transformations leaving iI fixed are all of the form

$$Z_1 = (AZ + B)(-BZ + A)^{-1},$$

i. e.,

$$\frac{Z_1 - iI}{Z_1 + iI} = (A + Bi)\left(\frac{Z - iI}{Z + iI}\right)(A' + B'i).$$

Since

$$I = AA' + BB' = (A + Bi)(A - Bi)',$$

we have $A + Bi$ unitary, and then the theorem follows from Theorem 5.

Proof of Theorem 8. 1) Without loss of generality, we may assume that the two points are

$$Ii \quad \text{and} \quad Di, \qquad D = [d_1, \cdots, d_n], \qquad d_\nu > 0.$$

Both points belong to the subspace $X = O$. Since

$$\begin{aligned}\sigma(Y^{-1}dZY^{-1}dZ) &= \sigma(Y^{-1}dXY^{-1}dX) + \sigma(Y^{-1}dY\,Y^{-1}dY)\\ &\geq \sigma(Y^{-1}dY\,Y^{-1}dY).\end{aligned}$$

and the equality holds only for $dX = O$, the geodesics connecting the two points all lie in the subspace $X = O$.

2) Further let

$$Y = M'\Lambda M$$

where

$$M=(\gamma_{ij}),\qquad (\gamma_{ij}=0 \quad \text{for} \quad i>j,\ \ \gamma_{ii}=1),$$

and

$$\Lambda=[q_1,\cdots,q_n]$$

are obtained according to Jacobi's reduction of positive definite quadratic forms.

(Notice that the space with positive definite Y is mapped topologically into the space with $q_\nu>O$ in the new variables (q,γ)).

Then we have

$$\dot Y=\dot M'\Lambda M+M'\dot\Lambda M+M'\Lambda\dot M,$$

$$Y^{-1}\dot Y=M^{-1}\Lambda^{-1}\dot\Lambda M+M^{-1}\Lambda^{-1}M'^{-1}\dot M'\Lambda M+M^{-1}\dot M,$$

and

$$\sigma((Y^{-1}\dot Y)^2)=\sigma((\Lambda^{-1}\dot\Lambda)^2)+2\sigma((\dot M^{-1}\dot M)^2)$$
$$+4\sigma(M^{-1}\Lambda^{-1}\dot\Lambda\dot M)+2\sigma(M^{-1}\Lambda^{-1}M'^{-1}\dot M'\Lambda M),$$

since

$$\sigma(AB)=\sigma(BA)\quad\text{and}\quad\sigma(A')=\sigma(A).$$

Further, let

$$M^{-1}=(n_{hk}),\qquad n_{hk}=0\quad\text{for}\quad h>k,$$

then

$$\sigma((M^{-1}\dot M)^2)=\sum_{i,j,k,l}n_{ij}\dot m_{jk}n_{kl}\dot m_{li}=\sum_{i\leq j\leq k\leq l\leq i}n_{ij}\dot m_{jk}n_{kl}\dot m_{li}=\sum(n_{ii}\dot m_{ii})^2=0,$$

since $\dot m_{ii}=0$; and

$$\sigma(M^{-1}\Lambda^{-1}\dot\Lambda\dot M)=\sum n_{ij}q_j^{-1}\dot q_j\dot m_{ji}=\sum_{i=j}n_{ij}q_j^{-1}\dot q_j\dot m_{ji}=0.$$

Thus

$$\sigma(Y^{-1}\dot YY^{-1}\dot Y)=\sigma((\Lambda^{-1}\dot\Lambda)(\Lambda^{-1}\dot\Lambda))+2\sigma((\Lambda^{\frac12}\dot MM^{-1}\Lambda^{-\frac12})(\Lambda^{\frac12}\dot MM^{-1}\Lambda^{-\frac12})'$$
$$\geq\sigma((\Lambda^{-1}\dot\Lambda)(\Lambda^{-1}\dot\Lambda)'),$$

and the equality holds for $\dot M=O$. Therefore the geodesics connecting the two points lie in the subspace with $\dot M=O$, i.e., the subspace of real and diagonal Z.

The subspace of real and diagonal Z is Euclidean, since

$$\sigma((\Lambda^{-1}d\Lambda)(\Lambda^{-1}d\Lambda)'=\sum_{i=1}^{n}(d\log q_i)^2,$$

and we have proved the theorem.

Evidently, we may deduce

THEOREM 10. *All geodesics are symplectic images of the curves*

$$Z = i[\lambda_1^s, \cdots, \lambda_n^s], \qquad \lambda_\nu > 0$$

and $\sum_{\nu=1}^{n} \log^2 \lambda_\nu = 1$.

THEOREM 11. *The equations of the geodesics of the space are given by*

$$d^2Z/ds^2 + i(dZ/ds)Y^{-1}dZ/ds = O.$$

The previous results give a sufficient indication of the algebraic treatment. We shall now give a general treatment which seems to be a "direct" attack.

8. A general type of Riemannian geometry. Let

$$S = S^{(n,m)} = (s_{ij}), \qquad T = T^{(m,n)} = (t_{ij}).$$

We consider the geometry with the Riemannian metric

$$\sigma((I^{(n)} - ST)^{-1}dS(I^{(m)} - TS)^{-1}dT).$$

Let

$$\Lambda_{ij} = (\partial/\partial s_{ij})S.$$

LEMMA. *If*

$$\sigma(\Lambda_{ij}S') = 0$$

for all i and j, then $S = O^{(n,m)}$.

The lemma is evident.

THEOREM 12. *The equations of the geodesics of the space are given by*

$$d^2I/ds^2 + 2(dT/ds)(I - ST)^{-1}(S - STS)(I - TS)^{-1}(dT/ds) = O,$$
$$d^2S/ds^2 + 2(dS/ds)(I - TS)^{-1}(T - TST)(I - ST)^{-1}(dS/ds) = O,$$

where I denotes, under evident circumstances, either $I^{(n)}$ *or* $I^{(m)}$.

Proof. We have

$$\begin{aligned}\partial\sigma/\partial s_{ij} &= \sigma\{(I - ST)^{-1}\Lambda_{ij}T(I - ST)^{-1}(dS/ds)(I - TS)^{-1}(dT/ds)\\ &\quad + (I - ST)^{-1}(dS/ds)(I - TS)^{-1}T\Lambda_{ij}(I - TS)^{-1}(dT/ds)\}\\ &= \sigma(\Lambda_{ij}[T(I - ST)^{-1}(dS/ds)(I - TS)^{-1}(dT/ds)(I - ST)^{-1}\\ &\quad + (I - TS)^{-1}(dT/ds)(I - ST)^{-1}(dS/ds)(I - TS)^{-1}T]).\end{aligned}$$

Next

$$(d/ds)(\partial\sigma/\partial s_{ij}) = (d/as)(\sigma(\Lambda_{ij}(I-TS)^{-1}(dT/ds)(I-ST)^{-1}$$
$$= \sigma(\Lambda_{ij}[(I-TS)^{-1}(TdS/ds + (dT/ds)S)(I-TS)^{-1}(dT/ds)(I-ST)^{-1}$$
$$+ (I-TS)^{-1}(dT/ds)(I-ST)^{-1}(SdT/ds + (dS/ds)T)(I-ST)^{-1}$$
$$+ (I-TS)^{-1}(d^2T/ds^2)(I-ST)^{-1}]).$$

Thus

$$(d/ds)(\partial\sigma/\partial s_{ij}) - \partial\sigma/\partial s_{ij} = \sigma(\Lambda_{ij}(I-TS)^{-1}M(I-ST)^{-1},$$

where

$$M = d^2T/ds^2 + (dT/ds)(I-ST)^{-1}SdT/ds + (dT/ds)S(I-TS)^{-1}(dT/ds),$$

since

$$T(I-ST)^{-1} = (I-TS)^{-1}T, \text{ etc.}$$

The equations of the geodesics are

$$(d/ds)(\partial\sigma/\partial s_{ij}) - \partial\sigma/\partial s_{ij} = 0$$

for all i and j. We have, then,

$$d^2T/ds^2 + 2(dT/ds)(I-ST)^{-1}(S-STS)(I-TS)^{-1}(dT/ds) = 0,$$

since

$$(I-ST)^{-1}S + S(I-TS)^{-1}$$
$$=(I-ST)^{-1}(S(I-TS) + (I-ST)S)(I-TS)^{-1}$$
$$= 2(I-ST)^{-1}(S-STS)(I-TS)^{-1}.$$

Interchanging S and T, we have the other differential matrix-equation.

THEOREM 13. *The Riemannian curvature tensor of the space is given by*

$$\sigma(K(U,Y)K(U,Y) - K(U,Y)K(X,V)$$
$$- K(X,Y)K(U,V) + K(X,V)K(X,V))$$

where (U,V) *and* (X,Y) *are two directions and*

$$K(X,Y) = (I-ST)^{-1}X(I-TS)^{-1}Y.$$

Proof. (The method is borrowed from Siegel's paper). We write

$$R(S,T) = (I-TS)^{-1}T = T(I-ST)^{-1}$$
$$= (I-TS)^{-1}(T-TST)(I-ST)^{-1}.$$

Then

$$
\begin{aligned}
dR(S,T) &= (I-TS)^{-1}dT + (I-TS)^{-1}(TdS+dTS)(I-TS)^{-1}T \\
&= (I-TS)^{-1}TdS(I-TS)^{-1}T + (I-TS)^{-1}dT(I+S(I-TS)^{-1}T) \\
(1) \qquad &= (I-TS)^{-1}(TdST + dT)(I-ST)^{-1} \\
&= R(S,T)dSR(S,T) + (I-TS)^{-1}dT(I-ST)^{-1}.
\end{aligned}
$$

Now we define two covariant differentials $(\delta_1 U, \delta_1 V)$, $(\delta_2 U, \delta_2 V)$ by

$$\delta_i U = -UR(S,T)\delta_i S - \delta_i SR(S,T)U, \qquad i = 1, 2,$$

and $\delta_i V$ is defined similarly by interchanging S and T formally. Then

$$
\begin{aligned}
\delta_1\delta_2 U &= -\delta_1 UR(S,T)\delta_2 S - U\delta_1 R(S,T)\delta_2 S \\
&\quad - \delta_2 S\delta_1 R(S,T)U - \delta_2 SR(S,T)\delta_1 U \\
&= (UR(S,T)\delta_1 S + \delta_1 SR(S,T)U)R(S,T)\delta_2 S \\
&\quad - U(R(S,T)\delta_1 SR(S,T) + (I-TS)^{-1}\delta_1 T(I-ST)^{-1})\delta_2 S \\
&\quad - \delta_2 S(R(S,T)\delta_1 SR(S,T) + (I-TS)^{-1}\delta_1 T(I-ST)^{-1})U \\
&\quad + \delta_2 SR(S,T)(UR(S,T)\delta_1 S + \delta_1 SR(S,T)U) \\
&= \delta_1 SR(S,T)UR(S,T)\delta_2 S + \delta_2 SR(S,T)UR(S,T)\delta_1 S \\
&\quad - U(I-TS)^{-1}\delta_1 T(I-ST)^{-1}\delta_2 S \\
&\quad - \delta_2 S(I-TS)^{-1}\delta_1 T(I-ST)^{-1}U.
\end{aligned}
$$

We have, then,

$$
\begin{aligned}
U^* = (\delta_1\delta_2 - \delta_2\delta_1)U &= U(P(\delta_1 S, \delta_2 T) - P(\delta_2 S, \delta_1 T)) \\
&\quad + (Q(\delta_1 S, \delta_2 T) - Q(\delta_2 S, \delta_1 T))U,
\end{aligned}
$$

where

$$
\begin{aligned}
P(A,B) &= (I-TS)^{-1}B(I-ST)^{-1}A, \\
Q(A,B) &= A(I-TS)^{-1}B(I-ST)^{-1}.
\end{aligned}
$$

Similarly, interchanging S and T, we have

$$
\begin{aligned}
V^* = (\delta_1\delta_2 - \delta_2\delta_1)V &= V(P^*(\delta_1 T, \delta_2 S) - P^*(\delta_2 T, \delta_1 S)) \\
&\quad + (Q^*(\delta_1 T, \delta_2 S) - Q^*(\delta_2 T, \delta_1 S))V,
\end{aligned}
$$

where P^* and Q^* have obvious definitions.

We introduce a further covariant vector (X, Y). We have to evaluate

$$2R = \sigma(P(U^*, Y) + P(X, V^*)),$$

which is

$$\begin{aligned}&\sigma\{(I-TS)^{-1}Y(I-ST)^{-1}U[P(\delta_1S,\delta_2T)-P(\delta_2S,\delta_1T)]\\&\quad+U(I-TS)^{-1}Y(I-ST)^{-1}[Q(\delta_1S,\delta_2T)-Q(\delta_2S,\delta_1T)]\\&\quad+(I-ST)^{-1}X(I-TS)^{-1}V[P^*(\delta_1T,\delta_2S)-P^*(\delta_2T,\delta_1S)]\\&\quad+V(I-ST)^{-1}X(I-TS)^{-1}[Q^*(\delta_1T,\delta_2S)-Q^*(\delta_2T,\delta_1S)]\\&=\sigma(P(U,Y)P(\delta_1S,\delta_2T)-P(U,Y)P(\delta_2S,\delta_1T)\\&\quad+P(\delta_1S,Y)P(U,\delta_2T)-P(\delta_2S,Y)P(U,\delta_1T)\\&\quad+P(\delta_2S,V)P(X,\delta_1T)-P(\delta_1S,V)P(X,\delta_2T)\\&\quad+P(\delta_2S,\delta_1T)P(X,V)-P(\delta_1S,\delta_2T)P(X,V)).\end{aligned}$$

Putting $\delta_1(S,T)=(U,V)$ and $\delta_2(S,T)=(X,Y)$, we obtain the Riemannian curvature tensor

$$\begin{aligned}R=\sigma[&P(U,Y)P(U,Y)-P(U,Y)P(X,V)\\&-P(X,Y)P(U,V)+P(X,V)P(X,V)].\end{aligned}$$

Changing P into K, we obtain the result stated in the theorem.

9. A specialization. In particular, we put

$$S=Z,\qquad T=\bar{Z}',\qquad V=\bar{U}',\qquad X=\bar{Y}'$$

in the formula of Theorem 13. By the lemma below we then have

THEOREM 14. *The Riemannian curvature tensor of the space with the metric*

$$\sigma(K(dZ,d\bar{Z}'))$$

is equal to

$$\sigma((K(U,\bar{X}')-K(X,\bar{U}'))^2),$$

where

$$K(A,B)=(I-Z\bar{Z}')^{-1}A(I-\bar{Z}'Z)^{-1}B.$$

LEMMA. *Under the hypothesis of the theorem*

$$\sigma(K(U,\bar{U}')K(X,\bar{X}'))=\sigma(K(X,\bar{U}')K(U,\bar{X}')).$$

Proof. 1) The trace of the product of two Hermitian matrices A and B is real, since

$$\sigma(AB)=\sum_{r,s}a_{rs}b_{sr}=\sum\bar{a}_{sr}\bar{b}_{rs}=\sigma(\bar{A}\bar{B}).$$

2) We have

$$\begin{aligned}&\sigma(K(U,\bar{U}')K(X,\bar{X}'))\\&=\sigma((I-\bar{Z}'Z)^{-1}\cdot\bar{X}'(I-Z\bar{Z}')^{-1}U(I-\bar{Z}'Z)^{-1}\bar{U}'(I-Z\bar{Z}')^{-1}X),\end{aligned}$$

which is the trace of the product of two Hermitian matrices. Thus it is real.

3) Since

$$\sigma((\overline{K(U,\bar{U}')K(X,\bar{X}')})') = \sigma(K(X,\bar{U}')K(U,\bar{X}')),$$

we have the theorem.

THEOREM 15. *The Riemannian curvatures of the three kinds of hyperbolic spaces are always non-positive.*

Proof. We have

$$K(U,\bar{X}') - K(X,\bar{U}')$$
$$= (I - Z\bar{Z}')^{-1}(U(I-\bar{Z}'Z)^{-1}\bar{X}' - X(I-\bar{Z}'Z)^{-1}\bar{U}') = (I-Z\bar{Z}')^{-1}M$$

where M is skew-Hermitian. Then the Riemannian curvature tensor is

$$R = -\sigma((I-Z\bar{Z}')^{-1}M(I-Z\bar{Z}')^{-1}\bar{M}').$$

For those points making $I - Z\bar{Z}'$ positive definite, we have a matrix P such that

$$(I - Z\bar{Z}')^{-1} = P\bar{P}';$$

then

$$R = -\sigma(T\bar{T}'),$$

where $T = \bar{P}'MP$. Thus R is non-positive at all points for all directions of the space.

We conclude consequently that passing through two points there is one and only one geodesic and that from a point we can draw a geodesic perpendicular to a given geodesic, etc.[6]

It may also be proved that the spaces are Einstein spaces, i. e., their Ricci tensors are proportional to the fundamental tensors. According to a result due to Schouten and Struik, the spaces cannot be conformal to Euclidean spaces.

10. A concluding remark. Theorem 14 and its consequences and the properties of the groups given in Cartan's paper lead to a neat generalization of the theory of automorphic functions.

NATIONAL TSING HUA UNIVERSITY.
INSTITUTE OF MATHEMATICS, ACAD. SINICA.

[6] E. Cartan, *Leçons sur la géométrie des espaces de Riemann*, Paris, 1925, Note 3.

Reprinted from the
AMERICAN JOURNAL OF MATHEMATICS
Vol. 66, pp. 531–563, 1944

ON THE THEORY OF AUTOMORPHIC FUNCTIONS OF A MATRIX VARIABLE, II—THE CLASSIFICATION OF HYPERCIRCLES UNDER THE SYMPLECTIC GROUP.*

By Loo-Keng Hua.

1. Introduction. The present paper is a continuation of the paper I with the same title,[1] which gives a brief account of the geometrical aspect of the theory.

Throughout the paper, capital Latin letters denote $n \times n$ matrices with complex elements unless the contrary is stated. A' denotes the transposed matrix of A and $\bar{A}$, the conjugate imaginary matrix of A. I denotes the unit matrix and O, the zero matrix.

We define a hypercircle to be the set of points (symmetric matrices) Z for which the Hermitian matrix

$$\bar{Z}H_1Z + LZ + \bar{Z}\bar{L}' + H_2$$

is positive definite, where H_1 and H_2 are Hermitian matrices.

The object of the present paper is to classify completely hypercircles under the (non-homogeneous) symplectic group $\mathfrak{G}$, which consists of all the symplectic transformations defined by:

$$Z_1 = (AZ + B)(CZ + D)^{-1}, \quad AB' = BA', \quad CD' = DC', \quad AD' - BC' = I.$$

The letter $\mathfrak{G}$ will be kept in this sense throughout the paper.

Our classification of hypercircles depends on the theory of pairs of Hermitian matrices. Because all the available treatments (or at least all the treatments available to the author in China, cf. 6) of the subject contain a mistake, we find it necessary to resume the theory.

2. Symmetric pairs of matrices. Let

$$\mathfrak{F} = \begin{pmatrix} O & I \\ -I & O \end{pmatrix},$$

Note.—Because of the poor mail service between the U. S. and China, a number of minor changes in this paper have been made here, with the consent of the editors, by Prof. Hua's friend Dr. Hsio-Fu Tuan.

* Received April 21, 1943.

[1] This *Journal*, vol. 66 (1944), pp. 470-488.

which is a $2n \times 2n$ skew symmetric matrix. This notation will be kept throughout.

DEFINITION 1. *A pair of matrices A and B is said to be symmetric to each other, or to form a symmetric pair (A, B), if $AB' = BA'$.*

Clearly (A, B) *is a symmetric pair if and only if*

$$(A, B)\mathfrak{F}(A, B)' = 0,$$

since the left hand side is equal to

$$(-B, A)(A, B)' = -BA' + AB'.$$

DEFINITION 2. *A pair of matrices (C, D) is said to be conjugate to another pair of matrices (A, B) if $AD' - BC' = I$.*

According to this definition, the conjugate relation is skew in the two pairs: if (C, D) is conjugate to (A, B), then $-(A, B) = (-A, -B)$ is conjugate to (C, D). In the following we shall often speak of conjugate pairs, when the order of the pairs is immaterial.

Clearly (C, D) *is conjugate to* (A, B) *if and only if*

$$(A, B)\mathfrak{F}(C, D)' = I,$$

since the left hand side is exactly $AD' - BC'$.

THEOREM 1. *The transformation*

$$Z_1 = (AZ + B)(CZ + D)^{-1}$$

with the matrix

$$\mathfrak{T} = \begin{pmatrix} A & B \\ C & D \end{pmatrix}$$

(in the following we often speak of the transformation $\mathfrak{T}$) *belongs to* $\mathfrak{G}$, *if and only if*

$$\mathfrak{T}\mathfrak{F}\mathfrak{T}' = \mathfrak{F}.$$

Proof. Since the left hand side of the equation to be proved is

$$\begin{pmatrix} AB' - BA', & AD' - BC' \\ CB' - DA', & CD' - DC' \end{pmatrix},$$

the result follows immediately.

Putting this result in another form, we have:

THEOREM 2. *The transformation*

$$Z_1 = (AZ + B)(CZ + D)^{-1}$$

with the matrix

$$\mathfrak{T} = \begin{pmatrix} A & B \\ C & D \end{pmatrix}$$

belongs to $\mathfrak{G}$ *if and only if* (A, B) *and* (C, D) *are two symmetric pairs such that* (C, D) *is conjugate to* (A, B).

THEOREM 3. *If* (A, B) *is a symmetric pair, then*

$$(A_1, B_1) = Q(A, B)\mathfrak{T}$$

is also a symmetric pair, where $\mathfrak{T}$ *is in* $\mathfrak{G}$.

Proof. We have

$$\begin{aligned}(A_1, B_1)\mathfrak{F}(A_1, B_1)' &= Q(A, B)\mathfrak{T}\mathfrak{F}\mathfrak{T}'(A, B)'Q' \\ &= Q(A, B)\mathfrak{F}(A, B)'Q' = O.\end{aligned}$$

THEOREM 4. *If* (A, B) *is conjugate to* (C, D), *and if*

$$(A_1, B_1) = Q(A, B)\mathfrak{T}, \qquad (C_1, D_1) = Q'^{-1}(C, D)\mathfrak{T},$$

then (A_1, B_1) *is conjugate to* (C_1, D_1), *where* Q *is non-singular and* $\mathfrak{T}$ *is in* $\mathfrak{G}$.

Proof. We have

$$(A_1, B_1)\mathfrak{F}(C_1, D_1)' = Q(A, B)\mathfrak{T}\mathfrak{F}\mathfrak{T}'(C, D)'Q^{-1} = I.$$

DEFINITION 3. *Two symmetric pairs of matrices* (A_1, B_1) *and* (A, B) *are said to be equivalent if we have a non-singular matrix* Q *and a transformation* $\mathfrak{T}$ *of* $\mathfrak{G}$ *such that*

$$(A_1, B_1) = Q(A, B)\mathfrak{T}.$$

This relation will be denoted by

$$(A_1, B_1) \sim (A, B).$$

THEOREM 5. *The relation "$\sim$" possesses the properties: determination, reflexivity, symmetry and transitivity.*

DEFINITION 4. *A pair of matrices* (A, B) *is said to be non-singular if the matrix* (A, B) *is of rank* n.

THEOREM 6. *Any two non-singular symmetric pairs of matrices are equivalent.*

Proof. It is sufficient to prove that

$$(A, B) \sim (I, O).$$

1) If A is non-singular, then $A^{-1}B = S$ is symmetric. Then

$$(A, B) = A(I, S) = A(I, O) \begin{pmatrix} I & S \\ O & I \end{pmatrix}.$$

The result follows, since

$$\begin{pmatrix} I & S \\ O & I \end{pmatrix}$$

belongs to $\mathfrak{G}$.

2) If A is singular, then we have two non-singular matrices P and Q such that

$$A_1 = PAQ = \begin{pmatrix} I^{(r)} & O \\ O & O^{(n-r)} \end{pmatrix}.$$

Let

$$(A_1, B_1) = P(A, B) \begin{pmatrix} Q & O \\ O & Q'^{-1} \end{pmatrix},$$

where

$$B_1 = PBQ'^{-1} = \begin{pmatrix} s^{(r)} & m \\ l & t^{(n-r)} \end{pmatrix}, \text{ say.}$$

Since

$$\begin{pmatrix} Q & O \\ O & Q'^{-1} \end{pmatrix}$$

belongs to $\mathfrak{G}$, (A_1, B_1) is a non-singular and symmetric pair. Consequently $s^{(r)}$ is symmetric and l is a null matrix.

Let

$$(A_2, B_2) = (A_1, B_1) \begin{pmatrix} I & -S \\ O & I \end{pmatrix},$$

where

$$S = \begin{pmatrix} s^{(r)} & O \\ O & I^{(n-r)} \end{pmatrix}.$$

Then

$$A_2 = A_1, \qquad B_2 = -A_1 S + B_1 = \begin{pmatrix} O & m \\ O & t \end{pmatrix}.$$

Since (A_2, B_2) is non-singular, so also is t. Let

$$(A_3, B_3) = (A_2, B_2) \begin{pmatrix} I & O \\ I & I \end{pmatrix};$$

then

$$A_3 = A_2 + B_2 = \begin{pmatrix} I^{(r)} & m \\ O & t \end{pmatrix}$$

which is non-singular. By 1), we have

$$(A_3, B_3) \sim (I, O).$$

The result follows.

THEOREM 7. *The subgroup which leaves a non-singular symmetric pair of matrices invariant is simply isomorphic to the group which consists of all transformations of the form*

$$Z_1 = Q'ZQ + S,$$

where Q is non-singular and S is symmetric.

Proof. It is sufficient to consider the group which leaves (O, I) invariant. In fact, we have Q and $\mathfrak{T}$ such that

$$Q(A, B)\mathfrak{T} = (O, I).$$

Let Q_0 and $\mathfrak{T}_0$ be such that

$$Q_0(O, I)\mathfrak{T}_0 = (O, I).$$

Then

$$Q^{-1}Q_0Q(A, B)\mathfrak{T}\mathfrak{T}_0\mathfrak{T}^{-1} = (A, B).$$

The isomorphism of the group whose elements leave (O, I) invariant and the group whose elements leave (A, B) invariant is evident.

Let

$$Q(O, I)\mathfrak{T} = (O, I), \qquad \mathfrak{T} = \begin{pmatrix} A & B \\ C & D \end{pmatrix}.$$

Then, we have

$$(QC, QD) = (O, I),$$

i. e., $C = O$, $D = Q^{-1}$. Then $A = Q'$ and $B = SQ^{-1}$.

The group is isomorphic to the group formed by the matrices

$$\begin{pmatrix} Q' & SQ^{-1} \\ O & Q^{-1} \end{pmatrix}.$$

The result is now evident.

COROLLARY. *The transformations leaving (O, I) invariant are of the form*

$$\begin{pmatrix} Q' & SQ^{-1} \\ O & Q^{-1} \end{pmatrix},$$

where Q is non-singular and S is symmetric.

THEOREM 8. *Given a non-singular symmetric pair of matrices (A, B), we have a non-singular symmetric pair of matrices (C, D) as its conjugate. The totality of all possible pairs (C, D) depends on $n(n+1)$ parameters.*

Proof. 1) First we consider the case $(A, B) = (0, I)$. Let (C, D) be a pair satisfying our requirement; then

$$I = AD' - BC' = -C'.$$

Thus the conjugate pairs of (A, B) are

$$(-I, S),$$

where the S are symmetric. The theorem is true for $(A, B) = (0, I)$.

2) By Theorem 6, we have Q and $\mathfrak{T}$ such that

$$Q(A, B)\mathfrak{T} = (0, I).$$

We define (C, D) by $Q'^{-1}(C, D)\mathfrak{T} = (-I, S)$. Then (C, D) satisfies our requirement.

Further let Q_1 and $\mathfrak{T}_1$ be matrices satisfying also

$$Q_1(A, B)\mathfrak{T}_1 = (0, I).$$

Then, we have

$$(C, D) = Q'_1(-I, S_1)\mathfrak{T}_1^{-1}.$$

We shall now prove that this is equal to $Q'(-I, S)\mathfrak{T}^{-1}$. Since

$$Q_1Q^{-1}(0, I)\mathfrak{T}^{-1}\mathfrak{T}_1 = (0, I)$$

by the corollary of Theorem 7, we have

$$\mathfrak{T}^{-1}\mathfrak{T}_1 = \begin{pmatrix} Q'^{-1}Q'_1 & S_2QQ_1^{-1} \\ 0 & QQ_1^{-1} \end{pmatrix}.$$

Then

$$\begin{aligned} Q'_1(-I, S_1)\mathfrak{T}^{-1} &= Q'_1(-I, S_1)\begin{pmatrix} Q'^{-1}Q'_1 & S_2QQ_1^{-1} \\ 0 & QQ_1^{-1} \end{pmatrix}\mathfrak{T}^{-1} \\ &= Q'_1(-I, S_1)\begin{pmatrix} Q_1'^{-1}Q' & -Q_1'^{-1}Q'S_2 \\ 0 & Q_1Q^{-1} \end{pmatrix}\mathfrak{T}^{-1} \\ &= Q'(-I, Q'^{-1}Q'_1S_1Q_1Q^{-1} + S_2)\mathfrak{T}^{-1}. \end{aligned}$$

Hence we have always the same collection of pairs of matrices conjugate to (A, B).

3. Hypercircles.

DEFINITION 1. *The transformation of symmetric pairs*

$$(W_1, W_2) = Q(Z_1, Z_2)\mathfrak{T}$$

for a non-singular matrix Q and a transformation $\mathfrak{T}$ belonging to $\mathfrak{G}$ is called a homogeneous representation of

$$W = (AZ + B)(CZ + D)^{-1}.$$

The group so obtained is called the group $\mathfrak{G}_H$.

DEFINITION 2. *A hypercircle is defined by the set of points corresponding to symmetric matrices Z such that the Hermitian matrix*

$$\bar{Z}H_1Z + LZ + \bar{Z}\bar{L}' + H_2$$

is positive definite, where H_1 and H_2 are Hermitian matrices. Or, in "homogeneous" coördinates, a hypercircle is defined by the set of points corresponding to symmetric pairs (W_1, W_2) such that the Hermitian matrix

$$\bar{W}_1H_1W'_1 + \bar{W}_2LW'_1 + \bar{W}_1\bar{L}'W'_2 + \bar{W}_2H_2W'_2 = (\overline{W_1, W_2})\mathfrak{H}(W_1, W_2)'$$

is positive definite, where

$$\mathfrak{H} = \begin{pmatrix} H_1 & \bar{L}' \\ L & H \end{pmatrix}.$$

$\mathfrak{H}$ is called the matrix of the hypercircle.

Remark. $\mathfrak{H}$ is a general $2n \times 2n$ Hermitian matrix. Thus the following results may be interpreted purely algebraically without reference to hypercircles.

THEOREM 9. *The transformation $(W_1, W_2) = Q(Z_1, Z_2)\mathfrak{T}$ carries a hypercircle with the matrix $\mathfrak{H}$ to a hypercircle with the matrix $\mathfrak{H}_1 = \mathfrak{T}\mathfrak{H}\mathfrak{T}'$.*

Proof. Since

$$(\bar{W}_1, \bar{W}_2)\mathfrak{H}(W_1, W_2)' = \bar{Q}(\bar{Z}_1, \bar{Z}_2)\bar{\mathfrak{T}}\mathfrak{H}\mathfrak{T}'(Z_1, Z_2)'Q',$$

the theorem follows.

DEFINITION 3. *If we have $\mathfrak{T}$ belonging to $\mathfrak{G}$ such that $\mathfrak{H}_1 = \bar{\mathfrak{T}}\mathfrak{H}\mathfrak{T}'$, we say that $\mathfrak{H}_1$ and $\mathfrak{H}$ are conjunctive under $\mathfrak{G}$.*

Evidently, "conjunctivity under $\mathfrak{G}$" possesses the properties: symmetry, reflexivity and transitivity. Naturally, this suggests the classification of hyper-

4

circles under $\mathfrak{G}$. This problem is by no means easy but it is solved completely. First of all, we introduce the following notion:

DEFINITION 4. *For a hypercircle with the matrix $\mathfrak{H}$, we define*

$$\mathfrak{H}'\mathfrak{F}\mathfrak{H} = \begin{pmatrix} H'_1\bar{L} - L'H_1, & H'_1H_2 - L'\bar{L}' \\ -H'_2H_1 + \bar{L}L, & \bar{L}H_2 - H'_2\bar{L}' \end{pmatrix}$$

to be the discriminantal matrix of the hypercircle. It will be denoted by $\mathfrak{D}(\mathfrak{H})$. Evidently $\mathfrak{D}(\mathfrak{H})$ is skew-symmetric.

THEOREM 10. *If $\mathfrak{H}_1$ and $\mathfrak{H}_2$ are conjunctive under $\mathfrak{G}$, then $\mathfrak{D}(\mathfrak{H}_1)$ and $\mathfrak{D}(\mathfrak{H}_2)$ are congruent under $\mathfrak{G}$. More precisely, if $\bar{\mathfrak{T}}\mathfrak{H}_1\mathfrak{T}' = \mathfrak{H}_2$, then*

$$\mathfrak{T}\mathfrak{D}(\mathfrak{H}_1)\mathfrak{T}' = \mathfrak{D}(\mathfrak{H}_2).$$

Proof. Since

$$\mathfrak{D}(\mathfrak{H}_2) = \mathfrak{H}'_2\mathfrak{F}\mathfrak{H}_2 = \mathfrak{T}\mathfrak{H}'_1\bar{\mathfrak{T}}'\mathfrak{F}\bar{\mathfrak{T}}\mathfrak{H}_1\mathfrak{T}' = \mathfrak{T}\mathfrak{H}'_1\mathfrak{F}\mathfrak{H}_1\mathfrak{T}' = \mathfrak{T}\mathfrak{D}(\mathfrak{H}_1)\mathfrak{T}',$$

we have the result.

4. The canonical form of the discriminantal matrix. The problem of congruence of $\mathfrak{D}(\mathfrak{H}_1)$ and $\mathfrak{D}(\mathfrak{H}_2)$ under $\mathfrak{G}$ is equivalent to the problem of congruence of the pairs of skew symmetric matrices $(\mathfrak{D}(\mathfrak{H}_1), \mathfrak{F})$ and $(\mathfrak{D}(\mathfrak{H}_2), \mathfrak{F})$. The latter problem is solved in most treatises on elementary divisors. For the sake of completeness, the author quotes the following results:

THEOREM 11. *Let $\mathfrak{B}$ and $\mathfrak{B}_1$ be two non-singular matrices. The pairs of skew symmetric matrices $(\mathfrak{A}, \mathfrak{B})$ and $(\mathfrak{A}_1, \mathfrak{B}_1)$ are congruent if and only if $\mathfrak{A} + \lambda\mathfrak{B}$ and $\mathfrak{A}_1 + \lambda\mathfrak{B}_1$ have the same invariant factors (or the same elementary divisors).*

(For the proof see, e. g., MacDuffee, *Theory of Matrices*, Theorems 35. 4 and 30. 1.)

THEOREM 12. *There exist pairs of skew symmetric matrices of degree $2n$, one of which is non-singular, having any given admissible invariant factors. More precisely, let*

$$h_{2i} = h_{2i-1} = g_i = (\lambda - \lambda_1)^{l_{i1}} \cdots (\lambda - \lambda_k)^{l_{ik}},$$
$$(1 \leq i \leq n, \quad l_{ij} \geq 0, \quad 1 \leq j \leq k)$$

be the given $2i$-th invariant factors (since in a skew symmetric matrix, the $2i$-th invariant factor is equal to the $(2i\text{-}1)$-th invariant factor), let g_i divide g_{i+1} and let $\Sigma l_{ij} = n$. We define τ_i to be the direct sum of matrices

$$\tau_i = \tau_{i1} \dotplus \tau_{i2} \dotplus \cdots \dotplus \tau_{ik}$$

where

$$\tau_{ij} = \begin{pmatrix} \lambda_j & 1 & 0 & \cdots & 0 \\ 0 & \lambda_j & 1 & \cdots & 0 \\ 0 & 0 & \lambda_j & \cdots & 0 \\ \cdot & \cdot & \cdot & \cdot & \cdot \\ 0 & 0 & 0 & \cdots & \lambda_j \end{pmatrix}$$

is of degree l_{ij} *and* $1 \leq j \leq h$. *Further we define* T *by the direct sum*

$$T = \tau_n \dotplus \tau_{n-1} \dotplus \cdots \dotplus \tau_{n-t}.$$

Then the pair of skew symmetric matrices $(\mathfrak{E}, \mathfrak{F})$ *with*

$$\mathfrak{E} = \begin{pmatrix} O & T \\ -T' & O \end{pmatrix}, \quad \mathfrak{F} = \begin{pmatrix} O & I \\ -I & O \end{pmatrix}$$

possesses the preassigned invariant factors.

Proof. Let δ_i be the greatest common divisor of the i-rowed minors of $T - \lambda I$. Then, evidently,

$$\delta_i = g_1 \cdots g_i.$$

Further let d_i be the greatest common divisor of the i-rowed minors of $\mathfrak{E} - \lambda\mathfrak{F}$. We need only find the d_i for any even i. It is evident that d_{2i} is the g. c. d. of $\delta_i^2, \delta_{i-1}\delta_{i+1}, \delta_{i-2}\delta_{i+2}, \cdots$. Since

$$\delta_i^2 = g_1^2 \cdots g_i^2, \qquad \delta_{i-t}\delta_{i+t} = g_1^2 \cdots g_{i-t}^2 g_{i-t+1} \cdots g_{i+t},$$

and g_i divides g_{i+1}, we have δ_i^2 dividing $\delta_{i-t}\delta_{i+t}$. Thus $d_{2i} = \delta_i^2$. Then

$$h_{2i}h_{2i-1} = \frac{d_{2i}}{d_{2i-1}} \frac{d_{2i-1}}{d_{2i-2}} = \frac{\delta_{i-1}^2}{\delta_i^2}.$$

Since $h_{2i} = h_{2i-1}$, we have

$$h_{2i} = h_{2i-1} = g_i.$$

Consequently we have

THEOREM 13. *Every discriminantal matrix is congruent under* $\mathfrak{G}$ *to a matrix of the form*

$$\begin{pmatrix} O & T \\ -T' & O \end{pmatrix}$$

where T *has the same meaning as given in Theorem 12. Consequently, every* *hypercircle is conjunctive under $\mathfrak{G}$ to a hypercircle with its discriminantal matrix of the prescribed form.*

Proof. By Theorems 11 and 12, we have $\mathfrak{T}$ such that

$$\mathfrak{T}\mathfrak{D}(\mathfrak{H})\mathfrak{T}' = \begin{pmatrix} O & T \\ -T' & O \end{pmatrix}$$

and

$$\mathfrak{T}\mathfrak{F}\mathfrak{T}' = \mathfrak{F}.$$

Let $\bar{\mathfrak{T}}^{-1}\mathfrak{H}\mathfrak{T}'^{-1} = \mathfrak{H}_1$; then $\mathfrak{H}_1$ has its discriminantal matrix in the described form.

5. Proof of the theorem that every hypercircle is conjunctive under $\mathfrak{G}$ to a "binomial" hypercircle.

THEOREM 14. *Every hypercircle is conjunctive under $\mathfrak{G}$ to a "binomial" hypercircle, or more precisely a hypercircle with the matrix*

$$\begin{pmatrix} H_1 & O \\ O & H_2 \end{pmatrix}, \quad H_1 = \begin{pmatrix} h_1^{(r)} & O \\ O & O \end{pmatrix}, \quad H_2 = \begin{pmatrix} h_2^{(r)} & O \\ O & O \end{pmatrix}, \quad \det(h_1^{(r)}) \neq 0.$$

Proof. 1) The theorem is well-known for $n = 1$. By Theorem 13, it is sufficient to consider a hypercircle with the matrix

$$\begin{pmatrix} H_1 & \bar{L}' \\ L & H_2 \end{pmatrix}$$

satisfying the condition

$$\begin{pmatrix} H_1 & \bar{L}' \\ L & H_2 \end{pmatrix} \begin{pmatrix} O & I \\ -I & O \end{pmatrix} \begin{pmatrix} H_1 & \bar{L}' \\ L & H_2 \end{pmatrix} = \begin{pmatrix} O & * \\ * & O \end{pmatrix},$$

i. e., $H'_1L = L'H_1$ and $\bar{L}H_2 = H'_2\bar{L}'$.

If H_1 is non-singular, then

$$S = LH_1^{-1} = \bar{H}_1^{-1}\bar{L}'$$

is symmetric. We have evidently that

$$\bar{Z}H_1Z + LZ + \bar{Z}\bar{L}' + H_2 = (\bar{Z} + S)H_1(Z + \bar{S}) + H_2 - SH_1\bar{S},$$

which is "binomial" in $Z + \bar{S}$. A similar result holds when H_2 is non-singular. The theorem is thus true for these cases.

2) Before going further, we require two lemmas.

LEMMA 1. *Any symmetric matrix* S *may be expressed as* $S = TT'$ *where* T *is a matrix with zeros above the main diagonal* (well-known).

LEMMA 2. *For any given matrix* Q, *we have a non-singular symmetric matrix* S *such that* QS *is symmetric.*

In fact, it is sufficient to find a non-singular solution of the matrix equation

$$QS = SQ'$$

where the symmetric matrix S is considered as an unknown. We have a non-singular matrix Γ such that

$$Q_1 = \Gamma^{-1} Q \Gamma$$

is of the Jordan's normal form, i. e. a direct sum of matrices of the form

$$J_i^{(j)} = \begin{pmatrix} \lambda_i & 1 & 0 & \cdots & 0 \\ 0 & \lambda_i & 1 & \cdots & 0 \\ 0 & 0 & \lambda_i & \cdots & 0 \\ \cdot & \cdot & \cdot & \cdot & \cdot \\ 0 & 0 & 0 & \cdots & \lambda_j \end{pmatrix}$$

Then

$$Q_1 S_1 = S_1 Q'_1$$

where $\Gamma^{-1} S \Gamma'^{-1} = S_1$. Therefore, it is sufficient to find a solution of the equation with $Q = J_i^{(j)}$. Evidently

$$S^{(j)} = \begin{pmatrix} 0 & 0 & \cdots & 0 & 1 \\ 0 & 0 & \cdots & 1 & 0 \\ \cdot & \cdot & \cdot & \cdot & \cdot \\ 0 & 1 & \cdots & 0 & 0 \\ 1 & 0 & \cdots & 0 & 0 \end{pmatrix}$$

is a solution, since $S = S'$ and

$$J_i S = \begin{pmatrix} 0 & 0 & \cdots & 1 & \lambda_i \\ 0 & 0 & \cdots & \lambda_i & 0 \\ \cdot & \cdot & \cdot & \cdot & \cdot \\ 1 & \lambda_i & \cdots & 0 & 0 \\ \lambda_i & 0 & \cdots & 0 & 0 \end{pmatrix} = SJ'_i.$$

3) We now consider all conjunctive hypercircles of $\mathfrak{H}$ under $\mathfrak{G}$

with "binomial" discriminantal matrices. Let $\mathfrak{H}$ be one of them with H_1 of the highest rank r. If $r = n$, this problem was solved in 1).

We have a non-singular matrix Q such that

$$\bar{Q}' H_1 Q = \begin{pmatrix} h^{(r)} & O \\ O & O \end{pmatrix}, \qquad \det(h) \neq 0.$$

Since

$$\mathfrak{T} = \begin{pmatrix} Q' & O \\ O & Q^{-1} \end{pmatrix}$$

carries a hypercircle with "binomial" discriminantal matrix into one of the same nature, we may assume, without loss of generality, that

$$H_1 = \begin{pmatrix} h & O \\ O & O \end{pmatrix}, \qquad \det(h) \neq 0, \qquad H_2 = \begin{pmatrix} g_{11} & g_{12} \\ g_{21} & g_{22} \end{pmatrix}.$$

We shall now establish that $r \neq 0$ and that we may assume $\det(g_{11}) \neq 0$. Let

$$\mathfrak{D}(\mathfrak{H}) = \begin{pmatrix} O & K \\ -K' & O \end{pmatrix}$$

be its discriminantal matrix. We may transform $\mathfrak{H}$ in such a way that K is symmetric. In fact, by Lemmas 1 and 2, we have a symmetric matrix S such that (i) SK is symmetric and (ii) $S = TT'$ where T is a matrix with zeros above the main diagonal. Let

$$\mathfrak{T} = \begin{pmatrix} T' & O \\ O & T^{-1} \end{pmatrix}$$

which belongs to $\mathfrak{G}$. Then

$$\mathfrak{T}\mathfrak{D}(\mathfrak{H})\mathfrak{T}' = \begin{pmatrix} O & T'KT'^{-1} \\ -T^{-1}K'T & O \end{pmatrix},$$

where $T^{-1}K'T$ is symmetric, since

$$TT'K = K'TT' \quad \text{implies} \quad T^{-1}K'T = T'KT'^{-1}.$$

Further the first element in $\bar{\mathfrak{T}}\mathfrak{H}\mathfrak{T}'$ is equal to

$$\bar{T}' H_1 T = \begin{pmatrix} * & * \\ O & * \end{pmatrix} \begin{pmatrix} h & O \\ O & O \end{pmatrix} \begin{pmatrix} * & O \\ * & * \end{pmatrix} = \begin{pmatrix} * & O \\ O & O \end{pmatrix}$$

of which the rank is still r. We may assume that

$$H_1=\begin{pmatrix} h & 0 \\ 0 & 0 \end{pmatrix}, \qquad \mathfrak{D}(\mathfrak{H})=\begin{pmatrix} 0 & S \\ -S & 0 \end{pmatrix},$$

where S is symmetric.

Let ρ be any number. We have

$$\begin{pmatrix} I & 0 \\ \rho I & I \end{pmatrix}\begin{pmatrix} 0 & S \\ -S & 0 \end{pmatrix}\begin{pmatrix} I & \rho I \\ 0 & I \end{pmatrix}=\begin{pmatrix} 0 & S \\ -S & 0 \end{pmatrix}$$

and

$$\overline{\begin{pmatrix} I & 0 \\ \rho I & I \end{pmatrix}}\begin{pmatrix} H_1 & \bar{L}' \\ L & H_2 \end{pmatrix}\begin{pmatrix} I & \rho I \\ 0 & I \end{pmatrix}=\begin{pmatrix} H_1 & * \\ * & H_0 \end{pmatrix},$$

where

$$H_0=|\bar{\rho}|^2 H_1+\bar{\rho}\bar{L}'=\rho L+H_2.$$

It is evident that for ρ arge, $r=0$ if and only if $H_1=L=H_2=0$. Let

$$H_0=\begin{pmatrix} k_{11}{}^{(r)} & k_{12} \\ k_{21} & k_{22} \end{pmatrix},$$

then

$$k_{11}=|\rho|^2 h+\rho^*+\bar{\rho}^*+*.$$

For ρ large, k_{11} is non-singular.

4) Now we may assume that

$$H_1=\begin{pmatrix} h^{(r)} & 0 \\ 0 & 0 \end{pmatrix}, \qquad H_2=\begin{pmatrix} g_{11} & g_{12} \\ g_{21} & g_{22} \end{pmatrix}, \qquad L=\begin{pmatrix} l_{11} & l_{12} \\ l_{21} & l_{22} \end{pmatrix}$$

where $\det(h)\neq 0$, $\det(g_{11})\neq 0$ and $r\neq 0$. Let

$$R=\begin{pmatrix} I & 0 \\ -g'_{12}\bar{g}_{11}{}^{-1} & I \end{pmatrix}, \qquad \mathfrak{T}=\begin{pmatrix} R'^{-1} & 0 \\ 0 & R \end{pmatrix}$$

which belongs to $\mathfrak{G}$, such that

$$\bar{\mathfrak{T}}\mathfrak{H}\mathfrak{T}'=\begin{pmatrix} \bar{R}'^{-1} & 0 \\ 0 & \bar{R} \end{pmatrix}\begin{pmatrix} H_1 & \bar{L}' \\ L & H_2 \end{pmatrix}\begin{pmatrix} R^{-1} & 0 \\ 0 & R' \end{pmatrix}=\begin{pmatrix} \begin{pmatrix} h & 0 \\ 0 & 0 \end{pmatrix} & * \\ * & \begin{pmatrix} g & 0 \\ 0 & g_0 \end{pmatrix} \end{pmatrix}$$

where $g=g_{11}$. (In fact

$$\bar{R}'H_1R^{-1}=\begin{pmatrix} I & * \\ 0 & I \end{pmatrix}\begin{pmatrix} h & 0 \\ 0 & 0 \end{pmatrix}\begin{pmatrix} I & 0 \\ * & I \end{pmatrix}=\begin{pmatrix} h & 0 \\ 0 & 0 \end{pmatrix},$$

$$\bar{R}H_2R'=\begin{pmatrix} I & 0 \\ -\bar{g}'_{12} & g_{11}{}^{1} \end{pmatrix}\begin{pmatrix} g_{11} & g_{12} \\ \bar{g}'_{12} & g_{22} \end{pmatrix}\begin{pmatrix} I & -g_{11}{}^{-1}g_{12} \\ 0 & I \end{pmatrix}=\begin{pmatrix} g_{11} & 0 \\ 0 & g_0 \end{pmatrix}.)$$

Since the rank of H_2 cannot be higher than r, $g_0 = O$.

Now we may write

$$H_1 = \begin{pmatrix} h & O \\ O & O \end{pmatrix}, \quad L = \begin{pmatrix} l_{11} & l_{12} \\ l_{21} & l_{22} \end{pmatrix}, \quad H_2 = \begin{pmatrix} g & O \\ O & O \end{pmatrix}$$

where both h and g are non-singular. Since $H'_1 L = L' H_1$ and $\bar{L} H_2 = H'_2 \bar{L}'$, we have

$$h' l_{11} = l'_{11} h, \qquad \bar{l}_{11} g = g' \bar{l}'_{11}, \qquad l_{12} = l_{21} = O.$$

As in 1), we may then assume that $l_{11} = O$ and $\det(h) \neq 0$, but now g may be singular. By induction, we have $a_1^{(n-r)}$, $b_1^{(n-r)}$, $c_1^{(n-r)}$ and $d_1^{(n-r)}$ such that

$$\overline{\begin{pmatrix} a_1 & b_1 \\ c_1 & d_1 \end{pmatrix}} \begin{pmatrix} O & l'_{22} \\ l_{22} & O \end{pmatrix} \begin{pmatrix} a_1 & b_1 \\ c_1 & d_1 \end{pmatrix}' = \begin{pmatrix} h_2^{(n-r)} & O \\ O & g_2^{(n-r)} \end{pmatrix},$$

$$\overline{\begin{pmatrix} a_1 & b_1 \\ c_1 & d_1 \end{pmatrix}} \begin{pmatrix} O & I^{(n-r)} \\ -I^{(n-r)} & O \end{pmatrix} \begin{pmatrix} a_1 & b_1 \\ c_1 & d_1 \end{pmatrix}' = \begin{pmatrix} O & I^{(n-r)} \\ -I^{(n-r)} & O \end{pmatrix},$$

and we may assume that the rank of h_2 is higher than that of g_2, for otherwise

$$\begin{pmatrix} O & I \\ I & O \end{pmatrix}' \begin{pmatrix} h_2 & O \\ O & g_2 \end{pmatrix} \begin{pmatrix} O & I \\ I & O \end{pmatrix} = \begin{pmatrix} g_2 & O \\ O & h_2 \end{pmatrix}.$$

Let

$$A = \begin{pmatrix} I & O \\ O & a_1 \end{pmatrix}, \quad B = \begin{pmatrix} O & O \\ O & b_1 \end{pmatrix}, \quad C = \begin{pmatrix} O & O \\ O & c_1 \end{pmatrix}, \quad D = \begin{pmatrix} I & O \\ O & d_1 \end{pmatrix}$$

then

$$\mathfrak{T} = \begin{pmatrix} A & B \\ C & D \end{pmatrix}$$

belongs to $\mathfrak{G}$ and H_1 of $\bar{\mathfrak{T}} \mathfrak{H} \mathfrak{T}'$ is equal to $\begin{pmatrix} h & O \\ O & h_2 \end{pmatrix}$. Since its rank cannot be higher than r, we have $h_2 = O$. Consequently, $g_2 = O$. Then $l_{22} = O$. The result is now proved.

6. A lemma. For reasons explained in the Introduction, we find it necessary first to discuss the theory of pairs of Hermitian matrices **(6-9)**[2] as a basis for the classification of hypercircles **(10-16)**.

---

[2] Cf. Dickson, *Modern algebraic theories*, p. 123, Theorem 10; MacDuffee, *Theory of matrices*, p. 63, Theorem 36.5; Turnbull and Aitken, *Theory of canonical matrices*, p. 131, Lemma III; and Logsdon, *American Journal of Mathematics*, vol. 44 (1922), pp. 247-260. An earlier paper of Muth, *Journ. für Math.*, vol. 128 (1905), pp. 302-321 should be mentioned as one of importance in this connection.

LEMMA. *If $q(x)$ is a polynomial, with real coefficients, which has no negative or zero root, then we have a real polynomial $\chi(x)$ such that $\chi^2(x)-x$ is divisible by $q(x)$.*

Proof. Let

$$q(x)=\lambda \prod_{i=1}^{s}(x-a_i)^{l_i}\prod_{j=1}^{t}((x-\alpha_i)(x-\bar{\alpha}_i)),$$

where $a_i>0$ and α_i is complex.

1) The theorem is true for

$$q(x)=(x-a)^l.$$

In fact the theorem is true for $l=1$, for then $\chi(x)=\sqrt{a}$ is a solution. Suppose that we have a real polynomial $\chi_{l-1}(x)$ such that

$$\chi^2_{l-1}(x)-x=(x-a)^{l-1}\lambda(x), \qquad l>1,$$

where $\lambda(x)$ is a polynomial with real coefficients. Evidently $\chi_{l-1}(a)\neq 0$. Then

$$\chi_l(x)=\chi_{l-1}(x)-\tfrac{1}{2}\frac{\chi(a)}{\chi_{l-1}(a)}(x-a)^{l-1}$$

satisfies our requirement, since

$$\begin{aligned}\chi^2_l(x)-x&\equiv\chi^2_{l-1}(x)-x-\frac{\lambda(a)}{\chi_{l-1}(a)}\chi_{l-1}(x)(x-a)^{l-1}\\ &\equiv\left(\lambda(x)-\frac{\lambda(a)}{\chi_{l-1}(a)}\chi_{l-1}(x)\right)(x-a^{l-1})\\ &\equiv 0 \pmod{(x-a)^l}.\end{aligned}$$

2) The theorem is true for

$$q(x)=((x-\alpha)(x-\bar{\alpha}))^l.$$

In fact

$$\chi(x)=\frac{1}{\sqrt{2\,|\alpha|+\alpha+\bar{\alpha}}}(x+|\alpha|)$$

satisfies our requirement for $l=1$, since

$$\begin{aligned}\chi^2(x)-x&=\frac{1}{2\,|\alpha|+\alpha+\bar{\alpha}}(x^2-2\,|\alpha|\,x+|\alpha|^2)-x\\ &=\frac{1}{2\,|\alpha|+\alpha+\bar{\alpha}}(x-\alpha)(x-\bar{\alpha})\\ &\equiv 0 \pmod{(x-\alpha)(x-\bar{\alpha})}\end{aligned}$$

and $2\,|\alpha|+\alpha+\bar{\alpha}>0$.

Let $\chi_{l-1}(x)$ be a real polynomial satisfying

$$\chi^2_{l-1}(x) - x = ((x-\alpha)(x-\alpha))^{l-1}\lambda(x), \qquad l > 1.$$

It may be verified directly that

$$\chi_l(x) = \chi_{l-1}(x) + ((x-\alpha)(x-\bar{\alpha}))^{l-1}(sx+t)$$

satisfies our requirement, where the real numbers s and t are given by

$$\lambda(\alpha) + 2(s\alpha + t)\chi_{l-1}(\alpha) = 0$$

(The existence of s and t is easily seen, since α is not real and $\chi_{l-1}(\alpha) \neq 0$).

3) Let $q_1(x)$ and $q_2(x)$ be two real polynomials without common divisor, and let $\chi_1(x)$ and $\chi_2(x)$ be two real polynomials satisfying

$$\chi_1^2(x) - x \equiv 0 \ (\mathrm{mod}\ q_1(x))$$

and

$$\chi_2^2(x) - x \equiv 0 \ (\mathrm{mod}\ q_2(x)).$$

It is well-known that we have two real polynomials $h_1(x)$ and $h_2(x)$ such that

$$h_1(x)q_1(x) + h_2(x)q_2(x) = 1.$$

Then on letting

$$\chi(x) = \chi_1(x)h_2(x)q_2(x) + \chi_2(x)h_1(x)q_1(x),$$

we have

$$\chi^2(x) - x \equiv 0 \ (\mathrm{mod}\ q_1(x)q_2(x)).$$

Applying the process repeatedly, we have the theorem.

7. A theorem on pairs of Hermitian forms.

THEOREM 15. *If H and K are two Hermitidn linear λ-matrices having the same elementary divisors, then we have two non-singular matrices Γ_1 and Γ_2 such that*

$$\begin{aligned} \bar{\Gamma}_1 H \Gamma'_1 &= h_1^{(r_1)} + h_2^{(r_2)}, \\ \bar{\Gamma}_2 K \Gamma'_2 &= k_1^{(r_1)} + k_2^{(r_2)}, \end{aligned} \qquad r_1 + r_2 = h, \quad r_1 \geq 0, \quad r_2 \geq 0\,[3]$$

and, we have two non-singular matrices $p_1^{(r_1)}$ and $p_2^{(r_2)}$ such that

$$\begin{aligned} \bar{p}_1 h_1 p'_1 &= k_1, \\ \bar{p}_2 h_2 p'_2 &= -k_2. \end{aligned}$$

[3] In case $r_1 = 0$, $h^{(r_1)}$ is left out.

Proof. 1) By the hypothesis we have two non-singular matrices P and Q such that

$$PHQ = K.$$

Since $PHQ = PH\bar{p}' \cdot \bar{p}'^{-1}Q$, we may assume that $P = I$.

Since H and K are both Hermitian we have

$$HQ = \bar{Q}'H = K.$$

We have a non-singular matrix T such that

$$T^{-1}QT = q_1^{(r_1)} \dotplus q_2^{(r_2)}$$

where q_1 has non-negative characteristic roots and q_2 has only negative characteristic roots. We may assume without loss of generality that

$$Q = q_1 \dotplus q_2.$$

Let

$$H = \begin{pmatrix} h_{11} & h_{12} \\ \bar{h}'_{12} & h_{22} \end{pmatrix}.$$

Since $HQ = \bar{Q}'H$, we have $h_{12}q_2 = \bar{q}'_1 h_{12}$. Since $\bar{q}'_1$, q_2 have no common characteristic root, then $h_{12} = O$. Thus

$$H = h_1^{(r_1)} \dotplus h_2^{(r_2)}.$$

Consequently

$$K = k_1^{(r_1)} \dotplus k_2^{(r_2)},$$

and

$$h_1 q_1 = \bar{q}'_1 h_1 = k_1, \qquad h_2 q_2 = \bar{q}'_2 h_2 = k_2.$$

2) In the lemma of **6** we take $q(x)$ to be the characteristic polynomial of q_1. Then we have a real polynomial $\chi(x)$ such that

$$\chi^2(q_1) = q_1.$$

Then, letting $p_1 = \chi(q_1)$, we have

$$k_1 = h_1 q_1 = h_1 \chi^2(q_1) = \chi(\bar{q}'_1) h_1 \chi(q_1) = \bar{p}'_1 h_1 p_1.$$

Next, in the lemma of **6**, we take $q(x)$ to be the characteristic polynomial of $-q_2$. Then we have a polynomial $\chi(x)$ such that

$$\chi^2(-q_2) = -q_2.$$

Let $p_2 = \chi(-q_2)$, then

$$k_2 = h_2 q_2 = -h_2 \chi^2(-q_2) = -\chi(-\bar{q}'_2) h_2 \chi(-q_2) = -\bar{p}'_2 h_2 p_2.$$

The theorem is then proved.

8. Canonical form of pairs of Hermitian forms. First of all, we introduce the following notations: Let

$$j^{(t)} = \begin{pmatrix} 0 & 0 & \cdots & 0 & 1 \\ 0 & 0 & \cdots & 1 & 0 \\ \cdot & \cdot & \cdot & \cdot & \cdot \\ 0 & 1 & \cdots & 0 & 0 \\ 1 & 0 & \cdots & 0 & 0 \end{pmatrix}$$

be a t-rowed square matrix (a_{ij}) with

$$a_{ij} = \begin{cases} 1 & \text{for } i+j=n+1, \\ 0 & \text{otherwise;} \end{cases}$$

and let

$$m^{(t)}(\lambda) = \begin{pmatrix} 0 & 0 & \cdots & 0 & \lambda \\ 0 & 0 & \cdots & \lambda & 1 \\ \cdot & \cdot & \cdot & \cdot & \cdot \\ 0 & \lambda & \cdots & 0 & 0 \\ \lambda & 1 & \cdots & 0 & 0 \end{pmatrix}$$

be a t-rowed square matrix (b_{ij}) with

$$b_{ij} = \begin{cases} \lambda & \text{for } i+j=n+1, \\ 1 & \text{for } i+j=n+2, \\ 0 & \text{otherwise.} \end{cases}$$

(In case $n=1$, then $b_{11}=\lambda$).

THEOREM 16. *Let (A, B) and (A_1, B_1) be two pairs of Hermitian matrices. Let* $\det(\lambda A + B) = 0$ *have no real root and let A and A_1 be non-singular. A necessary and sufficient condition for the pairs to be conjunctive is that they have the same elementary divisors. More definitely, given*

$$g_i = ((\lambda-\lambda_1)(\lambda-\bar\lambda_1))^{t_{i1}} \cdots ((\lambda-\lambda_k)(\lambda-\bar\lambda_k))^{t_{ik}},$$

where $1 \leq i \leq n$ *and g_i divides g_{i+1} and* $\Sigma t_{ij} = \frac{1}{2}n$. *Let*

$$J = \dot{\sum_i} \dot{\sum_j} \begin{pmatrix} 0 & j^{(t_{ij})} \\ j^{(t_{ij})} & 0 \end{pmatrix}$$

and

$$M = \dot{\sum_i} \dot{\sum_j} \begin{pmatrix} 0 & m^{(t_{ij})}(\lambda_i) \\ m^{(t_{ij})}(\bar\lambda_i) & 0 \end{pmatrix}$$

where the $\dot\Sigma$'s denote direct sums and, for $t_{ij}=0$, the corresponding term is to be left out. Then $\lambda J - M$ has the preassigned g_i as its i-th elementary *divisor. Further every pair of Hermitian matrices (A, B) with g_i as its i-th elementary divisor is conjunctive to (J, M).*

Proof. It is not difficult to verify that $\lambda J - M$ has g_i as its i-th elementary divisor.

1) In Theorem 15, we take

$$H = \lambda A - B, \qquad K = \lambda J - M.$$

If $r_2 = 0$, the theorem is evident. If $r_1 = 0$, then we have a non-singular matrix P such that $\bar{P}HP' = -K$. Let

$$Q = \dot{\sum_i} \dot{\sum_j} \begin{pmatrix} I^{(t_{ij})} & 0 \\ 0 & -I^{(t_{ij})} \end{pmatrix}.$$

Then $\bar{Q}JQ' = -J$ and $\bar{Q}MQ' = -M$. Thus $\bar{Q}\bar{P}HP'Q' = K$ and the theorem is true.

2) Consider first the particular case where we have

$$g_n = ((x - \alpha)(x - \bar{\alpha}))^{n/2}$$

and $g_{n-1} = \cdots = g_1 = 1$. $\lambda J - M$ cannot be conjunctive to a direct sum of two Hermitian matrices. For otherwise we would have two non-singular matrices P and Q such that

$$P(\lambda J - M)Q = \begin{pmatrix} p_1 & 0 \\ 0 & p_2 \end{pmatrix}$$

and p_1 and p_2 are Hermitian. Then either $(x - \alpha)^{n/2}$ or $(x - \bar{\alpha})^{n/2}$, and hence both, would divide $\det(p_1)$. This is impossible. Then we have either $r_1 = 0$ or $r_2 = 0$ in this case. The result is then true for this particular case.

3) If $r_1 \neq 0$, $r_2 \neq 0$, then we have to consider h_1 and h_2 in Theorem 15 separately. Applying induction on the number of the distinct invariant factors, we have the theorem.

THEOREM 17. *Every pair (A, B), $\det(A) \neq 0$, of Hermitian matrices is conjunctive to the following pair (J, M), where*

$$J = \dot{\sum_i} \dot{\sum_j} \epsilon_{ij} j^{(s_{ij})} \dotplus \dot{\sum_i} \dot{\sum_j} \begin{pmatrix} 0 & j^{(t_{ij})} \\ j^{(t_{ij})} & 0 \end{pmatrix},$$

$$M = \dot{\sum_i} \dot{\sum_j} \epsilon_{ij} m^{(s_{ij})}(c_i) \dotplus \dot{\sum_i} \dot{\sum_j} \begin{pmatrix} 0 & m^{(t_{ij})}(\lambda_i) \\ m^{(t_{ij})}(\bar{\lambda}_i) & 0 \end{pmatrix};$$

the first $\dot{\sum_i}$ *runs over all real roots of* $\det(\lambda A + B) = 0$ *and the second* $\dot{\sum_i}$ *runs over all pairs of complex roots of* $\det(\lambda A + B) = 0$, *and* $\epsilon_{ij} = \pm 1$.

The proof of this theorem is completely analogous to that of Theorem 16.

DEFINITION. *The pair of forms* (J, M) *obtained in Theorem* 17 *is called the canonical form of all the pairs conjunctive to it.*

For a fixed c, we may arrange s_{ij} as

$$\begin{aligned} s_{i1} = s_{i2} = \cdots = s_{i\alpha} > s_{i\alpha+1} = \cdots = s_{i\alpha+\beta} \\ > s_{i\alpha+\beta+1} = \cdots = s_{i\alpha+\beta+\gamma} \\ > \cdots = \cdots = s_{i\alpha+\beta+\cdots+\eta}. \end{aligned}$$

We set

$$\begin{aligned} \sigma_1^{(i)} &= \epsilon_{i1} + \cdots + \epsilon_{i\alpha}, \\ \sigma_2^{(i)} &= \epsilon_{i\alpha+1} + \cdots + \epsilon_{i\alpha+\beta}, \\ \sigma_3^{(i)} &= \epsilon_{i\alpha+\beta+1} + \cdots + \epsilon_{i\alpha+\beta+\gamma}. \end{aligned}$$

The constants $\sigma_1^{(i)}, \sigma_2^{(i)}, \cdots$ are called *the system of signatures of the pairs of forms with respect to the real root c.*

To each real root we have a system of signatures. The totality of all the elementary divisors and all the systems of signatures is called the system of elementary divisors with signatures.

9. Law of inertia.

THEOREM 18. *The system of elementary divisors with signatures characterize the conjunctivity of pairs of Hermitian matrices completely. More exactly, the elementary divisors and the systems of signatures are the same for all conjunctive pairs of Hermitian matrices (law of inertia); pairs with different elementary divisors or with the same elementary divisors but different systems of signatures are not conjunctive.*

Proof. 1) It is known that if two pairs of Hermitian matrices are conjunctive, then their elementary divisors are the same. Further, it is evident that two canonical pairs with the same elementary divisors and the same system of signatures are conjunctive.

Thus it is sufficient to establish the result by showing that any two canonical pairs of Hermitian matrices with the same elementary divisors but different systems of signatures are not conjunctive.

2) Let (J, M) and

$$J_1 = \dot{\sum_i} \dot{\sum_j} \epsilon'_{ij} j^{(s_{ij})} \dot{+} \dot{\sum_i} \dot{\sum_j} \begin{pmatrix} 0 & j^{(t_{ij})} \\ j^{(t_{ij})} & 0 \end{pmatrix}$$

$$M_1 = \dot{\sum_i} \dot{\sum_j} \epsilon_{ij} m^{(s_{ij})}(c_i) \dot{+} \dot{\sum_i} \dot{\sum_j} \begin{pmatrix} 0 & m^{(t_{ij})}(\lambda_i) \\ m^{(t_{ij})}(\bar{\lambda}_i) & 0 \end{pmatrix}$$

be two canonical pairs of Hermitian matrices with the same elementary divisors. If (J, M) and (J_1, M_1) are conjunctive, then we have a non-singular $n \times n$ matrix Γ such that

$$\bar{\Gamma}(J, M)\Gamma' = (J_1, M_1).$$

Then

$$\bar{\Gamma}(MJ^{-1}) = (MJ^{-1})\Gamma,$$

since $MJ^{-1} = M_1 J_1^{-1} = \bar{\Gamma}(MJ^{-1})\bar{\Gamma}^{-1}$. Since $J^2 = I$, and

$$MJ^{-1} = \dot{\sum_i} \dot{\sum_j} m^{(s_{ij})}(c_i) j^{(s_{ij})} \dot{+} \dot{\sum_i} \dot{\sum_j} \begin{pmatrix} 0 & m^{(t_{ij})}(\lambda_i) j^{(t_{ij})} \\ m^{(t_{ij})}(\bar{\lambda}_i) j^{(t_{ij})} & 0 \end{pmatrix}$$

we have

$$\Gamma = \dot{\sum_i} \Gamma_i \dot{+} \dot{\sum_i} \begin{pmatrix} \Gamma_{11}^{(i)} & \Gamma_{12}^{(i)} \\ \Gamma_{21}^{(i)} & \Gamma_{22}^{(i)} \end{pmatrix}$$

and

$$\bar{\Gamma}_i(\dot{\sum_j} m^{(s_{ij})}(c_i) j^{(s_{ij})}) = (\dot{\sum_j} m^{(s_{ij})}(c_i) j^{(s_{ij})})\bar{\Gamma}_i.$$

Also

$$\bar{\Gamma}_i(\dot{\sum} \epsilon_{ij} j^{(s_{ij})})\Gamma'_i = \dot{\sum} \epsilon'_{ij} j^{(s_{ij})}$$

$$\bar{\Gamma}_i(\dot{\sum} \epsilon_{ij} m^{(s_{ij})}(c_i))\Gamma'_i = \dot{\sum} \epsilon'_{ij} m^{(s_{ij})}(c_i).$$

Thus it is sufficient to prove the theorem for the case with a unique real root c.

3) We require a

LEMMA. *Let H^A denote the adjoint matrix of H.*

(i) *If H and K are two conjunctive non-singular Hermitian λ-matrices, then H^A and K^A are conjunctive also; furthermore, if we arrange H^A and K^A as polynomials in λ, then their corresponding coefficients (which are matrices) are conjunctive.*

(ii) *If* $\det(H) \neq 0$ *and*

$$H = h_1 \dot{+} h_2 \dot{+} \cdots \dot{+} h_t,$$

then

(iii)
$$\frac{H^A}{d(H)} = \frac{h_1^A}{d(h_1)} \dotplus \frac{h_2^A}{d(h_2)} \dotplus \cdots \dotplus \frac{h_t^A}{d(h_t)}.$$

$$(m^{(t)}(\lambda))^A = (-1)^{\frac{1}{2}(t-1)(t-2)} \begin{pmatrix} 1, & -\lambda, & \lambda^2, \cdots, & (-\lambda)^{t-1} \\ -\lambda, & \lambda^2, & -\lambda^3, \cdots, & 0 \\ \cdot & \cdot & \cdot & \cdot \\ (-\lambda)^{t-1}, & 0, & 0, \cdots, & 0 \end{pmatrix}$$

which is a t-rowed square matrix (a_{ij}) *with*

$$a_{ij} = \begin{cases} (-\lambda)^{i+j+2} & \textit{for} \quad i+j \leq t+1, \\ 0 & \textit{otherwise.} \end{cases}$$

All these results may be verified easily.

4) Since

$$\bar{\Gamma}(J, M)\Gamma' = (J_1, M_1),$$

we have

$$\bar{\Gamma}((\lambda - c)J + M)\Gamma' = (\lambda - c)J_1 + M_1$$

for any λ. We write, dropping the subscript i,

$$M(\lambda) = (\lambda - c)J + M = \dot{\sum_j} \epsilon_j m^{(s_j)}(\lambda)$$

and

$$M_1(\lambda) = (\lambda - c)J_1 + M_1 = \dot{\sum_j} \epsilon'_j m^{(s_j)}(\lambda).$$

They are conjunctive for any λ. Thus $\det(M(\lambda))$ and $\det(M_1(\lambda))$ have the same sign, i. e., $\prod_j \epsilon_j^{s_j} = \sum_j \epsilon'^{s_j}_j$, since

$$\det(\epsilon_j m^{(s_j)}(\lambda)) = (-1)^{\frac{1}{2}s_j(s_j-1)}(\epsilon_j\lambda)^{s_j}.$$

Further, let

$$\Pi\, \epsilon_j^{s_j}(-1)^{\frac{1}{2}s_j(s_j-1)} = \epsilon,$$

$$M(\lambda)^A = \epsilon\lambda^n \left(\dot{\sum_j} \frac{\epsilon_j(m^{(s_j)}(\lambda))^A}{\det(m^{(s_j)}(\lambda))} \right)$$
$$= \epsilon \dot{\sum_j} \epsilon_j(-1)^{\frac{1}{2}s_j(s_j-1)}(m^{(s_j)}(\lambda))^A \lambda^{n-s_j}.$$

The coefficient of λ^{n-s_1} is equal to

$$\epsilon(-1)^{(s_1-1)} \dot{\sum_{1 \leq j \leq a}} \epsilon_j \begin{pmatrix} 1 & 0 & 0 & \cdots & 0 \\ 0 & 0 & 0 & \cdots & 0 \\ \cdot & \cdot & \cdot & \cdot & \cdot \\ 0 & 0 & 0 & \cdots & 0 \end{pmatrix},$$

since $(-1)^{\frac{1}{2}s_1(s_1-1)}(-1)^{\frac{1}{2}(s_1-1)(s_1-2)}=(-1)^{(s_1-1)}$. By (i) of the lemma, the signature of this matrix is equal to that of the corresponding expression of $M_1(\lambda)$; hence

$$\sum_{j=1}^{a}\epsilon_j=\sum_{j=1}^{a}\epsilon'_j.$$

The coefficient of $\lambda^{n-s_{a+1}}$ is of the form

$$\epsilon\Big(\dot{\sum_{1\leq j\leq a}}\epsilon_jP_j\Big)\dotplus\epsilon(-1)^{(s_{a+1}-1)}\dot{\sum_{a+1\leq j\leq a+\beta}}\epsilon_j\begin{pmatrix}1&0&0&\cdots&0\\0&0&0&\cdots&0\\\cdot&\cdot&\cdot&\cdot&\cdot\\0&0&0&\cdots&0\end{pmatrix}.$$

The corresponding expression of $M_1(\lambda)$ may be written as

$$\epsilon\Big(\dot{\sum_{1\leq j\leq a}}\epsilon_jP_j\Big)\dotplus\epsilon(-1)^{(s_{a+1}-1)}\dot{\sum_{a+1\leq j\leq a+\beta}}\epsilon'_j\begin{pmatrix}1&0&0&\cdots&0\\0&0&0&\cdots&0\\\cdot&\cdot&\cdot&\cdot&\cdot\\0&0&0&\cdots&0\end{pmatrix}$$

(by arranging the first part such that $\epsilon_i=\epsilon'_i$ for $1\leq i\leq \alpha$). Thus we have

$$\sum_{j=a+1}^{a+\beta}\epsilon_j=\sum_{j=a+1}^{a+\beta}\epsilon'_j.$$

The result follows by induction.

10. Normal form of hypercircles.

THEOREM 19. *Every hypercircle is conjunctive under $\mathfrak{G}$ to a hypercircle with the matrix*

$$\begin{pmatrix}H_1&O\\O&H_2\end{pmatrix},\qquad H_1=\begin{pmatrix}h_1^{(r)}&O\\O&O\end{pmatrix},\qquad H_2=\begin{pmatrix}h_2^{(r)}&O\\O&O\end{pmatrix}$$

where h_1 and h_2 may be expressed as two direct sums

$$h_1=\dot{\sum_i}\dot{\sum_j}\epsilon_{ij}j^{(s_{ij})}\dotplus\dot{\sum_i}\dot{\sum_j}\begin{pmatrix}O&j^{(t_{ij})}\\j^{(t_{ij})}&O\end{pmatrix}$$

and

$$h_2=\dot{\sum_i}\dot{\sum_j}\epsilon_{ij}m^{(s_{ij})}(c_i)\dotplus\dot{\sum_i}\dot{\sum_j}\begin{pmatrix}O&m^{(t_{ij})}(\lambda_i)\\m^{(t_{ij})}(\bar{\lambda}_i)&O\end{pmatrix}$$

where the c's are real and the λ's are complex numbers.

Proof. By Theorem 14, we have only to consider the case with

5

$$H_1 = \begin{pmatrix} h_1 & 0 \\ 0 & 0 \end{pmatrix}, \qquad H_2 = \begin{pmatrix} h_2 & 0 \\ 0 & 0 \end{pmatrix}, \qquad \det(h_1) \neq 0.$$

Consider the pair of Hermitian matrices $(h_1^{-1}, \bar{h}_2)$.

By Theorem 17, we have a non-singular matrix γ such that

$$\bar{\gamma} h_1^{-1} \gamma' = \dot{\Sigma} \dot{\Sigma} \epsilon_{ij} j^{(s_{ij})} \dotplus \dot{\sum_i} \dot{\sum_j} \begin{pmatrix} 0 & j^{(t_{ij})} \\ j^{(t_{ij})} & 0 \end{pmatrix}$$

$$\bar{\gamma} \bar{h}_2 \gamma' = \dot{\Sigma} \dot{\Sigma} \epsilon_{ij} m^{(s_{ij})}(c_i) \dotplus \dot{\sum_i} \dot{\sum_j} \begin{pmatrix} 0 & m^{(t_{ij})}(\bar{\lambda}_i) \\ m^{(t_{ij})}(\lambda_i) & 0 \end{pmatrix}.$$

Let

$$A = \begin{pmatrix} \bar{\gamma}'^{-1} & 0 \\ 0 & 0 \end{pmatrix}, \quad D = \begin{pmatrix} \bar{\gamma} & 0 \\ 0 & 1 \end{pmatrix}, \quad B = C = 0, \quad \mathfrak{T} = \begin{pmatrix} A & B \\ C & D \end{pmatrix};$$

$\mathfrak{T}$ belongs to $\mathfrak{G}$. Then

$$\bar{\mathfrak{T}} \begin{bmatrix} \begin{pmatrix} h_1 & 0 \\ 0 & 0 \end{pmatrix} & \begin{pmatrix} 0 & 0 \\ 0 & 0 \end{pmatrix} \\ \begin{pmatrix} 0 & 0 \\ 0 & 0 \end{pmatrix} & \begin{pmatrix} h_2 & 0 \\ 0 & 0 \end{pmatrix} \end{bmatrix} \mathfrak{T}'$$

gives the required form. (Notice that

$$\left(\dot{\sum_{i,j}} \dot{\Sigma} \epsilon_{ij} j^{(s_{ij})} \dotplus \dot{\sum_i} \dot{\sum_j} \begin{pmatrix} 0 & j^{(t_{ij})} \\ j^{(t_{ij})} & 0 \end{pmatrix} \right)^2 = I \Bigg).$$

THEOREM 20. *Every hypercircle with a matrix of the form given in Theorem 19 has a canonical discriminantal matrix. Apart from ϵ_{ij}, all other quantities in the expression of the matrix of the hypercircle are completely determined by its discriminantal matrix.*

The proof of the theorem needs only a direct verification.

Thus for a given discriminantal matrix we have only *a finite number* of hypercircles, more exactly, the number of hypercircles is $\leqq 2n$. We have to consider further whether the forms given in Theorem 19 are equivalent. The answer will be given in **15.**

11. Complete reducibility.

DEFINITION. *A sub-set $\mathfrak{C}$ of $\mathfrak{G}$ is said to be completely reducible, if we have a transformation $\mathfrak{W}$ belonging to $\mathfrak{G}$ such that the elements of $\mathfrak{W}^{-1}\mathfrak{C}\mathfrak{W}$ are of the form*

$$\begin{pmatrix} A & B \\ C & D \end{pmatrix}$$

with

$$A=\begin{pmatrix} a_1 & O \\ O & a_2 \end{pmatrix}, \quad B=\begin{pmatrix} b_1 & O \\ O & b_2 \end{pmatrix}, \quad C=\begin{pmatrix} c_1 & O \\ O & c_2 \end{pmatrix}, \quad D=\begin{pmatrix} d_1 & O \\ O & d_2 \end{pmatrix}.$$

THEOREM 21. *Let $\mathfrak{H}$ and $\mathfrak{K}$ be two hypercircles with the same discriminantal matrix $\mathfrak{D}$, and let* $\det(\mathfrak{D}-\lambda\mathfrak{F})=0$ *have more than one distinct root. The transformations which carry $\mathfrak{H}$ to $\mathfrak{K}$ are completely reducible. In particular, if $\mathfrak{H}=\mathfrak{K}$, they form a completely reducible group.*

Proof. We may assume that

$$\mathfrak{D}=\begin{pmatrix} O & T \\ -T' & O \end{pmatrix};$$

where $T=t_1 \dotplus t_2$ and t_1 and t_2 have no common characteristic roots.

Suppose that $\bar{\mathfrak{T}}\mathfrak{H}\mathfrak{T}'=\mathfrak{K}$ where $\mathfrak{T}$ belongs to $\mathfrak{G}'$, then $\mathfrak{T}\mathfrak{D}\mathfrak{T}'=\mathfrak{D}$. Since $\mathfrak{T}\mathfrak{F}\mathfrak{T}'=\mathfrak{F}$ we have $\mathfrak{T}'^{-1}=-\mathfrak{F}\mathfrak{T}\mathfrak{F}$. Then $\mathfrak{T}\mathfrak{D}=-\mathfrak{D}\mathfrak{F}\mathfrak{T}\mathfrak{F}$.
Put

$$\mathfrak{F}=\begin{pmatrix} A & B \\ C & D \end{pmatrix},$$

then

$$\begin{pmatrix} A & B \\ C & D \end{pmatrix}\begin{pmatrix} O & T \\ -T' & O \end{pmatrix}=\begin{pmatrix} O & T \\ -T' & O \end{pmatrix}\begin{pmatrix} D & -C \\ -B & A \end{pmatrix}$$

i. e.,

$$T'C=CT, \qquad T'D=DT',$$
$$TA=AT, \qquad TB=BT'.$$

Since t_1, t_2 have no common characteristic root, we have

$$A=a_1\dotplus a_2, \qquad B=b_1\dotplus b_2,$$
$$C=c_1\dotplus c_2, \qquad D=d_1\dotplus d_2.$$

The theorem follows.

In order to investigate the conjunctivity under $\mathfrak{G}$ of the forms in Theorem 19, we need only investigate the conjunctivity under $\mathfrak{G}$ of

$$h_1=\dot{\sum}\epsilon_i j^{(t_i)}$$
$$h_2=\dot{\sum}\epsilon_i m^{(t_i)}(c)$$

where c is a real number. The solutions are quite different according to $c<0$, >0 or $=0$.

12. Conjunctivity under $\mathfrak{G}$ for $c > 0$.

THEOREM 22. *The hypercircle with the matrix*

$$\begin{pmatrix} j^{(t)} & 0 \\ 0 & m^{(t)}(c) \end{pmatrix}$$

is conjunctive under $\mathfrak{G}$ *to that with*

$$-\begin{pmatrix} j^{(t)} & 0 \\ 0 & m^{(t)}(c) \end{pmatrix}$$

provided $c < 0$.

Proof. We shall first establish the following preliminary result:

We have a real and symmetric matrix $s^{(t)}$ such that

$$sj^{(t)}s = -m(c),$$

if $c < 0$.

The result is true for $t = 1$, since

$$\sqrt{|c|}\cdot 1\cdot\sqrt{|c|} = -c, \text{ i. e., } s = \sqrt{-c}.$$

The result is also true for $t = 2$, since

$$\begin{pmatrix} 0 & \sqrt{-c} \\ \sqrt{-c}, & -\tfrac{1}{2}(\sqrt{-c})^{-1} \end{pmatrix}\begin{pmatrix} 0 & 1 \\ 1 & 0 \end{pmatrix}\begin{pmatrix} 0 & \sqrt{-c} \\ \sqrt{-c}, & -\tfrac{1}{2}(\sqrt{-c})^{-1} \end{pmatrix}$$

$$= -\begin{pmatrix} 0 & c \\ c & 1 \end{pmatrix}, \quad \text{i. e.,} \quad s = \begin{pmatrix} 0 & \sqrt{-c} \\ \sqrt{-c}, & -\tfrac{1}{2}(\sqrt{-c})^{-1} \end{pmatrix}$$

Suppose that the theorem is true for t, then we shall prove that it is also true for $t + 2$, i. e., suppose we have s such that

$$sjs = -m(c)$$

and

$$\det(sj + \sqrt{-c}\, I^{(t)}) \neq 0.$$

Then, we solve

$$\begin{pmatrix} 0 & 0 & z \\ 0 & s^{(t)} & w' \\ z & w & u \end{pmatrix}\begin{pmatrix} 0 & 0 & 1 \\ 0 & j^{(t)} & 0 \\ 1 & 0 & 0 \end{pmatrix}\begin{pmatrix} 0 & 0 & z \\ 0 & s & w' \\ s & w & u \end{pmatrix} = -m^{(t+2)}(c)$$

i. e., we find real numbers z, u and a t-dimensional vector w such that

$$z^2 = -c, \qquad w'z + sjw' = \begin{pmatrix} -1 \\ 0 \\ \cdot \\ \cdot \\ \cdot \\ 0 \end{pmatrix}, \qquad 2uz + wjw' = 0.$$

The first equation gives $z = \sqrt{-c}$, the second is then soluble in w if and only if

$$\det (sj + \sqrt{-c}\, I^{(t)}) \neq 0$$

which is true by assumption, and from the third we then have the value of u. Set

$$s^{(t+2)} = \begin{pmatrix} 0 & 0 & z \\ 0 & s^{(t)} & w' \\ s & w & u \end{pmatrix}$$

where z, w, u are determined in this way; then $s^{(t+2)}$ satisfies

$$s^{(t+2)} j^{(t+2)} s^{(t+2)} = -m^{(t+2)}(c)$$

and

$$\det (s^{(t+2)} j^{(t+2)} + \sqrt{-c}\, I^{(t+2)}) \\ = -4c \det (s^{(t)} j^{(t)} + \sqrt{-c}\, I^{(t)}) \neq 0.$$

The preliminary result is now proved. Let

$$\mathfrak{T} = \begin{pmatrix} O & s^{-1} \\ -s & O \end{pmatrix}$$

which belongs to $\mathfrak{G}$. Then

$$\bar{\mathfrak{T}} \begin{pmatrix} j^{(t)} & O \\ O & m^{(t)}(c) \end{pmatrix} \mathfrak{T}' = \begin{pmatrix} s^{-1} m(c) s^{-1} & O \\ O & sjs \end{pmatrix} = -\begin{pmatrix} j^{(t)} & O \\ O & m(c) \end{pmatrix}.$$

The theorem follows.

Consequently, the signs ϵ_{ij} corresponding to a negative c_i in Theorem 19 may be replaced by $+1$.

13. Conjunctivity under $\mathfrak{G}$ for $c > 0$.

THEOREM 23. *If $\mathfrak{H}_1$ and $\mathfrak{H}_2$ are conjunctive under $\mathfrak{G}$, then the two pairs of Hermitian matrices*

$$(\bar{\mathfrak{H}}_1, \mathfrak{F}\mathfrak{H}_1{}^A\mathfrak{F})$$

and

$$(\bar{\mathfrak{H}}_2, \mathfrak{F}\mathfrak{H}_2{}^A\mathfrak{F})$$

are also conjunctive under $\mathfrak{G}$.

Proof. Let $\mathfrak{T}$ be an element of $\mathfrak{G}$ and $\bar{\mathfrak{T}}\mathfrak{H}_1\mathfrak{T}' = \mathfrak{H}_2$. Since $\mathfrak{T}\mathfrak{F}\mathfrak{T}' = \mathfrak{F}$ and $\mathfrak{T}^{-1} = \mathfrak{T}^A$ we have

$$\mathfrak{H}_2{}^A = \mathfrak{T}'^A\mathfrak{H}_1{}^A\mathfrak{T}^A = \mathfrak{T}'^{-1}\mathfrak{H}_1{}^A\mathfrak{T}^{-1}$$

and

$$\mathfrak{F}\mathfrak{H}_2{}^A\mathfrak{F} = \mathfrak{F}\mathfrak{T}'^{-1}\mathfrak{H}_1{}^A\mathfrak{T}^{-1}\mathfrak{F} = \mathfrak{T}\mathfrak{F}\mathfrak{H}_1{}^A\mathfrak{F}\mathfrak{T}'.$$

Therefore

$$\mathfrak{T}(\lambda\bar{\mathfrak{H}}_1 + \mu\mathfrak{F}\mathfrak{H}_1{}^A\mathfrak{F})\bar{\mathfrak{T}}' = \lambda\bar{\mathfrak{H}}_2 + \mu\mathfrak{F}\mathfrak{H}_2{}^A\mathfrak{F}.$$

THEOREM 24. *Let* $c > 0$, *and*

$$h_1 = \dot{\sum} \epsilon_i j^{(s_i)}$$
$$h_2 = \dot{\sum} \epsilon_i m^{(s_i)}(c).$$

For different systems of signatures we have non-conjunctive hypercircles (under $\mathfrak{G}$*) with matrices*

$$\mathfrak{H} = \begin{pmatrix} h_1 & O \\ O & h_2 \end{pmatrix}$$

under $\mathfrak{G}$.

Proof. Let

$$\mathfrak{H} = \begin{pmatrix} h_1 & O \\ O & h_2 \end{pmatrix}, \qquad \mathfrak{K} = \begin{pmatrix} k_1 & O \\ O & k_2 \end{pmatrix}$$

be two such hypercircles with different systems of signatures. If they are conjunctive under $\mathfrak{G}$, then

$$(\mathfrak{H}, -\mathfrak{F}\mathfrak{H}^A\mathfrak{F}) = \left(\begin{pmatrix} h_1 & O \\ O & h_2 \end{pmatrix}, \begin{pmatrix} \det(h_1)h_2{}^A & O \\ O & \det(h_2)h_1{}^A \end{pmatrix}\right)$$

and

$$(\mathfrak{K}, -\mathfrak{F}\mathfrak{K}^A\mathfrak{F}) = \left(\begin{pmatrix} k_1 & O \\ O & k_2 \end{pmatrix}, \begin{pmatrix} \det(k_1)k_2{}^A & O \\ O & \det(k_2)k_1{}^A \end{pmatrix}\right)$$

are conjunctive.

We shall now prove that

$$\phi = \lambda h_1 + \mu \det(h_1)h_2{}^A$$

is conjunctive to

$$\psi = \lambda h_2 + \mu \det (h_2) h_1^A.$$

We have

$$h_1^A \phi h_2 = h_1^A (\lambda h_1 + \mu \det (h_1) h_2^A) h_2$$
$$= \det (h_1) (\lambda h_2 + \mu \det (h_2) h_1^A) = \det (h_1) \psi.$$

Then

$$h_1^A \phi \bar{h}_1'^A \cdot h_1 h_2 = (\det (h_1))^2 \psi.$$

Now $[\det (h_1)]^2$ is positive and $h_1 h_2$ is a matrix with a positive characteristic root c. Hence as in the proof of Theorem 15, we have a matrix p such that

$$p h_1^A \phi \bar{h}_1'^A \bar{p}' = \psi.$$

Thus ϕ and ψ have the same system of elementary divisors with the same systems of signatures. Thus if $(\mathfrak{H}, \mathfrak{F}\mathfrak{H}^A\mathfrak{F})$ and $(\mathfrak{K}, \mathfrak{F}\mathfrak{K}^A\mathfrak{F})$ are conjunctive, then

$$(h_1, \det (h_2) h_1^A), \qquad (k_2, \det (k_2) k_1^A$$

are conjunctive, then (since $h_1^{-1} = h_1$, $k_1^{-1} = k_1$),

$$(h_2, h_1), \qquad (k_2, k_1)$$

are conjunctive. By Theorem 16, they are conjunctive if and only if they have the same systems of signatures.

Consequently the signs ϵ_{ij} corresponding to a positive c_i in Theorem 19 are significant.

14. Conjunctivity under $\mathfrak{G}$ for $c = 0$.

Here we require a preliminary lemma.

LEMMA. Let

$$t^{(l)} = \begin{pmatrix} \lambda & 1 & 0 & \cdots & 0 \\ 0 & \lambda & 1 & \cdots & 0 \\ \cdot & \cdot & \cdot & \cdot & \cdot \\ 0 & 0 & 0 & \cdots & 1 \\ 0 & 0 & 0 & \cdots & \lambda \end{pmatrix}$$

be an l-rowed matrix. The solution of

$$x^{(l,m)} t^{(m)} = t^{(l)} x^{(l,m)}$$

is of the form

$$x^{(l,m)} = \begin{pmatrix} x_1, & x_2, & \cdots, & x_m \\ 0, & x_1, & \cdots, & x_{m-1} \\ \cdot & \cdot & \cdot & \cdot \\ 0, & 0, & \cdots, & x_1 \\ 0, & 0, & \cdots, & 0 \\ \cdot & \cdot & \cdot & \cdot \end{pmatrix} \quad \text{if } l > m,$$

$$x^{(l,m)} = \begin{pmatrix} 0 \cdots 0 & x_1, & x_2, & \cdots, & x_l \\ 0 \cdots 0 & 0, & x_1, & \cdots, & x_{l-1} \\ \cdot & \cdot & \cdot & \cdot & \cdot \\ 0 \cdots 0, & 0, & 0, & \cdots, & x_1 \end{pmatrix} \quad \text{if } l > m,$$

$$x^{(l,l)} = \begin{pmatrix} x_1, & x_2, & \cdots, & x_l \\ 0, & x_1, & \cdots, & x_{l-1} \\ \cdot & \cdot & \cdot & \cdot \\ 0, & 0, & \cdots, & x_1 \end{pmatrix}.$$

THEOREM 25. *Theorem 24 is also true for $c = 0$.*

Proof. Let

$$\mathfrak{H} = \begin{pmatrix} H_1 & 0 \\ 0 & H_2 \end{pmatrix}, \qquad \mathfrak{K} = \begin{pmatrix} K_1 & 0 \\ 0 & K_2 \end{pmatrix},$$

and let

$$T = H'_1 H_2 = K'_1 K_2 = \dot{\sum} j^{(s_i)} m^{(s_i)}(0),$$

where

$$j^{(s)} m^{(s)}(0) = \begin{pmatrix} 0 & 1 & 0 & \cdots & 0 \\ 0 & 0 & 1 & \cdots & 0 \\ \cdot & \cdot & \cdot & \cdot & \cdot \\ 0 & 0 & 0 & \cdots & 1 \\ 0 & 0 & 0 & \cdots & 0 \end{pmatrix}$$

(For $s = 1$, it is zero.)

Evidently $H_1{}^2 = K_1{}^2 = I$. Let

$$\bar{\mathfrak{T}} \mathfrak{H} \mathfrak{T}' = \mathfrak{K}, \qquad \mathfrak{T} = \begin{pmatrix} A & B \\ C & D \end{pmatrix}.$$

Since $\mathfrak{T}\mathfrak{F}\mathfrak{T}' = \mathfrak{F}$, $\mathfrak{F}^2 = -I^{(2n)}$ and $\mathfrak{T}\mathfrak{D}\mathfrak{T}' = \mathfrak{D}$, we have $\mathfrak{T}\mathfrak{D}\mathfrak{F} = \mathfrak{D}\mathfrak{F}\mathfrak{T}$. Now $\mathfrak{D} = \begin{pmatrix} 0 & T \\ -T' & 0 \end{pmatrix}$; consequently, we have

$$AT = TA, \qquad BT' = TB, \qquad CT = T'C, \qquad DT' = T'D.$$

Now we use Greek letters to denote matrices commutative with T. Then

$$A=\alpha,\qquad B=\beta H_1,\qquad C=H_1\gamma,\qquad D=H_1\delta H_1,$$

since $T'=H_1TH_1$. Since

$$\overline{\begin{pmatrix} A & B\\ C & D\end{pmatrix}}\begin{pmatrix} H_1 & O\\ O & H_2\end{pmatrix}\begin{pmatrix} A & B\\ C & D\end{pmatrix}'=\begin{pmatrix} K_1 & O\\ O & K_2\end{pmatrix},$$

we have

$$K_1=\bar{A}H_1A'+\bar{B}H_2B'=\bar{\alpha}H_1\alpha'+\bar{\beta}H_1H_2H_1\beta'=\bar{\alpha}H_1\alpha'+T\bar{\beta}H_1\beta'.$$

Write

$$K_1=(k_{ij})_{1\leq i,j\leq\kappa}$$

with

$$k_{ii}=\epsilon'_{ij}{}^{(s_i)},\qquad k_{ij}=0\quad\text{for}\quad i\neq j.$$

Similarly, we write

$$H_1=(h_{ij})_{1\leq i,j\leq\kappa}$$

with

$$h_{ii}=\epsilon_{ij}{}^{(s_i)},\qquad h_{ij}=0\quad\text{for}\quad i\neq j.$$

Further, we write

$$T=(t_{ij})$$

with

$$t_{ij}=j^{(s_i)}m^{(s_i)}(0),\qquad t_{ij}=0\quad\text{for}\quad i\neq j;$$

and finally, we write

$$\alpha=(a_{ij}),\qquad a_{ij}=a_{ij}{}^{(s_i,s_j)}.$$

Then

$$k_{ij}=\sum_{\lambda,\mu}\bar{a}_{i\lambda}h_{\lambda\mu}a'_{j\mu}+\sum_{\lambda\ldots}t_{i\lambda}\cdots.$$

Now we consider the element in the $(s_i,1)$-position. The contribution from k_{ij} is either ϵ'_i for $i=j$ or 0 for $i\neq j$. The contribution from $\sum_{\lambda\ldots}t_{i\lambda}\cdots$ is zero, since the last row of $t_{i\lambda}$ is zero.

By the lemma, since

$$a_{ik}t_{kk}=t_{ii}a_{ik},$$

we have

$$a_{ik}=\begin{pmatrix} 0 & \cdots & 0, & x_{ik} & * & \cdots & * \\ \cdot & \cdot & \cdot & \cdot & \cdot & \cdot & * \\ 0, & \cdot & \cdot & \cdot & \cdot & 0 & x_{ik}\end{pmatrix}\quad\text{for } s_i>s_k,$$

$$\text{or} = \begin{pmatrix} x_{ik} & * & \cdots & * \\ \cdot & \cdot & \cdot & \cdot \\ 0 & 0 & \cdots & x_{ik} \\ 0 & 0 & \cdots & 0 \\ \cdot & \cdot & \cdot & \cdot \end{pmatrix} \quad \text{for } s_i < s_k,$$

$$\text{or} = \begin{pmatrix} x_{ik} & * & \cdots & * \\ 0 & x_{ik} & \cdots & * \\ \cdot & \cdot & \cdot & \cdot \\ 0 & 0 & \cdots & x_{ik} \end{pmatrix} \quad \text{for } s_i = s_k.$$

The element in the $(s_i, 1)$-position of $\bar{a}_{i\lambda} h_{\lambda\mu} a'_{j\mu}$ is zero for $\lambda \neq \mu$; is zero for $s_i < s_\lambda$; is zero for $s_j > s_\mu$; and is

$$\sum_{s_\lambda = s_i = s_j} \bar{x}_{i\lambda} \epsilon_\lambda x_{j\lambda} \quad \text{for} \quad s_\lambda = s_\mu = s_i = s_j.$$

Thus we obtain

$$\sum_{s_\lambda = s_i = s_j} \bar{x}_{i\lambda} \epsilon_\lambda x_{j\lambda} = \begin{cases} \epsilon'_i & \text{if } i = j, \\ 0 & \text{if } i \neq j. \end{cases}$$

Let all the elements s_ν equal to s_μ be

$$s_{\eta+1}, \cdots, s_{\eta+\xi}.$$

Then

$$\begin{pmatrix} \epsilon'_{\eta+1} & 0 & \cdots & 0 \\ 0 & \epsilon'_{\eta+2} & \cdots & 0 \\ \cdot & \cdot & \cdot & \cdot \\ 0 & 0 & \cdots & \epsilon'_{\eta+\xi} \end{pmatrix} = (\overline{x_{ij}}) \begin{pmatrix} \epsilon_{\eta+1} & 0 & \cdots & 0 \\ 0 & \epsilon_{\eta+2} & \cdots & 0 \\ \cdot & \cdot & \cdot & \cdot \\ 0 & 0 & \cdots & \epsilon_{\eta+\xi} \end{pmatrix} (x_{ij})'.$$

Thus

$$\epsilon_{\eta+1} + \cdots + \epsilon_{\eta+\xi} = \epsilon'_{\eta+1} + \cdots + \epsilon'_{\eta+\xi}.$$

The result follows.

15. Canonical form of hypercircles. We now summarize the results of **10-14**.

THEOREM 26. *Every hypercircle is conjunctive under* $\mathfrak{G}$ *to a hypercircle with the matrix*

$$\begin{pmatrix} H_1 & 0 \\ 0 & H_2 \end{pmatrix}, \quad H_1 = \begin{pmatrix} h_1^{(r)} & 0 \\ 0 & 0 \end{pmatrix}, \quad H_2 = \begin{pmatrix} h_2^{(r)} & 0 \\ 0 & 0 \end{pmatrix},$$

where h_1 *and* h_2 *may be expressed as two direct sums*

$$h_2 = \sum_{c_i \geqq 0}\sum \epsilon_{ij} m^{(s_{ij})}(c_i) + \dot{\sum_{c_i<0}}\dot{\sum} m^{(s_{ij})}(c_i) + \dot{\sum_i}\dot{\sum_j}\begin{pmatrix} 0 & m^{(t_{ij})}(\lambda_i) \\ m^{(t_{ij})}(\lambda_i) & 0 \end{pmatrix}$$

and

$$h_1 = \dot{\sum_{c_i \geqq 0}}\dot{\sum} \cdot \epsilon_{ij} j^{(s_{ij})} + \dot{\sum_{c_i<0}}\dot{\sum} j^{(s_{ij})} + \dot{\sum_i}\dot{\sum_j}\begin{pmatrix} 0 & j^{(t_{ij})} \\ j^{(t_{ij})} & 0 \end{pmatrix},$$

where the first double summation runs over non-negative c's, the second runs over negative c's and the third runs over all complex λ's.

Moreover, to each non-negative c, we may define the system of signatures as we did for the pairs of Hermitian matrices. Elementary divisors and systems of signatures characterize completely the conjunctivity of hypercircles under $\mathfrak{G}$.

Thus the problem of the conjunctivity of hypercircles under $\mathfrak{G}$ is now solved completely.

16. A final remark.

The treatment is much simpler for the case of the group $\mathfrak{G}_{II}$ which consists of all transformations of the form

$$Z_1 = (AZ + B)(CZ + D)^{-1},$$
$$A\bar{B}' = B\bar{A}', \qquad C\bar{D}' = D\bar{C}' \qquad A\bar{D}' - B\bar{C}' = I.$$

It is evident that a transformation with the matrix

$$\mathfrak{T} = \begin{pmatrix} A & B \\ C & D \end{pmatrix}$$

belongs to $\mathfrak{G}_{II}$ if and only if

$$\bar{\mathfrak{T}}\mathfrak{F}\mathfrak{T}' = \mathfrak{F}.$$

Correspondingly, the transformation of hypercircles may be written as

$$\bar{\mathfrak{T}}\mathfrak{H}\mathfrak{T}' = \mathfrak{K}.$$

Thus, the pair $\mathfrak{H}, \mathfrak{K}$ are conjunctive under $\mathfrak{G}_{II}$ in the strict sense, if and only if the pairs of Hermitian matrices

$$(\mathfrak{H}, i\mathfrak{F}), \qquad (\mathfrak{K}, i\mathfrak{F})$$

are conjunctive.

The classification of the hypercircles under $\mathfrak{G}_{II}$ is thus simply a straightforward application of the preceding results on pairs of Hermitian forms.

NATIONAL TSING HUA UNIVERSITY OF CHINA,
INSTITUTE OF MATHEMATICS, ACADEMIA SINICA.

Annals of Mathematics
Vol. 47, No. 2, April, 1946

ON THE THEORY OF FUCHSIAN FUNCTIONS OF SEVERAL VARIABLES

By Loo-keng Hua

(Received August 20, 1945)

1. Introduction

The paper contains a part of the author's general treatment of the theory of Fuchsian functions of several complex variables which may be considered as the first approximation of the author's precise results concerning Fuchsian functions of a matrix variable. The hypothesis is comparatively simple and weak. Broadly speaking any discontinuous group of automorphs of a bounded transitive space will fulfil our requirements for most purposes. In reviewing the history of the theory of automorphic functions of several variables, we find either that the group is too specific [2][1] or that the hypotheses are too complicated [3]. The present treatment seems to be comparatively satisfactory in both respects.

The space is metrized in a way which seems to be simpler and more precise than the metric of S. Bergmann [4]. In §4, the author establishes that any discontinuous group of the space is properly discontinuous. The convergence of Poincaré theta series has been investigated and a criterion, which seems to be the best possible, has been obtained.

If we restrict ourselves to the case that the discontinuous group has a compact fundamental domain, we can go a good deal further. As an example, we give a generalization of Siegel's theorem concerning the dependence of automorphic forms.

As an illustration, the author gives a discussion for Picard's hyperabelian functions at the end of the paper.

2. Analytic (or pseudo-conformal) automorphs of a bounded domain

Let $\mathfrak{R}$ be a bounded domain of the $2n$-dimensional space

$$(z) = (z_1, \cdots, z_n), \qquad z_k = x_k + iy_k, \qquad 1 \leqq k \leqq n.$$

If there is no ambiguity, we use simply z to denote the complex vector (z). The domain $\mathfrak{R}$ will be referred to as space $\mathfrak{R}$. Without loss of generality we may assume that the origin

$$(0) = (0, \cdots, 0)$$

is an interior point of the space $\mathfrak{R}$. Let Γ be the group of analytic automorphs of the space $\mathfrak{R}$. Suppose that Γ is transitive, i.e. that any interior point of $\mathfrak{R}$ may be carried to (0) by a transformation of the group Γ.

[1] A list of references is given at the end of the paper. The booklet of Behnke and Thullen [1] will be referred to as B-T in the paper.

The analytic automorphs with the fixed point 0 form a subgroup Γ_0 of Γ. It is called the *group of stability*. It is known that the group Γ_0 is compact (H. Cartan [5]). More precisely, since every transformation of Γ_0 is uniquely determined by its linear terms, each transformation of Γ_0 may be expressed by a matrix which is formed by the coefficients of linear terms. Moreover, owing to the compactness, by a linear transformation, we may write a transformation of Γ_0 as

$$w = t_{0,U}(z)$$

where U denotes a unitary matrix.

More definitely, let

$$U = (u_{ij}),$$

then the transformation

$$w = t_{0,U}(z)$$

denotes the analytic mapping

$$w_i = \sum_{j=1}^{n} u_{ij} z_j + \text{terms of higher powers}$$

(Cf. B-T, Kapitel 6).

Consider the cosets of Γ/Γ_0. Let

$$w = t_{a,I}(z) \qquad \text{(I being identity)}$$

be the transformation carrying (0) to $(a) = (a_1, \cdots, a_n)$. As a tends to zero, it approaches the identity transformation. Then the elements of Γ may be written as

$$w = t_{a,U}(z) = t_{a,I}(t_{0,U}(z)), \tag{1}$$

where a runs over all interior points of $\mathfrak{R}$ and U runs over all admissible unitary matrices in the group of stability. The transformation given by (1), with fixed (a), form a coset of Γ/Γ_0, which carries (0) to (a). Such a formulation suggests that the geometry is not very far from the ordinary hyperbolic geometry.

It is known that a family of analytic functions bounded in the aggregate form a normal family. In particular

$$t_{a,U}(z)$$

form a normal family, that is, in any sequence of the mappings

$$t_{a_1,U_1}(z), \cdots, t_{a_n,U_n}(z), \cdots,$$

we can select a subsequence

$$t_{a_{n_i},U_{n_i}}(z)$$

such that each component approaches an analytic function as a limit.

Let

$$J_{a,U}(z)$$

be the functional determinant of the transformation

$$w = t_{a,U}(z).$$

Let b be a point on the boundary of the space then we have a sequence a_i such that

$$\lim_{a_i \to b} J_{a_i,U}(z) = 0.$$

In fact, by normality, we have a_i such that

$$t_{a_i U}(z)$$

converges uniformly either to an automorph of $\mathfrak{R}$ or to a degenerate transformation. Since the limiting transformation carries 0 to b, it cannot be an automorph of $\mathfrak{R}$. Then we have the assertion (Cf. B-T, Satz 55).

3. Metrization of the space

We write

$$w = t_{a,U}^{-1}(z) = f(z, a, U),$$

as the inverse mapping of

$$z = t_{a,U}(w).$$

Then

$$f(a, a, U) = 0.$$

We have

$$dw_s = \sum_{t=1}^{n} dz_t\, a_{ts}, \qquad a_{ts} = \frac{\partial w_z}{\partial z_t}.$$

Let $J_{a,U}(z)$ denote the matrix

$$(a_{ts}) = \left(\frac{\partial w_s}{\partial z_t}\right).$$

THEOREM 1. *The Hermitian differential form*

$$(\overline{dz})\overline{J_{z,U}(z)}J_{z,U}(z)'\,(dz)'$$

is positive definite for z belonging to the space $\mathfrak{R}$ and is invariant under the group Γ, where $\bar{M}$ and M' denote the conjugate and the transposed matrices of M respectively.

PROOF. Since in $\mathfrak{R}$, the function

$$d(J_{zU}(z)) \neq 0,$$

where $d(M)$ denote the determinant of the matrix M, the Hermitian form is evidently definite. Let

$$z = f(x, \beta, V)$$

and

$$w = f(f(x, \beta, V), \alpha, U) = f(x, \gamma, W). \tag{1}$$

Putting $x = \gamma$ in (1), we have

$$0 = f(f(\gamma, \beta, V), \alpha, U).$$

Consequently, we have

$$\alpha = f(\gamma, \beta, V). \tag{2}$$

We have

$$(dz) = (dx) J_{\beta,V}(x).$$

Further

$$\begin{aligned} J_{z,U}(z) &= (J_{\alpha,U}(z))_{\alpha=z} \\ &= (J_{\alpha,U}(f(x, \beta, V)))_{\alpha=f(x,\beta,V)} \\ &= (J_{\beta,V}(x))^{-1} \left(\frac{\partial f(x, \gamma, W)}{\partial x} \right)_{\gamma=x} \end{aligned}$$

by (2).

Further, from

$$t_{\gamma,W}(w) = t_{\gamma,U}(u),$$

we deduce that $u = 0$ implies $w = 0$, i.e.

$$u = t_{0,X}(w),$$

where X is a unitary matrix. Then

$$f^{-1}(w, \gamma, W) = f^{-1}(f^{-1}(w, 0, X), \gamma, U),$$

and consequently

$$f(f(w, \gamma, U), 0, X) = f(w, \gamma, W).$$

We have immediately that

$$\left(\frac{\partial (f(w, \gamma, W))}{\partial w} \right)_{w=\gamma} = \left(\frac{\partial f(w, \gamma, U)}{\partial w} \right)_{w=\gamma} \left(\frac{\partial f(x, 0, X)}{\partial x} \right)_{x=0}.$$

Thus, we obtain

$$\left(\frac{\partial f(x, \gamma, W)}{\partial x} \right)_{\gamma=x} = \left(\frac{\partial f(x, \gamma, U)}{\partial x} \right)_{\gamma=x} X,$$

and

$$J_{z,U}(z) = (J_{\beta,U}(x))^{-1} J_{x,U}(x) X.$$

Then, we deduce

$$\begin{aligned}&(\overline{dz})\ \overline{J_{z,U}(z)}\ J_{z,U}(z)'(dz)'\\ &= (\overline{dx})\ \overline{J_{\beta,V}(x)}\ \overline{J_{\beta V}(x)}^{-1}\ \overline{J_{z,U}(x)}\ \bar{X}X'(J_{z,U}(x))'(J_{\beta,V}(x))'^{-1}(J_{\beta,V}(x))'(dx)'\\ &= (\overline{dx})\ \overline{J_{z,U}(x)}\ J_{z,U}(x)'(dx)'.\end{aligned}$$

The theorem is now established.

Incidentally, we have established also the following theorem:

THEOREM 2. *The Hermitian differential form*

$$(\overline{dz})\overline{J_{z,U}(z)}J_{z,U}(z)'(dz)'$$

is independent of U.

Is the metric unique? We have the following answer.

THEOREM 3. *If*

$$(\overline{dz})H(z, \bar{z})(dz)',\ H = (h_{ij}),\ \bar{H}' = H$$

is a positive definite Hermitian form invariant under the group Γ, *then it is equivalent to the form given in Theorem* 1 *by a suitable choice of coordinate system. More precisely, we have a constant matrix* C *such that*

$$H(w, \overline{W}) = \overline{J_{z,U}(z)}J_{z,U}(z)'$$

where $(w) = (z)C$.

PROOF. Let

$$z = f(w, \beta, V)$$

Then

$$\overline{J_{\beta,V}(w)}H(z, \bar{z})J_{\beta,V}(w)' = H(w, \bar{w}).$$

In particular, for $z = 0$, we have

$$H(\beta, \bar{\beta}) = \overline{J_{\beta,V}(\beta)}H(0, 0)J_{\beta,V}(\beta)'. \tag{1}$$

For $\beta = 0$, we have

$$H(0, 0) = \bar{V}H(0, 0)V',$$

since

$$J_{0,V}(0) = V.$$

We may choose P such that

$$H(0, 0) = \bar{P}P'.$$

Then

$$\bar{P}P' = \bar{V}\bar{P}P'V'.$$

Now $P^{-1}VP$ is unitary. Let

$$H(z, \bar{z}) = \bar{P}K(\bar{z}, \bar{z})P'.$$

Then, from (1), we have

$$H(z, \bar{z}) = \overline{J_{z,V}(z)}\bar{P}P'J_{z,V}(z)',$$

i.e.

$$K(z, \bar{z}) = \overline{P^{-1}J_{z,V}(z)P}(P^{-1}J_{z,V}(z)P)'.$$

The theorem follows.

THEOREM 4. *If the group of stability is irreducible, then the nonsingular invariant differential form is unique up to a constant factor.*

With the notation of the proof of Theorem 3, we have

$$H(0, 0) = \bar{V}H(0, 0)V'.$$

The result follows easily.

Evidently we have

THEOREM 5. *The volume element*

$$| d(J_{z,U}(z)) |^2 dx_1 \cdots dx_n dy_1 \cdots dy_n$$

is invariant under the group Γ, *where* $d(M)$ *denotes the determinant of the matrix* M.

REMARK. The present result depends only on the property that the group of stability is compact. The boundedness of $\mathfrak{R}$ is not used in its full force. It seems to be true that the Riemannian curvature of the space $\mathfrak{R}$ is never positive. But the author failed to find a proof for this important theorem. Moreover the transitivity is also a non-essential assumption.

4. Discontinuous group

A subgroup G of the group Γ is called *discontinuous,* if every infinite sequence of the transformations of G does not converge to a transformation of Γ. G is called *properly discontinuous,* if the set of images of an inner point of $\mathfrak{R}$ has no limit point in $\mathfrak{R}$. Evidently every properly discontinuous group is discontinuous. Now we shall prove that

THEOREM 6. *Every discontinuous group of the space* $\mathfrak{R}$ *is properly discontinuous.*

PROOF. Supposing the contrary, without loss of generality, we assume that

$$a_1, a_2, \cdots, a_n, \cdots$$

is a sequence of points equivalent to zero under the group G, and that the sequence approaches a. The corresponding transformations are

$$t_{a_1,U_1}(z), \cdots, t_{a_n,U_n}(z), \cdots.$$

They form a normal family, we have then a limit transformation

$$t_{a,U}(z),$$

where U is a limit element of the aggregate of unitary matrices $U_1, \cdots, U_n, \cdots$. This contradicts our supposition.

THEOREM 7. *If $w = t_{0,U}(z)$ belongs to G, then U is of finite order. Consequently, there is a finite number of transformations of a discontinuous group having an interior fixed point.*

The theorem is evident, since an infinite set of unitary matrices has always a limit element.

DEFINITION. A transformation with a fixed point interior to $\mathfrak{R}$ is called an *elliptic transformation*. The corresponding unitary matrix is called *multiplier* of the transformation.

REMARK. In the proof of Theorem 6, we do not need the full force on the boundedness of $\mathfrak{R}$. That the group of stability be compact will meet our requirement. The geometries having hypercircle as absolute with a regular metric given in Hua [6] satisfy the requirement, so that our Theorem 6 is still true for them.

5. Distance

DEFINITION. Let a and b be two points of the space. We define the distance $\Delta(a, b)$ between two points a and b to be the greatest lower bound of the absolute value of the integral

$$\int_C \sqrt{(\overline{dz})H(\bar{z}, z)\,(dz)'}$$

for all possible rectifiable curves C in the space $\mathfrak{R}$ connecting a and b.

We may easily establish the following properties:

1) For any pair of points a and b in $\mathfrak{R}$, $\Delta(a, b)$ is finite.

2) $\Delta(a, b) = \Delta(b, a)$.

3) $\Delta(a, b) = 0$ if and only if $a = b$.

In fact, surrounding a we construct a sphere S of radius ρ which separates a and b, and lies entirely in $\mathfrak{R}$. Let $\lambda(z)$ be the least characteristic root of $H(z, \bar{z})$ which is continuous in z, and let q be the least value which $\lambda(z)$ takes on the closure of S. Evidently $q > 0$. Then

$$\int_C (\overline{dz})H\ (dz)' \geqq \int_C \lambda(q)\ (d\bar{z})\ (dz)'$$

where C' denotes the part of C lying in S. Then

$$\int_C (\overline{dz})H\ (dz)' \geqq q \int_{C'} (\overline{dz})\ (dz)' \geqq q\rho.$$

Then we have 3).

4) If a, b, c are three points of the space, we have

$$\Delta(a, c) \leqq \Delta(a, b) + \Delta(b, c).$$

Consequently, $\Delta(a, b)$ is a continuous function of a and b.

5) Let $w = t(z)$ be a transformation of Γ, we have

$$\Delta(a, b) = \Delta(t(a), t(b)).$$

6) As a tends to a boundary point c of $\mathfrak{R}$, we have

$$\lim_{a \to c} \Delta(a, b) = \infty.$$

In fact, we may assume $b = 0$. Let λ be the greatest lower bound of $\Delta(0, c)$ for c running over all boundary points of $\mathfrak{R}$. Given $\epsilon > 0$, we have a point c and a curve C connecting c and 0 lying in $\mathfrak{R}$ except the terminal point c, such that

$$\int_C \sqrt{\overline{dz} H (dz)'} \leqq \lambda + \epsilon.$$

Taking a point f on C, we have

$$\int_C = \int_{(0,f)} + \int_{(f,c)} \leqq \lambda + \epsilon$$

i.e.

$$\Delta(c, f) \leqq \lambda + \epsilon - \Delta(0, f).$$

We have a transformation $w = t(z)$ of Γ carrying f into 0, and

$$\Delta(t(c), 0) = \Delta(c, f) \leqq \lambda + \epsilon - \Delta(0, f)$$

Since $t(c)$ is also a boundary point, $\Delta(t(c), 0) \geqq \lambda$, and we have

$$\Delta(0, f) \leqq \epsilon$$

for any $\epsilon > 0$. By 3) this is impossible.

7) The points x satisfying

$$\Delta(a, x) \leqq \rho$$

form a compact set. It is called a non-Euclidean sphere with center a and radius ρ.

The topology defined in the Euclidean sense is equivalent to the topology defined by considering non-Euclidean spheres as a complete system of neighborhoods.

REMARK. If we can establish that the space $\mathfrak{R}$ is of non-negative Riemannian curvature, the distance between two points is equal to the length of the unique geodesic connecting both points. The theory will be more elegant than the present one.

6. Fundamental region

By the result of §4, G is enumerable. We may assume that 0 is not a fixed point. In fact, the fixed points of a transformation form a manifold of dimension $\leqq 2n - 1$ and there are an enumerable many of them, thus we may choose a point which is not a fixed point of all the transformations of G. By a transformation, we take it to be 0. Then we may omit the unitary matrix from the subscript, we arrange the transformations of the discontinuous group G in the following order

$$t_i(z) = t_{a_i, U_i}(z), \qquad i = 0, 1, 2, \cdots,$$

according to

$$0 = \Delta(a_0, 0) < \Delta(a_1, 0) \leqq \Delta(a_2, 0) \leqq \cdots \leqq \Delta(a_i, 0) \leqq \cdots$$

and $t_0(z) = z$ is the identity. We have then

$$\lim_{i \to \infty} \Delta(a_i, 0) = \infty.$$

DEFINITION. A subdomain F of $\mathfrak{R}$ is called a *fundamental region* of a discontinuous group G, if the images of F under G covers $\mathfrak{R}$ without gaps and overlappings.

THEOREM 8. *There exists a fundamental region. More definitely, let a be any point which is not a fixed point, the point z satisfying*

$$\Delta(z, a) \leqq \Delta(z, t_i(a)), \qquad \text{for } i = 1, 2, \cdots,$$

form a fundamental region which is denoted by $F(a)$. It is called the radial region with center a.

PROOF. In fact, to every point z, we have a nearest, in the non-Euclidean sense, point $t_k(a)$ in the aggregate $t_i(a)$ $(i = 1, 2, \cdots)$, or one of its nearest points. Then it belongs to $F(t_k(a))$ which is an image of $F(a)$. There is no point belonging to the interior of two $F(t_i(a))$, for otherwise

$$\Delta(z, t_k(a)) \leqq \Delta(z, t_j(a)) \qquad \text{for } j \neq k$$

and

$$\Delta(z, t_j(a)) \leqq \Delta(z, t_k(a)).$$

Consequently

$$\Delta(z, t_j(a)) = \Delta(z, t_k(a)),$$

i.e. z is on the boundary of $F(t_j(a))$.

THEOREM 9. *Every compact domain M in $\mathfrak{R}$ is covered by a finite number of images of $F(a)$ under G.*

PROOF. Let M_1 be the set of points obtained as the sum of a set of closed non-Euclidean spheres with center at any point m of M and radius $\Delta(m, a)$. Then M_1 is also compact. If the intersection of M and $F(t_k(a))$ is non-empty, then M_1 contains $t_k(a)$. In fact, let p be a point in the intersection, then

$$\Delta(p, t_k(a)) \leqq \Delta(p, a).$$

By definition of M_1, $t_k(a)$ belongs to M_1. Since

$$\lim_{k \to \infty} \Delta(0, t_k(a)) = \infty,$$

we have the theorem.

Consequently, we have

THEOREM 10. *If $F(a)$ is compact, the compactness is independent of the choice of a.*

PROOF. Let $F(b)$ be the radial region with center b. By Theorem 11, $F(a)$ is covered by a finite number of $F(t_i(b))$'s. Let them be

$$F(t_{\lambda_1}(b)), \cdots, F(t_{\lambda_s}(b)).$$

Then $F(b)$ is covered by

$$F(t_{\lambda_1}^{-1}(a)), \cdots, F(t_{\lambda_s}^{-1}(a)).$$

In fact, let P be a point of $F(b)$, there is one of

$$t_{\lambda_1}(P), \cdots, t_{\lambda_s}(P)$$

lying in $F(a)$. Then P belongs to $F(t_{\lambda_i}^{-1}(a))$. Since $F(t_{\lambda_i}^{-1}(a))$ is compact, we have the theorem.

DEFINITION. *The bisecting manifold of two points a and b of $\mathfrak{R}$ is defined by the points x satisfying*

$$\Delta(x, a) = \Delta(x, b).$$

Thus the fundamental region is bounded by a number, finite or infinite, of bisecting manifolds.

THEOREM 11. *If $F(a)$ is compact, it is bounded by a finite number of bisecting manifolds.*

PROOF. Let δ be the diameter of F, i.e. the least upper bound of the non-Euclidean distances between any two points of F. Suppose that the theorem is false. Let the bisecting manifolds be defined by

$$\Delta(a_\alpha, x) = \Delta(a, x), \qquad \alpha = i_1, i_2, \cdots$$

Then

$$\Delta(a_\alpha, a) \leqq 2\delta.$$

The fact that

$$\lim_{\alpha \to \infty} \Delta(a_\alpha, a) = \infty$$

contradicts our previous inequality.

7. Lemmas

LEMMA 1. *Let*

$$w_k = f_k(z_1, \cdots, z_n), \qquad 1 \leqq k \leqq n \tag{1}$$

be an analytic mapping. Let

$$z_k = x_k + iy_k, \qquad w_k = u_k + iv_k.$$

The mapping (1) *induces a transformation of 2n-dimensional real space. The Jacobian of the induced transformation is equal to the square of the absolute value of the Jacobian of* (1).

PROOF. We have

$$dw_i = \sum_{j=1}^{n} dz_j \frac{\partial f_i}{\partial z_j}$$

of which the determinant is the Jacobian of (1). Taking conjugate complexes, we have

$$d\bar{w}_i = \sum_{j=1}^{n} \overline{dz_j}\, \overline{\frac{\partial f_i}{\partial z_j}}.$$

Thus

$$\frac{\partial(w_1, \cdots, w_n, \bar{w}_1, \cdots, \bar{w}_n)}{\partial(z_1, \cdots, z_n, \bar{z}_1, \cdots, \bar{z}_n)} = \left| \frac{\partial w}{\partial z} \right|^2.$$

Further

$$(w_k, \bar{w}_k) = (u_k, v_k)\begin{pmatrix} 1 & 1 \\ i & -i \end{pmatrix}, \qquad (z_k, \bar{z}_k) = (x_k, y_k)\begin{pmatrix} 1 & 1 \\ i & -i \end{pmatrix},$$

we have the result.

LEMMA 2. *If* $f(z_1, \cdots, z_n)$ *is a function regular in and on the polycylinder* C *with center* $(z_1^0, \cdots, z_n^0)$ *and radius* ρ, i.e.

$$| z_i - z_i^0 | \leqq \rho, \qquad 1 \leqq i \leqq n,$$

we have

$$\int \cdots \int_C | f(z_1, \cdots, z_n) |^2 \, dx_1 \cdots dx_n \, dy_1 \cdots dy_n \geqq | f(z_1^0, \cdots, z_n^0) |^2 \pi^n \rho^{2n}.$$

The equality holds only when f *is a constant.*

PROOF. Let

$$f(z_1, \cdots, z_n) = \sum_{m_1, \cdots, m_n = 0}^{\infty} a_{m_1, \cdots, m_n} (z_1 - z_1^0)^{m_1} \cdots (z_n - z_n^0)^{m_n}.$$

The integral is equal to

$$\int_0^\rho \cdots \int_0^\rho \int_0^{2\pi} \cdots \int_0^{2\pi} | \sum a_{m_1, \cdots, m_n} \rho_1^{m_1} \cdots \rho_n^{m_n} e^{i(m_1\theta_1 + \cdots + m_n\theta_n)} |^2$$
$$\times \rho_1 \cdots \rho_n \, d\rho_1 \cdots d\rho_n \, d\theta_1 \cdots d\theta_n$$

$$= (2\pi)^n \int_0^\rho \cdots \int_0^\rho \sum | a_{m_1, \cdots, m_n} |^2 \rho_1^{2m_1+1} \cdots \rho_n^{2m_n+1} \, d\rho_1 \cdots d\rho_n$$

$$\geqq (2\pi)^n \int_0^\rho \cdots \int_0^\rho | a_{0, \cdots, 0} |^2 \rho_1 \cdots \rho_n \, d\rho_1 \cdots d\rho_n$$

$$= \pi^n \rho^{2n} | f(z_1^0, \cdots, z_n^0) |^2.$$

The last statement of the lemma follows by re-examining the inequality.

8. Poincaré theta series

We construct the theta series

$$\Theta_k(z) = \sum_{i=0}^{\infty} \left(\frac{\partial t_i(z)}{\partial z} \right)^k .$$

THEOREM 12. *The theta series converges absolutely (and uniformly) for $k \geqq 2\lambda$, if the integral*

$$\int_{\mathfrak{R}} d(H(z, \bar{z}))^{1-\lambda} \, dx_1 \cdots dx_n \, dy_1 \cdots dy_n$$

converges, where

$$H(z, \bar{z}) = \overline{J_{z,U}(z)} J_{z,U}(z)'.$$

PROOF. It is easy to see that we have a constant Ω such that

$$\Omega^{-1} \leqq d(H(z, \bar{z})) \leqq \Omega,$$

where $\Omega = \Omega(\mathfrak{R}^*) > 0$ for all z lying in a compact region $\mathfrak{R}^*$ in the interior of $\mathfrak{R}$.

We choose a polycylinder C with center z_0 and with radius ρ such that $t_i(C)$'s do not overlap. Then, by Lemma 2 of §5,

$$\pi^n \rho^{2n} \left| \frac{\partial(t_i(z))}{\partial z} \right|^{2\lambda}_{\text{at } z=z_0} \leqq \int_C \left| \frac{\partial t_i(z)}{\partial z} \right|^{2\lambda} dx \, dy$$

$$\leqq \Omega^{|1-\lambda|} \int_C \left| \frac{\partial t_i(z)}{\partial z} \right|^{2\lambda} d(H(z, \bar{z}))^{1-\lambda} \, dx \, dy$$

$$= \Omega^{|1-\lambda|} \int_{t_i(C)} d(H(w, \bar{w}))^{1-\lambda} \, du \, dv.$$

Thus we have

$$\pi^n \rho^{2n} \sum_{i=0}^{\infty} \left| \frac{\partial t_i(z)}{\partial z} \right|^{2\lambda}_{\text{at } z=z_0} \leqq \Omega^{|1-\lambda|} \int_{\mathfrak{R}} d(H(w, \bar{w}))^{1-\lambda} \, du \, dv.$$

This establishes the absolute convergence of the series. Similarly, we have the uniformity of convergence.

Evidently the integral converges for $\lambda = 1$, we have the following

THEOREM 13. *The theta series converges for $k \geqq 2$.*

Is the constant given in Theorem 12 the best possible? The answer seems to be affirmative.

"Verzerrungssatz." Let $\mathfrak{R}^*$ be a compact subdomain of $\mathfrak{R}$. There exists a constant Λ (> 0) depending only on $\mathfrak{R}^*$ so that, for any two points z and z^* of $\mathfrak{R}^*$, we have

$$\Lambda^{-1} \leqq J_{a,U}(z)/J_{a,U}(z^*) \leqq \Lambda.$$

In general is this statement true? The author cannot answer it. Nevertheless for the four main types of symmetric bounded spaces the statement holds. (Cf. the last section of the paper.)

THEOREM 14. *If the "Verzerrungssatz" is true and if the fundamental region is compact, the theta series diverges when the integral in Theorem 12 does.*

PROOF. Let F be the fundamental region. Then

$$\left|\frac{\partial t_i(z)}{\partial z}\right|^{2\lambda} \geqq \Lambda_n^{-1} \int_F \left|\frac{\partial t_i(z)}{\partial z}\right|^{2\lambda} d(H(z, \bar{z}))^{1-\lambda}\, dx\, dy \cdot \left(\int_F d(H(z, \bar{z}))^{1-\lambda}\, dx\, dy\right)^{-1}.$$

Let

$$\int_F d(H(z, \bar{z}))^{1-\lambda}\, dx\, dy = P,$$

which is finite. Then

$$\sum_{i=0}^{\infty} \left|\frac{\partial t_i(z)}{\partial z}\right|^{2\lambda} \geqq \Lambda^{-1} P^{-1} \sum_{i=0}^{\infty} \int_{t_i(F)} d(H(z, \bar{z}))^{1-\lambda}\, dx\, dy$$

$$= \Lambda^{-1} P^{-1} \int_{\mathfrak{R}} d(H(z, \bar{z}))^{1-\lambda}\, dx\, dy .$$

The theorem follows.

In case the fundamental region is compact we have many advantages. In fact the essential difficulty arising from the "parabolic vertices" disappears. Broadly speaking under this hypothesis the theory of automoprhic functions may be developed quite satisfactorily.

From the discussion of §3, we see that as we apply the transformation

$$w = t_{\alpha, V}(z),$$

we have

$$(dz) = (dw) J_{\alpha, V}(z)$$

and

$$J_{z, U}(z) = (J_{\alpha, V}(z))^{-1} J_{2, o}(z) W,$$

where W is a unitary matrix. By the method of adjugation, we may obtain invariant integrals of any dimension. By them, we may find some improvements of Theorem 12 for special groups.

§9. Dependence of Fuchsian forms

In the present section we shall give a generalization of a theorem due to Siegel [7].

DEFINITION. An analytic function $f(z)$ of n complex variables $(z) = (z_1, \cdots, z_n)$ is called a *Fuchsian form of weight k with multiplier system v of a group G in the space $\mathfrak{R}$*, if it is meromorphic in the space $\mathfrak{R}$ and if

$$f(t_i(z)) = v(t_i) \left(\frac{\partial f_i(z)}{\partial z}\right)^k f(z)$$

for all transformations $t_i(z)$ belonging to G, where k is a constant and $v(t_i)$ is a number depending only on t_i.

The Fuchsian forms of weight 0 with multiplier system 1 of a group G are called the *Fuchsian functions* in the space $\mathfrak{R}$ of the group G.

By a well-known method (See, e.g. Picard of [2] and [3] or Blumenthal of [4]), we may establish the following theorem.

THEOREM 15. *For K large enough, there are $n + 1$ algebraically (actually analytically) independent integral Fuchsian forms of dimension K with $v = 1$. Consequently there are n algebraically (actually analytically) independent Fuchsian functions of the space $\mathfrak{R}$ with respect to the group G.*

In order to establish the dependence of $n + 1$ Fuchsian functions, we use Siegel's quick method.

Let $L = L(G, k, v)$ denote the set of all integral Fuchsian forms of weight k and multiplier v. If f_1 and f_2 belong to L, so does $\lambda_1 f_1 + \lambda_2 f_2$, for any complex λ_1 and λ_2. Hence L is a vector space with certain dimension d (infinite or finite). Now we suppose that k is real and $v(t_i)$ is of absolute value 1.

Siegel found that d is finite for Fuchsian forms of a symmetric matrix-variable and for G having a compact fundamental domain. Now we shall extend the result to any circular region and then, by it establish the dependence of automorphic functions.

By the definition of circular region, we have, in Γ, the subgroup of transformations

$$(z_1, \cdots, z_n) = e^{i\theta}(w_1, \cdots, w_n).$$

Consequently

$$d(H(ze^{i\theta}, \bar{z}e^{i\theta})) = d(H(z, \bar{z})).$$

Thus, let λ be a scalar,

$$d(H(\lambda z), \bar{\lambda}\bar{z}))$$

is a function of $\lambda\bar{\lambda}$. Let $\lambda\bar{\lambda} = t$; it is a function of t.

In order to prove the result, we need, besides the original idea of Siegel, the following self-evident lemma:

LEMMA. *For brevity, we write*

$$h(z) = h(z, \bar{z}), = d(H(z, \bar{z})).$$

Let

$$b = \max_{z \in F} \left(\frac{\sum_{i=1}^{n} \left(z_i \frac{\partial h}{\partial z_i} + \bar{z}_i \frac{\partial h}{\partial \bar{z}_i} \right)}{h(z)} ; 1 \right)$$

where F denotes a compact region (later the fundamental region of the group G). Then, for $t > 0$, we have

$$\lim_{t \to 1} \frac{\log |h(t^{\frac{1}{2}} z)| - \log |h(z)|}{\log t} \leq b$$

for z belonging to F.

With this b instead of the b in Siegel's paper, we may establish that

THEOREM 16. *Suppose that $\mathfrak{R}$ is a circular region and that the fundamental domain is compact. The dimension d of the vector space $L(G, k, v)$ is 0 for $k < 0$ and 0 or 1 for $k = 0$. It is finite and $\leqq ck$ where $c = (n+1)b^n$.*

From Theorem 16, we may deduce, as Siegel that

THEOREM 17. *Suppose that $\mathfrak{R}$ is a circular region, and that G possesses a compact fundamental domain. Fuchsian functions of the space $\mathfrak{R}$ with the group G form an algebraic functional field with exactly n independent elements.*

REMARK. The condition that $\mathfrak{R}$ is a circular region may be abolished, if we adopt the method due to Poincaré and Blumenthal [8], which is very much more elaborate than that due to Siegel. It is very likely that every bounded transitive space is a circular region. Thus the author presents such a treatment.

10. An introduction of vectorial automorphic functions

The preceding generalization of the theory of automorphic functions seems to be not the most appropriate one. It may be considered as an introduction of a way along which the author has proceeded to a certain extent. Such an idea will be proved to be fruitful in the study of automorphic functions of a matrix variable. The main idea may be described, in short, by the following phrase: "functions of vectorial variables with values over a vectorial domain."

DEFINITION. Let

$$f_1((z)), \cdots, f_n(z)),$$

be n functions meromorphic of the n variables

$$(z) = (z_1, \cdots, z_r)$$

in $\mathfrak{R}$. For a transformation

$$t_\lambda: \qquad w_i = t_i^{(\lambda)}(z_1, \cdots, z_n) \qquad 1 \leqq i \leqq n,$$

of G, we have

$$(f_1, \cdots, f_n)(w) = (f_1, \cdots, f_n)(z)(J(t_\lambda))^{-k}$$

where $J(t_\lambda)$ is the Jacobian matrix

$$\left(\frac{\partial w_i}{\partial z_j}\right).$$

Then $(f_1, \cdots, f_n)$ is called a *vectorial Fuchsian form of weight k.*

A vectorial Fuchsian form of weight 0 is called a *vectorial Fuchsian function.* If the Jacobian of $f_1, \cdots, f_n$ is not identically zero, the vectorial form is said to be *non-degenerate.*

THEOREM 18. *To each Fuchsian function we can construct a vectorial Fuchsian form of weight -1.*

PROOF: Let $\chi(z_1, \cdots, z_n)$ be a Fucnsian function. Then

$$\left(\frac{\partial \chi}{\partial w_1}, \cdots, \frac{\partial \chi}{\partial w_n}\right)(w) = \left(\frac{\partial \chi}{\partial z_1}, \cdots, \frac{\partial \chi}{\partial z_n}\right)(z)\, J(t_\lambda)$$

is a vectorial Fuchsian form of weight -1.

As a consequence of Theorem 15, we have

THEOREM 19. *There exists a non-degenerate vectorial Fuchsian function.*

THEOREM 20. *Suppose that G has a compact fundamental region. A vectorial Fuchsian function takes each vector an equal number of times, except those values lying on manifolds of dimension $< 2n$.*

The theorem is a consequence of Blumenthal's result [9] concerning the theory of eliminations.

From a vectorial Fuchsian function

$$(f_1, \cdots, f_n)(z),$$

we construct

$$\left(\frac{\partial f_i}{\partial z_j}\right) = J(z_1, \cdots, z_n).$$

Then

$$J(z_1, \cdots, z_n) = J(w_1, \cdots, w_n)J(t_\lambda),$$

which may be described as a "matrix Fuchsian form." Suppose that the vectorial function is nonsingular, we have

$$d(J(z_1, \cdots, z_n) = d(J(w_1, \cdots, w_n)d(J(t_\lambda)).$$

Thus to each non-degenerate vectorial Fuchsian function, we can construct a Fuchsian form of weight -1.

Let χ be any Fuchsian form of dimension k. Let $w_1, \cdots, w_n$ be a nondegenerate vectorial Fuchsian function. Then

$$\chi d(J(w_1, \cdots, w_n))^k = g$$

is a Fuchsian function. By Theorem 17, J may be expressed as a polynomial of $w_1, \cdots, w_n$ and w_{n+1}, where w_{n+1} is a properly chosen Fuchsian function. Then, we have

$$\chi(J(w_1, \cdots, w_n)^k = A(w_1, \cdots, w_n, w_{n+1}).$$

From this formal footing, we may extend some of the classical results. Since the author finds no way to get rid of the condition that G possesses a compact fundamental domain, he will not proceed to give a full discussion of its development.

11. A digression

It was determined by E. Cartan [10], that there are six types of irreducible bounded transitive symmetric spaces. Among them there are four general types and two special types. By means of the notation of matrices, the four general types may be expressed easily as follows.

(i) *Geometry of symmetric matrices.* Let Z denote n-rowed symmetric matrices. The space $\mathfrak{R}$ is formed by the points Z satisfying

$$I - Z\bar{Z} > 0.$$

(For an Hermitian matrix H, $H > 0$ means that H is positive definite.) The group of motion of the space $\mathfrak{R}$ is constituted by all transformations of the form

$$Z_1 = (AZ + B)(\bar{B}Z + \bar{A})^{-1}$$

where

$$\begin{pmatrix} A & B \\ \bar{B} & \bar{A} \end{pmatrix}\begin{pmatrix} 0 & I \\ -1 & 0 \end{pmatrix}\begin{pmatrix} A & B \\ \bar{B} & \bar{A} \end{pmatrix}' = \begin{pmatrix} 0 & I \\ -1 & 0 \end{pmatrix}.$$

This is known as Siegel's symplectic geometry.

(ii) *Geometry of skew symmetric matrices.* The space $\mathfrak{R}$ is formed by the points, defined by skew symmetric matrices Z, satisfying

$$I - \bar{Z}Z' > 0.$$

The group of motion of the space $\mathfrak{R}$ is constituted by all transformations of the form

$$Z_1 = (AZ + B)(-\bar{B}Z + \bar{A})^{-1}$$

where

$$\begin{pmatrix} A & B \\ -\bar{B} & \bar{A} \end{pmatrix}\begin{pmatrix} 0 & I \\ I & 0 \end{pmatrix}\begin{pmatrix} A & B \\ -\bar{B} & \bar{A} \end{pmatrix}' = \begin{pmatrix} 0 & I \\ I & 0 \end{pmatrix}.$$

(iii) *Geometry of rectangular matrices.* Let $Z = Z^{(m,n)}$ be an $m \times n$-rowed matrices. The space $\mathfrak{R}$ is formed by the points Z satisfying

$$I^{(m)} - \bar{Z}Z' > 0.$$

The group of motion ϕ is constituted by all transformations of the form

$$Z_1 = (AZ + B)(CZ + D)^{-1}$$

where

$$\overline{\begin{pmatrix} A^{(m)} & B^{(m,n)} \\ C & D \end{pmatrix}}\begin{pmatrix} I^{(m)} & 0 \\ 0 & -I^{(n)} \end{pmatrix}\begin{pmatrix} A & B \\ C & D \end{pmatrix}' = \begin{pmatrix} I^{(m)} & 0 \\ 0 & -I^{(n)} \end{pmatrix}.$$

(iv) *Geometry of complex spheres* (given at §13)

The four types of geometries possess special non-Euclidean properties which enable us to improve our general results. The first three types are included in the discussion of the author's papers [6]. As an illustration of the present general treatment, we shall study the final case to a certain extent. Note that this is the case known as Picard's hyperabelian group.

12. A particular kind of geometry of matrices

The matrices of the present section are all real.

Let $X = X^{(2,n)}$ be a $2 \times n$ rowed real matrix. The space $\mathfrak{R}_0$ is formed by the points X satisfying

$$I^{(2)} - XX' > 0 \qquad (1)$$

The motion of the space is given by

$$X_1 = (AX + B)(CX + D)^{-1}, \tag{2}$$

where

$$\begin{pmatrix} A^{(2)} & B \\ C & D \end{pmatrix}\begin{pmatrix} I^{(2)} & 0 \\ 0 & -I^{(n)} \end{pmatrix}\begin{pmatrix} A & B \\ C & D \end{pmatrix}' = \begin{pmatrix} I^{(2)} & 0 \\ 0 & -I^{(n)} \end{pmatrix},$$

i.e.

$$AA' - BB' = I^{(2)}, \qquad AC' = BD', \qquad CC' - DD' = -I^{(n)}, \tag{3}$$

and the determinant of

$$\begin{pmatrix} A & B \\ C & D \end{pmatrix}$$

is equal to 1.

From (3), we have consequently

$$A'A - C'C = I^{(n)}, \qquad A'B = C'D, \qquad B'B - D'D = -I^{(2)} \tag{4}$$

and

$$X_1 = (XB' + A')^{-1}(XD' + C'). \tag{5}$$

Let dX denote the differential[2] of X. Then

$$\begin{aligned} dX_1 &= (XB' + A')^{-1}dXD' - (XB' + A')^{-1}dXB'(XB' + A')^{-1}(XD' + C') \\ &= (XB' + A')^{-1}dX(CX + D)^{-1}. \end{aligned} \tag{6}$$

Further

$$\begin{aligned} I - X_1X_1' &= (XB' + A'^{-1})[(XB' + A')(BX' + A) - (XD' + C') \\ &\qquad \cdot (DX' + C)](BX' + A)^{-1}. \\ &= (XB' + A')^{-1}(I - XX')(BX' + A)^{-1} \end{aligned} \tag{7}$$

and

$$I - X_1'X_1 = (CX + D)'^{-1}(I - X'X)(CX + D)^{-1}. \tag{7'}$$

Consequently

$$d(XB' + A')^2 = d(CX + D)^2,$$

since

$$d(I - X'X) = d(I - XX').$$

[2] Notice that $d(M)$ denotes the determinant value of M and dM denotes the differential of M.

Then

$$d(XB' + A') = d(CX + D), \tag{8}$$

since

$$d\begin{pmatrix} A & B \\ C & D \end{pmatrix} = 1,$$

and $\begin{pmatrix} A & B \\ C & D \end{pmatrix}$ form a continuous piece.

From (6), (7) and (7′), we obtain an invariant differential positive definite quadratic form

$$\sigma((I^{(2)} - XX')^{-1}dX(I^{(n)} - X'X)^{-1}dX'), \tag{9}$$

where $\sigma(M)$ denotes the trace of the matrix M.

The Jacobian of the transformation (2) is equal to

$$d(XB' + A')^{-2-n}$$

by (6). The volume element of the space is given by

$$d(I - XX')^{-(2+n)/2}\dot{X} \tag{10}$$

where $\dot{X} = \pi dx_{rs}$ (Here we use again the fact that

$$d(I - XX') = d(I - X'X)).$$

Given a point P of the space $\mathfrak{R}_0$. We have two matrices Q and R such that

$$Q(I - PP')Q' = I, \tag{11}$$

and

$$R(I - P'P)R' = I. \tag{12}$$

(Notice that each of $I - PP' > 0$ and $I - P'P > 0$ implies the other.)

Then

$$X_1 = Q(X - P)(-P'X + I)^{-1}R^{-1} \tag{13}$$

is a transformation of the space carrying P into zero. Thus the space $\mathfrak{R}_0$ is *transitive*.

The group of stability at 0 is evidently given by

$$X_1 = AXD^{-1}, \tag{14}$$

where A and D are orthogonal. Every transformation may be considered as a combination of (13) and (14).

The Jacobian of (13) is given by

$$(d(XP' - I)d(Q))^{-2}d(-P'X + I)^{-n}d(R)^{-n} = (d(XP' - I))^{-2-n} \cdot (d(I - P'P))^{(2+n)/2} \tag{15}$$

since $d(XP' - I^{(2)}) = d(P'X - I^{(n))})$.

Let $\mathfrak{R}^*$ be a compact region interior to $\mathfrak{R}_0$, to establish the "Verzerrungssatz," we have to prove that there exists a constant Ω (> 0) depending on $\mathfrak{R}^*$ such that

$$\Omega^{-1} \leqq \left| \frac{d(X_1 P' - I)}{d(X_2 P' - I)} \right|^{-2-n} \leqq \Omega, \tag{16}$$

for all X_1 and X_2 in $\mathfrak{R}^*$ and all P of $\mathfrak{R}_0$. This can be established easily by the argument of continuity. (Notice that (16) establishes the result for (13) and that for (14) it is evident.)

It may be verified by the method given previously (the first paper of Hua [6]) that the space possesses a non-positive Riemannian curvature.

13. Geometry of complex spheres

The space $\mathfrak{R}$ is formed by complex vectors z satisfying

$$|zz'|^2 + 1 - 2\bar{z}z' > 0 \tag{1}$$

and

$$|zz'| < 1. \tag{2}$$

The group of motion of the space is constituted by

$$z_1 = \left\{\left[\left(\tfrac{1}{2}(zz' + 1), \frac{i}{2}(zz' - 1)\right)A' + zB'\right]\begin{pmatrix}1\\ i\end{pmatrix}\right\}^{-1} \times \left\{\left(\tfrac{1}{2}(zz' + 1), \frac{i}{2}(zz' - 1)\right)C' + zD'\right\}, \tag{3}$$

where

$$\begin{pmatrix} A & B \\ C & D \end{pmatrix}\begin{pmatrix} I^{(2)} & 0 \\ 0 & -I^{(n)} \end{pmatrix}\begin{pmatrix} A & B \\ C & D \end{pmatrix}' = \begin{pmatrix} I^{(2)} & 0 \\ 0 & -I^{(m)} \end{pmatrix},$$

and A, B, C and D are real and

$$d\begin{pmatrix} A & B \\ C & D \end{pmatrix} = +1.$$

The relation between the present space $\mathfrak{R}$ and that the space $\mathfrak{R}_0$ given in §12 is given by the following one to one transformation

$$X = 2\begin{pmatrix} zz' + 1 & i(zz' - 1) \\ \overline{zz'} + 1 & -i(\overline{zz'} - 1) \end{pmatrix}^{-1}\begin{pmatrix} z \\ \bar{z} \end{pmatrix}, \tag{4}$$

which is real.

In fact, the matrix $I - XX'$ is evidently conjunctive to

$$\overline{\begin{pmatrix} zz' + 1 & i(zz - 1) \\ \overline{zz'} + 1 & -i(\overline{zz'} - 1) \end{pmatrix}}\begin{pmatrix} zz' + 1 & i(zz' - 1) \\ \overline{zz'} + 1 & -i(\overline{zz'} - 1) \end{pmatrix}' - 4\overline{\begin{pmatrix} z \\ \bar{z} \end{pmatrix}}\begin{pmatrix} z \\ \bar{z} \end{pmatrix}'$$

$$= 2\begin{pmatrix} 1 + (zz')^2 - 2\bar{z}z' & 0 \\ 0 & 1 + |zz'|^2 - 2\bar{z}z' \end{pmatrix}.$$

Thus to each z of $\mathfrak{R}$, by (1) there is a point X of $\mathfrak{R}_0$.

Conversely, to each point $X = \begin{pmatrix} x \\ x_0 \end{pmatrix}$ of $\mathfrak{R}_0$, we have

$$\begin{aligned} 2z &= (zz' + 1)x + i(zz' - 1)x_0 \\ &= zz'(x + ix_0) + (x - ix_0). \end{aligned}$$

Consequently, we have

(5) $4zz' = (zz')^2(x + ix_0)(x + ix_0)' + 2(xx' + x_0x')zz' + (x - ix_0)(x - ix_0)'$,

which is a quadratic equation in zz'. Since

$$\left| \frac{(x + ix_0)(x + ix_0)'}{(x - ix_0)(x - ix_0)'} \right| = 1,$$

the equation (5) has a unique solution

$$zz' = \frac{2 - (xx' + x_0 x_0') - 2\sqrt{(1 - xx')(1 - x_0 x_0') - (x_0 x')^2}}{(x + ix_0)(x + ix_0)'}$$

with $| zz' | < 1$ for X belonging to $\mathfrak{R}_0$. Therefore the mapping is one to one.

Now we are going to find the Jacobian of the transformation (4). Let $w = x + ix_0$. Then (4) takes the form

$$2z = zz'w + \overline{w}. \tag{6}$$

We have

$$2\frac{\partial z_i}{\partial w_j} = 2\sum_{k=1}^{n} \left(z_k \frac{\partial z_k}{\partial w_j} \right) w_i + zz' \delta_{ij}, \tag{7}$$

where $\delta_{ij} = 0$ for $i \neq j$ and $\delta_{ii} = 1$. We write (7) in the form of matrices:

$$(I - P)\left(\frac{\partial z}{\partial w}\right) = \tfrac{1}{2} zz' I, \tag{8}$$

where $P = w'z$. Similarly

$$(I - P)\left(\frac{\partial z}{\partial \overline{w}}\right) = \tfrac{1}{2}I. \tag{9}$$

Combining (8) and (9) and their conjugate equations, we have

$$\begin{aligned} \left| \frac{\partial(z, \bar{z})}{\partial(w, \overline{w})} \right| &= 2^{-2n} \,| d(I - P)|^{-2} (1 - | zz' |^2)^n \\ &= 2^{-2n} \,| 1 - wz' |^{-2} (1 - |zz' |^2)^n, \end{aligned} \tag{10}$$

since $d(I - P) = 1 - wz'$. From (6) and its conjugate equation, by eliminating $\overline{w}$, we have

$$2(\overline{zz'}\, z - \bar{z}) = (| zz' |^2 - 1)w,$$

and

$$wz' = \frac{2(\bar{z}z - |zz'|^2)}{1 - |zz'|^2}.$$

Therefore, we have

$$(11) \qquad \left|\frac{\partial(z, \bar{z})}{\partial(w, \bar{w})}\right| = \frac{(1 - |zz'|^2)^{n+2}}{2^{2n}(1 + |zz'|^2 - 2zz')^2}.$$

From (3), we have

$$\rho z_1 = \left(\tfrac{1}{2}(zz' + 1), \frac{i}{2}(zz' - 1)\right)C' + zD'$$

$$\rho = \left(\left(\tfrac{1}{2}(zz' + 1), \frac{i}{2}(zz' - 1)\right)A' + zB'\right)\begin{pmatrix}1\\ i\end{pmatrix}.$$

We define λ_1 and λ_2 by

$$\rho(\lambda_1, \lambda_2) = \left(\tfrac{1}{2}(zz' + 1), \frac{i}{2}(zz' - 1)\right)A' + zB'.$$

We have

$$(12) \qquad (\lambda_1, \lambda_2)\begin{pmatrix}1\\ i\end{pmatrix} = 1, \qquad \text{i.e. } \lambda_1 + i\lambda_2 = 1.$$

Then

$$\begin{pmatrix}\rho & 0\\ 0 & \bar{\rho}\end{pmatrix}\begin{pmatrix}\lambda_1, \lambda_2; z_1\\ \bar{\lambda}_1, \bar{\lambda}_2; \bar{z}_1\end{pmatrix} = \begin{pmatrix}\tfrac{1}{2}(zz' + 1), & \frac{i}{2}(zz' - 1); z\\ \tfrac{1}{2}(\bar{z}\bar{z}' + 1), & -\frac{i}{2}(\bar{z}\bar{z}' - 1); \bar{z}\end{pmatrix}\begin{pmatrix}A' & C'\\ B' & D'\end{pmatrix}.$$

Since

$$\overline{\begin{pmatrix}\tfrac{1}{2}(zz' + 1), & \frac{i}{2}(zz' - 1); z\\ \tfrac{1}{2}(\bar{z}\bar{z}' + 1); & -\frac{i}{2}(\bar{z}\bar{z}' - 1); \bar{z}\end{pmatrix}}\begin{pmatrix}I^{(2)} & 0\\ 0 & -I^{(n)}\end{pmatrix}\begin{pmatrix}\tfrac{1}{2}(zz' + 1), & \frac{i}{2}(zz' - 1); z\\ \tfrac{1}{2}(\bar{z}\bar{z}' + 1), & -\frac{i}{2}(\bar{z}\bar{z}' - 1); \bar{z}\end{pmatrix}'$$

$$= \begin{pmatrix}* & 0\\ 0 & *\end{pmatrix},$$

we have

$$\overline{\begin{pmatrix}\lambda_1, \lambda_2; z_1\\ \bar{\lambda}_1, \bar{\lambda}_2; \bar{z}_1\end{pmatrix}}\begin{pmatrix}I^{(2)} & 0\\ 0 & -I^{(n)}\end{pmatrix}\begin{pmatrix}\lambda_1, \lambda_2; z_1\\ \bar{\lambda}_1, \bar{\lambda}_2; \bar{z}_1\end{pmatrix}' = \begin{pmatrix}* & 0\\ 0 & *\end{pmatrix},$$

i.e.

$$(13) \qquad \lambda_1^2 + \lambda_2^2 - z_1 z_1' = 0.$$

Combining (12) and (13), we have

$$\lambda_1^2 - (1 - \lambda_1)^2 - z_1 z_1' = 0$$

i.e.

$$\lambda_1 = \tfrac{1}{2}(1 + z_1 z_1').$$

Consequently

$$\lambda_2 = \frac{i}{2}(z_1 z_1' - 1).$$

We have, therefore,

$$\begin{pmatrix} \rho & 0 \\ 0 & \bar{\rho} \end{pmatrix} \begin{pmatrix} \frac{1}{2}(z_1 z_1' + 1), & \frac{i}{2}(z_1 z_1' - 1), & z_1 \\ \frac{1}{2}(\overline{z_1 z_1'} + 1), & -\frac{i}{2}(\overline{z_1 z_1'} - 1), & \bar{z}_1 \end{pmatrix}$$

(14)

$$= \begin{pmatrix} \frac{1}{2}(zz' + 1), & \frac{i}{2}(zz' - 1); z \\ \frac{1}{2}(\bar{z}\bar{z}' + 1), & -\frac{i}{2}(\bar{z}\bar{z}' - 1); \bar{z} \end{pmatrix} \begin{pmatrix} A' & C' \\ B' & D' \end{pmatrix}.$$

Consequently

(15)
$$\begin{aligned} X_1 &= (A' + XB')^{-1}(C' + XD') \\ &= (AX + B)(CX + D)^{-1}. \end{aligned}$$

Now we have established the relationship between two kinds of geometries. This asserts again that the geometries of matrices play an essential role in the study of automorphic functions of several variables.

Translating the results of the last section, we have the following properties of the space $\mathfrak{R}$:

1) The space is transitive.

2) The volume element is equal to

$$\frac{\dot{z}}{(1 + |zz'|^2 - 2\bar{z}z')^n}, \quad \dot{z} = \pi\, dx_r\, dy_r, \quad z = x + iy,$$

by (10) of §12 and (11) of §13. The integral of Theorem 12 converges for $\lambda > 1 - \frac{1}{n}$.[3] In fact, for $\lambda \geqq 1$, the assertion is evident, and, for $\lambda \leqq 1$, we have

[3] The constant seems not to be a best possible one. Correspondingly, for the space $\mathfrak{R}_0$, we can evaluate exactly the value

$$\int_{\mathfrak{R}_0} d(I - XX')^{\lambda} \dot{X} = \pi^{2n} 2^n \Gamma(2\lambda + 1)/\Gamma(2\lambda + n + 2).$$

But for the present case the author is unable to evaluate the value of (16).

$$(16) \qquad \int_{\Re} (1 + |zz'|^2 - 2\bar{z}z')^{-n(1-\lambda)} \dot{z} \leqq \int_{1-\bar{z}z'>0} \frac{\dot{z}}{(1 - 2\bar{z}z')^{n(1-\lambda)}},$$

which converges for $\lambda > 1 - \frac{1}{n}$, since $|zz'| < 1$ and $1 + |zz'|^2 - 2\bar{z}z' > 0$ imply $\bar{z}z' < 1$.

3) The "Verzerrungssatz" is true.

4) The space has a non-positive Riemannian curvature.

$$\int_{\Re_0} d(I - \bar{x}\bar{x}')^\lambda \dot{x} = \pi^{2n} 2^n \Gamma(2\lambda + 1)/\Gamma(2\lambda + n + 2).$$

5) The group of stability at 0 is given by the transformations with $B = C = 0$ and orthogonal A and D. In particular

$$A = \begin{pmatrix} \cos\theta & \sin\theta \\ -\sin\theta & \cos\theta \end{pmatrix}, \qquad D = I,$$

we have

$$z_1 = e^{-i\theta} z,$$

since

$$\left(\tfrac{1}{2}(zz' + 1), \frac{i}{2}(zz' - 1)\right)\begin{pmatrix} \cos\theta & \sin\theta \\ -\sin\theta & \cos\theta \end{pmatrix}\begin{pmatrix} 1 \\ i \end{pmatrix} = e^{i\theta},$$

the space is a circular region.

List of References

[1] Behnke u. Thullen, *Theorie der Funktionen mehrerer komplexer Veränderlichen*, Ergebnisse der Math. u. ihrer Grenzgebiete, Bd. 3, Heft 3 (1934), Julius Springer, Berlin.

[2] 1. Hyper-Fuchsian group; Picard, Acta Math. **1** (1882), p. 297; **2** (1883), p. 114 and **5** (1884), p. 121. Cf. also, Picard, Jour. de Math., (4) **1** (1885), p. 357; Bull. de la Soc. math. de France, **15** (1887), p. 148; Alezias, Ann. scient. Ec. Norm. Sup. (3) **19** (1902), p. 201 (also Thèse, Paris, 1901) and Craig, Trans. Amer. Math. Soc. **11** (1910), p. 37.

2. Hyper-Abelian group: Picard, Jour. de Math. (4) **1** (1885), p. 87. Cf. also Bourget, Annales de la Faculté des Sciences de Toulouse (1898) (also Thèse, Paris, 1898), Cotty, ibid. (1911) (also Thèse, Paris, 1911) and Giraud, Ann. scient. Ec. Norm. Sup. (3), **32** (1915), p. 237 (also Thèse Paris, 1916); ibid. (3) **33** (1916), p. 303 and p. 331.

3. Linear group. Fubini, Annali di Mate. (3) **11** (1905), p. 159; **12** (1906), p. 347; **14** (1908), p. 33. Cf. also, A Hurwitz, Math. Annalen **61** (1905), p. 325.

4. Group over the algebraic field: Blumenthal, Math. Annalen, **56** (1902), p. 509; **58** (1904), p. 497.

5. Symplectic group: Siegel, Math. Annalen **116** (1939), 615–657; Amer. Jour. Math. **65** (1943), 1–86. Cf. also Sugawara, Annals of Math. (2) **41** (1940), 488–494. (note that in Sugawara's paper, there is an essential mistake, Cf. the lemma 2 of §7).

[3] FUBINI, see 3 of [2].
WIRTINGER, Wien. Berichte **108** (1899), p. 1239.
GIRAND, Leçons sur les fonctions automorphes (1920), Gauthier-Villars, Paris.
[4] BERGMANN, Jour. für. Math. **169** (1933).
[5] H. CARTAN, Jour. de Math. (9) **10** (1931), p. 1.
[6] HUA, Amer. Jour. of Math. **66** (1944), pp. 470–488; pp. 531–563.
[7] SIEGEL, Annals of Math. (2) **43** (1942), pp. 613–616.
[8] POINCARÉ, Acta Math. **26** (1902), 48–98. Blumenthal, see 4 of [2].
[9] BLUMENTHAL, Math. Annalen, **57** (1903).
[10] E. CARTAN, Abh. Math. Sem. Hamburg Univ. **11** (1935), pp. 116–162.
(Note that some of the references on the list are not available in China. The author found these titles through indirect sources.)

TSING HUA UNIVERSITY
KUNMING, CHINA

Reprinted from the
QUARTERLY JOURNAL OF MATHEMATICS
Oxford Series, Vol. 17, No. 68, pp. 214–222, 1946

ON THE EXTENDED SPACE OF SEVERAL COMPLEX VARIABLES (I): THE SPACE OF COMPLEX SPHERES

By L. K. HUA (*Peiping*)

[Received 1 October 1945]

IN the theory of functions of a complex variable we introduce the point at infinity to make the extended space (Cauchy plane) compact. The procedure of introducing point at infinity depends, in effect, on the group G of linear fractional transformations

$$z^* = (az+b)/(cz+d), \tag{1}$$

where $ad-bc \neq 0$. It is well known that the group G is 'complete' in the sense that any analytic automorph of the extended space is a transformation of the form (1). The aim of the present series of papers is to extend this discussion to the study of functions of several complex variables.

Let G be a group of transformations

$$z_i' = f_i(z_1,..., z_n) \tag{2}$$

of n complex variables. Suppose that G satisfies the assumptions given by Osgood.† The manifold at infinity is introduced by means of the group (2). The totality of finite points and the points at infinity form the extended space $\mathfrak{R} = \mathfrak{R}(G)$. Immediately, we have the problem: 'Is the group G which defines the extended space $\mathfrak{R}(G)$ complete?' More precisely, we seek the group G such that $\mathfrak{R}(G)$ admits no analytic automorph other than those of G.

This important problem has been answered, so far as I am aware, only for two very special cases: (i) the space of the theory of functions, in which the group G is given by

$$z_i^* = (\alpha_k z_k+\beta_k)/(\gamma_k z_k+\delta_k) \quad (\alpha_k \delta_k-\beta_k \gamma_k \neq 0;\ 1 \leqslant i,\, k \leqslant n), \tag{3}$$

and (ii) the extended projective space, in which the group is given by

$$z_i^* = (c_1^{(i)}z_1+...+c_n^{(i)}z_n+c_0^{(i)})/(c_1^{(0)}z_1+...+c_n^{(0)}z_n+c_0^{(0)}), \tag{4}$$

where

$$(c_i^j)_{0\leqslant i,\, j\leqslant n}$$

is a non-singular matrix.

Recently I established results for several other groups, and they seem to form a complete system in the sense of the structure of

† *Lehrbuch der Funktionentheorie*, II_1 (Teubner, 1929), 293–301.

groups. The present paper contains a proof for the space of complex spheres. I hope to give the other cases in succeeding papers.

In the Lie geometry of hyperspheres in the $(n-1)$-dimensional space, we introduce 'homogeneous coordinates' $(u_1,\dots, u_n, v_1, v_2)$ to represent a hypersphere with centre $(\xi_1,\dots, \xi_{n-1})$ and radius R by means of the relations:

$$u_1^2+\dots+u_n^2-v_1^2-v_2^2 = 0, \tag{5}$$

$$\left.\begin{aligned} u_i &= \rho\xi_i \quad (1 \leqslant i \leqslant n-1),\\ v_2 &= \rho R,\\ u_n &= \tfrac{1}{2}\rho\Big(1-\sum_{i=1}^{n-1}\xi_i^2+R^2\Big),\\ v_1 &= \tfrac{1}{2}\rho\Big(1+\sum_{i=1}^{n-1}\xi_i^2-R^2\Big), \end{aligned}\right\} \tag{6}$$

and, inversely, we have

$$\xi_i = \frac{u_i}{u_n+v_1}, \qquad R = \frac{v_2}{u_n+v_1}.$$

The elements with $u_n+v_1 = 0$ represent improper hyperspheres. The Lie group of the geometry is given by

$$(u_1^*,\dots, u_n^*, v_1^*, v_2^*) = \rho(u_1,\dots, u_n, v_1, v_2)F, \tag{7}$$

where F is an $(n+2)$-rowed matrix leaving the quadratic relation (5) invariant. In non-homogeneous coordinates we have

$$\left.\begin{aligned} \xi_i^* &= f_i(\xi_1,\dots, \xi_{n-1}, R) \quad (1 \leqslant i \leqslant n-1),\\ R^* &= f_n(\xi_1,\dots, \xi_{n-1}, R). \end{aligned}\right\} \tag{8}$$

If now we extend the geometry to the complex field, we obtain an extended space $\mathfrak{R}$ defined by n complex variables $(\xi_1,\dots, \xi_{n-1}, R)$, and the new group is obtained from (8) by varying F in the complex field and preserving the relation (5).

For the sake of convenience in the complex field, we can modify our notation slightly. Write

$$\xi_1 = z_1, \quad \dots, \quad \xi_{n-1} = z_{n-1}, \quad iR = z_n. \tag{9}$$

Now the homogeneous coordinates of the 'complex sphere' $(z_1,\dots, z_n)$ are given by

$$x_i = \rho z_i, \quad y_i = \rho \sum_{i=1}^{n} z_i^2, \quad y_2 = \rho. \tag{10}$$

The transformation takes the form

$$(x_1^*,\dots, x_n^*, y_1^*, y_2^*) = \rho(x_1,\dots, x_n, y_1, y_2)F, \tag{11}$$

where F leaves the quadratic relation

$$\sum_{i=1}^{n} x_i^2-y_1y_2 = 0 \tag{12}$$

invariant. Write F as

$$\begin{pmatrix} T & v_1' & v_2' \\ u_1 & a & b \\ u_2 & c & d \end{pmatrix}, \tag{13}$$

where T is an n-rowed matrix and u_1, u_2, v_1, v_2 denote four n-vectors. Now corresponding to (8), we have

$$(z_1^*,\ldots,z_n^*) = \frac{(z_1,\ldots,z_n)T+u_1\sum_{i=1}^n z_i^2+u_2}{(z_1,\ldots,z_n)v_2'+b\sum_{i=1}^n z_i^2+d}. \tag{14}$$

Using (14) instead of (2) I shall prove here that *the group G defined by* (14) *contains all analytic automorphs of the space* $\mathfrak{R}(G)$.

We begin by finding the Laguerre subgroup H of the Lie group G: that is the group of transformations carrying improper points into improper points. More definitely, we look for those (14) which have no variables in the denominator. Consequently $v_2 = 0$, $b = 0$. From

$$\begin{pmatrix} T & v_1' & 0 \\ u_1 & a & 0 \\ u_2 & c & d \end{pmatrix}\begin{pmatrix} I^{(n)} & 0 & 0 \\ 0 & 0 & -\frac{1}{2} \\ 0 & -\frac{1}{2} & 0 \end{pmatrix}\begin{pmatrix} T' & u_1' & u_2' \\ v_1 & a & c \\ 0 & 0 & d \end{pmatrix} = \rho^2\begin{pmatrix} I & 0 & 0 \\ 0 & 0 & -\frac{1}{2} \\ 0 & -\frac{1}{2} & 0 \end{pmatrix} \tag{15}$$

we deduce that

$$TT' = \rho^2 I, \qquad u_1 = 0,$$

$$v_1 = \frac{2}{d}u_2 T', \qquad ad = \rho^2, \qquad c = \frac{u_2 u_2'}{d}.$$

Therefore

$$F = \begin{pmatrix} \rho\Gamma & 2\rho\Gamma u_2'/d & 0 \\ 0 & \rho^2/d & 0 \\ u_2 & u_2 u_2'/d & d \end{pmatrix}, \tag{16}$$

where Γ is orthogonal. F is a product of

$$\begin{pmatrix} \rho & 0 & 0 \\ 0 & \rho^2 & 0 \\ 0 & 0 & 1 \end{pmatrix}, \tag{17}$$

$$\begin{pmatrix} \Gamma & 2u_2\Gamma & 0 \\ 0 & 1 & 0 \\ u_2 & u_2 u_2' & 1 \end{pmatrix}, \tag{18}$$

and

$$\begin{pmatrix} I & 0 & 0 \\ 0 & 1/d & 0 \\ 0 & 0 & d \end{pmatrix}. \tag{19}$$

Corresponding to (18), we have the mapping

$$(z_1^*, \ldots, z_n^*) = (x_1^*, \ldots, x_n^*)/y_2^*$$
$$= \{(x_1, \ldots, x_n)\Gamma + y_2 v_2\}/y_2 = (z_1, \ldots, z_n)\Gamma + u_2. \qquad (20)$$

Corresponding to (17) and (19), we have the mapping

$$(z_1^*, \ldots, z_2^*) = \rho(z_1, \ldots, z_n). \qquad (21)$$

The group H is generated by (20) and (21).

It follows that any two finite points are equivalent under the transformation (20). For points (i.e. improper complex spheres) at infinity, I make the assertion:

Every point at infinity is carried into a finite point by means of one of the following $n+1$ transformations:

$$z_i^* = -z_i \Big/ \Big(\sum_{j=1}^{n} z_j^2\Big) \quad (1 \leqslant i \leqslant n), \qquad (22)$$

and, for a fixed p $(1 \leqslant p \leqslant n)$,

$$\left.\begin{aligned} z_i^* &= z_i \Big/ \Big(1 - 2z_p + \sum_{j=1}^{n} z_j^2\Big) \quad (i \neq p), \\ z_p^* &= \Big(-z_p + \sum_{j=1}^{n} z_j^2\Big) \Big/ \Big(1 - 2z_p + \sum_{j=1}^{n} z_j^2\Big). \end{aligned}\right\} \qquad (23)$$

In fact, write $\quad \mathfrak{b} = (x_1, \ldots, x_n, y_1, y_2).$

Let $\mathfrak{a}$ be a vector $(a_1, \ldots, a_n, b_1, b_2)$. Then, evidently

$$\mathfrak{b}^* = -\frac{2\langle \mathfrak{a}, \mathfrak{b}\rangle}{\langle \mathfrak{a}, \mathfrak{a}\rangle}\mathfrak{a} + \mathfrak{b} \qquad (24)$$

is a transformation of the extended space, where

$$\langle \mathfrak{a}, \mathfrak{b}\rangle = \sum_{j=1}^{n} a_j x_j - \tfrac{1}{2}(b_1 y_2 + b_2 y_1). \qquad (25)$$

(Actually, this is known as Lie inversion in the geometry of spheres.) Suppose that $\mathfrak{b}$ is a point at infinity: that is, it has $y_2 = 0$. Then we have

$$y_2^* = -\frac{2\langle \mathfrak{a}, \mathfrak{b}\rangle}{\langle \mathfrak{a}, \mathfrak{a}\rangle} b_2. \qquad (26)$$

Evidently, at least one of the vectors

$$\mathfrak{a} = (0, \ldots, 0, 0, 1, 1) \qquad (27)$$

and

$$\mathfrak{a} = \Big(0, \ldots, \underset{p\text{th}}{1}, 0, \ldots, 0, 1\Big) \qquad (28)$$

makes (26) non-vanishing.

Corresponding to (27) we have the transformation

$$z_i^* = \frac{x_i^*}{y_2^*} = \frac{x_i}{-y_1} = -\frac{z_i}{\sum_{j=1}^{n} z_j^2}. \tag{29}$$

Corresponding to (28) we have

$$\mathfrak{b}^* = -(2x_p - y_1)(0,..., 1, 0,..., 0, 1) + \mathfrak{b},$$

i.e.

$$x_i^* = x_i \quad (i \neq p), \qquad x_p^* = -(2x_p - y_1) + x_p = -x_p + y_1,$$

$$y_1^* = y_1, \qquad y_2^* = -(2x_p - y_1) + y_2.$$

Then, when $i \neq p$,

$$z_i^* = \frac{x_i^*}{y_2^*} = \frac{x_i}{-2x_p + y_1 + y_2}$$

$$= \frac{z_i}{1 - 2z_p + \sum_{j=1}^{n} z_j^2}, \tag{30$_1$}$$

and

$$z_p^* = x_p^*/y_2^* = (-x_p + y_1)/(-2x_p + y_1 + y_2)$$

$$= \frac{-z_p + \sum_{j=1}^{n} z_j^2}{1 - 2z_p + \sum_{j=1}^{n} z_j^2}. \tag{30$_2$}$$

Notice that (30) may be obtained from (29) by means of the transformations

$$z_i = w_i \quad (i \neq p), \qquad z_p = 1 - w_p,$$

and

$$z_i^* = -w_i^* \quad (i \neq p), \qquad z_p^* = -(1 - w_p^*).$$

The Jacobian of the transformation (29) is equal to

$$\Big(\sum_{j=1}^{n} z_j^2\Big)^{-n} \tag{31}$$

and that of (30) is equal to

$$\Big(1 - 2z_p + \sum_{j=1}^{n} z_j^2\Big)^{-n}. \tag{32}$$

I now assume that $n \geqslant 3$.

Let

$$z_i^* = f_i(z_1,..., z_n) \quad (1 \leqslant i \leqslant n) \tag{33}$$

be an analytic mapping carrying the extended space of complex spheres into itself. By a theorem due to Osgood,† the mapping (33) is birational. Consequently, (33) can be written as

$$z_i^* = p_i(z_1,..., z_n)/q(z_1,..., z_n), \tag{34}$$

† Ibid. 299.

where p_i $(1 \leqslant i \leqslant n)$ and q are $n+1$ polynomials without common divisor other than constant.

1. There is a point $(z_1^{(0)},\dots,z_n^{(0)})$ satisfying

$$p_i(z_1^{(0)},\dots,z_n^{(0)}) \neq 0$$

and

$$q(z_1^{(0)},\dots,z_n^{(0)}) \neq 0.$$

The transformation

$$z_i = w_i + z_i^{(0)},$$

of the form (20), converts (34) into a new transformation in which

$$p_i(0,\dots,0) \neq 0, \qquad q(0,\dots,0) \neq 0. \tag{35}$$

The transformation

$$z_i^* = f_i\left(\frac{-z_1}{\sum\limits_j z_j^2},\dots,\frac{-z_n}{\sum\limits_j z_j^2}\right), \tag{36}$$

which is the product of (33) and (22), also maps the extended space on itself. Write

$$z_i^* = \left\{p_i\left(\frac{-z_1}{\sum z_j^2},\dots,\frac{-z_n}{\sum z_j^2}\right)(\sum z_j^2)^\lambda\right\}\Big/\left\{q\left(\frac{-z_1}{\sum z_j^2},\dots,\frac{-z_n}{\sum z_j^2}\right)(\sum z_j^2)^\lambda\right\},$$

where λ is the least integer that makes all the numerators and the denominator integral. On account of (35), we find that

$$p_i\left(\frac{-z_1}{\sum z_j^2},\dots,\frac{-z_n}{\sum z_j^2}\right)(\sum z_j^2)^\lambda$$

and

$$q\left(\frac{-z_1}{\sum z_j^2},\dots,\frac{-z_n}{\sum z_j^2}\right)(\sum z_j^2)^\lambda$$

are all of degree 2λ.

Consider the Jacobian of (36). Let Δ and Δ_1 be the respective inverses of the Jacobians of (33) and (36). Δ and Δ_1 are polynomials; for, otherwise, there would exist points making the Jacobian vanish. From (36) and (31) we have

$$\Delta_1(z_1,\dots,z_n) = \Delta\left(\frac{-z_1}{\sum z_j^2},\dots,\frac{-z_n}{\sum z_j^2}\right)(\sum z_j^2)^n. \tag{37}$$

Since $q(0,\dots,0) \neq 0$, we have $\Delta(0,\dots,0) \neq 0$. Consequently, $\Delta_1(z_1,\dots,z_n)$ is a polynomial of degree $2n$.

Now we may assume, without loss of generality, that p_j and q are polynomials of degree 2λ, that their terms of the highest degree are constant multiples of $(\sum z_j^2)^\lambda$, and that the Jacobian Δ of (34) is a polynomial of degree $2n$ and its terms of highest degree a constant multiple of $(\sum z_j^2)^n$.

2. We decompose the polynomial q into irreducible factors

$$q = q_1^{\lambda_1} \dots q_l^{\lambda_l}. \tag{38}$$

I am going to prove that q_1^n divides the inverse of the Jacobian Δ. Suppose firstly that q_1^2 does not divide $\sum_{j=1}^{n} p_j^2$. The inverse of the Jacobian of the transformations of (29) and (34) is equal to

$$\left\{\sum_{j=1}^{n} \left(\frac{p_j}{q}\right)^2\right\}^n \Delta(z_1,\dots,z_n).$$

It follows that q_1^n divides Δ.

Suppose that q_1^2 divides $\sum_{j=1}^{n} p_j^2$. Without loss of generality, we may assume that q_1 does not divide p_1. The inverse of the Jacobian of the product of (30) and (34) is equal to

$$\left\{1-2\frac{p_1}{q}+\sum_{j=1}^{n} \left(\frac{p_j}{q}\right)^2\right\}^n \Delta(z_1,\dots,z_n).$$

Evidently q_1^n divides $\Delta(z_1,\dots,z_n)$.

Since $\sum z_j^2$ is an irreducible polynomial when $n \geqslant 3$, and Δ is of degree $2n$, we find immediately that $l = 1$ and q_1 is of degree 2. We therefore have

$$q(z_1,\dots,z_n) = \left(a\sum_{j=1}^{n} z_j^2+\dots\right)^\lambda \tag{39}$$

and

$$\Delta(z_1,\dots,z_n) = \rho\left(\sum_{j=1}^{n} z_j^2\right)^n+\dots.$$

Consequently

$$\Delta(z_1,\dots,z_n) = \text{constant}\times\{q(z_1,\dots,z_n)\}^{n/\lambda}. \tag{40}$$

By means of a translation (if necessary), we can assume that

$$q(z_1,\dots,z_n) = \left(\sum_{j=1}^{n} z_j^2+c\right)^\lambda. \tag{41}$$

3. The product of (29) and (34) is equal to

$$z_i^* = \frac{p_i}{(\sum p_j^2)/q}. \tag{42}$$

If q does not divide $\sum p_j^2$, there exists a manifold which is mapped into the point $z_i^* = 0$. This is impossible. Therefore q divides $\sum p_j^2$. By the argument that gave (41), we have

$$\frac{\sum p_j^2}{q} = \Big(a\sum_{j=1}^{n} z_j^2 + \sum_{j=1}^{n} \beta_j z_j + \gamma\Big)^{\lambda}. \tag{43}$$

Applying the same argument to the product of (30) and (34), we have immediately

$$q - 2p_k + \frac{1}{q}\sum p_j^2 = \Big(\alpha_k \sum_{j=1}^{n} z_j^2 + \sum_{j=1}^{n} \beta_{kj} z_j + \gamma_k\Big)^{\lambda}. \tag{44}$$

We suppose that $\lambda > 1$. From (41), (43), (44) we obtain

$$2z_k^* = 1 + \left(\frac{\alpha \sum z_j^2 + \sum \beta_j z_j + \gamma}{\sum z_j^2 + c}\right)^{\lambda} - \left(\frac{\alpha_k \sum z_j^2 + \sum \beta_{kj} z_j + \gamma_k}{\sum z_j^2 + c}\right)^{\lambda}. \tag{45}$$

Then we have

$$2\frac{\partial z_k^*}{\partial z_l} = \lambda \left(\frac{\alpha \sum z_j^2 + \sum \beta_j z_j + \gamma}{\sum z_j^2 + c}\right)^{\lambda-1} \frac{\partial}{\partial z_l}\left(\frac{\alpha \sum z_j^2 + \sum \beta_j z_k + \gamma}{\sum z_j^2 + c}\right) -$$

$$-\lambda \left(\frac{\alpha_k \sum z_j^2 + \sum \beta_{kj} z_j + \gamma_k}{\sum z_j^2 + c}\right)^{\lambda-1} \frac{\partial}{\partial z_l}\left(\frac{\alpha_k \sum z_j^2 + \sum \beta_{kj} z_j + \gamma_k}{\sum z_j^2 + c}\right).$$

If there exists a point such that

$$\alpha \sum z_j^2 + \sum \beta_j z_j + \gamma = 0, \tag{46}$$

$$\alpha_k \sum z_j^2 + \sum \beta_{kj} z_j + \gamma_k = 0 \tag{47}$$

but

$$\sum_{j=1}^{n} z_j^2 + c \neq 0,$$

then the point will make the Jacobian vanish. This violates the one-to-one relationship. Thus (46) and (47) imply

$$\sum_{j=1}^{n} z_j^2 + c = 0. \tag{48}$$

Consequently, (48) is a linear combination of (46) and (47). Further, $\sum (p_i/q)^2$ cannot be a constant. Thus (47) is a linear combination of (46) and (48). This is impossible when $n \geqslant 3$ because of the independence of $z_1^*, \ldots, z_n^*$.

4. We therefore have $\lambda = 1$. From (41), (43), (44), with some slight modification, we may write (33) as

$$z_k^* = \frac{\sum \beta_{ij} z_j + \gamma_i}{\sum z_j^2 + c}. \tag{49}$$

Since q divides $\sum_{j=1}^{n} p_j^2$, we have

$$\sum_{i=1}^{n}\Big(\sum_{j=1}^{n} \beta_{ij} z_j + \gamma_i\Big)^2 = \rho\Big(\sum_{j=1}^{n} z_j^2 + c\Big). \tag{50}$$

It follows immediately that

$$\sum_{k=1}^{n} \beta_{ki}\beta_{kj} = p^2\delta_{ij}, \tag{51}$$

$$\sum_{k=1}^{n} \beta_{ki}\gamma_k = 0, \tag{52}$$

and

$$\sum_{k=1}^{n} \gamma_k^2 = c. \tag{53}$$

From (51) we find that

$$\frac{1}{\rho}(\beta_{ki}) = b_{ki}$$

is an orthogonal matrix. From (52) and (53) we get $\gamma_k = 0$, $c = 0$. Thus, by multiplying the transformations of the groups G, (33) now takes the form

$$z_i^* = \frac{\rho \sum_{j=1}^{n} b_{ij} z_j}{\sum_{j=1}^{n} z_j^2}. \tag{54}$$

It belongs evidently to the group G. Therefore, we have established the completeness of the group G when $n \geqslant 3$.

Consequently every element of G is a product of the transformations (20), (21), (22).

I should remark that when $n = 1$ the theorem is well known. When $n = 2$, it is not difficult to establish that the group is not simple, the space being a topological product of two Cauchy planes. More precisely, we use

$$x_1 x_2 - y_1 y_2 = 0$$

instead of (12). The group G may be obtained from

$$z_1^* = \frac{a_1 z_1 + b_1}{c_1 z_1 + d_1}, \qquad z_2^* = \frac{a_2 z_2 + b_2}{c_2 z_2 + d_2},$$

with an additional permutation

$$z_1^* = z_2, \qquad z_2^* = z_1.$$

This is the case known as 'the space of the theory of functions'. Thus we have solved the problem completely.

Sonderdruck aus:
Schriftenreihe des Instituts für Mathematik
Bei der Deutschen Akademie der Wissenschaften zu Berlin, Heft 1
Erschienen im Akademie-Verlag · Berlin 1957

On the RIEMANNian curvature in the space of several complex variables*)

By LOO-KENG HUA, Peking

§ 1. Introduction. Let $z = (z^1, \ldots, z^n)$ be a vector of n complex variables $z^k = x^k + iy^k$ where x^k and y^k are real numbers. Let $\mathfrak{D}$ be a bounded schlicht domain in the $2n$-dimensional space formed by x^k and y^k $(1 \leq k \leq n)$. It was established by S. BERGMANN [1] that the class $\mathfrak{L}^2 = \mathfrak{L}^2(\mathfrak{D})$ of functions $f(z)$, regular in $\mathfrak{D}$ and square integrable over $\mathfrak{D}$ (i. e.

$$\int_{\mathfrak{D}} |f(z)|^2 \dot{z} < \infty, \quad \dot{z} = \prod_{k=1}^{n} dx^k dy^k) \tag{1.1}$$

forms a HILBERT space. We can select from $\mathfrak{L}^2$ a complete orthonormal system

$$\varphi_0(z), \varphi_1(z), \ldots, \varphi_\alpha(z), \ldots. \tag{1.2}$$

By orthonormal we mean that

$$\int_{\mathfrak{D}} \varphi_\alpha(z) \overline{\varphi_\beta(z)} \dot{z} = \begin{cases} 0 & \text{if} \quad \alpha \neq \beta, \\ 1 & \text{if} \quad \alpha = \beta. \end{cases} \tag{1.3}$$

He introduces also the kernel function

$$K(z, \bar{z}) = \sum_{\alpha=0}^{\infty} \varphi_\alpha(z) \overline{\varphi_\alpha(z)}, \tag{1.4}$$

which is independent of the choice of the complete orthonormal system $\{\varphi_\alpha\}$.

We metricalize the space $\mathfrak{D}$ by introducing the Hermitian differential form, under tensor convention,

$$T_{i\bar{j}}\, dz^i\, d\bar{z}^j, \tag{1.5}$$

where

$$T_{i\bar{j}} = \frac{\partial^2}{\partial z^i\, \partial \bar{z}^j} \log K(z, \bar{z}). \tag{1.6}$$

The RIEMANNian curvature of the space $\mathfrak{D}$ is defined by

$$R = \frac{R_{\bar{k}ij\bar{l}}\, dz^i\, dz^j\, d\bar{z}^k\, d\bar{z}^l}{(T_{i\bar{k}}\, dz^i\, d\bar{z}^k)^2}, \tag{1.7}$$

where

$$R_{\bar{k}ij\bar{l}} = -\frac{\partial^2}{\partial z^j\, \partial \bar{z}^l} T_{i\bar{k}} + \left(\frac{\partial}{\partial z^j} T_{i\bar{p}}\right) \cdot T^{p\bar{q}} \left(\frac{\partial}{\partial \bar{z}^l} T_{q\bar{k}}\right) \tag{1.8}$$

and $(T^{i\bar{k}})$ is the inverse matrix of $(T_{i\bar{k}})$.

By means of variational method, FUCHS [1] proved that

$$R > 2. \tag{1.9}$$

In the first part of the paper, we shall establish a more general result by means of an elementary method. We shall assume neither the completeness, nor the orthonormality of the system φ_ν nor the boundedness of the domain $\mathfrak{D}$. In the proof, we only use the analytic property that if $f(z)$ is an analytic function, then

*) The materials of the paper are contained in two papers (HUA [1], [2]).

$\frac{\partial}{\partial \bar{z}^k} f(z) = 0$. Beside this property, all the other manipulations are purely algebraic. We express also $2 - R$ as a sum of squares.

In the second part, we estimate R from below. We shall prove that under certain restriction, we have

$$R \geqq -n. \tag{1.10}$$

The restriction does not exclude the interesting case that $\mathfrak{D}$ is a transitive domain.

In the third part, as a feable generalization of celebrated Riemann theorem on conformal mapping, we prove that every non-continuable bounded domain with constant curvature can be carried onto a unit sphere by means of an analytic mapping (or pseudo-conformal mapping). Such a result seems to be the first step toward the interesting classification problem of non-continuable domain under analytic mappings.

Part I

§ 2. Unitary geometry.

Let z be a variable over the domain $\mathfrak{D}$. For the time being we do not assume the boundedness on $\mathfrak{D}$. Let H be a non-singular Hermitian matrix with elements $H_{i\bar{j}}$ which are analytic functions of z and $\bar{z}$. The fundamental Hermitian form is given by

$$dz\, H\, \overline{dz'}\,, \tag{2.1}$$

where A' denotes the transposed matrix of A.

Let

$$z = t(w) \tag{2.2}$$

be an analytic mapping with Jacobian

$$J = J(w) = \frac{\partial(z^1, \ldots, z^n)}{\partial(w^1, \ldots, w^n)}, \tag{2.3}$$

which is a non-singular matrix.

Then (2.2) carries (2.1) into

$$dw\, K\, \overline{dw'}$$

where

$$K = J\, H\, \bar{J}', \tag{2.4}$$

since

$$dz = dw\, J. \tag{2.5}$$

Differentiating (2.4), we have

$$dK = (J\, dH + dJ\, H)\, \bar{J}'$$

and

$$dK \cdot K^{-1} = J\,(dH \cdot H^{-1})\, J^{-1} + dJ \cdot J^{-1}.$$

Differentiating conjugately, we have

$$\bar{d}\,(dK \cdot K^{-1}) = J\, \bar{d}\,(dH \cdot H^{-1})\, J^{-1},$$

that is

$$(\bar{d}\,dK - dK\cdot K^{-1}\,\bar{d}K)\,K^{-1} = J\,(\bar{d}\,dH - dH\cdot H^{-1}\,\bar{d}H)\,H^{-1}\,J^{-1}.$$

Notice that d and $\bar{d}$ are the operations $\frac{\partial}{\partial z^j}\,dz^j$ and $\frac{\partial}{\partial \bar{z}^j}\,d\bar{z}^j$ respectively.

Multiplying by (2.4), we have

$$\bar{d}\,dK - dK\cdot K^{-1}\,\bar{d}K = J\,(\bar{d}\,dH - dH\cdot H^{-1}\,\bar{d}H)\,\bar{J}'. \tag{2.6}$$

Therefore $\bar{d}\,dH - dH\cdot H^{-1}\,\bar{d}H$ is an Hermitian matrix which transforms covariantly with H. Multiplying by vector dw on the left and $\overline{dw'}$ on the right, we have

$$dw\,(\bar{d}\,dK - dK\cdot K^{-1}\,\bar{d}K)\,\overline{dw'} = dz\,(\bar{d}\,dH - dH\cdot H^{-1}\,\bar{d}H)\,\overline{dz'}. \tag{2.7}$$

In comparing with the definition (1.8), we obtain

$$R = -\frac{dz\,(\bar{d}\,dH - dH\,H^{-1}\,\bar{d}H)\,\overline{dz'}}{(dz\,H\,\overline{dz'})^2}. \tag{2.8}$$

Theorem 1. Let $q(z, \bar{z})$ be an analytic function of z and $\bar{z}$ and $\overline{q(z, \bar{z})} = q(z, \bar{z})$. Then

$$H^* = q(z, \bar{z})\,H$$

is again an Hermitian matrix. Let R^* be the RIEMANNian curvature with respect to the fundamental tensor H^*.

Then

$$q\,R^* = R - \frac{\bar{d}\,d\log q(z, \bar{z})}{dz\,H\,\overline{dz'}}.$$

Proof. We have

$$\begin{aligned}&\bar{d}\,dH^* - dH^*\cdot H^{*-1}\,\bar{d}H^*\\ &= \bar{d}\,(dq\,H + q\,dH) - (dq\,H + q\,dH)\,q^{-1}\,H^{-1}\,(\bar{d}q\,H + q\,\bar{d}H)\\ &= q\,(\bar{d}\,dH - dH\cdot H^{-1}\,\bar{d}H) + (\bar{d}\,dq - q^{-1}\,dq\,\bar{d}q)\,H.\end{aligned}$$

Since

$$\bar{d}\,d\log q = \bar{d}(q^{-1}\,dq) = q^{-1}\,(\bar{d}\,dq - q^{-1}\,dq\,\bar{d}q),$$

we have the theorem.

§ 3. Unitary geometry with fundamental tensor (1.5).

Now we take

$$H_{i\bar{j}} = T_{i\bar{j}} = \partial_i\partial_{\bar{j}}\log K = \frac{1}{K}\,\partial_i\partial_{\bar{j}}K - \frac{1}{K^2}\,\partial_i K\,\partial_{\bar{j}}K \tag{3.1}$$

where $\partial_i = \frac{\partial}{\partial z^i}$, $\partial_{\bar{j}} = \frac{\partial}{\partial \bar{z}^j}$. For the present, we assume neither the completeness nor the orthonormality of the sequence $\{\varphi_\nu\}$. For assuring the meaning of $\log K$ we assume only that the functions φ_ν have no common zero in $\mathfrak{D}$.

Substituting (1.4) into (3.1) we have

$$T_{i\bar{j}} = \frac{1}{K^2} \sum_{\alpha,\beta=0}^{\infty} (\varphi_\alpha \bar{\varphi}_\alpha \, \partial_i \varphi_\beta \, \partial_{\bar{j}} \bar{\varphi}_\beta - \partial_i \varphi_\alpha \bar{\varphi}_\alpha \varphi_\beta \, \partial_{\bar{j}} \bar{\varphi}_\beta)$$
$$= \frac{1}{K^2} \sum_{\alpha>\beta} (\varphi_\alpha \, \partial_i \varphi_\beta - \varphi_\beta \, \partial_i \varphi_\alpha) \overline{(\varphi_\alpha \, \partial_j \varphi_p - \varphi_\beta \, \partial_j \varphi_\alpha)}\,, \tag{3.2}$$

and

$$dz\, H\, d\bar{z}' = \frac{1}{K^2} \sum_{\alpha>\beta} | \varphi_\alpha \, d\varphi_\beta - \varphi_\beta \, d\varphi_\alpha |^2 . \tag{3.3}$$

We obtain at once

Theorem 2. Let

$$\varphi_0(z),\ \varphi_1(z),\ \ldots,\ \varphi_\alpha(z),\ \ldots$$

be a sequence of functions analytic in $\mathfrak{D}$ and without common zero in $\mathfrak{D}$. If the series

$$K(z, \bar{z}) = \sum_{\alpha=0}^{\infty} \varphi_\alpha(z) \overline{\varphi_\alpha(z)}$$

converges uniformly in any compact $\mathfrak{D}^*$ interior to $\mathfrak{D}$, the Hermitian differential form $d\bar{d} \log K$ is positive definite (including semi-definite).

If $d\bar{d} \log K = 0$, from (3.3), we have

$$\varphi_\alpha \, d\varphi_\beta - \varphi_\beta \, d\varphi_\alpha = 0$$

for all α and β. That is, we have a vector $\xi = (\xi^1, \ldots, \xi^n)$ such that

$$(\varphi_\alpha \, \partial_i \varphi_\beta - \varphi_\beta \, \partial_i \varphi_\alpha) \xi^i = 0\,, \tag{3.4}$$

that is, the rank of the $n \times \infty$ matrix

$$\begin{pmatrix} \varphi_\alpha & \partial_1 \varphi_\beta - \varphi_\beta & \partial_1 \varphi_\alpha \\ \cdots & \cdots & \cdots \\ \varphi_\alpha & \partial_n \varphi_\beta - \varphi_\beta & \partial_n \varphi_\alpha \end{pmatrix}_{\alpha>\beta\geq 0} \tag{3.5}$$

is $< n$. Notice that

$$\varphi_\alpha (\varphi_\beta \, \partial_i \varphi_\nu - \varphi_\nu \, \partial_i \varphi_\beta) + \varphi_\beta (\varphi_\nu \, \partial_i \varphi_\alpha - \varphi_\alpha \, \partial_i \varphi_\nu) + \varphi_\nu (\varphi_\alpha \, \partial_i \varphi_\beta - \varphi_\beta \, \partial_i \varphi_\alpha) = 0\,. \tag{3.6}$$

Then, if $\varphi_0(z)$ does not vanish at $z = z_0$, the rank of (3.5) at $z = z_0$ is equal to that of

$$\begin{pmatrix} \varphi_0 & \partial_1 \varphi_\beta - \varphi_\beta & \partial_1 \varphi_0 \\ \cdots & \cdots & \cdots \\ \varphi_0 & \partial_n \varphi_\beta - \varphi_\beta & \partial_n \varphi_0 \end{pmatrix}_{\beta=1,2,\ldots},$$

that is, that of

$$\begin{Bmatrix} \partial_1 \left(\dfrac{\varphi_\beta}{\varphi_0}\right) \\ \cdots\cdots \\ \partial_n \left(\dfrac{\varphi_\beta}{\varphi_0}\right) \end{Bmatrix}. \tag{3.7}$$

Theorem 3. If the linear closure of the sequence $\{\varphi_\nu\}$ contains the functions $1, z^1, \ldots, z^n$, then the fundamental tensor (3.3) is always strictly definite. Consequently, if $\{\varphi_\nu\}$ is a complete orthonormal system, the fundamental tensor is strictly definite. By linear closure, we mean the set formed by the functions $f(z)$ representable by series

$$f(z) = \sum_{\nu=0}^{\infty} a_\nu \varphi_\nu(z)$$

convergent uniformly in any $\mathfrak{D}^*$ interior to $\mathfrak{D}$.

Proof. From

$$1 = \sum_{\nu=0}^{\infty} a_{0\nu} \varphi_\nu(z)\,,$$

$$z^k = \sum_{\nu=0}^{\infty} a_{k\nu} \varphi_\nu(z)\,, \qquad 1 \leq k \leq n\,,$$

we deduce that

$$\begin{aligned} dz^k &= \sum_{\nu=0}^{\infty} a_{k\nu}\, d\varphi_\nu(z) \\ &= \sum_{\nu=0}^{\infty}\sum_{\mu=0}^{\infty} a_{k\nu}\, d\varphi_\nu \cdot a_{0\mu}\varphi_\mu - \sum_{\nu=0}^{\infty}\sum_{\mu=0}^{\infty} a_{k\nu}\varphi_\nu\, a_{0\mu}\, d\varphi_\mu \\ &= \sum_{\nu>\mu} a_{k\nu}\, a_{0\mu}\,(\varphi_\mu\, d\varphi_\nu - \varphi_\nu\, d\varphi_\mu)\,, \end{aligned}$$

i. e.

$$\delta_i^k = \sum_{\nu>\mu} a_{k\nu}\, a_{0\mu}\,(\varphi_\mu\, \partial_i\varphi_\nu - \varphi_\nu\, \partial_i\varphi_\mu)\,.$$

The rank of (3.5) is evidently equal to n. Theorem 3 is proved.

For the later usage, we prove the following corollary.

Theorem 4. If

$$K(z, \bar z) = |\xi(z)|^2 \left(1 + a \sum_{i=1}^{n} |\varphi_i(z)|^2\right)^b \quad (a,\ b \text{ real}) \tag{3.8}$$

is a kernel function of a Hilbert space defined in § 1, then the Jacobian of $\varphi_i(z)$ does not vanish in $\mathfrak{D}$.

Proof. We have

$$d\bar d \log K(z, \bar z) = b\, d\bar d \log\left(1 + a\sum_{i=1}^{n} |\varphi_i(z)|^2\right)$$

$$= \frac{b}{\left(1 + a\sum_{i=1}^{n} |\varphi_i(z)|^2\right)^2}\left[a\sum_{i=1}^{n} |d\varphi_i|^2 \left(1 + a\sum_{j=1}^{n} |\varphi_j|^2\right) - a^2 \sum_{i=1}^{n} d\varphi_i\, \bar\varphi_i \sum_{j=1}^{n} \varphi_j\, \overline{d\varphi_j}\right]$$

$$= \frac{b}{\left(1 + a\sum_{i=1}^{n} |\varphi_i(z)|^2\right)^2}\left[a\sum_{i=1}^{n} |d\varphi_i|^2 + a^2 \sum_{n\geq\nu>\mu\geq 1} |d\varphi_\nu\,\varphi_\mu - d\varphi_\mu\,\varphi_\nu|^2\right].$$

If the Jacobian of $\varphi_i(z)$, i. e.

$$\left|\partial_j \varphi_i(z)\right|_{1 \leq i,j \leq n}$$

vanishes, at a point $z = z_0$, then we have a non-zero vector $\xi = (\xi^1, \ldots, \xi^n)$ such that

$$\partial_j \varphi_i(z)\, \xi^j = 0 \,,$$

at $z = z_0$.

That is, the Hermitian form

$$d\bar{d} \log K(\bar{z}, z)$$

vanishes at $z = z_0$ for $dz^i = \xi^i$. Combining with theorem 3, we have theorem 4.

§ 4. Riemannian curvature ≤ 2.

We arrange (α, β) $(\alpha > \beta)$ according to the following order: If $\alpha < \alpha'$, we put (α, β) preceding (α', β') and the relation is denoted by $(\alpha, \beta) < (\alpha', \beta')$; and if $\alpha = \alpha'$ but $\beta < \beta'$, then we put (α, β) preceding (α, β') and also denoted by $(\alpha, \beta) < (\alpha, \beta')$. After the arrangement we numerate the indices by $\gamma = 0, 1, 2, 3, \ldots$; more precisely $\gamma = \frac{1}{2}\alpha(\alpha - 1) + \beta$. Our fundamental tensor can be written by

$$T_{i\bar{j}}\, dz^i\, d\bar{z}^j = \frac{1}{K^2} \sum_{\gamma=0}^{\infty} u_\gamma \bar{u}_\gamma \tag{4.1}$$

and

$$T_{i\bar{j}} = \frac{1}{K^2} \sum_{\gamma=0}^{\infty} u^i_\gamma \bar{u}^j_\gamma \,,$$

where

$$u_\gamma = \varphi_\alpha\, d\varphi_\beta - \varphi_\beta\, d\varphi_\alpha \tag{4.2}$$

and

$$u^i_\gamma = \varphi_\alpha\, \partial_i \varphi_\beta - \varphi_\beta\, \partial_i \varphi_\alpha \,. \tag{4.3}$$

Taking $q = K^2$ in theorem 1, and

$$T^*_{i\bar{j}} = \sum_{\gamma=0}^{\infty} u^i_\gamma \bar{u}^j_\gamma \,, \tag{4.4}$$

we have then

$$q\, R^* = R - \frac{\bar{d}\, d \log K^2}{\bar{d}\, d \log K} = R - 2 \,. \tag{4.5}$$

The aim of the present section is to prove that $R^* \leq 0$, that is

Theorem 5. Let $\{\varphi_\nu\}$, $\nu = 0, 1, 2, \ldots$, be a sequence of analytic functions in a domain $\mathfrak{D}$, and $\{\varphi_\nu\}$ have no common zero in $\mathfrak{D}$. Suppose that

$$K(z, \bar{z}) = \sum_{\nu=0}^{\infty} \varphi_\nu(z)\, \overline{\varphi_\nu(z)}$$

converges uniformly in any compact domain interior to $\mathfrak{D}$, and that $\{\varphi_\nu/\varphi_0\}$,

$\nu = 1, 2, \ldots,$ contains n independent functions. Then the RIEMANNian curvature of the space $\mathfrak{D}$ defined by the metric

$$\bar{d}\, d \log K(z, \bar{z}) \tag{4.6}$$

does not exceed 2.

Proof. Let $H^* = (T^*_{i\bar{j}})$ and

$$N = \begin{pmatrix} u^1_\gamma \\ \ldots \\ u^n_\gamma \end{pmatrix}_{\gamma = 0, 1, \cdots} .$$

Then, evidently we have

$$H^* = N\, \bar{N}' ,$$

and

$$\begin{aligned} &\bar{d}\, dH^* - dH^* \cdot H^{*-1}\, \bar{d}H^* \\ &= dN \left(I - \bar{N}' (N\, \bar{N}')^{-1} N\right) \overline{dN}' . \end{aligned}$$

Substituting into (2.8), we have

$$R^* = -\frac{dz\, dN\, (I - \bar{N}'\, (N\, \bar{N}')^{-1}\, N)\, \overline{dN}'\, d\bar{z}'}{((dz\, N)\, (\overline{dz\, N})')^2} . \tag{4.7}$$

Since

$$\begin{aligned} I - \bar{N}'(N\, \bar{N}')^{-1}\, N &= \left(I - \bar{N}'(N\, \bar{N}')^{-1}\, N\right)^2 \\ &= \left(I - \bar{N}'(N\, \bar{N}')^{-1}\, N\right) \overline{\left(I - \bar{N}'(N\, \bar{N}')^{-1}\, N\right)'} \end{aligned}$$

is a positive definite Hermitian matrix, we have therefore $R^* \leqq 0$.

Remark. Theorem 5 is „best possible". In fact for $n = 2$, and taking six functions

$$A,\ z^1,\ z^2,\ a_i(z^1)^2 + 2\, b_i\, z^1\, z^2 + c_i(z^2)^2 , \qquad i = 1, 2, 3,$$

as our $\{\varphi_\nu\}$, we can easily calculate that, at $z = 0$,

$$2 - R = 4\, A^2\, \frac{\sum_{i=1}^{3} \left|a_i(dz^1)^2 + 2\, b_i\, dz^1\, dz^2 + c_i(dz^2)^2\right|^2}{(|dz^1|^2 + |dz^2|^2)^2} . \tag{4.8}$$

We determined (HUA [1]) also the extremal case, i. e. $R = 2$, and proved (HUA [1]) that if the linear closure of the sequence $\{\varphi_\nu\}$ contains

$$1,\ z^1,\ \ldots,\ z^n,\ z^i\, z^j \quad (1 \leqq i \leqq j \leqq n) ,$$

then we have

$$R < 2 . \tag{4.9}$$

§ 5. Expression of $2 - R$ as a sum of squares.

Theorem 6. We use T^* to denote the determinant of $T^*_{i\bar{j}}$, and

$$x_\gamma = \partial_j u^i_\gamma\, dz^i\, dz^j . \tag{5.1}$$

Then

$$-T^* R^*_{\bar{k}ij\bar{l}}\, dz^i\, dz^j\, d\bar{z}^k\, d\bar{z}^l = \sum_{\gamma_1>\gamma_2>\cdots>\gamma_{n+1}} \left| P_{\gamma_1,\ldots,\gamma_{n+1}} \right|^2, \tag{5.2}$$

where

$$P_{\gamma_1,\ldots,\gamma_{n+1}} = \begin{vmatrix} u^1_{\gamma_1}, & \ldots, & u^n_{\gamma_1}, & x_{\gamma_1} \\ \cdots & \cdots & \cdots & \cdots \\ u^1_{\gamma_{n+1}}, & \ldots, & u^n_{\gamma_{n+1}}, & x_{\gamma_{n+1}} \end{vmatrix}. \tag{5.3}$$

Proof. 1. First let us consider

$$A = T^* \partial_j \partial_{\bar{l}} T^*_{i\bar{k}}\, dz^i\, dz^j\, d\bar{z}^k\, d\bar{z}^l$$

$$= \sum_{\alpha_1=0}^{\infty} \cdots \sum_{\alpha_n=0}^{\infty} \begin{vmatrix} u^1_{\alpha_1} \overline{u^1_{\alpha_1}}, & \ldots, & u^1_{\alpha_1} \overline{u^n_{\alpha_1}} \\ \cdots & \cdots & \cdots \\ u^n_{\alpha_n} \overline{u^1_{\alpha_n}}, & \ldots, & u^n_{\alpha_n} \overline{u^n_{\alpha_n}} \end{vmatrix} \sum_{\beta=0}^{\infty} x_\beta \bar{x}_\beta$$

$$= \sum_{\alpha_1=0}^{\infty} \cdots \sum_{\alpha_n=0}^{\infty} u^1_{\alpha_1} \cdots u^n_{\alpha_n} \begin{vmatrix} \overline{u^1_{\alpha_1}}, & \ldots, & \overline{u^n_{\alpha_1}} \\ \cdots & \cdots & \cdots \\ \overline{u^1_{\alpha_n}}, & \ldots, & \overline{u^n_{\alpha_n}} \end{vmatrix} \sum_{\beta=0}^{\infty} x_\beta \bar{x}_\beta$$

$$= \sum_{\alpha_1>\alpha_2>\cdots>\alpha_n} \begin{vmatrix} u^1_{\alpha_1} \ldots u^n_{\alpha_1} \\ \cdots\cdots \\ u^1_{\alpha_n} \ldots u^n_{\alpha_n} \end{vmatrix} \overline{\begin{vmatrix} u^1_{\alpha_1} \ldots u^n_{\alpha_1} \\ \cdots\cdots \\ u^1_{\alpha_n} \ldots u^n_{\alpha_n} \end{vmatrix}} \sum_{\beta=0}^{\infty} x_\beta \bar{x}_\beta, \tag{5.4}$$

by (4.4).

Write

$$P_{\gamma_1,\ldots\gamma_{n+1}} = \begin{vmatrix} u^1_{\gamma_1}, & \ldots, & u^n_{\gamma_1}, & x_{\gamma_1} \\ \cdots & \cdots & \cdots & \cdots \\ u^1_{\gamma_{n+1}}, & \ldots, & u^n_{\gamma_{n+1}}, & x_{\gamma_{n+1}} \end{vmatrix} = \sum_{i=1}^{n+1} (-1)^{n+i} P_i\, x_{\gamma_i}. \tag{5.5}$$

If

$$\alpha_1 > \alpha_2 > \cdots > \alpha_r > \beta > \alpha_{r+1} > \cdots > \alpha_n,$$

we take $\alpha_1 = \gamma_1, \ldots, \alpha_r = \gamma_r, \beta = \gamma_{r+1}, \alpha_{r+1} = \gamma_{r+2}, \ldots, \alpha_n = \gamma_{n+1}$, then

$$A = \sum_{\gamma_1>\gamma_2>\cdots>\gamma_{n+1}} \sum_{i=1}^{n+1} p_i \bar{p}_i\, x_{\gamma_i} \bar{x}_{\gamma_i} + B, \tag{5.6}$$

where

$$B = \sum_{r=1}^{n} \sum_{\alpha_1>\cdots>\alpha_n} \begin{vmatrix} u^1_{\alpha_1} \ldots u^n_{\alpha_1} \\ \cdots\cdots \\ u^1_{\alpha_n} \ldots u^n_{\alpha_n} \end{vmatrix} \overline{\begin{vmatrix} u^1_{\alpha_1} \ldots u^n_{\alpha_1} \\ \cdots\cdots \\ u^1_{\alpha_n} \ldots u^n_{\alpha_n} \end{vmatrix}}\, x_{\alpha_r} \bar{x}_{\alpha_r}. \tag{5.7}$$

2. Next we consider

$$C = -T^* \partial_j T_{i\bar{p}}^{*}\, T^{*p\bar{q}}\, \partial_{\bar{l}} T^*_{q\bar{k}}\, dz^i\, dz^j\, d\bar{z}^k\, d\bar{z}^l$$

$$= -\sum_{\beta_1=0}^{\infty} \sum_{\beta_2=0}^{\infty} \bar{u}^p_{\beta_1} \left(T^* T^{*p\bar{q}}\right) u^q_{\beta_2}\, x_{\beta_1} \bar{x}_{\beta_2}. \tag{5.8}$$

As in 1. we obtain

$$
\begin{aligned}
&T^* \, T^{*p\bar q} \\
&= (-1)^{p+q} \sum_{\alpha_1=0}^{\infty} \cdots \sum_{\alpha_{n-1}=0}^{\infty}
\begin{vmatrix}
u^1_{\alpha_1} \overline{u^1_{\alpha_1}}, & \ldots, & u^1_{\alpha_1} \overline{u^{p-1}_{\alpha_1}}, & u^1_{\alpha_1} \overline{u^{p+1}_{\alpha_1}}, & \ldots, & u^1_{\alpha_1} \overline{u^n_{\alpha_1}} \\
\cdots & \cdots & \cdots & \cdots & \cdots & \cdots \\
u^{q-1}_{\alpha_{q-1}} \overline{u^1_{\alpha_{q-1}}}, & \ldots, & u^{q-1}_{\alpha_{q-1}} \overline{u^{p-1}_{\alpha_{q-1}}}, & u^{q-1}_{\alpha_{q-1}} \overline{u^{p+1}_{\alpha_{q-1}}}, & \ldots, & u^{q-1}_{\alpha_{q-1}} \overline{u^n_{\alpha_{q-1}}} \\
u^{q+1}_{\alpha_q} \overline{u^1_{\alpha_q}}, & \ldots, & u^{q+1}_{\alpha_q} \overline{u^{p-1}_{\alpha_q}}, & u^{q+1}_{\alpha_q} \overline{u^{p+1}_{\alpha_q}}, & \ldots, & u^{q+1}_{\alpha_q} \overline{u^n_{\alpha_q}} \\
\cdots & \cdots & \cdots & \cdots & \cdots & \cdots \\
u^n_{\alpha_{n-1}} \overline{u^1_{\alpha_{n-1}}}, & \ldots, & u^n_{\alpha_{n-1}} \overline{u^{p-1}_{\alpha_{n-1}}}, & u^n_{\alpha_{n-1}} \overline{u^{p+1}_{\alpha_{n-1}}}, & \ldots, & u^n_{\alpha_{n-1}} \overline{u^n_{\alpha_{n-1}}}
\end{vmatrix} \\
&= (-1)^{p+q} \sum_{\alpha_1=0}^{\infty} \cdots \sum_{\alpha_{n-1}=0}^{\infty} u^1_{\alpha_1} \cdots u^{q-1}_{\alpha_{q-1}} u^{q+1}_{\alpha_q} \cdots u^n_{\alpha_{n-1}}
\begin{vmatrix}
\overline{u^1_{\alpha_1}}, & \ldots, & \overline{u^{p-1}_{\alpha_1}}, & \overline{u^{p+1}_{\alpha_1}}, & \ldots, & \overline{u^n_{\alpha_1}} \\
\cdots & \cdots & \cdots & \cdots & \cdots & \cdots \\
\overline{u^1_{\alpha_{n-1}}}, & \ldots, & \overline{u^{p-1}_{\alpha_{n-1}}}, & \overline{u^{p+1}_{\alpha_{n-1}}} & \ldots, & \overline{u^n_{\alpha_{n-1}}}
\end{vmatrix} \\
&= (-1)^{p+q} \sum_{\alpha_1 > \alpha_2 > \cdots > \alpha_{n-1}}
\begin{vmatrix}
u^1_{\alpha_1}, & \ldots, & u^{q-1}_{\alpha_1}, & u^{q+1}_{\alpha_1}, & \ldots, & u^n_{\alpha_1} \\
\cdots & \cdots & \cdots & \cdots & \cdots & \cdots \\
u^1_{\alpha_{n-1}}, & \ldots, & u^{q-1}_{\alpha_{n-1}}, & u^{q+1}_{\alpha_{n-1}}, & \ldots, & u^n_{\alpha_{n-1}}
\end{vmatrix} \times \\
&\qquad \times \overline{\begin{vmatrix}
u^1_{\alpha_1}, & \ldots, & u^{p-1}_{\alpha_1}, & u^{p+1}_{\alpha_1}, & \ldots, & u^n_{\alpha_1} \\
\cdots & \cdots & \cdots & \cdots & \cdots & \cdots \\
u^1_{\alpha_{n-1}}, & \ldots, & u^{p-1}_{\alpha_{n-1}}, & u^{p+1}_{\alpha_{n-1}}, & \ldots, & u^n_{\alpha_{n-1}}
\end{vmatrix}} \, .
\end{aligned} \tag{5.9}
$$

Therefore, we have

$$
\begin{aligned}
&\sum_{\beta_1, \beta_2=0}^{\infty} T^* \, T^{*p\bar q} \, \overline{u^p_{\beta_1}} \, u^q_{\beta_2} \, x_{\beta_1} \, \overline{x_{\beta_2}} \\
&= \sum_{\beta_1, \beta_2=0}^{\infty} \sum_{\alpha_1 > \alpha_2 > \cdots > \alpha_{n-1}}
\begin{vmatrix}
u^1_{\beta_2}, & \ldots, & u^n_{\beta_2} \\
u^1_{\alpha_1}, & \ldots, & u^n_{\alpha_1} \\
\cdots & \cdots & \cdots \\
u^1_{\alpha_{n-1}}, & \ldots, & u^n_{\alpha_{n-1}}
\end{vmatrix}
\overline{\begin{vmatrix}
u^1_{\beta_1}, & \ldots, & u^n_{\beta_1} \\
u^1_{\alpha_1}, & \ldots, & u^n_{\alpha_1} \\
\cdots & \cdots & \cdots \\
u^1_{\alpha_{n-1}}, & \ldots, & u^n_{\alpha_{n-1}}
\end{vmatrix}}
x_{\beta_1} \, \overline{x_{\beta_2}} \\
&= B + \sum_{\alpha_1 > \cdots > \alpha_{n-1}} \sum_{\beta_1 \neq \beta_2}
\begin{vmatrix}
u^1_{\beta_2}, & \ldots, & u^n_{\beta_2} \\
u^1_{\alpha_1}, & \ldots, & u^n_{\alpha_1} \\
\cdots & \cdots & \cdots \\
u^1_{\alpha_{n-1}}, & \ldots, & u^n_{\alpha_{n-1}}
\end{vmatrix}
\overline{\begin{vmatrix}
u^1_{\beta_1}, & \ldots, & u^n_{\beta_1} \\
u^1_{\alpha_1}, & \ldots, & u^n_{\alpha_1} \\
\cdots & \cdots & \cdots \\
u^1_{\alpha_{n-1}}, & \ldots, & u^n_{\alpha_{n-1}}
\end{vmatrix}}
x_{\beta_1} \, \overline{x_{\beta_2}} \, .
\end{aligned} \tag{5.10}
$$

If $\alpha_i = \beta_1$ or β_2, the corresponding term vanishes. Therefore we may assume that α_i equals neither β_1 nor β_2. Suppose $\beta_1 > \beta_2$, we arrange

$$
\alpha_1 > \cdots > \alpha_u > \beta_1 > \alpha_{u+1} > \cdots > \alpha_{u+v} > \beta_2 > \alpha_{u+v+1} > \cdots > \alpha_{n-1} \, .
$$

By taking

$$
\begin{aligned}
&\gamma_i = \alpha_i \ (1 \leqq i \leqq u) \, , \quad \gamma_{u+1} = \beta_1 \, , \quad \gamma_{u+1+j} = \alpha_{u+j} \ (1 \leqq j \leqq v) \, , \\
&\gamma_{u+v+2} = \beta_2 \, , \quad \gamma_{u+v+2+k} = \alpha_{u+v+k} \ (1 \leqq k \leqq n-u-v-1) \, ,
\end{aligned}
$$

the corresponding term can be written as

$$(-1)^{u+u+v}\, p_{u+1}\, \bar{p}_{u+v+2}\, x_{\gamma_{u+1}}\, \bar{x}_{\gamma_{u+v+2}}\,.$$

If $\beta_1 < \beta_2$, we have a similar result.

Therefore

$$\begin{aligned} A + C &= \sum_{\gamma_1>\gamma_2>\cdots>\gamma_{n+1}} \left(\sum_{i=1}^{n} p_i\, \bar{p}_i\, x_{\gamma_i}\, \bar{x}_{\gamma_i} + \sum_{j\neq k} (-1)^{j+k}\, p_j\, \bar{p}_k\, x_{\gamma_j}\, \bar{x}_{\gamma_k}\right) \\ &= \sum_{\gamma_1>\gamma_2>\cdots>\gamma_{n+1}} \left(\sum_{j=1}^{n} (-1)^{j-1}\, p_i\, x_{\gamma_j}\right) \left(\sum_{k=1}^{n} (-1)^{k-1}\, \bar{p}_k\, \bar{x}_{\gamma_k}\right) \\ &= \sum_{\gamma_1>\gamma_2>\cdots>\gamma_{n+1}} \begin{vmatrix} u^1_{\gamma_1}, & \ldots, & u^n_{\gamma_1}, & x_{\gamma_1} \\ \cdots & \cdots & \cdots & \cdots \\ u^1_{\gamma_{n+1}}, & \ldots, & u^n_{\gamma_{n+1}}, & x_{\gamma_{n+1}} \end{vmatrix} \overline{\begin{vmatrix} u^1_{\gamma_1}, & \ldots, & u^n_{\gamma_1}, & x_{\gamma_1} \\ \cdots & \cdots & \cdots & \cdots \\ u^1_{\gamma_{n+1}}, & \ldots, & u^n_{\gamma_{n+1}}, & x_{\gamma_{n+1}} \end{vmatrix}}\,; \end{aligned} \tag{5.11}$$

the theorem is proved.

§ 6. Ricci tensor.

Let the determinant of the Hermitian form $\bar{d}\, d \log K$ be denoted by

$$D = \left|\partial_i \partial_{\bar{j}} \log K\right|\,. \tag{6.1}$$

From (4.1), we have

$$\begin{aligned} D &= \frac{1}{K^{2n}} \sum_{\alpha_1=0}^{\infty} \cdots \sum_{\alpha_n=0}^{\infty} \begin{vmatrix} u^1_{\alpha_1}\, \overline{u^1_{\alpha_1}}, & \ldots, & u^1_{\alpha_1}\, \overline{u^n_{\alpha_1}} \\ \cdots & \cdots & \cdots \\ u^n_{\alpha_n}\, \overline{u^1_{\alpha_n}}, & \ldots, & u^n_{\alpha_n}\, \overline{u^n_{\alpha_n}} \end{vmatrix} \\ &= \frac{1}{K^{2n}} \sum_{\alpha_1>\alpha_2>\cdots>\alpha_n} \Phi_{\alpha_1,\ldots,\alpha_n}(z)\, \overline{\Phi_{\alpha_1,\ldots,\alpha_n}(z)}\,, \end{aligned} \tag{6.2}$$

where

$$\Phi_{\alpha_1,\ldots,\alpha_n}(z) = \begin{vmatrix} u^1_{\alpha_1}, & \ldots, & u^n_{\alpha_1} \\ \cdots & \cdots & \cdots \\ u^1_{\alpha_n}, & \ldots, & u^n_{\alpha_n} \end{vmatrix}\,. \tag{6.3}$$

We arrange the order of $(\alpha_1, \ldots, \alpha_n)$ $(\alpha_1 > \alpha_2 > \cdots > \alpha_n)$ in the following way: If $\alpha_1 = \beta_1, \ldots, \alpha_{r-1} = \beta_{r-1}$ but $\alpha_r > \beta_r$, we define that $(\beta_1, \ldots, \beta_n)$ precedes $(\alpha_1, \ldots, \alpha_n)$. We numerate this sequence by numbers $0, 1, 2, \ldots, \omega, \ldots$ Then we have

$$D = \frac{1}{K^{2n}} \sum_{\omega=0}^{\infty} \Phi_\omega(z)\, \overline{\Phi_\omega(z)}\,. \tag{6.4}$$

Similar to the method given in § 3, the Hermitian form defined by the Ricci tensor can be written as

$$\begin{aligned} -\partial_i\, \partial_{\bar{j}} \log D\, dz^i\, d\bar{z}^j &= -\sum_{\omega>\eta} \left|\Phi_\omega\, d\Phi_\eta - \Phi_\eta\, d\Phi_\omega\right|^2 \Big/ \left(\sum_{\omega=0}^{\infty} \left|\Phi_\omega\right|^2\right)^2 \\ &\quad + 2n \sum_{\alpha>\beta} \left|\varphi_\alpha\, d\varphi_\beta - \varphi_\beta\, d\varphi_\alpha\right|^2 \Big/ \left(\sum_{\alpha=0}^{\infty} \left|\varphi_\alpha\right|^2\right)^2. \end{aligned} \tag{6.5}$$

Part II

§ 7. Space with further restriction.

In part II we add a further restriction that

$$\sum_{\nu=0}^{\infty} \Phi_\nu(z)\,\overline{\Phi_\nu(z)} = c\left(\sum_{\nu=0}^{\infty} \varphi_\nu(z)\,\overline{\varphi_\nu(z)}\right)^{2n+1}, \tag{7.1}$$

where c is a constant and Φ_ν is defined in § 6 or Φ_ν runs over all the n-rowed minors of the infinite matrix

$$\begin{pmatrix} \varphi_\alpha \partial_1 \varphi_\beta - \varphi_\beta \partial_1 \varphi_\alpha \\ \cdots\cdots\cdots\cdots \\ \varphi_\alpha \partial_n \varphi_\beta - \varphi_\beta \partial_n \varphi_\alpha \end{pmatrix}_{\alpha>\beta,\ \alpha,\beta=0,1,2,\cdots} . \tag{7.2}$$

First of all, let us give some explanation about the condition (7.1). It is known [cf. (6.1)] that

$$D = \left|\partial_i \partial_{\bar{j}} \log K\right| = \frac{1}{K^{2n}} \sum_{\gamma=0}^{\infty} \left|\Phi_\gamma\right|^2$$

is the volume density of the RIEMANNian space. (7.1) can be written as

$$c\,K = D\,, \tag{7.3}$$

that is, the kernel function is proportional to the volume density. Consequently the RIEMANNian space so obtained is an EINSTEIN space.

If the space $\mathfrak{D}$ is transitive, let

$$z = t(w)$$

be a transformation carrying $\mathfrak{D}$ onto itself. Then the kernel function $K(z, \bar{z})$ of the space satisfies

$$K(z, \bar{z}) = K(w, \bar{w})\,\left|J(w)\right|^2,$$

and also

$$D(z, \bar{z}) = D(w, \bar{w})\,\left|J(w)\right|^2.$$

Since

$$\frac{K(z, \bar{z})}{D(z, \bar{z})} = \frac{K(w, \bar{w})}{D(w, \bar{w})}$$

for all w,

$$\frac{K(z, \bar{z})}{D(z, \bar{z})}$$

is a constant. Therefore, for the space $\mathfrak{D}$ admitting a transitive group of analytic transformations, the geometry defined in § 1 satisfies our condition (7.1).

Now let us go back to our original situation. We introduce several lemmas which are useful for simplification:

Theorem 7. Let $z = t(w)$ be an analytic mapping with JACOBIAN $J(w)$. Let

$$\psi_\nu(w) = \varphi_\nu\big(t(w)\big)\,J(w)\,. \tag{7.4}$$

We construct $\Psi_\nu(w)$ from $\psi_\nu(w)$ as $\Phi_\nu(z)$ from $\varphi_\nu(z)$. Then, the relation (7.1) holds, if we replace $\varphi_\nu(z)$ by $\psi_\nu(w)$ and $\Phi_\nu(z)$ by $\Psi_\nu(w)$.

Proof. Since

$$\psi_\alpha(w)\frac{\partial}{\partial w^i}\psi_\beta(w)-\psi_\beta(w)\frac{\partial}{\partial w^i}\psi_\alpha(w)=\frac{\partial z^j}{\partial w^i}\big(\varphi_\alpha(z)\,\partial_j\varphi_\beta(z)-\varphi_\beta(z)\,\partial_j\varphi_\alpha(z)\big)\big(J(w)\big)^2$$

and

$$\Psi_\nu(w)=\det\left(\frac{\partial z^j}{\partial w^i}\right)\Phi_\nu(z)\big(J(w)\big)^{2n}=\Phi_\nu(z)\big(J(w)\big)^{2n+1},$$

we have

$$\sum_{\nu=0}^{\infty}\Psi_\nu(w)\overline{\Psi_\nu(w)}=|J(w)|^{2(2n+1)}\sum_{\nu=0}^{\infty}\Phi_\nu(z)\overline{\Phi_\nu(z)}$$
$$=c\,|J(w)|^{2(2n+1)}\left(\sum_{\nu=0}^{\infty}\varphi_\nu(z)\overline{\varphi_\nu(z)}\right)^{2n+1}$$
$$=c\left(\sum_{\nu=0}^{\infty}\psi_\nu(w)\overline{\psi_\nu(w)}\right)^{2n+1}.$$

Theorem 8. If

$$\psi_\nu(z)=\sum_{\mu=0}^{\infty}a_{\nu\mu}\varphi_\mu(z),\qquad \nu=0,1,2,\ldots,$$

where $(a_{\nu\mu})=U$ is an unitary matrix, the conclusion of theorem 7 is again true.

Proof. From

$$\psi_\alpha(z)\,\partial_i\psi_\beta(z)-\psi_\beta(z)\,\partial_i\psi_\alpha(z)$$
$$=\sum_{\nu=0}^{\infty}\sum_{\mu=0}^{\infty}a_{\alpha\nu}a_{\beta\mu}(\varphi_\nu\,\partial_i\varphi_\mu-\varphi_\mu\,\partial_i\varphi_\nu)$$
$$=\sum_{\nu>\mu}(a_{\alpha\nu}a_{\beta\mu}-a_{\alpha\mu}a_{\beta\nu})(\varphi_\nu\,\partial_i\varphi_\mu-\varphi_\mu\,\partial_i\varphi_\nu),$$

we deduce that

$$\begin{pmatrix}\psi_\alpha\partial_1\psi_\beta-\psi_\beta\partial_1\psi_\alpha\\ \cdots\cdots\cdots\cdots\\ \psi_\alpha\partial_n\psi_\beta-\psi_\beta\partial_n\psi_\alpha\end{pmatrix}=\begin{pmatrix}\varphi_\nu\partial_1\varphi_\mu-\varphi_\mu\partial_1\varphi_\nu\\ \cdots\cdots\cdots\cdots\\ \varphi_\nu\partial_n\varphi_\mu-\varphi_\mu\partial_n\varphi_\nu\end{pmatrix}(A_{\nu\mu,\alpha\beta})\tag{7.5}$$

where
$$A_{\mu\nu,\alpha\beta}=a_{\alpha\nu}a_{\beta\mu}-a_{\alpha\mu}a_{\beta\nu}.$$

Since

$$\sum_{\alpha>\beta}A_{\nu\mu,\alpha\beta}\bar{A}_{\nu_1\mu_1,\alpha\beta}=\frac{1}{2}\sum_{\alpha=0}^{\infty}\sum_{\beta=0}^{\infty}(a_{\alpha\nu}a_{\beta\mu}-a_{\alpha\mu}a_{\beta\nu})(\bar{a}_{\alpha\nu_1}\bar{a}_{\beta\mu_1}-\bar{a}_{\alpha\mu_1}\bar{a}_{\beta\nu_1})$$
$$=\frac{1}{2}(\delta_{\nu\nu_1}\delta_{\mu\mu_1}-\delta_{\mu\nu_1}\delta_{\nu\mu_1}-\delta_{\nu\mu_1}\delta_{\mu\nu_1}+\delta_{\mu\mu_1}\delta_{\nu\nu_1})$$
$$=\begin{cases}1, & \text{if }\ \nu=\nu_1,\ \mu=\mu_1,\\ 0, & \text{otherwise},\end{cases}$$

the matrix $(A_{\nu\mu,\alpha\beta})$ is unitary.

Since the n-rowed minors of (7.5) have also unitary relation, we have

$$\sum_{\nu=0}^{\infty} \Psi_\nu(z) \overline{\Psi_\nu(z)} = \sum_{\nu=0}^{\infty} \Phi_\nu(z) \overline{\Phi_\nu(z)}$$

which establishes theorem 8.

§ 8. Reductions.

Now we are going to perform some simplifications which do not restrict the generality.

We may suppose that $\mathfrak{D}$ contains the origin, $\varphi_0(0) \neq 0$, and that the JACOBian of φ_i/φ_0 $(1 \leq i \leq n)$ does not vainsh at the origin. Consider the expansion

$$\varphi_\nu(z) = a_0^\nu + \sum_{i=1}^{n} a_i^\nu z^i + \sum_{i,j=1}^{n} a_{ij}^\nu z^i z^j + \cdots, \tag{8.1}$$

where $a_{ij}^\nu = a_{ji}^\nu$. Since $\sum_{\nu=0}^{\infty} |\varphi_\nu(0)|^2$ converges, we have a unitary matrix $U = (u_{\nu\mu})_{\nu,\mu=0,1,2,\cdots}$ such that

$$(a_0^0, a_0^1, \ldots, a_0^\nu, \ldots) U = (\varrho, 0, \ldots, 0, \ldots)$$

where

$$\varrho^2 = \sum_{\nu=0}^{\infty} |\varphi_\nu(0)|^2 .$$

Consider

$$\sum_{\nu=0}^{\infty} \varphi_\nu(z) u_{\nu\mu}$$

instead of $\varphi_\mu(z)$, we may assume, without loss of generality, that

$$\varphi_0(0) = a_0^0 \neq 0, \quad \varphi_\nu(0) = 0 \quad (\nu > 0).$$

Consider the transformation

$$\left.\begin{aligned} w^1 &= (a_0^0)^{-1} \int_0^{z^1} \varphi_0(t, z^2, \ldots, z^n)\, dt \\ w^k &= z^k \qquad \text{for} \quad 2 \leq k \leq n. \end{aligned}\right\} \tag{8.2}$$

The transformation carries the origin into the origin and its JACOBian

$$\frac{\partial w}{\partial z} = (a_0^0)^{-1} \varphi_0(z)$$

equals 1 at the origin. By theorem 7, with $\psi_0(w) = \varphi_0(z)\, J(w) = a_0^0$, we may assume without loss of generality that

$$\varphi_0(z) = A, \quad \varphi_\nu(0) = 0$$

for $\nu > 0$. Notice that the property "the JACOBian of φ_k/φ_0 $(1 \leq k \leq n)$ does not vanish at the origin" still holds.

We write

$$\varphi_\nu(z) = l_\nu(z) + q_\nu(z), \quad \nu > 0,$$

where $l_\nu(z)$ are linear terms of $\varphi_\nu(z)$ and $q_\nu(z)$ contain neither linear nor constant terms. Since the JACOBIAN of φ_k/φ_0 $(1 \leq k \leq n)$ does not vanish at the origin, therefore

$$l_k(z) \qquad (1 \leq k \leq n)$$

are linearly independent forms. There is a unitary matrix

$$V = (v_{\nu\,\mu})_{\nu,\,\mu = 1,2,\ldots}$$

such that

$$\sum_{\mu=1}^{\infty} \varphi_\nu(z)\, v_{\nu\,\mu}\,, \quad \mu = n+1\,, \quad n+2,\ \ldots .$$

have no linear terms. Therefore, from now on, we may assume that, for $\nu > n$,

$$\varphi_\nu(z) = q_\nu(z)$$

which are power series without constant and linear terms.

Hereafter, we use $q(z)$ to denote a power series without constant and linear terms and they are not necessarily the same at each occurrence.

Let $w^i = l_i(z) = \sum_{j=1}^{n} a^i_j z^j$. By theorem 7 we obtain

$$\psi_0(w) = Aa\,, \quad \psi_i(w) = \big(w^i + q_i(w)\big)\, a\,, \quad 1 \leq i \leq n\,,$$

$$\psi_\nu(w) = q_\nu(w)\, a\,, \quad \nu > n$$

where $a = \det\,(a^i_j)^{-1}$. If we replace $\varphi_\alpha(z)$ by $\varphi_\alpha(z)\, a$ $(\alpha = 0, 1, 2, \ldots)$, the relation (7.1) becomes

$$\sum_{\nu=0}^{\infty} \Phi_\nu(z)\, \overline{\Phi_\nu(z)} = c_1 \left(\sum_{\alpha=0}^{\infty} \varphi_\alpha(z)\, \overline{\varphi_\alpha(z)} \right)^{2n+1},$$

where c_1 is again a constant. Therefore without loss of generality, we may assume that

$$\varphi_0(z) = A\,, \quad \varphi_i(z) = z^i + q_i(z)\,, \quad 1 \leq i \leq n\,,$$

$$\varphi_\nu(z) = q_\nu(z)\,, \qquad \nu > n\,. \tag{8.3}$$

Next let us study the relation (7.1). The right hand side does not contain those terms which are functions of $\bar{z}$ and independent of z. Such terms may occur on the left if $\Phi_\nu(z)$ contain constant terms. That is, it can only happen in $|\Phi_0(z)|^2$, where

$$\Phi_0 = \begin{vmatrix} \varphi_0 \partial_1 \varphi_1 - \varphi_1 \partial_1 \varphi_0, & \ldots, & \varphi_0 \partial_1 \varphi_n - \varphi_n \partial_1 \varphi_0 \\ \cdots & \cdots & \cdots \\ \varphi_0 \partial_n \varphi_1 - \varphi_1 \partial_n \varphi_0, & \ldots, & \varphi_0 \partial_n \varphi_n - \varphi_n \partial_n \varphi_0 \end{vmatrix}$$

$$= A^n \begin{vmatrix} \partial_1 \varphi_1, & \ldots, & \partial_1 \varphi_n \\ \cdots & \cdots & \cdots \\ \partial_n \varphi_1, & \ldots, & \partial_n \varphi_n \end{vmatrix}.$$

Since the right hand side of (7.1) contains no term, different from constant and independent of z, we see that Φ_0 must be a constant. Consequently

$$\begin{vmatrix} \partial_1\varphi_1, & \dots, & \partial_1\varphi_n \\ \dots & \dots & \dots \\ \partial_n\varphi_1, & \dots, & \partial_n\varphi_n \end{vmatrix} = 1 .$$

Taking $\varphi_i(z) = w^i$ in theorem 7, we may assume without loss of generality that

$$\left.\begin{array}{l} \varphi_0(z) = A\,, \quad \varphi_i(z) = z_i\,, \quad i = 1, 2, \dots, n\,; \\ \varphi_\nu(z) = q_\nu(z), \quad \nu = n+1, \quad n+2, \dots . \end{array}\right\} \tag{8.4}$$

Applying theorem 7 again, with $z = A\,w$, the functions so obtained are all divided by A^{n+1}, then without loss of generality we may assume that $A = 1$. Comparing coefficients we may assume also that $c_1 = 1$.

§ 9. Lower bound of the curvature.

Theorem 9. Under the same assumption as in theorem 5, we assume further that $\{\varphi_\nu\}$ satisfies (7.1), we have

$$R \geqq -n\,. \tag{9.1}$$

Proof. We may assume that $\mathfrak{D}$ contains the origin and without loss of generality we shall prove (9.1) only at $z = 0$. Since the manifold defined by the JACOBian of

$$\varphi_j/\varphi_0\,, \quad 1 \leqq i \leqq n$$

vanishing is of lower dimension, it is enough to assume that the JACOBian does not vanish at the origin. By the argument used in § 8, we may assume that

$$\varphi_0(z) = 1, \quad \varphi_k(z) = z^k \qquad (1 \leqq k \leqq n)$$

$$\varphi_\nu(z) = q_\nu(z)\,, \qquad \nu = n+1,\ n+2, \dots .$$

We substitute into (7.1) and consider the terms in (7.1) linear both in z and $\bar{z}$, the right hand side gives

$$(2n+1)\sum_{i=1}^{n} |z^i|^2\,,$$

and the left gives

$$\sum_{i>j} (|\,z^i\,|^2 + |\,z^j\,|^2) + \sum_{i=1}^{n}\ \sum_{\nu=n+1}^{\infty} \left|\frac{\partial q_\nu^*}{\partial z^i}\right|^2 = (n-1) \sum_{i=1}^{n} |z^i|^2 + \sum_{i=1}^{n}\ \sum_{\nu=n+1}^{\infty} \left|\frac{\partial q_\nu^*}{\partial z^i}\right|^2,$$

where q_ν^* is the quadratic part of q_ν. Equating both members, we have

$$\sum_{i=1}^{n}\ \sum_{\nu=n+1}^{\infty} \left|\frac{\partial q_\nu^*}{\partial z^i}\right|^2 = (n+2)\sum_{i=1}^{n} |z^i|^2\,. \tag{9.2}$$

Now we study the value of the RIEMANNian curvature at the origin. First let us find the value of N at the origin. It is evidently

$$N = (I^{(n)}, 0)\,.$$

17*

Therefore

$$I - \overline{N}'(N\overline{N}')^{-1}N = \begin{pmatrix} 0^{(n)} & 0 \\ 0 & I^{(\infty)} \end{pmatrix}.$$

At $z = 0$, we have

$$R^* = -\frac{\sum_{\nu=n+1}^{\infty} |\varphi_0\, d\, d\varphi_\nu - \varphi_\nu\, d\, d\varphi_0|^2}{(dz\, \overline{dz}')^2}$$

$$= -\sum_{\nu=n+1}^{\infty} |d\, dq_\nu^*|^2 / (dz\, \overline{dz}')^2 . \tag{9.3}$$

Let

$$q_\nu^* = \sum_{i,j=1}^{n} b_{ij}^\nu z^i z^j, \quad b_{ij}^\nu = b_{ji}^\nu .$$

Then (9.2) becomes

$$4 \sum_{i=1}^{n} \sum_{\nu=n+1}^{\infty} \left| \sum_{j=1}^{n} b_{ij}^\nu z^j \right|^2 = (n+2) \sum_{i=1}^{n} |z^i|^2 .$$

Consequently

$$\sum_{\nu=n+1}^{\infty} \sum_{i=1}^{n} \left| \sum_{j=1}^{n} b_{ij}^\nu dz^j \right|^2 = \frac{(n+2)}{4} \sum_{i=1}^{n} |dz^i|^2 .$$

By Schwarz' inequality, we have

$$\sum_{\nu=n+1}^{\infty} |d\, dq_\nu^*|^2 = 4 \sum_{\nu=n+1}^{\infty} \left| \sum_{i=1}^{n} \sum_{j=1}^{n} b_{ij}^\nu dz^i dz^j \right|^2$$

$$\leq 4 \sum_{\nu=n+1}^{\infty} \left(\sum_{i=1}^{n} \left| \sum_{j=1}^{n} b_{ij}^\nu dz^j \right|^2 \cdot \sum_{i=1}^{n} |dz^i|^2 \right)$$

$$= (n+2) \left(\sum_{i=1}^{n} |dz^i|^2 \right)^2 .$$

Substituting into (9.3), we have

$$-R^* \leq (n+2) .$$

The theorem is now proved.

Remark. It can be shown (Hua [1]) that if the linear closure of $\{\varphi_\nu\}$ contains

$$z^i z^j, \qquad 1 \leq i, j \leq n ,$$

then we have the strict inequality

$$R > -n .$$

Part III

§ 10. Domain with constant curvature.

Theorem 10. Let $\mathfrak{D}$ be a bounded non-continuable domain with constant curvature. There is an analytic mapping carrying $\mathfrak{D}$ onto the unit sphere

$$|z^1|^2 + \cdots + |z^n|^2 < 1 .$$

Proof. 1. The kernel function of a domain with constant curvature can be expressed as

$$|\xi(z)|^2\left(1+a\sum_{\nu=1}^{n}|\varphi_\nu(z)|^2\right)^b \qquad (a,\, b \text{ real}) \tag{10.1}$$

(BOCHNER [1]). By theorem 4, the JACOBian never vanishes in $\mathfrak{D}$. We introduce new variables

$$w^i = \varphi_i(z)\,, \qquad i=1,\ \dots,\ n\,. \tag{10.2}$$

Now we may assume that

$$|\xi(z)|^2\,(1+a\,z\,\bar{z}')^b \tag{10.3}$$

is our kernel function, where $z\,\bar{z}' = \sum_{i=1}^{n}|z^i|^2$. Notice that $\xi(z)$ may not be that of $\xi(z)$ as before. From now on $\xi(z)$ may be not the same at each occurrence.

2. Expand (10.3) in the following way

$$|\xi(z)|^2\left(1+a\,b\,z\,\bar{z}'+\frac{1}{2}\,a^2\,b\,(b-1)\,(z\,\bar{z}')^2+\cdots\right),$$

and

$$(z\,\bar{z}')^m = \sum_{l_1+\cdots+l_n=m}\frac{m!}{l_1!\dots l_n!}\,|z^1|^{2l_1}\dots|z^n|^{2l_n}\,.$$

Let

$$\psi_l(z)=\psi_{l_1,\dots,l_n}(z)=\sqrt{\frac{m!}{l_1!\dots l_n!}}\,(z^1)^{l_1}\dots(z^n)^{l_n}\,.$$

Then (10.3) can be written as

$$|\xi(z)|^2\left(1+ab(|z^1|^2+\cdots+|z^n|^2)+\cdots+a^m\frac{b(b-1)\cdots(b-m+1)}{m!}\sum_l|\psi_l(z)|^2+\cdots\right). \tag{10.4}$$

3. Suppose that $\varphi_0(z),\,\varphi_1(z),\,\dots$ be the orthonormal system of the domain $\mathfrak{D}$, then we have

$$\sum_{\nu=0}^{\infty}\left|\frac{\varphi_\nu(z)}{\xi(z)}\right|^2 = 1+a\,b\,(|z^1|^2+\cdots+|z^n|^2)+\cdots$$
$$+\,a^m\frac{b\,(b-1)\cdots(b-m+1)}{m!}\sum|\psi_l(z)|^2+\cdots\,. \tag{10.5}$$

Since $\xi(z)$ never vanishes in $\mathfrak{D}$, the both sides of (10.5) converge uniformly in a neighborhood of $z=0$. Applying the operator $\partial_1^m\,\partial_{\bar{1}}^m$ and putting $z=0$, we deduce that

$$a^m\,b\,(b-1)\dots(b-m+1)>0$$

for all values of m. Consequently

$$a<0, \qquad b<0\,. \tag{10.6}$$

Performing a suitable transformation if necessary, we may without loss of generality, assume that $a=-1$, $b=-\lambda$, $\lambda>0$.

4. From

$$K(z,\bar{z}) = |\xi(z)|^2 (1 - z\bar{z}')^{-\lambda}$$

we have

$$\begin{aligned} K(z,\bar{w}) &= \xi(z)\overline{\xi(w)}(1 - z\bar{w}')^{-\lambda} \\ &= \xi(z)\overline{\xi(w)}\left(1 + \lambda z\bar{w}' + \cdots + \frac{\lambda(\lambda+1)\cdots(\lambda+m-1)}{m!}\sum_l \psi_l(z)\overline{\psi_l(w)} + \cdots\right), \end{aligned} \tag{10.7}$$

which converges uniformly in w as z sufficiently small. Multiplying both sides by

$$\xi(w)\psi_{p_1,\ldots,p_n}(w)$$

and integrating with respect to w over $\mathfrak{D}$, we have then

$$\begin{aligned} \xi(z)\psi_{p_1,\ldots,p_n}(z) &= \int_{\mathfrak{D}} K(z,\bar{w})\xi(w)\psi_{p_1,\ldots,p_n}(w)\dot{w} \\ &= \xi(z)\sum_{m=0}^{\infty}\sum_{l_1+\cdots+l_n=m} a_{l_1,\ldots,l_n}\psi_{l_1,\ldots,l_n}(z), \end{aligned} \tag{10.8}$$

where

$$a_{l_1,\ldots,l_n} = \frac{\lambda(\lambda+1)\ldots(\lambda+m-1)}{m!}\int_{\mathfrak{D}} |\xi(w)|^2\overline{\psi_{l_1,\ldots,l_n}(w)}\psi_{p_1,\ldots,p_n}(w)\dot{w}. \tag{10.9}$$

Therefore

$$\psi_{p_1,\ldots,p_n}(z) = \sum_{m=0}^{\infty}\sum_{l_1+\cdots+l_n=m} a_{l_1,\ldots,l_n}\psi_{l_1,\ldots,l_n}(z).$$

From the uniqueness of power series, we have

$$a_{l_1,\ldots,l_n} = \begin{cases} 0 & \text{if} \quad (l_1,\ldots,l_n) \neq (p_1,\ldots,p_n), \\ 1 & \text{if} \quad l_k = p_k \quad (1 \le k \le n). \end{cases} \tag{10.10}$$

Consequently,

$$\xi(z)\Phi_{l_1,\ldots,l_n}(z) = \sqrt{\frac{\lambda(\lambda+1)\ldots(\lambda+m-1)}{m!}}\,\xi(z)\psi_{l_1,\ldots,l_n}(z) \tag{10.11}$$

form a complete orthonormal system of the domain $\mathfrak{D}$.

5. By some calculation, we find that the curvature of the domain considered is equal to $-2/\lambda$.

6. A function belonging to $\mathfrak{L}^2$ has the expansion

$$f(z) = \xi(z)\sum_l a^{l_1,\ldots,l_n}\Phi_{l_1,\ldots,l_n}(z), \tag{10.12}$$

$$\sum_l |a^{l_1,\ldots,l_n}|^2 < \infty.$$

The power series

$$f_1(z) = \sum_l a^{l_1,\ldots,l_n}\Phi_{l_1,\ldots,l_n}(z) \tag{10.13}$$

form a set $\mathfrak{M}^2$. Corresponding to each $f(z)$ of $\mathfrak{L}^2$, we have a function $f_1(z)$ in $\mathfrak{M}^2$ and conversely. Since $\mathfrak{L}^2$ contains 1, therefore $\mathfrak{M}^2$ contains the function $\frac{1}{\xi(z)}$.

Now let us consider the coefficients satisfying $l_1 + \cdots + l_n = m$. If

$$z^i = \sum_{j=1}^{n} a_j^i w^j , \tag{10.14}$$

then we have

$$\psi_{l_1,\ldots,l_n}(z) = \sum_k A_{l_1,\ldots,l_n}^{k_1,\ldots,k_n} \psi_{k_1,\ldots,k_n}(w) .$$

From the relation

$$\sum_l a^l \psi_l(z) = \sum_k b^k \psi_k(w) ,$$

we deduce

$$a^{l_1,\ldots,l_n} = \sum_k B_{k_1,\ldots,k_n}^{l_1,\ldots,l_n} b^{k_1,\ldots,k_n} .$$

It is well known that if (a_j^i) is unitary, so are (A_l^k) and (B_k^l). Therefore we have

$$\sum_{l_1+\cdots+l_n=m} |b^{l_1,\ldots,l_n}|^2 = \sum_{l_1+\cdots+l_n=m} |a^{l_1,\ldots,l_n}|^2 . \tag{10.15}$$

Therefore, we have established that, for unitary (10.14), if $f_1(z)$ belongs to $\mathfrak{M}^2$ so does $f_1(w)$. Therefore $\xi(z)\, f_1(w)$ belongs to $\mathfrak{L}^2$. The corresponding series converges uniformly in any region interior to $\mathfrak{D}$.

Let r be the distance from origin to the farthest point on the boundary of $\mathfrak{D}$. From the previous consideration, $f_1(z)$ is analytic every-where in the sphere $z\bar{z}' < r^2$, in particular, $\frac{1}{\xi(z)}$ is analytic in the sphere. Therefore, every function which is analytic in $\mathfrak{D}$, is continuable over the sphere $z\bar{z}' < r^2$. The theorem is now proved.

References

BERGMANN, S. [1], Sur les fonctions orthogonales de plusieurs variables complexes avec les applications à la théorie des fonctions analytiques, Gauthier-Villiars, Paris 1947.

BOCHNER, S. [1], Curvature in Hermitian metric, Bull. Amer. math. Soc. 53 (1947), 179—195.

FUCHS, B. A. [1], Über geodätische Mannigfaltigkeiten einer bei pseudokonformen Abbildungen invarianten Riemannschen Geometrie, Recueil math. 2 (1937), 567—594.

HUA, L. K. [1], On the Riemann curvature of the non-euclidean space of several complex variables, Acta math. Sinica 4 (1954), 141—168.

[2] On non-continuable domain with constant curvature, Acta math. Sinica 4 (1954).

V/12/6 0,030

SCIENTIA SINICA

Vol. VIII, No. 10, 1959

MATHEMATICS

THEORY OF HARMONIC FUNCTIONS IN CLASSICAL DOMAINS*

L. K. HUA (华罗庚) AND K. H. LOOK (陸启鏗)

(Institute of Mathematics, Academia Sinica)

TABLE OF CONTENTS

* Received July 3, 1959.

INTRODUCTION

This paper contains the results presented in several notes (Hua [3—5]; Hua and Look [1—6]; Look [2]), which were or will be published in Chinese, or were sketched in a short English summary.

§1. *Heuristic statement*

Let $\mathfrak{M}$ be an m-dimensional Riemannian manifold of class C^1 with the fundamental tensor $g_{ij}(i,j=1,2,\cdots,m)$. Let $x=(x^1,\cdots,x^m)$ be an arbitrary local coordinate system. Then we have the Beltrami operator

$$\Delta=\sum_{i,j=1}^{m} g^{ij}\left(\frac{\partial^2}{\partial x^i\partial x^j}-\sum_{k=1}^{m}\left\{ {k \atop ij} \right\}\frac{\partial}{\partial x^k}\right),$$

where g^{ij} as usual denotes the contravariant tensor of g_{ij} and

$$\left\{ {k \atop ij} \right\}=\frac{1}{2}\sum_{l=1}^{m} g^{kl}\left(\frac{\partial g_{il}}{\partial x^j}+\frac{\partial g_{jl}}{\partial x^i}-\frac{\partial g_{ij}}{\partial x^l}\right)$$

denotes the Christoffel symbol.

A real-valued function $u(x)$ of class C^2 defined in $\mathfrak{M}$ is said to be harmonic in $\mathfrak{M}$ if

$$\Delta u(x)=0. \tag{1}$$

Now we assume that $\mathfrak{M}$ is not compact, and the matrix (g^{ij}) is positive definite in $\mathfrak{M}$ and it becomes semi-definite on the "boundary". Alternatively, it can be described with regard to the study of harmonic functions in the "whole" Riemannian manifold. Naturally it is related to the theory of linear partial differential equations of second order of elliptic type which degenerates on the "boundary" of the manifold.

Different from the case with two variables, the geometrical structure of the boundary plays an important role. According to the authors' limited knowledge, such an attempt on the theory of differential equations of degenerated elliptic type has not been achieved before. However, we do not aim at the general theory of differential equations of degenerated elliptic type, since there still remain difficulties to be overcome.

The aim of this paper is to treat those cases which are related to the theory of functions of several complex variables. More precisely, we are going to establish the theory of harmonic functions of classical domains.

According to the rank of (g^{ij}) on the boundary, we introduce the concept of slit space. The different ranks of the matrix (g^{ij}) on the "closure" of $\mathfrak{M}$ are denoted by

$$m = m_1 > m_2 > \cdots > m_s \geqslant 0.$$

The set of points with rank m_r is denoted by $\mathfrak{C}^{(m_r)}$. Each slit space excluding its slit is denoted by $\mathfrak{C}^{(m_r)}$ $(r = 1, \cdots, s-1)$. The cases with which we are concerned are that $\mathfrak{C}^{(m_r)}$ is homeomorphic to the topological product of the space X and Y where X is a manifold of class C^2 and Y is a compact set.

The Dirichlet problem is defined in the following: Given a function continuous on $\mathfrak{C} = \mathfrak{C}^{(m_s)}$, whether there exists a unique solution of (1) on the closure of $\mathfrak{M}$ (certainly in the heuristic sense).

The uniqueness is established under comparatively general condition. To establish the existence theorem, our treatment is based on the Poisson integral which depends on the transibility of the space.

Let $B(x, \xi)$ be the functional determinant on $\mathfrak{C}$ of such a transformation of the group of motion of $\mathfrak{M}$ that carries the point x into a fixed point 0. Then

$$P(x, \xi) = C|B(x, \xi)|$$

with a well-determined constant C is our "Poisson" kernel. The solution of our Dirichlet problem can be expressed by

$$u(x) = \int_{\mathfrak{C}} \varphi(\xi) P(x, \xi) \dot{\xi},$$

where $\varphi(\xi)$ is the given boundary value on $\mathfrak{C}$ and $\dot{\xi}$ the volume element of $\mathfrak{C}$.

§2. *Harmonic functions in the hyperbolic spaces of matrices*

Let $\mathfrak{R}_{\mathrm{I}}$ (or $\mathfrak{R}_{\mathrm{I}}(m, n)$) denote the domain formed by $m \times n$ complex matrices Z making the Hermitian matrix $I - Z\bar{Z}'$ positive definite, and let $\mathfrak{C}_{\mathrm{I}}$ denote the set formed by the $m \times n$ matrices U satisfying $U\bar{U}' = I$. The differential equation for this region is given by

$$\sum_{\alpha,\beta=1}^{n} \sum_{j,k=1}^{m} \left(\delta_{\alpha\beta} - \sum_{l=1}^{m} \bar{z}_{l\alpha} z_{l\beta}\right)\left(\delta_{jk} - \sum_{\gamma=1}^{n} \bar{z}_{j\gamma} z_{k\gamma}\right) \frac{\partial^2 u(Z)}{\partial \bar{z}_{j\alpha} \partial z_{k\beta}} = 0.$$

A function is said to be harmonic on the closure $\overline{\mathfrak{R}}_{\mathrm{I}}$ of $\mathfrak{R}_{\mathrm{I}}$, if it is continuous on $\overline{\mathfrak{R}}_{\mathrm{I}}$, and it satisfies, in the sense explained in the text, the equation on $\overline{\mathfrak{R}}_{\mathrm{I}} - \mathfrak{C}_{\mathrm{I}}$.

Given a continuous function $\varphi(U)$ on $\mathfrak{C}_{\mathrm{I}}$, then the Poisson integral

$$u(Z) = \frac{1}{\prod_{l=1}^{m} \omega_{2(n-l)+1}} \int_{\mathfrak{C}_{\mathrm{I}}} \varphi(U) \frac{\det (I - Z\bar{Z}')^{n}}{|\det (I - Z\bar{U}')|^{2n}} \dot{U}, \quad \omega_{p-1} = \frac{2\pi^{p/2}}{\Gamma\left(\frac{p}{2}\right)}$$

gives the unique solution of the Dirichlet problem.

Similar results are obtained for the hyperbolic spaces of symmetric and skew symmetric matrices.

The case $m=n$ for $\mathfrak{R}_{\mathrm{I}}$ has been considered by J. Mitchell[1], but her treatment was unsatisfactory, since the given boundary value was of a very special class of function and the uniqueness was not proved.

§3. *Harmonic functions in the hyperbolic space of Lie-sphere*

Let $\mathfrak{R}_{\mathrm{IV}}$ denote the domain formed by n complex variables $z = (z_1, \cdots, z_n)$ satisfying

$$1 + |zz'|^2 - 2\bar{z}z' > 0, \quad |zz'| < 1,$$

and $\mathfrak{C}_{\mathrm{IV}}$ denote the set of the vectors

$$\xi = e^{i\theta}x,$$

where x is a real n-vector satisfying $xx' = 1$.

The differential equation of the domain is given by

$$(1 + |zz'|^2 - 2z\bar{z}')\left(\sum_{\alpha=1}^{n} \frac{\partial^2 u(z)}{\partial z_\alpha \partial \bar{z}_\alpha} - 2 \sum_{\alpha,\beta=1}^{n} z_\alpha \bar{z}_\beta \frac{\partial^2 u(z)}{\partial z_\alpha \partial \bar{z}_\beta}\right)$$
$$+ 2 \sum_{\alpha,\beta=1}^{n} (\bar{z}_\alpha - \overline{zz'} z_\alpha)(z_\beta - zz'\bar{z}_\beta) \frac{\partial^2 u(z)}{\partial z_\alpha \partial \bar{z}_\beta} = 0. \qquad (1)$$

Similarly we define a function to be harmonic on the closure of $\mathfrak{R}_{\mathrm{IV}}$.

The Dirichlet problem is also solved completely. Given a continuous function $\varphi(\xi)$ on $\mathfrak{C}_{\mathrm{IV}}$, the Poisson integral

$$u(z) = \frac{2}{\omega_1 \omega_{n-1}} \int_{\mathfrak{C}_{\mathrm{IV}}} \varphi(\xi) \frac{(1 + |zz'|^2 - 2\bar{z}z')^{n/2}}{|1 + zz'\bar{\xi}\bar{\xi}' - 2\bar{\xi}z'|^{n}} \dot{\xi}$$

gives the unique solution of (1).

The result was published in a note (Hua and Look [2]) and a weaker result was obtained later by Lowdenslager[1]. He did not prove the uniqueness, and the path was a particular one and it approached only a part of the boundary.

§4. *Applications*

In Chapter IV, we shall talk briefly about some applications of the result of the present paper.

First, we obtain a convergence theorem on the theory of harmonic analysis on a unitary group. More precisely, we introduced the Abel summability, and proved that any continuous function on the unitary group can be expressed as the Abel sum of its Fourier expansion.

Furthermore, to each continuous function on $\mathfrak{C}_I$ we have a harmonic function in $\mathfrak{R}_I$, but the converse is not true. That is, there may exist a harmonic function in $\mathfrak{R}_I$, but the boundary values on $\mathfrak{C}_I$ may not exist. We define a generalized function on $\mathfrak{C}_I$ by means of a harmonic function on $\mathfrak{R}_I$. According to L. Schwartz, it can also be called "a distribution".

The theory of distributions on the unitary group can be established in this way.

Finally, we give an example to illustrate the possibility to extend our results to the theory of harmonic functions with real variables.

§5. *Supplementary remarks*

It is interesting to remark that though we start with a differential equation, yet the harmonic function so obtained satisfies a system of differential equations of second order.

Next the harmonic property we defined is invariant under pseudo-conformal mappings. A similar problem was considered before by S. Bergmann[1], but his extended classes may be altered by a pseudo-conformal mapping.

The degeneracy of the differential equation on the boundary suggests a study of the differential equation of mixed type, which will not be considered in this paper.

As the classical theory of harmonic functions was generalized successfully in certain sense by Hodge[1] to n-dimensional compact Riemannian manifold, it seems to be worth while to make an attempt to consider the open manifold. Though it is far from a complete general theory, yet through these concrete examples, we see the light to proceed forward. We shall also mention the results of Duff and Spencer[1], in which they considered the open subdomain of a Riemannian manifold, yet their differential equations are positive definite throughout the domain and also on its boundary.

Chapter I

Some Preliminary Theorems

§1.1. *Spaces with slits*

Definition. A compact metric space R is called a slit space or a space with slit S if S is a non-empty closed subset of R of which each point is an accumulating point of $R-S$ and if $R-S$ is homeomorphic to a topological product $X \times Y$, where X is a connected m-dimensional differential manifold of class C^2, called the base space, and Y is a compact set, called the side space, and the homeomorphic mapping $\varphi: X \times Y \to R-S$ is called the coordinate function.

Consequently, X cannot be compact, for otherwise so also will be $X \times Y$, which could not be homeomorphic to $R-S$.

Example 1. The closure R of any bounded domain in an n-dimensional Euclidean space E^n can be considered as a slit space with the boundary as its slit. Now Y is a set of a single point.

Example 2. The closed bi-cylinder R:

$$|z_1| \leqslant 1, \quad |z_2| \leqslant 1$$

can be considered as a space with the boundary B as slit. Moreover, the boundary B

$$|z_1| \leqslant 1, \ |z_2| = 1 \text{ and } |z_1| = 1, \ |z_2| \leqslant 1$$

can be also considered as a space with slit C:

$$|z_1| = |z_2| = 1.$$

In fact, the set $B-C$ is formed by two sets

$$|z_1| = 1, \quad |z_2| < 1$$

and

$$|z_1| < 1, \quad |z_2| = 1.$$

Each one is a product to the interior of a unit circle and a circumference of a unit circle. Therefore $B-C$ is homeomorphic to $X \times Y$ where X is the interior of a unit circle and Y the set of two circumferences.

Example 3. The space to be denoted by $\mathfrak{R}_{\text{IV}}$ is formed by the complex n-vectors z satisfying

$$1 + |zz'|^2 - 2z\bar{z}' > 0, \ 1 - |zz'| > 0. \qquad (1.1.1)$$

The closure of the space is denoted by $\overline{\mathfrak{R}}_{\text{IV}}$. Each complex vector z can be expressed as

$$z = (z_1, z_2, 0, \cdots, 0)\Gamma, \tag{1.1.2}$$

where Γ is a real orthogonal matrix. Substituting into (1.1.1), we have

$$1 + |z_1^2 + z_2^2|^2 - 2(|z_1|^2 + |z_2|^2) > 0, \ 1 - |z_1^2 + z_2^2| > 0.$$

Let

$$w_1 = z_1 - iz_2, \ w_2 = z_1 + iz_2.$$

It follows that

$$(1 - |w_1|^2)(1 - |w_2|^2) > 0, \ 1 - |w_1 w_2| > 0.$$

Consequently, we have

$$|w_1| < 1, \quad |w_2| < 1. \tag{1.1.3}$$

This is a bi-cylinder. The boundary $\mathfrak{B}_{IV}$ of $\mathfrak{R}_{IV}$ is therefore formed by the sets (1.1.2) with

$$|w_1| = 1, \ |w_2| \leqslant 1 \text{ and } |w_1| \leqslant 1, \ |w_2| = 1.$$

We use $\mathfrak{C}_{IV}$ to denote the set of points (1.1.2) with $|w_1| = |w_2| = 1$. We are going to prove that $\mathfrak{C}_{IV}$ is a slit of $\mathfrak{B}_{IV}$. More precisely, $\mathfrak{B}_{IV} - \mathfrak{C}_{IV}$ is homeomorphic to $X \times Y$ where X denotes the interior of a unit circle and Y denotes the cosets of $O(n)$ by $O(n-2)$, where $O(n)$ is the real orthogonal group of order n.

In fact, each point z of $\mathfrak{B}_{IV} - \mathfrak{C}_{IV}$ can be expressed as

$$e^{i\theta}(z_1, z_2, 0, \cdots, 0)\Gamma$$

with $|w_1| < 1$ and $w_2 = 1$ or

$$z = e^{i\theta}\left(\frac{1}{2}(1 + w_1), -\frac{i}{2}(1 - w_1), 0, \cdots, 0\right)\Gamma, \ |w_1| < 1. \tag{1.1.4}$$

Further, since

$$e^{i\theta}\left(\frac{1}{2}(1 + w_1), -\frac{i}{2}(1 - w_1)\right)$$

$$= \left(\frac{1}{2}(e^{i\theta} + e^{-i\theta}e^{2i\theta}w_1), -\frac{i}{2}(e^{i\theta} - e^{-i\theta}e^{2i\theta}w_1)\right)$$

$$= \left(\frac{1}{2}(1 + e^{2i\theta}w_1), -\frac{i}{2}(1 - e^{2i\theta}w_1)\right)\begin{pmatrix} \cos\theta & \sin\theta \\ -\sin\theta & \cos\theta \end{pmatrix},$$

each z of $\mathfrak{B}_{IV} - \mathfrak{C}_{IV}$ can be expressed as

$$z = \left(\frac{1}{2}(1+w), -\frac{i}{2}(1-w), 0, \cdots, 0\right)\Gamma, \ |w| < 1. \qquad (1.1.5)$$

If we have another expression

$$z = \left(\frac{1}{2}(1+w_0), -\frac{i}{2}(1-w_0), 0, \cdots, 0\right)\Gamma_0,$$

then we have, with $\Gamma_1 = \Gamma\Gamma_0^{-1}$,

$$\left(\frac{1}{2}(1+w), -\frac{i}{2}(1-w), 0, \cdots, 0\right)\Gamma_1$$
$$= \left(\frac{1}{2}(1+w_0), -\frac{i}{2}(1-w_0), 0, \cdots, 0\right).$$

It follows that

$$\Gamma_1 = \begin{pmatrix} \gamma & 0 \\ 0 & \Gamma^{(n-2)} \end{pmatrix},$$

where $\Gamma^{(n-2)}$ is an $(n-2)$–rowed real orthogonal matrix and

$$\gamma = \begin{pmatrix} \cos\psi & \sin\psi \\ -\sin\psi & \cos\psi \end{pmatrix}, \text{ or } \begin{pmatrix} \cos\psi & \sin\psi \\ \sin\psi & -\cos\psi \end{pmatrix}.$$

For the first expression of γ, from

$$\left(\frac{1}{2}(1+w), -\frac{i}{2}(1-w)\right)\begin{pmatrix} \cos\psi & \sin\psi \\ -\sin\psi & \cos\psi \end{pmatrix}$$
$$= \left(\frac{1}{2}(1+w_0), -\frac{i}{2}(1-w_0)\right)$$

we have immediately $\psi = 0$ and the second expression is impossible.

§1.2. *Partial differential equation on slit space*

Let $(x^1, \cdots, x^m)$ be a local coordinate system of X, and let φ denote the mapping from $X \times Y$ onto $R - S$. A continuous function defined on $R - S$ is said to possess continuous second derivatives with respect to the coordinate of X, if $f(\varphi(x, y))$ $(x \in X, y \in Y)$ has derivatives up to the second order with respect to $x^1, \cdots, x^m$ and these derivatives are continuous on $X \times Y$.

A positive definite operator defined on X is a differential operator

$$\Delta = \sum_{j,k=1}^{m} a_{jk} \frac{\partial^2}{\partial x^j \partial x^k} + \sum_{l=1}^{m} b_l \frac{\partial}{\partial x^l}$$

with continuous coefficients a_{jk} and b_l in X such that (a_{jk}) is a positive definite symmetric matrix and Δ is independent of the choice of the local coordinates.

Theorem 1.2.1. Let f be a function continuous in R and with the continuous second derivatives on $R-S$ with respect to the coordinate of X. If, for each $y \in Y$, f satisfies

$$\Delta f(\varphi(x, y)) = 0,$$

then f reaches its minimal and maximal values on the slit S.

Proof. Since R is a compact metric space, $f(P)$ reaches its maximal value at a point P_0 of R, that is,

$$f(P) \leqslant f(P_0)$$

for any P of R. If $P_0 \in S$, there is nothing to prove. If $P_0 \notin S$, we put $P_0 = \varphi(x_0, y_0)$ where $x_0 \in X$, $y_0 \in Y$. We define $\psi(x,y) = f(\varphi(x,y))$. Then we have $\psi(x,y) \leqslant \psi(x_0,y_0)$; in particular

$$\psi(x, y_0) \leqslant \psi(x_0, y_0).$$

In case $\psi(x,y_0) = \psi(x_0,y_0)$ for all points $x \in X$, since X is not compact, there is at least a sequence of points $x_1, \cdots, x_k, \cdots$ in X which has no accumulating point in X. But, since R is compact, the image points $P_k = \varphi(x_k, y_0)$ must have at least an accumulating point $Q \in R$. Q must lie on S. Without loss of generality, we can assume that P_k converges to Q and then $\lim f(P_k) = f(Q) = f(P_0)$.

In case $\psi(x,y_0) = \psi(x_0,y_0)$ not for all points $x \in X$, since X is connected, there is at least a point $x_1 \in X$ such that $\psi(x_1,y_0) = \psi(x_0,y_0)$, but in any neighbourhood of x_1 in X, there is at least a point $x \in X$ such that

$$\psi(x, y_0) < \psi(x_1, x_0).$$

We choose a connected neighbourhood $U(x_1)$ of x_1 in X, which is contained in a local coordinate neighbourhood of X and the closure of which is compact. Since

$$\Delta\psi(x, y_0) = 0,$$

we can apply the extremal principle of linear partial differential equations of elliptic type (see, for example, Miranda [1], p. 5) to the domain $U(x_1)$ and see that this case is impossible.

Consequently, we always have $f(P) \leqslant \max_{Q \in S} f(Q)$. By a similar method, we also have $f(P) \geqslant \min_{Q \in S} f(Q)$. The proof of the theorem is complete.

A sequence of spaces

$$R_1 \supset R_2 \supset \cdots \supset R_k$$

is called a chain of slit spaces, if each R_ν is a slit space with $R_{\nu+1}$ as its slit ($\nu = 1, \cdots, k-1$).

Theorem 1.2.2. Denote the slit of R_k by $\mathfrak{C}$. Let Δ_ν be a positive definite operator defined in X_ν, the base space of R_ν. Suppose that f is a continuous function defined in R_1 and on each $R_\nu - R_{\nu+1}$ ($\nu = 1, \cdots, k; R_{k+1} = \mathfrak{C}$), it possesses continuous second derivatives with respect to the coordinate of X_ν. If for each point y of Y_ν, the side space of R_ν, f satisfies

$$\Delta_\nu f(\varphi_\nu(x, y)) = 0, \quad (\nu = 1, \cdots, k)$$

for x on X_ν, where φ_ν is the coordinate function of R_ν, then f reaches its maximal and minimal values on $\mathfrak{C}$.

The theorem is an immediate consequence by successive applications of our previous theorem.

§1.3. *Several lemmas*

Theorem 1.3.1. Let m be an integer $\geqslant 2$ and $0 < r < 1$. Then we have

$$\int \cdots \int_{x_1^2 + \cdots + x_m^2 = 1} \frac{(1 - r^2)^{m/2}}{|1 - r(x_1 - ix_2)|^m} \dot{x} = \frac{2\pi^{m/2}}{\Gamma\left(\frac{m}{2}\right)},$$

where $\dot{x}$ denotes the volume element of the sphere $x_1^2 + \cdots + x_m^2 = 1$.

Proof. Changing variables

$$x_1 = \rho \cos\theta, \quad x_2 = \rho \sin\theta, \quad x_3 = x_3, \quad \cdots, \quad x_m = x_m,$$

where $\rho = \sqrt{1 - x_3^2 - \cdots - x_m^2}$, we have

$$dx_1^2 + \cdots + dx_m^2 = \rho^2\, d\theta^2 + d\rho^2 + \sum_{\nu=3}^{m} dx_\nu^2$$

$$= \rho^2 d\theta + \frac{1}{\rho^2}(x_3 dx_3 + \cdots + x_m dx_m)^2 + \sum_{\nu=3}^{m} dx_\nu^2.$$

Since

$$\det\left(I^{(m-2)} + \frac{1}{\rho^2}(x_3, \cdots, x_m)'(x_3, \cdots, x_m)\right)$$

$$= 1 + \frac{1}{\rho^2}(x_3^2 + \cdots + x_m^2) = \frac{1}{\rho^2},$$

we have

$$\dot{x} = d\theta\, dx_3 \cdots dx_m,$$

and

$$\int \cdots \int_{x_1^2+\cdots+x_m^2=1} \frac{\dot{x}}{|1 - r(x_1 - ix_2)|^m} = \int \cdots \int_{x_3^2+\cdots+x_m^2<1} dx_3 \cdots dx_m \int_0^{2\pi} \frac{d\theta}{|1 - r\rho e^{-i\theta}|^m}.$$

Using the development

$$(1 - r\rho e^{-i\theta})^{-\frac{m}{2}} = \sum_{k=0}^{\infty} \frac{\Gamma\left(\frac{1}{2}m + k\right)}{\Gamma\left(\frac{m}{2}\right)\Gamma(k+1)} (r\rho e^{-i\theta})^k,$$

we have

$$\int_0^{2\pi} \frac{d\theta}{|1 - r\rho e^{-i\theta}|^m} = 2\pi \sum_{k=0}^{\infty} \left(\frac{\Gamma\left(\frac{m}{2} + k\right)}{\Gamma\left(\frac{m}{2}\right)\Gamma(k+1)} \right)^2 (r\rho)^{2k}.$$

On the other hand, since

$$\int \cdots \int_{x_3^2+\cdots+x_m^2<1} (1 - x_3^2 - \cdots - x_m^2)^k dx_3 \cdots dx_m = \frac{\left(\Gamma\left(\frac{1}{2}\right)\right)^{m-2} \Gamma(k+1)}{\Gamma\left(\frac{m}{2} + k\right)},$$

we have consequently,

$$\int \cdots \int_{x_1^2+\cdots+x_m^2=1} \frac{\dot{x}}{|1 - r(x_1 - ix_2)|^m} = \frac{2\pi^{\frac{m}{2}}}{\Gamma\left(\frac{m}{2}\right)} \sum_{k=1}^{\infty} \frac{\Gamma\left(\frac{m}{2} + k\right)}{\Gamma\left(\frac{m}{2}\right)\Gamma(k+1)} r^{2k}$$

$$= \frac{2\pi^{\frac{m}{2}}}{\Gamma\left(\frac{m}{2}\right)} (1 - r^2)^{-\frac{m}{2}}.$$

Theorem 1.3.2. If $\varphi(x_1, \cdots, x_m)$ is a continuous function on the sphere $x_1^2 + \cdots + x_m^2 = 1$, then

$$\lim_{r \to 1} \int \cdots \int_{x_1^2+\cdots+x_m^2=1} \varphi(x_1, \cdots, x_m) \frac{(1 - r^2)^{\frac{m}{2}}}{|1 - r(x_1 - ix_2)|^m} \dot{x} = \frac{2\pi^{\frac{m}{2}}}{\Gamma\left(\frac{m}{2}\right)} \varphi(1, 0, \cdots, 0).$$

Proof. By the previous theorem we deduce that

$$\int_{x_1^2+\cdots+x_m^2=1}\cdots\int \varphi(1,0,\cdots,0)\frac{(1-r^2)^{\frac{m}{2}}}{|1-r(x_1-ix_2)|^m}\dot{x}=\omega_{m-1}\varphi(1,0,\cdots,0),$$

where $\omega_{m-1}=2\pi^{\frac{m}{2}}/\Gamma\left(\frac{m}{2}\right)$. It is sufficient to prove that, for any given $\varepsilon>0$, we can choose r sufficiently near to 1 such that

$$\left|\int_{x_1^2+\cdots+x_m^2=1}\cdots\int (\varphi(x_1,\cdots,x_m)-\varphi(1,0,\cdots,0))\frac{(1-r^2)^{\frac{m}{2}}}{|1-r(x_1-ix_2)|^m}\dot{x}\right|<\varepsilon. \tag{1.2.1}$$

We use spherical coordinates

$$x_1=\cos\theta_1,\ x_2=\sin\theta_1\cos\theta_2,\ \cdots,\ x_m=\sin\theta_1\cdots\sin\theta_{m-1},$$

and notice that when $\theta_1=0$, $(x_1,\cdots,x_m)=(1,0,\cdots,0)$. It is known that

$$\dot{x}=\sin^{m-2}\theta_1\sin^{m-3}\theta_2\cdots\sin\theta_{m-2}\,d\theta_1\cdots d\theta_{m-1},$$

and the integral in (1.2.1) becomes

$$\int_0^{2\pi}d\theta_{m-1}\int_0^{\pi}\sin\theta_{m-2}\,d\theta_{m-2}\cdots\int_0^{\pi}\sin^{m-3}\theta_2\,d\theta_2$$

$$\times\int_0^{\pi}(\varphi(x_1,\cdots,x_m)-\varphi(1,0,\cdots,0))\frac{(1-r^2)^{\frac{m}{2}}\sin^{m-2}\theta_1\,d\theta_1}{|1-r(\cos\theta_1-i\sin\theta_1\cos\theta_2)|^m}$$

$$=I_1+I_2,$$

where I_1 equals the part of the integral with θ_1 integrated from 0 to δ and I_2 the remaining part of the integral.

Since $\varphi(x_1,\cdots,x_m)$ is continuous on $x_1^2+\cdots+x_m^2=1$, we can choose $\delta(>0)$ so small that, for $0\leqslant\theta_1\leqslant\delta$, we have

$$|\varphi(x_1,\cdots,x_m)-\varphi(1,0,\cdots,0)|<\frac{1}{2\omega_{m-1}}\varepsilon.$$

Hence

$$|I_1|<\frac{\varepsilon}{2}\cdot\frac{1}{\omega_{m-1}}\int_0^{2\pi}d\theta_{m-1}\int_0^{\pi}\sin\theta_{m-2}\,d\theta_{m-2}\cdots\int_0^{\pi}\sin^{m-3}\theta_2\,d\theta_2$$

$$\times\int_0^{\delta}\frac{(1-r^2)^{\frac{m}{2}}\sin^{m-2}\theta_1\,d\theta_1}{|1-r(\cos\theta_1-i\sin\theta_1\cos\theta_2)|^m}$$

$$\leqslant \frac{\varepsilon}{2} \cdot \frac{1}{\omega_{m-1}} \int \cdots \int_{x_1^2+\cdots+x_m^2=1} \frac{(1-r^2)^{\frac{m}{2}}}{|1-r(x_1-ix_2)|^m} \dot{x} = \frac{\varepsilon}{2}.$$

For a fixed δ, we have, for $\delta \leqslant \theta_1 \leqslant \pi$,

$$|1 - r(\cos\theta_1 - i\sin\theta_1\cos\theta_2)|^m = |(1 - r\cos\theta_1)^2 + r^2\sin^2\theta_1\cos^2\theta_2|^{\frac{m}{2}}$$
$$\geqslant (1 - r\cos\theta_1)^m > (1 - \cos\delta)^m = 2^m \sin^{2m}\frac{\delta}{2}.$$

Let M be the upper bound of $|\varphi(x_1, \cdots, x_m)|$, and taking r sufficiently near to 1 such that

$$0 < (1 - r^2)^{\frac{m}{2}} < \frac{2^{m-1}\sin^{2m}\frac{\delta}{2}}{M\omega_{m-1}} \cdot \frac{\varepsilon}{2}.$$

Then obviously

$$|I_2| \leqslant 2M \int_0^{2\pi} d\theta_{m-1} \int_0^{\pi} \sin\theta_{m-2}\, d\theta_{m-2} \cdots \int_0^{\pi} \sin^{m-3}\theta_2\, d\theta_2$$
$$\times \int_{\delta}^{\pi} \frac{(1-r^2)^{\frac{m}{2}} \sin^{m-2}\theta_1}{|1 - r(\cos\theta_1 - i\sin\theta_1\cos\theta_2)|^m} d\theta_1$$
$$\leqslant \frac{2M(1-r^2)^{\frac{m}{2}}}{2^m \sin^{2m}\frac{\delta}{2}} \int_0^{2\pi} d\theta_{m-1} \int_0^{\pi} \sin\theta_{m-2}\, d\theta_{m-2} \cdots \int_0^{\pi} \sin^{m-3}\theta_2\, d\theta_2$$
$$\times \int_0^{\pi} \sin^{m-2}\theta_1\, d\theta_1$$
$$= \frac{M\omega_{m-1}(1-r^2)^{\frac{m}{2}}}{2^{m-1}\sin^{2m}\frac{\delta}{2}} < \frac{\varepsilon}{2}.$$

The inequality (1.2.1) follows from the estimations of I_1 and I_2.

Chapter II

The Dirichlet Problems in the Hyperbolic Spaces of Matrices

§2.1. *A differential operator of* $\mathfrak{R}_{\mathrm{I}}$

We use $\mathfrak{R}_{\mathrm{I}}$ to denote the domain formed by $m \times n$ matrices $Z = (z_{j\alpha})_{1\leqslant j\leqslant m,\, 1\leqslant \alpha\leqslant n}$ $(m \leqslant n)$ making the Hermitian matrices

$$I - Z\bar{Z}' \tag{2.1.1}$$

positive definite, where Z' and $\bar{Z}$ denote the transposed and complex conjugate matrices of Z respectively. For simplicity, we also use $H>0$ and $H\geqslant 0$ to denote that the Hermitian matrix H is positive definite and positive semi-definite respectively.

We shall consider the linear differential operator

$$\Delta_{\mathrm{I}}=\sum_{\alpha,\beta=1}^{n}\sum_{j,k=1}^{m}\left(\delta_{\alpha\beta}-\sum_{l=1}^{m}\bar{z}_{l\alpha}z_{l\beta}\right)\left(\delta_{jk}-\sum_{\gamma=1}^{n}\bar{z}_{j\gamma}z_{k\gamma}\right)\frac{\partial^2}{\partial z_{k\beta}\partial\bar{z}_{j\alpha}},\qquad(2.1.2)$$

where $\delta_{\alpha\beta}=0$ or 1 according as $\alpha\neq\beta$ or $\alpha=\beta$.

We introduce a matrix operator

$$\partial_Z=\begin{pmatrix}\frac{\partial}{\partial z_{11}},&\cdots,&\frac{\partial}{\partial z_{1n}}\\ \cdots&\cdots&\cdots\\ \frac{\partial}{\partial z_{m1}},&\cdots,&\frac{\partial}{\partial z_{mn}}\end{pmatrix}.\qquad(2.1.3)$$

Then the differential operator Δ_{I} can be expressed as

$$\Delta_{\mathrm{I}}=\operatorname{tr}((I-Z\bar{Z}')\bar{\partial}_Z(I-\bar{Z}'Z)\partial_Z'),$$

where $\operatorname{tr}M$ denotes the trace of a matrix M. Notice that the expression is performed as a formal multiplication; we are not applying $\bar{\partial}_Z$ as differential operator on $(I-\bar{Z}'Z)$. Such an agreement will be understood throughout this paper, i.e., if A is a matrix operator and B is any matrix, then AB means the formal product, and if u is a function, $A\circ u$ means that we apply the operator A on u.

We shall study first the covariant property of ∂_Z under the group Γ^{I} of motions of $\mathfrak{R}_{\mathrm{I}}$. It is known (Hua [1]) that the transformations of Γ^{I} can be written as

$$W=(AZ+B)(CZ+D)^{-1}=(Z\bar{B}'+\bar{A}')^{-1}(Z\bar{D}'+\bar{C}'),\qquad(2.1.4)$$

where $A=A^{(m)},B=B^{(m,n)},C=C^{(n,m)}$ and $D=D^{(n)}$ satisfy the relations:

$$A\bar{A}'-B\bar{B}'=I,\ A\bar{C}'=B\bar{D}',\ C\bar{C}'-D\bar{D}'=-I,\qquad(2.1.5)$$

or what is the same thing,

$$\bar{A}'A-\bar{C}'C=I,\ \bar{A}'B=\bar{C}'D,\ \bar{B}'B-\bar{D}'D=-I.\qquad(2.1.6)$$

By differentiating (2.1.4), we have

$$\begin{aligned}dW&=[AdZ-(AZ+B)(CZ+D)^{-1}C\,dZ](CZ+D)^{-1}\\&=(Z\bar{B}'+\bar{A}')^{-1}[(Z\bar{B}'+\bar{A}')A-(Z\bar{D}'+\bar{C}')C]dZ(CZ+D)^{-1}\end{aligned}$$

$$= (Z\bar{B}' + \bar{A}')^{-1} dZ(CZ + D)^{-1}. \tag{2.1.7}$$

From the relations

$$\frac{\partial}{\partial z_{j\alpha}} = \sum_{k=1}^{m} \sum_{\beta=1}^{n} \frac{\partial w_{k\beta}}{\partial z_{j\alpha}} \frac{\partial}{\partial w_{k\beta}}, \quad dw_{j\alpha} = \sum_{k=1}^{m} \sum_{\beta=1}^{n} \frac{\partial w_{j\alpha}}{\partial z_{k\beta}} dz_{k\beta},$$

we deduce easily

$$\partial_Z' = (CZ + D)^{-1}\partial_W'(Z\bar{B}' + \bar{A}')^{-1}. \tag{2.1.8}$$

From (2.1.4) and (2.1.6), we have

$$\begin{aligned} I - \bar{W}'W &= I - \overline{(CZ + D)}'^{-1}\overline{(AZ + B)}'(AZ + B)(CZ + D)^{-1} \\ &= \overline{(CZ + D)}'^{-1}[\overline{(CZ + D)}'(CZ + D) \\ &\quad - \overline{(AZ + B)}'(AZ + B)](CZ + D)^{-1} \\ &= (\bar{Z}'\bar{C}' + \bar{D}')^{-1}(I - \bar{Z}'Z)(CZ + D)^{-1}, \end{aligned} \tag{2.1.9}$$

and

$$I - W\bar{W}' = (Z\bar{B}' + \bar{A}')^{-1}(I - Z\bar{Z}')(B\bar{Z}' + A)^{-1}. \tag{2.1.10}$$

Combining with (2.1.8), we have

$$\begin{aligned} &(I - Z\bar{Z}')\bar{\partial}_Z(I - \bar{Z}'Z)\partial_Z' \\ &= (Z\bar{B}' + \bar{A}')(I - W\bar{W}')\bar{\partial}_W(I - \bar{W}'W)\partial_W'. \end{aligned} \tag{2.1.11}$$

Consequently we have the important identity

$$\begin{aligned} \Delta_{\mathrm{I}} &= \operatorname{tr}((I - Z\bar{Z}')\bar{\partial}_Z(I - \bar{Z}'Z)\partial_Z') \\ &= \operatorname{tr}((I - W\bar{W}')\bar{\partial}_W(I - \bar{W}'W)\partial_W'). \end{aligned} \tag{2.1.12}$$

That is, we have

Theorem 2.1.1. The differential operator Δ_{I} is invariant under the transformations (2.1.4) of the group Γ^{I}.

§2.2. *Harmonic functions in* $\mathfrak{R}_{\mathrm{I}}$

A real-valued function $u(Z)$ possessing continuous second derivatives is said to be harmonic in $\mathfrak{R}_{\mathrm{I}}$, if it satisfies

$$\Delta_{\mathrm{I}} \circ u(Z) = 0 \tag{2.2.1}$$

in $\mathfrak{R}_{\mathrm{I}}$.

From Theorem 2.1.1, we deduce immediately

Theorem 2.2.1. The property "harmonic" is invariant under the group Γ^{I}. More precisely, if $u(Z)$ is harmonic in $\mathfrak{R}_{\mathrm{I}}$, so is

$u((AZ+B)(CZ+D)^{-1})$ where A, B, C and D satisfy (2.1.5).

In $\mathfrak{R}_{\mathrm{I}}$, we have a Poisson kernel (Hua [4])

$$P_{\mathrm{I}}(Z, U) = \frac{1}{V(\mathfrak{C}_{\mathrm{I}})} \cdot \frac{\det(I - Z\bar{Z}')^{n}}{|\det(I - Z\bar{U}')|^{2n}}, \tag{2.2.2}$$

where Z belongs to $\mathfrak{R}_{\mathrm{I}}$ and $U(=U^{(m,n)})$ is an $m \times n$ matrix satisfying

$$U\bar{U}' = I^{(m)}, \tag{2.2.3}$$

and $\mathfrak{C}_{\mathrm{I}}$ consists of all matrices U satisfying (2.2.3). Further $V(\mathfrak{C}_{\mathrm{I}})$ denotes the total volume of $\mathfrak{C}_{\mathrm{I}}$, which is known to be[1)]

$$V(\mathfrak{C}_{\mathrm{I}}) = \frac{2^{m}\pi^{mn-\frac{1}{2}m(m-1)}}{(n-m)!\cdots(n-1)!} = \prod_{l=1}^{m} \frac{2\pi^{n-l+1}}{(n-l)!}$$
$$= \prod_{l=1}^{m} \omega_{2(n-l)+1}, \tag{2.2.4}$$

where ω_{2p-1} is the volume of the sphere $|z_1|^2 + \cdots + |z_p|^2 = 1$.

From (2.1.10), it follows that (2.1.4) carries the characteristic manifold onto itself. Let

$$V = (AU + B)(CU + D)^{-1}, \tag{2.2.5}$$

then we have

$$I - W\bar{V}' = (Z\bar{B}' + \bar{A}')^{-1}(I - Z\bar{U}')(B\bar{U}' + A)^{-1}, \tag{2.2.6}$$

and then

$$\frac{\det(I - W\bar{W}')^{n}}{|\det(I - W\bar{V}')|^{2n}} = \frac{\det(I - Z\bar{Z}')^{n}}{|\det(I - Z\bar{U}')|^{2n}} |\det(B\bar{U}' + A)|^{2n}. \tag{2.2.7}$$

Consequently we have the following

Theorem 2.2.2. Under the transformation (2.1.4), the Poisson kernel satisfies

$$P_{\mathrm{I}}(W, V) = P_{\mathrm{I}}(Z, U)\, |\det(B\bar{U}' + A)|^{2n}. \tag{2.2.8}$$

Theorem 2.2.3. The Poisson kernel $P_{\mathrm{I}}(Z, U)$ satisfies a system of partial differential equations

$$\bar{\partial}_Z (I - \bar{Z}'Z) \partial'_Z \circ P_{\mathrm{I}}(Z, U) = 0, \tag{2.2.9}$$

---

1) This volume differs from that given previously in a monograph of Hua[4], by a factor $2^{m(n-1)-\frac{1}{2}m(m-1)}$. This fact is due to the definition of the volume element $\dot{U}$.

or, what is the same thing,

$$\sum_{\alpha,\beta=1}^{n}\left(\delta_{\alpha\beta}-\sum_{k=1}^{m}\bar{z}_{k\alpha}z_{k\beta}\right)\frac{\partial^2 P_1(Z,U)}{\partial\bar{z}_{j\alpha}\partial z_{l\beta}}=0,\ (j,l=1,\cdots,m). \qquad (2.2.10)$$

Proof. (i) First we prove that at the point $Z=0$, $P_1(Z,U)$ satisfies (2.2.9).

We write $U=(u_{j\alpha})$. At $Z=0$, we have

$$\left[\sum_{\alpha,\beta=1}^{n}\left(\delta_{\alpha\beta}-\sum_{k=1}^{n}\bar{z}_{k\alpha}z_{k\beta}\right)\frac{\partial^2 P_1(Z,U)}{\partial\bar{z}_{j\alpha}\partial z_{l\beta}}\right]_{Z=0}$$

$$=\left[\sum_{\alpha=1}^{n}\frac{\partial^2 P_I(Z,U)}{\partial\bar{z}_{j\alpha}\partial z_{l\alpha}}\right]_{Z=0}$$

$$=\frac{1}{V(\mathfrak{C}_1)}\left[\sum_{\alpha=1}^{n}\frac{\partial^2}{\partial\bar{z}_{j\alpha}\partial z_{l\alpha}}\det(I-Z\bar{Z}')^n\det(I-Z\bar{U}')^{-n}\det(I-U\bar{Z}')^{-n}\right]_{Z=0}$$

$$=\frac{1}{V(\mathfrak{C}_1)}\left[\sum_{\alpha=1}^{n}\frac{\partial^2}{\partial\bar{z}_{j\alpha}\partial z_{l\alpha}}\left(1-n\sum_{\beta=1}^{n}\sum_{k=1}^{m}|z_{k\beta}|^2+\cdots\right)\right.$$

$$\left.\times\left(1+n\sum_{\beta=1}^{n}\sum_{k=1}^{m}z_{k\beta}\bar{u}_{k\beta}+\cdots\right)\left(1+n\sum_{\beta=1}^{n}\sum_{k=1}^{m}\bar{z}_{k\beta}u_{k\beta}+\cdots\right)\right]_{Z=0}$$

$$=\frac{1}{V(\mathfrak{C}_1)}\left[\sum_{\alpha=1}^{n}(-n)\delta_{jl}+n^2\sum_{\alpha=1}^{n}u_{j\alpha}\bar{u}_{l\alpha}\right]. \qquad (2.2.11)$$

In matrix form, (2.2.11) can be written as

$$[\bar{\partial}_Z(I-\bar{Z}'Z)\partial_Z'\circ P_1(Z,U)]_{Z=0}=\frac{1}{V(\mathfrak{C}_1)}[-n^2I+n^2U\bar{U}']=0,$$

since $U\bar{U}'=I$.

(ii) Let T be any point of $\mathfrak{R}_1$, then there exist matrices A and D such that

$$(I-T\bar{T}')^{-1}=\bar{A}'A$$

and

$$(I-\bar{T}'T)^{-1}=\bar{D}'D.$$

We put

$$B=-AT,\quad C=-D\bar{T}'.$$

Obviously A,B,C,D satisfy the conditions (2.1.6) and the corresponding transformation (2.1.4) takes the form

$$W=A(Z-T)(I-\bar{T}'Z)^{-1}D^{-1}, \qquad (2.1.12)$$

which carries the point $Z=T$ into $W=0$. According to (2.1.8), (2.1.9) and Theorem 2.2.2, we have

$$\begin{aligned}&[\partial_Z(I-\bar{Z}'Z)\partial_Z'\circ P_{\mathrm{I}}(Z,U)]_{Z=T}\\&=|\det(I-T\bar{U}')A|^{-2n}(I-T\bar{T}')^{-1}A^{-1}[\partial_W(I-\bar{W}'W)\partial_W'\circ P_{\mathrm{I}}(W,V)]_{W=0}\\&\quad\times\bar{A}'^{-1}(I-T\bar{T}')^{-1}=0.\end{aligned}$$

The theorem is now completely proved.

As a consequence, we have

Theorem 2.2.4. The Poisson kernel $P_{\mathrm{I}}(Z,U)$ is harmonic in $\mathfrak{R}_{\mathrm{I}}$ with respect to the variable Z.

By the process of differentiating under integral signs we have

Theorem 2.2.5. For any continuous function $\varphi(U)$ on $\mathfrak{C}_{\mathrm{I}}$, the Poisson integral

$$u(Z)=\int_{\mathfrak{C}_{\mathrm{I}}}\varphi(U)P_{\mathrm{I}}(Z,U)\dot{U} \tag{2.2.13}$$

gives a function harmonic in $\mathfrak{R}_{\mathrm{I}}$.

In fact,

$$\Delta_{\mathrm{I}}\circ u(Z)=\int_{\mathfrak{C}_{\mathrm{I}}}\varphi(U)\Delta_{\mathrm{I}}\circ P_{\mathrm{I}}(Z,U)\dot{U}=0.$$

§2.3. *The structure of the closure of* $\mathfrak{R}_{\mathrm{I}}$

More explicitly, sometimes we use $\mathfrak{R}_{\mathrm{I}}(m,n)$ and $\mathfrak{C}_{\mathrm{I}}(m,n)$ to denote $\mathfrak{R}_{\mathrm{I}}$ and $\mathfrak{C}_{\mathrm{I}}$ respectively.

Let $\mathfrak{B}_{\mathrm{I}}^{(m-r)}$ denote the set of matrices Z such that $I-Z\bar{Z}'\geqslant 0$ and of rank $\leqslant r$. Evidently $\mathfrak{B}_{\mathrm{I}}^{(0)}$ is the closure of $\mathfrak{R}_{\mathrm{I}}$ and $\mathfrak{B}_{\mathrm{I}}^{(m)}$ is $\mathfrak{C}_{\mathrm{I}}$.

Theorem 2.3.1. We let

$$\mathfrak{C}_{\mathrm{I}}^{(r)}=\mathfrak{B}_{\mathrm{I}}^{(m-r)}-\mathfrak{B}_{\mathrm{I}}^{(m-r+1)},\quad(r=0,1,\cdots,m-1).$$

Then $\mathfrak{C}_{\mathrm{I}}^{(r)}$ is invariant and forms a transitive set under the transformation group Γ^{I}. More precisely, there is a transformation of the group Γ^{I} to carry any given point of $\mathfrak{C}_{\mathrm{I}}^{(r)}$ into

$$\begin{pmatrix}I^{(m-r)} & 0^{(m-r,\,n-m+r)}\\ 0^{(r,\,m-r)} & 0^{(r,\,n-m+r)}\end{pmatrix}.$$

Proof. From (2.1.10) we have

$$(Z\bar{B}'+\bar{A}')(I-W\bar{W}')(BZ'+A)=I-Z\bar{Z}'.$$

It follows that

$$r(I - Z\bar{Z}') \leqslant r(I - W\bar{W}'),$$

where $r(X)$ denote the rank of X. The inversion establishes

$$r(I - Z\bar{Z}') \geqslant r(I - W\bar{W}').$$

Therefore $\mathfrak{C}_{\mathrm{I}}^{(r)}$ is a set invariant under Γ^{I}.

Let Z be any point on $\mathfrak{C}_{\mathrm{I}}^{(r)}$; there are unitary matrices $U = U^{(m)}$ and $V = V^{(n)}$ such that

$$UZV = \begin{pmatrix} \lambda_1 & 0 & \cdots & 0 & 0 & \cdots & 0 \\ 0 & \lambda_2 & \cdots & 0 & 0 & \cdots & 0 \\ \cdots & \cdots & \cdots & \cdots & \cdots & \cdots & \cdots \\ 0 & 0 & \cdots & \lambda_m & 0 & \cdots & 0 \end{pmatrix}, \quad \lambda_1 \geqslant \lambda_2 \geqslant \cdots \geqslant \lambda_m \geqslant 0.$$

Since the rank of $I - Z\bar{Z}'$ is equal to r, we have $\lambda_1 = \cdots = \lambda_{m-r} = 1$ and $\lambda_{m-r+1} < 1$. The theorem follows from the following

Theorem 2.3.2. Let

$$T_0 = U_0' \begin{pmatrix} 0 & 0 \\ 0 & T_1^{(r, n-m+r)} \end{pmatrix} V_0, \quad I^{(r)} - T_1\bar{T}_1' > 0.$$

The transformation

$$W = A(Z - T_0)(I - \bar{T}_0'Z)^{-1}D^{-1}$$

of Γ^{I}, where

$$A = \begin{pmatrix} I^{(m-r)} & 0 \\ 0 & A_1 \end{pmatrix} \bar{U}_0, \quad D = \begin{pmatrix} I & 0 \\ 0 & D_1 \end{pmatrix} V_0$$

and

$$\bar{A}_1'A_1 = (I - T_1\bar{T}_1')^{-1}, \ \bar{D}_1'D_1 = (I - \bar{T}_1'T_1)^{-1},$$

carries $T = U_0'\begin{pmatrix} I^{(m-r)} & 0 \\ 0 & T_1 \end{pmatrix} V_0$ into

$$\begin{pmatrix} I^{(m-r)}, & 0 \\ 0 & 0 \end{pmatrix},$$

and

$$Z = U_0'\begin{pmatrix} \rho I^{(m-r)} & 0 \\ 0 & Z_1 \end{pmatrix} V_0, \ 0 \leqslant \rho < 1, \ I - Z_1\bar{Z}_1' > 0$$

into

$$W = U_0'\begin{pmatrix} \rho I^{(m-r)} & 0 \\ 0 & W_1 \end{pmatrix} V_0$$

with

$$W_1 = A_1(Z_1 - T_1)(I - \bar{T}_1'Z_1)^{-1}D_{\mathrm{I}}^{-1}.$$

The present theorem is evident.

Theorem 2.3.3. The closure $\mathfrak{B}_{\mathrm{I}}^{(0)}$ of $\mathfrak{R}_{\mathrm{I}}$ has the following chain of slit spaces:

$$\mathfrak{B}_{\mathrm{I}}^{(0)} \supset \mathfrak{B}_{\mathrm{I}}^{(1)} \supset \cdots \supset \mathfrak{B}_{\mathrm{I}}^{(m-1)}.$$

More precisely, $\mathfrak{C}_{\mathrm{I}}^{(r)} = \mathfrak{B}_{\mathrm{I}}^{(m-r)} - \mathfrak{B}_{\mathrm{I}}^{(m-r+1)}$ (where $r = 1, \cdots, m$ and $\mathfrak{B}_{\mathrm{I}}^{(m)} = \mathfrak{C}_{\mathrm{I}}$ is the slit of $\mathfrak{B}_{\mathrm{I}}^{(m-1)}$) is homeomorphic to the topological product

$$\mathfrak{R}_{\mathrm{I}}(r, n-m+r) \times \mathfrak{M}_{\mathrm{I}}^{(m-r)},$$

where $\mathfrak{M}_{\mathrm{I}}^{(s)}$ is defined in the following way.

We use $\mathfrak{U}(m)$ to denote the m–rowed unitary group. Let s be an integer satisfying $0 < s < m$; we consider the pairs of matrices

$$(U, V), \quad U \in \mathfrak{U}(m), \quad V \in \mathfrak{U}(n).$$

Two pairs (U, V) and (U_1, V_1) are called equivalent, if there exist three unitary matrices $U^{(s)}$, $U^{(m-s)}$ and $V^{(n-s)}$ such that

$$U = \begin{pmatrix} U^{(s)} & 0 \\ 0 & U^{(m-s)} \end{pmatrix} U_1, \quad V = \begin{pmatrix} \bar{U}^{(s)} & 0 \\ 0 & V^{(n-s)} \end{pmatrix} V_1.$$

By the equivalence, we classify pairs into classes. Each class is considered as an element; the totality of elements is defined to be the set $\mathfrak{M}_{\mathrm{I}}^{(s)}$.

Proof. It is known that each element of $\mathfrak{C}_{\mathrm{I}}^{(r)}$ can be expressed as

$$Z = U'\begin{pmatrix} I^{(m-r)} & 0 \\ 0 & W \end{pmatrix} V,$$

where $U = U^{(m)}$ and $V = V^{(n)}$ are unitary matrices and $W = W^{(r, n-m+r)}$ satisfies

$$I^{(r)} - W\bar{W}' > 0.$$

If there is another expression

$$Z = U_1' \begin{pmatrix} I^{(m-r)} & 0 \\ 0 & W_1 \end{pmatrix} V_1,$$

we have then

$$U_2' \begin{pmatrix} I & 0 \\ 0 & W \end{pmatrix} = \begin{pmatrix} I & 0 \\ 0 & W_1 \end{pmatrix} V_2, \tag{2.3.1}$$

where

$$U_2 = UU_1^{-1}, \quad V_2 = V_1V^{-1}.$$

From (2.3.1), we have

$$U_2' \begin{pmatrix} I & 0 \\ 0 & W\overline{W}' \end{pmatrix} \overline{U}_2 = \begin{pmatrix} I & 0 \\ 0 & W_1 \end{pmatrix} V_2\overline{V}_2' \begin{pmatrix} I & 0 \\ 0 & \overline{W}_1' \end{pmatrix},$$

i.e.,

$$U_2' \begin{pmatrix} I & 0 \\ 0 & W\overline{W}' \end{pmatrix} = \begin{pmatrix} I & 0 \\ 0 & W_1\overline{W}_1' \end{pmatrix} U_2'.$$

Since none of the characteristic roots of $W\overline{W}'$ and $W_1\overline{W}_1'$ is equal to 1, consequently

$$U_2 = \begin{pmatrix} U^{(m-r)} & 0 \\ 0 & U^{(r)} \end{pmatrix}.$$

By argument similar to that of (2.3.1), we deduce

$$V_2 = \begin{pmatrix} \overline{U}^{(m-r)} & 0 \\ 0 & V^{(n-m+r)} \end{pmatrix}.$$

The theorem follows.

§2.4. *The boundary properties of the Poisson integral of* $\mathfrak{R}_{\mathrm{I}}$

Now we study the properties of the Poisson integral

$$u(Z) = \int_{\mathfrak{C}_{\mathrm{I}}} \varphi(U) P_{\mathrm{I}}(Z, U) \dot{U} \tag{2.4.1}$$

as Z approaches a boundary point of $\mathfrak{R}_{\mathrm{I}}$ from its interior.

Theorem 2.4.1. Let $m > 1$. Let $\varphi(U)$ be a real-valued function continuous on the characteristic manifold $\mathfrak{C}_{\mathrm{I}}(m,n)$ of $\mathfrak{R}_{\mathrm{I}}(m,n)$ and let

$$Q = \begin{pmatrix} 1 & 0 \\ 0 & Z_0^{(m-1,n-1)} \end{pmatrix}, \quad Z = \begin{pmatrix} \rho & 0 \\ 0 & Z_0^{(m-1,n-1)} \end{pmatrix}$$

with $0 \leqslant \rho < 1$ and $I - Z_0\bar{Z}_0' > 0$. Then

$$\lim_{\rho\to 1}\int_{\mathfrak{C}_{\mathrm{I}}(m,n)}\varphi(U)P_{\mathrm{I}}(Z,U)\dot{U}$$

$$=\int_{\mathfrak{C}_{\mathrm{I}}(m-1,n-1)}\varphi\left(\begin{pmatrix}1 & 0\\ 0 & U_1\end{pmatrix}\right)P_{\mathrm{I}}^{(m-1,n-1)}(Z_0,U_0)\dot{U}_0 \tag{2.4.2}$$

uniformly with respect to Z_0, where $U_0 \in \mathfrak{C}_{\mathrm{I}}(m-1, n-1)$ and $P_{\mathrm{I}}^{(m-1,n-1)}(Z_0, U_0)$ is the Poisson kernel of $\mathfrak{R}_{\mathrm{I}}(m-1, n-1)$, that is

$$P_{\mathrm{I}}^{(m-1,n-1)}(Z_0,U_0)=\frac{1}{V(\mathfrak{C}_{\mathrm{I}}(m-1,n-1))}\cdot\frac{\det(I-Z_0\bar{Z}_0')^{n-1}}{|\det(I-Z_0\bar{U}_0')|^{2(n-1)}}.$$

Proof. First let us consider the special case with $Z_0 = 0$ and

$$Z=\begin{pmatrix}\rho & 0\\ 0 & 0^{(m-1,n-1)}\end{pmatrix},\quad 0\leqslant\rho<1.$$

Then, we have

$$u(Z)=\frac{1}{V(\mathfrak{C}_{\mathrm{I}})}\int_{\mathfrak{C}_{\mathrm{I}}}\varphi(U)\frac{(1-\rho^2)^n}{|1-\rho\bar{u}_{11}|^{2n}}\dot{U}, \tag{2.4.3}$$

where

$$U=\begin{pmatrix}u\\ U_1\end{pmatrix},\quad u=(u_{11},\cdots,u_{1n}). \tag{2.4.4}$$

(2.4.3) can be written as

$$u(Z)=\frac{1}{V(\mathfrak{C}_{\mathrm{I}})}\int_{u\bar{u}'=1}\tau(u)\frac{(1-\rho^2)^n}{|1-\rho\bar{u}_{11}|^{2n}}\dot{u} \tag{2.4.5}$$

with

$$\tau(u)=\int_{U_1}\varphi(U)\dot{U}_1, \tag{2.4.6}$$

where U_1 runs over $(m-1)\times n$ matrices satisfying

$$\begin{pmatrix}u\\ U_1\end{pmatrix}\overline{\begin{pmatrix}u\\ U_1\end{pmatrix}}'=I^{(m)}. \tag{2.4.7}$$

Here u is considered as parameters satisfying $u\bar{u}'=1$.

For any fixed u, the set of U_1 satisfying (2.4.7) forms a manifold of which the volume element is denoted by $\dot{U}_1$. It is easy to see that $\tau(u)$ is a continuous function on $u\bar{u}'=1$. We apply Theorem 1.3.2

to (2.4.5), then we have

$$\lim_{\rho\to 1} u(Z) = \frac{\omega_{2n-1}}{V(\mathfrak{C}_{\mathrm{I}})}\tau(1, 0, \cdots, 0). \tag{2.4.8}$$

Notice that, by (2.2.4),

$$\frac{\omega_{2n-1}}{V(\mathfrak{C}_{\mathrm{I}}(m, n))} = \frac{1}{V(\mathfrak{C}_{\mathrm{I}}(m-1, n-1))}.$$

Substituting $u = (1, 0, \cdots, 0)$ into (2.4.7) we have

$$U_1 = (0, U_0), \quad U_0 = U_0^{(m-1, n-1)}$$

and $U_0\bar{U}_0' = I^{(m-1)}$. Consequently, we have, from (2.4.6) and (2.4.8),

$$\lim_{\rho\to 1} u(Z) = \frac{1}{V(\mathfrak{C}_{\mathrm{I}}(m-1, n-1))}\int_{\mathfrak{C}_{\mathrm{I}}(m-1, n-1)} \varphi\left(\begin{pmatrix} 1 & 0 \\ 0 & U_0\end{pmatrix}\right)\dot{U}_0. \tag{2.4.9}$$

If $Z_0 \neq 0$, then by Theorem 2.3.2 there is a transformation $W = \Phi(Z)$ of Γ^{I} carrying

$$\begin{pmatrix} 1 & 0 \\ 0 & Z_0\end{pmatrix} \text{ into } \begin{pmatrix} 1 & 0 \\ 0 & O\end{pmatrix} \text{ and } \begin{pmatrix} \rho & 0 \\ 0 & Z_1\end{pmatrix} \text{ into } \begin{pmatrix} \rho & 0 \\ 0 & W_1\end{pmatrix},$$

where

$$W_1 = A_1(Z_1 - Z_0)(I - \bar{Z}_0'Z)^{-1}D_1^{-1} \tag{2.4.10}$$

with

$$\bar{A}_1'A_1 = (I - Z_0\bar{Z}_0')^{-1}, \; \bar{D}_1'D_1 = (I - \bar{Z}_0'Z_0)^{-1}.$$

For $U \in \mathfrak{C}_{\mathrm{I}}$, we have

$$V = \Phi(U) \in \mathfrak{C}_{\mathrm{I}}$$

and

$$Z = \Phi^{-1}(W).$$

It is known that

$$\dot{V} = |\det(B\bar{U}' + A)|^{-2n}\dot{U}.$$

According to Theorem 2.2.2, we have, for $Z = \begin{pmatrix} \rho & 0 \\ 0 & Z_0\end{pmatrix}$,

$$\lim_{\rho\to 1}\int_{\mathfrak{C}_{\mathrm{I}}(m, n)} \varphi(U)P_{\mathrm{I}}(Z, U)\dot{U}$$

$$= \lim_{\rho\to 1}\frac{1}{V(\mathfrak{C}_{\mathrm{I}}(m, n))}\int_{\mathfrak{C}_{\mathrm{I}}(m, n)} \varphi(\Phi^{-1}(V))\frac{(1-\rho^2)^n}{|1-\rho\bar{v}_{11}|^{2n}}\dot{V}$$

$$= \frac{1}{V(\mathfrak{C}_\mathrm{I}(m-1, n-1))} \int_{\mathfrak{C}_\mathrm{I}(m-1,\, n-1)} \varphi\left(\Phi^{-1}\left(\begin{pmatrix} 1 & 0 \\ 0 & V_0 \end{pmatrix}\right)\right) \dot{V}_0.$$

Notice that the above formula holds uniformly with respect to Z_0, since in the proof of (2.4.9), for any given $\varepsilon > 0$ there is a ρ depending only on the upper bound of $|\varphi|$ such that (cf. Th. 1.3.2)

$$\left| \frac{1}{V(\mathfrak{C}_\mathrm{I}(m, n))} \int_{\mathfrak{C}_\mathrm{I}(m,\, n)} \varphi(\Phi^{-1}(V)) \frac{(1-\rho^2)^n}{|1-\rho \bar{v}_{11}|^{2n}} \dot{V} \right.$$
$$\left. - \frac{1}{V(\mathfrak{C}_\mathrm{I}(m-1, n-1))} \int_{\mathfrak{C}_\mathrm{I}(m-1,\, n-1)} \varphi\left(\Phi^{-1}\left(\begin{pmatrix} 1 & 0 \\ 0 & V_0 \end{pmatrix}\right)\right) \dot{V}_0 \right| < \varepsilon,$$

where ρ is independent of Z_0.

By (2.4.10), we have

$$\Phi^{-1}\left(\begin{pmatrix} 1 & 0 \\ 0 & V_0 \end{pmatrix}\right) = \begin{pmatrix} 1 & 0 \\ 0 & U_0 \end{pmatrix},$$

where

$$V_0 = A_1(U_0 - Z_0)(I - \bar{Z}_0' U_0)^{-1} D_1^{-1}.$$

Hence

$$\dot{V}_0 = |\det(I - \bar{Z}_0' U_0) A_1|^{-2(n-1)} \dot{U}_0 = \frac{\det(I - Z_0 \bar{Z}_0')^{n-1}}{|\det(I - Z_0 \bar{U}_0')|^{2(n-1)}} \dot{U}_0.$$

Finally we have

$$\lim_{\rho \to 1} \int_{\mathfrak{C}_\mathrm{I}(m,\, n)} \varphi(U) P_\mathrm{I}(Z, U) \dot{U}$$
$$= \frac{1}{V(\mathfrak{C}_\mathrm{I}(m-1, n-1))} \int_{\mathfrak{C}_\mathrm{I}(m-1,\, n-1)} \varphi\left(\begin{pmatrix} 1 & 0 \\ 0 & U_0 \end{pmatrix}\right) P_\mathrm{I}^{(m-1,\, n-1)}(Z_0, U_0) \dot{U}_0$$

uniformly with respect to Z_0.

Theorem 2.4.2. Let $\varphi(U)$ be a real-valued function continuous on the characteristic manifold $\mathfrak{C}_\mathrm{I}(m, n)$ of $\mathfrak{R}_\mathrm{I}(m, n)$.

(i) If $Q \in \mathfrak{C}_\mathrm{I}^{(r)}$ $(0 < r < m)$, the limit

$$\lim_{Z \to Q} \int_{\mathfrak{C}_\mathrm{I}} \varphi(U) P_\mathrm{I}(Z, U) \dot{U}$$

exists and defines a function continuous in $\mathfrak{C}_\mathrm{I}^{(r)} \cong \mathfrak{R}_\mathrm{I}(r, n-m+r) \times \mathfrak{M}_\mathrm{I}^{(m-r)}$; besides, it is harmonic with respect to the coordinate of $\mathfrak{R}_\mathrm{I}(r, n-m+r)$. More precisely, if we set

$$Q = U_0' \begin{pmatrix} I^{(m-r)} & 0 \\ 0 & Z_1 \end{pmatrix} V_0,$$

the function

$$u(Z_1, U_0, V_0) = \lim_{Z\to Q} \int_{\mathfrak{C}_I} \varphi(U) P_I(Z, U)\dot{U}$$

is harmonic in $\mathfrak{R}_I(r, n-m+r)$ with respect to the variable Z_1 for any $(U_0, V_0) \in \mathfrak{M}_I^{(m-r)}$.

(ii) For $Q \in \mathfrak{C}_I = \mathfrak{C}_I(m, n)$, then

$$\lim_{Z\to Q} \int_{\mathfrak{C}_I} \varphi(U) P_I(Z, U)\dot{U} = \varphi(Q).$$

Proof. (i) When $0 < r < m$, we let

$$Z = U_0' \begin{pmatrix} \Lambda^{(m-r)} & \mathbf{0} \\ \mathbf{0} & Z_1 \end{pmatrix} V_0,$$

where

$$\Lambda^{(m-r)} = [\lambda_1, \cdots, \lambda_{m-r}], \ 1 > \lambda_1 \geqslant \cdots \geqslant \lambda_{m-r} \geqslant 0.$$

Applying the previous theorem repeatedly, we see that

$$\begin{aligned} &\lim_{\lambda_{m-r}\to 1} \cdots \lim_{\lambda_1\to 1} \int_{\mathfrak{C}_I(m,n)} \varphi(U) P_I(Z, U)\dot{U} \\ &= \lim_{\lambda_{m-r}\to 1} \cdots \lim_{\lambda_1\to 1} \int_{\mathfrak{C}_I(m,n)} \varphi(U_0' U V_0) P_I\left(\begin{pmatrix} \Lambda^{(m-r)} & 0 \\ 0 & Z_1 \end{pmatrix}, U\right)\dot{U} \\ &= \frac{1}{V(\mathfrak{C}_I(r, n-m+r))} \int_{\mathfrak{C}_I(r, n-m+r)} \varphi\left(U_0' \begin{pmatrix} I^{(m-1)} & 0 \\ 0 & U_1 \end{pmatrix} V_0\right) \\ &\quad \times \frac{\det(I - Z_1\bar{Z}_1')^{n-m+r}}{|\det(I - Z_1\bar{U}_1')|^{2(n-m+r)}} \dot{U}_1 \end{aligned} \tag{2.4.11}$$

uniformly with respect to Z_1, U_0, V_0.

Obviously the above expression is a continuous function of Z_1, U_0, V_0 and for any U_0 and V_0 it is harmonic in $\mathfrak{R}_I(r, n-m+r)$.

Now we take in $\mathfrak{R}_I$ an arbitrary sequence of points $Z_1, Z_2, \cdots, Z_k, \cdots$, which approaches Q.

We write

$$u(Z) = \int_{\mathfrak{C}_I} \varphi(U) P_I(Z, U)\dot{U},$$

and $u_r(Q)$ as the limiting function of (2.4.11).

For any given $\varepsilon > 0$, since $u_r(Q)$ is continuous, there is a neighbourhood $\mathfrak{B}(Q)$ of Q in the space of the mn complex variables

$Z=(z_{j\alpha})$ such that for any point $P\in\mathfrak{B}(Q)\cap\mathfrak{C}_{\mathrm{I}}^{(r)}$, we always have

$$|u_r(P)-u_r(Q)|<\frac{\varepsilon}{2}.$$

Each point of Z_k has a representation

$$Z_k=U_k'\begin{pmatrix}\Lambda_k & 0\\ 0 & Z_{0k}\end{pmatrix}V_k,$$

where

$$\Lambda_k=[\lambda_1^{(k)},\cdots,\lambda_{m-r}^{(k)}],\qquad 1>\lambda_1^{(k)}\geqslant\cdots\geqslant\lambda_{m-r}^{(k)}\geqslant 0.$$

Since $Z_k\to Q$, we must have $\Lambda_k\to I^{(m-r)}$. This implies that when we take k sufficiently large, $\lambda_1^{(k)},\cdots,\lambda_{m-r}^{(k)}$ can be as near to 1 as we please. Since (2.4.11) holds uniformly for all $Q\in\mathfrak{C}_{\mathrm{I}}^{(r)}$, we see that for $Q_k=U_k'\begin{pmatrix}I & 0\\ 0 & Z_{0k}\end{pmatrix}V_k\in\mathfrak{C}_{\mathrm{I}}^{(r)}$,

$$|u(Z_k)-u_r(Q_k)|<\frac{\varepsilon}{2},$$

when k is sufficiently large.

Obviously $Q_k\to Q$, hence we can take k so large that $Q_k\in\mathfrak{B}(Q)\cap\mathfrak{C}_{\mathrm{I}}^{(r)}$. Then

$$|u(Z_k)-u_r(Q)|\leqslant|u(Z_k)-u_r(Q_k)|+|u_r(Q_k)-u_r(Q)|<\varepsilon.$$

This shows that for any sequence of points $Z_k\to Q$, the part (i) of our theorem holds.

(ii) When $Q\in\mathfrak{C}_{\mathrm{I}}$, we set

$$Q=U_0'(I^{(m)},\ 0\ )V_0,$$

and

$$Z=U_0'(\Lambda,\ 0\ )V_0,\ \Lambda=[\lambda_1,\cdots,\lambda_m].$$

Then, by Theorem 2.4.1 and Theorem 1.3.2,

$$\lim_{\lambda_m\to1}\cdots\lim_{\lambda_1\to1}\int_{\mathfrak{C}_{\mathrm{I}}(m,\,n)}\varphi(U)P_{\mathrm{I}}(Z,U)\dot U$$

$$=\lim_{\lambda_m\to1}\frac{1}{\omega_{2(n-m)+1}}\int_{u\bar u'=1}\varphi\left(U_0'\begin{pmatrix}I^{(m-1)} & 0\\ 0 & u\end{pmatrix}V_0\right)\frac{(1-\lambda_m^2)^{n-m+1}}{|1-\lambda_m\bar u_1|^{2(n-m+1)}}\dot u$$

$$=\varphi(U_0'(I^{(m)},\ 0\ )V_0)=\varphi(Q),$$

where $u=(u_1,\cdots,u_{n-m+1})$ with $u\bar u'=1$.

As in the proof of part (i), for any sequence of points Z_1, $Z_2, \cdots$ $(Z_k \to Q)$ in $\mathfrak{R}_{\mathrm{I}}$, we always have

$$\lim_{Z_k \to Q} \int_{\mathfrak{C}_{\mathrm{I}}} \varphi(U) P_{\mathrm{I}}(Z, U) \dot{U} = \varphi(Q).$$

The theorem is proved.

§2.5. *A Dirichlet problem of* $\mathfrak{R}_{\mathrm{I}}$

A real-valued function $u(Z)$ is said to be harmonic on the closure of $\mathfrak{R}_{\mathrm{I}}$, if it is continuous in $\mathfrak{B}_{\mathrm{I}}^{(0)}$, and on each $\mathfrak{C}_{\mathrm{I}}^{(r)} = \mathfrak{B}_{\mathrm{I}}^{(m-r)} - \mathfrak{B}_{\mathrm{I}}^{(m-r+1)}$ $(r = 1, \cdots, m)$ it is harmonic with respect to the coordinate of the base space $\mathfrak{R}_{\mathrm{I}}(r, n-m+r)$ of the slit space $\mathfrak{B}_{\mathrm{I}}^{(m-r)}$.

Notice that the differential equation on $\mathfrak{C}_{\mathrm{I}}^{(r)}$ can also be considered as a consequence of the differential equation

$$\operatorname{tr}((I - Z\bar{Z}')\bar{\partial}_Z(I - \bar{Z}'Z)\partial'_Z) \circ u = 0. \tag{2.5.1}$$

In fact, for $Z = \begin{pmatrix} I & 0 \\ 0 & Z_1 \end{pmatrix}$, the previous equation reduces to

$$\operatorname{tr}\left(\begin{pmatrix} 0 & 0 \\ 0 & I - Z_1\bar{Z}_1' \end{pmatrix}\begin{pmatrix} * & * \\ * & \bar{\partial}_{Z_1} \end{pmatrix}\begin{pmatrix} 0 & 0 \\ 0 & I - \bar{Z}_1'Z_1 \end{pmatrix}\begin{pmatrix} * & * \\ * & \partial'_{Z_1} \end{pmatrix}\right) \circ u = 0,$$

i.e.,

$$\begin{aligned}\operatorname{tr}&\left(\begin{pmatrix} 0 & 0 \\ * & (I - Z_1\bar{Z}_1')\bar{\partial}_{Z_1} \end{pmatrix}\begin{pmatrix} 0 & 0 \\ * & (I - \bar{Z}_1'Z_1)\partial'_{Z_1} \end{pmatrix}\right) \circ u \\ &= \operatorname{tr}((I - Z_1\bar{Z}_1')\bar{\partial}_{Z_1}(I - \bar{Z}_1'Z_1)\partial'_{Z_1}) \circ u = 0.\end{aligned}$$

Similarly for the case $Z = U'\begin{pmatrix} I & 0 \\ 0 & Z_1 \end{pmatrix}V$ with unitary U and V.

Now we formulate the following Dirichlet problem:

Given a continuous function $\varphi(U)$ on $\mathfrak{C}_{\mathrm{I}}$, whether there exists a unique function harmonic on the closure of $\mathfrak{R}_{\mathrm{I}}$, which takes the given boundary value $\varphi(U)$ on $\mathfrak{C}_{\mathrm{I}}$.

The answer is positive. We have

Theorem 2.5.1. Let $\varphi(U)$ be a real-valued continuous function on $\mathfrak{C}_{\mathrm{I}}$. Then

$$u(Z) = \int_{\mathfrak{C}_{\mathrm{I}}} \varphi(U) P_{\mathrm{I}}(Z, U) \dot{U} \tag{2.5.2}$$

is the unique function harmonic on the closure of $\mathfrak{R}_{\mathrm{I}}$, which takes the given boundary value $\varphi(U)$ on $\mathfrak{C}_{\mathrm{I}}$.

Proof. The existence of the problem follows from Theorem 2.4.2; we remain to prove the uniqueness of the theorem.

Suppose that there is another function $u_1(Z)$ which is harmonic on the closure of $\mathfrak{R}_{\mathrm{I}}$ and which takes the given boundary value $\varphi(U)$ on $\mathfrak{C}_{\mathrm{I}}$. Then $u_1(Z) - u(Z)$ is again a function which is harmonic on the closure of $\mathfrak{R}_{\mathrm{I}}$ and vanishes on $\mathfrak{C}_{\mathrm{I}}$. The principle of extremity, Theorem 1.2.2, asserts that it vanishes identically. Therefore we have the theorem.

Notice that the Poisson kernel, by Theorem 2.2.3, satisfies a system of m^2 partial differential equations

$$\bar{\partial}_Z(I - \bar{Z}'Z)\partial_Z' \circ P_{\mathrm{I}}(Z,U) = 0.$$

Consequently, the function represented by the Poisson integral satisfies also a system of m^2 partial differential equations.

§2.6. *Harmonic functions in* $\mathfrak{R}_{\mathrm{II}}$

Let Z denote an n–rowed symmetric matrix of the form

$$Z = \begin{pmatrix} \sqrt{2}\, z_{11}, & z_{12}, & \cdots, & z_{1n} \\ z_{12}, & \sqrt{2}\, z_{22}, & \cdots, & z_{2n} \\ \cdots & \cdots & \cdots & \cdots \\ z_{1n}, & z_{2n}, & \cdots, & \sqrt{2}\, z_{nn} \end{pmatrix}, \tag{2.6.1}$$

and let $\mathfrak{R}_{\mathrm{II}}$ denote the domain

$$I - Z\bar{Z} > 0. \tag{2.6.2}$$

It is a space of $\frac{1}{2}n(n+1)$ complex variables

$$z_{11}, z_{12}, \cdots, z_{1n}, z_{22}, \cdots, z_{2n}, \cdots, z_{nn}.$$

We introduce the operator

$$\partial_Z = \begin{pmatrix} \sqrt{2}\dfrac{\partial}{\partial z_{11}}, & \dfrac{\partial}{\partial z_{12}}, & \cdots, & \dfrac{\partial}{\partial z_{1n}} \\ \dfrac{\partial}{\partial z_{12}}, & \sqrt{2}\dfrac{\partial}{\partial z_{22}}, & \cdots, & \dfrac{\partial}{\partial z_{2n}} \\ \cdots & \cdots & \cdots & \cdots \\ \dfrac{\partial}{\partial z_{1n}}, & \dfrac{\partial}{\partial z_{2n}}, & \cdots, & \sqrt{2}\dfrac{\partial}{\partial z_{nn}} \end{pmatrix}, \tag{2.6.3}$$

and

$$\Delta_{\mathrm{II}} = \mathrm{tr}\,((I - \bar{Z}Z)\partial_Z(I - Z\bar{Z})\bar{\partial}_Z) \tag{2.6.4}$$

or

$$\Delta_{II} = \sum_{\alpha,\beta,\lambda,\mu=1}^{n} \left(\delta_{\lambda\mu} - \sum_{\sigma=1}^{n} P_{\lambda\sigma}P_{\mu\sigma}z_{\lambda\sigma}\bar{z}_{\mu\sigma}\right)$$
$$\times \left(\delta_{\alpha\beta} - \sum_{\gamma=1}^{n} P_{\alpha\gamma}P_{\beta\gamma}z_{\alpha\gamma}\bar{z}_{\beta\gamma}\right) P_{\lambda\alpha}P_{\mu\beta}\frac{\partial^2}{\partial z_{\lambda\alpha}\partial\bar{z}_{\mu\beta}},$$

where $z_{\alpha\beta} = z_{\beta\alpha}$ and

$$P_{\alpha\beta} = \begin{cases} \sqrt{2}, & \text{for } \alpha = \beta, \\ 1, & \text{for } \alpha \neq \beta. \end{cases} \tag{2.6.5}$$

A real-valued function $u(Z)$ possessing continuous second derivatives is said to be harmonic in $\mathfrak{R}_{II}$, if it satisfies in $\mathfrak{R}_{II}$ the differential equation

$$\Delta_{II} \circ u(Z) = 0. \tag{2.6.6}$$

Theorem 2.6.1. If $u(Z)$ is harmonic in $\mathfrak{R}_{II}$, it remains to be harmonic in $\mathfrak{R}_{II}$, after a transformation of the group Γ^{II} of the motions of $\mathfrak{R}_{II}$.

Proof. It is known that a transformation of $\mathfrak{R}_{II}$ is of the form

$$W = (AZ + B)(\bar{B}Z + \bar{A})^{-1}, \tag{2.6.7}$$

where

$$A'\bar{B} = \bar{B}'A, \quad \bar{A}'A - B'\bar{B} = I.$$

Now we have

$$dW = (\bar{B}Z + \bar{A})'^{-1}\, dZ(\bar{B}Z + \bar{A})^{-1}.$$

If we use $a_{\alpha\beta}(1 \leqslant \alpha, \beta \leqslant n)$ to denote the elements of the matrix $(\bar{B}Z + \bar{A})^{-1}$, then the above relation can be written as

$$P_{\alpha\beta}\, dw_{\alpha\beta} = \sum_{\lambda,\mu=1}^{n} a_{\lambda\alpha}P_{\lambda\mu}\, dz_{\lambda\mu}a_{\mu\beta}.$$

Since $z_{\lambda\mu} = z_{\mu\lambda}$, we have

$$\frac{\partial w_{\alpha\beta}}{\partial z_{\lambda\mu}} = \begin{cases} \dfrac{1}{P_{\alpha\beta}}(a_{\lambda\alpha}a_{\mu\beta} + a_{\mu\alpha}a_{\lambda\beta}), & (\lambda \neq \mu) \\ \dfrac{1}{P_{\alpha\beta}}\sqrt{2}\, a_{\lambda\alpha}a_{\mu\beta}, & (\lambda = \mu). \end{cases}$$

For $\lambda \neq \mu$, we have

$$\frac{\partial}{\partial z_{\lambda\mu}} = \sum_{\alpha\leqslant\beta} \frac{\partial w_{\alpha\beta}}{\partial z_{\lambda\mu}} \frac{\partial}{\partial w_{\alpha\beta}}$$

$$= \sum_{\alpha<\beta} (a_{\lambda\alpha}a_{\mu\beta} + a_{\mu\alpha}a_{\lambda\beta}) \frac{\partial}{\partial w_{\alpha\beta}} + \sum_{\alpha=1}^{n} \frac{1}{2} (a_{\lambda\alpha}a_{\mu\alpha} + a_{\mu\alpha}a_{\lambda\alpha}) \sqrt{2} \frac{\partial}{\partial w_{\alpha\alpha}}$$

$$= \sum_{\alpha<\beta} a_{\lambda\alpha}a_{\mu\beta}P_{\alpha\beta} \frac{\partial}{\partial w_{\alpha\beta}} + \sum_{\alpha>\beta} a_{\lambda\alpha}a_{\mu\beta}P_{\alpha\beta} \frac{\partial}{\partial w_{\alpha\beta}} + \sum_{\alpha=\beta} a_{\lambda\alpha}a_{\lambda\beta}P_{\alpha\beta} \frac{\partial}{\partial w_{\alpha\beta}}$$

$$= \sum_{\alpha,\beta=1}^{n} a_{\lambda\alpha}\left(P_{\alpha\beta} \frac{\partial}{\partial w_{\alpha\beta}}\right) a_{\mu\beta}.$$

For $\lambda=\mu$, we have

$$\sqrt{2} \frac{\partial}{\partial z_{\lambda\lambda}} = \sqrt{2} \sum_{\alpha\leqslant\beta} \frac{\partial w_{\alpha\beta}}{\partial z_{\lambda\lambda}} \frac{\partial}{\partial w_{\alpha\beta}}$$

$$= \sum_{\alpha<\beta} 2a_{\lambda\alpha}a_{\lambda\beta} \frac{\partial}{\partial w_{\alpha\beta}} + \sum_{\alpha=1}^{n} \sqrt{2}\, a_{\lambda\alpha}a_{\lambda\alpha} \frac{\partial}{\partial w_{\alpha\alpha}}$$

$$= \sum_{\alpha<\beta} a_{\lambda\alpha}a_{\lambda\beta} \frac{\partial}{\partial w_{\alpha\beta}} + \sum_{\alpha>\beta} a_{\lambda\alpha}a_{\lambda\beta}P_{\alpha\beta} \frac{\partial}{\partial w_{\alpha\beta}} + \sum_{\alpha=\beta} a_{\lambda\alpha}a_{\lambda\beta}P_{\alpha\beta} \frac{\partial}{\partial w_{\alpha\beta}}$$

$$= \sum_{\alpha,\beta=1}^{n} a_{\lambda\alpha}\left(P_{\alpha\beta} \frac{\partial}{\partial w_{\alpha\beta}}\right) a_{\lambda\beta}.$$

Consequently, we can put

$$P_{\lambda\mu} \frac{\partial}{\partial z_{\lambda\mu}} = \sum_{\alpha,\beta=1}^{n} a_{\lambda\alpha}\left(P_{\alpha\beta} \frac{\partial}{\partial w_{\alpha\beta}}\right) a_{\mu\beta}$$

for $\lambda, \mu = 1, \cdots, n$, or

$$\partial_Z = (\bar{B}Z + \bar{A})^{-1}\partial_W(\bar{B}Z + \bar{A})'^{-1}.$$

On the other hand

$$\begin{aligned} I - W\bar{W} &= I - (\bar{B}Z + \bar{A})'^{-1}(AZ + B)'(\bar{A}\bar{Z} + \bar{B})(B\bar{Z} + A)^{-1} \\ &= (\bar{B}Z + \bar{A})'^{-1}(I - Z\bar{Z})(B\bar{Z} + A)^{-1}. \end{aligned} \tag{2.6.8}$$

We have finally

$$\begin{aligned} &(I - \bar{Z}Z)\partial_Z(I - Z\bar{Z})\bar{\partial}_Z \\ &= (\bar{Z}B' + A')[(I - \bar{W}W)\partial_W(I - W\bar{W})\bar{\partial}_W](\bar{Z}B' + A')^{-1}. \end{aligned}$$

This leads to

$$\begin{aligned} &\operatorname{tr}((I - \bar{W}W)\partial_W(I - W\bar{W})\bar{\partial}_W) \\ &= \operatorname{tr}((I - \bar{Z}Z)\partial_Z(I - Z\bar{Z})\bar{\partial}_Z), \end{aligned}$$

which proves our theorem.

The Poisson kernel of $\mathfrak{R}_{II}$ is known (Hua [4]) to be

$$P_{II}(Z, S) = \frac{1}{V(\mathfrak{C}_{II})} \cdot \frac{\det(I - Z\bar{Z})^{\frac{1}{2}(n+1)}}{|\det(I - Z\bar{Z})|^{n+1}}, \tag{2.6.9}$$

where S is an n-rowed symmetric unitary matrix, i.e., $S\bar{S} = I$ and is expressed as

$$S = \begin{pmatrix} \sqrt{2}\, s_{11}, & s_{12}, & \cdots, & s_{1n} \\ s_{12}, & \sqrt{2}\, s_{22}, & \cdots, & s_{2n} \\ \cdots & \cdots & \cdots & \cdots \\ s_{1n}, & s_{2n}, & \cdots, & \sqrt{2}\, s_{nn} \end{pmatrix}. \tag{2.6.10}$$

Further, we have[1)]

$$V(\mathfrak{C}_{II}) = 2^{\frac{1}{2}n(n+1)}\pi^{\frac{1}{4}n(n+1)} \frac{\Gamma\left(\frac{1}{2}\right)}{\Gamma\left(\frac{n+1}{2}\right)} \prod_{\nu=1}^{n-1} \frac{\Gamma\left(\frac{1}{2}(n-\nu)+1\right)}{\Gamma(n-\nu+1)}.$$

Since $\Gamma(x)\,\Gamma\left(x + \frac{1}{2}\right) = \frac{\sqrt{\pi}}{2^{2x-1}}\Gamma(2x)$, we have

$$\frac{\Gamma\left(\frac{\mu}{2}+1\right)}{\Gamma(\mu+1)} = \frac{\Gamma\left(\frac{\mu+1}{2}\right)\Gamma\left(\frac{\mu+1}{2}+\frac{1}{2}\right)}{\Gamma\left(\frac{\mu+1}{2}\right)\Gamma(\mu+1)} = \frac{\sqrt{\pi}}{2^{\mu}\Gamma\left(\frac{\mu+1}{2}\right)},$$

and then

$$\begin{aligned} V(\mathfrak{C}_{II}) &= 2^{\frac{1}{2}n(n+1)}\pi^{\frac{1}{4}n(n+1)} \frac{\Gamma\left(\frac{1}{2}\right)}{\Gamma\left(\frac{n+1}{2}\right)} \prod_{\mu=1}^{n-1} \frac{\Gamma\left(\frac{\mu}{2}+1\right)}{\Gamma(\mu+1)} \\ &= 2^{\frac{1}{2}n(n+1)}\pi^{\frac{1}{4}n(n+1)} \frac{\pi^{\frac{n}{2}}}{2^{\frac{1}{2}n(n-1)}} \prod_{\mu=1}^{n} \frac{1}{\Gamma\left(\frac{\mu+1}{2}\right)} \\ &= \prod_{\mu=1}^{n} \frac{2\pi^{\frac{1}{2}(\mu+1)}}{\Gamma\left(\frac{\mu+1}{2}\right)} = \prod_{\mu=1}^{n} \omega_{\mu}. \end{aligned} \tag{2.6.11}$$

---

1) The constant differs from the original one by a factor 2^{n^2} (cf. the footnote on p. 1046).

Theorem 2.6.2. After the transformation (2.6.7), the Poisson kernel becomes

$$P_{\mathrm{II}}(Z, S) = P_{\mathrm{II}}(W, T)|\det(B\bar{S} + A)|^{-(n+1)},$$

where

$$T = (AS + B)(\bar{B}S + \bar{A})^{-1}. \tag{2.6.12}$$

Proof. Clearly

$$I - W\bar{T} = (\bar{B}Z + \bar{A})'^{-1}(I - Z\bar{S})(B\bar{S} + A)^{-1}.$$

Combining with (2.6.8), we have immediately

$$\frac{\det(I - Z\bar{Z})^{\frac{1}{2}(n+1)}}{|\det(I - Z\bar{S})|^{n+1}} = \frac{\det(I - W\bar{W})^{\frac{1}{2}(n+1)}|\det(\bar{B}Z + \bar{A})|^{n+1}}{|\det((I - W\bar{T})(\bar{B}Z + \bar{A})(B\bar{S} + A))|^{n+1}}$$

$$= \frac{\det(I - W\bar{W})^{\frac{1}{2}(n+1)}}{|\det(I - W\bar{T})|^{n+1}}|\det(B\bar{S} + A)|^{-(n+1)}.$$

Theorem 2.6.3. The Poisson kernel $P_{\mathrm{II}}(Z, S)$ is harmonic in $\mathfrak{R}_{\mathrm{II}}$ with respect to the variable Z.

Proof. According to Theorems 2.6.1 and 2.6.2 and the method used in the proof of Theorem 2.2.3, it is sufficient to prove that

$$[\Delta_{\mathrm{II}} \circ P_{\mathrm{II}}(Z, S)]_{Z=0} = 0.$$

By (2.6.4) and (2.6.9), we have

$$[\Delta_{\mathrm{II}} \circ P_{\mathrm{II}}(Z, S)]_{Z=0}$$

$$= \frac{1}{V(\mathfrak{C}_{\mathrm{II}})}\Bigg[\sum_{\alpha,\lambda=1}^{n} P_{\lambda\alpha}P_{\lambda\alpha}\frac{\partial^2}{\partial z_{\lambda\alpha}\partial\bar{z}_{\lambda\alpha}}\det(I - Z\bar{Z})^{\frac{n+1}{2}}\det(I - Z\bar{S})^{-\frac{n+1}{2}}$$

$$\times \det(I - \bar{Z}S)^{-\frac{n+1}{2}}\Bigg]_{Z=0}$$

$$= \frac{1}{V(\mathfrak{C}_{\mathrm{II}})}\Bigg[2\sum_{\lambda\leqslant\alpha}\frac{\partial^2}{\partial z_{\lambda\alpha}\partial\bar{z}_{\lambda\alpha}}\Big\{1 - (n + 1)\sum_{\beta\leqslant\gamma}|z_{\beta\gamma}|^2 + \cdots\Big\}$$

$$\times \Big\{1 + (n + 1)\sum_{\beta\leqslant\gamma} z_{\beta\gamma}\bar{S}_{\beta\gamma} + \cdots\Big\}\Big\{1 + (n+1)\sum_{\beta\leqslant\gamma}\bar{z}_{\beta\gamma}S_{\beta\gamma} + \cdots\Big\}\Bigg]_{Z=0}$$

$$= \frac{1}{V(\mathfrak{C}_{\mathrm{II}})}\Big[-2(n + 1)\sum_{\lambda\leqslant\alpha} 1 + 2(n + 1)^2\sum_{\lambda\leqslant\alpha}\bar{S}_{\lambda\alpha}S_{\lambda\alpha}\Big]$$

$$= \frac{1}{V(\mathfrak{C}_{\mathrm{II}})}\Big[-n(n + 1)^2 + (n + 1)^2\sum_{\lambda,\alpha=1}^{n}|P_{\lambda\alpha}S_{\lambda\alpha}|^2\Big]$$

$$= \frac{1}{V(\mathfrak{C}_{\mathrm{II}})}[-n(n + 1)^2 + (n + 1)^2 n] = 0.$$

Consequently, for any real-valued function $\varphi(S)$ continuous on $\mathfrak{C}_{\mathrm{II}}$, the Poisson integral

$$u(Z) = \int_{\mathfrak{C}_{\mathrm{II}}} \varphi(S) P_{\mathrm{II}}(Z, S) \dot{S}$$

defines a harmonic function in $\mathfrak{R}_{\mathrm{II}}$.

§2.7. *The boundary properties of the Poisson integral of* $\mathfrak{R}_{\mathrm{II}}$

More precisely, we use $\mathfrak{R}_{\mathrm{II}}(n)$ and $\mathfrak{C}_{\mathrm{II}}(n)$ to denote $\mathfrak{R}_{\mathrm{II}}$ and $\mathfrak{C}_{\mathrm{II}}$ respectively. Let $\mathfrak{B}_{\mathrm{II}}^{(n-r)}$ be the set of symmetric matrices Z such that $I - Z\bar{Z}$ is positive semi-definite and of rank $\leqslant r$ $(0 \leqslant r \leqslant n)$. Evidently $\mathfrak{B}_{\mathrm{II}}^{(0)}$ is the closure of $\mathfrak{R}_{\mathrm{II}}$ and $\mathfrak{B}_{\mathrm{II}}^{(n)}$ is the characteristic manifold $\mathfrak{C}_{\mathrm{II}}$ of $\mathfrak{R}_{\mathrm{II}}$.

Analogous to Theorem 2.3.1, we have

Theorem 2.7.1. We define $\mathfrak{C}_{\mathrm{II}}^{(r)} = \mathfrak{B}_{\mathrm{II}}^{(n-r)} - \mathfrak{B}_{\mathrm{II}}^{(n-r+1)}$. Then $\mathfrak{C}_{\mathrm{II}}^{(r)}$ is an invariant subspace of $\mathfrak{B}_{\mathrm{II}}^{(0)}$ and is transitive under the group Γ^{II}.

Further we have

Theorem 2.7.2. The sequence of sets

$$\mathfrak{B}_{\mathrm{II}}^{(0)} \supset \mathfrak{B}_{\mathrm{II}}^{(1)} \supset \cdots \supset \mathfrak{B}_{\mathrm{II}}^{(n-1)}$$

forms a chain of slit spaces. The slit of $\mathfrak{B}_{\mathrm{II}}^{(n-1)}$ is $\mathfrak{C}_{\mathrm{II}}$. More precisely, $\mathfrak{C}_{\mathrm{II}}^{(r)} = \mathfrak{B}_{\mathrm{II}}^{(n-r)} - \mathfrak{B}_{\mathrm{II}}^{(n-r+1)}$ is homeomorphic to the topological product

$$\mathfrak{R}_{\mathrm{II}}(r) \times \mathfrak{M}_{\mathrm{II}}^{(n-r)},$$

where the set $\mathfrak{M}_{\mathrm{II}}^{(s)}$ can be defined as the following. Two n-rowed unitary matrices U and V are said to be equivalent if there exist an s-rowed real orthogonal matrix $\Gamma^{(s)}$ and an unitary matrix $U^{(n-s)}$ such that

$$U = \begin{pmatrix} \Gamma^{(s)} & 0 \\ 0 & U^{(n-s)} \end{pmatrix} V.$$

By equivalence, we classify the unitary matrices into classes. The totality of classes defines the set $\mathfrak{M}_{\mathrm{II}}^{(s)}$.

Proof. It is known that each element Z of $\mathfrak{C}_{\mathrm{II}}^{(r)}$ can be expressed as

$$Z = U' \begin{pmatrix} I^{(n-r)} & 0 \\ 0 & W^{(r)} \end{pmatrix} U, \tag{2.7.1}$$

where $I - W\overline{W} > 0$.

From

$$Z = U_0'\begin{pmatrix} I & 0 \\ 0 & W_0 \end{pmatrix} U_0,$$

we deduce that

$$U_1'\begin{pmatrix} I & 0 \\ 0 & W \end{pmatrix} U_1 = \begin{pmatrix} I & 0 \\ 0 & W_0 \end{pmatrix},$$

where $U_1 = UU_0^{-1}$.

Consequently, we have

$$U_1 = \begin{pmatrix} \Gamma^{(n-r)} & 0 \\ 0 & U^{(r)} \end{pmatrix}.$$

The theorem follows.

Theorem 2.7.3. Let $\varphi(S)$ be a continuous real-valued function defined on the characteristic manifold $\mathfrak{C}_{II}(n)$ of $\mathfrak{R}_{II}(n)$.

(i) Let

$$Q = U_0'\begin{pmatrix} I^{(n-r)} & 0 \\ 0 & Z_1 \end{pmatrix} U_0$$

be an arbitrary point of $\mathfrak{C}_{II}^{(r)}(0 < r < n)$, then

$$\lim_{z \to Q} \int_{\mathfrak{C}_{II}(n)} \varphi(S) P_{II}(Z, S)\dot{S}$$

$$= \frac{1}{V(\mathfrak{C}_{II}(r))} \int_{\mathfrak{C}_{II}(r)} \varphi\left(U_0'\begin{pmatrix} I^{(n-r)} & 0 \\ 0 & S_1 \end{pmatrix} U_0\right) \frac{\det(I^{(r)} - Z_1\bar{Z}_1)^{\frac{r+1}{2}}}{|\det(I - Z_1\bar{S}_1)|^{r+1}} \dot{S}_1.$$

The last integral represents a function continuous in $\mathfrak{C}_{II}^{(r)}$ ($\cong \mathfrak{R}_{II}(r) \times \mathfrak{M}_{I}^{(n-r)}$) and harmonic with respect to Z_1 in $\mathfrak{R}_{II}(r)$.

(ii) If $S_0 \in \mathfrak{C}_{II}(n)$, then

$$\lim_{z \to S_0} \int_{\mathfrak{C}_{II}} \varphi(S) P_{II}(Z, S)\dot{S} = \varphi(S_0).$$

Proof. It is sufficient to prove the following special case for $r = n-1$ and $Z = \begin{pmatrix} \rho & 0 \\ 0 & O \end{pmatrix}$, that is

$$\lim_{\rho \to 1} \frac{1}{V(\mathfrak{C}_{II}(n))} \int_{\mathfrak{C}_{II}(n)} \varphi(S) \frac{(1-\rho^2)^{\frac{n+1}{2}}}{|1 - \rho\bar{s}_{11}|^{n+1}} \dot{S}$$

$$= \frac{1}{V(\mathfrak{C}_{II}(n-1))} \int_{\mathfrak{C}_{II}(n-1)} \varphi\left(\begin{pmatrix} 1 & 0 \\ 0 & S_1 \end{pmatrix}\right) \dot{S}_1, \tag{2.7.2}$$

since the remaining part of the proof can be carried out by the same method used in §2.4.

Let

$$S = \begin{pmatrix} s_{11} & s \\ s' & S_1 \end{pmatrix}, \tag{2.7.3}$$

where $S = (s_{12}, \cdots, s_{1n})$. Write

$$s_{11} = x_1 + ix_2,\ s_{12} = x_3 e^{i\theta_1},\ \cdots,\ s_{1n} = x_{n+1} e^{i\theta_{n-1}}. \tag{2.7.4}$$

From $|s_{11}|^2 + s\bar{s}' = 1$, it follows that $xx' = 1$, where $x = (x_1, \cdots, x_{n+1})$ is a real vector. Now we are going to prove that for a given x satisfying $xx' = 1$, we can construct a unitary symmetric matrix S, that is, there exists an $(n-1) \times (n-1)$ symmetric matrix S_1 such that

$$s_{11}\bar{s} + s\bar{S}_1 = 0, \quad s'\bar{s} + S_1\bar{S}_1 = I. \tag{2.7.5}$$

Without loss of generality we take $\theta_1 = \cdots = \theta_{n-1} = 0$. There is a real orthogonal matrix $\Gamma = \Gamma^{(n-1)}$ such that

$$s = (\lambda, 0, \cdots, 0)\Gamma, \quad \lambda = \sqrt{x_3^2 + \cdots + x_{n+1}^2}.$$

Equations of (2.7.5) become

$$s_{11}(\lambda, 0, \cdots, 0) + (\lambda, 0, \cdots, 0)\bar{T} = 0,$$

$$(\lambda, 0, \cdots, 0)'(\lambda, 0, \cdots, 0) + T\bar{T} = I,$$

where $T = \Gamma' S_1 \Gamma$. Then

$$T = [\bar{s}_{11}, 1, \cdots, 1]$$

is a solution.

Now we take $(x_1, \cdots, x_{n+1})$ as parameters of (2.7.5). For a fixed $(x_1, x_2, \cdots, x_{n+1})$, the manifold of (2.7.5) is denoted by Σ. Then the Poisson integral can be written as

$$\frac{1}{V(\mathfrak{C}_{II})} \int_{\mathfrak{C}_{II}} \varphi(S) \frac{(1-\rho^2)^{\frac{n+1}{2}}}{|1-\rho\bar{s}_{11}|^{n+1}} \dot{S} = \int_{xx'=1} \psi(x) \frac{(1-\rho^2)^{\frac{n+1}{2}}}{|1-\rho(x_1 - ix_2)|^{n+1}} \dot{x},$$

where

$$\psi(x) = \frac{1}{V(\mathfrak{C}_{II}(n))} \int_\Sigma \varphi(S) \dot{\Sigma}.$$

According to Theorem 1.3.2, we have

$$\lim_{\rho\to 1}\int_{xx'=1}\psi(x)\frac{(1-\rho^2)^{\frac{n+1}{2}}}{|1-\rho(x_1-ix_2)|^{n+1}}\dot{x}=\omega_n\psi(1,0,\cdots,0).$$

Therefore

$$\lim_{\rho\to 1}\frac{1}{V(\mathfrak{C}_{\rm II}(n))}\int_{\mathfrak{C}_{\rm II}(n)}\varphi(S)\frac{(1-\rho^2)^{\frac{n+1}{2}}}{|1-\rho\bar{s}_{11}|^{n+1}}\dot{S}$$

$$=\frac{1}{V(\mathfrak{C}_{\rm II}(n-1))}\int_{S_1\in\mathfrak{C}_{\rm II}(n-1)}\varphi\left(\begin{pmatrix}1 & 0\\ 0 & S_1\end{pmatrix}\right)\dot{S}_1.$$

This proves (2.7.2).

§ 2.8. *A Dirichlet problem of* $\mathfrak{R}_{\rm II}$.

A real-valued function $u(Z)$ is said to be harmonic on the closure of $\mathfrak{R}_{\rm II}$ if it is continuous on the characteristic manifold $\mathfrak{C}_{\rm II}(n)$ and on each $\mathfrak{C}_{\rm II}^{(r)}(r=1,\cdots,n)$ it is harmonic with respect to the coordinate of the base space $\mathfrak{R}_{\rm II}(r)$.

Similar to the proof of Theorem 2.5.1, we solve a corresponding Dirichlet problem for $\mathfrak{R}_{\rm II}$.

Theorem 2. 8. 1. Given a real-valued continuous function $\varphi(S)$ on the characteristic manifold $\mathfrak{C}_{\rm II}$, the Poisson integral

$$u(Z)=\int_{\mathfrak{C}_{\rm II}}\varphi(S)P_{\rm II}(Z,S)\dot{S}$$

gives the unique function which is harmonic on the closure of $\mathfrak{R}_{\rm II}$ and takes the given boundary value $\varphi(S)$ on $\mathfrak{C}_{\rm II}$.

Remark. It can be proved as in § 2.2 that the Poisson kernel satisfies a set of equations

$$\partial_Z(I-Z\bar{Z})\bar{\partial}_Z\circ P_{\rm II}(Z,U)=0,$$

and so does the function defined by the Poisson integral.

§ 2.9. *Harmonic functions in* $\mathfrak{R}_{\rm III}$.

Let Z be an n–rowed skew-symmetric matrix

$$Z=\begin{pmatrix}0 & z_{12} & \cdots & z_{1n}\\ -z_{12} & 0 & \cdots & z_{2n}\\ \cdots & \cdots & \cdots & \cdots\\ -z_{1n} & -z_{2n} & \cdots & 0\end{pmatrix}, \qquad (2.9.1)$$

and let $\mathfrak{R}_{\rm III}$ denote the domain

$$I + Z\bar{Z} > 0 \tag{2.9.2}$$

which is a domain of $\frac{1}{2}\, n(n-1)$ complex variables $z_{12}, \cdots, z_{1n}, z_{23}, \cdots, z_{2n}, \cdots, z_{n-1,n}$.

We introduce the operators

$$\partial_Z = \begin{pmatrix} 0\,, & \frac{\partial}{\partial z_{12}}, & \cdots, & \frac{\partial}{\partial z_{1n}} \\ -\frac{\partial}{\partial z_{12}}, & 0\,, & \cdots, & \frac{\partial}{\partial z_{2n}} \\ \cdots & \cdots & \cdots & \cdots \\ -\frac{\partial}{\partial z_{1n}}, & -\frac{\partial}{\partial z_{2n}}, & \cdots, & 0 \end{pmatrix} \tag{2.9.3}$$

and

$$\Delta_{\mathrm{III}} = \mathrm{tr}\,((I + \bar{Z}Z)\partial_Z(I + Z\bar{Z})\bar{\partial}_Z), \tag{2.9.4}$$

i.e.,

$$\Delta_{\mathrm{III}} = \sum_{\lambda,\mu,\alpha,\beta=1}^{n} \Big(\delta_{\lambda\mu} - \sum_{\sigma=1}^{n} z_{\lambda\sigma}\bar{z}_{\mu\sigma}\Big)\Big(\delta_{\alpha\beta} - \sum_{\gamma=1}^{n} z_{\alpha\gamma}\bar{z}_{\beta\gamma}\Big)\, q_{\lambda\alpha}q_{\mu\beta}\, \frac{\partial^2}{\partial z_{\lambda\alpha}\partial\bar{z}_{\mu\beta}},$$

where $z_{\alpha\beta} = -z_{\beta\alpha}$ and

$$q_{\alpha\beta} = \begin{cases} 0 & \text{for } \alpha = \beta, \\ 1 & \text{for } \alpha \neq \beta. \end{cases} \tag{2.9.5}$$

A real-valued function $u(Z)$ possessing continuous second derivatives is said to be harmonic in $\mathfrak{R}_{\mathrm{III}}$ if it satisfies, in $\mathfrak{R}_{\mathrm{III}}$, the differential equation

$$\Delta_{\mathrm{III}} \circ u(Z) = 0. \tag{2.9.6}$$

Theorem 2.9.1. If $u(Z)$ is harmonic in $\mathfrak{R}_{\mathrm{III}}$, it remains to be harmonic in $\mathfrak{R}_{\mathrm{III}}$ after any transformation of the group Γ^{III} of motions of $\mathfrak{R}_{\mathrm{III}}$.

Proof. It is known that the transformation of Γ^{III} is of the form

$$W = (AZ + B)(-\bar{B}Z + \bar{A})^{-1}, \tag{2.9.7}$$

where

$$A'\bar{B} = -\bar{B}'A, \quad A'\bar{A} - \bar{B}'B = I.$$

Differentiating (2.9.7), we have

$$dW = (-\bar{B}Z + \bar{A})'^{-1}\, dZ(-\bar{B}Z + \bar{A})^{-1}.$$

Let the elements of $(-\bar{B}Z + \bar{A})^{-1}$ be $b_{\alpha\beta}(\alpha, \beta = 1, \cdots, n)$. The previous equality can be written as

$$q_{\alpha\beta}\, dw_{\alpha\beta} = \sum_{\lambda,\mu=1}^{n} b_{\lambda\alpha} q_{\lambda\mu}\, dz_{\lambda\mu} b_{\mu\beta},$$

where $z_{\lambda\mu} = -z_{\mu\lambda}$. Then, we have

$$q_{\lambda\mu} \frac{\partial w_{\alpha\beta}}{\partial z_{\lambda\mu}} = q_{\alpha\beta}(b_{\lambda\alpha} b_{\mu\beta} - b_{\mu\alpha} b_{\lambda\beta}).$$

Hence

$$q_{\lambda\mu} \frac{\partial}{\partial z_{\lambda\mu}} = \sum_{\alpha<\beta} q_{\lambda\mu} \frac{\partial w_{\alpha\beta}}{\partial z_{\lambda\mu}} \frac{\partial}{\partial w_{\alpha\beta}}$$

$$= \sum_{\alpha<\beta} q_{\alpha\beta} b_{\lambda\alpha} b_{\mu\beta} \frac{\partial}{\partial w_{\alpha\beta}} - \sum_{\alpha<\beta} q_{\alpha\beta} b_{\lambda\beta} b_{\mu\alpha} \frac{\partial}{\partial w_{\alpha\beta}}$$

$$= \sum_{\alpha,\beta=1}^{n} b_{\lambda\alpha} \left(q_{\alpha\beta} \frac{\partial}{\partial w_{\alpha\beta}} \right) b_{\mu\beta},$$

that is,

$$\partial_Z = (-\bar{B}Z + \bar{A})^{-1} \partial_W (-\bar{B}Z + \bar{A})'^{-1}.$$

Applying the formula

$$I + W\bar{W} = (-\bar{B}Z + \bar{A})'^{-1}(I + Z\bar{Z})(-B\bar{Z} + A)^{-1},$$

we have

$$\operatorname{tr}((I + \bar{Z}Z)\partial_Z(I + Z\bar{Z})\bar{\partial}_Z) = \operatorname{tr}((I + \bar{W}W)\partial_W(I + W\bar{W})\bar{\partial}_W).$$

This proves the theorem.

The Poisson kernel of $\mathfrak{R}_{\mathrm{III}}$ is equal to

$$P_{\mathrm{III}}(Z, K) = \frac{1}{V(\mathfrak{C}_{\mathrm{III}})} \cdot \frac{\det(I + Z\bar{Z})^{a}}{|\det(I + Z\bar{K})|^{2a}} \tag{2.9.8}$$

(Hua [4], Hua and Look [1]), where

$$a = \begin{cases} \dfrac{n-1}{2} & \text{for even } n, \\[2ex] \dfrac{n}{2} & \text{for odd } n, \end{cases} \tag{2.9.9}$$

and[1)]

1) The constant differs from the original one by the fact given in the footnote on p. 1046.

$$V(\mathfrak{C}_{\mathrm{III}}(n)) = \begin{cases} \displaystyle\prod_{\nu=1}^{\frac{1}{2}n} \omega_{4\nu-3} & \text{for even } n, \\ \displaystyle\frac{1}{2\pi} \prod_{\nu=1}^{\frac{1}{2}(n+1)} \omega_{4\nu-3} = \prod_{\nu=2}^{\frac{1}{2}(n+1)} \omega_{4\nu-3} & \text{for odd } n. \end{cases} \tag{2.9.10}$$

Moreover, the matrix K in (2.9.8) is of the form

$$K = U'F^{(n)}U, \tag{2.9.11}$$

where U is an n-rowed unitary matrix and

$$F^{(n)} = \begin{cases} \begin{pmatrix} 0 & 1 \\ -1 & 0 \end{pmatrix} \dotplus \cdots \dotplus \begin{pmatrix} 0 & 1 \\ -1 & 0 \end{pmatrix} & \text{for even } n, \\ \begin{pmatrix} 0 & 1 \\ -1 & 0 \end{pmatrix} \dotplus \cdots \dotplus \begin{pmatrix} 0 & 1 \\ -1 & 0 \end{pmatrix} \dotplus 0 & \text{for odd } n. \end{cases} \tag{2.9.12}$$

Theorem 2.9.2. After the transformation (2.9.7) of Γ^{III}, the Poisson kernel becomes

$$P_{\mathrm{III}}(Z, K) = P_{\mathrm{III}}(W, J) | \det(B\bar{K} + \bar{A})|^{-a},$$

where

$$J = (AK + B)(-\bar{B}K + \bar{A})^{-1}. \tag{2.9.13}$$

The proof is similar to that of Theorem 2.6.2.

Theorem 2.9.3. The Poisson kernel $P_{\mathrm{III}}(Z, K)$ is harmonic in $\mathfrak{R}_{\mathrm{III}}$ with respect to the variable Z.

Proof. It is sufficient to prove that

$$[\Delta_{\mathrm{III}} \circ P_{\mathrm{III}}(Z, K)]_{Z=0} = 0.$$

In fact, let $K = (k_{\alpha\beta})$ and then

$$[\Delta_{\mathrm{III}} \circ P_{\mathrm{III}}(Z, K)]_{Z=0} = \left[\sum_{\lambda,\alpha=1}^{n} q_{\lambda\alpha}^2 \frac{\partial^2 P_{\mathrm{III}}(Z, K)}{\partial z_{\lambda\alpha} \partial \bar{z}_{\lambda\alpha}}\right]_{Z=0}$$

$$= \frac{1}{V(\mathfrak{C}_{\mathrm{III}})} \left[\sum_{\lambda,\alpha=1}^{n} q_{\lambda\alpha}^2 \frac{\partial^2}{\partial z_{\lambda\alpha} \partial \bar{z}_{\lambda\alpha}} \det(I + Z\bar{Z})^a \det(I + Z\bar{K})^{-a} \det(I + \bar{Z}K)^{-a}\right]_{Z=0}$$

$$= \frac{1}{V(\mathfrak{C}_{\mathrm{III}})} \left[\sum_{\lambda,\alpha=1}^{n} q_{\lambda\alpha}^2 \frac{\partial^2}{\partial z_{\lambda\alpha} \partial \bar{z}_{\lambda\alpha}} \left\{1 - 2a \sum_{\beta<\nu} |z_{\beta\nu}|^2 + \cdots\right\}\left\{1 + 2a \sum_{\beta<\nu} z_{\beta\nu}\bar{k}_{\beta\nu} + \cdots\right\}\right.$$

$$\left. \times \left\{1 + 2a \sum_{\alpha<\nu} \bar{z}_{\beta\nu} k_{\beta\nu} + \cdots\right\}\right]_{Z=0}$$

$$= \frac{1}{V(\mathfrak{C}_{\mathrm{III}})}\left[(-2a)\sum_{\lambda,\alpha=1}^{n} q_{\lambda\alpha}^2 + 4a^2\sum_{\lambda,\alpha=1}^{n} k_{\lambda\alpha}\bar{k}_{\lambda\alpha}\right]$$

$$= \frac{1}{V(\mathfrak{C}_{\mathrm{III}})}[-2an(n-1) + 4a^2 \operatorname{tr}(K\bar{K}')]. \tag{2.9.14}$$

From (2.9.11) and (2.9.12), we have

$$\operatorname{tr}(K\bar{K}') = \begin{cases} n & \text{for even } n, \\ n-1 & \text{for odd } n. \end{cases}$$

Then, by (2.9.9), it follows that (2.9.14) equals zero.

For any real-valued function $\varphi(K)$ continuous on $\mathfrak{C}_{\mathrm{III}}$, the Poisson integral

$$u(Z) = \int_{\mathfrak{C}_{\mathrm{III}}} \varphi(K) P_{\mathrm{III}}(Z, K)\dot{K}$$

defines a harmonic function in $\mathfrak{R}_{\mathrm{III}}$.

§ 2.10. *The boundary properties of the Poisson integral of* $\mathfrak{R}_{\mathrm{III}}$

More precisely, we use $\mathfrak{R}_{\mathrm{III}}(n)$ and $\mathfrak{C}_{\mathrm{III}}(n)$ to denote $\mathfrak{R}_{\mathrm{III}}$ and $\mathfrak{C}_{\mathrm{III}}$ respectively. Let $\mathfrak{B}_{\mathrm{III}}^{(2r)}$ be the set of all skew-symmetric matrices Z such that $I + Z\bar{Z}$ is positive semi-definite and of rank $\leqslant n-2r$ $\left(r = 1, 2, \cdots, \left[\frac{n}{2}\right]\right)$. Note that the rank of $I + Z\bar{Z}$ has the same parity as n. In fact, there is a unitary matrix U (Hua [1]) such that

$$UZU' = \begin{pmatrix} 0 & \lambda_1 \\ -\lambda_1 & 0 \end{pmatrix} \dotplus \begin{pmatrix} 0 & \lambda_2 \\ -\lambda_2 & 0 \end{pmatrix} \dotplus \cdots \dotplus \begin{pmatrix} 0 & \lambda_r \\ -\lambda_r & 0 \end{pmatrix} \dotplus 0^{(n-2r)}.$$

Then $U(I+Z\bar{Z})\bar{U}' = [1-|\lambda_1|^2, 1-|\lambda_1|^2, \cdots, 1-|\lambda_r|^2, 1-|\lambda_r|^2, 1, \cdots, 1]$. The assertion is therefore true.

Further $\mathfrak{B}_{\mathrm{III}}^{(0)}$ is the closure of $\mathfrak{R}_{\mathrm{III}}$ and $\mathfrak{B}_{\mathrm{III}}^{2[\frac{n}{2}]}$ is equal to $\mathfrak{C}_{\mathrm{III}}$.

We have analogously the following

Theorem 2. 10. 1. We define $\mathfrak{C}_{\mathrm{III}}^{(n-2r)} = \mathfrak{B}_{\mathrm{III}}^{(2r)} - \mathfrak{B}_{\mathrm{III}}^{(2r+2)}$. Then $\mathfrak{C}_{\mathrm{III}}^{(n-2r)}$ is an invariant subspace of $\mathfrak{B}_{\mathrm{III}}^{(0)}$ and is transitive under the transformations of the group Γ^{III}.

Theorem 2. 10. 2. We have a chain of slit spaces

$$\mathfrak{B}_{\mathrm{III}}^{(0)} \supset \mathfrak{B}_{\mathrm{III}}^{(2)} \supset \cdots \supset \mathfrak{B}_{\mathrm{III}}^{2[\frac{n}{2}]-2}.$$

The slit of $\mathfrak{B}_{\mathrm{III}}^{2[\frac{n}{2}]-2}$ is $\mathfrak{C}_{\mathrm{III}}$. More precisely, $\mathfrak{B}_{\mathrm{III}}^{(n-2r)}$ is homeomorphic to

$$\mathfrak{R}_{\text{III}}(n-2r) \times \mathfrak{M}_{\text{III}}^{(2r)},$$

where $\mathfrak{M}_{\text{III}}^{(2r)}$ is a set defined as follows:

Two n-rowed unitary matrices U and V are said to be equivalent if there exist a $2r$-rowed unitary symplectic matrix $P^{(2r)}$ and a unitary matrix $U^{(n-2r)}$ such that

$$U = \begin{pmatrix} P^{(2r)} & 0 \\ 0 & U^{(n-2r)} \end{pmatrix} V.$$

By equivalence, we classify the unitary matrices into classes. The totality of classes defines the set $\mathfrak{M}_{\text{III}}^{(2r)}$.

Proof. It is known (Hua [1]) that each element Z of $\mathfrak{B}_{\text{III}}^{(n-2r)}$ can be expressed as

$$Z = U' \begin{pmatrix} F^{(2r)} & 0 \\ 0 & W^{(n-2r)} \end{pmatrix} U,$$

where $I + W\overline{W} > 0$.

From

$$Z = U_0' \begin{pmatrix} F^{(2r)} & 0 \\ 0 & W_0 \end{pmatrix} U_0$$

we deduce that

$$U_1' \begin{pmatrix} F^{(2r)} & 0 \\ 0 & W \end{pmatrix} U_1 = \begin{pmatrix} F^{(2r)} & 0 \\ 0 & W_0 \end{pmatrix},$$

where $U_1 = UU_0^{-1}$.

Consequently, we have

$$U_1 = \begin{pmatrix} P^{(2r)} & 0 \\ 0 & U^{(n-2r)} \end{pmatrix}, \quad P'FP = F^{(2r)}.$$

The theorem follows.

Theorem 2.10.3. Let $\varphi(K)$ be a continuous real-valued function defined on the characteristic manifold $\mathfrak{C}_{\text{III}}(n)$ of $\mathfrak{R}_{\text{III}}(n)$.

(i) Let

$$Q = U_0' \begin{pmatrix} F^{(2r)} & 0 \\ 0 & Z_0 \end{pmatrix} U_0$$

be an arbitrary point on $\mathfrak{B}_{\text{III}}^{(n-2r)} \left(0 < r < \left[\frac{n}{2}\right]\right)$, then

$$\lim_{Z\to Q}\int_{\mathfrak{C}_{\mathrm{III}}(n)} \varphi(K)P_{\mathrm{III}}(Z, K)\dot{K}$$

$$= \frac{1}{V(\mathfrak{C}_{\mathrm{III}}(n-2r))}\int_{\mathfrak{C}_{\mathrm{III}}(n-2r)} \varphi\left(U_0'\begin{pmatrix} F^{(2r)} & 0 \\ 0 & K_0\end{pmatrix}U_0\right)\frac{\det(I+Z_0\bar{Z}_0)^{a-r}}{|\det(I+Z_0\bar{K}_0)|^{2(a-r)}}\dot{K}_0.$$

The last integral represents a function continuous in $\mathfrak{B}_{\mathrm{III}}^{(n-2r)}$ ($\cong \mathfrak{R}_{\mathrm{III}}(n-2r)\times\mathfrak{M}_{\mathrm{III}}^{(2r)}$) and harmonic with respects to Z_0 in $\mathfrak{R}_{\mathrm{III}}(n-2r)$.

(ii) if $K_0\in\mathfrak{C}_{\mathrm{III}}(n)$, then

$$\lim_{Z\to K_0}\int_{\mathfrak{C}_{\mathrm{III}}(n)} \varphi(K)P_{\mathrm{III}}(Z, K)\dot{K} = \varphi(K_0).$$

Proof. It is sufficient to prove the particular case with $r=1$ and

$$Z=\begin{pmatrix} 0 & \rho \\ -\rho & 0\end{pmatrix}\dotplus 0^{(n-2)},\quad Q=\begin{pmatrix} 0 & 1 \\ -1 & 0\end{pmatrix}\dotplus 0^{(n-2)},$$

that is,

$$\lim_{\rho\to 1}\frac{1}{V(\mathfrak{C}_{\mathrm{III}}(n))}\int_{\mathfrak{C}_{\mathrm{III}}(n)}\varphi(K)\frac{(1-\rho^2)^{2a}}{|1-\rho\bar{k}_{12}|^{4a}}\dot{K}$$

$$= \frac{1}{V(\mathfrak{C}_{\mathrm{III}}(n-2))}\int_{\mathfrak{C}_{\mathrm{III}}(n-2)}\varphi\left(\begin{pmatrix} F^{(2)} & O \\ 0 & K_1\end{pmatrix}\right)\dot{K}_1. \tag{2.10.1}$$

First we consider the case n being even. Now $a=\frac{1}{2}(n-1)$. Let $K\in\mathfrak{C}_{\mathrm{III}}(n)$ and write

$$K=\begin{pmatrix} 0 & k \\ -k' & L\end{pmatrix},\quad L=-L'. \tag{2.10.2}$$

From $K\bar{K}'=I$, we see that

$$k\bar{k}'=1,\ k\bar{L}'=0,\ k'\bar{k}+L\bar{L}'=I^{(n-1)}. \tag{2.10.3}$$

Conversely, for any given vector

$$k=(k_{12},\ \cdots,\ k_{1n})$$

such that $k\bar{k}'=1$, there is a skew-symmetric matrix L satisfying (2.10.3). In fact, there exists a unitary matrix $U=U^{(n-1)}$ such that

$$k=(1,0,\cdots,0)U.$$

The matrix

$$L = U'\begin{pmatrix} 0 & 0 \\ 0 & F^{(n-2)} \end{pmatrix} U$$

satisfies our requirement.

Now we have

$$\frac{1}{V(\mathfrak{C}_{\mathrm{III}}(n))}\int_{\mathfrak{C}_{\mathrm{III}}(n)} \varphi(K)\,\frac{(1-\rho^2)^{n-1}}{|1-\rho\bar{k}_{12}|^{2(n-1)}}\,\dot{K}$$
$$= \cdot \int_{k\bar{k}'=1} \psi(k)\,\frac{(1-\rho^2)^{n-1}}{|1-\rho\bar{k}_{12}|^{2(n-1)}}\,\dot{k},$$

where

$$\psi(k) = \frac{1}{V(\mathfrak{C}_{\mathrm{III}}(n))}\int_{\substack{k\bar{L}'=0 \\ k'\bar{k}+L\bar{L}'=I}} \varphi\left(\begin{pmatrix} 0 & k \\ -k' & L \end{pmatrix}\right)\dot{L}.$$

By Theorem 1.3.2, we have

$$\lim_{\rho\to 1}\frac{1}{V(\mathfrak{C}_{\mathrm{III}}(n))}\int_{\mathfrak{C}_{\mathrm{III}}(n)} \varphi(K)\,\frac{(1-\rho^2)^{n-1}}{|1-\rho\bar{k}_{12}|^{2(n-1)}}\,\dot{K}$$
$$= \omega_{2n-3}\psi(1, 0, \cdots, 0) = \frac{1}{V(\mathfrak{C}_{\mathrm{III}}(n-2))}\int_{\mathfrak{C}_{\mathrm{III}}(n-2)} \varphi\left(\begin{pmatrix} F^{(2)} & 0 \\ 0 & L \end{pmatrix}\right)\dot{L}.$$

Therefore we have (2.10.1) for even n.

Now we consider the case with odd n. The closure of $\mathfrak{R}_{\mathrm{III}}(n)$ can be imbedded into that of $\mathfrak{R}_{\mathrm{III}}(n+1)$ and $\mathfrak{C}_{\mathrm{III}}(n)$ is contained in $\mathfrak{C}_{\mathrm{III}}(n+1)$. In fact, any $U_1 \in \mathfrak{C}_{\mathrm{III}}(n+1)$ can be written in the form (Hua and Look [1])

$$K_1 = \begin{pmatrix} K & U'h' \\ -hU & 0 \end{pmatrix}, \quad K = U'F^{(n)}U,$$

$$h = (0, \cdots, 0, e^{i\theta}).$$

Now, for any given function $\varphi(K)$ continuous on $\mathfrak{C}_{\mathrm{III}}(n)$, we can define a function $\varphi^*(K_1)$ continuous on $\mathfrak{C}_{\mathrm{III}}(n+1)$ such that

$$\varphi^*(K_1) = \varphi(K).$$

According to (2.9.10), we have

$$V(\mathfrak{C}_{\mathrm{III}}(n)) = \frac{1}{2\pi} V(\mathfrak{C}_{\mathrm{III}}(n+1)).$$

Then

$$\frac{1}{V(\mathfrak{C}_{\rm III}(n))}\int_{\mathfrak{C}_{\rm III}(n)}\varphi(K)\frac{\det(I+Z\bar{Z})^{n/2}}{|\det(I+Z\bar{K})|^{n}}\dot{K}$$

$$=\frac{1}{2\pi V(\mathfrak{C}_{\rm III}(n))}\int_0^{2\pi}\int_{\mathfrak{C}_{\rm III}(n)}\varphi^*\left(\begin{pmatrix}K & U'h'\\ -hU & 0\end{pmatrix}\right)\frac{\det(I+Z\bar{Z})^{n/2}}{|\det(I+Z\bar{K})|^{n}}d\theta\dot{K}$$

$$=\frac{1}{V(\mathfrak{C}_{\rm III}(n+1))}\int_{\mathfrak{C}_{\rm III}(n+1)}\varphi^*(K_1)\frac{\det\left(I^{(n+1)}+\begin{pmatrix}Z&0\\0&0\end{pmatrix}\begin{pmatrix}\bar{Z}&0\\0&0\end{pmatrix}\right)^{n/2}}{\left|\det\left(I^{(n+1)}+\begin{pmatrix}Z&0\\0&0\end{pmatrix}\bar{K}_1\right)\right|^{n}}\dot{K}_1.$$

Since $n+1$ is even, we can apply the result just proved to obtain the formula (2.10.1) for odd n.

§2.11. *A Dirichlet problem of* $\mathfrak{R}_{\rm III}$

A real-valued function $u(Z)$ is said to be harmonic on the closure of $\mathfrak{R}_{\rm III}$ if it is continuous on the characteristic manifold $\mathfrak{C}_{\rm III}(n)$, and on each $\mathfrak{B}_{\rm III}^{(n-2r)}$ $\left(r=0,1,\cdots,\left[\frac{n}{2}\right]-1\right)$ it is harmonic with respect to the coordinate of the base space $\mathfrak{R}_{\rm III}(n-2r)$.

Similarly, the solution of the Dirichlet problem of $\mathfrak{R}_{\rm III}$ is given by

Theorem 2.11.1. Given a real-valued continuous function $\varphi(K)$ on the characteristic manifold $\mathfrak{C}_{\rm III}$, the Poisson integral

$$u(Z)=\int_{\mathfrak{C}_{\rm III}}\varphi(K)P_{\rm III}(Z,K)\dot{K}$$

gives the unique function which is harmonic on the closure of $\mathfrak{R}_{\rm III}$ and takes the given boundary value $\varphi(K)$ on $\mathfrak{C}_{\rm III}$.

Chapter III

The Dirichlet Problem in the Hyperbolic Space of Lie-Sphere

§3.1. *Harmonic functions in* $\mathfrak{R}_{\rm IV}$

Let $n\geqslant 2$ and let $\mathfrak{R}_{\rm IV}$ denote the domain

$$1+|zz'|^2-2z\bar{z}'>0,\quad 1-|zz'|>0 \tag{3.1.1}$$

in the space of n complex variables $z=(z_1,\cdots,z_n)$. Let $\mathfrak{C}_{\rm IV}$ denote the characteristic manifold of $\mathfrak{R}_{\rm IV}$, that is, the set of vectors

$$\xi=e^{i\theta}x,\quad xx'=1,\quad 0\leqslant\theta<\pi,$$

where x is a real n-vector. The total volume of $\mathfrak{C}_{\rm IV}$ is equal to

$$V(\mathfrak{C}_{\mathrm{IV}}) = \frac{2\pi^{\frac{n}{2}+1}}{\Gamma\left(\frac{n}{2}\right)}.$$

Since $\mathfrak{R}_{\mathrm{IV}}$ is a transitive domain, we proved previously (Look [1]) that it admits an invariant quadratic differential form

$$dzT^{\mathrm{IV}}\,\overline{dz'},$$

where

$$T^{\mathrm{IV}} = \frac{1}{(1+|zz'|^2-2z\bar{z}')^2}\left[(1+|zz'|^2-2z\bar{z}')I^{(n)}-2\begin{pmatrix} z \\ \bar{z}\end{pmatrix}'\begin{pmatrix} 1-2z\bar{z}' & \overline{zz'} \\ zz' & -1\end{pmatrix}\overline{\begin{pmatrix} z \\ \bar{z}\end{pmatrix}}\right]. \tag{3.1.2}$$

The inversion of T^{IV} is equal to

$$(T^{\mathrm{IV}})^{-1} = (1+|zz'|^2-2z\bar{z}')(I-2\bar{z}'z)+2(z'-zz'\bar{z}')(\bar{z}-\overline{zz'}z). \tag{3.1.3}$$

In fact, it follows from the formal identity about matrices that

$$\begin{aligned}(I-PQ\bar{P}')^{-1} &= I+\sum_{l=1}^{\infty}(PQ\bar{P}')^l = I+P\sum_{l=0}^{\infty}(Q\bar{P}'P)^lQ\bar{P}' \\ &= I+P(I-Q\bar{P}'P)^{-1}Q\bar{P}' = I+P(Q^{-1}-\bar{P}'P)^{-1}\bar{P}'.\end{aligned}$$

Taking

$$Q = \frac{2}{1+|zz'|^2-2z\bar{z}'}\begin{pmatrix} 1-2z\bar{z}' & \overline{zz'} \\ zz' & -1\end{pmatrix}, \quad P = \begin{pmatrix} z \\ \bar{z}\end{pmatrix}',$$

we have

$$\begin{aligned}(T^{\mathrm{IV}})^{-1} &= (1+|zz'|^2-2\bar{z}z')\left\{I^{(n)}+\begin{pmatrix} z \\ \bar{z}\end{pmatrix}'\left[\frac{1}{2}\begin{pmatrix} 1 & \overline{zz'} \\ zz' & -1+2z\bar{z}'\end{pmatrix}\right.\right. \\ &\qquad \left.\left.-\overline{\begin{pmatrix} z \\ \bar{z}\end{pmatrix}}\begin{pmatrix} z \\ \bar{z}\end{pmatrix}'\right]^{-1}\overline{\begin{pmatrix} z \\ \bar{z}\end{pmatrix}}\right\} \\ &= (1+|zz'|^2-2z\bar{z}')I+2\begin{pmatrix} z \\ \bar{z}\end{pmatrix}'\begin{pmatrix} 1 & -\overline{zz'} \\ -zz' & -1+2z\bar{z}'\end{pmatrix}\overline{\begin{pmatrix} z \\ \bar{z}\end{pmatrix}} \\ &= (1+|zz'|^2-2z\bar{z}')\left[I-2\begin{pmatrix} z \\ \bar{z}\end{pmatrix}'\begin{pmatrix} 0 & 0 \\ 0 & 1\end{pmatrix}\overline{\begin{pmatrix} z \\ \bar{z}\end{pmatrix}}\right] \\ &\quad +2\begin{pmatrix} z \\ \bar{z}\end{pmatrix}'\begin{pmatrix} 1 & -\overline{zz'} \\ -zz' & |zz'|^2\end{pmatrix}\overline{\begin{pmatrix} z \\ \bar{z}\end{pmatrix}} \\ &= (1+|zz'|^2-2z\bar{z}')(I-2\bar{z}'z)+2(z'-\overline{zz'}\bar{z}')(\bar{z}-\overline{zz'}z).\end{aligned}$$

Consequently, we have an invariant differential operator

$$\Delta_{\mathrm{IV}} = \bar{\partial}_z (T^{\mathrm{IV}})^{-1} \partial'_z,$$

where $\partial_z = \left(\frac{\partial}{\partial z_1}, \cdots, \frac{\partial}{\partial z_n}\right)$ and the convention of §2.1 holds also here. More precisely,

$$\begin{aligned}\Delta_{\mathrm{IV}} &= (1 + |zz'|^2 - 2z\bar{z}')\bar{\partial}_z(I - 2\bar{z}'z)\partial'_z \\ &\quad + 2\bar{\partial}_z(z' - zz'\bar{z}')(\bar{z} - \bar{z}\bar{z}'z)\partial'_z \\ &= (1 + |zz'|^2 - 2z\bar{z}')\left(\sum_{\alpha=1}^{n} \frac{\partial^2}{\partial z_\alpha \partial \bar{z}_\alpha} - 2\sum_{\alpha,\beta=1}^{n} z_\alpha \bar{z}_\beta \frac{\partial^2}{\partial z_\alpha \partial \bar{z}_\beta}\right) \\ &\quad + 2\sum_{\alpha,\beta=1}^{n} (\bar{z}_\alpha - \bar{z}\bar{z}' z_\alpha)(z_\beta - zz'\bar{z}_\beta)\frac{\partial^2}{\partial z_\alpha \partial \bar{z}_\beta}.\end{aligned} \tag{3.1.4}$$

A real-valued function $u(z)$ possessing continuous second derivatives is said to be harmonic in $\mathfrak{R}_{\mathrm{IV}}$ if it satisfies the partial differential equation

$$\Delta_{\mathrm{IV}} \circ u(z) = 0.$$

Obviously, we have

Thecrem 3.1.1. If $u(z)$ is harmonic in $\mathfrak{R}_{\mathrm{IV}}$, the function obtained from $u(z)$ by a transformation of the group Γ^{IV} of motions of $\mathfrak{R}_{\mathrm{IV}}$ remains to be harmonic.

It is known (Hua [2]) that the transformation of the group Γ^{IV} is of the form

$$w = \left\{\left[\left(\frac{zz'+1}{2}, i\frac{zz'-1}{2}\right)A' + zB'\right]\binom{1}{i}\right\}^{-1}\left\{\left(\frac{zz'+1}{2}, i\frac{zz'-1}{2}\right)C' + zD'\right\}, \tag{3.1.5}$$

where $A = A^{(2)}$, $B = B^{(2,n)}$, $C = C^{(n,2)}$, $D = D^{(n)}$ are real matrices satisfying

$$AA' - BB' = I^{(2)},\ CC' - DD' = -I^{(n)},\ AC' = BD',\ \det A > 0. \tag{3.1.6}$$

The transformation (3.1.5) which carries the point $t = (t_1, \cdots, t_n)$ in $\mathfrak{R}_{\mathrm{IV}}$ into the point $0 = (0, \cdots, 0)$ can be written as

$$w = \left\{\left[\left(\frac{zz'+1}{2}, i\frac{zz'-1}{2}\right) - zT'\right]A'\binom{1}{i}\right\}^{-1}\left\{z - \left(\frac{zz'+1}{2}, i\frac{zz'-1}{2}\right)T\right\}D', \tag{3.1.7}$$

where

$$A'A = (I - TT')^{-1},\ D'D = (I - T'T)^{-1},\ \det A > 0 \tag{3.1.8}$$

and

$$T = 2\begin{pmatrix} tt' + 1 & i(tt' - 1) \\ \bar{t}\bar{t}' + 1 & -i(\bar{t}\bar{t}' - 1) \end{pmatrix}^{-1}\begin{pmatrix} t \\ \bar{t} \end{pmatrix}. \tag{3.1.9}$$

We now consider the Poisson kernel $P_{\mathrm{IV}}(z, \xi)$ of $\mathfrak{R}_{\mathrm{IV}}$ under the transformation of Γ^{IV}. It is known that (Hua [4])

$$P_{\mathrm{IV}}(z, \xi) = \frac{1}{V(\mathfrak{C}_{\mathrm{IV}})} \cdot \frac{(1 + |zz'|^2 - 2z\bar{z}')^{n/2}}{|1 + zz'\bar{\xi}\bar{\xi}' - 2z\bar{\xi}'|^n}. \tag{3.1.10}$$

We at first prove

Theorem 3.1.2. $\mathfrak{C}_{\mathrm{IV}}$ is invariant under the group Γ^{IV}.

Proof. Evidently $\mathfrak{C}_{\mathrm{IV}}$ is invariant under the group of stability Γ_0^{IV}, the element of which is of the form

$$w = e^{i\psi} z\Gamma, \tag{3.1.11}$$

where Γ is an $n \times n$ real orthogonal matrix and ψ is real. Hence we can restrict ourselves to proving that $\mathfrak{C}_{\mathrm{IV}}$ is invariant under the transformation (3.1.7) where

$$T = \begin{pmatrix} \lambda_1 & 0 & 0 & \cdots & 0 \\ 0 & \lambda_2 & 0 & \cdots & 0 \end{pmatrix}, \quad 1 > \lambda_1 \geqslant \lambda_2 \geqslant 0 \tag{3.1.12}$$

and

$$A = \begin{pmatrix} \frac{1}{\sqrt{1-\lambda_1^2}} & 0 \\ 0 & \frac{1}{\sqrt{1-\lambda_2^2}} \end{pmatrix}, \quad D = \begin{pmatrix} \frac{1}{\sqrt{1-\lambda_1^2}} & 0 \\ 0 & \frac{1}{\sqrt{1-\lambda_2^2}} \end{pmatrix} \dotplus I^{(n-2)}. \tag{3.1.13}$$

Whenever $z = \xi = e^{i\theta} x$, the corresponding point of (3.1.7) is

$$\begin{aligned} w &= \left\{\left[\left(\frac{e^{2i\theta}+1}{2}, i\,\frac{e^{2i\theta}-1}{2}\right) - e^{i\theta} x T'\right] A' \begin{pmatrix} 1 \\ i \end{pmatrix}\right\}^{-1} \\ &\quad \times \left\{e^{i\theta} x - \left(\frac{e^{2i\theta}+1}{2}, i\,\frac{e^{2i\theta}-1}{2}\right) T\right\} D' \\ &= \left\{[(\cos\theta, -\sin\theta) - (\lambda_1 x_1, \lambda_2 x_2)] \begin{pmatrix} \frac{1}{\sqrt{1-\lambda_1^2}} \\ \frac{i}{\sqrt{1-\lambda_2^2}} \end{pmatrix}\right\}^{-1} \\ &\quad \times \{x - (\lambda_1 \cos\theta, -\lambda_2 \sin\theta, 0, \cdots, 0)\} D' \end{aligned}$$

$$= \left\{\frac{\cos\theta - \lambda_1 x_1}{\sqrt{1-\lambda_1^2}} - i\frac{\sin\theta - \lambda_2 x_2}{\sqrt{1-\lambda_2^2}}\right\}^{-1}\left(\frac{x_1 - \lambda_1\cos\theta}{\sqrt{1-\lambda_1^2}}, \frac{x_2 + \lambda_2\sin\theta}{\sqrt{1-\lambda_2^2}}, x_3, \cdots, x_n\right)$$

$$= e^{i\psi} y,$$

where

$$\psi = \arg\left\{\frac{\cos\theta - \lambda_1 x_1}{\sqrt{1-\lambda_1^2}} - i\frac{\sin\theta + \lambda_1 x_2}{\sqrt{1-\lambda_1^2}}\right\}$$

and

$$y = \left|\frac{\cos\theta - \lambda_1 x_1}{\sqrt{1-\lambda_1^2}} - i\frac{\sin\theta + \lambda_2 x_2}{\sqrt{1-\lambda_2^2}}\right|^{-1}\left(\frac{x_1 - \lambda_1\cos\theta}{\sqrt{1-\lambda_1^2}}, \frac{x_2 + \lambda_2\sin\theta}{\sqrt{1-\lambda_2^2}}, x_3, \cdots, x_n\right).$$

Obviously, y is a real vector. It remains to prove $yy' = 1$. In fact,

$$yy' = \frac{\dfrac{(x_1 - \lambda_1\cos\theta)^2}{1-\lambda_1^2} + \dfrac{(x_2 + \lambda_2\sin\theta)^2}{1-\lambda_2^2} + x_3^2 + \cdots + x_n^2}{\dfrac{(\cos\theta - \lambda_1 x_1)^2}{1-\lambda_1^2} + \dfrac{(\sin\theta + \lambda_2 x_2)^2}{1-\lambda_2^2}}.$$

$$= \{(1-\lambda_2^2)[(x_1 - \lambda_1\cos\theta)^2 - (1-\lambda_1^2)x_1^2]$$
$$+ (1-\lambda_1^2)[(x_2 + \lambda_2\sin\theta)^2 - (1-\lambda_2^2)x_2] + (1-\lambda_1^2)(1-\lambda_2^2)\}$$
$$\times \{(1-\lambda_2^2)(\cos\theta - \lambda_1 x_1)^2 + (1-\lambda_1^2)(\sin\theta + \lambda_2 x_2)^2\}^{-1} = 1.$$

The theorem is proved.

Theorem 3.1.3. After the transformation (3.1.7), the Poisson kernel becomes

$$P_{IV}(w, \zeta) = \frac{|1 + tt'\overline{\xi\xi'} - 2t\bar{\xi}'|^n}{(1 + |tt'|^2 - 2t\bar{t}')^{n/2}} P_{IV}(z, \xi),$$

where $\zeta \in \mathfrak{C}_{IV}$ is the point corresponding to the point $\xi \in \mathfrak{C}_{IV}$ under (3.1.7).

Proof. The Cauchy kernel $H_{IV}(z, \bar{\xi})$ of $\mathfrak{R}_{IV}$ is known to be (Hua [4])

$$H_{IV}(z, \bar{\xi}) = \frac{1}{V(\mathfrak{C}_{IV})} \cdot \frac{1}{(1 + zz'\overline{\xi\xi'} - 2z\bar{\xi}')^{n/2}}.$$

Hence the Poisson kernel and the Cauchy kernel satisfy the following relation

$$P_{IV}(z, \bar{\xi}) = \frac{H(z, \bar{\xi})H(\xi, \bar{z})}{H(z, \bar{z})}. \tag{3.1.14}$$

It is known that, after the transformation (3.1.7), the Cauchy kernel suffers

$$H_{\mathrm{IV}}(z, \bar{\xi}) = H_{\mathrm{IV}}(w, \bar{\zeta}) B^{\frac{1}{2}}(z, t) \overline{B^{\frac{1}{2}}(\xi, t)}, \tag{3.1.15}$$

where $B(z, t)$ is the functional determinant of (3.1.7). Hence,

$$P_{\mathrm{IV}}(z, \xi) = P_{\mathrm{IV}}(w, \zeta) |B(\xi, t)|. \tag{3.1.16}$$

If we take $z = t$ in (3.1.15), then the corresponding point is $w = 0$, and (3.1.15) becomes

$$H_{\mathrm{IV}}(t, \bar{\xi}) = H_{\mathrm{IV}}(0, \bar{\zeta}) B^{\frac{1}{2}}(t, t) \overline{B^{\frac{1}{2}}(\xi, t)}. \tag{3.1.17}$$

It is known (Hua [4]) that

$$B(t, t) = \frac{1}{(1 + |tt'|^2 - 2t\bar{t}')^{n/2}}$$

and $H_{\mathrm{IV}}(0, \bar{\zeta}) = \dfrac{1}{V(\mathfrak{C}_{\mathrm{IV}})}$. We obtain from (3.1.17)

$$|B(\xi, t)| = \frac{(1 + |tt'|^2 - 2t\bar{t}')^{n/2}}{|1 + tt'\overline{\xi\xi'} - 2t\bar{\xi}'|^n}.$$

Substituting the above value into (3.1.16), we get the required result.

Theorem 3.1.4. The Poisson kernel $P_{\mathrm{IV}}(z, \xi)$ is harmonic in $\mathfrak{R}_{\mathrm{IV}}$ with respect to the variable z.

Proof. After Theorems 3.1.1 and 3.1.3, it is sufficient to prove that

$$[\Delta_{\mathrm{IV}} \circ P_{\mathrm{IV}}(z, \xi)]_{z=0} = 0.$$

In fact, by (3.1.4), we have

$$[\Delta_{\mathrm{IV}} \circ P_{\mathrm{IV}}(z, \xi)]_{z=0} = \left[\sum_{a=1}^{n} \frac{\partial^2 P_{\mathrm{IV}}(z, \xi)}{\partial z_a \partial \bar{z}_a}\right]_{z=0}$$

$$= \frac{1}{V(\mathfrak{C}_{\mathrm{IV}})} \left[\sum_{a=1}^{n} \frac{\partial^2}{\partial z_a \partial \bar{z}_a} (1 + |zz'|^2 - 2z\bar{z}')^{n/2} (1 + zz'\overline{\xi\xi'} - 2z\bar{\xi}')^{-n/2} \right.$$

$$\left. \times (1 + \bar{z}\bar{z}'\xi\xi' - 2\bar{z}\xi')^{-n/2}\right]_{z=0}$$

$$= \frac{1}{V(\mathfrak{C}_{\mathrm{IV}})} \left[\sum_{a=1}^{n} \frac{\partial^2}{\partial z_a \partial \bar{z}_a} (1 - nz\bar{z}' + \cdots)(1 + nz\bar{\xi}' + \cdots) \right.$$

$$\left. \times (1 + n\bar{z}\xi' + \cdots)\right]_{z=0}$$

$$= \frac{1}{V(\mathfrak{C}_{\mathrm{IV}})} \left[-n \sum_{a=1}^{n} 1 + n^2 \sum_{a=1}^{n} \xi_a \bar{\xi}_a\right] = \frac{1}{V(\mathfrak{C}_{\mathrm{IV}})} [-n^2 + n^2 xx'] = 0.$$

This proves the theorem.

§ 3.2. *The boundary properties*

For the sake of convenience, we take a linear transformation

$$\overset{*}{z}_1 = z_1 - iz_2, \quad \overset{*}{z}_2 = z_1 + iz_2, \quad \overset{*}{z}_\alpha = z_\alpha \quad (\alpha = 3, \cdots, n), \tag{3.2.1}$$

which carries $\mathfrak{R}_{IV}$ onto $\mathfrak{R}_{IV}^*$, the domain defined by

$$1 + |\overset{*}{z}_1\overset{*}{z}_2 + \overset{*}{z}{}_3^2 + \cdots + \overset{*}{z}{}_n^2|^2 - [|\overset{*}{z}_1|^2 + |\overset{*}{z}_2|^2 + 2(|\overset{*}{z}_3|^2 + \cdots + |\overset{*}{z}_n|^2)] > 0, \quad 1 - |\overset{*}{z}_1\overset{*}{z}_2 + \overset{*}{z}{}_3^2 + \cdots + \overset{*}{z}{}_n^2| > 0. \tag{3.2.2}$$

Denote by $\overline{\mathfrak{R}}_{IV}^*$ the closure of $\mathfrak{R}_{IV}^*$ and by $\mathfrak{B}_{IV}^*$ its boundary. The characteristic manifold $\mathfrak{C}_{IV}$ is transformed to be $\mathfrak{C}_{IV}^*$, the points of which can be represented as

$$\overset{*}{\xi} = e^{i\theta}(x_1 - ix_2), \; \overset{*}{\xi}_2 = e^{i\theta}(x_1 + ix_2), \; \overset{*}{\xi}_\alpha = e^{i\theta}x_\alpha, \; (\alpha = 3, \cdots, n), \tag{3.2.3}$$

where θ and $x = (x_1, \cdots, x_n)$ are real numbers with $xx' = 1$.

Let $\overset{*}{\Gamma}{}^{IV}$ and $\overset{*}{\Gamma}{}_0^{IV}$ be the groups corresponding to the groups Γ^{IV} and Γ_0^{IV} respectively. Since any homeomorphic mapping carries the boundary into boundary and $\mathfrak{C}_{IV}^*$ is invariant under $\overset{*}{\Gamma}{}^{IV}$ by Theorem 3.1.2, obviously $\mathfrak{B}_{IV}^* - \mathfrak{C}_{IV}^*$ is invariant under $\overset{*}{\Gamma}{}^{IV}$. Moreover, $\mathfrak{B}_{IV}^* - \mathfrak{C}_{IV}^*$ is transitive under $\overset{*}{\Gamma}{}^{IV}$, i.e.,

Theorem 3.2.1. Any point of $\mathfrak{B}_{IV}^* - \mathfrak{C}_{IV}^*$ can be transformed into the point $(0, 1, 0, \cdots, 0)$ by $\overset{*}{\Gamma}{}^{IV}$.

Proof. Let t_0 be an arbitrary point of $\mathfrak{B}_{IV}^* - \mathfrak{C}_{IV}^*$. According to the Example 3 given in §1.1, we can assume without loss of generality that

$$t_0 = (t_1, 1, 0, \cdots, 0), \quad |t_1| < 1.$$

From (3.1.7) and (3.1.8), we know that the transformation of $\overset{*}{\Gamma}{}^{IV}$ is of the form

$$\overset{*}{w} = \left\{\left[\left(\frac{1}{2}(\overset{*}{z}Q^{-1}Q'^{-1}\overset{*}{z}{}' + 1), \frac{i}{2}(\overset{*}{z}Q^{-1}Q'^{-1}\overset{*}{z}{}' - 1)\right) - \overset{*}{z}Q^{-1}T'\right]A'\begin{pmatrix}1\\ i\end{pmatrix}\right\}^{-1} \times \left\{\overset{*}{z}Q^{-1} - \left(\frac{1}{2}(\overset{*}{z}Q^{-1}Q'^{-1}\overset{*}{z}{}' + 1), \frac{i}{2}(\overset{*}{z}Q^{-1}Q'^{-1}\overset{*}{z}{}' - 1)\right)T\right\}D'Q, \tag{3.2.4}$$

where $Q = \begin{pmatrix}1 & 1\\ -i & i\end{pmatrix} \dotplus I^{(n-2)}$ and

$$T = 2\begin{pmatrix} \overset{*}{t}Q^{-1}Q'^{-1}\overset{*}{t}{}' + 1 & i(\overset{*}{t}Q^{-1}Q'^{-1}\overset{*}{t}{}' - 1) \\ \hline \overline{\overset{*}{t}Q^{-1}Q'^{-1}\overset{*}{t}{}' + 1} & -i(\overline{\overset{*}{t}Q^{-1}Q'^{-1}\overset{*}{t}{}' - 1})\end{pmatrix}^{-1}\begin{pmatrix}\overset{*}{t}Q^{-1}\\ \hline \overline{\overset{*}{t}Q^{-1}}\end{pmatrix} \tag{3.2.5}$$

with $A'A=(I-TT')^{-1}$, $D'D=(I-T'T)^{-1}$, $\det A>0$.

Now we take $\overset{*}{t}=(t_1, 0, \cdots, 0)$. Then

$$T=\frac{1}{2}\begin{pmatrix} t_1+\bar{t}_1, & it_1-i\bar{t}_1, & 0, & \cdots, & 0\\ it_1-i\bar{t}_1, & -t_1-\bar{t}_1, & 0, & \cdots, & 0\end{pmatrix}.$$

Hence

$$TT'=\begin{pmatrix}|t_1|^2 & 0\\ 0 & |t_1|^2\end{pmatrix},$$

and

$$T'T=\begin{pmatrix}|t_1|^2 & 0\\ 0 & |t_1|^2\end{pmatrix}\dotplus O^{(n-2)}.$$

Besides we choose

$$A=\begin{pmatrix}\frac{1}{\sqrt{1-|t_1|^2}} & 0\\ 0 & \frac{1}{\sqrt{1-|t_1|^2}}\end{pmatrix} \text{ and } D=\begin{pmatrix}\frac{1}{\sqrt{1-|t_1|^2}} & 0\\ 0 & \frac{1}{\sqrt{1-|t_1|^2}}\end{pmatrix}\dotplus I^{(n-2)}.$$

After T, A and D are so chosen, we see that the transformation (3.2.4) at the point $\overset{*}{z}=t_0$ is equal to

$$\begin{aligned}\overset{*}{w}&=\left\{\left[\left(\frac{t_1+1}{2}, i\frac{t_1-1}{2}\right)-\frac{1}{2}(t_1+1, it_1-i, 0, \cdots, 0)T'\right]A'\begin{pmatrix}1\\ i\end{pmatrix}\right\}^{-1}\\ &\quad\times\left\{\frac{1}{2}(t_1+1, it_1-i, 0, \cdots, 0)-\frac{1}{2}(t_1+1, it_1-i)T\right\}D'Q\\ &=\left\{\left[(t_1+1, it_1-i)-\frac{1}{2}((t_1+1)(t_1+\bar{t}_1)\right.\right.\\ &\quad\left.\left.-(t_1-1)(t_1-\bar{t}_1), i(t_1+1)(t_1-\bar{t}_1)-i(t_1-1)(t_1+\bar{t}_1))\right]A'\begin{pmatrix}1\\ i\end{pmatrix}\right\}^{-1}\\ &\quad\times\left\{(t_1+1, it_1-i, 0, \cdots, 0)-\frac{1}{2}((t_1+1)(t_1+\bar{t}_1)\right.\\ &\quad-(t_1-1)(t_1-\bar{t}_1), i(t_1+1)(t_1-\bar{t}_1)\\ &\quad\left.-i(t_1-1)(t_1+\bar{t}_1), 0, \cdots, 0)\right\}D'Q\\ &=\left\{\frac{1}{\sqrt{1-|t_1|^2}}(1-|t_1|^2, -i+i|t_1|^2)\begin{pmatrix}1\\ i\end{pmatrix}\right\}^{-1}\\ &\quad\times\left\{\frac{1}{\sqrt{1-|t_1|^2}}(1-|t_1|^2, -i+i|t_1|^2, 0, \cdots, 0)Q\right\}\\ &=(0, 1, 0, \cdots, 0).\end{aligned}$$

The theorem is proved.

By Example 3 in §1.1, we have

Theorem 3.2.2. $\overline{\mathfrak{R}}_{\mathrm{IV}}^{*} \supset \mathfrak{C}_{\mathrm{IV}}^{*}$ form a chain of slit spaces. The slit of $\mathfrak{B}_{\mathrm{IV}}^{*}$ is $\mathfrak{C}_{\mathrm{IV}}^{*}$ and $\mathfrak{B}_{\mathrm{IV}}^{*} - \mathfrak{C}_{\mathrm{IV}}^{*}$ is homeomorphic to the topological product of $X = \{|t| < 1\}$ and $Y = O(n)/O(n-2)$.

Now we consider the Poisson integral of $\mathfrak{R}_{\mathrm{IV}}^{*}$

$$u(\overset{*}{z}) = \int_{\mathfrak{C}_{\mathrm{IV}}^{*}} \varphi(\overset{*}{\xi}) P_{\mathrm{IV}}^{*}(\overset{*}{z}, \overset{*}{\xi}) \overset{*}{\xi}, \tag{3.2.6}$$

where

$$P_{\mathrm{IV}}^{*}(\overset{*}{z}, \overset{*}{\xi}) = \frac{1}{V(\mathfrak{C}_{\mathrm{IV}}^{*})}[1 + |\overset{*}{z}_1\overset{*}{z}_2 + \overset{*}{z}{}_3^2 + \cdots + \overset{*}{z}{}_n^2|^2$$

$$- (|\overset{*}{z}_1|^2 + |\overset{*}{z}_2|^2 + 2|\overset{*}{z}_3|^2 + \cdots + 2|\overset{*}{z}_n|^2)]^{n/2} \times |1 + \{(\overset{*}{z}_1\overset{*}{z}_2 + \overset{*}{z}{}_3^2$$

$$+ \cdots + \overset{*}{z}{}_n^2)(\overline{\overset{*}{\xi}_1}\overline{\overset{*}{\xi}_2} + \overline{\overset{*}{\xi}{}_3^2} + \cdots + \overline{\overset{*}{\xi}{}_n^2}) - (\overset{*}{z}_1\overline{\overset{*}{\xi}_1} + \overset{*}{z}_2\overline{\overset{*}{\xi}_2} + 2\overset{*}{z}_3\overline{\overset{*}{\xi}_3} + \cdots + 2\overset{*}{z}_n\overline{\overset{*}{\xi}_n})|\}^{-n}$$

and

$$V(\mathfrak{C}_{\mathrm{IV}}^{*}) = \frac{2\pi^{\frac{n}{2}+1}}{\Gamma\left(\frac{n}{2}\right)}. \tag{3.2.7}$$

Theorem 3.2.3. Let $\varphi(\overset{*}{\xi})$ be a real-valued function continuous in the characteristic manifold $\mathfrak{C}_{\mathrm{IV}}^{*}$ of $\mathfrak{R}_{\mathrm{IV}}^{*}$.

(i) If $z_0 = (t_1, 1, 0, \cdots, 0)\ A$ is a point of $\mathfrak{B}_{\mathrm{IV}}^{*} - \mathfrak{C}_{\mathrm{IV}}^{*}$ $(A \in \overset{*}{\Gamma}{}_0^{\mathrm{IV}})$, then

$$\lim_{\overset{*}{z} \to z_0} \int_{\mathfrak{C}_{\mathrm{IV}}^{*}} \varphi(\overset{*}{\xi}) P_{\mathrm{IV}}^{*}(\overset{*}{z}, \overset{*}{\xi}) \dot{\overset{*}{\xi}}$$

$$= \frac{1}{\pi} \int_0^{\pi} \varphi((e^{2i\theta}, 1, 0, \cdots, 0)A) \frac{1 - |t_1|^2}{|1 - t_1 e^{-2i\theta}|^2} d\theta. \tag{3.2.8}$$

Notice that the last integral is a harmonic function in the usual sense with respect to the variable t_1.

(ii) If $\overset{*}{\zeta} \in \mathfrak{C}_{\mathrm{IV}}^{*}$, then

$$\lim_{\overset{*}{z} \to \overset{*}{\zeta}} \int_{\mathfrak{C}_{\mathrm{IV}}^{*}} \varphi(\overset{*}{\xi}) P_{\mathrm{IV}}^{*}(\overset{*}{z}, \overset{*}{\xi}) \dot{\overset{*}{\xi}} = \varphi(\overset{*}{\zeta}).$$

Proof. By a method previously used, we should only prove that when $z = (0, r, 0, \cdots, 0)$, $0 \leqslant r < 1$, and $z_0 = (0, 1, 0, \cdots, 0)$ we have

$$\lim_{r \to 1} \int_{\mathfrak{C}_{\mathrm{IV}}^{*}} \varphi(\overset{*}{\xi}) P_{\mathrm{IV}}^{*}((0, r, 0, \cdots, 0), \overset{*}{\xi}) \dot{\overset{*}{\xi}} = \frac{1}{\pi} \int_0^{\pi} \varphi(e^{2i\theta}, 1, 0, \cdots, 0)\, d\theta. \tag{3.2.9}$$

In fact,

$$\int_{\mathfrak{C}_{IV}^*} \varphi(\overset{*}{\xi}) P_{IV}^*((0, r, 0, \cdots, 0), \overset{*}{\xi}) \dot{\overset{*}{\xi}}$$

$$= \frac{1}{V(\mathfrak{C}_{IV}^*)} \int_{xx'=1} \int_0^{\pi} \varphi(\overset{*}{\xi}) \frac{(1 - r^2)^{n/2}}{|1 - re^{-i\theta}(x_1 - ix_2)|^n} d\theta \dot{x}.$$

We make the following change:

$$y_1 - iy_2 = e^{-i\theta}(x_1 - ix_2), \ y_3 = x_3, \cdots, y_n = x_n, \ \theta = \theta.$$

Then

$$\int_{\mathfrak{C}_{IV}^*} \varphi(\overset{*}{\xi}) P_{IV}^*((0, r, 0, \cdots, 0), \overset{*}{\xi}) \dot{\overset{*}{\xi}}$$

$$= \frac{1}{V(\mathfrak{C}_{IV}^*)} \int_{yy'=1} \psi(y) \frac{(1 - r^2)^{n/2}}{|1 - r(y_1 - iy_2)|^n} \dot{y},$$

where

$$\psi(y) = \int_0^{\pi} \varphi(e^{2i\theta}(y_1 - iy_2), y_1 + iy_2, e^{i\theta} y_3, \cdots, e^{i\theta} y_n) d\theta.$$

Applying Theorems 1.3.2, we have

$$\lim_{r \to 1} \frac{1}{V(\mathfrak{C}_{IV}^*)} \int_{yy'=1} \psi(y) \frac{(1 - r^2)^{n/2}}{|1 - r(y_1 - iy_2)|^n} \dot{y} = \frac{\omega_{n-1}}{V(\mathfrak{C}_{IV}^*)} \psi(1, 0, \cdots, 0).$$

This proves formula (3.2.9).

§ 3.3. *A Dirichlet problem of* $\mathfrak{R}_{IV}$

A real-valued function $\varphi(\xi)$ harmonic in $\mathfrak{R}_{IV}$ is said to be harmonic on the closure of $\mathfrak{R}_{IV}$, if it is continuous in $\overline{\mathfrak{R}}_{IV}$ and on $\mathfrak{B}_{IV} - \mathfrak{C}_{IV}$; it is harmonic in the usual sense with respect to the coordinate of $X = \{|t| < 1\}$.

Applying Theorems 3.2.3 and 1.2.2 we solve the corresponding Dirichlet problem of $\mathfrak{R}_{IV}$:

Theorem 3. 3. 1. If a real-valued continuous function $\varphi(\xi)$ is given in the characteristic manifold $\mathfrak{C}_{IV}$ of $\mathfrak{R}_{IV}$, then

$$u(z) = \int_{\mathfrak{C}_{IV}} \varphi(\xi) P_{IV}(z, \xi) \dot{\xi}$$

is the unique function harmonic on the closure of $\mathfrak{R}_{IV}$, which takes the given boundary value $\varphi(\xi)$ on $\mathfrak{C}_{IV}$.

Chapter IV

Applications

In this chapter we give a few applications and remarks which are not too lengthy to be included here.

§4.1. *A convergence theorem in the theory of representations*

Let $\mathfrak{U}(n)$ be the unitary group of order n and let $A_{f_1,\cdots,f_n}(U)$ defined for $U\in\mathfrak{U}(n)$ be the (unitary) representation with the signature $(f_1,f_2,\cdots,f_n)$, where $f_1,\cdots,f_n$ are integers satisfying $f_1\geqslant f_2\geqslant\cdots\geqslant f_n$. Sometimes, for simplicity, we use f to denote $(f_1,f_2,\cdots,f_n)$. Let $N(f)$ be the order of the matrix $A_f(U)$ and let

$$A_f(U)=(a^f_{ij}(U))_{1\leqslant i,j\leqslant N(f)}.$$

After normalization, we let

$$\varphi^f_{ij}(U)=\sqrt{\frac{N(f)}{C}}\,a^f_{ij}(U), \tag{4.1.1}$$

where $C=V(\mathfrak{C}_{\mathrm{I}}(n,n))$ is the total volume of $\mathfrak{U}(n)(=\mathfrak{C}_{\mathrm{I}}(n,n))$ (see §2.2).

We put, in Theorem 2.4.2(ii), $m=n$ and $Z=rV$, and have

$$\lim_{r\to1}\frac{1}{C}\int_{\mathfrak{U}(n)}\varphi(U)\,\frac{(1-r^2)^{n^2}}{|\det(I-rV\bar U')|^n}\,\dot U=\varphi(V). \tag{4.1.2}$$

It is known (Hua [4]) that

$$\det(I-Z\bar U')^{-n}=C\sum_{f_1\geqslant\cdots\geqslant f_n\geqslant0}\ \sum_{i,j=1}^{N(f)}\varphi^f_{ij}(Z)\overline{\varphi^f_{ij}(U)}$$

$$=\sum_{f_1\geqslant\cdots\geqslant f_n\geqslant0}N(f)\,\mathrm{tr}\,(A_f(Z\bar U')), \tag{4.1.3}$$

which converges uniformly in any compact subset of $\mathfrak{R}_{\mathrm{I}}$, in particular, in the closed set

$$rI-Z\bar Z'\geqslant0,\quad 0\leqslant r<1.$$

The expression

$$\varphi^f_{ij}(U)\overline{\varphi^g_{kl}(U)}$$

with $f_1\geqslant\cdots\geqslant f_n\geqslant0$ and $g_1\geqslant\cdots\geqslant g_n\geqslant0$ appears in the representation

$$A_f(U)\times\overline{A_g(U)}.$$

Therefore, $\varphi^f_{ij}(U) \times \overline{\varphi^g_{kl}(U)}$ can be expressed as a finite linear combination of the functions in the sequence (4.1.1). More precisely,

$$a^f_{ij}(U)\overline{a^g_{kl}(U)} = \sum \lambda^h_{st} a^h_{st}(U)$$

with

$$\sum |\lambda^h_{pq}|^2 = 1.$$

Then the Poisson kernel of $\mathfrak{R}_\mathrm{I}(n,n)$ equals

$$\begin{aligned}\frac{1}{C}\frac{\det(I - Z\bar{Z}')^n}{|\det(I - Z\bar{U}')|^{2n}} &= C\det(I - Z\bar{Z}')^n \\ &\quad\times \sum_{f\geqslant 0}\sum_{g\geqslant 0}\sum_{i,j}\sum_{s,t} \varphi^f_{ij}(Z)\overline{\varphi^f_{ij}(U)}\varphi^g_{st}(U)\overline{\varphi^g_{st}(Z)} \\ &= \sum_h \sum_{k,l} \Phi^h_{kl}(Z)\overline{\varphi^h_{kl}(U)}. \end{aligned} \tag{4.1.4}$$

This series converges uniformly in $rI - Z\bar{Z}' \geqslant 0$.

Multiplying (4.1.4) by $\varphi^f_{ij}(U)$ on both sides and integrating term by term, we obtain

$$\Phi^f_{ij}(Z) = \frac{1}{C}\int_{\mathfrak{U}(n)} \varphi^f_{ij}(U)\frac{\det(I - Z\bar{Z}')^n}{|\det(I - Z\bar{U}')|^{2n}}\dot{U}, \tag{4.1.5}$$

which by (4.1.2) becomes

$$\lim_{r\to 1}\Phi^f_{ij}(rU) = \varphi^f_{ij}(U). \tag{4.1.6}$$

Now, let $u(U)$ be an arbitrary continuous function defined in $\mathfrak{U}(n)$. We have formally the Fourier series of $u(U)$:

$$\sum_f \sum_{i,j} a^f_{ij}\varphi^f_{ij}(U), \quad a^f_{ij} = \int_{\mathfrak{U}(n)} u(U)\overline{\varphi^f_{ij}(U)}\dot{U}.$$

Since

$$\begin{aligned}\frac{1}{C}\int_{\mathfrak{U}} u(U)\frac{\det(I - Z\bar{Z}')^n}{|\det(I - Z\bar{U}')|^{2n}}\dot{U} &= \sum_f \sum_{i,j}\Phi^f_{ij}(Z)\int_{\mathfrak{U}} u(U)\,\overline{\varphi^f_{ij}(U)}\dot{U} \\ &= \sum_f \sum_{i,j} a^f_{ij}\Phi^f_{ij}(Z),\end{aligned}$$

which converges uniformly on any compact subset of $\mathfrak{R}_\mathrm{I}$ and

$$u(U) = \lim_{r\to 1}\Big(\sum_f \sum_{i,j} a^f_{ij}\Phi^f_{ij}(rU)\Big), \tag{4.1.7}$$

the right hand side is defined to be the Abel sum of the Fourier series $\sum_f \sum_{i,j} a_{ij}^f \varphi_{ij}^f(U)$.

Theorem 4.1.1. Every continuous function on $\mathfrak{A}(n)$ can be represented by its Fourier series in the sense of the Abel summability.

We shall express $\Phi_{ij}^f(rU)$ in terms of $\varphi_{ij}^f(U)$.

Let

$$B_f(rV) = \sqrt{\frac{C}{N(f)}}\,(\Phi_{ij}^f(rV))_{1\leqslant i,j\leqslant N(f)}. \tag{4.1.8}$$

By (4.1.5) we have

$$B_f(rV) = \frac{1}{C}\int_{\mathfrak{A}} \frac{(1-r^2)^{n^2}}{|\det(I - rV\bar{U}')|^{2n}} A_f(U)\dot{U}.$$

Let W be any unitary matrix. If we change V and U into WV and WU respectively in the above integral, then we have

$$B_f(rWV) = \frac{1}{C}\int_{\mathfrak{A}} \frac{(1-r^2)^{n^2}}{|\det(I - rV\bar{U}')|^{2n}} A_f(WU)\dot{U} = A_f(W)B_f(rV). \tag{4.1.9}$$

Similarly,

$$B_f(rVW) = B_f(rV)A_f(W). \tag{4.1.10}$$

From (4.1.9) and (4.1.10) with $V = I$, we have, for any unitary W,

$$A_f(W)B_f(rI) = B_f(rI)A_f(W).$$

By Schur's lemma, we have immediately

$$B_f(rI) = \rho^f(r)A_f(W),$$

where $\rho^f(r)$ is a function in r alone and

$$\rho^f(r) \to 1 \text{ for } r \to 1. \tag{4.1.11}$$

According to Theorem 4.1.1, we have

Theorem 4.1.2. Let $u(U)$ be a function continuous on $\mathfrak{A}(n)$. For any given $\varepsilon > 0$, there is a number $\delta > 0$ and a positive integer N_0 such that whenever $1-\delta < r < 1$, we have

$$\left| u(U) - \sum_{N_0 \geqslant f_1 \geqslant \cdots \geqslant f_n \geqslant -N_0} \sum_{i,j} a_{ij}^f \rho^f(r)\varphi_{ij}^f(U) \right| < \varepsilon.$$

To guarantee that the Poisson integral

$$\int_{\mathfrak{A}} \varphi(U)P_1(Z, U)\dot{U} \tag{4.1.12}$$

represents a harmonic function in $\mathfrak{R}_{\mathrm{I}}$, it is not necessary to assume that $\varphi(U)$ is continuous. In fact, if we assume that $\varphi(U)$ is integrable (or belongs to L), the integral (4.1.12) still exists in the interior of $\mathfrak{R}_{\mathrm{I}}$ and it has also the expansion

$$\sum_{f}\sum_{i,j} a_{ij}^{f}\Phi_{ij}^{f}(Z), \tag{4.1.13}$$

which converges uniformly for $rI - Z\bar{Z}' \geqslant 0$, where $0 \leqslant r < 1$. It is to be remarked that the assumption about the continuity can be replaced by a local one, i.e., if $\varphi(U)$ is a function continuous at a point U_0, then the Fouries series

$$\sum_{f}\sum_{i,j} a_{ij}^{f}\varphi_{ij}^{f}(U)$$

is Abel summable to $u(U_0)$.

§4.2. *The distribution on unitary group*

To each function which is harmonic in $\mathfrak{R}_{\mathrm{I}}$ (also in $\mathfrak{R}_{\mathrm{II}}$, $\mathfrak{R}_{\mathrm{III}}$ and $\mathfrak{R}_{\mathrm{IV}}$), we define a distribution. More precisely, if we start with a harmonic function

$$u(Z) = \sum_{f_1 \geqslant \cdots \geqslant f_n}\sum_{i,j=1}^{N(f)} a_{ij}^{f}\Phi_{ij}^{f}(Z),$$

the distribution is defined by the formal power series

$$u(U) = \sum_{f_1 \geqslant \cdots \geqslant f_n}\sum_{i,j} a_{ij}^{f}\varphi_{ij}^{f}(U),$$

which may converge or may not.

Given two distributions $u(U)$, $v(U)$, we define a convolution

$$(u(U), \overline{v(U)}) = \sum_{f_1 \geqslant \cdots \geqslant f_n}\sum_{i,j} a_{ij}^{f}\bar{b}_{ij}^{f}$$

if the series converges or is summable in the Abel sense, i.e.,

$$\lim_{r\to 1}\int_{\mathfrak{C}_{\mathrm{I}}} u(U)\overline{v(rU)}\dot{U} = \lim_{r\to 1}\left[\sum_{f}\sum_{i,j} a_{ij}^{f}\bar{b}_{ij}^{f}\rho^{f}(r)\right].$$

The distribution expressed by the Poisson kernel $P_{\mathrm{I}}(rV, U)$ is called the delta function on the unitary group and is denoted by $\delta_V(U)$. We have

$$(u(U), \delta_V(U)) = \lim_{r\to 1}\int_{\mathfrak{C}_{\mathrm{I}}} u(U)P_{\mathrm{I}}(rV, U)\dot{U} = u(V).$$

The detail of the study of the theory of distribution on a compact group and on the homogeneous space will not be given here.

§4.3. *A note to the harmonic functions of real variables*

Let $\mathfrak{R}(m,n)$ be the set of all real $m\times n$ matrices $X=(x_{i\alpha})_{1\leqslant i\leqslant m,\ 1\leqslant\alpha\leqslant n}$ satisfying

$$I - XX' > 0. \tag{4.3.1}$$

Without loss of generality, we always assume $m \leqslant n$.

$\mathfrak{R}=\mathfrak{R}(m,n)$ admits a group $\mathfrak{g}$ of motions, the element of which is the transformation

$$Y = (AX + B)(CX + D)^{-1}, \tag{4.3.2}$$

where $A=A^{(m)}$, $B=B^{(m,n)}$, $C=C^{(n,m)}$, $D=D^{(n)}$ are real matrices satisfying

$$A'A - C'C = I^{(m)},\ A'B = C'D,\ D'D - B'B = I^{(n)}. \tag{4.3.3}$$

It is not hard to see that $CX + D$ is non-singular whenever X belongs to the closure of $\mathfrak{R}$, and

$$(AX + B)(CX + D)^{-1} = (XB' + A')^{-1}(XD' + C'). \tag{4.3.4}$$

Moreover, $\mathfrak{R}$ is transitive under $\mathfrak{g}$, since for any point $X_0\in\mathfrak{R}$ there is a transformation of $\mathfrak{g}$

$$Y = A(X - X_0)(I - X_0'X)^{-1}D^{-1} \tag{4.3.5}$$

with

$$A'A = (I - X_0X_0')^{-1},\ D'D = (I - X_0'X_0)^{-1}, \tag{4.3.6}$$

which carries X_0 into O.

In $\mathfrak{R}$, we can introduce a Riemann metric

$$ds^2 = \operatorname{tr}(dX(I - X'X)^{-1}dX'(I - XX')^{-1}), \tag{4.3.7}$$

which is invariant under $\mathfrak{g}$. If we arrange the pairs of indices $(i\alpha)$ into the order

$$(11), (12), \cdots, (1n), (21), (22), \cdots, (2n), \cdots, (m1), (m2), \cdots, (mn),$$

then the contravariant tensor $g^{(i\alpha)(j\beta)}$ associated to the fundamental tensor $g_{(i\alpha)(j\beta)}$ of the Riemann metric (4.3.7) can be written as

$$g^{(i\alpha)(j\beta)} = \left(\delta_{ij} - \sum_{\gamma=1}^{n} x_{i\gamma}x_{j\gamma}\right)\left(\delta_{\alpha\beta} - \sum_{k=1}^{m} x_{k\alpha}x_{k\beta}\right). \tag{4.3.8}$$

We are going to evaluate the Beltrami operator

$$\Delta = g^{(i\alpha)(j\beta)}\left(\frac{\partial^2}{\partial x_{i\alpha}\partial x_{j\beta}} - \begin{Bmatrix} (k\gamma) \\ (i\alpha)\ (j\beta) \end{Bmatrix} \frac{\partial}{\partial x_{k\gamma}}\right). \tag{4.3.9}$$

Here we use the summation convention. The Latin letters $i, j, \cdots$ run from 1 to m and the Greek letters $\alpha, \beta, \cdots$ run from 1 to n.

For simplicity we denote

$$\begin{aligned} I - XX' &= (a_{ij}),\ (I - XX')^{-1} = (A_{ij}), \\ I - X'X &= (b_{\alpha\beta}),\ (I - X'X)^{-1} = (B_{\alpha\beta}). \end{aligned} \tag{4.3.10}$$

Since

$$\begin{aligned} \frac{\partial g_{(i\alpha)(j\beta)}}{\partial x_{l\lambda}} &= - g_{(i\alpha)(p\mu)} g_{(j\beta)(q\nu)} \frac{\partial g^{(p\mu)(q\nu)}}{\partial x_{l\lambda}} \\ &= g_{(i\alpha)(p\mu)} g_{(j\beta)(q\nu)}[(\delta_{lp}x_{q\lambda} + \delta_{lq}x_{p\lambda})(\delta_{\nu\mu} - x_{s\nu}x_{s\mu}) \\ &\quad + (\delta_{pq} - x_{p\sigma}x_{q\sigma})(\delta_{\lambda\mu}x_{l\nu} + \delta_{\lambda\nu}x_{l\mu})], \end{aligned}$$

we have

$$\begin{aligned} & g^{(i\alpha)(j\beta)} g^{(k\gamma)(l\lambda)} \frac{\partial g_{(i\alpha)(j\beta)}}{\partial x_{l\lambda}} \\ &= g^{(k\gamma)(l\lambda)} g_{(p\mu)(q\nu)}[(\delta_{lp}x_{q\lambda} + \delta_{lq}x_{p\lambda})b_{\mu\nu} + a_{pq}(\delta_{\lambda\mu}x_{l\nu} + \delta_{\lambda\nu}x_{l\mu})] \\ &= a_{kl}b_{\gamma\lambda}A_{pq}B_{\mu\nu}\, [(\delta_{lp}x_{q\lambda} + \delta_{lq}x_{p\lambda})b_{\mu\nu} + a_{pq}(\delta_{\lambda\mu}x_{l\nu} + \delta_{\lambda\nu}x_{l\mu})] \\ &= n(a_{kl}b_{\gamma\lambda}A_{pq}\delta_{lp}x_{q\lambda} + a_{k\lambda}b_{\gamma\lambda}A_{pq}\delta_{lq}x_{p\lambda}) \\ &\quad + m(a_{kl}b_{\gamma\lambda}B_{\mu\nu}\delta_{\lambda\mu}x_{l\nu} + a_{kl}b_{\gamma\lambda}B_{\mu\nu}\delta_{\lambda\nu}x_{l\mu}) \\ &= 2nb_{\gamma\lambda}x_{k\lambda} + 2m\, a_{kl}\, x_{l\gamma}. \end{aligned} \tag{4.3.11}$$

Similarly,

$$\begin{aligned} & 2g^{(i\alpha)(j\beta)} g^{(k\gamma)(l\lambda)} \frac{\partial g_{(i\alpha)(l\lambda)}}{\partial x_{j\beta}} \\ &= 2g^{(i\alpha)(j\beta)} g^{(k\gamma)(l\lambda)} g_{(i\alpha)(p\mu)} g_{(l\lambda)(q\nu)}\, [(\delta_{jp}x_{q\beta} + \delta_{jq}x_{p\beta})b_{\mu\nu} \\ &\quad + a_{pq}(\delta_{\beta\mu}x_{j\nu} + \delta_{\beta\nu}x_{j\mu})] \\ &= 2(m+1)x_{k\mu}b_{\mu\gamma} + 2(n+1)a_{pk}x_{p\gamma}. \end{aligned} \tag{4.1.12}$$

Substituting (4.3.11) and (4.3.12) into (4.3.9), we obtain

$$\begin{aligned} & g^{(i\alpha)(j\beta)} \begin{Bmatrix} (k\gamma) \\ (i\alpha)\ (j\beta) \end{Bmatrix} = (m+1)x_{k\mu}b_{\mu\gamma} + (n+1)a_{kp}x_{p\gamma} - nb_{\gamma\lambda}x_{k\lambda} - ma_{kl}x_{l\gamma} \\ &= (n - m + 1)a_{kl}x_{p\gamma} - (n - m - 1)x_{k\mu}b_{\mu\gamma} \\ &= (n - m + 1)(\delta_{kp} - x_{k\sigma}x_{p\sigma})x_{p\gamma} - (n - m - 1)x_{k\mu}(\delta_{\mu\gamma} - x_{l\mu}x_{l\gamma}) \\ &= 2(\delta_{kp} - x_{k\sigma}x_{p\sigma})x_{p\gamma}. \end{aligned}$$

Hence the Beltrami operator of $\mathfrak{R}$ becomes

$$\Delta = \sum_{i,j=1}^{m} \sum_{\alpha,\beta=1}^{n} \left(\delta_{ij} - \sum_{\gamma=1}^{n} x_{i\gamma}x_{j\gamma}\right)\left(\delta_{\alpha\beta} - \sum_{k=1}^{m} x_{k\alpha}x_{k\beta}\right) \frac{\partial^2}{\partial x_{i\alpha}\partial x_{j\beta}}$$
$$- 2 \sum_{k,p=1}^{m} \sum_{\gamma=1}^{n} x_{p\gamma}\left(\delta_{kp} - \sum_{\sigma=1}^{n} x_{k\sigma}x_{p\sigma}\right) \frac{\partial}{\partial x_{k\gamma}}. \qquad (4.3.9)'$$

If we introduce the matrix operator

$$\partial_X = \begin{Bmatrix} \frac{\partial}{\partial x_{11}} & \frac{\partial}{\partial x_{12}} & \cdots & \frac{\partial}{\partial x_{1n}} \\ \frac{\partial}{\partial x_{21}} & \frac{\partial}{\partial x_{22}} & \cdots & \frac{\partial}{\partial x_{2n}} \\ \cdots & \cdots & \cdots & \cdots \\ \frac{\partial}{\partial x_{m1}} & \frac{\partial}{\partial x_{m2}} & \cdots & \frac{\partial}{\partial x_{mn}} \end{Bmatrix}, \qquad (4.3.13)$$

then

$$\Delta = \operatorname{tr}\left[(I - XX')\partial_X(I - X'X)\partial'_X - 2(I - XX')X\partial'_X\right]. \qquad (4.3.9)''$$

Since Δ is invariant under $\mathfrak{g}$, obviously we have

Theorem 4.3.1. Any harmonic function in $\mathfrak{R}$ remains harmonic after the transformation of $\mathfrak{g}$.

Let $\mathfrak{B}^{(0)}$ be the closure of $\mathfrak{R}$ and $\mathfrak{B}^{(m-r)}$ be the set of matrices X such that $I-XX'$ is positive semi-definite and of rank $\leqslant r$. Then we have

Theorem 4.3.2. $\mathfrak{B}^{(0)} \supset \mathfrak{B}^{(1)} \supset \cdots \supset \mathfrak{B}^{(m-1)}$ form a chain of slit spaces. The slit of $\mathfrak{B}^{(m-1)}$ is $\mathfrak{C} = \mathfrak{C}(m, n)$ which is the set of real $m \times n$ matrices Γ such that

$$\Gamma\Gamma' = I^{(m)}. \qquad (4.3.14)$$

More precisely, the set $\mathfrak{C}^{(r)} = \mathfrak{B}^{(m-r)} - \mathfrak{B}^{(m-r+1)}$ is homeomorphic to the topological product

$$\mathfrak{R}(r, n - m + r) \times \mathfrak{M}^{(m-r)},$$

where $\mathfrak{M}^{(m-r)}$ is defined in the following way: A pair of matrices $(\Gamma_1, \Gamma_2), \Gamma_1 \in O(m), \Gamma_2 \in O(n)$, is said to be equivalent to the pair $(\overset{*}{\Gamma}_1, \overset{*}{\Gamma}_2)$, $\overset{*}{\Gamma}_1 \in O(m)$, $\overset{*}{\Gamma}_2 \in O(n)$, if

$$\Gamma_1 = \begin{pmatrix} A & O \\ O & B \end{pmatrix} \overset{*}{\Gamma}_1, \quad \Gamma_2 = \begin{pmatrix} A & O \\ O & C \end{pmatrix} \overset{*}{\Gamma}_2,$$

where A, B, C are $(m - r) \times (m - r), r \times r, (n - m + r) \times (n - m + r)$

real orthogonal matrices respectively. We identify the equivalent pairs and form a quotient space which we denote by $\mathfrak{M}^{(m-r)}$.

Proof. It should be noticed that the matrix X of $\mathfrak{C}^{(r)}$ can be written as

$$X = \Gamma_1' \begin{pmatrix} I^{(m-r)} & O \\ O & X_1 \end{pmatrix} \Gamma_2,$$

where $\Gamma_1 \in O(m)$, $\Gamma_2 \in O(n)$ and $I^{(r)} - X_1 X_1' > O$. The remaining proof of this theorem is analogous to that of Theorem 2.3.3.

Similar to Theorem 2.3.1, we can prove

Theorem 4.3.3. $\mathfrak{C}^{(r)}$ is invariant under $\mathfrak{g}$ and any point of $\mathfrak{C}^{(r)}$ can be transformed by $\mathfrak{g}$ into the point

$$\begin{pmatrix} I^{(m-r)} & O \\ O & O^{(r,n-m+r)} \end{pmatrix}.$$

A function $u(X)$ is said to be harmonic on the closure of $\mathfrak{R}$, if it is continuous in $\mathfrak{B}^{(0)}$ and on each $\mathfrak{C}^{(r)}(r=1,\cdots,m)$ it is harmonic with respect to the coordinate of $\mathfrak{R}(r, n-m+r)$.

A corresponding Dirichlet problem is solved:

Theorem 4.3.4. If a continuous function $\varphi(\Gamma)$ is given in $\mathfrak{C}$, then the function

$$u(X) = \frac{1}{V(\mathfrak{C}(m,n))} \int_{\mathfrak{C}(m,n)} \varphi(\Gamma) \frac{\det(I - XX')^{\frac{n-1}{2}}}{\det(I - X\Gamma')^{n-1}} \dot{\Gamma}, \tag{4.3.15}$$

$$V(\mathfrak{C}(m,n)) = \prod_{\nu=n-m+1}^{n} \omega_{\nu-1}$$

is the unique function harmonic on the closure of $\mathfrak{R}$, which takes the given boundary value $\varphi(\Gamma)$ on $\mathfrak{C}$.

Proof. (i) At first we prove that $u(X)$ is harmonic in $\mathfrak{R}(m,n)$. It is sufficient to prove that

$$\Delta \cdot \frac{\det(I - XX')^{\frac{n-1}{2}}}{\det(I - X\Gamma')^{n-1}} = 0$$

with respect to X.

Since after the transformation (4.3.2) we have

$$\frac{\det(I - YY')^{\frac{n-1}{2}}}{\det(I - Y\Gamma_1')^{n-1}} = \frac{\det(I - XX')^{\frac{n-1}{2}}}{\det(I - X\Gamma')^{n-1}} \det(B\Gamma' + A)^{n-1},$$

where

$$\Gamma_1 = (A\Gamma_1 + B)(C\Gamma_1 + D)^{-1},$$

it should remain only to prove

$$\left[\Delta\circ\frac{\det(I - XX')^{\frac{n-1}{2}}}{\det(I - X\Gamma')^{n-1}}\right]_{X=0} = 0.$$

In fact, suppose that $\Gamma = (\gamma_{i\alpha})$, then

$$\left[\Delta\circ\frac{\det(I - XX')^{\frac{n-1}{2}}}{\det(I - X\Gamma')^{n-1}}\right]_{X=0} = \left[\sum_{i=1}^{m}\sum_{\alpha=1}^{n}\frac{\partial^2}{\partial x_{i\alpha}\partial x_{i\alpha}}\cdot\frac{\det(I - XX')^{\frac{n-1}{2}}}{\det(I - X\Gamma')^{n-1}}\right]_{X=0}$$

$$= \left[\sum_{i=1}^{m}\sum_{\alpha=1}^{n}\frac{\partial^2}{\partial x_{i\alpha}^2}\left\{1 - \frac{n-1}{2}\sum_{i=1}^{m}\sum_{\alpha=1}^{n}x_{i\alpha}^2 + \cdots\right\}\right.$$

$$\times\left\{1+(n-1)\sum_{i=1}^{m}\sum_{\alpha=1}^{n}x_{i\alpha}\gamma_{i\alpha} - \frac{n-1}{2}\sum_{i,j=1}^{m}\sum_{\alpha,\beta=1}^{n}(x_{i\alpha}\gamma_{i\alpha}x_{i\beta}\gamma_{j\beta} - x_{i\alpha}\gamma_{j\alpha}x_{j\beta}\gamma_{i\beta})\right.$$

$$\left.\left.+\frac{(n-1)n}{2}\sum_{i,j=1}^{m}\sum_{\alpha,\beta=1}^{n}x_{i\alpha}\gamma_{i\alpha}x_{j\beta}\gamma_{j\beta} + \cdots\right\}\right]_{X=0}$$

$$= -\frac{n-1}{2}\sum_{i=1}^{m}\sum_{\alpha=1}^{n}2 - \frac{n-1}{2}\sum_{i=1}^{m}\sum_{\alpha=1}^{n}(\gamma_{i\alpha}\gamma_{i\alpha} - \gamma_{i\alpha}\gamma_{i\alpha})$$

$$+\frac{n(n-1)}{2}\sum_{i=1}^{m}\sum_{\alpha=1}^{n}2\gamma_{i\alpha}\gamma_{i\alpha}$$

$$= -(n-1)mn + n(n-1)m = 0.$$

Hence $u(X)$ is harmonic in $\mathfrak{R}$.

(ii) We want to prove that $u(X)$ in $\mathfrak{C}^{(r)}$ is harmonic with respect to the coordinate of $\mathfrak{R}(r, n-m+r)$; i.e., if $Q=\Gamma_1'\begin{pmatrix}I^{(m-r)} & 0\\ 0 & X_0\end{pmatrix}\Gamma_2$,

$$\lim_{X\to Q}u(X) = \frac{1}{V(\mathfrak{C}(r, n-m+r))}\int_{\mathfrak{C}(r,n-m+r)}\varphi\left(\Gamma_1'\begin{pmatrix}I^{(m-r)} & 0\\ 0 & \Gamma_0\end{pmatrix}\Gamma_2\right)$$

$$\times\frac{\det(I - X_0X_0')^{\frac{n-m+r-1}{2}}}{\det(I - X_0\Gamma_0')^{n-m+r-1}}\dot{\Gamma}_0. \qquad (4.3.16)$$

It is sufficient to prove the particular case that for $r = m-1$, $X_0 = 0$, $\Gamma_1 = I^{(m)}, \Gamma_2 = I^{(n)}$ and $X = \begin{pmatrix}\rho & 0\\ 0 & O\end{pmatrix}$, $(0 \leqslant \rho < 1)$,

$$\lim_{\rho\to1}\frac{1}{V(\mathfrak{C}(m, n))}\int_{\mathfrak{C}(m,n)}\varphi(\Gamma)\frac{(1-\rho^2)^{\frac{n-1}{2}}}{(1-\rho\gamma_{11})^{n-1}}\dot{\Gamma}$$

$$= \frac{1}{V(\mathfrak{C}(m-1, n-1))} \int_{\mathfrak{C}(m-1,n-1)} \varphi\left(\begin{pmatrix} 1 & 0 \\ 0 & \Gamma_0 \end{pmatrix}\right) \dot{\Gamma}_0.$$

Denote $\Gamma = \begin{pmatrix} \gamma \\ \Gamma_1 \end{pmatrix}$, where $\gamma = (\gamma_{11}, \cdots, \gamma_{1n})$ satisfies $\gamma\gamma' = 1$.

Let

$$\psi(\gamma) = \int_{\binom{\gamma}{\Gamma_1}\binom{\gamma}{\Gamma_1}'=I} \varphi\left(\begin{pmatrix} \gamma \\ \Gamma_1 \end{pmatrix}\right) \dot{\Gamma}_1.$$

According to a theorem analogous to Theorem 1.3.2, we can prove

$$\lim_{\rho\to 1} \frac{1}{V(\mathfrak{C}(m, n))} \int_{\mathfrak{C}(m,n)} \varphi(\Gamma) \frac{(1-\rho^2)^{\frac{n-1}{2}}}{(1-\rho\gamma_{11})^{n-1}} \dot{\Gamma}$$

$$= \lim_{\rho\to 1} \frac{1}{V(\mathfrak{C}(m, n))} \int_{\gamma\gamma'=1} \psi(\gamma) \frac{(1-\rho^2)^{\frac{n-1}{2}}}{(1-\rho\gamma_{11})^{n-1}} \dot{\gamma}$$

$$= \frac{\omega_{n-1}}{V(\mathfrak{C}(m, n))} \int_{\mathfrak{C}(m-1,n-1)} \varphi\left(\begin{pmatrix} 1 & 0 \\ 0 & \Gamma_0 \end{pmatrix}\right) \dot{\Gamma}_0$$

$$= \frac{1}{V(\mathfrak{C}(m-1, n-1))} \int_{\mathfrak{C}(m-1,n-1)} \varphi\left(\begin{pmatrix} 1 & 0 \\ 0 & \Gamma_0 \end{pmatrix}\right) \dot{\Gamma}_0.$$

Moreover, according to formula (4.3.16), we know that $u(X)$ takes the given boundary value $\varphi(\Gamma)$. Again, by Theorem 1.2.2, $u(X)$ is the unique solution satisfying the conditions of our theorem.

References

Bergmann, S. [1] 1953 Kernel function and extended class in the theory of functions of complex variables. *Colloque sur les fonc. plu. variables*, 135—157.

Duff, G. F. D. and Spencer, D. C. [1] 1952 *Ann. of Math.* **56**, 128—168.

Hodge, W. V. D. [1] 1952 *The theory and applications of harmonic integrals.* 2nd ed.

Hua, L. K.

[1] 1944 *Amer. J. Math.* **66**, 470—488.

[2] 1946 *Ann. of Math.* **47**, 167—191.

[3] 1957 *Science Record*, New Ser. **1**, 7—9.

[4] 1958 *Harmonic analysis of the classical domains of analytic functions of several complex variables*, Peking (in Chinese).

[5] 1958 *Science Record*, New Ser. **2**, 281—283.

Hua, L. K. and Look, K. H.

[1] 1958 *Science Record*, New Ser. **2,** 19—22.

[2] 1958 *Science Record*, New Ser. **2,** 77—80.

[3] 1958 *Acta Math. Sinica*, **8,** 531—547 (in Chinese).

[4] 1959 *Acta Math. Sinica*, **9** (in Chinese).

[5] 1959 *Acta Math. Sinica*, **9** (in Chinese).

[6] Theory of harmonic functions in classical domains IV. Harmonic functions in the hyperbolic space of Lie-sphere (to appear).

Look, K. H.

[1] 1956 *Unitary geometry in the theory of functions of several complex variables* (in Chinese).

[2] 1959 *Science Record*, New Ser. **3,** 289—294.

Lowdenslager, D. B. [1] 1958 *Ann. of Math.* **67,** 467—484.

Miranda, C. [1] 1955 *Equazion alle derivate parziali di tipo ellittico.*

Mitchell, J. [1] 1955 *Trans. Amer. Soc.* **79,** 401—422.

MISCELLANEOUS

Reprinted from the
JOURNAL OF MATHEMATICS AND PHYSICS
Vol. 15, pp. 249–263, 1936

ON FOURIER TRANSFORMS IN L^p IN THE COMPLEX DOMAIN

BY LOO-KENG HUA AND SHIEN-SIU SHÜ[1]

Introduction. In this paper we are concerned with functions belonging to H^pL^p as introduced by A. C. Offord.[2] In particular, when $p = 2$, the class H^2L^2 is identical with L^2. R. E. A. C. Paley and N. Wiener[3] have obtained some very beautiful results in the case of L^2 on the Fourier transform of a function vanishing exponentially, of a function analytic in a strip, of a function analytic in a half-plane, etc. The purpose of this paper is to extend those results to H^pL^p. A difficulty in making the generalization is that for H^pL^p we have no corresponding Plancherel theorem

$$\int_{-\infty}^{\infty} | F(x) |^2 \, dx = \int_{-\infty}^{\infty} | f(u) |^2 \, du$$

where $F(x)$ is the Fourier transform of $f(u)$. The main results may be summarized as following:

1) If $f(u)$ is a function belonging to L^p in any finite interval and

$$f(x) = \begin{cases} O(e^{-\mu u} K(u)) & (u \to \infty), \\ O(e^{\lambda u} K(u)) & (u \to -\infty) \end{cases}$$

where $K(u)$ is a constant or a function of u belonging to L^p $(-\infty, \infty)$ then $f(u)$ belongs to H^pL^p, $p \geqq 2$ (Theorems 1 and 2).

2) If $F(s)$ is analytic in a strip, if $F(s)$ belongs uniformly to L^p in the strip and if $F(s)$ belongs to H^p on the boundaries of the strip, then for any interior point s, $F(s)$ can be expressed by the associated Cauchy integrals, and $F(s)$ belongs to H^pL^p in the strip (Theorems 3 and 5).

[1] National Tsing Hua University, Peiping, China.

[2] A. C. Offord, On Fourier Transforms III, Trans. Amer. Math. Soc. 38 (1935), pp. 250–266.

[3] Paley and Wiener, Fourier Transforms in the Complex Domain, 1934, Amer. Math. Soc. Colloquium Publication, vol. 19.

3) The conclusions of 2) are valid if we replace the condition that $F(s)$ belongs uniformly to L^p in the strip by the requirement that

$$F(s) = O\left(e^{e^{\rho |t|}}\right)$$

(Theorem of Phragmén-Lindelöf type; Theorem 7).

4) If $F(s)$ is analytic over the right half-plane if $F(s)$ belongs uniformly to L^p or

$$\lim_{r\to\infty} \frac{1}{r} \log | F(re^{\theta i}) | = 0$$

uniformly over the right half-plane and if $F(it)$ belongs to H^p, then for any interior s, $F(s)$ can be expressed by an associated Cauchy integral and $F(s)$ belongs to H^pL^p. (Theorems 8 and 10.)

5) The two following classes of entire functions are identical:

(a) the class of entire functions $F(s)$ belonging to H^pL^p along the real axis and satisfying the condition

$$F(s) = O(e^{A |s|}),$$

(b) the class of all entire functions of the form $F(s) = \int_{-A}^{A} f(u)\, e^{ius}\, du$, where $f(u)$ belongs to L^p over $(-A, A)$ (Theorem 11).

6) If $F(z)$ is an entire function such that

$$\lim_{r\to\infty} \frac{1}{r} \overset{+}{\log} | F(re^{\theta i}) | = 0$$

and does not vanish identically, it can not belong to H^pL^p along any line (Theorem 12).

We are indebted to Prof. N. Wiener for his lectures on Fourier Transforms at Tsing Hua University and we must take this opportunity to express our deep gratitude for his many valuable suggestions.

§1. In this section let us assume $f(u)$ to be a function belonging to L^p in any finite interval, $p \geqq 2$, and

$$f(u) = \begin{cases} O(e^{-\mu u}\, K(u)) & (u \to \infty), \\ O(e^{\lambda u}\, K(u)) & (u \to -\infty) \end{cases} \tag{1.1}$$

where $K(u)$ is a constant or a function belonging to $L^p(-\infty, \infty)$.

THEOREM 1. Let $\delta > 0$ and $-\lambda + \delta \leqq \sigma \leqq \mu - \delta$. Then

$$F(\sigma, t) = \int_{-\infty}^{\infty} f(u)\, e^{\sigma u - itu}\, du \tag{1.2}$$

exists and is bounded uniformly over the strip, and

$$F(\sigma, -t) = F(z), \qquad z = \sigma + it,$$

is an analytic function of z.

Proof. It is not very difficult to prove that $f(u)e^{\sigma u}$ belongs to L^{p_1} where $0 < p_1 \leqq p$. In particular, $p_1 = 1$, we see that (1.2) exists. Furthermore, over $-\lambda + \delta \leqq \sigma \leqq \mu - \delta$, we have

$$\begin{aligned} | F(\sigma, t) | &\leqq \int_{-\infty}^{\infty} | f(u) | e^{\sigma u}\, du \\ &\leqq \text{const.} \left[\int_{-\infty}^{-N} | K(u) | e^{\delta u}\, du + \int_{N}^{\infty} | K(u) | e^{-\delta u}\, du \right] \\ &\quad + \int_{-N}^{0} | f(u) | e^{-\lambda u}\, du + \int_{0}^{N} | f(u) | e^{\mu u}\, du \\ &= \text{const. (Independent of } \sigma \text{ and } t). \end{aligned}$$

Next let us consider the function

$$\begin{aligned} &\int_{-\infty}^{\infty} f(u) \left(\frac{e^{(z+\Delta z)u} - e^{zu}}{\Delta z} \dot{-} ue^{zu} \right) du \\ &= \int_{-\infty}^{\infty} u f(u)\, e^{zu} \left(\frac{e^{\Delta zu} - 1}{\Delta zu} - 1 \right) du \\ (1.3) \quad &= \int_{-\infty}^{0} u f(u)\, e^{\left(z - \frac{\delta}{2}\right)u} e^{\frac{\delta}{2}u} \left(\frac{e^{\Delta zu} - 1}{\Delta zu} - 1 \right) du \\ &\quad + \int_{0}^{\infty} u f(u)\, e^{\left(z + \frac{\delta}{2}\right)u} e^{-\frac{\delta}{2}u} \left(\frac{e^{\Delta zu} - 1}{\Delta zu} - 1 \right) du. \end{aligned}$$

We are now going to prove that when $u < 0$, $| R(\Delta z) | < \frac{\delta}{2}$, we have

$$| G(u, \Delta z) | = \left| e^{\frac{\delta}{2}u} \left(\frac{e^{\Delta zu} - 1}{\Delta zu} - 1 \right) \right| \leqq 3.$$

In fact, when $| \Delta zu | \geqq 1$,

$$| G(u, \Delta z) | \leqq e^{R\left(\frac{\delta}{2} + \Delta z\right)u} + 2e^{\frac{\delta}{2}u} \leqq 3$$

when $|\Delta zu| < 1$

$$| G(u, \Delta z) | \leqq \left| \frac{u\Delta z}{2!} + \frac{(u\Delta z)^2}{3!} + \cdots \right|$$

$$\leqq \frac{1}{2!} + \frac{1}{3!} + \cdots \leqq e - 1 < 3.$$

Similarly when $u > 0$, $| R(\Delta z) | < \frac{\delta}{2}$, we have

$$\left| e^{-\frac{\delta}{2}u} \left(\frac{e^{\Delta zu} - 1}{\Delta zu} - 1 \right) \right| \leqq 3.$$

Using Lebesgue's convergence theorem we have that (1.3) tends to zero as $\Delta z \to 0$. Thus

$$\frac{d}{dz} \int_{-\infty}^{\infty} f(u)\, e^{zu}\, du = \int_{-\infty}^{\infty} f(u)\, u\, e^{zu}\, du.$$

THEOREM 2. Under the same hypotheses as before, we have

$$\int_{-\infty}^{\infty} | F(\sigma, t) |^p \, dt < C$$

where C is a constant independent of σ. Thus, by Offord's criterion of H^p, $f(u)e^{\sigma u}$ belongs to H^pL^p.[4] Consequently $F(\sigma, t)$ belongs to H^pL^p.[5]

Proof. Firstly, let us prove that there exists a number T such that $|t| > T$

$$| F(\sigma, t) | < 1$$

for any σ on $-\lambda + \delta \leqq \sigma \leqq \mu - \delta$. Suppose this be false when $t \to \infty$. Let $t_1 < t_2 < \cdots < t_n \cdots$ be a sequence which tends to infinity. Then for each t_i, there is at least one $\sigma_i(-\lambda + \delta \leqq \delta_i \leqq \mu - \delta)$ and one $\tau_i(> t_i)$ such that

$$| F(\sigma_i, \tau_i) | \geqq 1.$$

Let σ_0 be one of the limiting points of $\{\sigma_i\}$. (It does exist by the Bolzano-Weiertrass theorem.)

[4] Loc. cit. Th. 9.
[5] Loc. cit. Th. 12.

By the Riemann-Lebesque Theorem there exists a number t_0, such that when $t > t_0$, we have

$$| F(\sigma_0, t) | < \tfrac{1}{4}.$$

On the other hand, since $F(\sigma, -t)$ is an analytic function of $\sigma + it$

$$| F(\sigma + \Delta\sigma, t) - F(\sigma, t) | \leqq \left| \int_\sigma^{\sigma+\Delta\sigma} F'_\sigma(\sigma, t)\, dt \right|$$

$$\leqq \int_\sigma^{\sigma+\Delta\sigma} \left| \int_{-\infty}^{\infty} uf(u)\, e^{(\sigma-it)u}\, du \right| dt.$$

The integrand is uniformly bounded over the strip $-\lambda + \delta \leqq \sigma \leqq \mu - \delta$. Hence we can choose δ_0, such that whenever

$$| \sigma - \sigma_0 | < \delta_0$$

we obtain

$$| F(\sigma, t) - F(\sigma_0, t) | < \tfrac{1}{4}$$

for any value of t. Therefore

$$| F(\sigma, t) | < \tfrac{1}{2} \qquad (t > t_0).$$

We can then choose m great enough, such that $| \sigma_m - \sigma_0 | < \delta_0$ and $t_m > t_0$, thus we have

$$F(\sigma_m, \tau_m) | < \tfrac{1}{2}.$$

This is a contradiction.

Now let us consider

$$\int_{-\infty}^{\infty} | F(\sigma, t) |^p\, dt = \left[\int_{-\infty}^{-T} + \int_{-T}^{T} + \int_T^{\infty} \right] | F(\sigma, t) |^p\, dt$$

$$\leqq \int_{-T}^{T} | F(\sigma, t) |^p\, dt + \left[\int_{-\infty}^{-T} + \int_T^{\infty} \right] | F(\sigma, t) |^2\, dt.$$

The first term is easily seen to be bounded, since its integrand is bounded by theorem 1. The second term is uniformly bounded over the given strip by the Plancherel theorem in L^2.

§2. **Theorem 3.** If $F(s)$ is analytic over $-\lambda \leqq \sigma \leqq \mu$, $s = \sigma + it$ and

$$\int_{-\infty}^{\infty} | F(\sigma + it) |^p\, dt < \text{const.}$$

over this region, then when s is an interior point of this region, we have

$$(2.1)\quad F(s) = \frac{1}{2\pi}\int_{-\infty}^{\infty} \frac{F(\mu + iy)}{\mu + iy - s}\, dy - \frac{1}{2\pi}\int_{-\infty}^{\infty} \frac{F(-\lambda + iy)}{-\lambda + iy - s}\, dy.$$

The proof of this theorem is quite similar to that given in N. Wiener and Paley's book (pp. 3–5) this could be done merely by using Hölder inequality instead of Schwarz inequality.

Consequently we have

THEOREM 4. Under the hypotheses of theorem 3, $F(s)$ is bounded over any region $-\lambda + \delta \leqq \sigma \leqq \mu - \delta$.

By the Hölder inequality we obtain that

$$\left| \int_{-\infty}^{\infty} \frac{F(\mu + iy)}{\mu + iy - s}\, dy \right| \leqq \left\{ \int_{-\infty}^{\infty} | F(\mu + iy) |^p\, dy \right\}^{\frac{1}{p}} \left\{ \int_{-\infty}^{\infty} \frac{dy}{| \mu + iy - s |^q} \right\}^{\frac{1}{q}}$$

is bounded since $2 < p < \infty$ and $q = \dfrac{p}{p - 1} > 1$. Similarly, the other term of (2.1) is also bounded.

THEOREM 5. Besides the hypotheses of theorem 3, we assume further that $F(\sigma + it)$ belongs to H^p for $\sigma = \mu$ and $-\lambda$, then there exists a measurable function $f(x)$ such that

$$(2.2)\quad \int_{-\infty}^{\infty} | f(x) |^p e^{p\mu x}\, dx < M; \qquad \int_{-\infty}^{\infty} | f(x) |^p e^{-p\lambda x}\, dx < M,$$

and that over the open interval $-\lambda < \sigma < \mu$, $F(\sigma + it)$ and $f(x)e^{\sigma x}$ are Fourier transforms of each other in L^p. Moreover, for each σ on this open interval

$$\frac{1}{\sqrt{2\pi}} \int_{-\infty}^{\infty} f(u)\, e^{(\sigma+it)u}\, du$$

converges in the ordinary sense to the function $F(\sigma + it)$ everywhere on $-\infty < t < \infty$.

Proof. Since $F(\sigma + it)$ belongs to H^pL^p for $\sigma = \mu$ and $\sigma = -\lambda$,

$$\frac{1}{\sqrt{2\pi}} \int_{-\infty}^{\infty} F(\mu + it)\, e^{-itx}\, dt$$

is summable $(C, 1)$ almost everywhere to a function $f(\mu, x)$ which is the Fourier transform of $F(\mu + it)$ in L^p and therefore it belongs also to H^pL^p. Similarly we define $f(-\lambda, t)$. (Offord, Theorem 12.)

Let us put

$$G(x) = \begin{cases} 0 & x < 0; \\ e^{-\alpha x + itx} & x > 0;\ \alpha > 0. \end{cases}$$

We obtain

$$\frac{1}{\sqrt{2\pi}}\int_{-\infty}^{\infty} G(x)\, e^{ixy}\, dx = \frac{1}{\sqrt{2\pi}}\int_0^{\infty} e^{ixy-(\alpha-it)x}\, dx$$
$$= \frac{1}{\sqrt{2\pi}}\frac{1}{\alpha - it - iy} = g(y)$$

$g(y)$ is bounded in any finite range. By theorems 1 and 2, $G(x)$ belongs to H^qL^q, $q = \dfrac{p}{p-1}$. By Offord's theorem 1,

$$\frac{1}{\sqrt{2\pi}}\int_{-\infty}^{\infty} \frac{e^{-ixy}}{\alpha - i(t+y)}\, dy = \begin{cases} 0 & , x < 0, \\ e^{(-\alpha+it)x}, & x > 0. \end{cases} \tag{c, 1}$$

Again by the use of Offord's theorem 3, when $-\lambda + \delta \leqq \sigma \leqq \mu - \delta$, and for any value of t we have

$$\int_0^{\infty} f(\mu, x)\, e^{(\sigma-\mu)x}\, e^{itx}\, dx = \frac{1}{\sqrt{2\pi}}\int_{-\infty}^{\infty} \frac{F(\mu + iy)}{\mu + iy - (\sigma + it)}\, dy$$

by putting $\alpha = \mu - \sigma$. It is easily seen that the two integrals of both sides converge in the ordinary sense. Similarly,

$$\int_{-\infty}^{0} f(-\lambda, x)\, e^{(\sigma+\lambda)x}\, e^{itx}\, dx = \frac{1}{\sqrt{2\pi}}\int_{-\infty}^{\infty} \frac{F(-\lambda + iy)}{-\lambda + iy - (\sigma + it)}\, dy.$$

By theorem 3, we get

$$\begin{aligned} F(s) &= \frac{1}{\sqrt{2\pi}}\int_{-\infty}^{0} f(-\lambda, x)\, e^{(\sigma+\lambda)x}\, e^{itx}\, dx \\ &\quad + \frac{1}{\sqrt{2\pi}}\int_0^{\infty} f(\mu, x)\, e^{(\sigma-\mu)x}\, e^{itx}\, dx = \frac{1}{\sqrt{2\pi}}\int_{-\infty}^{\infty} f(x)\, e^{\sigma x}\, dx \end{aligned} \tag{2.3}$$

where

$$f(x) = \begin{cases} f(-\lambda, x)\, e^{\lambda x}, & (x < 0), \\ f(\mu, x)\, e^{-\mu x} & (x > 0) \end{cases} \tag{2.4}$$

and (2.3) converges everywhere in t in the ordinary sense. By theorem 1, for $-\lambda < \sigma < \mu$, $f(x)\, e^{\sigma x}$ and $F(s)$ belong to L^pH^p, Moreover

$$\begin{aligned} \int_{-\infty}^{\infty} |f(x)|^p\, e^{p\mu x}\, dx &= \int_{-\infty}^{0} |f(-\lambda, x)|^p\, e^{p(\mu+\lambda)x}\, dx + \int_0^{\infty} |f(\mu, x)|^p\, dx \\ &\leqq \int_{-\infty}^{0} |f(-\lambda, x)|^p\, dx + \int_0^{\infty} |f(\mu, x)|^p\, dx \\ &< \text{const.} \end{aligned}$$

Similarly

$$\int_{-\infty}^{\infty} | f(x) |^p e^{-p\lambda x} dx < \text{const.}$$

§3. Theorems of the Phragmén-Lindelöf type.

THEOREM 6. If $F(\sigma + it)$ be analytic over $-\lambda \leqq \sigma \leqq \mu$, and $F(\sigma + it)$ belongs to H^pL^p for $\sigma = -\lambda$ and $\sigma = \mu$, $p \geqq 2$, and if

$$| F(\sigma + it) | < M$$

uniformly in the strip $-\lambda \leqq \sigma \leqq \mu$, then the conclusions of theorems 3 and 5 are valid.

Proof. The conclusion of theorem 3 is easily verified, just as shown in Paley and Wiener's book, p. 9.

In the proof of theorem 5, we have only used the facts that the conclusions of theorem 3 is valid and that $F(\sigma + it)$ belongs to H^pL^p for $\sigma = -\lambda$ and $\sigma = \mu$. These are true in the present case.

THEOREM 7. If $F(s)$ is an analytic function of s over the strip $-\lambda \leqq \sigma \leqq \mu$, if $F(s)$ belongs to H^pL^p $(p \geqq 2)$ for $\sigma = -\lambda$ and $\sigma = \mu$, and if

$$| F(\sigma + it) | = O\left(e^{e^{\rho | t |}}\right) \qquad (-\lambda \leqq \sigma \leqq \mu)$$

where $\rho < \pi/(\lambda + \mu)$, then the conclusions of theorems 3, 4 and 5 are valid.

Proof. Let us consider the function

$$F_\epsilon(\sigma + it) = \frac{1}{\epsilon} \int_t^{t+\epsilon} F(\sigma + it)\, dt$$

which satisfies the following four properties:

(1) $$F_\epsilon(\sigma + it) = O\left(e^{e^{\rho' | t |}}\right) \qquad (-\lambda < \sigma < \mu),$$

(2) $$| F_\epsilon(\mu + it) | \leqq \frac{1}{\epsilon^{1-\frac{1}{q}}} \left[\int_t^{t+\epsilon} | F(\mu + it) |^p\, dt\right]^{\frac{1}{p}},$$

$$| F_\epsilon(-\lambda + it) \leqq \frac{1}{\epsilon^{1-\frac{1}{q}}} \left[\int_t^{t+\epsilon} | F(-\lambda + it) |^p\, dt\right]^{\frac{1}{p}}, \qquad q = \frac{p-1}{p}.$$

Thus, for each $\epsilon > 0$, $F_\epsilon(\sigma + it)$ is uniformly bounded over $-\lambda \leqq \sigma \leqq \mu$ by the classical Phragmén-Lindelöf theorem.

(3) $F_\epsilon(\mu + it)$ and $F_\epsilon(-\lambda + it)$ belong to L^p uniformly in ϵ. Since by the Hölder inequality, we obtain

$$\int_a^b \left| \frac{1}{\epsilon} \int_t^{t+\epsilon} F(\mu + iy)\, dy \right|^p dt \leqq \frac{1}{\epsilon^{p-\frac{p}{q}}} \int_a^b dt \int_t^{t+\epsilon} | F(\mu + iy) |^p\, dy$$

$$= \frac{1}{\epsilon} \int_{a\epsilon+}^{b} dy \int_{y-\epsilon}^{y} | F(\mu + iy) |^p\, dt + \frac{1}{\epsilon} \int_a^{a+\epsilon} dy \int_a^y | F(\mu + iy) |^p\, dt$$

$$+ \frac{1}{\epsilon} \int_b^{b+\epsilon} dy \int_{y-\epsilon}^{b} | F(\mu + iy) |^p\, dt < \int_a^{b+\epsilon} | F(\mu + iy) |^p\, dy$$

$$\leqq \int_{-\infty}^{\infty} | F(\mu + iy) |^p\, dy.$$

Similarly, $F_\epsilon(-\lambda + it)$ belongs to L^p uniformly in ϵ.

(4) For each $\epsilon > 0$, $F_\epsilon(\mu + it)$ and $F_\epsilon(-\lambda + it)$ belong to H^p. Let us put

$$F(\mu + it) = \frac{1}{\sqrt{2\pi}} \int_{-\infty}^{\infty} f(\mu, x)\, e^{ixt}\, dx \qquad (C, 1)$$

where $f(\mu, x)$ belongs to H^pL^p. Therefore

$$F(\mu + it) - \frac{1}{\sqrt{2\pi}} \int_{-w}^{w} \left(1 - \frac{|x|}{w}\right) f(\mu, x)\, e^{itx}\, dx = o(1)$$

as $w \to \infty$. Thus we have

$$F_\epsilon(\mu + it) - \frac{1}{\sqrt{2\pi}} \int_{-w}^{w} \left(1 - \frac{|x|}{w}\right) f(\mu, x)\, \frac{e^{i\epsilon x} - 1}{i\epsilon x}\, e^{ixt}\, dx = o(1)$$

almost everywhere in t. Hence

$$F_\epsilon(\mu + it) = \frac{1}{\sqrt{2\pi}} \int_{-\infty}^{\infty} f(\mu, x)\, \frac{e^{i\epsilon x} - 1}{i\epsilon x}\, e^{ixt}\, dx.$$

The right integral converges everywhere to a function which is finite everywhere and belongs to L in every finite range, then by Offord's theorems 8 and 5, we obtain that the function

$$\frac{f(\mu, x)}{i\epsilon x} (e^{i\epsilon x} - 1)$$

belongs to L_p^* and $F_\epsilon(\mu + it)$ belongs to H^p.

Then by the use of the previous theorem we have

$$F_\epsilon(s) = \frac{1}{2\pi}\int_{-\infty}^{\infty}\frac{F(\mu+iy)}{\mu+iy-s}\,dy - \frac{1}{2\pi}\int_{-\infty}^{\infty}\frac{F(-\lambda+iy)}{-\lambda+iy-s}\,dy,\ -\lambda<\sigma<\mu.$$

On account of the fact that both of the integrals converges uniformly in ϵ, let ϵ tend to zero we have

$$F(s) = \frac{1}{2\pi}\int_{-\infty}^{\infty}\frac{F(\mu+iy)}{\mu+iy-s}\,dy - \frac{1}{2\pi}\int_{-\infty}^{\infty}\frac{F(-\lambda+iy)}{-\lambda+iy-s}\,dy.$$

Thus, together with the conditions that $F(\mu+it)$ and $F(-\lambda+it)$ belong to H^pL^p, we obtain the conclusions of theorems 4 and 5.

§4. The Fourier transform of a function in a half-plane.

THEOREM 8. If $F(\sigma+it)$ is analytic for $\sigma \geqq 0$ and $F(it)$ belongs to H^p, and if

$$\int_{-\infty}^{\infty} |F(\sigma+it)|^p\,dt < \text{const.}$$

for $0 \leqq \sigma < \infty$, then the formula

$$F(s) = -\frac{1}{2\pi}\int_{-\infty}^{\infty}\frac{F(iy)}{iy-s}$$

holds for $R(s) > 0$. Moreover, there exists a function $f(x)$ which vanishes almost everywhere for $x > 0$ and which belongs to $L^p(-\infty, 0)$ such that for $\sigma > 0$

$$F(\sigma+it) = \int_{-\infty}^{0} f(x)\,e^{x(\sigma+it)}\,dx$$

where the integral converges everywhere in t in the ordinary sense. $F(\sigma+it)$ and $f(x)e^{\sigma x}$ are Fourier transforms of each other in L^p for each $\sigma > 0$, and both of them belong to H^pL^p.

Proof. By theorem 3, we establish the formula

$$F(s) = -\frac{1}{2\pi}\int_{-\infty}^{\infty}\frac{F(iy)}{iy-s}\,dy.$$

since

$$\left|\int_{-\infty}^{\infty}\frac{F(\mu+iy)}{\mu+iy-s}\,dy\right| \leqq \text{const.}\left\{\int_{-\infty}^{\infty}\frac{dy}{|\mu+iy-s|^q}\right\}^{\frac{1}{q}} = o(1)$$

as $\mu \to \infty$.

Now, just as we have done on theorem 5, we can prove that

$$F(s) = \frac{1}{\sqrt{2\pi}} \int_{-\infty}^{0} f(0, x)\, e^{\sigma x + itx}\, dx$$

where $f(0, x)$ is the Fourier transform of $F(it)$ in L^p.

We define

$$f(x) = \begin{cases} f(0, x) & (x < 0), \\ 0 & (x > 0). \end{cases}$$

Therefore

$$\int_{-\infty}^{\infty} | f(x) |^p \, dx = \int_{-\infty}^{0} | f(0, x) |^p \, dx < \infty.$$

And by theorems 1 and 2, $f(x)\, e^{\sigma x}$ belongs to $H^p L^p$. So is $F(\sigma + it)$ for each $\sigma > 0$.

THEOREM 9. If $F(s)$ is analytic over $0 \leqq \sigma < \infty$, if

$$| F(s) | < M$$

uniformly over $\sigma \geqq 0$, and if $F(it)$ belongs to $H^p L^p$, then the conclusions of theorem 8 are valid.

Proof. Just as we have done before, for $\sigma > 0$ we have

$$(4.1) \qquad F(s) = \frac{1}{2\pi} \lim_{B \to \infty} \int_{-B}^{B} \frac{F(\mu + iy)}{\mu + iy - s}\, dy - \frac{1}{2\pi} \int_{-\infty}^{\infty} \frac{F(iy)}{iy - s}\, dy$$

where μ is any number greater than σ. Let us consider

$$(4.2) \qquad \begin{aligned} \frac{1}{2\pi i} \int_{B}^{B+1} dA \Bigg[\int_{-Ai}^{\mu - Ai} + \int_{\mu - Ai}^{\mu + Ai} + \int_{\mu + Ai}^{Ai} + \int_{Ai}^{i(t+\rho)} \\ + \int_{i(t-\rho)}^{-Ai} \Bigg] \frac{F(z)}{z - it}\, dz + \frac{1}{2\pi i} \int_{-\frac{\pi}{2}}^{\frac{\pi}{2}} F(\rho e^{\theta i} + it)\, i d\theta = 0 \end{aligned}$$

where $\rho e^{\theta i} = z - it$. In the first place,

$$\begin{aligned} \frac{1}{2\pi} \int_{-\frac{\pi}{2}}^{\frac{\pi}{2}} F(\rho e^{\theta i} + it)\, d\theta = \frac{1}{2\pi} \int_{-\frac{\pi}{2}}^{\frac{\pi}{2}} F(it)\, dt \\ + \frac{\rho}{2\pi} \int_{-\frac{\pi}{2}}^{\frac{\pi}{2}} \frac{F(\rho e^{i\theta} + it) - F(it)}{\rho e^{i\theta}}\, e^{\theta i}\, d\theta = \frac{F(it)}{2} + O(\rho) \end{aligned}$$

as ρ tends to zero since $F(z)$ is analytic at $z = it$.

As before, let B tend to infinity then (4.2) becomes

$$\frac{1}{2\pi}\left\{\lim_{B\to\infty}\int_{-\infty}^{B}\frac{F(\mu+iy)}{\mu+iy-it}\,dy-\left[\int_{-B}^{t-\rho}+\int_{t+\rho}^{B}\right]\frac{F(iy)}{i(y-t)}\,dy\right\} + \tfrac{1}{2}F(iy)+O(\rho)=0. \tag{4.3}$$

Since $F(iy)$ belongs to L^p, by a theorem due to Riesz,[6] when ρ tends to zero, the integral

$$\frac{1}{2\pi}\int_{-\infty}^{\infty}\frac{F(iy)}{y-t}\,dy$$

converges almost everywhere in t to a function $g(t)$ which belongs to L^p (in the sense of the principal value of Cauchy). Again,

$$\lim_{B\to\infty}\int_{-B}^{B}\left[\frac{F(\mu+iy)}{\mu+iy-s}-\frac{F(\mu+it)}{\mu+iy-it}\right]dy \leqq \text{const.}\int_{-\infty}^{\infty}\frac{dy}{|\mu+iy-s\,|\,|\mu+iy-it|}=o(1) \tag{4.4}$$

as μ tends to infinity.

Together with (4.1), (4.3) and (4.4), we obtain

$$F(\sigma+it)=\text{const.}-\frac{1}{2\pi}\int_{-\infty}^{\infty}\frac{F(iy)\,dy}{iy-\sigma-it}=g(t)-\tfrac{1}{2}F(it) -\frac{1}{2\pi}\int_{-\infty}^{\infty}\frac{F(iy)}{iy-\sigma-it}\,dy. \tag{4.5}$$

Because $g(t)-F(it)$ belongs to L^p, the constant in (4.5) must vanish and we have the formula

$$F(\sigma+it)=-\frac{1}{2\pi}\int_{-\infty}^{\infty}\frac{F(iy)}{iy-\sigma-it}\,dy.$$

The other parts of the conclusion follow immediately as we have indicated in the preceding theorem.

By an argument similar to the one we have used in theorem 7, we obtain the following theorem of Phragmén-Lindelöf type:

THEOREM 10. If $F(s)$ is analytic over $0\leqq\sigma<\infty$, if

$$\lim_{r\to\infty}\frac{1}{r}\log|F(re^{\theta i})|=0$$

[6] M. Riesz, Sur les fonctions conjuguées, Math. Zeits. 27 (1928), pp. 218–244.

and if $F(it)$ belongs to H^pL^p, then the conclusions of theorem 8 are valid.

§5. Entire functions of exponential type.

THEOREM 11. The two following classes of entire functions are identical:

(1) the class of entire functions $F(z)$ belonging to H^pL^p along real axis and satisfying the condition

$$F(z) = O(e^{A|z|});$$

(2) the class of all entire functions of the form

$$F(z) = \int_{-A}^{A} f(u)\, e^{iuz}\, du$$

where $f(u)$ belongs to L^p over $(-A, A)$.

Proof. The class (2) is contained in the class (1). In the first place, $F(x)$ belongs to H^pL^p by theorems 1 and 2. In the second place

$$\begin{aligned} |F(z)| &\leqq \left\{\int_{-A}^{A} |f(u)|^p\, du\right\}^{\frac{1}{p}} \left\{\int_{-A}^{A} |e^{iquz}|\, du\right\}^{\frac{1}{q}} \\ &= \text{const.} \left\{\int_{-A}^{A} e^{qu|Iz|}\, du\right\}^{\frac{1}{q}} \\ &= \text{const.} \left\{\frac{e^{qA|Iz|} - e^{-qA|Iz|}}{q|Iz|}\right\}^{\frac{1}{q}} \\ &= O(e^{A|z|}). \end{aligned}$$

On the other hand, the class of (1) is of the form (2). In order to prove this, let us consider the function

$$G(z) = \frac{e^{-Az}}{\epsilon} \int_{z}^{z+i\epsilon} F(iw)\, dw$$

which is bounded over the imaginary axis and positive real axis, and which is at most of exponential growth. By the Phragmén-Lindelöf theorem, it is bounded on the right half-plane. As we have done for theorem 7, $G(it)$ belongs to H^pL^p. Hence by theorem 8 there exists $f_\epsilon(x)$ belonging to $L^p(-\infty, 0)$ such that

$$G(z) = \frac{1}{\sqrt{2\pi}} \int_{-\infty}^{0} f_\epsilon(x)\, e^{zx}\, dx \qquad R(z) > 0 \tag{5.1}$$

and $G(z)$ belongs to H^pL^p for each $x = R(z) > 0$. Thus we have

$$F_\epsilon(z) = \frac{1}{\epsilon}\int_z^{z+\epsilon} F(w)dw = iG(iz)\, e^{iAz}$$
$$= \int_{-\infty}^{A} if_\epsilon(x - A)\, e^{izx} \qquad (I(z) < 0).$$

Similarly, we obtain

$$\frac{1}{\epsilon}\int_z^{z+\epsilon} F(w)dw = \int_{-A}^{\infty} i\bar{f}_\epsilon(x + A)\, e^{izx}\, dx \tag{5.2}$$

for $I(z) > 0$.

Let $H(z) = F_\epsilon(-iz)$. Then

$$\begin{aligned} H(z) &= \int_{-A}^{\infty} i\bar{f}_\xi(x + A)\, e^{zx}\, dx && (R(z) > 0),\\ &= \int_{-\infty}^{A} if_\epsilon(x - A)\, e^{zx}\, dx && (R(z) < 0). \end{aligned} \tag{5.3}$$

Since $F(z)$ is at most of exponential growth, so is $H(z)$; because of (5.3), $H(z)$ belongs to H^pL^p for each $\sigma = Rz \neq 0$. Thus by theorem 7 $H(it)$ belongs to H^pL^p in t and there exists a function $g_\epsilon(x)$ such that

$$\frac{1}{\epsilon}\int_z^{z+\epsilon} F(w)\, dw = \int_{-\infty}^{\infty} g_\epsilon(u)\, e^{izu}\, du \qquad (C, 1) \tag{5.4}$$

for any z. Comparing with (2.4) and (5.3) we have

$$g_\epsilon(x) = \begin{cases} i\bar{f}_\epsilon(x + A) & \text{for} \quad -A \leqq x < 0,\\ if_\epsilon(x - A) & \text{for} \quad A \geqq x > 0,\\ 0 & \text{for} \quad |x| > A. \end{cases}$$

In particular,

$$\frac{1}{\epsilon}\int_x^{x+\epsilon} F(w)\, dw = \int_{-A}^{A} g_\epsilon(u)\, e^{ixu}\, du.$$

On the other hand since $F(x)$ belongs to H^pL^p, we have

$$F(x) = \int_{-\infty}^{\infty} f(u)\, e^{ixu}\, du, \qquad (C, 1)$$

and

$$\frac{1}{\epsilon}\int_x^{x+\epsilon} F(x)\, dx = \int_{-\infty}^{\infty} f(u)\, \frac{e^{i\epsilon u} - 1}{i\epsilon u}\, e^{ixu}\, du.$$

Hence comparing with (5.4) $f(u)$ must vanish outside $(-A, A)$. Therefore, as ϵ tends to zero, (5.4) becomes

$$F(z) = \int_{-A}^{A} f(u)\, e^{izu}\, du.$$

An immediate corollary is

THEOREM 12. If $F(z)$ is an entire function such that

$$\lim_{r \to \infty} \frac{1}{r} \overset{+}{\log} \mid F(re^{i\theta}) \mid = 0$$

and does not vanish identically, it can not belong to H^pL^p along any line.

If it does, we may take this line to be the real axis. Thus for every $A > 0$, the Fourier transform of $F(x)$ will vanish almost everywhere outside $(-A, A)$ by the preceding theorem. Hence $F(z)$ must vanish identically and this contradicts our hypothesis.

[*Extracted from the Journal of the London Mathematical Society, Vol.* 14, 1939.]

A REMARK ON THE MOMENT PROBLEM

LOO-KENG HUA*.

Fox's results† on the moment problem can be generalized in the following way.

Let (a, b) be a finite or infinite interval, and let $p(t)$ be a real-valued function such that $t^s p(t)$ is summable in (a, b) $(s = 0, 1, \ldots, 2n)$ for some $n \geqslant 0$. Let $\{P_i(t)\}$ $(i = 0, 1, \ldots, n)$ be a set of polynomials with real coefficients such that (i) the degree of any $P_i(t)$ is at most n, and (ii)

$$\int_a^b p(t)\, P_i(t)\, P_j(t)\, dt = \begin{cases} 0 & (i \neq j), \\ 1 & (i = j). \end{cases}$$

If $p(t)$ is non-negative and not equivalent to zero, such a set of polynomials always exists.

Let
$$P_i(t) = \sum_{j=0}^{n} a_{ij} t^j \quad (i = 0, 1, \ldots, n).$$

We note that the polynomials $P_i(t)$ are linearly independent. For, if

$$\sum_{j=0}^{n} \lambda_j P_j(t) = 0,$$

then
$$\sum_{j=0}^{n} \lambda_j \int_a^b p(t)\, P_i(t)\, P_j(t)\, dt = 0 \quad (i = 0, 1, \ldots, n),$$

i.e., $\lambda_i = 0$ $(i = 0, 1, \ldots, n)$. It follows that the matrix (a_{ij}) is non-singular and has an inverse matrix (b_{ij}), so that

$$t^i = \sum_{j=0}^{n} b_{ij} P_j(t) \quad (i = 0, 1, \ldots, n).$$

THEOREM. *For any set of numbers* $\{c_r\}$ $(r = 0, 1, \ldots, n)$, *the system of equations*

$$\int_a^b p(t) f(t)\, t^r\, dt = c_r \quad (r = 0, 1, \ldots, n) \tag{1}$$

* Received 31 November, 1938; read 19 January, 1939.

† C. Fox, "The solution of a moment problem", *Journal London Math. Soc.*, 13 (1938), 12–14.

has one and only one solution of the form

$$f(t)=\sum_{i=0}^{n} e_i P_i(t),$$

viz., that given by

$$e_i=\sum_{j=0}^{n} a_{ij} c_j \quad (i=0,\ 1,\ \ldots,\ n).$$

(i) Suppose that

$$f(t)=\sum_{i=0}^{n} e_i P_i(t)$$

is a solution of (1). Then

$$\sum_{j=0}^{n} a_{ij} c_j=\sum_{j=0}^{n} a_{ij}\int_a^b p(t) f(t) t^j\,dt=\int_a^b p(t) f(t) P_i(t)\,dt$$

$$=\sum_{k=0}^{n} e_k \int_a^b p(t) P_k(t) P_i(t)\,dt=e_i \quad (i=0,\ 1,\ \ldots,\ n).$$

(ii) Suppose that

$$f(t)=\sum_{i=0}^{n} e_i P_i(t),$$

where

$$e_i=\sum_{j=0}^{n} a_{ij} c_j \quad (i=0,\ 1,\ \ldots,\ n).$$

Then

$$\int_a^b p(t) f(t) t^r\,dt=\sum_{i,\,j,\,k=0}^{n} b_{rk} a_{ij} c_j \int_a^b p(t) P_i(t) P_k(t)\,dt$$

$$=\sum_{i,\,j=0}^{n} b_{ri} a_{ij} c_j=c_r \quad (r=0,\ 1,\ \ldots,\ n),$$

i.e., $f(t)$ is a solution of (1).

The following examples may be mentioned.

(I) If $p(t)=1$, $a=0$, $b=1$, we have Fox's result.

(II) Let $p(t)=e^{-t}$, $a=0$, $b=+\infty$. If $L_n(t)$ is the Laguerre polynomial of degree n, we can take

$$P_i(t)=\frac{1}{i!} L_i(t)=\sum_{j=0}^{i} (-1)^j \binom{i}{j} \frac{1}{j!} t^j.$$

Our results then state that the only solution $f(t)$ of

$$\int_0^\infty e^{-t} f(t) t^r\,dt=c_r \quad (r=0,\ 1,\ \ldots,\ n),$$

such that $f(t)$ is a polynomial of degree not more than n, is

$$f(t) = \sum_{i=0}^{n} \left\{ \sum_{j=0}^{n} (-1)^j \binom{i}{j} \frac{1}{j!} c_j \right\} \frac{1}{i!} L_i(t).$$

Many other examples can easily be constructed.

National Tsing Hua University,
Kunming,
Yunnan, China.

ESTIMATION OF AN INTEGRAL*

LOO-KENG HUA

[*Translated from the Chinese*]

Let $\omega(u)$ be a real function for $u \geqslant 1$, defined by

$$\begin{cases} \omega(u) = u^{-1}, & 1 \leqslant u \leqslant 2; \\ \dfrac{d}{du}(u\omega(u)) = \omega(u-1), & u > 2. \end{cases} \tag{1}$$

In 1937, the Soviet mathematician Buchstab [1] estimated

$$\lim_{u\to\infty} \omega(u) = e^{-\gamma}, \tag{2}$$

where γ is the Euler constant. This result is a consequence of his work on number theory. His original proof was obtained by Brun's method in number theory. Evidently, this is purely an analytical problem, since the definition of $\omega(u)$ and the conclusion are not related to number theory. Thus we ask whether we can use an analytic method to prove this proposition. In 1950, the Dutch mathematician De Bruijn [2] achieved this. But in 1951, Buchstab proved more precisely that

$$|\omega(u) - e^{-\gamma}| < e^{-u(\log u + \log\log u - 1) + O(u \log\log u/\log u)}. \tag{3}$$

by a number-theoretical method.

In this paper I use an analytical method to prove a result more precise than (3).

Lemma 1. *Let*

$$g(x) = \exp\left(-x + \int_0^x \frac{e^{-t}-1}{t}\,dt\right). \tag{4}$$

This function has the properties

(i) $$g(x) = \frac{d}{dx}(xe^x g(x));$$

(ii) $$\int_0^\infty g(x)\,dx = e^{-\gamma}.$$

* Published in Sci. Sinica **4** (1951), 393–402.

Proof. Differentiating

$$\frac{d}{dx}\left(xe^{x}g(x)\right)=\frac{d}{dx}\left(x\exp\int_0^x\frac{e^{-t}-1}{t}\,dt\right)$$

$$=\exp\left(\int_0^x\frac{e^{-t}-1}{t}\,dt\right)\left(1+x\frac{e^{x}-1}{x}\right)=g(x).$$

Integrating from 0 to ∞,

$$\int_0^\infty g(x)\,dx=\left[xe^{x}g(x)\right]_0^\infty=\left[x\exp\left(\int_0^x\frac{e^{-t}-1}{t}\,dt\right)\right]_0^\infty$$

$$=\lim_{x\to\infty}\exp\left(\int_0^x\frac{e^{-t}-1}{t}\,dt+\log x\right)=e^{-\gamma}.$$

Lemma 2. *Let*

$$h(u)=\int_0^\infty g(x)e^{-ux}\,dx. \tag{5}$$

This integral is absolutely convergent for $u\geqslant-1$, *and uniformly convergent for* $u\geqslant-1+\epsilon$. *Also,* $h(u)$ *has the following properties*:

(i) $$h(0)=e^{-\gamma};$$

(ii) $$\lim_{u\to\infty}uh(u)=1;$$

(iii) $$uh'(u-1)+h(u)=0,\qquad \textit{for}\quad u>0;$$

(iv) $$uh(u-1)+\int_{u-1}^{u}h(t)\,dt=1,\qquad \textit{for}\quad u>0.$$

Proof. (i) can be obtained from Abel's Lemma and Lemma 1, (ii). Putting $ux=y$, then

$$uh(u)=\int_0^\infty g\left(\frac{y}{u}\right)e^{-y}\,dy,$$

and (ii) follows.

Differentiating under the sign of integration, by (5)

$$uh'(u-1)=-u\int_0^\infty g(x)e^{-(u-1)x}x\,dx$$

$$=\int_0^\infty xe^{x}g(x)\,d(e^{-ux})$$

$$=xe^{x}g(x)e^{-ux}\Big]_0^\infty-\int_0^\infty e^{-ux}\,d(xe^{x}g(x))$$

$$=-\int_0^\infty e^{-ux}g(x)\,dx=-h(u).$$

(Lemma 1, (i)), and (iii) are proved.

Now we prove

$$q(u)=uh(u-1)+\int_{u-1}^{u}h(t)\,dt$$

is a constant. Differentiating $q(u)$, from (iii) we have

$$q'(u) = h(u-1) + uh'(u-1) + h(u) - h(u-1) = 0.$$

So $q(u)$ is a constant. Let $u \to \infty$. By (ii) we have

$$\lim_{u\to\infty} q(u) = 1.$$

Therefore (iv) is proved.

Lemma 3. *For $u \geqslant 2$, we have the identity*

$$\int_{u-1}^{u} \omega(t)h(t)\,dt + u\omega(u)h(u-1) = e^{-\gamma}.$$

Proof. Denote the left-hand side by $p(u)$. By (1) and Lemma 2, (ii)

$$p'(u) = \omega(u)h(u) - \omega(u-1)\,h(u-1) + \frac{d}{du}(u\omega(u))h(u-1)$$
$$+ u\omega(u)h'(u-1) = 0,$$

so $p(u)$ is a constant. By (1), we have

$$p(2) = \int_1^2 \omega(t)h(t)\,dt + 2\omega(2)h(1)$$
$$= \int_1^2 \frac{h(t)}{t}\,dt + h(1)$$
$$= -\int_1^2 h'(t-1)\,dt + h(1) \qquad \text{(by Lemma 2, (ii))}$$
$$= h(0) = e^{-\gamma} \qquad \text{(by Lemma 1, (i))}.$$

Lemma 4. *Let*

$$W(u) = \omega(u) - e^{-\gamma}; \tag{6}$$

then for $u > 0$ we have

$$W(u) = -\frac{1}{uh(u-1)}\int_{u-1}^{u} W(t)h(t)\,dt. \tag{7}$$

Proof. By Lemma 3 and Lemma 2 (iv),

$$\int_{u-1}^{u} \omega(t)h(t)\,dt + u\omega(u)h(u-1) = e^{-\gamma}\left(\int_{u-1}^{u} h(t)\,dt + uh(u-1)\right).$$

Hence

$$\int_{u-1}^{u} W(t)h(t)\,dt + uh(u-1)W(u) = 0.$$

This is (7).

Let

$$F(u) = |W(u)|.$$

By (7), we have

$$F(u) \leqslant \frac{1}{uh(u-1)} \int_{u-1}^{u} F(t)h(t)\,dt$$

$$= \frac{1}{uh(u-1)} \int_0^1 F(u-1+\vartheta)h(u-1+\vartheta)\,d\vartheta$$

$$\leqslant \frac{1}{u} \int_0^1 F(u-1+\vartheta)\,d\vartheta, \tag{8}$$

(since $h(u)$ is a monotone decreasing function).

Lemma 5. *Suppose $f(u)$ is a positive function, and $f(u)$ satisfies* (8) *for sufficiently large u; then, as $u \to \infty$, we have*

$$F(u) \leqslant e^{-u(\log u + \log\log u + \log\log u/\log u - 1) + O(u/\log u)}.$$

Proof.

1. Let $M(u) = \max_{u \leqslant x \leqslant \infty} F(x)$. From (8) we have

$$M(u) \leqslant \frac{M(u-1)}{u} \leqslant \frac{M(u-2)}{u(u-1)} \leqslant \cdots = O\left(\frac{1}{P(u)}\right),$$

where

$$\log P(u) = \log u + \log(u-1) + \cdots$$

Since $\log x$ is an increasing function,

$$\log P(u) \geqslant \int_1^u \log x\,dx = u(\log u - 1).$$

Hence

$$M(u) = O(e^{-u(\log u - 1)}). \tag{9}$$

2. Let

$$F_1(u) = F(u)e^{u(\log u - 1)}$$

and

$$M_1(u) = \max_{u \leqslant x \leqslant \infty} F_1(u).$$

From (8) we immediately obtain

$$M_1(u) \leqslant \frac{M_1(u-1)}{u} \int_0^1 \frac{e^{u(\log u - 1)}}{e^{(u-1+t)(\log(u-1+t)-1)}}\,dt$$

$$= \frac{M_1(u-1)}{u} \int_0^1 \exp(\Phi(t))\,dt, \tag{10}$$

where

$$\begin{aligned}\Phi(t) &= u(\log u - 1) - (u + t - 1)(\log(u + t - 1) - 1)\\ &\leqslant u(\log u - 1) - (u + t - 1)(\log(u - 1) - 1)\\ &= \log u - 1 + (u - 1)(\log u - \log(u - 1)) - t(\log(u - 1) - 1)\\ &\leqslant \log u - t(\log(u - 1) - 1).\end{aligned}$$

Substituting in (10), we have

$$\begin{aligned}M_1(u) &\leqslant M_1(u - 1)\int_0^1 e^{-t(\log(u-1)-1)}\,dt\\ &\leqslant \frac{M_1(u - 1)}{\log(u - 1) - 1}\,.\end{aligned}$$

Successively using this expression, we obtain

$$M_1(u) = O\left(\frac{1}{P_1(u)}\right),$$

where

$$\begin{aligned}\log P_1(u) &= \log(\log(u - 1) - 1) + \log(\log(u - 2) - 1) + \cdots\\ &\geqslant \int^{u-1} \log(\log t - 1)\,dt\\ &= \left[t\log(\log t - 1) - \int \frac{dt}{\log t - 1}\right]^{u-1}\\ &= u\log\log u - C_1\frac{u}{\log u}\,,\end{aligned}$$

and C_1 is a constant > 1.

Now we have proved

$$F(u) = O(e^{-u(\log u + \log\log u - 1) + C_1 u/\log u}). \tag{11}$$

This result is more accurate than Buchstab's, but we can get a much better result.

3. Let

$$F_2(u) = F_1(u)e^{u\log\log u - C_1 u/\log u},$$

and

$$M_2(u) = \max_{u \leqslant x \leqslant \infty} F_2(x).$$

From (8) we immediately obtain

$$M_2(u) \leqslant \frac{M_2(u - 1)}{u}\int_0^1 \exp(\Phi(t))\,dt, \tag{12}$$

where

$$\Phi(t) - \log u = -\log u + u(\log u + \log\log u - 1) - C_1 u/\log u - (u + t - 1)$$
$$\times (\log(u + t - 1) + \log\log(u + t - 1) - 1) + C_1 \frac{u + t - 1}{\log(u + t - 1)}$$
$$\leqslant (u - 1)\log u + u \log\log u - C_1 u/\log u$$
$$- (u + t - 1)(\log(u - 1) + \log\log(u - 1) - 1) + C_1 \frac{u + t - 1}{\log(u - 1)}$$
$$= (u - 1)\log \frac{u}{u - 1} + \log\log u + (u - 1)\log\left(\frac{\log u}{\log(u - 1)} \right)$$
$$- 1 - C_1\left(\frac{u}{\log u} - \frac{u - 1}{\log(u - 1)} \right)$$
$$- t\left(\log(u - 1) + \log\log(u - 1) - 1 - \frac{C_1}{\log(u - 1)} \right)$$
$$\leqslant \log\log u + (u - 1)\frac{1}{u - 1} + (u - 1)\frac{1}{(u - 1)\log(u - 1)} - 1$$
$$- \frac{C_1}{\log(u - 1)} - t\left(\log u + \log\log u - 1 - \frac{C_1}{\log(u - 1)} \right)$$

(since $\log(1 + x) \leqslant x$)

$$\leqslant \log\log u - t(\log u + \log\log u - 2) + \frac{1}{\log(u - 1)}$$

(for sufficiently large u).
Substituting in (12), we have

$$M_2(u) \leqslant M_2(u - 1)\frac{e^{1/\log(u-1)}\log u}{\log u + \log\log u - 2}$$
$$\leqslant M_2(u - 1)\exp\left(-\log\left(1 + \frac{\log\log u}{\log u} - \frac{2}{\log u} \right) + \frac{1}{\log(u - 1)} \right)$$
$$\leqslant M_2(u - 1)\exp\left(- \frac{\log\log u}{\log u} + \frac{C_2}{\log u} \right)$$

(since $\log(1 + x) \geqslant x - Cx^2$). But

$$\frac{\log\log u}{\log u} - \frac{C_2}{\log u} + \frac{\log\log(u - 1)}{\log(u - 1)} - \frac{C_2}{\log(u - 1)} + \cdots$$
$$\geqslant \int^{u-1} \frac{\log\log t}{\log t}\, dt - C_2 \int^{u} \frac{dt}{\log t}$$
$$\geqslant \frac{u\log\log u}{\log u} - C_3 \frac{u}{\log u}.$$

So we obtain

$$M_2(u) \leqslant e^{u \log\log u/\log u - C_3 u/\log u},$$

that is

$$F(u) \leqslant e^{-u(\log u + \log\log u + \log\log u/\log u - 1) + C_4 u/\log u}.$$

Remark. We can iterate this method, so as to obtain even sharper results.

Combining Lemma 5, Lemma 4 and (6), we have

Theorem:

$$|\omega(u) - e^{-\gamma}| \leqslant e^{-u(\log u + \log\log u + \log\log u/\log u - 1) + O(u/\log u)}.$$

If it suffices to prove merely

$$\lim_{u\to\infty} \omega(u) = e^{-\gamma}.$$

Then we have the following simple method, but need to quote some theorems on the Laplace integral. Let

$$f(s) = \int_0^\infty e^{-us}\, d\alpha(u) = \int_0^\infty e^{-us}\, dw(u+1), \tag{13}$$

where

$$\alpha(u) = \omega(u+1) - 1.$$

Integrating by parts

$$\begin{aligned} f(s) &= s\int_0^\infty e^{-us}\alpha(u)\, du = -1 + s\int_0^\infty e^{-us}\omega(u+1)\, du \\ &= -1 + s\int_0^\infty e^{-us}\, d((u+2)\omega(u+2)). \end{aligned}$$

Changing u to $u-1$, and using (1) we have

$$\begin{aligned} f(s) &= -1 + se^s\int_0^\infty e^{-us}\, d((u+1)\omega(u+1)) \\ &= -1 + se^s\left[\int_0^\infty e^{-us}(u+1)\, d\omega(u+1) + \int_0^\infty e^{-us}\omega(u+1)\, du\right]. \end{aligned} \tag{14}$$

Differentiating under the integral we have

$$\int_0^\infty ue^{-us}\, d\omega(u+1) = -f'(s),$$

and integrating by parts

$$\begin{aligned} \int_0^\infty e^{-us}\omega(u+1)\, du &= -\frac{e^{-us}}{s}\omega(u+1)\Big|_0^\infty + \frac{1}{s}\int_0^\infty e^{-us}\, d\omega(u+1) \\ &= \frac{1}{s} + \frac{1}{s}f(s). \end{aligned}$$

Putting these two expressions into (14), we have

$$f(s) = -1 + se^{s}\Big[f(s) - f'(s) + \frac{1}{s} + \frac{1}{s} f(s)\Big],$$

i.e. the ordinary differential equation of first order

$$f'(s) + \big(s^{-1}(e^{-s} - 1) - 1\big)f(s) = s^{-1}(1 - e^{-s}). \tag{15}$$

Since $\lim_{u \to 0}\alpha(u) = \lim_{u \to 0}(\omega(u+1) - 1) = 0$, it follows from Abel's Theorem (ex. [4], p. 183, Cor. 1c), that

$$\lim_{s \to \infty} f(s) = 0. \tag{16}$$

Solving the differential equation (15) and using condition (16), we obtain

$$\begin{aligned} f(s) &= -e^{s + \int_0^s t^{-1}(1-e^{-t})\,dt} \int_s^\infty e^{-u - \int_0^u t^{-1}(1-e^{-t})\,dt}\, \frac{1 - e^{-u}}{u}\, du \\ &= e^{s + \int_0^s t^{-1}(t-e^{-t})\,dt} \int_s^\infty e^{-u - \int_0^u t^{-1}(1-e^{-t})\,dt}\, d\Big(-u - \int_0^u t^{-1}(1 - e^{-1})\,dt\Big) \\ &\quad + c^{s + \int_0^s t^{-1}(1-e^{-t})\,dt} \int_s^\infty e^{-u - \int_0^u t^{-1}(1-e^{-t})\,dt}\, du \\ &= -1 + e^{s + \int_0^s t^{-1}(1-e^{-t})\,dt} \int_s^\infty e^{-u - \int_0^u t^{-1}(1-e^{-t})\,dt}\, du \\ &= -1 + e^{s + \int_0^s t^{-1}(1-e^{-t})\,dt}\Big[ue^{-\int_0^u t^{-1}(1-e^{-t})\,dt}\Big]_s^\infty \\ &= -1 - se^{s} + e^{-\gamma + s + \int_0^s t^{-1}(1-e^{-t})\,dt}. \end{aligned} \tag{17}$$

It is clear that

$$\lim_{s \to 0} f(s) = -1 + e^{-\gamma}.$$

Then, by a Tauberian Theorem, we obtain

$$\lim_{u \to \infty} \alpha(u) = \lim_{u \to \infty} (\omega(u+1) - 1) = -1 + e^{-\gamma}.$$

This is

$$\lim_{u \to \infty} \omega(u) = e^{-\gamma}.$$

But note that it is necessary to check the Tauberian Condition, i.e. $\int_0^t u\, d\alpha(u) = O(t)$. (ex. [4], p. 187, Thm. 36). Since

$$d\omega(u) = \frac{1}{u}(\omega(u-1) - \omega(u))\, du,$$

then

$$\begin{aligned} \int_0^t u\, d\alpha(u) &= \int_0^t u\, d\omega(u+1) = \int_1^t \big[\omega(u) - \omega(u+1)\big]\, du + O(1) \\ &= -\int_t^{t+1} \omega(u)\, du + O(1) = O(1) \end{aligned}$$

(it is easy to prove $\omega(u) = O(1)$). So the Tauberian Condition is satisfied. The Theorem is now completely proved.

Note that, incidentally, we have found (17) to be the Laplace transform of $\omega(u+1)$.

(Summary)

Let $\omega(u)$ be the function defined by (1) and (2), $f(s)$ be the Laplace transform of $\omega(u+1)$. We proved that $f(s)$ satisfies the differential equation (14) with the initial conditon $\lim_{s\to\infty} f(s) = 0$. Solving the differential equation, we obtain the explicit expression (17). By Tauberian theorem, we deduce that $\lim_{u\to\infty}\omega(u) = e^{-\gamma}$, where γ is the Euler constant. Further from (2), we have

$$\omega'(u) = -\frac{1}{u}\int_{u-1}^{u} \omega^{t}(t)\,dt,$$

we deduce then

$$\omega'(u) = O(e^{-u(\log u + \log\log u + \log\log u/\log u - 1) + O(u/\log u)}).$$

Integrating, we have

$$\omega(u) = e^{-\gamma} + O(e^{-u(\log u + \log\log u + \log\log u/\log u - 1) + O(u/\log u)})$$

which is sharper than a result due to Buchstab and answers a conjecture of De Bruijn. The proof in the text requires no knowledge beyond advanced calculus.

References

[1] Buchstab, A. A., 1937: An asymptotic estimation of a general number-theoretic function. Matem. Sb., 2 (144), 1239–1246.

[2] DeBruijn, N. G., 1950: On the number of uncancelled elements in the sieve of Eratosthenes, Nederl. Akad. Wetensch. Proc., **52**, 803–812; Indagationes Mathematicae, **12**, 247–256.

[3] Buchstab, A. A., 1951: On the estimation of the number of numbers in an arithmetical progression which cannot be divided by "relatively" small primes. Matem. Sb., 28 (70), 166–184.

[4] Widder, D. V., The Laplace Transform.

ON THE CALCULATION OF MINERAL RESERVES AND HILLSIDE AREAS ON CONTOUR MAPS*

HUA LOO-KENG AND WANG YUAN

§1. Introduction

Our geographers, mineralogists and geologists have presented many practical methods for calculating mineral reserves and hillside areas which enable us to engage in further investigations. In the present paper the authors attempt to compare these methods, to interpret their mutual relations and implicit errors, and to present some suggestions.

In mineral-body geometry (see [2]–[4]) one uses the Bauman formula, the frustum formula and the trapezoid formula for the approximate calculation of mineral reserves by divided layers. Suppose the mineral reserve volumes evaluated from these formulas are v, v_1 and v_2 respectively. In this paper we prove that they satisfy the inequality:

$$v \leqslant v_1 \leqslant v_2,$$

and we determine the conditions for taking the equal sign. The authors consider that the comparison of these three formulas should be based mainly on the number of dimensions, or degrees of freedom; consequently, we regard the Bauman formula as less limited than the others.

This paper presents a double-layer formula for the calculation of mineral reserves; this formula is due to our finding a new method for the proof of the Bauman formula. The proof is not only simple but also easy to improve still further. The advantages of our formula are in the consideration of more factors than in the Bauman formula without causing much complication; it also considers more factors than the Sobolevskiĭ formula (which generally uses a double layer formula for calculating the reserves, see [2]–[4]). We recommend it to technicians who engage in the calculation of mineral reserves.

For the calculation of hillside areas, the Volkov method (see [5]–[6]) is often used in geography and the Bauman method is often used in mineral-body geometry (see [1]–[2]). This paper points out that Bauman's method is more accurate than Volkov's, while both methods yield lower results than the exact value. Our paper completely determines those curved surfaces which can be handled with arbitrary accuracy by those two methods. In detail, the error depends on the change of the angle of inclination of the points on the surface. Only if the variation of the angle of

* Published in Acta Math. Sinica, **11** (1961) 1, 29–40. Translated from the Chinese by Hu Dihe.

inclination is small for all points on the surface can the Volkov method yield accurate results; and only if the angles of inclination of points between two adjacent contours differ from each other slightly can the Bauman method yield accurate results. Under other circumstances these two methods may give large errors. Therefore we suggest drawing several rays through the highest point on a contour map. If the surface is approximately a ruled surface, we may evaluate the individual surface areas between adjacent rays and then add them together. In case the variation of the angles of inclination for the surface between adjacent contours and adjacent rays is large, we may evaluate each individual surface area formed by rays and contours, and then add them together. The error of the results thus obtained will be relatively small.

§2. Calculation of mineral reserves

1. The Bauman method. Given a contour map of a mineral reserve with height difference h, the contour shown on the map represents essentially the *cross sectional* area at a fixed altitude. Let us evaluate the body volume between two such planes. The distance between these two planes is the height difference h. We denote the sections enclosed by the lower and the upper contours by A and B respectively (see figure 1; their areas are also represented by A and B). Bauman suggested evaluating the volume v between these two altitudes by

$$v=\left[\frac{1}{2}(A+B)-\frac{T(A,B)}{6}\right]h \tag{1}$$

where $T(A,B)$ is the area of the diagram shown below, and is called the Bauman correction number.

As shown in figure 2, draw the ray OP from the highest point O. The length of the section of the ray between A and B on the map is l. Construct another diagram, figure 3, take point O', then take $O'P'=l$ in the same direction as OP. When P moves one cycle around the contour, P' traces out a diagram whose area is called the Bauman correction number. Because of its dependence on sections A and B, we denote it by $T(A,B)$.

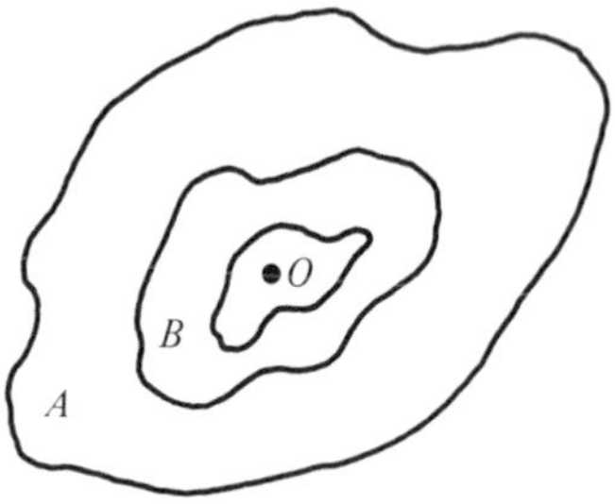

Figure 1

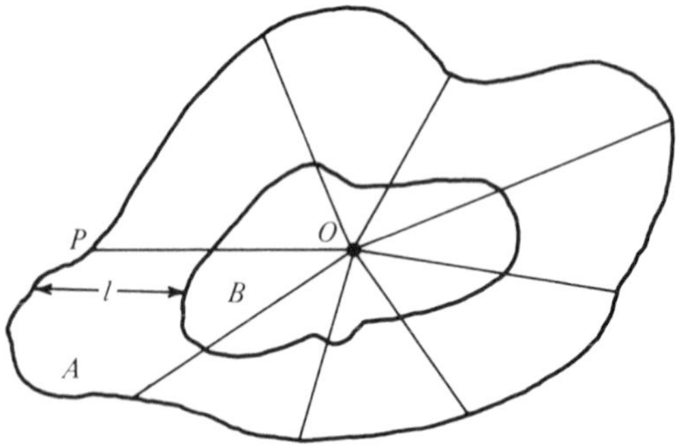

Figure 2

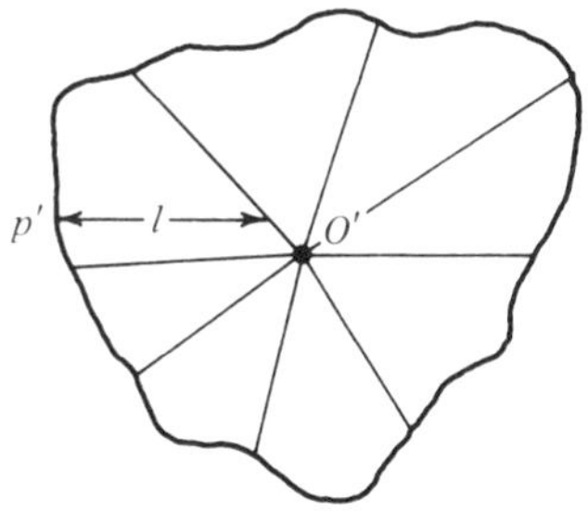

Figure 3

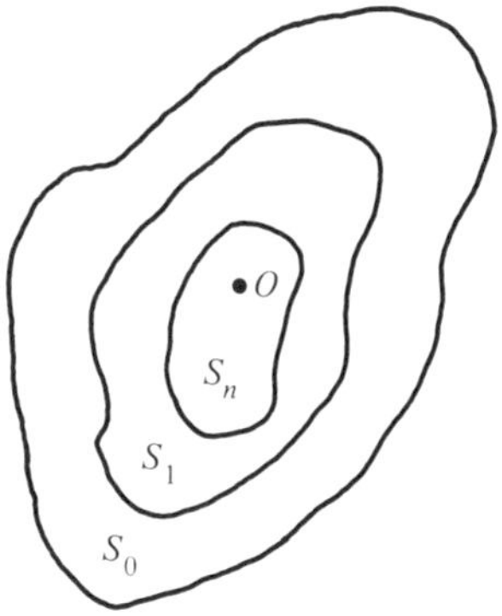

Figure 4

Adding up the computed volumes layer by layer, we obtain the mineral reserve volume V. In other words, supposing the areas enclosed by the $n + 1$ contours of the contour map to be $S_0, S_1, \ldots, S_n$, then the mineral reserve volume may be evaluated approximately by the formula

$$V = \left(\frac{S_0 + S_n}{2} + \sum_{m=1}^{n-1} S_m \right) h - \frac{h}{6} \sum_{m=0}^{n-1} T(S_m, S_{m+1}), \tag{2}$$

where h is contour distance (figure 4).

Theorem (BAUMAN). *Suppose that the lower surface A and the upper surface B (their areas are also represented by A and B) of a known body are planes, and A is parallel to B, h is the height between them, O is a point on B. If all the sections formed by the body and any arbitrary plane passing through O and perpendicular to B are quadrilateral, the body volume v is precisely as shown in formula* (1).

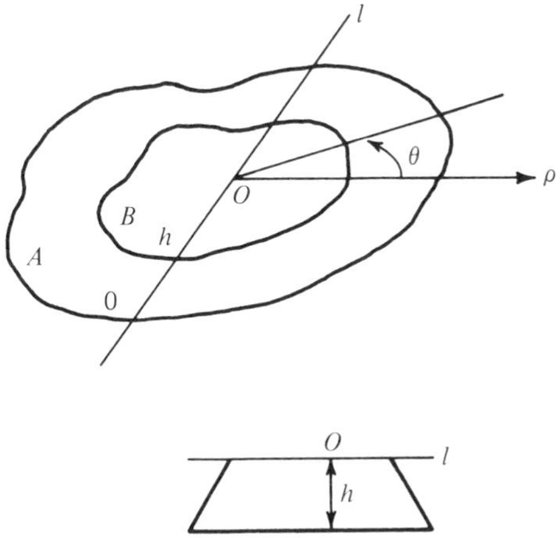

Figure 5

Proof. Introduce polar coordinates with origin O (see figure 5). Let the equation of the contour of altitude z in polar coordinates be

$$\rho = \rho(z, \theta), \qquad (0 \leqslant \theta \leqslant 2\pi),$$

where $\rho(z, 0) = \rho(z, 2\pi)$. Hereafter we will always assume that $\rho(z, \theta)$ $(0 \leqslant \theta \leqslant 2\pi, 0 \leqslant z \leqslant h)$ is continuous. We may assume the altitudes of A and B to be 0 and h. Also let

$$\rho_1(\theta) = \rho(0, \theta), \qquad \rho_2(\theta) = \rho(h, \theta).$$

By assumption we have

$$\rho(Z, \theta) = \frac{Z}{h}\rho_2(\theta) + \frac{h - Z}{h}\rho_1(\theta) \qquad (0 \leqslant Z \leqslant h).$$

Hence the body volume is

$$\begin{aligned}
\frac{1}{2}\int_0^h \int_0^{2\pi} \rho^2(Z, \theta) d\theta dZ &= \frac{1}{2}\int_0^{2\pi}\int_0^h \left(\frac{Z}{h}\rho_2(\theta) + \frac{h - Z}{h}\rho_1(\theta)\right)^2 dZ\, d\theta \\
&= \frac{h}{2}\int_0^{2\pi}\left(\frac{\rho_1^2(\theta)}{3} + \frac{\rho_2^2(\theta)}{3} + \frac{\rho_1(\theta)\rho_2(\theta)}{3}\right) d\theta \\
&= \frac{h}{2}\left[\frac{1}{2}\int_0^{2\pi}\rho_1^2(\theta)\, d\theta + \frac{1}{2}\int_0^{2\pi}\rho_2^2(\theta)\, d\theta\right] \\
&\quad - \frac{h}{6}\left[\frac{1}{2}\int_0^{2\pi}(\rho_1(\theta) - \rho_2(\theta))^2 d\theta\right] \\
&= \frac{h}{2}(A + B) - \frac{h}{6}T(A, B).
\end{aligned}$$

The theorem is proved.

2. *The relations between Bauman, frustum and trapezoid formulae.* If the bases of a body are in parallel planes, h is the altitude, O is a point on B, the following two formulae, in addition to the Bauman formula, are often used to evaluate the approximate volume of the body:

$$\text{Frustum formula:} \quad v_1 = \frac{h}{3}(A + B + \sqrt{AB}), \tag{3}$$

$$\text{Trapezoid formula:} \quad v_2 = \frac{h}{2}(A + B); \tag{4}$$

in general, it is appropriate to use formula (3) if $(A - B)/A > 40\%$, and formula (4) if $(A - B)/A < 40\%$.

Theorem 1. *The inequality*

$$v \leqslant v_1 \leqslant v_2 \tag{5}$$

holds for all cases. Also, $v = v_1$ *if and only if the body is the frustum of a pyramid or a cone, and the perpendicular line from the vertex to the base passes through point* O*; while* $v_1 = v_2$ *if and only if* $A = B$.

Proof. Use the same assumptions as in the Bauman theorem. From the Bauman formula and the Bunjakovskiĭ-Schwarz inequality, we have

$$v = \frac{h}{6}\int_0^{2\pi}\left(\rho_1^2(\theta) + \rho_2^2(\theta) + \rho_1(\theta)\rho_2(\theta)\right)d\theta$$

$$\leqslant \frac{h}{3}\left[\frac{1}{2}\int_0^{2\pi}\rho_1^2(\theta)\,d\theta + \frac{1}{2}\int_0^{2\pi}\rho_2^2(\theta)\,d\theta + \frac{1}{2}\sqrt{\int_0^{2\pi}\rho_1^2(\theta)\,d\theta\int_0^{2\pi}\rho_2^2(\theta)\,d\theta}\,\right]$$

$$= \frac{h}{3}\left[A + B + \sqrt{AB}\,\right] = v_1$$

if and only if $\rho_1(\theta) = C\rho_2(\theta)$ $(0 \leqslant \theta \leqslant 2\pi$, C is a constant); that is, $v = v_1$ if the body is a frustum and the perpendicular from the vertex to base A passes through the point O (figure 6).

Also, since

$$v_2 - v_1 = \frac{h}{2}(A + B) - \frac{h}{3}\left(A + B + \sqrt{AB}\right) = \frac{h}{6}\left(\sqrt{A} - \sqrt{B}\right)^2 \geqslant 0,$$

we have

$$v_1 \leqslant v_2,$$

with equality if and only if $A = B$. The theorem is proved.

These three formulae should be compared by looking at the number of dimensions. Since the dimension of a plane is 2, the formula obtained by considering this number (the number of degrees of freedom) to be 1 is less general.

Thus the trapezoid formula is obtained by regarding the middle section as the arithmetic mean of both bases, that is, regarding the number of degrees of freedom as 1.

The Bauman formula, however, regards the middle section to be of 2 degrees of freedom. To be precise, it assumes that $\rho(z,\theta)$ is obtained by a linear relationship of $\rho(0,\theta)$ and $\rho(h,\theta)$ (see 1).

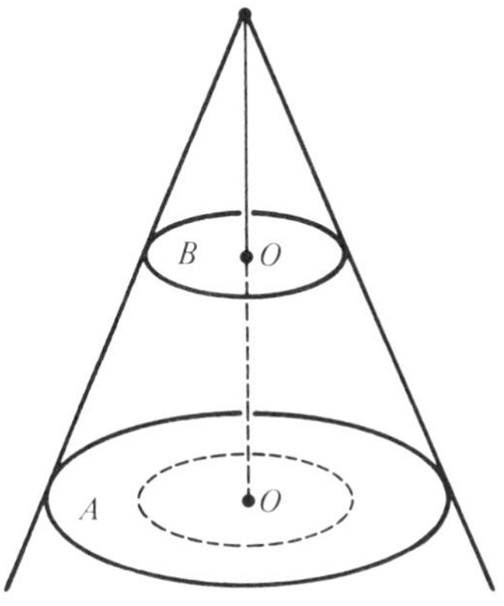

Figure 6

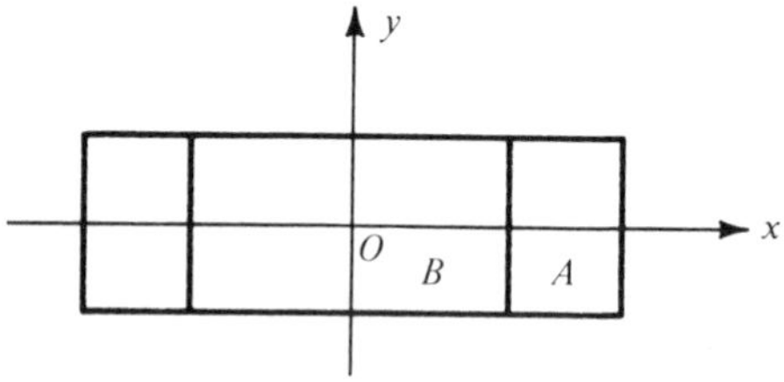

Figure 7

The frustum formula also regards the degrees of freedom of the middle section to be 2, but it further assumes that $\rho(0,\theta) = c\rho(h,\theta)$ $(0 \leqslant \theta \leqslant 2\pi)$, where c is a constant.

Therefore we think that the Bauman formula has more generality, and in general should give more accurate results for the approximate calculation of solid volumes. Nevertheless, we do not discard the possibility that the other two formulae may be more suitable for some specific bodies. For instance, consider a trapezoid with bases of equal width (as shown in figure 7). The trapezoid formula can yield its actual volume, while either the Bauman formula or the frustum formula will give a lower result. However, we should note that in this case the number of degrees of freedom of the section of the trapezoid is 1 (because the magnitude along the y-axis is constant).

We can also estimate the relative deviation of the trapezoid formula and the frustum formula with respect to Bauman formula.

For instance, when $(A - B)/A < 40\%$ $(or\ B > 3A/5)$, the relative deviation of the result computed by the trapezoid formula and the result computed by the Bauman formula is given by

$$\Delta = \frac{v_2 - v}{v} = \frac{\frac{1}{2}(A+B)h - \frac{1}{2}(A+B)h + (h/6)T(A,B)}{\frac{1}{2}(A+B)h - (h/6)T(A,B)}$$

$$= \frac{T(A,B)}{3(A+B) - T(A,B)}\,.$$

Since

$$T(A,B) \leqslant A - B$$

(or $\frac{1}{2}\int_0^{2\pi}(\rho_1(\theta) - \rho_2(\theta))^2 d\theta \leqslant \frac{1}{2}\int_0^{2\pi}\rho_1^2(\theta)\,d\theta - \frac{1}{2}\int_0^{2\pi}\rho_2^2(\theta)\,d\theta$, an inequality which obviously holds), therefore

$$\Delta \leqslant \frac{A-B}{2A+4B}\,.$$

By substituting the condition $B > \frac{3}{5}A$, we have

$$\Delta \leqslant \frac{A - \frac{3}{5}A}{2A + \frac{12}{5}A} = \frac{1}{11} < 10\%.$$

3. Suggestion of a formula for the calculation of mineral reserves. The Bauman formula is obtained by assuming the $\rho(z,\theta)$ to be a linear combination of $\rho(0,\theta)$ and $\rho(h,\theta)$. If we estimate two adjacent layers together, that is, if we know three adjacent contours $\rho(0,\theta)$, $\rho(h,\theta)$ and $\rho(zh,\theta)$, we obtain the approximation to the surface by joining the contours $\rho(0,\theta)$, $\rho(h,\theta)$ and $\rho(zh,\theta)$ with the surface constructed by parabolas; hence we suggest the following method of calculation.

Let A, B, C denote the sections enclosed by the three adjacent contours respectively (areas are also denoted by A, B, C), and let h be the distance between A and B as well as between B and C; then the total volume of these two layers can be approximately computed by the formula

$$v_3 = \frac{h}{3}(A + 4B + C) - \frac{h}{15}(2T(A,B) + 2T(B,C) - T(A,C)). \tag{6}$$

If we omit the second term, equation (6) will be reduced identically to the familiar Sobolevskiǐ formula. Adding up the volumes of every double-layer, we obtain the approximate formula for the volume V of the total mineral reserve. In other words, supposing the areas enclosed by the $2n + 1$ contours of a contour map to be $S_0, S_1, \cdots, S_{2n}$, and the contour distance to be h, the mineral reserve volume V may be approximately computed by the formula

$$V = \frac{h}{3}\left[S_0 + S_{2n} + 4\sum_{i=0}^{n-1} S_{2i+1} + 2\sum_{i=1}^{n-1} S_{2i}\right]$$
$$- \frac{h}{15}\left[2\sum_{i=0}^{n-1} T(S_{2i}, S_{2i+1}) + 2\sum_{i=0}^{n-1} T(S_{2i+1}, S_{2i+2}) - \sum_{i=0}^{n-1} T(S_{2i}, S_{2i+2})\right]. \tag{7}$$

Note. If the contour map contains an even number of contours, the uppermost layer can be estimated alone, and (7) can be used for the calculation of the rest of the layers.

Theorem 2. *Let the upper base C and the lower base A of a body be planes, let B be the middle section (areas are also denoted by C, A, B respectively), and A, C be parallel to B, let h be the distance between A and B and also between B and C, and let O be a point on C (figure 8). If the boundary of the section obtained in the plane*

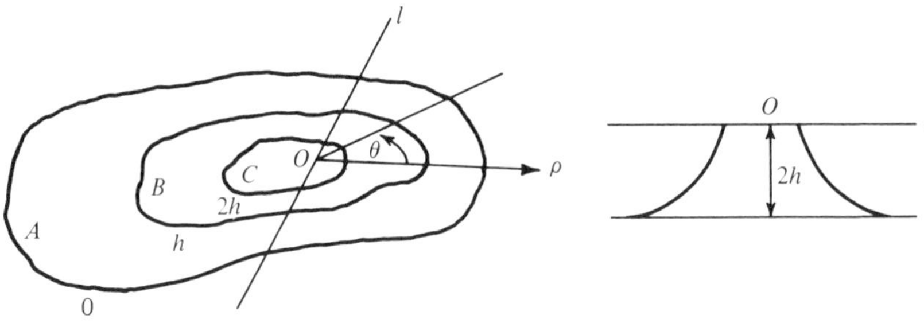

Figure 8

passing through O and perpendicular to C is composed of two straight lines and two parabolas, the volume v_3 of the body is precisely that shown in (6).

Proof. Introduce polar coordinates with origin O, let the equation of the contour of altitude z be

$$\rho = \rho(z,\theta) \quad (0 \leqslant \theta \leqslant 2\pi,\ \rho(z,0) = \rho(z,2\pi)).$$

We may also assume the altitudes of A, B, C to be 0, h, zh, and denote

$$\rho_1(\theta) = \rho(0,\theta),\ \rho_2(\theta) = \rho(h,\theta),\ \rho_3(\theta) = \rho(2h,\theta).$$

By assumption we have

$$\rho(z,\theta) = \frac{(z-h)(z-2h)}{2h^2}\rho_1(\theta) - \frac{z(z-2h)}{h^2}\rho_2(\theta) + \frac{z(z-h)}{2h^2}\rho_3(\theta). \tag{8}$$

Therefore the volume v_3 of the body is

$$\frac{1}{2}\int_0^{2h}\int_0^{2\pi}\rho^2(z,\theta)\,d\theta\,dz$$

$$= \frac{1}{2}\int_0^{2\pi}d\theta\int_0^{2h}\left[\frac{(z-h)(z-2h)}{2h^2}\rho_1(\theta) - \frac{z(z-2h)}{h^2}\rho_2(\theta) + \frac{z(z-h)}{2h^2}\rho_3(\theta)\right]^2 dz$$

$$= \frac{h}{2}\int_0^{2\pi}\left[\frac{4}{15}\rho_1^2(\theta) + \frac{16}{15}\rho_2^2(\theta) + \frac{4}{15}\rho_3^2(\theta) + \frac{4}{15}\rho_1(\theta)\rho_2(\theta)\right.$$
$$\left. + \frac{4}{15}\rho_2(\theta)\rho_3(\theta) - \frac{2}{15}\rho_1(\theta)\rho_3(\theta)\right]d\theta$$

$$= \frac{h}{2}\int_0^{2\pi}\left[\frac{\rho_1^2(\theta)}{3} + \frac{4\rho_2^2(\theta)}{3} + \frac{\rho_3^2(\theta)}{3} - \frac{2}{15}(\rho_1(\theta) - \rho_2(\theta))^2\right.$$
$$\left. - \frac{2}{15}(\rho_2(\theta) - \rho_3(\theta))^2 + \frac{1}{15}(\rho_1(\theta) - \rho_3(\theta))^2\right]d\theta$$

$$= \frac{h}{3}(A + 4B + C) - \frac{h}{15}(2T(A,B) + 2T(B,C) - T(A,C)).$$

The theorem is proved.

§3. Calculation of hillside areas

4. *The Bauman method and the Volkov method.* We first introduce methods often used by mineralogists and geographers. Suppose there is a map with contours of contour distance Δh; hereafter we will always assume there exists a summit and the contour is a closed curve (other conditions can easily be dealt with by extension from this case). Suppose that, starting from the highest point, we draw the contours (l_{n-1}), (l_{n-2}), $\cdots$, (l_0) one after the other (figure 9). Let the altitude of (l_0) be 0, and denote the altitude of (l_n) by h, the area between (l_i) and (l_{i+1}) by B_i (or the area of the projection).

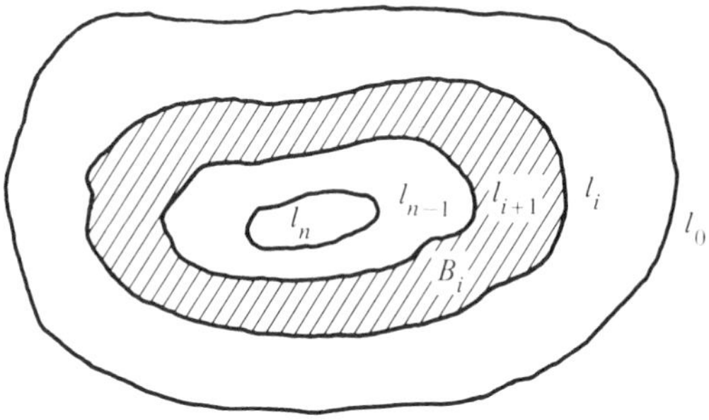

Figure 9

I. The procedure of the method frequently used in the mineral body geometry:

a. $C_i = \frac{1}{2}(l_i + l_{i+1})\Delta h$ (the area enclosed by the contour);

b. $\sum_{i=0}^{n-1}\sqrt{B_i^2 + C_i^2}$ is the asymptotic value of the hillside area (Bauman method).

II. The procedure of the method frequently used in geography:

a. $l = \sum_{i=0}^{n-1} l_i$ (the total length of the contours), $B = \sum_{i=0}^{n-1} B_i$ (the total area of the projections), $\operatorname{tg}\alpha = \Delta h \cdot l / B$ (the average angle of inclination);

b. $B \sec\alpha = \sqrt{B^2 + (\Delta h \cdot l)^2}$ is the asymptotic formula of the hillside area (Volkov method).

Note. $\sqrt{a^2 + b^2}$ can be obtained quickly by a graphical method (with the right-triangle formula).

Which one of these two methods is better? How close are the results obtained by these methods to the actual hillside area? In other words, as the number of the contours becomes infinite in such a way that $\Delta h \to 0$, what are the results given by these methods? Are they the actual hillside areas? In general, the answer is no. Only for some very special surfaces is the answer affirmative. Later we shall determine these specific surfaces and shall provide the relative differences between the results from these methods and the actual value, and at the same time indicate the procedures of computation to avoid large errors.

5. *Relation among* ***Ba, Vo*** *and S.* Introduce polar coordinates with the peak (l_n) as origin O. Let the equation of the contour of altitude z be

$$\rho = \rho(z, \theta) \quad (0 \leqslant \theta \leqslant 2\pi),$$

where $\rho(z, 0) = \rho(z, \pi)$. Hereafter we always assume that

$$\frac{\partial\rho(z,\theta)}{\partial\theta} \quad \text{and} \quad \frac{\partial\rho(z,\theta)}{\partial z} \qquad (0 \leqslant \theta \leqslant 2\pi, 0 \leqslant z \leqslant h)$$

are continuous. Let $z_i = h_i/n$; then the area enclosed by l_i is equal to

$$\frac{1}{2}\int_0^{2\pi}\rho^2(2_i,\theta)\,d\theta,$$

from which, by the mean-value theorem, we have

$$\begin{aligned}B_i &= \frac{1}{2}\int_0^{2\pi}\left[\rho^2(z_i,\theta)-\rho^2(z_{i+1},\theta)\right]d\theta\\ &= -\int_0^{2\pi}\rho(z_i',\theta)\frac{\partial\rho(z_i',\theta)}{\partial z_i'}\,d\theta\,\Delta h,\end{aligned}$$

where $z_i' \in [z_i, z_{i+1}]$, and $\Delta h = h/n$. The length of (l_i) is equal to

$$l_i = \int_{(l_i)} ds = \int_0^{2\pi}\sqrt{\rho^2(z_i,\theta)+\left(\frac{\partial\rho(z_i,\theta)}{\partial\theta}\right)^2}\,d\theta.$$

The result obtained from the Bauman method is

$$C_i = \int_0^{2\pi}\sqrt{\rho^2(z_i'',\theta)+\left(\frac{\partial\rho(z_i'',\theta)}{\partial\theta}\right)^2}\,d\theta\,\Delta h,$$

where the mean-value theorem has been applied with $z_i'' \in [z_i, z_{i+1}]$; hence as $\Delta h\to 0$, $\sum_{i=0}^{n-1}\sqrt{B_i^2+C_i^2}$ approaches

$$\mathbf{Ba} = \int_0^h\sqrt{\left(\int_0^{2\pi}\rho\frac{\partial\rho}{\partial z}\,d\theta\right)^2+\left(\int_0^{2\pi}\sqrt{\rho^2+\left(\frac{\partial\rho}{\partial\theta}\right)^2}\,d\theta\right)^2}\,dz. \tag{9}$$

This is the value of the hillside area calculated by the Bauman method as $\Delta h\to 0$.

It is also easy to see that

$$B = \frac{1}{2}\int_0^{2\pi}\rho^2(0,\theta)\,d\theta = \int_0^{2\pi}d\theta\int_0^h -\rho\frac{\partial\rho}{\partial z}\,dz$$

(note $\rho(h,\theta)=0$) and the limit of $\Delta h\cdot l$ must be equal to

$$\begin{aligned}&\lim_{n\to\infty}\sum_{i=0}^{n-1}\frac{h}{n}\int_0^{2\pi}\sqrt{\rho^2(z_i,\theta)+\left(\frac{\partial\rho(z_i,\theta)}{\partial\theta}\right)^2}\,d\theta\\ &\quad= \int_0^h dz\int_0^{2\pi}\sqrt{\rho^2+\left(\frac{\partial\rho}{\partial\theta}\right)^2}\,d\theta;\end{aligned}$$

therefore as $\Delta h\to 0$, the estimated value of the hillside area by the Volkov method approaches

$$\mathbf{Vo} = \sqrt{\left(\int_0^{2\pi}d\theta\int_0^h -\rho\frac{\partial\rho}{\partial z}\,dz\right)^2+\left(\int_0^{2\pi}d\theta\int_0^h\sqrt{\rho^2+\left(\frac{\partial\rho}{\partial\theta}\right)^2}\,dz\right)^2}\,. \tag{10}$$

Since

$$ds^2 = \left[\left(\frac{\partial \rho}{\partial \theta} \right)^2 + \rho^2 \right] d\theta^2 + 2 \frac{\partial \rho}{\partial \theta} \frac{\partial \rho}{\partial z} d\theta \, dz + \left(1 + \left(\frac{\partial \rho}{\partial z} \right)^2 \right) dz^2,$$

the area S of the hillside surface is

$$S = \int_0^{2\pi} d\theta \int_0^h \sqrt{\rho^2 + \left(\frac{\partial \rho}{\partial \theta} \right)^2 + \left(-\rho \frac{\partial \rho}{\partial z} \right)^2} \, d\theta. \tag{11}$$

For the comparison of **Ba**, **Vo** and S, we introduce a function of a complex variable

$$f(z,\theta) = \rho \frac{\partial \rho}{\partial z} + i \sqrt{\rho^2 + \left(\frac{\partial \rho}{\partial \theta} \right)^2}, \tag{12}$$

and thus obtain

$$\mathbf{Ba} = \int_0^h \left| \int_0^{2\pi} f(z,\theta)\, d\theta \right| dz, \tag{13}$$

$$\mathbf{Vo} = \left| \int_0^h \int_0^{2\pi} f(z,\theta)\, d\theta \, dz \right|, \tag{14}$$

and

$$S = \int_0^h \int_0^{2\pi} |f(z,\theta)| \, d\theta \, dz. \tag{15}$$

Hence the inequality

$$\mathbf{Vo} \leqslant \mathbf{Ba} \leqslant S \tag{16}$$

is obviously established.

Therefore we recognize: (i) the Bauman method is more accurate than the Volkov method; (ii) the obtained results are smaller than the actual value; (iii) because the Bauman method yields a smaller result, we can make the modification $C_i = l_i \Delta h$. Thus we not only simplify the computation but also increase the value.

Before investigating the surface for which $\mathbf{Vo} = S$ and $\mathbf{Ba} = S$, we introduce the following lemma:

Lemma. *If $f(x)$ is a complex function in the interval $[a,b]$, where a and b are real numbers, the necessary and sufficient condition for the equality*

$$\left| \int_a^b f(x)\, dx \right| = \int_a^b |f(x)| \, dx \tag{17}$$

is that the ratio of the imaginary part to the real part of $f(x)$ should be constant.

Proof. Let $f(x) = \rho(x) e^{i\theta(x)}$, $\rho(x) \geqslant 0$ and $\theta(x)$ is a real function. It is obvious that (17) is established provided that $\theta(x)$ is a constant with respect to x. Conversely,

since

$$\left(\left|\int_a^b f(x)\,dx\right|\right)^2 = \int_a^b\int_a^b f(x)\overline{f(y)}\,dx\,dy = \int_a^b\int_a^b \rho(x)\rho(y)e^{i(\theta(x)-\theta(y))}dx\,dy$$

$$= 2\iint_{a\leqslant x<y\leqslant b} \rho(x)\rho(y)\cos[\theta(x)-\theta(y)]\,dx\,dy,$$

$$\left(\int_a^b |f(x)|\,dx\right)^2 = 2\iint_{a\leqslant x<y\leqslant b} \rho(x)\rho(y)\,dx\,dy.$$

Therefore if (17) is established, there must exist

$$\cos(\theta(x)-\theta(y)) \equiv 1,$$

or $\theta(x) \equiv \theta(y)$, as the lemma asserts.

It is easy to see that the lemma holds for multiple integrals, too.

From the lemma we see that the necessary and sufficient condition for the establishment of

$$\mathbf{Vo} = \left|\int_0^{2\pi}\int_0^h f(z,\theta)\,dz\,d\theta\right| = \int_0^{2\pi}\int_0^h |f(z,\theta)|\,dz\,d\theta = S$$

is that the ratio of the imaginary part to the real part of $f(z,\theta)$ should be a constant c; hence we obtain the partial differential equation

$$\rho^2 + \left[\frac{\partial\rho}{\partial\theta}\right]^2 = c^2\left[\rho\frac{\partial\rho}{\partial z}\right]^2. \tag{18}$$

In other words, only for those functions $\rho = \rho(z,\theta)$ which satisfy this partial differential equation can the Volkov method yield an exact answer. These, of course, should meet the following conditions: $\rho(h,\theta) = 0$ and $\rho(0,\theta) = \rho_0(\theta)$ (the equation of the base of the surface).

We will not solve this partial differential equation; instead, we will try to see its geometric meaning. Regard θ and z as parameters, that is,

$$x = \rho\cos\theta, \qquad y = \rho\sin\theta, \qquad z = z,$$

and ρ as a function of θ and z. From

$$\frac{\partial x}{\partial\theta} = \frac{\partial\rho}{\partial\theta}\cos\theta - \rho\sin\theta, \qquad \frac{\partial y}{\partial\theta} = \frac{\partial\rho}{\partial\theta}\sin\theta + \rho\cos\theta, \qquad \frac{\partial z}{\partial\theta} = 0,$$

$$\frac{\partial x}{\partial z} = \frac{\partial\rho}{\partial z}\cos\theta, \qquad \frac{\partial y}{\partial z} = \frac{\partial\rho}{\partial z}\sin\theta, \qquad \frac{\partial z}{\partial z} = 1$$

we know that the direction of the normal at the point (θ,z) of the surface is

$$\left(\frac{\partial\rho}{\partial\theta}\sin\theta + \rho\cos\theta,\ -\frac{\partial\rho}{\partial\theta}\cos\theta + \rho\sin\theta,\ -\rho\frac{\partial\rho}{\partial z}\right).$$

From (18) we see that the cosine of the angle α between the normal and the z-axis (or the inclination at point (θ, z)) is equal to

$$\cos\alpha = \frac{-\rho(\partial\rho/\partial z)}{\sqrt{(\rho\partial\rho/\partial z)^2 + (\partial\rho/\partial\theta)^2 + \rho^2}} = \frac{1}{\sqrt{1+c^2}},$$

which is a constant. That is to say, the tangent plane meets the horizontal plane (xy-plane) at a fixed angle α. We will now interpret the geometric properties of such surfaces.

Construct an arbitrary vertical plane from the high-water mark to the xy-plane. Then tangent line of the curve at every point has the same intersection angle with the xy-plane. Hence it is a straight line.

Construct a base from any closed plane curve (l_0). Take any point (l_n) with projection inside the base as the high-water mark. The straight line passing through the high-water mark and perpendicular to the base is called the axis. Through any point A in (l_0), construct a straight line which lies in the plane formed by A and the axis, and meets the base at an angle α. The diagram constructed by such straight lines is the one which satisfies **Vo** $= S$.

Therefore, if there exists a peak and no steep angle downward, then these will be only surfaces whose bases are circles, or polygons formed by tangent lines of a circle, or a construction of some circular arcs and some tangent lines, and whose axis is the line passing through the center of the circle and perpendicular to the base (see figure 10).

Among well-known examples, only the Mongolian tent, the pyramid, and diagrams combined from these can be arbitrarily well approximated by the Volkov method.

But when does **Ba** $= S$? Of course, when **Vo** $= S$, **Ba** $= S$. Is there any surface other than those mentioned above? Yes, we demonstrate it as follows: from

$$\mathbf{Ba} = \int_0^h \left| \int_0^{2\pi} f(z,\theta)\, d\theta \right| dz = \int_0^h \int_0^{2\pi} |f(z,\theta)|\, d\theta\, dz = S$$

we obtain

$$\int_0^h \left(\int_0^{2\pi} |f(z,\theta)|\, d\theta - \left| \int_0^{2\pi} f(z,\theta)\, d\theta \right| \right) dz = 0.$$

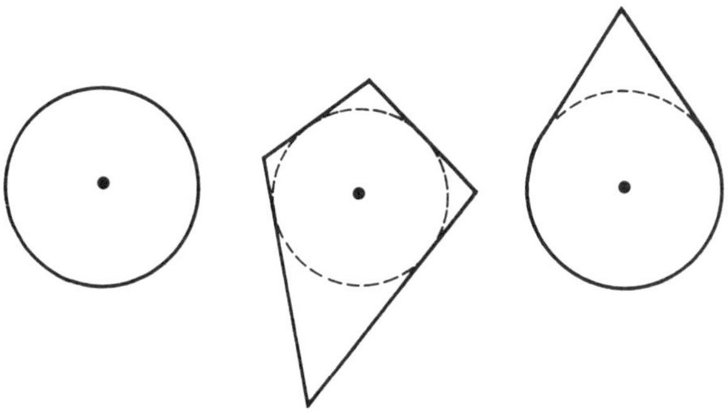

Figure 10

Since the integrand is not negative, we have for any z that

$$\int_0^{2\pi}|f(z,\theta)|\,d\theta=\left|\int_0^{2\pi}f(z,\theta)\,d\theta\right|.$$

Therefore, when we fix z, the ratio of the imaginary part to the real part of $f(z,\theta)$ is constant, that is, c in equation (18) is a function of z alone. Hence, only those surfaces with equal inclination for the same altitude can satisfy $\mathbf{Ba}=S$. Among well-known examples, only the Gourd and the White Pagoda (in the North Sea) can be approximated arbitrarily well by the Bauman method.

Now we estimate the error for these two methods. Suppose that the cosines of inclination of the points on the surface are between two positive constants ξ and η,

$$\xi \leqslant \cos\alpha \leqslant \eta,$$

or

$$\xi \leqslant \frac{-\rho(\partial\rho/\partial z)}{\sqrt{(\rho(\partial\rho/\partial z))^2+(\partial\rho/\partial\theta)^2+\rho^2}} \leqslant \eta.$$

From this we obtain

$$\frac{\rho^2+(\partial\rho/\partial\theta)^2}{(\rho(\partial\rho/\partial z))^2+(\partial\rho/\partial\theta)^2+\rho^2} \geqslant 1-\eta^2,$$

so that

$$\int_0^{2\pi}\int_0^h\sqrt{\rho^2+(\partial\rho/\partial\theta)^2}\,dz\,d\theta$$

$$\geqslant\sqrt{1-\eta^2}\int_0^{2\pi}d\theta\int_0^h\sqrt{(\rho(\partial\rho/\partial z))^2+(\partial\rho/\partial\theta)^2+\rho^2}\,dz=\sqrt{1-\eta^2}\,S,$$

$$\int_0^{2\pi}\int_0^h -\rho\frac{\partial\rho}{\partial z}\,dz\,d\theta \geqslant \xi S,$$

from which

$$\mathbf{Vo}\geqslant\sqrt{\xi^2S^2+(1-\eta^2)S^2}=\sqrt{1+\xi^2-\eta^2}\,S.$$

Also because $1>\eta\geqslant\xi>0$, hence

$$\frac{\xi}{\eta}\leqslant\sqrt{1+\xi^2-\eta^2},$$

(squaring both sides of this inequality, we have $(\eta^2-\xi^2)(1-\eta^2)\geqslant 0$) and we obtain

$$\mathbf{Vo}\geqslant\frac{\xi}{\eta}S.$$

Thus we have proved the following theorem.

Theorem 3. *If all the cosines of the inclinations α at any point on the surface $\rho=\rho(z,\theta)$ $(0\leqslant z\leqslant h,\ 0\leqslant\theta\leqslant 2\pi)$ satisfy $0<\xi\leqslant\cos\alpha\leqslant\eta$, then the inequality*

$$\frac{\xi}{\eta}S\leqslant Vo\leqslant\mathbf{Ba}\leqslant S \tag{19}$$

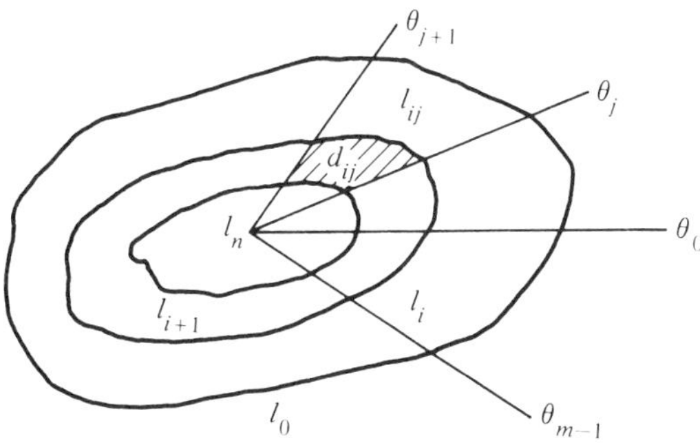

Figure 11

is established. The necessary and sufficient condition for **Vo** $= S$ *is that any point on the surface has equal inclination; the necessary and sufficient condition for* **Ba** $= S$ *is that the points at equal altitude have the same inclination.*

6. *Suggestions for methods of calculation.* From theorem 3 we can see that the Volkov method can give accurate results only when the variation of the inclination of points on the surface is small, and the Bauman method can give accurate results only when the difference of inclination of points between two adjacent altitudes is small. For other cases these methods may yield greater error.

Accordingly, we suggest the following methods of calculation: introduce several rays $\theta_0, \theta_1, \cdots, \theta_{m-1}$ from the high-water mark (l_n) on the contour map (see figure 11), where the amplitude of θ_j is equal to $2\pi j/m$. The area enclosed by rays θ_j, θ_{j+1} and contours l_i, l_{i+1} is denoted by d_{ij}; the length of the section of l_i intercepted by θ_j and θ_{j+1} is denoted by l_{ij}.

Method I.

a. $D_j = \sum_{i=0}^{n-1} d_{ij}$ (the area on the contour map between rays θ_j and θ_{j+1});
b. $E_j = [\sum_{i=0}^{n-1} l_{ij}]\Delta h$ (the sum of the areas of the vertical screen between two vertical walls);
c. $\sigma_1 = \sum_{j=0}^{m-1}\sqrt{D_j^2 + E_j^2}$ is the asymptotic value of the surface area.

Method II.

a. $e_{ij} = l_{ij}\Delta h$ (the area of the vertical screen between two vertical walls);
b. $\sigma_2 = \sum_{i=0}^{n-1}\sum_{j=0}^{m-1}\sqrt{d_{ij}^2 + e_{ij}^2}$ is the asymptotic value of the surface area.

In the same manner as in the preceding section we know that

$$K = \int_0^{2\pi}\sqrt{\left(\int_0^h -\rho\frac{\partial\rho}{\partial z}\,dz\right)^2 + \left(\int_0^h\sqrt{\rho^2 + (\partial\rho/\partial\theta)^2}\,dz\right)^2}\,d\theta$$

$$= \int_0^{2\pi}\left|\int_0^h f(z,\theta)\,dz\right|\,d\theta \tag{20}$$

and

$$S = \int_0^{2\pi} \int_0^h \sqrt{\rho^2 + (\partial\rho/\partial\theta)^2 + (\rho(\partial\rho/\partial z))^2}\, dz\, d\theta$$
$$= \int_0^{2\pi} \int_0^h |f(z,\theta)|\, dz\, d\theta \tag{21}$$

are the values which σ_1 and σ_2 approach as $n \to \infty$, $m \to \infty$ (see (12) for the definition of $f(z,\theta)$).

Evidently, $\mathbf{Vo} \leqslant K \leqslant S$ (see (10)), and in the same manner as in the preceding section we see that the necessary and sufficient condition for $K = S$ is that the surface is a ruled surface. Since σ_2 approaches the actual area, method II is the most accurate and reliable.

References

[1] V. I. Bauman, On the calculation of mineral reserves, *Gornyĭ žurnal*, December, 1908. (Russian)

[2] I. N. Ušakov, *Mineral reserve geometry*, Coal Industry Publishing House, 1957. (Chinese)

[3] P. A. Ryžov, *Mineral body geometry*, Geology Publishing House, 1957. (Chinese)

[4] S. S. Izakson, *The verification of calculation and the determination of calculation error of mineral reserves*, Coal Industry Publishing House, 1958. (Chinese)

[5] N. M. Volkov, *Principles and methods of the measurement of a diagram*, 1950. (Chinese)

[6] Lu Shu-fen, "Problems of the calculation of ground surface area on a contour map," *Journal of Mensuration* **4** (1960), no. 1. (Chinese)

ON UNIFORM DISTRIBUTION AND NUMERICAL ANALYSIS (I)

(NUMBER-THEORETIC METHOD)

HUA LOO-KENG (华罗庚) AND WANG YUAN (王 元)

Received June 13, 1973.

ABSTRACT

In this paper, the uniformly distributed sequences of sets defined by means of a real cyclotomic field have been dealt with. We have obtained the estimations of their discrepancies and applied them to the problem of numerical integration.

I. INTRODUCTION

Let G_s be a unit cube

$$0 \leqslant x_1 \leqslant 1, \cdots, 0 \leqslant x_s \leqslant 1$$

of s-dimensional space, and $n_1 < n_2 < \cdots$ be a sequence of positive integers. Let

$$P_{n_l}(j) = (x_1^{(n_l)}(j), \cdots, x_s^{(n_l)}(j)), \quad (1 \leqslant j \leqslant n_l)$$

be a set of points in G_s. For any $(\gamma_1, \cdots, \gamma_s) \in G_s$, let $N_{n_l}(\gamma_1, \cdots, \gamma_s)$ denote the number of points of $P_{n_l}(j)(1 \leqslant j \leqslant n_l)$ satisfying the inequalities

$$0 \leqslant x_1^{(n_l)}(j) < \gamma_1, \cdots, 0 \leqslant x_s^{(n_l)}(j) < \gamma_s.$$

If

$$\lim_{l \to \infty} \frac{N_{n_l}(\gamma_1, \cdots, \gamma_s)}{n_l} = \gamma_1 \cdots \gamma_s,$$

then the sequence of sets $(P_{n_l}(j))(n_1 < n_2 < \cdots)$ is called uniformly distributed on G_s. Futhermore, if we have sharper condition

$$\left| \frac{N_{n_l}(\gamma_1, \cdots, \gamma_s)}{n_l} - \gamma_1 \cdots \gamma_s \right| < \varphi(n_l),$$

where $\varphi(n_l) = o(1)$, then the sequence of sets $(P_{n_l}(j))(n_1 < n_2 < \cdots)$ is uniformly distributed with discrepency $\varphi(n)$. In case $n_l = l$, $x_1^{(l)}(j) = x_1(j), \cdots,$ and $x_s^{(l)}(j) = x_s(j)(l = 1, 2, \cdots)$, the sequence $P(j) = (x_1(j), \cdots, x_s(j))$ $(j = 1, 2, \cdots)$ will be called uniformly distributed on G_s.

Let p be a prime $\geqslant 5$. Set $r = \frac{1}{2}(p-1)$ and $q = r - 1 = \frac{1}{2}(p-3)$. The real cyclotomic field $\Re_r = R\left(2 \cos \frac{2\pi}{p}\right)$ is an algebraic field of degree r. Let

$$\omega_1 = 2 \cos \frac{2\pi}{p}, \quad \omega_2 = 2 \cos \frac{4\pi}{p}, \cdots, \quad \omega_q = 2 \cos \frac{2\pi q}{p}$$

and let

$$\left|\frac{h_i^{(l)}}{n_l}-\omega_i\right|\leqslant c(\Re_r)n_l^{-1-\frac{1}{q}}, \quad (1\leqslant i\leqslant q)$$

be the simultaneous approximations of ω's by rationals, where we use $c(f,\cdots,g)$ to denote the positive constants depending on $f,\cdots,g$ only, but not always with the same value.

Theorem 1. *Let*

$$P(j)=(\{\omega_1 j\},\cdots,\{\omega_q j\}) \quad (j=1,2,\cdots), \tag{1.1}$$

where $\{x\}$ *denotes the fractional part of* x. *Then the sequence* $P(j)(j=1,2,\cdots)$ *has discrepancy*

$$\varphi(n)=c(\Re_r,\ \varepsilon)n^{-1+\varepsilon},$$

ε *being any pre-assigned positive number.* (The convention will be adopted throughout the present paper).

Theorem 2. *Set*

$$P_{n_l}(j)=\left(\left\{\frac{j}{n_l}\right\},\ \left\{\frac{h_1^{(l)}j}{n_l}\right\},\cdots,\left\{\frac{h_q^{(l)}j}{n_l}\right\}\right) \quad (1\leqslant j\leqslant n_l). \tag{1.2}$$

Then the sequence of sets $(P_{n_l}(j))$ *has discrepancy*

$$\varphi(n)=c(\Re_r,\varepsilon)n^{-\frac{1}{2}-\frac{1}{2q}+\varepsilon}.$$

Let $E_s^\alpha(C)$ be the class of functions

$$f(x_1,\cdots,x_s)=\sum\overset{\infty}{\underset{-\infty}{\cdots}}\sum C(m_1,\cdots,m_s)e^{2\pi i(m_1x_1+\cdots+m_sx_s)},$$

in which the Fourier coefficients satisfy

$$|C(m_1,\cdots,m_s)|\leqslant\frac{C}{(\bar{m}_1\cdots\bar{m}_s)^\alpha},$$

where α and C are positive constants and $\bar{m}=\max(1,|m|)$.

Suppose $\alpha>1$ throughout the present paper.

Applying the sequences of sets (1.1) and (1.2) to the problem of numerical integration, we have

Theorem 3. *Let* l *be the least integer* $\geqslant\alpha$ *and let* $\mu_{n,l,j}$ *be a set of integers defined by*

$$\left(\sum_{j=-n}^{n}z^j\right)^l=\sum_{j=-n_l}^{n_l}\mu_{n,l,j}z^j.$$

Then

$$\sup_{f\in E_q^\alpha(C)}\left|\int_0^1\cdots\int_0^1 f(x_1,\cdots,x_q)dx_1\cdots dx_q-\frac{1}{(2n+1)^l}\sum_{j=-n_l}^{n_l}\mu_{n,l,j}f(\omega_1 j,\cdots,\omega_q j)\right|$$
$$\leqslant C\cdot c(\Re_r,\alpha,\varepsilon)n^{-\alpha+\varepsilon}. \tag{1.3}$$

Theorem 4. *We have*

$$\sup_{f\in E_r^\alpha(C)}\left|\int_0^1\cdots\int_0^1 f(x_1,\cdots,x_r)\,dx_1\cdots dx_r-\frac{1}{n_l}\sum_{j=1}^{n_l}f\left(\frac{j}{n_l},\frac{h_1^{(l)}j}{n_l},\cdots,\frac{h_q^{(l)}j}{n_l}\right)\right|$$

$$\leqslant C\cdot c(\mathfrak{R}_r,\alpha,\varepsilon)n_l^{-\frac{\alpha}{2}-\frac{\alpha}{2q}+\varepsilon}. \tag{1.4}$$

It is well known that under certain conditions, the integral of non-periodic function may be calculated by an integral of periodic function[14]. We may also use the following simple formula instead of formula (1.3) for the case $\alpha=2$.

$$\sup_{f\in E_q^2(C)}\left|\int_0^1\cdots\int_0^1 f(x_1,\cdots,x_q)\,dx_1\cdots dx_q-\frac{1}{n}\sum_{j=-n}^{n}\left(1-\frac{|j|}{n}\right)f(\omega_1 j,\cdots,\omega_q j)\right|$$

$$\leqslant C\cdot c(\mathfrak{R}_r,\varepsilon)n^{-2+\varepsilon}. \tag{1.5}$$

It is followed immediately by the Ω-result of Roth[9] on the discrepancy of uniformly distributed sequence of sets that the estimation given in Theorem 1 is the best possible one of its kinds apart from some possible improvements about the order n^ε. This type of distribution may be recognized as the best distributed sequence of sets. The other known best distributed sequences of sets are those proposed by Коробов[14] and Halton[2] in 1959 and 1960 respectively. We can prove also that the error term given in Theorem 3 does not allow essential improvements[15]. Perhaps, it is worth mentioning that only $c(\mathfrak{R}_r)\log n_l$ elementary operations are required for obtaining the sequence of integers $(h_1^{(l)},\cdots,h_q^{(l)};n_l)$ in (1.2).

The classical method in numerical integration is established by means of the sequence of sets

$$\left(\frac{j_1}{m},\cdots,\frac{j_s}{m}\right)\quad(0\leqslant j_1,\cdots,j_s\leqslant m-1). \tag{1.6}$$

That is, the integral over G_s is calculated approximately by the sum

$$\frac{1}{m^s}\sum_{j_1,\cdots,j_s}f\left(\frac{j_1}{m},\cdots,\frac{j_s}{m}\right). \tag{1.7}$$

The number of points of (1.6) is $n=m^s$. We can prove easily that the discrepancy of (1.6) is $\geqslant n^{-1/s}$ and the error term in classical quadrature formula of the functions belonging to $E_s^\alpha(C)$ over G_s is $\geqslant 2Cn^{-\alpha/s}$. (Take $f(x_1,\cdots,x_s)=C(e^{2\pi i m_1 x_1}+e^{-2\pi i m_1 x_1})/m^\alpha$. Then we have $\int_0^1\cdots\int_0^1 f(x_1,\cdots,x_s)dx_1\cdots dx_s=0$ and $\frac{1}{n}\sum_{j_1,\cdots,j_s}f\left(\frac{j_1}{m},\cdots,\frac{j_s}{m}\right)=\frac{2C}{n^{\alpha/s}}$.)

In 1959, Бахвалов[12] and the authors[6] proved independently the formula (1.4) for the case $q=1$ with the error term $O(F_l^{-\alpha}\log 3F_l)$ by means of the Fibonacci sequence (F_l), where

$$F_l=\frac{1}{\sqrt{5}}\left(\left(\frac{1+\sqrt{5}}{2}\right)^{l+1}-\left(\frac{1-\sqrt{5}}{2}\right)^{l+1}\right)\quad(l\geqslant 1).$$

Since $\mathfrak{R}_2 = R\left(2\cos\frac{2\pi}{5}\right) = R(\sqrt{5})$, the sequences of sets (1.1) and (1.2) may be recognized as the generalizations of the sequences of sets

$$\left\{\frac{\sqrt{5}-1}{2}j\right\} \ (j = 1, 2, \cdots) \text{ and } \left(\left\{\frac{j}{F_l}\right\}, \left\{\frac{F_{l-1}j}{F_l}\right\}\right) \ (1 \leqslant j \leqslant F_l) \tag{1.8}$$

obtained by Golden section and Fibonacci sequence respectively. Hence we suggested also the possibility to treat the high dimensional case of numerical integration by means of a real cyclotomic field and gave some numerical examples of the formula (1.4) for the case $2 \leqslant q \leqslant 10$[6–8]. Indeed, other totally real fields can be used as well instead of the cyclotomic field, for example, the Dirichlet field $R(\sqrt{P_1}, \cdots, \sqrt{P_h})$, where P's are h distinct primes, but it seems that the cyclotomic field often gives best result among the fields of the same degree. Another advantage of cyclotomic field is convenience for calculation.

The proofs of the above theorems depend on the following important result of Wolfgang M. Schmidt[10] concerning the simultaneous approximations of algebraic numbers by rationals:

Lemma 1.1. *Let* $\alpha_1, \cdots, \alpha_s$ *be a set of real algebraic numbers such that* $1, \alpha_1, \cdots, \alpha_s$ *are linearly independent over rational field* R. *Then we obtain*

$$\langle \alpha_1 m_1 + \cdots + \alpha_s m_s \rangle \geqslant c(\alpha_1, \cdots, \alpha_s, \varepsilon)(\bar{m}_1 \cdots \bar{m}_s)^{-1-\varepsilon},$$

$m_1, \cdots, m_s$ *denoting any set of integers which are not all equal to zeros and* $\langle x \rangle = \min(\{x\}, 1 - \{x\})$.

We have also used the number-theoretic methods in numerical analysis introduced by Бахвалов[12], Коробов[14], Haselgrove[3] and Hlawka[5].

If a similar result of A. Baker[1] is used instead of Lemma 1.1, we may also obtain the corresponding results on numerical analysis. For example, we may prove that the sequence

$$P(j) = (\{ej\}, \{e^2 j\}, \cdots, \{e^s j\}), \ (j = 1, 2, \cdots) \tag{1.9}$$

has discrepancy $\varphi(n) = c(s, \varepsilon)n^{-1+\varepsilon}$.

By Wang 520-calculator, we obtain the following example of the formula (1.5)

$$\sup_{f \in E_4^2(C)} \left| \int_0^1 \cdots \int_0^1 f(x_1, \cdots, x_4)dx_1 \cdots dx_4 - \frac{1}{3000} \sum_{j=-3000}^{3000} \left(1 - \frac{|j|}{3000}\right) \right.$$
$$\left. \times f\left(2j\cos\frac{2\pi}{11}, 2j\cos\frac{4\pi}{11}, 2j\cos\frac{6\pi}{11}, 2j\cos\frac{8\pi}{11}\right) \right| \leqslant 0.065C.$$

II. The Totally Real Algebraic Field

Let $\mathscr{F}_s$ denote a totally real algebraic field of the degree s. For a number η of $\mathscr{F}_s$, let $\eta^{(1)}(=\eta), \eta^{(2)}, \cdots, \eta^{(s)}$ be its conjugates. Assume that

$$\omega_1, \cdots, \omega_s$$

is an integral basis of $\mathscr{F}_s$. Form a matrix

$$\Omega = (\omega_j^{(i)}) \quad (1 \leqslant i, \ j \leqslant s).$$

The matrix

$$S = \Omega'\Omega = \left(\sum_{k=1}^{s} \omega_i^{(k)}\omega_j^{(k)}\right), \quad (1 \leqslant i, j \leqslant s)$$

is called the fundamental matrix of $\mathscr{F}_s$. Clearly, it is a symmetric matrix with rational integer elements. The invariants of the fundamental matrix under the modular groups are characteristic properties of the algebraic number field. The determinant $\det S$ of S is called the discriminant of the field.

Let η be a unit of $\mathscr{F}_s$ satisfying

$$|\eta| > 1, \quad |\eta^{(j)}| \leqslant c(\mathscr{F}_s)|\eta|^{-\frac{1}{s-1}} \quad (2 \leqslant j \leqslant s). \tag{2.1}$$

We express η as follows:

$$\eta = \sum_{i=1}^{s} k_i \omega_i, \tag{2.2}$$

where k's are rational integers. From (2.2) and its conjugates, it appears

$$(\eta^{(1)}, \cdots, \eta^{(s)}) = (k_1, \cdots, k_s)\Omega'. \tag{2.3}$$

Hence

$$(\eta^{(1)}, \cdots, \eta^{(s)})\Omega = (k_1, \cdots, k_s)S = (h_1, \cdots, h_s) \quad \text{(by definition)}$$

or

$$h_j = \sum_{i=1}^{s} \eta^{(i)} \omega_j^{(i)}, \quad (1 \leqslant j \leqslant s).$$

Therefore, we have

$$|\eta\omega_j - h_j| \leqslant \sum_{i=2}^{s} |\eta^{(i)}|\,|\omega_j^{(i)}| \leqslant c(\mathscr{F}_s)|\eta|^{-\frac{1}{s-1}}$$

by (2.1). Suppose

$$1 = \sum_{i=1}^{s} a_i\omega_i$$

and

$$n = \sum_{i=1}^{s} a_i h_i.$$

Then we deduce that

$$|\eta - n| = \left|\sum_{i=1}^{s} a_i\omega_i\eta - \sum_{i=1}^{s} a_i h_i\right| \leqslant \sum_{i=1}^{s} |a_i|\,|\omega_i\eta - h_i| \leqslant c(\mathscr{F}_s)|\eta|^{-\frac{1}{s-1}}.$$

Hence

$$\left|\omega_j - \frac{h_j}{n}\right| \leqslant c(\mathscr{F}_s)\,|n|^{-1-\frac{1}{s-1}}, \quad (1 \leqslant j \leqslant s). \tag{2.4}$$

In the case of $s > 2$, various classical methods can only prove the existence of an infinitely many sets of integers $(h_1, \cdots, h_s; n)$ satisfying (2.4), but this does not suggest effective way to finding $(h_1, \cdots, h_s; n)$. It is shown in this section that the problem for finding $(h_1, \cdots, h_s; n)$ is equivalent to the problem for finding a unit η in the totally real algebraic field $\mathscr{F}_s$ so that (2.1) is satisfied.

If a complete set of independent units of $\mathscr{F}_s$ is known, then by the following theorem, we can find a sequence of units (η_l) satisfying (2.1) and

$$|\eta_l| \to \infty \quad (\text{as } l \to \infty).$$

As a result, we have infinitely many sets of integers

$$(h_1, \cdots, h_s; n)$$

satisfying (2.4).

Theorem 2.1. *In the totally real algebraic field $\mathscr{F}_s$, we have a sequence of units η_l $(= \eta_l^{(1)})$ $(l = 1, 2, \cdots)$, whose conjugates satisfy*

$$|\eta_l^{(i)}| < e^{-(2l-1)c} \quad (2 \leqslant i \leqslant s)$$

and

$$e^{-2c}|\eta_l^{(j)}| \leqslant |\eta_l^{(i)}| \leqslant e^{2c}|\eta_l^{(j)}| \quad (2 \leqslant i, \ j \leqslant s),$$

where $c = c(\mathscr{F}_s) > 0$.

Proof. Let

$$\varepsilon_1, \cdots, \varepsilon_{s-1}$$

be a complete set of independent units of $\mathscr{F}_s$. Set

$$\xi^{(i)} = \varepsilon_1^{(i)l_1} \cdots \varepsilon_{s-1}^{(i)l_{s-1}} \quad (2 \leqslant i \leqslant s)$$

and

$$c = \max_{2 \leqslant i \leqslant s} \left(\sum_{j=1}^{s-1} |\log|\varepsilon_j^{(i)}|| \right).$$

Since

$$\det(\log|\varepsilon_j^{(i)}|) \neq 0, \quad (2 \leqslant i \leqslant s, \ 1 \leqslant j \leqslant s-1),$$

we denote the solution of the system of linear equations

$$\log|\xi^{(2)}| = \cdots = \log|\xi^{(s)}| = -2cl - 1$$

by

$$l_1 = l_1^{(l)}, \cdots, l_{s-1} = l_{s-1}^{(l)}.$$

Set

$$[l_i^{(l)}] = a_i^{(l)}, \quad (1 \leqslant i \leqslant s-1),$$

where $[x]$ denotes the integer part of x. Define

$$\eta_l = \varepsilon_1^{a_1^{(l)}} \cdots \varepsilon_{s-1}^{a_{s-1}^{(l)}}.$$

Then it follows

$$\log|\eta_l^{(i)}| = \sum_{j=1}^{s-1} a_j^{(l)} \log|\varepsilon_j^{(i)}| \leqslant \sum_{j=1}^{s-1} l_j^{(l)} \log|\varepsilon_j^{(i)}| + \sum_{j=1}^{s-1} |\log|\varepsilon_j^{(i)}||$$

$$= \log|\xi^{(i)}| + \sum_{j=1}^{s-1} |\log|\varepsilon_j^{(i)}|| < -2cl + c = -(2l-1)c \quad (2 \leqslant i \leqslant s)$$

and

$$|\log|\eta_l^{(i)}| - \log|\eta_l^{(j)}|| \leqslant |\log|\xi^{(i)}| - \log|\xi^{(j)}||$$
$$+ \sum_{k=1}^{s-1} (|\log|\varepsilon_k^{(i)}|| + |\log|\varepsilon_k^{(j)}||) \leqslant 2c \quad (2 \leqslant i, \ j \leqslant s).$$

The theorem is proved.

Remark. It follows immediately from Theorem 2.1 that there requires only $c(\mathscr{F}_s)\cdot \log|n|$ elementary operations for finding the set of integers $(h_1, \cdots, h_s; n)$ satisfying (2.4).

III. Some Examples

1) Let p be a prime $\geqslant 5$, $r = \frac{1}{2}(p-1)$ and $q = r - 1 = \frac{1}{2}(p-3)$. The real cyclotomic field $\Re_r = R\left(2\cos\frac{2\pi}{p}\right)$ is an algebraic field of degree r. The field has a set of integral basis

$$\omega_1 = 2\cos\frac{2\pi}{p}, \quad \omega_2 = 2\cos\frac{4\pi}{p}, \cdots, \quad \omega_r = 2\cos\frac{2\pi r}{p}. \tag{3.1}$$

Let σ be a cyclic permutation of $(\omega_1, \cdots, \omega_r)$ $(\omega_1 \to \omega_2, \cdots, \omega_r \to \omega_1)$. Then from a number η $(= \eta^{(1)})$ of $\Re_r$, we have its q conjugates $\eta^{(2)}, \cdots, \eta^{(r)}$ under the q transformations $\sigma, \sigma^2, \cdots, \sigma^q$. Hence

$$S = \Omega'\Omega = pI - 2M,$$

where I is the identical matrix and $M = (m_{ij})$, in which $m_{ij} = 1$.

Taking a set of complete independent units $\varepsilon_1, \cdots, \varepsilon_q$ of $\Re_r$, then by Theorem 2.1, we construct a unit

$$\eta = \varepsilon_1^{a_1} \cdots \varepsilon_q^{a_q},$$

satisfying

$$|\eta| > 1, \quad e^{-2c}|\eta^{(j)}| \leqslant |\eta^{(i)}| \leqslant e^{2c}|\eta^{(j)}|, \quad (2 \leqslant i, \ j \leqslant r),$$

where $c = c(\Re_r) > 0$.

If

$$\eta = \sum_{i=1}^{r} k_i \omega_i,$$

then from

$$(h_1, \cdots, h_r) = (k_1, \cdots, k_r)S = (k_1, \cdots, k_r)(pI - M),$$

we have

$$h_i = pk_i - 2\sum_{j=1}^{r} k_j, \quad (1 \leqslant i \leqslant r).$$

Since

$$\sum_{i=1}^{r} \omega_i = -1,$$

hence it is found that

$$n = -\sum_{j=1}^{r} h_j = -\sum_{j=1}^{r} k_j .$$

Therefore, we have the simultaneous Diophantine approximations

$$\left| \frac{h_i}{n} - \omega_i \right| \leqslant c(\mathfrak{R}_r) n^{-1-\frac{1}{q}}, \quad (1 \leqslant i \leqslant r). \tag{3.2}$$

It is well-known that

$$\rho_l = \frac{\sin \frac{\pi}{p} g^{l+1}}{\sin \frac{\pi}{p} g^l}, \quad (1 \leqslant l \leqslant q)$$

is a set of complete independent units of the cyclotomic field $\mathfrak{R}_r$, where g denotes a primitive root mod p. It is more convenient to use the set of units

$$\omega_l = 2 \cos \frac{2\pi l}{p}, \quad (1 \leqslant l \leqslant q).$$

However, they do not always form an independent set. In order that they form a set of complete independent units, it is necessary and sufficient that (i) 2 is a primitive root mod p, or (ii) 2 belongs to the exponent r mod p and $p \equiv 7 \pmod 8$[8].

2) Let $p_1, \cdots, p_h$ be h distinct primes, $m = 2^h$ and $l = m - 1$. The real Dirichlet field $\mathscr{D}_m = R(\sqrt{p_1}, \cdots, \sqrt{p_h})$ is an algebraic field of degree m. Let

$$\varepsilon_1, \cdots, \varepsilon_l$$

be a set of complete independent units of $\mathscr{D}_m$. Take ε's as the solutions of Pell's equations

$$x^2 - p_{i_1} \cdots p_{i_k} y^2 = \pm 4$$

with the least of $\frac{x}{2} + \frac{\sqrt{p_{i_1} \cdots p_{i_k}}}{2} y$, where $k \geqslant 1$ and $1 \leqslant i_1 < \cdots < i_k \leqslant h$ is any choice of $1, 2, \cdots, h$. Suppose

$$\varepsilon_i = \begin{cases} \frac{x_i}{2} + \frac{\sqrt{d_i} y_i}{2}, & x_i \equiv y_i \equiv 1 \pmod 2 \quad (1 \leqslant i \leqslant \tau), \\ x_i + \sqrt{d_i}\, y_i, & (\tau + 1 \leqslant i \leqslant l), \end{cases}$$

where x_i and y_i are rational integers.

The integral basis of $\mathscr{D}_m$ is

$$\omega_1 = 1, \ \omega_2 = \varepsilon_1, \cdots, \omega_{\tau+1} = \varepsilon_\tau, \ \omega_{\tau+2} = \sqrt{d_{\tau+1}}, \cdots, \omega_m = \sqrt{d_l}.$$

Consider l transformations

$$(\sigma_{i_1, \cdots, i_k}) \quad \sqrt{p_\nu} \to \begin{cases} -\sqrt{p_\nu} \text{ for } \nu = i_j \quad (1 \leqslant j \leqslant k) \\ \sqrt{p_\nu}, \text{ otherwise,} \end{cases}$$

where $k \geqslant 1$ and $1 \leqslant i_1 < \cdots < i_k \leqslant h$ is any choice of 1, 2, $\cdots$, h. Applying these transformations to any element η $(= \eta^{(1)})$ of $\mathscr{D}_m$, we have its l conjugates $\eta^{(2)}$, $\cdots$, $\eta^{(m)}$. Form a matrix

$$\Omega = (\omega_i^{(j)}) \quad (1 \leqslant i, \ j \leqslant m),$$

then we obtain

$$S = \Omega' \Omega = \begin{pmatrix} A & 0 \\ 0 & B \end{pmatrix},$$

where $A = (a_{ij})$ $(1 \leqslant i, \ j \leqslant \tau + 1)$ and $B = (b_{\mu\nu})$ $(1 \leqslant \mu, \ \nu \leqslant l - \tau)$ in which $a_{11} = 2^h$, $a_{ii} = 2^{h-2}(x_{i-1}^2 + d_{i-1} y_{i-1}^2)$ $(2 \leqslant i \leqslant \tau + 1)$, $a_{1j} = a_{j1} = 2^{h-1} x_{j-1}$ $(2 \leqslant j \leqslant \tau + 1)$, $a_{ij} = 2^{h-2} x_{i-1} x_{j-1} (2 \leqslant i, \ j \leqslant \tau + 1, \ i \neq j)$, $b_{\mu\mu} = 2^h d_{\tau+\mu}$ $(1 \leqslant \mu \leqslant l - \tau)$ and $b_{\mu\nu} = 0$ $(\mu \neq \nu)$.

Construct a unit

$$\eta = \varepsilon_1^{a_1} \cdots \varepsilon_l^{a_l}$$

of $\mathscr{D}_m$ according to Theorem 2.1 such that

$$|\eta| > 1, \quad e^{-2c} |\eta^{(j)}| \leqslant |\eta^{(i)}| \leqslant e^{2c} |\eta^{(j)}| \quad (2 \leqslant i, j \leqslant m),$$

where $c = c(\mathscr{D}_m) > 0$.

If

$$\eta = \sum_{i=1}^{m} k_i \omega_i,$$

then from

$$(k_1, \ \cdots, \ k_m) \, S = (h_1, \ \cdots, \ h_m)$$

we have

$$n = h_1 = 2^{h-1} \left(2k_1 + \sum_{i=1}^{\tau} x_i k_{i+1} \right),$$

$$h_j = 2^{h-2} \left(2x_{j-1} k_1 + d_{j-1} y_{j-1}^2 k_j + \sum_{i=2}^{\tau+1} x_{i-1} \, x_{j-1} \, k_i \right) \quad (2 \leqslant j \leqslant \tau + 1),$$

$$h_j = 2^h k_j d_{j-1} \quad (\tau + 2 \leqslant j \leqslant m).$$

Hence the simultaneous Diophantine approximations

$$\left| \omega_i - \frac{h_i}{n} \right| \leqslant c(\mathscr{D}_m) \, n^{-1-\frac{1}{l}}, \quad (1 \leqslant i \leqslant m).$$

IV. Uniform Distribution

In this paper, we use $\boldsymbol{\gamma} = (\gamma_1, \ \cdots, \ \gamma_s)$ to denote the vector with real components and $\mathbf{m} = (m_1, \ \cdots, \ m_s)$ the vector with integral components. We also use the notations $\|\boldsymbol{\gamma}\| = \bar{\gamma}_1 \cdots \bar{\gamma}_s$, $|\boldsymbol{\gamma}| = |\gamma_1 \cdots \gamma_s|$ and $(\boldsymbol{\alpha}, \ \boldsymbol{\beta}) = \sum_{i=1}^{s} \alpha_i \beta_i$ (the scalar product of $\boldsymbol{\alpha}$ and $\boldsymbol{\beta}$). In this section, we shall prove the formula of Erdös-Turan-Koksma.

Theorem 4.1. *Let η be a number satisfying $\frac{1}{6} > \eta > 0$ and let h be an integer $> \frac{1}{\eta}$. Then for any $\boldsymbol{\gamma} \in G_s$, we have*

$$\left| \frac{1}{n_l} N_{n_l}(\boldsymbol{\gamma}) - |\boldsymbol{\gamma}| \right| < 2^{s+2} \varphi(n_l),$$

where

$$\varphi(n_l) = \sum_{|m_j| \leqslant h}' \frac{1}{\|\pi \mathbf{m}\|} \left| \frac{1}{n} \sum_{j=1}^{n_l} e^{2\pi i (P_{n_l}(j), \mathbf{m})} \right| + \frac{\log^s 9h}{\eta h} + 2^{s+1} \eta,$$

in which Σ' denotes a sum with an exception $\mathbf{m} = \mathbf{o} = (0, \cdots, 0)$.

Lemma 4.1.[13] *Let r be a positive integer, α, β be real numbers and Δ satisfy*

$$0 < \Delta < \frac{1}{2}, \quad \Delta \leqslant \beta - \alpha \leqslant 1 - \Delta.$$

Then there exists a function $\psi(x)$ with periodic 1 such that

(i) $\psi(x) = 1$, *for* $\alpha + \frac{1}{2}\Delta \leqslant x \leqslant \beta - \frac{1}{2}\Delta$,

(ii) $0 \leqslant \psi(x) \leqslant 1$, *for* $\alpha - \frac{1}{2}\Delta \leqslant x \leqslant \alpha + \frac{1}{2}\Delta$ *and* $\beta - \frac{1}{2}\Delta \leqslant x \leqslant \beta + \frac{1}{2}\Delta$,

(iii) $\psi(x) = 0$, *for* $\beta + \frac{1}{2}\Delta \leqslant x \leqslant 1 + \alpha - \frac{1}{2}\Delta$,

(iv) $\psi(x)$ *possesses a Fourier expansion*

$$\psi(x) = \beta - \alpha + \Sigma' C(m) e^{2\pi i m x},$$

of which the coefficients satisfy

$$|C(m)| \leqslant \min\left(\beta - \alpha, \frac{1}{\pi|m|}, \left(\frac{1}{\pi|m|}\right)^{r+1} \left(\frac{r}{\Delta}\right)^r\right).$$

Lemma 4.2. *Set $0 < \delta \leqslant \varphi(n_l)$. If a uniformly distributed sequence of sets $(P_{n_l}(j))$ $(n_1 < n_2 < \cdots)$ satisfies*

$$\left| \frac{1}{n_l} N_{n_l}(\boldsymbol{\gamma}) - |\boldsymbol{\gamma}| \right| < \varphi(n_l)$$

for $\delta \leqslant \gamma_i \leqslant 1 - \delta$ $(1 \leqslant i \leqslant s)$, then

$$\left| \frac{1}{n_l} N_{n_l}(\boldsymbol{\gamma}) - |\boldsymbol{\gamma}| \right| < 2^{s+2} \varphi(n_l)$$

holds for all $\boldsymbol{\gamma} \in G_s$.

Proof. For simplicity, hereafter we omit the index l of n_l and $h^{(l)}$.

1) Set $\delta \leqslant \alpha_i < \beta_i \leqslant 1 - \delta$ $(1 \leqslant i \leqslant s)$ and let $N_n(\boldsymbol{\alpha}, \boldsymbol{\beta})$ be the number of points $P_n(j)$ $(1 \leqslant j \leqslant n)$ satisfying

$$\alpha_i \leqslant x_i^{(n)}(j) < \beta_i \quad (1 \leqslant i \leqslant s).$$

We can prove easily that

$$\left| \frac{1}{n} N_n(\boldsymbol{\alpha}, \boldsymbol{\beta}) - |\boldsymbol{\beta} - \boldsymbol{\alpha}| \right| < 2^s \varphi(n).$$

2) Let $\mathscr{D}$ be the domain

$$\delta \leqslant x_i < 1 - \delta \quad (1 \leqslant i \leqslant s)$$

and $\bar{\mathscr{D}}$ denote the complementary set of $\mathscr{D}$ in G_s. The measure of $\bar{\mathscr{D}}$ is $\leqslant 2^s\delta \leqslant 2^s\varphi(n)$. By 1), the number of points $P_n(j)$ belonging to $\mathscr{D}$ equals

$$(1 - 2\delta)^s n + \vartheta 2^s n\varphi(n),$$

where ϑ and ϑ's that will appear later are not the same, but with absolute value $\leqslant 1$. Therefore, the number of points of $P_n(j)$ belonging to $\bar{\mathscr{D}}$ does not exceed

$$(1 - (1 - 2\delta)^s)n + 2^s\varphi(n)n \leqslant 2^{s+1}\varphi(n)n.$$

3) Combining 1) with 2), we deduce that for an arbitrary $\boldsymbol{\gamma} \in G_s$ we have

$$\left| \frac{1}{n} N_n(\boldsymbol{\gamma}) - |\boldsymbol{\gamma}| \right| < (2^s + 2^s + 2^{s+1})\varphi(n) = 2^{s+2}\varphi(n).$$

The lemma is proved.

The Proof of Theorem 4.1. By Lemma 4.2, it is sufficient to prove that

$$\left| \frac{1}{n} N_n(\boldsymbol{\gamma}) - |\boldsymbol{\gamma}| \right| < \varphi(n)$$

under the conditions

$$3\eta < \gamma_i \leqslant 1 - 3\eta \quad (1 \leqslant i \leqslant s).$$

Introducing function

$$G_x(y) = \begin{cases} 1, & \text{for } 0 \leqslant y < x, \\ 0, & \text{for } x \leqslant y < 1, \end{cases}$$

we have evidently

$$\frac{1}{n} N_n(\mathbf{x}) = \frac{1}{n} \sum_{j=1}^{n} G_{x_1}(x_1^{(n)}(j)) \cdots G_{x_s}(x_s^{(n)}(j)). \tag{4.1}$$

For $3\eta \leqslant x \leqslant 1 - 3\eta$, we construct according to Lemma 4.1 two auxiliary functions $G_x^{(1)}(y)$ and $G_x^{(2)}(y)$. Function $G_x^{(1)}(y)$ satisfies

(i) $G_x^{(1)}(y) = 1$, for $2\eta \leqslant y \leqslant x - \eta$,

(ii) $0 \leqslant G_x^{(1)}(y) \leqslant 1$, for $\eta \leqslant y \leqslant 2\eta$ and $x - \eta \leqslant y \leqslant x$,

(iii) $G_x^{(1)}(y) = 0$, for $x \leqslant y \leqslant 1 + \eta$,

(iv) $G_x^{(1)}(y)$ has a Fourier expansion

$$G_x^{(1)}(y) = x - 2\eta + \Sigma' C_1(m) e^{2\pi i m y},$$

where

$$|C_1(m)| \leqslant \min\left(x - 2\eta, \frac{1}{\pi|m|}, \frac{1}{\eta\pi^2 m^2}\right).$$

Function $G_x^{(2)}(y)$ satisfies

(ⅰ)′ $G_x^{(2)}(y) = 1$, for $-\eta \leqslant y \leqslant x$,

(ⅱ)′ $0 \leqslant G_x^{(2)}(y) \leqslant 1$, for $-2\eta \leqslant y \leqslant -\eta$ and $x \leqslant y \leqslant x + \eta$,

(ⅲ)′ $G_x^{(2)}(y) = 0$, for $x + \eta \leqslant y \leqslant 1 - 2\eta$,

(ⅳ)′ $G_x^{(2)}(y)$ has a Fourier expansion

$$G_x^{(2)}(y) = x + 2\eta + \Sigma' C_2(m) e^{2\pi i m y},$$

where

$$|C_2(m)| \leqslant \min\left(x + 2\eta, \ \frac{1}{\pi |m|}, \ \frac{1}{\eta \pi^2 m^2}\right).$$

From (ⅳ) it follows

$$G_x^{(1)}(y) = x - 2\eta + \sum_{|m| \leqslant h}{}' C_1(m) e^{2\pi i m y} + \vartheta \sum_{|m| > h} \frac{1}{\eta \pi^2 m^2}$$

$$= x - 2\eta + \sum_{|m| \leqslant h}{}' C_1(m) e^{2\pi i m y} + \frac{2\vartheta}{\pi^2 \eta h},$$

where $\displaystyle\sum_{m > h} \frac{1}{m^2} \leqslant \int_h^\infty \frac{dt}{t^2} = h^{-1}$. Put $x = x_i$ and $y = y_i$ in the above formula and write $C_0^{(1)} = x_i - 2\eta$ for all i. Then we have s equations. Multiplying separatively the left hand sides and the right hand sides of these equations, we have

$$G_{x_1}^{(1)}(y_1) \cdots G_{x_s}^{(1)}(y_s) = \sum_{|m_i| \leqslant h} C_1(m_1) \cdots C_1(m_s) e^{2\pi i(m_1 y_1 + \cdots + m_s y_s)}$$

$$+ \frac{2\vartheta}{\pi^2 \eta h}\left(1 + \sum_{|m| \leqslant h} |C_1(m)|\right)^s = (x_1 - 2\eta) \cdots (x_s - 2\eta)$$

$$+ \sum_{|m_i| \leqslant h}{}' C_1(m_1) \cdots C_1(m_s) e^{2\pi i(m_1 y_1 + \cdots + m_s y_s)} + \frac{\vartheta \log^s 9h}{\eta h}$$

$$\cdot \left(1 + \sum_{|m| \leqslant h} |C_1(m)| \leqslant 2 + \frac{2}{\pi} + \frac{2}{\pi} \int_1^h \frac{dt}{t} < 2 + \log h < \log 9h\right).$$

Consequently, it is found that

$$G_{x_1}^{(1)}(y_1) \cdots G_{x_s}^{(1)}(y_s) = x_1 \cdots x_s + \sum_{|m_i| \leqslant h}{}' C_1(m_1) \cdots C_1(m_s) e^{2\pi i(m_1 y_1 + \cdots + m_s y_s)}$$

$$+ \vartheta\left(\frac{\log^s 9h}{\eta h} + 2^{s+1}\eta\right). \tag{4.2}$$

Similarly, we find

$$G_x^{(2)}(y_1) \cdots G_{x_s}^{(2)}(y_s) = x_1 \cdots x_s + \sum_{|m_i| \leqslant h}{}' C_2(m_1) \cdots C_2(m_s) e^{2\pi i(m_1 y_1 + \cdots + m_s y_s)}$$

$$+ \vartheta\left(\frac{\log^s 9h}{\eta h} + 2^{s+1}\eta\right). \tag{4.3}$$

From the definition of $G_x^{(1)}(y)$ and $G_x^{(2)}(y)$, we have

$$G_{x_1}^{(1)}(y_1) \cdots G_{x_s}^{(1)}(y_s) \leqslant G_{x_1}(y_1) \cdots G_{x_s}(y_s) \leqslant G_{x_1}^{(2)}(y_1) \cdots G_{x_s}^{(2)}(y_s). \tag{4.4}$$

Hence the theorem follows from (4.1), (4.2), (4.3) and (4.4).

V. The Proof of Theorem 1

Since 1, $\omega_1, \cdots, \omega_q$ are linearly independent over rational field R, Theorem 1 follows from Lemma 1.1 and the following

Theorem 5.1. *If for any vector* $\mathbf{m} \neq \mathbf{o}$, *the inequality*

$$\langle(\mathbf{m}, \boldsymbol{\gamma})\rangle > b\|\mathbf{m}\|^{-a} \tag{5.1}$$

holds, where a *and* b *are constants satisfying* $s+1 \geqslant a > 1$ *and* $1 \geqslant b > 0$, *then the sequence*

$$P(j) = (\{\gamma_1 j\}, \cdots, \{\gamma_s j\}) \quad (j = 1, 2\cdots)$$

has discrepancy

$$\varphi(n) = c(a, b, s) n^{-1+2s(a-1)} (\log 3n)^{1+s\delta_{1,a}},$$

where $\delta_{\alpha,\beta}$ *denotes Kronecker symbol.*

Lemma 5.1. *Let* δ *be any real number. Then*

$$\left|\sum_{j=1}^{n} e^{2\pi i\delta j}\right| \leqslant \min\left(n, \frac{1}{2\langle\delta\rangle}\right).$$

Proof. If δ is not an integer, then we have

$$\left|\sum_{j=1}^{n} e^{2\pi i\delta j}\right| = \left|\frac{e^{2\pi i\delta(n+1)} - e^{2\pi i\delta}}{e^{2\pi i\delta} - 1}\right| \leqslant \frac{1}{|\sin \pi\delta|} \leqslant \frac{1}{2\langle\delta\rangle}.$$

Hence we have the lemma.

Lemma 5.2. *Let* $g(\mathbf{m})$ *be a non-negative function of* $\mathbf{m}$. *Then*

$$\sum_{|m_i|\leqslant h}{}' \frac{g(\mathbf{m})}{\|\mathbf{m}\|} \leqslant \sum_{l=0}^{s} \sum_{\mathbf{i}} \frac{1}{h^l} \sum_{m_{i_{l+1}}=1}^{h} \cdots \sum_{m_{i_s}=1}^{h} \frac{1}{m_{i_{l+1}}^2 \cdots m_{i_s}^2}$$
$$\times \sum_{|k_{i_1}|\leqslant h+1} \cdots \sum_{|k_{i_l}|\leqslant h+1} \sum_{|k_{i_{l+1}}|\leqslant m_{i_{l+1}}} \cdots \sum_{|k_{i_s}|\leqslant m_{i_s}}{}' g(\mathbf{k}),$$

where $\sum_{\mathbf{i}}$ *denotes a sum in which* $\mathbf{i} = (i_1, \cdots, i_s)$ *runs over all permutations of* $(1, 2, \cdots, s)$.

Proof. Since

$$\sum_{|m|\leqslant h} \frac{g(m)}{\overline{m}} = g(0) + \sum_{m=1}^{h} \frac{1}{m}(g(m) + g(-m))$$
$$= g(0) + \sum_{m=1}^{h} \left(\frac{1}{m} - \frac{1}{m+1}\right) \sum_{1\leqslant k\leqslant m} (g(k) + g(-k))$$
$$+ \frac{1}{h+1} \sum_{|k|\leqslant h+1}{}' g(k) \leqslant \sum_{m=1}^{h} \frac{1}{m^2} \sum_{|k|\leqslant m} g(k) + \frac{1}{h} \sum_{|k|\leqslant h+1} g(k),$$

we have

$$\sum_{|m_i|\leqslant h}' \frac{g(\mathbf{m})}{\|\mathbf{m}\|} \leqslant \sum_{|m_i|\leqslant h}' \frac{1}{\bar{m}_1\cdots\bar{m}_{s-1}} \left(\sum_{m_s=1}^{h} \frac{1}{m_s^2} \sum_{|k_s|\leqslant m_s} g(m_1, \cdots, m_{s-1}, k_s) \right.$$

$$\left. + \frac{1}{h} \sum_{|k_s|\leqslant h+1} g(m_1, \cdots, m_{s-1}, k_s) \right) + \sum_{|m_i|\leqslant h} \frac{1}{\bar{m}_1\cdots\bar{m}_{s-1}}$$

$$\times \left(\sum_{m_s=1}^{h} \frac{1}{m_s^2} \sum_{|k_s|\leqslant m_s}' g(m_1, \cdots, m_{s-1}, k_s) + \frac{1}{h} \sum_{|k_s|\leqslant h+1}' g(m_1, \cdots, m_{s-1}, k_s) \right)$$

$$\leqslant \cdots \leqslant \sum_{l=0}^{s} \sum_{\mathbf{i}} \frac{1}{h^l} \sum_{m_{i_{l+1}}=1}^{h} \cdots \sum_{m_{i_s}=1}^{h} \frac{1}{m_{i_{l+1}}^2 \cdots m_{i_s}^2}$$

$$\times \sum_{|k_{i_1}|\leqslant h+1} \cdots \sum_{|k_{i_l}|\leqslant h+1} \sum_{|k_{i_{l+1}}|\leqslant m_{i_{l+1}}} \cdots \sum_{|k_{i_s}|\leqslant m_{i_s}}' g(\mathbf{k}).$$

The lemma is proved.

Lemma 5.3. *Set* $\mathbf{m} \neq \mathbf{o}$ *and* $Q = [2^{sa}\|\mathbf{m}\|^a b^{-1}] + 1$. *If* (5.1) *holds for any* $\mathbf{m} \neq \mathbf{o}$, *then in any interval* $(P, P + Q^{-1}]$, *there contains at most a point* $(\mathbf{k}, \boldsymbol{\gamma}) = \sum_{i=1}^{s} k_i\gamma_i$, *where* $\mathbf{k}$ *is a vector with integral components which satisfy* $|k_i| \leqslant |m_i|$ $(1 \leqslant i \leqslant s)$.

Proof. If there contain two points $(\mathbf{k}', \boldsymbol{\gamma})$ and $(\mathbf{k}'', \boldsymbol{\gamma})$ in the interval $(P, P + Q^{-1}]$, where $\mathbf{k}' \neq \mathbf{k}''$, $|k_i'| \leqslant |m_i|$ and $|k_i''| \leqslant |m_i|$ $(1 \leqslant i \leqslant s)$, then we obtain

$$\langle(\mathbf{k}' - \mathbf{k}'', \boldsymbol{\gamma})\rangle \leqslant Q^{-1}.$$

On the other hand, from (5.1), we have

$$\langle(\mathbf{k}' - \mathbf{k}'', \boldsymbol{\gamma})\rangle > b\|\mathbf{k}' - \mathbf{k}''\|^{-a} \geqslant 2^{-sa}b\|\mathbf{m}\|^{-a} > Q^{-1},$$

which leads to a contradiction. Hence we have the lemma.

Lemma 5.4. *Set* $\mathbf{m} \neq \mathbf{o}$ *and* $Q = [2^{sa}\|\mathbf{m}\|^a b^{-1}] + 1$. *If* (5.1) *holds for any* $\mathbf{m} \neq \mathbf{o}$, *then*

$$\sum_{|k_i|\leqslant|m_i|}' \frac{1}{\langle(\mathbf{k}, \boldsymbol{\gamma})\rangle} \leqslant 4Q \log 3Q.$$

Proof. Divide the interval $(0, 1]$ into Q subintervals

$$I_j = \left(\frac{j}{Q}, \frac{j+1}{Q}\right] \quad (j = 0, 1, \cdots, Q-1).$$

By (5.1), we deduce that none of the points $(\mathbf{k}, \boldsymbol{\gamma})$ lies in the interval I_0, where $\mathbf{k} \neq \mathbf{o}$ and $|k_i| \leqslant |m_i|$ $(1 \leqslant i \leqslant s)$. It follows by Lemma 5.3 that there contains at most a point $(\mathbf{k}, \boldsymbol{\gamma})$ in any interval I_j, where $j \geqslant 1$. Hence

$$\sum_{|k_i|\leqslant|m_i|}' \frac{1}{\langle(\mathbf{k}, \boldsymbol{\gamma})\rangle} \leqslant 4 \sum_{j=1}^{Q-1} \frac{Q}{j} \leqslant 4Q \log 3Q.$$

The lemma follows.

Lemma 5.5. *Let h be an integer $\geqslant 2$. If (5.1) holds for any $\mathbf{m} \neq \mathbf{o}$, then*

$$\sum_{|m_i| \leqslant h}' \frac{1}{\|\mathbf{m}\| \langle (\mathbf{m}, \boldsymbol{\gamma}) \rangle} \leqslant c(a, b, s) h^{s(a-1)} (\log h)^{1+s\delta_{1,a}}.$$

Proof. From Lemmas 5.2 and 5.4, we have

$$\sum_{|m_i| \leqslant h}' \frac{1}{\|\mathbf{m}\| \langle (\mathbf{m}, \boldsymbol{\gamma}) \rangle} \leqslant \sum_{l=0}^{s} \sum_{\mathbf{i}} \frac{1}{h^l} \sum_{m_{i_{l+1}}=1}^{h} \cdots \sum_{m_{i_s}=1}^{h} \frac{1}{m_{i_{l+1}}^2 \cdots m_{i_s}^2}$$

$$\times \sum_{|k_{i_1}| \leqslant h+1} \cdots \sum_{|k_{i_l}| \leqslant h+1} \sum_{|k_{i_{l+1}}| \leqslant m_{i_{l+1}}} \cdots \sum_{|k_{i_s}| \leqslant m_{i_s}}' \frac{1}{\langle (\mathbf{k}, \boldsymbol{\gamma}) \rangle}$$

$$\leqslant \sum_{l=0}^{s} \sum_{\mathbf{i}} \frac{1}{h^l} \sum_{m_{i_{l+1}}=1}^{h} \cdots \sum_{m_{i_s}=1}^{h} \frac{1}{m_{i_{l+1}}^2 \cdots m_{i_s}^2} c(a, b, s) h^{la} (m_{i_{l+1}} \cdots m_{i_s})^a \log h$$

$$\leqslant c(a, b, s) h^{s(a-1)} (\log h)^{1+s\delta_{1,a}}.$$

The lemma is proved.

The Proof of Theorem 5.1. By Lemmas 5.1 and 5.5, we have

$$\sum_{|m_i| \leqslant h}' \frac{1}{\|\mathbf{m}\|} \left| \frac{1}{n} \sum_{j=1}^{n} e^{2\pi i (\mathbf{m}, \boldsymbol{\gamma}) j} \right| \leqslant \frac{1}{n} \sum_{|m_i| \leqslant h}' \frac{1}{2\|\mathbf{m}\| \langle (\mathbf{m}, \boldsymbol{\gamma}) \rangle}$$

$$\leqslant c(a, b, s) n^{-1} h^{s(a-1)} (\log h)^{1+s\delta_{1,a}}.$$

Take $\eta = \dfrac{1}{7n}$ and $h = 8n^2$. Then we have the theorem by Theorem 4.1.

VI. The Proof of Theorem 2

We use P_M^s to denote the s-dimensional parallelopiped with edges parallel to coordinate axes and volume $\leqslant M$, and $\mathbf{u} = (u_0, \cdots, u_s)$ and $\mathbf{h} = (1, h_1, \cdots, h_s)$ the $(s+1)$-dimensional vectors with integral components.

Evidently, we may deduce Theorem 2 from Lemma 1.1 and the following two theorems.

Theorem 6.1. *Let n be an integer > 1 and M be a number $\geqslant 1$. Further let $\mathbf{a} = (a_1, \cdots, a_s)$ be a vector with integral components. If the congruence*

$$(\mathbf{a}, \mathbf{m}) = \sum_{i=1}^{s} a_i m_i \equiv 0 \pmod{n} \tag{6.1}$$

has no solution in the domain

$$\|\mathbf{m}\| \leqslant M, \quad \mathbf{m} \neq \mathbf{o}, \tag{6.2}$$

then the set

$$\left(\left\{ \frac{a_1 j}{n} \right\}, \cdots, \left\{ \frac{a_s j}{n} \right\} \right), \quad (1 \leqslant j \leqslant n),$$

has discrepancy

$$\varphi(n) = c(s, \varepsilon) M^{-1+\varepsilon}.$$

Theorem 6.2. *Let*

$$\left|\frac{h_i}{n} - \gamma_i\right| \leqslant dn^{-1-\frac{1}{s}}, \quad (1 \leqslant i \leqslant s), \tag{6.3}$$

be the simultaneous approximations of γ's by rationals, where d is a positive constant. If (5.1) holds for any $\mathbf{m} \neq \mathbf{o}$, then there exists constant $c(a, b, d, s)$ (< 1) such that the congruence

$$(\mathbf{h}, \mathbf{u}) = u_0 + \sum_{i=1}^{s} h_i u_i \equiv 0 \pmod{n} \tag{6.4}$$

has no solution in the domain

$$\|\mathbf{u}\| \leqslant c(a, b, d, s) n^{\left(1+\frac{1}{s}\right)/(a+1)}, \quad \mathbf{u} \neq \mathbf{o}. \tag{6.5}$$

Lemma 6.1. *Let l be an integer $\geqslant 1$. Then the s-dimensional domain*

$$\|\mathbf{m}\| < lM \tag{6.6}$$

can be covered by at most $c(\varepsilon)^s l^{1+\varepsilon} M^\varepsilon$ parallelopipeds of the type P_M^s.

Proof. Take

$$c(\varepsilon) = 2^{2+\varepsilon} \sum_{j=0}^{\infty} (\bar{j}^{-(1+\varepsilon)} + 2^{-\varepsilon j}).$$

1) For $s = 1$, since the domain (6.6) is the interval $(-lM, lM)$, it can be covered by at most

$$\frac{2lM}{M} = 2l$$

intervals of the type $[c, c + M]$, where c is a real number. Hence the lemma is true for $s = 1$.

2) Suppose k is a positive integer and assume that the lemma holds for $s = 1, \cdots, k$. Now we proceed to prove the validity of the lemma for $s = k + 1$.

Divide the domain

$$\bar{m}_1 \cdots \bar{m}_{k+1} < lM \tag{6.7}$$

into $2[\log_2 M] + 3$ sub-domains

(i) $m_{k+1} = j, \quad \bar{j} \leqslant l,$

(ii) $2^i l < |m_{k+1}| \leqslant 2^{i+1} l, \quad (i = 0, 1, \cdots, [\log_2 M])$

by the hyperplanes

$$m_{k+1} = 0, \quad \pm 2^i l, \quad (i = 0, 1, \cdots, [\log_2 M]).$$

3) Suppose that $m_{k+1} = j$. Then

$$\bar{m}_1 \cdots \bar{m}_k < \frac{lM}{\bar{j}} < \left(\left[\frac{l}{\bar{j}}\right] + 1\right) M.$$

Hence by our inductive hypothesis, we see that the above k-dimensional domain can be covered by at most

$$Q = c(\varepsilon)^k \left(\left[\frac{l}{\bar{j}}\right] + 1\right)^{1+\varepsilon} M^\varepsilon$$

parallelopipeds of the type P_M^k. Using these P_M^k as bases and 1 as height, we construct P_M^{k+1}. Hence the sub-domain $m_{k+1} = j$ can be covered by at most Q parallelopipeds of the type P_M^{k+1}. Consequently, the domain defind by (i) can be covered by at most

$$2 \sum_{j=0}^{l} c(\varepsilon)^k \left(\left[\frac{l}{j}\right] + 1\right)^{1+\varepsilon} M^\varepsilon \tag{6.8}$$

parallelopipeds of the type P_M^{k+1}.

4) Consider the sub-domain of (6.7)

$$2^i l < m_{k+1} \leqslant 2^{i+1} l. \tag{6.9}$$

Then for $m_{k+1} = 2^i l + 1$, we have

$$\bar{m}_1 \cdots \bar{m}_k < \frac{M}{2^i}.$$

Hence according to our inductive hypothesis, the above domain can be covered by at most

$$c(\varepsilon)^k \left(\frac{M}{2^i}\right)^\varepsilon \tag{6.10}$$

parallelopipeds of the type $P_{M/2^i}^k$. Using the $P_{M/2^i}^k$ as base and 2^i as height, we construct P_M^{k+1}. Since $2^i l + 2^i l + 1 > 2^{i+1} l$, the domain (6.9) can be covered by at most $c(\varepsilon)^k l \left(\frac{M}{2^i}\right)^\varepsilon$ parallelopipeds of the type P_M^{k+1}. Consequently, the domain defined by (ii) can be covered by at most

$$2c(\varepsilon)^k \sum_{i=0}^{[\log_2 M]} l \left(\frac{M}{2^i}\right)^\varepsilon \tag{6.11}$$

parallelopipeds of the type P_M^{k+1}.

5) It follows by (6.8) and (6.11) that the domain (6.7) can be covered by at most

$$c(\varepsilon)^k l^{1+\varepsilon} M^\varepsilon \sum_{j=0}^{\infty} \left(\frac{2^{2+\varepsilon}}{j^{1+\varepsilon}} + \frac{2}{2^{\varepsilon j}}\right) \leqslant c(\varepsilon)^{k+1} l^{1+\varepsilon} M^\varepsilon$$

parallelopipeds of the type P_M^{k+1}. Hence the lemma follows by mathematical induction.

Remark. The term $c(\varepsilon)^s l^{1+\varepsilon} M^\varepsilon$ in the above lemma may be replaced by $c(s) l \log^{s-1} \cdot 3lM$[12].

Lemma 6.2. *Let T_M^l be the number of solutions of the congruence* (6.1) *in the domain* (6.6). *If the congruence* (6.1) *has no solution in the domain* (6.2), *then we have*

$$T_M^l \leqslant c(\varepsilon)^s l^{1+\varepsilon} M^\varepsilon.$$

Proof. By Lemma 6.1, it is sufficient to prove that the congruence (6.1) has at most 1 solution in any parallelopiped of the type P_M^s. Suppose that the congruence (6.1) has two solutions $\mathbf{m}'$ and $\mathbf{m}''$ in certain P_M^s, where $\mathbf{m}' \neq \mathbf{m}''$. Let $\mathbf{m} = \mathbf{m}' - \mathbf{m}''$. Then $\|\mathbf{m}\| \leqslant M$ and

$$(\mathbf{a}, \mathbf{m}) = (\mathbf{a}, \mathbf{m}') - (\mathbf{a}, \mathbf{m}'') \equiv 0 \pmod{n}.$$

This leads to a contradiction. Hence we have the lemma.

The Proof of Theorem 6.1. Take $3\delta s = \varepsilon$. By Lemma 6.2, we have

$$\sum_{|m_i|\leqslant h}' \frac{1}{\|\mathbf{m}\|}\left|\frac{1}{n}\sum_{j=1}^{n} e^{2\pi i(\mathbf{a},\,\mathbf{m})j/n}\right| = \sum_{\substack{|m_i|\leqslant h\\ (\mathbf{a},\,\mathbf{m})\equiv 0 \pmod n}}' \frac{1}{\|\mathbf{m}\|} \leqslant \sum_{l=1}^{h^s} \frac{(T_M^{l+1}-T_M^l)}{lM}$$

$$= \frac{1}{M}\sum_{l=1}^{h^s} T_M^{l+1}\left(\frac{1}{l}-\frac{1}{l+1}\right) + \frac{T_M^{h^s+1}}{(h^s+1)M} \leqslant c(\delta)^s M^{-1+\delta} h^{s\delta}.$$

$(T_M^1 = 0)$. Take $\eta = \frac{1}{7M}$ and $h = 7([M]+1)^2$. Then the theorem follows by Theorem 4.1.

The Proof of Theorem (6.2). Let $\mathbf{u} \neq \mathbf{o}$ be a solution of the congruence (6.4). If $u_i = 0$ $(1 \leqslant i \leqslant s)$, we obtain $u_0 \neq 0$. From (6.4), we have $u_0 \equiv 0 \pmod n$. Hence the $\|\mathbf{u}\| \geqslant n$. Consequently, $\mathbf{u}$ is not belonging to the domain (6.5). Therefore we may suppose $(u_1, \cdots, u_s) \neq (0, \cdots, 0)$. If

$$\bar{u}_1\cdots\bar{u}_s \geqslant \left(\frac{b}{2ds}\right)^{\frac{1}{a+1}} n^{\left(1+\frac{1}{s}\right)/(a+1)},$$

then it appears

$$\|\mathbf{u}\| \geqslant \left(\frac{b}{2ds}\right)^{\frac{1}{a+1}} n^{\left(1+\frac{1}{s}\right)/(a+1)},$$

hence we have the theorem. Now suppose

$$\bar{u}_1\cdots\bar{u}_s < \left(\frac{b}{2ds}\right)^{\frac{1}{a+1}} n^{\left(1+\frac{1}{s}\right)/(a+1)}.$$

Since

$$\langle \alpha - \beta\rangle \geqslant \langle\alpha\rangle - \langle\beta\rangle,$$

from (5.1) and (6.3), we have

$$\frac{|u_0|}{n} \geqslant \left\langle \frac{u_0}{n}\right\rangle = \left\langle \frac{1}{n}\sum_{i=1}^{n} h_i u_i\right\rangle \geqslant \left\langle \sum_{i=1}^{s} \gamma_i u_i\right\rangle - \left\langle \sum_{i=1}^{s}\left(\frac{h_i}{n}-\gamma_i\right)u_i\right\rangle$$

$$\geqslant \frac{b}{(\bar{u}_1\cdots\bar{u}_s)^a} - \frac{ds}{n^{1+\frac{1}{s}}}(\bar{u}_1\cdots\bar{u}_s).$$

Therefore

$$|u_0|\bar{u}_1\cdots\bar{u}_s > \frac{1}{2}(2ds)^{\frac{a-1}{a+1}} b^{\frac{2}{a+1}} n^{\left(1+\frac{1}{s}\right)/(a+1)}.$$

The theorem follows.

VII. The Proof of Theorem 3

We use the notations

$$I(f) = \int_0^1\cdots\int_0^1 f(x_1, \cdots, x_s)\, dx_1\cdots dx_s$$

and

$$P_n(f) = I(f) - \frac{1}{(2n+1)^l} \sum_{j=-nl}^{nl} \mu_{n,l,j} f(j\boldsymbol{\gamma}).$$

Clearly, Theorem 3 is the consequence of Lemma 1.1 and the following

Theorem 7.1. *If* (5.1) *holds for any* $\mathbf{m} \neq \mathbf{o}$, *then*

$$\operatorname*{Sup}_{f \in E_s^\alpha(C)} |P_n(f)| \leqslant C \cdot c(a, b, \alpha, s)\, n^{(-\alpha+s\alpha(a-1))/(a-1)} (\log 3n)^{\alpha+s\alpha\delta_{1,\alpha}}.$$

Proof. Since

$\left|\frac{\sin h\pi\delta}{h \sin \pi\delta}\right| \leqslant 1$ (where h is a positive integer and δ is a real number but not an integer),

$$\frac{1}{(2n+1)^l} \sum_{j=-nl}^{nl} \mu_{n,l,j} f(j\boldsymbol{\gamma}) = \frac{1}{(2n+1)^l} \sum C(\mathbf{m}) \sum_{j=-nl}^{nl} \mu_{n,l,j}\, e^{2\pi i(\mathbf{m},\boldsymbol{\gamma})j}$$

$$= C(\mathbf{o}) + \frac{1}{(2n+1)^l} \sum{}' C(\mathbf{m}) \left(\sum_{j=-n}^{n} e^{2\pi i(\mathbf{m},\boldsymbol{\gamma})j} \right)^l$$

$$= C(\mathbf{o}) + \sum{}' C(\mathbf{m}) \left(\frac{\sin (2n+1)\pi(\mathbf{m},\boldsymbol{\gamma})}{(2n+1)\sin \pi(\mathbf{m},\boldsymbol{\gamma})} \right)^l$$

and

$$C(\mathbf{o}) = I(f),$$

then we have

$$\operatorname*{Sup}_{f \in E_s^\alpha(C)} |P_n(f)| \leqslant C \sum{}' \frac{1}{\|\mathbf{m}\|^\alpha} \left| \frac{\sin (2n+1)\pi(\mathbf{m},\boldsymbol{\gamma})}{(2n+1)\sin \pi(\mathbf{m},\boldsymbol{\gamma})} \right|^\alpha = C(\Sigma_1 + \Sigma_2), \tag{7.1}$$

where Σ_1 denotes a sum in which the $\mathbf{m}$'s satisfy the conditions $|m_i| \leqslant n^{\frac{a}{a-1}}$ $(1 \leqslant i \leqslant s)$ and $\mathbf{m} \neq \mathbf{o}$, while Σ_2 the remaining part.

For $\alpha > 0$, $a_i > 0$ and $\sum_i a_i < \infty$, we have

$$\sum_i a_i^\alpha = \sum_i \left(\frac{a_i}{\sum_j a_j} \right)^\alpha \left(\sum_k a_k \right)^\alpha \leqslant \sum_i \frac{a_i}{\sum_j a_j} \left(\sum_k a_k \right)^\alpha = \left(\sum_k a_k \right)^\alpha.$$

Hence by Lemmas 5.1 and 5.5 it is found that

$$\sum{}_1 \leqslant \frac{1}{2^\alpha (2n+1)^\alpha} \sum_{|m_i| \leqslant n^{\frac{a}{a-1}}}{}' \frac{1}{\|\mathbf{m}\|^\alpha \langle(\mathbf{m},\boldsymbol{\gamma})\rangle^\alpha}$$

$$\leqslant \frac{1}{2^\alpha (2n+1)^\alpha} \left(\sum_{|m_i| \leqslant n^{\frac{a}{a-1}}}{}' \frac{1}{\|\mathbf{m}\| \langle(\mathbf{m},\boldsymbol{\gamma})\rangle} \right)^\alpha$$

$$\leqslant c(a, b, \alpha, s)\, n^{-\alpha+s\alpha(a-1)/(a-1)} (\log 3n)^{\alpha+s\alpha\delta_{1,\alpha}}. \tag{7.2}$$

It is evidently

$$\sum\nolimits_2 \leqslant \sum_{i=1}^{s} \sum_{|m_i| > n^{\frac{\alpha}{\alpha-1}}} \frac{1}{|m_i|^{\alpha}} \sum \frac{1}{\|\mathbf{m}\|^{\alpha}} \leqslant c(\alpha, s)\, n^{-\alpha}. \tag{7.3}$$

Substituting (7.2) and (7.3) into (7.1), we have the theorem.

VIII. The Proof of Theorem 4

Use the notation

$$Q_n(f) = I(f) - \frac{1}{n} \sum_{j=1}^{n} f\left(\frac{j\mathbf{a}}{n}\right).$$

Theorem 4 follows immediately from Lemma 1.1, Theorem 6.2 and the following

Theorem 8.1. *If the congruence* (6.1) *has no solution in the domain* (6.2), *then we have*

$$\sup_{f \in E_s^{\alpha}(C)} |Q_n(f)| \leqslant C \cdot c(\alpha, \varepsilon)^s M^{-\alpha+\varepsilon}.$$

Proof. Clearly, we may suppose that $\varepsilon < \alpha - 1$. Since

$$\frac{1}{n} \sum_{j=1}^{n} f\left(\frac{j\mathbf{a}}{n}\right) = \frac{1}{n} \sum_{j=1}^{n} \sum C(\mathbf{m}) e^{2\pi i(\mathbf{a},\, \mathbf{m})j/n}$$

$$= C(\mathbf{o}) + \sum{}' C(\mathbf{m}) \frac{1}{n} \sum_{j=1}^{n} e^{2\pi i(\mathbf{a},\, \mathbf{m})j/n} = C(\mathbf{o}) + \sum_{(\mathbf{a},\mathbf{m}) \equiv 0 \pmod n}{}' C(\mathbf{m}),$$

from Lemma 6.2, it is known

$$\sup_{f \in E_s^{\alpha}(C)} |Q_n(f)| \leqslant C \sum_{(\mathbf{a},\mathbf{m}) \equiv 0 \pmod n}{}' \frac{1}{\|\mathbf{m}\|^{\alpha}} \leqslant C \sum_{l=1}^{\infty} \frac{(T_M^{l+1} - T_M^{l})}{(lM)^{\alpha}}$$

$$= C \sum_{l=1}^{\infty} T_M^{l+1} \left(\frac{1}{l^{\alpha}} - \frac{1}{(l+1)^{\alpha}}\right) \leqslant C \cdot c(\varepsilon)^s M^{-\alpha+\varepsilon}$$

$$\times \sum_{l=1}^{\infty} \frac{\alpha}{l^{\alpha-\varepsilon}} \leqslant C \cdot c(\alpha, \varepsilon)^s M^{-\alpha+\varepsilon}$$

$\left(\frac{1}{l^{\alpha}} - \frac{1}{(l+1)^{\alpha}} = \alpha \int_{l}^{l+1} x^{-\alpha-1} dx \leqslant \frac{\alpha}{l^{\alpha+1}}\right)$. The theorem is proved.

IX. Examples

Denote

$$P_n^*(f) = I(f) - \frac{1}{n} \sum_{j=-(n-1)}^{n-1} \left(1 - \frac{|j|}{n}\right) f(j\boldsymbol{\gamma}).$$

First of all, we prove the following

Theorem 9.1. *Let $\gamma_1, \cdots, \gamma_s$ be a set of real number such that $1, \gamma_1, \cdots, \gamma_s$ are linearly independent over the rational field R. Then we have*

$$\operatorname*{Sup}_{f \in E_s^2(C)} |P_n^*(f)| \leqslant C\left(\frac{\pi^2}{6}\right)^s (W(n; \gamma_1, \cdots, \gamma_s) - 1),$$

where

$$W(n; \gamma_1, \cdots, \gamma_s) = \frac{3^s}{n} + 2\frac{3^s}{n}\sum_{j=1}^{n-1}\left(1 - \frac{j}{n}\right)\prod_{\nu=1}^{s}(1 - 2\{\gamma_\nu j\})^2.$$

Lemma 9.1. *We have*

$$\sum_{m=-\infty}^{\infty} \frac{e^{2\pi i m x}}{\frac{\pi^2}{6} m^2} = 3(1 - 2\{x\})^2.$$

Proof. Since

$$3\int_0^1 (1 - 2x)^2 e^{2\pi i m x}\, dx = \begin{cases} 0, & \text{for } m = 0, \\ \dfrac{6}{\pi^2 m^2}, & \text{for } m \neq 0, \end{cases}$$

therefore we have the lemma.

The Proof of Theorem 9.1. Since

$$\sum_{j=-(n-1)}^{n-1} (n - |j|) = \sum_{k=0}^{n-1}\sum_{j=-k}^{k} 1 = n^2$$

and

$$\sum_{j=-(n-1)}^{n-1} (n - |j|) e^{2\pi i j \delta} = \sum_{k=0}^{n-1}\sum_{j=-k}^{k} e^{2\pi i j\delta}$$

$$= \frac{1}{\sin \pi\delta}\sum_{k=0}^{n-1} \sin(2k+1)\pi\delta = \left(\frac{\sin n\pi\delta}{\sin \pi\delta}\right)^2,$$

where δ is a real number but not an integer, we find

$$\frac{1}{n}\sum_{j=-(n-1)}^{n-1}\left(1 - \frac{|j|}{n}\right) f(j\boldsymbol{\gamma}) = \frac{1}{n^2}\sum_{j=-(n-1)}^{n-1}(n - |j|)\sum C(\mathbf{m}) e^{2\pi i(\mathbf{m}, \boldsymbol{\gamma})j}$$

$$= \frac{1}{n^2}\sum C(\mathbf{m})\sum_{j=-(n-1)}^{n-1}(n - |j|) e^{2\pi i(\mathbf{m}, \boldsymbol{\gamma})j}$$

$$= C(\mathbf{o}) + \frac{1}{n^2}\sum{}' C(\mathbf{m})\left(\frac{\sin n\pi(\mathbf{m}, \boldsymbol{\gamma})}{\sin \pi(\mathbf{m}, \boldsymbol{\gamma})}\right)^2.$$

Hence by Lemma 9.1, we have

$$\operatorname*{Sup}_{f \in E_s^2(C)} |P_n^*(f)| \leqslant \frac{C}{n^2}\sum{}' \frac{1}{\|\mathbf{m}\|^2}\left(\frac{\sin n\pi(\mathbf{m}, \boldsymbol{\gamma})}{\sin \pi(\mathbf{m}, \boldsymbol{\gamma})}\right)^2 = \frac{C}{n^2}\sum{}' \frac{1}{\|\mathbf{m}\|^2}$$

$$\times \sum_{k=0}^{n-1}\sum_{j=-k}^{k} e^{2\pi i(\mathbf{m}, \boldsymbol{\gamma})j} \leqslant \frac{C}{n^2}\left(\frac{2\pi}{6}\right)^s \sum_{k=0}^{n-1}\sum_{j=-k}^{k}\sum{}' \frac{e^{2\pi i(\mathbf{m}, \boldsymbol{\gamma})j}}{\left\|\frac{\pi^2}{6}\mathbf{m}\right\|}$$

$$= \frac{C}{n^2}\left(\frac{\pi^2}{6}\right)^s \sum_{k=0}^{n-1} \sum_{j=-k}^{k} \left(3^s \prod_{\nu=1}^{s} (1-2\{\gamma_\nu j\})^2 - 1\right)$$

$$= C\left(\frac{\pi^2}{6}\right)^s \left(\sum_{j=-(n-1)}^{n-1} (n-|j|)\frac{3^s}{n^2}\prod_{\nu=1}^{s} (1-2\{\gamma_\nu j\})^2 - 1\right)$$

$$= C\left(\frac{\pi^2}{6}\right)^s (W(n; \gamma_1, \cdots, \gamma_s) - 1).$$

The theorem is proved.

By Wang 520 calculator, we obtain the following two tables:

Table 1: $s = 3$

n	$W\left(n; \frac{\sqrt{5}-1}{2}, \sqrt{2}, \sqrt{10}\right)$	$W(n; e, e^2, e^3)$
100	1.08877	1.10689
500	1.01351	1.00914
1000	1.00572	1.00294

Table 2: $s = 4$

n	$W\left(n; 2\cos\frac{2\pi}{11}, 2\cos\frac{4\pi}{11}, 2\cos\frac{6\pi}{11}, 2\cos\frac{8\pi}{11}\right)$	$W(n; e, e^2, e^3, e^4)$
1000	1.03263	1.13899
1500	1.02139	1.11848
3000	1.00887	

References

[1] Baker, A. 1965 On some Diophantine inequalities involving the exponential function, *Can. J. Math.*, **17**(4), 616—626.

[2] Halton, J. H. 1960 On the efficiency of certain quasirandom sequences of points in evaluating multi-dimensional integrals, *Num. Math.*, **27**(2), 84—90.

[3] Haselgrove, C. B. 1961 A method for numerical integration, *Math. Comp.*, **15**(76), 323—337.

[4] Hlawka, E. 1962 Zur angenäherten Berechnung mehrfacher Integrale, *Mon. Math.*, **66**(2), 140—151.

[5] ——— 1964 Uniform distribution modulo 1 and numerical analysis, *Comp. Math.*, **16**(1—2), 92—105.

[6] Hua, Loo-keng & Wang Yuan 1960 Remarks concerning numerical integration, *Sci. Rec.*, **4**(1), 8—11.

[7] ——— 1964 On Diophantine approximations and numerical integrations (I) (II), *Sci. Sin.*, **13**(6), 1007—1010.

[8] ——— 1965 On numerical integration of periodic functions of several variables, *Sci. Sin.*, **14**(7), 964—978.

[9] Roth, K. F. 1954 On irregularities distribution, *Math.*, **1**(2), 73—79.

[10] Schmidt, Wolfgang M. 1970 Simultaneous approximation to algebraic numbers by rationals, *Acta Math.*, **125**, 189—201.
[11] Wyel, H. 1913 Über die Gleichverteilung von Zahlen mod Eins, *Math. Ann.*, **77**, 313—352.
[12] Бахбалов Н. С. 1959 О приближенном вычислении кратных интегралов, *Вес. Мос. Ун-та.*, **4**, 3—18.
[13] Виноградов И. М. 1971 Метод тригонометрических Сумм в теории чисел, *Физмат. Лит.*, Изд. «Наука».
[14] Коробов Н. М. 1963 Теоретико-числовые методы в приближенном анализе, *Физмат. Лит. Мос.*.
[15] Шарыгин И. Ф. 1963 Оценки снизу погрешиости квадратурных формул на классах функций, *Жур. Мат. и Мат. Физ.* **3**(2), 370—376.
[16] *Applications of Number Theory to Numerical Analysis*, Zaremba, S. K. (Ed.), Acad. Press, 1972.
[17] 津田孝夫 1973 多变数问题的数值解析，株式会社（日文）.

Note added on the 19th of May, 1973. A theorem similar to Theorem 5.1 was proved independently by H. Niederreiter and some results given in [8] were improved by S. Haber (See [16]). In the course of publication, we add two more new books, [16] and [17].

APPLIED MATHEMATICS

Applied Mathematics Group, Academia Sinica

In the past twenty years, besides his research in pure mathematics, Professor L. K. Hua also has contributed significantly to the application of mathematical methodology to the national economy. He led a group of applied mathematicians, technicians, and workers, and toured twenty-three provinces in China, visiting thousands of factories. In this way, he has succeeded in handing over directly to the ordinary workers a number of useful mathematical methods which have helped to bring about many effective industrial applications. The amount of time and energy he has spent in this endeavour is undoubtedly not less than that given to any one of his pure mathematical undertakings. For this reason, it is proper to include in the present selecta a list of his contributions in this respect. As these cover many different trades and professions, it is impossible to describe each of them in detail.

Textile Industry

1. To promote the quality of 2014 yarn Khaki cloth.
2. To solve the off-color problem for Meili silk fabric.
3. To promote the efficiency of the loom.
4. To boost the per spinning frame yield of fine yarn.
5. To promote the efficiency of heat setting for terylene cotton cloth.
6. To reduce the breakage of fine yarn.
7. To reduce the case of cotton winding by improving the surface condition of the rubber roll.
8. To make the dynamic balance of a cylinder.
9. To improve the dyeing quality and save raw material.

Electronics

1. To trial–produce the new 160V capacitor.
2. To salvage 1000 kilometers of molybdenum wire.
3. To improve the damp-proofing property of damp-proof lacquer for transistors.
4. To debug the power amplifier of an XD_1 signal generator.
5. To solve the source voltage fluctuation for a "BP-3 wide frequency spectrum analyser".

6. To recover the rare metal—tantalum.
7. To increase the tantalum powder utilization coefficient in making tantalum electrolytic capacitors.
8. To improve the quality of ground monocrystalline silicon wafers.
9. To control the aluminium film's thickness.
10. To anneal high purity aluminium foils.
11. To improve the quality of the point-etching process.

Metallurgy

1. To promote the efficiency of ballmill.
2. To improve the result of aluminium sealing from top of furnace during pouring and casting H80 welding-rod steel.
3. To overcome a technical barrier in the smelting of Si–Cr alloys.
4. To reduce the time for steelmaking in an arc furnace.
5. To improve the quality of 2Cr13 stainless steel.
6. To boost the production of cobalt.
7. To boost the production of thallium.
8. To improve the coating quality of silicon steel sheets.
9. To increase the production of a three-rolled cold rolling mill.
10. To reduce the waste products rate for a Ø 500 roll mill.
11. To increase the recovery rate of metal manganese (Mn).
12. To solve the problem of cavity due to contraction in steel ingots.
13. To prolong the furnace life (i.e., life span of furnace lining).

Coal Mining

1. To boost the coal production by rational arrangement.
2. To promote the efficiency of a coal mining combine by regulating its parameters.
3. To reduce the powder consumption and boost per unit area yield of a coal face.
4. To increase the recovery rate of fine coal.
5. To raise the breaking strength of loop anchor chains and connecting loops.

Electric Power

1. To restore the output of a turbo-generating set.
2. To promote the boiler's efficiency.
3. To revise the industrial water system.
4. To optimize the operation of a feed pump.
5. To realize the automatic frequency regulation during synchronization.
6. To lower the temperature of bearing in a steam turbine.
7. To improve the efficiency of water softening of a moving bed.

Communication and Transportation

1. To organize railway construction.
2. To increase the loading capacity of a railway station.
3. To economize fuel consumption of motor vehicles and ships.
4. To improve meteorological navigation.
5. To organize construction of the Yangtze River Bridge at Luzhou.

Construction and Building Material

1. The organization of a construction.
2. The budgeting of a construction.
3. The organization of bridge construction.
4. To boost the production of fibre board.
5. To reduce the cost of polyvinyl chloride glue mud.
6. To boost the production of cement.
7. To improve the quality of terrazzo products.
8. To improve the expansion coefficient of expanded pearl rock.
9. To reduce the cost of slag-concrete.
10. To trial-produce poly-calcium sulphide solution.
11. To improve the efficiency of making concrete tubes by the centrifugal process.

Foodstuff, Oil Products, Food Processing

1. To increase the rate of finished product rice in the rice-processing.
2. To increase the rate of finished product oil in the oil crops-processing.
3. To increase the rate of finished product flour in the wheat-processing.
4. To increase the rate of finished product wine in the wine-making.
5. To increase the reclamation rate of sugar.
6. To increase the production of malt.
7. To reduce the rate of reprocessing for fine dried noodles.
8. To increase the rate of finished product sugar from maltose.
9. To improve bean curd quality and reduce soybean consumption.
10. To improve the quality of confectionary.
11. To increase the reclamation rate of protein in pig bristle dissolving process.

Design

1. To design a radio network.
2. To design a filter.
3. To design a compensator.
4. To design an airfield for a given terrain.
5. To design an optical lens.
6. To design planetary gears.

7. To design a frequency band of a radio transmitter.
8. To design switches in a circuit.
9. To design a water sampling device.
10. To design the location of a multi-stage pumping station.

Chemical Engineering

1. To improve the sensitivity of liquid crystal to color-changing temperature.
2. To improve the quality of sebacic acid.
3. To increase the recovery rate of amine dicyanide.
4. To improve the production and quality of active carbon.
5. To prolong the life of caprylene with hydrogen catalyst.
6. To economize potassium silicofluoride in the production of potassium zirco-fluoride.
7. To increase the recovery rate of the oxygen-resist 1010.
8. To improve the separation efficiency of rectifying tower.
9. To increase the extraction rate of furfural.
10. To boost the recovery rate of caustic soda.
11. To increase the production of lysol.
12. To increase the production and save the raw materials in the sulphide soda processing.
13. To reduce electricity consumption and increase the production of calcium carbide.
14. To increase the production of the granule-forming tower.
15. To increase the production of the gas generator.
16. To increase the production of phosphate fertilizer.

Petroleum Industry

1. To increase the efficiency of the demulsification agent GP122.
2. To improve the dehydration quality of crude oil.
3. To increase the rate of total extraction for decompression tower in the ordinary and decompressing process.
4. To evaluate the maximal production capacity of an oil well.
5. To increase the wax-melting rate of the chemical dewaxing.
6. To find the optimal conditions in boiler operation.
7. To reduce the viscocity of thick oil.
8. To select the wave-filtering factor of the return transmitting apparatus at the base of seismological information.
9. To increase the production of tiny ball aluminium silicate (catalyst).
10. To trial-produce new oil specimens by platinum reforming.
11. To improve the quality of anti-solidification agent 605.

Light Industry

1. To improve the quality of the thermos bottle.
2. To increase the production of soap.

3. To improve the quality of paper.
4. To improve the quality of tanning and making leather.
5. To increase the production of leather shoes.
6. To improve the light efficiency of fluorescent lamps.
7. To raise the production of cigarettes.
8. To improve the quality of the inner coating of tin cans.
9. To raise the efficiency of the duck's down–separating machine.
10. To manufacture matches from withered wood.
11. To increase the production of transparent polyamide fibre thread.
12. To increase the production of special glycerine.
13. To improve the quality of TiF_2 glass.

Machine Manufacturing

1. To increase the processing efficiency and precision in various machine tools.
2. The optimal approximation problem for gear-pair meshing.
3. The static balancing of grinding wheels.
4. To improve the surface smooth finish of the mirror scale of a floor-type boring machine.
5. To improve the quality of nodular cast iron.
6. To improve the quality in flange making.
7. High-speed chrome plating.
8. To improve the quality of lacquer plating by electrophoresis.
9. Hardening process for all kinds of cutter products.
10. High-frequency hardening process for the plane surface of gears.
11. Heat treatment of vibrating films.
12. Automatic welding for single-layer and two-side submerged arc welding.
13. To manufacture drills with 35 Chromium-vanadium.
14. The neck-forming process of oxygen bottles.
15. The cyanideless zinc-plating process.
16. To reduce the coke–iron ratio of a cupola furnace.

Drug Manufacturing

1. To increase the production of paracetamol.
2. To increase the rate of iodine acquisition from the sea tangle.
3. To reduce the cost of sulphadiazine.
4. To increase the production of the dipterex.
5. To improve the tablet-forming process of tetracycline.
6. To boost the production of sodium methanolate.
7. To increase the reclamation rate for salt of hydrochloric acid of terramycin.
8. To save the raw materials in making furadantin.
9. To raise the fermentation index of tetracycline.

Loo-Keng Hua: Publications

Articles

1. Study on Sturm Theorem. *Science* **14**(1929), 545–548. (in Chinese)
2. On incorrectness of Shu Jia Ju's paper. *Science* **15**(1930), 307–309. (in Chinese)
3. Study on function $T^{-1}\{H(x)\}$. *Science* **15**(1931), 871–888, 1055–1062. (in Chinese)
4. A Theorem of integral calculus. *Science* **15**(1931), 1716–1719.
5. Generalization and trial of additive angles formula in trigonometry. *Science* **15**(1931), 1930–1945. (in Chinese).
6. A new function. *Science* **15**(1931), 1051–2058.
7. On pseudo-periodic functions. *Trans. Sci. Soc. China* **8**(1934), 15–18; also *Tohoku Math J.* **40**(1934), 27–33.
8. On the representation of integers by circulant. *Trans. Sci. Soc. China* **8**(1934), 19–21; also *Tohoku Math. J.* **39**(1934), 316–321.
9. On a theorem of Hermite. *Trans. Sci. Soc. China* **8**(1934), 157–158.
10. A note on Minkowski's theorem of homogeneous linear forms. *Trans. Sci. Soc. China* **8**(1934), 160–161.
11. On the hypergeometric functions of higher order. *Tohoku Math. J.* **39**(1934), 253–263.
12. Note on diophantine equation of two circulants. *Tohoku Math. J.* **40**(1934), 34–35.
13. Note on Pell's equation. *Tohoku Math. J.* **40**(1934), 36.
14. Waring's problem for cubes. *Bull. Calcutta Math. Soc.* **26**(1934), 139–140.
15. On a certain kind of operations connected with linear algebra. *Tohoku Math. J.* **41**(1935), 222–246.
16. A proof of Hadamard's theorem. *Tohoku Math. J.* **41**(1935), 247–248.
17. The representation of integers as sums of the cubic function $(x^3 + 5x)/6$. *Tohoku Math. J.* **4**(1935), 356–360.
18. On the representation of integers by the sums of seven cubic functions. *Tohoku Math. J.* **4**(1935), 361–366.
19. The representation of integers as sums of cubic function $(x^3 + 2x)/3$. *Tohoku Math. J.* **41**(1935), 367–370.
20. On an easier Waring–Kamke problem. *Sci. Repts. Tsing Hua Univ.* **A3**(1935), 247–260.
21. On Waring's theorems with cubic polynomial Summands. *Math. Ann.* **III**(1935), 622–628.
22. On Waring's problem with polynomial summands. *Amer. J. Math.* **58**(1936), 553–562; also *Chinese Math. Soc.* **1**(1936), 23–61.
23. Note on boundedly convergent power series. *Sci. Repts. Tsing Hua Univ.* **A3**(1936), 345–351.
24. A problem on the additive theory of number of several variables. *Math. Zeit.* **41**(1936), 708–712.
25. On Waring's problem. *Tohoku Math. J.* **42**(1936), 210–225.
26. An easier Waring–Kamke problem. *J. London Math. Soc.* **II**(1936), 4–5.
27. On Fourier transforms in L^p in the complex domain. *J. Math. Phys.* **15**(1936), 249–263. (with S. S. Shu)
28. A problem in the additive theory of numbers of several variables. *J. London Math. Soc.* **12**(1937), 257–261.
29. A generalization of an easier Waring-Kamke problem. *J. London Math. Soc.* **12**(1937), 262–264.
30. On a generalized Waring problem. *Proc. London Math. Soc.* (2) **43**(1937), 161–182.
31. On the representation of integers as the Sums of the k-th powers of primes. *Dokl. Akad. Nauk SSSR(N.S)* **17**(1937), 167–168.

32. Some results in the additive prime-number theory. *Quart. J. Math.* Oxford Ser. **9**(1938), 68–80.
33. Some results in the additive prime number theory. *Dokl. Akad. Nauk SSSR(N.S)* **18**(1938), 3.
34. Some results in the additive theory of numbers. *Dokl. Akad. Nauk SSSR(N.S)* **18**(1938), 4.
35. Some results in Waring's Problem for small powers. *Dokl. Akad. Nauk SSSR(N.S)* **18**(1938), 527–528.
36. On Waring's problem. *Quart. J. Math.* Oxford Ser. **9**(1938), 199–202.
37. On Tarry's problem. *Quart. J. Math.* Oxford Ser. **9**(1938), 315–320.
38. On an exponential sum. *J. London Math. Soc.* **13**(1938), 54–61; also *J. Chinese Math. Soc.* **2**(1940), 301–312. (with further progress.)
39. On the representation of numbers as the sums of the powers of primes. *Math. Zeit.* **44**(1938), 335–346.
40. A generalization of Lendesdorff's Theorem. *Proc. Indian Acad. Sci.* **7**(1938), 390–392.
41. On Waring's problem for fifth powers. *Proc. London Math. Soc.* (2) **45**(1939), 144–160.
42. A remark on the moment problem. *J. London Math. Soc.* **14**(1939), 84–86.
43. On a lemma due to Vinogradow. *Dokl. Akad. Nauk SSSR(N.S)* **24**(1939), 419–420.
44. On a system of diophantine equations. *Dold. Akad. Nauk SSSR(N.S)* **27**(1940), 312–313.
45. On a generalized Waring problem, II. *J. Chinese Math. Soc.* **2**(1940), 175–191.
46. Some "Anzahl" theorems for groups of prime-power orders. *J. Chinese Math. Soc.* **2**(1940), 313–319. (with H. F. Tuan).
47. On Waring's problem with cubic polynomial Summands. *J. Indian Math. Soc.* **4**(1940), 127–135; also *Sci. Repts. Tsing Hua Univ.* **A4**(1940), 55–83.
48. On a theorem due to Vinogradow. *Quart. J. Math.* Oxford Ser. **11**(1940), 161–176.
49. Sur une somme exponentielle. *C. R. Acad. Sci. Paris* **210**(1940), 520–523.
49A. On an exponential sum. *J. Chinese Math. Soc.* **2**(1940), 301–312.
50. Sur le probleme de Waring relatif a un polynome du troisieme degre. *C. R. Acad. Sci. Paris* **210**(1940).
51. On the number of solutions of certain congruences. *Sci. Repts. Tsing Hua Univ.* **A4**(1940), 113–133. (with S. H. Min)
52. Determination of the groups of odd-prime-power order p^n which contains a cyclic subgroup of index p^2. *Sci. Repts. Tsing Hua Univ.* **A4**(1940), 145–154. (with H. F. Tuan)
53. Some recent progress in theory of numbers. *Quart. Wu Han Univ.* **7**(1940). (with K. L. Chung) (in Chinese).
54. A note on the class number of ternary quadratic forms. *J. London Math. Soc.* **16**(1941), 82–83.
55. On diophantine approximation. *Dokl. Akad. Nauk SSSR(N.S)* **32**(1941), 395–396.
56. Some problems of the geometrical theory of numbers. *Sci. Record* **1**(1942), 19–21.
57. On character sums. *Sci. Record* **1**(1942), 21–23.
58. On a double exponential sum. *Sci. Record* **1**(1942), 23–25. (with S. H. Min)
59. An analogue of Tarry's problem. *Sci. Record* **1**(1942), 26–29. (with S. H. Min)
60. On the number of partitions of a number into unequal parts. *Trans. Amer. Math. Soc.* **51**(1942), 194–201.
61. On the least primitive root of a prime. *Bull. Amer. Math. Soc.* **48**(1942), 726–730.
62. On the least solution of Pell's equation. *Bull. Amer. Math. Soc.* **48**(1942), 731–735.
63. The lattice-points in a circle. *Quart. J. Math.* Oxford Ser. **13**(1942), 18–29.
64. On the distribution of quadratic non-residues and the Euclidean algorithm in real quadratic fields, I–II. *Trans. Amer. Math. Soc.* **56**(1944), 527–546, 547–569. (II, with S. H. Min)
65. On the lack of an Euclidean algorithm in $R(\sqrt{61})$. *Amer. J. Math.* **67**(1945), 209–211. (with W. T. Shih).
66. On the theory of automorphic functions of a matrix variable. I. Geometrical basis. II. The classification of hypercircles under the symplectic group. *Amer. J. Math.* **66**(1944), 470–488, 531–563.
67. Geometries of matrices. I. Generalizations of von Staudt's Theorem. I_1. Arithmetical construction. *Trans. Amer. Math. Soc.* **57**(1945), 441–481, 482–490.
68. A remark on a result due to Blichfeldt. *Bull. Amer. Math. Soc.* **51**(1945), 537–539.
69. Geometries of matrices. *Sci. Record.* **1**(1945), 262–267.
70. The theory of automorphic functions of a matrix variable. *Sci. Record* **1**(1945), 303–305.
71. On the Euclidean algorithm in the real quadratic fields. *Sci. Record* **1**(1945), 319 (with W. T. Shih).

72. Geometries of symmetric matrices over the real field. I–II. *Dokl. Akad. Nauk SSSR(N.S)* **53**(1946), 95–97, 195–196.
73. Automorphism of real symplectic group. *Dokl. Akad. Nauk SSSR(N.S)* **53**(1946), 303–306.
74. On the theory of Fuchsian functions of several variables. *Ann. of Math.* **47**(1946), 167–191.
75. On the extended space of several complex variables (1). The space of complex spheres. *Quart. J. Math.* Oxford Ser. **17**(1946), 214–222; also *Sci. Record* **2**(1947), 6–8.
76. Orthogonal classification of Hermitian matrices. *Trans. Amer. Math. Soc.* **59**(1946), 508–523.
77. Geometries of matrices. II. Study of involutions in the geometry of symmetric matrices. *Trans. Amer. Math. Soc.* **61**(1947), 193–228.
78. Geometries of matrices. III. Fundamental Theorems in the geometries of symmetric matrices. *Trans. Amer. Math. Soc.* **61**(1947), 229–255.
79. Theory of automorphic functions of several complex variables. *Akad. Nauk Gruzin SSR. Trudy Tbillis Mat. Inst. Razadze* **15**(1947), 143–273.
80. Some results on additive theory of numbers. *Proc. Nat. Acad. Sci. U.S.A.* **33**(1947), 136–137.
81. Some "Anzahl" theorems for groups of prime power orders. *Sci. Repts. Tsing Hua Univ.* **A4**(1947), 313–327.
82. On a double exponential sum. *Sci. Repts. Tsing Hua Univ.* **A4**(1947), 484–518. (with S. H. Min)
83. A theorem on matrices and its application to Grassmann space. *Sci. Repts. Tsing Hua Univ.* **A5**(1948), 150–181.
84. Introduction to the theory of vector modular forms. *Akad. Nauk Azerbaidzan SSR. Trudy Inst. Fiz. Mat.* **3**(1948), 32–43. (in Russian).
85. On the automorphisms of the symplectic group over any field. *Ann. of Math.* **49**(1948), 739–759.
86. On the existence of solutions of certain equations in a finite field. *Proc. Nat. Acad. Sci. U.S.A.* **34**(1948), 258–263. (with H. S. Vandiver).
87. Characters over certain types of rings with applications to the theory of equations in a finite field. *Proc. Nat. Acad. Sci. U.S.A.* **35**(1949), 94–99. (with H. S. Vandiver)
88. On the automorphisms of a sfield. *Proc. Nat. Acad. Sci. U.S.A.* **35**(1949), 386–389.
89. On the number of solutions of some trinomial equations in a finite field. *Proc. Nat. Acad. Sci. U.S.A.* **35**(1949), 477–481 (with H. S. Vandiver)
90. On the nature of the solutions of certain equations in a finite field. *Proc. Nat. Acad. Sci. U.S.A.* **35**(1949), 481–487 (with H. S. Vandiver).
91. Some properties of a sfield. *Proc. Nat. Acad. Sci. U.S.A.* **35**(1949), 533–537.
92. On the generators of symplectic modular group. *Trans. Amer. Math. Soc.* **65**(1949), 415–426. (with I. Reiner)
93. Geometry of symmetric matrices over any field with characteristic other than two. *Ann. of Math.* **50**(1949), 8–31.
94. Improvement of a result of Wright. *J. London Math. Soc.* **24**(1949), 157–159.
95. An improvement of Vinogradov's mean-value theorem and several applications. *Quart. J. Math.* **20**(1949), 48–61.
96. On the multiplicative group of a field. *Sci. Record* **3**(1950), 1–6.
97. On semi-homomorphisms of rings and their applications in projective geometry. *Sci. Sinica* **1**(1950), 1–6. (in Chinese); also *Uspehi Mat. Nauk* **8**(55) (1953), 143–148. (in Russian)
98. Fundamental theorem of the projective geometry on a line and geometry of matrices. *C. R. 1er Cong. Math. Hongrois* (1950), 317–325. Akad. Kiado, Budapest, 1952.
99. A theorem on matrices over a sfield and its applications. *Acta. Math. Sinica* **1**(1951), 109–163.
100. On exponential sums over an algebraic number field. *Canadian J. Math.* **3**(1951), 44–51.
101. Supplement to the paper of Dieudonné on the automorphisms of classical groups. *Memoirs Amer. Math. Soc.* **2**(1951), 96–122.
102. Automorphisms of the unimodular group. *Trans. Amer. Math. Soc.* **71**(1951), 331–348. (with I. Reiner)
103. Estimation of an integral. *Sci. Sinica* **4**(1951), 393–402. (in Chinese)
104. On the number of solutions of Tarry's problem. *Acta Sci. Sinica* **1**(1952), 1–76.
105. On the automorphism and isomorphisms of linear group. *Acta Sci. Sinica* **2**(1952–1953), 1–52. (with Wan Zhe Xian)
106. Automorphisms of the projective unimodular group. *Trans. Amer Math. Soc.* **72**(1952), 467–473. (with I. Reiner)

107. A note on the total matrix ring over a non-commutative field. *Ann. Soc. Polon. Math.* **25**(1952), 188–198.

108. A generalization of Hamiltonian matrices. *Acta. Sci. Sinica* **2**(1953), 1–58.

109. The present position of Mathematics in China, *Ke Xue Tong Bao* **2**(1953), 1–5. (in Chinese); also *Vestnik Akad. Nauk SSSR* **6**(1953), 14–20.

110. Theory of functions of several complex variables. I. A complete orthonormal system in the hyperbolic space of matrices. *Acta Sci. Sinica* **4**(1953), 288–323. (in Chinese)

111. On the theory of functions of several complex variables. I. A complete orthonormal system in the hyperbolic space of rectangular matrices. II. A complete orthonormal system in the hyperbolic space of hyperspheres. *Dokl. Akad. Nauk SSSR(N.S)* **93**(1953), 775–777, 983–984. (in Russian)

112. On the estimation of the unitary curvature of the space of several complex variables. *Sci. Sinica* **4**(1955), 1–26; *Acta Math. Sinica* **4**(1954), 143–168.

113. On non-continuable domain with constant curvature. *Acta Math. Sinica* **4**(1954), 317–332, (in Chinese); *Sci. Sinica* **4**(1955), 27–32.

114. Theory of functions of several complex variables. II. A complete orthonormal system in the hyperbolic space of hyperspheres. *Acta Math. Sinica* **5**(1955), 1–25. (in Chinese)

115. On the theory of functions of several complex variables, A complete orthonormal system in the hyperbolic space of symmetric and anti-symmetric matrices. *Dokl. Akad. Nauk SSSR*, **101**(1955), 29–30.

116. An inequality for determinants. *Acta Math. Sinica*. **5**(1955), 463–470. (in Chinese)

117. Some algebraic identities. *Izv. Nat. Inst. Bulgarian Akad. Nauk* **2**(1956), 3–12.

118. Some definite integrals. *Acta Math. Sinica* **6**(1956), 302–312.

119. On exponential sums. *Sci. Record (N.S)* **1**(1957), no. 1, 1–4.

120. On the major arcs of Waring Problem. *Sci. Record (N.S)* **1**(1957), no. 3, 17–18.

121. On a system of partial differential equations. *Sci. Record (N.S)* **1**(1957), 367–371.

122. On the Riemannian curvature in the space of several complex variables. *Schr. Forschungs-inst. Math.* **1**(1957), 245–263.

123. Geometry of rectangular matrices and their application to real projective and non-euclidean geometry. *Sci. Sinica* **6**(1957), 995–1011. (with B. A. Rosenfeld)

124. On Cauchy formula for the space of skew-symmetric matrices of odd order. *Sci. Record (N.S)* **2**(1958), 19–22. (with K. H. Look)

125. Boundary properties of the Poisson integral of Lie sphere. *Sci. Record (N.S)* **2**(1958), 77–80. (with K. H. Look)

126. A convergence theorem in the space of continuous functions on a compact group. *Sci. Record (N.S)* **2**(1958), 280–284.

127. A subgroup of the orthogonal group with respect to an indefinite quadratic form. *Sci. Record (N.S)* **2**(1958), 329–331.

128. Theory of harmonic functions of classical domain I. Harmonic functions in the hyperbolic space of matrices. *Acta Math. Sinica* **8**(1958), 531–547. II. Harmonic functions in the hyperbolic space of symmetric matrices. III. Harmonic functions in the hyperbolic space of skew symmetric matrices. *Acta Math. Sinica* **9**(1959), 295–305, 306–314. (all in Chinese) (with K. H. Look)

129. Theory of harmonic functions in classical domains. *Sci. Sinica* **8**(1959), 1031–1094. (with K. H. Look)

130. Research works in mathematics in China from 1949–1959. By mathematics group with L. K. Hua and others. *Sci. Sinica* **8**(1959), 1218–1228; also: Mathematical research in Communist China in the past ten years. New York, U.S.A. *Joint Publications Research Service*, (1960), 91; also: A brief review of mathematical investigations in China for the last decade. *Uspehi Mat. Nauk* **15**(1960), no. 3(93), 193–201.

131. Remarks concerning numerical integration. *Sci. Record (N.S)* **4**(1960), 8–11. (with Wang Yuan)

132. Calculation of deposit amount and hillside area on contour map. *Acta Math. Sinica* **10**(1959), 19–40. (in Chinese)

133. Application of mathematical methods to wheat harvest. *Acta Math. Sinica* **11**(1960), 63–75. (in Chinese)

133A. On the calculation of mineral reserves and hillside areas on contour maps. *Acta Math. Sinica* **11**(1961)1, 29–40. (in Chinese)

134. Symplectic similarity of symplectic matrices. *J. Zhong Shan Univ.* **4**(1962), 1–12.
135. Real congruence of matrices *J. Zhong Shan Univ.* **4**(1962), 13–31.
136. Finiteness and infinity, discreteness and continuity. *Ke Xue Tong Bao* (1963), 4–21. (with Wang Yuan)
137. Introduction to generalized functions. *Progress in Math.* **6**(1963), 391–409.
138. On diophantine approximations and numerical integrations. (I) (II). *Sci. Sinica* **13**(1964), 1007–1010.
139. Harmonic analysis on unitary group. *Peking symposium Gen.* **165**(1964), 15–32.
140. On an inequality of Opial. *Sci. Sinica* **14**(1965), 789–790.
141. On an inequality of Harnack's type. *Sci. Sinica* **14**(1965), 791.
142. On Lavrentiev's partial differential equation of the mixed type. *Sci. Sinica* **13**(1964), 1755–1762; also *Acta Math. Sinica* **15**(1965), 873–882.
143. On canonical form of system of differential equations of second order with constant coefficients in two variables and two unknown functions. *Ke Xue Tong Bao*, (1964), 1100–1103. (with Wu Zhi Qian and Lin Wei)
144. On uniqueness theorem of Dirichlet problem of system of partial differential equations of elliptic type of second order with constant coefficients. *Acta Math. Sinica* **15**(1965), 242–248. (with Wu Zhi Qian and Lin Wei)
145. On the classification of the system of differential equations of the second order. *Sci. Sinica* **14**(1965), 461–465.
146. Opening paper on partial differential equation of mixed type. *J. of Univ. of Sci. & Tech. China* **1**(1965), 1–27.
147. On numerical integration of periodic functions of several variables. *Sci. Sinica* **14**(1965), 964–977. (with Wang Yuan)
148. On uniform distribution and numerical analysis (Number-theoretic method). *Sci. Sinica* (I) **16**(1973), 483–505; (II) **17**(1974), 331–348; (III) **18**(1975), 184–198.
149. Optimum-seeking method of several variables. *Ke Xue Tong Bao.* (I) (II) **18**(1973), 165–166. (III) (IV) **19**(1974), 317–319.
150. A note on simultaneous diophantine approximations to algebraic integers *Sci. Sinica* **20**(1977), 563–567.

Books and Monographs

1. *Additive Theory of Prime Numbers* (in Russian). Trudy Inst. Math. Steklov 22 (1947), 1–179. Chinese translation (revised), Academic Press, Peking 1957; Hungarian translation, Académiai Kiadó, Budapest, 1959; German translation, Teubner, Leipzig, 1959; English translation, AMS, Providence, Rhode Island, 1965.
2. *Introduction to Theory of Numbers* (in Chinese). Academic Press, Peking, 1957.
3. *Harmonic Analysis of Functions of Several Complex Variables in the Classical Domains* (in Chinese). Academic Press, Peking, 1958. Revised edition, 1965; Russian translation, Izd. inostran. lit., Moskva, 1959; English translation, AMS, Providence, Rhode Island, 1963.
4. *Abschätzungen von Exponentialsummen und ihre Anwendung in der Zahlentheorie*. Teubner, Leipzig, 1959; Chinese translation, Academic Press, Peking, 1963; Russian translation, Mir, Moskva, 1964.
5. *Numerical Integration and its Applications* (revised *Numerical Calculation of Integral*) (in Chinese). Academic Press, Peking, 1963, (with Wang Yuan).
6. *Classical Groups* (in Chinese). Shanghai Sci. Press, Shanghai, 1963, (with Wan Zhe-Xian).
7. *Introduction to Higher Mathematics* (in Chinese). Vol. I, part 1, 2. Academic Press, Peking, 1963.
8. *Starting with Unit Circle* (in Chinese). Academic Press, Peking, 1977.
9. *Application of Number Theory to Numerical Analysis* (in Chinese). Academic Press, Peking, 1978, (with Wang Yuan).
10. *Systems of Partial Differential Equations of the Second Order of Two Unknown Functions and Two Independent Variables with Constant Coefficients* (in Chinese). Academic Press, Peking, 1978, (with Wu Zhi Qian and Lin Wei).

Popular Books (All in Chinese)

1. *Starting with Yang Hui Triangle*. Academic Press, Peking, 1956.
2. *To Young Mathematicians*. Chinese Youth Press, Peking, 1956.
3. *Property and Role of Mathematics*. Academic Press, Peking, 1959.
4. *Starting with Zhu Chung Zi's ratio of circumference*. Chinese Youth Press, Peking, 1962.
5. *Talks on Mathematical Problem Concerning Honeycomb Structure*. Academic Press, Peking, 1962.
6. *Mathematical Induction*. Shanghai Education Press, Shanghai, 1963.
7. *Starting with Sun Zi's "Magic Calculation" (Chinese remainder theorem)*. People's Education Press, Peking, 1964.
8. *Popular Talk on Overall Planning Method and its Supplement (revised)*. Chinese Industry Press, Peking, 1965.
9. *Popular Talk on Optimum-Seeking Method*. Academic Press, Peking, 1971.
10. *Popular Talk on Optimum-Seeking Method and its Supplement*. National Defence Industry Press, Peking, 1971.
11. *Talks on Optimum-Seeking Method*. Liaoning People's Press, Shenyang, 1973.

Springer-Verlag is the publisher of three monographs by Loo-keng Hua:

Starting with the Unit Circle
Background to Higher Analysis
1981

Starting With the Unit Circle is a useful introduction to a particularly active field of research: harmonic analysis in several variables. Professor Hua starts with the notion of a unit disc, and goes on to solve the Dirichlet problem on the unit ball in n-dimensions. Along the way he considers the basic facts of the Lorentz group, the axioms of special relativity, and partial differential equations of mixed type. The exposition is explicit and elementary, providing concrete explanations of advanced ideas, and presenting material and viewpoints not generally found in standard texts. *Starting With the Unit Circle* is a very stimulating book, introducing the reader to the author's personal perspective of an important area of mathematics.

Introduction to Number Theory
1982

Introduction to Number Theory is an encyclopedic account of modern number theory, accessible to the beginning graduate student. Consistent with Hua's other work, the presentation emphasises the close relationship between number theory and mathematics as a whole. The original edition of the work was published in 1957 in Chinese; the new edition contains recent results in number theory as well as new notes on results which will help the reader acquaint himself with current research literature. Problems are provided to help the student acquire a better grasp of the material. The recent notes were compiled by Professor Wang Yuan and by the translator, Dr. Peter Shiu.

Applications of Number Theory to Numerical Analysis (with Professor Wang Yuan, Institute of Mathematics, Academia Sinica)
1981

The subject of this book is the use of deep and advanced number theoretical techniques in numerical analysis, such as the determination of a series of uniformly distributed sets in the s-dimensional unit cube ($s \geqslant 2$) (subsequent approximation of definite integrals with minimal error). Several of the techniques explained in this book are significantly superior to older well-established methods of approximation. The Appendix contains a table of good lattice point sets. Basically, the book is self-contained and accessible to readers with a knowledge of elementary number theory. Deeper and complicated technical results are either derived or referenced in the text.